Handbook of Steel Construction

FIFTH EDITION

FIRST PRINTING OCTOBER 1991

Canadian Institute of Steel Construction

201 Consumers Road, Suite 300
Willowdale, Ontario M2J 4G8

Fifth Edition

First Printing October 1991

ISBN 0-88811-072-3

PRINTED IN CANADA

by

Universal Offset Limited
Markham, Ontario

R. van Dieren

CONTENTS

FOREWORD

The Canadian Institute of Steel Construction is a national industry organization representing the structural steel, open-web steel joist and steel plate fabricating industries in Canada. Formed in 1930 and granted a Federal charter in 1942, the CISC functions as a non-profit organization promoting the efficient and economic use of fabricated steel in construction.

As a member of the Canadian Steel Construction Council, the Institute has a general interest in all uses of steel in construction. CISC works in close co-operation with the Steel Structures Education Foundation (SSEF) to develop educational courses and programmes related to the design and construction of steel structures. The CISC supports and actively participates in the work of the Standards Council of Canada, the Canadian Standards Association, the Associate Committee on the National Building Code and numerous other organizations, in Canada and other countries, involved in research work and the preparation of codes and standards.

Preparation of engineering plans is not a function of the CISC. The Institute does provide technical information through its professional engineering staff, through the preparation and dissemination of publications, through the medium of seminars, courses, meetings, video tapes, and computer programs. Architects, engineers and others interested in steel construction are encouraged to make use of CISC information services.

CISC is located at 201 Consumers Road, Suite 300, Willowdale, Ontario, M2J 4G8, Telephone (416) 491-4552 and Fax (416) 491-6461.

PREFACE

This handbook has been prepared and published by the Canadian Institute of Steel Construction. It is an important part of a continuing effort to provide current, practical information to assist educators, designers, fabricators, and others interested in the use of steel in construction.

The First Edition of the CISC Handbook of Steel Construction was published in 1967 with subsequent editions following each new edition of the CSA structural steel design standard, now called CAN/CSA-S16.1-M89.

This Fifth Edition is based primarily on CSA Standard CAN/CSA-S16.1-M89. Part 1 is a reprint of CAN/CSA-S16.1-M89 in its entirety. To assist with understanding the requirements of this CSA Standard, Part 2 is a Commentary on CAN/CSA-S16.1-M89 prepared by CISC. Part 3 contains information on bolts and welds with tables for design and evaluation of various structural framing connections. Changes in Part 3 relate to slip-critical connections, prying action, and tension members. Part 4 contains information on compression members, and includes new tables for WWF sections in CSA G40.21 grade 350W steel, in addition to tables based on CSA G40.21 grade 300W steel. Tables for determining the appropriate class of sections used as beam-columns have been added as well as tables for double angle struts built-up from the imperial series of angles. Part 5 contains information on flexural members, and includes additional tables for WWF beams in CSA G40.21 grade 350W steel. The composite beam tables have been revised to reflect current construction conditions, and now extend to 40% shear connection. In Part 6, tables related to mechanical and chemical properties of steel and other information extracted from CSA Standards G40.20, G40.21, and G312.3 have been revised according to approved changes for 1991. Tables of properties and dimensions for the imperial series of angles have been added. New Canadian designations for imperial series angles and HSS sections are used. For the first time, tables of properties and dimensions are provided for cold-formed channels.

In this Edition, colour coding has been introduced to identify W-shapes which are likely to be more readily available on short notice, and a rolling code to provide guidance on availability of angles has been added.

The revised CISC Code of Standard Practice - Fifth edition, June 8, 1991, leads the information found in Part 7.

Permission to reprint portions of their publications, granted by the Canadian Standards Association and the American Institute of Steel Construction, Inc., is gratefully acknowledged. The contribution of many individuals who helped make publication of this book possible is sincerely appreciated.

Although no effort has been spared in an attempt to ensure that all data in this book is factual and that the numerical values are accurate to a degree consistent with current structural design practice, the Canadian Institute of Steel Construction does not assume responsibility for errors or oversights resulting from the use of the information contained herein. Anyone making use of the contents of this book assumes all liability arising from such use. All suggestions for improvement of this publication will receive full consideration for future printings.

M. I. Gilmor
Editor
September, 1991

DESIGNATIONS

Standard designations should always be used to identify structural steel products on drawings and other documents. In Canada, the official designation is the metric (SI) designation, and examples of correct designations for most of the commonly used steel products are provided below. These designations should be used on all design drawings, for detailing purposes and for ordering material.

Shape		Example
Welded Wide Flange Shapes		WWF 900 x 169
W Shapes		W 610 x 113
Miscellaneous M Shapes		M 200 x 9.7
Standard Beams (S Shapes)		S 380 x 64
Standard Channels (C Shapes)		C 230 x 30
Miscellanous Channels (MC Shapes)		MC 250 x 12.5
Structural Tees	- cut from WWF Shapes	WWT 250 x 138
	- cut from W Shapes	WT 155 x 43
	- cut from M shapes	MT 100 x 25.5
Bearing Piles (HP Shapes)		HP 130 x 110
Equal Leg angles	- Imperial Series	L 102 x 102 x 9.5
	- Metric series	L 100 x 100 x 10
Unequal Leg angles	- Imperial Series	L 127 x 89 x 9.5
	- Metric Series	L 125 x 90 x 10
Plates (thickness x width)		PL 8 x 500
Square bars (side, mm)		Bar 25⌷
Round bars (diameter, mm)		Bar 25 ϕ
Flat Bars (thickness x width)		Bar 5 x 60
Round Pipe (outside diameter x thickness)		Pipe 323.9 OD 9.53
Hollow Structural sections	- Rectaugular	HSS 152 x 102 x 9.5 Class C [#]
	- Square	HSS 152 x 152 x 9.5 Class C [#]
	- Round	HSS 141 x 9.5 Class C [#]
Cold Formed Channels		CFC 310 x 89 x 4.93

[#] *Class C or H*

GENERAL NOMENCLATURE

Explanations of the nomenclature used in many sections of this book appear in those specific sections. In addition, the following symbols are included here for convenience. See also pages 1-17 to 1-20, inclusive.

A	Area
A_b	Cross-sectional area of one bolt based on nominal diameter
A_f	Flange area
A_w	Web area; shear area; effective throat area of weld
a	Centre-to-centre distance between transverse web stiffeners, depth of concrete compression zone
a/h	Aspect ratio, ratio of distance between stiffeners to web depth
B_f	Bearing force in a member or component under factored load
B_r	Factored bearing resistance of a member or component
B_x	Bending factor with respect to axis x-x
B_y	Bending factor with respect to axis y-y
b	Width of stiffened or unstiffeneed compression elements; design effective width of concrete or cellular slab
C_e	Euler buckling load = $1\,970\,000\ A/(KL/r)^2$
C_f	Compressive force in a member or component under factored load; factored axial load
C_r	Factored compressive resistance of a member or component; factored compressive resistance of steel acting at the centroid of that part of the steel area in compression
C'_r	Compressive resistance of concrete acting at the centroid of the concrete area in compression
C_w	Warping torsional constant
C_y	Axial compressive load at yield stress
c	Distance from neutral axis to outer fibre of structural shape
D	Outside diameter of circular sections; diameter of rocker or roller; stiffener factor; fillet weld size (millimetres)
d	Depth; overall depth of a section; diameter of bolt or stud
E	Elastic modulus of steel (200 000 MPa assumed)
E_c	Elastic modulus of concrete
e	End distance; lever arm between the compressive resistance, C_r, and tensile resistance, T_r
e'	Lever arm between the compressive resistance, C'_r, of concrete and tensile resistance, T_r, of steel
F_{cr}	Critical plate buckling stress
F_s	Ultimate shear strength
F_u	Specified minimum tensile strength (Megapascals)
F_y	Specified minimum yield stress, yield point or yield strength
f'_c	Specified compressive strength of concrete at 28 days (Megapascals)
g	Transverse spacing between fastener gauge lines (gauge distance)
h	Clear depth of web between flanges; height of stud
I	Moment of inertia
I_x	Moment of inertia about axis x-x
I_y	Moment of inertia about axis y-y
J	St. Venant's torsion constant
K	Effective length factor
K_x	Effective length factor with respect to axis x-x
K_y	Effective length factor with respect to axis y-y

KL Effective length

k Distance from outer face of flange to web toe of fillet of rolled shapes

L Length

L_{cr} Maximum unbraced length adjacent to a plastic hinge

L_u Maximum unsupported length of compression flange for which no reduction in factored moment resistance, M_r, is required

L_x Unsupported length with respect to axis x-x

L_y Unsupported length with respect to axis y-y

M_f Bending moment in a member or component under factored load

M_{f1} Smaller factored end moment of a beam-column; factored bending moment at a point of concentrated load

M_{f2} Larger factored end moment of a beam-column

M_p Plastic moment = ZF_y

M_r Factored moment resistance of a member or component

M'_r Factored moment resistance of a member of a given unbraced length greater than L_u

M_{rc} Factored moment resistance of composite beam

M_y Yield moment = SF_y

m Number of faying surfaces or shear planes in a bolted joint, equal to 1 for bolts in single shear and 2 for bolts in double shear

N Length of bearing of an applied load

P Concentrated load

Q_r Sum of the factored resistances of all shear connectors between points of maximum and zero moment

q_r Factored resistance of a shear connector

R End reaction or concentrated transverse load applied to a flexural member

r Radius of gyration

r_t Radius of gyration, about its axis of symmetry, of a T section comprising the compression flange and 1/6 of the web

r_u Radius of gyration with respect to axis u-u

r_v Radius of gyration with respect to axis v-v

r_x Radius of gyration with respect to axis x-x

r_y Radius of gyration with respect to axis y-y

r_z Radius of gyration with respect to axis z-z

S Elastic section modulus

S_x Elastic section modulus with respect to axis x-x

S_y Elastic section modulus with respect to axis y-y

s Centre-to-centre spacing (pitch) between successive fastener holes in line of stress

T_f Tensile force in a member or component under factored load

T_r Factored tensile resistance of a member or component; factored tensile resistance of the steel acting at the centroid of that part of the steel area in tension

t Thickness

U Amplification factor for stability analysis of beam-columns

V_f Shear force in a member or component under factored load

V_r Factored shear resistance of a member or component

V_s Slip resistance of a bolted joint

W Total uniformly distributed load; concentrated load

w Web thickness; load per unit of length

Z Plastic section modulus of a steel section

α	Load factor
β	Value used in bolt interaction equation
γ	Importance factor
λ	Non-dimensional slenderness ratio in column formula
μ	Coefficient related to the slip resistance of a bolted joint
ϕ	Resistance factor
ψ	Load combination factor
ω	Coefficient used to determine equivalent uniform bending effect in beam-columns
Δ	Deflection of a point of a structure
ASTM	American Society for Testing and Materials
CISC	Canadian Institute of Steel Construction
CPMA	Canadian Paint Manufacturers' Association
CSCC	Canadian Steel Construction Council
CSA	Canadian Standards Association
SSRC	Structural Stability Research Council

PART ONE
CAN/CSA-S16.1-M89, "LIMIT STATES DESIGN OF STEEL STRUCTURES"

General

This Standard is reprinted with the permission of the Canadian Standards Association and contains all revisions approved at the time of printing.

CSA Standards are subject to periodical review and amendments will be published from time to time by CSA.

National Standard of Canada CAN/CSA-S16.1-M8

Limit States Design of Steel Structures

Prepared by
Canadian Standards Association

Approved by
Standards Council of Canada

Technical Editor: J. N. Carnegie
Managing Editor: Bernard Kelly

ISSN 0317-5669
Published in December 1989 by Canadian Standards Association
178 Rexdale Boulevard, Rexdale (Toronto), Ontario, Canada M9W 1R3.

Contents

Appendices

Technical Committee on Steel Structures

D.J.L. Kennedy	University of Alberta, Edmonton	*Chairman*
G.L. Kulak	University of Alberta, Edmonton	*Vice-Chairman*
M.I. Gilmor	Canadian Institute of Steel Construction, Willowdale, Ontario	*Secretary*
D.E. Allen	National Research Council of Canada, Ottawa, Ontario (resigned 1989)	
M. Archer-Shee	Canadian Welding Bureau, Dartmouth, Nova Scotia	
M. Aregawi	Department of Buildings and Inspections, Toronto, Ontario	
G.S. Baichwal	Ontario Hydro, Toronto	
W.W. Baigent	Canron Inc., Rexdale, Ontario	
P.C. Birkemoe	University of Toronto, Toronto, Ontario	
W. Blackwell	Blackwell Engineering Limited, Toronto, Ontario	
D. Calder	Calder Engineering Inc., North Vancouver, British Columbia	
M. Comeau	Campbell Comeau Engineering Ltd., Halifax, Nova Scotia	
J. Dawe	University of New Brunswick, Fredericton	
J.C. Draper	CP Rail Limited, Montreal, Quebec	
R.W. Dryden	Canadian Building Systems, Burlington, Ontario	
T. Gaul	Empire Iron Works Limited, Winterburn, Alberta	
J.E. Henderson	Henderson Engineering Services Milton, Ontario	
M.U. Hosain	University of Saskatchewan, Saskatoon	

R.G. Keen	Stelco Inc., Hamilton, Ontario	
J.K. Komocki	National Defence Department—Canada, Ottawa, Ontario (resigned 1989)	
C.J. Montgomery	The Cohos Evamy Partners, Edmonton, Alberta	
C.D. Nixon	National Capital Region, Ottawa, Ontario (resigned 1989)	
A. Picard	Université Laval, Québec, Québec	
R.G. Redwood	McGill University, Montreal, Quebec	
E.J. Rohacek	Quinn Dressel Associates, Toronto, Ontario	
R.M. Schuster	University of Waterloo, Waterloo, Ontario	*Associate*
J. Springfield	Carruthers and Wallace Limited, Toronto, Ontario	
D.C. Stringer	Dominion Bridge, Montreal, Quebec	
D.L. Tarlton	Canadian Sheet Steel Building Institute, Willowdale, Ontario	
R. Vincent	Le Groupe Canam Manac Inc., Montréal, Québec	
B.A. Wing	The Algoma Steel Corporation Limited, Mississauga, Ontario	
A.R. Zaki	CN Rail, Montreal, Quebec	
J. Carnegie	Canadian Standards Association, Rexdale, Ontario	*Standards Administrator, Nonvoting*

Preface

This is the fourth edition in Canada of a general limit states design Standard for steel structures. It is simply entitled *Limit States Design of Steel Structures* because it is appropriate for the design of a broad range of structures. The scope recognizes that the requirements for the design of specific structures, such as bridges for example, are given in other Standards and that supplementary requirements may be necessary for some particular structures. The Standard sets out minimum requirements for the design of steel structures and, it is expected, will only be used by engineers competent in this field.

The Standard is prepared in SI units and supersedes the SI edition published in 1984. The three limit states SI editions of 1989, 1984, and 1978 were based on the first limit states edition, written in imperial units, introduced in 1974. The 1974 edition has now been withdrawn as it was technically out of date. The four limit states design editions were preceded by seven working stress design editions published in 1969, 1965, 1961, 1954, 1940, 1930, and 1924. All of these have now been withdrawn.

The sequence of eleven Standards marks a progression in technology, a change in the system of units, and more importantly, a change in the philosophy of design from working stress to limit states.

Although the basic limit states format as set out in 1974 has proven itself in use and remains unaltered, a considerable number of technical changes reflecting the latest research developments have been incorporated in this edition. These are based on an increased understanding of the behaviour of structural materials, elements, and members and thus of the overall behaviour of structures. This increased understanding has itself been enhanced by the limit states approach as the designer explicitly recognizes the different possible modes of failure and designs against them.

Major changes have been introduced in the analysis and design of beam-columns. At the least, a second order elastic analysis of the structure, taking lateral deformations into account directly, must be performed. Beam-columns are checked for the three possible failure modes related to cross-sectional strength, overall member strength, and lateral torsional buckling. The design of tension members, now including provisions for shear lag and tear-out of a block of material, is related to the two basic failure modes of net-section fracture and gross-section yield. Shear transfer in composite beams has received special attention. Provisions for the use of snug-tight bolts have been introduced. An entirely new clause on seismic design gives requirements for different types of lateral-load-resisting systems to be used in conjunction with the seismic design requirements of the *National Building Code of Canada.* In addition to these and other technical changes, significant changes have been introduced in the clauses on drawings, materials, design requirements, open-web steel joists, fabrication, and erection to clarify their intent.

The clauses in the Standard relating to fabrication and erection serve to show that design cannot be considered in isolation but that it is part of the design and construction sequence.

This Standard has been adopted by the Associate Committee on the National Building Code as the reference Standard for steel structures in Section 4.6 of the *National Building Code* for 1990.

This Standard was prepared by the Technical Committee on Steel Structures for Buildings under the jurisdiction of the Standards Steering Committee on Structures, and was approved by these Committees. This Standard has been approved as a National Standard of Canada by the Standards Council of Canada.

December 1989.

Notes:

(1) *Use of the masculine gender in this Standard is not meant to exclude the feminine gender when applied to persons. Similarly, use of the singular does not exclude the plural (and vice versa) when the sense allows.*

(2) *Although the intended primary application of this Standard is stated in its Scope, it is important to note that it remains the responsibility of the user of the Standard to judge its suitability for his particular purpose.*
(3) *CSA Standards are subject to periodic review, and suggestions for their improvement will be referred to the appropriate committee.*
(4) *All enquiries regarding this Standard, including requests for interpretation, should be addressed to Canadian Standards Association, Standards Division, 178 Rexdale Boulevard, Rexdale, Ontario M9W 1R3.*
Requests for interpretation should
(a) define the problem, making reference to the specific clause, and, where appropriate, include an illustrative sketch;
(b) provide an explanation of circumstances surrounding the actual field condition; and
(c) be phrased where possible to permit a specific "yes" or "no" answer.
Interpretations are published in CSA Information Update. *For subscription details and a free sample copy, write to CSA Sales Promotions or telephone (416) 747-4116.*

CAN/CSA-S16.1-M89
Limit States Design of Steel Structures

1. Scope and Application

1.1

This Standard provides rules and requirements for the design, fabrication, and erection of steel structures where the design is based on limit states. The term "steel structures" refers to structural members and frames that consist primarily of structural steel components, including the detail parts, welds, bolts, or other fasteners required in fabrication and erection. This Standard also applies to structural steel components in structures framed in other materials.

1.2

Requirements for steel structures such as bridges, antenna towers, offshore structures, and cold-formed steel structural members are given in other CSA Standards.

1.3

This Standard applies unconditionally to steel structures, except that supplementary rules or requirements may be necessary for
(a) unusual types of construction;
(b) mixed systems of construction;
(c) steel structures that
(i) have great height or spans;
(ii) are required to be moveable or be readily dismantled;
(iii) are exposed to severe environmental conditions or possible severe loads such as those resulting from vehicle impact or chemical explosion;
(iv) are required to satisfy aesthetic, architectural, or other requirements of a nonstructural nature;
(v) employ materials or products not listed in Clause 5; or
(vi) have other special features that could affect design, fabrication, or erection; and
(d) tanks, stacks, other platework structures, poles, and piling.

1.4

A rational design based on theory, analysis, and engineering practice, acceptable to the regulatory authority, may be used in lieu of the formulae provided in this Standard. In such cases the design shall provide nominal margins (or factors) of safety at least equal to those intended in the provisions of this Standard (see Appendix E).

2. Definitions and Symbols

2.1 Definitions

The following definitions apply to this Standard:

Approved — approved by the regulatory authority.

Camber — the deviation from straightness of a member or any portion of a member with respect to its major axis. Frequently, camber is specified and produced in a member to compensate for deflections that will occur in the member when loaded. (See Clause 6.2.2.) Unspecified camber is sometimes referred to as bow.

Concrete — Portland cement concrete in accordance with CSA Standard A23.1.

Designer — the professional engineer responsible for the design.

Erection tolerances — tolerances related to the plumbness, alignment, and level of the piece as a whole. The deviations are determined by considering the locations of the ends of the piece. (See Clause 30.)

Fabrication tolerances — tolerances allowed from the nominal dimensions and geometry, such as the cutting to length, finishing of ends, cutting of bevel angles, and for fabricated members, out-of-straightness such as sweep and camber. (See Clause 28.)

Limit states — those conditions of a structure in which the structure ceases to fulfil the function for which it was designed. Those states concerning safety are called the ultimate limit states and include exceeding of load-carrying capacity, overturning, sliding, fracture, and fatigue. Those states that restrict the intended use and occupancy of the structure are called serviceability limit states and include deflection, vibration, and permanent deformation.

Factors

Load factor, α — a factor, given in Clause 7.2, applied to a specified load for the limit states under consideration that takes into account the variability of the loads and load patterns and the analysis of their effects.

Load combination factor, ψ — a factor, given in Clause 7.2, applied to factored loads other than dead load that takes into account the reduced probability of a number of loads from different sources acting simultaneously.

Importance factor, γ — a factor, given in Clause 7.2, applied to factored loads that takes into account the consequences of collapse as related to the use and occupancy of the structure.

Resistance factor, ϕ — a factor, given in the appropriate clauses in this Standard, applied to a specified material property or the resistance of a member, connection, or structure that, for the limit state under consideration, takes into account the variability of material properties, dimensions, workmanship, type of failure, and uncertainty in prediction of member resistance. To maintain simplicity of the design formulae in this Standard, the type of failure and the uncertainty in prediction of member resistance have been incorporated in the expressions of member resistance. (See Appendix E for a more detailed discussion.)

Loads

Gravity load — (newtons) is equal to the mass of the object (kilograms) being supported multiplied by the acceleration due to gravity, g (9.81 m/s^2).

Specified loads (D, L, Q, and T) — those loads prescribed by the regulatory authority (see Clause 7.1).

Factored load — the product of a specified load and its load factor.

Mill tolerances — variations allowed from the nominal dimensions and geometry with respect to cross-sectional area, nonparallelism of flanges, and out-of-straightness such as sweep or camber in the product as manufactured and given in CSA Standard G40.20.

Regulatory authority — a federal/provincial/municipal ministry, department, board, agency, or commission that has responsibility for regulating, by statute, the use of products, materials, or services.

Resistance

Resistance, R — the resistance of a member, connection, or structure as calculated in accordance with this Standard based on the specified material properties and nominal dimensions.

Factored resistance, ϕR — the product of the resistance and the appropriate resistance factor.

Sweep — the deviation from straightness of a member or any portion of a member with respect to its minor axis.

2.2 Symbols

The following symbols are used throughout this Standard. Deviations from them and additional nomenclature are noted where they appear.

A = area

A_b = cross-sectional area of a bolt based on its nominal diameter

A_c = transverse area of concrete between longitudinal shear planes; cross-sectional area of concrete in composite columns

A_{cv} = the critical area of two longitudinal shear planes, one on each side of the area A_c, extending from the point of zero moment to the point of maximum moment

A_f = flange area

A_g = gross area

A_m = area of fusion face

A_n = critical net area

A_{ne} = effective net area

A'_{ne} = effective net area reduced for shear lag

A_p = concrete pull-out area

A_r = area of reinforcing steel

A_s = area of steel section including cover plates; area of bottom (tension) chord of steel joist; area of stiffener or pair of stiffeners

A_{sc} = area of steel shear connector

A_w = web area; shear area; effective throat area of weld

a = centre-to-centre distance between transverse web stiffeners; depth of concrete compression zone

a' = length of cover plate termination

a/h = aspect ratio; ratio of distance between stiffeners to web depth

B = bearing force in a member or component under specified load

B_f = bearing force in a member or component under factored load

B_r = factored bearing resistance of a member or component

b = width of stiffened or unstiffened compression elements; design effective width of concrete or cover slab

C = compressive force in a member or component under specified load; axial load

C_e = Euler buckling strength = $\pi^2 EI/L^2$

C_{ec} = Euler buckling strength of a concrete-filled hollow structural section

C_f = compressive force in a member or component under factored load; factored axial load

C_r = factored compressive resistance of a member or component; factored compressive resistance of steel acting at the centroid of that part of the steel area in compression

C_{rc} = factored compressive resistance of a composite column

C'_r = compressive resistance of concrete acting at the centroid of the concrete area in compression

C_w = warping torsional constant (mm^6)

C_y = axial compressive load at yield stress
c_1 = coefficient used to determine slip resistance
D = outside diameter of circular sections; diameter of rocker or roller; stiffener factor; dead load
d = depth; overall depth of a section; diameter of bolt or stud
d_b = depth of beam
E = elastic modulus of steel (200 000 MPa assumed)
E_c = elastic modulus of concrete
e = end distance; lever arm between the compressive resistance, C_r, and the tensile resistance, T_r
e′ = lever arm between the compressive resistance, C_r', of concrete and tensile resistance, T_r, of steel
F = strength or stress
F_{cr} = critical plate-buckling stress in compression, flexure, or shear
F_{cre} = elastic critical plate-buckling stress in shear
F_{cri} = inelastic critical plate-buckling stress in shear
F_s = ultimate shear stress
F_{sr} = allowable stress range in fatigue
F_{st} = factored axial force in the stiffener
F_t = tension-field post-buckling stress
F_u = specified minimum tensile strength
F_y = specified minimum yield stress, yield point, or yield strength
F_y' = yield level including effect of cold-working
F_{yr} = specified yield strength of reinforcing steel
f_c' = specified compressive strength of concrete at 28 days
G = shear modulus of steel (77 000 MPa assumed)
g = transverse spacing between fastener gauge lines (gauge distance)
h = clear depth of web between flanges; height of stud; storey height
h_d = depth of steel deck
I = moment of inertia of cover-plated section
J = St. Venant torsion constant
K = effective length factor
KL = effective length
k = distance from outer face of flange to web-toe of fillet of I-shaped sections
k_b = buckling coefficient
k_s = mean slip coefficient
k_v = shear buckling coefficient
L = length; length of longitudinal weld; live load
L_c = length of channel shear connector
L_{cr} = maximum unbraced length adjacent to a plastic hinge
L_n = net length (that is, gross length less design allowance for holes within the length)
M = bending moment in a member or component under specified load
M_f = bending moment in a member or component under factored load

M_{fc} = bending moment in a girder, under factored load, at theoretical cut-off point

M_{fg} = first-order moment under factored gravity loads determined assuming that there is no lateral translation of the frame

M_{ft} = first-order translational moment under factored lateral loads, or the moment resulting from lateral translation of an unsymmetrical frame, or the moment resulting in an unsymmetrically loaded frame under factored gravity loading (see Clause 8.6.2)

M_{f1} = smaller factored end moment of a beam-column; factored bending moment at a point of concentrated load

M_{f2} = larger factored end moment of a beam-column

M_p = plastic moment = ZF_y

M_r = factored moment resistance of a member or component

M_{rc} = factored moment resistance of a composite beam; factored moment resistance of a column reduced for the presence of axial load

M_u = critical elastic moment of a laterally unbraced beam

M_y = yield moment = SF_y

m = number of faying surfaces or shear planes in a bolted joint, equal to 1.0 for bolts in single shear and 2.0 for bolts in double shear

N = length of bearing of an applied load

n = number of bolts; number of shear connectors required between the point of maximum positive bending moment and the adjacent point of zero moment

n' = number of shear connectors required between any concentrated load and nearest point of zero moment in a region of positive bending moment

P = force to be developed in a cover plate

Q = wind or earthquake load

Q_r = sum of the factored resistances of all shear connectors between points of maximum and zero moment

q_r = factored resistance of a shear connector

q_{rr} = factored resistance of a shear connector in a ribbed slab

q_{rs} = factored resistance of a shear connector in a solid slab

R = end reaction or concentrated transverse load applied to a flexural member; resistance of a member, connection, or structure; transition radius

r = radius of gyration

r_y = radius of gyration of a member about its weak axis

S = elastic section modulus of a steel section

s = centre-to-centre longitudinal spacing (pitch) of any two successive fastener holes

T = tensile force in a member or component under specified load; load due to temperature change, etc (see Clause 7.1.1)

T_f = tensile force in a member or component under factored load

T_r = factored tensile resistance of a member or component; in composite construction, factored tensile resistance of the steel acting at the centroid of that part of the steel area in tension

T_y = axial tensile load at yield stress

t = thickness; thickness of flange; average flange thickness of channel shear connector

t_c = concrete or cover slab thickness

U_1 = factor to account for moment gradient and for second-order effects of axial force acting on the deformed member

U_2 = amplification factor to account for second-order effects of gravity loads acting on the laterally displaced storey

V = shear force in a member or component under specified load

V_f = shear force in a member or component under factored load

V_h = total horizontal shear to be resisted at the junction of the steel section or joist and the slab or steel deck

V_r = factored shear resistance of a member or component

V_s = slip resistance of a bolted joint

V_{st} = factored shear force in column web to be resisted by stiffener

w = web thickness; width of plate

w_d = average width of flute of steel deck

w' = sum of thickness of column web plus doubler plates

w_n = net width (that is, gross width less design allowance for holes within the width)

X_u = ultimate strength as rated by the electrode classification number

x = subscript relating to strong axis of a member

y = subscript relating to weak axis of a member, distance from centroid of cover plate to neutral axis of cover-plated section

Z = plastic section modulus of a steel section

α = load factor

β = value used in bolt interaction equation

γ = importance factor

λ = nondimensional slenderness ratio in column formula

Δ_f = relative first-order lateral (translational) displacement of the storey due to factored loads (coincident with M_{ft})

ΣC_f = sum of factored axial compressive loads of all columns in the storey

ΣV_f = sum of factored lateral loads above the storey; the total first-order storey shear

κ = ratio of the smaller factored moment to the larger factored moment at opposite ends of the unbraced length, positive for double curvature and negative for single curvature

ϕ = resistance factor (see definition under Factors, in Clause 2.1)

ψ = load combination factor

ω_1 = coefficient used to determine equivalent uniform bending effect in beam-columns

ω_2 = coefficient to account for increased moment resistance of a laterally unsupported beam segment when subject to a moment gradient

ρ = density of concrete; slenderness ratio

2.3 Units

Equations and expressions appearing in this Standard are compatible with the following SI (metric) units:
force: N (newtons)
length: mm (millimetres)
moment: N•mm
strength or stress: MPa (megapascals)

3. Reference Publications

3.1

This Standard refers to the following publications and where such reference is made it shall be to the latest edition and revisions thereto, unless otherwise specified:

CSA Standards

CAN/CSA-A23.1-M90,
Concrete Materials and Methods of Concrete Construction;

CAN3-A23.3-M84,
Design of Concrete Structures for Buildings;

B95-1962,
Surface Texture (Roughness, Waviness, and Lay);

CAN/CSA-G40.20-M87,
General Requirements for Rolled or Welded Structural Quality Steel;

CAN/CSA-G40.21-M87,
Structural Quality Steels;

G164-M1981,
Hot Dip Galvanizing of Irregularly Shaped Articles;

CAN/CSA-S37-M86,
Antennas, Towers, and Antenna-Supporting Structures;

CAN/CSA-S136-M89,
Cold Formed Steel Structural Members;

CAN3-S304-M84,
Masonry Design for Buildings;

W47.1-1983,
Certification of Companies for Fusion Welding of Steel Structures;

W48.1-M1980,
Mild Steel Covered Arc Welding Electrodes;

W48.3-M1982,
Low-Alloy Steel Covered Arc Welding Electrodes;

W48.4-M1980,
Solid Mild Steel Filler Metals for Gas Shielded Arc Welding;

W48.5-M1990,
Carbon Steel Electrodes for Flux and Metal Cored Arc Welding;

W48.6-M1980,
Bare Mild Steel Electrodes and Fluxes for Submerged-Arc Welding;

W55.3-1965,
Resistance Welding Qualification Code for Fabricators of Structural Members Used in Buildings;

W59-M1989,
Welded Steel Construction (Metal Arc Welding).

ASTM Standards

A27/A27M-87,
Specification for Steel Castings, Carbon, for General Application;

A108-89a,
Specification for Steel Bars, Carbon, Cold Finished, Standard Quality;

A148/A148M-89a,
Specification for Steel Castings, High Strength, for Structural Purposes;

A307-89,
Specification for Carbon Steel Bolts and Studs, 60 000 psi Tensile Strength;

A325-89,
Specification for High-Strength Bolts for Structural Steel Joints;

A325M-89,
Specification for High-Strength Bolts for Structural Steel Joints [Metric];

A486/A486M-84,
Specification for Steel Castings for Highway Bridges;

A490-89,
Specification for Heat-Treated Steel Structural Bolts, 150 ksi MinimumTensile Strength;

A490M-89,
Specification for High-Strength Steel Bolts, Classes 10.9 and 10.9.3, for Structural Steel Joints [Metric];

A514/A514M-88,
Specification for High-Yield-Strength, Quenched and Tempered Alloy Steel Plate, Suitable for Welding;

A521-76 (Reapproved 1987),
Specification for Steel, Closed-Impression Die Forgings for General Industrial Use;

A570/A570M-88,
Specification for Steel, Sheet and Strip, Carbon, Hot-Rolled, Structural Quality;

A668-85a,
Specification for Steel Forgings, Carbon and Alloy, for General Industrial Use.

CGSB* Standards
CAN/CGSB-1.40-M89,
Primer, Structural Steel, Oil Alkyd Type;

1-GP-81M-1978,
Primer, Alkyd, Air Drying and Baking for Vehicles and Equipment;

CAN/CGSB-1.140-M89,
Oil Alkyd Type Red Lead Iron Oxide Primer;

1-GP-166M-1979,
Primer, Basic Lead Silico-Chromate, Oil Alkyd Type.

CISC/CPMA† Standards
1-73a,
A Quick-Drying One-Coat Paint for Use on Structural Steel;

2-75,
A Quick-Drying Primer for Use on Structural Steel.

CISC‡
Code of Standard Practice for Structural Steel, 1980.

National Building Code of Canada, 1990

SSPC§ Specifications
SP 2-82,
Hand Tool Cleaning;

SP 3-82,
Power Tool Cleaning;

SP 4-82,
Flame Cleaning of New Steel;

SP 5-85,
White Metal Blast Cleaning;

SP 6-85,
Commercial Blast Cleaning;

SP 7-85,
Brush-Off Blast Cleaning;

SP 10-85,
Near-White Blast Cleaning.

Structural Stability Research Council
*Guide to Stability Design Criteria for Metal Structures, Fourth Edition.***

Research Council on Structural Connections
*Guide to Design Criteria for Bolted and Riveted Joints, Second Edition.***

*Canadian General Standards Board.
†Canadian Institute of Steel Construction/Canadian Paint Manufacturers' Association.
‡Canadian Institute of Steel Construction.
§Steel Structures Painting Council.
**Wiley Interscience.

4. Drawings

4.1 Design Drawings

4.1.1

Design drawings shall be drawn to a scale adequate to convey the required information. The drawings shall show a complete design of the structure with members suitably designated and located, including such dimensions and detailed description as necessary to permit the preparation of fabrication and erection documents. Floor levels, column centres, and offsets shall be dimensioned. The term "drawings" may include computer output and other data.

4.1.2

Design drawings shall designate the design standards used, shall show clearly the type or types of construction (as defined in Clause 8) to be employed, shall show the category of the structural system used for seismic design (see Clause 27), and shall designate the material or product standards applicable to the members and details depicted (see Clause 5). Drawings shall give the governing combinations of shears, moments, and axial forces to be resisted by the connections.

4.1.3

Where high-strength bolted joints are required to resist shear between connected parts, design drawings shall indicate the type of joint, slip-critical or bearing-type, to be provided (see Clause 23).

4.1.4

The size and location of stiffeners, reinforcement, and bracing required to stabilize compression elements, and the camber of beams, girders, and trusses shall be shown on design drawings.

4.2 Fabrication and Erection Documents

4.2.1 Connection Design Details

Connection design details shall be prepared in advance of preparing shop details and submitted to the designer for confirmation that the intent of the design is met. Connection design details shall provide details of typical and special types of connections, and other data necessary for the preparation of shop details. Connection design details shall be referenced to the design drawings.

4.2.2 Shop Details

Shop details shall be prepared in advance of fabrication and submitted to the designer for review. Shop details shall provide complete information for the fabrication of various members and components of the structure, including the required material and product standards; the location, type, and size of all mechanical fasteners; bolt installation requirements; and welds.

4.2.3 Erection Diagrams

Erection diagrams shall be submitted to the designer for review. Erection diagrams are general arrangement drawings showing the principal dimensions of the structure, piece marks, sizes of the members, size and type of bolts, bolt installation requirements, elevations of column bases, all necessary dimensions and details for setting anchor bolts, and all other information necessary for the assembly of the structure.

4.2.4 Erection Procedures

Erection procedures shall outline the construction methods, erection sequence, temporary bracing requirements, and other engineering details necessary for shipping, erecting, and maintaining the stability of the steel frame. Erection procedures shall be supplemented by drawings and sketches to identify the location of stabilizing elements. Erection procedures shall be submitted for review when so specified.

4.2.5 Field Work Details

Field work details shall be submitted to the designer for review. Field work details shall provide complete information for modifying fabricated members in the shop or on the job site. All operations required to modify the member shall be shown on the field work details. If extra materials are necessary to make modifications, shop details shall be required.

5. Material: Standards and Identification

5.1 Standards

5.1.1 General

Acceptable material and product standards and specifications (latest editions) for use under this Standard are listed in Clauses 5.1.3 to 5.1.9, inclusive. Materials and products other than those listed also may be used if approved. Approval shall be based on published specifications that establish the properties, characteristics, and suitability of the material or product to the extent and in the manner of those covered in listed standards.

5.1.2 Strength Levels

The yield strength, F_y, and the tensile strength, F_u, used as the basis for design shall be the specified minimum values as given in the material and product standards and specifications. The levels reported on mill test certificates shall not be used as the basis for design.

5.1.3 Structural Steel

CAN/CSA G40.21-M,
Structural Quality Steels.

5.1.4 Sheet Steel

ASTM A570,
Specification for Steel, Sheet and Strip, Carbon, Hot-Rolled, Structural Quality.

Other standards for structural sheet are listed in Clause 2.4 of CSA Standard S136. Only structural-quality sheet standards that specify chemical composition and mechanical properties will be acceptable for use with this Standard. Mill test certificates that list the chemical composition and the mechanical properties shall be available, upon request, in accordance with Clause 5.2.1(a).

5.1.5 Cast Steel

ASTM A27,
Specification for Steel Castings, Carbon, for General Application;

ASTM A148,
Specification for Steel Castings, High Strength, for Structural Purposes;

ASTM A486,
Steel Castings for Highway Bridges.

5.1.6 Forged Steel

ASTM A521,
Specification for Steel, Closed-Impression Die Forgings for General Industrial Use;

ASTM A668,
Specification for Steel Forgings, Carbon and Alloy, for General Industrial Use.

5.1.7 Bolts

ASTM A307,
Specification for Carbon Steel Bolts and Studs, 60 000 psi Tensile Strength;

ASTM A325,
Specification for High-Strength Bolts for Structural Steel Joints;

ASTM A325M,
Specification for High-Strength Bolts for Structural Steel Joints [Metric];

ASTM A490,
Specification for Heat-Treated Steel Structural Bolts, 150 ksi Minimum Tensile Strength;

ASTM A490M,
Specification for High-Strength Steel Bolts, Classes 10.9 and 10.9.3, for Structural Steel Joints [Metric].

Note: *Before specifying metric bolts, the designer should check on their current availability in the quantities required.*

5.1.8 Welding Electrodes

CSA W48.1,
Mild Steel Covered Arc Welding Electrodes;

CSA W48.3,
Low-Alloy Steel Covered Arc Welding Electrodes;

CSA W48.4,
Solid Mild Steel Filler Metals for Gas Shielded Arc Welding;

CSA W48.5,
Carbon Steel Electrodes for Flux and Metal Cored Arc Welding;

CSA W48.6,
Bare Mild Steel Electrodes and Fluxes for Submerged-Arc Welding.

5.1.9 Studs

ASTM A108,
Specification for Steel Bars, Carbon, Cold Finished, Standard Quality [Grades 1015 and 1018].

5.2 Identification

5.2.1 Methods

The materials and products used shall be identified as to specification, including type or grade, if applicable, by one of the following means, except as provided in Clauses 5.2.2 and 5.2.3:
(a) mill test certificates or producer's certificates satisfactorily correlated to the materials or products to which they pertain; and
(b) legible markings on the material or product made by its producer in accordance with the applicable material or product standard.

5.2.2 Unidentified Structural Steel

Unidentified structural steel shall not be used unless approved by the building designer. If the use of unidentified steel is authorized, F_y shall be taken as 210 MPa and F_u shall be taken as 380 MPa.

5.2.3 Tests to Establish Identification

Unidentified structural steel may be tested to establish identification when permitted by the building designer. Testing shall be done by an approved testing agency in accordance with CSA Standard G40.20. The test results, taking into account both mechanical properties and chemical composition, shall form the basis for classifying the steel as to specification. Once classified, the specified minimum values for steel of that specification grade shall be used as the basis for design (see Clause 5.1.2).

5.2.4 Affidavit

The fabricator, if requested, shall provide an affidavit stating that the materials and products that have been used in fabrication conform to the applicable material or product standards called for by the design drawings or specifications.

6. Design Requirements

6.1 General

6.1.1 Limit States

As set out in this Standard, steel structures shall be designed to be serviceable during the useful life of the structure and safe from collapse during construction and during the useful life of the structure. Limit states define the various types of collapse and unserviceability that are to be avoided; those concerning safety are called the ultimate limit states (strength, overturning, sliding, and fatigue) and those concerning serviceability are called the serviceability limit states (deflections, vibration, and permanent deformation). The object of limit states design calculations is to keep the probability of a limit state being reached below a certain value previously established for the given type of structure. This is achieved in this Standard by the use of load factors applied to the specified loads (see Clause 7) and resistance factors applied to the specified resistances (see Clause 13).

The various limit states are set out in this Clause. Some of these relate to the specified loads and others to the factored loads. Camber, provisions for expansion and contraction, and corrosion protection are further design requirements related to serviceability and durability. All limit states shall be considered in the design.

6.1.2 Structural Integrity

The general arrangement of the structural system and the connection of its members shall be designed to provide resistance to widespread collapse as a consequence of local failure. The requirements of this Standard generally provide a satisfactory level of structural integrity for steel structures. Supplementary provisions may be required for structures where accidental loads such as vehicle impact or explosion are likely to occur (see Clause 1.3). (Further guidance is contained in Chapter 4, Commentary C of the Supplement to the *National Building Code of Canada*.)

6.2 Requirements Under Specified Loads

6.2.1 Deflection

6.2.1.1

Steel members and frames shall be proportioned so that deflections are within acceptable limits for the nature of the materials to be supported and for the intended use and occupancy.

6.2.1.2

In the absence of a more detailed evaluation, see Appendix I for recommended values for deflections.

6.2.1.3

Roofs shall be designed to withstand any additional loads likely to occur as a result of ponding. (See Clause 7.1.1.) (Further guidance is contained in Chapter 4, Commentary I of the Supplement to the *National Building Code of Canada*.)

6.2.2 Camber

6.2.2.1

Camber of beams, trusses, or girders, if required, shall be called for on the design drawings. Generally, trusses and crane girders of 25 m or greater span should be cambered for approximately the dead-plus-half-live-load deflection. (See Clause 16 for requirements for open-web joists and Clause 28.9.5 for fabrication tolerances.)

6.2.2.2

Any special camber requirements necessary to bring a loaded member into proper relation with the work of other trades shall be stipulated on the design drawings.

6.2.3 Dynamic Effects

6.2.3.1

Suitable provision shall be made in the design for the effect of live load that induces impact or vibration, or both. In severe cases, such as structural supports for heavy machinery that causes substantial impact or vibration when in operation, the possibility of harmonic resonance, fatigue, or unacceptable vibration shall be investigated.

6.2.3.2

Special consideration shall be given to floor systems susceptible to vibration, such as large open floor areas free of partitions, to ensure that such vibration is acceptable for the intended use and occupancy. (Guidance regarding floor vibrations is contained in Appendix G.)

6.2.3.3

Unusually flexible structures (generally those whose ratio of height to effective resisting width exceeds 4:1) shall be investigated for lateral vibrations under dynamic wind load. Lateral accelerations of the structure shall be checked to ensure that such accelerations are acceptable for the intended use and occupancy. (Information on lateral accelerations under dynamic wind loads can be found in Appendix H.)

6.2.4 Resistance to Fatigue

Structural steelwork shall be designed to resist the effects of fatigue under the specified loads in accordance with Clause 14.

6.2.5 Prevention of Permanent Deformation

6.2.5.1

For composite beams that are unshored during construction, the stress in the steel beam induced by the specified loads shall not exceed $0.90F_y$. The stress shall be calculated as the sum of the stresses due to any specified loads applied before the concrete strength reaches $0.75f'_c$, calculated on the steel section alone, plus the stresses at the same location, due to the remaining specified loads calculated on the composite section.

6.2.5.2

Slip-critical joints, in which the design load is assumed to be transferred by the slip resistance of the clamped faying surfaces, shall be proportioned using the provisions of Clause 13.12 to resist without slipping the moments and forces induced by the specified loads (see Clause 23).

6.3 Requirements Under Factored Loads

6.3.1 Strength

Structural steelwork shall be proportioned to resist moments and forces resulting from the application of the factored loads acting in the most critical combination, taking into account the resistance factors as specified in the appropriate clauses of this Standard.

6.3.2 Overturning

The building or structure shall be designed to resist overturning resulting from the application of the factored loads acting in the most critical combination, taking into account the importance of the building as specified in Clause 7, and taking into account the resistance factors as specified in the appropriate clauses of this Standard.

6.4 Expansion and Contraction

Suitable provision shall be made for expansion and contraction commensurate with the service and erection conditions of the structure.

6.5 Corrosion Protection

6.5.1

Steelwork shall have sufficient corrosion protection to minimize any corrosion likely to occur in the service environment.

6.5.2

Interiors of buildings conditioned for human comfort may be generally assumed to be noncorrosive environments; however, the need for corrosion protection shall be assessed and protection shall be furnished in those buildings where it is deemed to be necessary.

6.5.3

Corrosion protection of the inside surfaces of enclosed spaces permanently sealed away from any external source of oxygen is unnecessary.

6.5.4

The minimum required thickness of steelwork situated in a noncorrosive environment and therefore not requiring corrosion protection is governed by the provisions of Clause 11.

6.5.5

Corrosion protection shall be provided by means of suitable alloying elements in the steel, by protective coatings, or by other effective means, either singly or in combination.

6.5.6

Localized corrosion likely to occur from entrapped water, excessive condensation, or other factors shall be minimized by suitable design and detail. Where necessary, positive means of drainage shall be provided.

6.5.7

If the corrosion protection, specified for steelwork exposed to the weather or to other environments in which progressive corrosion can occur, is likely to require maintenance or renewal during the service life of the structure, the steelwork so protected shall have a minimum thickness of 4.5 mm (exclusive of fill plates and shims).

7. Loads and Safety Criterion

7.1 Specified Loads

7.1.1

Except as provided for in Clauses 7.1.2 and 7.1.3, the following loads and influences as specified by the regulatory authority shall be considered in the design of structural steelwork:

D — Dead loads, including the weight of steelwork and all permanent materials of construction, partitions, stationary equipment, and additional weight of concrete and finishes resulting from deflections of supporting members, and the forces due to prestressing;

L — Live loads, including loads due to intended use and occupancy of structures, moveable equipment, snow, rain, soil, or hydrostatic pressure, impact, and any other live loads stipulated by the regulatory authority;

Q — Wind or earthquake loads;

T — Influences resulting from temperature changes, shrinkage, or creep of component materials or from differential settlement.

7.1.2

If it can be shown by engineering principles or if it is known from experience that neglect of some or all of the effects due to T does not affect the structural safety or serviceability, they need not be considered in the calculations.

7.1.3

Suitable provision shall be made for loads imposed on the steel structure during its erection. During subsequent construction, suitable provision shall be made to support the construction loads on the steel structure with an adequate margin of safety.

7.2 Safety Criterion and Effect of Factored Loads

7.2.1

The structural steelwork shall be designed to have sufficient strength or stability, or both, such that

Factored resistance ≥ Effect of factored loads

where the factored resistance is determined in accordance with other clauses of this Standard and the effect of factored loads is determined in accordance with Clauses 7.2.2 to 7.2.5. In cases of overturning, uplift, and stress reversal, no positive anchorage is required if the stabilizing effect of dead load multiplied by a load factor of less than 1.00 given in Clause 7.2.3 is greater than the effect of loads tending to cause overturning, uplift, and stress reversal multiplied by load factors greater than 1.00 given in Clause 7.2.3.

7.2.2

Effect of factored loads, in force units, is the structural effect due to the specified loads multiplied by load factors, α, defined in Clause 7.2.3; a load combination factor, ψ, defined in Clause 7.2.4; and an importance factor, γ, defined in Clause 7.2.5. The factored load combinations shall be taken as follows:

$$\alpha_D D + \gamma\psi\,(\alpha_L L + \alpha_Q Q + \alpha_T T)$$

7.2.3

Load factors, α, shall be taken as follows:

α_D = 1.25; except that when the dead load resists overturning, uplift, or reversal of load effect, $\alpha_D = 0.85$;

α_L = 1.50;

α_Q = 1.50 for wind or 1.00 for earthquake;

α_T = 1.25.

7.2.4

The load combination factor, ψ, shall be taken as follows:

(a) when only one of L, Q, and T act, $\psi = 1.00$;
(b) when two of L, Q, and T act, $\psi = 0.70$;
(c) when all of L, Q, and T act, $\psi = 0.60$.

The most unfavourable effect shall be determined by considering L, Q, and T acting alone with $\psi = 1.00$, or in combination with $\psi = 0.70$ or 0.60.

7.2.5

The importance factor, γ, shall be not less than 1.00, except for those structures where it can be shown that collapse is not likely to cause injury or other serious consequences, it shall be not less than 0.80.

8. Analysis of Structure

8.1 General

8.1.1

In proportioning the structure to meet the various design requirements of Clause 6, the methods of analysis given in this Clause shall be used. The distribution of internal forces and bending moments shall be determined both under the specified loads to satisfy the requirements of serviceability and fatigue in Clause 6 and under the factored loads as required to satisfy strength and overturning requirements in Clause 7.

8.1.2

Two basic types of construction and associated design assumptions, designated "continuous" and "simple", are permitted for all or part of a structure under this Standard. The distribution of internal forces and bending moments throughout the structure will depend on the type or types of construction chosen and the forces to be resisted.

8.2 Continuous Construction

In continuous construction the beams, girders, and trusses are rigidly framed or are continuous over supports. Connections are generally designed to resist the bending moments and internal forces calculated by assuming that the original angles between intersecting members remain unchanged as the structure is loaded.

8.3 Simple Construction

8.3.1

Simple construction assumes that the ends of beams, girders, and trusses are free to rotate under load in the plane of loading. Resistance to lateral loads, including sway effects, shall be ensured by a suitable system of bracing or shear walls or by the design of part of the structure as continuous construction, except as provided in Clause 8.3.2.

8.3.2

A frame designed to support gravity loads on the basis of simple construction may be proportioned to resist lateral loads, including sway effects, by distributing the moments resulting from such loading among selected joints of a frame by a recognized empirical method provided that
(a) the connection and connected members are proportioned to resist the moments and forces caused by lateral loads;
(b) the connected members have solid webs;
(c) the beam or girder can support the full gravity load when assumed to act as a simple beam;
(d) the connection has adequate capacity for inelastic rotation when subjected to the factored gravity and lateral loads;
(e) the mechanical fasteners or welds of the connection are proportioned to resist 1.5 times the moments and forces produced by the factored gravity and lateral loads; and
(f) in assessing the stability of the structure, in accordance with Clause 8.6, the effect of the flexibility of the connection is taken into account.

8.4 Elastic Analysis

Under a particular loading combination, the forces and moments throughout all or part of the structure may be determined by an analysis that assumes that individual members behave elastically.

8.5 Plastic Analysis

Under a particular loading combination, the forces and moments throughout all or part of the structure may be determined by a plastic analysis provided that
(a) the steel used has $F_y \leq 0.80F_u$ and exhibits the load-strain characteristics necessary to achieve moment redistribution;
(b) the width-thickness ratios meet the requirements of Class 1 sections as given in Clause 11.2;
(c) the members are braced laterally in accordance with the requirements of Clause 13.7;
(d) web stiffeners are supplied on a member at a point of load application where a plastic hinge would form;
(e) splices in beams or columns are designed to transmit 1.1 times the maximum calculated moment under factored loads at the splice location or $0.25M_p$, whichever is greater;
(f) members are not subject to repeated heavy impact or fatigue; and
(g) the influence of inelastic deformation on the strength of the structure is taken into account. (See also Clause 8.6.)

8.6 Stability Effects

8.6.1

The analyses referred to in Clauses 8.4 and 8.5 shall include the sway effects in each storey produced by the vertical loads acting on the structure in its displaced configuration. These second-order effects due to the relative translational displacement (sway) of the ends of a member shall preferably be determined from a second-order analysis. Alternatively, the elastic second-order effects may be accounted for by
(a) amplifying the translational moments obtained from a first-order elastic analysis by the factor

$$U_2 = \frac{1}{1 - \left[\frac{\Sigma C_f \Delta_f}{\Sigma V_f h}\right]}$$

thus, $M_f = M_{fg} + U_2 M_{ft}$

(b) performing an analysis in accordance with Appendix J.

Note: *The amplification factor given in Clause 8.6.1(a) is the limit of the iterative procedure given in Appendix J.*

The second-order moment, M_f, shall be used for the design of both columns and beams.

When the elastic second-order (P Δ) effects exceed 40% of the primary effects, ie, $U_2 > 1.4$, either increase the stiffness of the frame to reduce Δ_f, or perform a second-order elasto-plastic analysis.

8.6.2

For the gravity load case, translational moments shall be taken as not less than the moment produced by pseudo-lateral loads, applied at each storey, equal to 0.005 times the factored gravity loads acting at each storey.

Note: *This is equivalent to obtaining moments from a first-order analysis of the structure based on the assumption of an initial displacement of the structure equal to 0.005 of the height.*

9. Design Lengths of Members

9.1 Simple Span Flexural Members

Beams, girders, and trusses may be designed on the basis of simple spans, whose length may be taken as the distance between the centres of gravity of supporting members. Alternatively, the span length of beams and girders may be taken as the actual length of such members measured between centres of end connections. The length of trusses designed as simple spans may be taken as the distance between the extreme working points of the system of triangulation employed. In all cases the design of columns or other supporting members shall provide for the effect of any significant moment or eccentricity arising from the manner in which a beam, girder, or truss may actually be connected or supported.

9.2 Continuous Span Flexural Members

Beams, girders, or trusses having full or partial end restraint due to continuity or cantilever action shall be proportioned to carry all moments, shears, and other forces at any section assuming the span, in general, to be the distance between the centres of gravity of the supporting members. Supporting members shall be proportioned to carry all moments, shears, and other forces induced by the continuity of the supported beam, girder, or truss.

9.3 Members in Compression

9.3.1 General

Members in compression shall be designed on the basis of their effective length, KL (the product of the effective length factor, K, and the unbraced length, L). Unless otherwise specified in this Standard the unbraced length, L, shall be taken as the length of the compression member between the centres of restraining members. The unbraced length may differ for different cross-sectional axes of a compression member. At the bottom storey of a multi-storey structure or for a single-storey structure, L shall be taken as the length from the top of the base plate to the centre of restraining members at the next higher level.

9.3.2 In Frames

The effective length factor, K, shall be taken as 1.0 for compression members of frames
(a) in which sway effects have been included in the analysis used to determine the moments and forces; or
(b) in which the sway effects in addition to the lateral loads are resisted by bracing or shear walls, unless the degree of rotational restraint afforded at the ends of the unbraced lengths shows that a value of K less than 1.0 is applicable (see Appendices B and C).

9.3.3 In Trusses

Unless otherwise specified in this Standard or unless analysis shows that a smaller value is applicable, the effective length factor, K, shall be taken as 1.0 for compression members in trusses (see Clause C8 of Appendix C).

10. Slenderness Ratios

10.1 General

The slenderness ratio of a member in compression shall be taken as the ratio of the effective length, KL, to the corresponding radius of gyration, r. The slenderness ratio of a member in tension shall be taken as the ratio of the unbraced length, L, to the corresponding radius of gyration.

10.2 Maximum Slenderness Ratio

10.2.1

The slenderness ratio of a member in compression shall not exceed 200.

10.2.2

The slenderness ratio of a member in tension shall not exceed 300. This limit may be waived if other means are provided to control flexibility, sag, vibration, and slack in a manner commensurate with the service conditions of the structure, or if it can be shown that such factors are not detrimental to the performance of the structure or of the assembly of which the member is a part.

11. Width-Thickness Ratios: Elements in Compression

11.1 Classification of Sections

11.1.1

For the purposes of this Standard, structural sections shall be designated as Class 1, 2, 3, or 4, depending on the maximum width-thickness ratios of their elements subject to compression, and as otherwise specified in Clause 11.1.2 and 11.1.3. The Classes are defined as follows:
(a) **Class 1** sections (plastic design sections) will permit attainment of the plastic moment and subsequent redistribution of the bending moment;
(b) **Class 2** sections (compact sections) will permit attainment of the plastic moment but need not allow for subsequent moment redistribution;
(c) **Class 3** sections (noncompact sections) will permit attainment of the yield moment; and
(d) **Class 4** sections will generally have local buckling of elements in compression as the limit state of structural capacity.

11.1.2

Class 1 sections, when subject to flexure, shall have an axis of symmetry in the plane of loading and, when subject to axial compression, shall be doubly symmetric.

11.1.3

Class 2 sections, when subject to flexure, shall have an axis of symmetry in the plane of loading unless the effects of asymmetry of the section are included in the analysis.

11.2 Maximum Width-Thickness Ratios of Elements Subject to Compression

The width-thickness ratio of elements subject to compression shall not exceed the limits given in Table 1 for the specified section classification.

11.3 Width and Thickness

11.3.1

For elements supported along only one edge parallel to the direction of compressive force, the width shall be taken as follows:
(a) for plates, the width, b, is the distance from the free edge to the first row of fasteners or line of welds;

(b) for legs of angles, flanges of channels and zees, and stems of tees, the width, b, is the full nominal dimension; and
(c) for flanges of beams and tees, the width, b, is one-half of the full nominal dimension.

11.3.2

For elements supported along two edges parallel to the direction of compressive force, the width shall be taken as follows:
(a) for flange or diaphragm plates in built-up sections, the width, b, is the distance between adjacent lines of fasteners or lines of welds;
(b) for flanges of rectangular hollow structural sections, the width, b, is the clear distance between webs less the inside corner radius on each side;
(c) for webs of built-up sections, the width, h, is the distance between adjacent lines of fasteners or the clear distance between flanges when welds are used; and
(d) for webs of hot-rolled sections, the width, h, is the clear distance between flanges.

11.3.3

The thickness of elements is the nominal thickness. For tapered flanges of rolled sections, the thickness is the nominal thickness halfway between a free edge and the corresponding face of the web.

Table 1
Maximum Width-Thickness Ratios: Elements in Compression

	Section Classification			
Description of element	**Class 1 plastic design**	**Class 2 compact**	**Class 3 noncompact**	**Class 4* slender**
Legs of angles and elements supported along one edge, except as noted	–	–	$\frac{b}{t} \leq \frac{200}{\sqrt{F_y}}$	See Clause 13
Angles in continuous contact with other elements; plate-girder stiffeners	–	–	$\frac{b}{t} \leq \frac{200}{\sqrt{F_y}}$	See Clause 13
Stems of T-sections	$\frac{b}{t} \leq \frac{145}{\sqrt{F_y}}$ †	$\frac{b}{t} \leq \frac{170}{\sqrt{F_y}}$ †	$\frac{b}{t} \leq \frac{340}{\sqrt{F_y}}$	See Clause 13
Flanges of I- or T-sections; plates projecting from compressive elements; outstanding legs of pairs of angles in continuous contact ‡	$\frac{b}{t} \leq \frac{145}{\sqrt{F_y}}$	$\frac{b}{t} \leq \frac{170}{\sqrt{F_y}}$	$\frac{b}{t} \leq \frac{200}{\sqrt{F_y}}$	See Clause 13
Flanges of channels	–	–	$\frac{b}{t} \leq \frac{200}{\sqrt{F_y}}$	See Clause 13
Flanges of rectangular hollow structural sections	$\frac{b}{t} \leq \frac{420}{\sqrt{F_y}}$	$\frac{b}{t} \leq \frac{525}{\sqrt{F_y}}$	$\frac{b}{t} \leq \frac{670}{\sqrt{F_y}}$	See Clause 13
Flanges of box sections, flange cover plates and diaphragm plates, between lines of fasteners or welds	$\frac{b}{t} \leq \frac{525}{\sqrt{F_y}}$	$\frac{b}{t} \leq \frac{525}{\sqrt{F_y}}$	$\frac{b}{t} \leq \frac{670}{\sqrt{F_y}}$	See Clause 13

(Continued)

Table 1 (Concluded)
Maximum Width-Thickness Ratios: Elements in Compression

	Section Classification			
Description of element	**Class 1 plastic design**	**Class 2 compact**	**Class 3 noncompact**	**Class 4* slender**
Perforated cover plates	–	–	$\frac{b}{t} \leq \frac{840}{\sqrt{F_y}}$	–
Webs in axial compression	$\frac{h}{w} \leq \frac{670}{\sqrt{F_y}}$	$\frac{h}{w} \leq \frac{670}{\sqrt{F_y}}$	$\frac{h}{w} \leq \frac{670}{\sqrt{F_y}}$	See Clause 13
Webs in flexural compression	$\frac{h}{w} \leq \frac{1100}{\sqrt{F_y}}$	$\frac{h}{w} \leq \frac{1700}{\sqrt{F_y}}$	$\frac{h}{w} \leq \frac{1900}{\sqrt{F_y}}$	See Clause 13
Webs in combined flexure and axial compression	$\frac{h}{w} \leq \frac{1100}{\sqrt{F_y}}\left(1 - 0.39\frac{C_f}{C_y}\right)$	$\frac{h}{w} \leq \frac{1700}{\sqrt{F_y}}\left(1 - 0.61\frac{C_f}{C_y}\right)$	$\frac{h}{w} \leq \frac{1900}{\sqrt{F_y}}\left(1 - 0.65\frac{C_f}{C_y}\right)$	See Clause 13
Circular hollow sections	$\frac{D}{t} \leq \frac{13\,000}{F_y}$	$\frac{D}{t} \leq \frac{18\,000}{F_y}$	$\frac{D}{t} \leq \frac{23\,000}{F_y}$	–

* *Class 4 includes all sections not otherwise specified.*
† *See Clause 11.1.3.*
‡ *Can be considered as Class 1 or Class 2 sections if angles are continuously connected by adequate mechanical fasteners or welds, and if there is an axis of symmetry in the plane of loading.*

12. Gross and Net Areas

12.1 Application

Members in tension shall be proportioned on the basis of the areas associated with the potential failure modes. Members in compression shall be proportioned on the basis of the gross area. (For beams and girders, see Clause 15.)

12.2 Gross Area

Gross area shall be calculated by summing the products of the thickness and the gross width of each element (flange, web, leg, plate), as measured normal to the axis of the member.

12.3 Effective Net Area

12.3.1

The effective net area, A_{ne}, shall be determined by summing the critical net areas, A_n, of each segment along a potential path of minimum resistance. The net areas shall be calculated as follows:

(a) for a segment normal to the force (ie, in direct tension)
$A_n = w_n t$

(b) for a segment parallel to the force (ie, in shear)
$A_n = 0.6\ L_n t$

(c) for a segment inclined to the force
$A_n = w_n t + s^2 t/4g$

12.3.2

In calculating w_n and L_n, the width of bolt holes shall be taken as 2 mm larger than the specified hole diameter. Where it is known that drilled holes will be used, this allowance may be waived.

12.3.3 Effective Net Area Reduction — Shear Lag

12.3.3.1

When fasteners transmit load to each of the cross-sectional elements of a member in tension in proportion to their respective areas, the reduced effective net area is equal to the effective net area:
$A'_{ne} = A_{ne}$

12.3.3.2

When bolts transmit load to some but not all of the cross-sectional elements and only when the critical net area includes the net area of unconnected elements, the reduced effective net area shall be taken as follows:

(a) for WWF, W, M or S shapes with flange widths not less than two-thirds the depth, and for structural tees cut from these shapes, when only the flanges are connected with three or more transverse lines of fasteners:
$A'_{ne} = 0.90\ A_{ne}$

(b) for all other structural shapes connected with three or more transverse lines of fasteners:
$A'_{ne} = 0.85\ A_{ne}$

(c) for all members connected with two transverse lines of fasteners:

$A'_{ne} = 0.75\ A_{ne}$

where A_n is calculated in accordance with Clause 12.3.1(a) and (c).

12.3.3.3

When transverse welds transmit load to some of the cross-sectional elements, the effective net area shall be taken as the area of the connected elements.

12.3.3.4

When longitudinal welds transmit load, the reduced effective net area of connected parts shall be taken as follows:

(a) when $L \geq 2w$, $A'_{ne} = 1.00\ A_g$
(b) when $2w > L \geq 1.5w$, $A'_{ne} = 0.87\ A_g$
(c) when $1.5w > L \geq w$, $A'_{ne} = 0.75\ A_g$

provided $L \not< w$
where w = distance between welds
Note: *See also Clause 12.3.1.*

12.3.3.5

Larger values of the reduced effective net area may be used if justified by test or rational analysis.

12.3.4

For angles, the gross width shall be the sum of the widths of the legs minus the thickness. The gauge for holes in opposite legs shall be the sum of the gauges from the heel of the angle minus the thickness.

12.3.5

In calculating the net area across plug or slot welds, the weld metal shall not be taken as adding to the net area.

12.4 Pin-Connected Members in Tension

12.4.1

In pin-connected members in tension, the net area, A_n, across the pin hole, normal to the axis of the member, shall be at least 1.33 times the cross-sectional area of the body of the member. The net area of any section on either side of the axis of the member, measured at an angle of 45° or less to the axis of the member, shall be not less than 0.9 times the cross-sectional area of the body of the member.

12.4.2

The distance from the edge of the pin hole to the edge of the member, measured transverse to the axis of the member, shall not exceed four times the thickness of the material at the pin hole.

12.4.3

The diameter of a pin hole shall be not more than 1 mm larger than the diameter of the pin.

13. Member and Connection Resistance

13.1 General

To meet the strength requirements of this Standard all factored resistances, as determined in this Clause, shall be greater than or equal to the effect of factored loads determined in accordance with Clause 7.2, and ϕ shall be taken as 0.90 unless otherwise specified.

13.2 Axial Tension

The factored tensile resistance, T_r, developed by a member subjected to an axial tensile force shall be taken as

(a) the least of

- (i) $T_r = \phi A_g F_y$
- (ii) $T_r = 0.85 \phi A_{ne} F_u$
- (iii) $T_r = 0.85 \phi A'_{ne} F_u$

(b) for pin connections,

$T_r = 0.75 \phi A_n F_y$

13.3 Axial Compression

13.3.1

The factored axial compressive resistance for W shapes and for hollow structural sections manufactured according to CSA Standard G40.20, Class C (cold-formed non-stress-relieved) and conforming to the requirements of Clause 11 of this Standard for Class 1, 2, or 3 sections shall be taken as*

(a) $0 \leq \lambda \leq 0.15$ $\quad C_r = \phi A F_y$

(b) $0.15 < \lambda \leq 1.0$ $\quad C_r = \phi A F_y (1.035 - 0.202\lambda - 0.222\lambda^2)$

(c) $1.0 < \lambda \leq 2.0$ $\quad C_r = \phi A F_y (-0.111 + 0.636\lambda^{-1} + 0.087\lambda^{-2})$

(d) $2.0 < \lambda \leq 3.6$ $\quad C_r = \phi A F_y (0.009 + 0.877\lambda^{-2})$

(e) $3.6 < \lambda$ $\quad C_r = \phi A F_y \lambda^{-2} = \phi A \left[\frac{1\,970\,000}{(KL/r)^2}\right]$

where $\lambda = \frac{KL}{r}\sqrt{\frac{F_y}{\pi^2 E}}$

**These expressions are applicable to the W shapes normally rolled in Canada and can be assumed to be valid for other doubly symmetric Class 1, 2, or 3 sections, except for solid round non-stress-relieved cold-straightened bars greater than 50 mm diameter (refer to CSA Standard S37). Welded H shapes should have flange edges flame-cut. Singly symmetric, asymmetric, or cruciform sections should be checked as to whether torsional-flexural buckling is critical.*

13.3.2

The factored axial compressive resistance for hollow structural sections manufactured according to CSA Standard G40.20, Class H (hot-formed or cold-formed stress-relieved) and conforming with the requirements of Clause 11 of this Standard for Class 1, 2, or 3 sections, shall be taken as

(a) $0 \leq \lambda \leq 0.15$ $\quad C_r = \phi A F_y$

(b) $0.15 < \lambda \leq 1.2$ $\quad C_r = \phi A F_y (0.990 + 0.122\lambda - 0.367\lambda^2)$

(c) $1.2 < \lambda \leq 1.8$ $\quad C_r = \phi A F_y (0.051 + 0.801\lambda^{-2})$

(d) $1.8 < \lambda \leq 2.8$ $\quad C_r = \phi A F_y (0.008 + 0.942\lambda^{-2})$

(e) $2.8 < \lambda$ $\quad C_r = \phi A F_y \lambda^{-2} = \phi A \left[\frac{1\,970\,000}{(KL/r)^2}\right]$

where $\lambda = \frac{KL}{r}\sqrt{\frac{F_y}{\pi^2 E}}$

13.3.3

The factored compressive resistance, C_r, developed by a member subject to an axial compressive force and designated as a Class 4 section according to Clause 11 shall be determined as follows:
(a) For sections that are less than or equal to 4.5 mm thick and that are not hollow structural sections, the compressive resistance shall be calculated in accordance with CSA Standard S136.
(b) For hollow structural sections and sections greater than 4.5 mm thick that can be shown to be not critical in torsional buckling or not subject to torsional-flexural buckling, the compressive resistance shall be calculated in accordance with the requirements of Clause 13.3.1 or 13.3.2, as applicable. The area, A, shall be taken as the effective area determined in accordance with CSA Standard S136. The slenderness ratio shall be calculated using gross section properties.
(c) Singly symmetric sections, which may be subject to torsional-flexural buckling, and channels with unstiffened flanges shall be designed by rational analysis.

13.4 Shear

13.4.1 Webs of Flexural Members with Two Flanges

13.4.1.1 Elastic Analysis

Except as noted in Clause 13.4.1.2, the factored shear resistance, V_r, developed by the web of a flexural member shall be taken as

$V_r = \phi A_w F_s$

where

A_w = shear area (dw for rolled shapes and hw for girders); and F_s is as follows:

(a) $\frac{h}{w} \le 439\sqrt{\frac{k_v}{F_y}}$ $\qquad F_s = 0.66F_y$

(b) $439\sqrt{\frac{k_v}{F_y}} < \frac{h}{w} \le 502\sqrt{\frac{k_v}{F_y}}$ $\qquad F_s = F_{cri} = 290\frac{\sqrt{F_y k_v}}{(h/w)}$

(c) $502\sqrt{\frac{k_v}{F_y}} < \frac{h}{w} \le 621\sqrt{\frac{k_v}{F_y}}$ $\qquad F_s = F_{cri} + F_t$

$$F_{cri} = \frac{290\sqrt{F_y k_v}}{(h/w)}$$

$$F_t = (0.50\,F_y - 0.866\,F_{cri})\left(\frac{1}{\sqrt{1 + (a/h)^2}}\right)$$

(d) $621\sqrt{\frac{k_v}{F_y}} < \frac{h}{w}$ $\qquad F_s = F_{cre} + F_t$

$$F_{cre} = \frac{180\,000 k_v}{(h/w)^2}$$

$$F_t = (0.50\,F_y - 0.866\,F_{cre})\left(\frac{1}{\sqrt{1 + (a/h)^2}}\right)$$

where

k_v = shear buckling coefficient

$k_v = 4 + \frac{5.34}{(a/h)^2}$ when $a/h < 1$

No STIFF USE 5.34

$$k_v = 5.34 + \frac{4}{(a/h)^2} \quad \text{when } a/h \geq 1$$

a/h = aspect ratio, the ratio of the distance between stiffeners to web depth.

13.4.1.2 Plastic Analysis

In structures designed on the basis of a plastic analysis as defined in Clause 8.5, the factored shear resistance, V_r, developed by the web of a flexural member subjected to shear shall be taken as

$V_r = 0.55\phi w d F_y$

13.4.1.3 Maximum Slenderness

The slenderness ratio (h/w) of a web shall not exceed $83\ 000/F_y$

where

F_y = specified minimum yield point of the compressive flange steel.

This limit may be waived if analysis indicates that buckling of the compressive flange into the web will not occur at factored load.

13.4.1.4 Combined Shear and Moment in Girders

Transversely stiffened girders depending on tension-field action to carry shear (ie, with $h/w > 502\sqrt{k_v/F_y}$) shall be proportioned in such a way that the following limits are observed:

$$\frac{V_f}{V_r} \leq 1.0$$

$$\frac{M_f}{M_r} \leq 1.0$$

$$0.727\frac{M_f}{M_r} + 0.455\frac{V_f}{V_r} \leq 1.0$$

where

V_r is established according to Clause 13.4 and M_r is established according to Clause 13.5 or 13.6, as applicable.

13.4.2 Webs of Flexural Members not Having Two Flanges

The factored shear resistance for cross sections not having two flanges (eg, solid rectangles, rounds, tees) shall be determined by rational analysis. The factored shear stress at any location in the cross section shall be taken as not greater than $0.66\phi F_y$ and shall be reduced where shear buckling is a consideration.

13.4.3 Gusset Plates

The factored tensile resistance or combined tension and shear resistance of gusset plates shall be calculated in accordance with Clause 13.2(a) based on an effective net area as given in Clause 12.3.

13.4.4 Connecting Elements

The factored shear resistance of connecting elements such as gusset plates, framing angles, and shear tabs shall be taken as

$V_r = 0.50\phi L_n t F_u$

13.4.5 Pins

The total factored shear resistance of the nominal area of pins shall be taken as

$V_r = 0.66\phi A F_y$

13.5 Bending — Laterally Supported Members

The factored moment resistance, M_r, developed by a member subjected to bending moments and where continuous lateral support is provided to the compressive flange shall be taken as

(a) for Class 1 and Class 2 sections:

$M_r = \phi Z F_y = \phi M_p$

(b) for Class 3 sections:

$M_r = \phi S F_y = \phi M_y$

(c) for Class 4 sections:

(i) when both the web and the compressive flange fall within Class 4 of Table 1, the value of M_r shall be determined in accordance with CSA Standard S136. The calculated value, F_y', applicable to cold-formed members shall be determined by using only the values for F_y and F_u that are specified in the relevant structural steel material standard;

(ii) for beams or girders whose flanges meet the requirements of Class 3 and whose webs exceed the limits for Class 3, see Clause 15;

(iii) for beams or girders whose webs meet the requirements of Class 3 and whose flanges exceed the limits for Class 3, the moment resistance shall be calculated in accordance with CSA Standard S136.

Alternatively, the moment resistance may be calculated as

$M_r = \phi S_e F_y$

where

S_e = the effective section modulus determined using an effective flange width of $670t/\sqrt{F_y}$ for flanges supported along two edges parallel to the direction of stress and an effective width of $200t/\sqrt{F_y}$ for flanges supported along one edge parallel to the direction of stress. For flanges supported along one edge, in no case shall b/t exceed 60.

13.6 Bending — Laterally Unsupported Members

Where continuous lateral support is not provided to the compression flange of a member subjected to bending, the factored moment resistance, M_r, may be taken as

(a) for doubly symmetric Class 1 and 2 sections

(i) when $M_u > 0.67\, M_p$

$$M_r = 1.15\phi M_p \left(1 - \frac{0.28 M_p}{M_u}\right) \text{ but not greater than } \phi M_p$$

(ii) when $M_u \leq 0.67\, M_p$

$M_r = \phi M_u$

where the critical elastic moment of the unbraced member is given by

$$M_u = \frac{\omega_2 \pi}{L} \sqrt{E I_y G J + \left(\frac{\pi E}{L}\right)^2 I_y C_w}$$

where

L = length of unbraced portion of beam, mm

ω_2 = $1.75 + 1.05\kappa + 0.3\kappa^2 \leq 2.5$, for unbraced lengths subject to end moments

= 1.0 when the bending moment at any point within the unbraced length is larger than the larger end moment or when there is no effective lateral support for the compression flange at one of the ends of the unsupported length

C_w = 0.0 for hollow structural sections

κ = ratio of the smaller factored moment to the larger factored moment at opposite ends of the unbraced length, positive for double curvature and negative for single curvature

(b) for doubly symmetric Class 3 and 4 sections and for channels

(i) when $M_u > 0.67\ M_y$

$$M_r = 1.15\phi M_y \left(1 - \frac{0.28 M_y}{M_u}\right)$$

but not greater than ϕM_y for Class 3 sections and the value given in Clause 13.5 (c) (iii) for Class 4 sections

(ii) when $M_u \leq 0.67\ M_y$

$M_r = \phi M_u$

where M_u and ω_2 are defined in Clause 13.6 (a)(ii)

(c) For cantilever beams, a rational method of analysis taking into account the lateral support conditions at the support and tip of the cantilever such as that given in the Structural Stability Research Council's Guide to Stability Design Criteria for Metal Structures should be used.

(d) For monosymmetric shapes, a rational method of analysis such as that given in the Structural Stability Research Council's Guide to Stability Design Criteria for Metal Structures should be used.

13.7 Lateral Bracing for Members in Structures Analysed Plastically

Members in structures or portions of structures in which the distributions of moments and forces have been determined by a plastic analysis shall be braced to resist lateral and torsional displacement at all hinge locations. The laterally unsupported distance, L_{cr}, from such braced hinge locations to the nearest adjacent point on the frame similarly braced shall not exceed

$L_{cr} = 550 r_y / \sqrt{F_y}$ for $\kappa < -0.5$

$L_{cr} = 980 r_y / \sqrt{F_y}$ for $\kappa \geq -0.5$

Both bracing requirements should be checked and the more severe shall govern the location of the braced point. Bracing is not required at the location of the last hinge to form in the failure mechanism assumed as the basis for proportioning the structure. Except for the aforementioned regions, the maximum unsupported length of members in structures analysed plastically need not be less than that permitted for the same members in structures analysed elastically.

13.8 Axial Compression and Bending*

**More detailed methods of determining the resistance of columns subject to biaxial bending are given in Appendix F.*

13.8.1 Member Strength and Stability — All Classes of Sections Except Class 1 Sections of I-Shaped Members

Members required to resist both bending moments and an axial compressive force shall be proportioned so that

$$\frac{C_f}{C_r} + \frac{U_{1x} M_{fx}}{M_{rx}} + \frac{U_{1y} M_{fy}}{M_{ry}} \leq 1.0$$

where

M_f is the maximum end moment as defined in Clause 8.6.

The capacity of the member shall be examined for the following cases:

(a) cross-sectional strength, in which case

C_r is as defined in Clause 13.3 with the value $\lambda = 0$

M_r is as defined in Clause 13.5 (for the appropriate class of section)

U_{1x} and U_{1y} are taken as 1.0

(b) overall member strength, in which case

C_r is as defined in Clause 13.3 with the value $K = 1.0$, and is based on the maximum slenderness

ratio for biaxial bending. For uniaxial strong-axis bending, $C_r = C_{rx}$. (See also Clause 9.3.2.)
M_r is as defined in Clause 13.5 (for the appropriate class of section)
U_{1x} and U_{1y} are as defined in Clause 13.8.3
(c) lateral torsional buckling strength, when applicable, in which case
C_r is as defined in Clause 13.3 with the value K = 1.0, and is based on weak-axis buckling (see also Clause 9.3.2)
M_{rx} is as defined in Clause 13.6 (for the appropriate class of section)
M_{ry} is as defined in Clause 13.5 (for the appropriate class of section)
U_{1x} is as defined in Clause 13.8.3, but not less than 1.0
U_{1y} is as defined in Clause 13.8.3.

13.8.2 Member Strength and Stability — Class 1 Sections of I-Shaped Members

Members required to resist both bending moments and an axial compressive force shall be proportioned so that

$$\frac{C_f}{C_r} + \frac{0.85\,U_{1x}\,M_{fx}}{M_{rx}} + \frac{0.60U_{1y}\,M_{fy}}{M_{ry}} \leq 1.0$$

All terms in this expression are defined in Clause 13.8.1.

The capacity of the member shall be examined for
(a) cross-sectional strength;
(b) overall member strength; and
(c) lateral torsional buckling strength.

In addition, the member shall meet the following criterion:

$$\frac{M_{fx}}{M_{rx}} + \frac{M_{fy}}{M_{ry}} \leq 1.0$$

where
M_{rx} and M_{ry} are defined in Clause 13.5 or 13.6, as appropriate.

13.8.3 Value of U_1

In lieu of a more detailed analysis, the value of U_1, accounting for the second-order effects due to the deformation of a member between its ends, shall be taken as

$$U_1 = \left[\frac{\omega_1}{1-\frac{C_f}{C_e}}\right]$$

where
ω_1 is defined in Clause 13.8.4.

13.8.4 Values of ω_1

Unless otherwise determined by analysis, the following values shall be used for ω_1:
(a) for members not subjected to transverse loads between supports
$\omega_1 = 0.6 - 0.4\kappa \geq 0.4$
(b) for members subjected to distributed loads or a series of point loads between supports
$\omega_1 = 1.0$
(c) for members subjected to a concentrated load or moment between supports
$\omega_1 = 0.85$

For the purpose of design, members subjected to concentrated load or moment between supports (eg, crane columns) may be considered to be divided into two segments at the point of load (or moment) application. Each segment shall then be treated as a member that depends on its own flexural stiffness to prevent side-sway in the plane of bending considered, and ω_1 shall be taken as 0.85. In calculating the slenderness ratio for use in Clause 13.8, the total length of the member shall be used.

13.9 Axial Tension and Bending

Members required to resist both bending moments and an axial tensile force shall be proportioned so that

(a) $\frac{T_f}{T_r} + \frac{M_f}{M_r} \le 1.0$

where

$M_r = \phi M_p$ for Class 1 and Class 2 sections

$M_r = \phi M_y$ for Class 3 and Class 4 sections

(b) $\frac{M_f}{M_r} - \frac{T_f Z}{M_r A} \le 1.0$ for Class 1 and Class 2 sections;

$\frac{M_f}{M_r} - \frac{T_f S}{M_r A} \le 1.0$ for Class 3 and Class 4 sections

where

M_r is defined in Clause 13.5 or 13.6

13.10 Load Bearing

13.10.1

The factored bearing resistance, B_r, developed by a member or portion of a member subjected to bearing shall be taken as follows:

(a) on the contact area of machined, accurately sawn, or fitted parts,

$B_r = 1.50\phi F_y A$

(b) on expansion rollers or rockers,

$B_r = 0.00013\phi D L F_y^2$

where

B_r is in newtons

ϕ is taken as 0.90

D and L are the diameter and length, respectively, of roller or rocker

F_y is the specified minimum yield point of the weaker part in contact

(c) in bolted connections, the lesser of

(i) $B_r = 3\phi t d n F_u$

(ii) $B_r = \phi t n e F_u$

where

ϕ is taken as 0.67

F_u is the tensile strength of the plate

Note: *If the member is a tension member designed in accordance with Clause 13.2(a), then only Clause 13.10.1(c)(i) applies.*

13.10.2

The ratio of end distance to bolt diameter shall meet the requirements of Clause 22.8.

13.11 Bolts in Bearing-Type Connections

13.11.1 General

For bolts in bearing-type connections, ϕ_b shall be taken as 0.67 to ensure that the connection will not fail before the member.

13.11.2 Bolts in Shear

The factored resistance developed by a bolted joint subjected to shear shall be taken as the lesser of

(a) the factored bearing resistance, B_r, given in Clause 13.10.1(c); or

(b) the factored shear resistance of the bolts,* which shall be taken as

$V_r = 0.60\phi_b n m A_b F_u$

When the bolt threads are intercepted by any shear plane, the factored shear resistance of any joint shall be taken as 70% of V_r.

For joints longer than 1300 mm, the shearing resistance shall be taken as 80% of the aforementioned values.

**For A325M bolts, F_u is 830 MPa. For A490M bolts, F_u is 1040 MPa. For A325 bolts 1 inch or less in diameter, F_u is 825 MPa. For A325 bolts greater than 1 inch in diameter, F_u is 725 MPa. For A490 bolts, F_u is 1035 MPa.*

13.11.3 Bolts in Tension

The factored tensile resistance developed by a bolted joint, T_r, subjected to tension, T_f, shall be taken as

$T_r = 0.75\phi_b n A_b F_u$

13.11.4 Bolts in Combined Shear and Tension

A bolt in a joint that is required to develop resistance to both tension and shear shall be proportioned so that the following relationship is satisfied:

$$\frac{V_f^2}{m^2} + \beta T_f^2 \le 0.56\phi_b^2 \beta (A_b F_u)^2$$

where

β = an interaction factor derived from test results
= 0.69 for A325 bolts, shear plane through shank
= 0.41 for A325 bolts, shear plane through threads
= 0.56 for A490 bolts, shear plane through shank
= 0.30 for A490 bolts, shear plane through threads

except that V_f shall not exceed V_r as given in Clause 13.11.2.

13.12 Bolts in Slip-Critical Connections

13.12.1 General

The requirement for a slip-critical connection is that under the forces and moments produced by specified loads, slip of the assembly shall not occur. In addition, the effect of factored loads shall not exceed the resistances of the connection as given in Clause 13.11.

13.12.2 Shear Connections

The slip resistance, V_s, of a bolted joint, subjected to shear, V, shall be taken as:

$V_s = 0.53 c_1 k_s m n A_b F_u$

where k_s is the mean slip coefficient as determined from Table 2 or by tests carried out in accordance with *Method to Determine the Slip Coefficient for Coatings used in Bolted Connections* of the Research Council on Structural Connections, and c_1 is a coefficient that relates the specified initial tension and mean slip to a 5% probability of slip for bolts installed by turn-of-nut procedures. Table 2 gives c_1.

Table 2
Values of c_1 and k_s

Contact surface of bolted parts		k_s	c_1	c_1
Class	**Description**		**A325 Bolts and A325M Bolts**	**A490 Bolts and A490M Bolts**
A	Clean mill scale, or blast-cleaned with Class A coatings	0.33	0.82	0.78
B	Blast-cleaned or blast-cleaned with Class B coatings	0.50	0.89	0.85
C	Hot-dip galvanized with wire brushed surfaces	0.40	0.90	0.85

Notes:

(1) *Class A and Class B coatings are defined as those coatings that provide a mean slip coefficient of not less than 0.33 and 0.50, respectively.*

(2) *Values of c_1 for 5% probability of slip for values of k_s other than those listed in Table 2 may be found in the Research Council on Structural Connections'* Guide to Design Criteria for Bolted and Riveted Joints, *Second Edition, listed therein as values of the Slip Factor D.*

13.12.3 Connections in Combined Shear and Tension

A bolt in a joint that is required to develop resistance to both tension and shear shall be proportioned so that the following relationship is satisfied for the specified loads:

$$\frac{V}{V_s} + 1.9\frac{T}{nA_bF_u} \leq 1.0$$

where

V_s = slip resistance as defined in Clause 13.12.2.

13.13 Welds

13.13.1

The factored resistance for weld metal and base metal of welded joints shall be as shown in Table 3(A).

13.13.2

The factored compressive resistance of joints that use partial joint penetration groove welds shall be based on the effective throat area of the welds plus the area of the base metal fitted in contact bearing.

Note: *Joints in compression that depend on contact bearing, when assembled during fabrication, should have at least 75% of the entire contact area in full bearing, defined as a separation not exceeding* 0.5 mm, *and the separation at the edges of the joint should not exceed* 1.0 mm, *unless otherwise stipulated by the designer.*

13.13.3

The vector sum of factored longitudinal and transverse shear loads shall not exceed the factored resistances given in Table 3(A), unless an ultimate strength analysis is used that is acceptable to the designer.

13.13.4

Plug and slot welds shall be considered to provide shear resistance in the plane of the connected parts only.

Table 3(A)
Factored Resistance of Welds

Type of weld	Load	Matching condition: Requirements for matching*	Matching condition: Factored resistance	
Complete joint penetration groove welds	Tension or compression parallel to axis of weld	Matching condition; or if electrode classification is lower or higher than matching (see Note (2))	Same as calculated for base metal	
	Compression normal to axis of weld	Matching condition is mandatory	Same as calculated for base metal	
	Tension normal to axis of weld	Matching condition is mandatory	Same as calculated for base metal	
	Shear	Matching condition; or	Same as calculated for base metal	
		If electrode classification is lower than matching (see Note (1))	The smaller of (a) base metal $V_r = 0.67\phi A_m F_y$ (b) weld metal $V_r = 0.67\phi_w A_w X_u$	(5)
Partial joint penetration groove welds (3)	Tension or compression parallel to axis of weld	Matching condition; or if electrode classification is lower or higher than matching (see Note (2))	Same as calculated for base metal	
	Compression normal to axis of weld	Matching condition is mandatory	See Note (4)	
	Tension normal to axis of weld	Matching condition; or if electrode classification is lower than matching	$T_r = \phi_w A_n F_u$	(6)
	Shear	Matching condition; or if electrode classification is lower than matching	The smaller of (a) base metal $V_r = 0.67\phi A_m F_y$ (b) weld metal $V_r = 0.67\phi_w A_w X_u$	(5)

(Continued)

Table 3(A) (Continued)

Type of Weld	Load	Matching condition	
		Requirements for matching *	**Factored resistance**
Fillet welds	Tension or compression parallel to axis of weld	Matching condition; or if electrode classification is lower or higher than matching (see Note (2))	Same as calculated for base metal
	Tension or compression normal to axis of weld	For matching condition, use X_u of matching electrode classification	(5)
	Shear	For electrode classification lower than matching, use X_u of lower electrode	The smaller of (a) base metal $V_r = 0.67 \phi A_m F_y$ (b) weld metal $V_r = 0.67 \phi_w A_w X_u$
Plug and slot welds	Shear	For electrode classification one designation higher than matching, use X_u of matching electrode	

For G40.21-M steels and their matching electrode classifications refer to Table 3(B).

Notes:

(1) *Electrodes of strengths lower than that shown for the corresponding base metal grades of Table 3(B) may be used for complete joint penetration groove welds between webs and flanges of girders transferring shear loads.*

(2) *If shear is transferred between components, the design value is the smaller of that for the base metal or that for the weld metal.*

(3) *For hollow structural sections, the provisions of Appendix L of CSA Standard W59 may be used.*

(4) *Factored resistances of joints are calculated as for base metal, but on an area calculated as $A = A_m$ + area of base metal in contact bearing.*

(5) *The following symbols are used in this Table:*

A_w (a) *for all types of welds except plug and slot welds,*
A_w = *size of effective throat of weld × length of weld; and*
(b) *for plug and slot welds,*
A_w = *nominal area of the hole or slot, in the plane of the faying surface.*

A_m (a) *for fillet welds,*
A_m = *effective size × length of weld;*
(b) *for complete joint penetration groove welds,*
(i) *for butt joints,*
A_m = *thickness of base metal × length of weld;*
(ii) *for T-joints,*

Table 3(A) (Concluded)

A_m = size of fusion face in base metal × length of weld;

(c) for partial joint penetration groove welds,

A_m = size of fusion face in base metal × length of weld; however,

$A_m \leq$ thickness of base metal × length of weld; and

(d) for plug and slot welds,

A_m = nominal area of the hole or slot, in the plane of the faying surface.

A_n = applicable area of base metal normal to tensile load for joints in tension using partial joint penetration groove welds

F_u = specified ultimate tensile strength of base metal

F_y = specified minimum yield stress, yield point, or yield strength of base metal

V_r = factored shear resistance

X_u = ultimate strength as given by electrode classification number

ϕ_w = resistance factor (which is equal to 0.67 for weld metal)

(6) For members designed for tension and when overall ductile behaviour is desired (member yielding before weld fracture), $A_n/A_g > F_y/F_u$.

Table 3(B)
Base Metal and Matching Electrode Ultimate Strengths

Matching electrode ultimate strengths* MPa	G40.21-M Grades						
	260	300	350	380	400	480	700
410	X	X†					
480	X	X	X‡	X			
550					X‡		
620						X	
820							X

**Ultimate strengths have been determined from electrode classification numbers.*
†*For HSS only.*
‡*For unpainted applications using "A" or "AT" steels where the deposited weld metal shall have similar atmospheric corrosion resistance or similar colour characteristics to the base metal, or both, the requirements of Clauses 5.2.1.4 and 5.2.1.5 of CSA Standard W59 shall apply.*
Note: *For matching condition of ASTM steels, see Table 11-1 or Table 12-1 of CSA Standard W59.*

14. Fatigue

14.1 General

14.1.1

In addition to meeting the requirements of Clause 14 for fatigue, any member or connection shall also meet the requirements for the static load conditions using the factored loads.

14.1.2

Members and connections subjected to fatigue loading shall be designed, detailed, and fabricated so as to minimize stress concentrations and abrupt changes in cross-section.

14.1.3

Specified loads for the design of members or connections shall be used for all fatigue calculations.

14.1.4

A specified load less than the maximum specified load but acting with a greater number of cycles may govern and shall be considered.

14.1.5

Plate girders with $h/w > 3150 \sqrt{F_y}$ shall not be used under fatigue conditions.

14.1.6

Slotted holes shall not be used in bolted connections in members subject to fatigue.

14.2 Life

For guidance in determining the number of cycles, the life of the structure shall be assumed to be not less than 50 years unless otherwise stated.

14.3 10 000 Cycles of Load

When a load is expected to be applied not more than 10 000 times during the life of the structure, no special considerations beyond those in Clause 14.1.2 need apply.

14.4 More Than 10 000 Cycles of Load

When a load is expected to be applied more than 10 000 times in the life of the structure, the loaded members, connections, bolts, and welds shall be proportioned so that the probability of fatigue failure is acceptably small. In such cases the design shall be based on the best available information on the fatigue characteristics of the materials and components to be used. In the absence of more specific information, Clause 14.5 provides guidance in proportioning members and parts. Fatigue resistance shall be provided only for those loads considered to be repetitive and hence contributing to fatigue. Often the magnitude of a repeated load is less than the maximum static load for which the member or part is designed.

14.5 Allowable Stress Range in Fatigue

When this Clause is used as the basis for design, the members, connections, bolts, and welds shall be proportioned so that the calculated stress range does not exceed the allowable stress range, F_{sr}, given in Table 4(A) for the appropriate type and location of material shown in Table 4(B). The stress range is defined as the algebraic difference between the maximum stress and the minimum stress. Tensile stress is considered to have the opposite algebraic sign to compressive stress. Members subject to a stress range involving only compression need not be designed for fatigue. The information in Tables 4(A) and 4(B) is shown in Appendix K, Figure K2.

14.6 Secondary Effects

Secondary stresses, stresses due to deformations, and stresses due to out-of-plane movements are potential sources of fatigue cracks. Caution is therefore advised in detailing structures that are subjected to repetitive loads and in which these sources of stresses may be present.

14.7 Single Load Paths

Structures in which the failure of a single element could result in collapse or catastrophic failure should receive special attention when fatigue cracks are a possibility. In such a case, if the rules prescribed in Clause 14 are used, a system of periodic inspection and maintenance shall be in force during the life of the structure. Alternatively, the permissible stress ranges shall be limited to 0.80 times those given in Table 4(A).

Table 4(A)
Allowable Stress Ranges in Fatigue*

Category	F_{sr} (MPa) For 100 000 cycles	For 500 000 cycles	For 2 000 000 cycles	Endurance Limit ‡
a	435	255	165	165
b	338	200	124	110
b-1	269	159	100	83
c	245	145	90	70†
d	193	110	70	48
e	152	90	55	31
f	110	63	40	18
w	103	83	62	55
y	145	105	80	65

**See Table 4(B) and Appendix K, Figure K1.*

†*Except for transverse stiffener welds on girder webs or flanges, where* 83 MPa *should be used.*

‡*The lower limiting value as shown in Appendix K, Figure K1.*

Table 4(B)
Stress Range Categories for Various Applications

General condition	Situation	Stress range category (see Table 4(A))	Illustrative example (see Figure K2)
Plain material	Base metal with rolled or cleaned surfaces; flame-cut edges with a surface roughness not exceeding 1000 (25 μm) as defined by CSA Standard B95	a	1, 2
Built-up members	Base metal and weld metal in members (without attachments) built-up of plates or shapes connected by continuous complete penetration groove welds (backing bars removed), or by continuous fillet welds parallel to the direction of applied stress	b	3, 4, 5, 7
	Base metal and weld metal in members (without attachments) built-up of plates or shapes connected by continuous complete penetration groove welds (backing bars not removed) or by continuous partial penetration groove welds parallel to the direction of applied stress	b-1	3, 4, 5, 7
	Base metal at toe of transverse stiffener welds on girder webs or flanges	c	6

(Continued)

Table 4(B) (Continued)

General condition	Situation	Stress range category (see Table 4(A))	Illustrative example (see Figure K2)
Built-up members	Base metal at ends of partial-length welded cover plates narrower than the flange and having square or tapered ends (with or without welds across the ends), or wider than the flange with welds across the ends:		
	(a) Flange thickness ≤ 20 mm	e	7
	(b) Flange thickness > 20 mm	f	7
	Base metal at ends of partial-length welded cover plates wider than the flange, without welds across the ends	f	7
Groove-welded attachments: longitudinally loaded*	Base metal adjacent to details attached by complete or partial penetration groove welds when the detail length in the direction of stress, L, is less than 50 mm	c	6, 18
	Base metal adjacent to details attached by complete or partial penetration groove welds when the detail length in the direction of stress, L, is between 50 mm and 12 times the plate thickness but less than 100 mm	d	18
	Base metal adjacent to details attached by complete or partial penetration groove welds when the detail length in the direction of stress, L, is greater than 12 times the plate thickness or greater than 100 mm:		
	(a) Detail thickness ≤ 25 mm	e	18
	(b) Detail thickness > 25 mm	f	18
	Base metal adjacent to details attached by complete or partial penetration groove welds with a transition radius, R, as below, regardless of the detail length:		
	With the end welds ground smooth		12
	(a) R ≥ 600 mm	b	
	(b) 600 mm > R ≥ 150 mm	c	
	(c) 150 mm > R ≥ 50 mm	d	
	(d) 50 mm > R	e	
	For all transition radii, R, without end welds ground smooth	e	

(Continued)

Table 4(B) (Continued)

General condition	Situation	Stress range category (see Table 4(A))	Illustrative example (see Figure K2)
Groove-welded connections	Base metal and weld metal in or adjacent to complete penetration groove-welded splices of rolled or welded sections having similar profiles when welds are ground flush, with grinding in direction of applied stress and with weld soundness established by nondestructive examination	b	8, 9
	Base metal and weld metal in or adjacent to complete penetration groove-welded splices with 600 mm radius transitions in width when welds are ground flush, with grinding in direction of applied stress and with weld soundness established by nondestructive examination	b	11
	Base metal and weld metal in or adjacent to complete penetration groove-welded splices at transitions in width or thickness, with welds ground to provide slopes no steeper than 4 in 10, with grinding in direction of applied stress, and with weld soundness established by nondestructive examination:		
	(a) G40.21M-700Q and 700 QT base metal	b-1	10, 10A
	(b) All other base metals	b	10, 10A
	Base metal and weld metal in or adjacent to complete joint penetration groove-welded splices, with or without transitions, involving equal width and/or thicknesses of material or involving transitions having slopes no greater than 4 in 10 when in either case reinforcement is not removed and when weld soundness is established by nondestructive examination	c	8, 9, 10, 10A
Fillet-welded connections	Base metal at intermittent fillet welds	e	
	Base metal at details connected with transversely loaded welds, with the welds perpendicular to the direction of stress:		
	(a) Detail thickness ≤ 12 mm	c	19
	(b) Detail thickness > 12 mm	See Note	

(Continued)

Table 4(B) (Continued)

General condition	Situation	Stress range category (see Table 4(A))	Illustrative example (see Figure K2)
Fillet-welded attachments: longitudinally loaded * †	Base metal adjacent to fillet-welded attachments in which the detail length in the direction of stress, L, is less than 50 mm; stud-type shear connectors	c	13, 14, 18, 20
	Base metal adjacent to details attached by fillet welds when the detail length in the direction of stress, L, is between 50 mm and 12 times the plate thickness, but less than 100 mm	d	18, 20
	Base metal adjacent to details attached by fillet welds when the length in the direction of stress, L, is greater than 12 times the plate thickness or greater than 100 mm:	See tabulation in Example 12 of Figure K2	12
	(a) Detail thickness ≤ 25 mm	e	7,16,18, 20
	(b) Detail thickness > 25 mm	f	7, 16, 18
	Base metal adjacent to details attached by fillet welds with a transition radius, R, regardless of the detail length:		
	With the end welds ground smooth		12
	(a) R ≥ 50 mm	d	
	(b) 50 mm > R ≥ 0 mm	e	
	For all transition radii without the end welds ground smooth	e	12
Groove-welded attachments: transversely loaded*‡	Detail base metal attached by complete penetration groove welds with a transition radius, R, regardless of the detail length and with weld soundness transverse to the direction of stress established by nondestructive examination:		
	With equal plate thickness and reinforcement removed		12
	(a) R ≥ 600 mm	b	
	(b) 600 mm > R ≥ 150 mm	c	
	(c) 150 mm > R ≥ 50 mm	d	
	(d) 50 mm > R	e	
	With equal plate thickness and reinforcement not removed		12
	(a) R ≥ 150 mm	c	
	(b) 150 mm > R ≥ 50 mm	d	
	(c) 50 mm > R	e	
	With unequal plate thickness and reinforcement removed		12
	(a) R ≥ 50 mm	d	
	(b) 50 mm > R	c	
	For all transition radii, R, with unequal plate thickness and reinforcement not removed	e	12

(Continued)

Table 4(B) (Continued)

General condition	Situation	Stress range category (see Table 4(A))	Illustrative example (see Figure K2)
Groove-welded attachments: transversely loaded*‡	With the end welds ground smooth (a) R ≥ 50 mm (b) 50 mm > R	 d e	12
	For all transition radii, R, without the end welds ground smooth	e	12
Fillet-welded attachments: transversely loaded with weld in direction of principal stress * †	Detail base metal attached by fillet welds with a transition radius, R, regardless of the detail length (shear stress on the throat of fillet welds is governed by Category w):		
	With the ends ground smooth (a) R ≥ 50 mm (b) 50 mm > R	 d e	12
	For all transition radii, R, without the end welds ground smooth	e	12
Fillet welds	Shear stress on throat of fillet welds	w	16
Stud-type shear connectors	Shear stress on the nominal area of stud-type shear connectors	y	13
Mechanically fastened connections§	Base metal at gross section of high-strength bolted slip-critical connections, except axially loaded joints that induce out-of-plane bending in connected material	b	17
	Base metal at net section of high-strength bolted bearing-type connections	b	17
	Base metal at net section of bolted connections other than high-strength bearing-type	d	17
Anchor bolts and threaded parts	Tensile stress range on the tensile stress area of the threaded part, including effects of bending	e	
Fillet-welded HSS to base plate	Shear stress on fillet weld	f	21

*"Longitudinally loaded" signifies that the direction of applied stress is parallel to the longitudinal axis of the weld. "Transversely loaded" signifies that the direction of applied stress is perpendicular to the longitudinal axis of the weld.
†Gusset plates attached to girder flange surfaces with only transverse fillet welds are prohibited.
‡Transversely loaded partial penetration groove welds are prohibited.
§Bolts in connections subject to fatigue shall be pretensioned.

(Continued)

Table 4(B) (Concluded)

Note: *Allowable fatigue stress range on throat of fillet welds transversely loaded is a function of the effective throat and plate thickness. (Ref. Frank and Fisher,* Journal of the Structural Division, *ASCE, Vol. 105, No. ST9, September 1979.)*

$$S_r = S_r^c\left[(0.06 - 0.79\,H/t_p)/(1.1\,t_p^{\,1/6})\right]$$

where

S_r^c equal to the allowable stress range for Category c as given in Table 4(A). This assumes no penetration at the weld root.

15. Beams and Girders

15.1 Proportioning

Beams and girders consisting of rolled shapes (with or without cover plates), hollow structural sections, or fabricated sections shall be proportioned on the basis of the properties of the gross section or the modified gross section. No deduction need be made for fastener holes in webs or flanges unless the reduction of flange area by such holes exceeds 15% of the gross flange area, in which case the excess shall be deducted. The effect of openings other than holes for fasteners shall be considered in accordance with Clause 15.10.

15.2 Rotational Restraint at Points of Support

Beams and girders shall be restrained against rotation about their longitudinal axes at points of support.

15.3 Copes

15.3.1

The effect of copes on the lateral torsional buckling resistance of a beam or girder shall be taken into account.

15.3.2

The effect of copes in reducing the net area of the web available to resist transverse shear and the effective net area of potential paths of minimum resistance shall be taken into account (see Clause 12.3.1).

15.4 Reduced Moment Resistance of Girders with Thin Webs

When the web slenderness ratio, h/w, exceeds $1900/\sqrt{M_f/\phi S}$, the flange shall meet the width-thickness ratios of Class 3 sections in accordance with Clause 11 and the factored moment resistance of the beam or girder, M_r', shall be determined by

$$M_r' = M_r\left[1 - 0.0005\frac{A_w}{A_f}\left(\frac{h}{w} - 1900/\sqrt{M_f/\phi S}\right)\right]$$

where

M_r = factored moment resistance as determined by Clause 13.5 or 13.6 but not to exceed ϕM_y

15.5 Flanges

15.5.1

Flanges of welded girders preferably shall consist of a single plate or a series of plates joined end-to-end by complete penetration groove welds.

15.5.2

Flanges of bolted girders shall be proportioned so that the total cross-sectional area of cover plates does not exceed 70% of the total flange area.

15.5.3

Fasteners or welds connecting flanges to webs shall be proportioned to resist horizontal shear forces due to bending combined with any loads that are transmitted from the flange to the web other than by direct bearing. Spacing of fasteners or intermittent welds in general shall be in proportion to the intensity of the shear force and shall not exceed the maximum for compression or tension members as applicable, in accordance with Clause 19.

15.5.4

Partial-length flange cover plates shall be extended beyond the theoretical cut-off point and the extended portion shall be connected with sufficient fasteners or welds to develop a force in the cover plate at the theoretical cut-off point not less than

$$P = \frac{A M_{fc} y}{I_g}$$

where

P = required force to be developed in cover plate

A = area of cover plate

M_{fc} = moment due to factored loads at theoretical cut-off point

y = distance from centroid of cover plate to neutral axis of cover-plated section

I_g = moment of inertia of cover-plated section

Additionally, for welded cover plates the welds connecting the cover-plate termination to the beam or girder shall be designed to develop the force, P, within a length, a′, measured from the actual end of the cover plate, determined as follows:

(a) a′ = the width of cover plate when there is a continuous weld equal to or larger than three-fourths of the cover-plate thickness across the end of the plate and along both edges in the length a′;

(b) a′ = 1.5 times the width of cover plate when there is a continuous weld smaller than three-fourths of the cover-plate thickness across the end of the plate and along both edges in the length a′; and

(c) a′ = 2 times the width of cover plate when there is no weld across the end of the plate but there are continuous welds along both edges in the length a′.

15.6 Bearing Stiffeners

15.6.1

Pairs of bearing stiffeners on the webs of single-web beams and girders shall be required at points of concentrated loads and reactions wherever the bearing resistance on the web is exceeded (see Clause 15.9). Bearing stiffeners shall also be required at unframed ends of single-web girders having web depth-thickness ratios greater than $1100/\sqrt{F_y}$. Box girders may employ diaphragms designed to act as bearing stiffeners.

15.6.2

Bearing stiffeners shall bear against the flange or flanges through which they receive their loads and shall extend approximately to the end of the flange plates or flange angles. They shall be designed as columns in accordance with Clause 13.3, assuming the column section to consist of the pair of stiffeners and a centrally located strip of the web equal to not more than 25 times its thickness at interior stiffeners, or a strip equal to not more than 12 times its thickness when the stiffeners are located at the end of the web. The effective column length, KL, shall be taken as not less than three-fourths of the length of the stiffeners in calculating the ratio KL/r. Only that portion of the stiffeners outside of the angle fillet or the flange-to-web welds shall be considered effective in bearing. Angle bearing stiffeners shall not be crimped. Bearing stiffeners shall be connected to the web so as to develop the full force required to be carried by the stiffener into the web or vice versa.

15.7 Intermediate Transverse Stiffeners

15.7.1

Intermediate transverse stiffeners, when used, shall be spaced to suit the shear resistance determined in accordance with Clause 13.4, except that at girder end panels or at panels adjacent to large openings, the tension-field component shall be taken as zero unless means are provided to anchor the tension field.

15.7.2

The maximum distance between stiffeners, when stiffeners are required, shall not exceed the values shown in Table 5. Closer spacing may be required in accordance with Clause 15.7.1.

Table 5
Maximum Intermediate Transverse Stiffener Spacing

Web depth-thickness ratio (h/w)	Maximum distance between stiffeners, a, in terms of clear web depth, h
Up to 150	3h
More than 150	$\dfrac{67\,500h}{(h/w)^2}$

15.7.3

Intermediate transverse stiffeners may be furnished singly or in pairs. Width-thickness ratios shall conform to Clause 11. The moment of inertia of the stiffener, or pair of stiffeners if so furnished, shall be not less than $(h/50)^4$ taken about an axis in the plane of the web. The gross area of intermediate stiffeners, or pairs of stiffeners if so furnished, shall be given by the expression

$$A_s \geq \frac{aw}{2}\left[1 - \frac{a/h}{\sqrt{1 + (a/h)^2}}\right] CYD$$

where

a = centre-to-centre distance of adjacent stiffeners (ie, panel length)
w = web thickness
h = web depth
$C = \left[1 - \dfrac{310\,000 k_v}{F_y (h/w)^2}\right]$ but not less than 0.10

Y = ratio of specified minimum yield point of web steel to specified minimum yield point of stiffener steel
D = stiffener factor
= 1.0 for stiffeners furnished in pairs
= 1.8 for single angle stiffeners
= 2.4 for single plate stiffeners
k_v = shear buckling coefficient (see Clause 13.4.1)
F_y = specified minimum yield point of web steel

When the greatest shear, V_f, in an adjacent panel is less than that permitted by Clause 13.4.1, this gross area requirement may be reduced in like proportion by multiplying by the ratio V_f/V_r.

15.7.4

Intermediate transverse stiffeners shall be connected to the web for a shear transfer per pair of stiffeners (or per single stiffener when so furnished), in newtons per millimetre of web depth, h, not less than $1 \times 10^{-4} h F_y^{1.5}$, except that when the largest calculated shear, V_f, in the adjacent panels is less than V_r as calculated by Clause 13.4.1, this shear transfer may be reduced in the same proportion. However, the total shear transfer shall in no case be less than the value of any concentrated load or reaction required to be transmitted to the web through the stiffener. Fasteners connecting intermediate transverse stiffeners to the web shall be spaced not more than 300 mm from centre to centre. If intermittent fillet welds are used, the clear distance between welds shall not exceed 16 times the web thickness or 4 times the weld length.

15.7.5

When intermediate stiffeners are used on only one side of the web, the stiffeners shall be attached to the compression flange. Intermediate stiffeners used in pairs shall have at least a snug fit against the compression flange. When stiffeners are cut short of the tension flange, the distance cut short shall be equal to or greater than four times but not greater than six times the girder web thickness. Stiffeners preferably shall be clipped to clear girder flange-to-web welds.

15.8 Lateral Forces

The flanges of beams and girders supporting cranes or other moving loads shall be proportioned to resist any lateral forces produced by such loads.

15.9 Web Crippling and Yielding

Bearing stiffeners shall be provided where the factored concentrated loads or reactions exceed the factored compressive resistances of webs of rolled beams and welded plate girders. The factored compressive resistance of the web shall be calculated as follows:

(a) for interior loads (concentrated load applied a distance from the member end greater than the member depth), the smaller of

(i) $B_r = 1.10\phi w (N + 5k) F_y$

(ii) $B_r = 300\phi w^2 \left[1 + 3\left(\frac{N}{d}\right)\left(\frac{w}{t}\right)^{1.5}\right] \sqrt{F_y t / w}$

(b) for end reactions, the smaller of

(i) $B_r = 1.10\phi w (N + 2.5k) F_y$

(ii) $B_r = 150\phi w^2 \left[1 + 3\left(\frac{N}{d}\right)\left(\frac{w}{t}\right)^{1.5}\right] \sqrt{F_y t / w}$

where
w = web thickness

N = length of bearing (N shall be not less than k for end reactions)
k = distance from outer face of flange to web toe of flange-to-web fillet
t = flange thickness

15.10 Openings

15.10.1

Except as provided in Clause 15.1, the effect of all openings in beams and girders shall be considered in the design. At all points where the factored shear or moments at the net section would exceed the capacity of the member, adequate reinforcement shall be added to the member at that point to provide the required strength and stability.

15.10.2

Unreinforced circular openings may be located in the web of unstiffened prismatic compact beams or girders without considering net section properties provided that
(a) the specified design load for the member is uniformly distributed;
(b) the section has an axis of symmetry in the plane of bending;
(c) the openings are located within the middle third of the depth and the middle half of the span of the member;
(d) the spacing between the centres of any two adjacent openings, measured parallel to the longitudinal axis of the member, is a minimum of 2.5 times the diameter of the larger opening; and
(e) the factored maximum shear at the support does not exceed 50% of the factored shear resistance of the section.

15.10.3

If the forces at openings are determined by an elastic analysis, the procedure adopted shall be in accordance with published, recognized principles. The forces determined by such elastic analysis shall not exceed the factored resistances given in Clause 13 and, if applicable, Clause 14.

15.10.4

The strength and stability of the member in the vicinity of openings may be determined on the basis of assumed locations of plastic hinges, such that the resulting force distributions satisfy the requirements of equilibrium, provided that the analysis is carried out in accordance with Clauses 8.5(a), (b), and (f). However, for I-type members the width-thickness ratio of the flanges may meet the requirements of Class 2 sections, provided that the webs meet the width-thickness limit of Class 1 sections.

15.11 Torsion

15.11.1

Beams and girders subjected to torsion shall have sufficient strength and rigidity to resist the torsional moment and forces in addition to other moments or forces. The connections and bracing of such members shall be adequate to transfer the reactions to the supports.

15.11.2

The factored resistance of I-shaped members subject to combined flexure and torsion may be determined from moment-torque interaction diagrams that take into account the normal stress distribution due to flexure and warping torsion and the St. Venant torsion. Assumed normal stress distributions shall be consistent with the class of section.

15.11.3

Members subject to torsional deformations required to maintain compatibility of the structure need not be designed to resist the associated torsional moments provided that the structure satisfies the requirements of equilibrium.

15.11.4

For all members subject to loads causing torsion, the torsional deformations under specified loads shall be limited in accordance with the requirements of Clause 6.2.1.1. For members subject to torsion or to combined flexure and torsion, the maximum combined normal stress, as determined by an elastic analysis, arising from warping torsion and bending due to the specified loads shall not exceed F_y.

16. Open-Web Steel Joists

16.1 Scope

Clause 16 provides requirements for the design, manufacture, transportation, and erection of open-web steel joists used in the construction of buildings. Joists intended to act compositely with the deck shall be designed using the requirements of Clause 17 in conjunction with the requirements of this Clause. This Clause shall not be used for the design of joists not having an axis of symmetry in the plane of the joist.

16.2 General

Joists are steel trusses of relatively low mass with parallel or slightly pitched chords and triangulated web systems proportioned to span between masonry walls or structural supporting members, or both, and to provide direct support for floor or roof deck. In general joists are manufactured on a production-line basis employing jigs, with certain details of the members being standardized by the individual manufacturer.

16.3 Definitions

The following definitions apply to Clause 16:

Deck or decking — the structural floor or roof element spanning between adjacent joists and directly supported thereby. The terms deck and decking include cast-in-place or precast concrete slabs, profiled metal deck, wood plank or plywood, and other relatively rigid elements suitable for floor or roof construction.

Span of an open-web steel joist — the centre-to-centre distance of joist bearings.

Special open-web steel joists or special joists —
(a) joists subjected to the loads stipulated in Clause 16.6.2;
(b) cantilever joists, continuous joists, and joists having special support conditions; and
(c) joists having other special requirements such as lateral-force-resisting braces/ties for columns and beams.

Standard open-web joists — simply supported flexural members whose design is governed by the loading given in Clause 16.6.1. This definition does not include primary trusses that support joists, other secondary members, and special joists.

Tie joists — joists that have at least one end connected to a column to facilitate erection and that are designed to resist gravity loads only, unless otherwise specified.

16.4 Materials

Steel for joists shall be of a structural quality, suitable for welding, meeting the requirements of Clause 5.1.1. Structural members cold-formed to shape may use the effect of cold-forming in accordance with Clause 5.2 of CSA Standard S136. The calculated value of F_y' shall be determined using only the values for F_y and F_u that are specified in the relevant structural steel material standard. Yield levels reported on mill test certificates or determined according to Clause 9.3 of CSA Standard S136 shall not be used as the basis for design.

16.5 Drawings

16.5.1 Building Design Drawings

The building design drawings prepared by the building designer shall show
(a) the uniformly distributed specified live and dead gravity loads, the unbalanced loading condition and the concentrated load conditions given in Clause 16.6.1 or 16.6.2, and any special loading conditions such as nonuniform snow loads, horizontal loads, end moments, net uplift, and allowances for mechanical equipment;
(b) maximum joist spacing and, where necessary, camber, maximum joist depth, and shoe depth;
(c) where joists are not supported on steel members, maximum bearing pressures, or sizes of bearing plates;
(d) anchorage requirements in excess of the requirements of Clause 16.6.13;
(e) bracing as may be required by Clauses 16.6.7.2 or 16.9.1; and
(f) method and spacing of attachments of steel deck to the top chord.

Note: *It is recommended that the building design drawings include a note warning that attachments for mechanical, electrical, and other services shall be made by using approved clamping devices or u-bolt-type connectors and that no drilling or cutting shall be done unless approved by the building designer.*

16.5.2 Joist Design Drawings

Joist design drawings prepared by the joist manufacturer shall show, as a minimum, the specified loading, factored member loads, material specification, member sizes, dimensions, spacers, welds, shoes, anchorages, bracing, bearings, field splices, bridging locations, and camber.

16.6 Design

16.6.1 Loading for Standard Open-Web Steel Joists

Unless otherwise specified by the building designer (in accordance with Clause 16.6.2), the factored moment and shear resistances of an open-web steel joist at every section shall be not less than the moment and shear due to the following factored load conditions, considered separately:
(a) a uniformly distributed load equal to the total dead and live loads;
(b) an unbalanced load with 100% of the total dead and live loads on any continuous portion of the joist and 25% of total dead and live loads on the remainder to produce the most critical effect on any component; and
(c) a concentrated factored load applied at any panel point of 13.5 kN for floor joists for office or similar occupancy or 2 kN for roof joists.

16.6.2 Loading for Special Open-Web Steel Joists

The factored moment and shear resistances of special open-web steel joists at every section shall be not less than the moment and shear due to the loading conditions specified by the building designer in Clause 16.5.1(a) or those due to the factored dead load plus the following factored live load conditions, (a), (b), (c), or (d), considered separately:

(a) for floor joists, an unbalanced live load applied on any continuous portion of the joist to produce the most critical effect on any component;
(b) for roof joists, an unbalanced loading condition with 100% of the snow load plus other live loads applied on any continuous portion of the joist and 50% of the snow load on the remainder of the joist to produce the most critical effect on any component;
(c) for roof joists, wind uplift; and
(d) the appropriate factored concentrated load from Table 4.1.6. B of the *National Building Code of Canada,* applied at any one panel point to produce the most critical effect on any component.

16.6.3 Design Assumptions

Open-web steel joists shall be designed for loads acting in the plane of the joist applied to the compression chord, which is assumed to be prevented from lateral buckling by the deck.

For the purpose of determining axial forces in all members, the loads may be replaced by statically equivalent loads applied at the panel points.

16.6.4 Verification of Joist Manufacturer's Design

When the adequacy of the design of a joist cannot be readily demonstrated by a rational analysis based on accepted theory and engineering practice, the joist manufacturer may elect to verify the design by test. The test shall be carried out to the satisfaction of the building designer. The test loading shall be 1.10/0.90 times the factored loads used in the design.

16.6.5 Member and Connection Resistance

Member and connection resistance shall be calculated in accordance with the requirements of Clause 13 except as otherwise specified in Clause 16.

16.6.6 Width-Thickness Ratios

16.6.6.1

Width-thickness ratios of compressive elements of hot-formed sections shall be governed by Clause 11. Width-thickness ratios of compressive elements of cold-formed sections shall be governed by CSA Standard S136.

16.6.6.2

For purposes of determining the appropriate width-thickness ratio of compressive elements supported along one edge, any stiffening effect of the deck or the joist web shall be neglected.

16.6.7 Tension Chord

16.6.7.1

The tension chord shall be continuous and may be designed as an axially loaded tension member unless subject to eccentricities in excess of those permitted under Clause 16.6.11.4 or to applied load between panel points. The governing radius of gyration of the tension chord or any component thereof shall be not less than 1/240 of the corresponding unsupported length. For joists with the web in the y-plane, the unsupported length of chord for computing L_x/r_x shall be taken as the panel length centre-to-centre of panel points, and the unsupported length of chord for calculating L_y/r_y shall be taken as the distance between bridging lines connected to the tension chord. Joist shoes, when anchored, may be assumed to be equivalent to bridging lines. Moments due to concentrated loads shall be included in the design.

16.6.7.2

The chord shall be designed for the resulting compressive forces when net uplift is specified, when joists are made continuous or cantilevered, or when end moments are specified. Bracing, when required, shall be provided in accordance with the requirements of Clause 20.3.

When net uplift due to wind forces is a design requirement, a single line of bottom-chord bracing shall also be provided at each end of the joists near the first bottom-chord panel points.

16.6.8 Compression Chord

16.6.8.1

The compression chord shall be continuous and may be designed for axial compressive force alone when the panel length does not exceed 610 mm, when concentrated loads are not applied between the panel points, and when not subject to eccentricities in excess of those permitted under Clause 16.6.11.4. When the panel length exceeds 610 mm, the compression chord shall be designed as a continuous member subject to combined axial and bending forces.

16.6.8.2

The slenderness ratio, KL/r, of the compression chord or of its components shall not exceed 90 for interior panels nor 120 for end panels, where the governing KL/r shall be the maximum value determined by the following:

(a) for x-x (horizontal) axis, L_x shall be the centre-to-centre distance of panel points, and K = 0.9;

(b) for y-y (vertical) axis, L_y shall be the centre-to-centre distance of the attachments of the deck. The spacing of attachments shall be not more than the design slenderness ratio of the top chord times the radius of gyration of the top chord about its vertical axis and not more than 1000 mm, and K = 1.0;

(c) for z-z (skew) axis of individual components, L_z shall be the centre-to-centre distance of panel points or spacers, or both and K = 0.9. Decking shall not be considered to fulfil the function of batten plates or spacers for top chords consisting of two separated components.

and where

r = the appropriate radius of gyration.

16.6.8.3

Compression chords of joists in panel lengths exceeding 610 mm shall be proportioned such that

$$\frac{C_f}{C_r} + \frac{M_f}{M_r} \leq 1.0$$

where

M_r is given in Clause 13.5 and C_r is given in Clause 13.3.

At the panel point, C_r may be taken as ϕAF_y and Clause 13.5(a) may be used to determine M_r provided that the chord meets the requirements of a Class 2 section and $M_f/M_p < 0.25$.

The chord shall be assumed to be pinned at the joist supports.

16.6.8.4

When welding is used to attach steel deck to the chord of a joist for the transfer of load due to uplift or diaphragm action, any flat width of the chord component to be in contact with the deck shall be at least 5 mm larger than the nominal design dimensions of the welds, measured transverse to the longitudinal axis of the chord.

16.6.9 Webs

16.6.9.1

Webs shall be designed in accordance with the requirements of Clause 13 to resist the shear at any point due to the factored loads given in Clause 16.6.1 or 16.6.2. Particular attention shall be paid to possible reversals of shear.

16.6.9.2

The length of a web member shall be taken as the distance between the intersections of the axes of the web and the chords. For buckling in the plane of the web, the effective length factor shall be taken as 0.9 if the web consists of individual members. For all other cases, the effective length factor shall be taken as 1.0.

16.6.9.3

The slenderness ratio of a web member in tension need not be limited.

16.6.9.4

The slenderness ratio of a web member in compression shall not exceed 200.

16.6.10 Spacers and Battens

Compression members consisting of two or more sections shall be interconnected so that the slenderness ratio of each section calculated using its least radius of gyration is less than or equal to the design slenderness ratio of the built-up member. Spacers or battens shall be an integral part of the joist.

16.6.11 Connections and Splices

16.6.11.1

Component members of joists shall be connected by welding, bolting, or other approved means.

16.6.11.2

Connections and splices shall develop the factored loads without exceeding the factored member resistances given in Clause 16. Butt-welded splices shall develop the factored tensile resistance, T_r, of the member.

16.6.11.3

Splices may occur at any point in chord or web members.

16.6.11.4

Members connected at a joint preferably shall have their centroidal axes meet at a point. Where this is impractical and eccentricities are introduced, such eccentricities may be neglected if they do not exceed

(a) for continuous web members, the greater of the two distances measured from the neutral axis of the chord member to the extreme fibres of the chord member; and

(b) for noncontinuous web members, the distance measured from the neutral axis to the back (outside face) of the chord member.

When the eccentricity exceeds these limits, provision shall be made for the effects of total eccentricity.

Eccentricities assumed in design shall be those at maximum fabrication tolerances, which shall be stated on the shop details.

16.6.12 Bearings

16.6.12.1

Bearings at ends of joists shall be proportioned so that the factored bearing resistance of the supporting material is not exceeded.

16.6.12.2

Where a joist bears, with or without a bearing plate, on solid masonry or concrete support, the bearing shall meet the requirements of CSA Standards S304 and A23.3, respectively.

16.6.12.3

Where a joist bears on a member of the structural steel frame, the end of the bearing shall extend at least 65 mm beyond the face of the support, except that when the available bearing area is restricted this distance may be reduced provided that the bearing is adequately anchored to the support. In any case the factored bearing resistance shall not be exceeded.

16.6.12.4

The bearing detail and the end panels of the joist shall be proportioned to include the effect of the eccentricity between the centre of the bearing and the intersection of the axes of the chord and the end diagonal.

16.6.13 Anchorage

16.6.13.1

Joist ends shall be properly anchored to withstand the effect of factored loads, including net uplift, as follows:

(a) in no case shall the anchorage to masonry be less than the following:

(i) for floor joists, a 10 mm diameter rod at least 300 mm long embedded horizontally;

(ii) for roof joists, a 20 mm diameter anchor bolt 300 mm long embedded vertically with a 50 mm, 90° hook;

(b) the anchorage to steel shall be a connection capable of withstanding a horizontal load of not less than 10% of the end reaction of the joist but not less than one 20 mm diameter bolt or a pair of fillet welds satisfying the minimum size and length requirements of CSA Standard W59.

16.6.13.2

Tie joists may have their top and bottom chords connected to a column. Unless otherwise specified, tie joists shall have top and bottom chord connections each at least equivalent to those required by Clause 16.6.13.1. Either the top or bottom connection shall utilize a mechanical fastener.

16.6.13.3

Where joists are used as a part of a frame, the joist-to-column connections shall be designed to carry the moments and forces due to the factored loads (see Clause 7.2).

16.6.14 Deflection

16.6.14.1

Steel joists shall be proportioned so that deflection due to specified loads is within acceptable limits for the nature of the materials to be supported and the intended use and occupancy. Such deflection limits shall be as given in Clause 6.2.1 unless otherwise specified by the building designer.

16.6.14.2

The deflection may be established by test or may be calculated by assuming a moment of inertia equal to the gross moment of inertia of the chords about the centroidal axis of the joist and multiplying the calculated deflection derived on this basis by 1.10.

16.6.15 Camber

Unless otherwise specified by the building designer, the nominal camber, in millimetres, shall be equal to 0.07 times the square of the span, expressed in metres. For tolerances see Clause 16.11.9.

16.6.16 Vibration

The building designer shall give special consideration to floor systems where unacceptable vibration may occur. When requested, the joist manufacturer shall supply joist properties and details to the building designer. (See Appendix G.)

16.6.17 Welding

16.6.17.1

Arc welding design and practice shall conform to CSA Standard W59.

16.6.17.2

The resistance of resistance-welded joints shall be taken as established in CSA Standard W55.3, and the related welding practice shall conform to welding standards approved by the Canadian Welding Bureau under the same CSA Standard.

16.6.17.3

Fabricators and erectors of welded construction covered by this Standard shall be certified by the Canadian Welding Bureau in Division 1 or Division 2.1 to the requirements of CSA Standard W47.1 or W55.3, or both, as applicable.

16.6.17.4

The factored resistances of welds shall be equal to those given in Table 3(A).

16.6.17.5

When field welding joists to supporting members, surfaces to be welded shall be free of coatings that are detrimental to achieving an adequate weldment.

16.6.17.6

Flux and slag shall be removed from all welds.

16.7 Stability During Construction

Means shall be provided to support joist chords against lateral movement and to hold the joist in the vertical or specified plane during construction.

16.8 Bridging

16.8.1 General

Bridging transverse to the span of joists may be used to meet the requirements of Clause 16.7 and also to meet the slenderness ratio requirements for chords. Bridging is not to be considered "bracing" as defined in Clause 20.3.1.

16.8.2 Installation

All bridging and bridging anchors shall be completely installed before any construction loads, except for the weight of the workers necessary to install the bridging, are placed on the joists.

16.8.3 Types

Unless otherwise specified or approved by the building designer, the joist manufacturer shall supply bridging that may be either of the diagonal or horizontal type.

16.8.4 Diagonal Bridging

Diagonal bridging consisting of crossed members running from top chord to bottom chord of adjacent joists shall have a slenderness ratio, L/r, of not more than 200, where L is the length of the diagonal bridging member or one-half of this length when crossed members are connected at their point of intersection, and r is the least radius of gyration. All diagonal bridging shall be connected adequately to the joists by bolts or welds.

16.8.5 Horizontal Bridging

A line of horizontal bridging shall consist of a continuous member attached to either the top chord or the bottom chord. Horizontal bridging members shall have a slenderness ratio of not more than 300.

16.8.6 Attachment of Bridging

Attachment of diagonal and horizontal bridging to joist chords shall be by welding or mechanical means capable of resisting an axial load of at least 3 kN in the attached bridging member. These welds should meet the minimum length requirements stipulated in CSA Standard W59.

16.8.7 Anchorage of Bridging

Each line of bridging shall be adequately anchored at each end to sturdy walls or to main components of the structural frame, if practicable. If not practicable, diagonal and horizontal bridging shall be provided in combination between adjacent joists near the ends of bridging lines.

The ends of joists designed to bear on their bottom chords shall be held adequately in position by attachments to the walls or to the structural frame or by lines of bridging located at the ends except where such ends are built into masonry or concrete walls.

16.8.8 Bridging Systems

Bridging systems, including sizes of bridging members and all necessary details, shall be shown on the erection diagrams. If a specific bridging system is required by the design, the design drawings shall show all information necessary for the preparation of shop details and erection diagrams.

16.8.9 Spacing of Bridging

Diagonal and horizontal bridging, whichever is furnished, shall be spaced so that the unsupported length of the chord between bridging lines or between laterally supported ends of the joist and adjacent bridging lines does not exceed

(a) for compression chords, 170r; and
(b) for tension chords, 240r;

where
r = the applicable chord radius of gyration about its axis in the plane of the web.

Ends of joists anchored to supports may be assumed to be equivalent to bridging lines. If ends of joists are not so anchored before deck is installed, the distance from the face of the support to the nearest bridging member in the plane of the bottom chord shall not exceed 120r. In no case shall there be less than one line of horizontal or diagonal bridging attached to each joist spanning 4 m or more. If only a single line of bridging is required, it shall be placed at the centre of the joist span. If bridging is not used on joists less than 4 m in span, the ends of such joists shall be anchored to the supports so as to prevent overturning of the joist during placement of the deck.

16.9 Decking

16.9.1 Decking to Provide Lateral Support

Decking shall bear directly on the top chord of the joist and shall be sufficiently rigid to provide lateral support to the compression chord of the joist. In special cases where the decking is incapable of furnishing the required lateral support, the compression chord of the joist shall be braced laterally in accordance with the requirements of Clause 20.3.

16.9.2 Attachments

Attachments of decking considered to provide lateral support shall be capable of staying the top chords laterally. Attachments shall be deemed to fulfil this requirement when the attachments as a whole are adequate to resist a force in the plane of the decking of not less than 5% of the maximum force in the top chord and assumed to be uniformly distributed along the length of the top chord. The spacing of attachments shall be not more than the design slenderness ratio of the top chord times the radius of gyration of the top chord about its vertical axis, and not more than 1 m.

16.9.3 Diaphragm Action

Where decking is used in combination with joists to form a diaphragm for the purpose of transferring lateral applied loads to vertical bracing systems, special attachment requirements shall be fully specified on the building design drawings.

16.9.4 Cast-In-Place Slabs

Cast-in-place slabs used as decking shall have a minimum thickness of 50 mm. Forms for cast-in-place slabs shall not cause lateral displacement of the top chords of joists during installation of the forms or the placing of the concrete. Nonremovable forms shall be positively attached to top chords by means of clips, ties, wedges, fasteners, or other suitable means at intervals not exceeding 1 m; however, there shall be at least two attachments in the width of each form at each joist. Forms and their method of attachment shall be such that the cast-in-place slab, after hardening, is capable of furnishing lateral support to the joist chords.

16.9.5 Installation of Steel Deck

16.9.5.1

To facilitate installation of the steel deck, the location of the top chord of the joist shall be confirmed by marking the deck at suitable intervals, or by other suitable means.

16.9.5.2

(a) The installer of steel deck to be fastened to joists by arc spot welding shall be certified by the Canadian Welding Bureau to the requirements of CSA Standard W47.1, Division 1 or Division 2.1. Part of the work may be sublet to a Division 3 firm providing that the Division 1 or Division 2.1 firm retains responsibility for the sublet work.
(b) The installation welding procedures shall be approved by the Canadian Welding Bureau.
(c) The welders shall have current qualifications for arc spot welding issued by the Canadian Welding Bureau.

16.10 Shop Painting

Joists shall have one shop coat of protective paint of a type standard with the manufacturer unless otherwise specified.

16.11 Manufacturing Tolerances

16.11.1

The tolerance on the specified depth of the manufactured joist shall be ±7 mm.

16.11.2

The maximum deviation from the design location of a panel point measured along the length of a chord shall be 13 mm. In joists in which an individual end diagonal is attached to the bottom chord or in which the end diagonal is a continuation of an upturned bottom chord, the gravity axes of the members in such a joint should meet at a point. (See Clause 16.6.11.4.)

16.11.3

The maximum deviation from the design location of a panel point measured perpendicular to the longitudinal axis of the chord and in the plane of the joist shall be 7 mm.

16.11.4

The connections of web members to chords shall not deviate laterally more than 3 mm from that assumed in the design.

16.11.5

The maximum sweep of a joist or any portion of the length of the joist upon completion of manufacture shall be not greater than 1/500 of the length on which the sweep is measured.

16.11.6

The maximum tilt of bearing shoes shall be 1 in 50 measured from a plane perpendicular to the plane of the web and parallel to the longitudinal axis of the joist.

16.11.7

The tolerance on the specified shoe depth shall be ±3 mm.

16.11.8

The tolerance on the specified length of the joist shall be ±7 mm. The connection holes in a joist shall not vary from the detailed location by more than 2 mm for members 10 m or less in length or by more than 3 mm for members more than 10 m in length.

16.11.9

The tolerance in millimetres on the nominal or specified camber shall be

$$\pm\left(6 + \frac{\text{span, in metres}}{4}\right)$$

The resulting actual minimum camber in a joist shall be +3 mm, except that the maximum range in camber for joists of the same span shall be limited to 20 mm.

16.12 Inspection and Quality Control

16.12.1 Inspection

Material and workmanship shall be accessible for inspection at all times by qualified inspectors representing the building designer. Random in-process inspection shall be carried out by the manufacturer, and all joists shall be thoroughly inspected by the manufacturer before shipping.

16.12.2 Identification and Control of Steel

Steel used in the manufacture of joists shall, at all times in the manufacturer's plant, be marked to identify its specification (and grade, where applicable). This shall be done by suitable markings or by recognized colour-coding or by any system devised by the manufacturer that will ensure to the satisfaction of the building designer that the correct material is being used.

16.12.3 Quality Control

Upon request by the building designer, the manufacturer shall provide evidence of having suitable quality control measures to ensure that the joists meet all specified requirements. When testing is part of the manufacturer's normal quality control program, the loading criteria shall be 1.0/0.9 times the factored loads for the materials used in the joists.

For resistance welding, the quality control procedures outlined in CSA Standard W55.3 shall be met. For arc welding, the quality control requirements of CSA Standard W59 shall be met.

16.13 Handling and Erection

16.13.1 General

Care shall be exercised to avoid damage during strapping, transport, unloading, site storage and piling, and erection. Dropping of joists shall not be permitted. Special precautions shall be taken when erecting long, slender joists and hoisting cables shall preferably not be released until the member is stayed laterally by at least one line of bridging. Joists shall have all bridging attached and be permanently fastened into place before the application of any loads. Heavy construction loads shall be adequately distributed so as not to exceed the capacity of any joist. Field welding shall not cause damage to joists, bridging, deck, and supporting steel members.

16.13.2 Erection Tolerances

16.13.2.1

The maximum sweep of a joist or a portion of the length of a joist upon completion of erection shall not exceed the limit given in Clause 16.11.5 and shall be in accordance with the general requirements of Clause 30.

16.13.2.2

All members shall be free from twists, sharp kinks, and bends.

16.13.2.3

When joists are finally fastened in position in the field, the maximum deviation from the location shown on the erection diagrams shall be 15 mm.

16.13.2.4

The deviation normal to the specified plane of the web of a joist shall not exceed 1/50 of the depth of the joist.

17. Composite Beams

17.1 Application

The provisions of Clause 17 apply to composite beams consisting of steel sections, trusses, or joists interconnected with either a reinforced concrete slab or a steel deck with a concrete cover slab.

17.2 Definitions

The following definitions apply in this Clause:

Cover slab — the concrete above the flutes of the steel deck. All flutes shall be filled with concrete so as to form a ribbed slab.

Effective cover slab thickness, t — the minimum thickness of concrete measured from the top of the cover slab to the top of the steel deck. This thickness shall not be less than 65 mm unless the adequacy of a lesser thickness has been established by appropriate tests.

Effective slab thickness, t — the overall slab thickness, provided that
(a) the slab is cast with a flat underside;
(b) the slab is cast on corrugated steel forms having a height of corrugation not greater than 0.25 times the overall slab thickness; or
(c) the slab is cast on fluted steel forms whose profile meets the following requirements. The minimum concrete rib width shall be 125 mm; the maximum rib height shall be 40 mm but not more than 0.4 times the overall slab thickness; the average width between ribs shall not exceed 0.25 times the overall slab thickness nor 0.2 times the minimum width of concrete ribs.

In all other cases, effective slab thickness means the overall slab thickness minus the height of flute corrugation.

Flute — the portion of the steel deck that forms a valley.

Rib — the portion of the concrete slab that is formed by the steel deck flute.

Slab — a reinforced cast-in-place concrete slab at least 65 mm in effective thickness. The area equal to the effective width times the effective slab thickness shall be free of voids or hollows except for those specifically permitted in the definition of effective slab thickness.

Steel deck — a load-carrying steel deck, consisting of either
(a) a single fluted element (noncellular deck); or
(b) a two-element section consisting of a fluted element in conjunction with a flat sheet (cellular deck).

The maximum depth of the deck shall be 80 mm and the average width of the minimum flute shall be 50 mm. A steel deck may be of a type intended to act compositely with the cover slab in supporting applied load.

Steel joist — an open-web steel joist suitable for composite design.

Steel section — a steel structural section with a solid web or webs suitable for composite design. Web openings are permissible only on condition that their effects are fully investigated and accounted for in the design.

17.3 General

17.3.1 Deflections

Calculation of deflections shall take into account the effects of creep of concrete, shrinkage of concrete, and increased flexibility resulting from partial shear connection and from interfacial slip. These effects shall be established by test or analysis, where practicable. Consideration shall also be given to the effects of full or partial continuity in the steel beams and concrete slabs in reducing calculated deflections.

In lieu of tests or analysis, the effects of partial shear connection and interfacial slip, creep, and shrinkage may be assessed as follows:

(a) for increased flexibility resulting from partial shear connection and interfacial slip, calculate the deflections using an effective moment of inertia given by

$$I_e = I_s + 0.85\,(p)^{0.25}\,(I_t - I_s)$$

where

I_s = moment of inertia of steel beam

I_t = transformed moment of inertia of composite beam

p = fraction of full shear connection (use $p = 1.00$ for full shear connection);

(b) for creep, increase elastic deflections caused by dead loads and long-term live loads, as calculated in Item (a), by 15%; and

(c) for shrinkage of concrete, calculate deflection using a selected shrinkage strain and assuming the composite beam to be bent in single curvature by equal end moments. The shrinkage strain is affected by such factors as age of concrete, ratio of slab volume to surface area, concrete properties (water/cement ratio, per cent fines, entrained air, cement content), and the restraint provided by the steel beam and deck. (See Appendix L for methods of calculating deflections due to shrinkage strain.)

17.3.2 Vertical Shear

The web area of steel sections or web system of steel trusses and joists shall be proportioned to carry the total vertical shear, V_f.

17.3.3 End Connections

End connections of steel sections, trusses, and joists shall be proportioned to transmit the total end reaction of the composite beam.

17.4 Design Effective Width of Concrete

17.4.1

Slabs or cover slabs extending on both sides of the steel section or joist shall be deemed to have a design effective width, b, equal to the lesser of
(a) 0.25 times the composite beam span; or
(b) the average distance from the centre of the steel section, truss, or joist to the centres of adjacent parallel supports.

17.4.2

Slabs or cover slabs extending on one side only of the supporting section or joist shall be deemed to have a design effective width, b, not greater than the width of the top flange of the steel section or top chord of the steel joist or truss plus the lesser of
(a) 0.1 times the composite beam span; or
(b) 0.5 times the clear distance between the steel section, truss, or joist and the adjacent parallel support.

17.5 Slab Reinforcement

17.5.1 General

Slabs shall be adequately reinforced to support all loads and to control both cracking transverse to the composite beam span and longitudinal cracking over the steel section or joist. Reinforcement shall not be less than that required by the specified fire-resistance design of the assembly.

17.5.2 Parallel Reinforcement

Reinforcement parallel to the span of the beam in regions of negative bending moment of the composite beam shall be anchored by embedment in concrete that is in compression. The reinforcement of slabs that are to be continuous over the end support of steel sections or joists fitted with flexible end connections shall be given special attention. In no case shall such reinforcement at the ends of beams supporting ribbed slabs perpendicular to the beam be less than two 15 M bars.

17.5.3 Transverse Reinforcement, Solid Slabs

Unless it is known from experience that longitudinal cracking caused by composite action directly over the steel section or joist is unlikely, additional transverse reinforcement or other effective means shall be provided. Such additional reinforcement shall be placed in the lower part of the slab and anchored so as to develop the yield strength of the reinforcement. The area of such reinforcement shall be not less than 0.002 times the concrete area being reinforced and shall be uniformly distributed.

17.5.4 Transverse Reinforcement, Ribbed Slabs

17.5.4.1

Where the ribs are parallel to the beam span, the area of transverse reinforcement shall be not less than 0.002 times the concrete cover slab area being reinforced and shall be uniformly distributed.

17.5.4.2

Where the ribs are perpendicular to the beam span, the area of transverse reinforcement shall not be less than 0.001 times the concrete cover slab area being reinforced and shall be uniformly distributed.

17.6 Interconnection

17.6.1

Except as permitted by Clauses 17.6.2 and 17.6.4, interconnection between steel sections, trusses, or joists and slabs or steel decks with cover slabs shall be attained by the use of shear connectors as prescribed in Clause 17.7.

17.6.2

Unpainted steel sections, trusses, or joists that support slabs and are totally encased in concrete do not require interconnection by means of shear connectors provided that
(a) a minimum of 50 mm of concrete covers all portions of the steel section, truss, or joist except as noted in Item (c);
(b) the cover in Item (a) is reinforced to prevent spalling; and
(c) the top of the steel section, truss, or joist is at least 40 mm below the top and 50 mm above the bottom of the slab.

17.6.3

Studs may be welded through a maximum of two steel sheets in contact, each not more than 1.71 mm in overall thickness including coatings (1.52 mm in nominal base steel thickness plus zinc coating not greater than nominal 275 g/m^2). Otherwise, holes for placing studs shall be made through the sheets as necessary. Welded studs shall meet the requirements of CSA Standard W59.

17.6.4

Other methods of interconnection that have been adequately demonstrated by test and verified by analysis may be used to effect the transfer of forces between the steel section, truss, or joist and the slab or steel deck with cover slab. In such cases the design of the composite member shall conform to the design of a similar member employing shear connectors, insofar as practicable.

17.6.5

The diameter of a welded stud shall not exceed 2.5 times the thickness of the part to which it is welded, unless test data satisfactory to the designer are provided to establish the capacity of the stud as a shear connector.

17.7 Shear Connectors

17.7.1 General

The resistance factor, ϕ_{sc}, to be used with the shear resistances given in this Clause shall be taken as 0.80. The factored shear resistance, q_r, of other shear connectors shall be established by tests acceptable to the designer.

17.7.2 End-Welded Studs

End-welded studs shall be headed or hooked with $h/d \geq 4$. The projection of a stud in a ribbed slab, based on its length prior to welding, shall be at least two stud diameters above the top surface of the steel deck.

17.7.2.1

In solid slabs,
$q_{rs} = 0.50\phi_{sc} A_{sc}\sqrt{f_c' E_c} \leq \phi_{sc} A_{sc} F_u$
where F_u for commonly available studs is 415 MPa and q_{rs} is in newtons.

17.7.2.2

In ribbed slabs with ribs parallel to the beam,
(a) when $w_d/h_d \geq 1.50$
$q_{rr} = q_{rs}$
(b) when $w_d/h_d < 1.50$
$q_{rr} = [0.6 (w_d/h_d)(h/h_d - 1)]q_{rs} \leq 1.0\, q_{rs}$

17.7.2.3

In ribbed slabs with ribs perpendicular to the beam,
(a) when $h_d = 75$ mm
$q_{rr} = 0.35\,\phi_{sc}\,\rho\,A_p\sqrt{f_c'} \leq q_{rs}$
(b) when $h_d = 38$ mm
$q_{rr} = 0.61\,\phi_{sc}\,\rho\,A_p\sqrt{f_c'} \leq q_{rs}$
where
A_p is the concrete pull-out area taking the deck profile and stud burn-off into account. For a single stud, the apex of the pyramidal pull-out area, with four sides sloping at 45°, is taken as the centre of the top surface of the head of the stud. For a pair of studs, the pull-out area has a ridge extending from stud to stud.
ρ = 1.0 for normal-density concrete (2150 to 2500 kg/m^3)
= 0.85 for semi-low-density concrete (1850 to 2150 kg/m^3)

17.7.3 Channel Connectors

In solid slabs of normal-density concrete with $f_c' \geq 20$ MPa and a density of at least 2300 kg/m^3,
$q_{rs} = 36.5\,\phi_{sc}\,(t + 0.5w)L_c\sqrt{f_c'}$

17.8 Ties

Mechanical ties shall be provided between the steel section, truss, or joist and the slab or steel deck to prevent separation. Shear connectors may serve as mechanical ties if suitably proportioned. The maximum spacing of ties shall not exceed 1000 mm, and the average spacing in a span shall not exceed 600 mm or be greater than that required to achieve any specified fire-resistance rating of the composite assembly.

17.9 Design of Composite Beams with Shear Connectors

17.9.1

The composite beam shall consist of steel section, truss or joist, shear connectors, ties, and slab or steel deck with cover slab.

17.9.2

The properties of the composite section shall be calculated neglecting any concrete area that is in tension within the maximum effective area (equal to effective width times effective thickness). If a steel truss or joist is used, the area of its top chord shall be neglected in determining the properties of the composite section and only Clause 17.9.3(a) is applicable.

17.9.3

The factored moment resistance, M_{rc}, of the composite section with the slab or cover slab in compression shall be calculated as follows, where ϕ = 0.90 and ϕ_c, the resistance factor for concrete, = 0.60:

(a) **Case 1** — Full shear connection and plastic neutral axis in the slab; that is, $Q_r \geq \phi A_s F_y$ and $\phi A_s F_y \leq 0.85\phi_c b t f_c'$, where Q_r equals the sum of the factored resistances of all shear connectors between points of maximum and zero moment

$$M_{rc} = T_r e' = \phi A_s F_y e'$$

where e′ is the lever arm and is calculated using

$$a = \frac{\phi A_s F_y}{0.85\phi_c b f_c'}$$

(b) **Case 2** — Full shear connection and plastic neutral axis in the steel section; that is, $Q_r \geq 0.85\phi_c b t f_c'$ and $0.85\phi_c b t f_c' < \phi A_s F_y$

$$M_{rc} = C_r e + C_r' e'$$

$$C_r' = 0.85\phi_c b t f_c'$$

$$C_r = \frac{\phi A_s F_y - C_r'}{2}$$

(c) **Case 3** — Partial shear connection; that is, $Q_r < 0.85\phi_c b t f_c'$ and $< \phi A_s F_y$

$$M_{rc} = C_r e + C_r' e'$$

$$C_r' = Q_r$$

$$C_r = \frac{\phi A_s F_y - C_r'}{2}$$

where e′ is the lever arm and is calculated using

$$a = \frac{C_r'}{0.85\phi_c b f_c'}$$

17.9.4

No composite action shall be assumed in calculating flexural strength when Q_r is less than 0.4 times the lesser of $0.85\phi_c b t f_c'$ and $\phi A_s F_y$. No composite action shall be assumed in calculating deflections when Q_r is less than 0.25 times the lesser of $0.85\phi_c b t f_c'$ and $\phi A_s F_y$.

17.9.5

For full shear connection, the total horizontal shear, V_h, at the junction of the steel section, truss, or joist and the concrete slab or steel deck, to be resisted by shear connectors distributed between the point of maximum bending moment and each adjacent point of zero moment shall be

$$V_h = \phi A_s F_y$$

$$V_h = 0.85\phi_c b t f_c'$$

for Cases 1 and 2 as defined in Clause 17.9.3(a) and (b), respectively, and $Q_r \geq V_h$.

17.9.6

For partial shear connection the total horizontal shear, V_h, as defined in Clause 17.9.3(c) shall be

$$V_h = Q_r$$

17.9.7

Composite beams employing steel sections and concrete slabs may be designed as continuous members. The factored moment resistance of the composite section, with the concrete slab in the tension area of the composite section, shall be the factored moment resistance of the steel section alone, except that when sufficient shear connectors are placed in the negative moment region, suitably anchored concrete slab reinforcement parallel to the steel sections and within the design effective width of the concrete slab may be included in calculating the properties of the composite section. The total horizontal shear, V_h, to be resisted by shear connectors between the point of maximum negative bending moment and each adjacent point of zero moment shall be taken as $\phi A_r F_{yr}$.

17.9.8

The number of shear connectors to be located on each side of the point of maximum bending moment (positive or negative, as applicable), distributed between that point and the adjacent point of zero moment, shall be not less than

$$n = \frac{V_h}{q_r}$$

Shear connectors may be spaced uniformly, except that in a region of positive bending the number of shear connectors, n′, required between any concentrated load applied in that region and the nearest point of zero moment shall be not less than

$$n' = n\left(\frac{M_{f1} - M_r}{M_f - M_r}\right)$$

where

M_{f1} = positive bending moment under factored load at concentrated load point
M_r = factored moment resistance of the steel section alone
M_f = maximum positive bending moment under factored load

17.9.9 Longitudinal Shear

The longitudinal shear of composite beams with solid slabs or with cover slabs and steel deck parallel to the beam shall be taken as

$$V_u = \Sigma q_r - 0.85\phi_c f_c' A_c - \phi A_r F_{yr}$$

where A_r is the area of longitudinal reinforcement within the concrete area, A_c

For normal-weight concrete, the factored shear resistance along any potential longitudinal shear surfaces in the concrete slab shall be taken as

$$V_r = (0.80\,\phi A_r F_{yr} + 2.76\,\phi_c A_{cv}) \leq 0.50\,\phi_c f_c' A_{cv}$$

where A_r is the area of transverse reinforcement crossing shear planes, A_{cv}

17.10 Design of Composite Beams without Shear Connectors

17.10.1

Unpainted steel sections or joists supporting concrete slabs and encased in concrete in accordance with Clause 17.6.2 may be proportioned on the basis that the composite section supports the total load.

17.10.2

The properties of the composite section for determination of load carrying capacity shall be calculated by ultimate strength methods, neglecting any area of concrete in tension.

17.10.3

As an alternative method of design, encased simple-span steel sections or joists may be proportioned on the basis that the steel section, truss, or joist alone supports 0.90 times the total load.

17.11 Unshored Beams

For composite beams that are unshored during construction, the stresses in the tension flange of the steel section, truss, or joist due to the loads applied before the concrete strength reaches $0.75f'_c$ plus the stresses at the same location due to the remaining specified loads considered to act on the composite section, shall not exceed $0.90F_y$.

17.12 Beams During Construction

The steel section, truss, or joists alone shall be proportioned to support all factored loads applied prior to hardening of the concrete without exceeding its calculated capacity under the conditions of lateral support or shoring, or both, to be furnished during construction.

18. Composite Columns (Concrete-Filled Hollow Structural Sections)

18.1

The provisions of Clause 18 apply to composite columns consisting of steel hollow structural sections completely filled with concrete.

18.2

Hollow structural sections designated as Class 1, 2, or 3 sections that are completely filled with concrete may be assumed to carry compressive load as composite columns. Class 4 hollow structural sections that are completely filled with concrete may also be designed as composite columns provided that the width-thickness ratios of the walls of rectangular sections do not exceed $1350/\sqrt{F_y}$, and the outside diameter-to-thickness ratios of circular sections do not exceed $28\,000/F_y$.

18.3

The axial load assumed to be carried by the concrete shall be only that portion applied by direct bearing on the concrete. Other assumptions regarding the amount of load applied to the concrete may be employed if their adequacy has been demonstrated by test.

18.4

The factored compressive resistance of a composite column shall be taken as

$$C_{rc} = \tau C_r + \tau' C'_r$$

where

$$C'_r = 0.85\phi_c f'_c A_c \lambda_c^{-2}\left[\sqrt{1+0.25\lambda_c^{-4}} - 0.5\lambda_c^{-2}\right]$$

in which $\lambda_c = \frac{KL}{r_c}\sqrt{\frac{f'_c}{\pi^2 E_c}}$

r_c = radius of gyration of the concrete area, A_c

E_c = initial elastic modulus for concrete, considering the effects of long-term loading. For normal-weight concrete, with f'_c expressed in megapascals, this may be taken as $(1 + S/T)\,2500\sqrt{f'_c}$

where
S is the short-term load and T is the total load on the column.

For all rectangular hollow structural sections and for circular hollow structural sections with a height-to-diameter ratio of 25 or greater, $\tau = \tau' = 1.0$

Otherwise $\tau = \dfrac{1}{\sqrt{1 + \rho + \rho^2}}$

and $\tau' = 1 + \left(\dfrac{25\,\rho^2\tau}{(D/t)}\right)\left(\dfrac{F_y}{0.85\,f'_c}\right)$

where
$\rho = 0.02\,(25 - L/D)$

18.5

Where bending as well as axial compression is to be resisted, the bending shall be assumed to be resisted by the steel section alone. The steel section shall be proportioned as a beam-column to carry the total bending plus axial compression equal to the difference between the total axial compression and that portion that can be sustained by the concrete:

$M_f \le \tau\, M_r$

and if $C_f > \tau' C'_r$

$$\frac{C_f - \tau' C'_r}{\tau\, C_r} + \frac{\omega_1 M_f}{\tau\, M_r\left(1 - \dfrac{C_f - \tau' C'_r}{C_e}\right)} \le 1.0$$

Alternatively, the contribution of both the steel and the concrete to the bending resistance may be based on a rational analysis substantiated by tests, in which case the following equation shall be satisfied:

$$\frac{C_f}{C_{rc}} + \frac{\omega_1 M_f}{M_{rc}\left(1 - \dfrac{C_f}{C_{ec}}\right)} \le 1.0$$

19. General Requirements for Built-Up Members

19.1 Members in Compression

19.1.1

All components of built-up compression members and the transverse spacing of their lines of connecting bolts or welds shall meet the requirements of Clauses 10 and 11.

19.1.2

All component parts that are in contact with one another at the ends of built-up compression members shall be connected by bolts spaced longitudinally not more than four diameters apart for a distance equal to 1.5 times the width of the member or by continuous welds having a length of not less than the width of the member.

19.1.3

Unless closer spacing is required for transfer of load or for sealing inaccessible surfaces, the longitudinal spacing in-line between intermediate bolts or the clear longitudinal spacing between intermittent welds in built-up compression members shall not exceed the following, as applicable:

(a) $330t/\sqrt{F_y}$, but not more than 300 mm for the outside component of the section consisting of a

plate when the bolts on all gauge lines or intermittent welds along the component edges are not staggered, where t = thickness of the outside plate;
(b) $525t/\sqrt{F_y}$, but not more than 450 mm for the outside component of the section consisting of a plate when the bolts or intermittent welds are staggered on adjacent lines, where t = thickness of the outside plate;
(c) for compression members composed of two or more rolled shapes in contact or separated from one another by intermittent fillers, the slenderness ratio of any shape between points of interconnection shall not exceed the slenderness ratio of the built-up member. The least radius of gyration of each component part shall be used in computing the slenderness ratio of that part between points of interconnection with other component parts. (See also Clause 27.4.3.1.)

19.1.4

For compression members composed of two or more rolled shapes in contact or separated from one another by intermittent fillers, the equivalent slenderness ratio shall be taken as

$$\rho_e = \sqrt{\rho_o^{\ 2} + \rho_i^{\ 2}}$$

ρ_e = equivalent slenderness ratio of built-up member
ρ_o = slenderness ratio of built-up member acting as a unit
ρ_i = maximum slenderness ratio of component part of a built-up member between fasteners based on an effective length factor of 1.0 when the interconnectors are fastened with snug-tight bolts and 0.65 when fastened with welds or pretensioned bolts.

19.1.5

For starred angle compression members interconnected at least at the one-third points, Clause 19.1.4 need not apply.

19.1.6

The fasteners and interconnecting parts, if any, shall be proportioned to resist a force equal to one per cent of the total force in the built-up member.

19.1.7

The spacing requirements of Clauses 19.1.3, 19.2.3, and 19.2.4 may not always provide a continuous tight fit between components in contact. When the environment is such that corrosion could be a serious problem, the spacing of bolts or welds may need to be less than the specified maximum.

19.1.8

Open sides of compression members built up from plates or shapes shall be connected to each other by lacing, batten plates, or perforated cover plates.

19.1.9

Lacing shall provide a complete triangulated shear system and may consist of bars, rods, or shapes. The spacing of connections of lacing to a main component shall be such that the slenderness ratio of a main component between these points of connection does not exceed the governing slenderness ratio of the member as a whole. Lacing shall be proportioned to resist a shear normal to the longitudinal axis of the member of not less than 2.5% of the total axial load on the member plus the shear from transverse loads, if any.

19.1.10

The slenderness ratio of lacing members shall not exceed 140. The effective length for single lacing shall be the distance between connections to the main components; for double lacing connected at the intersections, the effective length shall be 70% of that distance.

19.1.11

Lacing members shall preferably be inclined to the longitudinal axis of the built-up member at an angle of not less than 45°.

19.1.12

Lacing systems shall have diaphragms in the plane of the lacing and as near to the ends as practicable, and at intermediate points where lacing is interrupted. Such diaphragms may be plates (tie plates) or shapes.

19.1.13

End tie plates used as diaphragms shall have a length not less than the distance between the lines of bolts or welds connecting them to the main components of the member. Intermediate tie plates shall have a length of not less than one-half of that prescribed for end tie plates. The thickness of tie plates shall be at least 1/60 of the width between lines of bolts or welds connecting them to the main components, and the longitudinal spacing of the bolts or clear longitudinal spacing between welds shall not exceed 150 mm. At least three bolts shall connect the tie plate to each main component or, alternatively, a total length of weld not less than one-third the length of tie plate shall be used.

19.1.14

Shapes used as diaphragms shall be proportioned and connected to transmit from one main component to the other a longitudinal shear equal to 5% of the axial compression in the member.

19.1.15

Perforated cover plates may be used in lieu of lacing and tie plates on open sides of built-up compressive members. The net width of such plates at access holes shall be assumed to be available to resist axial load, provided that

(a) the width-thickness ratio conforms to Clause 11;
(b) the length of the access hole does not exceed twice its width;
(c) the clear distance between access holes in the direction of load is not less than the transverse distance between lines of bolts or welds connecting the perforated plate to the main components of the built-up member; and
(d) the periphery of the access hole at all points has a minimum radius of 40 mm.

19.1.16

Battens consisting of plates or shapes may be used on open sides of built-up compression members that do not carry primary bending in addition to axial load. Battens shall be provided at the ends of the member, at locations where the member is laterally supported along its length, and elsewhere as determined by the following spacing requirements:

(a) when the slenderness ratio of the built-up member with respect to the axis perpendicular to the battens is equal to or less than 80% of the slenderness ratio with respect to the axis parallel to the battens, the spacing between battens shall be such that the slenderness ratio of a main component between ends of adjacent batten plates shall not exceed 50, and shall not exceed 70% of the slenderness ratio of the built-up member with respect to the axis parallel to the battens;

(b) when the slenderness ratio of the built-up member with respect to the axis perpendicular to the battens is more than 80% of the slenderness ratio with respect to the axis parallel to the battens, the spacing between battens shall be such that the slenderness ratio of a main component between ends of adjacent batten plates shall not exceed 40, and shall not exceed 60% of the slenderness ratio of the built-up member with respect to the axis perpendicular to the battens.

19.1.17

Battens shall have a length of not less than the distance between lines of bolts or welds connecting them to the main components of the member, and shall have a thickness of not less than 1/60 of this distance, if the batten consists of a flat plate. Battens and their connections shall be proportioned to resist, simultaneously, a longitudinal shear force,

$$V_f = \frac{0.025\,C_f d}{na}$$

and a moment,

$$M_f = \frac{0.025\,C_f d}{2n}$$

where

d = longitudinal centre-to-centre distance between battens, mm

a = distance between lines of bolts or welds connecting the batten to each main component, mm

n = number of parallel planes of battens

19.2 Members in Tension

19.2.1

Members in tension composed of two or more shapes, plates, or bars separated from one another by intermittent fillers shall have the components interconnected at fillers spaced so that the slenderness ratio of any component between points of interconnection shall not exceed 300.

19.2.2

Members in tension composed of two plate components in contact or a shape and a plate component in contact shall have the components interconnected so that the spacing between connecting bolts or clear spacing between welds does not exceed 36 times the thickness of the thinner plate nor 450 mm (see Clause 19.1.3).

19.2.3

Members in tension composed of two or more shapes in contact shall have the components interconnected so that the spacing between connecting bolts or the clear spacing between welds does not exceed 600 mm, except where it can be determined that a greater spacing would not affect the satisfactory performance of the member (see Clause 19.1.3).

19.2.4

Members in tension composed of two separated main components may have either perforated cover plates or tie plates on the open sides of the built-up member. Tie plates, including end tie plates, shall have a length of not less than two-thirds of the transverse distance between bolts or welds connecting them to the main components of the member and shall be spaced so that the slenderness ratio of any component between the tie plates does not exceed 300. The thickness of tie plates shall be at least 1/60 of the transverse distance between the bolts or welds connecting them to the main components and the longitudinal spacing of the bolts or welds shall not exceed 150 mm. Perforated cover plates shall comply with the requirements of Clause 19.1.14(b), (c), and (d).

19.3 Open Box-Type Beams and Grillages

Two or more rolled beams or channels used side-by-side to form a flexural member shall be connected together at intervals of not more than 1500 mm. Through-bolts and separators may be used provided that, in beams having a depth of 300 mm or more, no fewer than two bolts shall be used at each separator location. When concentrated loads are carried from one beam to the other or distributed between the beams, diaphragms having sufficient stiffness to distribute the load shall be bolted or welded between the beams. The design of members shall provide for torsion resulting from any unequal distribution of loads. Where beams are exposed, they shall be sealed against corrosion of interior surfaces or spaced sufficiently far apart to permit cleaning and painting.

20. Stability of Structures and Individual Members

20.1 General

20.1.1

In the design of a steel structure, care shall be taken to ensure that the structural system is adequate to resist the forces caused by the factored loads and to ensure that a complete structural system is provided to transfer the factored loads to the foundations, particularly when there is a dependence on walls, floors, or roofs acting as shear-resisting elements or diaphragms. (See also Clause 8.6.)

Note: *The structure should also be checked to ensure that adequate resistance to torsional deformations has been provided.*

20.1.2

Design drawings shall indicate all load-resisting elements essential to the integrity of the completed structure and shall show details necessary to ensure the effectiveness of the load-resisting system. Design drawings shall also indicate the requirements for roofs and floors used as diaphragms.

20.1.3

Erection diagrams shall indicate all load-resisting elements essential to the integrity of the completed structure. Permanent and temporary load-resisting elements essential to the integrity of the partially completed structure shall be clearly specified on the erection diagrams.

20.1.4

Where the portion of the structure under consideration does not provide adequate resistance to lateral forces, provision shall be made for transferring the forces to adjacent lateral-load-resisting elements.

20.2 Stability of Columns

Beam-to-column connections shall have adequate strength to transfer the lateral forces produced by possible out-of-plumbness as specified in Clause 30.7.2. These forces shall be calculated for the loading cases of Clause 7.2.4 using the appropriate load combination factors.

20.3 Stability of Beams, Girders, and Trusses

20.3.1

Bracing members assumed to provide lateral support to the compression flange of beams and girders or to the compression chord of trusses, and the connections of such bracing members, shall be proportioned to resist a force equal to 1% of the force in the compression flange or chord at the point of support.

20.3.2

When bracing of the compression flange or chord is effected by a slab or deck, the slab or deck and the means by which the calculated bracing forces are transmitted between the flange or chord and the slab or deck shall be adequate to resist a force in the plane of the slab or deck. This force shall be considered to be uniformly distributed along the length of the compression flange or chord and shall be taken as at least 5% of the maximum force in the flange or chord, unless a lesser amount can be justified by analysis.

20.3.3

Consideration shall be given to the probable accumulation of forces when a bracing member must transfer forces from one braced member to another.

20.3.4

Members that restrain beams and girders designed to resist loads causing torsion shall be proportioned according to the requirements of Clause 15.11. Special consideration shall be given to the connection of asymmetric sections such as channels, angles, and zees.

21. Connections

21.1 Alignment of Members

Axially loaded members that meet at a joint shall have their gravity axes intersect at a common point if practicable; otherwise, the results of bending due to the joint eccentricity shall be provided for.

21.2 Unrestrained Members

Except as otherwise indicated on the design drawings, all connections of beams, girders, and trusses shall be designed and detailed as flexible and ordinarily may be proportioned for the reaction shears only. Flexible beam connections shall accommodate end rotations of unrestrained (simple) beams. To accomplish this, inelastic action at the specified load levels in the connection is permitted.

21.3 Restrained Members

When beams, girders, or trusses are subject to both reaction shear and end moment due to full or partial end restraint or to continuous or cantilever construction, their connections shall be designed for the combined effect of shear, bending, and axial load. When beams are rigidly framed to the flange of an H-type column, stiffeners shall be provided on the column web as follows:

(a) opposite the compression flange of the beam when

$$B_r = \phi w_c (t_b + 5k) F_{yc} < \frac{M_f}{d_b}$$

except that for members with Class 3 or 4 webs,

$$B_r = \phi \frac{640\,000}{(h_c/w_c)^2} w_c (t_b + 5k)$$

(b) opposite the tension flange of the beam when

$$T_r = 7\phi(t_c)^2 F_{yc} < \frac{M_f}{d_b}$$

where
w_c = thickness of column web
t_b = thickness of beam flange
k = distance from outer face of column flange to web-toe of fillet, or to web-toe of flange-to-web weld in a welded column
F_{yc} = specified yield point of column
d_b = depth of beam
h_c = clear depth of column web
t_c = thickness of column flange

The stiffener or pair of stiffeners opposite either beam flange must develop a force equal to

$$F_{st} = \left(\frac{M_f}{d_b}\right) - B_r$$

Stiffeners shall also be provided on the web of columns, beams, or girders if V_r calculated from Clause 13.4.2 is exceeded, in which case the stiffener or stiffeners must transfer a shear force equal to

$$V_{st} = V_f - 0.55\phi w d F_y$$

In all cases, the stiffeners shall be connected so that the force in the stiffener is transferred through the stiffener connection. When beams frame to one side of the column only, the stiffeners need not be longer than one-half of the depth of the column.

21.4 Connections of Tensile or Compression Members

The connections at ends of tensile or compression members not finished to bear shall develop the force due to factored loads. However, the connection shall be designed for not less than 50% of the resistance of the member based on the condition (tension or compression) that governs the selection of the member.

21.5 Bearing Joints in Compression Members

21.5.1

Where columns bear on bearing plates or are finished to bear at splices, there shall be sufficient fasteners or welds to hold all parts securely in place.

21.5.2

Where other compression members are finished to bear, the splice material and connecting fasteners or welds shall be arranged to hold all parts in place and shall be proportioned for 50% of the calculated load.

21.6 Lamellar Tearing

Corner or T-joint details of rolled structural members or plates involving transfer of tensile forces in the through-thickness direction resulting from shrinkage due to welding executed under conditions of restraint shall be avoided where possible. If this type of connection cannot be avoided, measures shall be taken to minimize the possibility of lamellar tearing.

21.7 Placement of Fasteners and Welds

Except in members subject to repeated loads (as defined in Clause 14), disposition of fillet welds to balance the forces about the neutral axis or axes for end connections of single-angle, double-angle, or similar types of axially loaded members is not required. Eccentricity between the gravity axes of such members and the gauge lines of bolted end connections may also be neglected. In axially loaded members subject to repeated loads, the fasteners or welds in end connections shall have their centre of gravity on the gravity axis of the member unless provision is made for the effect of the resulting eccentricity.

21.8 Fillers

21.8.1

When load-carrying fasteners pass through fillers with a total thickness greater than 6 mm, the fillers shall be extended beyond the splice material and the filler extension shall be secured by sufficient fasteners to distribute the total force in the member at the ultimate limit state uniformly over the combined section of the member and the filler. Alternatively, an equivalent number of fasteners shall be included in the connection.

21.8.2

In welded construction, any filler with a total thickness greater than 6 mm shall extend beyond the edges of the splice plate and shall be welded to the part on which it is fitted with sufficient weld to transmit the splice plate load, applied at the surface of the filler, as an eccentric load. Welds that connect the splice plate to the filler shall be sufficient to transmit the splice plate load and shall be long enough to avoid overloading the filler along the toe of the weld. Any filler that is 6 mm or less in thickness shall have its edges made flush with the edges of the splice plate and the required weld size shall be equal to the thickness of the filler plate plus the size necessary to transmit the splice plate load.

21.9 Welds in Combination

If two or more of the general types of weld (groove, fillet, plug, or slot) are combined in a single connection, the effective capacity of each shall be calculated separately with reference to the axis of the group in order to determine the factored resistance of the combination.

21.10 Fasteners and Welds in Combination

21.10.1

When approved by the designer, high-strength bolts in slip-critical connections may be considered as sharing the specified load with welds in new work, provided that the factored resistance either of the high-strength bolts, or of the welds, is equal to or greater than the effect of the factored loads. At the specified load level, the load sharing shall be on the basis of the proportional capacities of the bolts in the slip-critical connection and 0.70 times the factored resistance of the welds.

21.10.2

In making alterations to structures, existing rivets and high-strength bolts may be used to carry forces resulting from existing dead loads and welding may be proportioned to carry all additional loads.

21.11 High-Strength Bolts (in Slip-Critical Joints) and Rivets in Combination

In making alterations, rivets and high-strength bolts in slip-critical joints may be considered as sharing forces due to specified dead and live loads.

21.12 Connections Using Bolts

21.12.1 Connections Using Snug-Tightened High-Strength Bolts

Snug-tightened high-strength bolts may be used in all connections except those specified in Clause 21.12.2.

21.12.2 Connections Using Pretensioned High-Strength Bolts

Pretensioned high-strength bolts (ASTM A325M, A490M, A325, A490) shall be used in
(a) slip-critical connections where slippage cannot be tolerated (such connections include those subject to fatigue or to frequent load reversal, or those in structures sensitive to deflection);
(b) connections proportioned in accordance with the requirements of Clause 27;
(c) all elements resisting crane loads;
(d) connections for supports of running machines or other live loads that produce impact or cyclic load;
(e) connections where the bolts are subject to tensile loadings (see Clause 23.1.4); and
(f) connections using oversize or slotted holes unless specifically designed to accommodate movement.

21.13 Welds

The use of welded connections is not restricted.

21.14 Special Fasteners

Fasteners of special types may be used when approved by the designer.

22. Bolting Details

22.1 High-Strength Bolts

A325M, A490M, A325, and A490 high-strength bolts and their usage shall conform to Clause 23.

22.2 A307 Bolts

Nuts on A307 bolts shall be tightened to an amount corresponding to the full effort of a person using a spud wrench. When so specified, nuts shall be prevented from working loose by the use of lock washers, lock nuts, jam nuts, thread burring, welding, or other methods approved by the designer.

22.3 Effective Bearing Area

The effective bearing area of bolts shall be the nominal diameter multiplied by the length in bearing. For countersunk bolts, half of the depth of the countersink shall be deducted from the bearing length.

22.4 Long Grips

A307 bolts that carry calculated loads and for which the grip exceeds five diameters shall have their number increased by 0.6% for each additional 1 mm in the grip.

22.5 Minimum Pitch

The minimum distance between centres of bolt holes should not be less than 3 bolt diameters and shall in no case be less than 2.7 diameters.

22.6 Minimum Edge Distance

The minimum distance from the centre of a bolt hole to any edge shall be that given in Table 6.

Table 6
Minimum Edge Distance for Bolt Holes

Bolt diameter		Minimum edge distance, mm	
Inch*	mm	At sheared edge	At rolled, sawn, or gas-cut edge †
5/8	–	28	22
–	16	28	22
3/4	–	32	25
–	20	34	26
7/8	–	38‡	28
–	22	38	28
–	24	42	30
1	–	44‡	32
–	27	48	34
1-1/8	–	51	38
–	30	52	38
1-1/4	–	57	41
–	36	64	46
Over 1-1/4	Over 36	1.75 × diameter	1.25 × diameter

**ASTM Standards A325 and A490 are written in inch-pound units. Accordingly, bolt diameters are shown in inches for these bolts only.*

†*Gas-cut edges shall be smooth and free from notches. Edge distance in this column may be decreased by* 3 mm *when the hole is at a point where calculated stress under factored loads is not more than 0.3 of the yield stress.*

‡*At ends of beam-framing angles, this distance may be* 32 mm.

22.7 Maximum Edge Distance

The maximum distance from the centre of any bolt to the nearest edge of parts in contact shall be 12 times the thickness of the outside connected part, but not greater than 150 mm.

22.8 Minimum End Distance

In the connection of tension members having more than two bolts in a line parallel to the direction of load, the minimum end distance (from centre of end fastener to nearest end of connected part) shall be governed by the edge distance values given in Table 6. In members having either one or two bolts in the line of load, the end distance shall be not less than 1.5 bolt diameters.

22.9 Slotted Holes

Maximum and minimum edge distance for bolts in slotted or oversize holes (as permitted in Clause 23.3.2) shall conform to the requirements given in Clauses 22.6, 22.7, and 22.8, assuming that the fastener can be placed at any extremity of the slot or hole.

23. Structural Joints Using ASTM A325M, A490M, A325, or A490 Bolts

23.1 General

23.1.1

Clause 23 deals with the design, assembly, and inspection of structural joints using ASTM A325M, A490M, A325, or A490 bolts, or equivalent fasteners. The bolts may or may not be required to be installed to a specific minimum tension, depending on the type of connection.

23.1.2

Design, fabrication, and erection drawings shall show the type(s) of bolts to be used and shall specify whether or not pretensioning is required. Joints required to resist shear between connected parts shall be designated as either bearing-type or slip-critical. For bearing-type joints, the design drawings shall indicate whether or not the bolts shall be tightened to a specified minimum tension.

23.1.3

In joints where the bolts are subject to shear, the presence or absence of threads in the shear planes of the joint shall be considered.

Determination of the strength of bearing-type joints shall take into account both the shear capacity of the fasteners and the bearing capacity of the connected materials.

23.1.4

Bolts used in joints that are required to support load by direct tension shall be proportioned so that the tensile load on the bolt area, independent of the initial tightening force, shall not exceed the factored tensile resistance as given in Clause 13.11.3. The applied load shall be taken as the sum of the external load plus any tension caused by a prying action due to deformation of the connected parts. If the connection is subject to repeated loading, prying forces shall be avoided and the bolts shall be pretensioned.

23.1.5

Joints subject to repeated loads shall be proportioned in accordance with Clause 14.

23.2 Bolts, Nuts, and Washers*

23.2.1

Except as provided in Clause 23.2.4, bolts, nuts, and washers shall conform to ASTM Standards A325, A325M, A490, and A490M.

**Before specifying metric bolts, the designer should check on their current availability in the quantities required.*

23.2.2

The length of bolts shall be such that the point of the bolt will be flush with or outside the face of the nut when completely installed.

23.2.3

If required, A325M and A325 bolts, nuts, and washers may be galvanized in accordance with the requirements of ASTM Standards A325M and A325. When a galvanized nut is installed on a galvanized bolt in a solid steel connection and with three to five threads in the grip, it shall be capable of producing a tensile-type fracture of the bolt and of rotating one full turn from snug before failure.

23.2.4

Other fasteners that meet the chemical and mechanical requirements of ASTM Standards A325M, A490M, A325, or A490 and that have body diameter and bearing areas under the head and nut given in those standards, may be used. Such fasteners may differ in other dimensions and their use shall be subject to the approval of the designer.

23.2.5

If necessary, washers may be clipped on one side to a point not closer than 7/8 of the bolt diameter from the centre of the washer hole.

23.3 Bolted Parts

23.3.1

Bolted parts shall fit together solidly when assembled and shall not be separated by gaskets or any other interposed compressible material.

23.3.2

Holes may be punched, sub-punched or sub-drilled and reamed, or drilled, as permitted by Clause 28.5. The nominal diameter of a hole shall be not more than 2 mm greater than the nominal bolt size, except that for pretensioned bolted joints, where shown on the design drawings, and at other locations approved by the designer, enlarged or slotted holes may be used with high-strength bolts 16 mm in diameter and larger. Joints that use enlarged or slotted holes shall be proportioned in accordance with the requirements of Clause 23 and Clauses 13.11 and 13.12, and shall meet the following conditions:

(a) Oversize holes are 4 mm larger than bolts 22 mm and less in diameter, 6 mm larger than bolts 24 mm in diameter, and 8 mm larger than bolts 27 mm and greater in diameter. Oversize holes shall not be used in bearing-type connections but may be used in any or all plies of slip-critical connections. Hardened washers shall be used under heads or nuts adjacent to the plies containing oversize holes.

(b) Short slotted holes are 2 mm wider than the bolt diameter and have a length that does not exceed the oversize diameter provisions of Item (a) by more than 2 mm. They may be used in any or all plies of slip-critical or bearing-type connections. Such slots may be used without regard to direction of loading in slip-critical connections but shall be normal to the direction of the load in bearing-type connections. Hardened washers shall be used under heads or nuts adjacent to the plies containing the slotted holes.

(c) Long slotted holes are 2 mm wider than the bolt diameter and have a length greater than that allowed in Item (b) but not more than 2.5 times the bolt diameter and may be used

(i) in slip-critical connections without regard to direction of loading. One-third more bolts shall be provided than would be needed to satisfy the requirements of Clause 13.12.

(ii) in bearing-type connections with the long dimension of the slot normal to the direction of loading. No increase in the number of bolts specified in Clause 13.11 is required.

(iii) in only one of the connected parts of either a slip-critical or bearing- type connection at an individual faying surface.

(iv) provided that structural plate washers or a continuous bar not less than 8 mm in thickness cover long slots that are in the outer plies of joints. These washers or bars shall have a size sufficient to completely cover the slot after installation.

(d) When A490 or A490M bolts greater than 26 mm in diameter are used in oversize or slotted

holes, hardened washers shall be at least 16 mm in thickness.
(e) The requirement for the nominal diameter of hole may be waived to permit the use of the following bolt diameters and hole combinations in bearing-type or slip-critical connections:

(i) either a 3/4 inch diameter bolt or an M20 bolt in a 22 mm hole;
(ii) either a 7/8 inch diameter bolt or an M22 bolt in a 24 mm hole; and
(iii) either a 1 inch diameter bolt or an M24 bolt in a 27 mm hole.

23.3.3

When assembled, all joint surfaces including those adjacent to bolt heads, nuts, and washers shall be free of scale (tight mill scale excepted), burrs, dirt, and foreign material that would prevent solid seating of the parts.

23.3.4

The condition of the contact surfaces for slip-critical connections as described in Table 2 shall be as follows:
(a) For clean mill scale, the surfaces shall be free of oil, paint, lacquer, or any other coating for all areas within the bolt pattern and for a distance beyond the edge of the bolt hole that is the greater of 25 mm or the bolt diameter.
(b) For Classes A and B, the blast-cleaning and the coating application must be the same as that used in the tests to determine the mean slip coefficient.
(c) For Class C, hot-dip galvanizing must be done in accordance with CSA Standard G164 and the surface subsequently roughened by means of hand wire-brushing. Power wire-brushing is not permitted.
(d) For all other coatings, the surface preparation and coating application for the joint must be the same as that used in the tests to determine the mean slip coefficient.

Coated joints shall not be assembled before the coatings have cured for the minimum time used in the tests to determine the mean slip coefficient.

23.4 Installation

23.4.1 Bolt Tension

High-strength bolts that are not required to be pretensioned shall be installed in properly aligned holes, to a snug-tight condition. This is defined as the tightness that exists when all plies in a joint are in firm contact. (For slotted holes, see Clause 23.3.2.)

Pretensioned bolts shall be tightened to at least the minimum bolt tension given in Table 7, in accordance with Clause 23.5 or 23.6.

Table 7
Bolt Tension

Bolt diameter		Minimum bolt tension* kN	
Inch	mm	A325M and A325 Bolts	A490M and A490 Bolts
1/2	–	53	67
5/8	–	85	107
–	16	91	114
3/4	–	125	157
–	20	142	178
7/8	–	174	218
–	22	176	220
–	24	205	257
1	–	227	285
–	27	267	334
1-1/8	–	249	356
–	30	326	408
1-1/4	–	316	454
1-3/8	–	378	538
–	36	475	595
1-1/2	–	458	658

**Equal to 70% of the specified minimum tensile strength given in the appropriate ASTM Standard, soft-converted where appropriate and rounded to the nearest kilonewton.*

23.4.2 Hardened Washers

23.4.2.1

Hardened washers are required under the head or nut when turned
(a) as required by Clause 23.7.4; and
(b) for pretensioned A490 and A490M bolts.

23.4.2.2

Hardened washers are required
(a) for oversize or slotted holes (see Clause 23.3.2).
(b) under the head and nut for A490 and A490M bolts when used with steel having a specified minimum yield point of less than 280 MPa.
(c) of at least 16 mm in thickness, when A490 and A490M bolts greater than 26 mm in diameter are used in oversize and slotted holes.

23.4.3 Bevelled Washers

Bevelled washers shall be used to compensate for lack of parallelism where, in the case of A325M and A325 bolts, an outer face of bolted parts has more than a 5% slope with respect to a plane normal to the bolt axis. In the case of A490M and A490 bolts, bevelled washers shall be used to compensate for any lack of parallelism due to slope of outer faces.

23.5 Turn-of-Nut Tightening

23.5.1

After aligning the holes in a joint, sufficient bolts shall be placed and brought to a snug-tight condition to ensure that the parts of the joint are brought into full contact with each other.

"Snug-tight" is the tightness attained by a few impacts of an impact wrench or the full effort of a person using a spud wrench.

23.5.2

Following the initial snugging operation, bolts shall be placed in any remaining open holes and brought to snug-tightness. Re-snugging may be necessary in large joints.

23.5.3

When all bolts are snug-tight, each bolt in the joint shall then be tightened additionally by the applicable amount of relative rotation given in Table 8, with tightening progressing systematically from the most rigid part of the joint to its free edges. During this operation there shall be no rotation of the part not turned by the wrench, unless the bolt and nut are match-marked to enable the amount of relative rotation to be determined.

23.6 Tightening by Use of a Direct Tension Indicator

Tightening by means of a direct tension indicator is permitted provided that it can be demonstrated by an accurate direct measurement procedure that the bolt has been tightened in accordance with Table 7.

23.7 Inspection

23.7.1

The inspector shall determine whether the requirements of Clauses 23.2, 23.3, 23.4, and 23.5 or 23.6 are met. Installation of bolts shall be observed to ascertain that a proper tightening procedure is employed. For those cases in which bolt pretension is required, the turned element of all bolts shall be visually examined for evidence that they have been tightened. For snug-tight connections, the inspection need only ensure that the bolts have been tightened sufficiently to bring the connected elements into full contact.

23.7.2

Tensions in bolts installed by the "turn-of-nut" method exceeding those given in Table 7 shall not be cause for rejection.

23.7.3

When bolts are installed in accordance with Clause 23.6, the verification that the bolt has been properly tightened is determined by the direct tension indicator.

23.7.4

For bolts in slip-critical connections and for bolts in pretensioned bearing-type connections, when there is disagreement concerning the results of inspection of bolt tension in the turn-of-nut method, the following arbitration inspection procedure shall be used unless a different procedure has been specified:
(a) The inspector shall use an inspection wrench that is a manual or power torque wrench capable of indicating a selected torque value.
(b) Three bolts of the same grade and diameter as those under inspection and representative of the lengths and conditions of those in the structure shall be placed individually in a calibration device capable of indicating bolt tension. There shall be a washer under the part turned if washers are so used in the structure or, if no washer is used, the material abutting the part turned shall be of the same specification as that in the structure.
(c) When the inspection wrench is a manual wrench, each bolt specified in Item (b) shall be tightened in the calibration device by any convenient means to an initial tension of approximately

15% of the required fastener tension and then to the minimum tension specified for its size in Table 7. Tightening beyond the initial condition shall not produce greater nut rotation than that permitted in Table 8. The inspection wrench shall then be applied to the tightened bolt, and the torque necessary to turn the nut or head 5° in the tightening direction shall be determined. The average torque measured in the tests of three bolts shall be taken as the job inspection torque to be used in the manner specified in Item (e). The job inspection torque shall be established at least once each working day.

(d) When the inspection wrench is a power wrench, it shall first be applied to produce an initial tension of approximately 15% of the required fastener tension and then adjusted so that it will tighten each bolt specified in Item (b) to a tension of at least 5% but not more than 10% greater than the minimum tension specified for its size in Table 7. This setting of the wrench shall be taken as the job inspection torque to be used in the manner specified in Item (e). Tightening beyond the initial condition must not produce greater nut rotation than that permitted in Table 8. The job inspection torque shall be established at least once each working day.

(e) Bolts, represented by the sample prescribed in Item (b), that have been tightened in the structure shall be inspected by applying, in the tightening direction, the inspection wrench and its job inspection torque to 10% of the bolts, but not less than 2 bolts, selected at random in each connection. If no nut or bolt head is turned by this application of the job inspection torque, the connection shall be accepted as properly tightened. If any nut or bolt head is turned by the application of the job inspection torque, this torque shall be applied to all bolts in the connection and all bolts whose nut or head is turned by the job inspection torque shall be tightened and reinspected. Alternatively, the fabricator or erector, at his option, may retighten all the bolts in the connection and then resubmit the connection for the specified inspection.

Table 8
Nut Rotation* from Snug-Tight Condition

Disposition of outer faces of bolted parts	Bolt length‡	Turn
Both faces normal to bolt axis or one face normal to axis and other face sloped 1:20 (bevelled washer not used)†	Up to and including 4 diameters	1/3
	Over 4 diameters and not exceeding 8 diameters or 200 mm	1/2
	Exceeding 8 diameters or 200 mm	2/3
Both faces sloped 1:20 from normal to bolt axis (bevelled washers not used)†	All lengths of bolts	3/4

**Nut rotation is rotation relative to a bolt regardless of whether the nut or bolt is turned. Tolerance on rotation: 30° over or under. This Table applies to coarse-thread heavy-hex structural bolts of all sizes and lengths used with heavy-hex semi-finished nuts.*

†Bevelled washers are necessary when A490M or A490 bolts are used.

‡Bolt length is measured from the underside of the head to the extreme end of point.

24. Welding

24.1 Arc Welding

Arc welding design and practice shall conform with CSA Standard W59.

24.2 Resistance Welding

The resistance of resistance-welded joints shall be taken as given in CSA Standard W55.3. Related welding practice shall be in conformance with welding standards approved by the Canadian Welding Bureau as given in CSA Standard W55.3.

24.3 Fabricator and Erector Qualification

Fabricators and erectors responsible for making welds for structures fabricated or erected under this Standard shall be certified by the Canadian Welding Bureau to the requirements of CSA Standard W47.1 (Division 1 or Division 2.1), or CSA Standard W55.3, or both, as applicable. Part of the work may be sublet to a Division 3 fabricator; however, the Division 1 or Division 2.1 fabricator or erector shall retain responsibility for the sublet work.

25. Column Bases

25.1 Loads

Suitable provision shall be made to transfer column loads and moments to footings and foundations.

25.2 Resistance

25.2.1 Compressive Resistance of Concrete

The compressive resistance of concrete shall be determined in accordance with Clause 10.15 of CSA Standard A23.3. When compression exists over the entire base plate area, the bearing pressure on the concrete may be assumed to be uniform over an area equal to the width of the base plate multiplied by a depth equal to $d - 2e$, where e is the eccentricity of the column load.

25.2.2 Resistance to Pull-Out

Anchor bolts subject to tensile forces shall be anchored to the foundation unit in such a manner that the required factored tensile force can be developed. Full anchorage is obtained when the factored pull-out resistance of the concrete is equal to or larger than the factored tensile resistance of the bolts. For methods of transferring the tensile forces from the anchors to the concrete, see CSA Standard A23.3.

25.2.3 Resistance to Transverse Loads

25.2.3.1

Shear resistance may be developed by friction between the base plate and the foundation unit or by bearing of the anchor bolts or shear lugs against the concrete. When shear acts toward a free edge, the requirements of CSA Standard A23.3 shall be met.

25.2.3.2

When loads are transferred by friction, the requirements of CSA Standard A23.3 shall be met.

25.2.3.3

When shear is transmitted by bearing of the anchor bolts on the concrete, the factored bearing resistance shall be taken as

$B_r = 1.4\phi_c n A f_c'$

where

ϕ_c = 0.60

n = number of anchor bolts in shear

A = bearing area, taken as the product of the bolt diameter, d, and an assumed depth of 5d

25.2.3.4

For methods of transmitting shear by bearing of shear lugs in the concrete, see CSA Standard A23.3.

25.2.4 Moment Resistance

Moment resistance shall be taken as the couple formed by the tensile resistance of the anchor bolts determined in accordance with Clause 26.2.1 or 26.2.3, as applicable, and by the concrete compressive resistance determined in accordance with Clause 10.15 of CSA Standard A23.3.

25.3 Finishing

Column bases shall be finished in accordance with the following requirements:

(a) The bottom surfaces of bearing plates and column bases that rest on masonry or concrete foundations and are grouted to ensure full bearing need not be planed.

(b) Steel-to-steel contact bearing surfaces of rolled steel bearing plates shall be finished in such a manner that the requirements of Clauses 28.8, 28.9.7, and 30.7.6 are satisfied. In general, rolled steel bearing plates 50 mm or less in thickness may be used without machining, provided that satisfactory contact is obtained. Rolled steel bearing plates more than 50 mm but not more than 100 mm thick may be straightened by pressing, or may be machined at bearing locations to obtain a satisfactory contact. Rolled steel bearing plates more than 100 mm thick, and other column bases, shall be machined at bearing locations.

26. Anchor Bolts

26.1 General

Anchor bolts shall be designed to resist the effect of factored uplift forces, bending moments, and shears determined in accordance with Clause 7.2. The anchorage of the anchor bolts in the foundation unit shall be such that the required load capacity can be developed. Forces present during construction as well as those present in the finished structure shall be resisted.

26.2 Bolt Resistance

26.2.1 Tension

The factored tensile resistance of an anchor bolt shall be taken as

$T_r = \phi_b A_n F_u$

where

ϕ_b = 0.67

A_n = the tensile stress area of the bolts

$= \frac{\pi}{4}(D - 0.938\,P)^2$ for metric bolts

$= \frac{\pi}{4}\left(D - \frac{0.974}{n}\right)^2$ for imperial bolts

where

P = the pitch of thread, mm, and

n = number of threads per inch

26.2.2 Shear

The factored shear resistance of the anchor bolts shall be taken as

$V_r = 0.60\phi_b n A_b F_u$

but not greater than the lateral bearing resistance given in Clause 25.2.3.3.

When the bolt threads are intercepted by the shear plane, the factored shear resistance shall be taken as 70% of V_r.

26.2.3 Shear and Tension

An anchor bolt required to develop resistance to both tension and shear shall be proportioned so that

$$\left(\frac{V_f}{V_r}\right)^2 + \left(\frac{T_f}{T_r}\right)^2 \leq 1$$

where

V_f is the portion of the total shear per bolt transmitted by bearing of the anchor bolts on the concrete (see Clause 25.2.3.3).

26.2.4 Tension and Bending

An anchor bolt required to develop resistance to both tension and bending shall be proportioned to meet the requirements of Clause 13.9(a). The tensile and moment resistances, T_r and M_r, shall be based on the properties of the cross section at the critical section; M_r shall be taken as $\phi_b S F_y$.

27. Seismic Design Requirements

27.1 General

27.1.1

Clause 27 provides requirements for the design of members and connections in the lateral-load-resisting systems of steel-framed buildings for which ductile response is required under seismic loading. This Clause is to be applied in conjunction with the requirements of Clause 4.1.9 of the *National Building Code of Canada,* 1990.

27.1.2

In calculating second-order effects at ultimate limit states under seismic load (see Clause 4.1.9.2(8) of the *National Building Code of Canada,* 1990), the seismic load shall be combined with gravity loads likely to be present under the action of the earthquake, and the limit of 1.4 on the value of U_2 given in Clause 8.6.1 need not be applied.

27.1.3

If structural or nonstructural elements not considered to form part of the lateral-load-resisting system have a significant effect on the structural response to earthquake motions, they shall be considered in the analysis.

27.1.4

Structural members and their connections that are not considered to form part of the lateral-load-resisting system shall be capable of maintaining their resistance when subject to seismically induced deformations.

27.1.5

Steel used in lateral-load-resisting systems shall conform to Clauses 5.1.3 and 8.5(a). Material other than this may be used if approved by the regulatory authority.

27.2 Ductile Moment-Resisting Frames

27.2.1 General

27.2.1.1

Ductile moment-resisting frames have the capacity to form plastic hinges in flexural members and, where necessary, in columns or in joints, and to maintain the member resistance at these hinges. They shall meet the specific requirements of this section.

27.2.1.2

Beams, columns, and joint panel zones shall be proportioned and braced to enable them to undergo large plastic deformations unless it can be demonstrated that the element being considered will remain elastic while one or more of the other elements at the joint is undergoing large plastic deformations.

The element or elements at a joint that may undergo large plastic deformations are termed critical elements, and they shall be identified.

An element undergoing large plastic deformations shall be assumed to apply relevant loads to the other elements at the joint equal to 1.2 times its unfactored yield resistance. An element may be considered as responding elastically if the loading on it does not exceed its factored resistance.

The resistance at the ends of beams shall be determined from Clause 13.5 or 13.6. At the ends of columns, the resistance shall be determined from Clause 13.8.1 or 13.8.2, and the resistance of the joint panel zone shall be determined from Clause 27.2.4.1.

27.2.1.3

For the purposes of Clauses 27.2.2, 27.2.3, and 27.2.4,

(a) in calculating the flexural resistance and section class of a beam-column, the gravity load contribution to the axial load shall be based on specified load levels; and

(b) the contribution of a composite slab shall be estimated in calculating the flexural resistance of, and the loading produced by, the girders.

27.2.2 Beams

27.2.2.1

When a beam is a critical element at a joint, it shall

(a) conform to Class 1 section requirements, and
(b) be laterally braced according to the requirements of Clause 13.7.
When a beam is not critical, it shall have
(c) flanges conforming to Class 1 section requirements, and
(d) webs conforming to Classes 1 or 2.

27.2.2.2

Abrupt changes in beam flange cross sections shall be avoided in regions where plastic hinges may occur.

27.2.2.3

When horizontal framing members are trusses, the columns shall be the critical element at the joint and shall conform to Clause 27.2.3.1. Where such buildings exceed one storey, the ratio of unbraced column height to least radius of gyration shall not exceed 60. The connections of the truss chord to the column shall have factored resistances at least equal to the lesser of
(a) the factored yield resistances of the chord sections; and
(b) the forces required to impose column moments equal to 1.2 times the column factored flexural resistance.

27.2.3 Columns (Including Beam-Columns)

27.2.3.1

When a column is a critical element at a joint, it shall
(a) conform to Class 1 section requirements;
(b) have a factored axial load not exceeding $0.60\phi AF_y$ for all load combinations; and
(c) be laterally braced according to the requirements of Clause 13.7.
When a column is not critical, it shall conform to Class 1 or Class 2 section requirements.

27.2.3.2

Splices that incorporate partial joint penetration groove welds shall be located at least one-fourth of the clear distance between beams but not less than one metre from the beam-to-column joints.

27.2.4 Column Joint Panel Zone

27.2.4.1

The horizontal shearing resistance of the column joint panel zone shall be taken as

$$V_r = 0.55\phi d_c w' F_{yc}\left[1 + \frac{3b_c t_c^2}{d_c d_b w'}\right]$$

where the subscripts b and c denote the beam and the column, respectively, and b_c is the width of the column flange.

27.2.4.2

When the joint panel zone is one of the critical elements at a joint, the following shall apply:
(a) The ratio of the sum of panel zone depth and width to panel zone thickness shall not exceed 90. Doubler plate thickness may be included with web thickness in calculating this ratio only if the doubler plate is plug-welded or bolted to the column web.
(b) Doubler plates, if present, shall be placed against the column web. They shall be groove- or fillet-welded to develop their full shear yield resistance.
(c) The column flange shall be braced, either directly, or indirectly, through the column web at

the level of both beam flanges by members proportioned according to Clause 20.3.1.
When the joint panel zone is not critical, the panel zone details shall be selected according to Clauses 21.3 and 27.2.4.3.

27.2.4.3

When beams are rigidly framed to the flange of an H-shaped column, the capacity of the column web to resist the tensile load due to seismic forces shall be taken as 0.6 times the value of T_r given in Clause 21.3.

27.2.5 Beam-to-Column Connections

27.2.5.1

The moment-resisting connection of a beam to a column shall have factored flexural resistance at least equal to the lesser of
(a) the factored bending resistance of the beam; and
(b) the moment that will induce in the column joint panel zone a shearing force equal to the factored resistance given in Clause 27.2.4.1.
Partial joint penetration groove welds shall not be used in these connections.

27.2.5.2

The beam web connection shall have a resistance adequate to carry shears induced by yielding of critical members at adjacent joints.

27.2.5.3

Flanges and connection plates in bolted connections of beam-to-column joints shall have a factored net section ultimate resistance, Clause 13.2(a) (ii) or (iii), at least equal to the factored gross area yield resistance, Clause 13.2(a)(i).

27.2.5.4

Plastic hinges shall be avoided at locations where the flange area has been reduced, such as at bolt holes, unless $F_y/F_u \leq 0.67$.

27.2.6 Bracing

Beams, columns, and beam-to-column connections shall be braced by members proportioned according to Clause 20.3.1. At beam-to-column connections, the compression force to be stabilized shall be that associated with complete yielding of the critical element at the connection. The possibility of complete load reversals shall be considered.

27.2.7 Fasteners

Fasteners connecting the separate elements of built-up flexural members that are critical shall have resistance adequate to support full yielding at potential plastic hinge locations.

27.3 Moment-Resisting Frames with Nominal Ductility

27.3.1

Moment-resisting frames with nominal ductility can sustain limited amounts of inelastic deformation through flexural action, joint panel zone shearing, or connection deformations. They shall meet the specific requirements of this Clause.

27.3.2

Members shall be Class 1 or 2, unless the connection components are the critical elements in the joints. If the joint panel zone is critical, the slenderness ratio of both the panel zone web and the doubler plate, if any, shall permit the attainment of the shear yield stress. If connection plates are critical, they shall exhibit ductile behaviour that allows significant rotation of the connection.

27.3.3

Either the beam-to-column moment-resisting connections shall satisfy the requirements of Clause 27.2.5.1, or the frame shall be designed to resist the factored load effects specified in Clause 7.2, with the earthquake loads multiplied by 1.25, and the resistance of the connections shall be associated with a ductile mode of failure.

27.4 Ductile Braced Frames

27.4.1 General

Ductile braced frames with concentric bracing have the capacity to absorb energy through yielding of braces. They shall meet the specific requirements of this Clause.

27.4.2 Bracing Systems

27.4.2.1

Diagonal braces shall be oriented such that, at each level in any planar frame, at least 30% of the horizontal shear carried by the bracing system shall be carried by tension braces and at least 30% shall be carried by compression braces.

27.4.2.2

Frames in which seismic load resistance is provided by any of the following shall not be considered as frames in this category:
(a) V- or chevron bracing, in which pairs of braces are located either above or below a beam and meet the beam at a single point within the middle half of the span;
(b) K- bracing, in which pairs of braces meet a column on one side near its mid-height between floors; or
(c) systems that do not meet the requirements of Clause 27.4.2.1.

27.4.3 Diagonal Bracing Members

27.4.3.1

Bracing members shall have a slenderness ratio, L/r, less than $1900/\sqrt{F_y}$. Symmetrical sections shall have width-thickness ratios for flanges no greater than 80% of the limits for Class 1 sections and for webs no greater than 100% of the limit for Class 1 sections. Legs of angles and flanges of channels shall have width-to-thickness ratios not greater than $115/\sqrt{F_y}$. Built-up bracing members shall have slenderness ratios of the individual parts no greater than 0.5 times the slenderness ratio of the member as a whole.

27.4.3.2

The factored compressive resistance of a brace shall be determined as the product of C_r, given in Clause 13.3, and a reduction factor equal to $1/(1 + 0.35\lambda)$. This factor need not be applied if the tension braces acting at the same level and in the same plane as the compression brace have sufficient reserve capacity to compensate for the reduction.

27.4.4 Bracing Connections

27.4.4.1

Eccentricities in bracing connections shall be minimized.

27.4.4.2

Welded brace connections shall have a factored resistance at least equal to the lesser of
(a) the axial tensile yield strength of the brace (A_gF_y); and
(b) 2.0 times the member force due to the factored seismic load plus the brace force due to specified gravity load.

Partial joint penetration groove welds shall not be used in these connections.

27.4.4.3

Bolted brace connections shall have a factored resistance based on ultimate tensile strength, Clause 13.2(a)(ii) or (iii), at least equal to the lesser of the loads defined in Clause 27.4.4.2 (a) and (b).

27.4.4.4

Brace connections including gusset plates shall be detailed to avoid brittle failures due to rotation of the brace when it buckles. This ductile rotational behaviour shall be allowed for, either in the plane of the frame or out of it, depending on the slenderness ratios.

27.4.4.5

Fasteners that connect the separate elements of built-up bracing members shall, if the overall buckling mode induces shear in the fastener, have resistances adequate to support one-half of the yield load of the smaller component being joined, with this force assumed to act at the centroid of the smaller member.

27.4.5 Other Connections

27.4.5.1

Beam-to-column connections and column splices that participate in the lateral-load-resisting system shall have factored resistances adequate to support the effects of the bracing connection loads given in Clause 27.4.4.2 while also supporting the gravity loads. They shall also support the effects of the factored load combinations detailed in Clause 7.2.

27.4.5.2

Column splices made with partial penetration groove welds and subject to net tension forces due to overturning effects shall have factored resistances equal to 150% of the calculated factored load effects, but not less than 50% of the flange yield load of the smaller column.

27.4.5.3

Connections shall resist forces arising from load redistribution following brace buckling or yielding.

27.4.6 Columns and Beams

Columns and beams shall be proportioned to resist the specified gravity loads together with the forces induced by the lesser of the braced forces given in Clause 27.4.4.2(b) and the brace resistances based on $\phi = 1.0$. The brace compressive resistance shall not include the reduction factor defined in Clause 27.4.3.2. Redistributed loads due to brace buckling or yielding shall be considered. The beam and column forces need not exceed 2.0 times the member force due to the factored seismic load, plus the specified gravity load.

27.5 Braced Frames with Nominal Ductility

27.5.1

Concentrically braced frames with nominal ductility have the capacity to absorb limited amounts of energy through inelastic bending or extension of bracing members. They shall meet the specific requirements of this Clause.

27.5.2

Diagonal compression bracing members shall be Class 2 sections, or shall have cross-section elements that can undergo limited straining while sustaining the yield stress. Columns, beams, and connections in the lateral-load-resisting system shall resist the specified gravity load together with the forces induced by the lesser of the forces given in Clause 27.5.3(b) and the brace resistances based on $\phi = 1.0$. The beam, column, and connection forces need not exceed 1.33 times the member forces due to factored seismic load plus the specified gravity loads.

27.5.3

Welded brace connections shall have a factored resistance at least equal to the lesser of
(a) the axial tensile yield strength of the brace (A_gF_y); and
(b) 1.33 times the member force due to the factored seismic load plus the brace force due to specified gravity load, except that for tension-only bracing the load so selected shall be multiplied by an additional factor of 1.10.

27.5.4

Bolted brace connections shall have a factored resistance based on ultimate tensile strength, Clause 13.2(a) (ii) or (iii), at least equal to the lesser of the loads defined in Clause 27.5.3 (a) and (b).

27.5.5

The beam attached to chevron or V-braces shall be continuous between columns. When a beam is supported from below by chevron braces, the beam shall have adequate resistance, with $\phi = 1.0$, to support its specified tributary gravity loads without the support provided by the braces.

27.6 Special Framing Systems

27.6.1

Frames in which ductile behaviour is to be provided by girder yielding in eccentrically braced frames shall be proportioned according to the requirements of Appendix D.

27.6.2

Other framing systems and frames that incorporate special bracing, base isolation, or energy-absorbing devices shall be designed on the basis of published research results, observed performance in past earthquakes or special investigation.

28. Fabrication

28.1 General

Unless otherwise specified, the provisions of Clause 28 shall apply to both shop and field fabrication.

28.2 Straightness of Material

Prior to layout or fabrication, rolled material shall be straight within established rolling mill tolerances. If straightening is necessary, it shall be done by means that will not injure the material. When heat is applied locally, the temperature of the heated area shall not exceed the limits given in CSA Standard W59. Sharp kinks and bends shall be cause for rejection.

28.3 Gas Cutting

Gas cutting shall be done by machine where practicable. Gas-cut edges shall conform to CSA Standard W59. Re-entrant corners shall be free from notches and shall have the largest practical radii, with a minimum radius of 14 mm.

28.4 Sheared or Gas-Cut Edge Finish

28.4.1

Planing or finishing of sheared or gas-cut edges of plates or shapes shall not be required, unless specifically noted on the drawings or included in a stipulated edge preparation for welding.

28.4.2

The use of sheared edges in the tension area shall be avoided in locations subject to plastic hinge rotation at factored loading. If used, such edges shall be finished smooth by grinding, chipping, or planing. These requirements shall be noted on design drawings and shop details where applicable.

28.4.3

Burrs shall be removed
(a) as required in Clause 23.3.3;
(b) when required for proper fit-up for welding; and
(c) when they create a hazard during or after construction.

28.5 Holes for Bolts or Other Mechanical Fasteners

28.5.1

Unless otherwise shown on design drawings or as specified in Clause 23.3.2, holes shall be made 2 mm larger than the nominal diameter of the fastener. Holes may be punched when the thickness of the material is not greater than the nominal fastener diameter plus 4 mm. For greater thicknesses, holes shall be either drilled from the solid or sub-punched or sub-drilled and reamed. The die for all sub-punched holes or the drill for all sub-drilled holes shall be at least 4 mm smaller than the required diameter of the finished hole. Holes in CSA Standard G40.21 (Type 700Q) or ASTM Standard A514 steels more than 13 mm thick shall be drilled.

28.5.2

In locations subject to plastic hinge rotation at factored loading, fastener holes in the tension area shall be either sub-punched and reamed or drilled full size.

28.5.3

The requirements of Clause 28.5.2 shall be noted on design drawings and shop details where applicable.

28.6 Bolted Construction

28.6.1

Drifting done during assembly to align holes shall not distort the metal or enlarge the holes. Holes in adjacent parts shall match sufficiently well to permit easy entry of bolts. If necessary, holes, except oversize or slotted holes, may be enlarged to admit bolts by a moderate amount of reaming; however, gross mismatch of holes shall be cause for rejection.

28.6.2

Assembly of high-strength bolted joints shall be in accordance with Clause 23.

28.7 Welded Construction

Workmanship and technique in arc-welded fabrication shall conform to those prescribed by CSA Standard W59. The welding practice in resistance-welded fabrication shall conform to that required by CSA Standard W55.3, and shall be approved by the Canadian Welding Bureau.

28.8 Finishing of Bearing Surfaces

Joints in compression that depend on contact bearing shall have the bearing surfaces prepared to a common plane by milling, sawing, or other suitable means. Surface roughness shall have a roughness height rating not exceeding 500 (12.5 μm) as defined in CSA Standard B95, unless otherwise specified.

28.9 Tolerances

28.9.1

Structural members consisting primarily of a single rolled shape shall be straight within the tolerances allowed by CSA Standard G40.20, except as specified in Clause 28.9.4.

28.9.2

Built-up bolted structural members shall be straight within the tolerances allowed for rolled wide-flange shapes by CSA Standard G40.20, except as specified in Clause 28.9.4.

28.9.3

Dimensional tolerances of welded structural members shall be those prescribed by CSA Standard W59, unless otherwise specified.

28.9.4

Fabricated compression members shall not have a deviation from straightness of more than one-thousandth of the axial length between points that are to be laterally supported.

28.9.5

Beams with bow within straightness tolerances shall be fabricated so that, after erection, the bow due to rolling or fabrication shall be upward.

28.9.6

All completed members shall be free from twists, bends, and open joints. Sharp kinks or bends shall be cause for rejection.

28.9.7

Joints in compression that depend on contact bearing when assembled during fabrication shall have at least 75% of the entire contact area in full bearing. (Full bearing shall be defined as a separation not exceeding 0.5 mm.) The separation of any remaining portion shall not exceed 1 mm. A gap of up to 3 mm may be packed with nontapered steel shims in order to meet the requirements of this Clause. Shims need not be other than mild steel, regardless of the grade of the main material.

28.9.8

A variation of 1 mm is permissible in the overall length of members with both ends finished for contact bearing.

28.9.9

Members without ends finished for contact bearing that are to be framed to other steel parts of the structure may have a variation from the detailed length not greater than 2 mm for members 10 m or less in length and not greater than 4 mm for members more than 10 m in length.

29. Cleaning, Surface Preparation, and Priming

29.1 General Requirements

29.1.1

All steelwork, except as exempted in Clauses 29.1.2, 29.1.3, and 29.2, or unless otherwise noted on design drawings or in the job specifications, shall be given one coat of primer or one-coat paint (see Clause 29.5) applied in the shop. The primer or one-coat paint shall be applied thoroughly and evenly to dry, clean surfaces by suitable means.

29.1.2

Steelwork that will subsequently be concealed by interior building finish need not be given a coat of primer, unless otherwise specified (see Clause 6.5).

29.1.3

Steelwork that will be encased in concrete need not be given a coat of primer. Steelwork that is designed to act compositely with reinforced concrete and that depends on natural bond for interconnection shall not be given a coat of primer.

29.1.4

Steelwork that will be shop-primed shall be cleaned of all loose mill scale, loose rust, weld slag and flux deposit, dirt, other foreign matter, and excessive weld spatter prior to application of the primer. Oil and grease shall be removed with a solvent. The fabricator shall be free to use any satisfactory method to clean the steel and prepare the surface for painting, unless a specific method of surface preparation is called for.

29.1.5

Primer shall be dry before loading primed steelwork for shipment.

29.1.6

Steelwork that will not be shop-primed after fabrication shall be cleaned of oil and grease with solvent cleaners and shall be cleaned of dirt and other foreign matter.

29.2 Requirements for Special Surfaces

29.2.1

Surfaces that will be inaccessible after assembly shall be cleaned or cleaned and primed, as required by Clause 29.1, prior to assembly. Inside surfaces of enclosed spaces that will be entirely sealed off from any external source of oxygen need not be primed.

29.2.2

In members in compression, surfaces that are finished to bear and assembled during fabrication shall be cleaned before assembly but shall not be primed unless otherwise specified.

29.2.3

Surfaces that are finished to bear and not assembled during fabrication shall be protected by a corrosion-inhibiting coating. The coating shall be of a type that can be readily removed prior to assembly or shall be of a type that makes such removal unnecessary.

29.2.4

Faying surfaces of high-strength bolted slip-critical joints shall not be primed or otherwise coated except as permitted by Clause 23.

29.2.5

Joints to be field-welded and surfaces to which shear connections are to be welded shall be kept free of primer and/or any other coating that could be detrimental to achieving a sound weldment, except that sheet steel decks may be welded to clean, primed steelwork.

29.3 Surface Preparation

Unless otherwise specified or approved, surface preparation shall be in conformance with one of the following applicable specifications of the Steel Structures Painting Council:
SP 2; SP 3; SP 4; SP 5; SP 6; SP 7; or SP 10.

29.4 Primer

Unless otherwise specified or approved, shop primer shall conform to one of the following standards of the Canadian General Standards Board:
1.40; 1-GP-81; 1.140; or 1-GP-166; or to CISC/CPMA Standard 2-75.

29.5 One-Coat Paint

Unless otherwise specified or approved, one-coat paint intended to withstand exposure to an essentially noncorrosive atmosphere for a period of time not exceeding six months shall conform to CISC/CPMA Standard 1-73a.

30. Erection

30.1 General

The steel framework shall be erected true and plumb within the specified tolerances. Temporary bracing shall be employed wherever necessary to withstand all loads to which the structure may be subject during erection and subsequent construction, including loads due to wind, equipment, and equipment operation. Temporary bracing shall be left in place undisturbed as long as required for the safety and integrity of the structure (see also Clause 26). The erector shall ensure during erection that an adequate margin of safety exists in the uncompleted structure and members using the factored member resistances calculated in accordance with Clause 13. (See also Clause 20.1.3.)

30.2 Temporary Loads

Wherever piles of material, erection equipment, or other loads are carried during erection, suitable provision shall be made to ensure that the loads can be safely sustained during their duration and without permanent deformation or other damage to any member of the steel frame and other building components supported thereby.

30.3 Adequacy of Temporary Connections

As erection progresses, the work shall be securely bolted or welded to take care of all dead, wind, and erection loads and to provide structural integrity as required.

30.4 Alignment

No permanent welding or bolting shall be done until as much of the structure as will be stiffened thereby has been suitably aligned.

30.5 Surface Preparation for Field Welding

The portions of surfaces that are to receive welds shall be thoroughly cleaned of all foreign matter, including paint film.

30.6 Field Painting

Unless otherwise specified, the cleaning of steelwork in preparation for field painting, touch-up of shop primer, spot-painting of field fasteners, and general field painting shall not be considered to be a part of the erection work.

30.7 Erection Tolerances

30.7.1 Elevation of Base Plates

Column base plates shall be considered to be at their proper elevation if the following tolerances are not exceeded:

(a) for single and multi-storey buildings designed as simple construction as provided in Clause 8.3,
± 5 mm from the specified elevation; and

(b) for single and multi-storey buildings designed as continuous construction as provided in Clause 8.2,
± 3 mm from the specified elevation.

30.7.2 Plumbness of Columns

Unless otherwise specified, columns shall be considered plumb if their verticality does not exceed

the following tolerances:
(a) for exterior columns of multi-storey buildings,
1 to 1000, but not more than 25 mm toward or 50 mm away from the building line in the first 20 storeys plus 2 mm for each additional storey, up to a maximum of 50 mm toward or 75 mm away from the building line over the full height of the building;
(b) for columns adjacent to elevator shafts,
1 to 1000, but not more than 25 mm in the first 20 storeys plus 1 mm for each additional storey, up to a maximum of 50 mm over the full height of the elevator shaft; and
(c) for all other columns,
1 to 500.

Column plumbness is measured from the actual column centreline at the base of the column. The location of the centreline of the column at the base shall fall within a 10 mm diameter circle of the location established on the drawings.

30.7.3 Horizontal Alignment of Members

Unless otherwise specified, spandrel beams shall be considered aligned when the offset of one end relative to the other from the alignment shown on the drawings does not exceed L/1000; however, the offset need not be less than 3 mm and shall not exceed 6 mm. For all other members, the corresponding offsets are L/500, 3 mm, and 12 mm.

30.7.4 Elevations of Members

Elevations of the ends of members shall be within 10 mm of the specified member elevation. Allowances shall be made for initial base elevation, column shortening, differential deflections, temperature effects, and other special conditions, but the maximum deviation from the specified slope shall not exceed L/500. The difference from the specified elevation between member ends that meet at a joint shall not exceed 6 mm.

30.7.5 Members with Adjustable Connections

Members specified to have adjustable connections (such as shelf angles, sash angles, and lintels) shall be considered to be within tolerances when the following conditions are met:
(a) Each piece is level within L/1000; however, the difference in elevation of the ends need not be less than 3 mm and shall not exceed 6 mm.
(b) Adjoining ends of members are aligned vertically and horizontally within 2 mm.
(c) The location of these members both vertically and horizontally is within 10 mm of the location established by the dimensions on the drawings.

30.7.6 Column Splices

Column splices and other compression joints that depend on contact bearing as part of the splice resistance shall, after alignment, have a maximum allowable separation of 6 mm. Any gap exceeding 1.5 mm shall be packed with nontapered steel shims. Shims need not be other than mild steel, regardless of the grade of the main material.

30.7.7 Joint Fit-Up

The fit-up of joints that are to be field-welded shall be within the tolerances shown on the erection diagrams and shall in no case, before welding is begun, exceed the tolerances specified in CSA Standard W59.

30.7.8 Special Tolerances

Unless otherwise required by operational characteristics of the crane, crane girders and monorail

beams shall be erected within the following tolerances:
(a) The slope of a member shall not exceed L/1000. However the difference in elevation of the ends need not be less than 3 mm and shall not exceed 6 mm. The difference in elevation of opposite points on two parallel girders shall not exceed 1/1000 of the distance between the girders, and shall not exceed 6 mm.
(b) The alignment of a horizontal member shall not exceed L/500. However, the difference in alignment of the ends need not be less than 3 mm and shall not exceed 8 mm.
(c) The distance between the ends of two parallel girders shall not deviate by more than 1 in 500 of the span of the girder. However the difference in the distances between the girder ends need not be less than 3 mm and shall not exceed 10 mm.

31. Inspection

31.1 General

Material and workmanship shall at all times be subject to inspection by qualified inspectors representing and responsible to the designer. The inspection shall cover shop work and field erection work to ensure compliance with this Standard.

31.2 Cooperation

Insofar as possible, all inspections shall be made in the fabricator's shop and the fabricator shall cooperate with the inspector, permitting access for inspection to all places where work is being done. The inspector shall cooperate in avoiding undue delay in the fabrication or erection of the steelwork.

31.3 Rejection

Material or workmanship not conforming to the provisions of this Standard may be rejected at any time during the progress of work when nonconformance to these provisions is established.

31.4 Inspection of High-Strength Bolted Joints

The inspection of high-strength bolted joints shall be performed in accordance with the procedures prescribed in Clause 23.7.

31.5 Inspection of Welding

The inspection of welding shall be in accordance with the applicable clause in CSA Standard W59.

31.6 Identification of Steel by Marking

In the fabricator's plant, steel used for main components shall at all times be marked to identify its specification (and grade, if applicable). This shall be done by suitable markings or by recognized colour coding, except that cut pieces identified by piece mark and contract number need not continue to carry specification identification markings when it has been satisfactorily established that such cut pieces conform to the required material specifications.

Appendix A
Standard Practice for Structural Steel

Note: *This Appendix is not a mandatory part of this Standard.*

A1.

Matters concerning standard practice not covered by this Standard but pertinent to the fabrication and erection of structural steel (such as a definition of structural steel items, the computation of weights, etc) are to be in accordance with the *Code of Standard Practice for Structural Steel* published by the Canadian Institute of Steel Construction, unless otherwise clearly specified in the plans and specifications issued to the bidders.

Appendix B
Effective Lengths of Members in Compression

Note: *This Appendix is not a mandatory part of this Standard.*

B1.

The slenderness ratio of a member in compression is defined as the ratio of the effective length to the applicable radius of gyration.

The effective length, KL, may be thought of as the actual unbraced length, L, multiplied by a factor, K, such that the product, KL, is equal to the length of a pin-ended compression member of equal capacity to the actual member. The effective length factor, K, of a column of finite unbraced length therefore depends on the conditions of restraint afforded to the column at its braced locations.

B2.

A variation in K between 0.65 and 2.0 would apply to the majority of cases likely to be encountered in actual structures. Figure B1 illustrates six idealized cases in which joint rotation and translation are either fully realized or nonexistent.

	(a)	(b)	(c)	(d)	(e)	(f)
Buckled shape of column is shown by dashed line	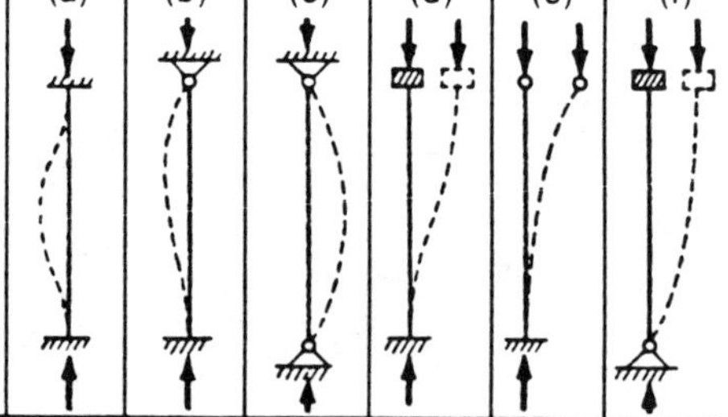					
Theoretical K value	0.5	0.7	1.0	1.0	2.0	2.0
Recommended design value when ideal conditions are approximated	0.65	0.80	1.0	1.2	2.0	2.0

End condition code

Rotation fixed	Translation fixed
Rotation free	Translation fixed
Rotation fixed	Translation free
Rotation free	Translation free

Figure B1

Appendix C
Criteria for Estimating Effective Column Lengths in Continuous Frames

Note: *This Appendix is not a mandatory part of this Standard.*

C1.

Because the Standard (see Clause 8.6) requires that the sway effects produced by the vertical loads acting on the structure in its displaced configuration be included in the analysis, this Appendix covers only cases in which there is no relative translational displacement of the ends of the column.

C2.

Figure C1 is a nomograph applicable to cases in which the equivalent I/L of adjacent girders that are rigidly attached to the columns is known; it is based on the assumption that all columns in the portion of the framework considered reach their individual critical load simultaneously.

In the usual building frame, not all columns would be loaded so as to simultaneously reach their buckling loads; thus some conservatism is introduced in the interest of simplification.

C3.

The equation on which this nomograph is based is as follows:

$$\frac{G_U G_L}{4}(\pi/K)^2 + \frac{G_U + G_L}{2}\left(1 - \frac{\pi/K}{\tan \pi/K}\right) + 2\left[\frac{\tan \pi/2K}{\pi/K}\right] = 1$$

C4.

Subscripts U and L refer to the joints at the two ends of the column section being considered. G is defined as

$$G = \frac{\Sigma I_c / L_c}{\Sigma I_g / L_g}$$

where Σ indicates a summation for all members rigidly connected to that joint and lying in the plane in which buckling of the column is being considered,

where
I_c is the moment of inertia and
L_c is the unsupported length of a column section;
I_g is the moment of inertia and
L_g is the unsupported length of a girder or other restraining member.
I_c and I_g are taken about axes perpendicular to the plane of buckling being considered.

C5.

For column ends supported by, but not rigidly connected to, a footing or foundation, G may be taken as 10 for practical designs. If the column end is rigidly attached to a properly designed footing, G may be taken as 1.0. Smaller values may be used if justified by analysis.

C6.

Refinements in girder I_g/L_g may be made when conditions at the far end of any particular girder are known definitely or when a conservative estimate can be made. For the case with no side-sway, multiply girder stiffnesses by the following factors:
(a) 1.5 if the far end of the girder is hinged; and
(b) 2.0 if the far end of the girder is fixed against rotation (ie, rigidly attached to a support that is itself relatively rigid).

C7.

Having determined G_U and G_L for a column section, the effective length factor, K, is determined at the intersection of the straight line between the appropriate points on the scales for G_U and G_L with the scale for K.

C8.

The nomograph may be used to determine the effective length factors for compressive members of trusses with rigid joints, provided that all members of the truss that meet at the joint under consideration do not reach their ultimate loads simultaneously (yielding in tension or buckling in compression), in which case the effective length factors should be taken as one. This latter case typically occurs when the members are sized individually for one dominating load combination.

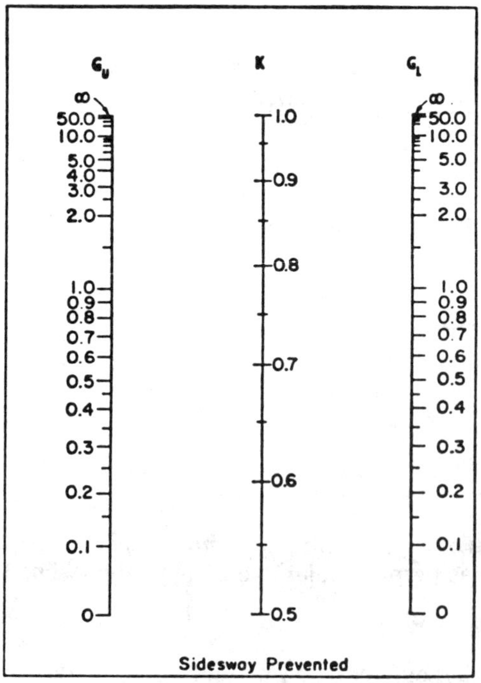

Figure C1
Nomograph for Effective Lengths of Columns in Continuous Frames

Appendix D
Seismic Design Requirements for Eccentrically Braced Frames

Note: *This Appendix is a mandatory part of this Standard.*

Members in the braced bays of eccentrically braced frames shall be designed in accordance with the following requirements.

D1. Link Beam

The link beam is the segment of a beam in an eccentrically braced frame that is designed to yield, either in flexure or in shear, prior to yield of other parts of the structure. Flanges shall conform to Class 1 section requirements and webs to Class 2. A link beam shall be provided at least at one end of each brace.

D2. Link Beam Resistance

The shear resistance of a link beam shall be taken as given in Clause 13.4.1.2 and the flexural resistance as given in Clause 13.5. Where the link beam resistance is controlled by shear, the flexural and axial resistances within the link shall be calculated neglecting the contribution of the web.

The reduced flexural resistance for use in Clauses D7 and D12 shall be taken as $1.18\phi M_p\ [1 - C_f/\phi AF_y]$, but not greater than ϕM_p.

D3. Link Beam Rotation

The rotation of the link segment relative to the rest of the beam, at a total frame drift of R times the drift determined for factored loading, shall not exceed the following:
(a) 0.060 radians for link segments having a clear length of $1.6M_r/V_r$ or less;
(b) 0.015 radians for link segments having a clear length of $2.6M_r/V_r$ or greater; and
(c) a value obtained by linear interpolation between the above limits for link segments having clear lengths between the above limits.
Note: *R is defined in Clause 4.1.9 of the* National Building Code of Canada, *1990.*

D4. Link Beam Web

Doubler plate reinforcement shall not be used on the web of a link beam and no openings shall be placed in it.

D5. Brace-to-Beam Connections

Brace-to-beam connections shall develop the compression resistance of the brace and transfer this force to the beam web. No part of the brace-to-beam connection shall extend into the web area of a link beam.

D6. Link Beam Stiffeners

Link beams shall have full-depth web stiffeners on both sides of the beam web at the brace end of the link beam. In addition, for link beams with a clear length within the limits of Clause D3 (c), full-depth stiffeners shall be placed a distance b from each end of the link. The stiffeners shall have a combined width of not less than b – 2w and a thickness of not less than 0.75w nor 10 mm.

D7. Intermediate Stiffeners

Full-depth intermediate web stiffeners shall be provided in either of the following conditions:
(a) where the link beam resistance is controlled by V_r; or
(b) where the link beam resistance is controlled by flexure and the shear determined by applying the reduced flexural resistance exceeds $0.45\phi dwF_y$.

D8. Web Stiffener Spacing

The spacing of link beam intermediate web stiffeners, when required, shall conform to the requirements given below:
(a) for link beams with a rotation angle of 0.06 radians, the spacing shall not exceed 38w – 0.2d;
(b) for link beams with a rotation angle of 0.03 radians or less, the spacing shall not exceed 56w – 0.2d; and
(c) for values of rotation angle between these limits, linear interpolation shall be used.

D9. Web Stiffener Location

Full-depth intermediate web stiffeners are required on only one side of the web for link beams less than 650 mm in depth and on both sides of the web for beams 650 mm or greater in depth. The thickness of one-side stiffeners shall not be less than 10 mm and the width shall not be less than 0.5b – w.

D10. Stiffener Welds

Fillet welds connecting the stiffener to the beam web shall develop a stiffener force of A_sF_y. Fillet welds connecting the stiffener to the flanges shall develop a stiffener force of $0.25A_sF_y$.

D11. Link Beam-to-Column Connection

D11.1

Where a link beam is adjacent to the column, the following requirements shall be met:
(a) the beam flanges shall have complete joint penetration groove welds to the column; and
(b) where the link beam resistance is controlled by shear (Clause D7), the web connection shall be welded to develop the full shear resistance of the beam web.

D11.2

Where the link beam is connected to the column web, the beam flanges shall have complete joint penetration groove welds to the connection plates and the web connection shall be welded to develop the shear resistance of the beam web. The rotation between the link beam and the column shall not exceed 0.015 radians at R times the drift due to factored loading.

D12. Brace Resistance

Each brace shall have a compressive resistance at least 1.5 times the axial force corresponding to the controlling resistance of the link beam. The controlling link beam resistance is either the shear resistance, V_r, or the reduced flexural resistance, M_r, whichever results in the lesser force in the brace.

D13. Column Resistance

The column resistance shall not be less than that required to sustain a load 1.25 times the load corresponding to the controlling resistance of the link beam.

D14. Roof Link Beam

A link beam is not required in roof beams of frames over five storeys in height.

D15. Concentric Brace in Combination

The first storey of a frame over five storeys in height may be concentrically braced if this storey can be shown to have a resistance of at least 1.5 times the loading associated with yielding of any other storey of the structure.

D16. Axial Forces

Axial forces in beams due to forces from the braces and due to transfer of seismic force to the end of the frames shall be considered in the design.

D17. Beam Flanges

Top and bottom flanges of beams shall be laterally braced at the ends of link beams and at intervals not exceeding $200/\sqrt{F_y}$ times the beam flange width. Lateral bracing at the ends of link beams shall be designed to resist 0.015 times the beam flange resistance. Intermediate bracing shall be designed to resist 0.01 times the beam flange force at the brace point for the loading condition corresponding to yielding of the link beam (see Clause D12). The resulting lateral displacement at the brace points shall not exceed 2.5 mm.

D18. Beam-to-Column Connections

Beam connections to columns may be designed to resist transverse shear only if the link beam is not adjacent to the column. Such connections must have capacity to resist a torsional moment of $0.01btdF_y$.

Appendix E
Margins of Safety

Note: *This Appendix is not a mandatory part of this Standard.*

E1.

An advantage of limit states design is that the probability of failure for different loading conditions is made more consistent, by the use of distinct load factors for the different loads to which the structure is subject, than in working stress design, where a single factor of safety is used. Furthermore, different resistance factors can, in a parallel manner, be applied to determine member resistances with a uniform reliability. The combination of the load factor and the inverse of the resistance factor gives a number comparable to the traditional factor of safety. In this Standard, a resistance factor of 0.90 is generally used.

E2.

For live loads, the load factor of 1.50 multiplied by the inverse of the resistance factor (1/0.90) equals 1.67, which is comparable to the safety factor contained in former working stress design standards. By using a load factor of 1.25 for dead load, probabilistic studies indicate that consistent probabilities of failure are determined over all ranges of dead-to-live-load ratios. The same probabilistic studies also show that load combination factors of 0.70 and 0.60 (depending on the number of loads taken in combination) applied only to live, wind, or earthquake and temperature loads, and a factor of 0.85 applied to dead load when it is counteractive to live loads, also result in a consistent probability of failure.

E3.

Resistance factors (see Clause 2.1) generally allow for underrun in the member or connection resistance as compared to that predicted. The underrun may arise from variability in material properties, dimensions, and workmanship, as well as from simplifications in the mathematical derivation of the resistance equations.

For simplicity in some cases in this Standard, uncertainty in the formulation of the theoretical member resistance has been incorporated directly into the expression for member resistance rather than using a lower value for the resistance factor. This is the case for the column curve, where the curve predicting the ultimate strengths as a function of slenderness ratio has been derived statistically taking into account residual stresses and initial out-of-straightness.

E4.

For bolts, a resistance factor of 0.67 is used to ensure that connector failures will not occur before general failure of the member as a whole. For long bolted joints and for cases in which shear planes intersect the threads, reduction factors are applied to the resistance formulations. As with bolts, a resistance factor of 0.67 is used for welds.

E5.

The expressions for resistance for welds are based on CSA Standard W59, and therefore maintain the margins of safety given in that Standard.

Appendix F
Columns Subject to Biaxial Bending

Notes:
(1) *This Appendix is not a mandatory part of this Standard.*
(2) *The precise design of beam-columns to resist biaxial bending is extremely complex. More refined design expressions than those given in Clause 13 are available, but these are shape-dependent. The Clauses in this Appendix provide design expressions for wide-flange shapes and square hollow structural sections.*

F1.

The bending moments under factored loads to be used in Clauses F2 and F3, M_{fx} and M_{fy}, are as defined in Clause 8.6.

F2.

Class 1 and 2 wide-flange shapes may be proportioned so that

(a) $$\left(\frac{M_{fx}}{M_{rcx}}\right)^{\zeta} + \left(\frac{M_{fy}}{M_{rcy}}\right)^{\zeta} \le 1.0$$

(b) $$\left(\frac{U_{1x}M_{fx}}{M_{ox}}\right)^{\eta} + \left(\frac{U_{1y}M_{fy}}{M_{oy}}\right)^{\eta} \le 1.0$$

where
U_{1x} and U_{1y} are as defined in Clause 13.8.3; however, U_{1x} shall not be less than 1.0

M_{rcx} and M_{rcy} are the factored moment resistance of the section, reduced for the presence of axial load, and may be taken as

Class 1:

$$M_{rcx} = 1.18\ M_{rx}\left[1 - \frac{C_f}{C_y}\right] \le M_{rx}$$

$$M_{rcy} = 1.19 M_{ry}\left[1 - \left(\frac{C_f}{C_y}\right)^2\right] \le M_{ry}$$

Class 2:

$$M_{rcx} = M_{rx}\left(1 - \frac{C_f}{C_y}\right)$$

$$M_{rcy} = M_{ry}\left(1 - \frac{C_f}{C_y}\right)$$

where M_{rx} and M_{ry} are defined in Clause 13.5(a)

M_{ox} and M_{oy} = maximum factored moment resistance of the column in the presence of the axial load but in the absence of the other orthogonal moment, may be taken as

$$M_{ox} = M_{rx}\left(1 - \frac{C_f}{C_r}\right)$$

$$M_{oy} = M_{ry}\left(1 - \frac{C_f}{C_r}\right)$$

$$\zeta = 1.6 - \frac{C_f/C_y}{2\log_e(C_f/C_y)}$$

For $b/d \geq 0.3$, $\eta = 0.4 + \frac{C_f}{C_y} + \frac{b}{d}$

For $b/d < 0.3$, $\eta = 0.1$

For M_{rx}, use M_r as defined in Clause 13.6(a); for M_{ry}, use M_r as defined in Clause 13.5(a) and C_r as defined in Clause 13.3.1 with K = 1.0 (see Clause 9.3).

Note: *For values of $C_f/C_y < 0.3$, the value of ζ may be taken as equal to the value of η.*

F3.

Class 1 and Class 2 square hollow structural sections (rolled* or fabricated) may be proportioned so that

(a) $\frac{M_{fx}}{M_{rx}} + 0.5\frac{M_{fy}}{M_{ry}} \leq 1.0$

(b) $\frac{C_f}{C_r} + 0.85\left(\frac{M_{fx}}{M_{rx}} + 0.5\frac{M_{fy}}{M_{ry}}\right) \leq 1.0$

where

M_{fx} = the numerically larger moment

M_{rx} and M_{ry} are as defined in Clause 13.5(a)

$C_r = \phi A F_y$

**Hot-rolled or stress-relieved such that residual stresses do not exceed $0.3F_y$.*

(c) $\frac{C_f}{C_r} + \upsilon\left(\frac{U_{1x} M_{fx}}{M_{rx}} + \frac{U_{1y} M_{fy}}{M_{ry}}\right) \leq 1.0$

where $\upsilon = \frac{\sqrt{(\omega_{1x}M_{fx})^2 + (\omega_{1y}M_{fy})^2}}{\omega_{1x}M_{fx} + \omega_{1y}M_{fy}}$

C_r is as defined in Clause 13.3.2 with K = 1.0 (see Clause 9.3)

M_{rx} and M_{ry} are as defined in Clause 13.5(a)

ω_1 is as defined in Clause 13.8.4

U_{1x} and U_{1y} are as defined in Clause 13.8.3

Note: *Design of Class 1 and Class 2 sections in accordance with the above requirements takes advantage of the redistribution of stress after initiation of yielding under the specified loads. Consideration should be given to this aspect of design if yielding under the specified loads would induce undesirable lateral deformations of a structure.*

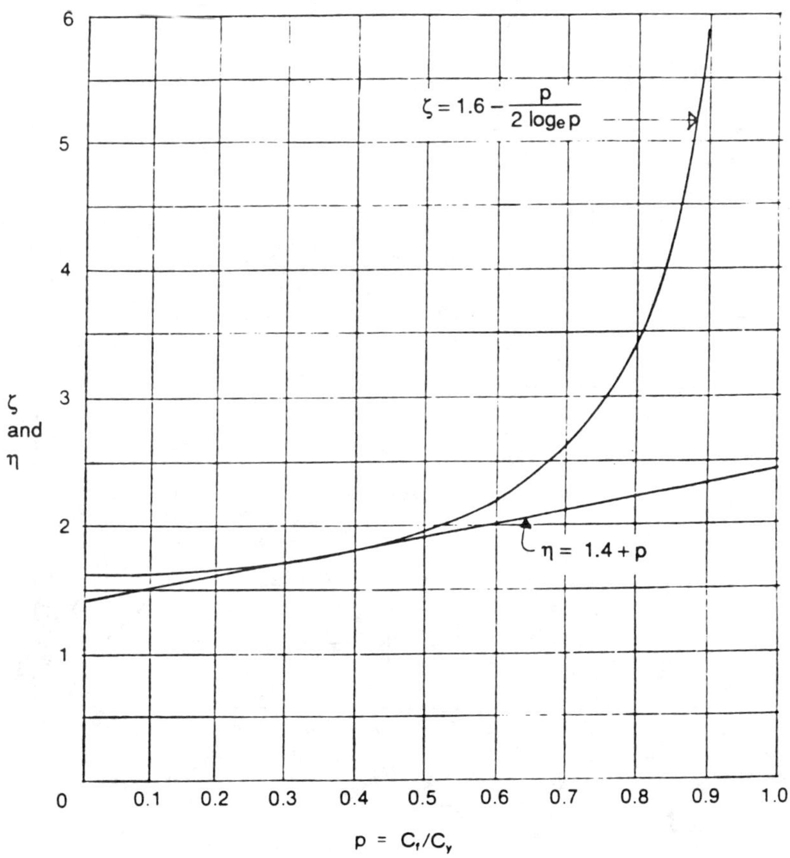

Figure F1
Plot of ζ and η Versus C_f/C_y for b/d = 1

$$\left(\frac{M_x}{M_{ox}}\right)^{\zeta} + \left(\frac{M_y}{M_{oy}}\right)^{\zeta} = 1$$

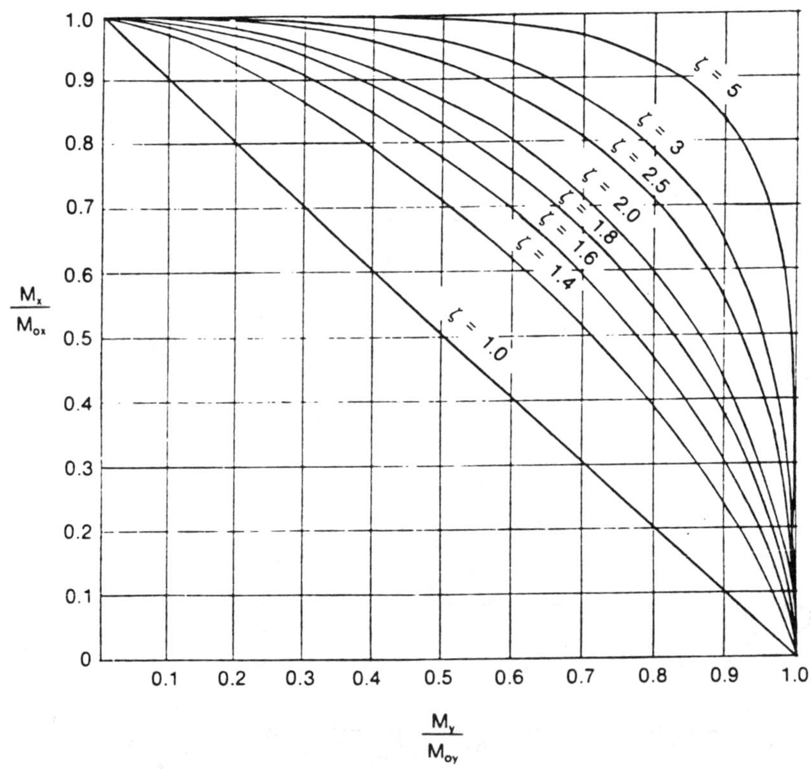

Figure F2
Plot of the Interaction Equation

Appendix G
Guide for Floor Vibrations

Note: *This Appendix is not a mandatory part of this Standard.*

G1.

The development of floors of lighter construction, longer spans, and less inherent damping has sometimes resulted in problems of objectionable floor vibrations during normal human activity. Fatigue or overloading of floor structures due to vibration is not covered in this Appendix, nor are floors subject to vibrations caused by rhythmic activities such as aerobic classes; some guidance on this is given in Reference (1) for assembly occupancies.

G2.

Two types of vibration problems arise in floor construction. Continuous vibrations arise due to the periodic forces of machinery, vehicles, or certain human activities such as dancing. These vibrations can be considerably amplified when the periodic forces are synchronized with a floor frequency — a condition called resonance. Transient vibrations, which decay as shown in Figure G1, arise due to footsteps or other impact.

G3.

The most important floor characteristics affecting vibration problems are the natural frequency in hertz (cycles per second), which usually corresponds to the lowest mode of vibration, and damping. The relation between damping, expressed as a percentage of critical damping (see Reference (2)), and decay of free vibration is shown in Figure G2. Other characteristics affecting transient vibration problems are mass, especially for heavy, long-span floors, and stiffness under point load, especially for light, short-span floors.

G4. Annoyance Criteria

G4.1

Generally people do not like floors to vibrate. For a continuous sinusoidal vibration lasting more than about ten cycles, an average threshold of definite perception is shown in Figure G3 in terms of peak acceleration. The threshold levels for different people range from about one-half to twice the level shown. In the frequency range 2–8 Hz, where people are most sensitive to vibration, the threshold corresponds to 0.5% g approximately, where g is the acceleration due to gravity. The threshold of definite perception shown in Figure G3 can be used to approximate a design criterion for residential, school, and office occupancies; the design level will be lower for sensitive occupancies (eg, operating rooms and special laboratories) and greater for industrial occupancies.

G4.2

For transient vibrations, as shown in Figure G1, the design threshold in terms of initial peak acceleration of a decaying vibration increases with an increase in damping as shown in Figure G3. People find a continuous vibration much more annoying than a vibration that dies out quickly.

G4.3

Design criteria for walking (footstep) vibrations equivalent to that for continuous vibration are shown in Figure G3 for different levels of damping (see Reference (3)). The criteria are related to acceleration from a heel impact test.

G5. Continuous Vibrations — Resonance

G5.1

Continuous vibrations caused by machines can be reduced by special design provisions (see References (2) and (4)) such as vibration isolation. Care should be taken at the planning stage to locate such machinery away from sensitive occupancies such as offices.

G5.2

Floor vibrations can also arise from heavy street traffic on bumpy pavement over soft subgrade. The annoyance increases considerably when repetitive vehicles such as buses create ground vibrations that synchronize with the floor frequency.

G5.3

Continuous vibrations caused by human activities may be a problem for light residential floors or for long-span floors used for special purposes such as dancing, concerts, or gymnastics. People alone or in unison can create periodic forces in the frequency range of approximately 1–4 Hz; therefore, for such occupancies natural frequencies lower than 5 Hz should be avoided. For very repetitive activities such as jumping exercises and dancing, it is recommended that the frequency of floors be 8 Hz or higher to avoid very noticeable vibration. See Reference (1) for more specific guidance on vibrations due to rhythmic human activities.

G6. Walking Vibrations

G6.1

Objectionable vibration due to footstep impact can occur in floor systems with light damping in residential, school, office, and similar occupancies. Since this is the most common source of annoyance, the remainder of this guide will be concerned with this problem. Types of construction that may give walking-vibration problems include open-web steel joists or steel beams with concrete deck and light wood deck floors using steel joists.

G7. Performance Test for Floor Vibration

G7.1

The vibration acceptability of a floor system to human activity can be evaluated by a performance test. Partitions, rugs, and furnishings, finishes, etc, contribute to reducing vibration annoyance and should therefore be considered in setting up the test floor. A measuring device that filters out frequencies greater than approximately 1.5 times the fundamental frequency should be located near midspan. A person who will give a subjective evaluation of the floor should also be sitting close to the measuring device.

G7.2

One test involves having a person of average weight with soft-soled shoes rise up on his toes and drop on his heels near the location of measurement. Fundamental frequency, damping from decay record (see Figure G2), and peak acceleration are obtained from the measurement, and the peak acceleration is plotted on Figure G3 to see how it compares with the criteria. Another test involves checking floor comfort when different persons walk down the floor; the average peak acceleration can then be compared with the criterion for steady motion given in Figure G3.

G8. Long-Span Steel Floors with Concrete Deck

G8.1

Walking vibrations may be a problem for open-web steel joists or steel beams with concrete deck, composite or noncomposite, generally of spans 7–20 m with frequencies in the range 4–15 Hz. For such floors, partitions, if properly located, provide more than enough damping to avoid excessive vibrations. On the other hand, walking vibrations may be serious for bare floors with very low inherent damping, as is the case for fully composite construction. Figure G3 shows that the criterion is roughly 10 times greater for 12% damping than for 3% damping.

G8.2

To assess vibration acceptability, a knowledge of frequency, damping, and the peak acceleration from heel impact is required. If design by performance testing is not feasible, these parameters should be estimated by calculation as follows:
(a) The frequency can be estimated by assuming full composite action, even for noncomposite construction. For a simply supported one-way system, the frequency f_1 is given by

$$f_1 = 156\sqrt{\frac{EI_T}{wL^4}} \qquad (1)$$

where
E is the modulus of elasticity of steel (200 000 MPa)
I_T is the moment of inertia (mm^4) of the transformed T-section (concrete transformed to steel) assuming a concrete flange equal in width to the spacing of steel joists or beams
L is the span in mm, and
w is the dead load of the T-section in N/mm of span.

When one-way systems are supported on steel girders, the frequency may be reduced. In this case the frequency can be approximated by

$$\frac{1}{f^2} = \frac{1}{f_1^{\,2}} + \frac{1}{f_2^{\,2}} \qquad (2)$$

where f_2 is the frequency of floor supported on steel girders, perpendicular to joists.

A continuous beam of equal spans on flexible supports should be treated as simply supported, since adjacent spans vibrate in opposite directions. For other conditions of span and restraint, the dynamically equivalent simply supported span is less than the full span and can be estimated from the fundamental mode shape.
(b) Damping is generally more difficult to estimate than frequency. A bare steel and concrete deck floor has a damping of approximately 3–4% critical for noncomposite construction and about 2% for fully composite construction. The addition of such components as floor finishing, rugs, furnishings, ceiling, fire-proofing, and ducts increases the damping by about 3% or more. Partitions, either above or below the floor, provide the most effective damping, especially when they are located in both directions. Even light partitions that do not extend to the ceiling provide considerable damping. Partitions along support lines or parallel to the floor joists and farther apart than approximately 6 m may not be effective, however, because the nodal lines of vibration form under the partitions. Human beings also provide damping, but this is less effective for heavy, long-span floors than for lighter, short-span floors. The following values are suggested for design calculation (see Reference (5)):

	Damping in per cent critical
Bare floor	3
Finished floor — ceiling, ducts, flooring, and furniture	6
Finished floor with partitions	12

(c) The peak acceleration from heel impact for floors of greater than 7 m span with frequencies lower than about 10 Hz can be estimated by assuming an impulse of 70 N•s suddenly applied to a simple spring and mass system whose mass gives the same response as that of the floor system represented as a simply supported beam vibrating in the fundamental mode. The peak acceleration, a_o, in per cent *g*, can be approximated by (see Reference (3))

$$a_o = 0.9\left[\frac{2\pi f \times \text{impulse}}{\text{equivalent mass}}\right]\left[\frac{100}{g}\right] = \frac{60f}{wBL} \quad (3)$$

where
f is the frequency, in hertz
w is the weight of the floor plus contents, in kilopascals
L is the span, in metres
B is the width of the equivalent beam, in metres

For steel joist or beam and concrete deck systems on stiff supports, L is the joist span and B can be approximated as 40 t_c, where t_c is the thickness of the concrete deck as determined from the average mass of concrete, including ribs. For joists or beams and concrete deck supported on flexible girders, where the girder frequency is much lower than the joist frequency and therefore girder vibration predominates, L is the girder span and B can be approximated as the width of floor supported by the girder. For cases where both frequencies are similar, Equation (2) can be used to determine frequency and BL can be estimated as follows:

$$BL = \left(\frac{f}{f_1}\right)^2 B_1L_1 + \left(\frac{f}{f_2}\right)^2 B_2L_2 \quad (4)$$

where the subscript 1 refers to the joist or beam system on rigid supports and the subscript 2 refers to the girder system.

G8.3

For floor spans of less than 7 m, the deflection limits given in Clause 6.2.1.2 in this Standard are recommended, where, for noncomposite construction, stiffness should be based on noncomposite action. In any case, care should be taken to avoid low damping.

G9. Light Wood Deck Floors Using Steel Joists

G9.1

Walking vibrations may be objectionable for light wood deck floors using steel joists with small rolled or cold-formed sections, generally with frequencies in the range 10 – 25 Hz. Although the same principles applying to long-span floors can be used for lighter floors with higher frequencies, the motion can no longer be represented by a simple impulse applied to the floor system. This is because the persons involved — the one causing and the one receiving the motion — interact with the floor to damp out the motion of the floor.

G9.2

Research carried out so far on steel joist floors with wood deck indicates that, in general, their characteristics for vibration acceptability are similar to those for wood joist floors. Evaluation tests of wood floors indicate that stiffness under point loading (0.5 to 1 mm maximum deflection under 1 kN) is the most important parameter affecting vibration comfort (see Reference (6)). Such a stiffness requirement also helps prevent cabinet swaying, china rattling, etc. A joist deflection limit of L/360 under 2 kPa loading provides sufficient stiffness if sufficient lateral stiffness is provided either in the deck or by cross-bridging.

G9.3

Floor damping is less important for light floors than for long-span floors, since the main source of damping is provided by the persons on the floor. Also, adding mass does not improve vibration comfort, since an increase in mass corresponds to a decrease in effective damping. Spans continuous over a support that is a party wall between housing units should be avoided, since people are more annoyed by vibrations originating outside their units than from within. For cold-formed C-joists, ceiling boards or straps should be attached to the bottom flange to prevent annoying high-frequency torsional vibrations in the joists.

G10. Corrective Measures for Unacceptable Floors

G10.1

Measures for correcting floors with annoying vibrations will depend on whether the vibrations are continuous or transient.

G10.2

For transient vibrations, the most effective measure is usually to increase the damping. This can be done by adding partitions or damper posts in the floor below. If these methods are not suitable, special devices such as vibration absorbers or damping materials can be incorporated into the floor system (see References (7) and (8)). For light floors, a rug is effective in reducing walking impact as well as in cushioning the sway of china cabinets.

G10.3

Corrective measures for continuous vibrations include vibration isolation, smoothing of road surface, and alteration of floor frequency to reduce resonance.

G11. References

(1) Supplement to the *National Building Code of Canada,* 1990, Commentary on Serviceability Criteria for Deflections and Vibrations.

(2) Thomson, W.T., *Vibration Theory and Applications.* Prentice-Hall.

(3) Allen, D.E. and Rainer, H., *Vibration Criteria for Long Span Steel Floors. Canadian Journal of Civil Engineering,* Vol. 3, No. 2, June 1976.

(4) Steffens, R.J., *Some Aspects of Structural Vibration.* Building Research Current Paper Engineering Series 37, Building Research Station, Ministry of Technology, Great Britain.

(5) Allen, D.L., *Vibrational Behaviour of Long-Span Floor Slabs.* Proceedings of the Canadian Structural Engineering Conference,1974.

(6) Onysko, D.M., *Performance of Wood-Joist Floor Systems — A Literature Review.* Forest Products Laboratory Information Report OP-X-24, Canadian Forestry Service, Department of Fisheries and Forestry, January 1970.

(7) Nelson, F.C., *The Use of Viscoelastic Material to Damp Vibrations in Buildings and Large Structures. AISC Engineering Journal,* Vol. 5, No. 2, April 1968, p. 72.

(8) Allen, D.E. and Pernica, G., *A Simple Absorber for Walking Vibrations. Canadian Journal of Civil Engineering,* Vol. 11, No. 1, March 1984.

(9) Lenzen, K.H., *Vibration of Steel Joists-Concrete Slab Floors. AISC Engineering Journal,* Vol. 3, No. 3, July 1966, p. 133.

(10) Wright, D.T. and Green, R., *Human Sensitivity to Vibrations.* Department of Civil Engineering, Report No. 7, Queen's University, February 1959.

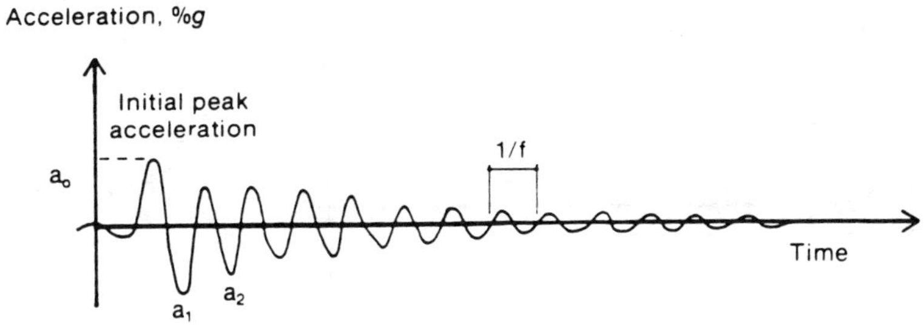

Figure G1
Typical Transient Vibration from Heel Drop
(High Frequencies Filtered Out)

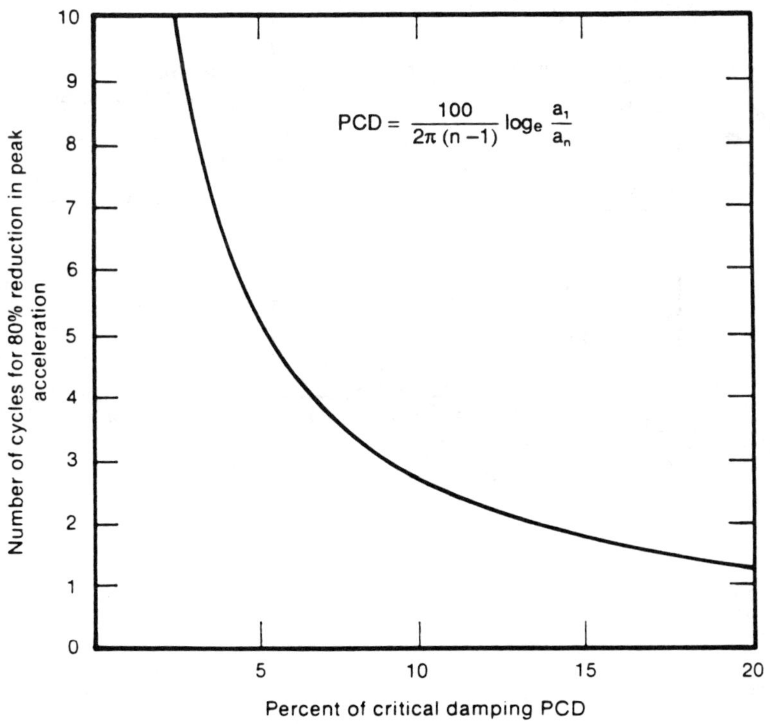

Figure G2
Relation Between Damping and Decay

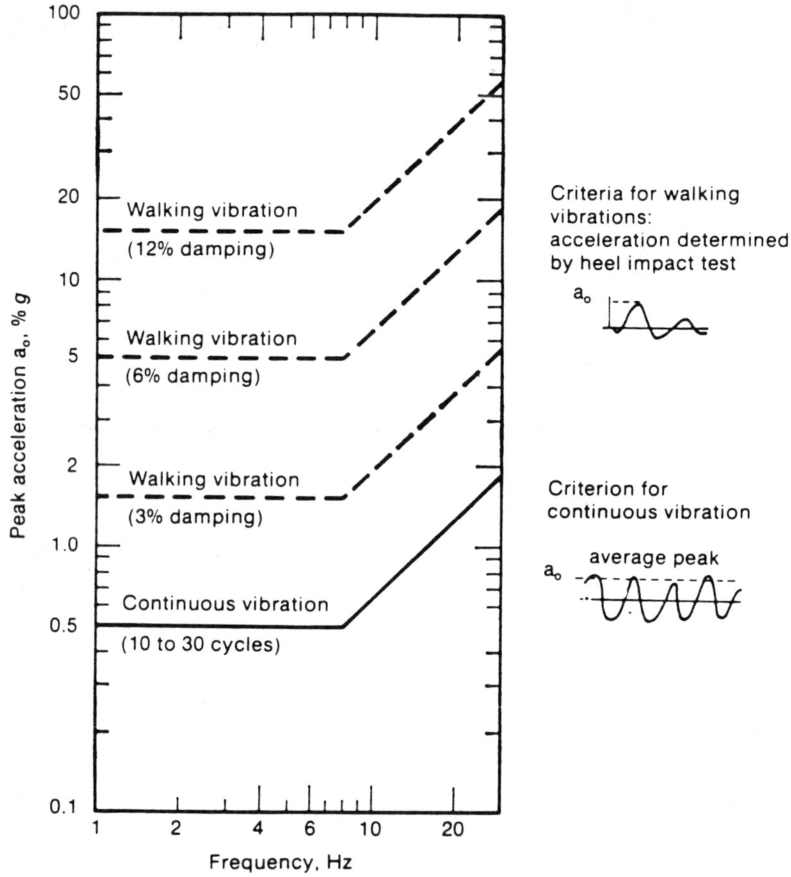

Figure G3
Annoyance Criteria for Floor Vibrations
(Residential, School, Office Occupancies)

Appendix H
Wind Sway Vibrations

Note: *This Appendix is not a mandatory part of this Standard.*

H1.

Wind motion of tall buildings or other flexible structures may create annoyance for human occupants unless measures are taken at the design stage. The main source of annoyance is lateral acceleration, although noise (grinding and wind howl) and visual effects can also cause concern.

H2.

For a given wind speed and direction, the motion of a building, which includes vibration parallel and perpendicular to the wind direction and twist, is best predicted by a wind tunnel test. Approximate calculation rules are, however, given in Reference (1).

H3.

In cases where wind motion is significant in design, the following should be considered:
(a) education of occupants that, although high winds may occasionally cause motion, the building is safe;
(b) minimization of noises through, for example detailing of building joints to avoid grinding and design of elevator guides to avoid scraping due to sway;
(c) minimization of twist through symmetry of layout, bracing, or outer walls (tube concept). Twist vibration also creates a magnified visual effect of relative motion of adjacent buildings;
(d) possible introduction of mechanical damping to reduce wind vibration.

H4. References

(1) Supplement to the *National Building Code of Canada,* 1990, Chapter 4, Commentary B on Wind Loads.

(2) Hansen, R.J., Reed, J.W., and Van Marcke, E.H., *Human Response to Wind-Induced Motion of Buildings. Journal of the Structural Division,* ASCE, Vol. 99, No. ST7, July 1973, pp. 1589–1605.

(3) Chen, P.W. and Robertson, L.E., *Human Perception Thresholds of Horizontal Motion. Journal of the Structural Division,* ASCE, Vol. 98, No. ST8, August 1972, pp. 1681–1695.

(4) Reed, J.W., *Wind-Induced Motion and Human Discomfort in Tall Buildings.* Department of Civil Engineering Research Report R71-42. Massachusetts Institute of Technology, November 1971.

(5) Hogan, M., *The Influence of Wind on Tall Building Design.* Faculty of Engineering Science Research Report BLWT-4-71, University of Western Ontario, March 1971.

(6) Council on Tall Buildings and Urban Habitat, Monograph on the Planning and Design of Tall Buildings. Volumes PC and SB. American Society of Civil Engineers, 1981.

Appendix I
Recommended Maximum Values for Deflections For Specified Design Live and Wind Loads

Note: *This Appendix is not a mandatory part of this Standard.*

I1.

Although this Appendix refers specifically to specified design live and wind loads, when setting forth deflection criteria, the designer should consider the inclusion of specified dead loads in some instances. For example, nonpermanent partitions, which are classified by the *National Building Code* as dead load, should be part of the loading considered under Appendix I if they are likely to be applied to the structure after the completion of finishes susceptible to cracking. Because some building materials augment the rigidity provided by the steelwork, the wind load assumed to be carried by the steelwork for calculating deflections can be somewhat reduced from the design wind load used in strength and stability calculations. The more common structural elements that contribute to the stiffness of a building are masonry walls, certain types of curtain walls, masonry partitions, and concrete around steel members. The **maximum** suggested amount of this reduction is 15%. In tall and slender structures (height greater than **4 times** the width), it is recommended that the wind effects be determined by means of dynamic analysis or wind tunnel tests.

		Design load	Application	Maximum deflection
Industrial type buildings	Vertical deflection	Live load	Simple span members supporting inelastic roof coverings	1/240 of span
		Live load	Simple span members supporting elastic roof coverings	1/180 of span
		Live load	Simple span members supporting floors	1/300 of span
		Maximum wheel loads (no impact)	Simple span crane runway girders for crane capacity of 225 kN and over	1/800 of span
		Maximum wheel loads (no impact)	Simple span crane runway girders for crane capacity under 225 kN	1/600 of span
	Lateral deflection	Crane lateral force	Simple span crane runway girders	1/600 of span
		Crane lateral force or wind	Building column sway *	1/400 to 1/200 of height

(Continued)

		Design load	Application	Maximum deflection
All other buildings	Vertical deflection	Live load	Simple span members of floors and roofs supporting construction and finishes susceptible to cracking	1/360 of span
		Live load	Simple span members of floors and roofs supporting construction and finishes not susceptible to cracking	1/300 of span
	Lateral deflection	Wind	Building sway, due to all effects	1/400 of building height
		Wind	Storey drift (relative horizontal movement of any two consecutive floors due to shear effects) in buildings with cladding and partitions without special provision to accommodate building frame deformation	1/500 of storey height
		Wind	Storey drift (as above), with special provision to accommodate building frame deformation	1/400 of height

*Permissible sway of industrial buildings varies considerably and depends on such factors as wall construction, building height, effect of deflection on the operation of crane, etc. Where the operation of the crane is sensitive to lateral deflections, a permissible lateral deflection of less than 1/400 of the height may be required.

Appendix J
Guide to Calculation of Stability Effects

Note: *This Appendix is not a mandatory part of this Standard.*

J1. General

This Appendix gives one approach to the calculation of the additional bending moments and forces generated by the vertical loads acting through the deflected shape of the structure. In this approach, the moments and forces are incorporated into the results of the analysis of the structure. Alternatively, a second-order analysis, which formulates equilibrium on the deformed structure, may be used to include the stability effects.

J2. Combined Loading Case

Step 1. Apply the factored load combination to the structure (see Clause 7.2.2).

Step 2. Calculate the lateral deflections at each floor level, Δ_i, by first-order elastic analysis.

Step 3. Calculate the artificial storey shears, V_i', due to the sway forces.

From Figure J1

$V_i' = \frac{\Sigma P_i}{h_i}\left[\Delta_{i+1} - \Delta_i\right]$ = artificial shear in storey i due to the sway forces

ΣP_i = sum of the column axial loads in storey i

h_i = height of storey i

Δ_{i+1}, Δ_i = displacements of storey i + 1 and i, respectively

Step 4. Calculate the artificial lateral loads, H_i', as follows:

$H_i' = V_{i-1}' - V_i'$

Step 5. Repeat Step 1, applying the artificial lateral loads, H_i', in addition to the factored load combination.

Step 6. Repeat Steps 2 through 5 until satisfactory convergence is achieved. Lack of convergence within 5 cycles may indicate an excessively flexible structure.

J3. Gravity Load Case

For the gravity load case, translational moments shall be taken as not less than the moment produced by pseudo-lateral loads, applied at each storey, equal to 0.005 times the factored gravity loads acting at each storey.

Note: *This is equivalent to obtaining moments from a first-order analysis of the structure based on the assumption of an initial displacement of the structure equal to 0.005 of the height.*

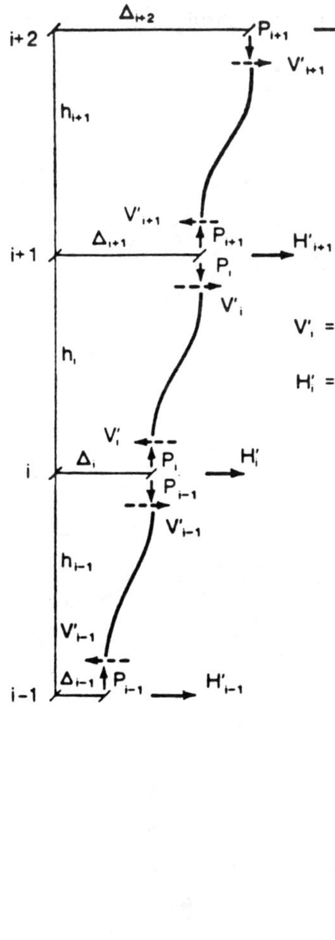

$$V'_i = \frac{\Sigma P_i}{h_i}(\Delta_{i+1} - \Delta_i)$$

$$H'_i = V'_i - V'_{i-1}$$

Figure J1
Sway Forces Due to Vertical Loads

Appendix K
Fatigue

Note: *This Appendix is not a mandatory part of this Standard.*

K1.

Figure K1 is a plot of the design curves for the allowable stress range for categories a to y in Tables 4(A) and 4(B) (Clause 14).

K2.

Figure K2 gives illustrative examples of the various fatigue categories described in Table 4(B).

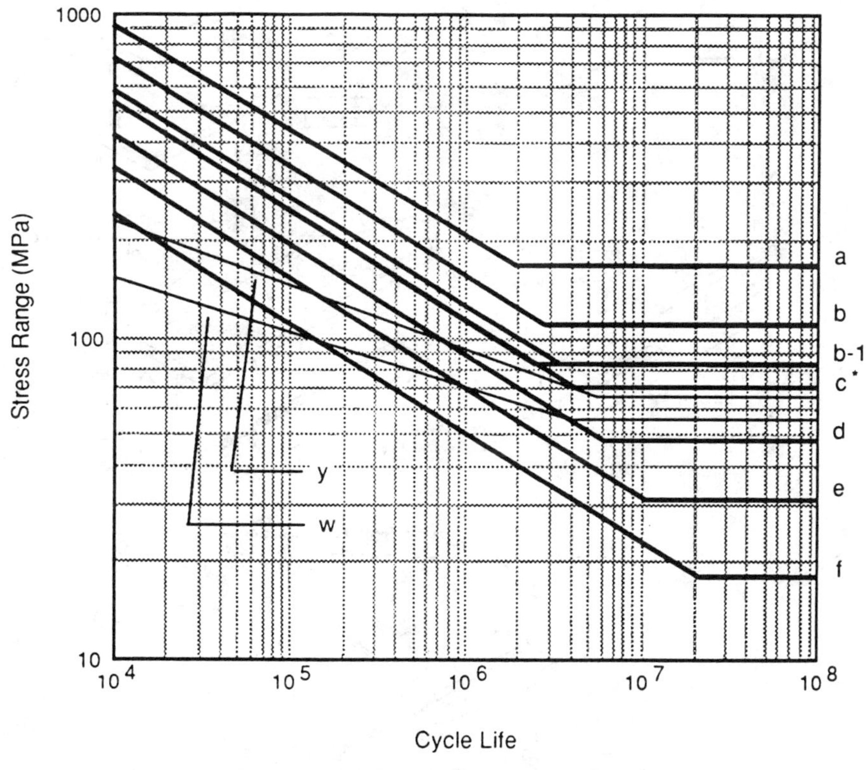

*For transverse stiffener welds on girder webs or flanges use 83 MPa instead of 70 MPa for the endurance limit.

Figure K1
Design Curves for the Allowable Stress Range for Categories a to y

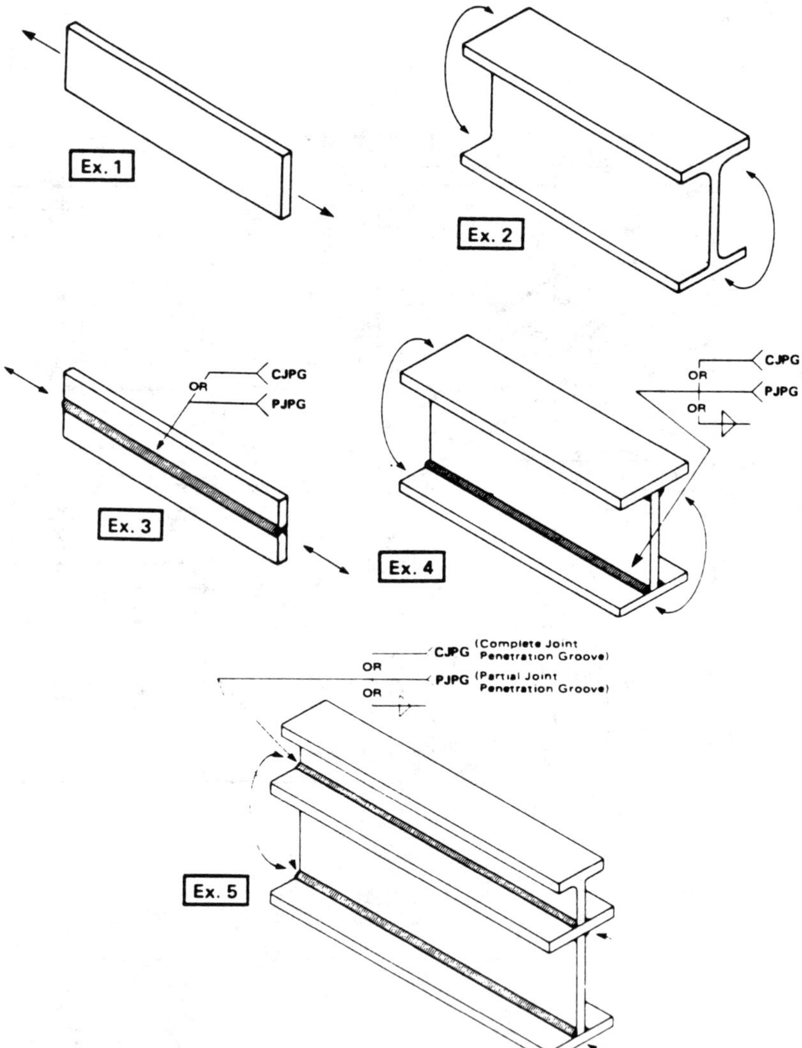

Note: *The example numbers are referenced in Table 4(B).*

Figure K2
Illustrative Examples of Various Details Representing Stress Range Categories

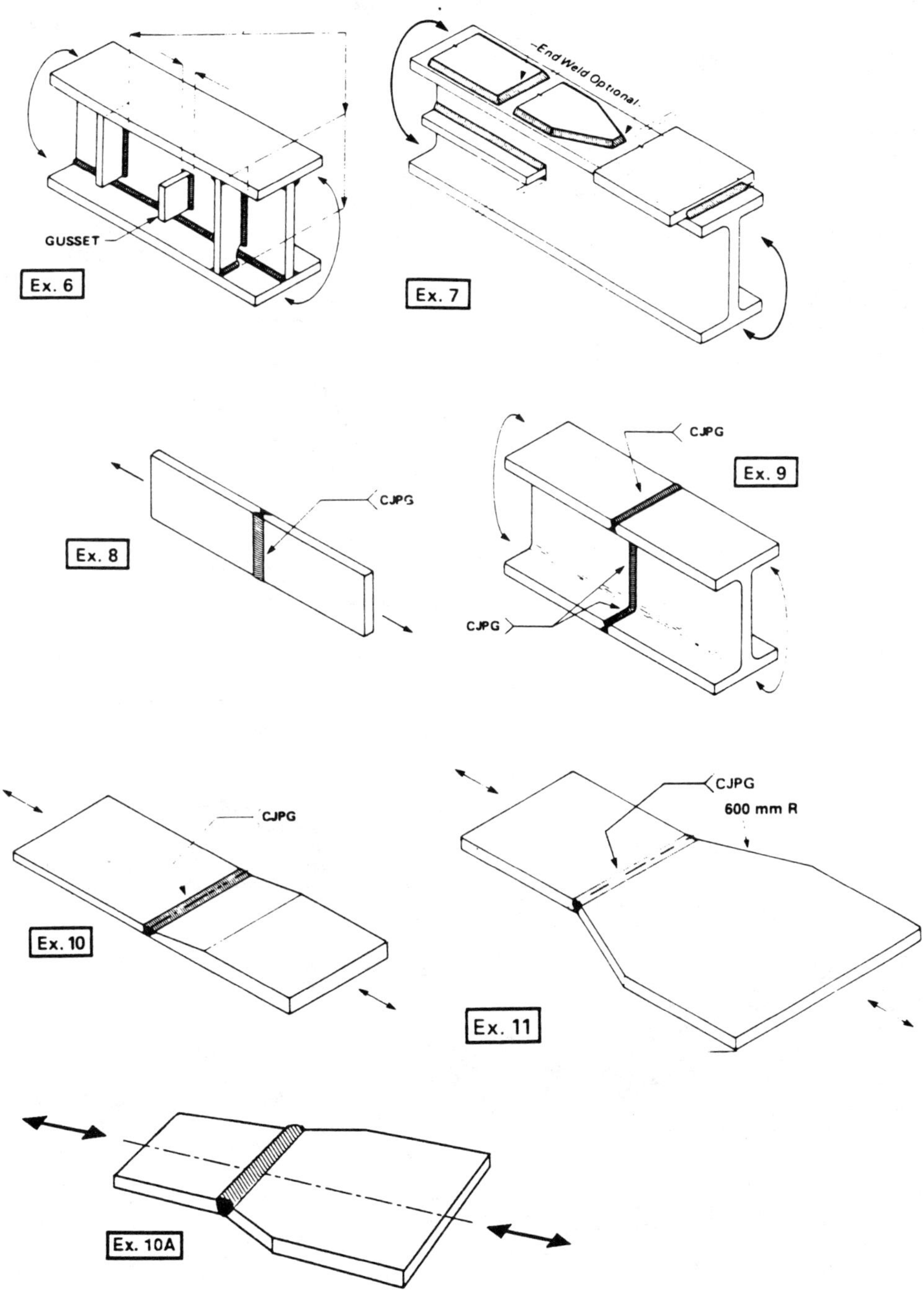

Note: *The example numbers are referenced in Table 4(B).*

Figure K2 (Continued)

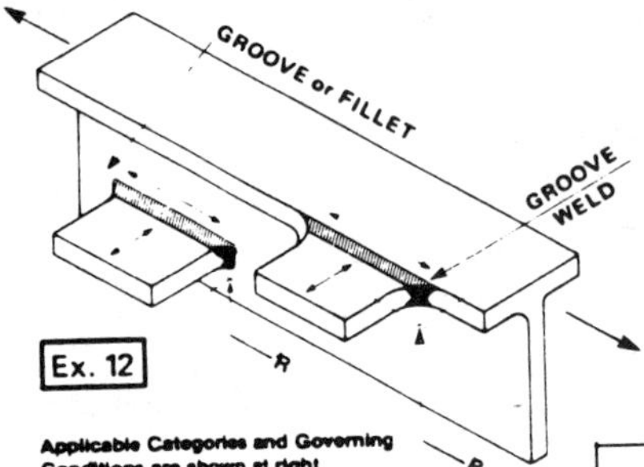

Ex. 12

Applicable Categories and Governing Conditions are shown at right.

TRANSITION RADIUS "R"	Fillet Connections	Groove Connections				
	To Web	To Web or to Flange	To Web (3)	To Flange (3)		
	Longitudinal Loading/Transverse Loading	Longitudinal Loading	Transverse Loading			
	Stress Range Category			Stress Range Category based on Condition of Joint (2)		
				1	2	3,4
50 mm > R ⩾ 0 mm	e	e(1)	e	e	e	e
150 mm > R ⩾ 50 mm (4)	d	d	d	d	d	e
600 mm > R ⩾ 150 mm (4)	d	c	c	c	c	e
R ⩾ 600 mm (4)	d	b	c	b	c	e

(1) *For longitudinal loading only, use Category d if detail length is between* 50 mm *and* 12 *times the plate thickness, but less than* 100 mm.

(2) *Condition of Joint:*

- *(1) Equal thickness of parts joined — reinforcement removed.*
- *(2) Equal thickness of parts joined — reinforcement not removed.*
- *(3) Unequal thickness of parts joined — reinforcement removed.*
- *(4) Unequal thickness of parts joined — reinforcement not removed.*

(3) *Weld soundness to be established by nondestructive examination.*

(4) *Terminal ends of welded joints to be ground smooth.*

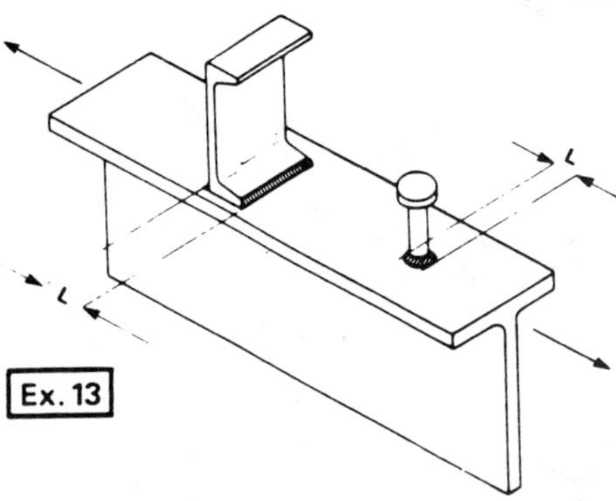

Ex. 13

Ex. 14

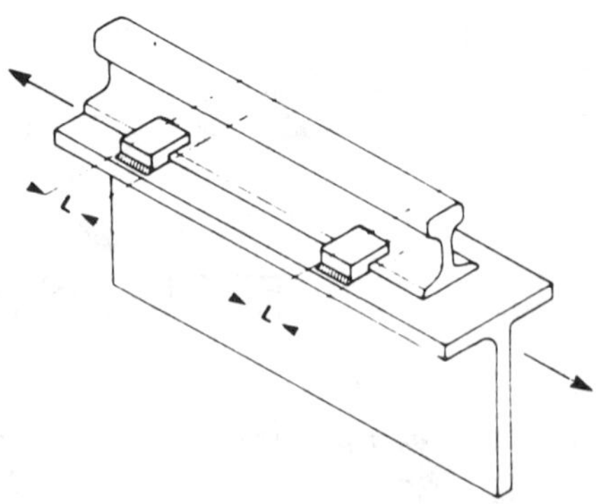

Note: *The example numbers are referenced in Table 4(B).*

Figure K2 (Continued)

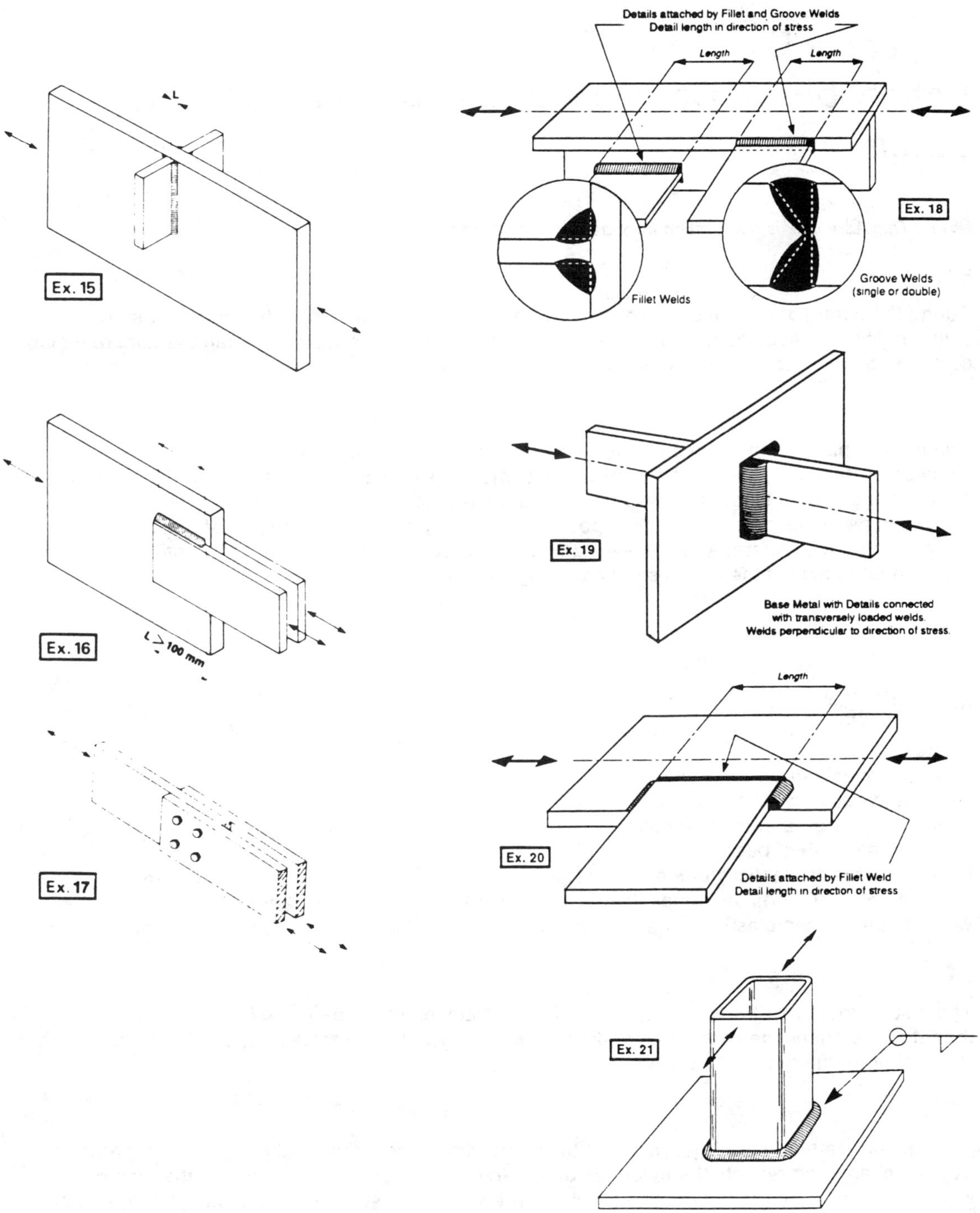

Note: *The example numbers are referenced in Table 4(B)*

Figure K2 (Concluded)

Appendix L
Deflections of Composite Beams Due to Shrinkage Strain

Note: *This Appendix is not a mandatory part of this Standard.*

L1.

During the curing of the concrete cover slab, the volumetric reduction (shrinkage strain) of the concrete induces additional flexural deflections in the composite member when the concrete slab contracts but the steel shape does not.

L2.

Two models have been suggested (see References (1) and (2)) to estimate the deflection due to shrinkage. Basically, these models assume a value of shrinkage strain acting on the effective area of the concrete slab. It is assumed that the effect of the shrinkage strain is equivalent to equal end moments acting on the composite beam. An appropriate modulus of elasticity is selected for the concrete, and the lever arm of the force is taken from the centre of the slab to the elastic neutral axis of the composite beam (Figure L1).

L3.

The resulting shrinkage deflection is

$$\Delta_{sh} = \frac{\varepsilon_{sh} E_c A_c L^2}{8EI} \cdot y_c$$

where

ε_{sh} = shrinkage strain
E_c = modulus of elasticity of concrete
A_c = effective area of concrete slab
L = span length of beam
E = modulus of elasticity of steel beam
I = moment of inertia of composite beam as given in Clause L4 or Clause L5
y_c = distance from elastic neutral axis to assumed line of action of the shrinkage force

L4.

The first method (see Reference (1)) uses the restrained shrinkage value for ε_{sh}, which is lower than the free shrinkage strain, the modulus of elasticity of the concrete, E_c, and the transformed moment of inertia of the composite beam.

L5.

The second method (see Reference (2)) takes the shrinkage strain as the strain appropriate to free shrinkage and reflects the influence on the free shrinkage of the following: the time over which shrinkage occurs, the relative humidity, the volume-to-surface ratio, slump, fines, and air content and cement content of the concrete mix. An age-adjusted modulus similar to that used in calculations of concrete creep is used for E_c. The moment of inertia of the beam is taken as the effective moment of inertia, I_e, given in Clause 17.3.1(a).

L6.

For both methods, care should be taken in selecting values for the quantities in the shrinkage deflection equation so as not to overestimate deflections, especially when used to determine the camber requirements. Many of the quantities will be influenced by site conditions.

L7. References

(1) Chien, E.Y.L. and Ritchie, J.K., *Design and Construction of Composite Floor Systems*, Canadian Institute of Steel Construction, August 1984.

(2) Montgomery, C.J., Kulak, G.L., and Shwartsburd, G., *Deflection of a Composite Floor System, Canadian Journal of Civil Engineering,* Vol. 10, No. 2, June 1983.

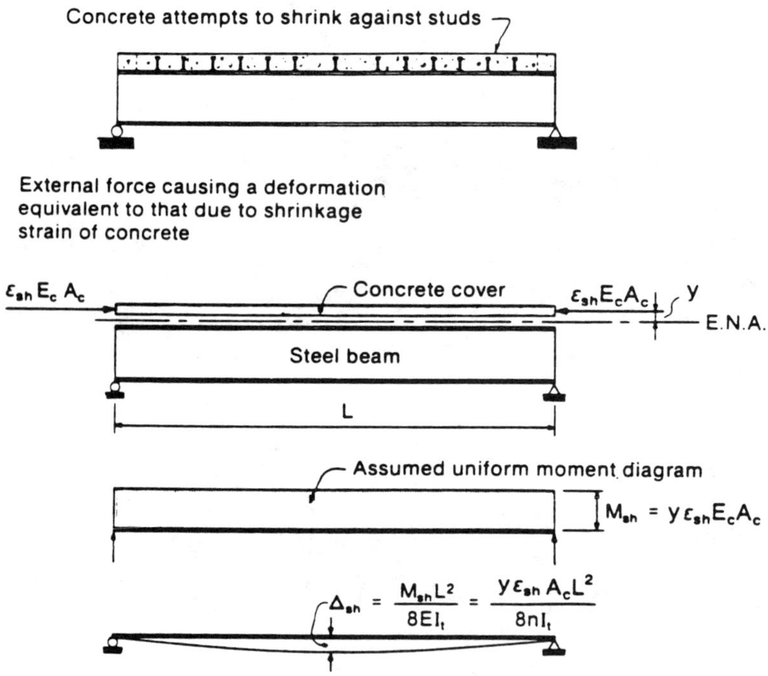

Figure L1
Shrinkage Strain Model

PART TWO
CISC COMMENTARY ON CAN/CSA-S16.1-M89

Preface

This Commentary has been prepared by the Canadian Institute of Steel Construction in order to provide guidance as to the intent of various provisions of the National Standard of Canada, written by CSA, CAN/CSA-S16.1-M89, "Limit States Design of Steel Structures". This Commentary and the information contained in the references cited provide an extensive background to the development of the Standard, its technical requirements, and the new or revised sections of the M89 edition.

The Institute gratefully acknowledges the efforts of P. C. Birkemoe, E. Chien, M. U. Hosain, G. L. Kulak, A. Picard, D. L. T. Oakes, R. G. Redwood, D. C. Stringer, and R. Vincent for their valuable contributions to the commentary. J. E. Henderson and D. J. L. Kennedy, Chairman of the S16 Committee, worked with the Editor on a clause-by-clause basis in preparation of this Commentary.

The information contained in the Commentary is provided by the Institute. It is not to be considered the opinion of the CSA Committee nor to detract from that Committee's responsibility and authority insofar as interpretation and revision of the Standard is concerned.

This Commentary is provided by the Institute as apart of its commitment to the education of those interested in the use of steel in construction. The Institute assumes no responsibility for errors or oversights resulting from the use of the information contained herein. Anyone making use of the contents of this Commentary assumes all liability arising from such use. All suggestions for improvements of this Commentary will receive full consideration for future printings.

M. I. Gilmor

Editor

Introduction

Since 1974 when the Canadian Standards Association first introduced the limit states design standard for structural steel, S16.1-1974 "Steel Structures for Buildings—Limit States Design", the Standard has undergone a number of technical improvements but its major requirements have remained virtually unchanged. However, with the introduction of the 1989 edition, a number of more significant changes have been introduced, in part reflecting the maturing of the Standard but also the acquisiton of more detailed information on behaviour during the past 15 years.

Specifically the following changes have been adopted for CAN/CSA-S16.1-M89:

1) The words "for buildings" have been removed from the title and scope of the Standard;
2) The clause on drawings has been substantially revised;
3) Explicit clauses now reinforce the fact that strength levels reported on mill test certificates are not to be used for design calculations;
4) A clause on structural integrity has been added;
5) Consideration of possible ponding of concrete or other finishes is to be included in dead load calculations;
6) Stability effects are to be accounted for directly. Several methods are permitted in Clause 8.6. Consistent with this, effective lengths of columns in frames are taken as the actual lengths;
7) An increased minimum sway effect under gravity loads only has been stipulated;
8) The width–thickness ratio limits for webs of beams and beam–columns have been increased;
9) Net area computations reflect planes of shear rupture as well as tensile failure and now include shear lag effects;
10) Yielding in axial tension is based on the gross section while rupture of the net section is based on the ultimate tensile strength;
11) Shear on the webs of members without two flanges is addressed as is shear on connecting elements, such as shear tabs or framing angles;
12) The term, ω, which accounted for the moment gradient on an unbraced length of a beam, has been moved to the numerator in the expression for M_u, and called ω_2. The effect remains unchanged. A note has been added for cantilevers;
13) The requirements for beam–columns have been condensed into a single expression which must be evaluated under several circumstances;
14) The bearing resistance against fasteners connecting a tension member is no longer determined by edge distance. Instead, potential failure modes along effective net sections (consisting of tension and shear segments) are evaluated;
15) The term "Slip-critical" replaces "Slip-resistant" (friction) for that category of bolted joints. The number of slip categories has been reduced to the three most common cases. In addition, the formulation has changed such that the slip coefficient is identified;
16) The resistance of partial joint penetration groove welds with tension normal to the axis of the weld, has been increased;
17) The fatigue stress ranges and categories have been revised to be consistent with those for bridges in the Ontario Highway Bridge Design Code;
18) A clause dealing with copes has been added;

19) The clause on web crippling has been revised and retitled to web crippling and yielding, and the clause on stability of thin webs has been removed;

20) The clause on torsion has been revised;

21) The requirements for open-web steel joists related to uplift, bracing, chord size, attachment, and installation of steel deck have been revised;

22) For composite beams, the effective width limit of 16t + b has been removed, slab reinforcement requirements have been re-written, shear connector resistances have been revised, a provision for 40% shear connection has been included, and a clause on longitudinal shear has been added;

23) For composite HSS columns, the contribution of both the steel and the concrete to the bending resistance may be considered;

24) For built-up compression members, an equivalent slenderness ratio dependent on the type of interconnection is prescribed along with the shear force to be resisted by the interconnection;

25) High strength bolts installed only to a snug-tightened condition are permitted;

26) Conditions under which pretensioned high strength bolts are required have been revised;

27) A new Clause 27, Seismic Design Requirements, to be used in conjuction with the requirements of the 1990 edition of the National Building Code of Canada, has been added;

28) The tolerances for compression joints which depend on contact bearing, either fabricated in the shop or erected in the field, have been changed;

29) Erection tolerances have been extensively rewritten and expanded;

30) Appendix B has been shortened consistent with change (6);

31) In Appendix C, the sidesway permitted nomograph has been removed; and,

32) Appendix D has been replaced with the seismic design requirements for eccentrically braced frames.

Standards based on working (allowable) stress, with the exception of a few clauses on plastic design, required the calculation of stresses at various points in the structure and a comparison of these stresses with allowable stresses. The allowable stresses were usually established as some portion of the yield point of the material. Elements or members subjected to compression were examined for stability, usually expressed as an allowable compressive stress. A limit states standard takes a different approach. To serve their intended purposes, all building structures must meet the following requirement—to reduce the occurrence of various types of collapse or unserviceability to a sufficiently small value. Limit states are those conditions of the structure corresponding to the onset of the various types of collapse or unserviceability. The conditions associated with collapse are called ultimate limit states; those associated with unserviceability are called serviceability limit states.

In limit states design, the capacity or performance of the structure or its components is checked against the various limit states at certain load levels. For the ultimate limit states of strength and stability, for example, the structure must retain its load carrying capacity up to factored load levels. For serviceability limit states, the performance of the structure at specified load levels must be satisfactory. (Specified loads are those prescribed by the Regulatory Authority. A factored load is the product of a specified load and its load factor.) Examples of the serviceability requirement include prevention of damage to non-structural elements, and restrictions on deflections, permanent deformations, slip in slip-critical connections, and acceleration under vibratory motion.

Because both the loads acting on a structure and the resistance of a member can only be defined statistically, the "factor of safety", commonly used in working stress standards, is divided, in limit states standards, into two parts—a load factor and a resistance factor. A load factor (α) is applied to the specified load to take into account the fact that loads higher than those anticipated may exist and also to take into account approximations in the analysis of the load effects. A resistance factor (ϕ) is applied to the nominal member strengths, or resistances (R), to take into account that the resistance of the member due to variability of the material properties, dimensions and workmanship may be less than anticipated, and also to take into account the type of failure and uncertainty in the prediction of the resistance. An advantage, therefore, of limit states design is that the factors assigned to loads arising from different sources can be related to their uncertainty of prediction, and the factors assigned to different members can be related to their reliability and to the different types of failure. Thus, a greater degree of consistency against failure can be obtained (Kennedy 1974; Allen 1975; Kennedy *et al.* 1976).

For the failure of structural steel members by yielding, the resistance factor is taken to be 0.90 (Kennedy and Gad Aly 1980). To maintain simplicity in design, the resistance formulas for buckling or other types of member failure have been adjusted so that a uniform resistance factor, $\phi = 0.90$, can be used and yet provide the necessary safety required in the definition of the resistance factor. For example, the resistance formulas for tension, clauses 13.2(a)(ii) and (iii), provide a higher safety factor against fracture across the net section than against yielding in the gross section. The only exceptions to $\phi = 0.90$ are:

- Bolts in bearing-type connections, $\phi_b = 0.67$, to ensure that the connectors will be stronger than the members being joined;
- Crushing resistance of concrete (for composite construction), $\phi_c = 0.60$, which is consistent with CAN/CSA-A23.3-M89 and takes into account the greater strength variability and the brittle type of failure associated with concrete;
- Shear connectors (for composite construction), $\phi_{sc} = 0.80$; and,
- Welds, to ensure that the weld is stronger than the member joined, $\phi_w = 0.67$.

Probabilistic studies (Allen 1975) show that consistent probabilities of failure are determined for all dead-to-live load ratios when a dead load factor of 1.25 and a live load factor of 1.50 are used. For certain types of industrial structures, if there is a high degree of uncertainty in the loads, the designer may elect to use larger load factors. Those given in S16.1 (taken from the NBCC, 1990) are minimums. However, in situations where the dead load and the live loads are counteractive it is as important as ever that α_D be taken as 0.85, or less as appropriate.

Kennedy (1974) and Allen (1975) provide considerably more information on the type of probabilistic, calibration, and design studies that were performed while developing the limit states standard. The National Building Code of Canada (NBCC 1990a) contains a more extensive discussion on limit states design. Kennedy and Gad Aly (1980) and Baker and Kennedy (1984) provide information on the statistical determination of the resistance factors (ϕ).

In the Commentary clauses that follow, the numbers and headings used refer to the relevant clause numbers and headings of National Standard of Canada CAN/CSA-S16.1-M89. This will be referred to simply as S16.1-M89.

1. SCOPE AND APPLICATION

This Standard applies generally to steel structures, and structural steel components in other structures. The analysis, design, detailing, fabrication, and erection requirements contained in the Standard normally provide a satisfactory level of structural integrity for most steel structures.

Clause 1.2 states that requirements for some specific types of structures and members are given in other CSA Standards. Situations where additional requirements may be necessary are given in Clause 1.3. Commentary C of the "Commentaries on Part 4 of the National Building Code of Canada" provides numerous references to the technical literature on the topic of structural integrity.

Clause 1.4 notes that the designer has the freedom (subject to approval from the Regulatory Authority) to use methods of design or analyses in lieu of the formulas given in the Standard. It is required, of course, that the structural reliability provided by the alternative (as measured by the reliability index, for example) be equal to, or greater, than those in the Standard. An example of such a rational method would be the design of stub-girders using the method set out by Chien and Ritchie (1984) based on tests (Bjorhovde and Zimmerman 1980; Kullman and Hosain 1985; Ahamd *et al.* 1990). Clause 13.13.3 is another example. Clause 1.4 does *not* permit the substitution of any other structural steel design standard (eg. AISC or CSA S16 1969) for S16.1-M89.

2. DEFINITIONS AND SYMBOLS

2.3 Units

All coefficients appearing in equations and expressions in this Standard are consistent with forces measured in Newtons and lengths in millimetres.

3. REFERENCE PUBLICATIONS

The Standards listed are the latest editions at the time of printing. When reference is made to undated publications in specific clauses of this Standard, it is intended that the latest edition and revisions of these publications be used.

4. DRAWINGS

4.1 Design Drawings

4.1.1 Structural steel design drawings, by themselves, should show all member designations, axis orientations, and dimensions needed to describe the complete steel structure. It should not be necessary, in order to ascertain information on structural steel components, to refer to drawings produced for the use of other trades.

4.1.2 The development of adequate connections for structural members requires that the design engineer determine the shears, moments and axial forces resulting from the governing load combinations for which the connection must be designed. For complex combinations, a useful presentation of this information may be to list the maximum value of each (eg. shear, moment, and axial force), along with the values of the others which coincide with that maximum. The principle is to provide co-existant sets of forces so that free body diagrams can be identified to ensure that governing forces are transmitted through connections and panels.

4.1.4 Structural stability, a fundamental consideration of design, extends to the behaviour of elements within a member as well as to the functioning of members in total. Stabilizing components are needed to achieve both the correct local behaviour and the correct overall behaviour anticipated by the design. Therefore, the design engineer must define bracing, stiffeners, and reinforcement that are required to prevent failure from instability. An example is web reinforcement in the vicinity of moment connections.

4.2 Fabrication and Erection Documents

Although five types of documents are identified in the Standard, many structures which use pre-engineered connections from company or industry sources require only *shop details* and *erection diagrams*.

4.2.1 Connection Design Details

Connection design details which often take the form of design brief sheets typically show the configuration and details of non-standard connections developed for specific situations. They are submitted to the design engineer for review to confirm that the structural intent has been understood and met, and they may be stamped by a professional engineer when appropriate. Drafting technicians use *connection design details* to prepare *shop details*.

4.2.2 Shop Details

Shop details frequently take the form of traditional shop drawings and are used to provide the fabrication shop with all the specific information required to produce the member. They are submitted to the design engineer for review to confirm that the structural intent has been understood and met. *Shop details* are not stamped by a professional engineer because they generally do not contain original engineering.

4.2.3 Erection Diagrams

Erection diagrams convey information about the permanent structure that is required by field personnel in order to assemble it. They are submitted to the design engineer for review, but are not stamped by a professional engineer because original engineering is generally not added by the fabricator.

4.2.4 Erection Procedures

Erection procedures outline methods and equipment, such as falsework and temporary guying cables, employed by the steel erector to assemble the structure safely. They may be submitted to the design engineer for review, and may be stamped by a professional engineer when appropriate.

4.2.5 Field Work Details

Field work details are drawings which describe modifications required to fabricated members. The work may be done either in the shop or at the job site depending on circumstances. When extra material is involved, *field work details* effectively become *shop details*. They are submitted to the design engineer for review.

5. MATERIALS: STANDARDS AND IDENTIFICATION

The design requirements contained in S16.1-M89 have been developed on the assumption that the materials and products which will be used are those listed in Clause 5. These materials and products are all covered by standards prepared by the Canadian Standards Association (CSA) or the American Society for Testing and Materials (ASTM).

The standards listed provide controls over manufacture and delivery of the materials and products which are necessary to ensure that the materials and products will have the characteristics assumed when the design provisions of S16.1-M89 were prepared. The use of materials and products other than those listed is permitted, provided that approval, based on published specifications, is obtained. In this case, the designer should assure himself that the materials and products have the characteristics required to perform satisfactorily in the structure. In particular, ductility is often as important as the strength of the material. Weldability and toughness may also be required in many structures.

The values for yield and tensile strength reported on mill test reports are not to be used for design. Only the specified minimum values published in product standards and specifications may be used. This has always been implicit in the requirements of the Standard by definition of the terms F_y and F_u but is now explicitly stated. Furthermore, when tests are done to identify steel, the specified minimum values of the steel, once classified, shall be used as the basis for design.

6. DESIGN REQUIREMENTS

This clause clearly distinguishes between those requirements which must be checked using specified loads (fatigue and the serviceability limit states) and those which must be checked using factored loads (the ultimate limit states, other than fatigue). Many of the serviceability requirements (deflections, vibrations, etc.) are stipulated qualitatively and guidance, in quantitative form, is provided in Appendices. Thus, the designer is permitted to use the best information available to him in order to satisfy the serviceability requirements, but is also provided with information that the S16.1 Committee considers to be generally suitable, when used with competent engineering judgement.

6.1.2 A structural intregrity clause has been added as a reminder that measures may be necessary to guard against progressive general failure resulting from a local incident. Being inherently ductile, steel structures have generally had an excellent record of behaviour when subjected to unusual or unexpected loadings. However, connection details are particularly important in achieving this ductile behaviour. Details which rely solely on friction due to gravity to provide nominal lateral force resistance may have little or no resistance to unanticipated lateral loads if subjected to abnormal uplift conditions and should be carefully evaluated for such an eventuality or completely avoided.

6.2.1.3 Even though deflections are checked under the actions of specified loads, additional loading may result from ponding of rain on roofs, or the ponding of finishes or concrete, while in the fluid state, on floors or roofs. Such additional loads are to be included in the design of the supporting members under ultimate limit states as required by Clause 7. More information on ponding is available in the National Building Code of Canada (NBCC 1990b).

6.2.3.2 Additional useful information on vibrations of floor systems may be found in the published technical literature (Allen 1974; Murray 1975; Allen and Rainer 1976; Rainer 1980; Allen *et al.* 1985).

7. LOADS AND SAFETY CRITERION

This clause sets forth the fundamental safety criterion that must be met, namely:

Factored Resistance ≥ Effect of Factored Loads,

or

$$\phi R \geq \alpha_D D + \gamma \psi (\alpha_L L + \alpha_Q Q + \alpha_T T)$$

The expression for the *effect of factored loads* is identical with that given in Part 4 of the

National Building Code of Canada (NBCC 1990c) as are the values given for the various load factors (α), load combination factors (ψ) and importance factors (γ). This data has been included in Clause 7 for the convenience of designers using S16.1-M89 and to emphasize the fact that the load and resistance factors are inextricably linked and must be considered or developed in conjunction with one another.

Dead loads are to include the additional mass of construction materials that will be built into a structure as a result of deflections of supporting members, such as a concrete floor slab placed to a level plane but supported by members that were not cambered and that deflect under the weight of the concrete.

The *factored resistance* is given by the product ϕ R where ϕ is the resistance factor and R is the nominal member strength, or resistance. The *factored resistances* of various types of members are given in Clauses 13, 15, 16, 17, 18, 21, 26 and 27.

8. ANALYSIS OF STRUCTURE

Clause 8 permits the use of two basic types of construction—"continuous" and "simple" —both of which are defined. In recognition of previous successful practice, a special form of "simple" construction is permitted. In this form of construction, a building frame may be designed to support gravity loads on the basis of "simple" construction and to support lateral loads due to wind or earthquake through the provision of moment-resisting joints. A number of limitations are imposed in Clause 8.3.2 if this method is used. In particular, the requirement for solid webs, (8.3.2(b)), eliminates the use of open-web steel joists as connected members of the frame (Nixon 1981).

The limitations are intended to ensure that the moment-resisting joints designed nominally for wind or earthquake moments alone have both the strength and ductility necessary to accommodate the "overload" which will result if factored gravity and lateral loads act concurrently. It is assumed that, if the connection has adequate capacity for inelastic rotation when subjected to the first application of factored gravity and lateral loading, under subsequent loading cycles the connection will behave elastically, although it will have a permanent inelastic deformation (Sourochnikoff 1950, Disque 1964). Such an assumption is valid except in joints where load fluctuation would create alternating plasticity in the connection (Popov and Pinkney 1969).

Clause 8 also permits the use of the two general methods of analysis—elastic and plastic analysis. Methods of elastic analysis are familiar to most designers.

8.5 Plastic Analysis

The use of plastic analyses at the factored load levels to determine the forces and moments throughout a structure implies that the structure achieves its limiting load capacity when sufficient plastic hinges have developed to transform the frame into a mechanism. As successive plastic hinges form, the structure develops an increased load-carrying capacity above that corresponding to the formation of the initial plastic hinge until a mechanism develops. To achieve this, the members in which the hinges form before the mechanism develops must be sufficiently stocky (Class 1 sections) and well braced so that inelastic rotations can occur without loss of moment capacity.

Deflections at the specified load level are, of course, limited in accordance with Clause 6.2.1. Plastically designed structures are usually "elastic" at specified load levels so that the deflections would generally be computed on the basis of an elastic analysis.

8.5(a) Material

The plastic method relies on certain basic assumptions for its validity (ASCE 1971). This Standard, therefore, imposes restrictions to preserve the applicability of the plastic theory. The basic restriction (Clause 8.5(a)) that the steel exhibits significant amounts of strain-hardening is required to ensure that satisfactory moment redistribution will occur (Adams and Galambos 1969). This behaviour should exist at the temperatures to which the structure will be subjected in service. Also, although not explicitly stated, plastically designed structures usually entail welded fabrication, and therefore the steel specified should also be weldable. At normal temperatures all the steels referred to in Clause 5.1.2 should be satisfactory except for CSA G40.21-M, 700 Q and 700 QT steels ($F_y > 0.80\ F_u$).

8.5(b) Width–Thickness Ratios

In order to preclude premature local buckling, and thus ensure adequate hinge rotation, compression elements in regions of plastic moment must have width–thickness ratios no greater than those specified for Class 1 (plastic design) sections in Clause 11.2. Although both Class 1 and Class 2 sections can attain the fully plastic moment, only Class 1 sections will maintain this moment through the rotation necessary for redistribution of moments implicit in the plastic method of analysis.

8.5(c) Lateral Bracing.

The lateral bracing requirements are considerably more severe than those for structures designed on the basis of an elastic moment distribution because of the rotation needed at the location of the plastic hinges. Two values of the critical length between braces, L_{cr}, are specified, one for the case where the moment gradient is pronounced, the other (more stringent) for the case of uniform, or near uniform moment. The moment gradient dividing the two cases has been selected as $\kappa = -0.5$ based on the test results of Lay and Galambos (1967). Both criteria should be applied, and the more severe requirement governs the determination of L_{cr}.

Since the final hinge in the failure mechanism does not require rotation capacity, the bracing spacing limitations of this clause do not apply, and the elastic bracing requirements of Clause 13.6(a) may be used.

Lateral bracing is required to prevent both lateral movement and twisting at a braced point. Lateral bracing is usually provided by floor beams or purlins which frame into the beam to be braced. These bracing members must have adequate axial strength and axial stiffness to resist the tendency to lateral deflection. These requirements are indicated in Clause 20.3. Further information on the design of bracing members is given in Lay and Galambos (1966). When the bracing member is connected to the compression flange of the braced member, the brace should possess bending stiffness to resist twisting of the braced member. Currently, there are insufficient experimental results to indicate the magnitude of stiffness required.

A concrete slab into which the compression flange is embedded or to which the compression flange is mechanically connected, as in composite construction, or metal decks welded to the top flange of the beam in the positive moment region, generally provide sufficient restraint to lateral and torsional displacements. When the lateral brace is connected to the tension flange, provision must be made for maintaining the shape of the cross-section and for preventing lateral movement of the compression flange. This can be accomplished with either diagonal struts to the compression flange or adequately designed web stiffeners.

8.5(d) Web Crippling.

Web stiffeners are required on a member at a point of load application where a plastic hinge would form. Stiffeners are also required at beam-to-column connections where the forces developed in the beam flanges would either cripple the column web or, in the case of tension loads, distort the column flange with incipient weld fracture. The rules for stiffener design are given in Clause 21.3. See ASCE (1971) for further details of stiffeners and Fisher *et al.* (1963) for special requirements pertaining to tapered and curved haunches.

When the shear force is excessive, additional stiffening may be required to limit shear deformations. The capacity of an unreinforced web to resist shear is taken to be that related to an average shear yield stress based on the Huber-Henckey-von Mises criterion of $F_y/\sqrt{3}$. For an effective depth of the web of a rolled shape of about 95% of the section depth, Clause 13.4.1.2 gives

$$V_r = 0.95\,\phi\,wd\,F_y/\sqrt{3} = 0.55\,\phi\,wd\,F_y$$

At beam-to-column connections, when the shear force exceeds that permitted above, the excess may be carried by providing doubler plates to increase the web thickness or by providing diagonal stiffeners (Figure 2–1). The force in the beam flange that is transferred into the web as a shear is closely

$$V = M/d_b$$

Equating this to the shear resistance as given in Clause 13.4.1.2 (where now, $w = w_c$ and $d = d_c$), and solving for the required web thickness,

$$w_c \geq \frac{M}{0.55\phi\, d_c\, d_b\, F_y}$$

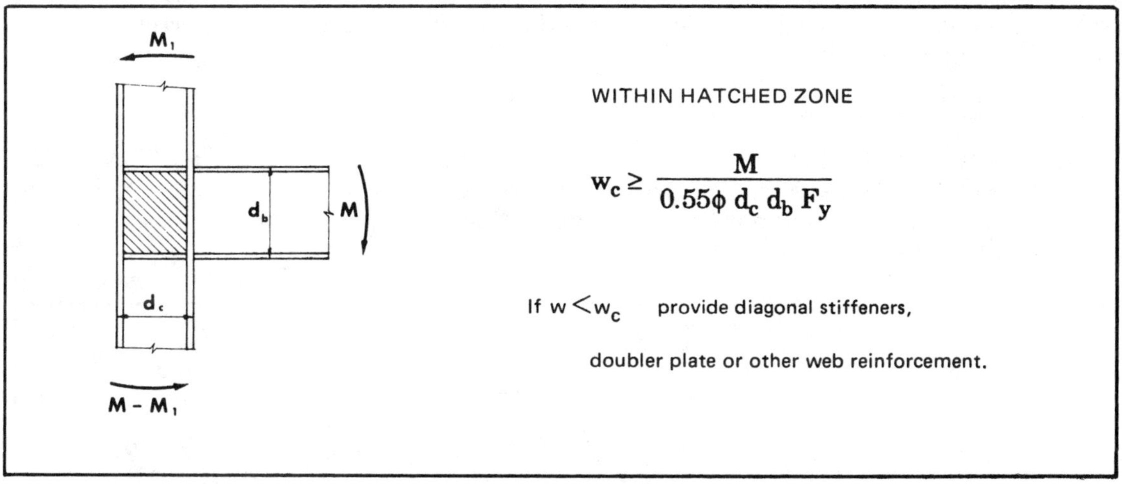

FIGURE 2–1
WEB THICKNESS AT BEAM-TO-COLUMN CONNECTIONS

If the actual web thickness is less than w_c, the required area of diagonal stiffeners may be obtained by considering the equilibrium of forces at the point where the top flange of the beam frames into the column. Using a lower bound approach, the total force to be transmitted ($V = M/d_b$) is assumed to be taken by the web and the horizontal component of the force in the diagonal stiffener:

$$V = M/d_b = 0.55\,\phi\,w_c\,d_c\,F_y + \phi\,F_y\,A_s\cos\theta$$

where

A_s = cross sectional area of diagonal stiffeners

$\theta = \tan^{-1}(d_b/d_c)$

The required stiffener area is therefore

$$A_s = \frac{1}{\cos\theta}\left[\frac{M}{\phi\, F_y\, d_b} - 0.55\, w_c\, d_c\right]$$

8.5(e) Splices

The bending moment diagram corresponding to the failure mechanism is the result of moment redistribution that occurred during the plastic hinging process. For example, points of inflection in the final bending moment distribution may have been required to resist significant moments to enable the failure mechanism to have developed (Hart and Milek 1965). To ensure that splices have sufficient capacity to enable the structure to reach its ultimate load capacity, a minimum connection requirement of 0.25 M_p is specified in Clause 8.5(e). Also, at any splice location, the moments corresponding to various factored loading conditions must be increased by 10% above the computed value. The splice is then designed either for the larger of the moments so increased or for the minimum requirement of 0.25 M_p.

8.5(f) Impact and Fatigue

The use of moment redistribution to develop the strength of the structure corresponding to a failure mechanism implies ductile behaviour. Members which may be repeatedly subjected to heavy impact and members which may be subject to fatigue should not be designed on the basis of a plastic analysis because ductile behavior cannot be anticipated under these conditions. Such members, at least for the present, are best proportioned on the basis of an elastic bending moment distribution.

8.5(g) Inelastic Deformations

For continuous beams, inelastic deformations may have a negligible effect on the strength of the structure. For other types of structures, in particular multi-storey frames, these secondary effects may have a significant influence on the strength of the structure (ASCE 1971).

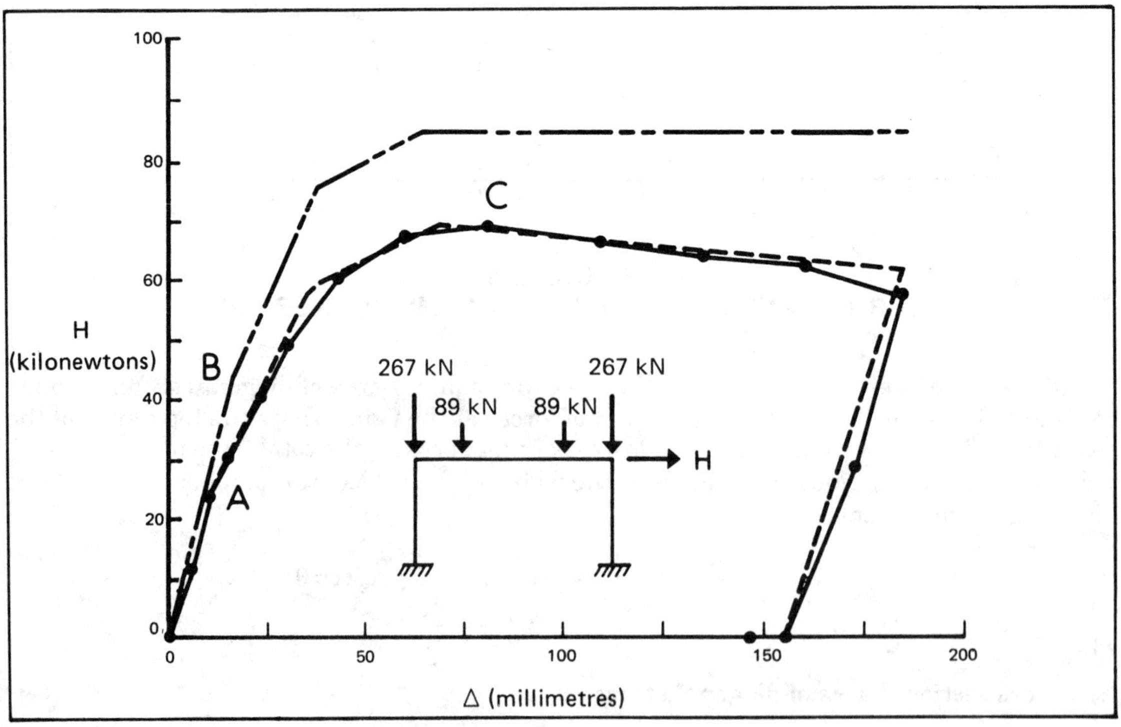

FIGURE 2–2
OBSERVED AND PREDICTED LOAD–DEFLECTION RELATIONSHIPS

In the structure shown in the inset of Figure 2–2, the secondary effects have reduced the lateral load carrying capacity (while maintaining the same vertical load) by approximately 25% (ASCE 1971; Adams 1974). The first plastic hinge formed at stage *A* in this structure while the ultimate strength (considering moment redistribution) was not attained until stage *C*. The inelastic deformations between these two stages have reduced the overall strength of the structure. Clause 8.6 requires that the sway effects produced by the vertical loads be accounted for in design. Therefore Clause 8.5(g) requires that, in a structure analyzed on the basis of a plastic moment distribution, the additional effects produced by inelastic sway deformations are accommodated. In most cases the actual strength of the structure can only be predicted by tracing the complete load–deflection relationship for the structure or for selected portions (Beedle *et al.* 1969). Methods are available to perform this type of design. For braced multi-storey frames, however, simpler techniques have also been developed (AISI 1968).

8.6 Stability Effects

Clause 8.6 recognizes that all building structures are subjected to sway deformations. The vertical loads acting through the deformed shape of the structure produce secondary bending moments, in the case of a moment resisting frame, or additional forces, in a vertical bracing system. These additional moments or forces (the stability effects) reduce the strength of the structure, as shown for a moment-resistant frame in Figure 2–2. In addition, bending moments and deflections, which exceed those predicted by a first order analysis, are produced at all stages of loading (Adams 1974). Similar effects are produced in structures containing a vertical bracing system, as shown in Figure 2–3 where the steel frame is linked to a shear wall (Adams 1974).

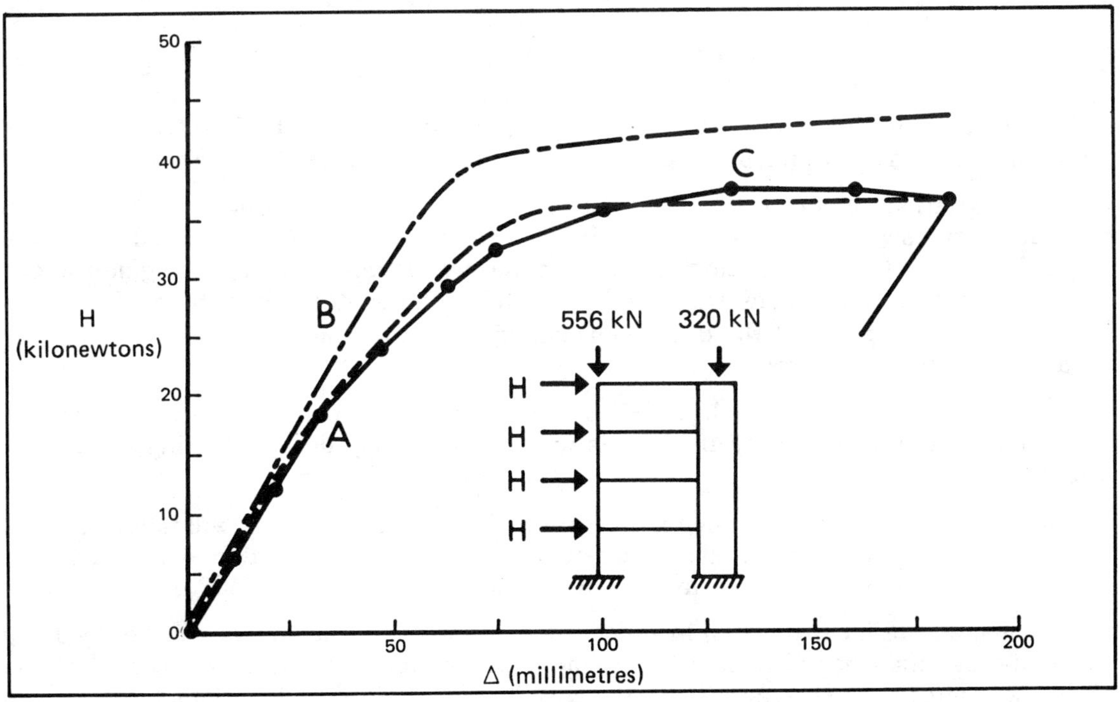

FIGURE 2–3
LOAD–DEFLECTION RELATIONSHIPS

8.6.1

In S16.1-M89, for the first time, the designer must account for the sway effects directly. This is done by (1) performing a second order geometric analysis for the moments and forces, or (2) accounting for these effects by (a) amplifying the first order elastic translational moments by the factor U_2, or (b) performing an anlysis according to Appendix J. In all cases, the lateral deflections of the structure under the translational loads must be determined. It is no longer acceptable to base the design of a moment resisting frame on a portal or cantilever type analysis to determine the moments without regard to the lateral deflections, although such methods are valuable for preliminary design.

Computer programs are available to perform analyses based on equilibrium of the deformed structure (Logcher *et al.* 1969; Galambos 1968). With this type of program, the additional moments or forces generated by the vertical loads acting on the displaced structure (the so-called PΔ effect) are taken into account. In addition, most second-order programs also account for the reduction in column stiffnesses, caused by their axial loads (Galambos 1968).

The second approach is simply to amplify the results of a first-order analysis to include the PΔ effects. It is implied in this approach that the reduction in individual member stiffnesses will be negligible. Adams (1974) suggested a simple check to ensure that this assumption is justified. The new Standard, however, limits this approach to structures for which the elastic second-order effects do not exceed 40% of the first-order effects. If the amplification factor exceeds this value, the designer may either increase the lateral stiffness of the structure to limit the amplification factor to 40%, or carry out a second-order elasto-plastic analysis.

In amplifying the elastic, first-order moments, two equivalent methods are given. In the first method, the first-order moments or forces are amplified by the factor U_2. The value of

$$U_2 = \frac{1}{1 - \frac{\Sigma C_f \Delta}{\Sigma V_f h}}$$

is, in fact, the limit of the so called PΔ/h shears as determined by the iterative procedure, the second method given in Appendix J of the Standard (Kennedy *et al.* 1990).

As a variation on the method of Appendix J, estimated deflections may be used to calculate initial sway forces (Wood *et al.* 1976). The structure is then analyzed under the lateral load caused by wind or earthquake plus the sway forces. If the resulting deflections are less than those assumed for the initial estimate of the PΔ effects then these effects have been over-estimated and (if the designer is satisfied that the situation is acceptable) the iterative process is not required.

It has been observed (Galambos 1988) that most regular structures meeting reasonable deflection limits under the specified loads will be adequately designed using the PΔ approach.

Because of its inherent simplicity, it is expected that most designers will follow the first method, using the amplification factor, U_2, to amplify the translational moments or forces. The moments and forces due to gravity loads are, of course, not amplified.

Because the designer has now included the sway effects in the analysis, either by doing a second-order analysis or by one of the two approximate methods, no further consideration of the sway effects is necessary. Thus, the checking of interaction equations for beam-columns can be legitimately based on the sway-prevented case. For this reason, sidesway-permitted nomographs have been deleted from Appendix C.

8.6.2

A structure subjected to gravity loads only may have insignificant lateral deflections, and may only fail by elastic or inelastic sway buckling. Figure 2–4 shows a frame subject to vertical loads only. As the loads are increased, the effects of the vertical loads, acting on whatever initial imperfections exist due to fabrication and erection, lead to failure through instability, much the same as for the combined loading case shown in Figures 2-2 and 2-3. To simulate this condition, Clause 8.6.2 requires that translational effects shall be introduced, acting in conjuction with the gravity loads, that are not less than those caused by pseudo-lateral loads equal to 0.005 times the factored gravity loads acting at each storey. For the case, where either the gravity loads or the structure or both are non-symmetric and the structure therefore sways sideways under vertical loads alone, the greater of the resulting translational effect and that due to the psuedo-lateral loads would govern. The psuedo-lateral loads are only considered for the gravity load case and are not considered in conjuction with lateral loading due to wind or earthquake.

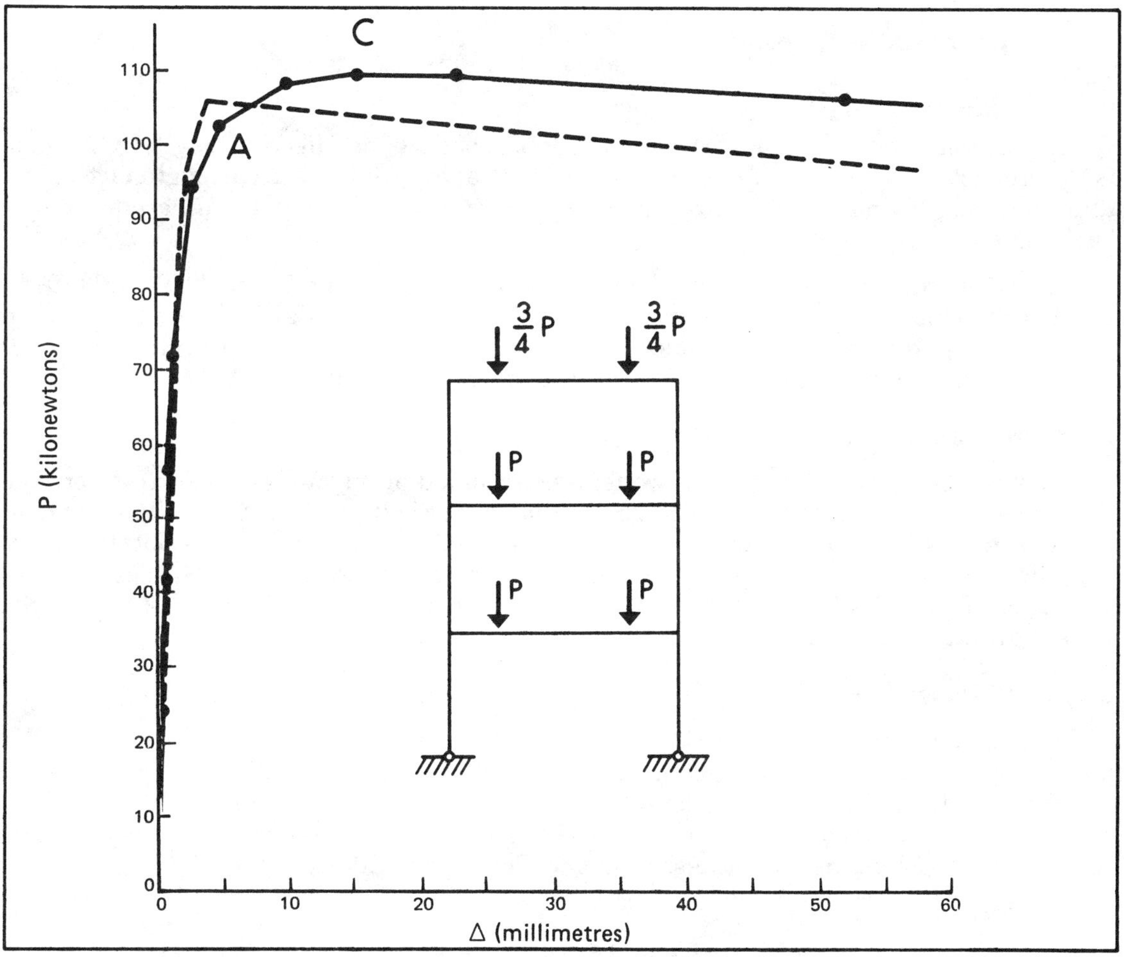

FIGURE 2–4
LOAD-DEFLECTION RELATIONSHIP—VERTICAL LOAD ONLY

Thus, the lateral instability or bifurcation problem has been replaced by a strength design check against lateral movements. Cheong Sait-Moy (1991) has shown that a ratio of the pseudo-lateral load of 0.0075 to the factored gravity load is adequate for frames with a U_2 factor of up to 3.

9. DESIGN LENGTH OF MEMBERS

9.1 Simple Span Flexural Members

For design purposes, it is usually convenient to consider the length of a member as equal to the distance between centres of gravity of supporting members. In most instances the difference resulting from considering a member to be that length rather than its actual length, centre-to-centre of end connections, is small. In some cases, however, there is sufficient difference to merit computing the actual length. Regardless of the length used for design, the actual connection detail may cause an eccentric load, or moment, to act on the supporting member and this effect must be taken into account.

9.3 Members in Compression

9.3.1 General

Compression members may be designed on the basis of an effective length (KL) which is the product of the actual unbraced length and the appropriate effective length factor. The unbraced length and the effective length factor may be different for different axes of buckling.

Information about effective lengths is given in Galambos (1988) and Tall *et al.*(1974). Further guidance is provided in Appendices B and C of the Standard. The CSP, a column selection computer program available to designers (CISC 1987), contains routines for computing effective lengths, based on the principles outlined in Appendix C.

9.3.2 In Frames

Clause 8.6 requires that the sway effects be included in the analysis used to determine the moments and forces. Therefore, having included these effects, the effective length factor in such frames is taken as 1.0. This also applies to frames in which the lateral loads, including sway effects, are resisted by direct acting bracing or shear walls. The designer has the option of using K factors less than 1.0 in this latter case if the rotational restraint at the ends show that such a value is applicable.

9.3.3 In Trusses

The effective length factor of 1.0 are to be used unless an analysis shows that a smaller value is appropriate. This approach is valid for trusses designed such that all members fail simultaneously, and conservative for others. For trusses in which some members are oversize, trusses for example with constant size chords, all members do not fail simultaneously and the effective lengths may be less than one. See Clause C8, Appendix C.

10. SLENDERNESS RATIOS

The maximum slenderness ratio of 200 for compression members, stipulated as long ago as the 1974 Standard, has been retained in S16.1-M89 for the reason that strength, or resistance, of a compression member becomes quite small as the slenderness ratio increases and the member becomes relatively inefficient.

For considerations of strength, no limiting slenderness ratio is required for a tension member and, indeed, none is applied to wire ropes and cables. However, a slenderness ratio limit of 300 is given with permission to waive this limit under specified conditions. The limit does assist in the handling of members and may help prevent flutter under oscillating loads such as those induced in wind bracing designed for tension loads only.

Members whose design is governed by earthquake loadings may be subject to more stringent slenderness ratios depending on the ductility requirements of the lateral load resisting system. See commentary on Clause 27.

WIDTH - THICKNESS RATIOS FOR COMPRESSION ELEMENTS			
Detail	**Class 1**	**Class 2**	**Class 3**
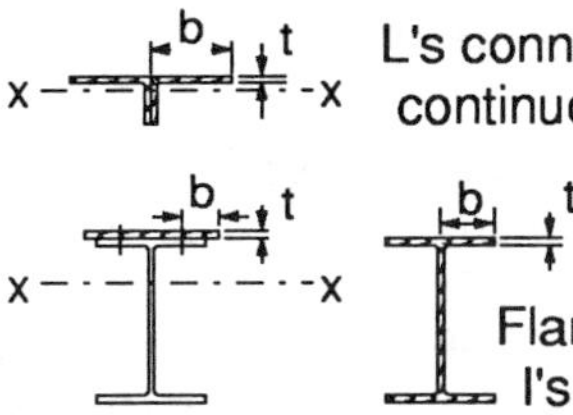 L's connected continuously; Flanges of I's or T's	$\frac{b}{t} \le \frac{145}{\sqrt{F_y}}$ ✢ ✢ 2-L's or cover PLs. symmetric abt. plane of bending - abt. x-x axis.	$\frac{b}{t} \le \frac{170}{\sqrt{F_y}}$ ✢ ✢ 2-L's or cover PLs. symmetric abt. plane of bending - abt. x-x axis.	$\frac{b}{t} \le \frac{200}{\sqrt{F_y}}$
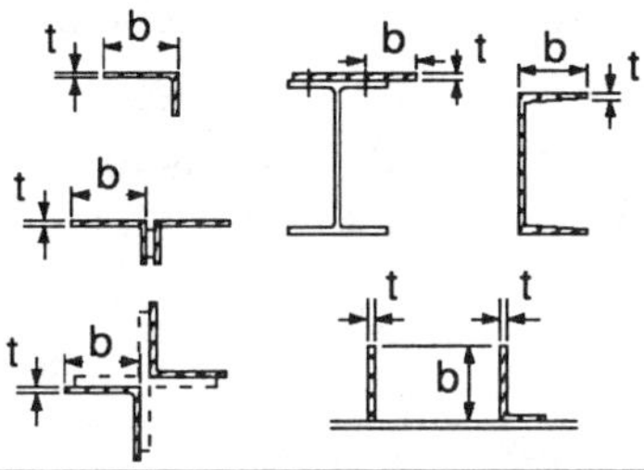	——	——	$\frac{b}{t} \le \frac{200}{\sqrt{F_y}}$ L's not continuously connected, flange of C's, asymmetric cover PL, plate gdr. stiffeners.
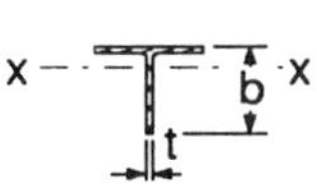 Stems of T's; ♣ symmetric abt. plane of bending, or incl. asymmetry effect in analysis.	$\frac{b}{t} \le \frac{145}{\sqrt{F_y}}$ ♣	$\frac{b}{t} \le \frac{170}{\sqrt{F_y}}$ ♣	$\frac{b}{t} \le \frac{340}{\sqrt{F_y}}$
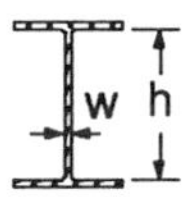	bending only $\frac{h}{w} \le \frac{1100}{\sqrt{F_y}}$ axial compression $\frac{h}{w} \le \frac{670}{\sqrt{F_y}}$	bending only $\frac{h}{w} \le \frac{1700}{\sqrt{F_y}}$ axial compression $\frac{h}{w} \le \frac{670}{\sqrt{F_y}}$	bending only $\frac{h}{w} \le \frac{1900}{\sqrt{F_y}}$ axial compression $\frac{h}{w} \le \frac{670}{\sqrt{F_y}}$
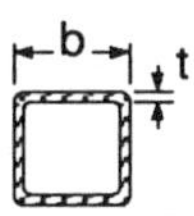 HSS	$\frac{h}{w} \le \frac{420}{\sqrt{F_y}}$	$\frac{h}{w} \le \frac{525}{\sqrt{F_y}}$	$\frac{h}{w} \le \frac{670}{\sqrt{F_y}}$
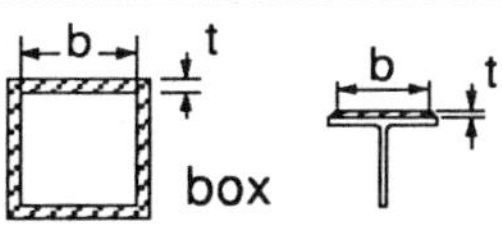 box	$\frac{h}{w} \le \frac{525}{\sqrt{F_y}}$	$\frac{h}{w} \le \frac{525}{\sqrt{F_y}}$	$\frac{h}{w} \le \frac{670}{\sqrt{F_y}}$
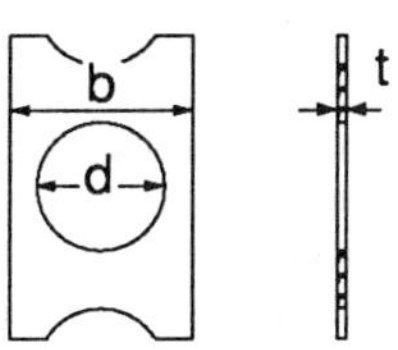	——	——	$\frac{b}{t} \le \frac{840}{\sqrt{F_y}}$
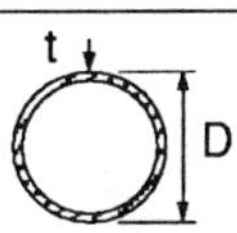	$\frac{D}{t} \le \frac{13\,000}{F_y}$	$\frac{D}{t} \le \frac{18\,000}{F_y}$	$\frac{D}{t} \le \frac{23\,000}{F_y}$

FIGURE 2–5
WIDTH–THICKNESS RATIOS FOR COMPRESSION ELEMENTS

11. WIDTH–THICKNESS RATIOS: ELEMENTS IN COMPRESSION

Clause 11.1 identifies four categories of cross-sections, Class 1 through Class 4, based upon the width-to-thickness ratios of the elements of the cross-section in compression. With the ratios given in Table 1 of Clause 11 for Classes 1, 2, or 3, the respective ultimate limit states will be attained prior to local buckling of the plate elements. These ultimate limit states are: Class 1—maintenance of the plastic moment capacity (beams), or the plastic moment capacity reduced for the presence of axial load (beam-columns), through sufficient rotation to fulfill the assumption of plastic analysis; Class 2—attainment of the plastic moment capacity for beams, and the reduced plastic moment capacity for beam-columns, but with no requirement for rotational capacity; Class 3—attainment of the yield moment for beams, or the yield moment reduced for the presence of axial load for beam–columns. Class 4—have plate elements which buckle locally before the yield strength is reached.

The distinction based on moment capacity does not apply to axially loaded members. Thus, Table 1 simply shows the same limit for each of Classes 1, 2, and 3 for webs in axial compression. The flange limit for columns should be taken as that for Class 3 beam flanges, that is, $200/\sqrt{F_y}$, (Dawe and Kulak 1984).

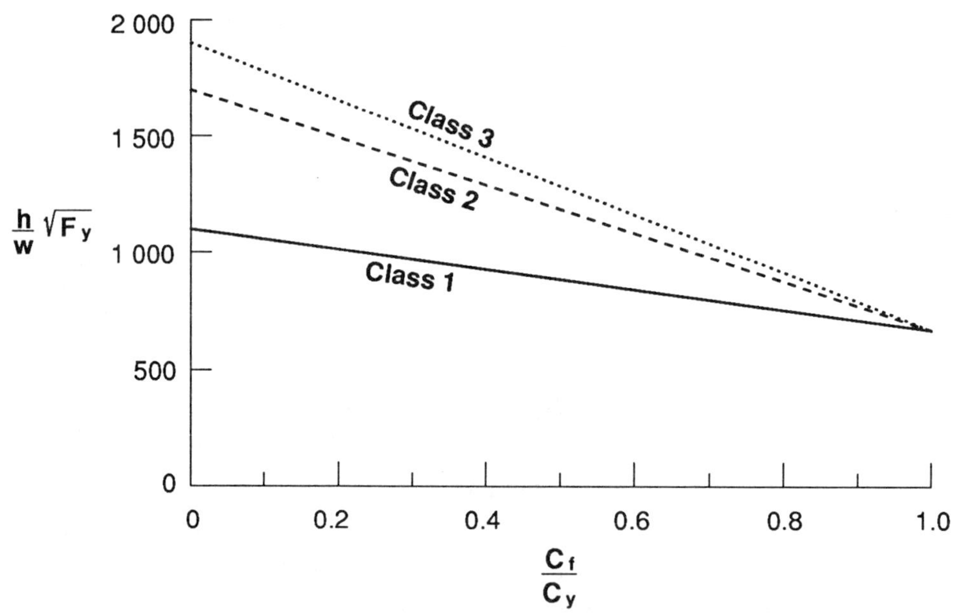

FIGURE 2–6
WEB SLENDERNESS RATIOS

Figure 2–5 illustrates the requirements for Classes 1, 2 and 3 sections.

The bases of the requirements, particularly those for W-shapes, come from a background of both experimental and theoretical studies. For example, the restrictions on flanges have both a theoretical basis (Kulak *et al.* 1990; ASCE 1971; Galambos 1988) and an extensive experimental background (Haaijer and Thurlimann 1958; Lay 1965; Lukey and Adams 1969). The restrictions for webs in flexural compression come from both theory and tests on Class 1 sections (Haaijer and Thurlimann 1958) but mostly from test results for Class 2 and 3 sections (Holtz and Kulak 1973 and 1975).

In Figure 2–6, the requirements are plotted for webs under axial compression and compression due to bending. Because the amount of web under compression varies from complete (columns) to one-half (beams), the depth-to-thickness limits vary as a function of the amount of axial load. The results presented here reflect the latest research results, particularly those of Dawe and Kulak (1986), and are significantly more liberal than previous limits.

Sections used for columns, beams, or beam–columns may be composed of elements whose width-to-thickness ratios exceed those prescribed for Class 3 provided that the resistance equations are adjusted accordingly. These sections, called Class 4, should be evaluated according to the rules given in Clause 13.

The requirement for Class 3 circular hollow sections, $D/t \leq 23\,000/\sqrt{F_y}$, is based on tests (Winter 1970) which indicates that tubes that satisfy this requirement can reach yield stress without local buckling. The more conservative requirements for Class 1 and Class 2 sections are based on research referred to in Stelco (1973).

12. GROSS AND NET AREAS

12.1 Application

The design and behaviour of tension members is integrally related to the proportioning and detailing of connections. Consequently, Clauses 12 and 13.2 are related. The two potential failure modes for tension members are yielding on the gross section and fracture through a net section.

12.3 Effective Net Area

This clause defines areas used to determine tension member resistances and incorporates the results of new research. The requirements apply to both bolted and welded connections.

12.3.1 The fracture path may consist of a series of segments, some loaded primarily in tension, including inclined paths, others primarily in shear. The critical path must be identified and evaluated. By determining the effective net area by summing the net area of each segment along the critical path, it is assumed, as has been demonstrated, (Birkemoe and Gilmor 1978; Ricles and Yura 1983; Hardash and Bjorhovde, 1985) that all segments fracture simultaneously.

The net area of segments normal to the force, in direct tension, and segments inclined to the force are determined as in previous standards. For the latter, the net area is increased by $s^2/4g$ for each inclined segment. Segments parallel to the force are in shear.

The Standard, therefore, now considers the failure mode of block tear-out that occurs, for example, when a block of material at the end of a tension member tears out through the fastener holes, as shown in Figure 2-7. According to Birkemoe and Gilmor (1978), the strength of the failure surface in shear is taken as 0.60 of that in tension. Thus, the load at which tear-out occurs is equal to the ultimate tensile strength times the net area in tension (including the $s^2/4g$ allowance for staggered holes) plus 0.60 times the ultimate tensile strength times the net area in shear. This load is equal to ultimate tensile strength times the net area in tension plus 0.60 of the area in shear.

Block tear-out may also occur at the end of a coped beam as shown in Figure 2-8. Here the applied force is the end shear on the beam. This force causes tension on horizontal planes and shear on vertical ones. Tests by Ricles and Yura (1983) on coped beams indicate that the Birkemoe-Gilmor model derived originally for this case is somewhat liberal, but those

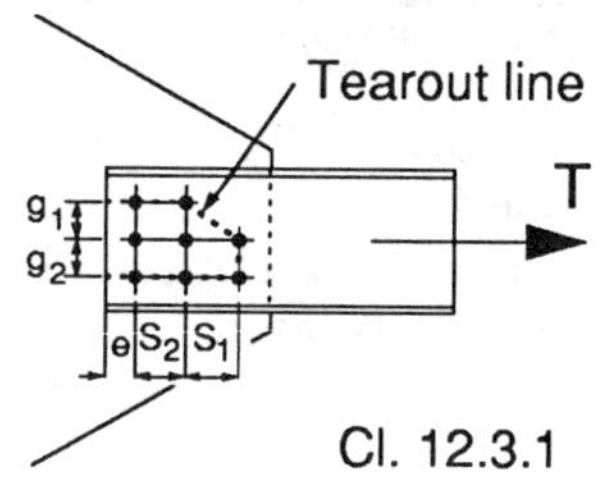

FIGURE 2-7
BLOCK TEAR-OUT AT END OF TENSION MEMBER

of Hardash and Bjorhovde (1985) on gusset plate connections indicate the reverse when the net areas in shear and tension are used.

The tear-out resistance, based on these net areas and the requirements of Clause 13.2, is therefore

$$0.85\ \phi\ F_u\ t\left[(g_2 - d) + \left(g_1 - d + \frac{s_1^2}{4\ g_1}\right) + 0.60\ (2s_2 + s_1 + 2e - 4d)\right]$$

for the paths shown in Figure 2-7 and

$$0.85\ \phi\ F_u\ t\left[(s + e_2 - 1.5d) + 0.60(2g + e_1 - 2.5d)\right]$$

for the paths shown in Figure 2-8.

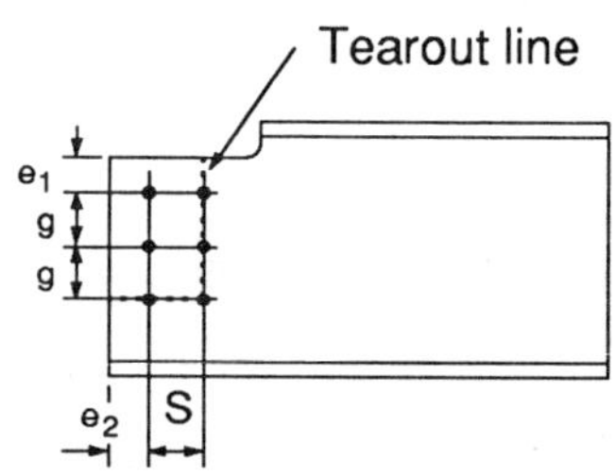

FIGURE 2-8
BLOCK TEAR-OUT AT END OF COPED BEAM

12.3.2 This allows for distortion or local material damage that may occur in forming the hole.

Clause 12.3.4 of S16.1-M84 limiting net areas, depending on the ratio of F_y/F_u, has been deleted because the new requirements, recognizing yielding on the gross section as a failure mode, greatly restricted its applicability.

12.3.3 Effective Net Area Reduction – Shear Lag When the critical net section fracture path crosses unconnected cross-sectional elements, the directly connected elements tend to reach their ultimate strength before the complete net section strength is reached due to shear lag. When all cross-sectional elements are directly connected, shear lag does not occur and the effective net area is the total net area.

The loss in efficiency due to shear lag can be expressed as a reduction in the net area. Munse and Chesson (1963) suggested that this reduction could be taken as $1 - \bar{x}/L$ where $\bar{x}$ is the distance from the shear plane to the centroid of that portion of the cross-section being developed and L is the connected length.

Because the connected length is usually not known at the time of tension member design, reduction factors have been derived for specific cases, as given in Clause 12.3.3.2, based on an extensive examination of the results of over 1000 tests (Kulak *et al* 1987). The reduction factor depends on the cross-sectional shape and the number of bolts (2, 3 or more) in the direction of the tensile load.

Similar behaviour has been observed in welded connections (Kulak *et al* 1987) when welds parallel to the tensile load in the member only are used. When the weld length is less than the distance between welds, it is likely that the weld is critical.

When block tear-out occurs in those elements which are directly connected, shear lag is not a factor. Shear lag is only considered when a potential failure path crosses unconnected elements.

12.3.3.5 The direct use of the Munse and Chesson reduction factor, $1 - \bar{x}/L$, is permitted by this clause. Figure 2-9 gives examples of dimensions used in calculating this shear lag reduction factor.

12.4 Pin-Connected Tension Members

The net area increase across the pin hole is to allow for the significant non-uniform distribution of stress (Johnston 1939). The minimum net area within the 45° arc each side of the longitudinal axis is to avoid end splitting.

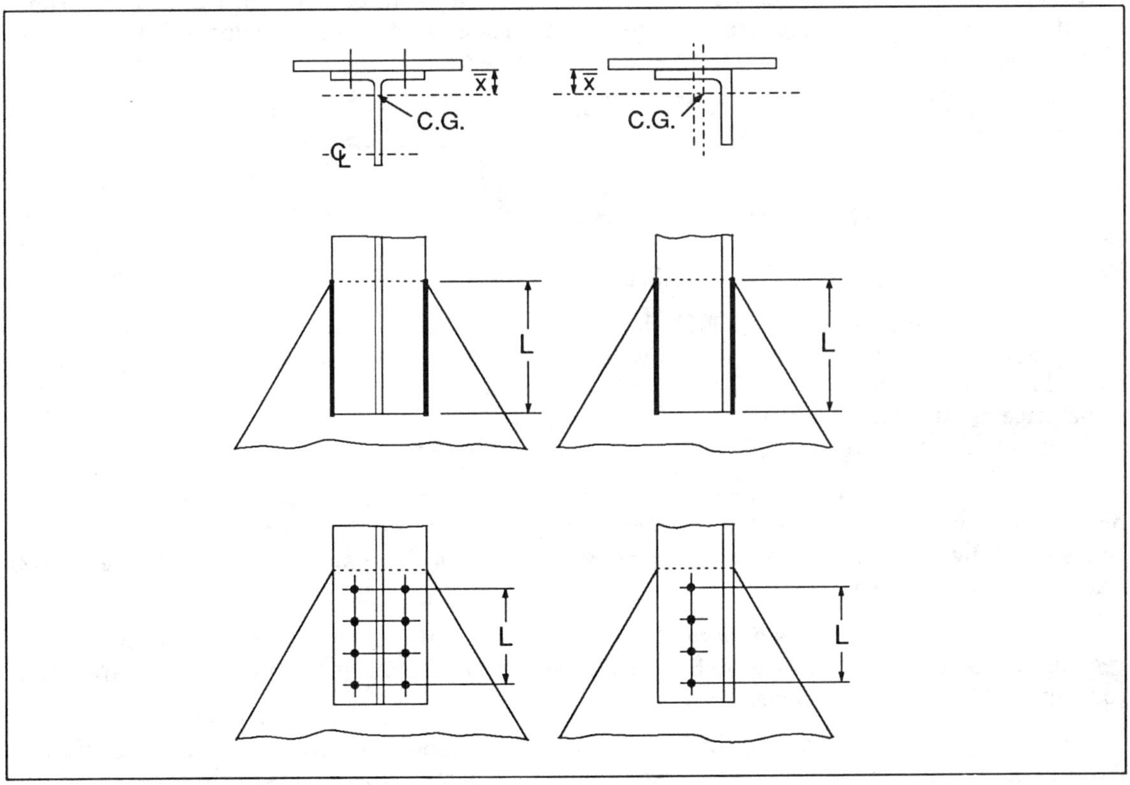

FIGURE 2-9
DIMENSIONS USED FOR SHEAR LAG CALCULATIONS

13. MEMBER AND CONNECTION RESISTANCE

13.1 General

The value of ϕ of 0.90 provides consistent and adequate values of the reliability index when used with the load factors of Clause 7.2 (Kennedy and Gad Aly 1980; Baker and Kennedy 1984).

13.2 Axial Tension

The two potential failure modes for tension members are yielding of the gross section and fracture of the net section. See Clause 12 for net section area calculations. Yielding on the gross section is a ductile mode of failure for which a resistance factor of 0.90 is appropriate. Because net section fracture occurs with little deformation and therefore little warning of failure, the resistance factor is effectively reduced by multiplying by 0.85. This philosophy is consistent with the reduced resistance factor used for connectors (bolts, welds, and shear connectors).

The reduction factor of 0.75 for pin connectors recognizes the greater non-uniformity of stress that occurs around a hole that is large relative to the material in which it is formed.

13.3 Axial Compression

Steel columns are conveniently classified as short, intermediate, or long members, and each category has an associated characteristic type of behaviour. A short column is one which can resist a load equal to the yield load ($C_y = AF_y$). A long column fails by elastic buckling. The maximum load depends only on the bending stiffness (EI) and length of the member. Columns in the intermediate range are most common in steel buildings. Failure is characterized by inelastic buckling and is greatly influenced by the magnitude and pattern of residual stresses that are present and the magnitude and shape of the initial imperfections or out-of-straightness. These effects lessen for both shorter and longer columns. The expressions in this clause account for these effects (Bjorhovde, 1972).

Figure 2–10 indicates the variations in strengths for three different set of columns with the same out-of-straightness patterns and different residual stress.

The resistance expressions of Clause 13.3.1 are based on Column Curve 2 of the Structural Stability Research Council (Galambos 1988) for W-shapes, and is used as the basis for the description of resistance of W-shapes and for cold-formed non-stress relieved, Class H, hollow structural sections (Clause 13.3.1), based on Bjorhovde and Birkemoe (1979).

Chernenko and Kennedy (1991) show that in the intermediate range, in particular, these expressions are quite conservative for welded H-shapes, such as WWF produced in Canada, when the flange edges have been flame cut. Because the expressions for column resistance are for x-axis or y-axis buckling only, singly symmetric, asymmetric, or cruciform sections should also be checked for lateral-torsional buckling.

Because of a more favourable residual stress pattern and out-of-straightness, hot-formed or cold-formed stress relieved hollow structural shapes (Class H) are assigned to the resistance given by Column Curve 1 of the Structural Stability Research Council (Clause 13.3.2) (Kennedy and Gad Aly, 1980).

For many heavy sections and welded sections fabricated from universal mill plate, it is recommended that Column Curve 3 be used, with resistance less than those corresponding to Column Curve 2 (Galambos, 1988).

Because the column strengths are influenced by the magnitude and distribution of residual stresses, care should be exercised in the use of the expressions of this Standard, for example, in determining the capacity of an existing column which is being reinforced in such a manner that there is an increase in compressive residual stresses in the fibres most remote

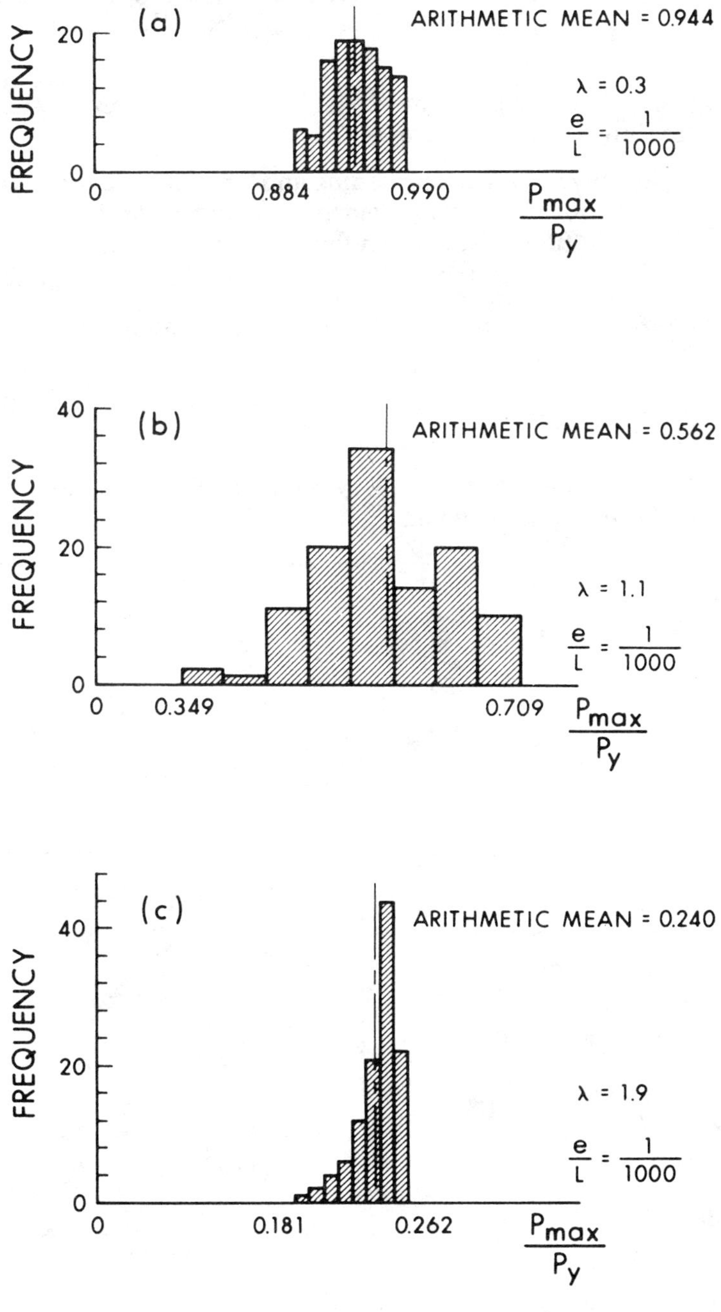

FIGURE 2-10
TYPICAL FREQUENCY DISTRIBUTION HISTOGRAMS FOR THE MAXIMUM STRENGTH OF 112 COLUMN CURVES (e/L = 1/1000)

from the centroid. In such a situation, adding material to reduce the slenderness ratio may be advantageous.

13.4 Shear

13.4.1.1 Elastic Analysis

The expressions for shear strength are given for stiffened plate girders; unstiffened plate girders and rolled beams are simply special cases for which the shear buckling coefficient, $k_v = 5.34$.

The four ranges of resistance based on Basler (1961), correspond to the following modes of behaviour and are illustrated in Figure 2–11:

(a) Full yielding followed by strain hardening and large deformation. The limiting stress of $0.66\ F_y$ corresponds to excessive deformation and is higher than that derived from von Mises criterion ($0.577\ F_y$), which forms the basis of Clause 13.4.1.2 for plastic analysis. The limit is consistent with the previous allowable stress of $0.4\ F_y$;

(b) A transition curve between strain hardening and inelastic buckling at full shear yielding. ($F_s = 0.577 F_y$);

(c) Inelastic buckling, F_{cri}, accompanied by post-buckling strength, F_t, due to tension field action, if the web is stiffened; and,

(d) Elastic buckling, F_{cre}, accompanied by post-buckling strength, F_t, due to tension field action, if the web is stiffened.

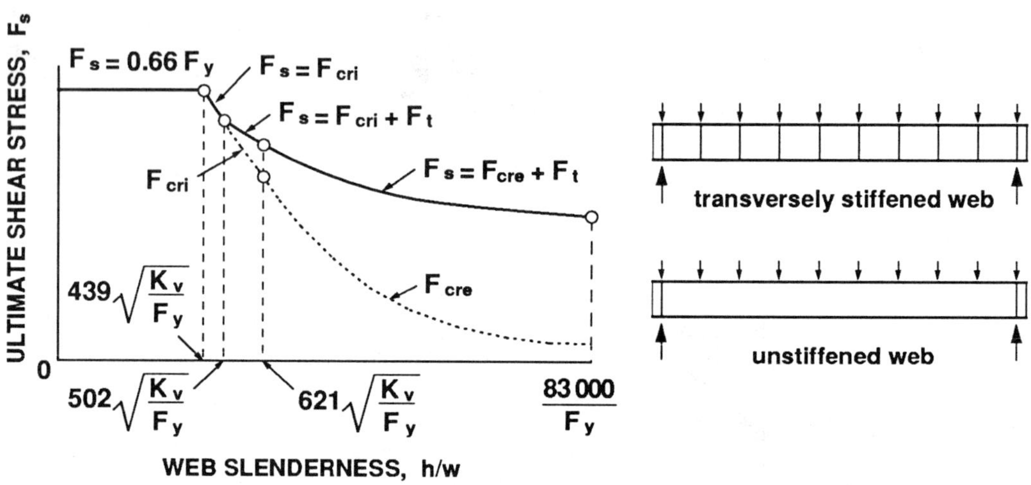

$$F_{cri} = \frac{290\sqrt{F_y k_v}}{h/w} \qquad k_v = 4 + \frac{5.34}{(a/h)^2} \text{ when } \frac{a}{h} < 1$$

$$F_{cre} = \frac{180\,000\,k_v}{(h/w)^2} \qquad k_v = 5.34 + \frac{4}{(a/h)^2} \text{ when } \frac{a}{h} \geq 1$$

F_t - See Cl. 13.4.1.1(c) and (d).

FIGURE 2-11
ULTIMATE SHEAR STRESS — WEBS OF FLEXURAL MEMBERS

For ranges (c) and (d), the expressions have been reformulated to separate buckling and tension field contributions.

In computing the shear resistance, it is assumed that the shear stress is distributed uniformly over the depth of the web. The web area (A_w) is the product of web thickness (w) and web depth (h) except for rolled shapes where it is customary to use the overall beam depth (d) in place of the web depth (h).

13.4.1.2 Plastic Analysis

For structures analysed plastically, high shears and moments may occur simultaneously at a hinge location. For this reason, the maximum shear stress is limited to the von Mises value, and when 95% of the depth of the web is assumed to be effective, the numerical coefficient 0.55 is obtained. Experimental evidence shows that this shear can be carried without reducing the flexural resistance below M_p (Yang and Beedle 1951).

13.4.1.3 Maximum Slenderness

This limit prevents the web from buckling under the action of the vertical components of the flange force arising as a result of the curvature of the girder, (Kulak *et al*, 1990).

13.4.1.4 Combined Shear and Moment in Girders

This requirement recognizes the limit state of the web yielding by the combined action of flexural stress and the post buckling components of the tension field development in the web near the flange (Basler, 1961b).

Figure 2–12 illustrates the interaction formula provided in Clause 13.4.1.4.

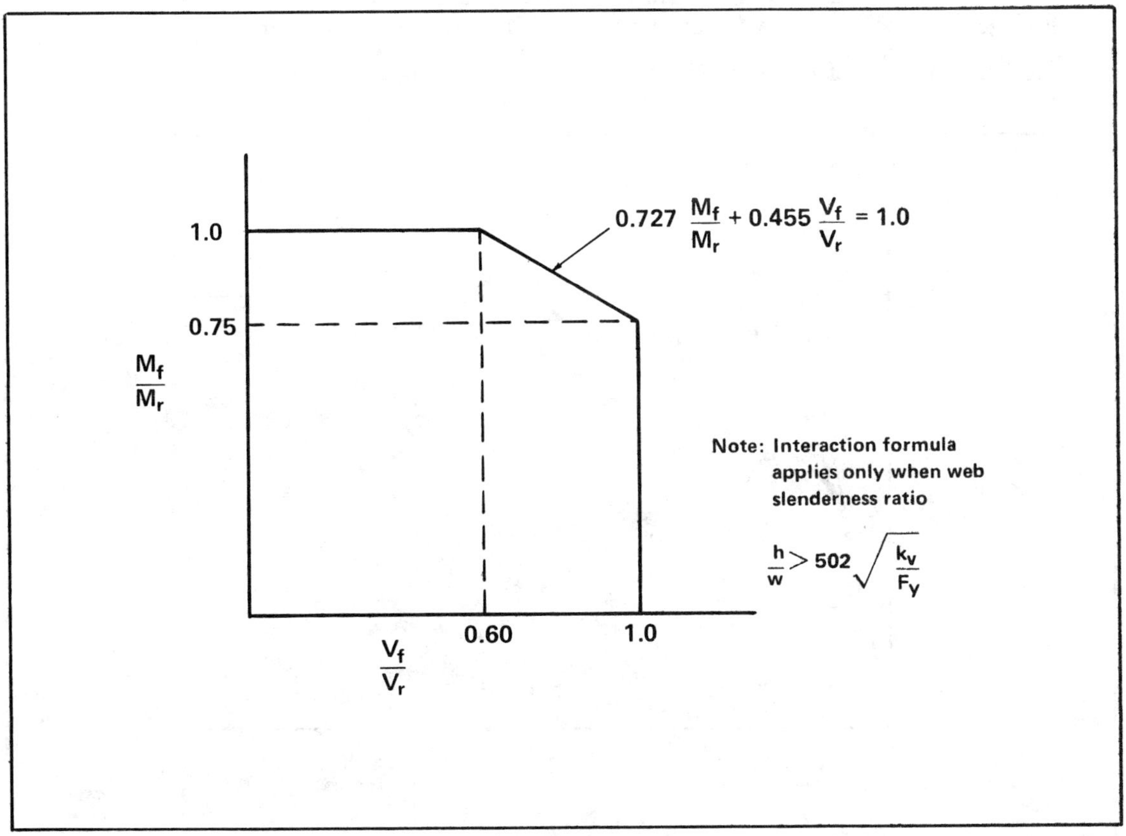

FIGURE 2-12
COMBINED SHEAR AND MOMENT IN WEBS OF TRANSVERSLY STIFFENED THIN WEB PLATE GIRDERS

13.4.2 Webs Of Flexural Members Not Having Two Flanges

When cross-sections do not have two flanges, the shear stress distribution can no longer be assumed to be uniform. The maximum shear stress can be based on strain-hardening provided shear buckling does not occur. (Not applicable to framed webs of beams - use Clause 13.4.4.)

13.4.3 Gusset Plates

This clause applies where gusset plate failure is the result of block tear-out, (Hardash and Bjorhovde 1985).

13.4.4 Connecting Elements

This clause applies when the elements are loaded primarily in shear and is somewhat conservative when compared with segments in shear in a block tear-out mode.

13.5 Bending: Laterally Supported Members

The factored moment resistances are consistent with the classification of cross-sections given in Clause 11, as illustrated by moment-rotation curves given in Figure 2-13.

The fully plastic moment, M_p, attained by Class 1 and 2 sections, implies that all fibres of the section are completely yielded but neglects the additional resistance developed due to strain-hardening.

The stress distribution for Class 3 sections is linear with a maximum stress equal to the yield stress.

Class 4 sections reach their maximum moment resistance when a plate element (flange or web) buckles locally. Class 4 sections are divided into three categories.

The first consists of those sections with Class 4 flanges and webs. This type of section

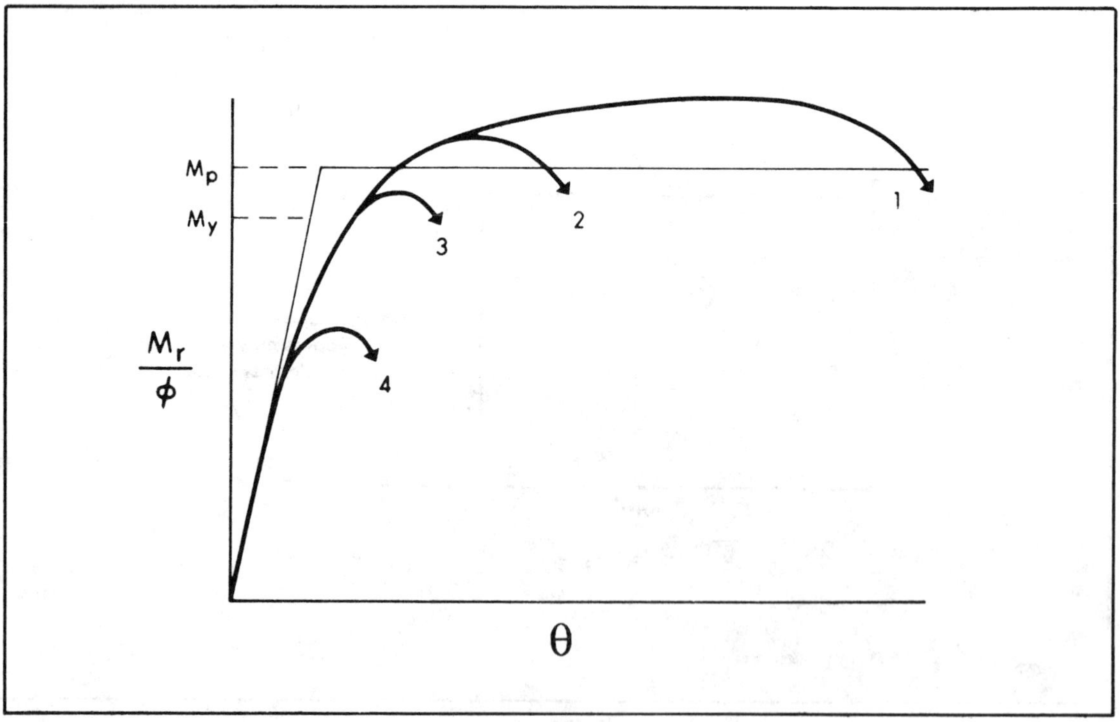

FIGURE 2-13
MOMENT-ROTATION CURVES

is designed to the requirements of CSA Standard S136 (CSA, 1989) using the material properties appropriate to the structural steel specified.

The second category consists of those sections with Class 3 flanges and Class 4 webs. Clause 13.5(c)(ii) requires that these sections are designed in accordance with Clause 15.

The third category, with Class 4 flanges and Class 3 webs, has moment resistances governed by local buckling of the compression elements as is given by CSA Standard CAN/CSA-S136. Alternatively, the designer can ignore the portion of the flange in excess of the b/t limit and compute a reduced, effective section modulus (Kalyanaraman *et al.*, 1977).

13.6 Bending: Laterally Unsupported Members

Laterally unsupported beams may fail by lateral torsional buckling at applied moments significantly less than the full cross-section strength (M_p or M_y). Even when the top flange is laterally supported, under some circumstances, for example, a roof beam subject to uplift, the laterally unsupported bottom flange may be in compression.

Two factors contribute to the lateral torsional buckling resistance, the St. Venant torsional resistance (related to the product GJ) developed by the shear stress in the cross-section and the warping resistance (related to C_w) developed by the cross-bending of the flanges.

Analogous to columns, the lateral torsional moment resistance depends on the unsupported length. Beams may be considered to be of short, intermediate, or long length depending on whether their moment resistance is their full strength, the inelastic lateral torsional buckling strength, or the elastic lateral torsional buckling strength, respectively, as shown in Figure 2-14, for Class 1 and 2 shapes. The curve for Class 3 sections is similar except that the maximum moment resistance is ϕM_y and for Class 4 sections, the maximum resistance is based on local buckling considerations.

Without the factor ω_2, the expression for M_u is that for a doubly symmetric beam subject to uniform moment. The factor ω_2, when expressed as a ratio of the end moments, κ, and

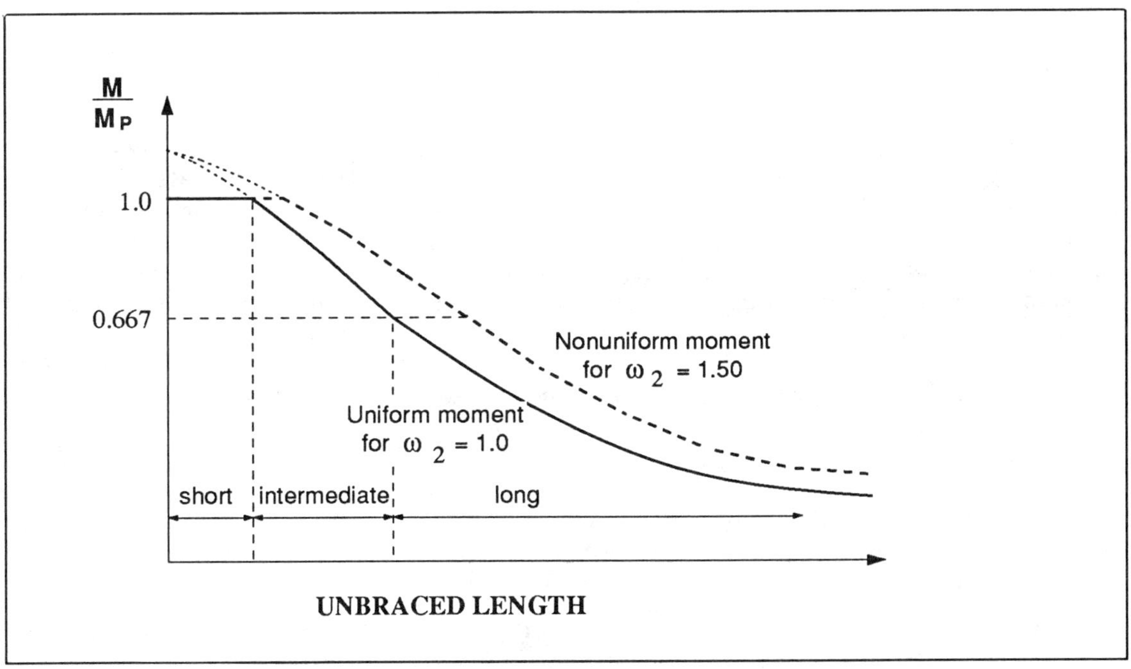

FIGURE 2-14
VARIATION OF UNIFORM AND NON-UNIFORM MOMENT RESISTANCE

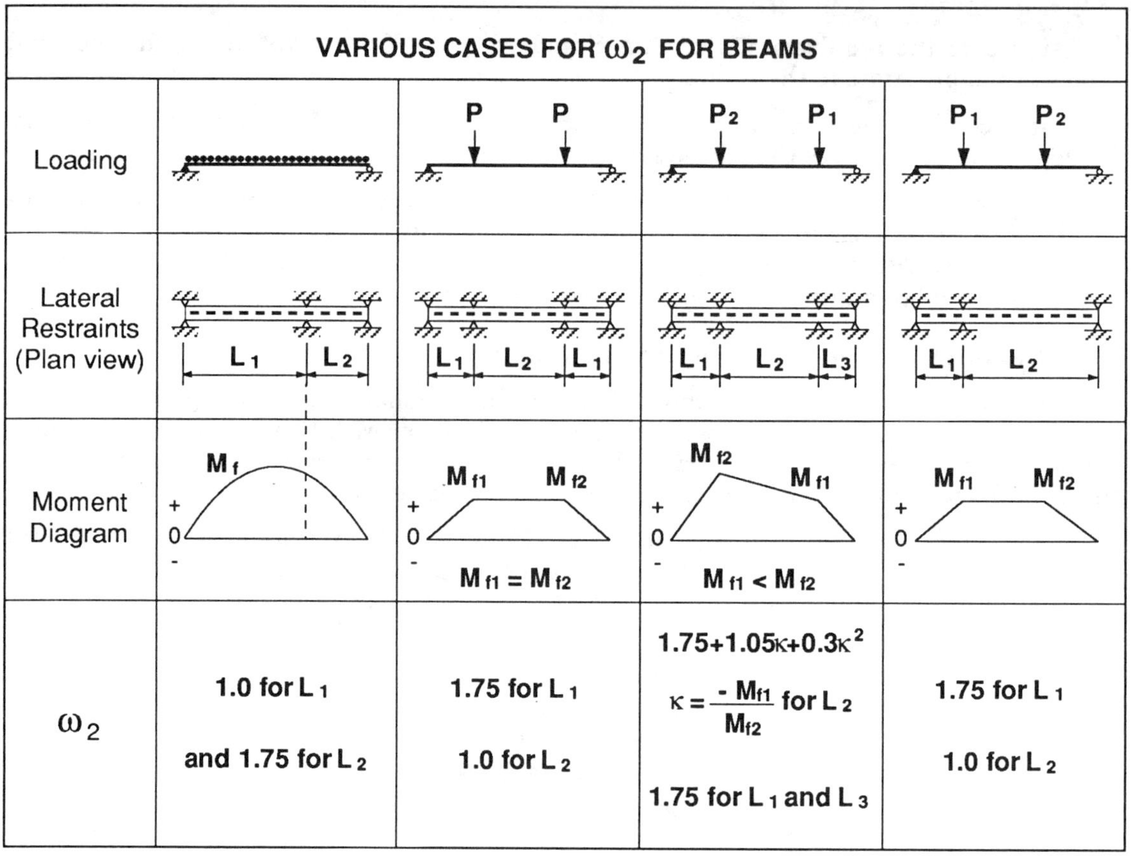

FIGURE 2-15
VARIOUS CASES OF ω_2 FOR BEAMS

having values ranging from 1.0 to 2.5, takes into account the fact that a varying moment is less critical than a uniform moment. The use of $\omega_2 = 1.0$ applied to the maximum moment in an unbraced length is somewhat conservative. Figure 2-15 illustrates ω_2 for beams with various loading and lateral constraint conditions. Also plotted on Figure 2-14 is the moment resistance for a beam for which $\omega_2 = 1.5$. It is seen that in the elastic region ($M_r < \frac{2}{3} M_p$) the full value of ω_2 is realized. In the inelastic region, the increase in M_r due to non-uniform moments gradually decreases to zero as the moment approaches ϕM_p. For laterally unsupported Class 2 and 3 beams, Baker and Kennedy (1984) give resistance factors ranging from 0.93 to 0.96 and therefore the value of 0.90 used in this Standard is conservative. The length, L, is generally taken as the distance between lateral supports. When beams are continuous through a series of lateral supports, interaction buckling (Trahair, 1968) occurs. The segment which tends to buckle laterally first is restrained by the ajoining segments (Nethercot and Trahair, 1976a, 1976b). Kirby and Nethercot (1978) and Schmitke and Kennedy (1985) give methods of computing effective lengths under these circumstances.

Significant information on lateral torsional buckling is summarized in Chen and Lui (1987).

The expression for M_u assumes that the beam is loaded at the shear centre. For other positions of the load, for monosymmetric beams, cantilever beams, unusual loading, and other support conditions, the SSRC Guide (Galambos 1988) may be consulted.

For structural systems utilizing the gerber girder system of cantilever girders and suspended spans Chien (1989) contains additional information.

13.7 Lateral Bracing for Members in Structures Analysed Plastically

This item has been covered in the commentary on Clause 8.5(c).

13.8 Axial Compression and Bending

For a general discussion of all aspects of these provisions and worked examples, see Kennedy *et al* (1990).

In S16.1-M89 the term "I-shaped" means any doubly-symmetric shape having two flanges and one web. This includes, for instance, both W-shapes and S-shapes.

The requirements in S16.1-M89 for beam-columns have been extensively revised from those in S16.1-M84. However, in most cases the resulting beam-column size is not significantly changed, if at all. The form of the interaction expressions has been retained but the meaning and usage of some of the terms have changed. The value each term takes is prescribed in the three sub-clauses (a), (b), and (c) depending on the particular mode of failure: cross-sectional strength, overall member strength, and lateral torsional buckling strength, respectively. Clause 13.8.1 is applicable to all classes of section while Clause 13.8.2 is applicable only to Class 1 sections of I-shaped members.

The interaction expressions must account for the following:

- a laterally supported member fails when it reaches its in-plane moment strength reduced for the presence of axial load;
- a laterally unsupported member may fail by lateral torsional buckling or a combination of weak axis buckling and lateral buckling;
- a relatively short member can reach its full cross-sectional strength whether it is laterally supported or not;
- when subjected to axial load only, the axial compressive resistance, C_r, depends on the maximum slenderness ratio;
- members bent about the weak axis do not exhibit out-of-plane behaviour,
- a constant moment has the most severe effect on in-plane behaviour. Other moment diagrams can be replaced by equivalent moment diagrams of reduced but uniform intensity;
- a constant moment has the most severe effect on the lateral torsional buckling behaviour. (See commentary on Clause 13.6). This effect disappears if the member is short enough, in which case, cross-sectional strength controls; and
- moments may be amplified by axial loads increasing the deflections.

Four modes of failure are addressed, as follows:

1) local bucking of an element

According to Clause 11.2 of the Standard, the compression elements of the cross-section must be proportioned such that local buckling does not occur prior to attainment of the cross-sectional strength. (See Commentary for Clause 11).

2) strength of the cross-section

As in previous standards, the cross-sectional strength of a shape used as a beam-column cannot be exceeded. Clause 13.8.2(a) gives the cross-sectional strength requirements for Class 1 sections of I-shaped members and Clause 13.8.1(a) for all other members. The cross-sectional strength is also the limiting strength of short members.

The cross-sectional strength of a Class 1 I-shaped section comprised of relatively stocky plate elements can be derived from the fully plastic stress distribution of the cross-section as shown in Figure 2-16. For uniaxial bending about the x-x axis and the y-y axis, expressions are respectively, using the limit states notation of this Standard:

$$M_{fx} = 1.18\,\phi\,M_{px}\,(1 - C_f/\phi\,C_y) \leq \phi\,M_{px}$$

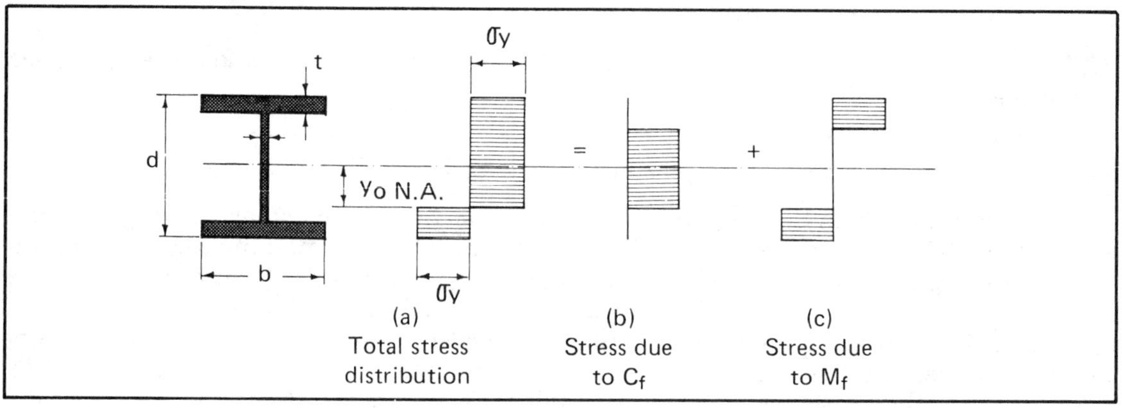

FIGURE 2–16
IDEALIZED STRESSES IN PLASTIFIED SECTION OF BEAM–COLUMN

$$M_{fy} = 1.19\,\phi\,M_{py}\,[1 - (C_f/\phi\,C_y)^2] \leq \phi\,M_{py}$$

Transposing the terms in the first expression and using a linear approximation for the second expression gives

$$\frac{C_f}{\phi\,C_y} + 0.85\,\frac{M_{fx}}{\phi\,M_{px}} \leq 1.0\,; \quad \frac{M_{fx}}{M_{px}} \leq 1.0$$

$$\frac{C_f}{\phi\,C_y} + 0.6\,\frac{M_{fy}}{\phi\,M_{py}} \leq 1.0\,; \quad \frac{M_{fy}}{M_{py}} \leq 1.0$$

as shown in Figure 2-17. For biaxial bending it is conservative to combine these expressions linearly to give, using the limit states notation of this Standard:

$$\frac{C_f}{\phi\,C_y} + 0.85\,\frac{M_{fx}}{\phi\,M_{px}} + 0.6\,\frac{M_{fy}}{\phi\,M_{py}} \leq 1.0\,; \quad \frac{M_{fx}}{\phi\,M_{px}} + \frac{M_{fy}}{\phi\,M_{py}} \leq 1.0$$

This is identical to the two expressions in Clause 13.8.2 when, in the latter, U_{1x} and U_{1y} are set equal to 1.0, as appropriate for the cross-sectional strength, C_r for $\lambda = 0$ equals ϕAF_y, and M_{rx} and M_{ry} equal ϕM_{px} and ϕM_{py} respectively.

A significant change from previous standards, recently confirmed by tests (Kulak and Dawe 1991), is that for uniaxial bending of Class 2 sections, the appropriate interaction expression is

$$\frac{C_f}{\phi\,C_y} + \frac{M_{fx}}{\phi\,M_{px}} \leq 1.0$$

Extending this linear expression to bending about the weak axis gives, for biaxial bending

$$\frac{C_f}{\phi\,C_y} + \frac{M_{fx}}{\phi\,M_{px}} + \frac{M_{fy}}{\phi\,M_{py}} \leq 1.0$$

which is identical to the expression given in Clause 13.8.1 when, in the latter, the appropriate values of the various cross-sectional resistance quantities are used.

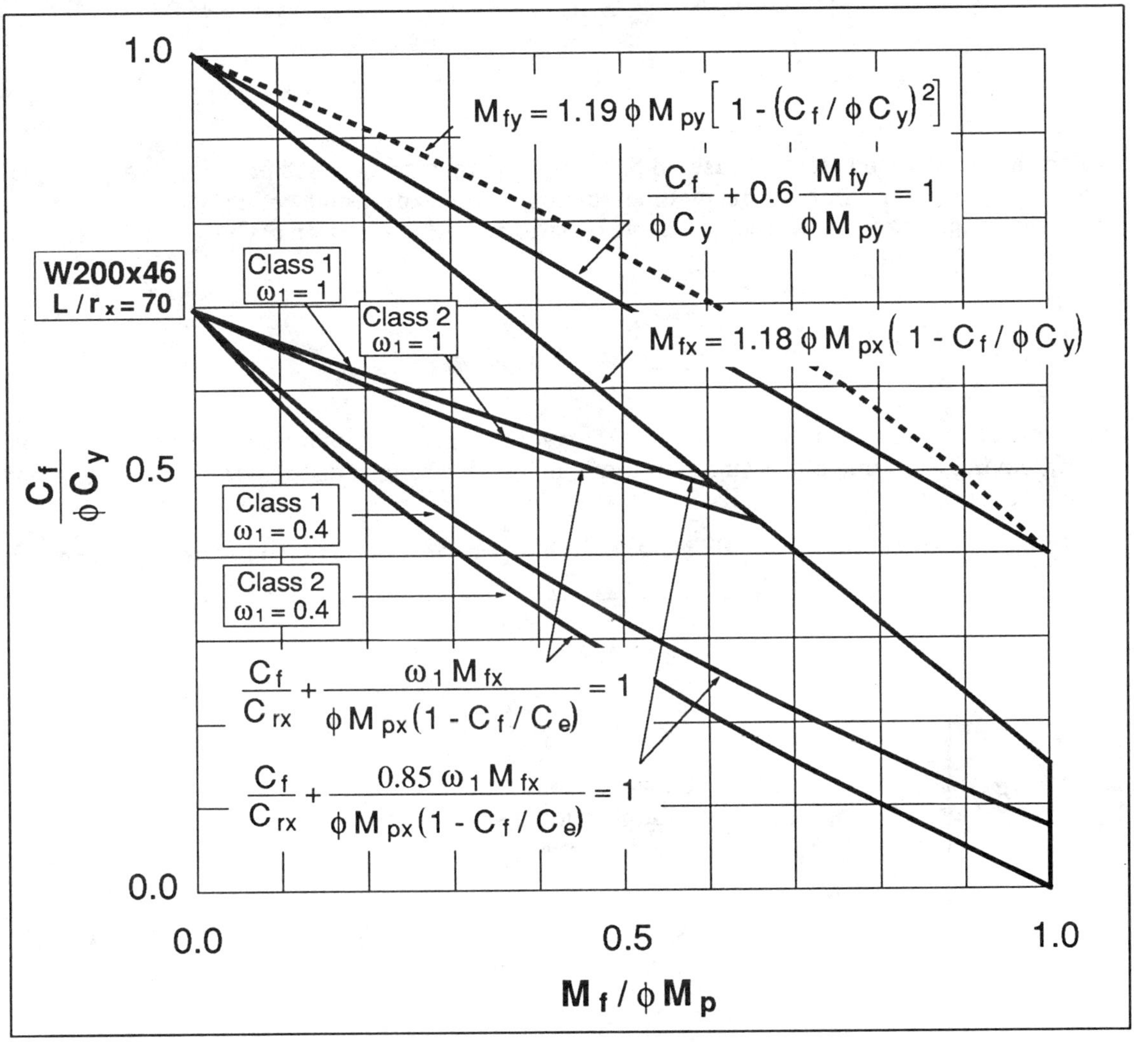

FIGURE 2–17
INTERACTION EXPRESSIONS FOR CLASS 1 AND 2 W-SHAPES

The expression of Clause 13.8.1 is also appropriate for Class 3 sections when the moment resistances are expressed in terms of M_y and for Class 4 sections when those resistances, C_r, M_{rx}, and M_{ry}, are based on local buckling.

3) overall member strength

The overall strength of a member depends on its slenderness. As an actual beam–column has length, the axial compressive resistance, C_r, depends on its slenderness ratio and will be less than or equal to the yield load. For any particular beam–column this fraction of the yield load can be established and is illustrated in Figure 2–17 for a W200x46 shape.

In Figure 2–18 the variation in moment resistance in terms of M/M_p as a function of the slenderness L/r_x is plotted schematically as a solid line for a particular laterally supported Class 1 section subject to a uniform moment about the x axis and carring an axial load of 0.35 C_y. An appropriate interaction expression for the in-plane strength of a Class 1 section is

$$\frac{C_f}{C_{rx}} + \frac{0.85\,\omega_1 M_f}{\phi M_p \left(1 - \frac{C_f}{C_e}\right)} = 1$$

which can be deduced from Clause 13.8.2 when the terms in that expression are appropriately defined. Note that if the member is short the expression reduces to that for the cross-sectional strength. The compressive resistance, C_{rx}, is a fuction of the slenderness ratio L/r_x.

The term

$$\omega_1 = 0.6 - 0.4\,\kappa \geq 0.4$$

multiplied by the maximum non-uniform moment, M_f, gives an equivalent uniform moment, $\omega_1 M_f$, having the same effect on the in-plane member strength as the non-uniform moment (Ketter 1961).

In order to account for the Pδ effects, the amplification of the moments caused by the

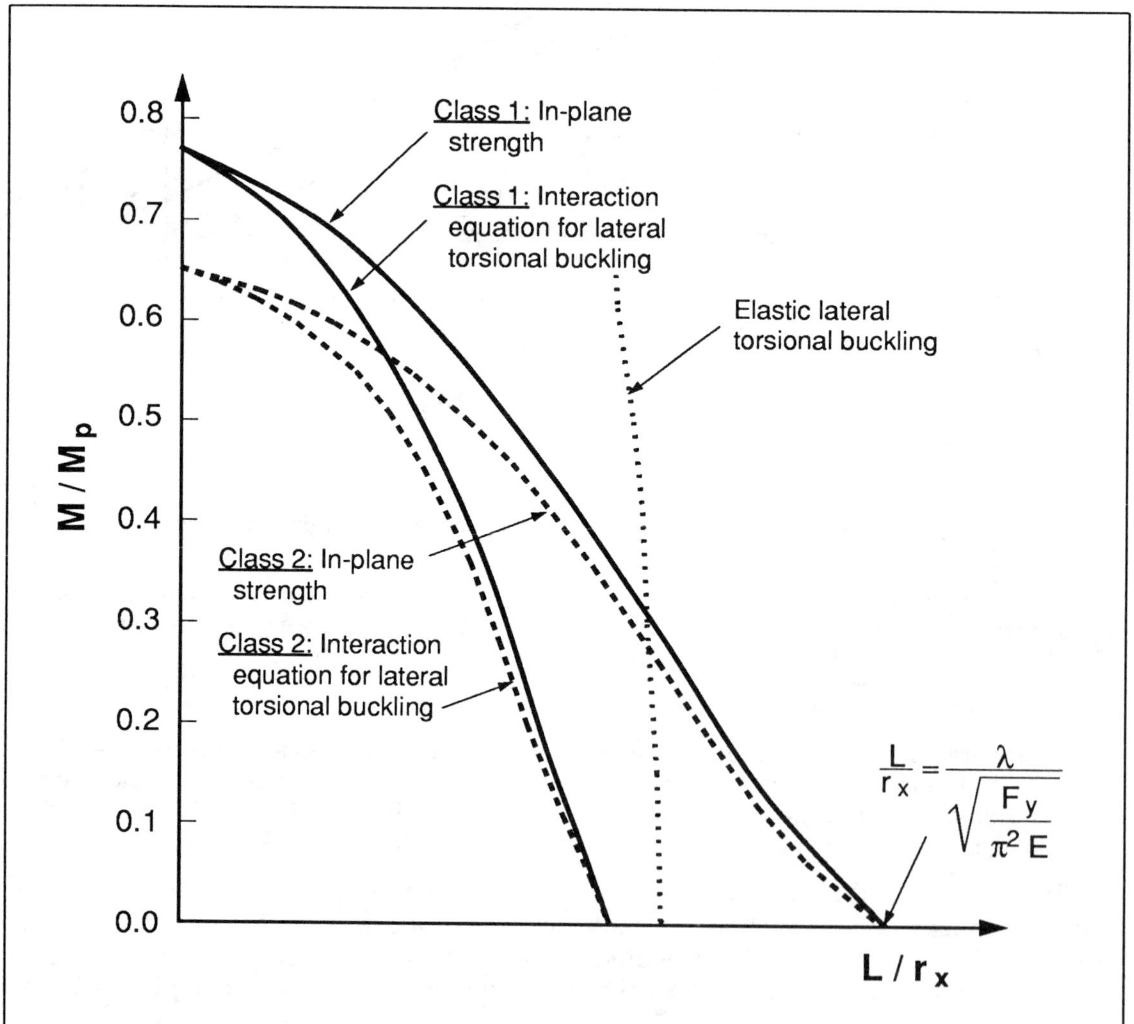

FIGURE 2-18
VARIATIONS OF MOMENT RESISTANCE WITH SLENDERNESS RATIO

axial loads acting on the deformed shape, the equivalent uniform moment, $\omega_1 M_f$, is amplified by the factor $\frac{1}{1-C_f/C_e}$. The in-plane strength of a Class 1 section is shown in Figure 2–17 for a particular section with $L/r_x = 70$. When $L/r_x = 0$ and $\omega_1 = 1$, the in-plane strength expressions 13.8.1(b) and 13.8.2(b) become the cross-sectional strength expressions 13.8.1(a) and 13.8.2(a) respectively.

In Figure 2-18, the curve of moment resistance versus slenderness ratio for the in-plane strength of a Class 2 section of equivalent cross-sectional strength to the Class 1 section is also given. It is similar to that for a Class 1 section except that, because the cross-sectional strength expression for Class 2 sections does not have the 0.85 factor that is appropriate for Class 1, the curve for Class 2 for zero slenderness ratio reaches only 0.65 of M_p and not 0.765 of M_p as for the Class 1 section.

For biaxial bending C_r is based conservatively on the maximum slenderness ratio. It could be argued that for biaxial bending the value used for C_r be interpolated between C_{rx} and C_{ry} on the basis of the proportion of the interaction fractions for bending about two axes. In other words, if a beam-column carries only a small portion of bending about the x axis, the decrease in C_r from C_{rx} toward C_{ry} should likewise be small.

In Figure 2–17, the envelope of in-plane strength interaction expressions is shown for $\omega_1 < 1$. When $\omega_1 < 1$, the limiting strength for low ratios of axial load is the cross-sectional strength expression. The dotted line in Figure 2-17 is the sole line for a Class 2 section. It is the in-plane strength interaction for a Class 2 section for $\omega_1 = 1.0$ for which the end points are C_f/C_r and M/M_p.

4) lateral torsional buckling strength

Building beam-columns are usually laterally unsupported for their full length and, even though they are subject to strong axis bending moments, failure occurs when the column, after bending about the strong axis, buckles about the weak axis and twists simultaneously. For such columns, the lateral torsional buckling strength is likely to be the least of the cross-sectional strength, the overall member strength, and the lateral torional buckling strength.

The curves in Figure 2-18 for a beam-column subject to uniform moment for Class 1 and 2 sections marked *interaction equation for lateral torsional buckling* demonstrate this effect. They are much below those for in-plane strength and only reach the full cross-sectional strength when the slenderness ratio is zero. The moment resistance is zero for laterally unsupported beam-columns when weak axis buckling occurs. Thus, for these members the axial compressive resistance is based on L/r_y and M_{rx} is based on the resistance of a laterally unsupported beam. When subjected to weak axis bending, members do not exhibit out-of-plane buckling behaviour and therefore the weak axis moment resistance is based on the full cross-sectional strength, the plastic moment or yield moment capacity as appropriate for the Class of the section.

In computing M_{rx} from Clause 13.6, the effect of non-uniform moments is included. Therefore, in the interaction expressions when lateral torsional buckling is being investigated, the factored moment, M_{fx}, must also be a non-uniform moment, and not be replaced by an *equivalent* lesser moment. It is for this reason that the value of U_{1x} cannot be less than 1.

13.8.4 This clause gives generally conservative values of ω_1, the factor by which the maximum value of the non-uniform moment is multiplied to give an equivalent uniform moment having the same effect as the applied non-uniform moment on the overall strength of the member. For further discussion on ω_1, see Galambos (1988) where it is called C_m. Figure 2-19, based on AISC (1986a), gives values of ω_1 for some special cases of transverse bending.

Case	ω_1	Case	ω_1
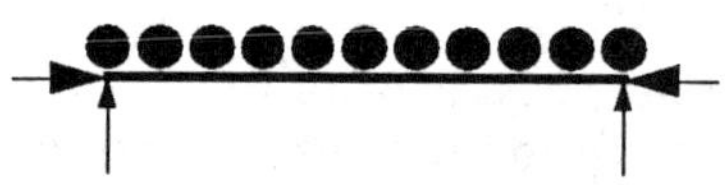	1.0	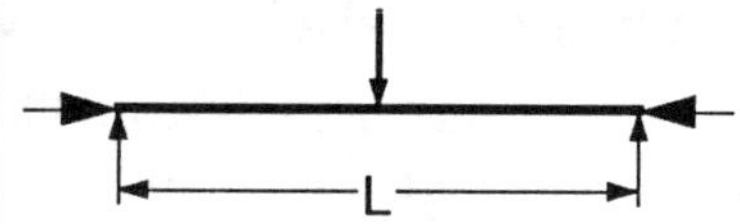	$1 - \frac{0.2\,C_f}{C_e}$
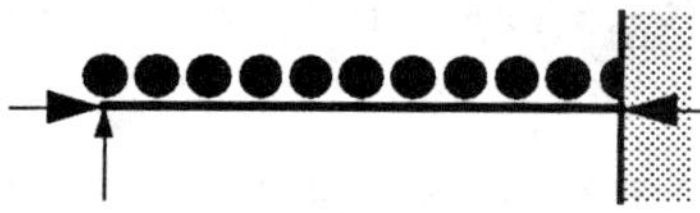	$1 - \frac{0.4\,C_f}{C_e}$	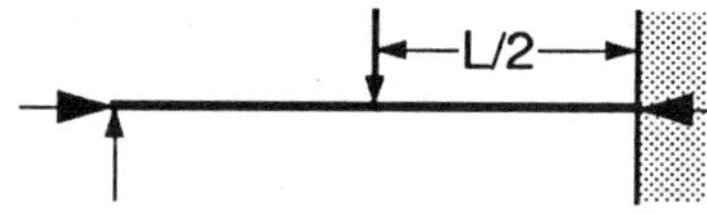	$1 - \frac{0.3\,C_f}{C_e}$
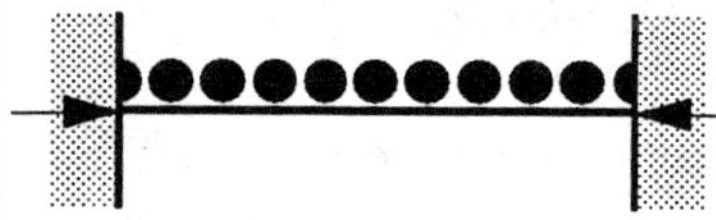	$1 - \frac{0.4\,C_f}{C_e}$	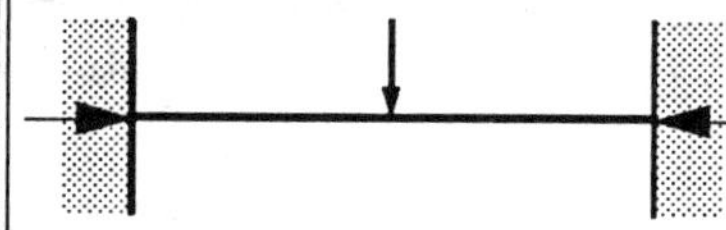	$1 - \frac{0.2\,C_f}{C_e}$

FIGURE 2-19
VALUES OF ω_1 FOR SPECIAL CASES OF LATERALLY LOADED BEAM-COLUMNS

Figures 2-20 and 2-21 give guidance for the design of beam-columns subjected to various bending moment effects.

13.9 Axial Tension and Bending

The linear interaction expression of Clause 13.9(a) is a cross-sectional strength check. Conservatively, it does not take into account the fact that the bending resistance for Class 1 sections does not vary linearly with axial force, for which case a factor of 0.85 in front of the moment term would appear to be appropriate (see Clause 13.8.2).

For members subjected predominantly to bending, i.e. when the tensile force is relatively small, failure may still occur by lateral torsional buckling. The expressions of Clause 13.9(b) result from that of Clause 13.9(a) when a negative sign is assigned to the tension interaction component and when M_r is based on the overall member behaviour taking lateral torsional buckling into account.

13.10 Load Bearing

The bearing resistance given for machined, accurately sawn, or fitted parts in contact, Clause 13.10.1(a), reflects the fact that a triaxial compressive stress state, restricting yielding of the parts in contact, generally exists. The value given is based on earlier working stress design standards, which have given satisfactory results.

For a cylindrical roller or rocker in contact with a flat surface, Clause 13.10.1(b), the maximum shearing stress developed due to a line load of q kN/mm, (Seeley and Smith, 1957) is,

$$\tau_{max} = 0.27\sqrt{\frac{qE}{\pi D(1-\nu^2)}}$$

<table>
<tr><th>Conditions **</th><th>Design Criteria</th></tr>
<tr><td>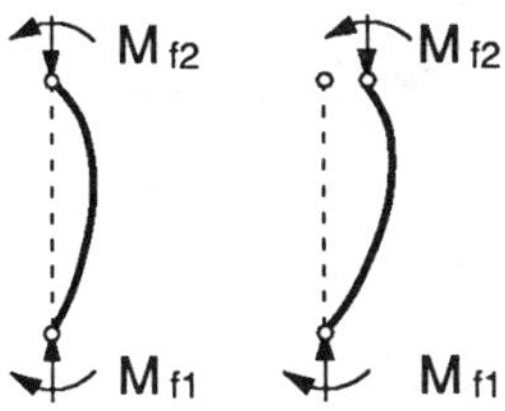
single curvature bending
$M_{f2} \geq M_{f1}$
$\omega_1 = 0.6 + 0.4 \dfrac{M_{f1}}{M_{f2}}$</td><td><u>Beam-Column Design Expressions</u>
P-Δ (frame sway effects), if any, are included in analysis.
(I) Class 1 Sections of I-Shapes
$\dfrac{C_f}{C_r} + \dfrac{0.85\, U_{1x} M_{fx}}{M_{rx}} + \dfrac{0.6\, U_{1y} M_{fy}}{M_{ry}} \leq 1.0$
$\dfrac{M_{fx}}{M^*_{rx}} + \dfrac{M_{fy}}{M^*_{ry}} \leq 1.0$
(II) All Classes Except Class 1 Sections of I-Shapes
$\dfrac{C_f}{C_r} + \dfrac{U_{1x} M_{fx}}{M_{rx}} + \dfrac{U_{1y} M_{fy}}{M_{ry}} \leq 1.0$</td></tr>
<tr><td>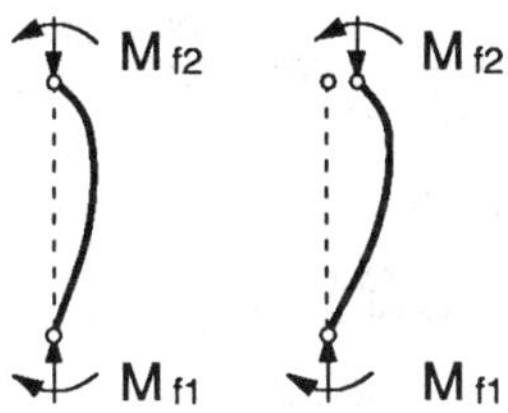
single curvature bending
$M_{f1} = 0$
$\omega_1 = 0.6$</td><td rowspan="2"><u>Three Member Strength Checks</u>
(a) Cross-sectional strength (use actual M_f at each location)
$C_r = \phi A F_y$
$M_r = \phi Z F_y$ for class 1 and class 2 sections
$= \phi S F_y$ for class 3 sections
= see Cl. 13.5 (c) of S16.1-M89. for class 4 sections
M^*_{rx} and M^*_{ry} as for M_r in Cl. 13.5 or 13.6, as appropriate
U_{1x} and U_{1y} are taken as 1.0

(b) Overall strength (use M_{f2} for M_f)
C_r = Factored compressive resistance (K=1) as in Cl.13.3 use C_{rx} for uniaxial strong-axis bending and max. slenderness for biaxial bending, or Interpolated between C_{rx} and C_{ry} on the basis of the proportion of interaction fractions for bending about two axes.
$M_r = \phi Z F_y$ for class 1 and class 2 sections
$= \phi S F_y$ for class 3 sections
= see Cl. 13.5 (c) of S16.1-M89. for class 4 sections
$U_1 = \omega_1 / (1 - C_f / C_e)$

(c) Lateral torsional buckling strength (use M_{f2} for M_f)
$C_r = C_{ry}$ = Factored comp. resistance (K=1) as in Cl.13.3
M_{rx} = value given by Cl. 13.6 of S16.1-M89
M_{ry} = as noted for M_r in (b) above
U_{1x} = as noted for U_1 in (b) above, but not less than 1.0
U_{1y} = as noted for U_1 in (b) above</td></tr>
<tr><td>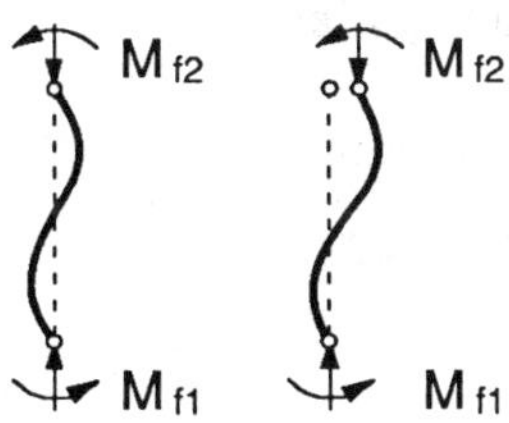
double curvature bending
$M_{f2} \geq M_{f1}$
$\omega_1 = 0.6 - 0.4 \dfrac{M_{f1}}{M_{f2}} \geq 0.4$</td></tr>
<tr><td colspan="2">C_f = Factored compressive load
C_r = Factored compressive resistance
M_f = Fac. bending moment (x-x or y-y axis)
M_r or M^*_r = Fac. moment resistance (x-x or y-y axis)
ω_1 = Coefficient used to determine equivalent uniform column bending effect (x-x or y-y axis)
U_1 = factor to account for moment gradient & member curvature second order effects</td></tr>
</table>

** Moments M_{f1} and M_{f2} may be applied about one or both axes.

FIGURE 2-20
PRISMATIC BEAM-COLUMNS–MOMENT AT ENDS–NO TRANSVERSE LOADS

FIGURE 2-21 — Prismatic Beam-Columns–Transverse Loads

Conditions**

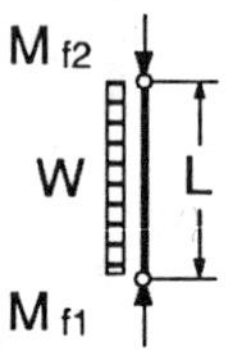

Loaded with UDL

$M_{f1} = M_{f2} = 0$

$M_{f3} = \frac{WL}{8}$ (max.)

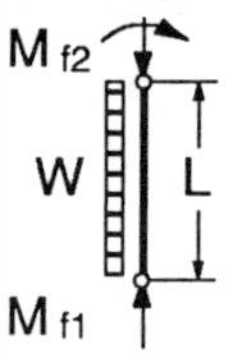

Loaded with UDL

$M_{f1} = 0$

$M_{f2} = \frac{WL}{8}$ (max.)

$M_{f3} = \frac{9\,WL}{128}$

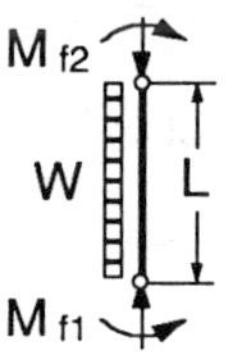

Loaded with UDL

$M_{f1} = M_{f2} = \frac{WL}{12}$ (max.)

$M_{f3} = \frac{WL}{24}$

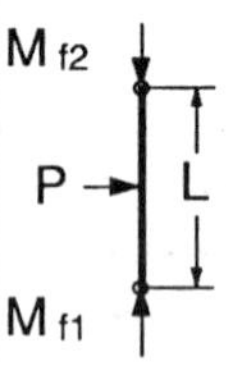

Loaded with PL

$M_{f1} = M_{f2} = 0$

$M_{f3} = \frac{PL}{4}$ (max.)

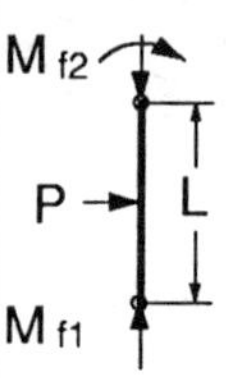

Loaded with PL

$M_{f1} = 0$

$M_{f2} = \frac{3PL}{16}$ (max.)

$M_{f3} = \frac{5\,PL}{32}$

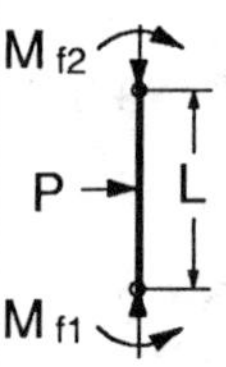

Loaded with PL

$M_{f1} = M_{f2} = \frac{PL}{8}$

$M_{f3} = \frac{PL}{8}$ (max.)

Design Criteria

Beam-Column Design Expressions

P-Δ (frame sway effects), if any, are included in analysis.

(I) Class 1 Sections of I-Shapes

$$\frac{C_f}{C_r} + \frac{0.85\,U_{1x} M_{fx}}{M_{rx}} + \frac{0.6\,U_{1y} M_{fy}}{M_{ry}} \leq 1.0$$

$$\frac{M_{fx}}{M^*_{rx}} + \frac{M_{fy}}{M^*_{ry}} \leq 1.0$$

(II) All Classes Except Class 1 Sections of I-Shapes

$$\frac{C_f}{C_r} + \frac{U_{1x} M_{fx}}{M_{rx}} + \frac{U_{1y} M_{fy}}{M_{ry}} \leq 1.0$$

Three Member Strength Checks

(a) Cross-sectional strength (use actual M_f at each location)

$C_r = \phi A F_y$

$M_r = \phi Z F_y$ for class 1 and class 2 sections

$= \phi S F_y$ for class 3 sections

= see Cl. 13.5 (c) of S16.1-M89. for class 4 sections

M^*_{rx} and M^*_{ry} as for M_r in Cl. 13.5 or 13.6, as appropriate

U_{1x} and U_{1y} are taken as 1.0

(b) Overall strength (use M_{max} for M_f)

C_r = Factored compressive resistance (K=1) as in Cl.13.3 use C_{rx} for uniaxial strong-axis bending and max. slenderness for biaxial bending, or Interpolated between C_{rx} and C_{ry} on the basis of the proportion of interaction fractions for bending about two axes.

$M_r = \phi Z F_y$ for class 1 and class 2 sections

$= \phi S F_y$ for class 3 sections

= see Cl. 13.5 (c) of S16.1-M89. for class 4 sections

$U_1 = \omega_1 / (1 - C_f / C_e)$

(c) Lateral torsional buckling strength (use M_{max} for M_f)

$C_r = C_{ry}$ = Factored comp. resistance (K=1) as in Cl.13.3

M_{rx} = value given by Cl. 13.6 of S16.1-M89

M_{ry} = as noted for M_r in (b) above

U_{1x} = as noted for U_1 in (b) above, but not less than 1.0

U_{1y} = as noted for U_1 in (b) above

C_f = Factored compressive load

C_r = Factored compressive resistance

M_f = Fac. bending moment (x-x or y-y axis)

M_r or M^*_r = Fac. moment resistance (x-x or y-y axis)

ω_1 = Coefficient used to determine equivalent uniform column bending effect (x-x or y-y axis). See Fig. 2-19 for design formulae.

U_1 = factor to account for moment gradient & member curvature second order effects

** Moments M_{f1} and M_{f2} may be applied about one or both axes.

FIGURE 2-21
PRISMATIC BEAM-COLUMNS–TRANSVERSE LOADS

where ν is Poisson's ratio. From this, the bearing resistance is then

$$\frac{B_r}{\phi} = qL = \frac{\pi DL(1-\nu^2)(\tau_{max})^2}{0.27^2 E}$$

Calibrating this resistance to that given in S16-1969 at F_y = 300MPa gives, τ_{max} = 0.77 F_y, and

$$\frac{B_r}{\phi} = 0.000\ 13\ DLF_y^2$$

The "Hertz" solution, as reported by Manniche and Ward-Hall (1975) gives the allowable load as

$$2.86\ DL\frac{(2.7\ F_y)^2}{E} = 0.000\ 10\ DLF_y^2$$

and indicates that the value of 0.000 13 obtained by calibration with the existing standard for a yield stress of about 300 MPa is somewhat conservative.

This is confirmed by Kennedy and Kennedy (1987) who reported that at this load no permanent deformation resulted and recommended that this value be used as a serviceability limit. They also reported that the rolling resistance of rollers varied as the fourth power of the unit normal load in kN/mm.

In bearing-type connections, Clause 13.10.1(c), excessive deformation in front of the loaded edge of the bolt hole may occur. Tests have shown (Munse 1959; Jones, 1958; de Back and de Jong 1968; Hirano 1970) that the ratio of the bearing stress (B_r/dt) to the ultimate tensile strength of the plate (F_u) is in the same ratio as the end distance of the bolt (e) to its diameter (d). Thus,

$$\frac{B_r}{\phi\, dt} = \frac{e}{d} F_u$$

Or, for n fasteners, $B_r = \phi\ tneF_u$

Because the test results do not provide data for e/d greater than 3, an upper limit of e = 3d is imposed. That is,

$$B_r \leq 3\ \phi\ tdnF_u$$

As for connections in general, the value of ϕ in Clause 13.10.1(c) is to be taken as 0.67.

The Standard states that 13.10.1(c)ii does not apply to tension members. This is because the potential for block tear-out, when the end distance, e, is small, is covered by Clause 12.3. In fact 13.10.1(c)(ii) is superfluous, as it is not applicable to compression members either. Here the bearing force is directed away from the end of the member and e > 3.

13.11 Bolts In Bearing-Type Connections

13.11.2 Bolts in Shear

Based on extensive testing, it has been established that the shear strength of high-strength bolts is approximately 0.60 times the tensile strength of the bolt material. However, if the bolt thread is intercepted by a shear plane, there is less shear area available. The ratio of the area through the thread root of a bolt to its shank area is about 0.70 for the usual structural sizes. Moreover, in long joints, the average resistance per fastener is less. The step reduction for joints longer than 1300 mm is a reasonable approximation of the true behaviour. In this context, "joint length" refers to a connection, such as an axially loaded lap splice, whose length is measured parallel to the direction of applied force. This clause does not apply to a shear connection at the end of a girder web where the load is distributed reasonably uniformly to the fasteners.

13.11.3 Bolts in Tension

The ultimate resistance of a single high-strength bolt loaded in tension is equal to the product of its tensile stress area (a value between the gross bolt area and the area at the root of the thread) and the ultimate tensile strength of the bolt. The tensile stress area is taken as 0.75 of the gross area of the bolt.

In calculating the tensile force acting on a bolt, prying action shall be taken into account. See Clause 23.1.4.

13.11.4 Bolts in Combined Shear and Tension

The expression given in this clause is an elliptical interaction equation developed directly from test results. The value 0.56 is the square of 0.75, the ratio of the tensile stress area to the nominal bolt area. The parameter, β, determined emperically, depends upon bolt type and location of the shear plane with respect to the threads.

13.12 Bolts in Slip-Critical Connections

13.12.2 Different installation procedures may result in different probabilities of slip, see Kulak *et al*, (1987).

Both the slip coefficient and the initial clamping force have considerable variation about their mean values. The coefficients of friction for coatings can vary as a function of the specific coating constituents and, therefore, values of k_s may differ from one coating specification to another. The value of k_s intended for use on a project should be specified.

The resistance expression given in this edition of S16.1 replaces that used in previous editions in that the mean slip coefficient, k_s is specifically included. The clamping force is due to the pretensioning of the bolts to an initial tension, T_i, which is a minimum of 70% of the tensile strength ($0.70\ A_s F_u$) where $A_s = 0.75 A_b$. Thus, the clamping force per bolt is

$$0.70 \times 0.75\ A_b\ F_u \quad \text{or} \quad 0.53\ A_b\ F_u$$

The value of c_1 establishes the probability level of slip for specific grades of bolts and the instalation method. Table 2 of S16.1-M89 gives values of c_1 for bolts installed by turn-of-nut procedures and for a probability level of 5%. Table 2 also provides mean slip coefficients for the three most common cases of surface conditions. Values of k_s and c_1 for many other common situations are given by Kulak *et al.* (1987).

The use of slip-critical connections should be the exception rather than the rule. They are the preferred solution only where cyclic loads or frequent load reversals are present, or where the use of the structure is such that the small one-time slips that may occur cannot be tolerated.

13.12.3 The resistance to slip is reduced as tensile load is applied and reaches zero when the parts are on the verge of separation, as no clamping force then remains. The interaction relationship is linear.

The term $1.9/(n\ A_b\ F_u)$ is the reciprocal of the initial bolt tension, $0.53\ n\ A_b\ F_u$.

13.13 Welds

When matching electrodes (electrodes with ultimate strengths equal to or greater than that of the base metal) are used, the resistance of complete penetration groove welds for any direction of loading is the same as for the base metal. This applies on a unit area basis, to fillet welds or partial joint penetration groove (PJPG) welds loaded in tension or compression parallel to the weld axis.

Gagnon and Kennedy (1989) established that the net area tensile resistance, i.e. on a unit area basis, transverse to the axis of a PJPG weld, is the same as for the base metal when matching electrodes are used. The previous conservative practice of assigning shear resistances to these welds has been replaced with tensile resistances, consistent with the tensile resistance of complete penetration welds equalling the full tensile resistance of the member.

The shear resistance of a weld is evaluated on the basis of both the resistance of the weld itself and of the base metal adjacent to the weld. Thus, fillet welds loaded at any angle, slot, plug, and PJPG welds loaded in shear have resistances equal to the lesser of weld throat or fusion face shear strength.

The resistance of the base metal ($V_r = 0.67\ \phi\ A_m\ F_y$), greater than that attributable to the shear yield strength $0.577\ F_y$, recognizes the beneficial effects of strain-hardening. The value of ϕ in this expression is taken as 0.90.

In the expression for the shear strength of the weld the factor 0.67 relates the shear strength of the weld to the electrode tensile strength, as given by the electrode classification number. Lesik and Kennedy (1990) give 0.75 for this factor, based on 126 tests reported in the literature.

In the expression ϕ_w is taken as 0.67 so that, as for bolts, the weld has a larger reliability index and reduced probability of failure, as compared to the member itself.

13.13.3 In addition to the conservative traditional approach of summing vectorially the longitudinal and transverse shear loads, this clause permits an ultimate strength analysis for welds. The latter, using the instantaneous shear centre concept, forms the basis of the eccentric load tables given in Part 3 (Butler and Kulak 1971, Butler *et al* 1972, Miazga and Kennedy 1989, Lesik and Kennedy 1990, Kennedy *et al* 1990). This ultimate strength analysis, recognizing the true behaviour of the weldments, results in much more consistent strength predictions than the traditional approach and is recommended for use. The use of this clause is also consistent with Clause 1.4.

14. FATIGUE

Clause 14 provides the requirements for the design of members and connections subject to fatigue. The limit state of fatigue is one for which the requirements are checked at specified load levels. In addition, Clause 14.1.1 requires that factored resistances be sufficient for the factored static loads.

A substantial amount of experimental data, developed on steel beams since 1967 under the sponsorship of the National Co-operative Highway Research Program (NCHRP 1970, 1974; Fisher 1974) of the USA, has shown that the most important factors governing fatigue resistance are the stress range and the type of detail.

Details are assigned to one of nine stress range categories (“a” to “y”), as compared with seven in the previous edition of the Standard. The two new categories, “b-1” and “y”, reflect more information on the fatigue behaviour of the details assigned to those categories.

The stress range categories, and details assigned to them, are identical to those in the Ontario Highway Bridge Design Code (MTO 1991a), with the exception that S16.1-M89 also includes a detail of a hollow structural section fillet welded to a base plate. An additional detail, based on work by Frank (1980), has been included for the fatigue strength of anchor bolts. The Commentary to the OHBDC (MTO 1991b) provides an extensive background to the development of the stress range concept and the categorization of individual details.

Only variable loads, live loads and impact, contribute to the stress range. Fatigue design is considered only for regions subjected to tensile stresses or stress reversal. A crack may appear in a region subject only to compression stresses if a zone of tensile residual stress exists there. Such a crack may propagate but usually stops once it leaves the tensile residual stress zone and enters the region with net compressive stresses.

Appendix K illustrates various details assigned to the stress range categories in Table 4(b).

A detail is considered satisfactory in fatigue provided that the stress range is less than that corresponding to the expected number of cycles of loading for the design life of the structure.

14.1.5 The maximum web slenderness permitted ($3150/\sqrt{F_y}$) is based on research conducted by Toprac and Natarajan (1971).

14.6 Secondary Effects

Secondary stress due to deformations and out-of-plane movements can also be a source of fatigue failures (Fisher 1978 and 1984). Crane girders, their attachments and supports, require careful design and attention to details to minimize fatigue cracks (Griggs 1976).

14.7 Single Load Paths

In a single load path system, the development of fatigue cracks in some structural elements may lead to collapse, or catastrophic failure of the structure. Unless there is a programme of periodic inspection and maintenance in force during the life of the structure, the permissible range of stress is limited to 0.80 of the values given in Table 4(a). This requirement is similar to that found in OHBDC (MTO 1991a) and recognizes that most fatigue cracks can be detected by periodic inspection and corrective actions taken.

15. BEAMS AND GIRDERS

15.1 Proportioning

Lilley and Carpenter (1940) have shown that reductions of flange area up to 15% can be disregarded due to the limited inelastic behaviour near the holes.

15.2 Rotational Restraint at Points of Support

A severe stability problem may exist when a beam or girder is continuous over the top of a column. The compression flange of the beam tends to buckle sideways and simultaneously, the beam-column junction tends to buckle sideways because of the compression in the column. Three mechanisms exist for providing lateral restraint: direct acting bracing, such as provided by bottom chord extensions of joists, beam web stiffeners welded to the bottom flange, or the distortional stiffness of the web. In the latter two cases, the connection of the beam flange to the column cap plate must have strength and stiffness (Chien, 1989). The restraint offered by the distortion of the web requires very careful assessment and is frequently neglected for simplicity. See also commentary on Clause 20.2.

15.3 Copes

Flanges are coped to permit beams to be connected to girder webs with simple connections while maintaining the tops of the flanges at the same elevation. Long copes may

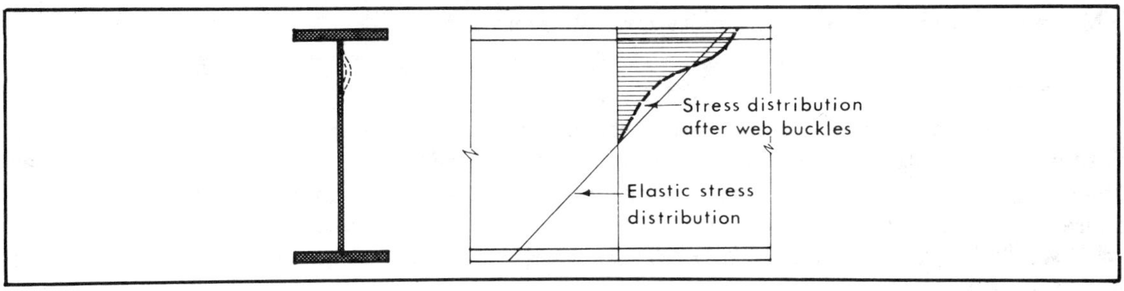

FIGURE 2-22
APPROXIMATE STRESS DISTRIBUTION IN GIRDERS WITH BUCKLED WEB

seriously effect the lateral torsional buckling resistance of a beam (Cheng and Yura, 1986). The reduced shear and moment resistance at the coped cross-section should be examined.

15.4 Reduced Moment Resistance of Girders with Thin Webs

A plate girder with Class 3 flanges and Class 4 webs has a maximum moment resistance less than ϕM_y because the Class 4 web buckles prematurely due to the compressive bending stresses. The reduction in moment resistance is based on Basler and Thurlimann (1961). Figure 2-22 shows an approximate stress distribution in a girder with a buckled web. The reduction in moment resistance is generally small, as shown in Figure 2-23.

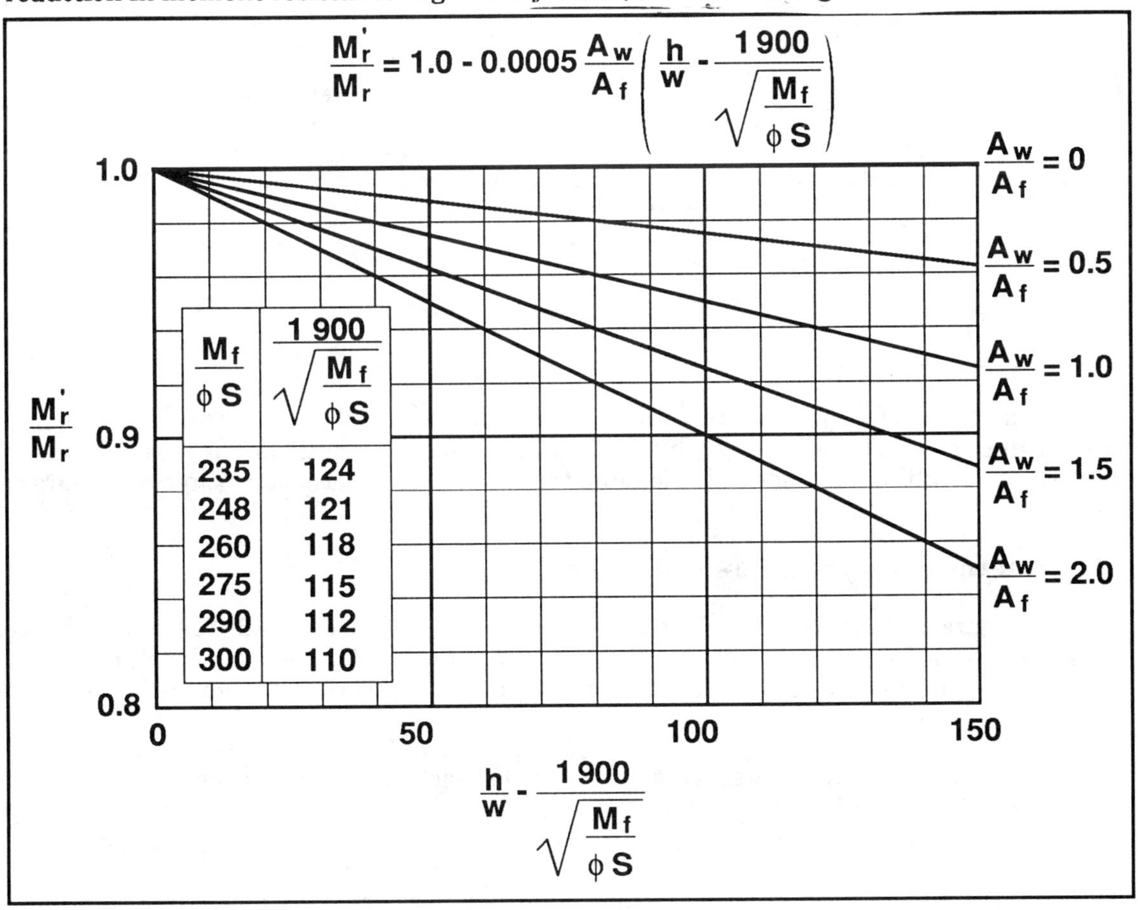

FIGURE 2-23
REDUCED MOMENT RESISTANCE IN GIRDERS WITH THIN WEBS

The limit of $1900/\sqrt{F_y}$ for the slenderness of a Class 3 web is replaced in this clause by $1900/\sqrt{M_f/(\phi S)}$ to account for the possibility that the factored moment may be less than ϕM_y thereby reducing the propensity for web buckling.

15.5 Flanges.

The theoretical cut-off point is the location where the moment resistance of the beam without cover plates equals the factored moment (Figure 2-24). The distance a′ increases as shear lag becomes more significant, as is the case when the weld size is smaller, or when there is no weld across the end of the plate. Theoretical and experimental studies of girders with welded cover plates (ASCE, 1967) shows that the cover plate load can be developed within length (a′).

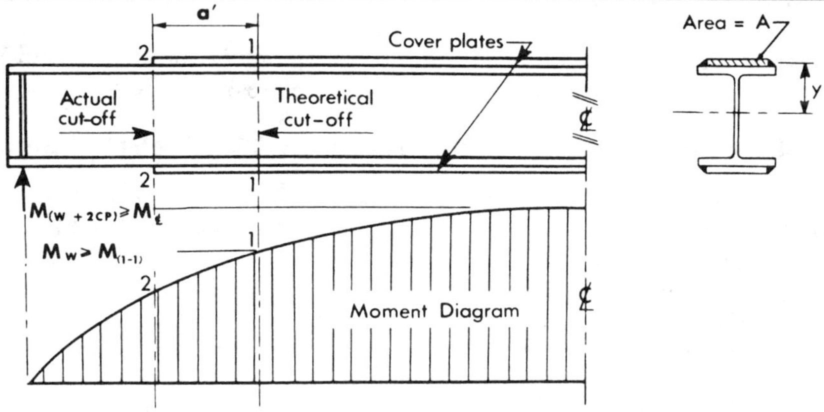

Force to be developed in cover plate within distance a′ is $\frac{A M_{(1\text{-}1)} y}{I_{(w+2CP)}}$

For welded cover plates a' is not to exceed lengths specified in clause 15.5.4 CAN/CSA-S16.1-M89. In certain cases this will necessitate an increase in weld size or an extension of cover plate so that force at distance a' from its end equals that which terminal welds will support.

FIGURE 2–24
COVER PLATE DEVELOPMENT

15.6 Bearing Stiffeners

The inclusion of a portion of the web in the column section resisting the direct load, and the assumption of an effective length of 75% of the stiffener's length, are approximations to the behaviour of the web under edge loading that have proved satisfactory in many years of use.

15.7 Intermediate Transverse Stiffeners

15.7.1 Figure 2-25 illustrates the action of a thin girder web under load. Tension fields are developed in the interior panels but not in the unanchored end panels, for which the maximum shear stress is, therefore, either the elastic or inelastic critical plate buckling stress in shear.

15.7.2 The limits on stiffener spacing are based on practical considerations. When $a/h > 3$, the tension field contribution is reduced. When $h/w > 150$ the maximum stiffener spacing is reduced for ease in fabrication and handling.

15.7.3 Clause 15.7.3 requires that intermediate transverse stiffeners have both a minimum moment of inertia and a minimum area. The former provides the required stiffness when web panels are behaving in an elastic manner; the latter ensures that the stiffener can

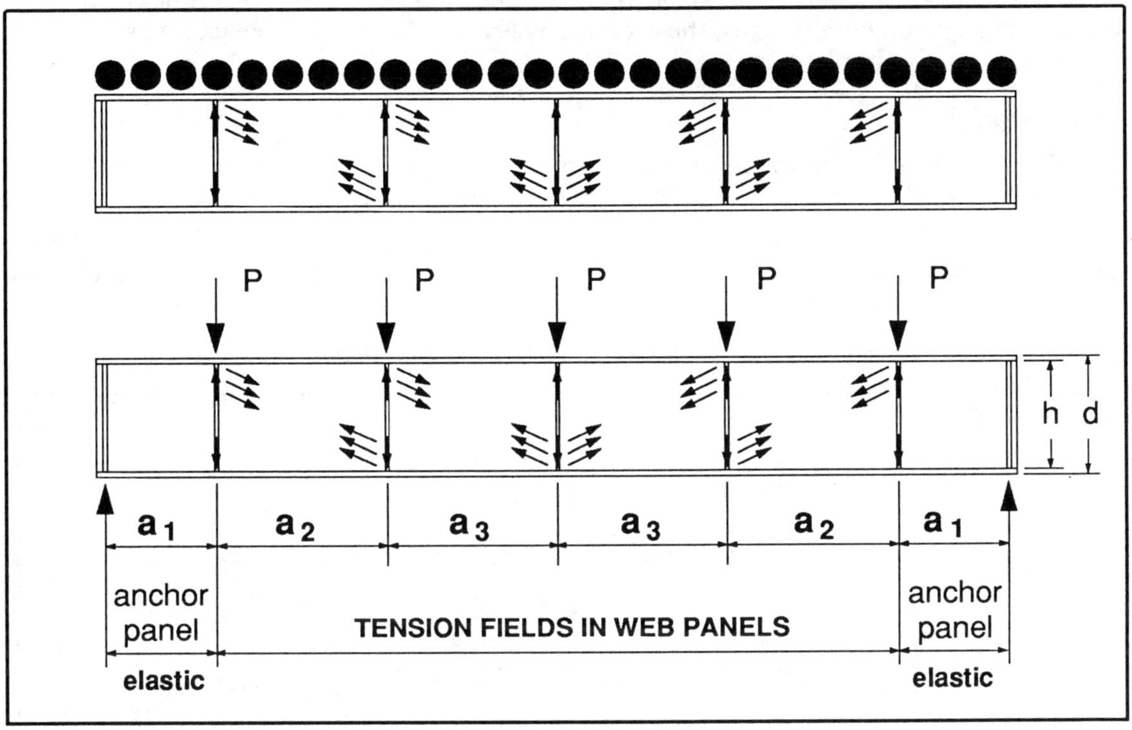

FIGURE 2–25
ACTION OF A THIN WEB PLATE GIRDER UNDER LOAD

sustain the compression, to which it is subjected, when the web panel develops a tension field. Since stiffeners subject to compression act as columns, stiffeners placed only on one side of the web are loaded eccentrically and are less efficient. The stiffener factor (D) is included in the formula for stiffener area to account for the lowered efficiency of stiffeners furnished singly, rather than in pairs.

15.7.4 The minimum shear to be transferred between the stiffener and the web is based on Basler (1961c).

15.7.5 The requirement of attaching single intermediate stiffeners to the compression flange is to prevent tipping of the flange under loading.

15.9 Web Crippling and Yielding

Unstiffened webs of beams and girders carrying concentrated loads applied normal to one flange can fail either from web yielding or from web crippling (a localized wrinkling, folding, or buckling of the web).

If the web is relatively thick, yielding will occur prior to crippling and expressions 15.9(a)(i) and 15.9(b)(i) govern web resistance. These expressions are based on test results (Graham *et al.* 1959) which show that a conservative estimate of strength is obtained by assuming the web yields over a length of N + 5k. The coefficient 1.10 was obtained by calibration to the previous working stress standard.

Relatively thin webs, cripple before yielding and expressions 15.9(a)(ii) and 15.9(b)(ii) govern the web resistance. The tests of Roberts (1981) and Roberts and Chong (1981) indicate that the important parameters governing the crippling load are the square of the web thickness, the length of the loaded area, the flange stiffness, and the yield strength of the web. The resistance expressions are empirically derived, and the coefficients 300 and 150 are in units of $\sqrt{MPa}$.

For unstiffened portions of webs where concentrated compressive loads are applied directly opposite on *both* flanges, the compressive resistance of the web acting as a column should also be investigated.

15.10 Openings

The conditions under which unreinforced circular openings may be used are based on work by Redwood and McCutcheon (1968) and are illustrated in Figure 2–26.

Elastic and plastic analysis to determine the effect of openings in a member are given in Bower *et al* (1971), Redwood (1973), and Redwood (1971, 1972, 1973), respectively. See Part 5 for worked examples.

A combination of vertical and horizontal intersecting stiffeners (particularly on both

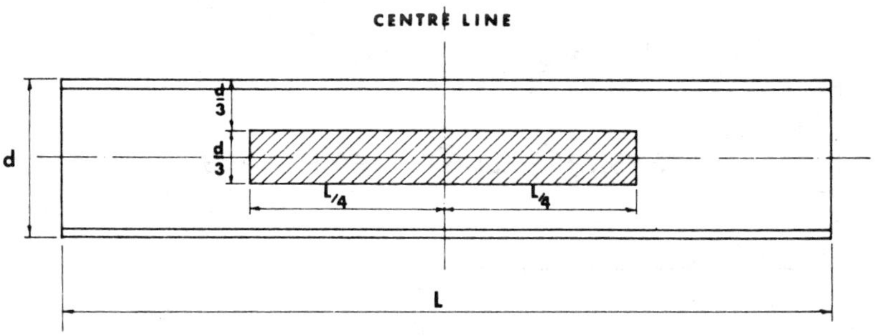

Unreinforced circular holes may be placed anywhere within the hatched zone without affecting the strength of the beam for design purposes, provided:

1. Beam supports uniformly distributed load only.
2. Beam section has an axis of symmetry in plane of bending.
3. Spacing of holes meets the requirements shown below.

SPACING OF HOLES

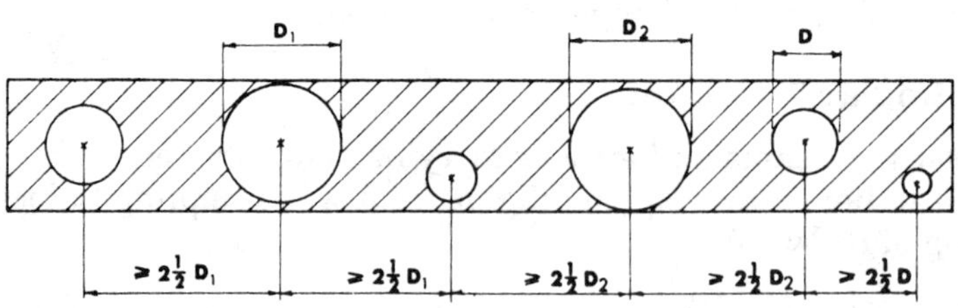

Spacing must be ⩾ 2½ times diameter of the larger opening of any two adjacent openings.

FIGURE 2–26
UNREINFORCED CIRCULAR WEB OPENINGS IN BEAMS

sides of a web) is seldom justified, and is quite expensive to fabricate. Generally, horizontal stiffeners alone are adequate. When both vertical and horizontal stiffeners are necessary, the horizontal stiffeners should be on one side of the web, and vertical stiffeners on the other, in order to achieve economy.

15.11 Torsion

In many cases, beams are not subject to torsion because of the restraint provided by slabs, bracing or other framing members.

Information on moment-torque interaction diagrams for I-shaped members is given in Driver and Kennedy (1989).

The torsional resistance of open sections having two flanges consists of the St. Venant torsional resistance and the warping torsional resistance.

Serviceability criteria will often govern the design of a beam subject to torsion. Limiting the maximum stress due to bending and warping, at the specified load level, to the yield strength guards against inelastic deformation. For elastic analyses, see Bethlehem Steel Corporation (1967), and Brockenbrough and Johnston (1974).

16. OPEN-WEB STEEL JOISTS

16.1 Scope

Open-web steel joists (OWSJ or joists), as described in Clause 16.2, are generally proprietary products whose design, manufacture, transport, and erection are covered by the requirements of Clause 16. The Standard clarifies the information to be provided by the building designer (user-purchaser) and the joist manufacturer (joist designer-fabricator).

16.3 Definitions

There are many variations of the simply supported joist which has given rise to the definition for *special open-web steel joists or special joists*. These joists are those subjected to specific loading conditions, cantilever joists, continuous joists, joists having special support conditions, and joists used as part of the frame (see Clause 16.6.13.3). As defined, tie joists are designed to resist gravity loads only and any connection to a column is to facilitate erection. Span is the distance centre-to-centre of the joist bearings.

16.4 Materials

The use of yield strength levels reported on mill test certificates for the purposes of design is prohibited, as this practice may significantly lower the margin of safety by not properly accounting for the statistical distribution of yield levels. Historically, all design rules have been, and still are, based on the use of the *specified* minimum yield point or yield strength. For structural members cold-formed to shape, the increase in yield strength due to cold forming, as given in Clause 5.2 of CAN/CSA-S136, may be taken into account provided that the increase is based on the specified minimum values in the relevant stuctural steel material standard.

16.5.1 Building Design Drawings

The Standard recognizes that the building designer may not be the joist designer; therefore, building design drawings are required to provide specific information for the design of the joists.

Loads such as unbalanced, non-uniform, concentrated, and net uplift, are to be shown by the building designer. A joist schedule (Figure 2–27) can be used to record all loads on joists.

JOIST SCHEDULE							
Mark	Depth (mm)	Spacing (mm)	Specified Dead Load	Specified Live Load	Specified Snow Load	Specified Wind Load	Remarks
J1	600	1 300	2.4 kPa	2.6 kPa			$\Delta_{live} = \frac{span}{320}$ Suggested I_x for vibration = ________
J2	700	2 000	8.9 kN; 1.5 kN/m; 3 m; 12 000		4.38 kN/m; 10.2 kN/m; 3 m; 12 000	-2.4 kN/m; 12 000	$\Delta_{live} = \frac{span}{240}$

FIGURE 2-27

All heavy concentrated loads such as those resulting from partitions, large pipes, mechanical, and other equipment to be supported by OWSJ, should be shown on the structural drawings. Small concentrated loads may be allowed for in the uniform dead load.

When the importance factor, γ, (see Clause 7.2.2) is not equal to 1.0, it should be specified by the building designer.

Options, such as attachments for deck when used as a diaphragm, special camber and any other special requirements should also be provided. Where vibration of a floor system is a consideration, it is recommended that the building designer give a suggested moment of inertia I_x. Because the depth of the joist supplied from different joist manufacturers may vary slightly, the depth should be specified when it is critical.

Although steel joist manufacturers may indicate the maximum clear openings for ducts, etc. which can be accommodated through the webs of each depth of their OWSJs, building designers should, in general, show on the building design drawings the size, location and elevation of openings required through OWSJs (Figure 2–28). Large ducts may be accommodated by special design. Ducts which require open panels and corresponding reinforcement of the joist should, where possible, be located within the middle half of the joist. This information is required prior to the time of tendering to permit inclusion of an appropriate cost.

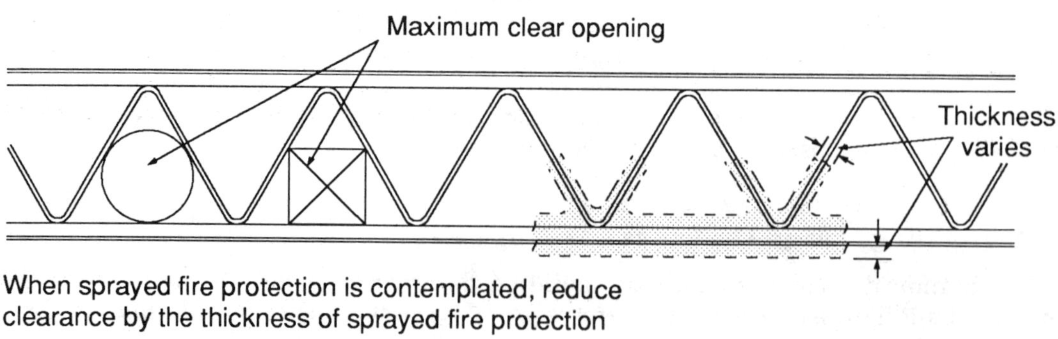

FIGURE 2-28
SIZES OF OPENINGS FOR ELECTRICAL AND MECHANICAL EQUIPMENT

Specific joist designations from a manufacturer's catalogue or from the AISC and Steel Joist Institute of the U.S.A. are not appropriate and should *not* be specified.

16.5.2 Joist Design Drawings

A joist manufacturer's design information may come in varying forms such as: design sheets, computer printout, and tables. Not all joist manufacturers make "traditional" detail drawings.

16.6.1 Loading for Standard Open-Web Steel Joists

To accommodate load tables in standard OWSJ catalogues, for such purposes as estimating, the Standard provides for specific unbalanced and concentrated loading conditions. The concentrated factored loads given for floor and roof joists are total loads to be applied at a panel point. In such load tables, only the total factored load is shown because the ratio of dead-to-live load of the floor system is not known.

Three loading conditions, each considered separately, are specified to guard against overloading of joist components due to unbalanced or concentrated loads.

16.6.2 Loading for Special Open-Web Steel Joists

The four factored live load combinations are consistent with Section 4.1 of the National Building Code of Canada (1990). In particular, as required by the National Building Code of Canada, roofs and the joists supporting them may be subject to uplift loads due to wind.

16.6.3 Design Assumptions

The loads may be replaced by statically equivalent loads applied at the panel points for the purpose of determining axial forces in all members. It is assumed that any moments induced in the joist chord by direct loading do not influence the magnitude of the axial forces in the members. Tests on trusses (Aziz 1972) have shown that the secondary moments induced at rigid joints due to joint rotations do not affect the ultimate axial forces determined by a pin-jointed truss analysis.

16.6.4 Verification of Joist Manufacturer's Design

When there is difficulty in analysing the effect of certain specific conditions, for example a particular web-chord connection, or a geometric configuration of a cold formed chord, a joist manufacturer may elect to verify the design assumption by a test. The numerical factor of 1.10/0.90, stipulated as a multiplier for the factored loads, may be explained as follows:

1) The factor of 1/0.90 is the reciprocal of the resistance factor.
2) The factor of 1.10 provides that the results of limited number of tests bear a similar statistical relationship to the entire series of joists that the average yield strength has to the specified minimum yield strength, F_y.

16.6.7 Tension Chord

A minimum radius of gyration is specified for tension chord members tc provide a minimum stiffness for handling and erection.

Under certain loading conditions, net compression forces may occur in segments of tension chords and must be considered. Bracing of the chord, for compression, may be provided by regular bridging only if the bridging meets requirements of Clause 20.3. As a minimum, lines of bracing are specifically required near the ends of tension chords in order to enhance stability when the wind causes a net uplift.

Bottom chord bracing may be required for continuous and cantilever joists as shown in Figure 2-29.

In those cases, where the bottom chord has little or no net compression, bracing is not

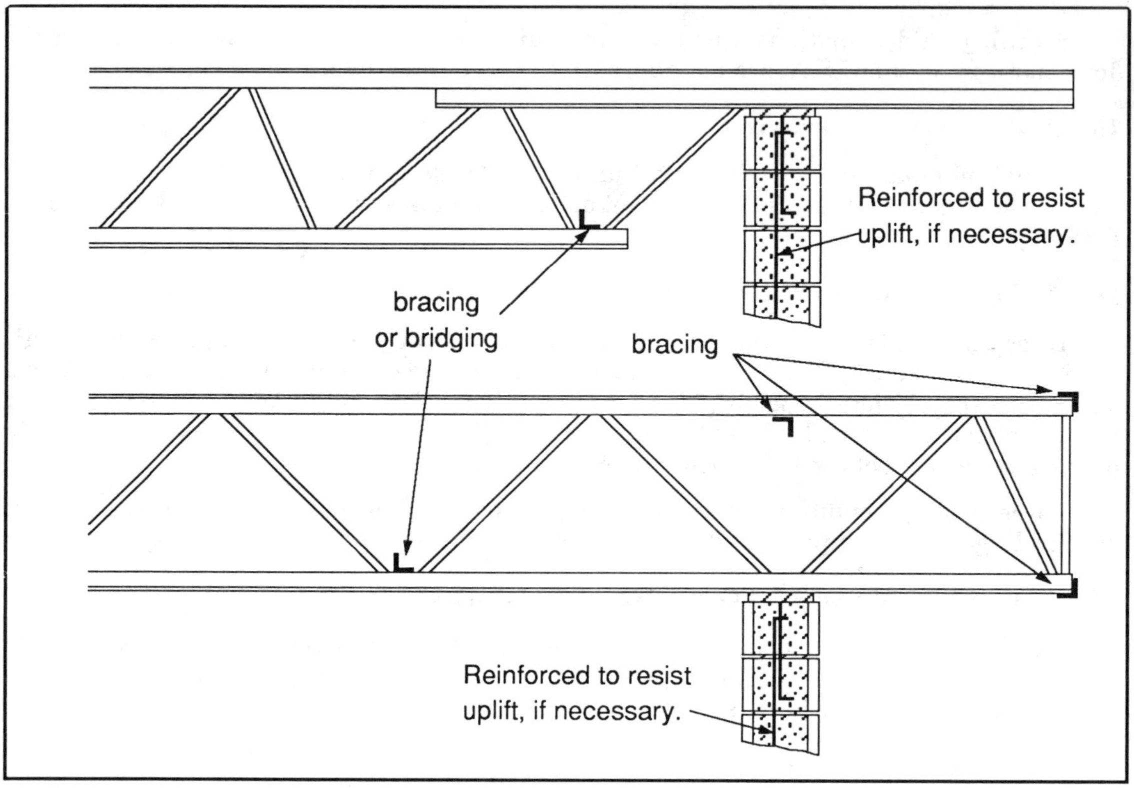

FIGURE 2-29
BRACING AND BRIDGING OF CANTILEVER JOISTS

required for cantilever joists. However, it is generally considered good practice to install a line of bridging at the first bottom chord panel point as shown in Figure 2-29.

16.6.8 Compression Chord

When the conditions set out in Clause 16.6.8.1 are fulfilled, only axial force need be considered when the panel length is less than 610 mm (Kennedy and Rowan 1964). In these cases, the stiffness of the floor or roof structure tends to help transfer loads to the panel points of the joist, thus offsetting the reduction in chord capacity due to local bending. When the panel length exceeds 610 mm, a simplified form of the beam–column interaction formula is used, based on the larger shape factor common to joist chords. When calculating bending moments in the end panel, it is customary to assume the end of the chord to be pinned, even though the joist bearing is welded to its support. The stiffening effect of supported deck or of the web is to be neglected when determining the appropriate width–thickness ratio (Clause 16.6.6.1) of the compression top chord.

The requirement in Clause 16.6.8.4, that the flat width of the chord component be at least 5 mm larger than the nominal dimension of the weld, should be considered an absolute minimum. Increasing the dimension may improve workmanship. See Clauses 16.9.5.1 and 16.9.5.2 regarding workmanship requirements when laying and attaching deck to joists.

16.6.9 Webs

The length of web members for design purposes are shown in Figure 2-30. With the exception of web members made of individual members, the effective length factor is always taken as 1.0. For individual members this factor is 0.9 for buckling in the plane of the web (see Clause C8 of Appendix C), but is 1.0 for buckling perpendicular to the plane of the web.

Web members in tension are not required to meet a limiting slenderness ratio. This is

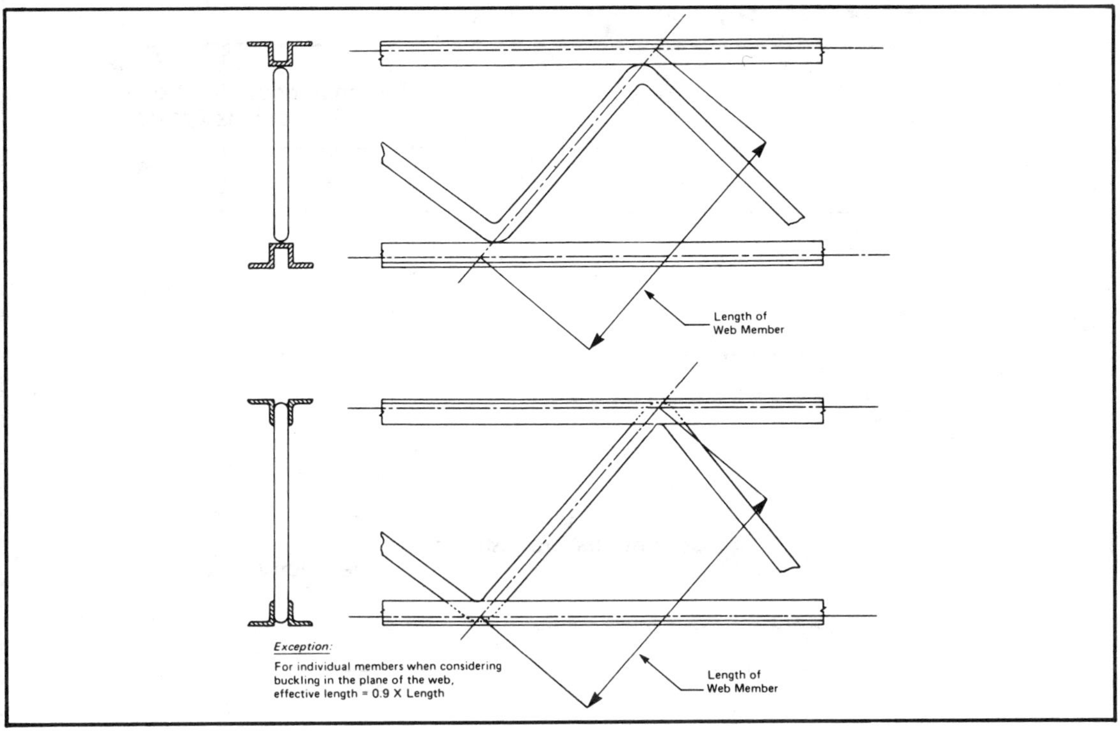

FIGURE 2-30
LENGTH OF JOIST WEB MEMBER

significant when flats are used as tension members; however, attention should be paid to those loading cases where the possibility of shear reversal exists. Under these circumstances, it is likely that diagonals (except for end diagonals) may have to resist compression forces.

16.6.10 Spacers and Battens

Spacers and battens must be an intregal part of the joist and (see Clause 16.6.8.2(c)) steel deck cannot be considered to act spacers or battens.

16.6.11 Connections and Splices

Although splices are permitted at any point in chord or web members, the splices must be capable of carrying the factored loads without exceeding the factored resistances of the members. Butt-welded splices are permitted provided they develop the factored tensile resistance of the member. As a general rule, it is preferable to have the gravity axes of members meet at a common point within a joint. However, when this is not practical, eccentricities may be neglected if they do not exceed those described in Clause 16.6.11.4; see Figure 2-31. Kaliandasani, et al (1977) have shown that the effect of small eccentricities is of minor consequence, except for eccentricities at the end bearing and the intersection of the end diagonal and bottom chord. (See also Clause 16.6.12.4.)

16.6.12 Bearings

16.6.12.1 As required by Clause 16.5.1(c), the factored bearing resistance of the supporting material or the size of the bearing plates must be given on the building design drawings.

16.6.12.2 It is likely that the centre of bearing will be eccentric with respect to the intersection of the axes of the chord and the end diagonal (Figure 2-32). Because the location of the centre of bearing is dependent on the field support conditions, and their construction

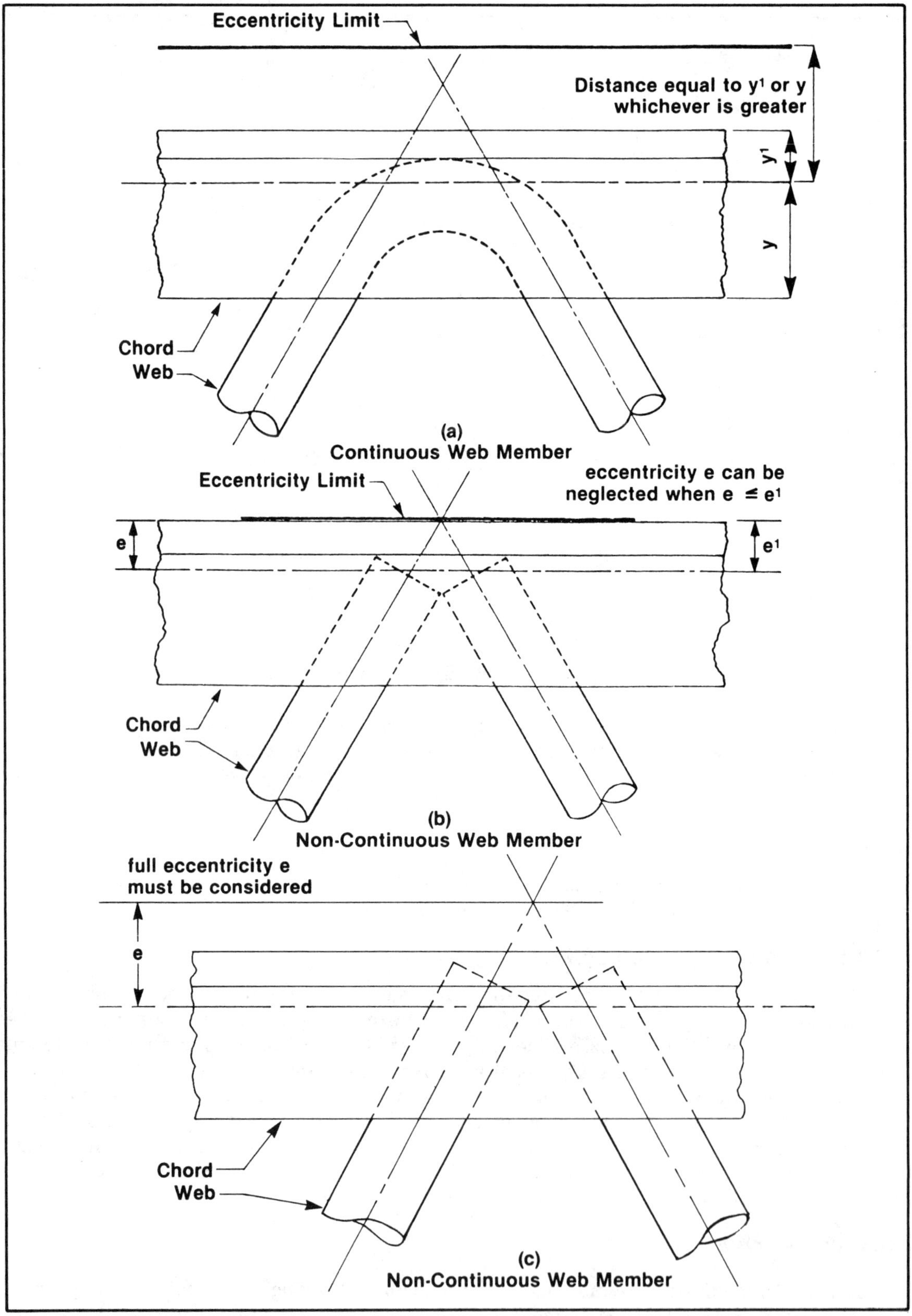

FIGURE 2-31
ECCENTRICITY LIMITS AT PANEL POINTS OF JOISTS

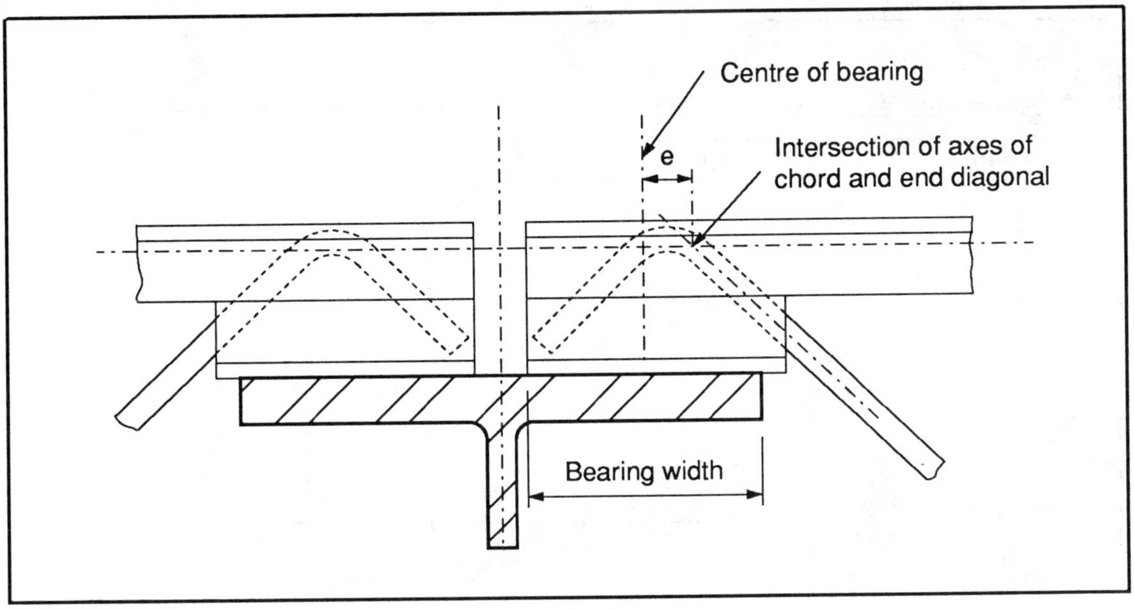

FIGURE 2-32
JOIST END BEARING ECCENTRICITY

tolerances, it may be wise to assume a maximum eccentricity when designing the bearing detail. In lieu of specific information, a reasonable assumption might be an eccentricity of 32 mm because the minimum bearing on a steel support is 65 mm. When detailing joists, care must be taken to provide clearance between the end diagonal and the supporting member or wall (Figure 2-33). A maximum 25 mm is suggested to minimize eccentricities. One solution, to obtain proper bearing, is to increase the depth of the bearing shoe.

For spandrel beams and other beams on which joists frame from one side only, good practice suggests that the centre of the bearing shoe be located within the middle third of the flange of the supporting beam (Figure 2-34(a)). As the depth of bearing shoes vary, the building designer should check with the manufacturer in setting top of steel elevations. By using a deep shoe, interference between the support and the end diagonal will be avoided (Figure 2-34(b)).

If the support is found to be improperly located, such that the span of the joist is increased, the resulting eccentricity may be greater than that assumed. Increasing the

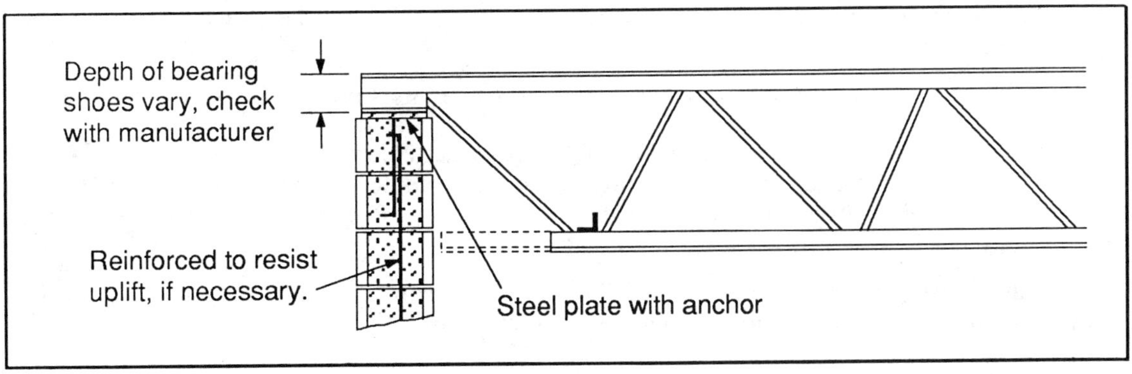

FIGURE 2-33
JOISTS BEARING ON STEEL PLATE ANCHORED TO CONCRETE AND MASONRY

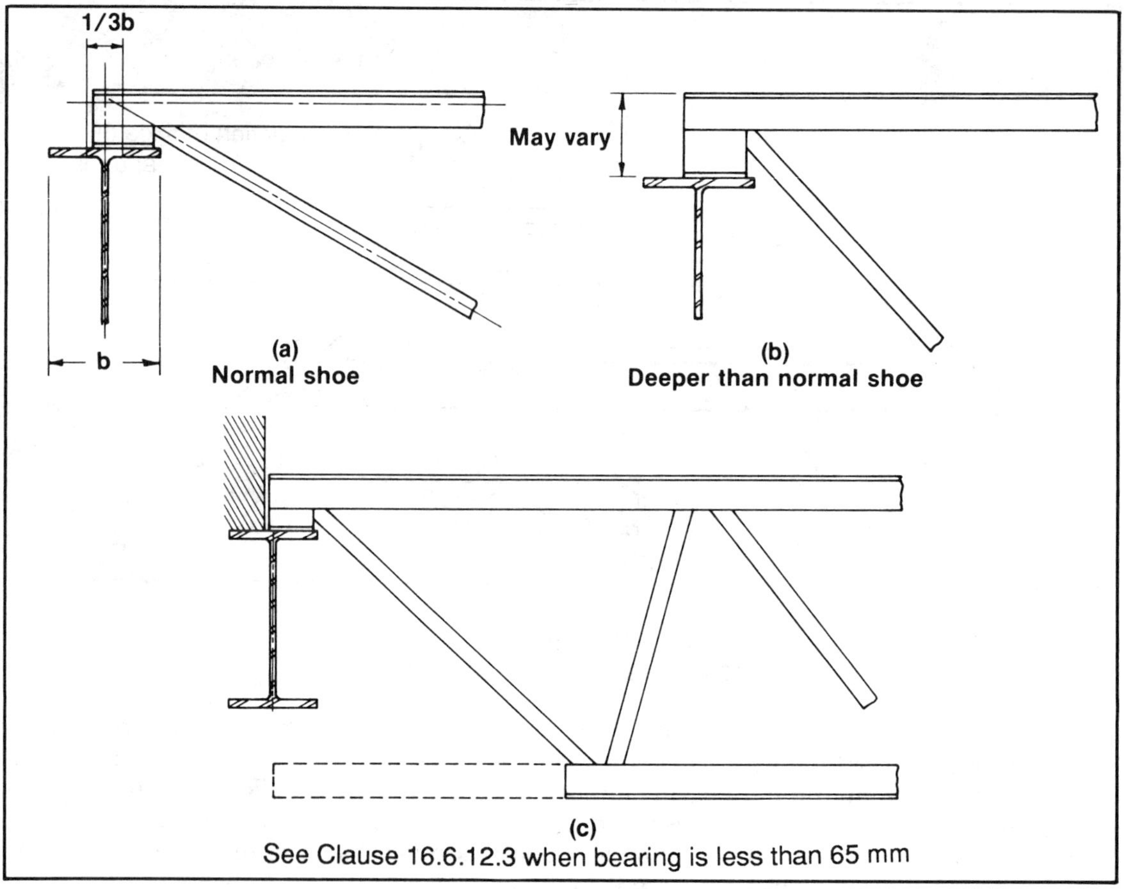

FIGURE 2-34
JOISTS BEARING ON STEEL

length of the bearing shoe to obtain proper bearing may create the more serious problem of increasing the amount of eccentricity.

16.6.13.2 The function of tie joists is to assist in the erection and plumbing of the steel frame. Either the top or bottom chord is connected by bolting and, after plumbing the columns, the other chord is usually welded (Figure 2-35). In most buildings, tie joists remain as installed with both top and bottom chords connected; however, current practices vary throughout Canada with, in some cases, the bottom chord connections to the columns being made with slotted holes. Shrivastava *et al.* (1979) studied the behaviour of tie joist connections and concluded that they may be insufficient to carry lateral loads which could result from rigid bolting.

The designation *tie joist* is not intended to be used for joists participating in frame action.

16.6.13.3 When joists are used as part of a frame to brace columns, or to resist lateral forces on the finished structure, the appropriate moments and forces are to be shown on the building design drawings to enable the joists and the joist-to-column connections to be designed by the joist manufacturer.

In cantilever girder suspended span (Gerber) roof framing, joists may also be used to provide stability for girders passing over columns (Chien 1989).

16.6.14 Deflection

The method of computing deflections given in Clause 16.6.14.2 has been verified by tests (Kennedy and Rowan 1964).

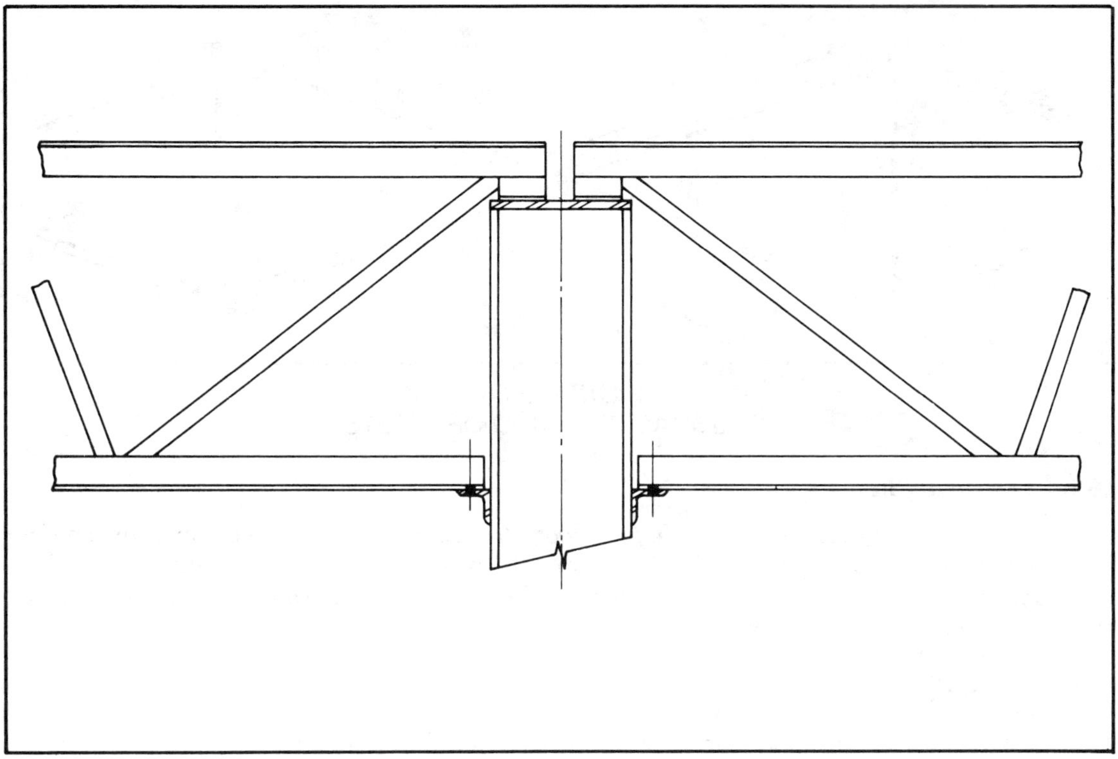

FIGURE 2-35
TIE JOISTS

16.6.15 Camber

The nominal camber based on Clause 16.6.15 is tabulated in Table 2–1 rounded to the nearest millimetre. Manufacturing tolerances are covered in Clause 16.11.9. The maximum difference in camber of 20 mm, set to limit the differential camber between two adjacent joists, is reached at a span of 17 000 mm.

Table 2–1

Camber (mm)			
Span	Nominal Camber	Minimum Camber	Maximum Camber
Up to 6 000	–	3	10
7 000	3	3	11
8 000	4	3	12
9 000	6	3	14
10 000	7	3	16
11 000	8	3	17
12 000	10	3	19
13 000	12	3	21
14 000	14	4	23
15 000	16	6	26
16 000	18	8	28

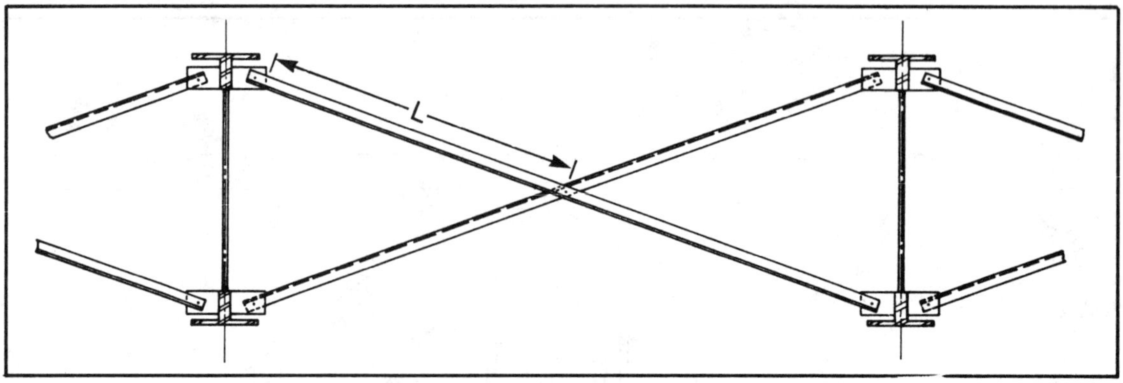

FIGURE 2-36
DIAGONAL BRIDGING OF JOISTS

16.6.16 Vibration

Appendix G of S16.1-M89, *Guide for Floor Vibrations*, contains recommendations for floors using steel joists. By increasing the floor thickness (mass), both the frequency and the peak acceleration are reduced, thus reducing the annoyance more efficiently than by increasing the moment of inertia (I_x) of the joists.

16.6.17 Welding

16.6.17.3 A fabricator or erector certified in Division 3 may meet the requirement of the Standard by having the work done under the supervision of a fabricator certified in Division 1 or 2.1. Many welded joints used in joists are not prequalified under CSA W59, therefore the certified fabricator must have all these welded joints qualified by the Canadian Welding Bureau.

16.6.17.6 Flux and slag are removed from all welds to assist in the inspection of the welds, as well as to increase the life of the protective coatings applied to the joists.

16.7 Stability During Construction

A distinction is made between bridging, put in to meet the slenderness ratio requirements for top and bottom chords, and the temporary support required by Clause 16.7 to hold joists against movement during construction. Permanent bridging, of course, can be used for both purposes.

16.8 Bridging

Figures 2-36 to 2-38 provide illustrations of bridging and details of bridging connections.

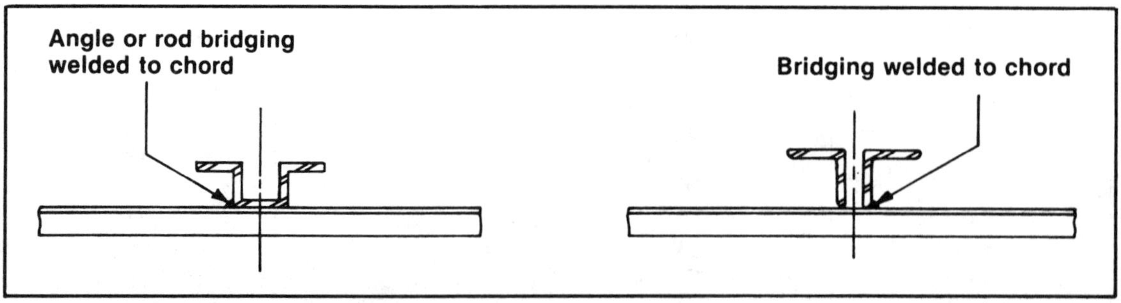

FIGURE 2-37
HORIZONTAL BRIDGING CONNECTIONS TO THE JOIST'S TOP CHORD

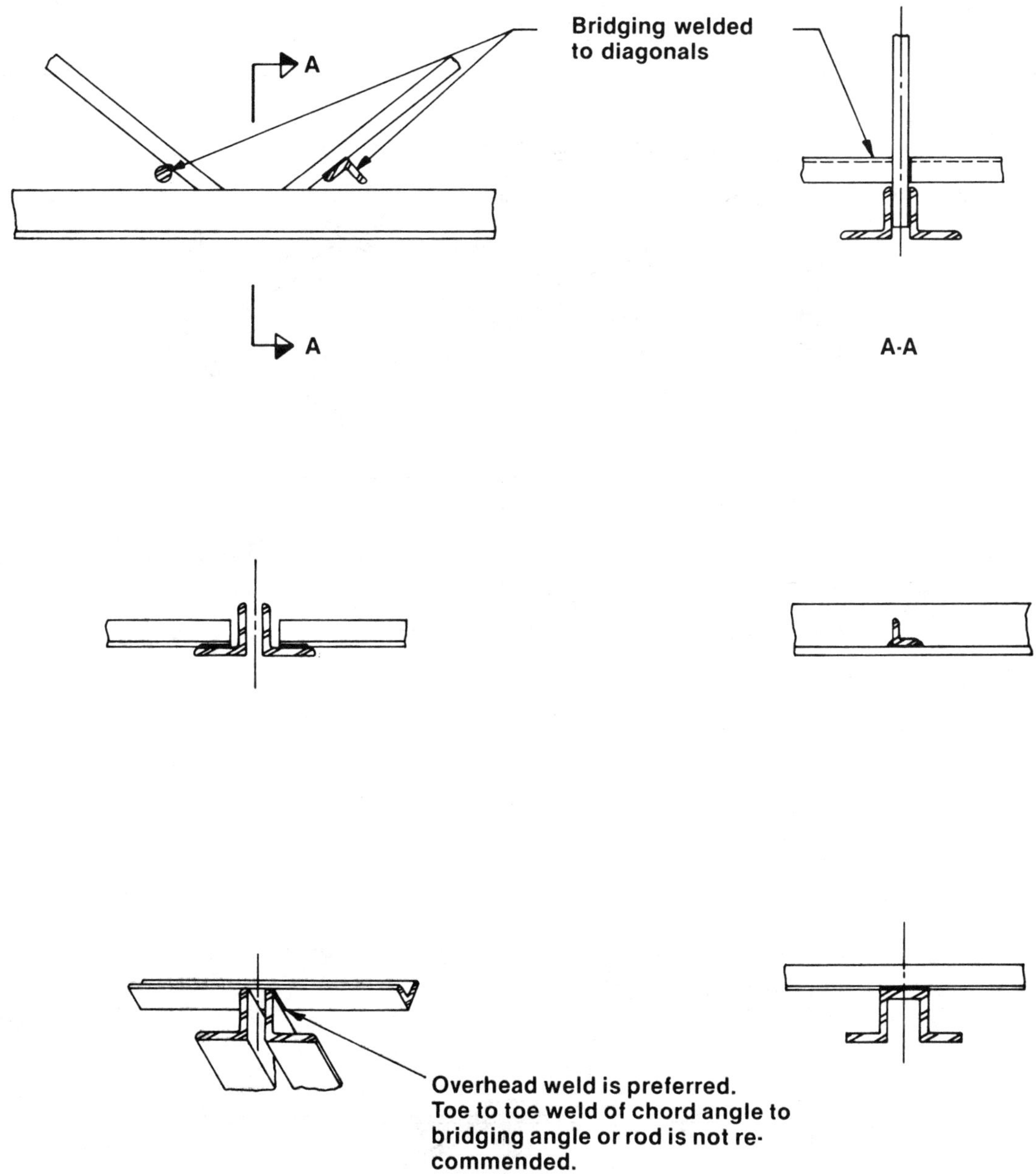

FIGURE 2-38
HORIZONTAL BRIDGING CONNECTIONS TO JOIST BOTTOM CHORD

16.8.7 Anchorage of Bridging

Ends of bridging lines may be anchored to the adjacent steel frame or adjacent concrete or masonry walls as shown in Figure 2-39.

JOISTS - ANCHORAGE OF BRIDGING

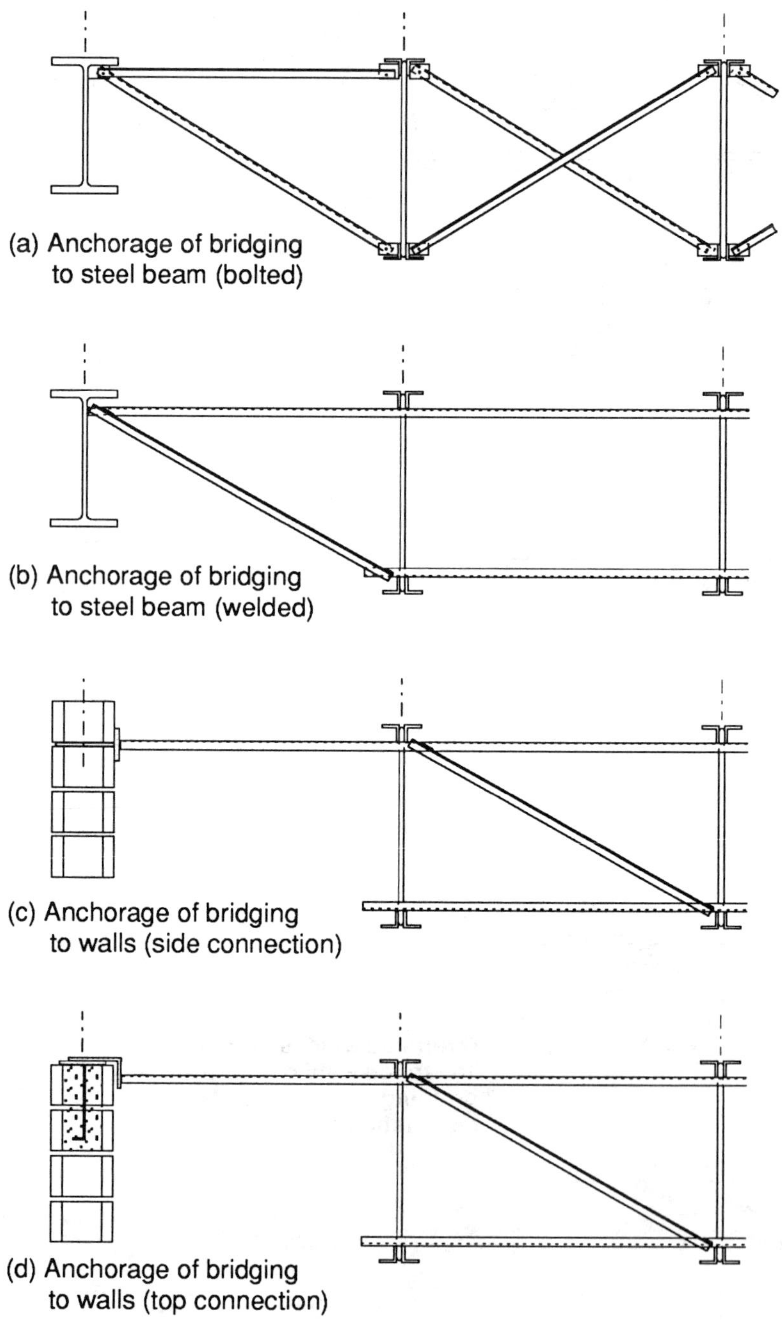

FIGURE 2-39
ANCHORAGE OF JOIST BRIDGING

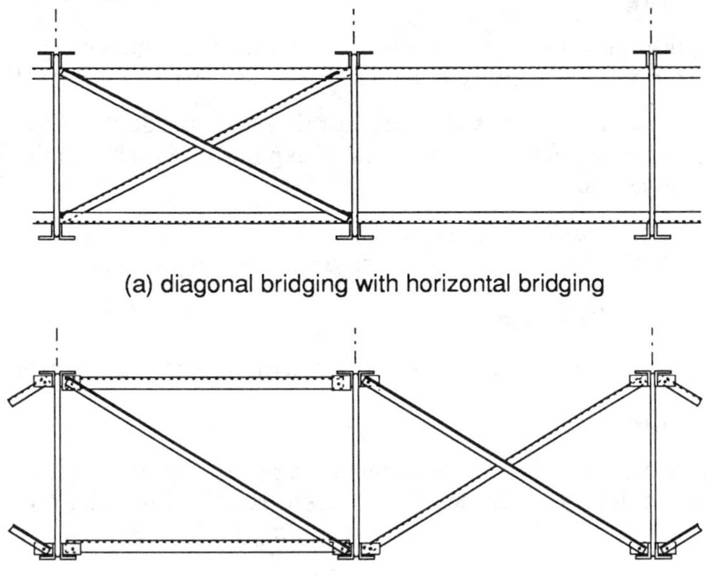

(a) diagonal bridging with horizontal bridging

(b) horizontal bridging with diagonal bridging

FIGURE 2-40
BRACING OF JOIST BRIDGING

Where attachment to the adjacent steel frame or walls is not practicable, diagonal and horizontal bridging shall be provided in combination between adjacent joists near the ends of bridging lines as shown in Figure 2-40. Joists bearing on the bottom chord will require bridging at the ends of the top chord.

16.8.9 Spacing of Bridging

Either horizontal or diagonal bridging is acceptable, although horizontal bridging is generally recommended for shorter spans, up to about 15 m, and is usually attached by welding. Diagonal bridging is recommended for longer spans and is usually attached by bolting. Bridging need not be attached at panel points and may be fastened at any point along the length of the joists. When horizontal bridging is used, bridging lines will not necessarily appear in pairs as the requirements for support of tension chords are not the same as those for compression chords. Because the ends of joists are anchored, the supports may be assumed to be equivalent to bridging lines.

16.9.1 Decking to Provide Lateral Support

When the decking complies with Clause 16.9 and is sufficiently rigid to provide lateral support to the top (compression) chord, the top chord bridging may be removed when it is no longer required. Bottom (tension) chord bridging is permanently required such that the unsupported length of the chord does not exceed 240r, as defined in Clause 16.8.9.

16.9.5 Installation of Steel Deck

16.9.5.1 Workmanship is of concern when decking is to be attached by arc-spot welding to top chords of joists. When the joist location is marked on the deck as the deck is positioned, the welders will be more likely to position the arc-spot welds correctly.

16.9.5.2 Arc-spot welds for attaching the deck to joists are structural welds and require proper welding procedures.

16.10 Shop Painting

Interiors of buildings conditioned for human comfort are generally assumed to be of a non-corrosive environment and therefore do not require corrosion protection.

Joists normally receive one coat of paint suitable for a production line application. This paint is generally adequate for three months of exposure, which should be ample time to enclose, or paint, the joists.

Special coatings, and paints with special preparations, are expensive because they have to be applied individually to each joist by spraying or other means.

16.11 Manufacturing Tolerances

Figure 2-41 illustrates many of the manufacturing tolerance requirements.

16.12.3 Quality Control

When testing forms part of the manufacturers normal quality control programme, the test shall follow steps 1 to 4 of the loading procedure given in Part 5 of *Steel Joist Facts* (CISC 1980).

16.13.1 Erection Tolerances

Figure 2-42 illustrates many of the erection tolerance requirements.

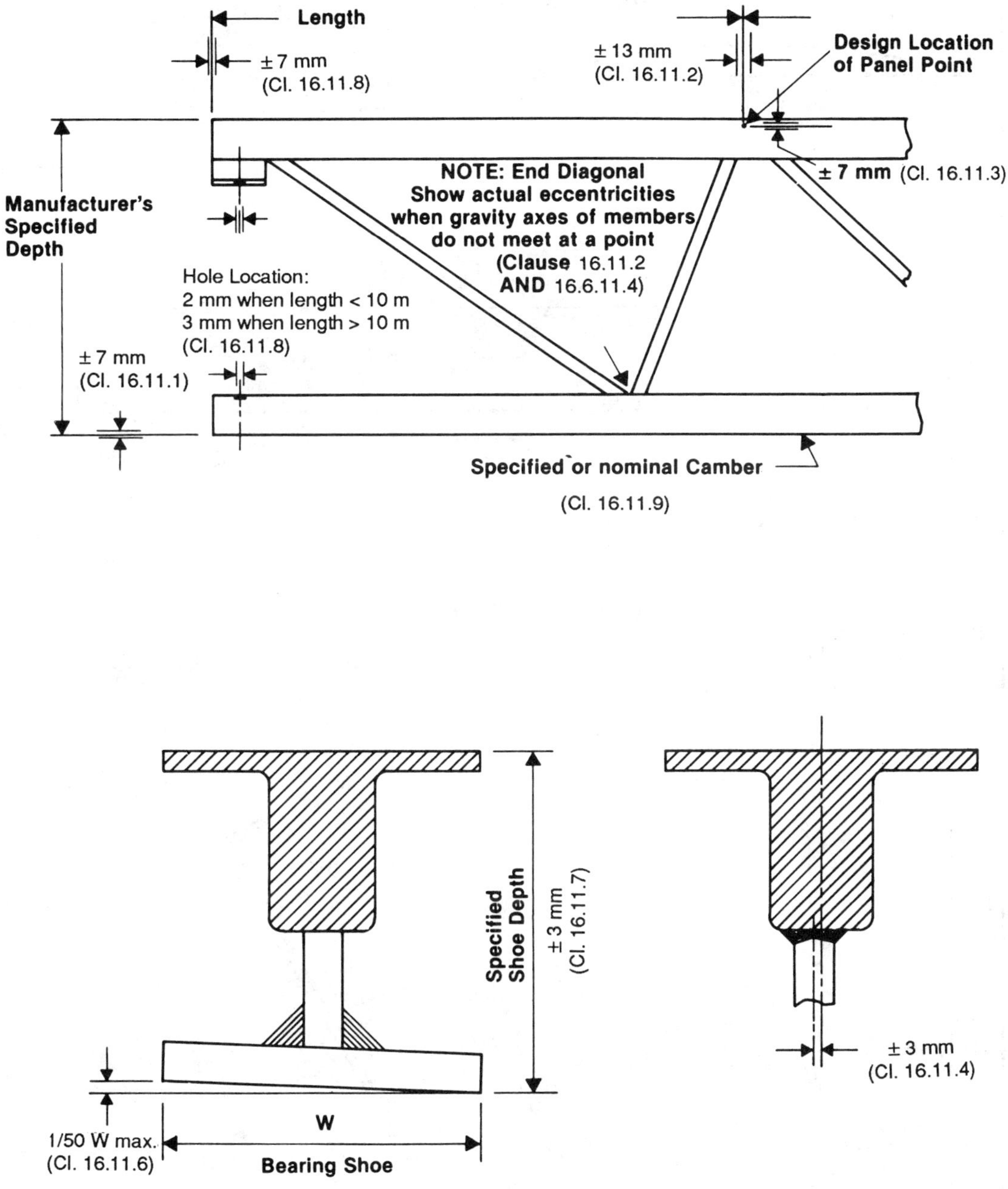

FIGURE 2-41
JOIST MANUFACTURING TOLERANCES

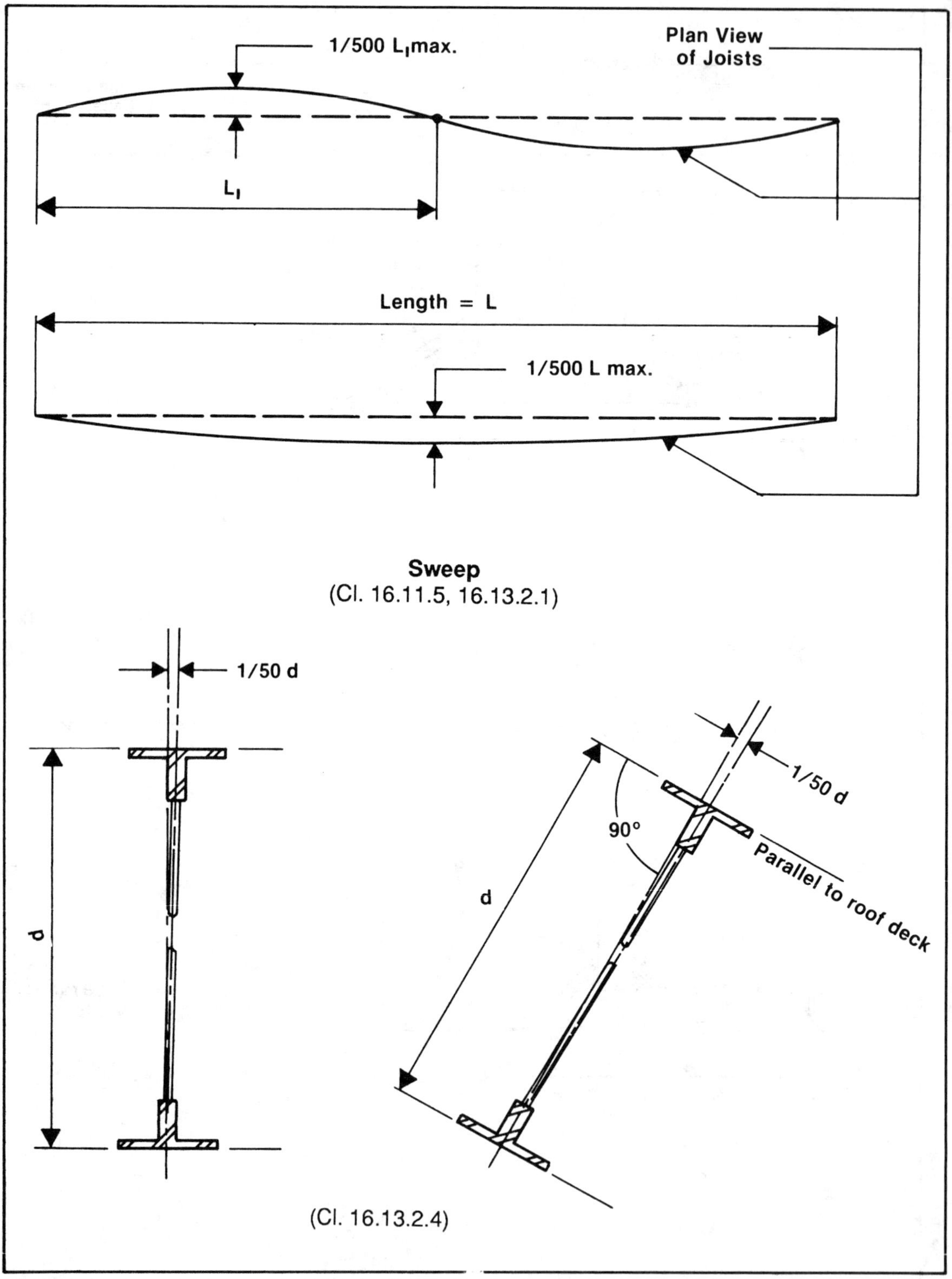

FIGURE 2-42
JOIST ERECTION TOLERANCES

17. COMPOSITE BEAMS

17.2 Definitions

Figure 2–43 illustrates various cases of effective slab and cover slab thickness.

17.3.1 Deflections

The moment of inertia is reduced from the transformed value to account for the increased flexibility resulting from partial shear connection, and for interfacial slip, similar to that coefficient proposed by Grant *et al.* (1977). However, the 0.85 accounts for the loss in stiffness due to interfacial slip, even with full shear connection.

The increase of the elastic deflection of 15% for creep is an arbitrary but reasonable value. For information on shrinkage induced deflection, see Chien and Ritchie (1984), Montgomery *et al.* (1983), Brattland and Kennedy (1986), and Jent (1989).

17.3.2 Vertical Shear

17.3.3 End Connections

These clauses follow from the assumption that the concrete does not carry any vertical shear.

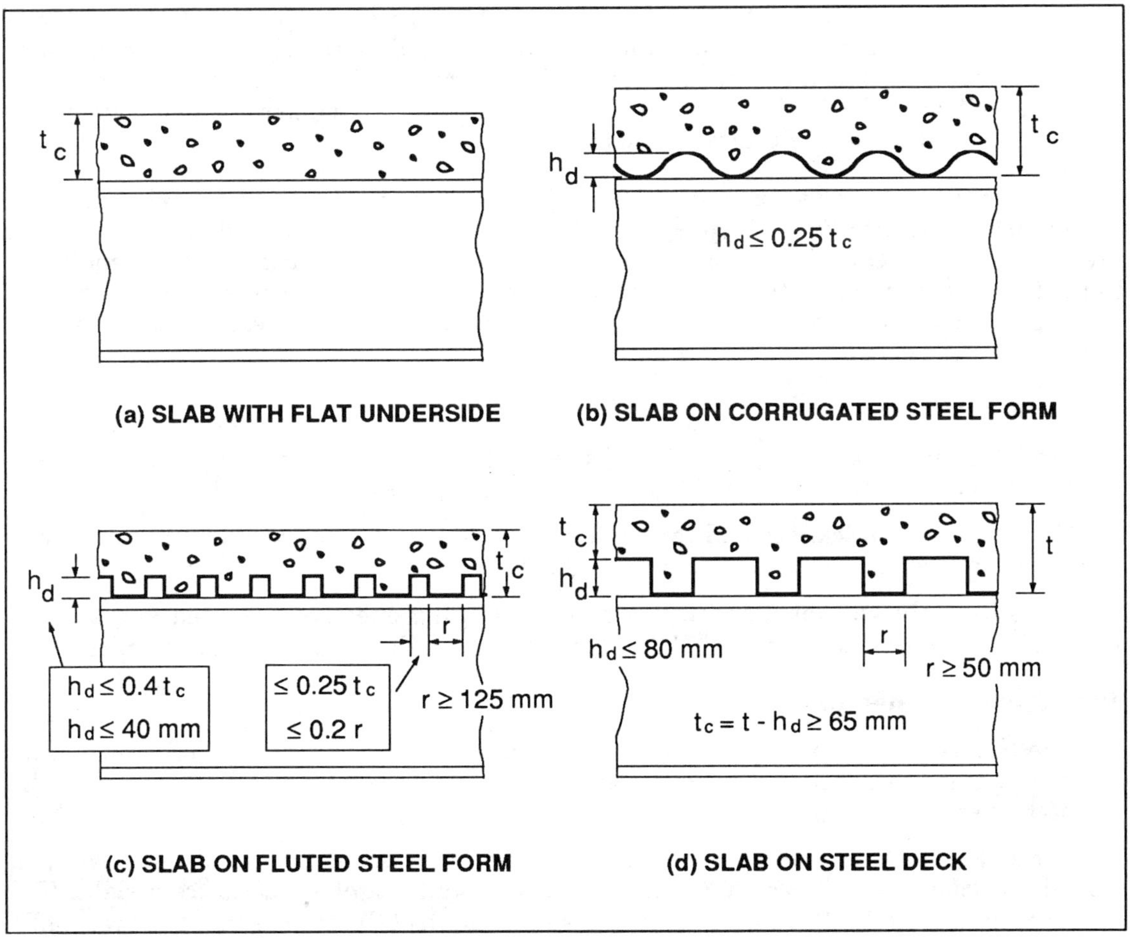

FIGURE 2-43
EFFECTIVE SLAB THICKNESS FOR COMPOSITE BEAMS

17.4 Design Effective Width of Concrete

As there was no basis in fact, the effective width limit based on slab thickness has been removed, as was the case with the AISC (1986b) and Eurocode No. 4 (1987) specifications. Although the effective width rules were formulated on the basis of elastic conditions (Robinson and Wallace, 1973, Adekola, 1968), the differences at ultimate load do not significantly affect the moment resistance of the composite beam (Elkelish and Robinson, 1986; Hagood *et al.*, 1968; Johnson, 1975; Heins and Fan, 1976).

17.5 Slab Reinforcement

The requirements for slab reinforcements have been revised and expanded.

17.5.2 The minimum requirement of two 15M bars at the ends of beams supporting ribbed slabs perpendicular to the beam is new. Jent (1989) has verified, experimentally, the effectiveness of the detail proposed by Ritchie and Chien (1980).

17.5.3 The longitudinal shear forces generated by interconnecting solid concrete slabs to steel sections, trusses, or joists by means of shear connectors may cause longitudinal cracking of the slab directly over the steel. This effect is independent of any flexural cracking which may occur due to the slab spanning continuously over supports, although the two effects may combine. Longitudinal shear cracking is more apt to start from the underside of the solid slab, whereas flexural cracking is more apt to start at the top surface of the slab. Investigations by Johnson (1970), El-Ghazzi *et al.* (1976), and Davies (1969) have shown that a minimum area of transverse reinforcing steel is required to improve the longitudinal shear capacity of a solid slab composite beam. The minimum reinforcement ratio is the same as that specified in CAN/CSA Standard A23.3 (CSA 1984) for temperature and shrinkage reinforcement in reinforced concrete slabs.

17.5.4 For the same reasons as for Clause 17.5.3, a minimum transverse reinforcement ratio of 0.002 has also been specified for composite beams with ribbed slab when the ribs are parallel to the beam span. This ratio is reduced to 0.001 when the ribs are perpendicular to the beam span, because the steel deck provides a measure of transverse reinforcement. Reinforcement of the cover slab may also be necessary for flexure, fire resistance, shrinkage, or temperature effects.

17.6 Interconnection

When unpainted sections, trusses, or joists are totally encased in concrete as specified, effective interconnection is obtained and no shear connectors are required.

The total sheet thickness and the total amount of zinc coating are limited in order to achieve sound welds.

Tests have shown that a shear connector is not fully effective if welded to a support which is too thin or flexible (Gobel 1968). For this reason the stud diameter is limited.

17.7 Shear Connectors

The factored resistance of end welded studs in a solid slab is treated differently from those in a ribbed slab, for which different provisions are specified, depending upon the orientation of the deck ribs.

For end welded studs in a solid slab, the values given in Clause 17.7.2.1 are based on work by Olgaard *et al.* (1971) in both normal and lightweight solid concrete slabs. The limiting value of $\phi_{sc} A_{sc} F_u$ represents the tensile strength of the stud as the stud eventually bends over and finally fails in tension.

According to Clause 17.7.2.2(a), the factored shear resistance for studs in solid slabs is applicable to ribbed slabs within the limitations given (Johnson 1975). The stud resistance is reduced in narrow flutes based on Grant *et al.* (1977). In order to minimize excessive localized stresses in concrete, it is recommended that the lateral spacing centre-to-centre of studs used in pairs be not less than four stud diameters. The minimum recommended longitudinal spacing of connectors, in both solid slabs and ribbed slabs with ribs parallel to the beam, is six diameters (Olgaard *et al.* 1971).

New provisions have been incorporated for ribbed slabs, with ribs perpendicular to the beam (Jayas and Hosain, 1988 and 1989). Push-out tests, as well as full size beam tests, indicated that failure in this type of composite beam would likely occur due to concrete pull-out. The equations of Clause 17.7.2.3, similar to those suggested by Hawkins and Mitchell (1984), provide better correlation to test results than those using the reduction factor method adopted by AISC (1986) and Figure 2-44. The maximum spacing limits specified for mechanical ties in Clause 17.8 is applicable to headed studs, as they function in this capacity.

Pullout area for specific deck profiles and studs are given in Part 5. Further information on end welded studs can be found in Johnson (1970), Chien and Ritchie (1984), and Robinson (1988).

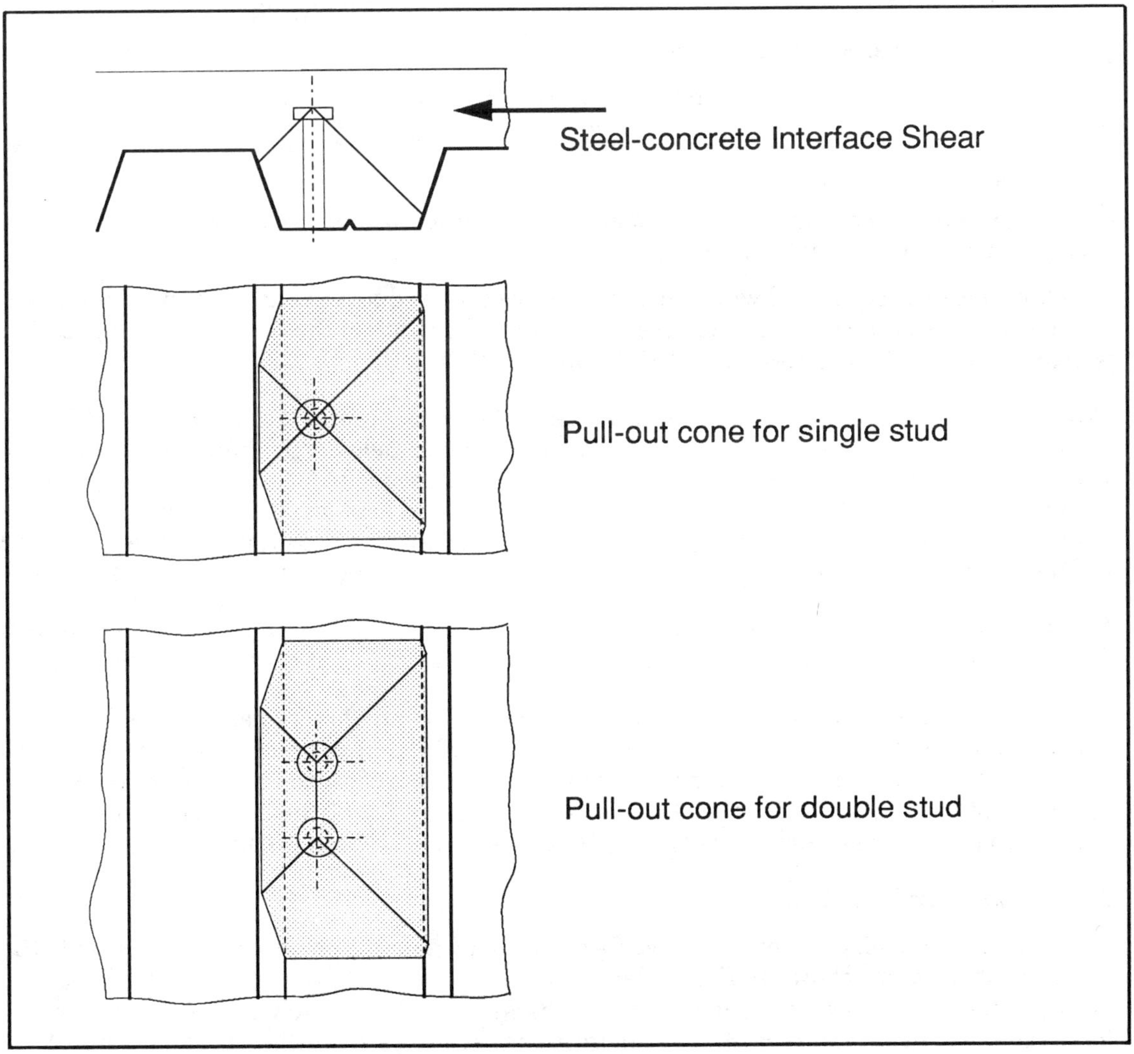

FIGURE 2-44
PULLOUT SURFACE AREAS WITH RIBBED METAL DECK

17.7.3 The shear value of channel connectors is based on Slutter and Driscoll (1965).

17.9 Design of Composite Beams with Shear Connectors.

The factored moment resistance of a composite flexural member is computed based on the ultimate capacity of the cross-section (Robinson, 1969; Vincent, 1969; Hansell and Viest, 1971; Robinson and Wallace, 1973; Tall *et al.*, 1974) where the following assumptions are made:

- Concrete in tension is neglected;
- Only the lower chord of a steel joist or truss is considered effective when computing the moment resistance;
- The internal couple consists of equal tension and compression forces;
- The forces are obtained as the product of a limit states stress (ϕF_y for steel and $0.85\ \phi_c\ f'_c$ for concrete) times the respective effective areas; and,
- To take into account the greater variability of concrete elements, the resistance factor is taken as 0.60 for concrete as compared to 0.90 for steel.

Three design cases are considered:

- Case 1 representing full shear connection with the plastic neutral axis in the slab;
- Case 2 representing full shear connection with the plastic neutral axis in the steel section; and,
- Case 3 representing partial shear connection for which the plastic neutral axis is always in the steel section.

Only Case 1 is permitted when joists or trusses are used to prevent buckling of top chord and overloading of the shear connectors. For Case 3, the depth of the concrete in compression is determined by the expression for "a" (Robinson 1969).

17.9.4 Robinson (1988) and Jayas and Hosain (1989) show that a lower limit of 40% of full shear connection is acceptable for strength calculations. Below this value, the interfacial slip is such that integral composite action cannot be assured. A lower limit is given for deflection designs as deflections are computed at specified load levels. This latter provision is used where the flexural strength is based on the bare steel beam; but, the increased stiffness due to the concrete is considered for deflection calculations.

17.9.5 Between the point of zero and maximum moment, a horizontal force associated with the internal resisting couple must be transmitted across the steel-concrete interface.

17.9.8 Uniform spacing of shear connectors is generally satisfactory because the flexibility of the connectors provides a redistribution of the interface shear among them. However, to ensure that sufficient moment capacity is achieved at points of concentrated load, the second provision of this clause is invoked. As the moment capacity of the steel section does not depend on shear connectors, this capacity is subtracted from both M_f and M_{f1}.

17.9.9 Longitudinal Shear

In order to develop the compressive force in the portion of the concrete slab outside the potential shear plane shown in Figure 2-45, net shear forces, totalling V_u, must be developed on these planes. The expressions for shear resistance are based on Mattock (1974). Values for semi-low density and low density concrete are given by Mattock *et al.* (PCI) and Chien and Ritchie (1984).

17.10 Design of Composite Beams Without Shear Connectors

This conservative approach assumes that the composite section is about 10% stronger

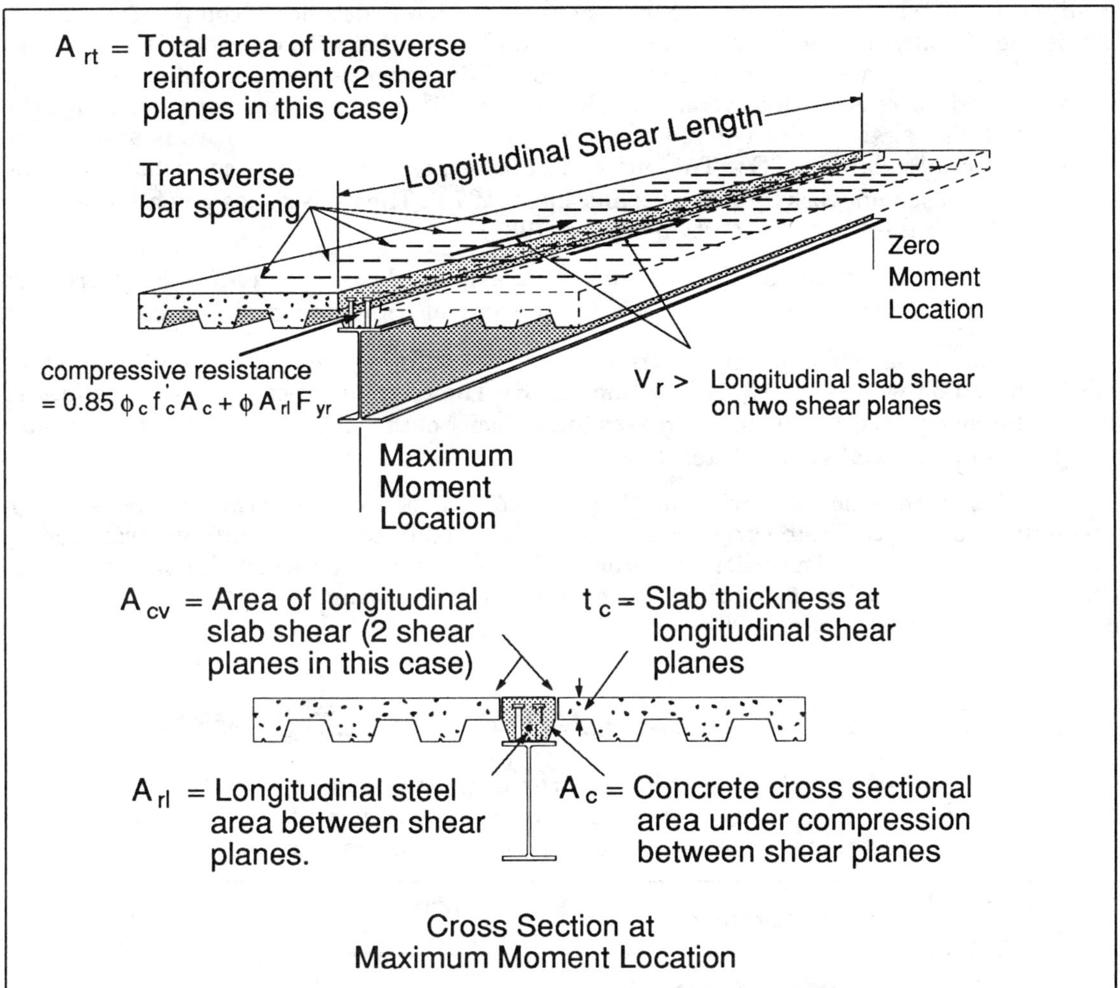

$$\mathbf{V_u = \sum q_r - 0.85\, \phi_c f'_c A_c - \phi A_{rl} F_{yr}}$$

$$\mathbf{V_r = 0.80\ \phi A_{rt} F_{yr} - 2.76\ \phi_c A_{cv} \le 0.5\ \phi_c f'_c A_{cv}}$$

FIGURE 2-45
POTENTIAL LONGITUDINAL SHEAR PLANES

than the bare steel member, although the moment resistance computed according to Clause 17.10.2 typically gives a larger value.

17.11 Unshored Beams

This provision guards against permanent deformations under specified loads by limiting the total stress in the bottom fibre of the steel section. The ultimate strength of the composite beam, which exhibits ductile behaviour is not affected by the stress state at the specified load level.

18 COMPOSITE COLUMNS (Concrete-Filled Hollow Structural Sections)

CIDECT (1970), Knowles and Park (1970), Wakabayashi (1977), STELCO (1981), and

Budijgnto (1983) have demonstrated that the compression resistance of composite columns, consisting of hollow structural sections (HSS) completely filled with concrete, is due to both the steel and the concrete core. The contribution of the concrete core is greatest for stocky members and decreases with increasing slenderness. The hollow steel section carries the same axial load as if it were not filled with concrete, and the concrete core is assumed to carry additional axial load, C'_r. This method of superposition and the contribution of the concrete are based on work by Knowles and Park (1970). The resistances given are in close agreement with the CIDECT solution (Stelco, 1981).

The value of E_c is modified to account for creep and leads to a reduced compressive resistance of the concrete core by increasing the value of λ and hence decreasing C'_r.

The triaxial load effect on the concrete due to the confining effect of the walls of circular HSS is based on work by Virdi & Dowling (1976). The triaxial effects increase the failure load of the concrete ($\tau' > 1.0$) and decreases the capacity of the steel section ($\tau' < 1.0$) because the steel is in a biaxial stress state.

The first interaction equation in Clause 18.5 is based on the lower bound solution assuming that the concrete carries its portion of the axial load and bending is resisted by the steel section alone. The second interaction expression is used when the moment resistance of the section is based on fully plastic stress blocks ignoring the portion of the concrete core in tension (Barber *et al*, 1987).

19. GENERAL REQUIREMENTS FOR BUILT-UP MEMBERS

The term *built-up* member refers to any structural member assembled from two or more components. Such members may be used to resist compression, tension, or bending and the requirements for fastening together the various components vary accordingly.

Tension Members	Requirements	Tension Members	Requirements
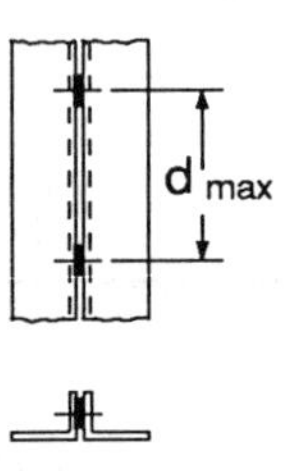	**TWO ROLLED SHAPES NOT IN CONTACT** d_{max} = 300 x Least radius of gyration of one component.	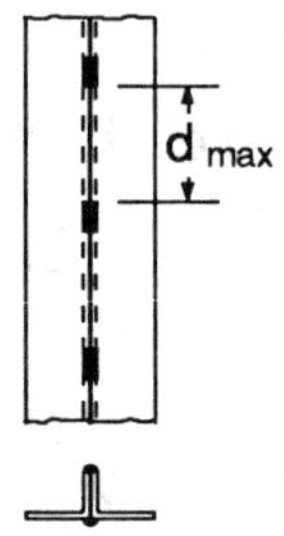	**TWO ROLLED SHAPES NOT IN CONTACT** d_{max} = 600 mm d_{max} may be increased, when justified.
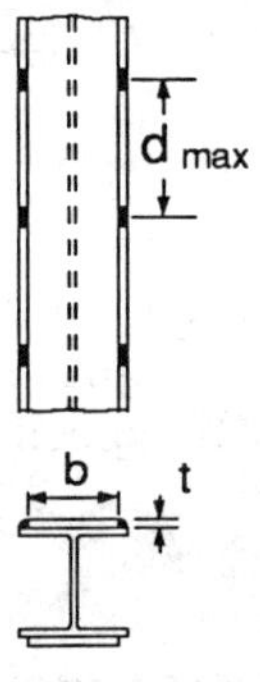	**SHAPE AND PLATE IN CONTACT** d_{max} = 36 t or 450 mm whichever is lesser.	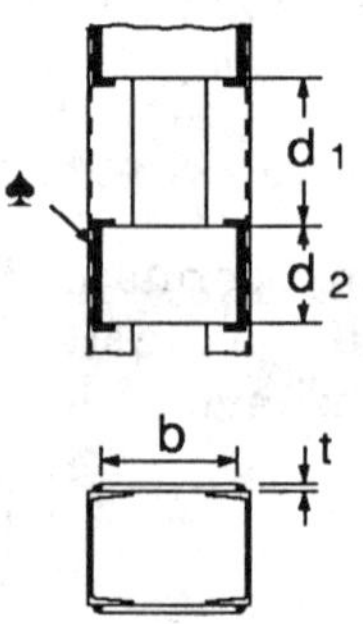	**BATTENS** $b \leq 60\,t$ $\quad d_2 \geq \frac{2b}{3}$ d_1 = 300 x Least radius of gyration of one component. ♠ For intermittent welds or fasteners maximum longl. pitch = 150 mm

FIGURE 2-46(a)
BUILT-UP MEMBER DETAILS

Compression Members	Requirements
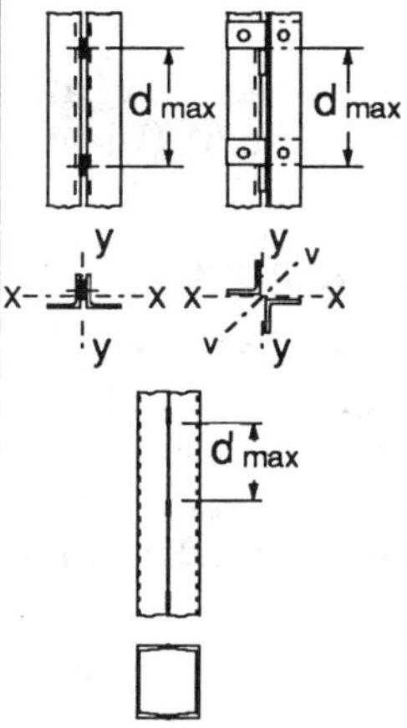 	**ROLLED SHAPES** $d^{*}_{max} = \left(\frac{KL}{r}\right) r_{min}$ KL/r = design slenderness of member as a whole. r_{min} = Least radius of gyration of one component * For starred angles, 1/3 points. Cl. 19.1.5 19.1.3
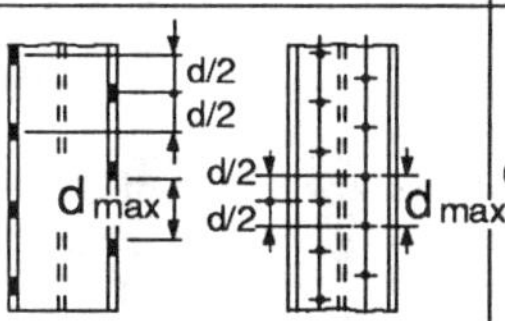 	**STAGGERED FASTENERS OR WELDS** $d_{max} = \frac{525\,t}{\sqrt{F_y}}$ or 450 mm t = Outside plate thickness 19.1.3
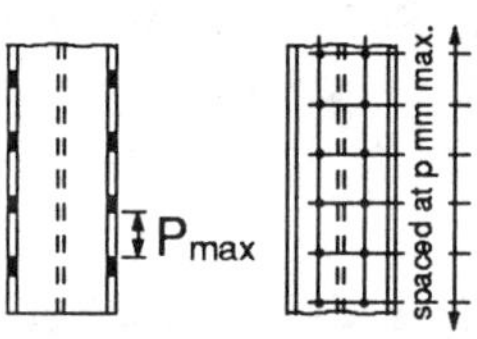 	**FASTENERS OR WELDS NOT STAGGERED** $P_{max} = \frac{330\,t}{\sqrt{F_y}}$ or 300 mm t = Outside plate thickness 19.1.3
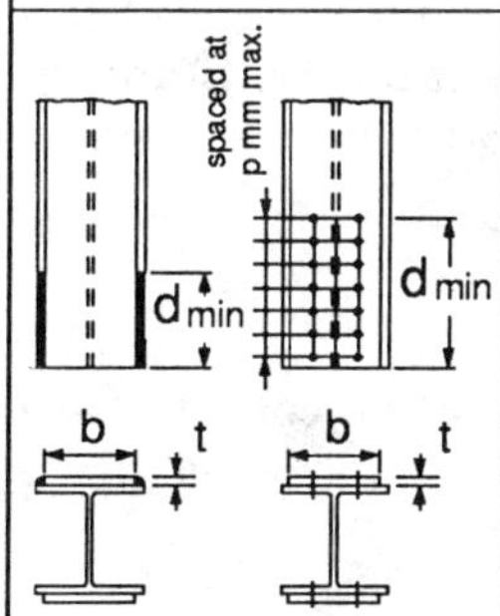 	**ENDS OF BUILT-UP COLUMNS** Welded Connection : $d_{min} = b$ Bolted Connection : $d_{min} = 1.5\,b$ P_{max} = 4 x diameter of fastener 19.1.2
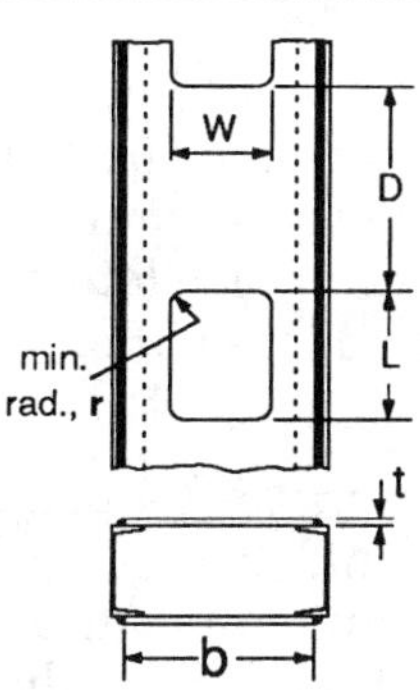 	**PERFORATED COVER PLATES** $b \leq \frac{840\,t}{\sqrt{F_y}}$ $L \leq 2W$ $D \geq b$ $r \geq 40$ mm Note: For bolted fabrication b≥ 400 mm is preferred. 19.1.15

Compression Members	Requirements
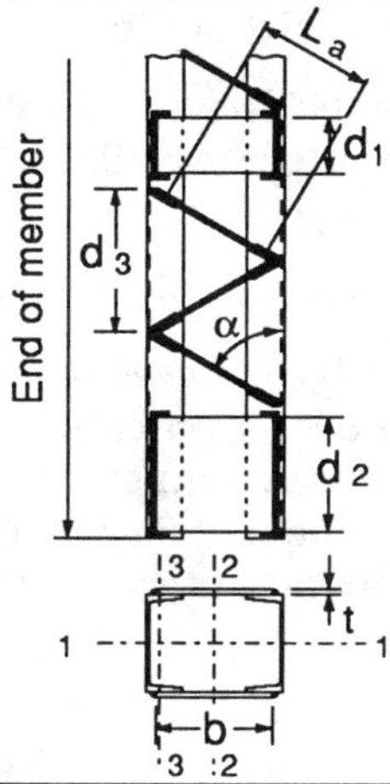 	**LACING AND TIE PLATES** $b \leq 60\,t$ $d_1 \geq b/2$ $d_2 \geq b$ $d_3 \leq \left(\frac{KL}{r_1}\right) r_3 \leq \left(\frac{KL}{r_2}\right) r_3$ $L_a \leq$ 140 x rad. of gyration of lacing member $\alpha \geq 45°$ KL = Eff. length of member referred to appropriate axis. 19.1.(9 - 13)
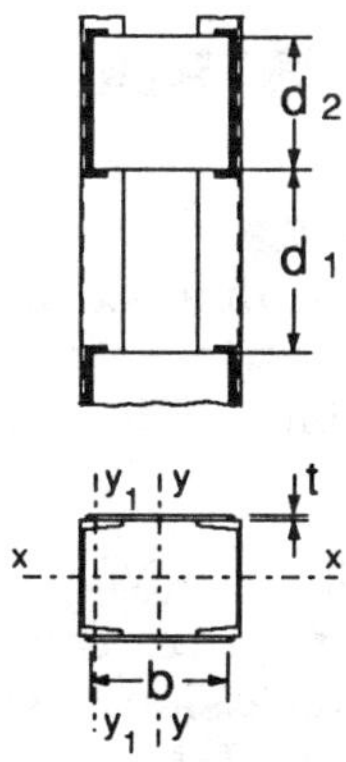 	**BATTENS** $b \leq 60\,t$ $d_2 \geq b$ 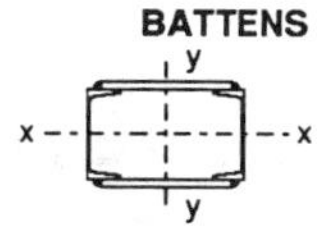 when $\frac{K_y L_y}{r_y} \leq \frac{0.8\,K_x L_x}{r_x}$ $\frac{d_1}{r_{y1}} \leq 50 \leq \frac{0.7\,K_x L_x}{r_x}$ 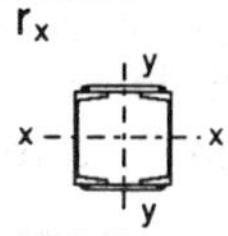 when $\frac{K_y L_y}{r_y} > \frac{0.8\,K_x L_x}{r_x}$ $\frac{d_1}{r_{y1}} \leq 40 \leq \frac{0.6\,K_y L_y}{r_y}$ 19.1.16

Beams and Grillages	Requirements
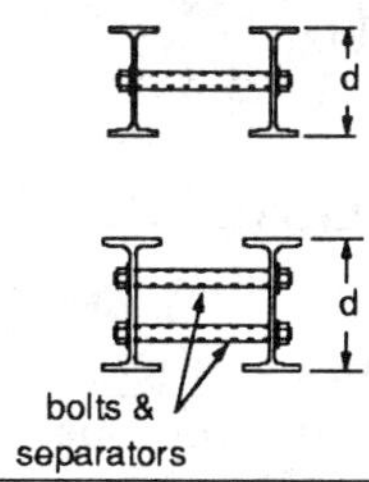 	**NON-LOAD-SHARING BEAMS** Not less than one bolt : d < 300 mm Two or more bolts : d ≥ 300 mm Centres of separator groups ≤ 1500 mm 19.3
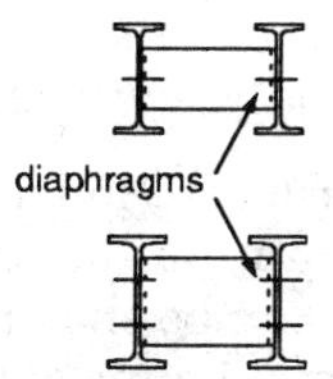 	**LOAD-SHARING BEAMS** Diaphragm shall have sufficient stiffness to distribute required loads. Centres of diaphragms ≤ 1500 mm 19.3

FIGURE 2-46(b)
BUILT-UP MEMBER DETAILS

The diagrams of Figure 2-46 illustrate the main provision of Clause 19. Many of the provisions are based on long established practice and have proven satisfactory.

Tension members are stitched together sufficiently to work in unison and to minimize vibration. For exposed members, components in contact should be fitted tightly together to minimize corrosion problems (Brockenbrough 1983).

When a built-up column buckles, shear is introduced in lacing bars (Clause 19.1.9) and battens and their connections (Clause 19.1.17), in addition to any transverse shears (Bleich, 1952).

Further discussion on columns with lacing and battens is given in Galambos (1988).

For compression members composed of two or more rolled shapes connected at intervals, Clause 19.1.4 requires the use of an equivalent slenderness ratio, increased to take into account the flexibility of the interconnector. This increase is applied to the axis of buckling where the buckling mode of the member involves relative deformation that produces shear forces (see Clause 19.1.6) in the interconnectors between the individual shapes (Duan & Chen, 1988).

The requirements for starred angles are based on work by Temple et al (1986), who showed that with fewer interconnectors the buckling strength was reduced.

20. STABILITY OF STRUCTURES AND INDIVIDUAL MEMBERS

20.1 General

Increased emphasis continues to be placed in this edition of the Standard on the designer's responsibility to ensure stability of the structure and of the individual members (see also Commentary on Clause 8.6). Clause 8.6 requires the structure as a whole to resist the PΔ effects.

The stability of the column-girder assembly and the girder web, when a girder is continuous over a column, requires careful assessment. The column, girder web, and the girder flange are all in compression, creating a condition of inherent instability. Stability can be achieved by providing lateral support to the girder-column joint or by properly designed web stiffeners restraining the rotation of the joint.

20.2 Stability of Columns

This clause addresses the local lateral forces that must be resisted when columns are erected out-of-plumb. The flexibility of the resisting elements should be considered when determining the lateral forces to be resisted.

20.3 Stability of Beams, Girders and Trusses

The requirements of Clause 20.3 provide reasonable bracing forces for most situations. Winter (1958) shows that the bracing force required to maintain stability depends on the initial out-of-straightness of the member, the deflection that occurs at the brace point in developing the brace force, and the number of braced points. Massey (1962) examined lateral bracing forces for beams while Zuk (1956) and Lay and Galambos (1966) considered bracing requirements for structures analysed plastically.

When an element in a structure must resist the "bracing forces" from more than one member, the average maximum out-of-straightness of the members should be used to compute the bracing forces. It can be shown statistically that the average maximum out-of-straightness is equal to the maximum out-of-straightness of one member divided by the square root of the number of members. Beaulieu and Adams (1980) provide more guidance in selected cases.

20.3.4 Because the shear centre of an asymmetric section does not coincide with the centroid, this section may be unintentionally loaded so as to produce torsion and biaxial bending. Both the connections and the members providing reactions should be checked.

21. CONNECTIONS

21.3 Restrained Members

When the compressive or tensile force transmitted by a beam flange to a column (approximately equal to the factored moment divided by the depth of the beam) exceeds the factored bearing or tensile resistance of the web of the column, stiffeners are required to develop the load in excess of the bearing or tensile resistance.

Taking the length of the column web resisting the compressive force as the thickness of the beam flange plus 5k (Graham *et al.* 1959) results in the first equation given in Clause 21.3 for the bearing resistance of columns with Class 1 and 2 webs.

For members with Class 3 and 4 webs, the bearing resistance of the web is limited by its buckling strength. The second equation given in Clause 21.3(a) for these members results in a bearing resistance, when the same loaded length (t_b + 5k) is used as in the first equation, about equal to the minimum of the AISC LRFD Specification (1986b) where the loaded length is taken as the clear depth of the column. Although not stated, the bearing resistance computed from the second equation should not exceed the first. In both expressions, if the compression flange is applied at the end of a column, the loaded length should be reduced to t_b + 2.5k.

Graham *et al.* (1959) also show, based on a yield line analysis, that the column flange bending resistance, when subject to a tensile load from the beam flange, can be taken conservatively to be $7\,t_c^{\,2}\,F_{yc}$. Tests have shown that connections proportioned in accordance with this equation have carried the plastic moment of the beam satisfactorily.

21.4 Connections of Tension or Compression Members.

Obviously, the end connections must transmit the factored loads. The 50% rule which has existed in structural design standards for many years guards against providing a connection inconsistent with the member it connects when the member size has been selected for some criterion other than strength.

21.6 Lamellar Tearing

In cases where shrinkage results as a consequence of welding under highly restrained conditions, very large tensile strains may be set up. If these are transferred across the through-thickness direction of rolled structural members or plates, lamellar tearing may result. Thornton (1973) and AISC (1973) give methods of minimizing lamellar tearing . Figure 2-47 illustrates one such case.

21.7 Placement of Fasteners and Welds

Gibson and Wake (1942) have shown that, except for cases of repeated loads, end welds on tension angles, and other similar members need not be placed so as to balance the forces about the neutral axis of the member.

21.8 Fillers

In bearing-type shear connections development of the filler before the splice material diminishes bending of the bolt. In slip-critical joints, tests with fillers up to 1 inch in thickness and with surface conditions comparable to other joint components show that the fillers act integrally with the remainder of the joint and they need not be developed before the splice material (Kulak *et al.* 1987).

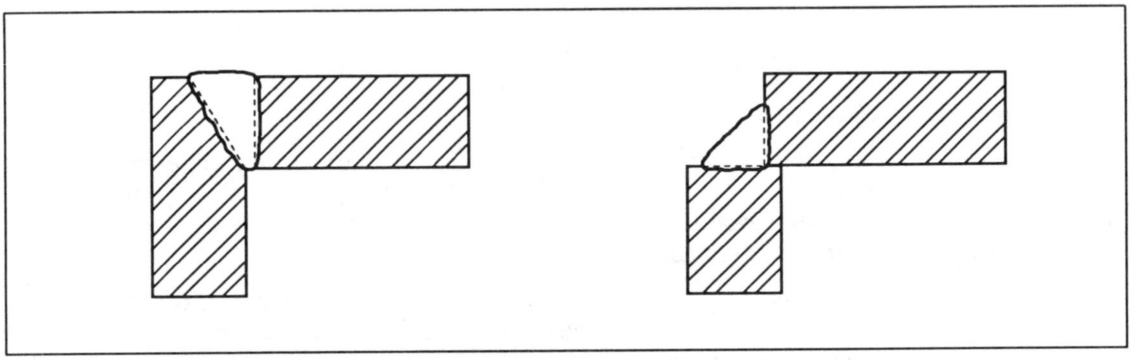

FIGURE 2-47
DETAILS TO MINIMIZE LAMELLAR TEARING

21.10 Fasteners and Welds in Combination

21.10.1 Because the relative movement in a slip-critical connection at the specified load level is small, the welds and the bolts will share in carrying the load. However, as the bolt capacity is based on a slip criterion at the specified load level and the welds are usually proportioned for factored loads, the weld strengths are discounted to the specified load level by multiplying by the factor 0.70. Moreover, the factored resistance of the connection must be equal to or greater than the effect of the factored loads. As an accurate prediction of the factored resistance involves the use of the load–deformation response of the weld group and of the bolts (Holtz and Kulak 1970), the Standard adopts the conservative approach of using the larger of the bolt or weld group factored resistance.

21.12 Connections Using Bolts

Connections using bolts may either have the bolts pretensioned or not depending on the joint type, loads to which it is subject, the type of holes used, as detailed in Clause 21.12.2.

Bolts which are not pretensioned must be installed to a snug-tightened condition. These may be A307, A325, or A490 bolts. Previous editions of the Standard permitted the use of A307 bolts snug-tightened but implied that all A325 and A490 bolts were pretensioned. Because the ultimate limit states of shear through the bolt and bearing on the plate material are not significantly affected by the level of pretension (Kulak *et al.* 1987), it is only logical to permit bolts of higher strength than the A307 bolt to also be installed snug-tight in similar connections. The previous edition, S16.1-M84, recognized this, in part, in its Clause 22.7.1 which relaxed the inspection requirements of certain bearing-type joints.

Only high-strength bolts, such as A325 and A490 and their metric equivalents, may be used in joints requiring pretensioned high-strength bolts.

21.12.2 Connections Using Pretensioned High-Strength Bolts

Snug-tightened bolts may be used, except for the specific cases given in Clause 21.12.2 where the use of pretensioned high-strength bolts is required.

(a) Pretensioning of the bolts provides the clamping force in slip-critical connections and hence the slip resistance at the specified load level.

(b) Pretensioning of the bolts provides energy dissipation under cyclic earthquake loading in connections proportioned in accordance with the seismic requirements given in Clause 27, although these connections are proportioned as bearing-type connections for the ultimate limit state.

(c) and (d) Pretensioning in both these connections, ensures that the bolts don't work loose and, of course, is necessary to ensure adequate fatigue behaviour.

(e) An example of such a connection is a tee hanger connection. (Such connections are not slip-resistant nor bearing-type.) Pretensioning reduces the prying action and the stress range.

(f) In these connections pretensioning prevents gross movement within the joint. See also Clause 23.3.2 to determine in which cases slip-critical connections are required.

21.13 Welds

Although the use of welded connections is not restricted, the use of certain weld details may not be permitted in fatigue or seismic loading. See Clauses 14 and 27.

Welding of high strength bolts should not be done as it will adversely affect the strength of the bolt.

22. BOLTING DETAILS

The details given reflect good practice and, for the most part, have remained unchanged over several editions of this Standard. Kulak *et al.* (1987) contains a comprehensive summary of bolt requirements.

23. STRUCTURAL JOINTS USING ASTM A325M, A490M, A325, OR A490 BOLTS

23.1 General

A325M, A490M, A325 and A490 bolts are produced by quenching and tempering (ASTM 1990a, 1990b, 1990c, 1990d). A325 bolts are not as strong as A490 bolts but have greater ductility. For this reason and reasons of availability, the use of A490 bolts is subject to restrictions as discussed subsequently.

The behaviour of the joint depends both on how the bolts are loaded and installed. For the first time, in this Standard, the use of snug-tightened high strength bolts is permitted. As there are four basic types of connections, three with bolts in shear and one with bolts in tension, it is absolutely essential that the drawings specify the type of connections used.

Bolts subject to tension must be pretensioned. Bolts subject to shear in slip-critical connections must also be pretensioned. Bolts subject to shear in bearing-type connections may either be pretensioned or snug-tightened.

Kulak *et al.* (1987) shows that the ultimate shear and bearing resistance of a bolted connection are not dependent on the pretension in the bolt. As the number of situations (Clause 21.12.2) where pretensioning is required is very limited, the norm for building construction is that snug-tightened bearing-type connections are used. Departures from the norm are only to be made with due consideration. Few joints in building construction are subject to frequent load reversal nor are there many situations where a one-time slip into bearing cannot be tolerated.

As a result of normal fabrication practice, minor misalignment of bolt holes may occur in connections with two or more bolts. Such misalignment, if anything, has a beneficial effect (Kulak *et al.* 1987) resulting in a stiffer joint, improved slip resistance and decreased rigid body motion.

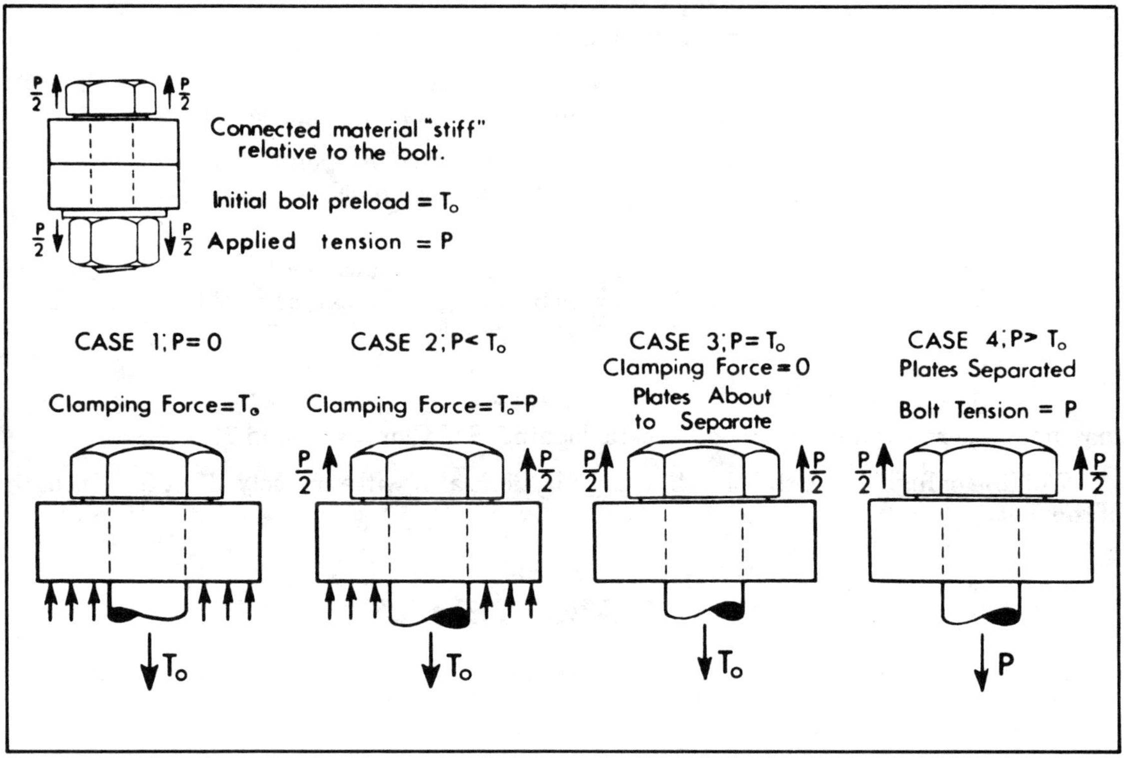

FIGURE 2–48
EFFECT OF APPLIED TENSION ON TIGHTENED HIGH-STRENGTH BOLTS

23.1.4 In Figure 2–48, where it is assumed that the connected material is relatively stiff with respect to the bolt, it is shown that, when an external tensile force is applied to the connected parts, the applied external force and the internal clamping force in the bolt are not additive. Measurement of actual bolt forces in connections of practical sizes has shown that there is an increase in bolt force over that assumed but it is usually only about 5 to 10%. Thus, the bolt needs to be proportioned only for the factored tensile load.

The effect of prying action on bolt tension, Figure 2–49, is an important design consideration. Kulak *et al.* (1987) summarizes the criteria for the design of tension connections when prying action is present. For load reversal or repeated load situations these connections must be proportioned so that prying is avoided.

23.2 Bolts, Nuts and Washers

The normal bolt assembly consists of a A325 or A490 bolt, with a heavy hex head, restricted thread length, and coarse threads, and a heavy hex nut. Proprietary versions are available which differ from the normal in various aspects, and, in some cases, may offer one or more advantages. Their use is permissible under the conditions set forth in Clause 23.2.4.

Designers are cautioned to ascertain the availability of A325M and A490M bolts before specifying their use. Alternatively, by basing the design on the bolt with the lower resistance either the Imperial series or the metric series bolt could be used depending upon availability.

Galvanized A325 bolts are permitted; however, galvanized A490 bolts are not permitted as they are especially susceptible to stress corrosion and hydrogen stress cracking (Kulak *et al.* 1987). The rotation requirement of this clause provides a means of testing the galvanized assembly for proper fit and for proper thread lubrication.

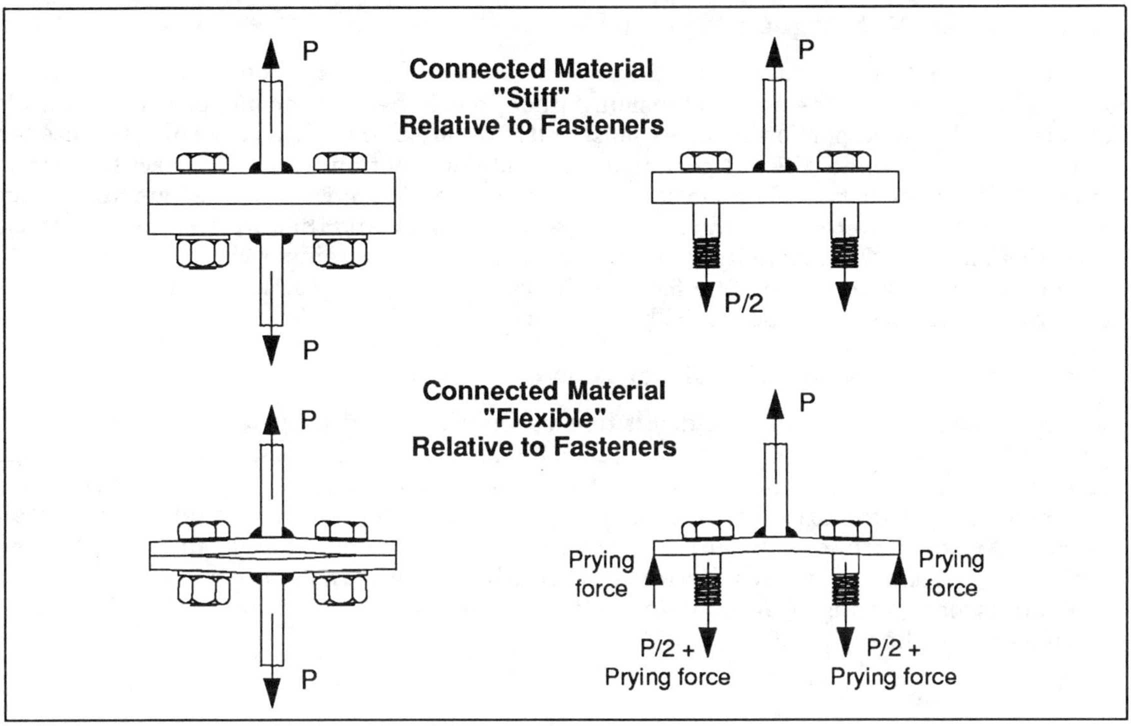

FIGURE 2–49
EFFECT OF PRYING ACTION ON BOLT TENSION

23.3 Bolted Parts

Details on the sizes and types of holes (standard, oversize, or slotted) permitted for bearing-type and slip-critical connections are given. While the Standard permits several hole making methods, punching and drilling are the most common. Incremental punching is sometimes used in fabricating slotted holes, especially long slots. Clause 23.3.2(e) allows selected Imperial bolts in metric holes as the industry phases from one series to the other. A hardened washer, when required, is intended to cover the hole (or bridge the slot) if it occurs in an outer ply. One-third more bolts are required in slip-critical connections using long sloted holes to account for the reduced clamping force that otherwise would be present (Kulak *et al.* 1987)

The treatment of the faying surfaces within the plies of slip-critical joints shall be consistent with the mean slip coefficient chosen for design (Clause 13.12). For clean mill scale, the surfaces must be free of substances which would reduce the slip coefficient. For other coatings, the surface preparation, coating application and curing should be similar to those used in the tests to obtain the slip coefficient. The Steel Structures Painting Council provides specifications (SSPC 1989) for cleaning and coating of steel structures. Kulak *et al.* (1987) provide information on slip for various surface conditions and coating types.

23.4 Installation

Bolts required to be pretensioned must be tightened to tensions of at least 70% of their specified minimum tensile strength. All other bolts need only be tightened sufficiently to ensure that all the plies are in firm contact, ie. snug-tightened.

Except when galvanized, A325 bolts may be reused once or twice, providing that proper control on the number of reuses can be established (Kulak *et al.* 1987; RCSC, 1987). A490 bolts should not be reused.

23.5 "Turn-of-Nut" Tightening.

Any installation procedure used for pretensioning high strength bolts involves elongating the bolt to produce the desired tension. Although the shank of the bolt probably remains elastic, the threaded portion behaves plastically. Because the bolt as a whole is tightened into the inelastic range (the flat portion of the load-deformation curve) the exact location of "snug-tight" is not critical. For the same reason, application of the specified amount of nut rotation results in pretensions that are not greatly variable and that are greater than those prescribed in Table 7. Although there is a reasonable margin against twist-off, the tolerance on nut rotation prescribed in the footnote to Table 8 is good practice, particularly when galvanized A325 bolts or black A490 bolts are used.

23.6 Tightening by Use of a Direct Tension Indicator

The Standard permits use of direct tension indicator bolting systems. All of these are proprietary in nature relying on a physical change in a part of the bolt system. This change attempts to indicate that the minimum bolt tension has been achieved. Systems that rely on irreversable deformations or fracture of a part serve only to indicate that during installation, a force or torque sufficient to deform or fracture the part had been reached. Even with such a system, reliable results are unlikely to be achieved unless the installation follows procedures for snugging of the joint and patterned tightening operations as given in that for "turn-of-nut" tightening (RCSC, 1987).

23.7 Inspection

Bolts, nuts and washers are normally received with a light residual coating of oil. This coating is not detrimental, in fact it is desirable, and should not be removed. Galvanized bolts and/or nuts may be coated with a special lubricant to facilitate tightening. Obviously, this should not be removed.

The inspection procedures used depend on whether the bolts are specified to be snug-tightened or pretentioned. In all cases, the inspector shall observe that the procedure for the installation of the bolts conforms with the requirements of this Standard.

When snug-tightening is specified, the tightening is deemed satisfactory when all of the connected elements are in full contact. Inadvertent pretensioning of snug-tightened bolts is normally not a cause for concern.

When pretensioning is specified, the tightening is deemed satisfactory when all of the elements are in full contact and observation of the sides of the nuts or bolt heads shows that they have been slightly peened by the wrench.

When bolts are tightened by the turn-of-nut method and when there is rotation of the part not turned by the wrench, the outer face of the nut may be match-marked with the bolt point before final tightening, thus affording the inspector visual means of noting nut rotation. Such marks may be made with crayon or paint by the wrench operator after the bolts have been snugged.

Should disagreement arise concerning the results of inspection of bolt tension of bolts specified to be pretensioned, arbitration procedures as given in Clause 23.7.4 are to be followed. The use of inspection torque values other than those established according to the requirements of Clause 23.7.4 is invalid because of the variability of the torque-tension relationship. The inspection procedure given in Clause 23.7 is the same as that recommended by the Research Council on Structural Connections (RCSC 1987) and places its emphasis on the need to observe the installation for the proper tightening procedures.

Regardless of the installation procedure or the type of direct tension indicator used, it is important to have all of the plies drawn up tight before starting the specific tightening procedure. This is particularly so for stiff joints that require pattern tightening (RCSC 1987).

24. WELDING

Consistent with CSA policy that the requirements of one standard are not repeated in another, the user of this Standard is referred to CSA Standards W59 and W55.3 for the requirements for arc and resistance welding respectively.

24.3 Fabricator and Erector Qualifications

The intent of Clause 24.3 is simply that the responsibility for structural welding shall lie with the fabricators and erectors certified by the Canadian Welding Bureau as stated specifically in the clause. Such certification should ensure that the fabricators and erectors have the capability to make structural welds of the quality assumed by S16.1-M89.

25. COLUMN BASES

The designer is referred to appropriate clauses of CAN/CSA-A23.3 for the various resistances of the reinforced concrete elements. The compressive resistance of concrete is based on a rectangular stress block as given in CAN/CSA A23.3.

In general, the use of base plates bearing directly on grout is preferred to the use of levelling plates interposed between the base plate and the grout. The latter condition may lead to uneven bearing.

26. ANCHOR BOLTS

In general, an anchor bolt is not a high-strength bolt, but rather an item which may be fabricated from a reinforcing bar or a plain bar of A36 or G40.21-300W steel, or other steel bar stock. The expressions for the tensile, shear, and combined shear and tensile resistance of anchor bolts are similar to those for high strength bolts. The tensile resistance is based directly on the tensile stress area, A_n, rather than 0.75 times the shank area as given in Clause 13.11. The basic elliptical interaction diagram is used for combined shear and tension. For tension and bending, the factored moment resistance is limited to the factored yield moment because the ductility of the steel used may be limited.

27. SEISMIC DESIGN REQUIREMENTS

This is the first time that specific seismic design requirements have been given in a steel design standard in Canada. While the requirements represent the best available knowledge, designers should be alert to possible changes.

27.1 General

The NBCC (1990) assigns force modification (i.e., load reduction) factors, R, to various structural systems in relation to their capacity to absorb energy by undergoing inelastic deformations. The more energy absorbing the structure, the higher is the assigned value of R, which is used as a divisor to reduce the magnitude of the seismic forces. Values of R greater than 1.0 can be justified only if the structure has the ability to undergo inelastic deformations without loss of resistance. The degree of redundancy is also considered because alternative load paths reduce the possibility of serious consequences arising from failure of a member.

The objective of Clause 27 is to provide details with ductility consistent with the R values assumed in the analysis. No reference is made to seismic zones, and the clause therefore applies to all steel structures in Canada for which an $R \geq 2$ is used. Ordinary steel structures (not meeting the special seismic requirements of this Clause) have sufficient inherent ductility to be assigned an R value of 1.5. They are especially appropriate for, but not restricted to, zones of low seismicity.

Clause 27 defines the requirements for four classes of frames:

- ductile moment resisting frames,
- moment resisting frames with nominal ductility,
- ductile braced frames, and
- braced frames with nominal ductility.

In addition, requirements for eccentrically braced frames are given in Appendix D. For special framing systems, see Clause 27.6.

Properly detailed moment resisting frames exhibit highly ductile behaviour and are highly redundant. Two categories of moment resisting frames are recognized: first, ductile moment resisting frames, in which members and connections are selected and braced to ensure that severe inelastic straining can take place; second, moment resisting frames with nominal ductility in which the proportions of members are adequate to provide the limited inelastic straining demanded under a greater design load and the connections are adequate to resist extreme forces.

Concentrically braced frames are those in which the centrelines of diagonal braces, beams, and columns are approximately concurrent with little or no joint eccentricity. Such frames usually have limited redundancy. Inelastic straining must take place in bracing members subjected principally to axial load. Compression members can absorb considerable energy by inelastic bending after buckling and in subsequent straightening after load reversal but the amount is small for slender members. Local buckling or buckling of components of built-up members also limits energy absorption.

Two categories of (concentrically) braced frames are considered: first, ductile braced frames, which exclude some bracing systems that have not behaved well under seismic load; second, braced frames with nominal ductility, designed for higher loads and of which less inelastic behaviour is expected. When braces are selected to satisfy drift or slenderness requirements, and have resistances greater than required to carry the seismic loads, the connections are required to be designed for increased loads commensurate with the capacity of the braces.

Because the behaviour of connections will often be critical for good performance under severe earthquake loading, the engineer's reponsibility for a seismically critical structure includes not only provision of connection design loads but also specification of connection type and details. SEAOC (1988), Krawinkler and Popov (1982), and Astaneh *et al.* (1986) indicate preferred types of connections.

27.1.2 In the computation of second-order effects, it is the gravity loads that are likely to be present that should be combined with the seismic loads. Thus, in computing ΣC_f, α_D is taken as 1.0. As the deflections are already increased by the R factor, the limit on U_2 is waived.

27.1.4 A simple end connection maintaining its shear capacity in the displaced configuration is an example of this situation.

27.2 Ductile Moment Resisting Frames

The NBCC force modification factor, R, is 4.0.

27.2.1.1 Under severe lateral load, a ductile moment resisting frame responds by inelastic deformation at beam-to-column joints and plastic hinges. The key locations are the beam-to-column joints, where inelasticity may develop in the girder, the column or, in the case of H-shaped columns (bent about the strong axis) the joint panel zone comprised of the column web and flanges, and column web stiffeners, if any. It is required that these elements maintain their ultimate strength while undergoing significant deformations. The panel zone provides excellent ability to absorb energy by means of cyclic plastic shearing deformations (Popov *et al.* 1986).

At plastic hinges, members absorb energy by undergoing inelastic cyclic bending while maintaining their flexural and shear resistances. The member should be prismatic in the region of the hinge in order to avoid excessively large strains and to ensure that as large a volume of steel as practicable is strained plastically.

27.2.1.2 The requirement that non-critical elements withstand 1.2 times the unfactored resistance of critical elements ensures that the former remain elastic. Thus, the non-critical elements need only meet the bracing and width-thickness requirements for members analysed elastically, as provided by Clauses 27.2.2, 27.2.3 and 27.2.4.

To determine the critical element at a joint, the resistances of the elements are ranked. Then 1.2 times the unfactored resistance of the single most critical element is applied as a load to the joint. Any of the remaining elements whose resistance exceeds the effect of the applied load is considered to be non-critical. It may be possible for one, two, or all three of the elements to be critical. All such elements must be braced to enable them to undergo large plastic deformations.

The Standard gives provisions for any of the three elements to be critical. To avoid a column sidesway mode of failure in a multi-storey structure that demands high ductility, it is preferable that the columns not be critical. It is advantageous to have both the beams and panel zones critical so that both absorb energy.

27.2.1.3 Clause (a) recognizes that the probable gravity load acting on the columns under this severe loading condition coincides with the specified load level. As well, depending on the use and occupancy, only a portion of the specified live load should be considered.

In evaluating the relative strengths of the structural components at the joint, an estimate should be made of the contribution of the slab. Under positive bending moment composite slabs bear against the column flange (in the case of a column bending about its strong axis) and, due to confinement of the concrete, ultimate compressive resistance of the material can reach values of 1.3 f_c'.

27.2.3 Columns (Including Beam-Columns)

27.2.3.1 The width–thickness requirements here follow from Clause 27.2.1.2. The axial load in the column is also restricted because the rapid deterioration of beam-column flexural strength when high axial loads are acting limits the ductility.

27.2.3.2 This requirement is to avoid plastic hinges near a splice in which there are partial penetration welds.

27.2.4 Column Joint Panel Zone

27.2.4.1 The column panel zone has a shear strength greater than the von Mises shear yield value on the web because of the contribution of the flexure of the column flanges necessary for the panel to yield in shear (Krawinkler and Popov 1982).

27.2.4.2 These requirements ensure that the panel zone can undergo cyclic plastic straining. The entire perimeter of the doubler plates shall be welded to contiguous elements, unless the doubler plate extends beyond the stiffeners. Lateral bracing to the column flange may be provided through transverse beams connected to the column web.

27.2.4.3. Note that the column-to-beam flange welds are subject to forces induced by both beam bending and local bending of the column and its flanges at the corners of the panel zone (Popov *et al.* 1986). The tensile resistance to a normal load on the column flange is reduced to 60% of the value given in Clause 21.3 to account for the highly non-uniform stresses in a beam flange when welded to a column with unstiffened flanges.

27.2.5 Beam-to-column Connections

27.2.5.1 To ensure yielding over as large a volume of material as possible, partial joint penetration groove welds are prohibited as they lead to confined zones of yielding.

27.2.5.3 and 27.2.5.4 These requirements ensure that extensive yielding on the gross section takes place before fracture on the net section and that extensive yielding takes place at plastic hinge locations.

27.2.6 Bracing of both top and bottom beam flanges as well as column flanges shall be considered.

27.2.7 Consideration should be given to the fact that plastic hinge locations will not be predicted by an elastic analysis of the frame. A few welded wide flange column shapes, with webs thicker than 20 mm, do not have complete penetration welds between their web and flanges. These are identified in the tables of properties and dimensions, Part 6.

27.3 Moment Resisting Frames With Nominal Ductility

The NBCC force modification factor, R, is 3.0.

27.3.2 These requirements are to ensure that the critical elements will undergo sufficient inelastic deformation.

27.3.3 Unless the design of the beam-to-column connections meets the requirements for ductile moment resisting frames, albeit with R = 3, both the members and connections shall be sized for seismic forces increased by 1.25 times (effective R = 2.4). Although a ductile mode of failure is still required of the connections, the overall ductility is less than that for a ductile moment-resisting frame.

27.4 Ductile Braced Frames

The NBCC force modification factor, R, is 3.0.

27.4.2.1 This requirement ensures some redundancy and also similarity between load-deflection characteristics in the two opposite directions. A significant proportion of the shear is carried by tension braces so that compression brace buckling will not cause a catastrophic loss in shear capacity. This is required in each planar braced frame because a loss of capacity in any one frame would change the torsional resistance of the structure.

27.4.2.2 The use of V- or chevron bracing is prohibited in this category because such bracing has been found to perform rather poorly under severe seismic shaking. V- or chevron bracing is allowed, with some restrictions, in braced frames with nominal ductility (See Clause 27.5) where the lateral load resisting system is designed for increased seismic forces. V-bracing alternating with chevron bracing in adjacent floors, with bracing forces transmitted directly between the braces on adjacent floors is in effect X-bracing and may be designed under this clause.

K-bracing is proscribed for the same reasons as V- or chevron bracing. Tension only bracing is not acceptable in view of its severely limited ability to absorb energy and the lack of redundancy.

27.4.3 Diagonal Bracing Members

27.4.3.1 The slenderness ratio is restricted because the energy absorbed by plastic bending of the braces diminishes with increased slenderness. Because of high curvatures in a member undergoing overall buckling, local buckling is exacerbated, with concomitant loss in energy absorbing capacity. To control local buckling, the limits on width-thickness ratios are more severe than Class 1 limits. Filling tubes with concrete (Liu and Goel 1988) or providing stiffeners are examples of inhibiting local buckling in critical regions. Built-up members may deteriorate rapidly under cyclic loading and therefore the spacing of stitch fasteners is restricted.

27.4.3.2 The reduction factor, based on SEAOC (1988), takes into account the fact that the compressive resistance diminishes under cyclic loading with slenderness ratio. This reduction stabilizes after a few cycles. It is sufficient that the sum of the resistances of the tension and compression braces acting at the same level and in the same plane exceed the factored load effect.

27.4.4 Bracing Connections

27.4.4.1 Eccentricities which are normally considered negligible (for example at the ends of bolted or welded angle members) may influence the failure mode of connections subjected to cyclic load (Astaneh *et al.* 1986).

27.4.4.2 A connection designed to meet the requirements of Clause (a) ensures that the member is capable of yielding on the gross section. The upper limit of 2.0 times the member force due to the factored seismic load plus the specified gravity loads restricts the force when the bracing members are oversized to facilitate connection, to limit deflections, to meet a slenderness or a b/t limit. For structures in severe seismic zones, however, it is recommended that in any case the bracing for the upper two or three stories be designed for nominal yield load of the brace because the dynamic response associated with higher modes of the inelastic structure may cause severe shear loading in the upper storeys.

These connection resistance requirements ensure that the maximum compressive load to which the brace is subjected during the initial cycles can be carried.

27.4.4.3 When the load of clause 27.4.4.2(a) applies, the net section factored resistance exceeds the gross section factored yield resistance by $1/\phi = 1.11$. This promotes ductile behaviour. For Grade 300W steel, the net section must be at least 0.88 A_g to meet this requirement. When Clause 27.4.4.2(b) applies, considerable reserve connection strength is provided and compensates for the lack of ductility when longitudinal yielding is not mobilized.

27.4.4.4 A brace which buckles out-of-plane will form a plastic hinge at mid-length and hinges in the gusset plate at each end. When braces attached to a single gusset plate buckle out-of-plane, there is a tendency for the plate to tear if it is restrained by its attachment to the adjacent frame members (Astaneh *et al.* 1986). Provision of a clear distance, approximately twice the plate thickness, between the end of the brace and the adjacent members allows the plastic hinge to form in the plate and eliminates the restraint. When in-plane buckling of the brace may occur, ductile rotational behaviour should be possible either in the brace or in the joint.

27.4.4.5 Buckling of double angle braces (legs back-to-back) about the axis of symmetry leads to transfer of load from one angle to the other, thus imposing significant loading on the stitch fastener (Astaneh *et al.* 1986).

27.4.5.1 This ensures that other connections in the lateral load resisting system sustain, at the factored resistance level, the nominal brace resistance force in addition to the specified gravity load.

27.4.6 The limit, Clause 27.4.4.2(b), on the column and beam forces of 2.0 times the member force due to the factored seismic load plus the specified gravity loads restricts these forces when the bracing members are oversized to facilitate connection, to limit deflections, or to meet slenderness or width-thickness limits. However, as recommended in Clauses 27.4.4.2 and 27.4.4.3, the bracing for the upper two or three stories of multi-storey structures in severe seismic zones should be designed for the nominal yield load of the brace. By using a resistance factor of 1.0 on the brace resistance, the other components can sustain, at the factored resistance level, the nominal brace resistance force in addition to the specified gravity load.

The reduced brace compressive resistance is not used as in the first cycles of load it does not apply. However, redistributed loads resulting from buckled compressive brace loads being transferred to the tension brace must be considered in beams and columns as well as in connections. The requirement for twice the member force may then be controlled by the redistributed force. The reduced brace resistance may affect these loads.

27.5 Braced Frames With Nominal Ductility

The NBCC force modification factor, R, is 2.0.

27.5.2 The first requirement is to control local buckling in compression braces under the seismic loads. See also the Commmentary on Clause 27.4.6. Whereas in Clause 27.4.4.2 the factor used is 2.0 times the member force, for the braced frame with nominal ductility, the factor is taken as 1.33.

27.5.3 and 27.5.4 See the Commentary on Clause 27.5.2. The additional factor of 1.10 for tension-only bracing systems is to ensure, for the slender members used in this case, that the impact resulting when slack is taken up not cause connection failure. Details leading to confined zones of yielding, such as occur at partial joint penetration groove welds should be avoided.

27.5.5 When the compression member of a chevron-braced system buckles, the tension brace can carry no additional load unless the beam resists the additional load in bending. This clause attempts to ensure that should one of the braces buckle, the beam can at least support the specified gravity load without collapse. This check involves comparing the beam nominal resistance ($\phi = 1.0$) with the specified gravity load effect.

Therefore the overall approach of the designer would include:

(1) designing the beam and bracing to carry the required factored combinations of dead, live, and earthquake loads. Under these circumstances the beam is supported by chevron bracing and its factored resistance is used. ($\phi = 0.90$).

(2) in order to meet Clause 27.5.5, design the beam, unpropped (i.e. not supported by the chevron bracing), to carry the specified (unfactored) gravity loads. The resistance factor of the beam is to be taken as $\phi = 1.0$.

27.6 Special Framing Systems

27.6.1 Design requirements to ensure ductile behaviour of eccentrically braced frames are based on extensive studies of their behaviour (SEAOC 1988). These requirements have been adopted in the Uniform Building Code (UBC 1988). Appendix D is based on the requirements of SEAOC (1988) with modifications to provide consistency with this Standard and with the National Building Code of Canada. SEAOC (1988) contains a commentary on these requirements.

27.6.2 Numerous other systems have been proposed to enhance the seismic response of structural steel frames, eg. Steimer (1986).

28. FABRICATION

This clause and the clauses on erection and inspection serve to show that design cannot be considered in isolation, but are part of the design and construction sequence. The resistance factors used in this Standard and the methods of analysis are related to fabrication, erection, and inspection tolerances and practices.

28.4.2 The use of sheared edges is restricted because the micro-cracking induced may reduce the ductility.

28.5.1 The thickness of 700Q steels that can be punched is restricted because of the excessive damage that occurs at the edge of the hole.

28.5.2 The restriction of this clause is similar to that of Clause 28.4.2

28.9 Tolerances

The resistance factors given in this Standard are consistent with the distribution of out-of-straightness of members produced to the straightness tolerances given here (Kennedy and Gad Aly 1980; Chernenko and Kennedy 1991).

28.9.7 Milling techniques will realistically result in some measurable deviation. Tests by Popov and Stephen (1977a) on columns with intentionally introduced gaps at milled splice joints indicated that the compressive resistance of spliced columns is similar to unspliced columns. Local yielding reduces the gap. While in these tests column splice gaps of 1.6 mm were left unshimmed, the Standard is more restrictive and defines full contact as a separation not exceeding 0.5 mm. Because shims will be subjected to either biaxial or triaxial stress fields, mild steel shims may be used regardless of the grade of main material.

29. CLEANING, SURFACE PREPARATION, AND PRIMING

There are four instances where steelwork need not be or should not be coated:

a) steelwork concealed by an interior building finish;

b) steelwork encased in concrete;

c) faying surfaces of slip-critical joints,except as permitted by Clause 23; and,

d) surfaces finished to bear unless otherwise specified.

30. ERECTION

Clauses 30.7.3, 30.7.4, 30.7.5 and 30.7.8 are written in a parallel manner, in that the offset of one end relative to the other, or the elevation of one end relative to the other, is expressed as a function of the length but with upper and lower limits. The lower limit represents a realistic assessment of adequate positioning and the upper limit is a maximum not to be exceeded by the largest members, as illustrated in Figure 2-50 for horizontal alignment of spandrel beams.

30.7 Erection Tolerances

This clause has been re-written and expanded to provide a more helpful definition of tolerances for the location of the ends of members with respect to their theoretical locations. Tolerances have now been given for column base plates, for alignment and elevations of horizontal or sloping members, and for crane or monorail beams and girders. Column splice tolerances have been redefined based on new research (see Commentary on Clause 28.9.7).

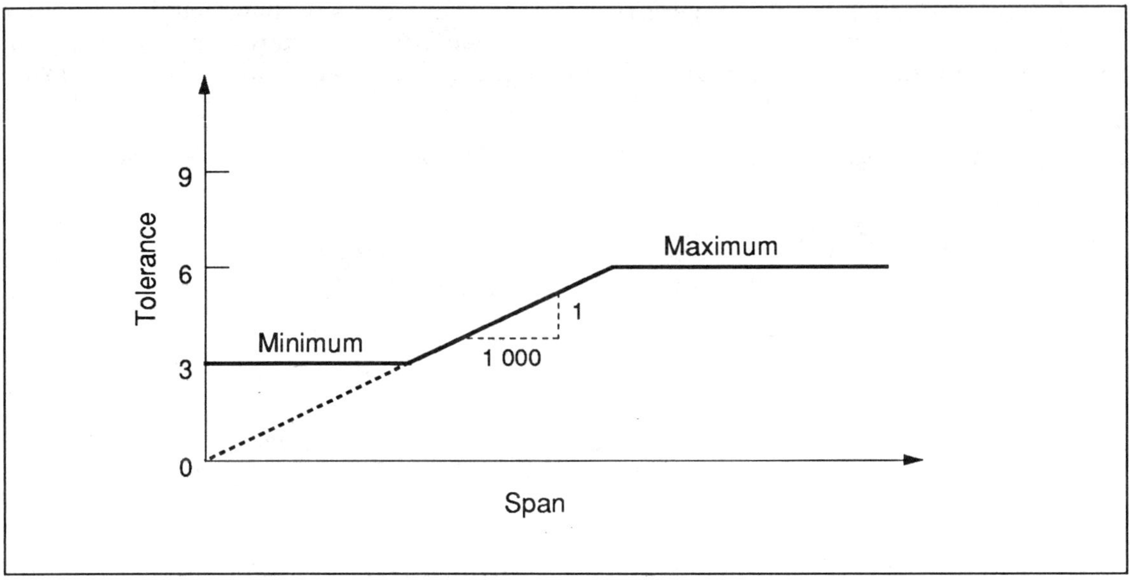

FIGURE 2-50
HORIZONTAL ALIGNMENT TOLERANCE OF SPANDREL BEAMS

APPENDIX D

Seismic Design Requirements for Eccentrically Braced Frames

The value of the NBCC force modification factor, R, is 3.5.

Material in Appendix D is essentially from SEAOC (1988) and the SEAOC commentary on eccentrically braced frames.

APPENDIX F

Columns Subject to Biaxial Bending

This appendix provides non-linear interaction expressions for Class 1 and 2 sections of W-shapes and HSS members, as a design alternative to the linear interaction expressions of Clause 13.8. In the commentary to Clause 13.8, the linearity of the interactions expressions of that clause was discussed. In fact, interaction curves relating axial load and the two orthogonal moments are not linear.

The simplest demonstration of non-linearity is that used by Ellis and Pillai (1971) for HSS. Consider a circular tube subject to eccentric loading about the x and y axes. Since the section properties are uniform about the polar axis, the interaction expressions of clause 13.8 would sum the moment effects as if a moment of $P(e_x + e_y)$ were applied, whereas the actual moment is $P\sqrt{(e_x^2 + e_y^2)}$. Thus, the interaction curve for a circular tube is a quadrant of a circle. Finding a similar effect applied to square HSS, Pillai determined that the available test data fitted the expression of clause F3(c) if the sum of the moment terms was modified by the expression

$$\gamma = \frac{e_x^2 + e_y^2}{e_x + e_y}$$

The value of γ is equivalent to ν in clause F3(c).

It has been found (Chen and Atsuta 1972 and 1973; Tebedge and Chen 1974, and Ross and Chen 1976) that a similar non-linear effect applied to W-shapes (Clause F1) where the exponents ζ and η have been evaluated from rigorously calculated interaction curves.

The expressions for M_{ox} and M_{oy} have been determined from Clause 13.8, transposed and applied for uniaxial bending only.

Essentially, the interaction expressions of Clause F1 define an interaction surface relating the boundary conditions on the three perpendicular axes (i.e. axial resistance and moment resistance about each axis). Therefore, the better these values are determined, the better will be the interaction surface.

Unlike W-shapes, significant biaxial bending is imposed on HSS whenever beams frame into the column about two perpendicular axes. The design expressions given in Clause F3 for square HSS were developed by Pillai (1970 and 1974) and Pillai and Kurian (1977).

Limitations

Significant economy may result in designs according to the expressions of this Appendix, but before adopting these, some thought should be given to the extent to which yielding under service load is likely, and whether further restrictions on the design are necessary. Springfield (1975) has proposed additional precautions.

REFERENCES

AASHTO. 1977. Standard specification for highway bridges. American Association of State Highway and Transportation Officials, Washington, DC.

ADAMS, P. F. 1974. The design of steel beam–columns. Canadian Steel Industries Construction Council, Willowdale, Ont.

ADAMS, P. F., and GALAMBOS, T. V. 1969. Material considerations in plastic design. International Association for Bridge and Structural Engineering, **29-II**.

ADEKOLA, A. O. 1968. Effective widths of composite beams of steel and concrete. The Structural Engineer, **46**(9): 285–289.

AHAMD, M.; CHIEN, E. Y. L., and HOSAIN, M. U. 1990. Modified stub-girder floor system: full scale tests. ASCE Structures Congress. Baltimore, MA.

AISC. 1973. Commentary on highly restrained welded connections. Engineering Journal, American Institute of Steel Construction, Third Quarter.

——— 1986a. Manual of steel construction. Load & resistance factor design. American Institute of Steel Construction. Chicago, IL.

——— 1986b. Load and resistance factor design specification for structural steel buildings. American Institute of Steel Construction. Chicago, IL.

AISI. 1968. Plastic design of braced multi-storey steel frames. American Iron and Steel Institute, Washington, DC.

ALLEN, D. L. 1974. Vibrational behaviour of long-span floor slabs. Canadian Journal of Civil Engineering, **1**(1).

ALLEN, D. E. 1975. Limit states design—a probabilistic study. Canadian Journal of Civil Engineering, **2**(1).

ALLEN, D. E., and RAINER, J. H. 1976. Vibration criteria for long-span floors. Canadian Journal of Civil Engineering, **3**(2).

ALLEN, D. E., RAINER, J. H., and PERNICA, G. 1985. Vibration criteria for assembly occumancies. Canadian Journal of Civil Engineering, **12**(3).

ASCE. 1967. Commentary on welded cover-plated beams. Subcommittee on Cover Plates, Task Committee on Flexural Members, ASCE Journal of the Structural Division, **93**(ST4).

——— 1971. Commentary on plastic design in steel. Manual of Engineering Practice, No. 41, American Society of Civil Engineers.

——— 1979. Structural design of tall steel buildings. pp. 624–628.

ASTANEH, A., GOEL, S.C., and HANSON, R.D. 1986. Earthquake-resistant design of double-angle bracings. Engineering Journal, AISC, **23**(4), 133–147.

ASTM. 1990a. Standard specification for high-strength bolts for structural steel joints [metric]. Standard A325M-90, American Society for Testing and Material.

——— 1990b. Standard specification for high-strength steel bolts, Class 10.9 and 10.9.3, for structural steel joints [metric]. Standard A490M-90, American Society for Testing and Material.

——— 1990c. Standard specification for high-strength bolts for structural steel joints. Standard A325-90, American Society for Testing and Material.

——— 1990d. Standard specification for heat-treated, steel structural bolts, 150 ksi tensile strength. Standard A490-90, American Society for Testing and Material.

AZIZ, T.S.A. 1972. Inelastic nonlinear behaviour of steel triangulated planar frames. Thesis, Carleton University, Ottawa, Ont.

de BACK, J., and de JONG, A. 1968. Measurement on connections with high strength bolts, particularly in view of the permissible arithmetical bearing stress. Report 6-68-3, Stevin Laboratory, Delft University of Technology, The Netherlands.

BAKER, K. A. and KENNEDY, D.J.L.. 1984. Resistance factors for laterally unsupported steel beams and biaxially loaded steel beam-columns. Canadian Journal of Civil Engineering, **11**(4): p. 1008-1019.

BARBER, E., KENNEDY, S.J., KENNEDY, D.J.L., and MacGREGOR. 1987. Flexural strength of concrete filled steel hollow sections. Proceedings, CSCE Centenial Conference. May 19-22. Montreal, Que., 493-512

BASLER, K. 1961a. New provisions for plate girder design. Proceedings, AISC National Engineering Conference.

——— 1961b. Strength of plate girders under combined bending and shear. ASCE Journal of the Structural Division, **87**(ST7).

——— 1961c. Strength of plate girders in shear. ASCE Journal of the Structural Division, **87**(ST7).

BASLER, K., and THURLIMANN, B. 1961. Strength of plate girders in bending. ASCE Journal of the Structural Division, **87**(ST6).

BEAULIEU, D., PERLYNN, M., DUNBAR, A., ADAMS, P. F., and KELLER, D. 1976. The Effects of column out-of-plumbs on the stability of core-braced buildings. Canadian Journal of Civil Engineering, **3**(3).

BEAULIEU, D., and ADAMS, P. F. 1980. Significance of structural out-of-plumb forces and recommendations for design. Canadian Journal of Civil Engineering, **7**(1).

BEEDLE, L. S. 1958. Plastic design of steel frames. John Wiley & Sons, New York, NY.

BEEDLE, L. S., LU, L. W., and LIM, L. C. 1969. Recent developments in plastic design practice. ASCE Journal of the Structural Division, **95**(ST9).

BEEDLE, L. S., and TALL, L,. 1960. Basic column strength. ASCE Journal of the Structural Division, **86**(ST7).

BETHLEHEM STEEL CORPORATION. 1967. Torsional analysis of rolled steel sections. Handbook 1963C, Bethlehem Steel Corp., Bethlehem, PA.

BIRKEMOE, P. C., and GILMOR, M. I. 1978. Behaviour of bearing critical, double-angle beam connections. Engineering Journal, AISC, Fourth Quarter.

BJORHOVDE, R. A. 1972. A probabilistic approach to maximum column strength. Procedings, ASCE Conference on Safety and Reliability of Metal Structures.

BJORHOVDE, R., and BIRKEMOE, P. C. 1979. Limit States design of HSS columns. Canadian Journal of Civil Engineering, **6**(2).

BJORHOVDE, R., and ZIMMERMAN, T. J.1980. Some aspects of stub-girder design. Proceedings of the Canadian Structural Engineering Conference. Canadian Steel Construction Council. Willowdale, Ont.

BLEICH, F. 1952. Buckling strength of metal structures. McGraw-Hill. New York., NY.

BOWER, J. E., et al. 1971. Suggested design guide for beams with web holes. ASCE Journal of the Structural Division, **97**(11).

BRATTLAND, A., and KENNEDY, D.J.L. 1986. Test of a full-scale composite truss. Proceedings, Canadian Structural Engineering Conference. pp. 8–1 to 8–50, C.I.S.C.

BROCKENBROUGH, R. L. 1983. Considerations in the design of bolted steel joints for weathering steel. AISC Engineering Journal, **20**(1).

BROCKENBROUGH, R. L., and JOHNSTON, B. G. 1974. Steel design manual.United States Steel Corp., Pittsburgh, PA.

BUDIJGNTO, P. 1983. Design methods for composite columns. Project Report G83-6, Department of Civil Engineering, McGill University.

BUTLER, L.J. and KULAK, G.L. 1971. Strength of fillet welds as a function of direction of load. Welding Research Supplement, Welding Journal. Welding Research Council, **36**(5), 2315-2345

BUTLER, L.J., PAL, S. and KULAK, G.L. 1972. Eccentrically loaded welded connections. ASCE Journal of the Structural Division, **98**(ST5), 989-1005

CHEN, W. F., and ATSUTA, T. 1972. Interaction equations for biaxially loaded sections. ASCE Journal of the Structural Division, **98**(ST5).

CHEN, W. F., and ATSUTA, T. 1973. Ultimate strength of biaxially loaded steel H-columns. ASCE Journal of the Structural Division, **99**(ST3).

CHEN, W. F., and LUI, E.M..1987. Structural stability, theory and implementation. Elsevier. New York, NY.

CHEONG SAIT-MOY, F. 1991. Column design in gravity loaded frames. ASCE Journal of Structural Engineering, **117**(ST5)

CHENG, J-J. R., and YURA, J. A. 1986. Local web buckling of coped beams. ASCE Journal of Structural Engineering, **112**(10).

CHERNENKO, D.E. and KENNEDY, D.J.L. 1991. An analysis of the performance of welded wide flange columns. Canadian Journal of Civil Engineering, **18**(4).

CHIEN, E.Y.L., and RITCHIE, J. K. 1984. Design and construction of composite floor systems. Canadian Institute of Steel Construction, Willowdale, Ont.

CHIEN, E.Y.L. 1989. Roof framing with cantilever (Gerber) girders & open web steel joists. Canadian Institute of Steel Construction, Willowdale, Ont.

CIDECT. 1970. Concrete filled hollow section steel columns design manual. Monograph No. 1, International Committee for the Study and Development of Tubular Structures. Whitefriars Press Ltd., London.

CISC. 1987. Column selection program. Canadian Institute of Steel Construction, Willowdale, Ont.

——— 1980. Steel Joist Facts, 2nd Edition. Canadian Institute of Steel Construction, Willowdale, Ont.

CSA. 1969. Steel structures for buildings. CSA Standard S16-1969, Canadian Standards Association, Rexdale, Ont.

——— 1984. Design of concrete structures for buildings. CSA Standard CAN3-A23.3-M84, Canadian Standards Association, Rexdale, Ont.

——— 1989. Cold formed steel structural members. CSA Standard S136-M89. Canadian Standards Association, Rexdale, Ont.

DAVIES, C. 1969. Tests on half-scale steel concrete composite beams with welded stud connectors. The Structural Engineer, **47**(1): 29–40.

DAWE, J. L., and KULAK, G. L. 1984. Local buckling of W shape columns and beams. ASCE Journal of Structural Engineering, **110**(6).

——— 1986. Local buckling behavior of beam–columns. ASCE Journal of Structural Engineering, **112**(11).

DISQUE, R. O. 1964. Wind connections with simple framing. Engineering Journal, AISC, July.

DRIVER, R. G., and KENNEDY, D. J. L. 1989. Combined flexure and torsion of I-shaped steel beams. Canadian Journal of Civil Engineering, **16**(2).

DUAN, L. and CHEN, W-F. 1988. Design rules of built-up members in Load and Resistance Factor design. ASCE Journal of Structural Engineering, **114**(11), 2544-2554.

EL-GHAZZI, M. N., ROBINSON, H., and ELKHOLY, I.A.S. 1976. Longitudinal shear capacity of slabs of composite beams. Canadian Journal of Civil Engineering, **3**(4).

ELKELISH, S., and ROBINSON, H. 1986. Effective widths of composite beams with ribbed metal deck. Canadian Journal of Civil Engineering, **13**(5): 575–582.

EUROCODE No. 4. 1987. Common unified rules for composite steel and concrete structures. Report EUR 9886 EN, Commission of the European Communities, Luxembourg.

FISHER, J. W. 1970. Design of composite beams with formed metal deck. Engineering Journal, AISC, July.

——— 1974. Guide to 1974 AASHTO fatigue specifications. AISC, Chicago, IL.

——— 1978. Fatigue cracking in bridges from out-of-plane displacements. Canadian Journal of Civil Engineering, **5**(4).

——— 1984. Fatigue and fracture in steel bridges—case studies. John Wiley & Sons.

FISHER, J. W., GALAMBOS, T. V., and KULAK, G. L. 1978. Load and resistance factor design criteria for connectors. ASCE. Journal of the Structural Division, **104**(ST9).

FISHER, J, W., LEE, G. C, YURA, J. A., and DRISCOLL, G. C. 1963. Plastic analysis and tests of haunched corner connections. Bulletin No. 91, Welding Research Council.

FRANK, K. H. 1980. Fatigue strength of anchor bolts. ASCE. Journal of the Structural Division, **106**(ST6).

GAGNON, D. G., and KENNEDY, D. J. L. 1989. Behaviour and ultimate tensile strength of partial joint penetration groove welds. Canadian Journal of Civil Engineering, **16**(3).

GALAMBOS, T. V. 1968. Structural members and frames. Prentice-Hall Inc., Englewood Cliffs, NJ.

GALAMBOS, T. V. 1988. Guide to stability design criteria for metal structures (4th ed.). Structural Stability Research Council, John Wiley & Sons, Inc., New York, NY.

GIBSON, G. T., and WAKE, B. T. 1942. An investigation of welded connections for angle tension members. Welding Journal, American Welding Society.

GOBEL, G. 1968. Shear strength of thin flange composite specimens. Engineering Journal, AISC, April.

GRAHAM, J. D., SHERBOURNE, A. N., KHABBAZ, R. N., and JENSEN, C. D. 1959. Welded interior beam-to-column connections. American Institute of Steel Construction.

GRANT, J. A., FISHER, J. W., and SLUTTER, R. G. 1977. Composite beams with formed steel deck. Engineering Journal, AISC, First Quarter.

GRIGGS, P. H. 1976. Mill Building structures. Proceedings, Canadian Structural Engineering Conference, Canadian Steel Construction Council, Willowdale, Ont.

HAAIJER, G., and THURLIMANN, B. 1958. On inelastic buckling in steel. ASCE Journal of the Engineering Mechanics Division, April.

HAGOOD, T. A., Jr., GUTHRIE, L., and HOADLEY, G. 1968. An investigation of the effective concrete slab width for composite construction. Engineering Journal, AISC, **5**(1): 20–25.

HANSELL, W. C., and VIEST, I. M. 1971. Load factor design for steel highway bridges. Engineering Journal, AISC, **8**(4).

HARDASH, S. and BJORHOVDE, R. 1985. New design criteria for gusset plates in tension. Engineering Journal, AISC, **21**(2), 77-94

HART, W. H., and MILEK, W. A. 1965. Splices in plastically designed continuous structures. Engineering Journal, AISC, April.

HAWKINS, N. M., and MITCHELL, D. 1984. Seismic response of composite shear connections. ASCE Journal of Structural Engineering, **110**(9), 2120-2136.

HEINS, C. P., and FAN, H. M.. 1976. Effective composite beam width at ultimate load. ASCE Journal of the Structural Division, **102**(ST11).

HIRANO, N. 1970. Bearing stresses in bolted joints. Society of Steel Construction of Japan, **6**(58), Tokyo.

HOLTZ, N. M., and KULAK, G. L. 1970. High strength bolts and welds in load-sharing systems. Studies in Structural Engineering, No. 8, Technical University of Nova Scotia.

——— 1973. Web slenderness limits for compact beams. SER 43, Department of Civil Engineering, University of Alberta.

——— 1975. Web slenderness limits for non-compact beams. SER 51, Department of Civil Engineering, University of Alberta.

JENT, K. A. 1989. Effects of shrinkage, creep and applied loads on continuous deck-slab composite beams. M.Sc. thesis, Queen's University, Kingston, Ont.

JAYAS, B. S., and HOSAIN, M. U. 1988. Behaviour of headed studs in composite beams: push-out tests. Canadian Journal of Civil Engineering, **15**(2), 240-253.

——— 1989. Behaviour of headed studs in composite beams: full-size tests. Canadian Journal of Civil Engineering, **16**(5), 712-724.

JOHNSTON, B. G. 1939. Pin connected plate links. Transactions, American Society of Civil Engineers.

JOHNSON, R. P. 1970. Longitudinal shear strength of composite beams. ACI Journal Procedings, **67**.

———. 1975. Composite structures of steel and concrete. Volume 1: beams, columns, frames and applications in buildings. Crosby Lookwood Staples, London, England. p. 210.

JOHNSON, R. P., and SMITH, D.G.E. 1975. Design rules for the control of deflections in composite beams. The Structural Engineer, **53**(9).

JONES, J. 1958. Bearing-ratio effect on strength of riveted joints. Transactions, American Society of Civil Engineers, **123**: 964–972.

KALIANDASANI, R. A., SIMMONDS, S. H., and MURRAY, D. W. 1977. Behaviour of open web steel joists. Report No. 62, Department of Civil Engineering, University of Alberta.

KALYANARAMAN, V., PEKOZ, T., and WINTER, G. 1977. Unstiffened compression elements. ASCE. Journal of the Structural Division, **103**(ST9).

KENNEDY, D.J.L. 1974. Limit states design—an innovation in design standards for steel structures. Canadian Journal of Civil Engineering, **1**(1).

KENNEDY, D.J.L., and ROWAN, W.H.D. 1964. Behaviour of compression chords of open web steel joists. Report to CISC.

KENNEDY, D.J.L., ALLEN, D. E., ADAMS, P. F., KULAK, G. L., TURNER, D. K., and TARLTON, D. L. 1976. Limit states design. Procedings of the Canadian Structural Engineering Conference, Canadian Steel Industries Construction Council, Willowdale, Ont.

KENNEDY, D.J.L., and GAD ALY, M. 1980. Limit states design of steel structures—performance factors. Canadian Journal of Civil Engineering, **7**(1).

KENNEDY, S. J., and KENNEDY, D. J. L. 1987. The performance and strength of hardened steel test roller assemblies. Proceedings, CSEC Centenial Conference, May 19-22. Montreal, Que.: 513-531.

KENNEDY, D.J.L., MIAZGA, G.S. and LESIK, D.F. 1990. Discussion of Evaluation of fillet weld shear strength of FCAW electrodes by McClellan. R.W. Welding Journal, August 1989, Welding Journal Reference, 44-46

KENNEDY, D.J.L., PICARD, A., and BEAULIEU, D. 1990. New Canadian provisions for the design of steel beam_columns. Canadian Journal of Civil Engineering, **17**(6)

KETTER, R. L. 1961. Further studies of the strength of beam-columns. ASCE, Journal of the Structural Division, **87**(ST6): 135–152

KIRBY, P. A. and NETHERCOT, D. A. 1978. Design for structural stability. Granada. London, England.

KNOWLES, R. B., and PARK, R. 1970. Axial load design for concrete filled steel tubes. ASCE, Journal of the Structural Division, **96**(ST10).

KRAWINKLER, H., and POPOV, E.P. 1982. Seismic behaviour of moment connections and joints. ASCE, Journal of the Structural Division, **108**(ST2), 373–391.

KULAK, G. L., ADAMS, P. F., and GILMOR, 1990. Limit states design in structural steel. Canadian Institute of Steel Construction, Willowdale, Ont.

KULAK, G. L., and DAWE, J. L. 1991. Discussion of design interaction equations for steel members. Sohai, I. S., Duan, L. and Chen W-F. ASCE, Journal of the Structural Division, **117**(ST7): 2191-2193.

KULAK, G. L., FISHER, J. W., and STRUIK, J.H.A. 1987. Guide to design criteria for bolted and riveted joints (2nd. ed.). John Wiley & Sons, New York, NY.

KULLMAN R. B. and HOSAIN, M. U. 1985. Shear capacity of stub-girders: full scale tests. ASCE, Journal of the Structural Division, **111**(ST1): 56-75.

LAY, M. G. 1965. Flange local buckling in wide-flange shapes. ASCE Journal of the Structural Division, **91**(ST6).

LAY, M. G., and GALAMBOS, T. V. 1966. Bracing requirements for inelastic steel beams. ASCE Journal of the Structural Division, **92**(ST2).

——— 1967. Inelastic beams under moment gradient. ASCE Journal of the Structural Division, **93**(ST1)

LESIK, D.F. and KENNEDY, D.J.L. 1990. Ultimate strength of fillet welded connections loaded in plane. Canadian Journal of Civil Engineering **17**(1), 55-67

LILLEY, S. B., and CARPENTER, S. T. 1940. Effective moment of inertia of a riveted plate girder. Transactions, American Society of Civil Engineers.

LIU, Z., and GOEL, S.C. 1988. Cyclic load behavior of concrete-filled tubular braces. ASCE Journal of Structural Engineering, **114**(ST7), 1488–1506.

LOGCHER, R. D., et al. 1969. ICES STRUDL II engineering users manual. Department of Civil Engineering, Massachusetts Institute of Technology, Cambridge, Mass.

LUKEY, A. F., and ADAMS, P. F. 1969. Rotation capacity of beams under moment gradient. ASCE Journal of the Structural Division, **95**(ST6).

MANNICHE, K., and WARD-HALL, G. 1975. Mission bridge—design and construction of the steel box girder. Canadian Journal of Civil Engineering, **2**(2).

MASSEY, C. 1962. Lateral bracing forces of steel I-beams. ASCE Engineering Mechanics Division, **88**(EM6).

MATTOCK, A. H. 1974. Shear transfer on concrete having reinforcement at an angle to the shear plane. Special Publication 42, Shear in Reinforced Concrete, American Concrete Institute, pp 17–42.

MATTOCK, A. H., LI, W. K., and WANG, T.C. Shear transfer in lightweight reinforced concrete. PCI Journal, **21**(1), 20-39.

MCGUIRE, W. 1968. Steel structures. Prentice Hall, Englewood Cliffs, NJ. p. 682.

MIAZGA, G.S. and KENNEDY, D.J.L. 1989. Behaviour of fillet welds as a function of the angle of loading. Canadian Journal of Civil Engineering Vol 16 pp 583-599

MONTGOMERY, C.J., KULAK, G. L., and SHWARTSBURD, G. 1983. Deflection of a composite floor system. Canadian Journal of Civil Engineering, **10**(2).

MTO. 1991a. Ontario Highway Bridge Design Code . Ministry of Transportation, Downsview, Ont.

——— 1991b. Ontario Highway Bridge Design Code — Commentary. Ministry of Transportation, Downsview, Ont.

MUNSE, W. H. 1959. The effect of bearing pressure on the static strength of riveted connections. Bulletin No. 454, Engineering Experimental Station, University of Illinois, Urbana, IL.

MUNSE, W. H. and CHESSON, E. 1963. Riveted and bolted joints: net section design. ASCE. Journal of the Structural Division, **89**(ST1), Part 1.

MURRAY, T. M. 1975. Design to prevent floor vibrations. Engineering Journal, AISC, third quarter.

NCRHP. 1970. Effect of weldments on the fatigue strength of steel beams. Report 102, National Cooperative Highway Research Program, Transportation Research Board, National Academy of Sciences, Washington, DC.

——— 1974. Fatigue strength of steel beams with welded stiffeners and attachments. Report 147, National Cooperative Highway Research Program, Transportation Research Board, National Academy of Sciences, Washington, DC.

NASH, D. S., and KULAK, G. L. 1976. Web slenderness limits for non-compact beam-columns. Report No. 53, Department of Civil Engineering, University of Alberta.

NBCC. 1990a. National building code of Canada. Supplement, Chapter 4, Commentary F. National Research Council of Canada, Ottawa, Ont.

——— 1990b. National building code of Canada. Supplement, Chapter 4, Commentary I, National Research Council of Canada, Ottawa, Ont.

——— 1990c. National building code of Canada. Clause 4.1.4. National Research Council of Cananda, Ottawa, Ont.

——— 1990d. National building code of Canada. Clause 4.1.6. National Research Council of Cananda, Ottawa, Ont.

NETHERCOT, D. A. and TRAHAIR, N. S. 1976a. Inelastic lateral buckling of determinate beams. ASCE Journal of the Structural Division, **102**(ST4):701-717.

——— 1976b. Lateral buckling approximations for elastic beams. ISE. The Stuctural Engineer. **54**(6): 197-204.

NIXON, D. 1981. The use of frame action to resist lateral loads in simple construction. Canadian Journal of Civil Engineering, **8**(4).

OLGAARD, J. G,, SLUTTER, R. G., and FISHER, J. W. 1971. Shear strength of stud connectors in light-weight and normal-weight concrete. Engineering Journal, AISC, April

PERLYNN, M. J. and KULAK, G. L. 1974. Web slenderness limits for compact beam-columns. SER 50, Department of Civil Engineering, University of Alberta.

PILLAI, U. S. 1970. Review of Recent Research on the Behaviour of Beam-Columns under Biaxial Bending. Civil Engineering Research Report No. CE 70-1, Royal Military College of Canada, Kingston, Ont.

——— 1974. Beam-columns of hollow structural sections. Canadian Journal of Civil Engineering, **1**(2).

PILLAI, U. S., and ELLIS, J. S. 1971. Hollow Tubular Beam-Columns in Biaxial Bending. ASCE Journal of the Structural Division, **97**(ST5).

PILLAI, U. S., and KURIAN, V. J. 1977. Tests on hollow structural section beam-columns. Canadian Journal of Civil Engineering, **4**(2).

POPOV, E. P., and PINKNEY, R. B. 1969. Cyclic yield reversal in steel building connections. ASCE Journal of the Structural Division, **95**(ST3).

POPOV, E. P., and STEPHEN, R. M. 1977a. Capacity of columns with splice imperfections. Engineering Journal, AISC, **14**(1).

——— 1977b. Tensile capacity of partial penetration groove welds. ASCE. Journal of the Structural Division, **103**(ST9).

POPOV, E.P., AMIN, N.R., LOUIE, J.J.C., and STEPHEN, R.M. 1986. Cyclic behaviour of large beam–column assemblies. Engineering Journal, AISC, **23**(1), 9–23.

RAINER, J. H. 1980. Dynamic tests on a steel-joist concrete-slab floor. Canadian Journal of Civil Engineering, **7**(2).

RCSC. 1987. Specification for structural joints using ASTM A325 or A490 bolts. Research Council on Riveted and Bolted Structural Joints.

REDWOOD, R. G. 1971. Simplified plastic analysis for reinforced web holes. Engineering Journal, AISC, **8**(4).

——— 1972. Tables for plastic design of beams with rectangular holes. Engineering Journal, AISC, **9**(1).

——— 1973. Design of beams with web holes. Canadian Steel Industries Construction Council, Willowdale, Ont.

REDWOOD, R., and MCCUTCHEON, J. 1968. Beam tests with unreinforced web openings. ASCE Journal of the Structural Division, **94**(ST1).

RICLES, J.M. and YURA, J.A. 1983. Strength of double-row bolted-web connections. ASCE Journal of the Structural Division, **109**(ST1), 126-142.

RITCHIE, J.K., and CHIEN, E.Y.L. 1980. Composite structural systems—design, construction and cost considerations. Proceedings, Canadian Structural Engineering Conference, C.I.S.C.

ROBERTS, T.M. 1981. Slender plate girders subjected of edge loading. Proceedings of Institute of Civil Engineers, Part 2, 71, September.

ROBERTS, T.M., and CHONG, C.K. 1981. Collapse of plate girders under edge loading. ASCE Journal of the Structural Division, **107**(8).

ROBINSON, H. 1969. Composite beam incorporating cellular steel decking. ASCE Journal of the Structural Division, **95**(ST3).

——— 1988. Multiple stud shear connections in deep ribbed metal deck. Canadian Journal of Civil Engineering, **15**(4).

ROBINSON, H., and WALLACE, I. W. 1973. Composite beams with 1-1/2 inch metal deck and partial and full shear connection. Transactions, Canadian Society for Civil Engineering, **16**(A-8), published in the Engineering Journal, Engineering Institute of Canada.

ROSS, D. A., and CHEN, W. F. 1976. Design criteria for steel I-columns under axial load and biaxial bending. Canadian Journal of Civil Engineering, **3**(3).

SEAOC. 1988. Recommended lateral force requirements and tentative commentary. Structural Engineers Association of California, Seismology Committee.

SEELEY, F. B., and SMITH, J. O. 1957. Advanced mechanics of materials (2nd ed.). John Wiley & Sons, Inc., New York, NY. pp. 365–367.

SHRIVASTAVA, S. C., REDWOOD, R. G., HARRIS, P. J., and ETTEHADIEH, A. A. 1979. End moments in open web steel tie joists. McGill University, June.

SLUTTER, R. G., and DRISCOLL, G. C. 1965. Flexural strength of steelconcrete composite beams. ASCE Journal of the Structural Division, **95**(ST2).

SCHMITKE, C.D. and KENNEDY, D.J.L. 1985. Effective lengths of laterally continuous, laterally unsupported steel beams. Canadian Journal of Civil Engineering. **12**(3), 603-616

SOUROCHNIKOFF, B. 1950. Wind-stresses in semi-rigid connections of steel framework. Transactions, American Society of Civil Engineers.

SPRINGFIELD, J., and ADAMS, P. F. 1972. Aspects of column design in tall steel buildings. ASCE Journal of the Structural Division, **98**(ST5).

SPRINGFIELD, J. 1975. Design of columns subject to biaxial bending. Engineering Journal, AISC, **12**(3), Third Quarter.

STELCO. 1973. Hollow structural sections—design manual for columns and beams. The Steel Company of Canada Limited, Hamilton, Ont.

——— 1981. Hollow structural suctions—design manual for concrete-filled HSS columns. The Steel Company of Canada Limited, Hamilton, Ont.

STIEMER, S. F. 1986. Innovative earthquake resistant steel design. Proceedings, Canadian Structural Engineering Conference. Canadian Steel Construction Council. Willowdale, Ont.

TALL, L., et al. 1974. Structural steel design (2nd ed.). The Ronald Press Company, New York, NY.

TEBEDGE, N., and CHEN, W. F. 1974. Design criteria for H-columns under biaxial loading. ASCE Journal of the Structural Division, **100**(ST3).

TEMPLE, M.C., SCHEPERS, J.A., and KENNEDY, D.J.L.1986. Interconnection of starred angle compression members. Canadian Journal of Civil Engineering, **13**(6).

THORNTON, C. H. 1973. Quality control in design and supervision can eliminate lamellar tearing. Engineering Journal, AISC, Fourth Quarter.

TOPRAC, A., and NATARAJAN, M. 1971. Fatigue strength of hybrid plate girders. ASCE Journal of the Structural Division, **97**(ST4).

TRAHAIR, N.S. 1968. Interaction buckling of narrow rectangular continuous beams. Civil Engineering Trans. Inst. of Eng. Australia.

UBC. 1988. Uniform building code. International Conference of Building Officials, Whittier, California.

VINCENT, G. S. 1969. Tentative criteria for load factor design of steel highway bridges. Steel Research Construction Bulletin No. 15, AISI, Washington, DC.

VIRDI, K. S., and DOWLING, P. J. 1976. A unified design method for composite columns. Memoires, IABSE, No. 36-11.

WAKABAYASHI, M. 1977 A new design method of long composite beam-columns. Proceedings, ASCE International Colloquium on Stability of Structures under Static and Dynamic Loads, Washington, DC., May.

WINTER, G. 1958. Lateral bracing of columns and beams. ASCE Journal of the Structural Division, **84**(ST2).

——— 1970. Commentary on the 1968 edition of the specification for the design of cold-formed steel structural members. American Iron and Steel Institute, Washington, DC.

WOOD, B. R., BEAULIEU, D., and ADAMS, P. F. 1976. Column design by P-delta method. ASCE Journal of the Structural Division, **102**(ST2).

——— 1976. Further aspects of design by P-delta method. ASCE Journal of the Structural Division, **102**(ST3).

YANG, C.H. and BEEDLE, L.S. 1951. Behavior of I and WF beams in shear. Fritz Engineering Lab. Report No. 205B21, Lehigh Univ., Bethlehem, Pa.

YURA, J. A., GALAMBOS, T. V., and Ravindra, M. K. 1978. The bending resistance of steel beams. ASCE Journal of the Structural Division, **104**(ST9).

ZUK, W. 1956. Lateral bracing forces on beams and columns. ASCE Engineering Mechanics Division, **82**(EM3).

PART THREE
CONNECTIONS AND TENSION MEMBERS

GENERAL INFORMATION

Part Three contains tables, examples, dimensions and general information of assistance to designers, detailers and others concerned with the design and detailing of connections and tension members according to the requirements of Clauses 12, 13.2, 13.9, 13.10.1(c), 13.11, 13.12, 13.13, 21.0, 22.0 and 23.0 of CAN/CSA-S16.1-M89. Information is included for both metric series and Imperial series bolts, though all design data is given in SI units. For convenience, Part Three is divided into seven main sections.

Bolt Data

Pages 3–5 to 3–37 contain information on diameter, area and strength of bolts, including bolt resistances and unit resistances, for evaluating bolts in bearing type connections, slip-critical connections, and bolts subjected to tension and prying action. Tables are also included for evaluating eccentric loads on various bolt groups.

Weld Data

Pages 3–38 to 3–53 contain information on factored resistance of welds, including values for various sizes of fillet welds with a comparison between Imperial and metric sized fillet welds. Tables are included for evaluating eccentric loads on various weld groups and configurations.

Framed Beam Shear Connections

Pages 3–54 to 3–67 contain information on common types of beam shear connections traditionally considered standard in the industry. Included are double angle beam connections, simple end plate connections, single angle beam connections and tee-type beam connections.

Seated Beam Shear Connections

Pages 3–68 to 3–73 contain information on unstiffened and stiffened seated beam shear connections of a type commonly used in practice, where direct framing of the supported beam is either not desirable or possible.

Moment Connections

Pages 3–74 to 3–82 contain examples of welded and welded / bolted moment connections, and information for the design of stiffeners on supporting columns.

Hollow Structural Section Connections

Pages 3–83 to 3–91 contain information regarding the connecting of HSS sections.

Tension Members

Pages 3–92 to 3–99 contain tables and examples for calculating net effective areas and for evaluating the unit tensile resistance of bolted and welded tension members for various grades of steel.

CONNECTION LOADS

Connections are designed and detailed for the member reactions and loads given on the structural steel design drawings by the designer. Most connections are designed for factored loads, however specified loads are used for calculating the slip resistance of slip-critical connections, such as those subjected to dynamic loadings.

In evaluating member loads and forces it is preferable to keep different types of load separate to facilitate application of the different load factors and load combination factors specified in Clause 7.2 of CAN/CSA-S16.1-M89 and other governing codes. However, this may not always be convenient, and Fig. 3–1 is included to permit an approximate evaluation of either the total specified load or the total factored load when either one is known, and the ratio of the specified dead load to specified live load is known or the unit specified dead load and unit specified live load are both known. The curve is based on load factors α given in Clause 7.2.3 and makes no allowance for possible live load reductions permitted by the applicable building codes.

Example

Given:

The total factored dead and live load reaction at the end of a beam to be designed as a slip-critical connection is 235 kN and the specified unit dead and live loads are 4.35 kPa and 2.16 kPa respectively. What is the reaction under specified load.

Solution:

Ratio D/L = 4.35/2.16 = 2.0

From Fig. 3–1, for D/L = 2.0, $(L + D) / (\alpha_L L + \alpha_D D) = 0.75$

Therefore specified load reaction is $0.75 \times 235 = 176$ kN

FIGURE 3–1

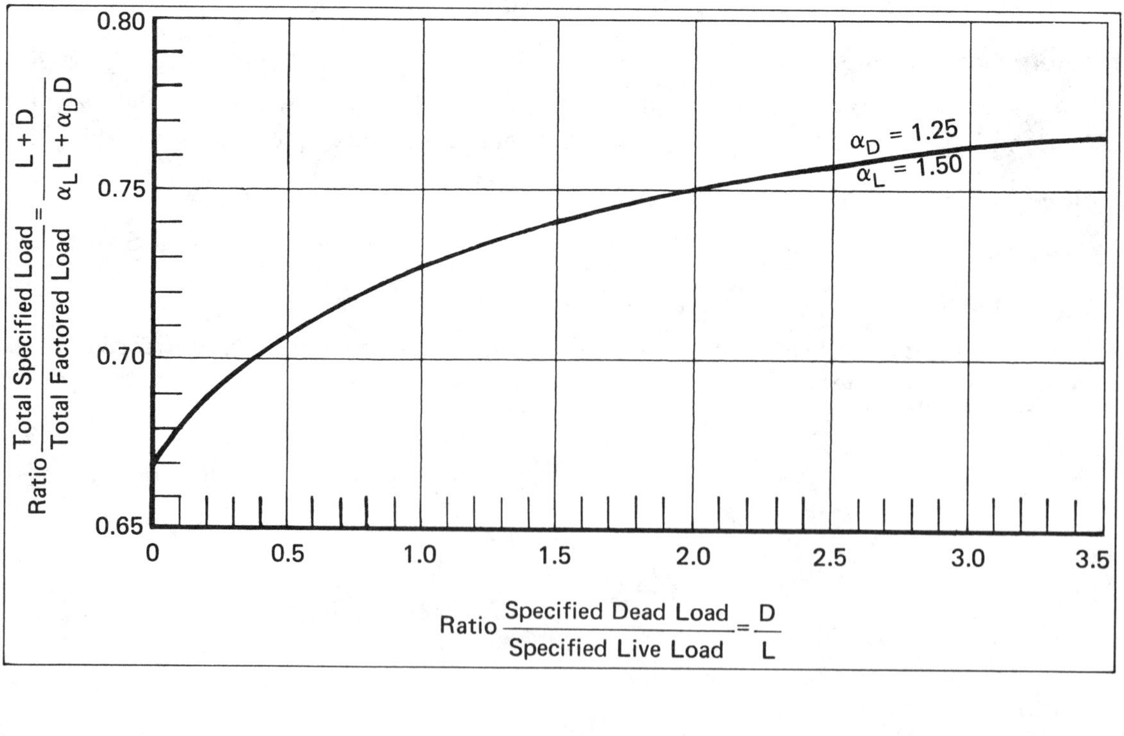

BOLT DATA

General

Tables in this section are based on CAN/CSA-S16.1-M89, and include information for both metric series and Imperial series bolts. Data for the metric series bolts is based on ASTM Specifications A325M and A490M and for the Imperial series bolts on ASTM Specifications A325, A490 and A307. Values are tabulated in ascending order of nominal cross-sectional area A_b (mm^2) to facilitate comparison between metric and Imperial sizes. Bold type is used for metric series bolts when both metric sizes and Imperial sizes appear on the same table. This section includes the following:

Bolt Data Metric and Imperial Series

Table 3–1 on page 3–9 lists the size, nominal diameter (mm), nominal area (mm^2) and values of A_bF_u for bolt sizes from M16 to M36 and 1/2 inch to 1-1/2 inch diameter.

Bolts in Bearing-Type Connections

Tables 3–2 to 3–7 on pages 3–7 to 3–11 list values of bearing and bolt resistances computed in accordance with Clauses 13.10.1(c) and 13.11. Tables 3–8 and 3–9 on page 3–13 assist in evaluating combined shear and tension on bolts.

Bolts in Slip-Critical Connections

Tables 3–10 and 3–11 on page 3–15 list resistances, for use with bolts in slip-critical connections, computed in accordance with Clause 13.12.

Bolts in Tension and Prying Action

Tables and design aids on pages 3–19 to 3–25 assist in evaluating the effects of prying action on bolts loaded in tension.

Eccentric Loads on Bolt Groups

Tables for evaluating eccentric loads on bolts in bearing-type and slip-critical connections for various bolt group configurations are given on pages 3–26 to 3–35.

Availability of Metric Bolts

Before specifying metric bolts, the designer should check on their current availability in the quantities required. To facilitate substitution, use the lower of the metric or Imperial values for similar sizes.

BOLT DATA
Metric and Imperial Series

Table 3-1

Bolt Size		Nominal Diameter of Bolt (mm)	Nominal Area (A_b) (mm^2)	$A_b F_u$** (kN)				
Metric*	Imperial			A325M	A490M	A325	A490	A307
	1/2	12.70	127			104	131	
	5/8	15.88	198			163	205	81.9
M16		**16.00**	**201**	**167**	**209**			
	3/4	19.05	285			235	295	118
M20		**20.00**	**314**	**261**	**327**			
M22		**22.00**	**380**	**316**	**395**			
	7/8	22.23	388			320	402	161
M24		**24.00**	**452**	**375**	**470**			
	1	25.40	507			418	524	
M27		**27.00**	**573**	**475**	**595**			
	1–1/8	28.58	641			465	664	
M30		**30.00**	**707**	**587**	**735**			
	1–1/4	31.75	792			574	819	
M36		**36.00**	**1018**	**845**	**1060**			
	1–1/2	38.10	1140			827	1180	

*The number following the letter M is the nominal bolt diameter in millimetres.
**See Table 3–3 page 3–8 for specified minimum tensile strengths, F_u.

BOLTS IN BEARING-TYPE CONNECTIONS

General

Connections are generally detailed as bearing-type unless specified otherwise by the designer. Bearing-type connections are designed for factored loads, and Tables 3–2 to 3–9 inclusive on the following pages assist in evaluating the requirements of Clause 13.11 of CAN/CSA-S16.1-M89. Clause 23.3.2 lists the size and type of holes permitted with bearing type connections.

Table 3–2, page 3–7, summarises the requirements of Clause 13.11.2 for bolts in shear and Clause 13.11.3 for bolts in tension, and lists expressions for factored resistance and unit factored resistance of bolts in bearing-type cornections.

Table 3–3, page 3–8, lists, for A325M, A490M, A325, A490 and A307 bolts, values of the specified minimum tension, F_u, values of unit factored shear resistances, 0.4 F_u and 0.28 F_u, and values of unit factored tensile resistances, 0.5 F_u.

Table 3–4, page 3–8, lists factored shear and tensile resistances in kN/bolt for both metric series and Imperial series bolts.

Table 3–5, page 3–9, lists values of the specified minimum tensile strength, F_u, for the common grades of structural steel, and values of unit factored bearing resistances, 0.67 F_u and 2.0 F_u.

Tables 3–6 and 3–7, pages 3–9 to 3–11 inclusive, list factored bearing resistance in kN/bolt for five different values of F_u for the connected material. Bearing resistances in these tables are given in terms of the material thickness, t, and end distance, e, and are independent of the diameter, d, and grade of the bolt. Maximum bearing values are reached when e = 3d. The shaded areas on each table give the approximate range of end distances for the bolt sizes listed, from the minimum end distance to that of e =3d, for which the tabulated factored bearing resistances are applicable. The shear values in Table 3–4 govern if less than the bearing values given in Tables 3–6 and 3–7.

For connections of G40.21-M 300W steel (F_u = 450 MPa), see Table 3–6, page 3–9.

Tables 3–8 and 3–9 on page 3–13 assist in evaluating bolts in combined shear and tension according to Clause 13.11.4.

Table 3–2

CAN/CSA-S16.1-M89 SUMMARY Bearing-Type Connections

Bolt Situation In Joint	Factored Resistance	Unit Factored Resistance (n = m = A_b = 1)	Clause Reference
BOLTS IN SHEAR			13.11.2
Shear on bolts with threads excluded from shear plane	$V_r = 0.60 \phi_b n m A_b F_u$	$V_r = 0.40 F_u$	13.11.2(b)
Shear on bolts with threads intercepted by shear plane	$V_r = 0.42 \phi_b n m A_b F_u$	$V_r = 0.28 F_u$	
For joints longer than 1300 mm	Shear = 0.8 V_r		
Bearing on main material (the lesser of)	$B_r = 3 \phi t d n F_u$ $B_r = \phi t n e F_u$	$B_r = 2.0 t d F_u$ $B_r = 0.67 t e F_u$	13.10.1(c)
BOLTS IN TENSION	$T_r = 0.75 \phi_b n A_b F_u$	$T_r = 0.50 F_u$	13.11.3

Notes: 1. Oversize holes are not permitted in bearing-type connections (see Clause 23.3.2(a) of CAN/CSA-S16.1-M89.
2. See Clause 23.3.2(b), (c) of CAN/CSA-S16.1-M89 re use of slotted holes in bearing-type connections.
3. See Clause 23.3.2(e) of CAN/CSA-S16.1-M89 for hole diameters permitted with M20 or 3/4-inch diameter, M22 or 7/8-inch diameter, and M24 or 1-inch diameter bolts.

UNIT FACTORED SHEAR AND TENSILE RESISTANCES**

Table 3–3

$\phi_b = 0.67$

Bolt Grade	Specified Minimum Tensile Strength, F_u (MPa)	Unit Factored Shear Resistance		Unit Factored Tensile Resistance $0.50\ F_u$ (MPa)
		Threads Excluded $0.40\ F_u$ (MPa)	Threads Intercepted $0.28\ F_u$ (MPa)	
A325M	**830**	**334**	**234**	**417**
A490M	**1040**	**418**	**293**	**523**
A325 (d ⩽ 1")	825	332	232	415
A325 (d ⩾ 1–1/8")	725	291	204	364
A490	1035	416	291	520
A307*	414	166	117	208

* Use of A307 bolts in connections is covered in Clauses 22.2 and 22.4 of CAN/CSA-S16.1-M89.
**Values for Imperial series bolts are based on ASTM specifications A325 and A490 soft converted to SI units.

FACTORED SHEAR AND TENSILE RESISTANCES (kN PER BOLT)

Table 3–4

Bolt Size		Nominal Area A_b (mm^2)	Factored Shear Resistance† – Single Shear**(kN/bolt)						Factored Tensile Resistance, T_r (kN/bolt)		
			Threads Excluded			Threads Intercepted ††					
Metric*	Imperial		A325 A325M	A490 A490M	A307	A325 A325M	A490 A490M	A307	A325 A325M	A490 A490M	A307
	1/2	127	42.2	52.8		29.5	37.0		52.7	66.0	
	5/8	198	65.7	82.4	32.9	45.9	57.6	23.2	82.2	103	41.2
M16		**201**	**67.1**	**84.0**		**47.3**	**58.9**		**83.8**	**105**	
	3/4	285	94.6	119	47.3	66.1	82.9	33.4	118	148	59.3
M20		**314**	**105**	**131**		**73.5**	**92.0**		**131**	**164**	
M22		**380**	**127**	**159**		**88.9**	**111**		**158**	**199**	
	7/8	388	129	161	64.4	90.0	113	45.4	161	202	80.7
M24		**452**	**151**	**189**		**106**	**132**		**188**	**238**	
	1	507	168	211		118	148		210	264	
M27		**573**	**191**	**240**		**134**	**168**		**239**	**300**	
	1–1/8	641	187	267		131	187		233	333	
M30		**707**	**236**	**296**		**165**	**207**		**295**	**370**	
	1–1/4	792	230	329		162	230		288	412	
M36		**1018**	**340**	**426**		**238**	**298**		**425**	**532**	
	1–1/2	1140	332	474		233	322		415	593	

* The number following the letter M is the nominal bolt diameter in millimetres.
† For joints longer than 1 300 mm, use 80% of the factored shear resistance.
** For double shear, multiply tabulated values by 2.0.
††Treads are intercepted if thin material next to the nut is combined with detailing for minimum bolt stick-through (the nut).

Table 3-5 UNIT FACTORED BEARING RESISTANCE

	Material Standard and Grade	Specified Minimum Tensile Strength, F_u (MPa)	0.67 F_u (MPa)	Maximum 2.0 F_u (MPa)
CAN/CSA G40.21-M	230G	380	255	760
	260W, 260WT	410	275	820
	300W for HSS only	410	275	820
	300W, 350W, 300WT	450	302	900
	350G, 350WT, 350R, 350A, 350AT	480	322	960
	380W and 380WT for HSS only	480	322	960
	400W, 400WT, 400A, 400AT	520	348	1040
	480W, 480WT, 480A, 480AT	590	395	1180
ASTM	A36	400	268	800
	A572 Grade 42	415	278	830
	Grade 50	450	302	900
	A441 F_y = 42	435	291	870
	F_y = 46	460	308	920
	F_y = 50	485	325	970

Table 3-6 FACTORED BEARING RESISTANCE, B_r* (kN/bolt)

Material Thickness t (mm)	CSA G40.21-M 300W, 350W, 300WT (F_u = 450 MPa)											
	End Distance**, e (mm)											
	20	25	30	35	40	45	50	60	70	80	90	100
4	24.1	30.1	36.2	42.2	48.2	54.3	60.3	72.4	84.4	96.5	109	121
4.5	27.1	33.9	40.7	47.5	54.3	61.1	67.8	81.4	95.0	109	122	136
5	30.1	37.7	45.2	52.8	60.3	67.8	75.4	90.4	106	121	136	151
6	36.2	45.2	54.3	63.3	72.4	81.4	90.4	109	127	145	163	181
7	42.2	52.8	63.3	73.9	84.4	95.0	106	127	148	169	190	211
8	48.2	60.3	72.4	84.4	96.5	109	121	145	169	193	217	241
9	54.3	67.8	81.4	95.0	109	122	136	163	190	217	244	271
10	60.3	75.4	90.4	106	121	136	151	181	211	241	271	301
11	66.3	82.9	99.5	116	133	149	166	199	232	265	298	332
12	72.4	90.4	109	127	145	163	181	217	253	289	326	362
13	78.4	98.0	118	137	157	176	196	235	274	314	353	392
14	84.4	106	127	148	169	190	211	253	295	338	380	422
15	90.4	113	136	158	181	204	226	271	317	362	407	452
16	96.5	121	145	169	193	217	241	289	338	386	434	482
17	103	128	154	179	205	231	256	308	359	410	461	513
18	109	136	163	190	217	244	271	326	380	434	488	543
19	115	143	172	200	229	258	286	344	401	458	516	573
20	121	151	181	211	241	271	301	362	422	482	543	603
21	127	158	190	222	253	285	317	380	443	507	570	633
22	133	166	199	232	265	298	332	398	464	531	597	663
23	139	173	208	243	277	312	347	416	485	555	624	693
24	145	181	217	253	289	326	362	434	507	579	651	724
25	151	188	226	264	301	339	377	452	528	603	678	754
28		211	253	295	338	380	422	507	591	675	760	844
30		226	271	317	362	407	452	543	633	724	814	904
32		241	289	338	386	434	482	579	675	772	868	965
35		264	317	369	422	475	528	633	739	844		

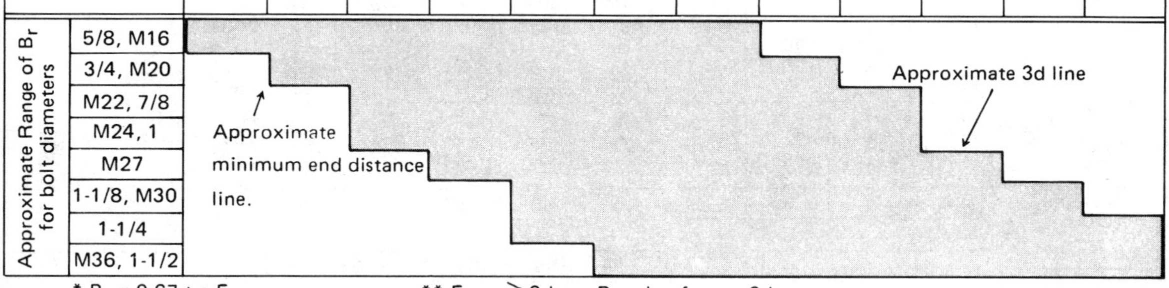

* B_r = 0.67 t e F_u ** For e > 3d use B_r value for e = 3d.

Notes: 1. Bolt shear governs if less than bearing. See Table 3–4.
2. Bearing values below dashed horizontal lines exceed single shear bolt values.

FACTORED BEARING RESISTANCE, B_r* (kN/bolt)

ASTM A36 (F_u = 400 MPa)

Material Thickness t (mm)	End Distance**, e (mm) 20	25	30	35	40	45	50	60	70	80	90	100
4	21.4	26.8	32.2	37.5	42.9	48.2	53.6	64.3	75.0	85.8	96.5	107
4.5	24.1	30.1	36.2	42.2	48.2	54.3	60.3	72.4	84.4	96.5	109	121
5	26.8	33.5	40.2	46.9	53.6	60.3	67.0	80.4	93.8	107	121	134
6	32.2	40.2	48.2	56.3	64.3	72.4	80.4	96.5	113	129	145	161
7	37.5	46.9	56.3	65.7	75.0	84.4	93.8	113	131	150	169	188
8	42.9	53.6	64.3	75.0	85.8	96.5	107	129	150	172	193	214
9	48.2	60.3	72.4	84.4	96.5	109	121	145	169	193	217	241
10	53.6	67.0	80.4	93.8	107	121	134	161	188	214	241	268
11	59.0	73.7	88.4	103	118	133	147	177	206	236	265	295
12	64.3	80.4	96.5	113	129	145	161	193	225	257	289	322
13	69.7	87.1	105	122	139	157	174	209	244	279	314	348
14	75.0	93.8	113	131	150	169	188	225	263	300	338	375
15	80.4	100	121	141	161	181	201	241	281	322	362	402
16	85.8	107	129	150	172	193	214	257	300	343	386	429
17	91.1	114	137	159	182	205	228	273	319	364	410	456
18	96.5	121	145	169	193	217	241	289	338	386	434	482
19	102	127	153	178	204	229	255	306	356	407	458	509
20	107	134	161	188	214	241	268	322	375	429	482	536
21	113	141	169	197	225	253	281	338	394	450	507	563
22	118	147	177	206	236	265	295	354	413	472	531	590
23	123	154	185	216	247	277	308	370	431	493	555	616
24	129	161	193	225	257	289	322	386	450	515	579	643
25	134	167	201	234	268	301	335	402	469	536	603	670
28	150	188	225	263	300	338	375	450	525	600	675	750
30	161	201	241	281	322	362	402	482	563	643	724	804
32		214	257	300	343	386	429	515	600	686	772	858
35		234	281	328	375	422	469	563	657	750	844	938

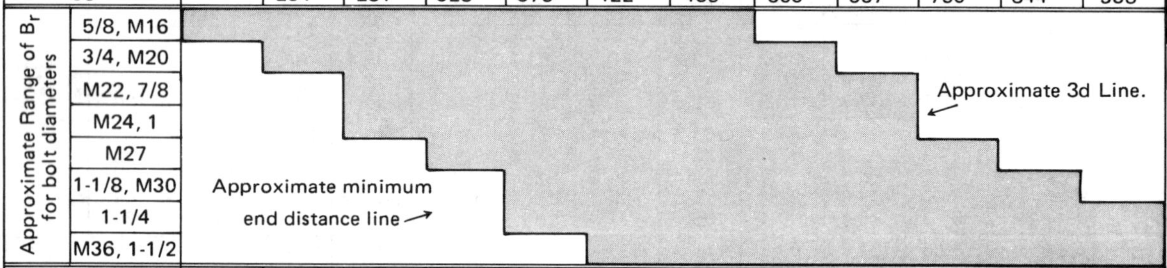

Material Thickness t (mm)	20	25	30	35	40	45	50	60	70	80	90	100
4	22.0	27.5	33.0	38.5	44.0	49.4	54.9	65.9	76.9	87.9	98.9	110
4.5	24.7	30.9	37.1	43.3	49.4	55.6	61.8	74.2	86.5	98.9	111	124
5	27.5	34.3	41.2	48.1	54.9	61.8	68.7	82.4	96.1	110	124	137
6	33.0	41.2	49.4	57.7	65.9	74.2	82.4	98.9	115	132	148	165
7	38.5	48.1	57.7	67.3	76.9	86.5	96.1	115	135	154	173	192
8	44.0	54.9	65.9	76.9	87.9	98.9	110	132	154	176	198	220
9	49.4	61.8	74.2	86.5	98.9	111	124	148	173	198	223	247
10	54.9	68.7	82.4	96.1	110	124	137	165	192	220	247	275
11	60.4	75.5	90.7	106	121	136	151	181	212	242	272	302
12	65.9	82.4	98.9	115	132	148	165	198	231	264	297	330
13	71.4	89.3	107	125	143	161	179	214	250	286	321	357
14	76.9	96.1	115	135	154	173	192	231	269	308	346	385
15	82.4	103	124	144	165	185	206	247	288	330	371	412
16	87.9	110	132	154	176	198	220	264	308	352	396	440
17	93.4	117	140	163	187	210	233	280	327	374	420	467
18	98.9	124	148	173	198	223	247	297	346	396	445	494
19	104	130	157	183	209	235	261	313	365	418	470	522
20	110	137	165	192	220	247	275	330	385	440	494	549
21	115	144	173	202	231	260	288	346	404	461	519	577
22	121	151	181	212	242	272	302	363	423	483	544	604
23	126	158	190	221	253	284	316	379	442	505	569	632
24	132	165	198	231	264	297	330	396	461	527	593	659
25	137	172	206	240	275	309	343	412	481	549	618	687
28	154	192	231	269	308	346	385	461	538	615	692	769
30	165	206	247	288	330	371	412	494	577	659	742	824
32	175	220	264	308	352	396	440	527	615	703	791	879
35	192	240	288	337	385	433	481	577	673	769	865	961

Material Thickness t (mm) — End Distance**, e (mm)

CSA G40.21-M 260W, 260WT (F_u = 410 MPa)

* $B_r = 0.67\ t\ e\ F_u$ ** For e > 3d use B_r value for e = 3d.

Notes: 1. Bolt shear governs if less than bearing. See Table 3—4.
2. Bearing values below dashed horizontal lines exceed single shear bolt values.

Table 3-7 FACTORED BEARING RESISTANCE, B_r* (kN/bolt)

CSA G40.21-M 350G, 350WT, 350R, 350A, 380W, 380WT (F_u = 480 MPa)

Material Thickness t (mm)	End Distance **, e (mm)											
	20	25	30	35	40	45	50	60	70	80	90	100
4	25.7	32.2	38.6	45.0	51.5	57.9	64.3	77.2	90.0	103	116	129
4.5	28.9	36.2	43.4	50.7	57.9	65.1	72.4	86.8	101	116	130	145
5	32.2	40.2	48.2	56.3	64.3	72.4	80.4	96.5	113	129	145	161
6	38.6	48.2	57.9	67.5	77.2	86.8	96.5	116	135	154	174	193
7	45.0	56.3	67.5	78.8	90.0	101	113	135	158	180	203	225
8	51.5	64.3	77.2	90.0	103	116	129	154	180	206	232	257
9	57.9	72.4	86.8	101	116	130	145	174	203	232	260	289
10	64.3	80.4	96.5	113	129	145	161	193	225	257	289	322
11	70.8	88.4	106	124	142	159	177	212	248	283	318	354
12	77.2	96.5	116	135	154	174	193	232	270	309	347	386
13	83.6	105	125	146	167	188	209	251	293	334	376	418
14	90.0	113	135	158	180	203	225	270	315	360	405	450
15	96.5	121	145	169	193	217	241	289	338	386	434	482
16	103	129	154	180	206	232	257	309	360	412	463	515
17	109	137	164	191	219	246	273	328	383	437	492	547
18	116	145	174	203	232	260	289	347	405	463	521	579
19	122	153	183	214	244	275	306	367	428	489	550	611
20	129	161	193	225	257	289	322	386	450	515	579	643
21	135	169	203	236	270	304	338	405	473	540	608	675
22	142	177	212	248	283	318	354	425	495	566	637	708
23	148	185	222	259	296	333	370	444	518	592	666	740
24	154	193	232	270	309	347	386	463	540	617	695	772
25	161	201	241	281	322	362	402	482	563	643	724	804
28		225	270	315	360	405	450	540	630	720	810	900
30		241	289	338	386	434	482	579	675	772	868	
32		257	309	360	412	463	515	617	720	823	926	
35			338	394	450	507	563	675	788	900		

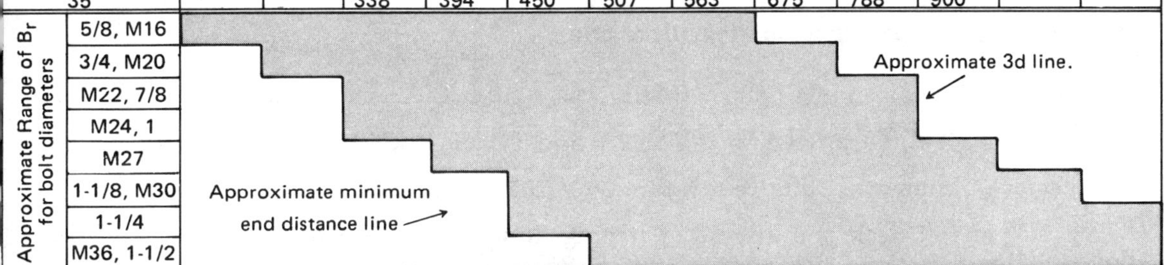

Material Thickness t (mm)	End Distance**, e (mm)											
	20	25	30	35	40	45	50	60	70	80	90	100
4	27.9	34.8	41.8	48.8	55.7	62.7	69.7	83.6	97.6	111	125	139
4.5	31.4	39.2	47.0	54.9	62.7	70.6	78.4	94.1	110	125	141	157
5	34.8	43.5	52.3	61.0	69.7	78.4	87.1	105	122	139	157	174
6	41.8	52.3	62.7	73.2	83.6	94.1	105	125	146	167	188	209
7	48.8	61.0	73.2	85.4	97.6	110	122	146	171	195	219	244
8	55.7	69.7	83.6	97.6	111	125	139	167	195	223	251	279
9	62.7	78.4	94.1	110	125	141	157	188	219	251	282	314
10	69.7	87.1	105	122	139	157	174	209	244	279	314	348
11	76.6	95.8	115	134	153	172	192	230	268	307	345	383
12	83.6	105	125	146	167	188	209	251	293	334	376	418
13	90.6	113	136	159	181	204	226	272	317	362	408	453
14	97.6	122	146	171	195	219	244	293	341	390	439	488
15	105	131	157	183	209	235	261	314	366	418	470	523
16	111	139	167	195	223	251	279	334	390	446	502	557
17	118	148	178	207	237	267	296	355	415	474	533	592
18	125	157	188	219	251	282	314	376	439	502	564	627
19	132	165	199	232	265	298	331	397	463	530	596	662
20	139	174	209	244	279	314	348	418	488	557	627	697
21	146	183	219	256	293	329	366	439	512	585	658	732
22	153	192	230	268	307	345	383	460	537	613	690	766
23	160	200	240	280	321	361	401	481	561	641	721	801
24		209	251	293	334	376	418	502	585	669	753	836
25		218	261	305	348	392	435	523	610	697	784	871
28		244	293	341	390	439	488	585	683	780	878	
30		261	314	366	418	470	523	627	732	836	941	
32			334	390	446	502	557	669	780	892		
35			366	427	488	549	610	732	854			

CSA G40.21-M 400W, 400WT, 400A, 400AT (F_u = 520 MPa)

* $B_r = 0.67\ t\ e\ F_u$ ** For $e > 3d$ use B_r value for $e = 3d$.

Notes: 1. Bolt shear governs if less than bearing. See Table 3–4.
2. Bearing values below dashed horizontal lines exceed single shear bolt values.

Bolts in Combined Shear and Tension — Bearing-Type Connections

Clause 13.11.4 of CAN/CSA-S16.1-M89 requires bolts subjected to shear and tension to satisfy the expression $V_f^2 + \beta T_f^2 \le \beta (0.75 \phi_b A_b F_u)^2$, where V_f is the factored shear load on the bolt and T_f is the factored tensile load on the bolt including prying effects. This expression can be written as $V_f^2 + \beta T_f^2 = \beta T_r^2$, where T_r is the factored tensile bolt resistance (See Table 3–4, page 3–8). If the shear tension ratio V_f / T_f on the bolts is X, then solving for V_f and T_f gives, $V_f = X T_f$ and $T_f = [\beta T_r^2/(X^2 + \beta)]^{1/2}$.

Combined shear and tension usually occurs for the threads not intercepted case, since a plate or flange thin enough to include threads in the shear plane (about 10 mm) has little capacity to transmit tension. Tables 3–8 and 3–9 give values of V_f and T_f for various shear-tension ratios, X, for M20, M22 and M24 A325M and A490M bolts and 3/4, 7/8 and 1-inch A325 and A490 bolts with threads excluded from the shear plane, where $\beta = 0.69$ for A325M and A325 bolts and $\beta = 0.56$ for A490M and A490 bolts.

Example

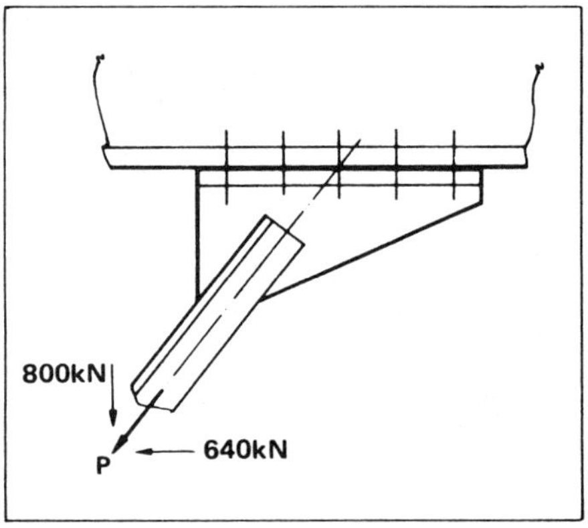

Given: A bracing connection for an inclined factored load P with a tension component T_f of 800 kN and a shear component V_f of 640 kN uses a tee section cut from a W410x85. Check the number of M20, A325M bolts required, assumning G40.21-M 300W steel with a 90 mm gauge on the tee flange and a 90 mm bolt pitch.

Solution: Ignore prying action for initial trial.

Shear-tension ratio is 640/800 = 0.80.

From Table 3–8, permitted V_f = 75.5 kN and T_f = 94.4 kN per bolt.

Therefore, number of bolts required = 640/75.5 or 800/94.4 = 8.48.

Try 10 bolts. Check the effects of prying action upon the bolt tension, and the connection capacity (see pages 3–19 to 3–25).

Applied factored tensile load per bolt = 800/10 = 80 kN = P_f.

For W410x85, t = 18.2 mm, w = 10.9 mm, and flange width = 181 mm.

$b = (90 - 10.9)/2 = 39.6$; $b' = 39.6 - 10 = 29.6$ mm; $1.25\,b = 49.5$

$a = (181 - 90)/2 = 45.5 < 1.25\,b$; $a' = 45.5 + 10 = 55.5$.

$$K = (4 \times 29.6 \times 10^3)/(0.9 \times 90 \times 300) = 4.87 \qquad \text{(Eqn. 1)}$$

$$\delta = 1 - 22/90 = 0.76 \qquad \text{(Eqn. 2)}$$

Use maximum allowed T_f in Eqn. 4, rather than T_r:

$$\alpha = \left(\frac{4.87 \times 94.4}{18.2^2} - 1\right) \times \frac{55.5}{0.76\,(55.5 + 29.6)} = 0.333 \text{ (for bolt } T_f\text{)};\ \delta\,\alpha = 0.253 \qquad \text{(Eqn. 4)}$$

Connection capacity = $(18.2^2/4.87)(1.253)10 = 852$ kN > 800 ∴ OK (Eqn. 5)

$$\alpha = \left(\frac{4.87 \times 80}{18.2^2} - 1\right) \times \frac{1}{0.76} = 0.232 \text{ (for applied load)};\ \delta\alpha = 0.176 \qquad \text{(Eqn. 6)}$$

$$T_f \approx 80\left[1 + \left(\frac{29.6}{55.5} \times \frac{0.176}{1.176}\right)\right] = 86.4 \text{ kN (bolt tension with prying)} \qquad \text{(Eqn. 7)}$$

Table 3-8

For A325M and A325 Bolts

Shear Tension Ratio $X = V_f/T_f$		Bolt Size											
		3/4		M20		M22		7/8		M24		1	
X	$1/X$	V_f	T_f	V_f	T_f	V_f	T_f	V_f	T_f	V_f	T_f	V_f	T_f
0	T_r	0	118	0	131	0	158	0	161	0	188	0	210
0.10		11.7	117	13.0	130	15.7	157	16.0	160	18.7	187	20.8	208
0.20		23.0	115	25.4	127	30.8	154	31.4	157	36.6	183	40.8	204
0.30		33.3	111	36.9	123	44.7	149	45.3	151	53.1	177	59.4	198
0.40		42.4	106	47.2	118	56.8	142	58.0	145	67.6	169	75.6	189
0.50		50.5	101	56.0	112	67.5	135	69.0	138	80.5	161	90.0	180
0.60		57.4	95.7	63.6	106	76.8	128	78.6	131	91.2	152	102	170
0.70		63.1	90.2	70.0	100	84.7	121	86.1	123	101	144	113	161
0.80		68.0	85.0	75.5	94.4	91.2	114	92.8	116	108	135	121	151
0.90		72.0	80.0	79.9	88.8	96.3	107	98.1	109	115	128	128	142
1.00	1.00	75.4	75.4	83.7	83.7	101	101	103	103	120	120	134	134
1.11	0.90	78.6	70.7	87.1	78.4	105	94.6	107	96.4	126	113	140	126
1.25	0.80	81.6	65.3	90.6	72.5	109	87.4	111	89.1	130	104	145	116
1.43	0.70	84.7	59.3	94.0	65.8	113	79.4	116	80.9	135	94.5	151	106
1.67	0.60	87.7	52.6	97.3	58.4	118	70.5	120	71.8	140	83.9	156	93.7
2.00	0.50	90.6	45.3	100	50.2	121	60.6	124	61.8	144	72.1	161	80.5
2.50	0.40	93.0	37.2	103	41.3	125	49.8	127	50.8	148	59.3	166	66.2
3.33	0.30	94.6	28.5	105	31.7	127	38.2	129	38.9	151	45.5	168	50.8
5.00	0.20	94.6	19.3	105	21.5	127	25.9	129	26.4	151	30.8	168	34.4
10.0	0.10	94.6	9.8	105	10.8	127	13.3	129	13.3	151	15.6	168	17.4
V_r	0	94.6	0	105	0	127	0	129	0	151	0	168	0

$V_f = XT_f$, $T_f = \sqrt{\beta T_r^2/(X^2+\beta)}$ where $\beta = 0.69$ for threads excluded.

Table 3-9

For A490M and A490 Bolts

Shear Tension Ratio $X = V_f/T_f$		Bolt Size											
		3/4		M20		M22		7/8		M24		1	
X	$1/X$	V_f	T_f	V_f	T_f	V_f	T_f	V_f	T_f	V_f	T_f	V_f	T_f
0	T_r	0	148	0	164	0	199	0	202	0	238	0	264
0.10		14.7	147	16.3	163	19.7	197	20.0	100	23.6	236	26.2	262
0.20		28.6	143	31.6	158	38.4	192	39.0	195	46.0	230	51.0	255
0.30		41.1	137	45.6	152	55.5	185	56.1	187	66.3	221	73.5	245
0.40		52.4	131	58.0	145	70.4	176	71.2	178	84.0	210	93.2	233
0.50		61.5	123	68.0	136	82.5	165	84.0	168	99.0	198	110	220
0.60		69.0	115	76.8	128	93.0	155	94.8	158	112	186	124	206
0.70		75.6	108	84.0	120	102	145	104	148	122	174	135	193
0.80		80.8	101	89.6	112	109	136	110	138	130	163	144	180
0.90		85.1	94.6	94.5	105	114	127	116	129	137	152	152	169
1.00	1.00	88.7	88.7	98.3	98.3	119	119	121	121	143	143	158	158
1.11	0.90	91.9	82.7	102	91.6	123	111	126	113	148	133	163	147
1.25	0.80	95.0	76.0	105	84.2	128	102	130	104	153	122	170	136
1.43	0.70	98.1	68.7	109	76.1	132	92.3	134	93.7	157	110	176	123
1.67	0.60	101	60.6	112	67.2	136	81.5	138	82.7	163	97.5	180	108
2.00	0.50	104	51.9	115	57.5	139	69.7	142	70.8	167	83.4	185	92.5
2.50	0.40	106	42.4	118	47.0	143	57.1	145	57.9	171	68.2	189	75.7
3.33	0.30	108	32.4	120	35.9	145	43.6	147	44.2	174	52.1	193	57.8
5.00	0.20	110	21.9	122	24.3	148	29.5	150	29.9	176	35.2	196	39.1
10.00	0.10	110	11.0	122	12.2	149	14.9	151	15.1	178	17.8	197	19.7
V_f	0	119	0	131	0	159	0	161	0	189	0	211	0

$V_f = XT_f$, $T_f = \sqrt{\beta T_r^2/(X^2+\beta)}$ where $\beta = 0.56$ for thread excluded.

BOLTS IN SLIP-CRITICAL CONNECTIONS

General

The name change from slip-resistant to slip-critical emphasizes that this type of connection is required only when the consequences of slip are critical to the performance of the structure. Clause 21.12.2(a) of CAN/CSA-S16.1-M89 requires slip-critical connections where slippage into bearing cannot be tolerated, such as structures sensitive to deflection, or subject to fatigue or frequent load reversals. Slip-critical shear joints transfer the specified loads by the slip resistance (friction) of the clamped faying surfaces in accordance with Clause 13.12.

In addition to the slip resistance, the strength (factored resistance) of the joint under factored loads must also be checked.

Tables

Tables 3–10 and 3–11 on page 3–15 are based on Clause 13.12.2 of S16.1-M89 for bolts in slip-critical connections.

Table 3–10 lists values c_1 for a 5% probability of slip, and values of unit slip resistance ($0.53\ c_1\ k_s\ F_u$) for A325M, A490M, A325 and A490 bolts for the contact surfaces (Class A, Class B and Class C) given in Table 2 of S16.1-M89.

Table 3–11 lists slip-resistance values ($V_s = 0.53\ c_1\ k_s\ m\ n\ A_b\ F_u$) for bolted joints with a single faying surface (m = 1) for Class A and Class B contact surface for M16 to M36 A325M and A490M bolts, and 1/2 to 1-1/2 inch A325 and A490 bolts.

Example

Given:

A single shear connection is subject to 370 kN at specified load level and 550 kN at factored load level. Select the number of M20 A325M bolts required for a slip-critical connection. Steel is G40.21-M 300W 6mm thick, and the surface is clean mill scale (Class A).

Solution:

(a) For specified loads

From Table 3–11, V_s = 37.4 kN (M20 A325M bolt for clean mill scale).

Number of bolts required = 370/37.4 = 9.9 — use 10

(b) Check connection at factored loads

From Table 3–4 (page 3–8), V_r = 73.5 kN (M20 A325M, threads intercepted)

Factored shear resistance of joint is $10 \times 73.5 = 735$ kN > 550 — OK

The factored bearing resistance of the joint for factored loads also needs to be checked, based upon thicknesses and end distances of the connected material:

For 6 mm thick material and an end distance of 30 mm, Table 3–6 (page 3–9) lists a bearing value of 54.3 kN per bolt which is less than 55.0 kN required for the 550 kN load.

Increase the edge distance, possibly to 35mm (B_r = 63.3 kN).

Table 3–10 UNIT SLIP RESISTANCE, $0.53 c_1 k_s F_u$* (MPa)

For 5% Probability of Slip

Contact Surface of Bolted Parts	Class	A	B	C
	Description	Clean mill scale, or blast-cleaned with Class A coatings $k_s = 0.33$	Blast-cleaned, or blast-cleaned with Class B coatings $k_s = 0.50$	Hot-dip galvanized with wire brushed surfaces $k_s = 0.40$
A325M	c_1	0.82	0.89	0.90
F_u = 830 MPa	**$0.53 c_1 k_s F_u$ (MPa)**	**119**	**196**	**158**
A490M	c_1	0.78	0.85	0.85
F_u = 1040 MPa	**$0.53 c_1 k_s F_u$ (MPa)**	**142**	**234**	**187**
A325 (d ≤ 1 inch)	c_1	0.82	0.89	0.90
F_u = 825 MPa	**$0.53 c_1 k_s F_u$ (MPa)**	**118**	**195**	**157**
A325 (d ≥ 1–1/8 inch)	c_1	0.82	0.89	0.90
F_u = 725 MPa	**$0.53 c_1 k_s F_u$ (MPa)**	**104**	**171**	**138**
A490	c_1	0.78	0.85	0.85
F_u = 1035 MPa	**$0.53 c_1 k_s F_u$ (MPa)**	**141**	**233**	**187**

* See CAN/CSA-S16.1-M89, Clause 13.12.2 for values of c_1 and k_s.

Table 3–11 SLIP RESISTANCE, V_s,* (kN/BOLT)

For Single Shear ** (m = 1)

Bolt Size		Nominal Area	Class A Surfaces		Class B Surfaces	
Metric+	Imperial	A_b (mm^2)	A325M, A325	A490M, A490	A325M, A325	A490M, A490
	1/2	127	15.0	17.9	24.7	29.6
	5/8	198	23.4	28.0	38.5	46.2
M16		**201**	**23.9**	**28.5**	**39.3**	**47.1**
	3/4	285	33.7	40.2	55.5	66.4
M20		**314**	**37.4**	**44.5**	**61.5**	**73.6**
M22		**380**	**45.2**	**53.9**	**74.4**	**89.0**
	7/8	388	45.9	54.8	75.5	90.5
M24		**452**	**53.8**	**64.1**	**88.5**	**106**
	1	507	60.0	71.6	98.7	118
M27		**573**	**68.2**	**81.3**	**112**	**134**
	1–1/8	641	66.6	90.5	110	149
M30		**707**	**84.2**	**100**	**138**	**166**
	1–1/4	792	82.4	112	135	185
M36		**1018**	**121**	**144**	**199**	**238**
	1–1/2	1140	119	161	195	266

* These resistances are for use with specified loads in accordance with Clause 13.12 of CAN/CSA-S16.1-M89.
** For double shear (m = 2) multiply tabulated values by 2.0.
+ The number following the letter M is the nominal bolt diameter in millimetres.

Bolts in Combined Shear and Tension — Slip-Critical Connections

Clause 13.12.3 of S16.1-M89 requires that bolts subject to both shear and tension in a slip-critical connection satisfy the following relationship for specified loads:

$$\frac{V}{V_s} + 1.9\frac{T}{n\,A_b\,F_u} \leq 1.0$$

The above relationship can conservatively be expressed (see Commentary page 2–38) as:

$$\frac{V}{V_s} + \frac{T}{T_i} \leq 1.0$$

If the shear-tension ratio V/ T on the bolts is X, solving for V and T gives V = X T, and $T = V_s / (X + V_s / T_i)$

Table 3–12 lists values of V and T for various shear-tension ratios X for Class A contact surfaces (clean mill scale or blast cleaned with Class A coatings, $k_s = 0.33$) using A325M or A325 bolts in single shear. This table can be used to establish directly the number of bolts required to satisfy the interaction equation for slip-critical connections subjected to a combination of shear and tension.

Clause 13.12.1 requires that the factored tension T_f, including any prying forces, not exceed the bolt tensile resistance T_r. See page 3–19 for a method of evaluating the prying effects.

Example

Given:

Find the number of M20 A325M bolts required in a slip-critical connection to resist a specified tension force of 320 kN and a specified shear force of 400 kN. The single faying surface consists of clean mill scale.

Solution:

Prying is not a factor when making the specified shear vs. specified tension interaction check. Within permitted loadings, prying is only a redistribution of the contact forces between the material surfaces, having no significant affect upon resistance to slipping.

However, the connection also has to be confirmed for strength as a bearing-type connection at factored loads.

Shear-tension ratio is 400 / 320 = 1.25

From Table 3–12, for M20 bolts and V/ T = 1.25,

permitted V = 30.9 kN and T = 24.7 kN per bolt

Therefore, number of bolts required is 400 / 30.9 or 320 / 24.7 = 12.9

Try 14 bolts

The bearing-type connection check, which includes bolt prying and flange strength using factored tension, would be done using Eqns. 4 and 5 as for the example on page 3–12.

Table 3–12

C_1 = 0.82

k_s = 0.33

SPECIFIED SHEAR AND TENSION (kN/BOLT)

Slip-Critical Connections, Class A Surfaces

A325M and A325 Bolts

Shear-Tension Ratio X = V/ T		3/4		**M20**		**M22**		7/8		**M24**		1	
X	1/ X	V	T	V	T	V	T	V	T	V	T	V	T
0.5		21.9	43.8	**24.5**	**49.0**	**29.9**	**59.7**	30.0	60.1	**35.3**	**70.6**	39.3	78.5
0.6		23.3	38.8	**26.0**	**43.3**	**31.7**	**52.8**	31.9	53.1	**37.4**	**62.4**	41.7	69.4
0.7		24.3	34.8	**27.2**	**38.8**	**33.1**	**47.2**	33.3	47.6	**39.1**	**55.9**	43.6	62.2
0.8		25.2	31.5	**28.1**	**35.2**	**34.2**	**42.8**	34.5	43.1	**40.5**	**50.6**	45.1	56.4
0.9		25.9	28.8	**28.9**	**32.1**	**35.2**	**39.1**	35.5	39.4	**41.7**	**46.3**	46.4	51.5
1.0	1.0	26.5	26.5	**29.6**	**29.6**	**36.0**	**36.0**	36.3	36.3	**42.6**	**42.6**	47.5	47.5
1.11	0.9	27.1	24.4	**30.2**	**27.2**	**36.7**	**33.0**	37.1	33.4	**43.5**	**39.2**	48.5	43.6
1.25	0.8	27.7	22.2	**30.9**	**24.7**	**37.5**	**30.0**	37.9	30.3	**44.5**	**35.6**	49.5	39.6
1.43	0.7	28.3	19.8	**31.6**	**22.1**	**38.3**	**26.8**	38.7	27.1	**45.5**	**31.8**	50.6	35.4
1.67	0.6	29.0	17.4	**32.3**	**19.4**	**39.2**	**23.5**	39.6	23.8	**46.5**	**27.9**	51.8	31.1
2.00	0.5	29.7	14.8	**33.0**	**16.5**	**40.1**	**20.0**	40.6	20.3	**47.6**	**23.8**	53.0	26.5
2.50	0.4	30.4	12.2	**33.8**	**13.5**	**41.0**	**16.4**	41.5	16.6	**48.7**	**19.5**	54.3	21.7
3.33	0.3	31.2	9.4	**34.7**	**10.4**	**42.0**	**12.6**	42.5	12.8	**49.9**	**15.0**	55.6	16.7
5.00	0.2	32.0	6.4	**35.5**	**7.1**	**43.0**	**8.6**	43.6	8.7	**51.1**	**10.2**	57.0	11.4
10.00	0.1	32.8	3.3	**36.4**	**3.6**	**44.1**	**4.4**	44.7	4.5	**52.4**	**5.2**	58.5	5.8
Vs	0	33.7	0	**37.4**	**0**	**45.2**	**0**	45.9	0	**53.8**	**0**	60.0	0

$V = XT, \quad T = \dfrac{V_s}{X + V_s / T_i}$

NOTES

BOLTS IN TENSION AND PRYING ACTION

General

Connections with fasteners loaded in tension occur in many common situations, such as hanger and bracing connections with tee-type gussets, and end plate moment connections. Clause 23.1.4 of CAN/CSA-S16.1-M89 requires that when bolts are loaded in direct tension, the effects of prying action be taken into account in proportioning the bolts and connected parts. This clause also requires that the connection be proportioned to avoid prying forces when subjected to repeated loading.

The actual stress distribution in the flange of a tee-type connection is extremely complex as it depends upon the bolt size and arrangement, and upon the strength and dimensions of the connecting flange. Consequently, various design methods have been proposed in the technical literature for proportioning such connections. The procedures given in this section are based on the recommendations contained on page 285 of *Guide to Design Criteria for Bolted and Riveted Joints*, 2nd edition, by Kulak, Fisher and Struik.

The procedures include a set of seven equations for selecting a trial section and for evaluating the bolt forces and flange capacity. Based on these equations, Table 3–13 on page 3–24 and Figure 3–2 on page 3–25 are provided for preliminary design and checking purposes.

The material in this section has been modified slightly from that in the previous edition of the Handbook. The determination of α by equilibrium Eqn. 4 for use in the connection capacity Eqn. 5 uses the full tensile resistance T_r of the bolts rather than the factored load per bolt including prying T_f (amplified bolt force). This provides a value for the maximum capacity of the connection. The design procedure has been rewritten to take direct advantage of Table 3–13 and Figure 3–2, and to reduce the calculations.

Table 3–13 and Figure 3–2 demonstrate the effect of applied factored tensile load per bolt and flange geometry upon flange thicknesses for various bolt sizes, assuming static loads. In general, prying effects can be minimized by dimensioning for minimum practical gauge distance and for maximum permissible edge distance. For repeated loading the flange must be made thick enough and stiff enough so that deformation of the flange is virtually eliminated. In addition, special attention must be paid to bolt installation to ensure that the bolts are properly pretensioned to provide the required clamping force.

The expressions for prying effects are based on tests carried out on tees. For angles, assuming the distribution of moment shown, the moment equilibrium equation can be derived from statics as

$$P_f\, b = Q\, a. \quad \text{Therefore } Q / P_f = b/a$$

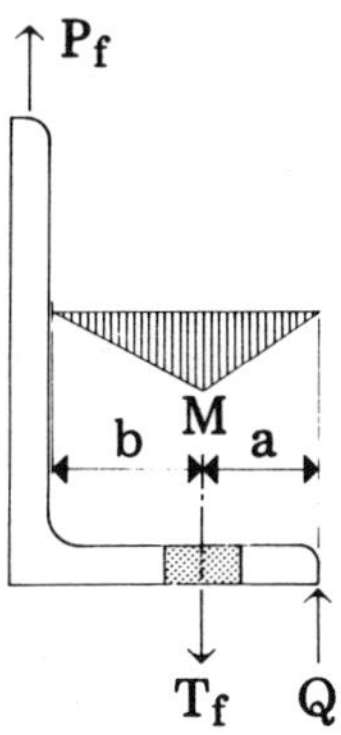

References

Douty, R. T., and McGuire, W. 1965. High strength bolted moment connections. Journal of the Structural Division, April, ASCE.

Kulak, G.L., Fisher, J.W., and Struik, J.H.A. 1987. Guide to design criteria for bolted and riveted joints, 2nd edition. John Wiley & Sons, New York, N.Y.

Nair, R. S., Birkemoe, P. C., and Munse, W. H. 1969. High strength bolts subject to tension and prying. Structural Research Series 353, September, Department of Civil Engineering, U. of Ill., Urbana.

Equations

$$K = 4\, b'\, 10^3 / (\phi\, p\, F_y) \qquad (1)$$

$$\delta = 1 - d'/p \qquad (2)$$

$$\text{Range of } t = \left(\frac{K P_f}{(1+\delta\alpha)}\right)^{1/2} \qquad \begin{matrix} t_{min} \text{ when } \alpha = 1.0 \\ t_{max} \text{ when } \alpha = 0.0 \end{matrix} \qquad (3)$$

$$\alpha = \left(\frac{K T_r}{t^2} - 1\right) \times \frac{a'}{\delta\,(a' + b')}, \quad 0 \le \alpha \le 1.0 \qquad (4)$$

$$\text{Connection capacity} = (t^2 / K)\,(1 + \delta\alpha)\, n \qquad (5)$$

$$\alpha = \left(\frac{K P_f}{t^2} - 1\right) \times \frac{1}{\delta} \qquad (6)$$

$$T_f \approx P_f\left[1 + \left(\frac{b'}{a'} \times \frac{\delta\alpha}{1+\delta\alpha}\right)\right] \le T_r \qquad (7)$$

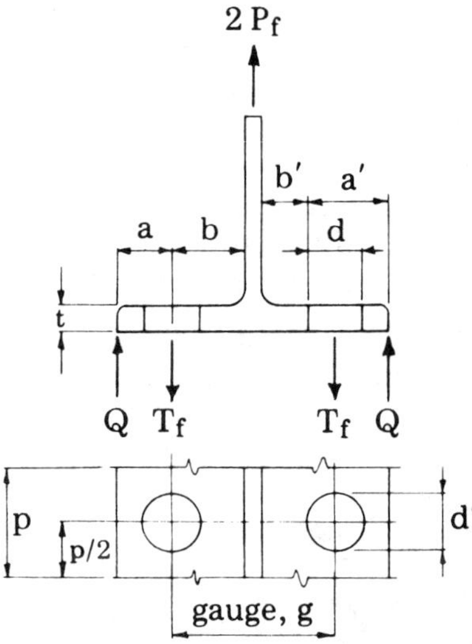

Nomenclature

K = Parameter as defined in Eqn. 1

P_f = Applied factored tensile load per bolt, kN

Q = Prying force per bolt at factored load ($Q = T_f - P_f$), kN

T_f = Factored load per bolt including prying (amplified bolt force), kN

T_r = Factored tensile resistance of bolts ($\phi_b\, 0.75\, A_b\, F_u$), kN

F_y = Yield strength of flange material, MPa

a = Distance from bolt line to edge of tee flange, not more than 1.25 b (mm)

a' = $a + d/2$

b = Distance from bolt line (gauge line) to face of tee stem (mm)

b' = $b - d/2$

d = Bolt diameter (mm)

d' = Nominal hole diameter (mm)

n = Number of flange bolts in tension

p = Length of flange tributary to each bolt, or bolt pitch (mm)

t = Thickness of flange (mm)

α = Ratio of sagging moment at bolt line to hogging moment at stem of tee

δ = Ratio of net area (at bolt line) to gross area at face of tee stem

ϕ = Resistance factor for the tee material (0.9)

Preliminary Design Tables

Table 3–13 on page 3–24 lists the maximum and minimum values of flange thickness *t* calculated with Eqn. 3 using $\alpha = 0.0$ and $\alpha = 1.0$, for a range of values of P_f. Results are tabulated for various flange bolt patterns and bolt sizes.

The maximum and minimum values of *t* indicate a range of flange thickness within which the bolts and flange are in equilibrium for the particular flange geometry, and in which the effects of flange flexure and prying reduce the effective tension capacity of the bolts. At the maximum thickness, P_f is the applied force per bolt, and at the minimum thickness P_f is the amplified bolt force including prying effects. When the flange thickness is greater than the larger value of t ($\alpha = 0.0$), the flange is generally sufficiently thick and stiff to virtually eliminate prying action, and the connection capacity will be limited by the tensile resistance of the bolts. When the flange thickness is less than the smaller value of t ($\alpha = 1.0$), the flange thickness will govern the connection capacity and the bolts will usually have excess capacity to resist the applied tension load in spite of prying effects. Within the range of flange thickness for $0.0 \leq \alpha \leq 1.0$, with the bolts and flange in equilibrium, the ratio T_f / P_f will increase from unity for t_{max} to a maximum value for t_{min}. In this range, the bolts control the capacity with the flange strength being increasingly utilized as the flange thickness decreases. (It can be helpful to note that the typical ratio of maximum to minimum flange thickness is about 1.33, and that at the mimimum flange thickness, the prying ratio T_f / P_f is about the same. When the maximum flange thickness is used, there is essentially no prying and the ratios t_{max} / t_{used} and T_f / P_f are both 1.0. Thus, a rough guide to the ratio t_{max} / t_{reqd} is the available prying ratio T_r / P_f of the desired bolt size, since T_r must be $\geq T_f$.)

The bolt pitch p should be approximately 4 to 5 times the bolt size ($4d \leq p \leq 5d$) and the gauge g should be kept as small as practicable. Also, dimension a for design purposes must not exceed 1.25 b.

Figure 3–2 on page 3–25 graphs the amplified bolt force T_f for various applied loads P_f, flange thicknesses t, and four different values of b (40mm, 45mm, 50mm and 55mm) with M20, M22 and M24, A325M bolts. These graphs can be used to evaluate the effects of flange thickness, gauge distance and bolt size on the amplified bolt force, and to establish reasonable trial connection parameters.

Design Procedure

Trial Section

1) Select an intended bolt size as a function of the applied factored tensile load per bolt P_f and the anticipated prying ratio.

2) With P_f, the bolt size, and trial values of b and p, use Eqns. 1, 2 and 3 (with $\alpha = 0.0$ and $\alpha = 1.0$) to identify a range of acceptable flange thicknesses. (Alternatively, use Table 3–13 on page 3–24.)

3) Identify an intended flange thickness.

Figure 3–2 (page 3–25) may also be used to identify an intended bolt size and flange geometry, based on the amplified bolt force being less than the bolt tensile resistance.

Design Check

1) Calculate α by using Eqn. 4, for use in Eqn. 5.

2) Calculate the connection capacity with Eqn. 5. If α from Eqn. 4 < 0.0, use $\alpha = 0.0$, and if $\alpha > 1.0$, use $\alpha = 1.0$.

3) Equations 6 and 7 can be used, if desired, to determine the total bolt tension including prying (amplified bolt force), that results from the applied load.

Note:

S16.1-M89, Clause 21.12.2(e) requires that all bolts subject to tensile loadings be pretensioned when installed, and Clause 23.1.4 requires that connections with repeated tension loads on bolts be proportioned to avoid prying forces.

Example 1

Given:

Design a tension tee connection with 4 ASTM A325M bolts in tension for a factored static load of 500 kN assuming the bolts on a 100 mm gauge, at a pitch of 110 mm, with the tee connected to rigid supports. Use G40.21-M 300W steel.

Solution:

Trial Section

Applied load per bolt = 500/4 = 125 kN = P_f

Assume M22 bolts, 24 mm hole diameter, 15 mm web. T_r = 158 kN

Available prying ratio T_r / P_f is 158/125 = 1.26 ∴ OK for near minimum flange

$b = (100 - 15) / 2 = 42.5\text{ mm}; \quad b' = 42.5 - 11 = 31.5$

$K = 4 \times 31.5 \times 10^3 / (0.9 \times 110 \times 300) = 4.25$ (Eqn. 1)

$\delta = 1 - (24/110) = 0.782$ (Eqn. 2)

$$t_{min} = \left(\frac{4.25 \times 125}{1.782}\right)^{1/2} = 17.3\text{ mm}; \quad t_{max} = \left(\frac{4.25 \times 125}{1.0}\right)^{1/2} = 23.0\text{ mm} \qquad \text{(Eqn. 3)}$$

(Alternatively, the range of t, by rough interpolation from Table 3–13 on page 3–24, could be seen to be about 17 to 23 mm.)

Two possible solutions are W460x97, t = 19.0 mm, and W530x138, t = 23.6 mm, the choice being determined by the maximum available prying ratio T_r / P_f for the bolt size finally selected.

Design Check

Try W460x97 with M22 bolts, d = 22 mm, d′ = 24 mm

t = 19.0, w = 11.4, flange width = 193

$b = (100 - 11.4) / 2 = 44.3\text{ mm}; \quad b' = 44.3 - 11 = 33.3; \quad 1.25\,b = 55.4$

$K = 4 \times 33.3 \times 10^3 / (0.9 \times 110 \times 300) = 4.48$ (Eqn. 1)

$\delta = 0.782$ (as above, by Eqn. 2)

$a = (193 - 100) / 2 = 46.5\text{ mm} < 1.25\,b; \quad a' = 46.5 + 11 = 57.5; \quad a' + b' = 90.8$

$$\alpha = \left(\frac{4.48 \times 158}{19.0^2} - 1\right) \times \frac{57.5}{0.782 \times 90.8} = 0.778 \qquad \text{(Eqn. 4)}$$

$\delta\alpha = 0.608$

Connection capacity = $(19.0^2 / 4.48)(1.608)\,4 = 518$ kN > 500 (Eqn. 5)

To find actual bolt tension, if desired:

$$\alpha = \left(\frac{4.48 \times 125}{19.0^2} - 1\right) \times \frac{1}{0.782} = 0.705 \qquad \text{(Eqn. 6)}$$

$\delta\alpha = 0.551$

$$T_f \approx 125\left[1 + \left(\frac{33.3}{57.5} \times \frac{0.551}{1 + 0.551}\right)\right] = 151\text{ kN}, \quad < 158 \qquad \text{(Eqn. 7)}$$

Tee stem capacity is $0.9(2 \times 110)\,11.4 \times 300 / 1\,000 = 677$ kN > 500 ∴ OK

(With M20 bolts, T_r / P_f = 131 /125 = 1.05, and the heavier W530x138 is required.)

Example 2

Given:

Use Table 3–13 to select bolt sizes and trial dimensions for a factored tensile load of 480 kN, assuming tees cut from a W460x106 section (G40.21-M 300W), and a gauge dimension of 90 mm.

Solution:

Since Table 3–13 is intended only for the selection of a trial section which must be checked with Eqns. 1, 2, 4 and 5 on page 3–20 (illustrated in the previous example), precise interpolation is not necessary. A comparison of the available ratio T_r / P_f for the bolts chosen, to the ratio of t_{max} to flange thickness, for the value of t_{max} selected from Table 3–13, will indicate whether a combination of bolt size, flange thickness and geometry may be a suitable choice.

For W460x106, t = 20.6 mm and w = 12.6 mm

For g = 90 mm, b = (90 – 12.6) / 2 = 38.7 (Use b = 40 for Table 3–13)

With 4 bolts, P_f = 480/4 = 120 kN, and T_r / P_f = 131/120 = 1.09 for M20 bolts, and 159/120 = 1.32 for M22 bolts.

Table 3–13, with b = 40, M20 bolts and P_f = 120, p = 100 gives t_{max} = 23.1 mm

$t_{max} / t = 23.1 / 20.6 = 1.12 \quad > T_r / P_f = 1.09$ Likely not OK

Table 3–13, with b = 40, M22 bolts and P_f = 120, p = 90 gives t_{max} = 23.9 mm

$t_{max} / t = 23.9 / 20.6 = 1.16 \quad < T_r / P_f = 1.32$ ∴ OK, check the design

Alternatively, Fig. 3–2 can be used to select the bolt size based on the flange thickness and the amplified bolt force.

Use graph for b = 40 mm (b = 38.7, see above)

Enter graph at applied load per bolt of 120 kN and flange thickness t = 20.6 mm

With M20 bolts, amplified bolt force ≈ 138 kN $> T_r$ = 131 kN No good

With M 22 bolts, amplified bolt force ≈ 125 kN $< T_r$ = 158 kN ∴ OK

Proceed with design check using M22 bolts; p = 5d = 110 mm.

b = 38.7; b′ = 38.7 – 11 = 27.7; 1.25 b = 48.4

$$K = 4 \times 27.7 \times 10^3 / (0.9 \times 110 \times 300) = 3.73 \quad \text{(Eqn. 1)}$$

$$\delta = 1 - (24 / 110) = 0.782 \quad \text{(Eqn. 2)}$$

$$a = (194 - 90) / 2 = 52.0 \text{ mm} \quad > 1.25\,b$$

Use a = 1.25 b = 48.4; a′ = 48.4 + 11 = 59.5; a′ + b′ = 87.2

$$\alpha = \left(\frac{3.73 \times 158}{20.6^2} - 1 \right) \times \frac{59.5}{0.782 \times 87.2} = 0.339 \quad \text{(Eqn. 4)} \qquad \delta\alpha = 0.265$$

Connection capacity = (20.6 ^{2} / 3.73) (1.265) 4 = 576 kN > 480 (Eqn. 5)

Check bolt tension (amplified bolt force)

$$\alpha = \left(\frac{3.73 \times 120}{20.6^2} - 1 \right) \times \frac{1}{0.782} = 0.070 \quad \text{(Eqn. 6)} \qquad \delta\alpha = 0.055$$

$$T_f \approx 120 \left[1 + \left(\frac{27.7}{59.5} \times \frac{0.055}{1 + 0.055} \right) \right] = 123 \text{ kN}, \quad < 158 \quad \text{(Eqn. 7)}$$

RANGE OF t

$$t = \left[\frac{K P_f}{(1 + \delta\alpha)} \right]^{1/2}$$

Table 3-13

t_{min} when $\alpha = 1.0$ and t_{max} when $\alpha = 0.0$

b (mm)	Bolt Size	P_f = 60 kN			P_f = 80 kN			P_f = 100 kN			P_f = 120 kN		
		pitch p (mm)			pitch p (mm)			pitch p (mm)			pitch p (mm)		
		80	90	100	80	90	100	80	90	100	80	90	100
35	3/4	12.8 16.8	11.9 15.9	11.2 15.0	14.7 19.4	13.8 18.3	13.0 17.4	16.5 21.7	15.4 20.5	14.5 19.4			
	M20	**12.7 16.7**	**11.9 15.7**	**11.2 14.9**	**14.7 19.2**	**13.7 18.1**	**12.9 17.2**	**16.4 21.5**	**15.3 20.3**	**14.4 19.2**	**17.9 23.6**	**16.8 22.2**	**15.8 21.1**
40	3/4	14.0 18.4	13.1 17.3	12.3 16.5	16.1 21.2	15.1 20.0	14.2 19.0	18.0 23.8	16.9 22.4	15.9 21.2			
	M20	**13.9 18.3**	**13.0 17.2**	**12.2 16.3**	**16.1 21.1**	**15.0 19.9**	**14.1 18.9**	**17.9 23.6**	**16.8 22.2**	**15.8 21.1**	**19.7 25.8**	**18.4 24.3**	**17.3 23.1**
45	3/4	15.1 19.9	14.1 18.7	13.3 17.8	17.4 22.9	16.3 21.6	15.3 20.5	19.4 25.6	18.2 24.2	17.1 22.9			
	M20	**15.0 19.7**	**14.0 18.6**	**13.2 17.6**	**17.3 22.8**	**16.2 21.5**	**15.3 20.4**	**19.4 25.5**	**18.1 24.0**	**17.1 22.8**	**21.2 27.9**	**19.8 26.3**	**18.7 24.9**
50	3/4	16.1 21.2	15.0 20.0	14.2 19.0	18.6 24.5	17.4 23.1	16.4 21.9	20.8 27.4	19.4 25.8	18.3 24.5			
	M20	**16.1 21.1**	**15.0 19.9**	**14.1 18.9**	**18.5 24.3**	**17.3 23.0**	**16.3 21.8**	**20.7 27.2**	**19.4 25.7**	**18.2 24.3**	**22.7 29.8**	**21.2 28.1**	**20.0 26.7**
55	3/4	17.1 22.5	15.9 21.2	15.0 20.1	19.7 26.0	18.4 24.5	17.4 23.2	22.0 29.0	20.6 27.4	19.4 26.0			
	M20	**17.0 22.4**	**15.9 21.1**	**15.0 20.0**	**19.7 25.8**	**18.4 24.3**	**17.3 23.1**	**22.0 28.9**	**20.5 27.2**	**19.4 25.8**	**24.1 31.6**	**22.5 29.8**	**21.2 28.3**

b (mm)	Bolt Size	P_f = 100 kN			P_f = 120 kN			P_f = 140 kN			P_f = 160 kN		
		pitch p (mm)			pitch p (mm)			pitch p (mm)			pitch p (mm)		
		90	100	110	90	100	110	90	100	110	90	100	110
40	**M22**	**16.6 21.8**	**15.6 20.7**	**14.8 19.8**	**18.2 23.9**	**17.1 22.7**	**16.2 21.6**	**19.6 25.9**	**18.5 24.5**	**17.5 23.4**			
	7/8	16.6 21.8	15.6 20.7	14.8 19.7	18.2 23.9	17.1 22.7	16.2 21.6	19.6 25.8	18.5 24.5	17.5 23.3	21.0 27.6	19.7 26.2	18.7 24.9
45	**M22**	**18.0 23.7**	**16.9 22.4**	**16.0 21.4**	**19.7 25.9**	**18.5 24.6**	**17.6 23.4**	**21.3 28.0**	**20.0 26.6**	**19.0 25.3**			
	7/8	18.0 23.6	16.9 22.4	16.0 21.4	19.7 25.9	18.5 24.5	17.5 23.4	21.2 27.9	20.0 26.5	18.9 25.3	22.7 29.9	21.4 28.3	20.3 27.0
50	**M22**	**19.2 25.3**	**18.1 24.0**	**17.2 22.9**	**21.1 27.8**	**19.8 26.3**	**18.8 25.1**	**22.8 30.0**	**21.4 28.4**	**20.3 27.1**			
	7/8	19.2 25.3	18.1 24.0	17.2 22.9	21.1 27.7	19.8 26.3	18.8 25.1	22.8 29.9	21.4 28.4	20.3 27.1	24.3 32.0	22.9 30.4	21.7 28.9
55	**M22**	**20.4 26.9**	**19.2 25.5**	**18.2 24.3**	**22.4 29.5**	**21.1 28.0**	**20.0 26.7**	**24.2 31.8**	**22.8 30.2**	**21.6 28.8**			
	7/8	20.4 26.9	19.2 25.5	18.2 24.3	22.4 29.4	21.1 27.9	20.0 26.6	24.2 31.8	22.8 30.2	21.6 28.8	25.8 34.0	24.3 32.3	23.1 30.8

b (mm)	Bolt Size	P_f = 140 kN			P_f = 160 kN			P_f = 180 kN			P_f = 200 kN		
		pitch p (mm)			pitch p (mm)			pitch p (mm)			pitch p (mm)		
		100	110	120	100	110	120	100	110	120	100	110	120
40	**M24**	**18.3 24.1**	**17.3 23.0**	**16.5 22.0**	**19.5 25.8**	**18.5 24.6**	**17.6 23.5**	**20.7 27.3**	**19.6 26.1**	**18.7 24.9**			
	1	18.1 23.8	17.1 22.7	16.3 21.7	19.4 25.4	18.3 24.3	17.4 23.2	20.5 27.0	19.4 25.7	18.5 24.6	21.6 28.4	20.5 27.1	19.5 26.0
45	**M24**	**19.8 26.2**	**18.8 24.9**	**17.9 23.9**	**21.2 28.0**	**20.1 26.7**	**19.1 25.5**	**22.5 29.7**	**21.3 28.3**	**20.3 27.1**			
	1	19.7 25.9	18.7 24.7	17.8 23.6	21.1 27.7	19.9 26.4	19.0 25.3	22.3 29.3	21.1 28.0	20.1 26.8	23.5 30.9	22.3 29.5	21.2 28.2
50	**M24**	**21.3 28.1**	**20.2 26.8**	**19.2 25.6**	**22.8 30.0**	**21.5 28.6**	**20.5 27.4**	**24.1 31.8**	**22.9 30.4**	**21.8 29.1**			
	1	21.2 27.8	20.0 26.5	19.1 25.4	22.6 29.7	21.4 28.4	20.4 27.1	24.0 31.5	22.7 30.1	21.6 28.8	25.3 33.2	24.0 31.7	22.8 30.3
55	**M24**	**22.6 29.9**	**21.4 28.5**	**20.4 27.3**	**24.2 31.9**	**22.9 30.4**	**21.8 29.1**	**25.7 33.9**	**24.3 32.3**	**23.1 30.9**			
	1	22.5 29.6	21.3 28.2	20.3 27.0	24.1 31.7	22.8 30.2	21.7 28.9	25.6 33.6	24.2 32.0	23.0 30.7	26.9 35.4	25.5 33.8	24.3 32.3

$K = 4b' \, 10^3 / (\phi p F_y)$ where $\phi = 0.90$ and $F_y = 300$ MPa

Figure 3.2 AMPLIFIED BOLT FORCE, T_f (kN)

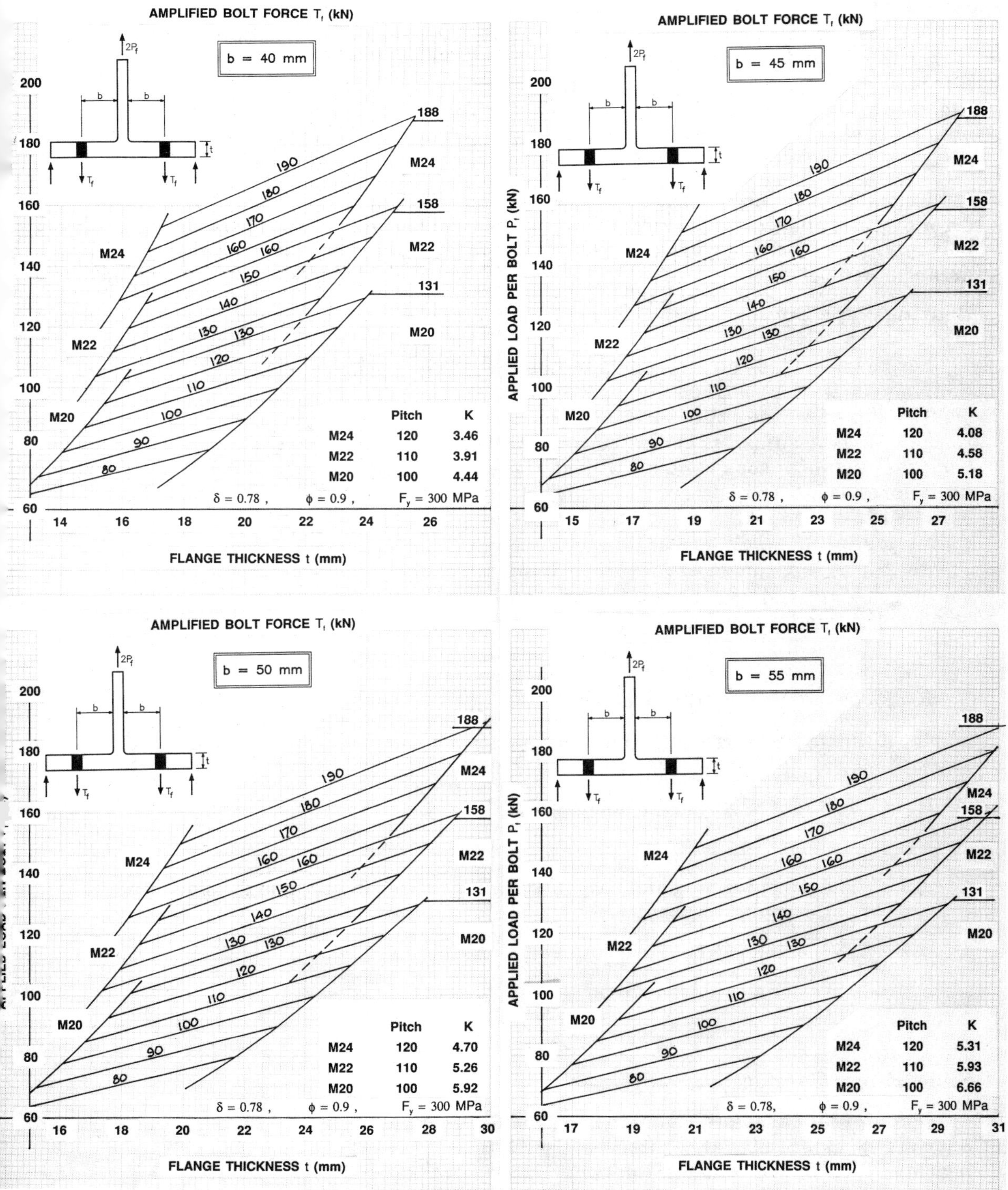

ECCENTRIC LOADS ON BOLT GROUPS

General

A bolted connection is eccentrically loaded when the line of action of the applied load passes outside the centroid of the bolt group. When the bolts are subjected to shear forces only, the effect of this eccentricity is to cause rotation about a single point called the instantaneous centre of rotation. The location of the instantaneous centre is obtained when the connection satisfies the three equilibrium equations for statics, $\Sigma F_x = 0$, $\Sigma F_y = 0$ and $\Sigma M = 0$ about the instantaneous centre.

Calculation of the instantaneous centre described in the references is a trial and error process, and the tables included in this section permit rapid evaluation of common bolt groups subjected to various eccentricities. All tables are based on symmetrical arrangements of bolts.

Bearing-Type Connections

For bearing type connections, a method of analysis is that described by Kulak *et al.* (1987). At the time the ultimate load is reached, it is assumed that the bolt furthest from the instantaneous centre will just reach its failure load. The resistance of each bolt is assumed to act on a line perpendicular to the radius joining the bolt to the instantaneous centre, and, Δ is assumed to vary linearly with the length of the radius. The resistance of each bolt is calculated according to the load deformation relationship:

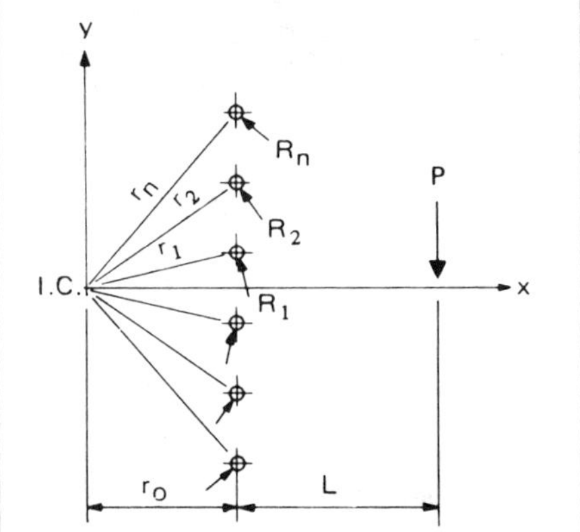

Forces on Eccentrically Loaded Connection

$$R = R_u\,(1 - e^{-\mu\Delta})^{\lambda} \text{..... Kulak } et\ al. \text{ (1987)}$$

and the ultimate load is reached when $\Delta = \Delta_{max}$ for the bolt furthest from the instantaneous centre, where

- R = bolt load at any given deformation
- R_u = ultimate bolt load
- Δ = shearing, bending and bearing deformation of the bolt, and local deformation of the connecting material
- μ, λ = regression coefficients
- e = base of natural logarithms

Slip-Critical Connections

For slip-critical connections, the method of anlaysis is essentially the same as that for bearing-type, except that the limiting slip resistance of the joint is reached when the maximum slip resistance of each individual bolt is reached as expressed by the relationship, $R = V_s = 0.53\, c_1\, k_s\, mn\, A_b\, F_u$ and the slip resistance of each bolt is assumed to be equal.

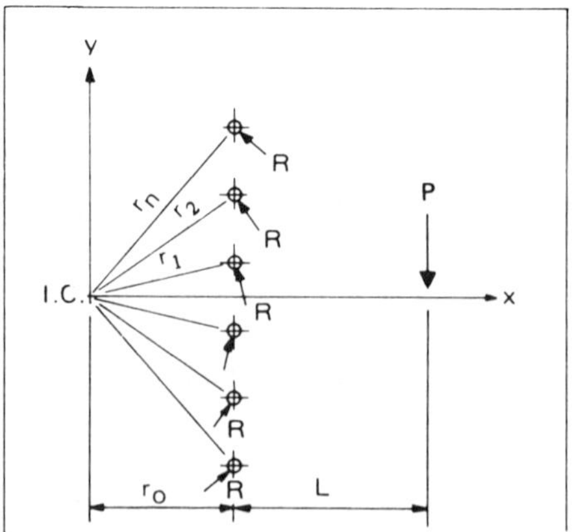

Forces on Eccentrically Loaded Slip-Resistant Connection

Tables

Tables 3–14 to 3–20 have been developed using the method described for bearing-type connections. Values tabulated are non-dimensional coefficients C and may be used for bolts of any diameter. In determining the coefficients C, the following values were used. R_u = 74 kips (329 kN), μ = 10.0, λ = 0.55, Δ_{max} = 0.34 inches (8.64 mm). These values were obtained experimentally for 3/4 inch diameter A325 bolts and are reported by Crawford and Kulak (1971).

The ultimate load P for each bolt group and eccentricity was computed and then divided by the maximum value of R (when $\Delta = \Delta max$) to obtain the values of C.

The tables may thus be used to obtain the factored resistance, expressed as a vertical load P, of a connection by multiplying the coefficient C, for any particular bolt group and eccentricity, by the factored shear resistance of a single bolt. i.e. $P_f = CV_r$.

Coefficients were developed in a similar way for slip-critical connections, except that the individual bolt resistances for all bolts in the group were assumed to be equal. The coefficients calculated in this way were from 5% to 10% higher than those for bearing-type connections. Thus only one set of tables, based on the bearing-type connections is provided for use with both bearing-type and slip-critical connections.

Use of Tables

Bearing-Type Connections

1) To obtain the coefficient C required for a given geometry of bolts and eccentricity of load, divide the factored load, P_f, by the factored shear resistance, V_r, of a single bolt for the appropriate shear condition. i.e. $C = P_f / V_r$.
2) To determine the capacity of a given connection, multiply the coefficient C for the bolt group and eccentricity, by the appropriate bolt shear resistance value, V_r, of a single bolt. i.e. $P_f = CV_r$.

V_r is the factored shear resistance of the bolt from Table 3–4. Used in this way these tables provide a margin of safety which is consistant with bolts in joints less than 1300 mm long and subjected to shear produced by concentric loads only.

Slip-Critical Connections

Although developed using the method for bearing-type connections, these tables can also be used for slip-resistant connections using the *specified* load, P, and the appropriate slip resistance value, V_s, for the bolt size and condition of the faying surface.

1) Required $C = P/V_s$
2) Capacity $P = CV_s$

V_s is the slip resistance determined from Tables 3–10 and 3–11.

References

CRAWFORD, S.F., and KULAK, G.L. 1971. Eccentrically loaded bolted connections. Journal of the Structural Division, ASCE, Vol. 97, No. ST3, March.

KULAK, G.L. 1975. Eccentrically loaded slip resistant connections. Engineering Journal, AISC, Vol. 12, No. 2, Second Quarter.

KULAK, G.L., ADAMS, P.F., and GILMOR, M.I. 1990. Limit states design in structural steel, CISC.

KULAK, G.L., FISHER, J.W., and STRUIK, J.H.A. 1987. Guide to design criteria for bolted and riveted joints; second edition. John Wiley and Sons.

SHERMER, C.L. 1971. Plastic behaviour of eccentrically loaded connections. Engineering Journal, AISC, Vol. 8, No. 2, April.

Example

1. Given:

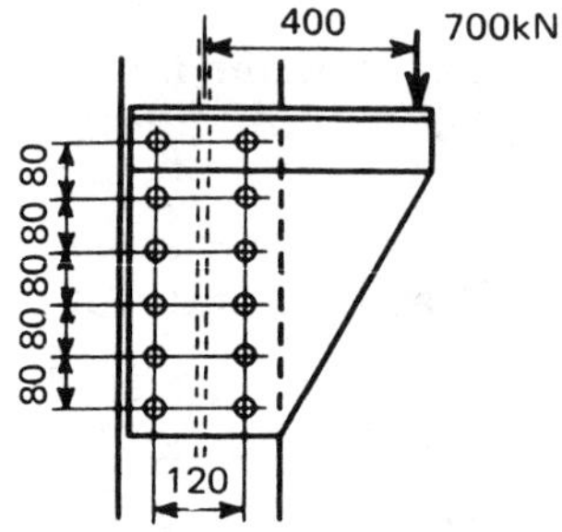

A double column bracket must be designed to support a factored load of 700 kN at an eccentricity of 400 mm. Find the number of M20 A325M bolts per flange required for a gauge dimension of 120 mm and a pitch of 80 mm assuming a bearing type connection.

Solution:

$P_f = 700/2 = 350$ kN, L = 400 mm

$V_r = 105$ kN (Table 3–4 page 3–8, single shear, threads excluded)

Required C = 350/105 = 3.33

From Tables 3–15 and 3–16 for 2 lines of bolts at 80 mm pitch

6 rows at 80 mm gauge, C = 3.49

at 320 mm gauge, C = 4.77

Interpolating for 120 mm gauge $C = 3.49 + (4.77 - 3.49) \times 40/240 = 3.70$

Use 6 rows of bolts (total 12 bolts)

Capacity is $3.70 \times 105 = 389$ kN per side

The bearing capacity of the connected material would also be checked for the various material thicknesses and end distances involved.

2. Given:

Find the number of M22 bolts required for a similar bracket assuming a slip-critical connection with clean mill scale and a specified load of 500 kN.

Solution:

P = 500/2 = 250 kN, L = 400 mm

$V_s = 45.2$ kN (Table 3–11, page 3–15)

Required C = 250/45.2 = 5.53

From Tables 3–15 and 3–16 for 2 lines of bolts at 80 mm pitch

8 rows at 80 mm gauge, C = 5.89

at 320 mm gauge, C = 7.16

Interpolating for 120 mm gauge, $C = 5.89 + (7.16 - 5.80) \times 40/240 = 6.10$

Use 8 rows of bolts (total 16 bolts)

Capacity is $6.10 \times 45.2 = 276$ kN per side

The ultimate strength of the joint would also be checked in bearing and shear for factored loads.

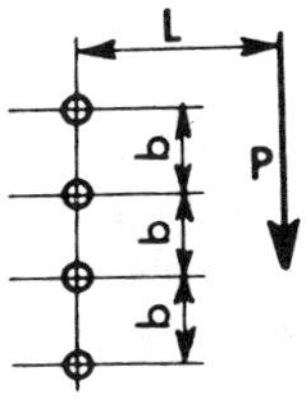

$$C = \frac{P_f}{V_r}, \text{ or } \frac{P}{V_s}$$

ECCENTRIC LOADS ON BOLT GROUPS
Coefficients C*

Table 3–14

Moment Arm, L, Millimetres												Number of Bolts	Pitch b mm
75	100	125	150	175	200	225	250	300	400	500	600		
0.94	0.74	0.60	0.51	0.44	0.39	0.35	0.31	0.26	0.20	0.16	0.13	2	
1.86	1.49	1.24	1.05	0.90	0.79	0.71	0.64	0.53	0.40	0.32	0.27	3	
2.95	2.50	2.15	1.85	1.62	1.44	1.30	1.18	0.99	0.75	0.60	0.51	4	
4.07	3.56	3.12	2.76	2.44	2.19	1.97	1.79	1.51	1.15	0.92	0.77	5	
5.17	4.67	4.21	3.77	3.37	3.06	2.77	2.53	2.16	1.65	1.34	1.12	6	80
6.25	5.79	5.30	4.82	4.38	4.01	3.66	3.37	2.89	2.23	1.80	1.51	7	
7.34	6.89	6.39	5.91	5.43	5.01	4.62	4.28	3.69	2.88	2.34	1.97	8	
8.40	7.99	7.51	7.02	6.54	6.07	5.64	5.24	4.57	3.60	2.93	2.48	9	
9.45	9.06	8.61	8.12	7.62	7.15	6.69	6.26	5.50	4.37	3.59	3.03	10	
10.5	10.1	9.71	9.23	8.73	8.25	7.75	7.30	6.49	5.20	4.30	3.65	11	
11.5	11.2	10.8	10.3	9.85	9.36	8.85	8.40	7.51	6.09	5.06	4.30	12	
1.01	0.82	0.67	0.57	0.49	0.43	0.39	0.35	0.29	0.22	0.18	0.15	2	
2.00	1.65	1.37	1.17	1.01	0.89	0.79	0.71	0.60	0.45	0.36	0.30	3	
3.11	2.68	2.33	2.04	1.80	1.60	1.44	1.31	1.11	0.84	0.68	0.57	4	
4.22	3.79	3.36	3.00	2.69	2.41	2.19	2.00	1.69	1.29	1.04	0.87	5	
5.31	4.89	4.45	4.06	3.66	3.34	3.05	2.81	2.40	1.85	1.50	1.26	6	90
6.40	5.99	5.58	5.13	4.73	4.35	4.00	3.70	3.20	2.48	2.02	1.69	7	
7.46	7.10	6.68	6.24	5.82	5.40	5.02	4.66	4.06	3.20	2.61	2.20	8	
8.51	8.18	7.77	7.34	6.90	6.48	6.07	5.69	5.00	3.98	3.27	2.76	9	
9.56	9.24	8.87	8.46	8.02	7.57	7.15	6.74	6.00	4.83	3.99	3.39	10	
10.6	10.3	9.95	9.54	9.12	8.69	8.25	7.81	7.02	5.73	4.76	4.06	11	
11.6	11.3	11.0	10.6	10.2	9.80	9.34	8.93	8.07	6.66	5.59	4.78	12	
1.09	0.89	0.74	0.63	0.55	0.48	0.43	0.39	0.33	0.25	0.20	0.16	2	
2.14	1.79	1.50	1.28	1.12	0.98	0.88	0.79	0.66	0.50	0.40	0.33	3	
3.25	2.85	2.50	2.21	1.95	1.76	1.59	1.45	1.22	0.94	0.75	0.63	4	
4.34	3.97	3.57	3.21	2.90	2.63	2.39	2.19	1.87	1.43	1.15	0.96	5	
5.44	5.07	4.69	4.29	3.92	3.60	3.32	3.05	2.63	2.04	1.66	1.39	6	100
6.51	6.16	5.78	5.40	5.01	4.64	4.31	4.01	3.49	2.73	2.23	1.88	7	
7.56	7.25	6.89	6.51	6.11	5.72	5.36	5.01	4.41	3.50	2.88	2.43	8	
8.60	8.32	7.98	7.61	7.22	6.81	6.44	6.07	5.40	4.35	3.59	3.05	9	
9.64	9.37	9.06	8.71	8.31	7.93	7.52	7.15	6.43	5.25	4.37	3.73	10	
10.7	10.4	10.1	9.79	9.42	9.03	8.64	8.24	7.50	6.20	5.21	4.46	11	
11.7	11.5	11.2	10.9	10.5	10.1	9.74	9.35	8.58	7.19	6.09	5.24	12	
1.25	1.02	0.85	0.74	0.65	0.57	0.51	0.46	0.39	0.29	0.24	0.20	2	
2.32	2.00	1.72	1.50	1.31	1.16	1.05	0.95	0.79	0.60	0.48	0.40	3	
3.44	3.11	2.80	2.51	2.25	2.04	1.85	1.69	1.45	1.11	0.90	0.75	4	
4.53	4.22	3.91	3.57	3.27	3.00	2.76	2.54	2.19	1.69	1.37	1.15	5	
5.60	5.32	5.02	4.68	4.37	4.05	3.77	3.50	3.05	2.41	1.96	1.66	6	120
6.65	6.40	6.11	5.78	5.45	5.13	4.82	4.53	4.01	3.20	2.63	2.23	7	
7.69	7.46	7.19	6.89	6.57	6.25	5.91	5.60	5.02	4.07	3.38	2.88	8	
8.72	8.51	8.26	7.98	7.68	7.36	7.01	6.69	6.06	5.00	4.20	3.59	9	
9.75	9.56	9.33	9.06	8.77	8.45	8.12	7.80	7.16	6.00	5.09	4.37	10	
10.8	10.6	10.4	10.1	9.85	9.55	9.24	8.91	8.25	7.02	6.01	5.21	11	
11.8	11.6	11.4	11.2	10.9	10.6	10.3	10.0	9.35	8.09	6.99	6.09	12	
1.45	1.24	1.07	0.94	0.83	0.74	0.66	0.60	0.51	0.39	0.31	0.26	2	
2.57	2.33	2.08	1.86	1.67	1.50	1.36	1.24	1.05	0.79	0.64	0.53	3	
3.66	3.44	3.21	2.95	2.73	2.51	2.32	2.15	1.85	1.45	1.18	0.99	4	
4.72	4.53	4.31	4.07	3.83	3.58	3.35	3.14	2.77	2.19	1.80	1.52	5	
5.77	5.60	5.40	5.17	4.93	4.68	4.44	4.21	3.77	3.06	2.54	2.16	6	160
6.80	6.65	6.47	6.26	6.03	5.79	5.54	5.30	4.83	4.01	3.37	2.89	7	
7.82	7.69	7.52	7.33	7.12	6.89	6.66	6.40	5.92	5.02	4.28	3.70	8	
8.84	8.72	8.57	8.40	8.20	7.98	7.75	7.51	7.02	6.07	5.25	4.58	9	
9.86	9.75	9.61	9.45	9.26	9.06	8.84	8.61	8.12	7.15	6.27	5.51	10	
10.9	10.8	10.6	10.5	10.3	10.1	9.92	9.70	9.23	8.25	7.31	6.50	11	
11.9	11.8	11.7	11.5	11.4	11.2	11.0	10.8	10.3	9.36	8.40	7.51	12	

* See page 3–26 for more details regarding these tables.

ECCENTRIC LOADS ON BOLT GROUPS
Coefficients C*

$$C = \frac{P_f}{V_r}, \text{ or } \frac{P}{V_s}$$

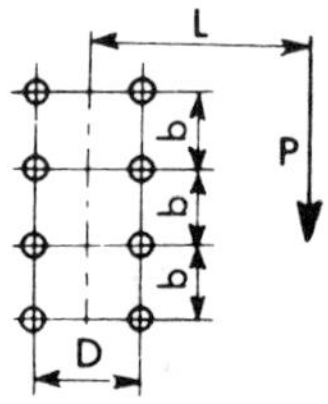

Table 3–15

Pitch b mm	Number of Bolts in Each Row	D = 80 mm											
		Moment Arm, L, Millimetres											
		75	100	125	150	175	200	225	250	300	400	500	600
	1	0.69	0.56	0.47	0.41	0.36	0.32	0.29	0.28	0.23	0.18	0.15	0.12
	2	2.15	1.76	1.49	1.28	1.12	1.03	0.93	0.84	0.71	0.54	0.43	0.36
	3	3.86	3.20	2.76	2.40	2.11	1.89	1.71	1.56	1.32	1.03	0.83	0.69
	4	5.89	5.09	4.46	3.90	3.51	3.14	2.84	2.59	2.20	1.70	1.38	1.15
	5	8.06	7.18	6.36	5.62	5.06	4.55	4.12	3.80	3.24	2.49	2.02	1.71
80	6	10.2	9.34	8.40	7.57	6.87	6.22	5.72	5.25	4.49	3.49	2.83	2.38
	7	12.4	11.5	10.5	9.68	8.85	8.06	7.44	6.85	5.92	4.62	3.75	3.16
	8	14.6	13.7	12.8	11.8	10.9	10.1	9.33	8.63	7.50	5.89	4.81	4.07
	9	16.7	15.9	14.9	14.0	13.0	12.1	11.3	10.5	9.22	7.29	5.99	5.06
	10	18.8	18.1	17.1	16.2	15.2	14.3	13.4	12.6	11.1	8.85	7.28	6.18
	11	20.9	20.2	19.3	18.4	17.4	16.5	15.5	14.6	13.0	10.5	8.70	7.40
	12	23.0	22.3	21.5	20.6	19.6	18.6	17.7	16.7	15.0	12.2	10.2	8.70
	2	2.21	1.87	1.59	1.37	1.20	1.07	0.99	0.90	0.76	0.58	0.46	0.39
	3	4.05	3.44	2.97	2.58	2.28	2.04	1.84	1.68	1.43	1.11	0.90	0.75
	4	6.18	5.46	4.76	4.23	3.76	3.38	3.10	2.83	2.41	1.85	1.51	1.26
	5	8.36	7.58	6.79	6.09	5.46	4.97	4.52	4.14	3.56	2.74	2.23	1.87
90	6	10.6	9.75	8.91	8.11	7.40	6.79	6.21	5.71	4.94	3.85	3.13	2.64
	7	12.7	12.0	11.1	10.2	9.44	8.71	8.08	7.46	6.49	5.08	4.16	3.51
	8	14.9	14.1	13.3	12.5	11.6	10.8	10.0	9.37	8.21	6.49	5.34	4.51
	9	17.0	16.3	15.5	14.7	13.8	12.9	12.2	11.4	10.1	8.05	6.63	5.62
	10	19.1	18.4	17.7	16.8	16.0	15.1	14.3	13.5	12.0	9.72	8.07	6.85
	11	21.1	20.5	19.8	19.0	18.2	17.3	16.4	15.6	14.1	11.5	9.60	8.20
	12	23.2	22.6	22.0	21.2	20.4	19.5	18.7	17.8	16.1	13.3	11.2	9.63
	2	2.34	1.99	1.68	1.46	1.28	1.14	1.05	0.96	0.81	0.61	0.49	0.41
	3	4.29	3.67	3.14	2.77	2.44	2.19	1.98	1.81	1.54	1.18	0.96	0.81
	4	6.45	5.74	5.09	4.50	4.06	3.66	3.32	3.07	2.62	2.01	1.63	1.38
	5	8.65	7.89	7.16	6.48	5.89	5.34	4.90	4.50	3.85	2.99	2.43	2.05
100	6	10.8	10.1	9.32	8.58	7.89	7.27	6.68	6.20	5.35	4.19	3.43	2.89
	7	13.0	12.3	11.5	10.7	10.0	9.30	8.66	8.03	7.03	5.56	4.55	3.85
	8	15.1	14.4	13.7	12.9	12.2	11.4	10.7	10.1	8.87	7.09	5.83	4.96
	9	17.2	16.6	15.9	15.2	14.4	13.6	12.9	12.1	10.8	8.75	7.26	6.17
	10	19.2	18.7	18.1	17.4	16.6	15.8	15.0	14.3	12.9	10.5	8.81	7.52
	11	21.3	20.8	20.2	19.5	18.8	18.0	17.2	16.4	15.0	12.4	10.5	8.98
	12	23.3	22.9	22.3	21.7	21.0	20.2	19.5	18.6	17.1	14.4	12.2	10.5
	2	2.57	2.15	1.88	1.63	1.44	1.28	1.16	1.05	0.91	0.69	0.56	0.47
	3	4.61	4.06	3.55	3.11	2.75	2.49	2.25	2.06	1.75	1.35	1.09	0.93
	4	6.84	6.21	5.59	5.04	4.58	4.15	3.82	3.51	3.00	2.34	1.90	1.59
	5	9.02	8.39	7.75	7.16	6.57	6.05	5.56	5.16	4.47	3.47	2.84	2.39
120	6	11.1	10.6	9.98	9.35	8.68	8.11	7.53	7.04	6.16	4.88	4.01	3.39
	7	13.2	12.7	12.2	11.5	10.9	10.3	9.63	9.06	8.02	6.44	5.33	4.53
	8	15.3	14.9	14.3	13.7	13.1	12.4	11.8	11.2	10.0	8.18	6.81	5.82
	9	17.4	17.0	16.5	15.9	15.3	14.7	14.0	13.4	12.1	10.0	8.44	7.23
	10	19.5	19.1	18.6	18.1	17.5	16.9	16.2	15.5	14.3	12.0	10.2	8.79
	11	21.5	21.1	20.7	20.2	19.6	19.0	18.4	17.8	16.5	14.0	12.0	10.5
	12	23.5	23.2	22.8	22.3	21.8	21.2	20.6	20.0	18.7	16.2	14.0	12.2
	2	2.89	2.51	2.22	1.95	1.76	1.58	1.43	1.30	1.11	0.86	0.69	0.58
	3	5.11	4.63	4.19	3.77	3.38	3.07	2.79	2.55	2.19	1.69	1.38	1.16
	4	7.29	6.85	6.35	5.89	5.44	5.03	4.67	4.32	3.77	2.96	2.42	2.05
	5	9.41	9.02	8.58	8.08	7.60	7.15	6.70	6.29	5.54	4.43	3.65	3.09
160	6	11.5	11.2	10.8	10.3	9.83	9.35	8.84	8.41	7.55	6.14	5.12	4.37
	7	13.6	13.3	12.9	12.5	12.0	11.5	11.1	10.6	9.63	8.03	6.78	5.82
	8	15.6	15.3	15.0	14.6	14.2	13.7	13.3	12.8	11.8	10.0	8.57	7.42
	9	17.7	17.4	17.1	16.8	16.4	15.9	15.5	15.0	14.0	12.1	10.5	9.16
	10	19.7	19.5	19.2	18.9	18.5	18.1	17.6	17.2	16.2	14.3	12.5	11.0
	11	21.7	21.5	21.3	20.9	20.6	20.2	19.8	19.4	18.4	16.5	14.6	13.0
	12	23.7	23.5	23.3	23.0	22.7	22.3	21.9	21.5	20.6	18.7	16.8	15.0

* See page 3–26 for more details regarding these tables.

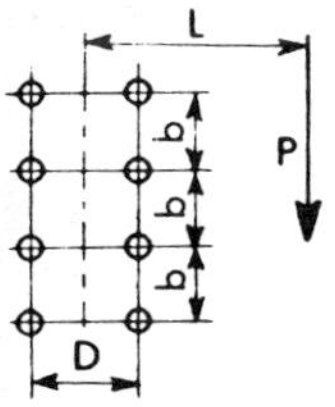

$$C = \frac{P_f}{V_r}, \text{ or } \frac{P}{V_s}$$

ECCENTRIC LOADS ON BOLT GROUPS
Coefficients C*

Table 3–16

D = 320 mm												Number of Bolts in Each Row	Pitch b mm
Moment Arm, L, Millimetres													
75	100	125	150	175	200	225	250	300	400	500	600		
1.36	1.23	1.12	1.03	0.95	0.88	0.83	0.78	0.69	0.57	0.48	0.42	1	
2.89	2.62	2.41	2.22	2.05	1.91	1.78	1.68	1.49	1.22	1.04	0.89	2	
4.56	4.17	3.81	3.53	3.26	3.03	2.84	2.66	2.38	1.95	1.65	1.43	3	
6.35	5.87	5.38	5.00	4.63	4.34	4.06	3.81	3.41	2.78	2.36	2.04	4	
8.31	7.67	7.10	6.60	6.16	5.74	5.41	5.07	4.52	3.72	3.16	2.72	5	
10.3	9.64	8.95	8.34	7.81	7.34	6.87	6.46	5.80	4.77	4.02	3.49	6	80
12.4	11.6	10.9	10.2	9.58	9.02	8.46	8.00	7.18	5.89	5.00	4.34	7	
14.5	13.7	12.9	12.2	11.5	10.8	10.2	9.64	8.67	7.16	6.08	5.25	8	
16.6	15.8	15.0	14.2	13.4	12.7	12.0	11.4	10.3	8.53	7.22	6.27	9	
18.6	17.9	17.1	16.3	15.4	14.7	13.9	13.2	12.0	9.97	8.49	7.38	10	
20.7	20.0	19.2	18.4	17.5	16.7	15.9	15.2	13.8	11.5	9.85	8.54	11	
22.8	22.1	21.3	20.5	19.7	18.8	18.0	17.1	15.6	13.2	11.3	9.82	12	
2.91	2.66	2.43	2.25	2.08	1.93	1.81	1.71	1.51	1.23	1.05	0.90	2	
4.63	4.23	3.89	3.61	3.34	3.10	2.90	2.72	2.44	1.99	1.69	1.46	3	
6.52	5.99	5.54	5.15	4.78	4.48	4.19	3.94	3.53	2.88	2.44	2.11	4	
8.48	7.87	7.34	6.85	6.36	5.97	5.58	5.28	4.70	3.87	3.26	2.84	5	
10.6	9.88	9.25	8.65	8.10	7.62	7.19	6.76	6.07	5.00	4.22	3.67	6	90
12.6	12.0	11.3	10.6	9.99	9.41	8.89	8.37	7.52	6.21	5.28	4.56	7	
14.7	14.1	13.3	12.6	11.9	11.3	10.7	10.1	9.12	7.59	6.42	5.58	8	
16.8	16.2	15.4	14.7	14.0	13.3	12.6	12.0	10.8	9.03	7.69	6.69	9	
18.9	18.3	17.6	16.8	16.1	15.3	14.6	13.9	12.7	10.6	9.05	7.87	10	
21.0	20.4	19.7	19.0	18.2	17.4	16.6	15.9	14.6	12.3	10.5	9.17	11	
23.0	22.5	21.8	21.1	20.3	19.5	18.8	18.0	16.5	14.0	12.1	10.5	12	
2.95	2.69	2.46	2.28	2.11	1.96	1.83	1.73	1.54	1.25	1.06	0.92	2	
4.71	4.32	3.98	3.66	3.42	3.18	2.97	2.79	2.50	2.04	1.73	1.50	3	
6.64	6.13	5.70	5.27	4.93	4.59	4.33	4.07	3.62	2.97	2.53	2.18	4	
8.66	8.08	7.56	7.04	6.60	6.20	5.80	5.45	4.89	4.03	3.40	2.95	5	
10.7	10.1	9.55	8.93	8.39	7.90	7.46	7.03	6.31	5.21	4.43	3.84	6	100
12.8	12.2	11.6	10.9	10.4	9.78	9.26	8.74	7.87	6.54	5.53	4.81	7	
14.9	14.3	13.7	13.0	12.4	11.8	11.1	10.6	9.57	7.98	6.80	5.91	8	
17.0	16.4	15.8	15.1	14.5	13.8	13.1	12.5	11.4	9.54	8.17	7.08	9	
19.1	18.5	17.9	17.3	16.6	15.9	15.2	14.5	13.3	11.2	9.64	8.40	10	
21.2	20.6	20.1	19.4	18.7	18.0	17.3	16.6	15.3	13.0	11.2	9.78	11	
23.2	22.7	22.2	21.5	20.9	20.2	19.4	18.7	17.3	14.9	12.9	11.3	12	
3.02	2.76	2.53	2.35	2.17	2.02	1.89	1.78	1.58	1.29	1.10	0.94	2	
4.87	4.47	4.13	3.84	3.55	3.33	3.11	2.92	2.60	2.14	1.82	1.57	3	
6.86	6.40	5.95	5.55	5.20	4.88	4.58	4.30	3.86	3.18	2.68	2.33	4	
8.95	8.44	7.94	7.44	7.02	6.58	6.21	5.88	5.25	4.33	3.68	3.18	5	120
11.0	10.5	10.0	9.48	8.93	8.48	8.00	7.59	6.84	5.70	4.82	4.19	6	
13.1	12.7	12.1	11.5	11.0	10.5	9.93	9.45	8.58	7.17	6.11	5.29	7	
15.2	14.8	14.2	13.7	13.1	12.5	12.0	11.4	10.4	8.81	7.53	6.56	8	
17.3	16.9	16.4	15.8	15.3	14.7	14.1	13.5	12.4	10.5	9.07	7.91	9	
19.3	18.9	18.5	17.9	17.4	16.8	16.2	15.6	14.5	12.4	10.7	9.40	10	
21.4	21.0	20.6	20.1	19.5	18.9	18.4	17.8	16.6	14.4	12.5	11.0	11	
23.4	23.1	22.7	22.2	21.7	21.1	20.5	19.9	18.7	16.3	14.4	12.7	12	
3.16	2.90	2.66	2.47	2.29	2.15	2.01	1.89	1.69	1.38	1.17	1.01	2	
5.14	4.78	4.46	4.16	3.86	3.62	3.39	3.20	2.85	2.35	1.99	1.73	3	
7.22	6.84	6.44	6.09	5.73	5.40	5.10	4.81	4.33	3.58	3.04	2.63	4	
9.32	8.96	8.56	8.14	7.74	7.33	6.96	6.61	6.00	4.97	4.23	3.69	5	160
11.4	11.1	10.7	10.3	9.82	9.40	8.97	8.57	7.82	6.61	5.65	4.93	6	
13.5	13.2	12.8	12.4	12.0	11.5	11.1	10.7	9.81	8.37	7.22	6.30	7	
15.5	15.3	14.9	14.5	14.1	13.7	13.2	12.8	11.9	10.3	8.94	7.84	8	
17.6	17.3	17.0	16.7	16.3	15.8	15.4	15.0	14.0	12.3	10.8	9.53	9	
19.6	19.4	19.1	18.8	18.4	18.0	17.6	17.1	16.2	14.4	12.7	11.3	10	
21.7	21.4	21.2	20.8	20.5	20.1	19.7	19.3	18.4	16.5	14.7	13.2	11	
23.7	23.5	23.2	22.9	22.6	22.2	21.8	21.4	20.5	18.7	16.8	15.2	12	

* See page 3–26 for more details regarding these tables.

ECCENTRIC LOADS ON BOLT GROUPS Coefficients C*

$$C = \frac{P_f}{V_r}, \text{ or } \frac{P}{V_s}$$

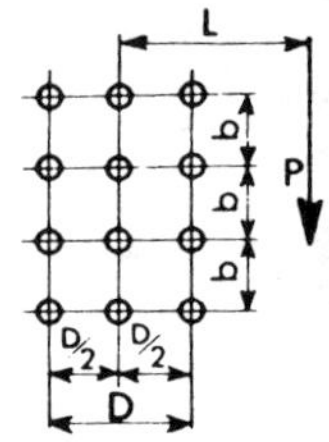

Table 3–17

Pitch b mm	Number of Bolts in Each Row	D = 160 mm											
		Moment Arm, L, Millimetres											
		75	100	125	150	175	200	225	250	300	400	500	600
	1	1.49	1.26	1.11	0.96	0.84	0.77	0.69	0.62	0.52	0.39	0.31	0.26
	2	3.52	3.05	2.63	2.30	2.09	1.87	1.69	1.54	1.30	1.02	0.82	0.69
	3	6.06	5.18	4.61	4.07	3.63	3.27	3.03	2.77	2.35	1.84	1.48	1.24
	4	8.94	7.86	7.00	6.22	5.65	5.11	4.65	4.26	3.70	2.85	2.31	1.97
	5	12.1	10.9	9.72	8.77	7.90	7.17	6.63	6.10	5.25	4.11	3.34	2.81
80	6	15.3	14.0	12.6	11.5	10.5	9.68	8.88	8.19	7.13	5.56	4.57	3.85
	7	18.5	17.2	15.9	14.6	13.4	12.4	11.4	10.6	9.18	7.25	5.94	5.05
	8	21.8	20.5	19.0	17.7	16.4	15.2	14.1	13.2	11.5	9.12	7.53	6.37
	9	25.0	23.7	22.4	20.9	19.6	18.3	17.1	15.9	14.1	11.2	9.25	7.88
	10	28.1	27.0	25.6	24.2	22.8	21.4	20.1	18.9	16.8	13.5	11.2	9.55
	11	31.3	30.2	28.9	27.5	26.0	24.6	23.3	22.0	19.6	15.9	13.2	11.3
	12	34.4	33.4	32.1	30.8	29.4	27.9	26.5	25.1	22.6	18.5	15.5	13.3
	2	3.64	3.16	2.73	2.39	2.18	1.95	1.76	1.61	1.36	1.07	0.86	0.72
	3	6.26	5.46	4.77	4.29	3.84	3.46	3.15	2.93	2.50	1.92	1.58	1.32
	4	9.30	8.31	7.35	6.63	5.96	5.47	4.99	4.58	3.92	3.07	2.49	2.09
	5	12.6	11.4	10.3	9.30	8.48	7.72	7.07	6.52	5.67	4.41	3.63	3.06
90	6	15.7	14.6	13.4	12.2	11.2	10.4	9.52	8.87	7.68	6.06	4.96	4.21
	7	19.0	17.8	16.6	15.4	14.3	13.2	12.2	11.4	10.0	7.88	6.51	5.50
	8	22.2	21.1	19.9	18.6	17.4	16.3	15.2	14.2	12.5	9.99	8.23	7.01
	9	25.4	24.3	23.1	21.9	20.6	19.4	18.2	17.1	15.2	12.3	10.2	8.65
	10	28.5	27.5	26.4	25.1	23.9	22.7	21.4	20.2	18.1	14.7	12.3	10.5
	11	31.6	30.7	29.6	28.5	27.2	25.9	24.6	23.4	21.1	17.4	14.5	12.5
	12	34.7	33.9	32.9	31.7	30.5	29.2	27.9	26.6	24.3	20.1	17.0	14.6
	2	3.77	3.28	2.84	2.49	2.27	2.03	1.84	1.68	1.42	1.12	0.90	0.76
	3	6.54	5.72	5.00	4.44	4.04	3.66	3.33	3.05	2.65	2.03	1.67	1.40
	4	9.62	8.66	7.78	7.03	6.34	5.75	5.32	4.89	4.20	3.29	2.68	2.25
	5	12.9	11.8	10.8	9.81	8.98	8.19	7.58	6.99	6.10	4.76	3.92	3.30
100	6	16.1	15.1	14.0	12.9	11.9	11.0	10.2	9.48	8.29	6.56	5.38	4.57
	7	19.3	18.3	17.2	16.1	15.0	14.0	13.0	12.2	10.7	8.54	7.07	5.99
	8	22.5	21.6	20.5	19.4	18.3	17.1	16.1	15.1	13.5	10.8	8.97	7.65
	9	25.7	24.8	23.8	22.6	21.5	20.4	19.3	18.2	16.3	13.3	11.0	9.44
	10	28.8	28.0	27.0	26.0	24.8	23.7	22.5	21.4	19.3	15.9	13.3	11.4
	11	31.9	31.1	30.2	29.2	28.1	27.0	25.8	24.7	22.4	18.7	15.8	13.6
	12	35.0	34.2	33.4	32.4	31.4	30.3	29.1	27.9	25.6	21.6	18.4	15.9
	2	4.02	3.45	3.06	2.70	2.40	2.21	2.00	1.83	1.56	1.19	0.99	0.83
	3	6.99	6.18	5.50	4.88	4.39	4.04	3.69	3.39	2.90	2.27	1.84	1.57
	4	10.2	9.29	8.48	7.66	7.01	6.45	5.92	5.46	4.76	3.71	3.05	2.57
	5	13.4	12.5	11.7	10.7	9.89	9.16	8.50	7.87	6.88	5.45	4.47	3.80
120	6	16.6	15.8	14.9	14.0	13.1	12.2	11.4	10.7	9.41	7.51	6.22	5.27
	7	19.8	19.1	18.2	17.2	16.3	15.3	14.5	13.7	12.1	9.81	8.16	6.94
	8	22.9	22.2	21.4	20.5	19.6	18.6	17.7	16.8	15.1	12.3	10.4	8.88
	9	26.0	25.4	24.6	23.8	22.9	21.9	21.0	20.0	18.2	15.1	12.8	11.0
	10	29.1	28.5	27.8	27.0	26.1	25.2	24.3	23.3	21.4	18.0	15.3	13.3
	11	32.2	31.7	31.0	30.2	29.4	28.5	27.5	26.6	24.6	21.1	18.1	15.7
	12	35.3	34.8	34.1	33.4	32.6	31.7	30.8	29.9	27.9	24.2	21.0	18.4
	2	4.38	3.87	3.45	3.07	2.79	2.52	2.30	2.11	1.83	1.41	1.15	0.98
	3	7.60	6.92	6.30	5.72	5.17	4.75	4.35	4.01	3.50	2.73	2.25	1.89
	4	10.9	10.2	9.54	8.82	8.22	7.59	7.06	6.61	5.76	4.58	3.78	3.20
	5	14.1	13.5	12.8	12.1	11.4	10.7	10.0	9.45	8.41	6.74	5.59	4.77
160	6	17.2	16.7	16.1	15.4	14.7	14.0	13.3	12.6	11.3	9.27	7.79	6.68
	7	20.3	19.8	19.3	18.7	18.0	17.2	16.6	15.8	14.5	12.1	10.2	8.79
	8	23.4	23.0	22.5	21.9	21.2	20.6	19.9	19.1	17.7	15.1	12.9	11.2
	9	26.4	26.1	25.6	25.1	24.5	23.8	23.2	22.4	21.0	18.2	15.8	13.8
	10	29.5	29.2	28.7	28.2	27.7	27.1	26.4	25.7	24.3	21.4	18.8	16.6
	11	32.5	32.2	31.8	31.4	30.9	30.3	29.7	29.0	27.6	24.7	21.9	19.5
	12	35.6	35.3	34.9	34.5	34.0	33.5	32.9	32.2	30.9	28.0	25.1	22.5

* See page 3–26 for more details regarding these tables.

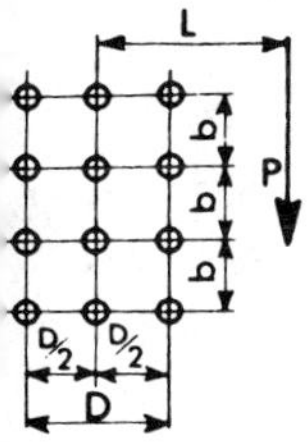

$$C = \frac{P_f}{V_r}, \text{ or } \frac{P}{V_s}$$

ECCENTRIC LOADS ON BOLT GROUPS
Coefficients C*

Table 3–18

D = 320 mm												Number of Bolts in Each Row	Pitch b mm
Moment Arm, L, Millimetres													
75	100	125	150	175	200	225	250	300	400	500	600		
1.97	1.78	1.63	1.49	1.37	1.28	1.18	1.11	0.98	0.77	0.62	0.53	1	
4.22	3.79	3.47	3.17	2.90	2.71	2.50	2.32	2.05	1.63	1.35	1.14	2	
6.66	6.04	5.49	5.07	4.66	4.30	4.03	3.75	3.32	2.66	2.18	1.87	3	
9.41	8.57	7.86	7.20	6.69	6.19	5.74	5.41	4.74	3.81	3.18	2.69	4	
12.3	11.3	10.4	9.63	8.88	8.24	7.74	7.22	6.42	5.18	4.33	3.67	5	
15.4	14.2	13.1	12.2	11.4	10.6	9.89	9.33	8.23	6.67	5.58	4.79	6	80
18.5	17.3	16.1	15.0	14.0	13.1	12.3	11.6	10.3	8.39	7.04	6.00	7	
21.7	20.4	19.2	18.0	16.9	15.9	14.9	14.0	12.5	10.2	8.59	7.40	8	
24.8	23.6	22.4	21.1	19.8	18.8	17.6	16.7	14.9	12.3	10.4	8.88	9	
27.9	26.8	25.6	24.3	23.0	21.7	20.5	19.5	17.5	14.4	12.2	10.5	10	
31.1	30.0	28.8	27.5	26.1	24.8	23.6	22.4	20.2	16.8	14.2	12.3	11	
34.2	33.2	32.0	30.7	29.4	28.0	26.7	25.4	23.1	19.3	16.4	14.2	12	
4.24	3.85	3.52	3.21	2.94	2.74	2.54	2.36	2.08	1.66	1.38	1.16	2	
6.81	6.18	5.61	5.14	4.78	4.42	4.14	3.85	3.41	2.74	2.25	1.93	3	
9.61	8.77	8.06	7.46	6.87	6.42	5.96	5.56	4.93	3.97	3.31	2.81	4	
12.6	11.7	10.8	9.97	9.27	8.60	8.02	7.56	6.66	5.44	4.51	3.87	5	
15.7	14.7	13.6	12.8	11.9	11.1	10.4	9.75	8.69	7.06	5.92	5.05	6	90
18.9	17.8	16.7	15.7	14.7	13.8	13.0	12.2	10.9	8.87	7.45	6.42	7	
22.1	21.0	19.9	18.8	17.6	16.7	15.7	14.8	13.3	10.9	9.21	7.89	8	
25.2	24.2	23.1	21.9	20.7	19.7	18.6	17.6	15.8	13.1	11.1	9.57	9	
28.3	27.4	26.3	25.1	24.0	22.8	21.6	20.5	18.6	15.5	13.1	11.3	10	
31.5	30.5	29.5	28.4	27.2	25.9	24.7	23.7	21.6	18.0	15.3	13.3	11	
34.6	33.7	32.7	31.5	30.4	29.2	28.0	26.8	24.5	20.7	17.7	15.4	12	
4.29	3.90	3.53	3.26	2.99	2.79	2.58	2.39	2.12	1.69	1.40	1.18	2	
6.91	6.28	5.75	5.27	4.90	4.53	4.21	3.96	3.47	2.79	2.32	1.99	3	
9.83	9.04	8.33	7.65	7.11	6.59	6.18	5.77	5.13	4.14	3.46	2.93	4	
12.9	12.0	11.1	10.3	9.60	8.97	8.36	7.83	6.98	5.66	4.74	4.07	5	
16.1	15.1	14.1	13.2	12.4	11.5	10.9	10.2	9.09	7.46	6.22	5.35	6	100
19.2	18.3	17.2	16.2	15.3	14.4	13.6	12.8	11.5	9.41	7.93	6.79	7	
22.4	21.4	20.5	19.4	18.4	17.4	16.4	15.6	14.0	11.6	9.78	8.44	8	
25.5	24.6	23.7	22.6	21.6	20.5	19.5	18.5	16.8	13.9	11.8	10.2	9	
28.6	27.8	26.8	25.8	24.8	23.6	22.6	21.5	19.7	16.5	14.1	12.2	10	
31.7	30.9	30.0	29.0	28.0	26.9	25.8	24.7	22.7	19.2	16.4	14.3	11	
34.8	34.1	33.2	32.3	31.2	30.1	29.0	27.9	25.8	22.0	19.0	16.5	12	
4.41	4.02	3.64	3.36	3.08	2.88	2.66	2.47	2.19	1.75	1.44	1.23	2	
7.17	6.58	5.99	5.53	5.11	4.77	4.44	4.14	3.68	2.96	2.47	2.09	3	
10.2	9.46	8.79	8.16	7.55	7.07	6.58	6.21	5.48	4.44	3.71	3.19	4	
13.4	12.6	11.7	11.0	10.3	9.66	9.03	8.51	7.56	6.21	5.18	4.45	5	
16.5	15.7	14.9	14.1	13.2	12.5	11.8	11.1	9.95	8.20	6.91	5.92	6	120
19.7	18.9	18.1	17.2	16.4	15.5	14.7	13.9	12.6	10.4	8.82	7.63	7	
22.8	22.1	21.3	20.4	19.5	18.7	17.8	17.0	15.4	12.9	11.0	9.49	8	
25.9	25.3	24.5	23.7	22.8	21.9	21.0	20.1	18.4	15.5	13.3	11.5	9	
29.0	28.4	27.7	26.9	26.0	25.1	24.2	23.3	21.5	18.3	15.8	13.8	10	
32.1	31.5	30.9	30.1	29.3	28.4	27.5	26.5	24.7	21.3	18.5	16.2	11	
35.2	34.6	34.0	33.3	32.5	31.6	30.7	29.8	27.9	24.3	21.3	18.7	12	
4.64	4.23	3.89	3.55	3.30	3.05	2.86	2.66	2.33	1.87	1.56	1.32	2	
7.62	7.05	6.55	6.06	5.64	5.23	4.88	4.61	4.07	3.30	2.77	2.38	3	
10.8	10.2	9.57	8.99	8.42	7.92	7.46	7.00	6.26	5.11	4.30	3.70	4	
14.0	13.4	12.7	12.1	11.5	10.9	10.3	9.72	8.77	7.21	6.10	5.27	5	
17.1	16.6	16.0	15.3	14.7	14.0	13.3	12.7	11.5	9.68	8.23	7.13	6	160
20.2	19.7	19.2	18.6	17.9	17.2	16.6	15.9	14.6	12.4	10.6	9.21	7	
23.3	22.9	22.4	21.8	21.2	20.5	19.8	19.1	17.7	15.2	13.2	11.5	8	
26.4	26.0	25.5	25.0	24.4	23.7	23.1	22.4	21.0	18.3	16.0	14.1	9	
29.4	29.1	28.6	28.1	27.6	27.0	26.3	25.6	24.2	21.4	18.9	16.8	10	
32.5	32.2	31.7	31.3	30.7	30.2	29.5	28.9	27.5	24.7	22.0	19.6	11	
35.5	35.2	34.8	34.4	33.9	33.4	32.7	32.1	30.8	27.9	25.1	22.6	12	

* See page 3–26 for more details regarding these tables.

ECCENTRIC LOADS ON BOLT GROUPS
Coefficients C*

$$C = \frac{P_f}{V_r}, \text{ or } \frac{P}{V_s}$$

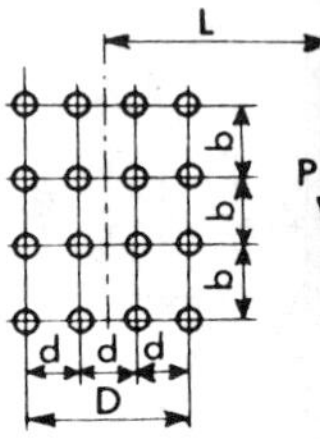

d = D/3

Table 3–19

Pitch b mm	Number of Bolts in Each Row	D = 240 mm Moment Arm, L, Millimetres: 75	100	125	150	175	200	225	250	300	400	500	600
	1	2.31	2.02	1.79	1.56	1.41	1.26	1.14	1.05	0.89	0.69	0.56	0.47
	2	5.12	4.52	3.97	3.58	3.20	2.89	2.63	2.46	2.10	1.62	1.31	1.13
	3	8.33	7.41	6.57	5.96	5.36	4.93	4.50	4.13	3.60	2.78	2.26	1.90
	4	12.1	10.8	9.77	8.78	8.05	7.33	6.71	6.27	5.40	4.26	3.47	2.92
	5	16.1	14.6	13.3	12.1	11.0	10.1	9.37	8.66	7.57	5.91	4.88	4.12
80	6	20.4	18.7	17.0	15.6	14.4	13.3	12.3	11.4	10.0	7.94	6.50	5.55
	7	24.6	22.9	21.2	19.5	18.1	16.8	15.5	14.5	12.7	10.1	8.40	7.11
	8	28.9	27.2	25.4	23.6	22.1	20.6	19.1	17.9	15.8	12.7	10.5	8.94
	9	33.1	31.5	29.7	27.9	26.2	24.5	22.9	21.6	19.1	15.4	12.8	11.0
	10	37.4	35.8	34.0	32.1	30.4	28.6	27.0	25.4	22.7	18.3	15.3	13.1
	11	41.6	40.1	38.4	36.5	34.7	32.9	31.1	29.4	26.4	21.6	18.1	15.5
	12	45.7	44.3	42.7	40.9	39.0	37.1	35.3	33.5	30.2	25.0	21.0	18.1
	2	5.23	4.55	4.07	3.61	3.29	2.98	2.71	2.48	2.17	1.67	1.36	1.16
	3	8.63	7.67	6.83	6.20	5.59	5.07	4.71	4.32	3.70	2.91	2.37	2.00
	4	12.5	11.2	10.1	9.24	8.38	7.74	7.10	6.55	5.73	4.46	3.69	3.11
	5	16.7	15.3	13.9	12.7	11.7	10.7	9.88	9.24	8.01	6.35	5.19	4.42
90	6	20.9	19.3	17.9	16.4	15.2	14.1	13.0	12.2	10.7	8.49	7.03	5.95
	7	25.2	23.7	22.1	20.6	19.2	17.8	16.6	15.5	13.7	11.0	9.06	7.73
	8	29.5	28.0	26.4	24.7	23.3	21.8	20.4	19.2	17.0	13.7	11.4	9.70
	9	33.7	32.3	30.8	29.1	27.5	25.9	24.4	23.0	20.6	16.6	13.9	11.9
	10	37.9	36.6	35.1	33.5	31.8	30.1	28.6	27.1	24.3	19.9	16.7	14.3
	11	42.0	40.8	39.4	37.8	36.1	34.5	32.8	31.2	28.3	23.3	19.7	17.0
	12	46.2	45.0	43.7	42.1	40.6	38.9	37.2	35.5	32.4	27.0	22.9	19.8
	2	5.35	4.66	4.17	3.71	3.39	3.06	2.79	2.56	2.24	1.73	1.40	1.18
	3	8.83	7.84	7.09	6.36	5.83	5.30	4.92	4.52	3.88	3.06	2.49	2.10
	4	12.9	11.7	10.6	9.59	8.82	8.06	7.49	6.92	6.06	4.74	3.92	3.30
	5	17.2	15.8	14.5	13.3	12.2	11.3	10.5	9.71	8.53	6.78	5.55	4.74
100	6	21.4	20.0	18.7	17.2	16.0	14.9	13.9	12.9	11.4	9.11	7.56	6.41
	7	25.7	24.4	22.9	21.5	20.0	18.8	17.5	16.5	14.6	11.7	9.78	8.35
	8	29.9	28.7	27.2	25.7	24.3	22.9	21.6	20.4	18.1	14.7	12.3	10.5
	9	34.1	32.9	31.6	30.1	28.6	27.1	25.8	24.4	21.8	18.0	15.1	12.9
	10	38.3	37.2	35.9	34.5	33.0	31.5	30.0	28.6	25.9	21.4	18.1	15.6
	11	42.4	41.4	40.2	38.8	37.4	35.8	34.3	32.8	30.0	25.1	21.3	18.4
	12	46.5	45.6	44.4	43.1	41.7	40.2	38.7	37.2	34.2	28.9	24.7	21.4
	2	5.52	4.90	4.40	3.92	3.53	3.25	2.97	2.73	2.38	1.84	1.50	1.26
	3	9.31	8.42	7.59	6.85	6.22	5.75	5.28	4.87	4.25	3.35	2.73	2.30
	4	13.6	12.4	11.4	10.4	9.57	8.88	8.19	7.59	6.66	5.29	4.34	3.70
	5	17.9	16.7	15.5	14.4	13.3	12.4	11.5	10.7	9.48	7.58	6.29	5.33
120	6	22.1	21.0	19.8	18.6	17.4	16.3	15.3	14.4	12.8	10.3	8.56	7.33
	7	26.3	25.3	24.2	23.0	21.7	20.5	19.4	18.3	16.3	13.3	11.1	9.55
	8	30.5	29.6	28.5	27.3	26.0	24.8	23.6	22.4	20.2	16.7	14.0	12.1
	9	34.6	33.8	32.8	31.6	30.4	29.2	27.9	26.7	24.3	20.3	17.2	14.8
	10	38.8	38.0	37.0	35.9	34.7	33.5	32.2	31.0	28.5	24.1	20.6	17.9
	11	42.9	42.1	41.2	40.2	39.1	37.9	36.7	35.4	32.8	28.1	24.3	21.2
	12	47.0	46.3	45.4	44.5	43.4	42.2	41.0	39.8	37.2	32.3	28.0	24.6
	2	5.95	5.32	4.80	4.31	3.96	3.60	3.30	3.08	2.66	2.10	1.71	1.44
	3	10.1	9.25	8.49	7.76	7.07	6.56	6.06	5.62	4.93	3.91	3.20	2.73
	4	14.4	13.6	12.7	11.8	11.0	10.3	9.55	8.97	7.92	6.35	5.24	4.48
	5	18.7	17.9	17.0	16.1	15.2	14.3	13.5	12.7	11.3	9.20	7.66	6.57
160	6	22.9	22.2	21.4	20.5	19.5	18.6	17.7	16.8	15.2	12.5	10.5	9.07
	7	27.0	26.4	25.6	24.8	23.9	23.0	22.0	21.1	19.3	16.2	13.8	11.9
	8	31.1	30.6	29.9	29.1	28.3	27.3	26.4	25.4	23.6	20.1	17.3	15.0
	9	35.2	34.7	34.1	33.4	32.5	31.7	30.8	29.8	27.9	24.2	21.1	18.5
	10	39.3	38.8	38.2	37.6	36.8	36.0	35.1	34.2	32.3	28.5	25.1	22.1
	11	43.4	42.9	42.4	41.8	41.1	40.3	39.5	38.6	36.7	32.8	29.2	26.0
	12	47.4	47.0	46.5	45.9	45.3	44.5	43.7	42.9	41.1	37.3	33.4	30.0

* See page 3–26 for more details regarding these tables.

* See p. 3—24 for more details regarding these tables.

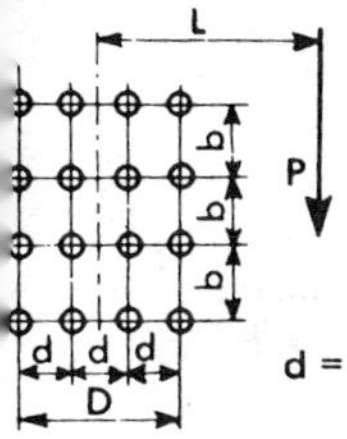

d = D/3

$$C = \frac{P_f}{V_r}, \text{ or } \frac{P}{V_s}$$

ECCENTRIC LOADS ON BOLT GROUPS
Coefficients C*

Table 3–20

D = 480 mm												Number of Bolts in Each Row	Pitch b mm
Moment Arm, L, Millimetres													
75	100	125	150	175	200	225	250	300	400	500	600		
2.91	2.69	2.51	2.33	2.18	2.04	1.90	1.79	1.58	1.26	1.05	0.89	1	
6.00	5.56	5.17	4.79	4.49	4.18	3.93	3.71	3.27	2.67	2.21	1.89	2	80
9.32	8.62	7.99	7.48	6.96	6.54	6.17	5.78	5.15	4.18	3.50	2.99	3	
12.8	11.9	11.1	10.3	9.72	9.08	8.58	8.12	7.20	5.92	4.98	4.26	4	
16.6	15.5	14.5	13.5	12.7	11.9	11.2	10.6	9.54	7.80	6.57	5.67	5	
20.5	19.2	18.1	16.9	15.9	15.0	14.2	13.4	12.0	9.95	8.41	7.21	6	
24.6	23.1	21.8	20.6	19.4	18.2	17.3	16.4	14.8	12.2	10.4	8.95	7	
28.7	27.3	25.8	24.3	23.0	21.7	20.6	19.6	17.7	14.8	12.6	10.9	8	
32.9	31.4	29.8	28.3	26.8	25.4	24.1	23.0	20.8	17.5	14.9	12.9	9	
37.0	35.6	34.0	32.4	30.8	29.3	27.9	26.6	24.1	20.3	17.4	15.2	10	
41.2	39.8	38.2	36.6	34.9	33.3	31.8	30.4	27.6	23.3	20.1	17.5	11	
45.4	44.0	42.4	40.8	39.1	37.4	35.8	34.2	31.4	26.6	22.9	20.0	12	
6.05	5.59	5.21	4.83	4.52	4.21	3.97	3.71	3.30	2.67	2.24	1.91	2	90
9.41	8.70	8.10	7.59	7.07	6.65	6.22	5.88	5.26	4.27	3.58	3.05	3	
13.1	12.1	11.3	10.6	9.95	9.31	8.80	8.26	7.41	6.10	5.09	4.39	4	
16.9	15.9	14.8	13.8	13.0	12.2	11.6	11.0	9.80	8.10	6.84	5.90	5	
20.9	19.7	18.6	17.4	16.4	15.5	14.6	13.9	12.5	10.3	8.74	7.56	6	
25.1	23.8	22.5	21.2	20.1	19.0	17.9	17.0	15.4	12.8	10.9	9.45	7	
29.2	27.9	26.5	25.1	23.8	22.6	21.5	20.4	18.5	15.5	13.2	11.5	8	
33.4	32.1	30.7	29.2	27.8	26.5	25.2	24.0	21.8	18.4	15.7	13.7	9	
37.6	36.3	34.9	33.4	32.0	30.6	29.1	27.8	25.4	21.4	18.5	16.1	10	
41.7	40.5	39.1	37.7	36.1	34.6	33.2	31.8	29.2	24.7	21.3	18.6	11	
45.9	44.7	43.4	41.9	40.4	38.9	37.4	35.9	33.0	28.2	24.4	21.4	12	
6.09	5.64	5.21	4.87	4.56	4.25	4.00	3.74	3.33	2.70	2.26	1.93	2	100
9.53	8.83	8.22	7.65	7.19	6.77	6.33	5.99	5.36	4.36	3.66	3.12	3	
13.3	12.4	11.6	10.8	10.1	9.54	8.96	8.49	7.61	6.23	5.25	4.54	4	
17.3	16.2	15.1	14.3	13.3	12.6	11.9	11.3	10.1	8.41	7.11	6.09	5	
21.3	20.2	19.0	17.9	16.9	16.0	15.2	14.3	12.9	10.8	9.14	7.92	6	
25.5	24.3	23.1	21.8	20.7	19.7	18.6	17.6	16.0	13.4	11.4	9.89	7	
29.6	28.5	27.2	25.9	24.7	23.5	22.3	21.3	19.3	16.2	13.9	12.1	8	
33.8	32.7	31.4	30.1	28.8	27.5	26.2	25.0	22.9	19.3	16.6	14.5	9	
38.0	36.9	35.6	34.3	33.0	31.6	30.2	29.0	26.7	22.6	19.5	17.1	10	
42.1	41.1	39.9	38.6	37.2	35.8	34.5	33.1	30.5	26.1	22.6	19.8	11	
46.3	45.2	44.1	42.8	41.5	40.1	38.7	37.3	34.6	29.8	25.9	22.7	12	
6.19	5.74	5.30	4.95	4.64	4.33	4.08	3.82	3.41	2.77	2.32	1.98	2	120
9.79	9.06	8.43	7.89	7.43	6.95	6.57	6.17	5.53	4.55	3.83	3.27	3	
13.7	12.8	12.0	11.3	10.6	9.95	9.43	8.88	7.99	6.62	5.59	4.80	4	
17.7	16.8	15.8	14.9	14.1	13.3	12.6	11.9	10.8	8.98	7.61	6.59	5	
21.9	20.9	19.9	18.9	17.9	16.9	16.1	15.3	13.9	11.6	9.91	8.60	6	
26.1	25.1	24.1	23.0	21.9	20.9	19.9	18.9	17.3	14.6	12.5	10.8	7	
30.3	29.3	28.3	27.2	26.0	24.9	23.9	22.8	20.9	17.8	15.3	13.3	8	
34.4	33.5	32.5	31.4	30.3	29.2	28.0	26.9	24.8	21.2	18.3	16.0	9	
38.5	37.7	36.8	35.7	34.6	33.4	32.3	31.1	28.8	24.9	21.6	19.0	10	
42.7	41.9	41.0	40.0	38.9	37.7	36.5	35.3	32.9	28.7	25.1	22.1	11	
46.8	46.0	45.2	44.2	43.1	42.0	40.8	39.6	37.3	32.7	28.7	25.5	12	
6.38	5.92	5.51	5.12	4.80	4.52	4.23	4.00	3.58	2.92	2.46	2.10	2	160
10.2	9.54	8.94	8.39	7.88	7.41	7.02	6.62	5.96	4.93	4.16	3.60	3	
14.3	13.6	12.9	12.1	11.5	10.9	10.3	9.77	8.84	7.38	6.27	5.43	4	
18.5	17.8	17.0	16.2	15.4	14.6	13.9	13.3	12.1	10.1	8.71	7.57	5	
22.7	22.0	21.2	20.4	19.5	18.7	17.9	17.1	15.7	13.3	11.5	10.0	6	
26.8	26.2	25.4	24.7	23.8	22.9	22.0	21.2	19.6	16.8	14.5	12.8	7	
31.0	30.4	29.7	28.9	28.1	27.2	26.4	25.4	23.7	20.6	17.9	15.8	8	
35.1	34.5	33.9	33.1	32.3	31.5	30.6	29.7	27.9	24.5	21.6	19.1	9	
39.2	38.7	38.1	37.4	36.6	35.8	34.9	34.1	32.2	28.6	25.4	22.7	10	
43.2	42.8	42.2	41.6	40.8	40.1	39.3	38.4	36.6	32.9	29.4	26.4	11	
47.3	46.9	46.3	45.7	45.1	44.3	43.5	42.7	40.9	37.2	33.6	30.3	12	

* See page 3–26 for more details regarding these tables.

ECCENTRIC LOAD ON BOLT GROUPS — SPECIAL CASE

High Strength Bolts

For connections where the eccentric load causes both shear and tension in the bolts, the following design method may be used when the fasteners are high strength bolts which have been tightened to the specified minimum initial tension.

A bracket connected by means of bolts with an initial tension, T_i, is shown below. Using an elastic analysis, and assuming that the connection material is rigid, the moment effect on the connection can be represented by a series of coupled forces whose neutral axis passes through the centroidal axis of the group.

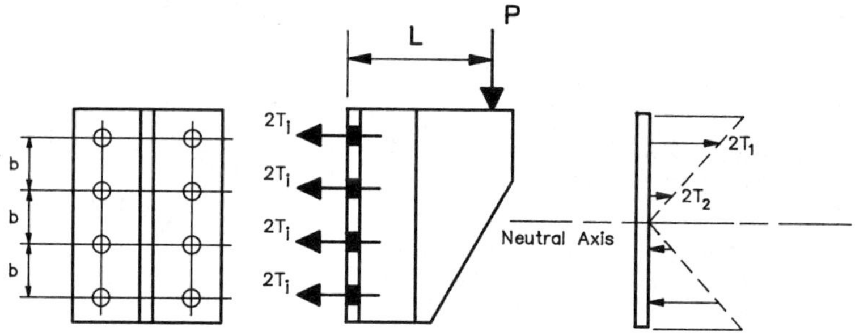

Fasteners in the top half of the connection are subject to applied shear and tension forces. The connection should be proportioned so that the maximum bolt tension, T_1, due to the moment, PL, when combined with the shear on the bolt meets the requirements of CAN/CSA-S16.1-M89 for bolts subject to combined shear and tension, Clause 13.11.4 for bearing-type connections, or Clause 13.12.3 for slip-critical connections.

Example 1

Given:

Check the adequacy of eight M20, A325M bolts (2 rows of 4, at 80 mm pitch) for the connection shown above for a factored load P of 300 kN at an eccentricity L of 150 mm. Assume the material thickness is adequate so that prying action on the bolts is not significant.

Solution:

Determine the maximum factored tension T_1 in one bolt by taking ΣM about the centre of the bolt group.

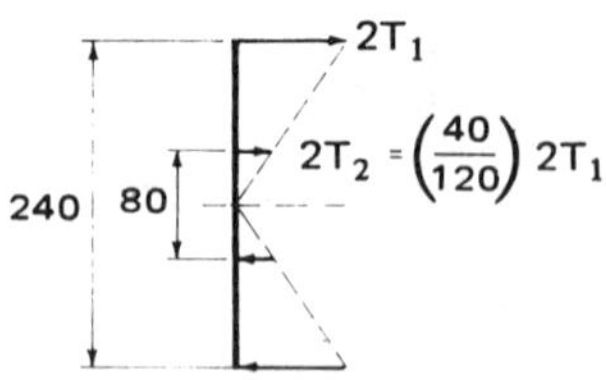

$(2T_1 \times 240) + (2T_1 \times (40/120) \times 80) = 300\ \text{kN} \times 150$

$480\ T_1 + 53.3\ T_1 = 45\ 000$

$T_1 = 84.4\ \text{kN} \quad < 131\ \text{kN} \quad$ (Table 3–4)

Factored shear per bolt = 300/8

= 37.5 kN $\quad$ < 105 kN $\quad$ (Table 3–4)

Check combined shear and tension for $V_f / T_f = 37.5 / 84.4 = 0.44$

From Table 3–8, page 3–13, for bearing-type connections,

maximum $V_f = 51$ kN (by interpolation) $\quad$ > 37.5 kN $\quad$ — OK

and maximum $T_f = 115$ kN (by interpolation) $\quad$ > 84.4 $\quad$ — OK

Example 2

Given:

Determine the number of M20, A325M bolts required to design the connection in Example 1 as a slip-critical connection for a specified load of 200 kN. Assume clean mill scale faying surfaces.

Solution:

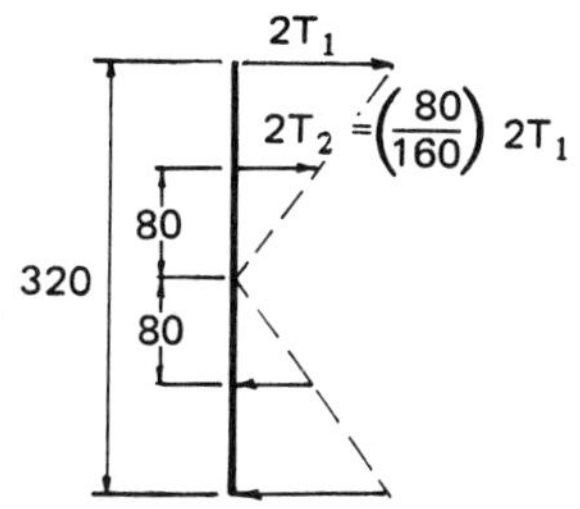

Try 10 bolts (2 rows of 5, at 80 mm pitch)

Taking Σ M about centre of bolt group gives

$(2T_1 \times 320) + (T_1 \times 160) = 200\ \text{kN} \times 150$

$640\ T_1 + 160\ T_1 = 30\ 000$

$T_1 = 37.5$ kN

$T_f = 1.5 \times 37.5 = 56.3$ kN < 131 (Table 3–4)

Specified shear per bolt = 200/10 = 20 kN < 37.4 (Table 3–11)

Check combined shear and tension for V/T = 20/37.5 = 0.53

From Table 3–12, page 3–17, for V/T of 0.60,

maximum V = 26.0 kN > 20 — OK

and maximum T = 43.3 kN > 37.5 — OK

WELD DATA

FACTORED RESISTANCE OF WELDS**

Table 3-21

Type of Weld	Load	Factored Resistance
Complete joint penetration groove welds	Tension or compression parallel to axis of weld*	Same as for base metal
	Tension normal to axis of weld	
	Compression normal to axis to weld	
	Shear	The smaller of (a) base metal $V_r = 0.67\ \phi\ A_m\ F_y$ or (b) weld metal $V_r = 0.67\ \phi_w\ A_w\ X_u$
Partial joint penetration groove welds	Tension or compression parallel to axis of weld*	Same as for base metal
	Compression normal to axis of weld	Same as for base metal
	Tension normal to axis of weld	$T_r = \phi_w A_n F_u$
	Shear	The smaller of (a) base metal $V_r = 0.67\ \phi\ A_m\ F_y$ or (b) weld metal $V_r = 0.67\ \phi_w\ A_w\ X_u$
Fillet Welds	Tension or compression parallel to axis of weld*	Same as for base metal
	Tension or compression normal to axis of weld	The smaller of (a) base metal $V_r = 0.67\ \phi\ A_m\ F_y$ or (b) weld metal $V_r = 0.67\ \phi_w\ A_w\ X_u$
	Shear	
Plug and slot welds	Shear	

* No load transfer occurs.

X_u = ultimate strength as rated by the Electrode Classification number.

** Factored resistance of welds based upon Clause 13.13.1, CAN/CSA-S16.1-M89 for matching conditions. More detailed information on the factored resistance of welds refer to CSA W59. See Clause 21.9 for the capacity of welds in combination, and Clause 21.10 for welds in combination with other fasteners.

ELECTRODE CLASSIFICATION[1] AND UNIT FACTORED WELD RESISTANCE[2], $\phi_w = 0.67, \phi = 0.90$

Table 3-22

WELD METAL				BASE METAL			
Metric Electrode Classification Number	Imperial Electrode Classification Number	Unit Factored Resistance On Weld Metal (MPa)		Specified Minimum Tensile Strength of Base Metal F_u (MPa)	Base Metal Specification and Grade CSA G40.21-M or ASTM	F_y (MPa)	Unit Factored Resistance on Base Metal $0.67\phi F_y$
		Shear on Effective Throat, A_w $0.67\ \phi_w\ X_u$	Shear per Millimetre of Filled Weld Size $0.67\ \phi_w\ X_u/\sqrt{2}$				
E410XX	E60XX	184	130	410 400	260W, 260WT A36	260 250	157 151
E480XX	E70XX	216	153	450[3] 480 480 415 450 435 460 485	300W, 300WT 350W, 350WT 350R, 350A 380W, 380WT A572 Gr 42 A572 Gr 50 A441 F_y = 42 A441 F_y = 46 A441 F_y = 50	300 350 380 290 345 290 315 345	181 211 229 175 208 175 190 208
E550XX	E80XX	248	175	520	400W, 400WT 400A, 400AT	400	241
E620XX	E90XX	279	197	590	480W, 480WT 480A, 480AT	480	289
E690XX	E100XX	310	220				
E760XX	E110XX	342	242	800	700Q, 700QT	700	422
E830XX	E120XX	374	264				

1. For complete information concerning electrode classification and strength matching of base metals, refer to CSA W59.
2. Factored weld resistance (kN) = Tabulated unit resistances x (A_w or A_m)/10^3.
3. F_u = 410 for 300W HSS.

WELD DATA

Table 3-23 FACTORED SHEAR RESISTANCE ON EFFECTIVE THROAT PER MILLIMETRE OF WELD LENGTH (kN)

Electrode Classification	Unit Shear Resist. (MPa)	Effective Throat Thickness (mm).												
		5	6	7	8	10	12	16	20	25	30	35	40	50
E 410 XX	184	.920	1.10	1.29	1.47	1.84	2.21	2.94	3.68	4.60	5.52	6.44	7.36	9.20
E 480 XX	**216**	**1.08**	**1.30**	**1.51**	**1.73**	**2.16**	**2.59**	**3.46**	**4.32**	**5.40**	**6.48**	**7.56**	**8.64**	**10.8**
E 550 XX	248	1.24	1.49	1.74	1.98	2.48	2.98	3.97	4.96	6.20	7.44	8.68	9.92	12.4
E 620 XX	279	1.40	1.67	1.95	2.23	2.79	3.35	4.46	5.58	6.98	8.37	9.76	11.2	14.0
E 690 XX	310	1.55	1.86	2.17	2.48	3.10	3.72	4.96	6.20	7.75	9.30	10.8	11.4	15.5
E 760 XX	342	1.71	2.05	2.39	2.74	3.42	4.10	5.47	6.84	8.55	10.3	12.0	13.7	17.1
E 830 XX	374	1.87	2.24	2.62	2.99	3.74	4.49	5.98	7.48	9.35	11.2	13.1	15.0	18.7

Table 3-24 FACTORED SHEAR RESISTANCE* OF FILLET WELDS PER MILLIMETRE OF WELD LENGTH, kN

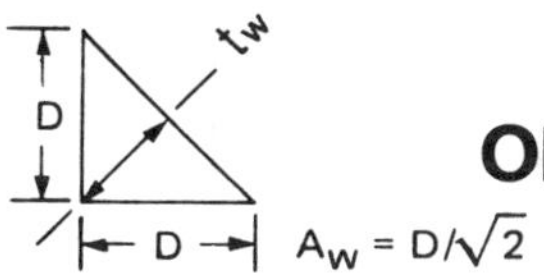

Metric Size Fillet Welds				
Electrode Classification				Fillet Weld Size, D
E410XX	**E480XX**	E550XX	E620XX	mm
.650	**.765**	.875	.985	**5**
.780	**.918**	1.05	1.18	**6**
1.04	**1.22**	1.40	1.58	**8**
1.30	**1.53**	1.75	1.97	**10**
1.56	**1.84**	2.10	2.36	**12**
1.82	**2.14**	2.45	2.76	**14**
2.08	**2.45**	2.80	3.15	**16**
2.34	**2.75**	3.15	3.55	**18**
2.60	**3.06**	3.50	3.94	**20**

Imperial Size Fillet Welds				
Fillet Weld Size, D	Electrode Classification			
inches	E410XX	E480XX	E550XX	E620XX
3/16	.619	.729	.833	.938
1/4	.826	.972	1.11	1.25
5/16	1.03	1.21	1.39	1.56
3/8	1.24	1.46	1.67	1.88
7/16	1.44	1.70	1.94	2.19
1/2	1.65	1.94	2.22	2.50
5/8	2.06	2.43	2.78	3.13
3/4	2.48	2.91	3.33	3.75

E410XX	E480XX	E550XX	E620XX	
184	**216**	248	279	Unit Factored Shear Resistance on Effective Throat (MPa)
130	**153**	175	197	Unit Factored Shear Resistance per millimetre of fillet weld size (MPa)

*Tabulated resistances for both metric and Imperial size fillet welds are based on X_u for the metric electrode classification.

ECCENTRIC LOADS ON WELD GROUPS

When the line of action of a load on a weld group does not pass through the centre of gravity of the group, the connection is eccentrically loaded. The traditional method of analysis of these weld groups has been elastic. Work reported by Butler *et al.* (1972) showed that the margins of safety for eccentrically loaded weld groups analysed elastically were both high and variable. They suggested a method of analysis based upon the load-deformation characteristics of the weld and the instantaneous centre of rotation analogy similar to that for eccentrically loaded bolt groups. For this method of analysis, the weld group is considered to be divided into a discrete number of finite weld elements. The resistance of the weld group to the external eccentric load is provided by the combined resistances of the weld elements.

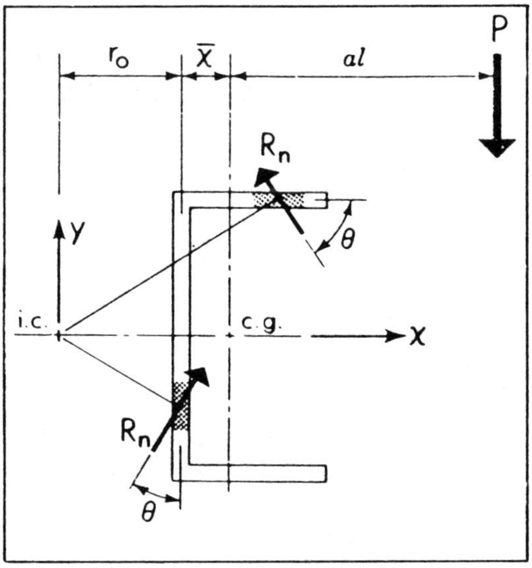

The resistance of each weld element can then be assumed to act on a line perpendicular to the radius, that radius being from the instantaneous centre of rotation to the centroid of the weld element, and can be expressed as follows:

$$R_n = R_{ULT}\,(1 - e^{-\mu\Delta})^{\lambda} \quad \text{(Butler } et\ al.\text{)}$$

where

R_{ULT}, μ, Δ and λ will depend on θ, the angle between the orientation of the weld and the resistance R_n, and have been determined from test specimens. (For definition of symbols see page 3–26, but applied to a weld element).

The ultimate load is obtained when the ultimate strength and deformation Δ of some weld element is reached. The resistance of the remaining weld elements is then computed from the above expression in which Δ is assumed to vary linearly with the distance from the instantaneous centre.

The correct location of the instantaneous centre is assured when the connection is in equilibrium, that is, when the three equations of statics, $\Sigma F_x = 0$, $\Sigma F_y = 0$ and $\Sigma M = 0$ are simultaneously satisfied.

Recent analyses by Lesik and Kennedy (1990) resulted in values of coefficients which are generally within 10% of those in Tables 3–25 to 3–32 which are based on the work of Butler *et al.* (1972).

References:

Butler, L.J., Pal, S., and Kulak, G.L. 1972. Eccentrically loaded welded connections. Journal of the Structural Division, May, ASCE, **98**(ST5).

Kulak, G.L., and Timler, P.A. 1984. Tests on eccentrically loaded fillet welds. SER No, 124, December, University of Alberta.

Lesik, D.F., and Kennedy, D.J.L. 1990. Ultimate strength of fillet welded connections loaded in plane. Canadian Journal of Civil Engineering, **17**(1), February.

Swannell, P., and Skewes, I.C. 1977. Design of welded brackets loaded in-plane: general theoretical ultimate load techniques and experimental programme. Australian WRA, RC #46, December, University of Queensland.

Tables

1. General

The values listed in Tables 3–25 to 3–32 inclusive were computed using the instantaneous centre of rotation method outlined above in the following manner.

For the various values of *a*, the eccentricity parameter, a weld length and corresponding eccentricity were selected for the appropriate weld configuration. The ultimate capacity was then computed based on the weld length and eccentricity for ¼ inch (6.35 mm) weld size and E70XX (E480XX) electrode. Butler *et al.* (1972) give, for the ¼ inch weld, 1 inch (25.4 mm) long, made with E60XX (E410XX) electrode, the following empirical equation for R_{ULT}, μ, λ and Δ_{max} :

$$R_{ULT} = \frac{10 + \theta}{0.92 + 0.0603\,\theta}$$

$$\mu = 75e^{0.0114\,\theta}$$

$$\lambda = 0.4e^{0.0146\,\theta}$$

$$\Delta_{max} = 0.225\,(\theta + 5)^{-0.47}$$

where θ is the angle (degrees) between the resultant force and the axis of the weld.

R_{ULT} was modified by the ratio of the electrode strengths because E410XX and E480XX have nearly the same specified ultimate elongations.

The ultimate capacities were reduced for a base weld leg size of 1 mm by division and then multiplied by the product of 0.67 ϕ ($\phi = 0.67$). The tabulated coefficient *C* is this reduced capacity divided by the weld length *l* in millimetres.

Values of the coefficient *C* above the horizontal line in each table are based on the resistance of the weld metal given in Table 3(A) of CSA/CAN-S16.1-M89 for the weld group concentrically loaded.

2. Use of Tables

The coefficients listed in Tables 3–25 to 3–32 inclusive are based on the use of E480XX electrodes. For E410XX electrodes, use 0.85 of the tabulated values.

(a) To determine the capacity *P* of the eccentrically loaded weld group in kilonewtons, multiply the appropriate coefficient *C* by the number of millimetres of weld size *D* and the length of the weld *l*, in millimetres.

(b) To determine the required number of millimetres of weld size *D*, divide the factored load *P*, in kilonewtons, by the appropriate coefficient *C* and the length of the weld *l*, in millimetres.

Values of *C* above the solid line are governed by the factored resistance of the weld group considered to be loaded concentrically.

3. Other Weld Configurations

For situations not covered by the tables of Eccentric Loads on Weld Groups, the method of analysis in which the vector sum of the factored longitudinal and transverse shear loads does not exceed the factored resistances of the weld is recommended as being convenient to use as it can readily be computed. Alternatively, interpolating between weld configurations in the tables which "bracket" the situation being evaluated will often be sufficient to confirm adequacy.

Example

For an example on the use of these tables, turn to page 3–50.

ECCENTRIC LOADS ON WELD GROUPS

Coefficients C*

Table 3-25

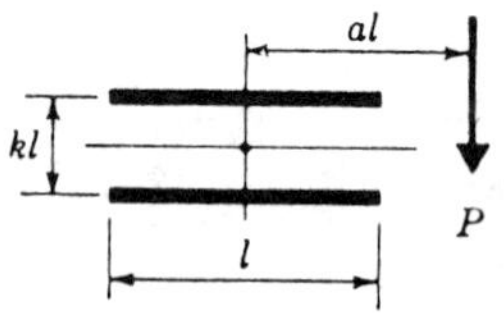

P = Factored eccentric load in kilonewtons.
l = Length of each weld in millimetres.
D = Number of millimetres in fillet weld size.
C = Coefficients tabulated below.

$$P = CDl$$

Required Minimum $C = \frac{P}{Dl}$

Required Minimum $D = \frac{P}{Cl}$

Required Minimum $l = \frac{P}{CD}$

a	k															
	0.0	0.1	0.2	0.3	0.4	0.5	0.6	0.7	0.8	0.9	1.0	1.2	1.4	1.6	1.8	2.0
0.2	0.278	0.282	0.294	0.305	0.305	0.305	0.305	0.305	0.305	0.305	0.305	0.305	0.305	0.305	0.305	0.305
0.3	0.230	0.234	0.245	0.260	0.277	0.294	0.305	0.305	0.305	0.305	0.305	0.305	0.305	0.305	0.305	0.305
0.4	0.194	0.198	0.208	0.222	0.238	0.255	0.272	0.291	0.305	0.305	0.305	0.305	0.305	0.305	0.305	0.305
0.5	0.168	0.171	0.180	0.193	0.207	0.224	0.239	0.256	0.273	0.290	0.305	0.305	0.305	0.305	0.305	0.305
0.6	0.147	0.150	0.158	0.170	0.183	0.198	0.214	0.228	0.244	0.260	0.276	0.305	0.305	0.305	0.305	0.305
0.7	0.130	0.133	0.140	0.151	0.164	0.177	0.192	0.206	0.220	0.235	0.250	0.280	0.304	0.305	0.305	0.305
0.8	0.117	0.119	0.126	0.136	0.148	0.161	0.174	0.187	0.201	0.214	0.228	0.257	0.282	0.304	0.305	0.305
0.9	0.106	0.108	0.114	0.123	0.134	0.147	0.159	0.171	0.184	0.197	0.209	0.236	0.263	0.285	0.304	0.305
1.0	0.096	0.098	0.105	0.113	0.123	0.134	0.146	0.158	0.169	0.182	0.193	0.217	0.245	0.267	0.287	0.305
1.2	0.082	0.084	0.089	0.097	0.105	0.115	0.125	0.136	0.146	0.156	0.168	0.189	0.213	0.235	0.255	0.274
1.4	0.071	0.073	0.077	0.084	0.092	0.100	0.109	0.119	0.129	0.138	0.147	0.167	0.187	0.208	0.229	0.247
1.6	0.063	0.064	0.068	0.074	0.081	0.089	0.097	0.105	0.114	0.123	0.132	0.149	0.167	0.185	0.206	0.224
1.8	0.056	0.058	0.061	0.067	0.073	0.080	0.087	0.095	0.102	0.111	0.119	0.134	0.151	0.167	0.185	0.203
2.0	0.051	0.052	0.055	0.060	0.066	0.072	0.079	0.086	0.093	0.100	0.108	0.123	0.138	0.152	0.168	0.185
2.2	0.047	0.048	0.051	0.055	0.060	0.066	0.072	0.079	0.085	0.092	0.099	0.113	0.126	0.140	0.154	0.170
2.4	0.043	0.044	0.046	0.051	0.055	0.061	0.067	0.073	0.079	0.085	0.091	0.104	0.116	0.129	0.142	0.156
2.6	0.040	0.040	0.043	0.047	0.051	0.056	0.062	0.067	0.073	0.079	0.085	0.097	0.108	0.120	0.132	0.145
2.8	0.037	0.038	0.040	0.043	0.048	0.052	0.058	0.063	0.068	0.074	0.079	0.090	0.101	0.112	0.124	0.135
3.0	0.034	0.035	0.037	0.041	0.045	0.049	0.054	0.059	0.064	0.069	0.074	0.084	0.095	0.105	0.116	0.126

*Coefficients in table are for E480XX electrodes.
The effect of eccentricity has been neglected for cases above the solid horizontal line.

Coefficients C*

Table 3-26

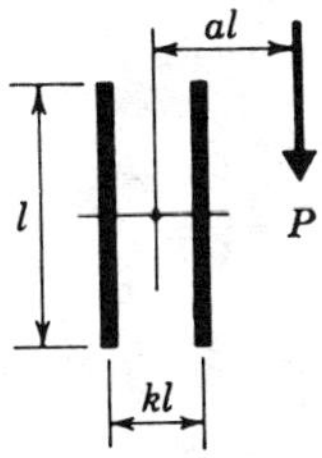

P = Factored eccentric load in kilonewtons.
l = Length of each weld in millimetres.
D = Number of millimetres in fillet weld size.
C = Coefficients tabulated below.

$$P = CDl$$

Required Minimum $C = \frac{P}{Dl}$

Required Minimum $D = \frac{P}{Cl}$

Required Minimum $l = \frac{P}{CD}$

a	k															
	0.0	0.1	0.2	0.3	0.4	0.5	0.6	0.7	0.8	0.9	1.0	1.2	1.4	1.6	1.8	2.0
0.2	.305	.305	.305	.305	.305	.305	.305	.305	.305	.305	.305	.305	.305	.305	.305	.305
0.3	.276	.276	.278	.281	.284	.287	.289	.291	.293	.295	.296	.298	.300	.301	.302	.303
0.4	.229	.230	.234	.240	.245	.250	.255	.259	.263	.266	.268	.273	.277	.280	.282	.284
0.5	.192	.194	.200	.207	.214	.220	.226	.232	.237	.241	.245	.251	.256	.261	.264	.267
0.6	.165	.167	.173	.181	.188	.196	.203	.209	.214	.219	.224	.232	.238	.243	.248	.252
0.7	.144	.146	.151	.160	.168	.176	.183	.190	.196	.201	.206	.215	.222	.228	.233	.238
0.8	.127	.129	.135	.143	.151	.159	.167	.174	.180	.186	.191	.200	.208	.214	.220	.225
0.9	.114	.116	.121	.129	.137	.145	.153	.160	.166	.172	.177	.187	.195	.202	.209	.214
1.0	.103	.105	.110	.118	.126	.133	.141	.148	.154	.160	.166	.176	.184	.191	.198	.204
1.2	.086	.088	.093	.100	.107	.115	.122	.128	.135	.141	.146	.156	.165	.173	.179	.186
1.4	.074	.076	.080	.086	.093	.100	.107	.113	.119	.125	.131	.140	.149	.157	.164	.170
1.6	.065	.067	.071	.076	.083	.089	.095	.101	.107	.113	.118	.128	.136	.144	.151	.157
1.8	.058	.059	.063	.068	.074	.080	.086	.092	.097	.102	.107	.117	.125	.133	.140	.146
2.0	.052	.053	.057	.062	.067	.073	.078	.084	.089	.094	.099	.108	.116	.123	.130	.136
2.2	.048	.049	.052	.056	.061	.066	.072	.077	.082	.087	.091	.100	.108	.115	.122	.128
2.4	.044	.045	.048	.052	.056	.061	.066	.071	.076	.080	.085	.093	.101	.108	.114	.120
2.6	.040	.041	.044	.048	.052	.057	.061	.066	.071	.075	.079	.087	.094	.101	.108	.113
2.8	.037	.038	.041	.044	.048	.053	.057	.062	.066	.070	.074	.082	.089	.095	.102	.107
3.0	.035	.036	.038	.041	.045	.049	.054	.058	.062	.066	.070	.077	.084	.090	.096	.102

*Coefficients in table are for E480XX electrodes.
The effect of eccentricity has been neglected for cases above the solid horizontal line.

ECCENTRIC LOADS ON WELD GROUPS

Coefficients C*

Table 3-27

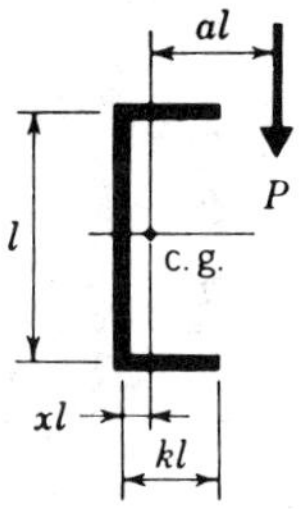

P = Factored eccentric load in kilonewtons.
l = Length of weld parallel to load P in millimetres.
D = Number of millimetres in fillet weld size.
C = Coefficients tabulated below.
xl = Distance from vertical weld to center of gravity of weld group.

$$P = CDl$$

Required Minimum $C = \frac{P}{Dl}$

Required Minimum $D = \frac{P}{Cl}$

Required Minimum $l = \frac{P}{CD}$

a	k															
	0.0	0.1	0.2	0.3	0.4	0.5	0.6	0.7	0.8	0.9	1.0	1.2	1.4	1.6	1.8	2.0
0.2	.153	.183	.214	.244	.275	.305	.336	.366	.397	.428	.458	.519	.580	.641	.702	.763
0.3	.138	.180	.214	.244	.275	.305	.336	.366	.397	.428	.458	.519	.580	.641	.702	.763
0.4	.114	.152	.188	.221	.254	.286	.320	.353	.387	.422	.455	.519	.580	.641	.702	.763
0.5	.096	.129	.161	.192	.222	.252	.282	.313	.345	.376	.409	.474	.544	.615	.687	.761
0.6	.082	.111	.140	.169	.196	.223	.251	.280	.308	.338	.368	.431	.496	.563	.632	.704
0.7	.072	.097	.122	.149	.174	.199	.225	.251	.278	.306	.334	.393	.455	.519	.586	.654
0.8	.063	.085	.109	.133	.156	.179	.203	.227	.252	.278	.305	.360	.419	.480	.545	.610
0.9	.057	.077	.098	.119	.141	.163	.184	.207	.231	.255	.280	.333	.388	.447	.508	.572
1.0	.051	.069	.088	.108	.129	.149	.169	.190	.212	.235	.259	.308	.360	.417	.476	.537
1.2	.043	.058	.074	.091	.109	.126	.144	.163	.182	.203	.223	.268	.317	.368	.422	.479
1.4	.037	.050	.064	.079	.094	.110	.126	.142	.159	.177	.196	.238	.282	.329	.378	.430
1.6	.033	.044	.056	.069	.083	.097	.111	.126	.141	.158	.176	.213	.253	.296	.342	.391
1.8	.029	.039	.050	.062	.074	.087	.100	.113	.127	.142	.159	.192	.230	.269	.312	.356
2.0	.026	.035	.045	.055	.067	.078	.090	.102	.115	.130	.144	.176	.210	.247	.286	.328
2.2	.024	.032	.041	.050	.061	.072	.082	.093	.106	.119	.132	.162	.193	.228	.264	.303
2.4	.022	.029	.037	.046	.056	.066	.076	.086	.098	.110	.122	.150	.179	.211	.246	.282
2.6	.020	.027	.035	.043	.052	.061	.070	.080	.091	.102	.113	.139	.167	.197	.229	.263
2.8	.019	.025	.032	.040	.048	.057	.065	.074	.085	.095	.106	.130	.156	.184	.215	.247
3.0	.017	.023	.030	.037	.045	.053	.061	.070	.079	.089	.099	.122	.147	.173	.202	.232
x	0	.008	.029	.056	.089	.125	.164	.204	.246	.289	.333	.424	.516	.610	.704	.800

*Coefficients in table are for E480XX electrodes.
The effect of eccentricity has been neglected for cases above the solid line.

Coefficients C*

Table 3-28

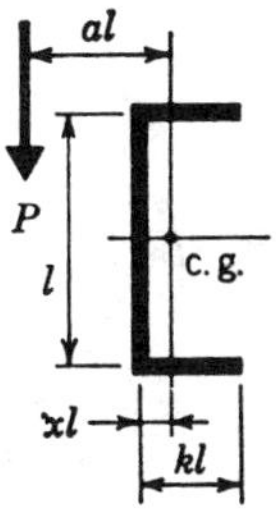

P = Factored eccentric load in kilonewtons.
l = Length of weld parallel to load P in millimetres.
D = Number of millimetres in fillet weld size.
C = Coefficients tabulated below.
xl = Distance from vertical weld to center of gravity of weld group.

$P = CDl$

Required Minimum $C = \frac{P}{Dl}$

Required Minimum $D = \frac{P}{Cl}$

Required Minimum $l = \frac{P}{CD}$

a	k															
	0.0	0.1	0.2	0.3	0.4	0.5	0.6	0.7	0.8	0.9	1.0	1.2	1.4	1.6	1.8	2.0
0.2	.153	.183	.214	.244	.275	.305	.336	.366	.397	.428	.458	.519	.580	.641	.702	.763
0.3	.138	.180	.214	.244	.275	.305	.336	.366	.397	.428	.458	.519	.580	.641	.702	.763
0.4	.114	.152	.189	.226	.263	.297	.331	.365	.397	.428	.458	.519	.580	.641	.702	.763
0.5	.096	.128	.161	.195	.229	.264	.296	.328	.360	.392	.424	.488	.555	.622	.691	.761
0.6	.082	.110	.138	.166	.199	.231	.264	.295	.325	.356	.387	.449	.513	.578	.644	.711
0.7	.072	.095	.118	.147	.177	.207	.237	.266	.296	.326	.355	.415	.475	.537	.601	.666
0.8	.063	.084	.105	.131	.157	.186	.215	.244	.272	.299	.327	.384	.443	.502	.563	.626
0.9	.057	.075	.095	.117	.143	.169	.195	.222	.249	.276	.303	.357	.412	.470	.529	.589
1.0	.051	.067	.086	.106	.130	.154	.179	.205	.231	.256	.281	.333	.387	.441	.497	.556
1.2	.043	.056	.072	.090	.110	.131	.153	.176	.199	.223	.246	.293	.341	.392	.444	.498
1.4	.037	.048	.062	.078	.094	.114	.133	.153	.175	.196	.217	.260	.305	.351	.399	.449
1.6	.033	.042	.055	.069	.083	.100	.117	.136	.155	.174	.194	.234	.274	.318	.362	.409
1.8	.029	.038	.048	.061	.074	.089	.105	.122	.138	.157	.175	.211	.250	.289	.331	.374
2.0	.026	.034	.044	.055	.067	.080	.095	.110	.125	.142	.159	.193	.229	.266	.304	.345
2.2	.024	.031	.040	.050	.061	.073	.086	.100	.114	.129	.144	.178	.210	.245	.282	.320
2.4	.022	.029	.036	.046	.056	.067	.079	.091	.105	.119	.133	.163	.195	.228	.262	.297
2.6	.020	.026	.034	.042	.052	.062	.073	.085	.097	.109	.123	.151	.182	.212	.245	.278
2.8	.019	.025	.031	.039	.048	.058	.068	.079	.090	.101	.114	.140	.169	.199	.229	.261
3.0	.017	.023	.029	.037	.045	.054	.064	.073	.084	.095	.107	.131	.158	.187	.215	.246
x	0	.008	.029	.056	.089	.125	.164	.204	.246	.289	.333	.424	.516	.610	.704	.800

*Coefficients in table are for E480XX electrodes.
The effect of eccentricity has been neglected for cases above the solid horizontal line.

ECCENTRIC LOADS ON WELD GROUPS

Coefficients C*

Table 3-29

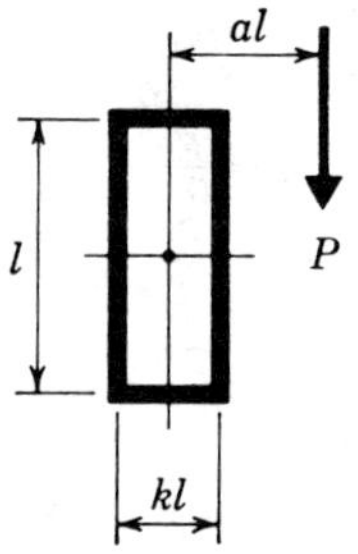

P = Factored eccentric load in kilonewtons.
l = Length of longer welds in millimetres.
D = Number of millimetres in fillet weld size.
C = Coefficients tabulated below.
Note: When load P is perpendicular to longer side l use table on facing page.

$P = CDl$

Required Minimum $C = \frac{P}{Dl}$

Required Minimum $D = \frac{P}{Cl}$

Required Minimum $l = \frac{P}{CD}$

a	k										
	0.0	0.1	0.2	0.3	0.4	0.5	0.6	0.7	0.8	0.9	1.0
0.2	.305	.336	.366	.397	.428	.458	.489	.519	.550	.580	.611
0.3	.276	.318	.360	.397	.428	.458	.489	.519	.550	.580	.611
0.4	.229	.267	.307	.346	.383	.417	.451	.486	.520	.555	.590
0.5	.192	.226	.263	.300	.337	.370	.403	.436	.469	.503	.536
0.6	.165	.194	.228	.262	.297	.330	.362	.394	.425	.457	.490
0.7	.144	.170	.200	.232	.265	.297	.327	.357	.388	.418	.449
0.8	.127	.150	.178	.208	.238	.269	.297	.326	.355	.384	.414
0.9	.114	.135	.159	.187	.216	.245	.272	.299	.327	.355	.383
1.0	.103	.122	.145	.170	.197	.224	.250	.276	.302	.329	.357
1.2	.086	.102	.122	.144	.167	.191	.216	.239	.264	.288	.313
1.4	.074	.088	.105	.124	.145	.167	.190	.211	.233	.255	.278
1.6	.065	.077	.092	.109	.128	.148	.169	.188	.208	.229	.250
1.8	.058	.069	.082	.098	.115	.133	.152	.170	.188	.208	.227
2.0	.052	.062	.074	.088	.104	.120	.137	.155	.172	.190	.207
2.2	.048	.056	.067	.081	.095	.110	.125	.142	.158	.174	.191
2.4	.044	.051	.062	.074	.087	.101	.116	.131	.146	.161	.177
2.6	.040	.047	.057	.069	.081	.093	.107	.122	.136	.150	.165
2.8	.037	.044	.053	.064	.075	.087	.100	.114	.127	.140	.154
3.0	.035	.041	.049	.060	.070	.081	.094	.106	.119	.132	.145

*Coefficients in table are for E480XX electrodes.
The effect of eccentricity has been neglected for cases above the solid horizontal line.

ECCENTRIC LOADS ON WELD GROUPS

Coefficients C*

Table 3-30

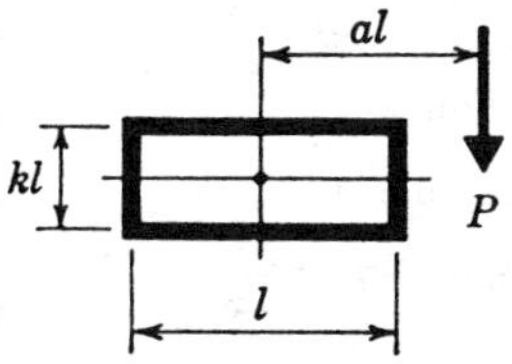

P = Factored eccentric load in kilonewtons.
l = Length of longer welds in millimetres.
D = Number of millimetres in fillet weld size.
C = Coefficients tabulated below.
Note: When load P is parallel to longer side l use table on facing page.

$$P = CDl$$

Required Minimum $C = \frac{P}{Dl}$

Required Minimum $D = \frac{P}{Cl}$

Required Minimum $l = \frac{P}{CD}$

a	k										
	0.0	0.1	0.2	0.3	0.4	0.5	0.6	0.7	0.8	0.9	1.0
0.2	.278	.309	.349	.388	.428	.458	.489	.519	.550	.580	.611
0.3	.230	.258	.297	.335	.375	.417	.463	.508	.550	.580	.611
0.4	.194	.220	.256	.291	.330	.367	.410	.453	.499	.543	.590
0.5	.168	.191	.223	.256	.293	.329	.366	.406	.449	.492	.536
0.6	.147	.168	.198	.228	.263	.296	.332	.367	.407	.448	.490
0.7	.130	.149	.177	.205	.237	.268	.302	.336	.371	.409	.449
0.8	.117	.134	.160	.186	.216	.245	.276	.308	.343	.376	.414
0.9	.106	.122	.145	.170	.197	.225	.255	.284	.317	.349	.383
1.0	.096	.111	.133	.156	.182	.207	.236	.264	.294	.324	.357
1.2	.082	.095	.114	.133	.157	.180	.205	.230	.257	.283	.313
1.4	.071	.083	.099	.117	.138	.158	.181	.203	.228	.252	.278
1.6	.063	.073	.088	.104	.123	.141	.162	.182	.204	.226	.250
1.8	.056	.065	.079	.093	.110	.127	.146	.165	.185	.205	.227
2.0	.051	.059	.072	.085	.100	.116	.133	.150	.169	.187	.207
2.2	.047	.054	.065	.077	.092	.106	.122	.138	.155	.173	.191
2.4	.043	.050	.060	.071	.085	.098	.113	.127	.144	.160	.177
2.6	.040	.046	.056	.066	.079	.091	.105	.118	.134	.149	.165
2.8	.037	.043	.052	.062	.073	.085	.098	.111	.125	.139	.154
3.0	.034	.040	.049	.058	.069	.079	.092	.104	.117	.130	.145

*Coefficients in table are for E480XX electrodes.
The effect of eccentricity has been neglected for cases above the solid horizontal line.

ECCENTRIC LOADS ON WELD GROUPS

Coefficients C*

Table 3-31

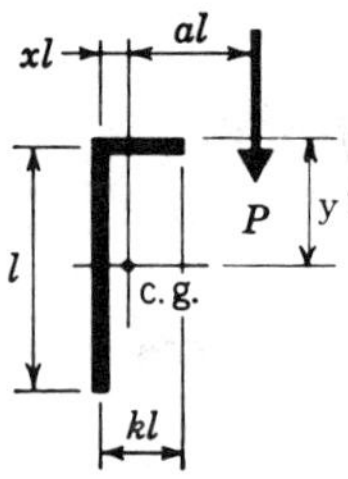

P = Factored eccentric load in kilonewtons.
l = Length of weld parallel to load P in millimetres.
D = Number of millimetres in fillet weld size.
C = Coefficients tabulated below.
xl = Distance from vertical weld to center of gravity of weld group.

$$P = CDl$$

$$\text{Required Minimum } C = \frac{P}{Dl}$$

$$\text{Required Minimum } D = \frac{P}{Cl}$$

$$\text{Required Minimum } l = \frac{P}{CD}$$

a	k 0.0	0.1	0.2	0.3	0.4	0.5	0.6	0.7	0.8	0.9	1.0	1.2	1.4	1.6	1.8	2.0
0.2	.153	.168	.183	.199	.214	.229	.244	.260	.275	.290	.305	.336	.366	.397	.428	.458
0.3	.138	.156	.175	.192	.208	.223	.238	.254	.270	.287	.304	.336	.366	.397	.428	.458
0.4	.114	.130	.147	.162	.177	.192	.206	.220	.235	.250	.266	.299	.335	.373	.411	.451
0.5	.096	.110	.124	.137	.151	.166	.179	.192	.205	.220	.234	.266	.300	.336	.373	.412
0.6	.082	.094	.107	.118	.130	.143	.157	.169	.181	.195	.209	.239	.271	.305	.341	.378
0.7	.072	.082	.093	.103	.114	.126	.138	.150	.162	.174	.188	.216	.246	.279	.313	.349
0.8	.063	.073	.082	.092	.101	.112	.123	.135	.146	.158	.170	.197	.226	.257	.290	.324
0.9	.057	.065	.074	.082	.091	.101	.111	.123	.133	.144	.155	.180	.208	.238	.269	.302
1.0	.051	.059	.067	.074	.082	.091	.101	.112	.122	.132	.143	.166	.193	.221	.251	.283
1.2	.043	.050	.056	.062	.069	.077	.085	.095	.104	.113	.122	.144	.168	.194	.221	.250
1.4	.037	.043	.048	.054	.060	.067	.074	.082	.090	.098	.107	.127	.148	.172	.197	.224
1.6	.033	.037	.042	.047	.053	.058	.065	.072	.080	.087	.095	.113	.133	.154	.177	.202
1.8	.029	.033	.038	.042	.047	.052	.058	.065	.072	.079	.086	.102	.120	.140	.161	.184
2.0	.026	.030	.034	.038	.042	.047	.052	.058	.065	.071	.078	.093	.109	.127	.147	.168
2.2	.024	.027	.031	.035	.038	.043	.048	.053	.059	.065	.071	.085	.100	.117	.136	.155
2.4	.022	.025	.028	.032	.035	.039	.044	.049	.055	.060	.066	.079	.093	.108	.126	.144
2.6	.020	.023	.026	.029	.033	.036	.041	.045	.051	.056	.061	.073	.086	.101	.117	.134
2.8	.019	.021	.024	.027	.030	.034	.038	.042	.047	.052	.057	.068	.081	.094	.109	.126
3.0	.017	.020	.023	.025	.028	.032	.035	.039	.044	.049	.053	.064	.075	.088	.103	.118
x	0	.005	.017	.035	.057	.083	.112	.144	.178	.213	.250	.327	.408	.492	.579	.667
y	.500	.455	.417	.385	.357	.333	.312	.294	.278	.263	.250	.227	.208	.192	.179	.167

*Coefficients in table are for E480XX electrodes.
The effect of eccentricity has been neglected for cases above the solid horizontal line.

Coefficients C*

Table 3-32

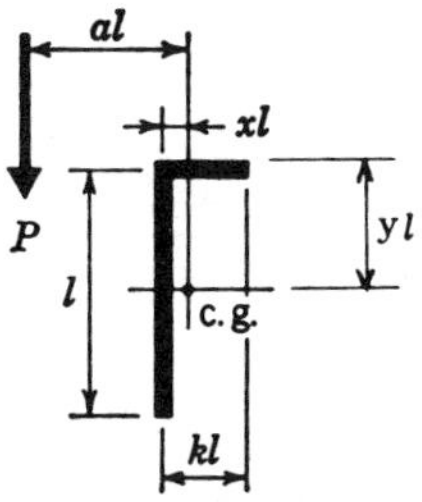

P = Factored eccentric load in kilonewtons.
l = Length of weld parallel to load P in millimetres.
D = Number of millimetres in fillet weld size.
C = Coefficients tabulated below.
xl = Distance from vertical weld to center of gravity of weld group.

$$P = CDl$$

Required Minimum $C = \frac{P}{Dl}$

Required Minimum $D = \frac{P}{Cl}$

Required Minimum $l = \frac{P}{CD}$

a	k															
	0.0	0.1	0.2	0.3	0.4	0.5	0.6	0.7	0.8	0.9	1.0	1.2	1.4	1.6	1.8	2.0
0.2	.153	.168	.183	.199	.214	.229	.244	.260	.275	.290	.305	.336	.366	.397	.428	.458
0.3	.138	.156	.174	.189	.205	.219	.234	.250	.265	.282	.300	.336	.366	.397	.428	.458
0.4	.114	.130	.146	.159	.173	.188	.202	.217	.233	.249	.266	.303	.338	.372	.407	.443
0.5	.096	.110	.122	.135	.149	.162	.176	.191	.206	.221	.237	.273	.310	.343	.377	.413
0.6	.082	.094	.105	.117	.129	.142	.154	.169	.183	.198	.213	.248	.284	.318	.351	.386
0.7	.072	.082	.092	.102	.114	.125	.138	.151	.164	.179	.193	.226	.260	.295	.327	.360
0.8	.063	.072	.081	.091	.101	.112	.124	.136	.149	.162	.176	.207	.240	.275	.305	.338
0.9	.057	.064	.073	.082	.091	.102	.112	.123	.135	.148	.161	.191	.222	.255	.286	.318
1.0	.051	.058	.066	.074	.082	.092	.102	.113	.124	.136	.149	.176	.207	.238	.269	.300
1.2	.043	.049	.055	.062	.070	.078	.087	.096	.106	.117	.128	.153	.180	.209	.239	.268
1.4	.037	.042	.048	.054	.060	.067	.075	.083	.092	.102	.112	.134	.159	.186	.215	.242
1.6	.033	.037	.042	.047	.053	.059	.066	.073	.082	.090	.099	.120	.142	.167	.193	.220
1.8	.029	.033	.037	.042	.047	.053	.059	.066	.073	.081	.089	.108	.129	.151	.176	.201
2.0	.026	.030	.034	.038	.042	.048	.053	.059	.066	.073	.081	.098	.117	.138	.161	.185
2.2	.024	.027	.031	.034	.039	.043	.048	.054	.060	.067	.074	.089	.107	.127	.148	.170
2.4	.022	.025	.028	.031	.035	.040	.044	.050	.055	.062	.068	.082	.099	.117	.137	.157
2.6	.020	.023	.026	.029	.033	.037	.041	.046	.051	.057	.063	.076	.092	.109	.127	.146
2.8	.019	.021	.024	.027	.030	.034	.038	.043	.048	.053	.059	.071	.086	.101	.118	.136
3.0	.017	.020	.022	.025	.028	.032	.036	.040	.044	.049	.055	.067	.080	.095	.110	.127
x	0	.005	.017	.035	.057	.083	.112	.144	.178	.213	.250	.327	.408	.492	.579	.667
y	.500	.455	.417	.385	.357	.333	.312	.294	.278	.263	.250	.227	.208	.192	.179	.167

*Coefficients in table are for E480XX electrodes.
The effect of eccentricity has been neglected for cases above the solid horizontal line.

Example

Given:

A column bracket supports a factored load of 500 kN. The width of the bracket is 300 mm. Welds are made using E480XX electrodes. For the weld configuration shown, find the required weld size. Steel is G40.21-M, 300W.

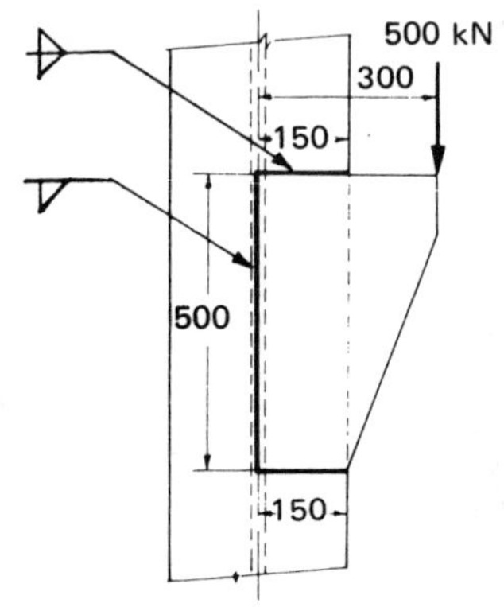

Solution:

Referring to Table 3–27, page 3–44:

$$D = \frac{P}{Cl}$$

= number of millimetres of fillet weld leg size

$k = 150 / 500 = 0.3$

From the bottom line of Table 3–27, for $k = 0.3$ $x = 0.056$

Referring to the figure in Table 3–27, $al + xl = 300$

For $l = 500$, $500\,a + 0.056(500) = 300$, $a = 0.544$

For $a = 0.544$ and $k = 0.3$, C = 0.18 by interpolation

Therefore, $D = \frac{500}{0.18 \times 500} = 5.5$, say 6 mm

Use 6 mm fillet welds made with E480XX electrodes.

Notes:

1. The final choice of the fillet weld size to be used in an actual connection will also be dependent upon the minimum and maximum sizes required by a) the physical thickness of the parts joined and b) the requirements of CSA Standard W59.
2. The strength of an actual connection will also be dependent upon the resistances of the connected parts.

ECCENTRIC LOADS ON WELD GROUPS SHEAR AND MOMENT

For the case of the eccentrically loaded fillet weld group shown below in Fig. 3–3, an analysis similar to that for the in-plane eccentricity is used but modified to account for the bearing of the plate at ultimate load. The magnitude and location of the horizontal force, H_B, due to plate bearing, is derived assuming a triangular stress block at ultimate load. (See Fig. 3–4). According to Dawe and Kulak (1974), the triangular stress block assumption provides a better prediction of the ultimate load than either a rectangular or parabolic distribution. The vertical force, V_B, below the neutral axis is taken to be the shear resistance of that portion of the weld below the neutral axis. This method of analysis accounts for the load-deformation characteristics of the weld.

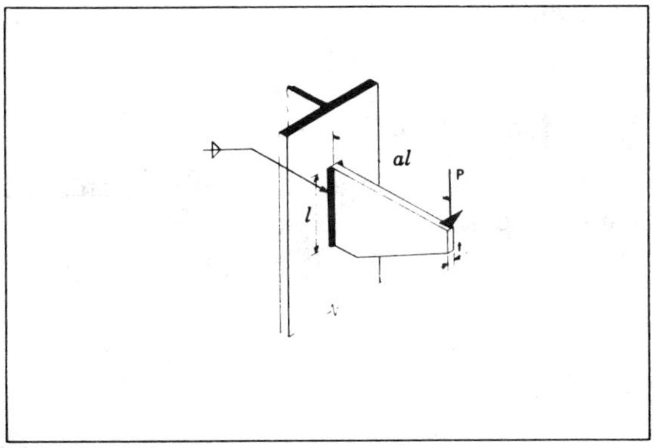

Figure 3–3

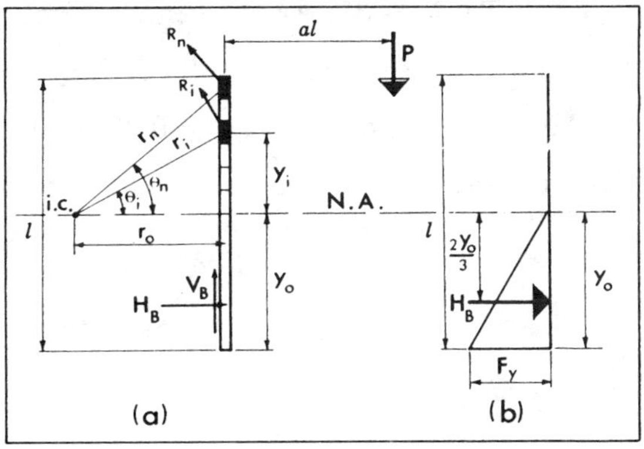

Figure 3–4

Table

Table 3–33 has been computed based upon the use of E480XX electrodes and F_y = 300 MPa following the method outlined by Dawe and Kulak. The coefficient C′ (P/*l*) is tabulated for various plate thicknesses *t* and fillet weld sizes. For joints using lower strength materials the value of C′ may be multiplied by F_y /300.

The tabulated values are conservative when used with plate material in excess of 300 MPa yield. The maximum value of C′ is determined by shear in the plate and is accounted for in the tabulated values.

Reference:

DAWE, J.L., and KULAK, G.L. 1974. Welded connections under combined shear and moment. Journal of the Structural Division, ASCE, **100**(ST4), April.

Example

Given:

A 12 mm plate carrying a 180 kN factored load is welded to a column with a pair of fillet welds 250 mm long. Find the fillet weld size required if the 180 kN load acts at a 130 mm eccentricity.

Solution:

$l = 250$ mm

$al = 130$ mm; therefore $a = 130/250 = 0.52$

C′ required is $P/l = 180/250 = 0.72$

Try 6 mm fillet weld

From Table 3–33, for t = 12 mm, and 6 mm weld size:

C′ = 0.754 for $a = 0.50$, and 0.629 for $a = 0.60$

Therefore, for $a = 0.52$, C′ = 0.73 (by interpolation) > 0.72 — OK

The minimum weld size based upon the thickness of the materials joined and the resistance of the connected parts must also be checked.

ECCENTRIC LOADS ON WELD GROUPS

Coefficients C′*

Table 3-33

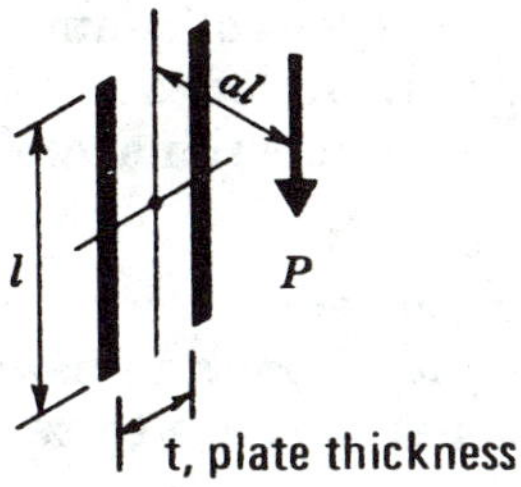

P = Factored eccentric load (kN)

l = Length of fillet welds parallel to load P (mm)

C′ = Coefficients tabulated below

$P = C' l$

Required minimum $C' = P/l$

Required minimum $l = P/C'$

Plate Thickness	8 mm	10 mm		12 mm			16 mm		
Weld Size	**5**	**5**	**6**	**5**	**6**	**8**	**6**	**8**	**10**
a = 0.2	1.30	1.47	1.60	1.55	1.76	2.00	1.84	2.34	2.60
0.3	0.889	1.04	1.10	1.16	1.24	1.36	1.47	1.66	1.78
0.4	0.671	0.784	0.828	0.885	0.941	1.02	1.13	1.25	1.34
0.5	0.537	0.630	0.665	0.710	0.754	0.818	0.911	1.01	1.07
0.6	0.448	0.526	0.555	0.594	0.629	0.683	0.762	0.839	0.895
0.7	0.385	0.450	0.475	0.509	0.540	0.586	0.653	0.720	0.767
0.8	0.336	0.394	0.416	0.446	0.473	0.512	0.573	0.631	0.673
0.9	0.298	0.350	0.369	0.396	0.420	0.455	0.509	0.560	0.598
1.0	0.269	0.316	0.332	0.357	0.379	0.410	0.458	0.506	0.537
1.2	0.223	0.263	0.276	0.297	0.316	0.342	0.382	0.421	0.448
1.4	0.191	0.225	0.238	0.255	0.270	0.292	0.327	0.360	0.384
1.6	0.168	0.197	0.207	0.224	0.236	0.256	0.286	0.315	0.336
1.8	0.150	0.175	0.184	0.198	0.210	0.228	0.255	0.280	0.299
2.0	0.134	0.157	0.166	0.178	0.188	0.205	0.229	0.251	0.269
2.2	0.122	0.143	0.151	0.162	0.172	0.186	0.208	0.229	0.244
2.4	0.111	0.132	0.137	0.149	0.158	0.170	0.191	0.210	0.224
2.6	0.103	0.121	0.127	0.136	0.145	0.157	0.176	0.194	0.206
2.8	0.096	0.113	0.118	0.127	0.135	0.146	0.162	0.180	0.192
3.0	0.089	0.105	0.111	0.119	0.126	0.137	0.152	0.168	0.179

Plate Thickness	20 mm			25 mm				40 mm			
Weld Size	**8**	**10**	**12**	**8**	**10**	**12**	**14**	**10**	**12**	**14**	**16**
a = 0.2	2.44	2.93	3.21	2.44	3.06	3.56	3.89	3.06	3.67	4.28	4.90
0.3	1.89	2.07	2.20	2.10	2.37	2.55	2.69	2.85	3.23	3.54	3.79
0.4	1.45	1.57	1.66	1.65	1.81	1.93	2.03	2.31	2.55	2.74	2.90
0.5	1.17	1.26	1.33	1.33	1.46	1.55	1.63	1.89	2.07	2.21	2.33
0.6	0.974	1.05	1.11	1.12	1.22	1.29	1.36	1.59	1.74	1.85	1.95
0.7	0.836	0.900	0.949	0.959	1.04	1.11	1.16	1.37	1.49	1.59	1.67
0.8	0.732	0.789	0.833	0.840	0.916	0.973	1.02	1.20	1.31	1.39	1.46
0.9	0.651	0.701	0.739	0.747	0.813	0.865	0.906	1.07	1.16	1.24	1.30
1.0	0.586	0.632	0.664	0.673	0.733	0.779	0.815	0.965	1.05	1.11	1.17
1.2	0.488	0.526	0.554	0.562	0.610	0.649	0.679	0.806	0.873	0.930	0.977
1.4	0.418	0.451	0.475	0.481	0.523	0.557	0.582	0.690	0.749	0.797	0.836
1.6	0.367	0.394	0.415	0.421	0.458	0.487	0.510	0.604	0.655	0.696	0.733
1.8	0.326	0.351	0.369	0.373	0.407	0.432	0.452	0.537	0.581	0.620	0.652
2.0	0.293	0.316	0.333	0.336	0.366	0.390	0.408	0.484	0.528	0.558	0.586
2.2	0.266	0.287	0.302	0.306	0.333	0.353	0.370	0.440	0.476	0.507	0.533
2.4	0.244	0.263	0.277	0.281	0.305	0.324	0.339	0.403	0.437	0.465	0.489
2.6	0.226	0.242	0.256	0.259	0.282	0.300	0.313	0.371	0.404	0.429	0.451
2.8	0.209	0.225	0.237	0.240	0.261	0.278	0.291	0.345	0.374	0.399	0.419
3.0	0.195	0.210	0.221	0.224	0.244	0.260	0.272	0.322	0.349	0.372	0.390

*Coefficients in table are for E480XX electrodes and plate with F_y = 300 MPa.

FRAMED BEAM SHEAR CONNECTIONS

General

This section of Part Three contains information on four common types of beam shear connections traditionally considered standard in the industry. Included are double angle connections, simple end plate connections, single angle connections and tee connections.

These types of connections are generally designed for strength requirements only, under the effects of factored loads, and the capacities of welds and of bolts in bearing-type connections are based on their factored resistances. The capacities are to be used with factored loads.

Tabulated bolt capacities for bearing-type connections are based on threads being excluded from the shear planes. When threads intercept a shear plane, these capacities should be reduced by 30%, i.e. use 70% of the tabulated values (CAN/CSA-S16.1-M89, Clause 13.11.2). Previous editions of S16.1 implied that threads were excluded from the shear plane when the material thickness adjacent to the nut was equal to 10 mm or more. This reference to 10 mm thickness has been removed from S16.1-M89 since a thickness of 10 mm may not guarantee that threads are excluded, unless special precautions are taken. For practical reasons, it is suggested that bearing-type shear connections be designed on the assumption of intercepted threads when combinations of thin material and detailing for minimum bolt stick-through (the nuts) are expected.

Bolt capacities of slip-critical joints for double angle and end plate connections have also been included for designing shear connections such as those subjected to fatigue or frequent load reversal. Values are based on Class A (clean mill scale or blast cleaned with Class A coatings) contact surfaces (k_s = 0.33). The capacities are to be used with specified loads only.

Tables of bolt and weld capacities are based on M20 and M22 A325M bolts and 3/4 and 7/8-inch diameter A325 bolts for bolted connections, and on E480XX electrodes for welded joints. They assume the use of detail material with a specified minimum yield strength F_y of 300 MPa.

Although based on specific arrangements of bolts and welds, the tables are general in nature and are intended to facilitate the design of all types of shear connections and not to preclude the use of other types of connections. These tables can be used by steel fabricators to prepare drawing office and shop standards, by design authorities to check fabricator standards, and by educational institutions to teach structural steel design and detailing.

The standard connections of individual fabricators will depend upon fabrication methods and material sources. They may differ from those shown in the tables regarding length and size of angles and other detail material, and gauge and pitch of bolts.

Minimum Material Thickness

Associated with the tables listing capacities of welds and capacities of bolts in bearing-type connections, is information concerning the minimum required thickness of supporting and supported material to develop the full connector capacities. These minimum thicknesses have been developed for three different values of minimum specified material yield strength (F_y = 250, 300 and 350 MPa), and were determined in the following manner.

For welded connections, for each weld size and for each value of F_y, the two equations for V_r were equated to the factored shear resistance of the web of the supported beam, and solved for the web material thickness t. The equations for V_r (Table 3(A), CSA/CAN-S16.1-M89) are $V_r = 0.67\,\phi\,A_m\,F_y$ for the base metal and

$V_r = 0.67\,\phi_w A_w X_u$ for the weld metal (with X_u = 480 MPa), and the web resistance equation is $V_r = \phi A_w F_s$ (with $F_s = 0.66\ F_y$).

For bolts in bearing-type connections, the minimum material thickness was derived by equating the bearing capacity of the material to the shear capacity of the bolts while assumming that supported beams are not coped. For webs of beams (both supporting and supported) and for webs and flanges of columns (supporting) the maximum factored bearing resistance has been assumed, namely $B_r = \phi\, t\, n\, 3d\, F_u$ (Clause 13.10.1(c)), while for framing angles, end plates and tees, an end distance *e* equal to 1.5 bolt diameters has been assumed in computing the factored bearing resistance $B_r = \phi\, t\, n\, e\, F_u$. The factored bearing resistance of the framing angles, end plates and tees is thus one-half the factored bearing resistance for the beam and column webs or the column flanges. Bolt shear values are either double or single shear as appropriate; thus for bearing type connections, the following conditions arise for the tabulated values of minimum thickness.

(a) For double angle connections the minimum thickness is that required for the web of the supported beam, for the connection angle, and for the supporting material when beams frame from two sides. When the beams frame from one side only, one-half the value listed is required for the thickness of supporting material.

(b) For end plate connections, the minimum thickness is that required for the supporting material when beams frame in from one side only. Double the value listed is required for the end plate and also for the supporting material when beams frame from two sides.

(c) For single angle connections the minimum thickness is that required for the web of the supported beam and for the supporting material when beams frame from one side only. Double the value listed is required for the connection angle and for the supporting material when beams frame from two sides.

(d) For tee connections the minimum thickness is that required for the web of the supported beam, for the flange of the tee and for the supporting material when beams frame from two sides. Double the value listed is required for the web of the tee. One-half the value listed is required for the supporting material when beams frame from one side.

DOUBLE ANGLE BEAM CONNECTIONS

Tables 3–34 and 3–35 on pages 3–60 and 3–61, respectively, list capacities of bolted and welded double angle beam connections. At the bottom of each table are values for minimum material thickness required to develop the connector capacities listed in the corresponding columns. For material thickness less than those listed, the corresponding connector capacities must be reduced by the ratio of the thickness of material supplied to the thickness of material listed. Any combination of welded or bolted legs can be selected from the tables.

Bolt capacities are based on concentric loading as tests have shown that eccentricity does not influence the ultimate strength of the bolts in connections using a single line of bolts in the web framing leg. Weld capacities include the effect of eccentricity for connection angles up to 310 mm in length. For longer connection angles the weld capacity is not reduced by the effect of eccentricity.

The connection angle length, L, has been based on a bolt pitch of 80 mm assuming an end distance of 35 mm. For connections angles with both legs welded, the angle lengths can be adjusted and capacities interpolated in accordance with the length used. Nominal minimum and maximum depths of supported beams appropriate to each length of connection angle are included. The suggested maximum depth assumes a connection length not less than half the beam depth to provide some measure of stiffness and stability. It should be recognized that these depths may not always be appropriate for a particular structure.

Table 3–34 for bolted connections lists bolt capacities for both bearing-type and slip-critical connections for three sizes of angle (width of leg and gauge dimension) and includes values for 2 to 13 bolts per vertical line based on a bolt pitch of 80 mm. For web framing legs, bolt capacities are based on the "double shear" condition and for outstanding legs on the "single shear" condition. Thus two vertical lines of bolts in the outstanding legs (one line in each angle leg) have the same capacity as one vertical line in the web framing leg. When beams are connected to both sides of the supporting material, the total bolt capacity in the outstanding legs is double that listed, provided the thickness of the supporting material is equal to or greater than that listed for the web of the supported beam.

For connection angles, the minimum required thickness to develop the bolt capacities (based on an assumed end distance of 1.5 bolt diameters) is the same as that listed for the web of the supported beam. This thickness will also provide adequate shear capacity for the angles. To ensure connection flexibility the angle thickness selected should not be greater than necessary, with a minimum thickness of 6 mm for practical reasons. When connection angles are less than about 12 mm thick, bolt threads may intercept a shear plane, in which case the bolt capacity must be reduced to 70% of the tabulated value. (See Clause 23.1.3 and Clause 13.11.2 of CSA/CAN-S16.1-M89).

Table 3–35 for welded connections lists weld capacities for the weld configurations shown, for web framing legs with welds, and outstanding legs with welds. Values are tabulated for four sizes of fillet weld and are based on the length and size (angle width W) of connection angles listed. The weld capacities for the outstanding leg assumes an angle thickness equal to the weld size plus 1 mm in determining the capacity under eccentric load.

Design of bearing-type connections for types and sizes of bolts other than those shown in Table 3–34 will be facilitated by the resistance tables on pages 3–7 to 3–11 inclusive, and for slip-critical connections by the tables on page 3–15.

Supported Beams with Copes.

Tests conducted at the Universities of Toronto and Texas have shown the importance of bolt spacing in double angle end connections when used to support beams with the top flange coped. High bearing stresses at the bolts can result in a web failure termed "Block shear" by Birkemoe and Gilmor (1978), and it may be necessary to increase the bolt pitch and/or the edge distance for bolts in the beam web.

Clause 13.4.2 of S16.1-M89 draws attention to the reduced shear resistance of coped beams (see also comments under "Factored Shear & Bearing Resistance" in Part Five), and Clause 12.3 of S16.1-M89 defines the effective net area for determining the block shear resistance.

Tests cited by Yura *et al.* (1980) have shown that the capacity of single row bolted connections computed assuming shear along the "block shear line" are conservative, but when two rows of bolts are used in the web framing leg, the effects of eccentricity should be taken into account.

References

BIRKEMOE, P.C. and GILMOR, M.I. 1978. Behaviour of bearing critical double angle beam connections. Engineering Journal, Fourth Quarter, AISC.

YURA, J.A., BIRKEMOE, P.E. and RICLES, J.M. 1980. Beam web shear connections –an experimental study. Beam-to-Column Building Connections: State of the Art, Preprint 80-179, April, ASCE.

Example 1

Bolted to beam web, bearing-type, welded to column flange.

Given:

W530x92 beam connected to flange of W250x73 column

Reaction due to factored loads = 580 kN.

Beam web thickness = 10.2 mm; column flange thickness = 14.2 mm.

G40.21-M 300W material, M20 A325M bolts, E480XX electrodes.

Solution:

Web framing legs — bolted (Table 3–34)

Vertical line with four bolts provides a nominal capacity of 840 kN.

Web thickness required for $F_y = 300$ ($F_u = 450$ MPa) steel

$= 11.7 \times 580 / 840 = 8.1$ mm < 10.2 — OK

Angle thickness required is same as required web of supported beam
= 8.1 mm — use 10

Minimum angle length required = 310 mm.

Actual bearing capacity of the web is $840 \times 10.2 / 11.7 = 732$ kN

Threads may be included in a shear plane; therefore apply 30% reduction in bolt shear capacity.

Bolt shear capacity is $0.7 \times 840 = 588$ kN > 580 — OK

Outstanding legs — welded (Table 3–35)

With L = 310 mm, W = 75 or 90 mm,

8 mm fillet welds provide a capacity of 754 kN > 580 — OK

10 mm thick angle required for bolting is also good for 8 mm fillet welds.

Use:

75x75x10 connection angles 310 mm long, four M20 A325M bolts in web framing leg and 8 mm fillet welds on outstanding legs.

Example 2

Welded to beam web, bolted to column flange, bearing-type.

Given:

Same as example 1

Solution:

Web framing legs — welded (Table 3–35)

6 mm fillet welds provide a capacity of 806 kN with angle length L = 310 mm and W = 75 mm

Web thickness required for 6 mm fillet welds, L = 310 mm

$= 10.3 \times 580 / 806 = 7.4$ mm < 10.2 — OK

Outstanding legs — bolted (Table 3–34)

Try 10 mm angle thickness (threads may be included in the shear plane).

For L = 310 mm, W = 75 or 90 mm, four bolts per vertical line,

Bearing capacity of angles is $840 \times 10 / 11.7 = 718$ kN > 580 — OK

Bolt shear capacity is $0.7 \times 840 = 588$ kN > 580 — OK

Required thickness of supporting material, beams framing from one side, is one half thickness of connection angles; therefore, column flange thickness of 14.2 mm is good.

Use:

90x75x10 connection angles, 310 mm long, 90 mm outstanding legs, g = 130 mm with eight M20 A325M bolts (2 rows of 4) and 6 mm fillet welds.

Example 3

Bolted to beam web and bolted to both sides of supporting member, supported beams not coped, bearing-type.

Given:

W530x92 beam, factored reaction 580 kN framing to both sides of web of WWF800x154 girder, web thickness 10.0 mm.

G40.21-M 300W material, M20 A325M bolts.

Solution:

Web framing legs — same as example 1

Outstanding legs — bolted to both sides of supporting member (Table 3–34).

Total reaction on girder web is $2 \times 580 = 1\ 160$ kN

For beams connected to both sides of supporting member, bolt capacities are double those listed in table, and required web thickness of supporting member is the same as that given for web thickness of supported beam.

For angle L = 310 mm, W = 75 or 90 mm, web thickness of 10 mm, four M20 A325M bolts per vertical line, bearing capacity of web

$= 2 \times 840 \times 10 / 11.7 = 1\ 440$ kN $> 1\ 160$ — OK

Use:

90x90x10 connection angles, 310 mm long, four M20 A325M bolts per vertical line in both leg framing and outstanding legs.

BOLTED DOUBLE ANGLE[1] BEAM CONNECTIONS

Table 3-34

M20, M22 A325M Bolts
3/4, 7/8 A325 Bolts

BOLT CAPACITY — EITHER LEG WITH BOLTS

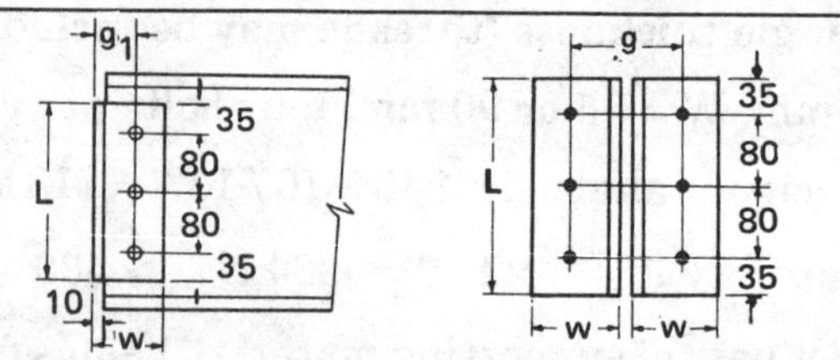

WEB FRAMING LEG   OUTSTANDING LEGS

Angle Width and Gauge		
W	g	g_1
100	140	65
90	130	60
75	100	45

Nominal depth of supported beam (mm) min.	Nominal depth of supported beam (mm) max.	Conn. Angle Length L (mm)	Bolts per Vertical Line	BEARING-TYPE CONNECTION[2] Factored Load Resistance (kN) Bolt Size 3/4	M20	M22	7/8	SLIP-CRITICAL CONNECTION[4] Specified Load Resistance (kN) Bolt Size 3/4	M20	M22	7/8
				W = 90 mm		W = 100 mm		W = 90 mm		W = 100 mm	
200	310	150	2	378	420	508	516	135	150	183	184
310	460	230	3	568	630	762	774	202	224	271	275
380	610	310	4	757	840	1020	1030	270	299	362	367
460	760	390	5	946	1050	1270	1290	337	374	452	459
530	920	470	6	1140	1260	1520	1550	404	449	542	551
610	1100	550	7	1320	1470	1780	1810	472	524	633	643
690	1200	630	8	1510	1680	2030	2060	539	598	723	734
800		710	9	1700	1890	2290	2320	607	673	814	826
900		790	10	1890	2100	2540	2580	674	748	904	918
920		870	11	2080	2310	2790	2840	741	823	994	1010
1100		950	12	2270	2520	3050	3100	809	898	1090	1100
1200		1030	13	2460	2730	3300	3350	876	972	1180	1190
				W = 75 mm		W = 90 mm		W = 75 mm		W = 90 mm	
200	310	150	2	378	420	508	516	135	150	183	184
310	460	230	3	568	630	762	774	202	224	271	275
380	610	310	4	757	840	1020	1030	270	299	362	367
460	760	390	5	946	1050	1270	1290	337	374	452	459
530	920	470	6	1140	1260	1520	1550	404	449	542	551
610	1100	550	7	1320	1470	1780	1810	472	524	633	643
690	1200	630	8	1510	1680	2030	2060	539	598	723	734
800		710	9	1700	1890	2290	2320	607	673	814	826
900		790	10	1890	2100	2540	2580	674	748	904	918
920		870	11	2080	2310	2790	2840	741	823	994	1010
1100		950	12	2270	2520	3050	3100	809	898	1090	1100
1200		1030	13	2460	2730	3300	3350	876	972	1180	1190

Specified Minimum Yield Strength of Material (MPa)	Minimum Required Web Thickness of Supported Beam[3] (mm)			
F_y = 250 (F_u = 400 MPa)	12.4	13.1	14.4	14.5
F_y = 300 (F_u = 450 MPa)	11.0	11.7	12.8	12.9
F_y = 350 (F_u = 480 MPa)	10.3	10.9	12.0	12.1

1. Connection angles are assumed to be material with F_y = 300 MPa.
2. When threads intercept a shear plane, use 70% of the values tabulated for bearing-type connections.
3. For connection angles, and for supporting material with beams framing from both sides, minimum required thickness is equal to tabulated values for web thickness of supported beam, and for supporting material with beams framing from one side is one-half the tabulated values.
4. Tabulated values for slip-critical connections assume Class A contact surfaces with k_s = 0.33.

WELDED DOUBLE ANGLE BEAM CONNECTIONS [a]

Table 3-35

E480XX Fillet Welds

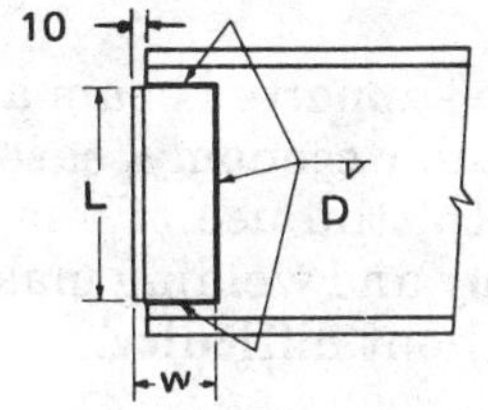

WEB FRAMING LEG WITH WELDS

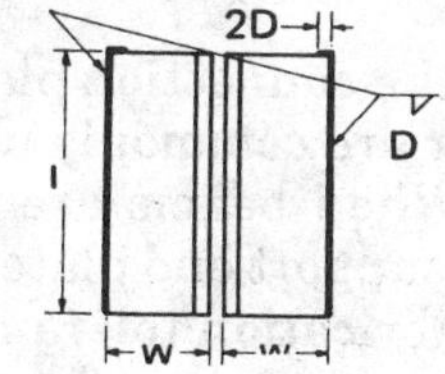

OUTSTANDING LEG WITH WELDS

WELD CAPACITY Factored Load Resistance (kN)				WELD CAPACITY Factored Load Resistance (kN)				Conn. Angle Length L	Nominal Depth of Supported Beam (mm)	
Fillet Size D (mm)				Fillet Size D (mm)						
5	6	8	10	5	6	8	10	(mm)	min.	max.
Angle Width W = 75 mm				Angle Width W = 90 mm						
394	473	631	789	135	157	202	246	150	200	310
550	660	880	1100	303	356	460	563	230	310	460
672	806	1080	1340	473	568	754	933	310	380	610
794	953	1270	1590	596	715	953	1190	390	460	760
916	1100	1470	1830	718	861	1150	1440	470	530	920
1040	1250	1660	2080	840	1010	1340	1680	550	610	1100
1160	1390	1860	2320	962	1150	1540	1920	630	690	1200
1280	1540	2050	2560	1080	1300	1740	2170	710	800	
1400	1690	2250	2810	1210	1450	1930	2410	790	900	
1530	1830	2440	3050	1330	1600	2130	2660	870	920	
1650	1980	2640	3300	1450	1740	2320	2900	950	1100	
1770	2130	2830	3540	1570	1890	2520	3150	1030	1200	
Angle Width W = 65 mm				Angle Width W = 75 mm						
382	459	612	765	146	170	218	265	150	200	310
519	623	831	1040	322	380	491	602	230	310	460
641	770	1030	1280	473	568	758	947	310	380	610
764	916	1220	1530	596	715	953	1190	390	460	760
886	1060	1420	1770	718	861	1150	1440	470	530	920
1010	1210	1610	2020	840	1010	1340	1680	550	610	1100
1130	1360	1810	2260	962	1150	1540	1920	630	690	1200
1250	1500	2000	2500	1080	1300	1740	2170	710	800	
1370	1650	2200	2750	1210	1450	1930	2410	790	900	
1500	1800	2390	2990	1330	1600	2130	2660	870	920	
1620	1940	2590	3240	1450	1740	2320	2900	950	1100	
1740	2090	2780	3480	1570	1890	2520	3150	1030	1200	
Minimum Required Web Thickness of Supported Beam (mm)				Minimum Thickness of Supporting Material with Beam Attached One Side[b]				Specified Minimum Yield Strength of Material (MPa)		
10.0	12.0	16.0	20.0	5.0	6.0	8.0	10.0	F_y = 250 (MPa)		
8.6	10.3	13.8	17.2	4.3	5.2	6.9	8.6	F_y = 300 (MPa)		
7.4	8.8	11.8	14.7	3.7	4.4	5.9	7.4	F_y = 350 (MPa)		

a. Connection angles are assumed to be material with F_y = 300 MPa.

b. For supporting material with beams framing from both sides, use double the tabulated value.

END PLATE CONNECTIONS

End plate connections with the connection plate welded to the supported beam and bolted to the supporting member are commonly used because of their economy, ease of fabrication, and performance. When beams are saw cut to length, the use of simple jigging procedures to locate and support end plates during assembly and welding makes it possible to meet the tighter fabrication tolerances required without difficulty.

Research on simple beam end plate shear connections has shown that their strength and flexibility compare favourably with double angle shear connections for similar material thickness, depth of connection, and arrangement of bolts (gauge and pitch). For practical reasons it is suggested that the minimum thickness of end plate be 6 mm, and for adequate flexibility that the maximum thickness be limited to 10 mm. The gauge dimension g should preferably be between 100 mm and 150 mm for plates up to 10 mm thick, but may be as low as 80 mm for minimum thickness plates with F_y not greater than 300 MPa.

Table 3–36 lists the capacities of bolts and welds for typical end plate connections with from 2 to 8 bolts per vertical line, together with the minimum thickness of supporting and supported material to develop the full capacity of the bolts and welds respectively. The table also includes reduction factors for bolt capacity in a bearing-type connection for end plate thicknesses of 6, 7, 8 and 10 mm, assuming an end distance of 1.5 bolt diameters.

For added safety during erection, clipped end plates with one upper corner of the end plate removed may be used. Tests at Queen's University demonstrated that clipped end plate connections have similar moment-rotation characteristics to unclipped end plate connections. Therefore weld capacities in Table 3-36 may be used directly for design, but tabulated bolt values must be reduced by the value of a single bolt.

Table 3-36 includes bolt capacities for slip-critical joints for those situations where bearing-type connections are not suitable.

References

VAN DALEN, K., and MACINTYRE, J.R. 1988. The rotational behaviour of clipped end plate connections. Canadian Journal of Civil Engineering, **15**(1), February.

VAN DALEN, K., and MACINTYRE, J.R. 1987. The shear strength of clipped end plate connections. Queen's University.

Example

Given:

W410x60 beam framing into web of WWF700x152 girder

Factored reaction 325 kN

Beam web thickness = 7.7 mm, girder web thickness = 11.0 mm, G40.21-M 300W steel, M20 A325M bolts, E480XX electrodes

Solution:

Assume 7 mm thick end plate; effective bolt capacity (bolts bearing on end plate) for 3 bolts per vertical line is $630 \times 0.60 = 378$ kN > 325 — OK

For 230 mm long end plate, weld capacity for 5 mm fillet welds made with E480XX electrodes is 350 kN > 325 — OK

Minimum length of end plate required for 7.7 mm web thickness (300W steel)

$= 230 \times 8.6 / 7.7 = 257$ mm. Use 260 mm

If beams were framing from both sides, the required web thickness for the girder would be twice the listed value, pro-rated for the actual load:

$2 \times 5.8 \times 325 / 630 = 6.0$ mm < 11.0 — O.K.

Use:

End plate 160x7 x 260 mm long connected to web of the supported beam with 5 mm E480XX fillet welds, and six M20 A325 bolts (2 rows of 3 at 100 mm gauge) with increased end distance and /or increased pitch.

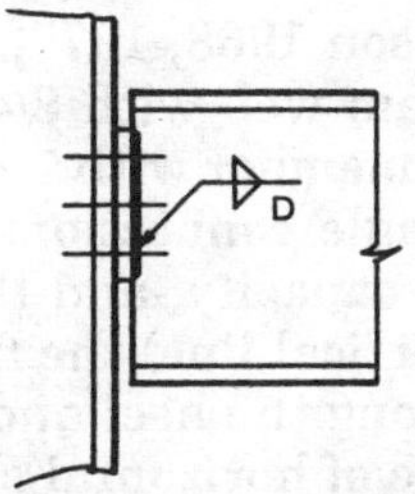

END PLATE BEAM CONNECTIONS
Table 3-36

M20, M22, A325M Bolts
3/4, 7/8, A325 Bolts
E480XX Fillet Welds

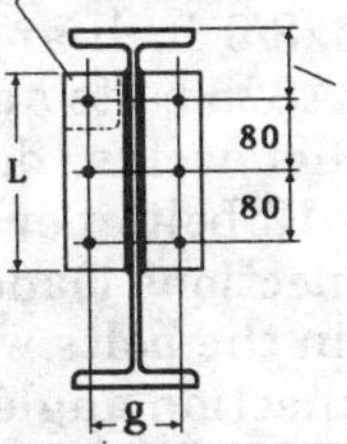

Bolts per Vertical Line	BEARING-TYPE CONNECTION [1] Factored Load Resistance [2] (kN)				WELD CAPACITY Factored Load Resistance (kN)			Connection Plate Length (mm)
	Bolt Size				Fillet Size D (mm)			
	3/4	M20	M22	7/8	5	6	8	
2	378	420	508	516	228	276	366	150
3	568	630	762	774	350	423	561	230
4	757	840	1020	1030	471	570	756	310
5	946	1050	1270	1290	593	718	952	390
6	1140	1260	1520	1550	714	865	1150	470
7	1320	1470	1780	1810	836	1010	1340	550
8	1510	1680	2030	2060	958	1160	1450	630

Material F_u (MPa)	Minimum Required Thickness of Supporting Material with Beams Framing From One Side [3] (mm)				Minimum Required Web Thicknessof Supported Beam (mm)			Material F_y (MPa)
400	6.2	6.6	7.2	7.3	10.0	12.0	16.0	250
450	5.5	5.8	6.4	6.4	8.6	10.3	13.8	300
480	5.2	5.5	6.0	6.0	7.4	8.8	11.8	350

End Plate Thickness (mm)	Reduction Factors for Bolt Capacities To Develop Bearing Resistance of End Plate (F_y = 300 MPa)			
6	0.55	0.52	0.47	0.47
7	0.64	0.60	0.55	0.55
8	0.70	0.69	0.63	0.63
10	0.91	0.86	0.78	0.78

Bolts per Vertical Line	SLIP-CRITICAL CONNECTION [1] Specified Load Resistance [4] (kN)			
	Bolt Size			
	3/4	M20	M22	7/8
2	135	150	183	184
3	202	224	271	275
4	270	299	362	367
5	337	374	452	459
6	404	449	542	551
7	472	524	633	643
8	539	598	723	734

1 For clipped end plate connections reduce tabulated bolt capacities by the value of one bolt.

2 When threads intercept a shear plane, use 70% of the values tabulated for bearing connections.

3 Minimum required thickness of supporting material with beams framing from both sides is double that listed.

4 Tabulated values for slip-critical connections assume Class A contact surfaces (k_s = 0.33).

SINGLE ANGLE BEAM CONNECTIONS

For some applications, single angle connections provide a satisfactory alternative to double angle or end plate connections. They are particularly suitable where limited access prevents the erection of beams with double angle or end plate connections, and where speed of erection is a primary consideration.

The connection angle may be either bolted or welded to the supporting and supported members; however, usual practice involves shop fillet welding to the supporting member and field bolting to the web of the supported beam.

Tests carried out at the University of British Columbia (Lipson 1968, 1977, 1980) using 4x3x3/8 inch angles with the 4-inch leg bolted to the beam web with 3/4 inch diameter A325 bolts and the 3-inch leg welded to the supporting member with 1/4-inch E70XX fillet welds, demonstrated that welded-bolted single angle connections with from 2 to 12 bolts per vertical line possess adequate rotational capacity, and that in those connections loaded to ultimate capacity (2 to 8 bolts per vertical line) the failure occurred in the bolts when the weld pattern included welding along the heel and ends of the connection angle. The tests also demonstrated that the use of horizontal slotted holes in the connection angle reduced the moment at the bolts without affecting the ultimate capacity of the connection.

Table 3–37 is based on this research, and assumes the use of 100x75x10 connection angles with the 75 mm leg welded to the supporting member and the 100 mm leg bolted to the supported web. Bolt capacities for bearing-type connections are provided for M20 and M22 A325M bolts and 3/4 and 7/8 inch A325 bolts based on their factored shear resistance for the appropriate number of bolts. The weld capacities have been established by assuming that a connection with 1/4-inch fillet welds has the same shear capacity as the 3/4-inch A325 bolts and then pro-rating for the three sizes of flllet welds shown in the table (i.e. weld capacity = V_r (3/4-inch bolts) x D/6.35, where D is the fillet weld size in mm).

References

LIPSON, S.L. 1980. Single-angle welded-bolted beam connections. Canadian Journal of Civil Engineering, **7**(2), June.

LIPSON, S.L. 1977. Single-angle welded-bolted connections. Structural Division Journal, ASCE, March.

LIPSON, S.L. 1968. Single-angle and single-plate beam framing connections. Proceedings, Canadian Structural Engineering Conference, C.I.S.C., Willowdale, Ontario, February: 141–162.

Example

Single angle welded-bolted beam connection (Table 3–37)

Given:

W410x60 beam, factored reaction 290 kN.

Beam web thickness 7.7 mm, 100x75x10 connection angle.

G40.21-M 300W material, M20 A325M bolts, E480XX electrodes.

Solution:

With threads included in the shear plane, bolt capacity with four M20 A325M bolts

is $0.7 \times 420 = 294$ kN > 290 — OK

Web thickness required is $5.8 \times 290 / 420 = 4.0$ mm < 7.7 — OK

Angle thickness required is $2 \times 4.0 = 8.0$ mm < 10 — OK

Angle length required for 4 bolts is 310 mm, and weld capacity using 5 mm E480XX fillet welds is 298 kN > 290 — OK

Use:

100x75x10 connection angle, 310 mm long, 75 mm leg welded to supporting member with 5 mm E480XX fillet welds; 100 mm leg bolted to web of supported beam with four M20 A325M bolts.

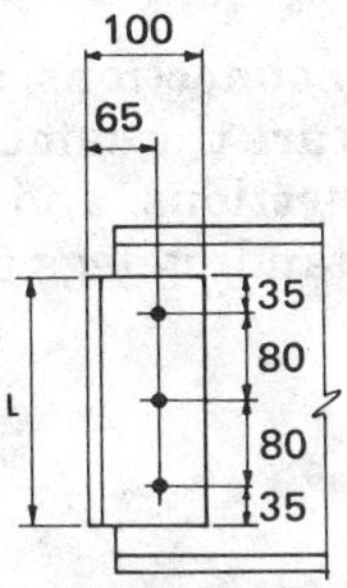

WEB FRAMING LEG
(Bolted to supported web)

SINGLE ANGLE BEAM CONNECTIONS

Table 3-37

M20, M22 A325M Bolts
3/4, 7/8 A325 Bolts
E480XX Fillet Welds

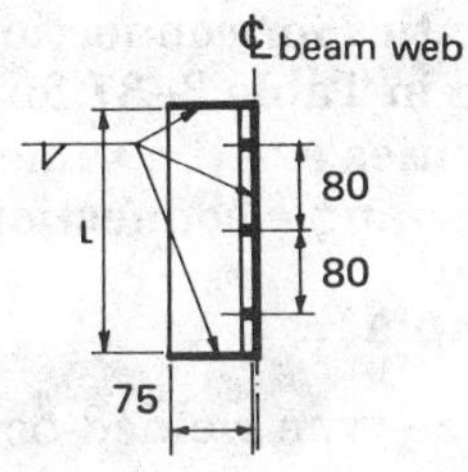

OUTSTANDING LEG
(Welded to supporting Material)

Bolts per Vertical Line	BEARING-TYPE CONNECTION Factored Load Resistance (kN)[1]				WELD CAPACITY Factored Load Resistance (kN)			Connection Angle Length L (mm)
	Bolt Size				Fillet Size D (mm)			
	3/4	**M20**	**M22**	7/8	5	6	8	
2	189	**210**	**254**	258	149	179	238	150
3	284	**315**	**381**	387	224	268	358	230
4	378	**420**	**508**	516	298	357	476	310
5	473	**525**	**635**	645	372	447	596	390
6	568	**630**	**762**	774	447	537	716	470
7	662	**735**	**889**	903	521	626	834	550
8	756	**840**	**1020**	1030	595	714	952	630
Material F_u (MPa)	Minimum Required Web Thickness of Supported Beam[2] (mm)				Minimum Required Thickness of Supporting Material With Beams Framing From One Side (mm)			Material F_y (MPa)
400	6.2	6.6	7.2	7.3	5.0	6.0	8.0	250
450	5.5	5.8	6.4	6.4	4.3	5.2	6.9	300
480	5.2	5.5	6.0	6.0	3.7	4.4	5.9	350

1. When threads intercept a shear plane, use 70% of the values tabulated.
2. For connection angles minimum required thickness is double the tabulated values.

TEE-TYPE BEAM CONNECTIONS

Tee-type beam connections combine some of the characteristics of single angle connections with the web-framing leg bolted in single shear, and of double-angle connections with the outstanding legs welded to the supporting member.

Their main advantage is speed and ease of erection. They are also commonly used where hole making in the supporting member is undesirable (connections to HSS columns), and to avoid coping the bottom flange of the supported beam for erection purposes.

Costs are generally higher than for other types of simple beam connections because of the higher costs of fabricating the tee-sections.

Table 3–38 lists bolt capacities for bearing-type web-framing connections, and weld capacities for connections of outstanding legs. The bolt capacities are the same as those listed in Table 3–37 for the web-framing legs of single angle connections, and the weld capacities are the same as those listed in Table 3–34 for the outstanding legs of welded double-angle connections.

Example

Tee-type welded-bolted beam connections (Table 3–38)

Given:

W410x60 beam, Factored reaction 325 kN.

Column — HSS 254 x 254 x 11

Beam web thickness 7.7 mm

G40.21-M 300W material, M20 A325M bolts, E480XX electrodes.

Solution:

Try tee cut from W200x59 beam; web thickness = 9.1 mm.

These thicknesses of beam web and tee web will result in threads intercepting the shear plane.

Five bolts per vertical line provide a capacity of $0.7 \times 525 = 367$ kN.

Beam web thickness required for F_y of 300 MPa is $5.8 \times 367 / 525$

$= 4.1$ mm < 7.7 — OK

Tee web thickness required is $2 \times 4.1 = 8.2$ mm < 9.1 mm — OK

Length of tee required for 5 bolts is 390 mm, and weld capacity for 5 mm fillet welds is 596 kN > 325 — OK

Use:

Tee cut from W200 x 59, 390 mm long, five M20 A325M bolts connecting webs of beam and tee, and 5 mm E480XX fillet welds to supporting material.

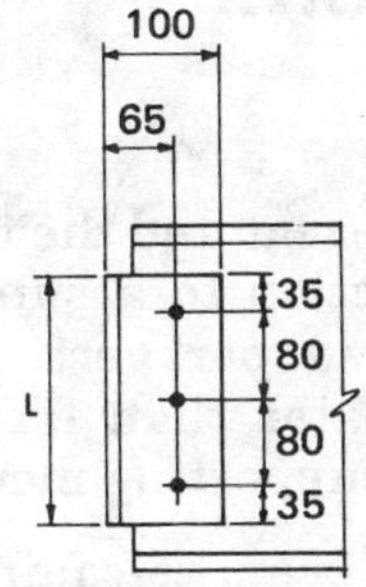

WEB FRAMING LEG
(Bolted to supported Web)

TEE-TYPE BEAM CONNECTIONS
Table 3-38

M20, M22 A325M Bolts
3/4, 7/8 A325 Bolts
E480XX Fillet Welds

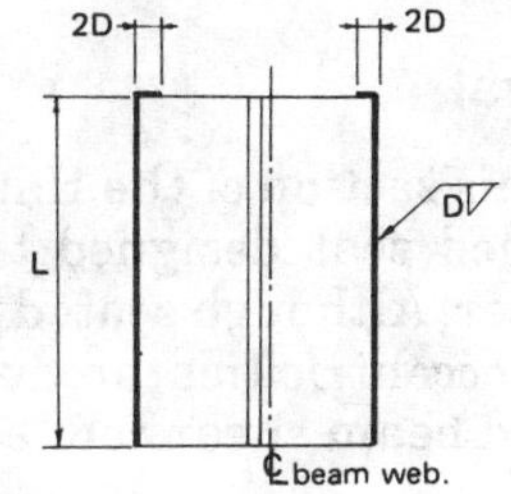

OUTSTANDING LEG
(Welded to supporting Material)

Bolts per Vertical Line	BEARING-TYPE CONNECTIONS Factored Load Resistance (kN)[1]				WELD CAPACITY Factored Load Resistance (kN)			Connection Tee Length L (mm)
	Bolt Size				Fillet Size D (mm)			
	3/4	M20	M22	7/8	5	6	8	
2	189	210	254	258	135	157	202	150
3	284	315	381	387	303	356	460	230
4	378	420	508	516	473	568	754	310
5	473	525	635	645	596	715	953	390
6	568	630	762	774	718	861	1150	470
7	662	735	889	903	840	1010	1340	550
8	756	840	1020	1030	962	1150	1540	630
Material F_u (MPa)	Minimum Required Web Thickness of Supported Beam[2] (mm)				Minimum Required Thickness of Supporting Material With Beams Framing from One Side (mm)			Material F_y (MPa)
400	6.2	6.6	7.2	7.3	5.0	6.0	8.0	250
450	5.5	5.8	6.4	6.4	4.3	5.2	6.9	300
480	5.2	5.5	6.0	6.0	3.7	4.4	5.9	350

1. When threads intercept a shear plane, use 70% of the values tabulated.
2. Minimum required thickness of Tee-web is double the tabulated values.

SEATED BEAM SHEAR CONNECTIONS

General

This section of the Handbook deals with the unstiffened angle seat and the tee-type stiffened seat designed to provide a simple beam shear connection to a supporting member. Although seated beam shear connections are designed to support vertical loads only, eccentricities produced by these connections may be greater than for simple framed beam shear connections and can influence the design of supporting members.

Seated beam shear connections are most commonly used at beam-to-column supports. If used at beam-to-girder supports, the girder web must be checked for adequate local stability and resistance. Economy with seated beam shear connections results from simple shop fabrication, together with ease and speed of field erection.

The unstiffened angle seat consists of a relatively thick angle either shop welded or bolted to the supporting member. When the supporting member is a column web, access for welding may be restricted. Load capacity of an unstiffened angle seat is limited by the angle thickness. This capacity can be increased by stiffening the angle, however stiffened angle seats are more expensive to fabricate, and stiffened seats using tee-stubs built up from plate are usually more economical. Stiffened seats designed for large loads are generally referred to as brackets and are beyond the scope of this section.

A seated beam must be stabilized with a flexible clip angle attached either to the top flange of the beam or to the beam web near the top of the beam. The clip angle must be thin enough to permit end rotation of the beam. Either welds or bolts can be used to connect the clip angle to the beam and supporting member. When welds are used, the fillet welds should be located along the toes of the angle.

Unstiffened Angle Seats

The capacity of unstiffened angle seats depends on the bending capacity of the seat angle, and is governed by the web thickness and effective bearing length of the supported beam. When the vertical leg of the seat angle is welded to the supporting member, the top of the angle is restrained by the welds, so that the capacity of the angle seat is assumed to be limited by the bending capacity of the outstanding leg. When the vertical leg is bolted to the supporting member, the top of the angle is not restrained by the bolts, so it is assumed that bending in the vertical leg, rather than the outstanding leg, controls the bending capacity of the angle seat.

Tables 3–39 and 3–40 list capacities for welded and bolted unstiffened angle seats of various thickness for seat lengths of 180 mm and 230 mm, assuming beams and seat angles of G40.21-M 300W material (F_y = 300 MPa) and welds made with E480XX electrodes. Capacities are based on the design models illustrated in the tables, with no allowance made for possible restraint provided by any connection between the seat and the bottom flange of the supported beam.

Tabulated values assume a seat angle fillet radius of 10 mm, and are based on a design clearance of 10 mm between the end of the supported beam and the face of the supporting member for seat angles of 10 mm or less in thickness, and a design clearance of 20 mm for seat angles greater than 10 mm in thickness.

It should be noted that beam web bearing capacities in Tables 3–39 and 3–40 are based on yielding capacity of the beam webs, Clause 15.9(b)(i) of CAN/CSA-S16.1-M89, which considers web thickness but not depth of the section. Web crippling of individual beams should be checked according to Clause 15.9(b)(ii) as this could govern the beam bearing capacity.

Vertical Leg Welded to Supporting Member

Table 3–39 for welded seats, tabulates the beam web bearing capacity for various angle thicknesses and beam web thicknesses, and the vertical leg connection capacity for various weld sizes and angle vertical leg lengths. The beam web bearing capacity was calculated with Clause 15.9(b)(i) by assuming $k = 3w_b$ and equating the value to the plastic bending capacity of the angle outstanding leg. The vertical leg capacity was determined with the model illustrated at the bottom of the table. The welds are assumed to carry the vertical load and a tension force computed from the effective eccentricity. Compression from the eccentric moment is assumed carried by contact at the middle of the angle. Vertical bending capacity of the angle is also checked.

Vertical Leg Bolted to Supporting Member

Table 3–40 for bolted seats tabulates the beam web bearing capacity and the seat angle capacity based on the model illustrated at the top of the table. Values below the solid and broken lines are governed by the seat angle capacity for seat angle lengths of 180 mm and 230 mm respectively. Table 3–40 also tabulates the bolt joint capacity for four sizes of bolt, for two or four bolts per seat angle for the conditions of threads excluded and threads intercepted.

Example

Given:

W530x82 beam, factored reaction 190 kN, G40.21-M 300W steel.

Beam web thickness 9.5 mm, flange width 209 mm.

Solution:

(a) *Unstiffened angle seat welded to supporting member*

Seat angle thickness for beam web bearing capacity:

Assume the end clearance is 20 mm (therefore angle thickness required is more than 10 mm). Interpolate in Table 3–39 for beam web thickness of 9.5 mm with L = 230 (to permit the 209 mm flange to be welded to the seat). A 13 mm thick angle provides a beam web bearing capacity of

(213 + 256) / 2 = 234 kN > 190 — OK

Vertical leg connection:

For an angle thickness of 13 mm with a vertical leg of 150 mm and a conservatively assumed eccentricity of 40 mm, 8 mm fillet welds provide connection capacity of

229 kN > 190 — OK

Use 150x100x13 seat angle 230 mm long with 150 mm leg welded to supporting member with 8 mm E480XX fillet welds on each side of the vertical leg.

(b) *Unstiffened angle seat bolted to supporting member*

Seat angle thickness and beam web bearing capacity:

From Table 3–40, a 16 mm seat angle 230 mm long provides an angle leg flexural capacity of 190 kN and a beam web bearing capacity of

(228 + 276) / 2 = 252 kN > 190 — OK

4 – M20 bolts threads excluded will provide a bolt joint capacity of 333 kN

Use 150x100x16 seat angle 230 mm long with 150 mm leg bolted to supporting member with 4 – M20 A325M bolts.

WELDED UNSTIFFENED ANGLE SEATS
TABLE 3-39

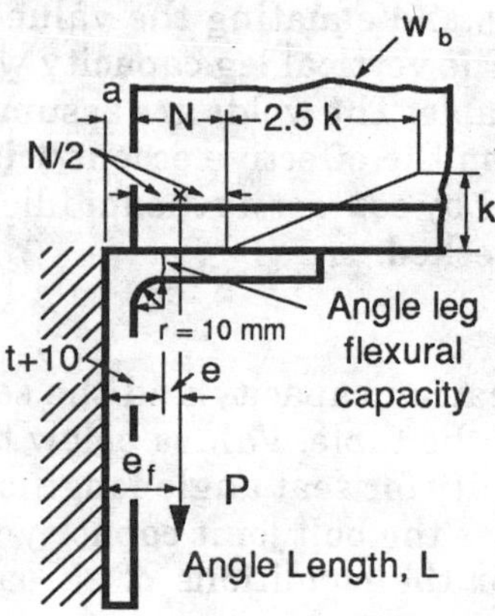

E480XX Fillet Welds
G40.21-M 300W Steel

Web bearing capacity, $P = 1.1\,\phi\, w_b\,(N+2.5k)\,F_y$

Angle leg flexural capacity, $P = \dfrac{(L\,t^2/4)\,\phi\,F_y}{N/2 + a - t - r}$

where, N is solved by equating web bearing capacity to angle leg flexural capacity, but $N \geq 2(r+t-a)$

For $t \leq 10$ then $a = 10$ | $k = 3\,w_b$
For $t > 10$ then $a = 20$ | $r = 10$ mm

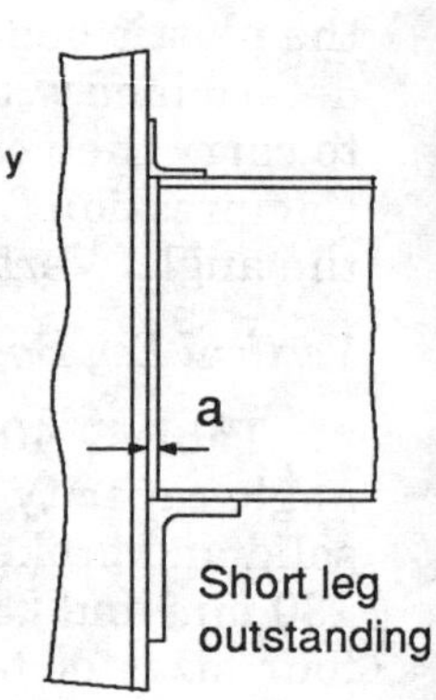

BEAM WEB BEARING CAPACITY - FACTORED LOAD RESISTANCE (kN)

Angle Thickness (mm)		8		10		13		16		20	
Angle Length (mm)		180	230	180	230	180	230	180	230	180	230
Beam Web Thickness (mm)	5	102	107	116	123	117	126	140	151	170	185
	6	130	135	146	152	142	152	168	180	202	218
	7	162	167	179	186	171	182	199	212	236	253
	8	199	204	217	223	205	215	234	247	274	291
	9			259	265	242	252	273	287	316	333
	10			306	312	284	294	317	330	362	379
	11					330	340	364	377	412	429
	12					381	390	417	429	466	484

VERTICAL LEG CONNECTION (kN)

Fillet Weld, D (mm)		6		6		8		8		10	
Angle Length (mm)		180	230	180	230	180	230	180	230	180	230
Angle Vertical Leg Length (mm)	90	94.2	73.7	115	90.2	149	116				
	100	118	92.5	141	114	174	144				
	125	181	141	184	174	231	221	218	218		
	150	235	209	229	229	289	289	274	274	326	326
	200			318	318	406	406	390	390	469	469
Assumed Eccentricity		25 mm		30 mm		40 mm		50 mm		60 mm	

Tensile Resistance $= 2\left\{0.67\,\phi_w\left[\dfrac{(l-a)\,D}{\sqrt{2}}\right]0.48\right\} = \dfrac{P\,e_f}{l/2}$

Shear Resistance $= 2\left\{0.67\,\phi_w\left[\dfrac{a\,D}{\sqrt{2}}\right]0.48\right\} = P$

Solving, $P = \dfrac{0.3047\,l^2\,D}{e_f + l}$ but $P \leq \dfrac{4\,\phi S_x F_y}{L}$ based on smallest L

Bearing Resistance $= 25\,a\,[\,1.5\,\phi F_y] > \dfrac{P\,e_f}{l/2}$

All variables in the expression are in (kN) and (mm) units

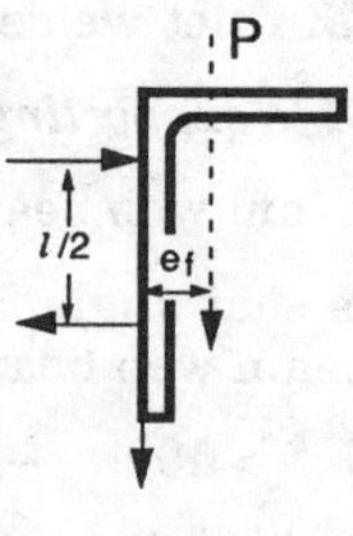

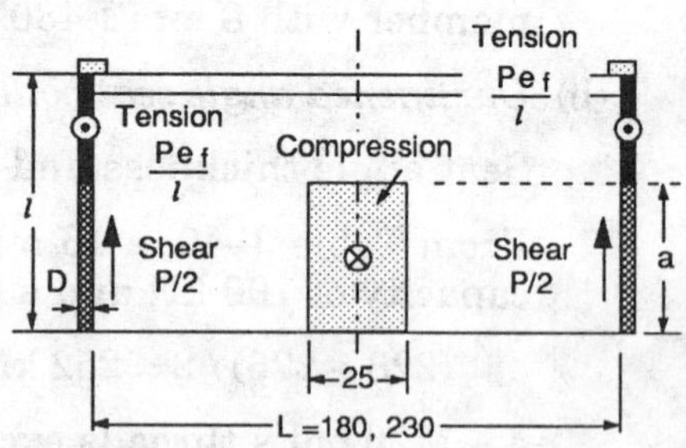

BOLTED UNSTIFFENED ANGLE SEATS

Table 3-40

M20, M22 A325M Bolts
3/4, 7/8 A325 Bolts
G40.21-M 300W Steel

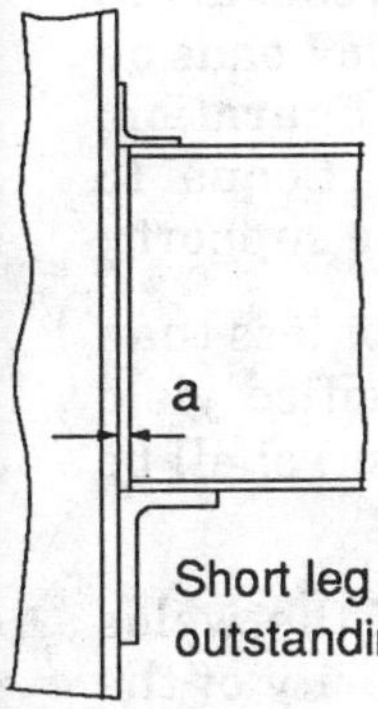

Short leg outstanding

Web bearing capacity, $P = 1.1\ \phi\ w_b\ (N+2.5k)\ F_y$

Angle leg flexural capacity, $P = \dfrac{(L t^2/4)\ \phi F_y}{N/2 + a - t/2}$

where, $N+a = \sqrt{2}\left[\sqrt{2}\,(t+r) - r\right]$

For $t \leq 10$ then $a = 10$ | $k = 3\ w_b$
For $t > 10$ then $a = 20$ | $r = 10$ mm

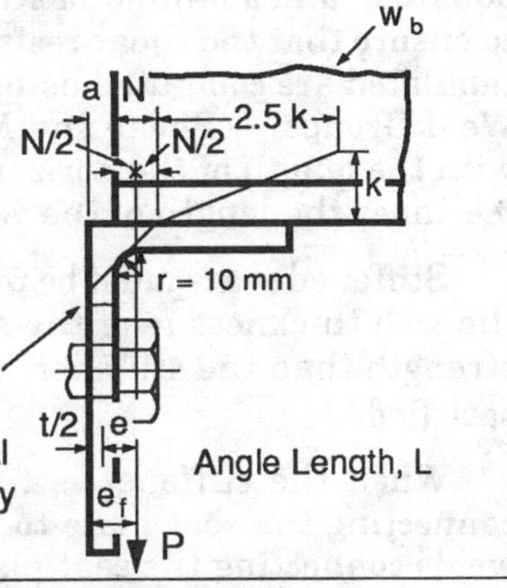

BEAM WEB BEARING CAPACITY - FACTORED LOAD RESISTANCE (kN)

Angle Thickness (mm)		8	10	13	16	20
Beam Web Thickness (mm)	5	73.3	79.2	73.3	82.2	94.1
	6	101	108	101	112	126
	7	134	142	134	146	163
	8		180	171	185	204
	9			212	228	250
	10				276	300
	11					354
	12					
Bearing Length, N (mm)		11.9	15.9	11.9	17.9	25.9
Vert. Leg Eccentricity, e		11.9	12.9	19.4	20.9	22.9

Angle Length (mm)	180	230	180	230	180	230	180	230	180	230
Angle leg flexural Capacity (kN)	**65.2**	**83.3**	**94**	**120**	**106**	**135**	**149**	**190**	**212**	**271**

VERTICAL LEG CONNECTION (kN)

Bolt Size		3/4		M20		M22		7/8	
No. of Bolts		2	4	2	4	2	4	2	4
Bolt Joint Capacity (kN)	threads excluded	150	301	167	333	200	401	206	411
	threads intercepted	127	254	141	281	169	338	174	347

Angle length, L

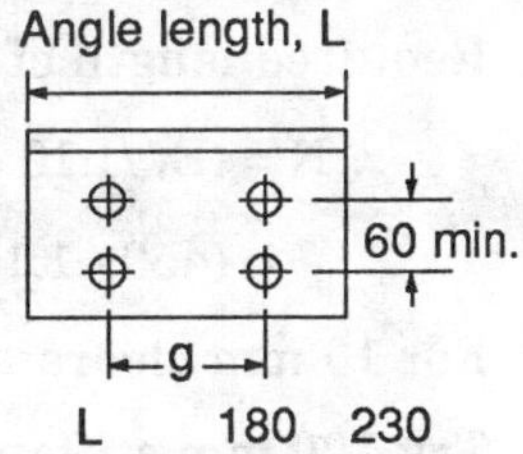

L	180	230
g	100	130

Bolt Joint Capacity = $n\sqrt{\beta(0.5\ A_b F_u)^2/(1+\beta)}$ where, n = number of bolts, $\beta = 0.69$ for threads excluded case, $\beta = 0.41$ for threads intercepted case. Design expression assumes that shear-tension ratio of bolts is 1.0. See Table 3-1 for values of $A_b F_u$

Values below the solid and broken lines are governed by seat angle capacity for seat angle lengths of 180 mm and 230 mm respectively.

Table 3–41 lists factored resistances of stiffened seats for the T-shaped weld configuration shown. Capacities are based on the use of E480XX electrodes, steel F_y = 300 MPa, and a minimum stiffener thickness *t* equal to 1.7 times the fillet weld leg size to ensure that the shear resistance of the stiffener is not exceeded. Factored resistances tabulated are computed using the theory outlined on page 3–51 for "Eccentric Loads on Weld Groups — Shear and Moment", modified for the tee-shaped weld configuration, with the length of the horizontal weld connecting the seat plate to the support equal to 0.4 times the length of the vertical weld connecting the stiffener plate to the support.

Stiffened seats must be proportioned so that the stiffener thickness *t* is not less than the web thickness *w* of the supported beam. If the beam has a higher specified yield strength than the stiffener, the relationship, $t \times F_y$ (stiffener) = $w \times F_y$ (beam) shall be satisfied.

When the stiffener is fitted to bear against the seat, the capacity of the welds connecting the seat plate to the stiffener shall be at least equal to the capacity of the welds connecting the seat plate to the supporting member. Welds or bolts may be used to connect the supported beam to the seat and for attachment of the clip angle required to stabilize the beam. See the sketches in Table 3–41 for the general arrangement.

When stiffened seats are in line on opposite sides of a column web, the size of the vertical fillet welds (for E480XX electrodes and F_y =300 MPa material) shall not exceed F_y / 515 times the thickness of the column web, so as not to exceed the shear resistance of the column web. As an alternative to limiting the weld size, a longer seat may be used to reduce the shear stresses in the column web.

Example

Given:

W530x82 beam with a factored reaction of 450 kN

Web thickness = 9.5 mm, flange width = 209 mm, k distance = 29 mm

Connected to web of W310x118 column, web thickness = 11.9 mm

Design stiffened welded seat for beams connected to both sides of column web.

G40.21-M 300W steel, E480XX electrodes

Solution:

(a) *Vertical stiffener*

Required length of bearing: $B_r = 1.10\,\phi\, w\,(N + 2.5k)\,F_y$ (Clause 15.9(b)(i))

$$\therefore N = (B_r / 1.10\,\phi\, w\, F_y) - 2.5\,k$$

$$= (450 / 1.10 \times 0.9 \times 9.5 \times 0.300) - (2.5 \times 29) = 87 \text{ mm}$$

For 10 mm clearance, minimum stiffener width = 87 + 10 = 97 mm

Try 100 mm stiffener width

For stiffeners both sides of column web, maximum effective weld size so that shear resistance of column web is not exceeded is

$$11.9 \times 300 / 515 = 6.93 \text{ mm}$$

Minimum stiffener thickness for shear is $1.7 \times 6.93 = 11.9$ mm

Try 12 mm stiffener with 6 mm fillet welds.

From Table 3–41, with w = 100 mm, 6 mm fillet welds and L = 275 mm, capacity provided is 500 kN > 450 — OK

Check b/t = 100/12 = 8.3 $< 200/\sqrt{F_y} = 11.6$ — OK (S16.1-M89, Table 1)

Use 12x100 stiffener x 275 mm long welded to column web with 6 mm fillet welds.

(b) *Horizontal Seat Plate*

Try 12 mm plate and 6 mm fillet welds (same as vertical stiffener)

Minimum length of weld required to attach seat plate to column web is

$0.4\,L = 0.4 \times 275 = 110$ mm

Minimum seat plate length, including clearances for horizontal stiffener welds is

$110 + (2 \times 6) + 12 = 134$ mm

Minimum length of seat plate assuming beam bolted to seat is

beam flange width = 209 min

Minimum length of seat plate assuming beam welded to seat is

$209 + 2\,(2 \times 6) = 233$ mm

Use 12x100 seat plate x 210 mm long welded to the column web with 6 mm fillet welds on underside of seat, and bolted to bottom flange of the beam with two M20 A325M bolts.

(c) *Weld between stiffener and seat plate*

Minimum length of weld required = 110 mm (same as between seat plate and column web) for 6 mm fillets

Length available is $2 \times 100 = 200$ mm — OK

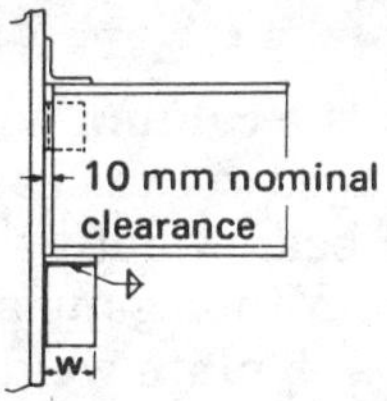

STIFFENED SEATED BEAM CONNECTIONS

Table 3-41

E480XX Electrodes

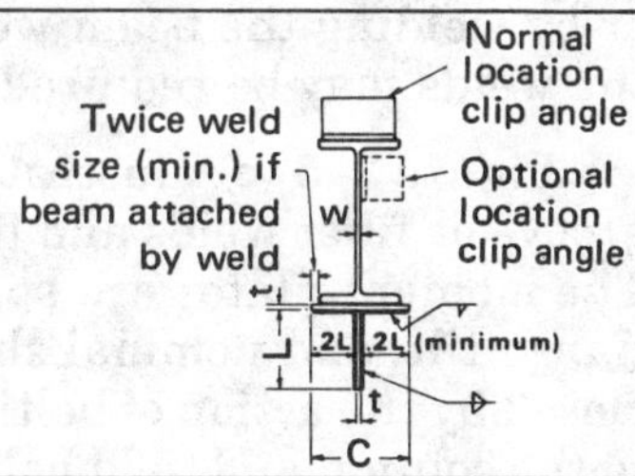

FACTORED RESISTANCE OF WELDS (kN)

Seat Width W (mm)	Fillet Size* D (mm)	Length of Stiffener L (mm) 150	175	200	225	250	275	300	325	350	400	450
100 $e_f = 60$	6	161	220	286	361	444	500	545	591	636	727	818
	8	212	288	378	480	590	666	727	788	848	969	1090
	10	260	356	468	593	731	833	909	985	1060	1210	1360
	12	307	422	556	704	869	1000	1090	1180	1270	1450	1640
125 $e_f = 72.5$	6	133	182	239	301	372	449	533	591	636	727	818
	8	175	240	316	397	490	594	707	788	848	969	1090
	10	215	297	388	493	611	738	878	985	1060	1210	1360
	12	254	351	461	586	727	879	1050	1180	1270	1450	1640
150 $e_f = 85$	6	114	155	203	258	320	385	458	537	621	727	818
	8	149	205	269	342	421	509	607	710	820	969	1090
	10	183	252	332	422	521	631	752	880	1020	1210	1360
	12	217	299	395	501	621	755	900	1050	1220	1450	1640
200 $e_f = 110$	6	88	120	158	200	247	299	357	417	484	632	796
	8	115	158	208	264	327	397	473	553	642	838	1060
	10	142	195	256	327	406	492	585	687	798	1040	1310
	12	168	231	305	389	483	585	697	820	950	1240	1570

*Minimum plate thickness, t = 1.7D. Material F_y = 300 MPa

MOMENT CONNECTIONS

General

Continuous construction requires moment resisting beam-to-column connections that will maintain, virtually unchanged, the original angles between intersecting members, at specified loads.

Rigid moment connections can be provided by using welds, bolts or combinations of welds and bolts. Numerous configurations and details are possible; Figure 3–5 shows four possible arragements.

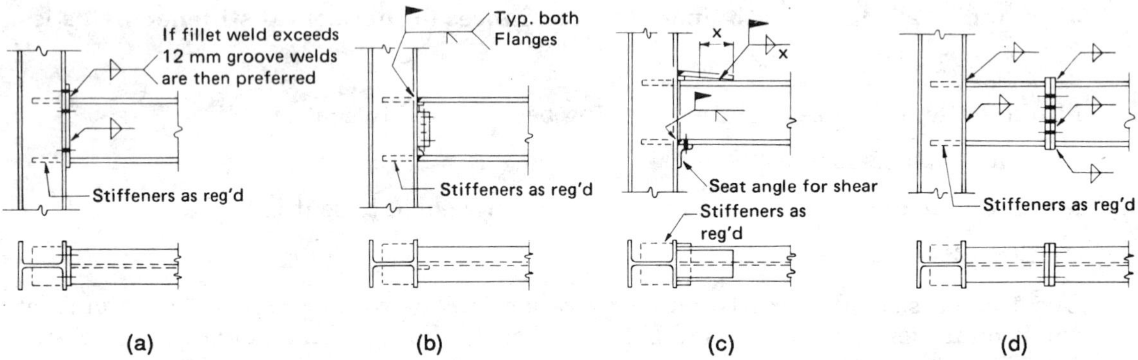

Figure 3–5

Figure 3–5 (a) illustrates a heavy plate shop welded to the end of the beam and field bolted to the column. The end plate distributes flange forces over a greater length of column web than does a fully welded joint, but prying action must be considered.

Figure 3–5 (b) illustrates beam flanges field welded directly to the column with groove welds. Shear capacity is developed by a seat angle, web framing angle or plate, or by welding the beam web directly to the column. Backing bars and run off tabs for the welds may be required.

Figure 3–5 (c) illustrates the use of moment plates shop welded to the column with groove or fillet welds and fillet-welded, or preferably bolted, to the flanges of the beam. The moment plates are spaced to accommodate rolling tolerances for beam depth and flange tilt, and nominal shims are provided to fill any significant gap. Minor gaps are closed by the action of bolting. Shear capacity is usually provided by a web plate welded to the column and field bolted to the beam. An alternative, shown in the sketch, is field groove welding of the bottom flange directly to the column along with the use of an angle as an erection seat and backing bar. Shear capacity may be provided by the seat angle.

Figure 3–5 (d) illustrates the use of short beam sections shop welded to the column, and field bolted to the beam near a point of contraflexure. An end plate connection is shown but lapping splice plates for the flanges and web may be more economical depending on the forces to be transmitted and the relative ease of achieving field fit-up. Another option is a field welded connection.

With the introduction of seismic design criteria (Clause 27) in CAN/CSA-S16.1-M89, restrictions are imposed on the type and arrangement of details permitted for moment connections controlled by seismic requirements. The restrictions, outlined in Clause 27, depend on the level of ductility assumed in the analysis.

Generally, in designing moment connections it is important to ensure that the connection provided will have adequate strength and stiffness, combined with sufficient rotation capacity to permit inelastic deformations to occur, as assumed or inherent in the analysis and design procedure used to proportion the structural frame. Since the welding. In particular, it is important to provide details with adequate ductility when framing into the weak axis of a column.

When applying the governing moments, shears and axial forces at a moment connection, a simplified approach may be used for design. Research (Chen and Lui 1988) has shown that the end moment and axial force in the beam can be assumed to be resisted by the beam flanges alone, with the end shear resisted by the beam web alone.

To ensure that the connection provided is consistent with the design assumptions used to proportion members of a structure, it is important that the designer provide the fabricator with governing maximum and coincident moments, shears and axial forces to be developed at the connection. See Clause 4.1.2 in S16.1-M89 and the Commentary.

Column Stiffeners

Where rigid connections are required, the resistance of a column section to local deformation is important. With relatively small beams connected to heavy columns, the columns will provide the degree of fixity assumed in the design of beams. With large beams, however, the columns will usually have to be strengthened locally by means of stiffeners, doubler plates or both.

Column stiffeners are provided opposite tension flanges of the connected beams to minimize curling of the column flanges with resultant overstressing of the central portion of the weld connecting the beam flange (or moment plate) to the column. Opposite compression flanges of the connected beams, column stiffeners are provided to prevent buckling of the column web. Horizontal plate stiffeners are the most common type of stiffener used. When different depth beams frame into opposite flanges of the column, either inclined stiffeners or horizontal plate stiffeners opposite the flange of each beam may be used. If shear generated in the column web at the moment connection exceeds the column shear capacity, "doubler" plates or diagonal plate stiffeners are used to locally increase the column web shear capacity.

Clause 21.3 of S16.1-M89 specifies requirements for web stiffeners on H-type columns when a beam is rigidly framed to the column flange.

References

The following references contain more detailed information on the design of moment connections. Some refer to allowable stress rules and must be interpreted for limit states applications.

ASCE. 1971. Commentary on plastic design in steel. American Society of Civil Engineers, New York, N.Y.

Blodgett, O.W. 1966. Design of welded structures. The James F. Lincoln Arc Welding Foundation.

Chen, W.F., and Lui, E.M. 1988. Static flange moment connections. Journal of Constructional Steel Research, Elsevier Science Publishers, New York, N.Y., Vol. 10: 65–66.

Chen, Haung and Beedle. 1974. Recent results on connection research at Lehigh. Regional Conference on Tall Buildings, Bangkok, pages 799–813.

Graham, Sherbourne and Khabbaz. 1959. Welded interior beam-to-column connections. The American Institute of Steel Construction, Chicago, Illinois.

Haung, Chen and Beedle. 1973. Behavior and design of steel beam-to-column moment connections. Welding Research Council Bulletin.

Krishnamurthy, N. 1978. A fresh look at bolted end plate behavior and design. Engineering Journal, A.I.S.C., **15**(2).

KRISHNAMURTHY, N. 1978. A fresh look at bolted end plate behavior and design. Engineering Journal, A.I.S.C., **15**(2).

PACKER, J.A. 1977. A limit state design method for the tension region of bolted beam-column connections. The Structural Engineer, **5**(10), October.

REGEE, HAUNG and CHEN. 1973. Test of a fully-welded beam-to-column connection. Welding Research Council Bulletin.

TALL L. (editor). 1964. Structural steel design (2nd edition). The Ronald Press Company, New York, N.Y.

WIC. 1989. Principles of welding design, module 32 — moment connections. Welding Institute of Canada, Oakville, Ontario.

Examples

Note: *In the following examples, the solution chosen in each case is intended to illustrate only one of several satisfactory solutions that could be used. In any given situation, the design will be influenced by the individual fabricator's experience, fabrication methods and erection procedures.*

Example 1

Given:

Design an interior beam-to-column connection for the following coincident forces and moments due to factored loads, assuming the unit specified dead and live loads are 3.8 kPa and 4.8 kPa respectively.

Factored beam moments = 240 kN·m and 320 kN·m

Factored beam shears = 110 kN and 130 kN

Steel = CSA G40.21-M 300W, Use E480XX electrodes for welds.

W310 x 86 Column

t_c = 16.3 mm
w_c = 9.1 mm
k_c = 33 mm
b = 254 mm
d = 310 mm
k_1 = 20 mm
T = 244 mm

W410 x 60 Beam

t = 12.8 mm
w = 7.7 mm
d = 407 mm
b = 178 mm
Class 1 (in bending)

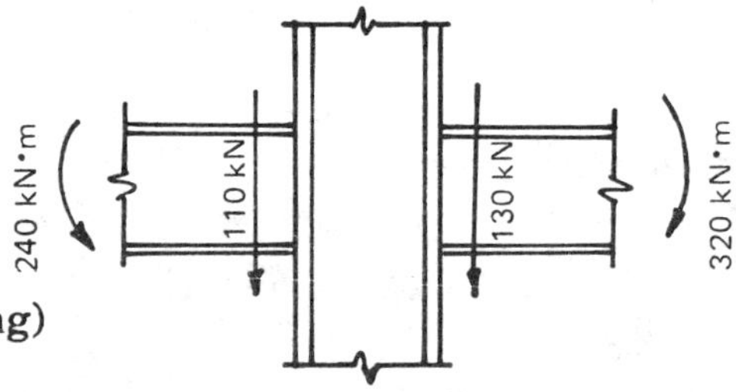

Solution:

(a) Web Connection

The design of the connection for the beam web to the column flange need only account for the vertical shear, neglecting eccentricity. (Design for 130 kN shear.)

Two alternatives are shown to illustrate a field bolted and a field welded condition.

(i) *Alternative 1*

Single plate field welded to beam web, shop welded to column flange, holes for 2 – M20 erection bolts.

To resist factored shear, try 5 mm fillet on 6 mm plate.

Required weld length = 130/0.765 = 170 mm (Table 3–24, page 3–39)

Use 230 mm length of plate — for a nominal beam depth of 410 mm (Table 3-35, page 3-61).

Check plate for factored shear capacity (Clause 13.4.4, S16.1-M89).

Net length is $230 - 2(20 + 2 + 2) = 182$ mm

V_r of plate is $0.5 \times 0.9 \times 182 \times 6 \times 0.450 = 221$ kN > 130 — OK

Use 6x75 x 230 mm plate with 5 mm E480XX fillet weld.

(ii) *Alternative 2*

Single plate field bolted to beam web, M20 A325M bolts, bearing-type, shop welded to column flange.

From Table 3–4, page 3–8, factored shear resistance, single shear, threads intercepted, for M20 A325M bolts = 73.5 kN

Therefore, 2 bolts $= 2 \times 73.5 = 147$ kN > 130 — OK

Check factored bearing resistance on beam web, w = 7.7 mm

From Table 3–6, page 3–9, bearing on 300W steel

= 73.9 kN per bolt for end distance = 35 mm and t = 7 mm

For 2 bolts, $= 2 \times 73.9 = 148$ kN > 130 — OK

Try 7 mm plate, 230 mm long, 2 bolts at 160 mm pitch, and check plate thickness for shear.

Net length = 182 mm, as for alternative 1

Required thickness of plate is $130 / (0.5 \times 0.9 \times 182 \times 0.450) = 3.5$ mm < 7 — OK

Use 7x80 x 230 plate and two M20 A325M bolts

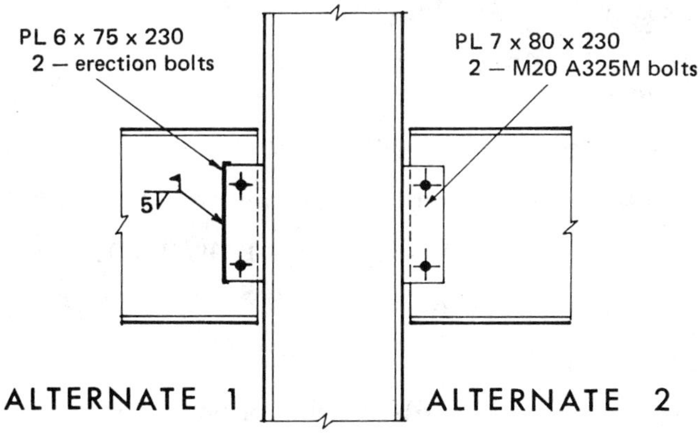

Alternative 2 replaces the two erection bolts with permanent high strength bolts, and eliminates vertical field welding (likely a better solution).

(b) Flange Connection

Two alternatives are shown to illustrate field bolted and field welded conditions.

(i) *Alternative 1*

Top and bottom moment plates shop welded to column, field bolted to beam flanges with A325M bolts in slip-critical connection.

The number of bolts required is determined on the basis of specified loads and all other strength checks are based on factored loads and factored resistances.

Determine the specified load moment using Figure 3–1 on page 3–4

Ratio D/L = 3.8 / 4.8 = 0.79 giving $(L + D) / (\alpha_L L + \alpha_D D) = 0.72$

Therefore, specified load moment is $0.72 \times 320 = 230$ kN·m

Flange force due to specified loads is $230 \times 1\,000 / 407 = 565$ kN

From Table 3–11, page 3–15, assuming M22 A325M bolts and Class A contact surfaces, number of bolts required

= 565 / 45.2 = 12.5 — use 14 bolts (2 rows of 7)

Assuming 80 mm pitch, 40 mm end distance, 130 mm clear to column flange, plate length required is

$(6 \times 80) + 40 + 130 = 650$ mm

Flange force due to factored loads is $320 \times 1\,000 / 407 = 786$ kN

From S16.1-M89, Clause 13.2(a), gross area of required plate is

$786 / (0.9 \times 0.300) = 2\,910 \text{ mm}^2$

and required net effective area is

$786 / (0.85 \times 0.9 \times 0.450) = 2\,280 \text{ mm}^2$

Try 200x16 plate (first preference thickness), and check areas.

Gross area is $200 \times 16 = 3\,200 \text{ mm}^2 > 2\,910$ — OK

Net area is $(200 - 52)16 = 2\,370 \text{ mm}^2 > 2\,280$ — OK

Ultimate strength for bolt shear and bearing on material thicknesses is OK by inspection.

(ii) *Alternative 2*

Moment plate field welded to column flange and top flange of beam, bottom flange of beam welded directly to column flange with groove weld.

As in alternative 1, the moment plate is designed to transmit the factored beam flange force of 786 kN.

Plate area required (gross) is $786 / (0.9 \times 0.300) = 2\,910 \text{ mm}^2$

Select plate width narrower than beam flange width to permit downhand welding.

Try 150 mm plate. (Maximum weld size would be $(178 - 4 - 150) / 2 = 12$ mm.)

Plate thickness required is $2\,910 / 150 = 19.4$ mm

Use 20 mm plate (first preference thickness)

From Table 3–24, for E480XX electrode, 12 mm fillet weld = 1.84 kN/mm

Weld length required is 786 / 1.84 = 427 mm. — use 430 mm

End weld length is 150 mm, therefore length each side is $(430 - 150) / 2 = 140$ mm

It is generally recommended that an unwelded length of plate equal to at least 1.2 times the plate width be provided.

Therefore minimum plate length is $140 + (1.2 \times 150) = 320$ mm

Use 20x150 x 320 plate welded to column flange with full penetration groove weld and welded to top flange of beam with 430 mm of 12 mm fillet welds.

A possible third alternative would be to field weld the top and bottom flanges of the beam directly to the column flange with full penetration groove welds using backing bars fitted against the column flange.

(c) Column Shear Capacity

The column will be subject to a shear force due to the unbalanced moment. S16.1-M89 Clause 21.3 requires stiffening of the column web if this shear exceeds $V_r = 0.55\,\phi\,w\,d\,F_y$

$= 0.55 \times 0.9 \times 9.1 \times 310 \times 0.300 = 419$ kN

Shear force is $(320 - 240) \times 1\,000 / 407 = 197$ kN < 419 — OK

Thus, no reinforcing of the web is required for shear. (Shear forces from the column, above and below the moment connections, are ignored for simplicity.)

(d) Column Stiffeners

Design the column stiffeners to S16.1-M89 Clause 21.3.

Clause 21.3(a): $B_r = 0.9 \times 9.1\,(12.8 + (5 \times 33))\,0.300 = 437$ kN < 786

Therefore, stiffeners are required opposite the compression flange

for capacity of $786 - 437 = 349$ kN

Clause 21.3(b): $T_r = 7 \times 0.9 \times 16.3^2 \times 0.300 = 502$ kN < 786

Stiffeners are also required opposite the tension flange

for capacity of $786 - 502 = 284$ kN

Total stiffener area required at compression flange

is $349 / (0.9 \times 0.300) = 1\,290$ mm^2

Maximum b/t ratio is $145 / \sqrt{300} = 8.37$ (to match the Class 1 beam)

Try 90 mm wide stiffener each side of column web (beam flange is 178 mm wide).

Minimum $t = 90 / 8.37 = 10.8$ mm — try 12 mm

Effective stiffener width to clear column k_1 distance is $(178 / 2) - 20 = 69$ mm

Effective stiffener area is $2 \times 69 \times 12 = 1\,660$ mm^2 $> 1\,290$ mm^2 — OK

Use 12x90 stiffener each side of column web opposite compression flange.

Use same stiffeners opposite tension flange.

(e) Stiffener Welds

Welds connecting stiffeners to column flange must be sufficient to develop force in stiffener of 349 kN

For double fillet welds at stiffener ends (length 69 mm), weld resistance required

is $349 / (2 \times 69) = 2.52$ kN/mm

From Table 3–24, page 3–39, 10 mm E480XX fillet welds provide

$2 \times 1.53 = 3.06$ kN/mm — OK

Welds connecting stiffeners to column web must transfer shear forces due to unbalanced beam moment of $197 / 2 = 98.5$ kN per side.

Approximate weld length available is T distance of 244 mm — assume 230 mm

Weld resistance required is $98.5 / 230 = 0.43$ kN/mm (one-sided weld will do)

Use single 5 mm fillet weld on each stiffener for 0.77 kN/mm (Table 3–24).

Example 2

Given:

Design an exterior beam-to-column connection for an elastically analysed frame, in which the column size is the same as example 1 and the beam is a W460x74 having a factored end moment of 320 kN·m and a factored end shear of 130 kN.

W310x86 column	W460x74 beam
See example 1	t = 14.5 mm
for dimensions	w = 9.0 mm
	d = 457 mm
	b = 190 mm
	Class 1 (in bending)

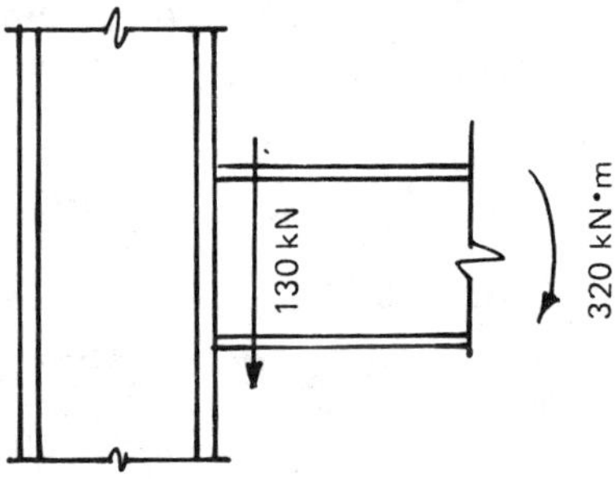

Solution

This example is basically an extension of Example 1 and the solutions given are intended only to provide information on other possibilities.

(a) Web Connection

Use an unstiffened seat angle shop welded to the column to carry the beam shear and to support the beam during erection.

From Table 3–39, page 3–70, for a beam web of 9 mm and a seat length of 230 mm, a 10 mm thick angle will provide a beam web bearing capacity of 265 kN, > 130 kN. Also a vertical leg of 125 mm with 6 mm fillet welds provides a vertical leg connection capacity of 174 kN, > 130 kN.

Use 125x90x10 angle x 230 mm long with 125 mm leg vertical, welded to column flange with 6 mm E480XX fillet welds.

(b) Flange Connection

Assume field welded connection with full penetration groove welds connecting top and bottom flanges of the beam directly to the column flange (suggested alternative 3 in Example 1). The seat angle would serve as the backing bar for the bottom flange weld.

(c) Column Shear Capacity

Shear force is $320 \times 1\,000 / 457 = 700\,\text{kN}$ (shears from column ignored)

Diagonal stiffeners will be used to carry shear in excess of the 419 kN shear capacity of the column web (see Example 1).

Therefore, horizontal component of stiffener force is $700 - 419 = 281$ kN

If θ is angle between stiffener and horizontal plane,

$$\cos\theta = 310 / (310^2 + 457^2)^{1/2} = 0.561$$

Force in stiffener is $281 / \cos\theta = 281 / 0.561 = 501$ kN

Total stiffener area required is $501 / (0.9 \times 0.300) = 1\,860\ \text{mm}^2$

For 90 mm wide stiffener, effective width is 69 mm (see Example 1)

Stiffener thickness required is $1\,860 / (2 \times 69) = 13.5$ mm — try 14 mm

Maximum b/t = 8.37 — $145 / \sqrt{F_y}$ (to match the Class 1 beam)

b/t supplied is 90/14 = 6.4 — OK

Use one 14 x 90 diagonal stiffener each side of column web.

(d) Horizontal Column Web Stiffeners

$B_r = 0.9 \times 9.1\,(14.5 + (5 \times 33))\,0.300 = 441\,\text{kN}$ < 700 (Clause 21.3(a))

Stiffeners required at compression flange for 700 – 441 = 259 kN

$T_r = 7 \times 0.9 \times 16.3^2 \times 0.300 = 502\,\text{kN}$ < 700 kN (Clause 21.3(b))

Stiffeners required at tension flange for 700 – 502 = 198 kN

Design stiffeners for 259 kN; area required is $259 / (0.9 \times 0.300) = 959\,\text{mm}^2$

Use two 12x90 stiffeners (see Example 1).

(e) Stiffener Welds

Diagonal Stiffeners.

Welds connecting stiffeners to column flanges must be sufficient to develop force in stiffener of 501 kN.

For double fillet welds at ends of stiffeners (length = 69 mm) weld resistance required is $501 / (2 \times 69) = 3.63\,\text{kN/mm}$

From Table 3–24, 12 mm E480XX fillet welds provide $2 \times 1.84 = 3.68\,\text{kN/mm}$. Use 12 mm E480XX fillet welds top and bottom at each end of stiffeners, and nominal 5 mm stitch welds between stiffener and column web.

Note that the angles formed by the diagonal stiffener with the column flange and the horizontal stiffener often are not suitable for fillet welds. However, by providing the size of a fillet weld, the designer indicates the required weld resistance. Shop detail drawings used by the fabricator must show an appropriate weld configuration for developing that resistance.

Horizontal Stiffeners

End weld must develop force in stiffener of 259 kN

Weld resistance required is $259 / (2 \times 69) = 1.88\,\text{kN/mm}$

Double 8 mm fillets provide $2 \times 1.22 = 2.44\,\text{kN/mm}$ (Table 3–24)

Welds connecting horizontal stiffeners to column web are required to transfer only a portion of the total horizontal shear force to the column web as the rest is transferred directly through the flange connection. Although the actual shear distribution is not known, a conservative approach is to design these welds to transfer the total design force in the stiffener. For an approximate weld length of 230 mm (see Example 1), weld resistance required is

$259 / (2 \times 230) = 0.56\,\text{kN/mm}$

Use single 5 mm E480XX fillet weld on each stiffener for 0.765 kN/mm.

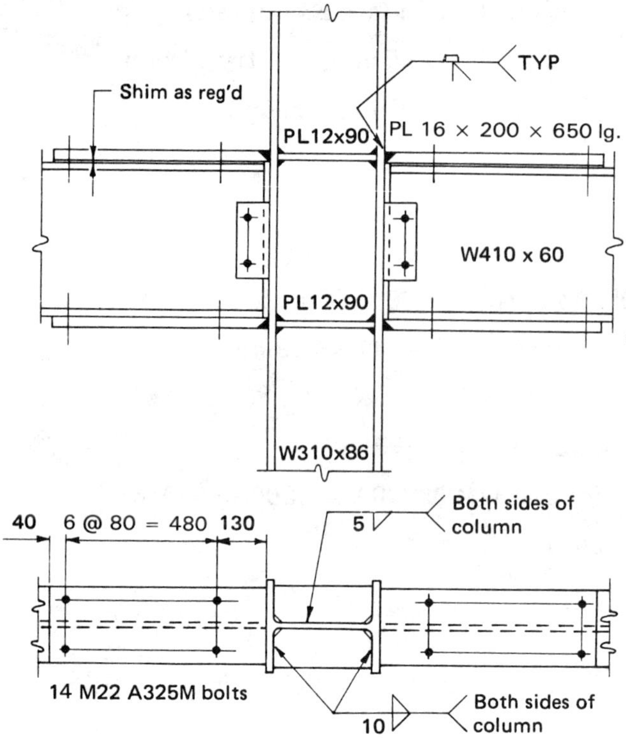

shop welded, field bolted.
web connection Alternate 2 .
flange connection Alternate 1.

Example 1

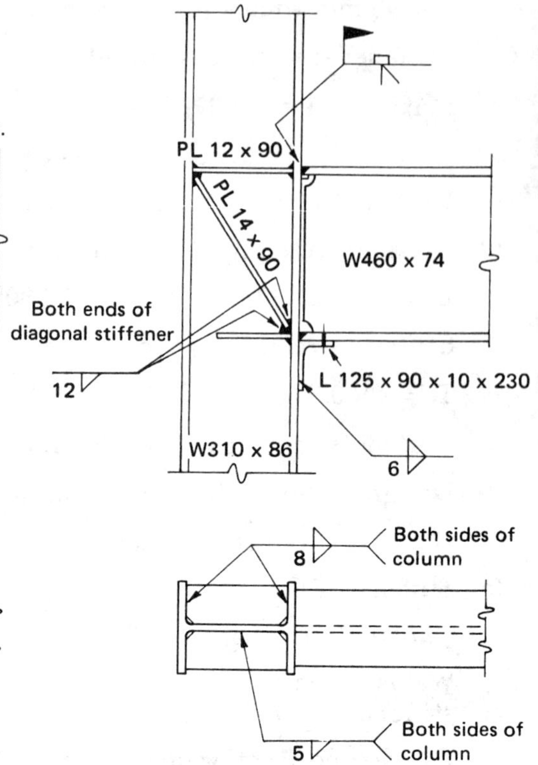

both shop and field welded.

Example 2

HOLLOW STRUCTURAL SECTION CONNECTIONS

General

The use of hollow structural sections as columns and truss components is quite common. HSS members may be selected for various reasons including, aesthetics and/or weight reduction for compression members. This section of the Handbook shows sketches of some of the more commonly used details, and includes information on weld details appropriate to HSS connections. Details include:

Beam to HSS column connections	Figure 3–6
Truss to column and girder joints	Figure 3–7
HSS truss connections	Figure 3–8
Connections for moment and shear	Figure 3–9
Other truss configurations	Figure 3–10
Welding details for hollow structural sections	Figure 3–11

The connections illustrated in Figures 3–6 and 3–7 are simple shear connections designed in a conventional manner; however, for the tee connection sufficient rotation of the joint must be allowed. Fang and White (1966) suggest a minimum width-to-thickness ratio for the tee flange of 10, where the width is the full flange width.

Many of the design recommendations for connections have resulted from research sponsored by the International Committee for the Study and Development of Tubular Structures (CIDECT). The International Institute of Welding (IIW) has also been instrumental in evaluating the research and publishing "state-of-the-art" design guidance. Further information on connection design is contained in the publications listed as references. The handbook by Packer and Henderson (1992), in particular is a practical and comprehensive guide dedicated to the Canadian market with design examples conforming to CAN/CSA-S16.1-M89.

Truss Connections

Since a prime application of HSS members is in architecturally exposed areas, careful attention must be given to aesthetics of the connections in addition to cost and fabrication techniques. The following points should be kept in mind when using HSS:

(1) Gap connections are usually the most economical connection type. Overlap connections require additional profiling of web members and extra welding. Reinforced connections are usually the most expensive.

(2) If fatigue is a design consideration, careful attention should be paid to the connection details. It is suggested that overlap connections with an overlap of 50% be used for trusses subjected to fatigue loading.

(3) Connection resistance increases as the width of truss web members approaches the width of the HSS chord and is a maximum when web width equals the chord width. However, welding costs increase when welds are placed on the corner radii of the chord; therefore to obtain optimum strength and economy with a rectangular HSS connection, the web width should be as wide as possible, but not greater than the width of the chord, minus about five times the chord wall thickness (since the outer corner radius is generally between two and three times the wall thickness).

(4) Connection resistance can also be improved by selecting chord members having a large wall thickness combined with web members having a small wall thickness. Frequently, a smaller size chord of thicker material with about the same mass per metre as a larger, thinner wall HSS will not require connection reinforcing or the more expensive overlap connections.

(5) Primary bending moments due to eccentricity *e* (Figure 3–8) may be neglected, with regard to connection design, provided the intersection of the centre lines of the web members lies within the following range measured from the centre line of the chord: 25% of the chord depth towards the outside of the truss, and 55% of the chord depth towards the inside of the truss.

(6) Profiling of round members is generally required when they are joined to other members. If aesthetics allow the web members to have the ends flattened instead of profiled, cost savings may be achieved.

(7) The effectiveness of load transfer from one HSS section to another is a function of dimensional parameters of the sections rather than the amount of welding. Packer and Henderson (1992) outline methods of calculating connection efficiency and weld effectiveness.

(8) Recent research by Frater and Packer (1990) has established that welding for HSS truss connections can be related to the loads rather than the member resistances; however, when web members of a K connection are inclined at more than 60° to the chord, welds along the heel of the web members must be considered ineffective.

E480XX electrodes are normally used for HSS supplied with 350 MPa minimum yield. Table 3–42 on page 3–90 gives the size of fillet welds necessary to develop the strength of the parent material in either shear or tension in accordance with Clause 13.13.1 of S16.1-M89. Keep in mind that such welds are not necessarily effective in transferring connection loads, as noted in item 7 above. (S16.1-M89 sizes are likely conservative in that Eurocode 3 incorporates International Institute of Welding 1989 design recommendations which consider the strength of an HSS to be fully developed when it is connected by a fillet weld whose throat is 1.10 times its wall thickness; however, Eurocode 3 also advocates partial or full penetration welds when the wall thicknesses exceeds 8 mm.)

In HSS connections the individual members are usually welded all around. Table 3–43 on page 3–91 gives the length of welds for web members connected to chord members at various angles θ, calculated in accordance with AWS D1.1 (1990).

References:

AWS. 1990. Structural welding code – steel, D1.1, ANSI /AWS. American Welding Society, Miami, Florida.

Eurocode 3. 1991. Design of steel structures: part 1 – general rules and rules for buildings.

Fang and White. 1966. Simple framing connections for square structural tubing. Structural Division Journal, American Society of Civil Engineers, April.

Frater, G.S., and Packer, J.A. 1990. Design of fillet weldments for hollow structural section trusses. CIDECT Report No. 5AN/2-90/7, Dept. of Civil Eng., University of Toronto, Ontario.

Packer, J.A., and Henderson, J.E. 1992. Handbook for the design of hollow structural section connections. Canadian Institute of Steel Construction, Willowdale, Ontario.

Stelco Inc. 1983. Proceedings of Canadian symposium on HSS.

Stelco Inc. 1981. Hollow structural sections – design manual for connections. (Out of print).

Wardenier, J., Kurobane, Y., Packer, J.A., Dutta, D., and Yeomans, N. 1991. Design guide for circular hollow section (CHS) joints under predominantly static loading. CIDECT.

FIGURE 3-6

BEAM TO HSS COLUMN CONNECTIONS

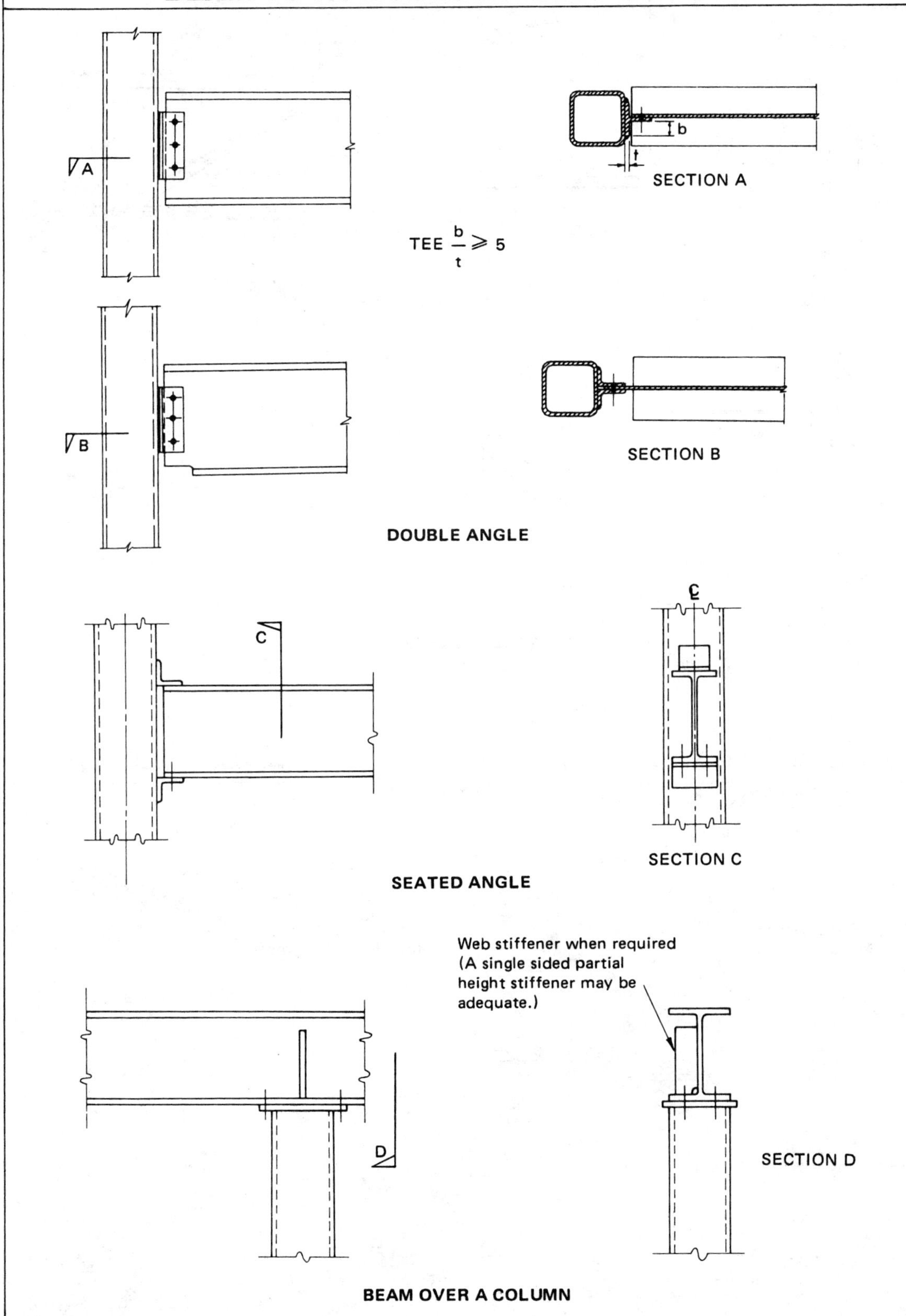

FIGURE 3-7

TRUSS TO COLUMN AND GIRDER JOINTS

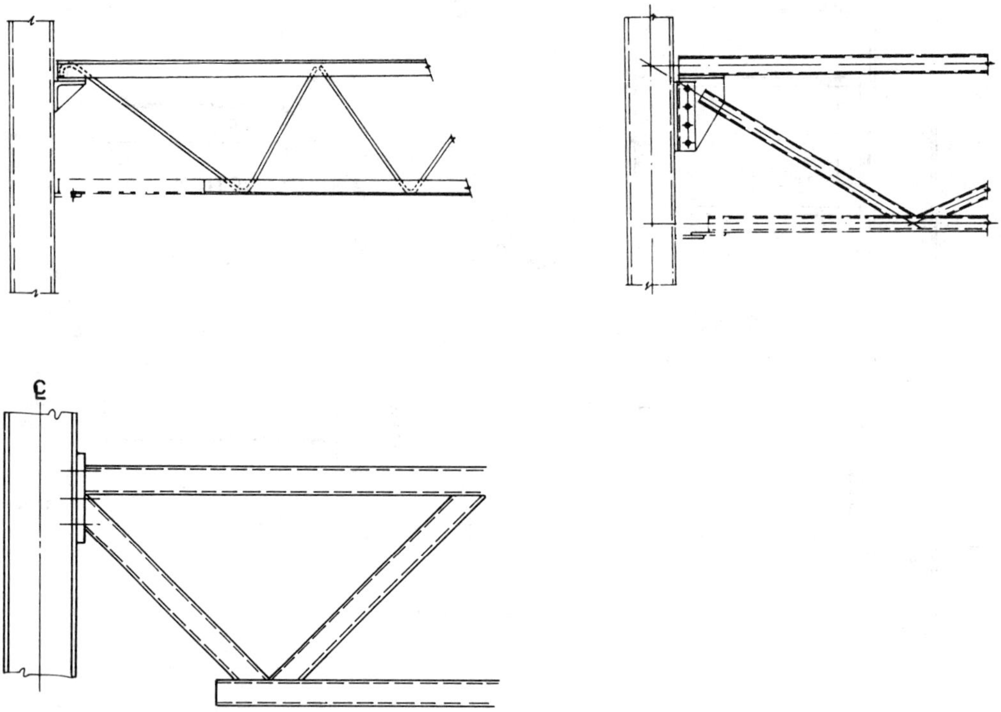

TRUSS TO COLUMN

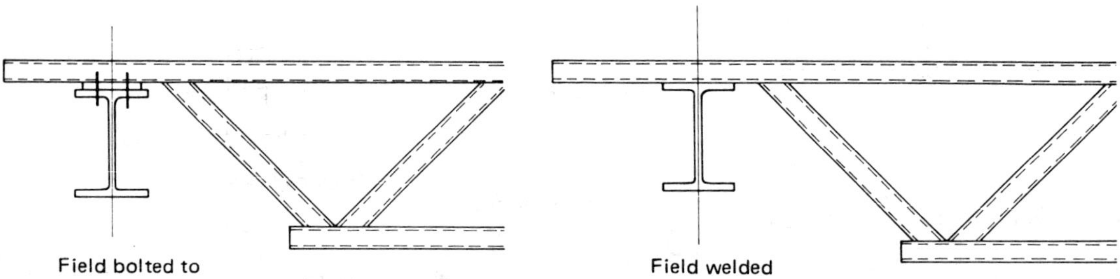

Field bolted to
Supporting member

Field welded
to supporting member

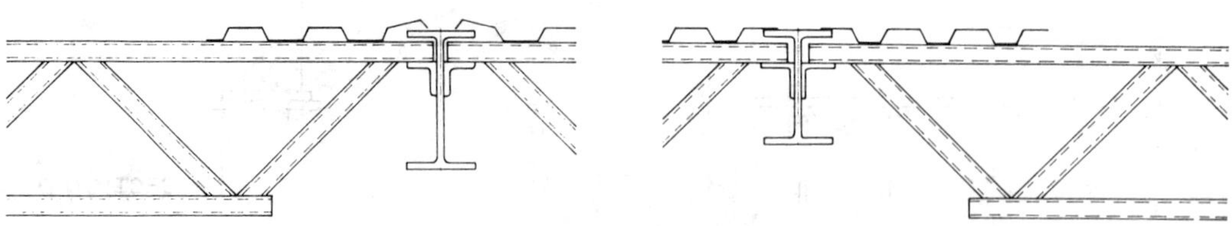

TRUSS TO GIRDER

FIGURE 3-8

HSS TRUSS CONNECTIONS

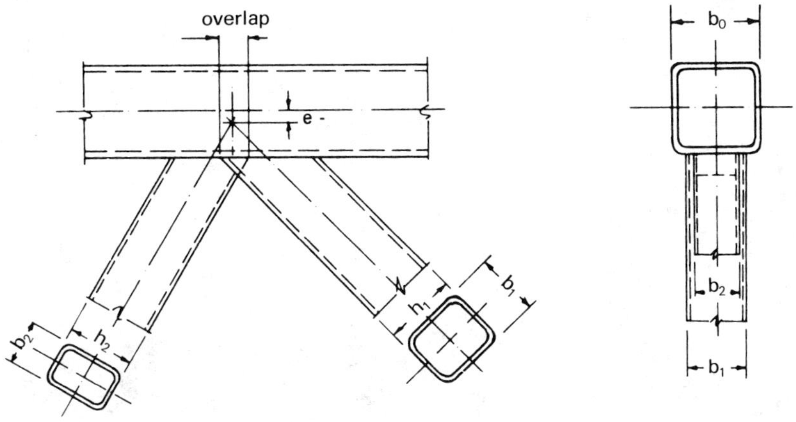

(a) OVERLAP CONNECTION

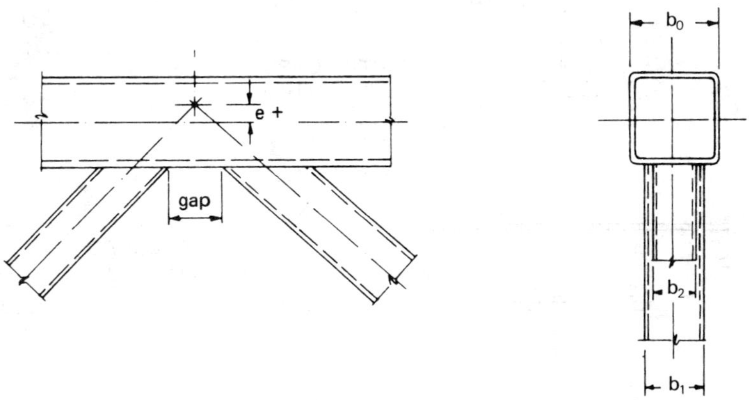

(b) GAP CONNECTION

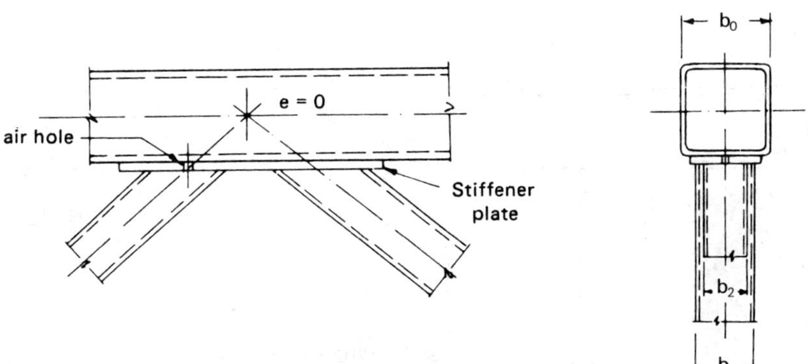

(c) STIFFENED GAP CONNECTION

FIGURE 3-9

CONNECTIONS FOR MOMENT AND SHEAR

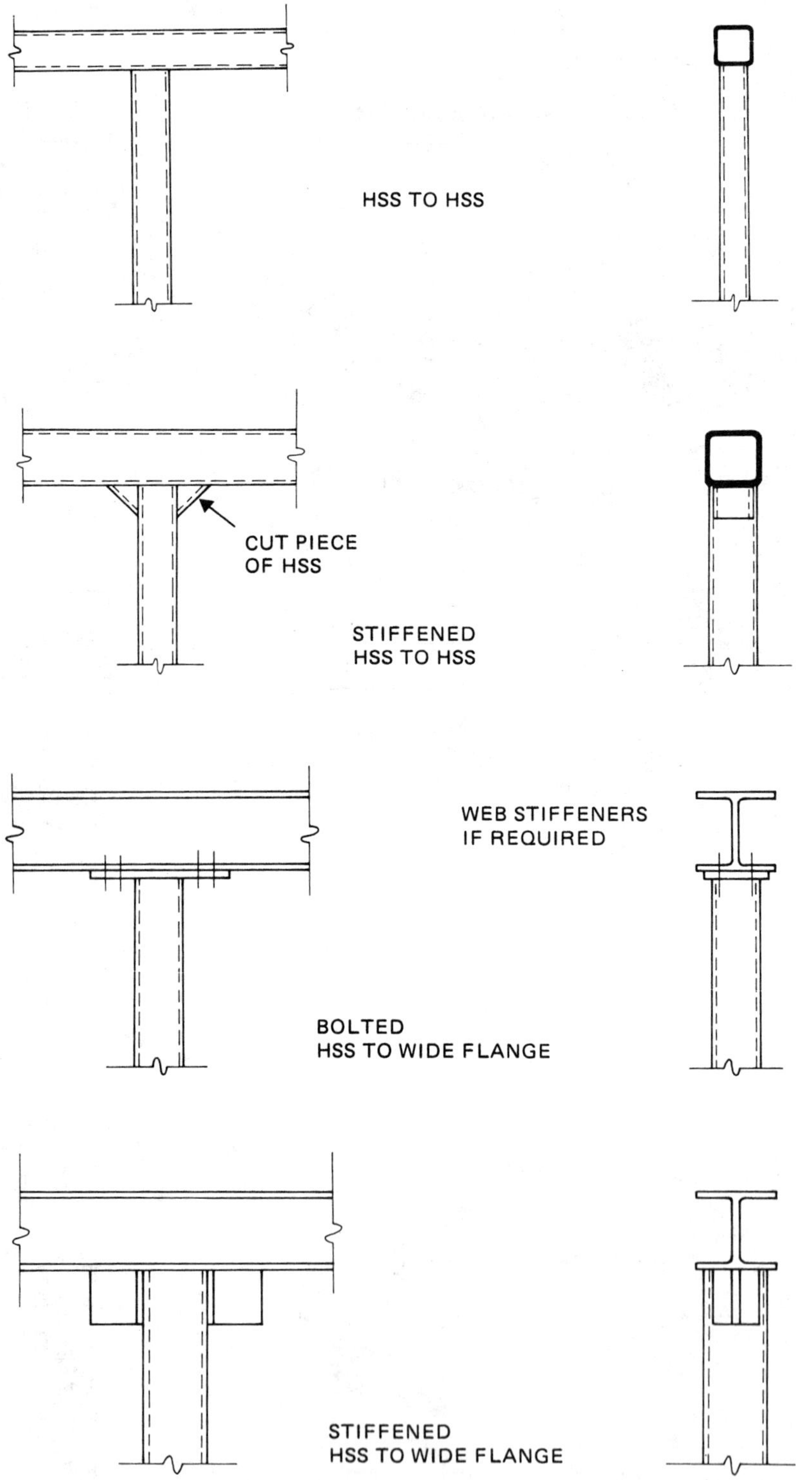

FIGURE 3-10

OTHER TRUSS CONFIGURATIONS

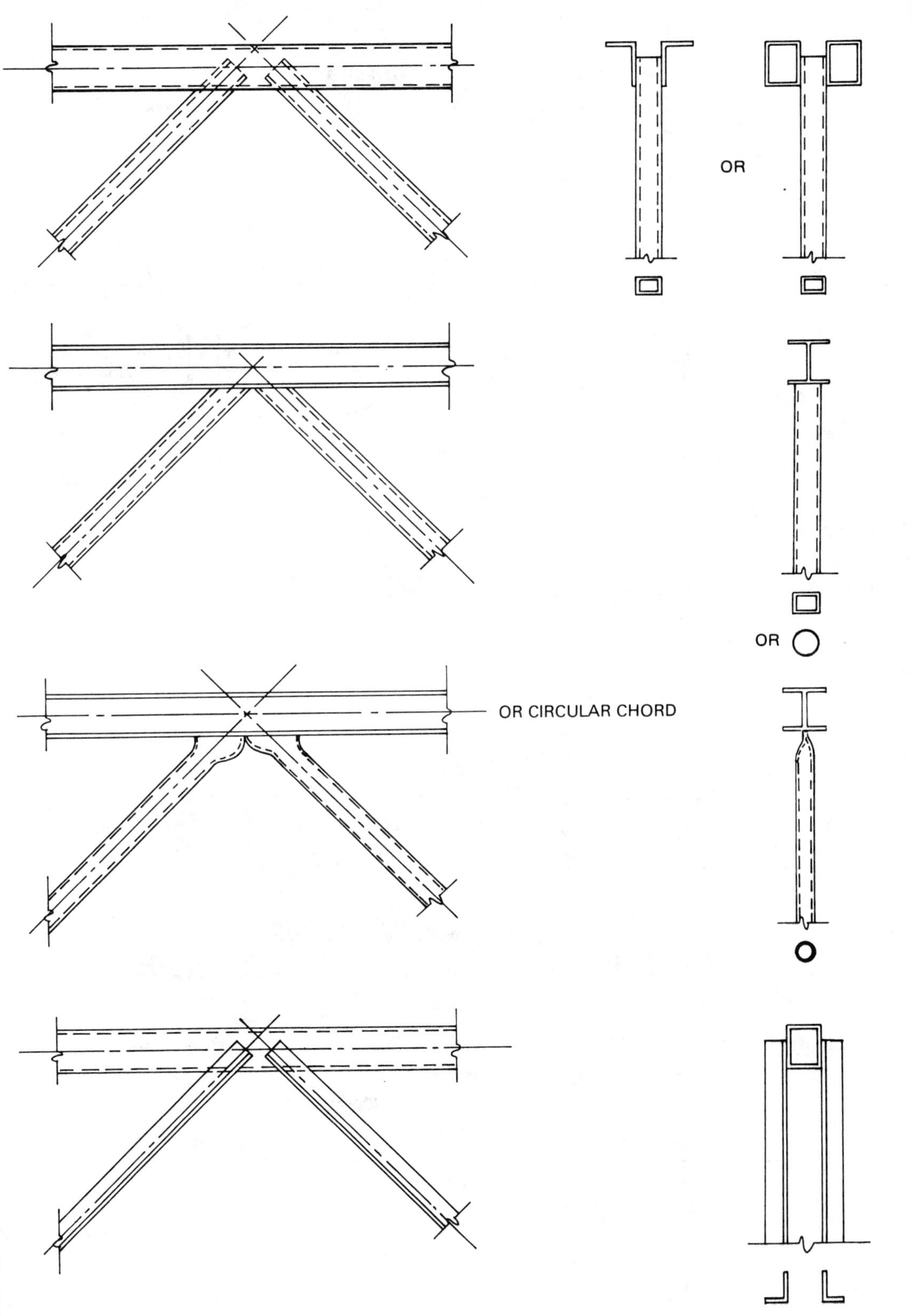

FIGURE 3-11

WELDING DETAILS FOR HOLLOW STRUCTURAL SECTIONS

Effective Throat:–

T – 5 mm for θ = 30° to 44°
T – 3 mm for θ = 45° to 59°

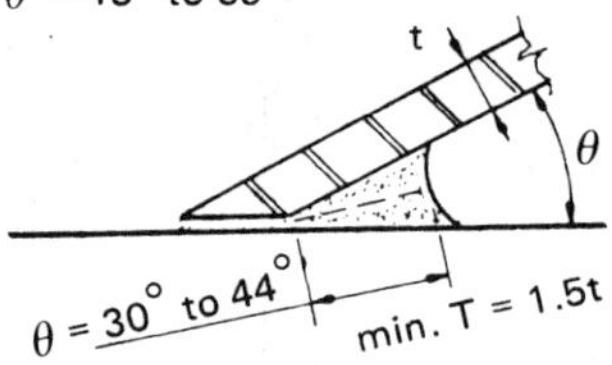

Detail A

θ = 30° to 59°

Effective Throat:–

T>0.707S
as per Table 4.2
in W59-M89

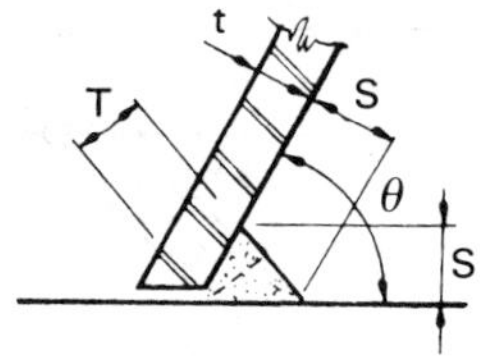

Detail B

θ = 60° to 90°

Effective Throat: — T = 0.707S

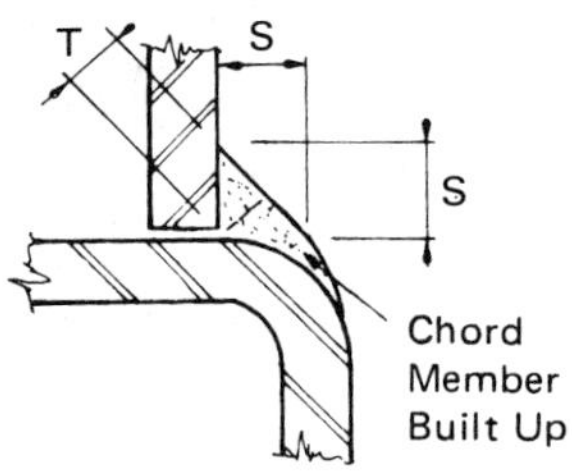

Detail C

θ = 90°

Effective Throat = T

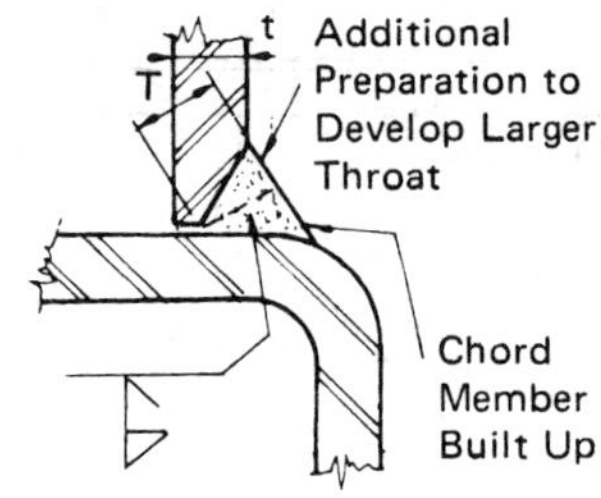

Detail D

θ = 90°

Effective Throat: — T = 0.707 x F x S

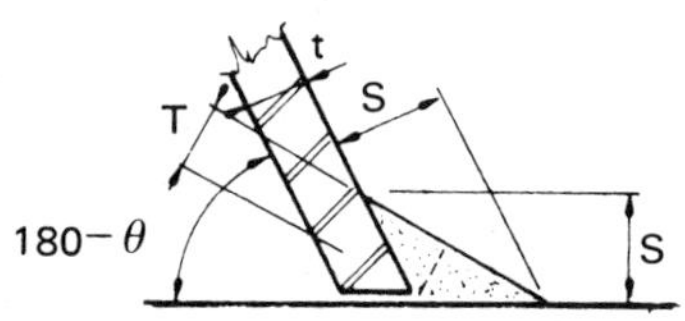

θ	91-100	101-106	107-113	114-120
F	0•95	0•90	0•85	0•80

Detail E

θ = 91° to 120°

Effective Throat: — T = t for θ = 135°
T<t for θ = 136° to 150°
T>t for θ = 121° to 134°

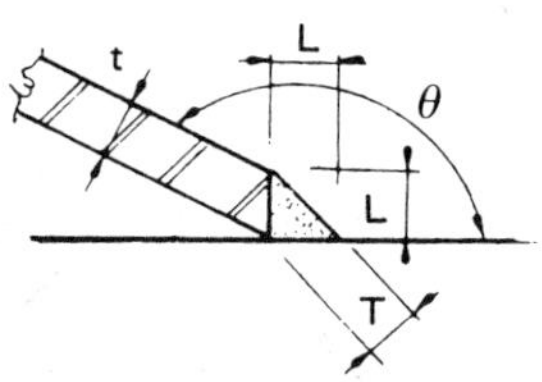

Detail F

θ = 121° to 150°

HSS CONNECTIONS
90° Fillet Size to Develop Wall Strength
Table 3-42

E480XX Fillet Welds **F_y = 350 MPa**

Wall Thickness (mm)	Fillet Leg Size (mm)	
	Wall in Shear	Wall in Tension
3.81	6	8
4.78	8	10
6.35	10	14
7.95	12	18
9.53	14	20
11.13	16	24
12.70	18	26

Table 3–43

HSS Web Members

HSS b x h x t (mm)	Angle θ (degrees) Between Web and Chord Member									
	30	35	40	45	50	55	60	65	70	90
25 x 25 x 3.8	133	121	113	107	102	98	95	93	91	89
32 x 32 x 3.8	171	157	146	138	132	127	123	120	118	114
38 x 38 x 4.8	204	187	174	164	157	151	147	143	140	136
51 x 51 x 6.4	272	249	232	219	209	201	195	191	187	181
64 x 64 x 6.4	348	319	297	280	268	258	250	244	240	232
76 x 76 x 8.0	416	381	355	335	320	308	299	292	286	278
89 x 89 x 9.5	484	443	413	390	372	359	348	340	333	323
102 x 102 x 13	544	498	464	438	418	403	391	382	374	363
127 x 127 x 13	697	637	593	561	535	516	500	488	479	464
152 x 152 x 13	849	776	723	683	652	628	610	595	584	566
178 x 178 x 13	1 000	916	853	806	770	741	719	702	689	668
203 x 203 x 13	1 150	1 055	983	928	887	854	829	809	794	769
254 x 254 x 13	1 460	1 330	1 240	1 170	1 120	1 080	1 050	1 020	1 000	972
305 x 305 x 13	1 760	1 610	1 500	1 420	1 360	1 310	1 270	1 240	1 210	1 180
51 x 25 x 4.8	181	170	161	155	150	146	143	141	139	136
25 x 51 x 4.8	227	203	186	174	164	156	150	145	142	136
76 x 51 x 8.0	317	294	277	264	254	247	241	236	233	227
51 x 76 x 8.0	363	328	302	283	268	257	248	241	235	227
102 x 51 x 8.0	370	346	329	316	306	298	292	287	283	278
51 x 102 x 8.0	463	415	380	354	334	318	306	297	289	278
89 x 64 x 8.0	393	363	342	325	313	303	295	289	285	278
64 x 89 x 8.0	439	398	367	345	327	313	303	294	288	278
102 x 76 x 9.5	461	426	400	380	365	353	344	337	332	323
76 x 102 x 9.5	507	460	425	399	379	364	351	342	335	323
127 x 51 x 9.5	415	391	374	361	351	343	337	332	329	323
51 x 127 x 9.5	554	494	451	418	393	374	359	347	338	323
127 x 64 x 9.5	464	435	413	396	384	374	366	360	356	348
64 x 127 x 9.5	580	521	477	444	419	400	384	372	363	348
127 x 76 x 13	499	464	438	419	404	393	384	377	372	363
76 x 127 x 13	590	531	489	457	432	413	398	386	377	363
152 x 102 x 13	650	602	568	541	521	505	493	484	476	464
102 x 152 x 13	743	672	619	580	549	526	507	493	482	464
178 x 127 x 13	802	741	697	664	638	618	602	590	581	566
127 x 178 x 13	896	811	749	703	667	639	617	600	587	566
203 x 102 x 13	755	706	671	644	624	608	595	585	578	566
102 x 203 x 13	943	847	776	722	681	649	624	605	590	566
203 x 152 x 13	954	880	827	786	755	731	712	697	686	668
152 x 203 x 13	1 050	951	880	826	784	752	727	707	692	668
254 x 152 x 13	1 060	984	929	889	857	833	814	799	788	769
152 x 254 x 13	1 250	1 130	1 040	968	916	875	844	819	800	769
305 x 203 x 13	1 360	1 260	1 190	1 130	1 090	1 060	1 030	1 010	997	972
203 x 305 x 13	1 560	1 410	1 300	1 210	1 150	1 100	1 060	1 030	1 010	972

Notes:
1. Outside corner radius assumed equal to 2t.
2. Perimeters shown in table are for the thickest wall HSS of each size; therefore perimeters are conservative for smaller wall thicknesses.
3. Perimeters calculated by: $K_a\,[4\pi t + 2(b-4t) + 2(h-4t)]$, where $K_a = ((h/\sin\theta) + b)/(h+b)$.

TENSION MEMBERS

General

Members subject to axial tension (i.e. when the resultant tensile load on the member is coincident with the longitudinal centroidal axis of the member) can be proportioned assuming a uniform stress distribution. The factored tensile resistance is calculated on the basis of yielding on the gross area and fracture on the effective net area (or effective net area reduced for shear lag) as per Clause 13.2 of CAN/CSA-S16.1-M89. Effective net area and effective net area reduced for shear lag are defined in Clause 12.3. These requirements, confirmed by tests and discussed in the Commentary, are substantially revised from those in S16.1-M84 to more accurately reflect behaviour. Tables 3–46 and 3–47 in the previous 4th edition of the Handbook are irrelevant and have been removed.

Fracture on the net section, rather than yielding, is the limit state because, with designs based on factored loads, yielding is unlikely under the expressions of Clause 13.2(a)(ii) or 13.2(a)(iii). Further, any deformation that might occur will be minimal since the volume of material affected is small and the beneficial effects of strain-hardening will be present.

Effective Net Area

Tables 3–44, 3–45 and 3–46 on the following pages are intended to simplify the calculation of effective net area according to the requirements of Clause 12.3.

Hole Diameters for Effective Net Area

Table 3–44 on page 3–98 lists the specified hole diameter for various bolt sizes according to Clause 23.3.2, and the diameter of holes for calculating effective net area according to Clause 12.3.2.

Reduction of Area for Holes

Table 3–45 on page 3–98 lists values for the reduction of area for holes of different diameter in material of various thicknesses.

Staggered Holes in Tension Members

Table 3–46 on page 3–99 lists values of $s^2/4g$ required to calculate the net width of any diagonal or zig-zag line of holes according to the requirements of Clause 12.3.1(c) for various pitches from 25 to 240 mm and for various gauges from 25 to 320 mm. Values of $s^2/4g$ for pitches and gauges between those listed can be interpolated.

Effective Net Area – Reduced for Shear Lag

Clause 12.3.3 of S16.1-M89 contains provisions for specific cases to determine loss of efficiency due to shear lag when tension members are not connected by all their elements. An alternative, with general application, is discussed in the Commentary on page 2–21. Table 3–47 on page 3–99 will assist with that general approach.

Example 1

Given:

The hole pattern shown occurs at a bolted splice in a built-up tension member consisting of two C310x45* sections. Determine the net area and the tensile resistance of the member for M20 A325M bolts, and material conforming to G40.21-M grades 300W, 350W and 480W. (* no longer available from Canadian sources)

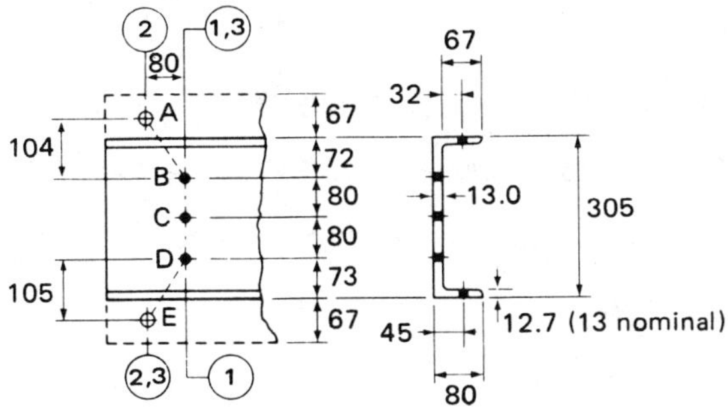

Solution:

(a) Gross area

Gross Area, A_g, from 2 – C310x45 is $2 \times 5\,690 = 11\,380\ mm^2$

Gross width is $305 + 2(80 - 13) = 439$ mm (S16.1-M89, Clause 12.3.4)

(b) Deduction for net section

Web width deduction for M20 bolts is 24 mm (Table 3–44)

Section 1—1

Less 3 holes, $3 \times 24 = -72$ mm

Section 2—2

Less 5 holes, 5×24	$= -120$	
Plus $s^2/4g$ for AB (s = 80, g = 104)	$= +\ 16$	(Table 3–46)[#]
Plus $s^2/4g$ for DE (s = 80, g = 105)	$= +\ 16$	
	$-\ 88$ mm	(governs)

Section 3—3

Less 4 holes, 4×24	$= -\ 96$
Plus $s^2/4g$	$= +\ 16$
	$-\ 80$ mm

[#]Values of $s^2/4g$ are obtained from Table 3–46 for s = 80 and interpolating by inspection for g = 104 and 105.

Net width is $439 - 88 = 351$ mm

Net area is $2 \times 351 \times 13 = 9\,130\ mm^2$

(c) Tensile Resistances

300W — (F_u = 450 MPa, from Table 3–5 on page 3–9)

$T_r = \phi\, A_g\, F_y$ (Clause 13.2(a)(i))

$= 0.9 \times 11\,380 \times 0.300 =$ **3 070 kN** (governs)

$T_r = 0.85\ \phi\ A_{ne}\ F_u$ (Clause 13.2(a)(ii))

$= 0.85 \times 0.9 \times 9\,130 \times 0.450 =$ **3 140** kN

350W — (F_u = 450 MPa)

$T_r = 0.9 \times 11\,380 \times 0.350 =$ **3 580** kN

$T_r = 0.85 \times 0.9 \times 9\,130 \times 0.450 =$ **3 140** kN (governs)

480W — (Fu = 590 MPa)

$T_r = 0.9 \times 11\,380 \times 0.480 =$ **4 920** kN

$T_r = 0.85 \times 0.9 \times 9\,130 \times 0.590 =$ **4 120** kN (governs)

(More bolts than shown would be needed to develop these resistances.)

Example 2

Given:

A C380x50 tension member carries factored axial tension of 850 kN, and has been connected through only the web with ten M22 A325 bolts as illustrated. Confirm the adequacy of the connection for 300W material.

d = 381 mm

w = 10.2 mm

A = 6 430 mm^2

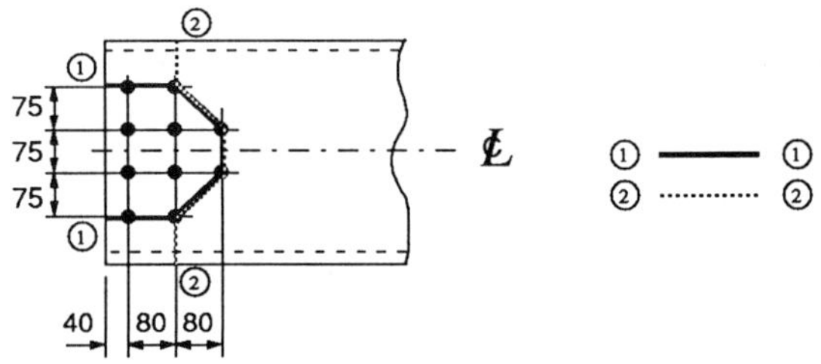

Solution:

Bolt V_r = 88.9 kN (Table 3–4, page 3–8, threads intercepted)

Bolt shear resistance is $10 \times 88.9 =$ **889** kN > 850 — OK

Gross section resistance is $\phi\ A_g\ F_y$

$= 0.9 \times 6\,430 \times 0.300 =$ **1 740** kN > 850 — OK

Web width deduction for each bolt hole is 26 mm (Table 3–44)

Net section 1—1 resistance

Tension element, $A_n = w_n t$ (Clause 12.3.1(a))

$= (75 - 26)10.2 = 500\ mm^2$

Shear elements, $A_n = 0.6\ L_n\ t$ (Clause 12.3.1(b))

$= 0.6 \times 2\ (40 + 80 - (1.5 \times 26))\ 10.2 = 991$

Inclined elements, $A_n = (w_n + s^2/4g)t$ (Clause 12.3.1(c))

$= 2\ (75 - 26 + 21.3)\ 10.2 = 1\,430$ (s = 80, g = 75)

$A_{ne} = 2\,920\ mm^2$

T_r of section 1—1 is $0.85\ \phi\ A_{ne}\ F_u$ (Clause 13.2(a)(ii))

$= 0.85 \times 0.9 \times 2\,920 \times 0.450 =$ **1 010 kN** > 850 — OK

Net section 2—2 resistance

Gross cross section $= + 6\,430\ mm^2$

Less 4 holes, $4 \times 26 \times 10.2 = - 1\,060$

Plus $2\,(s^2\,t/4g) = 2\,(80^2)\,10.2 / (4 \times 75) = + \;\; 435$

$A_{ne} = 5\,810\ mm^2$

T_r of section 2—2 is $0.85 \times 0.9 \times 5\,810 \times 0.450 =$ **2 000** kN > 850 — OK

Therefore, the connection is adequate and is governed by the bolt shear resistance, 889 kN, and the net-section resistance through section 1—1, 1010 kN.

Example 3

Given:

The flanges of a W200x52 tension member are connected to 16 mm gusset plates by 12 – M22 A325 bolts as shown. Material is 300W. Determine the connection resistance.

$t = 12.6$ mm

$A = 6\,660\ mm^2$

$d = 206$ mm

$b = 204$ mm

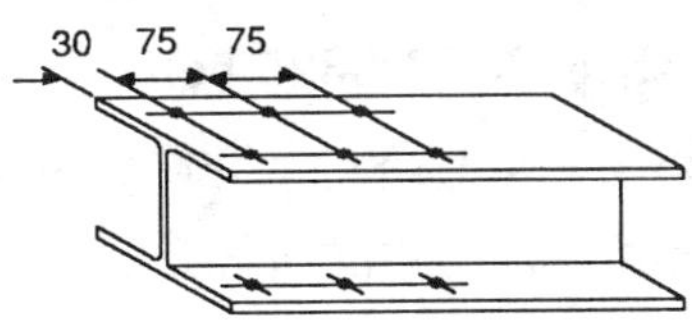

Solution:

Bolt $V_r = 127$ kN (Table 3–4, page 3–8, threads excluded)

Bolt shear resistance is $12 \times 127 =$ **1 520** kN

Note that the full shear resistance of the end row of bolts is available in spite of the 30 mm end distance. Clause 13.10.1(c)(ii), which would have limited their resistance to 114 kN (Table 3-6, page 3-9) for 12.6 mm thick flanges, is not applicable here (as per the note at the end of Clause 13.10.1). This recognizes that the end bolts cannot tear out in isolation, but that the entire bolt group must fail.

T_r (gross area) is $\phi\, A_g\, F_y$

$= 0.9 \times 6\,660 \times 0.300 =$ **1 800** kN

Net areas

Flange width deduction for each bolt hole is 26 mm (Table 3–44)

Gross area $= 6\,660\ mm^2$

Less 4 holes, $4 \times 26 \times 12.6 = - 1\,310$

$A_{ne} = 5\,350\ mm^2$

Reduction for shear lag is required because the web is not connected, and it will form part of the critical section.

As per S16.1-M89, Clause 12.3.3.2(a), the reduction coefficient is 0.90 when there are at least 3 transverse rows of bolts, and the flange width is not less than 2/3 the member depth.

$\therefore\ A'_{ne} = 0.90 \times 5\,350 = 4\,820\ \text{mm}^2$

T_r (net, reduced for shear lag) is $0.85\ \phi\ A'_{ne}\ F_u$ (Clause 13.2(a)(iii))

$= 0.85 \times 0.9 \times 4\,820 \times 0.450 = \mathbf{1\,660}$ kN

Therefore, the connection resistance is 1 520 kN (bolt shear).

Example 4

Given:

An HSS152x152x6.4 of 350W material carries 500 kN tension and is connected by a single plate welded into slots in the HSS walls as shown. Design the plate using 300W steel.

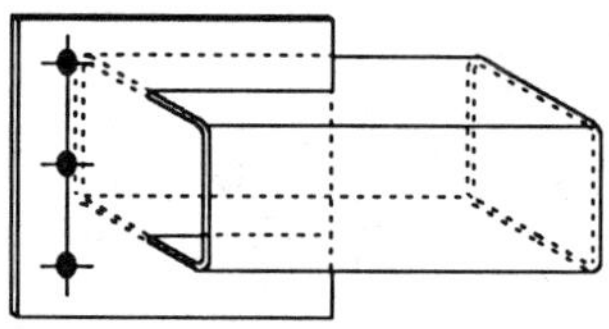

Solution:

Resistance of three M22 A325 bolts in double shear (threads intercepted) is

$= 3 \times 2 \times 88.9 = \mathbf{533}$ kN > 500 — OK

Try a PL12x230; $A = 2\,760\ \text{mm}^2$

Check plate bearing for a bolt carrying $500/3 = 167$ kN

Table 3–6 on page 3–9 shows that an edge distance of 50 mm on a 12 mm plate provides 181 kN bearing resistance — OK.

$T_r = \phi\ A_g\ F_y$ (S16.1-M89, Clause 13.2(a)(i))

$= 0.9 \times 2\,760 \times 0.300 = \mathbf{745}$ kN > 500 — OK

If holes are known to be drilled rather than punched, the specified hole diameter can be used for the net section. (Clause 12.3.2)

$\therefore\ A_{ne} = 12\,(230 - (3 \times 24)) = 1\,900\ \text{mm}^2$

$T_r = 0.85\ \phi\ A_{ne}\ F_u$ (Clause 13.2(a)(ii))

$= 0.85 \times 0.9 \times 1\,900 \times 0.450 = \mathbf{654}$ kN > 500 — OK

Shear Lag for the Plate

Clause 12.3.3.4 in S16.1-M89 provides for shear lag in plates that are connected by a pair of welds parallel to the load, and the length of the welds should generally not be less than the distance w between them. The effective net area is reduced if the length of the welds is less than 2w.

Distance between welds, $w = 152$ mm (use weld length = w)

$\therefore\ A'_{ne} = 0.75 \times 152 \times 12 = 1\,370\ \text{mm}^2$ (Clause 12.3.3.4(c))

Plate area beyond the welds is $(230 - 152)\ 12 = 936\ \text{mm}^2$

Resistance of the plate at the welds is $0.85\ \phi\ A'_{ne}\ F_u$ (Clause 13.2(a)(iii))

$= 0.85 \times 0.9\ (1\,370 + 936)\ 0.45 = \mathbf{794}$ kN > 500 — OK

Welds

Length must be $\geq w = 152$ mm

Fillet weld size is 500 kN / (4 × 152 × 0.153) = 5.4 mm — use 6

Shear Lag for the HSS

The distance between the welds measured along the developed wall of the HSS is about 300 mm, which appears to violate the requirement of Clause 12.3.3.4 that the length of welds must be at least equal to the distance between them. However, this requirement applies when unconnected elements of a shape must be included in the critical cross section (see Clause 12.3.3.2 in S16.1-M89 — the principle applies to elements connected by welding or by bolting). In this example, the unconnected walls of the HSS are not required for strength. Resistance provided by the connected walls alone is

$$0.85\,\phi\,(A_g/2 - (2 \times 12 \times 6.35))F_u = 0.765\,((3610/2) - 152)\,0.450$$

$$= \mathbf{569}\text{ kN} \quad > 500 \quad \text{— OK}$$

and shear lag need not be considered in this instance for the HSS member.

Therefore a PL12x230 x 250 mm long, slotted 152 mm into the HSS, meets the requirements of the standard.

If the load were 600 kN so that the unconnected walls of the HSS were needed for strength, a very conservative solution could be obtained by ignoring the distance between the welds in excess of the weld lengths. For example, the effective net area, reduced for shear lag, of the HSS if 200 mm weld lengths were used could be computed thus,

$$2\,(0.75 \times 200 \times 6.35) = 1\,910\text{ mm}^2$$

and the tension resistance, reduced for shear lag, would be

$$0.85 \times 0.9 \times 1\,910 \times 0.450 = \mathbf{658}\text{ kN} \quad > 600 \quad \text{— OK}$$

A somewhat more liberal assessment of this tension resistance is possible by a more involved analysis.

Eventually with shorter welds, the capacity of the welds or the shear resistance of the plate along the welds governs.

HOLE DIAMETERS FOR EFFECTIVE NET AREA

Table 3-44

Bolt Size		Specified Hole Diameter Clause 23.3.2		Diameter for Net Area Clause 12.3.2	
Metric	Imperial	Nominal Size (mm)	Oversize (mm)	Nominal Size (mm)	Oversize (mm)
M16		18	20	20	22
	3/4	22*	—	24	—
M20		22*	24	24	26
M22		24*	26	26	28
	7/8	24*	—	26	—
M24		27*	30	29	32
	1	27*	—	29	—
M27		29	35	31	37
M30		32	38	34	40
M36		38	44	40	46

Note: Diameter for net area is same as specified hole diameter if hole is drilled (Clause 12.3.2, S16.1-M89).
* For nominal diameter of hole, see Clause 23.3.2(e).

REDUCTION OF AREA FOR HOLES*

Table 3-45

Material Thickness (mm)	Diameter of Hole (mm)										
	20	22	24	26	28	29	31	32	34	37	40	44
5	100	110	120	130	140	145	155	160	170	185	200	220
6	120	132	144	156	168	174	186	192	204	222	240	264
7	140	154	168	182	196	203	217	224	238	259	280	308
8	160	176	192	208	224	232	248	256	272	296	320	352
9	180	198	216	234	252	261	279	288	306	333	360	396
10	200	220	240	260	280	290	310	320	340	370	400	440
11	220	242	264	286	308	319	341	352	374	407	440	484
12	240	264	288	312	336	348	372	384	408	444	480	528
13	260	286	312	338	364	377	403	416	442	481	520	572
14	280	308	336	364	392	406	434	448	476	518	560	616
16	320	352	384	416	448	464	496	512	544	592	640	704
18	360	396	432	468	504	522	558	576	612	666	720	792
20	400	440	480	520	560	580	620	640	680	740	800	880
22	440	484	528	572	616	638	682	704	748	814	880	968
25	500	550	600	650	700	725	775	800	850	925	1 000	1 100
28	560	616	672	728	784	812	868	896	952	1 040	1 120	1 230
30	600	660	720	780	840	870	930	960	1 020	1 110	1 200	1 320
32	640	704	768	832	896	928	992	1 020	1 090	1 180	1 280	1 410
35	700	770	840	910	980	1 020	1 080	1 120	1 190	1 300	1 400	1 540
38	760	836	912	988	1 060	1 100	1 180	1 220	1 290	1 410	1 520	1 670
40	800	880	960	1 040	1 120	1 160	1 240	1 280	1 360	1 480	1 600	1 760
45	900	990	1 080	1 170	1 260	1 300	1 400	1 440	1 530	1 660	1 800	1 980
50	1 000	1 100	1 200	1 300	1 400	1 450	1 550	1 600	1 700	1 850	2 000	2 200
55	1 100	1 210	1 320	1 430	1 540	1 600	1 700	1 760	1 870	2 040	2 200	2 420
60	1 200	1 320	1 440	1 560	1 680	1 740	1 860	1 920	2 040	2 220	2 400	2 640
70	1 400	1 540	1 680	1 820	1 960	2 030	2 170	2 240	2 380	2 590	2 800	3 080
80	1 600	1 760	1 920	2 080	2 240	2 320	2 480	2 560	2 720	2 960	3 200	3 520
90	1 800	1 980	2 160	2 340	2 520	2 610	2 790	2 880	3 060	3 330	3 600	3 960
100	2 000	2 200	2 400	2 600	2 800	2 900	3 100	3 200	3 400	3 700	4 000	4 400

*Area (mm^2) = Diameter of hole (mm) times material thickness (mm).

STAGGERED HOLES IN TENSION MEMBERS

Table 3–46

Values of $s^2/4g$

Pitch "s" (mm)	Gauge "g" (mm)															
	25	30	35	40	45	50	60	70	80	100	120	160	200	240	280	320
25					3	3	3	2	2	2	1	1	1	1	1	0
30				6	5	4	4	3	3	2	2	1	1	1	1	1
35			9	8	7	6	5	5	4	3	3	2	2	1	1	1
40		13	11	10	9	8	7	6	5	4	3	2	2	2	1	1
45	20	17	14	13	11	10	8	7	6	5	4	3	3	2	2	2
50	25	21	18	16	14	12	10	9	8	6	4	4	3	3	2	2
55	30	25	22	19	17	15	13	11	9	8	6	5	4	3	3	2
60	36	30	26	22	20	18	15	13	11	9	7	6	4	4	3	3
65	42	35	30	26	23	21	18	15	13	11	9	7	5	4	4	3
70	49	41	35	31	27	24	20	17	15	12	10	8	6	5	4	4
75		47	40	35	31	28	23	20	18	14	12	9	7	6	5	4
80			46	40	36	32	27	23	20	16	13	10	8	7	6	5
90				51	45	40	34	29	25	20	17	13	10	8	7	6
100						50	42	36	31	25	21	16	12	10	9	8
110							50	43	38	30	25	19	15	13	11	9
120									45	36	30	22	18	15	13	11
130										42	35	26	21	18	15	13
140										49	41	31	24	20	17	15
150											47	35	28	23	20	18
160												40	32	27	23	20
170												45	36	30	26	23
180												51	40	34	29	25
190													45	38	32	28
200													50	42	36	31
210														46	39	34
220														50	43	38
230															47	41
240																45

SHEAR LAG

Table 3–47

Values of $1 - \bar{x}/L$ *

L (mm)	Distance $\bar{x}$															
	10	15	20	25	30	35	40	45	50	55	60	65	70	80	90	100
40	0.75	0.63	0.50	0.38	0.25	0.13										
80	0.88	0.81	0.75	0.69	0.63	0.56	0.50	0.44	0.38	0.31	0.25	0.19	0.13			
120	0.92	0.88	0.83	0.79	0.75	0.71	0.67	0.63	0.58	0.54	0.50	0.46	0.42	0.33	0.25	0.17
160	0.94	0.91	0.88	0.84	0.81	0.78	0.75	0.72	0.69	0.66	0.63	0.59	0.56	0.50	0.44	0.38
200	0.95	0.93	0.90	0.88	0.85	0.83	0.80	0.78	0.75	0.73	0.70	0.68	0.65	0.60	0.55	0.50
240	0.96	0.94	0.92	0.90	0.88	0.85	0.83	0.81	0.79	0.77	0.75	0.73	0.71	0.67	0.63	0.58
280	0.96	0.95	0.93	0.91	0.89	0.88	0.86	0.84	0.82	0.80	0.79	0.77	0.75	0.71	0.68	0.64
320	0.97	0.95	0.94	0.92	0.91	0.89	0.88	0.86	0.84	0.83	0.81	0.80	0.78	0.75	0.72	0.69
360		0.96	0.94	0.93	0.92	0.90	0.89	0.88	0.86	0.85	0.83	0.82	0.81	0.78	0.75	0.72
400		0.96	0.95	0.94	0.93	0.91	0.90	0.89	0.88	0.86	0.85	0.84	0.83	0.80	0.78	0.75
440			0.95	0.94	0.93	0.92	0.91	0.90	0.89	0.88	0.86	0.85	0.84	0.82	0.80	0.77
480			0.96	0.95	0.94	0.93	0.92	0.91	0.90	0.89	0.88	0.86	0.85	0.83	0.81	0.79

* See Commentary for Clause 12.3.3, S16.1-M89 on page 2–20 for more information.

PART FOUR
COMPRESSION MEMBERS

GENERAL INFORMATION

Limits on Width-Thickness Ratios

See page 4–5.

Unit Factored Compressive Resistances for Compression Members, C_r/A

See page 4–10.

Stability Effects

Pages 4–16 to 4–20 describe the procedures given in Clause 8 of CAN/CSA-S16.1-M89 to account for stability effects.

Bending Factors for Beam-Columns

See page 4–21.

ω_1 — Equivalent Uniform Bending Coefficients

Table 4–7, page 4–24, lists values of ω_1 for various ratios, M_{f1}/M_{f2}, of factored end bending moments applied to beam-columns. The values of ω_1 are computed in accordance with the requirements of Clause 13.8.4, S16.1-M89. See the Design of Beam-Columns, page 4–100, for more information.

C_e/A — Euler Buckling Load per Unit Area

Table 4–8, page 4–25, lists values of C_e/A for KL/r ratios varying from 1 to 200. The values of C_e/A have been computed in accordance with the definition in Clause 2.2, S16.1-M89.

Amplification Factor U

Table 4–9, page 4–26, has been prepared to facilitate the design of beam-columns in accordance with the requirements of Clause 13.8, S16.1-M89 which incorporates variable U in the factor U_1. Values of the amplification factor corresponding to various combinations of C_f/C_e are listed.

Factored Axial Compressive Resistances of Columns

See page 4–27.

Design of Beam-Columns

See page 4–100.

Factored Moment Resistances of Columns

See page 4–111.

Double Angle Struts

The tables of factored axial compressive resistances for double angle struts, Imperial series starting on page 4–117 and Metric series starting on page 4–132, are based on the requirements of Clause 13.3.1, S16.1-M89, except that for slender angles the factored axial compressive resistance is computed based on the requirements of Clause 13.3.3, S16.1-M89.

Factored axial compressive resistances with respect to various effective lengths, in millimetres, relative to both the X–X and Y–Y axis, and the U–U and V–V axis for starred angles, are listed for angles made from CSA G40.21-M 300W. F_y for G40.21-M 300W steel angles is 300 MPa for all thicknesses listed.

The factored axial compressive resistances pertaining to effective lengths based on the Y-Y axis have been computed for angles spaced 10 mm back-to-back. To obtain factored compressive resistances for different spacing for Class 1, 2 and 3 angle struts, multiply the actual effective length by the ratio r_y/r_y'

where

r_y = value listed at the bottom of the tables, for angles 10 mm back-to-back

r_y' = r_y listed in tables in Part Six of this Handbook for actual distance back-to-back.

Using the modified effective length, the factored axial compressive resistance value can be selected directly from the tables.

The value r_z tabulated by itself and as part of the following ratios, r_z/r_x, r_z/r_y, r_z/r_u, and r_z/r_v, which appear with the properties of double angle struts, is the radius of gyration for a single angle about the Z–Z axis of that single angle. Values for r_x, r_y, r_u and r_v are those for the double angle strut. See Part Six of this Handbook for a more comprehensive list of angle properties.

The expressions in Clause 13.3.1, defining the axial load, and hence the resistance tables for double angle struts, are based on column strength for rolled W shapes. Double angles struts should be checked as to whether torsional-flexural buckling is critical. For values of KL = 0 for Class 4 angles, C_{ry} may be taken equal to C_{rx} provided that the strut is braced against torsional buckling.

New, more conservative requirements were added to S16.1-M89 (Clause 19.1.4) for determining the effective slenderness ratio of built-up compresssion members based on the number of interconnectors and method of interconnection.

Starred angles must be interconnected at least at the one-third points (Clause 19.1.5, S16.1-M89).

Column Base Plates

See page 4–154.

Anchor Bolts

See page 4–159.

LIMITS ON WIDTH-THICKNESS RATIOS

Table 4–1 below lists the particular width-thickness (b/t, h/w or D/t) ratio limits for various material yield strengths, for each general value given in Table 4–2.

Table 4–2 (page 4–6) which is taken from Clause 11 of CAN/CSA-S16.1-M89 lists the width-thickness ratios for Class 1, 2, and 3 sections for various elements in compression. All sections not meeting these requirements are Class 4.

Table 4–3 lists on page 4–7 WWF members in G40.21-M grade 350W, and on pages 4–8 and 4-9 WWF and W members in G40.21-M grade 300W, which are not Class 1 sections when used as beam-columns. The class for webs in combined flexural and axial compression is a function of the ratio of the factored axial load to the axial compressive load at yield stress, C_f/C_y, as is shown in Table 4–2. Values of C_f/C_y at which the webs change class are tabulated in Table 4–3. Some members with webs that are always Class 1 are controlled by flanges which are not Class 1. Therefore, these members and their flange classification are also included in the table.

Table 4–1
WIDTH-THICKNESS LIMITS

General Value	F_y (MPa)									
	248	280	290	300	320	330	350	380	400	480
$145/\sqrt{F_y}$	9.2	8.7	8.5	8.4	8.1	8.0	7.8	7.4	7.2	6.6
$170/\sqrt{F_y}$	10.8	10.2	10.0	9.8	9.5	9.4	9.1	8.7	8.5	7.8
$200/\sqrt{F_y}$	12.7	12.0	11.7	11.5	11.2	11.0	10.7	10.2	10.0	9.1
$340/\sqrt{F_y}$	21.6	20.3	20.0	19.6	19.0	18.7	18.2	17.4	17.0	15.5
$420/\sqrt{F_y}$	26.7	25.1	24.7	24.2	23.5	23.1	22.4	21.5	21.0	19.2
$525/\sqrt{F_y}$	33.3	31.4	30.8	30.3	29.3	28.9	28.1	26.9	26.2	24.0
$670/\sqrt{F_y}$	42.5	40.0	39.3	38.7	37.4	36.9	35.8	34.4	33.5	30.6
$840/\sqrt{F_y}$	53.3	50.2	49.3	48.5	47.0	46.2	44.9	43.1	42.0	38.3
$1100/\sqrt{F_y}$	69.8	65.7	64.6	63.5	61.5	60.6	58.8	56.4	55.0	50.2
$1700/\sqrt{F_y}$	108.0	101.6	99.8	98.1	95.0	93.6	90.9	87.2	85.0	77.6
$1900/\sqrt{F_y}$	120.7	113.5	111.6	109.7	106.2	104.6	101.6	97.5	95.0	86.7
$13\,000/F_y$	52.4	46.4	44.8	43.3	40.6	39.4	37.1	34.2	32.5	27.1
$18\,000/F_y$	72.6	64.3	62.1	60.0	56.2	54.5	51.4	47.4	45.0	37.5
$23\,000/F_y$	92.7	82.1	79.3	76.7	71.9	69.7	65.7	60.5	57.5	47.9

Table 4–2
WIDTH-THICKNESS RATIOS: COMPRESSION ELEMENTS

Description of Element	Section Classification		
	Class 1 Plastic Design	Class 2 Compact	Class 3 Non-compact
Legs of angles and elements supported along one edge, except as otherwise listed	—	—	$\frac{b}{t} \leq \frac{200}{\sqrt{F_y}}$
Angles in continuous contact with other elements; plate girder stiffeners	—	—	$\frac{b}{t} \leq \frac{200}{\sqrt{F_y}}$
Stems of T-sections	$\frac{b}{t} \leq \frac{145}{\sqrt{F_y}}$ *	$\frac{b}{t} \leq \frac{170}{\sqrt{F_y}}$ *	$\frac{b}{t} \leq \frac{340}{\sqrt{F_y}}$
Flanges of I or T sections; plates projecting from compresion elements; outstanding legs of pairs of angles in continuous contact **	$\frac{b}{t} \leq \frac{145}{\sqrt{F_y}}$	$\frac{b}{t} \leq \frac{170}{\sqrt{F_y}}$	$\frac{b}{t} \leq \frac{200}{\sqrt{F_y}}$
Flanges of channels	—	—	$\frac{b}{t} \leq \frac{200}{\sqrt{F_y}}$
Flanges of rectangular hollow structural sections	$\frac{b}{t} \leq \frac{420}{\sqrt{F_y}}$	$\frac{b}{t} \leq \frac{525}{\sqrt{F_y}}$	$\frac{b}{t} \leq \frac{670}{\sqrt{F_y}}$
Flanges of box sections, flange cover plates and diaphragm plates, between lines of fasteners or welds	$\frac{b}{t} \leq \frac{525}{\sqrt{F_y}}$	$\frac{b}{t} \leq \frac{525}{\sqrt{F_y}}$	$\frac{b}{t} \leq \frac{670}{\sqrt{F_y}}$
Perforated cover plates	—	—	$\frac{b}{t} \leq \frac{840}{\sqrt{F_y}}$
Webs in axial compression	$\frac{h}{w} \leq \frac{670}{\sqrt{F_y}}$	$\frac{h}{w} \leq \frac{670}{\sqrt{F_y}}$	$\frac{h}{w} \leq \frac{670}{\sqrt{F_y}}$
Webs in flexural compression	$\frac{h}{w} \leq \frac{1100}{\sqrt{F_y}}$	$\frac{h}{w} \leq \frac{1700}{\sqrt{F_y}}$	$\frac{h}{w} \leq \frac{1900}{\sqrt{F_y}}$
Webs in combined flexural and axial compression	$\frac{h}{w} \leq \frac{1100}{\sqrt{F_y}}\left(1 - 0.39\frac{C_f}{C_y}\right)$	$\frac{h}{w} \leq \frac{1700}{\sqrt{F_y}}\left(1 - 0.61\frac{C_f}{C_y}\right)$	$\frac{h}{w} \leq \frac{1900}{\sqrt{F_y}}\left(1 - 0.65\frac{C_f}{C_y}\right)$
Circular hollow sections	$\frac{D}{t} \leq \frac{13\,000}{F_y}$	$\frac{D}{t} \leq \frac{18\,000}{F_y}$	$\frac{D}{t} \leq \frac{23\,000}{F_y}$

* See Clauses 11.1.2 and 11.1.3 of CAN/CSA-S16.1-M89 (page 1-34).

**Can be considered as Class 1 or Class 2 sections if angles are continuously connected by adequate mechanical fasteners or welds, and there is an axis of symmetry in the plane of loading.

For Class 4 sections, see Clause 11 of CAN/CSA-S16.1-M89.

Table 4–3 CLASS OF SECTIONS FOR BEAM-COLUMNS *

G40.21-M 350W

Designation	Web			Flange
	1 $C_f/C_y \le$	2 $C_f/C_y \le$	3 $C_f/C_y \le$	
WWF2000X732	–	–	0.099	1
X648	–	–	0.084	1
X607	–	–	0.077	2
X542	–	–	0.069	2
WWF1800X700	–	0.106	0.251	1
X659	–	0.097	0.243	1
X617	–	0.088	0.236	1
X575	–	0.079	0.228	2
X510	–	0.070	0.221	2
WWF1600X622	–	–	0.118	1
X580	–	–	0.109	1
X538	–	–	0.099	1
X496	–	–	0.090	2
X431	–	–	0.080	2
WWF1400X597	–	0.174	0.308	1
X513	–	0.151	0.289	1
X471	–	0.140	0.279	2
X405	–	0.128	0.270	2
X358	–	0.128	0.270	2
WWF1200X487	–	0.376	0.478	1
X418	–	0.365	0.469	1
X380	–	0.354	0.459	2
X333	–	0.354	0.459	1
X302	–	0.343	0.450	2
X263	–	0.343	0.450	1
WWF1100X458	–	0.325	0.435	1
X388	–	0.312	0.424	1
X351	–	0.299	0.413	2
X304	–	0.299	0.413	1
X273	–	0.286	0.402	2
X234	–	0.286	0.402	1
WWF1000X447	–	0.454	0.543	1
X377	–	0.441	0.532	1
X340	–	0.428	0.521	2
X293	–	0.428	0.521	1
X262	–	0.415	0.511	2
X223	–	0.415	0.511	1
X200	–	0.402	0.500	1
WWF900X417	–	0.294	0.409	1
X437	–	0.278	0.395	1
X309	–	0.262	0.382	2
X262	–	0.262	0.382	1
X231	–	0.245	0.368	2
X192	–	0.245	0.368	1
X169	–	0.229	0.354	1
WWF800X339	–	0.442	0.533	1
X300	–	0.426	0.519	2
X253	–	0.426	0.519	1
X223	–	0.409	0.506	2
X184	–	0.409	0.506	1
X161	–	0.393	0.492	1
WWF700X245	0.027	0.590	0.657	1
X214	–	0.573	0.643	2
X196	–	0.563	0.635	3
X185	–	0.573	0.643	1
X152	–	0.557	0.630	1
WWF650X499	1.0	–	–	2
X400	1.0	–	–	4
WWF600X369	1.0	–	–	3
WWF550X280	1.0	–	–	4
WWF500X306	1.0	–	–	2
X276	1.0	–	–	2
X254	1.0	–	–	3
X223	1.0	–	–	4
X197	0.740	0.885	0.905	4
WWF450X248	1.0	–	–	2
X228	1.0	–	–	2
X201	1.0	–	–	3
X177	0.939	0.967	0.974	4
WWF400X202	1.0	–	–	2
X178	1.0	–	–	3
X157	1.0	–	–	3
WWF350X155	1.0	–	–	2
X137	1.0	–	–	2

See Table 4–2, page 4–6, for width-thickness criteria.

* Sections not listed are always Class 1.

– Indicates web is never that class.

CLASS OF SECTIONS FOR BEAM-COLUMNS *

G40.21-M 300W

Table 4–3

Designation	Web			Flange
	1 $C_f/C_y \le$	2 $C_f/C_y \le$	3 $C_f/C_y \le$	
WWF2000X732	–	0.053	0.206	1
X648	–	0.036	0.192	1
X607	–	0.028	0.185	1
X542	–	0.019	0.178	1
WWF1800X700	–	0.220	0.346	1
X659	–	0.211	0.339	1
X617	–	0.203	0.332	1
X575	–	0.195	0.325	1
X510	–	0.186	0.318	1
WWF1600X622	–	0.073	0.224	1
X580	–	0.063	0.215	1
X538	–	0.053	0.206	1
X496	–	0.042	0.197	1
X431	–	0.032	0.189	1
WWF1400X597	–	0.282	0.399	1
X513	–	0.261	0.381	1
X471	–	0.251	0.373	1
X405	–	0.241	0.364	1
X358	–	0.241	0.364	1
WWF1200X487	–	0.470	0.557	1
X418	–	0.460	0.548	1
X380	–	0.449	0.539	1
X333	–	0.449	0.539	1
X302	–	0.439	0.530	1
X263	–	0.439	0.530	1
WWF1100X458	–	0.422	0.517	1
X388	–	0.411	0.507	1
X351	–	0.399	0.497	1
X304	–	0.399	0.497	1
X273	–	0.387	0.487	1
X234	–	0.387	0.487	1
WWF1000X447	–	0.542	0.617	1
X377	–	0.530	0.607	1
X340	–	0.518	0.597	1
X293	–	0.518	0.597	1
X262	–	0.506	0.587	1
X223	–	0.506	0.587	1
X200	–	0.494	0.577	1
WWF900X417	–	0.394	0.493	1
X437	–	0.379	0.480	1
X309	–	0.364	0.467	1
X262	–	0.364	0.467	1
X231	–	0.349	0.455	1
X192	–	0.349	0.455	1
X169	–	0.334	0.442	1
WWF800X339	–	0.531	0.608	1
X300	–	0.516	0.595	1
X253	–	0.516	0.595	1
X223	–	0.501	0.582	1
X184	–	0.501	0.582	1
X161	–	0.485	0.569	1

Designation	Web			Flange
	1 $C_f/C_y \le$	2 $C_f/C_y \le$	3 $C_f/C_y \le$	
WWF700X245	0.215	0.668	0.722	1
X214	0.178	0.652	0.710	1
X196	0.156	0.643	0.702	2
X185	0.178	0.652	0.710	1
X152	0.142	0.637	0.697	1
WWF650X400	1.0	–	–	3
WWF600X369	1.0	–	–	3
WWF550X280	1.0	–	–	3
WWF500X276	1.0	–	–	2
X254	1.0	–	–	3
X223	1.0	–	–	3
X197	0.876	0.941	0.952	4
WWF450X228	1.0	–	–	2
X201	1.0	–	–	3
X177	1.0	–	–	3
WWF400X178	1.0	–	–	2
X157	1.0	–	–	3
WWF350X137	1.0	–	–	2
W1000X488	0.942	0.968	0.975	1
X444	0.788	0.905	0.922	1
X399	0.592	0.823	0.853	1
X363	0.483	0.778	0.815	1
X328	0.483	0.778	0.816	1
X286	0.483	0.778	0.816	3
W1000X443	0.976	0.982	0.987	1
X412	0.788	0.905	0.922	1
X371	0.593	0.824	0.854	1
X321	0.293	0.700	0.750	1
X296	0.294	0.700	0.750	1
W1000X272	0.293	0.700	0.750	1
X249	0.293	0.700	0.750	1
X222	0.223	0.671	0.725	1
W920X387	0.957	0.975	0.980	1
X365	0.879	0.942	0.953	1
X342	0.790	0.905	0.922	1
W920X313	0.913	0.956	0.965	1
X289	0.768	0.896	0.915	1
X271	0.670	0.856	0.881	1
X253	0.550	0.806	0.839	1
X238	0.452	0.766	0.805	1
X223	0.372	0.733	0.777	1
X201	0.272	0.691	0.742	1

See Table 4–2, page 4–6, for width-thickness criteria.

* Sections not listed are always Class 1.

– Indicates web is never that class.

Table 4–3 CLASS OF SECTIONS FOR BEAM-COLUMNS *

G40.21-M 300W

Designation	Web 1 $C_f/C_y \le$	Web 2 $C_f/C_y \le$	Web 3 $C_f/C_y \le$	Flange
W840X329	0.930	0.963	0.971	1
X299	0.797	0.908	0.925	1
W840X226	0.564	0.812	0.844	1
X210	0.474	0.775	0.812	1
X193	0.376	0.734	0.778	1
X176	0.265	0.688	0.740	1
W760X284	0.952	0.972	0.978	1
X257	0.816	0.916	0.931	1
W760X196	0.703	0.869	0.892	1
X185	0.616	0.834	0.862	1
X173	0.549	0.806	0.838	1
X161	0.459	0.769	0.807	1
X147	0.365	0.730	0.775	1
X134	0.125	0.630	0.691	2
W690X217	0.872	0.939	0.951	1
W690X170	0.766	0.895	0.914	1
X152	0.574	0.816	0.847	1
X140	0.460	0.769	0.808	1
X125	0.337	0.718	0.765	1
W610X174	0.912	0.956	0.965	1
X155	0.742	0.886	0.906	2
W610X140	0.799	0.909	0.925	1
X125	0.621	0.835	0.863	1
X113	0.497	0.784	0.820	1
X101	0.360	0.728	0.773	1
X91	0.168	0.648	0.706	2
X84	0.001	0.579	0.648	2
W610X92	0.442	0.762	0.801	1
X82	0.249	0.682	0.734	1
W530X150	0.967	0.979	0.984	1
W530X109	0.819	0.917	0.932	1
X101	0.704	0.870	0.892	1
X92	0.578	0.818	0.849	1
X82	0.433	0.758	0.798	1
X72	0.286	0.697	0.747	2
W530X85	0.596	0.825	0.855	1
X74	0.475	0.775	0.813	1
X66	0.286	0.697	0.747	1
W460X113	0.963	0.977	0.982	1
W460X89	0.920	0.959	0.967	1
X82	0.819	0.917	0.932	1
X74	0.644	0.845	0.872	1
X67	0.528	0.797	0.831	1
X61	0.431	0.757	0.798	2
W460X68	0.664	0.853	0.879	1
X60	0.402	0.745	0.787	1
X52	0.288	0.698	0.748	1
W410X74	0.978	0.983	0.988	1
X67	0.815	0.916	0.931	1
X60	0.564	0.812	0.844	1
X54	0.512	0.790	0.826	1
W410X46	0.369	0.731	0.776	1
X39	0.158	0.644	0.703	1
W360X162	1.0	–	–	2
X147	1.0	–	–	2
X134	1.0	–	–	3
W360X64	0.886	0.945	0.956	1
W360X57	0.871	0.939	0.950	1
X51	0.706	0.871	0.893	1
X45	0.619	0.835	0.863	2
W360X39	0.504	0.787	0.823	1
X33	0.253	0.683	0.736	1
W310X107	1.0	–	–	2
X97	1.0	–	–	3
W310X79	1.0	–	–	2
W310X45	0.786	0.904	0.921	1
X39	0.541	0.802	0.836	2
X31	0.188	0.657	0.713	3
W310X33	0.782	0.902	0.919	1
X28	0.605	0.829	0.858	1
X24	0.462	0.770	0.808	1
X21	0.256	0.684	0.737	2
W250X73	1.0	–	–	2
W250X49	1.0	–	–	2
W250X33	0.977	0.983	0.987	1
X24	0.604	0.829	0.858	3
W250X25	0.974	0.982	0.986	1
X22	0.892	0.948	0.958	1
X18	0.542	0.803	0.836	2
W200X46	1.0	–	–	2
W200X21	1.0	–	–	3
W200X15	0.784	0.903	0.920	2
W150X22	1.0	–	–	3
W150X14	1.0	–	–	2

See Table 4–2, page 4–6, for width-thickness criteria.

* Sections not listed are always Class 1.

– Indicates web is never that class.

UNIT FACTORED COMPRESSIVE RESISTANCES FOR COMPRESSION MEMBERS, C_r/A

General

Table 4–4 on pages 4–11 to 4–14 lists in MPa the unit factored compressive resistance, C_r/A, calculated in accordance with the requirements of Clause 13.3.1, CAN/CSA-S16.1-M89 for members with F_y varying from 248 to 700 MPa, for values of KL /r from 1 to 200. The values for F_y = 300 MPa are printed in boldface type. For hollow structural sections manufactured according to G40.20-M, Class C with F_y = 350 MPa, use values given in Table 4–4 for F_y = 350.

Table 4–5 on page 4–15 lists the unit factored compressive resistance, C_r /A, for compression members consisting of HSS manufactured according to G40.20-M, Class H with F_y = 350 MPa. The resistances have been calculated, for values of KL /r from 1 to 200, in accordance with the requirements of Clause 13.3.2, S16.1-M89.

Use

To obtain the factored compressive resistance, C_r, for doubly symmetric Class 1, 2 or 3 sections, multiply the unit factored compressive resistance, C_r /A, for the appropriate F_y and KL /r ratio, by the cross-sectional area, A, of the column section.

Examples

1. Given:

Find the factored compressive resistance of a W250x131 column of ASTM A36 steel (F_y = 248 MPa) for a KL /r ratio of 89.

Solution:

From page 6–58, for W250x131, A = 16 700 mm^2.

From Table 4–4, page 4–12, with KL /r = 89 and F_y = 248 MPa, C_r /A = 137 MPa

Therefore, $C_r = 137\ \text{MPa} \times 16\ 700\ \text{mm}^2 = 2\ 290 \times 10^3\ \text{N} = 2\ 290\ \text{kN}$

2. Given:

Find the factored compressive resistance of an HSS 254x152x11 Class H column for F_y = 350 MPa and KL /r = 89.

Solution:

From page 6–98 for HSS 254x152x11, A = 8 230 mm^2

From Table 4–5, page 4–15, with KL /r = 89 and F_y = 350 MPa, C_r /A = 195 MPa

Therefore, $C_r = 195\ \text{MPa} \times 8\ 230\ \text{mm}^2 = 1\ 600 \times 10^3\ \text{N} = 1600\ \text{kN}$

Notes:

1. Tables of C_r, factored axial compressive resistance, for columns in CSA G40.21-M grade 300W steels are given on pages 4–30 to 4–42, and for grade 350W on pages 4-44 to 4-50.
2. Tables of C_r for HSS Class C columns in G40.21-M grade 350W steel are given on pages 4–52 to 4–75.
3. Tables of C_r for HSS Class H columns in G40.21-M grade 350W steel are given on pages 4–76 to 4–99.
4. For columns not manufactured in Canada, heavy sections, and built-up sections, see Clause 13.3 of the CISC Commentary in Part Two of this Handbook for more information on compressive resistance.

UNIT FACTORED COMPRESSIVE RESISTANCES, C_r/A (MPa)* FOR COMPRESSION MEMBERS

$\frac{KL}{r}$ = 1 to 50

$\phi = 0.90$

Table 4–4

KL/r	F_y (MPa)									
	248	260	280	290	**300**	350	380	400	480	700
1	223	234	252	261	**270**	315	342	360	432	630
2	223	234	252	261	**270**	315	342	360	432	630
3	223	234	252	261	**270**	315	342	360	432	630
4	223	234	252	261	**270**	315	342	360	432	630
5	223	234	252	261	**270**	315	342	360	432	630
6	223	234	252	261	**270**	315	342	360	432	630
7	223	234	252	261	**270**	315	342	360	432	630
8	223	234	252	261	**270**	315	342	360	432	630
9	223	234	252	261	**270**	315	342	360	432	626
10	223	234	252	261	**270**	315	342	360	431	623
11	223	234	252	261	**270**	315	342	359	429	620
12	223	234	252	261	**270**	314	340	358	427	616
13	223	234	252	260	**269**	313	339	356	425	613
14	223	233	251	260	**268**	312	338	355	423	609
15	222	233	250	259	**267**	311	336	353	421	605
16	221	232	249	258	**266**	309	335	352	419	601
17	221	231	248	257	**265**	308	333	350	417	597
18	220	230	247	256	**264**	307	332	349	415	593
19	219	229	246	255	**263**	305	330	347	413	589
20	218	229	246	254	**262**	304	329	345	411	584
21	218	228	245	253	**261**	303	327	344	408	580
22	217	227	244	252	**260**	301	326	342	406	575
23	216	226	243	251	**259**	300	324	340	403	571
24	215	225	242	250	**258**	299	323	338	401	566
25	214	224	241	249	**257**	297	321	337	399	561
26	214	223	240	248	**256**	296	319	335	396	556
27	213	223	239	247	**255**	294	317	333	393	551
28	212	222	238	246	**253**	293	316	331	391	546
29	211	221	237	244	**252**	291	314	329	388	541
30	210	220	235	243	**251**	289	312	327	385	536
31	209	219	234	242	**250**	288	310	325	383	530
32	208	218	233	241	**249**	286	308	323	380	525
33	208	217	232	240	**247**	285	306	321	377	519
34	207	216	231	239	**246**	283	304	319	374	513
35	206	215	230	237	**245**	281	303	317	371	507
36	205	214	229	236	**243**	279	301	314	368	501
37	204	213	228	235	**242**	278	298	312	365	495
38	203	212	226	234	**241**	276	296	310	362	489
39	202	211	225	232	**239**	274	294	308	359	483
40	201	210	224	231	**238**	272	292	305	355	477
41	200	208	223	230	**237**	270	290	303	352	470
42	199	207	221	228	**235**	269	288	301	349	464
43	198	206	220	227	**234**	267	286	298	345	457
44	197	205	219	226	**232**	265	283	296	342	451
45	196	204	217	224	**231**	263	281	293	339	444
46	195	203	216	223	**229**	261	279	291	335	437
47	194	202	215	221	**228**	259	277	288	332	430
48	192	200	213	220	**226**	257	274	286	328	423
49	191	199	212	218	**225**	255	272	283	324	416
50	190	198	211	217	**223**	253	270	280	321	408

*Calculated in accordance with Clause 13.3.1 of CAN/CSA-S16.1-M89
For Class H hollow structural sections, see Table 4-5, page 4-15

$\frac{KL}{r}$ = 51 to 100

Table 4–4

UNIT FACTORED COMPRESSIVE RESISTANCES, C_r/A (MPa)* FOR COMPRESSION MEMBERS

$\phi = 0.90$

$\frac{KL}{r}$	F_y (MPa)									
	248	260	280	290	**300**	350	380	400	480	700
51	189	197	209	215	**221**	251	267	278	317	401
52	188	195	208	214	**220**	248	265	275	313	393
53	187	194	206	212	**218**	246	262	272	309	386
54	186	193	205	211	**217**	244	260	269	306	377
55	184	192	203	209	**215**	242	257	267	302	368
56	183	190	202	208	**213**	240	254	264	298	359
57	182	189	200	206	**212**	237	252	261	294	351
58	181	188	199	204	**210**	235	249	258	290	343
59	180	186	197	203	**208**	233	247	255	286	335
60	178	185	196	201	**206**	231	244	252	282	328
61	177	184	194	199	**205**	228	241	249	277	320
62	176	182	193	198	**203**	226	238	246	273	313
63	174	181	191	196	**201**	223	236	243	269	307
64	173	179	190	194	**199**	221	233	240	265	300
65	172	178	188	193	**197**	219	230	237	260	294
66	171	177	186	191	**195**	216	227	234	254	288
67	169	175	185	189	**194**	214	224	231	249	282
68	168	174	183	187	**192**	211	221	227	245	276
69	167	172	181	186	**190**	209	218	224	240	271
70	165	171	179	184	**188**	206	215	221	235	266
71	164	169	178	182	**186**	203	212	217	231	260
72	162	168	176	180	**184**	201	209	213	227	255
73	161	166	174	178	**182**	198	206	209	222	251
74	160	165	173	176	**180**	195	202	206	218	246
75	158	163	171	174	**178**	193	199	202	214	241
76	157	161	169	172	**176**	190	195	198	211	237
77	155	160	167	170	**174**	186	192	195	207	232
78	154	158	165	168	**172**	183	188	192	203	228
79	152	157	163	167	**169**	180	185	188	200	224
80	151	155	162	165	**167**	177	182	185	196	220
81	149	153	160	163	**165**	174	179	182	193	216
82	148	152	158	160	**163**	171	176	179	190	213
83	146	150	156	158	**160**	169	173	176	187	209
84	145	148	154	156	**158**	166	171	173	184	205
85	143	147	152	154	**155**	163	168	171	181	202
86	142	145	149	151	**153**	161	165	168	178	198
87	140	143	147	149	**151**	158	163	165	175	195
88	138	141	145	147	**148**	156	160	163	172	192
89	137	139	143	144	**146**	154	158	160	170	189
90	135	137	141	142	**144**	151	155	158	167	186
91	133	135	139	140	**142**	149	153	155	164	183
92	131	133	137	138	**140**	147	151	153	162	180
93	129	131	135	136	**138**	145	148	151	159	177
94	127	129	133	134	**136**	143	146	149	157	174
95	126	128	131	132	**134**	141	144	146	155	171
96	124	126	129	130	**132**	139	142	144	152	168
97	122	124	127	129	**130**	137	140	142	150	166
98	121	122	125	127	**128**	135	138	140	148	163
99	119	121	124	125	**126**	133	136	138	146	161
100	117	119	122	123	**125**	131	134	136	144	158

*Calculated in accordance with Clause 13.3.1 of **CAN/CSA-S16.1-M89**
For Class H hollow structural sections, see Table **4-5**, page **4-15**

UNIT FACTORED COMPRESSIVE RESISTANCES, C_r/A (MPa)* FOR COMPRESSION MEMBERS

$\phi = 0.90$

$\frac{KL}{r} = 101$ to 150

Table 4–4

$\frac{KL}{r}$	F_y (MPa)									
	248	260	280	290	**300**	350	380	400	480	700
101	116	118	120	122	**123**	129	132	134	142	156
102	114	116	119	120	**121**	127	131	133	140	154
103	113	114	117	119	**120**	126	129	131	138	151
104	111	113	116	117	**118**	124	127	129	136	149
105	110	112	114	115	**117**	122	125	127	134	147
106	108	110	113	114	**115**	121	124	126	132	145
107	107	109	111	113	**114**	119	122	124	130	142
108	106	107	110	111	**112**	118	120	122	128	139
109	104	106	108	110	**111**	116	119	121	127	137
110	103	105	107	108	**109**	115	117	119	125	134
111	102	103	106	107	**108**	113	116	117	123	132
112	101	102	104	106	**107**	112	114	116	122	130
113	99.4	101	103	104	**105**	110	113	114	120	128
114	98.2	99.7	102	103	**104**	109	111	113	118	126
115	97.0	98.5	101	102	**103**	108	110	112	117	123
116	95.9	97.3	99.5	101	**102**	106	109	110	115	121
117	94.8	96.1	98.3	99.4	**100**	105	107	109	114	119
118	93.7	95.0	97.2	98.2	**99.2**	104	106	107	112	118
119	92.6	93.9	96.0	97.0	**98.0**	102	105	106	111	116
120	91.5	92.8	94.9	95.9	**96.8**	101	103	105	110	114
121	90.4	91.8	93.8	94.8	**95.7**	99.9	102	104	108	112
122	89.4	90.7	92.7	93.7	**94.6**	98.7	101	102	107	110
123	88.4	89.7	91.6	92.6	**93.5**	97.6	99.7	101	106	109
124	87.4	88.7	90.6	91.5	**92.4**	96.4	98.5	99.8	104	107
125	86.4	87.7	89.6	90.5	**91.4**	95.3	97.3	98.6	103	105
126	85.5	86.7	88.6	89.5	**90.3**	94.2	96.2	97.4	102	104
127	84.5	85.7	87.6	88.4	**89.3**	93.1	95.1	96.3	100	102
128	83.6	84.8	86.6	87.5	**88.3**	92.0	93.9	95.1	99.1	101
129	82.7	83.8	85.6	86.5	**87.3**	91.0	92.9	94.0	97.5	99.3
130	81.8	82.9	84.7	85.5	**86.3**	89.9	91.8	92.9	96.1	97.9
131	80.9	82.0	83.8	84.6	**85.4**	88.9	90.7	91.8	94.7	96.5
132	80.0	81.1	82.8	83.6	**84.4**	87.9	89.7	90.8	93.3	95.1
133	79.2	80.3	81.9	82.7	**83.5**	86.9	88.6	89.7	92.0	93.7
134	78.3	79.4	81.1	81.8	**82.6**	85.9	87.6	88.7	90.7	92.4
135	77.5	78.6	80.2	81.0	**81.7**	85.0	86.6	87.7	89.4	91.2
136	76.7	77.7	79.3	80.1	**80.8**	84.0	85.7	86.7	88.1	89.9
137	75.9	76.9	78.5	79.2	**79.9**	83.1	84.7	85.7	86.9	88.7
138	75.1	76.1	77.7	78.4	**79.1**	82.2	83.8	84.7	85.7	87.5
139	74.3	75.3	76.8	77.6	**78.2**	81.3	82.8	83.8	84.5	86.3
140	73.6	74.5	76.0	76.7	**77.4**	80.4	81.9	82.8	83.4	85.2
141	72.8	73.8	75.2	75.9	**76.6**	79.5	81.0	81.6	82.3	84.0
142	72.1	73.0	74.5	75.1	**75.8**	78.7	80.1	80.5	81.2	82.9
143	71.3	72.3	73.7	74.4	**75.0**	77.8	79.2	79.4	80.1	81.9
144	70.6	71.5	72.9	73.6	**74.2**	77.0	78.4	78.4	79.0	80.8
145	69.9	70.8	72.2	72.8	**73.4**	76.1	77.2	77.3	78.0	79.8
146	69.2	70.1	71.4	72.1	**72.7**	75.3	76.2	76.3	77.0	78.8
147	68.5	69.4	70.7	71.3	**71.9**	74.5	75.2	75.3	76.0	77.8
148	67.9	68.7	70.0	70.6	**71.2**	73.7	74.2	74.4	75.0	76.8
149	67.2	68.0	69.3	69.9	**70.5**	73.0	73.3	73.4	74.1	75.8
150	66.5	67.3	68.6	69.2	**69.8**	72.2	72.3	72.5	73.1	74.9

*Calculated in accordance with Clause 13.3.1 of **CAN/CSA-S16.1-M89**
For Class H hollow structural sections, see Table **4-5**, page **4-15**

$\frac{KL}{r}$ = 151 to 200

Table 4–4

UNIT FACTORED COMPRESSIVE RESISTANCES, C_r/A (MPa)* FOR COMPRESSION MEMBERS

$\phi = 0.90$

$\frac{KL}{r}$	F_y (MPa)									
	248	260	280	290	**300**	350	380	400	480	700
151	65.9	66.7	67.9	68.5	**69.1**	71.2	71.4	71.6	72.2	74.0
152	65.2	66.0	67.2	67.8	**68.4**	70.3	70.5	70.7	71.3	73.1
153	64.6	65.4	66.6	67.1	**67.7**	69.4	69.6	69.8	70.4	72.2
154	64.0	64.7	65.9	66.5	**67.0**	68.5	68.8	68.9	69.6	71.4
155	63.4	64.1	65.3	65.8	**66.3**	67.7	67.9	68.1	68.7	70.5
156	62.8	63.5	64.6	65.2	**65.7**	66.9	67.1	67.3	67.9	69.7
157	62.2	62.9	64.0	64.5	**65.0**	66.0	66.3	66.4	67.1	68.9
158	61.6	62.3	63.4	63.9	**64.4**	65.2	65.5	65.7	66.3	68.1
159	61.0	61.7	62.8	63.3	**63.7**	64.5	64.7	64.9	65.5	67.3
160	60.4	61.1	62.2	62.7	**63.1**	63.7	63.9	64.1	64.7	66.5
161	59.8	60.5	61.6	62.1	**62.5**	62.9	63.2	63.3	64.0	65.8
162	59.3	60.0	61.0	61.5	**61.9**	62.2	62.4	62.6	63.3	65.0
163	58.7	59.4	60.4	60.9	**61.1**	61.5	61.7	61.9	62.5	64.3
164	58.2	58.8	59.8	60.3	**60.4**	60.8	61.0	61.2	61.8	63.6
165	57.7	58.3	59.3	59.7	**59.7**	60.1	60.3	60.5	61.1	62.9
166	57.1	57.8	58.7	58.9	**59.0**	59.4	59.6	59.8	60.4	62.2
167	56.6	57.2	58.1	58.2	**58.3**	58.7	58.9	59.1	59.8	61.5
168	56.1	56.7	57.5	57.6	**57.6**	58.0	58.3	58.4	59.1	60.9
169	55.6	56.2	56.8	56.9	**57.0**	57.4	57.6	57.8	58.4	60.2
170	55.1	55.7	56.2	56.3	**56.3**	56.7	57.0	57.2	57.8	59.6
171	54.6	55.1	55.5	55.6	**55.7**	56.1	56.4	56.5	57.2	59.0
172	54.1	54.6	54.9	55.0	**55.1**	55.5	55.7	55.9	56.6	58.3
173	53.6	54.1	54.3	54.4	**54.5**	54.9	55.1	55.3	55.9	57.7
174	53.1	53.7	53.7	53.8	**53.9**	54.3	54.5	54.7	55.3	57.1
175	52.6	53.0	53.1	53.2	**53.3**	53.7	54.0	54.1	54.8	56.5
176	52.2	52.4	52.6	52.6	**52.7**	53.1	53.4	53.5	54.2	56.0
177	51.7	51.8	52.0	52.1	**52.2**	52.6	52.8	53.0	53.6	55.4
178	51.3	51.3	51.4	51.5	**51.6**	52.0	52.3	52.4	53.1	54.8
179	50.6	50.7	50.9	51.0	**51.1**	51.5	51.7	51.9	52.5	54.3
180	50.1	50.2	50.4	50.4	**50.5**	50.9	51.2	51.3	52.0	53.8
181	49.6	49.7	49.8	49.9	**50.0**	50.4	50.6	50.8	51.4	53.2
182	49.0	49.1	49.3	49.4	**49.5**	49.9	50.1	50.3	50.9	52.7
183	48.5	48.6	48.8	48.9	**49.0**	49.4	49.6	49.8	50.4	52.2
184	48.0	48.1	48.3	48.4	**48.4**	48.9	49.1	49.3	49.9	51.7
185	47.5	47.6	47.8	47.9	**48.0**	48.4	48.6	48.8	49.4	51.2
186	47.0	47.1	47.3	47.4	**47.5**	47.9	48.1	48.3	48.9	50.7
187	46.6	46.7	46.8	46.9	**47.0**	47.4	47.6	47.8	48.4	50.2
188	46.1	46.2	46.3	46.4	**46.5**	46.9	47.2	47.3	48.0	49.8
189	45.6	45.7	45.9	46.0	**46.0**	46.5	46.7	46.9	47.5	49.3
190	45.2	45.3	45.4	45.5	**45.6**	46.0	46.2	46.4	47.0	48.8
191	44.7	44.8	45.0	45.1	**45.1**	45.5	45.8	45.9	46.6	48.4
192	44.3	44.4	44.5	44.6	**44.7**	45.1	45.3	45.5	46.2	48.2
193	43.8	43.9	44.1	44.2	**44.3**	44.7	44.9	45.1	45.7	47.7
194	43.4	43.5	43.7	43.7	**43.8**	44.2	44.5	44.6	45.3	47.2
195	43.0	43.1	43.2	43.3	**43.4**	43.8	44.1	44.2	44.9	46.7
196	42.6	42.7	42.8	42.9	**43.0**	43.4	43.6	43.8	44.4	46.2
197	42.2	42.3	42.4	42.5	**42.6**	43.0	43.2	43.4	44.0	45.8
198	41.8	41.8	42.0	42.1	**42.2**	42.6	42.8	43.0	43.6	45.3
199	41.4	41.4	41.6	41.7	**41.8**	42.2	42.4	42.6	43.2	44.9
200	41.0	41.1	41.2	41.3	**41.4**	41.8	42.0	42.2	42.8	44.4

*Calculated in accordance with Clause 13.3.1 of **CAN/CSA-S16.1-M89**
For Class H hollow structural sections, see Table **4-5**, page **4-15**

UNIT FACTORED COMPRESSIVE RESISTANCES, C_r/A (MPa)* FOR CLASS H HOLLOW STRUCTURAL SECTIONS

HSS CLASS H
F_y = 350 MPa

$\phi = 0.90$

Table 4–5

$\frac{KL}{r}$	$\frac{C_r}{A}$ (MPa)	$\frac{KL}{r}$	$\frac{C_r}{A}$ (MPa)	$\frac{KL}{r}$	$\frac{C_r}{A}$ (MPa)	$\frac{KL}{r}$	$\frac{C_r}{A}$ (MPa)	$\frac{KL}{r}$	$\frac{C_r}{A}$ (MPa)
1	315	41	298	81	219	121	113	161	67.1
2	315	42	297	82	216	122	112	162	66.3
3	315	43	296	83	213	123	110	163	65.5
4	315	44	295	84	210	124	109	164	64.7
5	315	45	293	85	207	125	107	165	64.0
6	315	46	292	86	204	126	106	166	63.3
7	315	47	291	87	201	127	104	167	62.5
8	315	48	289	88	198	128	103	168	61.8
9	315	49	288	89	195	129	102	169	61.1
10	315	50	286	90	192	130	100	170	60.4
11	315	51	285	91	188	131	99.0	171	59.8
12	315	52	283	92	184	132	97.7	172	59.1
13	315	53	281	93	181	133	96.5	173	58.4
14	315	54	280	94	177	134	95.3	174	57.8
15	315	55	278	95	174	135	94.1	175	57.2
16	315	56	276	96	170	136	93.0	176	56.5
17	315	57	274	97	167	137	91.7	177	55.9
18	314	58	273	98	164	138	90.4	178	55.3
19	314	59	271	99	161	139	89.1	179	54.7
20	314	60	269	100	158	140	87.9	180	54.2
21	314	61	267	101	156	141	86.7	181	53.6
22	313	62	265	102	153	142	85.5	182	53.0
23	313	63	263	103	150	143	84.4	183	52.5
24	312	64	261	104	148	144	83.2	184	51.9
25	312	65	259	105	145	145	82.1	185	51.4
26	311	66	256	106	143	146	81.0	186	50.9
27	311	67	254	107	140	147	80.0	187	50.4
28	310	68	252	108	138	148	78.9	188	49.9
29	309	69	250	109	136	149	77.9	189	49.4
30	309	70	247	110	134	150	76.9	190	48.9
31	308	71	245	111	132	151	75.9	191	48.4
32	307	72	242	112	130	152	75.0	192	47.9
33	306	73	240	113	128	153	74.0	193	47.4
34	306	74	237	114	126	154	73.1	194	47.0
35	305	75	235	115	124	155	72.2	195	46.5
36	304	76	232	116	122	156	71.3	196	46.1
37	303	77	230	117	120	157	70.4	197	45.6
38	302	78	227	118	118	158	69.6	198	45.2
39	301	79	224	119	117	159	68.7	199	44.8
40	300	80	222	120	115	160	67.9	200	44.4

* Calculated in accordance with Clause 13.3.2 of CAN/CSA-S16.1-M89. For Class C hollow structural sections, use Table 4–4, pages 4–11 to 4–14.

STABILITY EFFECTS

General

Clause 8.6 of CSA standard CAN/CSA-S16.1-M89 stipulates that the sway effects produced by the vertical loads acting on the structure in its displaced configuration be accounted for directly in the analysis for forces and moments. This can be done by 1) performing a second-order geometric analysis or 2) accounting for the sway effects by a) amplifying the first-order elastic translational moments by a factor U_2, or b) performing a $P\Delta$ analysis according to Appendix J. The approximate method of accounting for the sway effects by using K factors larger than 1.0 in determining the factored resistance of a beam-column is no longer an acceptable alternative.

While some software exists for a second-order geometric analysis, most designers will probably be more accustomed to using a first order elastic analysis. Thus, the methods of the second option given by S16.1-M89 to account for sway effects will be preferred, and are illustrated here.

Amplification Factor U_2

Clause 8.6.1 of S16.1 provides the designer with the option of multiplying the first order translational moments, ie. the moments due to the lateral loads, by the factor U_2 and adding these amplified moments to those caused by the gravity loads to obtain the moments used in the design of the beam-column. The amplification factor U_2 is taken as

$$U_2 = \frac{1}{1 - \left(\dfrac{\Sigma C_f \Delta_f}{\Sigma V_f h}\right)}$$

For situations where gravity loads predominate, a minimum pseudo-lateral load equal to 0.005 times the gravity loads acting at each storey is required by Clause 8.6.2.

$P\Delta$ Method of Analysis

Appendix J, Guide to Calculation of Stability Effects, of S16.1-M89 describes an iterative procedure whereby artificial storey shears are computed and added to the primary horizontal loads in computing the moments and forces in the structure. This method accounts directly for the sway effects, the stiffness and the actual deflected shape of the lateral load resisting system.

Since, for hand computation and preliminary analysis the iterative method can become onerous, several one-step $P\Delta$ methods of analysis can be found in the literature. One such method is the maximum deflection method in which $P\Delta$ artificial storey shears are computed assuming that sway deflections will be limited to a known value. Thus the $P\Delta$ effects will be estimated based upon a maximum deflection. The designer can then control the deflections by the appropriate selection of member stiffness.

References

Additional information on stability effects can be obtained from the following references:

PART TWO of this Handbook.

KENNEDY, D.J.L., PICARD, A., and BEAULIEU, D. 1990. New Canadian provisions for the design of steel beam-columns. Canadian Journal of Civil Engineering, **17**(6).

SSRC. 1988. Guide to stability design criteria for metal structures (4th ed). Structural Stability Research Council, John Wiley & Sons, Toronto.

KULAK, G. L., ADAMS, P.F., and GILMOR, M.I. 1990. Limit states design in structural steel, chapter 9, overall stability. CISC.

PICARD, A., and BEAULIEU, D. 1991. Calcul des charpentes d'acier. CISC

WOOD, B.R., BEAULIEU, D., ADAMS, P.F. 1976. Further aspects of design by P-delta method. Proceedings ASCE Journal of the Structural Division, **102**(ST3).

Example

1. Given :

For the three storey, three bay structure shown below, the roof and floor systems are relatively rigid diaphragms and hence the stability for the entire building is provided by the exterior rigid frames. Determine the preliminary design parameters for columns 1C2 and 2C2 with sway effects included in the analysis. Use Appendix J, CAN/CSA-S16.1-M89, assuming the load combination of dead plus full wind governs the design of the rigid frame column, 1C2.

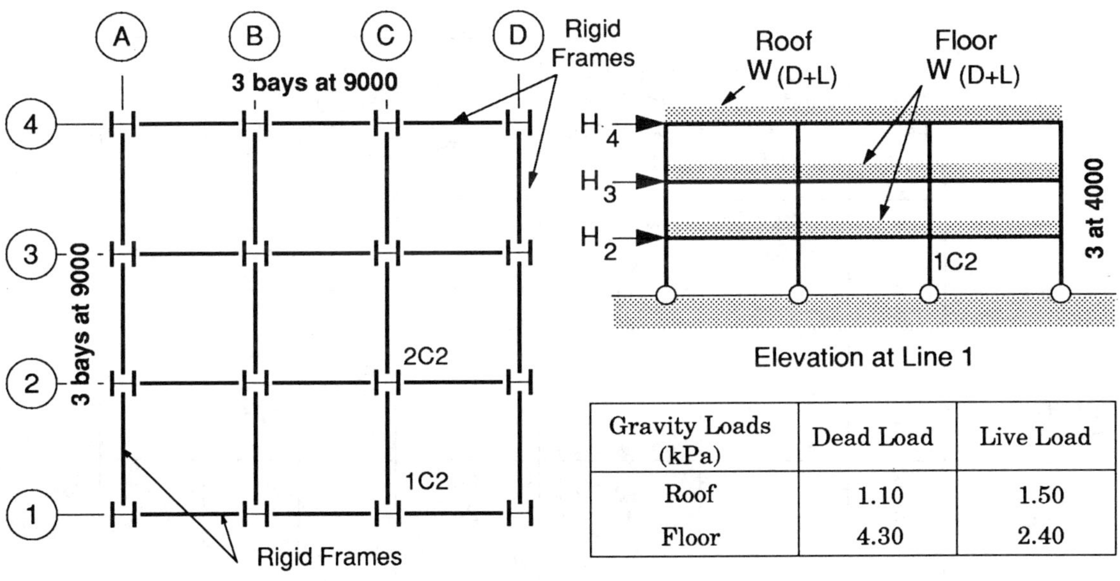

Gravity Loads (kPa)	Dead Load	Live Load
Roof	1.10	1.50
Floor	4.30	2.40

Spandrel load = 7.5 kN/m

Wind loads: 1/10 year wind, q_{10} = 0.36 kPa; 1/30 year wind, q_{30} = 0.43 kPa

Factored 1/30 wind loads are taken as H_4 = 43.4 kN; H_3 = H_2 = 83.6 kN

Solution :

a) Deflections

The deflections due to the factored 1/30 wind plus dead load combination are 26 mm, 21 mm and 14 mm for Δ_4 , Δ_3 and Δ_2 respectively and where determined from an elastic first order frame analysis based upon assumed member stiffness.

b) Gravity Loads

Since four of the interior columns "lean" (for stability) upon each rigid frame, the gravity loads used in the PΔ analysis must include the loads supported by these interior columns in addition to the gravity loads supported directly by the columns of the rigid frame.

Tributary area = 13.5 m x 27.0 m = 365 m^2 ; one-half perimeter = 27 + 27 = 54 m
Therefore, for the third storey, ΣP_3 = 1.25 (365 x 1.1 + 54.0 x 7.5) = 1 010 kN,
for the second storey, $\Sigma P_2 = \Sigma P_3$ + 1.25 (365 x 4.3 + 54.0 x 7.5) = ΣP_3 + 2 465
= 3 470 kN, for first storey, $\Sigma P_1 = \Sigma P_2$ + 2 465 = 5 940 kN

c) PΔ Analysis

Using the first order deflections due to the factored 1/30 year wind, the PΔ shears, V', artificial forces H' and the PΔ forces, H + H', are computed & tabulated below.

e.g., $V_2' = \dfrac{(\Delta_{f3} - \Delta_{f2})\Sigma P_2}{h}$ (from Appendix J, CAN/CSA-S16.1-M89)

Therefore, $V_2' = \dfrac{(21 - 14)\,3\,470}{4\,000} = 6.07$ kN

$V_3' = V_2' - H_2' = 6.07 - 1.26 = 4.81$ kN

The trial frame is sufficiently stiff that convergence is achieved in 2 cycles.

Iteration	Floor Level	Factored Wind Load, H (kN)	Sum of Col. Factored Axial Load, ΣP (kN)	Deflection, Δ (mm)	PΔ Shear, V' (kN)	PΔ Force, H' (kN)	H + H' (kN)
	4 3 2 1	43.4 83.6 83.6	1010 3470 5940	26 21 14	1.26 6.07 20.79	1.26 4.81 14.72	44.7 88.4 98.3
First	4 3 2 1			28 23 16	1.26 6.07 23.76	1.26 4.81 17.69	44.7 88.4 101.3
Second	4 3 2 1			28 24 16			

d) Preliminary Design Parameters

i) Interior Column, 2C2

Tributary area = 9 x 9 = 81 m^2
Live load reduction factor = $0.3 + \sqrt{9.8/81}$ = 0.648 (See Part 4, NBCC, 1990)
Total factored axial load,

C_f = 81 x [1.25(1.1+4.3)+1.5(1.5+2.4 x 0.648)] = 918 kN

Since column 2C2 has simple connections, assume $K_x = K_y = 1.0$

ii) Rigid Frame Column, 1C2

Using the forces determined from PΔ analysis, after convergence, the design moments for the beams and columns can be determined from an elastic frame analysis.

For column 1C2, they are taken as:

C_f = 450 kN, M_{f1} = -47.3 kN·m, M_{f2} = -163 kN·m, κ = (-47.3)/(-163) = 0.290

ω_1 = 0.48 (Table 4-7, double curvature)

Since PΔ forces are included in M_{f1} and M_{f2} , K may be less than or equal to 1.0. Use K_x = 0.9 and K_y = 1.0 for trial values.

2. Given :

Same as example 1, except that the storey deflection limit, h/500, is used to estimate the PΔ forces.

Solution :

Since the wind load used to compute the deflection under specified loads is the 1/10 year wind, the deflection must be increased to that due to the 1/30 year wind before computing the PΔ effect.

Therefore, lateral deflection under factored loads,

$$\Delta_f = \Delta_s \times \alpha_Q \times \frac{q_{30}}{q_{10}} \qquad \text{where, } \alpha_Q = 1.5 \quad \text{(Clause 7.2.3, S16.1-M89)}$$

$$= \Delta_s \times 1.5 \times \frac{0.43}{0.36} = 1.79\ \Delta_s$$

The following table summarizes the PΔ forces.

Floor Level	Factored Wind Load, H (kN)	Sum of Col. Factored Axial Load, ΣP (kN)	Deflection, Δ (mm)	PΔ Shear, V' (kN)	PΔ Force, H' (kN)	H + H' (kN)
4	43.4		43		3.54	46.9
		1010		3.54		
3	83.6		29		9.47	93.1
		3470		13.01		
2	83.6		14		7.78	91.4
		5940		20.79		
1						

In this example, the PΔ forces are over-estimated for levels 4 and 3 and under-estimated for level 2 when compared to the more accurate iterative method results

Preliminary Design Parameters

i) Interior Column, 2C2

Same as in example 1

ii) Rigid Frame Column, 2C2

For above PΔ analysis and frame analysis,

C_f = 450 kN, M_{f1} = -52.9 kN·m, M_{f2} = -167 kN·m, κ = (-52.9)/(-167) = 0.317

ω_1 = 0.47 (Table 4-7, double curvature)

3. Given :

Using gravity loads, 1/30 year wind and the first order frame deflections as given in example 1, and assuming the load combination of dead+full wind governs the design of the rigid frame column, 1C2, determine the preliminary design loads by using U_2 expression as in Clause 8.6.1 of CAN/CSA-S16.1-M89, for columns 1C2 with sway effects included in the analysis.

Solution :

The amplification factor, U_2 given by Clause 8.6.1 may be expressed as,

$$U_2 = \frac{1}{1 - \left(\frac{\Delta_f \Sigma P_f}{h \Sigma V_f}\right)}$$

where,
Δ_f = interstorey deflection due to factored 1/30 year wind
$\Sigma P_f = \Sigma$ factored axial loads (all columns) in the storey
ΣV_f = factored total first order storey shear

Floor Level	Factored Wind Load, H (kN)	Factored 1st. Order Storey-shear, $\Sigma V_f'$ (kN)	Sum of Col. Factored Axial Load, ΣP (kN)	Total Δ (mm)	Storey Δ_f (mm)	U_2	Total Fac. Storey Shear, $\Sigma V_f'$ (kN)	H + H' (kN)
4	43.4			26				44.7
		43.4	1010		5	1.03	44.7	
3	83.6			21				88.3
		127	3470		7	1.05	133	
2	83.6			14				101
		211	5940		14	1.11	234	
1								

In this example, the PΔ forces are as accurate as the iterative method results in example 1.

Preliminary Design Parameters

i) Interior Column, 2C2
Same as in example 1

ii) Rigid Frame Column, 2C2
Same as in example 1

While, for the sake of comparison, the factored storey shears were amplified by U_2, for which an analysis for moments may then be performed. The factor U_2 can be used to amplify the translational moments and shears directly.

BENDING FACTORS FOR BEAM-COLUMNS

The initial selection of a suitable steel section subject to axial load and moment can be simplified by replacing the applied axial load and bending moments with an equivalent total axial load. This procedure is discussed in more detail on page 4–100.

To simulate the effect of the bending moments, the applied moment (larger of the two end moments for columns with unequal end moments) can be multiplied by a "bending factor" which is determined for each structural steel shape from the properties of that shape. Table 4–6 lists average bending factors for each nominal size group of WWF column shapes, W shapes and hollow structural sections. This average bending factor is sufficiently accurate for an initial shape selection.

The bending factors are computed as follows:

for Class 1 sections, $B_x = 0.85 \times 10^3\ A / Z_x$

for Class 2 sections, $B_x = 10^3\ A / Z_x$

for Class 3 and Class 4 sections, $B_x = 10^3\ A / S_x$

for Class 1 sections, $B_y = 0.60 \times 10^3\ A / Z_y$

for Class 2 sections, $B_y = 10^3\ A / Z_y$

for Class 3 and Class 4 sections, $B_y = 10^3\ A / S_y$

where

A = total cross sectional area of the structural shape

S = elastic section modulus, about the X–X or Y–Y axis, as appropriate

Z = plastic section modulus, about the X–X or Y–Y axis, as appropriate

The bending factors are in m^{-1} units, and multiplying a bending moment in kilonewton metres by a bending factor from Table 4–6 results in an equivalent axial load in kilonewtons.

In cases where the factored moment resistance must be reduced when the unsupported member length exceeds L_u, the bending factors may be used to estimate the size of column required (which usually will require revision), or the bending factors can be increased to compensate for the reduction in M_r.

APPROXIMATE BENDING FACTORS
WWF and W Shapes

Table 4–6

Nominal Size	B_x (m^{-1})			B_y (m^{-1})		
	Class 1 Sections	Class 2 Sections	Class 3 & Class 4 Sections	Class 1 Sections	Class 2 Sections	Class 3 & Class 4 Sections
WWF650	3.18	3.74	4.21	4.60	7.67	11.7
WWF600	3.46	4.07	4.61	4.97	8.28	12.6
WWF550	3.79	4.46	5.05	5.36	8.93	13.6
WWF500	4.14	4.88	5.47	5.98	9.97	15.1
WWF450	4.59	5.40	6.04	6.57	11.0	16.6
WWF400	5.21	6.13	6.93	7.37	12.3	18.6
WWF350	6.04	7.10	8.12	8.36	13.9	21.1
920 x 420	2.37	2.79	3.15	8.86	14.8	22.8
920 x 305	2.52	2.97	3.42	14.3	23.9	37.7
840 x 400	2.54	2.99	3.39	9.35	15.6	24.1
840 x 290	2.73	3.21	3.70	15.0	25.1	39.3
760 x 380	2.79	3.28	3.69	9.62	16.0	24.7
760 x 265	3.04	3.57	4.12	16.6	27.7	43.5
690 x 355	3.10	3.65	4.10	10.3	17.1	26.4
690 x 255	3.32	3.91	4.47	16.7	27.9	43.5
610 x 325	3.48	4.09	4.61	11.2	18.6	28.6
610 x 230	3.72	4.38	5.00	18.3	30.5	47.5
530 x 315	3.91	4.60	5.18	11.2	18.7	28.7
530 x 210	4.21	4.95	5.65	19.5	32.4	50.6
460 x 285	4.54	5.34	5.99	12.2	20.4	31.3
460 x 190	4.87	5.73	6.52	21.0	35.0	54.4
410 x 260	5.02	5.91	6.63	12.9	21.5	32.9
410 x 180	5.41	6.37	7.21	21.8	36.4	56.3
410 x 140	5.73	6.74	7.74	31.9	53.1	83.3
360 x 430	4.56	5.36	6.81	6.50	10.8	16.7
360 x 405	5.01	5.90	7.08	7.03	11.7	18.0
360 x 395	5.38	6.33	7.23	7.47	12.4	18.9
360 x 370	5.60	6.58	7.29	8.19	13.6	20.7
360 x 255	5.82	6.85	7.67	12.8	21.3	32.6
360 x 205	6.02	7.09	7.90	17.0	28.3	43.4
360 x 170	6.16	7.25	8.15	22.5	37.4	57.6
310 x 330	5.73	6.74	8.22	8.86	14.8	22.7
310 x 310	6.34	7.46	8.48	9.78	16.3	24.9
310 x 255	6.64	7.82	8.65	12.5	20.8	31.6
310 x 200	6.83	8.04	8.96	16.4	27.3	41.9
310 x 165	6.81	8.02	8.96	21.7	36.1	55.4
250 x 255	7.77	9.14	10.4	11.7	19.5	29.6
250 x 200	8.23	9.69	10.8	15.9	26.5	40.3
250 x 145	8.19	9.64	10.8	24.2	40.4	62.1
200 x 200	9.74	11.5	12.9	14.8	24.6	37.5
200 x 165	10.2	12.0	13.4	19.4	32.3	49.2
150 x 150	13.3	15.6	17.5	21.0	35.0	53.3
130 x 125	16.2	19.1	21.6	24.0	40.0	61.1
100 x 100	20.4	24.0	27.8	31.0	51.6	79.4

Table 4-6

APPROXIMATE BENDING FACTORS
Hollow Structural Sections

Nominal Size	$B_x (m^{-1})$		$B_y (m^{-1})$	
	Class 1 & Class 2 Sections	Class 3 & Class 4 Sections	Class 1 & Class 2 Sections	Class 3 & Class 4 Sections
305 x 203	9.81	11.9	13.0	14.9
354 x 152	12.2	15.1	17.3	20.1
203 x 152	14.7	17.9	17.9	20.9
203 x 102	15.9	20.4	26.0	30.5
178 x 127	17.1	21.1	21.6	25.5
152 x 102	20.3	25.4	27.0	32.1
127 x 76	25.1	32.0	36.1	43.2
127 x 64	26.1	34.1	42.9	51.8
127 x 51	27.1	36.2	53.0	64.3
102 x 76	30.3	38.0	37.1	44.9
102 x 51	32.4	42.0	53.1	63.6
89 x 64	34.8	43.5	44.0	52.8
76 x 51	42.3	54.7	56.4	69.8
51 x 25	66.1	87.1	109	133
305 x 305	9.10	10.6	9.10	10.6
254 x 254	11.0	12.9	11.0	12.9
203 x 203	13.9	16.6	13.9	16.6
178 x 178	16.0	19.0	16.0	19.0
152 x 152	18.9	22.7	18.9	22.7
127 x 127	22.8	27.6	22.8	27.6
102 x 102	28.8	35.2	28.8	35.2
89 x 89	33.1	40.6	33.1	40.6
76 x 76	38.4	46.9	38.4	46.9
64 x 64	46.2	56.5	46.2	56.5
51 x 51	58.8	73.1	58.8	73.1
38 x 38	79.7	100	79.7	100
32 x 32	96.3	122	96.3	122
25 x 25	126	165	126	165

VALUES OF ω_1 *
Table 4–7

Single Curvature				Double Curvature			
$\frac{M_{f_1}}{M_{f_2}}$	ω_1	$\frac{M_{f_1}}{M_{f_2}}$	ω_1	$\frac{M_{f_1}}{M_{f_2}}$	ω_1	$\frac{M_{f_1}}{M_{f_2}}$	ω_1
1.00	1.00	0.50	0.80	0.00	0.60	0.55	0.40
0.95	0.98	0.45	0.78	0.05	0.58	0.60	0.40
0.90	0.96	0.40	0.76	0.10	0.56	0.65	0.40
0.85	0.94	0.35	0.74	0.15	0.54	0.70	0.40
0.80	0.92	0.30	0.72	0.20	0.52	0.75	0.40
0.75	0.90	0.25	0.70	0.25	0.50	0.80	0.40
0.70	0.88	0.20	0.68	0.30	0.48	0.85	0.40
0.65	0.86	0.15	0.66	0.35	0.46	0.90	0.40
0.60	0.84	0.10	0.64	0.40	0.44	0.95	0.40
0.55	0.82	0.05	0.62	0.45	0.42	1.00	0.40
		0.00	0.60	0.50	0.40		

* See Clause 13.8.4, CAN/CSA-S16.1-M89.

The value of ω_1 is used to modify the bending term in the beam-column interaction expression to account for various end moment and transverse loading conditions of the columns.

For columns of a frame not subject to transverse loads between supports, use the values of ω_1 shown in Table 4–7.

For members subjected to distributed loads or a series of point loads between supports, $\omega_1 = 1.0$, and for members subjected to a concentrated load or moment between supports, $\omega_1 = 0.85$.

The values of ω_1 given in Table 4–7 are derived from

$$\omega_1 = 0.6 - 0.4\,\kappa \geq 0.4$$

where

$\kappa = M_{f1} / M_{f2}$ for moments at opposite ends of the unbraced column length, positive for double curvature, and negative for single curvature in which,

M_{f1} = the smaller factored end moment, and

M_{f2} = the larger factored end moment.

C_e/A EULER BUCKLING LOAD PER UNIT OF AREA, MPa

Table 4-8

$\frac{KL}{r}$	$\frac{C_e}{A}$ (MPa)	$\frac{KL}{r}$	$\frac{C_e}{A}$ (MPa)	$\frac{KL}{r}$	$\frac{C_e}{A}$ (MPa)	$\frac{KL}{r}$	$\frac{C_e}{A}$ (MPa)	$\frac{KL}{r}$	$\frac{C_e}{A}$ (MPa)
1	1 970 000	41	1 172	81	300	121	135	161	76.0
2	492 500	42	1 117	82	293	122	132	162	75.1
3	218 889	43	1 065	83	286	123	130	163	74.1
4	123 125	44	1 018	84	279	124	128	164	73.2
5	78 800	45	973	85	273	125	126	165	72.4
6	54 722	46	931	86	266	126	124	166	71.5
7	40 204	47	892	87	260	127	122	167	70.6
8	30 781	48	855	88	254	128	120	168	69.8
9	24 321	49	820	89	249	129	118	169	69.0
10	19 700	50	788	90	243	130	117	170	68.2
11	16 281	51	757	91	238	131	115	171	67.4
12	13 681	52	729	92	233	132	113	172	66.6
13	11 657	53	701	93	228	133	111	173	65.8
14	10 051	54	676	94	223	134	110	174	65.1
15	8 756	55	651	95	218	135	108	175	64.3
16	7 695	56	628	96	214	136	107	176	63.6
17	6 817	57	606	97	209	137	105	177	62.9
18	6 080	58	586	98	205	138	103	178	62.2
19	5 457	59	566	99	201	139	102	179	61.5
20	4 925	60	547	100	197	140	101	180	60.8
21	4 467	61	529	101	193	141	99.1	181	60.1
22	4 070	62	512	102	189	142	97.7	182	59.5
23	3 724	63	496	103	186	143	96.3	183	58.8
24	3 420	64	481	104	182	144	95.0	184	58.2
25	3 152	65	466	105	179	145	93.7	185	57.6
26	2 914	66	452	106	175	146	92.4	186	56.9
27	2 702	67	439	107	172	147	91.2	187	56.3
28	2 513	68	426	108	169	148	89.9	188	55.7
29	2 342	69	414	109	166	149	88.7	189	55.1
30	2 189	70	402	110	163	150	87.6	190	54.6
31	2 050	71	391	111	160	151	86.4	191	54.0
32	1 924	72	380	112	157	152	85.3	192	53.4
33	1 809	73	370	113	154	153	84.2	193	52.9
34	1 704	74	360	114	152	154	83.1	194	52.3
35	1 608	75	350	115	149	155	82.0	195	51.8
36	1 520	76	341	116	146	156	81.0	196	51.3
37	1 439	77	332	117	144	157	79.9	197	50.8
38	1 364	78	324	118	141	158	78.9	198	50.2
39	1 295	79	316	119	139	159	77.9	199	49.7
40	1 231	80	308	120	137	160	77.0	200	49.3

To obtain C_e, in kilonewtons, multiply the tabular value by the cross sectional area, **A** , in square millimetres, and divide by 1 000.

AMPLIFICATION FACTOR *

Table 4–9

$$U = \frac{1}{1 - \frac{C_f}{C_e}}$$

$\frac{C_f}{C_e}$	U	$\frac{C_f}{C_e}$	U	$\frac{C_f}{C_e}$	U	$\frac{C_f}{C_e}$	U
.01	1.01	.26	1.35	.51	2.04	.76	4.17
.02	1.02	.27	1.37	.52	2.08	.77	4.35
.03	1.03	.28	1.39	.53	2.13	.78	4.55
.04	1.04	.29	1.41	.54	2.17	.79	4.76
.05	1.05	.30	1.43	.55	2.22	.80	5.00
.06	1.06	.31	1.45	.56	2.27	.81	5.26
.07	1.08	.32	1.47	.57	2.33	.82	5.56
.08	1.09	.33	1.49	.58	2.38	.83	5.88
.09	1.10	.34	1.52	.59	2.44	.84	6.25
.10	1.11	.35	1.54	.60	2.50	.85	6.67
.11	1.12	.36	1.56	.61	2.56	.86	7.14
.12	1.14	.37	1.59	.62	2.63	.87	7.69
.13	1.15	.38	1.61	.63	2.70	.88	8.33
.14	1.16	.39	1.64	.64	2.78	.89	9.09
.15	1.18	.40	1.67	.65	2.86	.90	10.0
.16	1.19	.41	1.69	.66	2.94	.91	11.1
.17	1.20	.42	1.72	.67	3.03	.92	12.5
.18	1.22	.43	1.75	.68	3.13	.93	14.3
.19	1.23	.44	1.79	.69	3.23	.94	16.7
.20	1.25	.45	1.82	.70	3.33	.95	20.0
.21	1.27	.46	1.85	.71	3.45	.96	25.0
.22	1.28	.47	1.89	.72	3.57	.97	33.3
.23	1.30	.48	1.92	.73	3.70	.98	50.0
.24	1.32	.49	1.96	.74	3.85	.99	100.0
.25	1.33	.50	2.00	.75	4.00		

* See Clause 13.8.3, CAN/CSA-S16.1-M89.

FACTORED AXIAL COMPRESSIVE RESISTANCES OF COLUMNS

Tables

The tables on pages 4–30 to 4–75 inclusive list the factored axial compressive resistances, C_r, in kilonewtons for WWF and W shapes and Class C HSS, produced to the requirements of CSA Standard G40.20-M. The resistances have been computed, for effective lengths with respect to the least radius of gyration varying from 0 mm to 12 000 mm, in accordance with the requirements of Clauses 13.3.1 and 13.3.3, CAN/CSA-S16.1-M89

The tables on pages 4–76 to 4–99 inclusive list the factored axial compressive resistances, C_r, in kilonewtons for Class H HSS produced to the requirements of G40.20-M, in accordance with Clauses 13.3.2 and 13.3.3, S16.1-M89.

In all, four sets of tables are provided:

Set 1 — WWF and W shapes conforming to CSA G40.20-M

— CSA G40.21-M, grade 300W

— only shapes available from Canadian mills are included, with those highlighted in white being commonly used sizes which are generally readily available.

Set 2 — WWF shapes conforming to CSA G40.20-M

— CSA G40.21-M, grade 350W.

Set 3 — HSS conforming to CSA G40.20-M, Class C

— CSA G40.21-M grade 350W.

Set 4 — HSS conforming to CSA G40.20-M, Class H

— CSA G40.21-M grade 350W.

In each set of tables, sections which are either Class 3 or 4, in the grade of steel for which the loads have been computed, are identified. In Sets 1 and 2, those welded sections in which the flange-to-web welds do not develop the entire web strength are identified.

The factored axial compressive resistances for Class 4 sections have been computed in accordance with the requirements of Clause 13.3.3 of S16.1-M89 and are so identified in the tables.

The applicable steel grade is listed at the top of each table, and the metric designation of each shape is given at the top of the columns, while the equivalent imperial size and mass are listed at the bottom of the tables. Properties and design data are included at the bottom of the tables as follows:

Area = Total cross-sectional area, mm^2.

Z_x = Plastic section modulus for bending about X–X axis, 10^3 mm^3

S_x = Elastic section modulus for bending about X–X axis, 10^3 mm^3. (S_e is given for Class 4 sections; see Clause 13.5(c), CSA-S16.1-M89.)

r_x = Radius of gyration about the strong, X–X, axis, mm.

Z_y = Plastic section modulus for bending about Y–Y axis, 10^3 mm^3

S_y = Elastic section modulus for bending about Y–Y axis, 10^3 mm^3. (S_e is given for Class 4 sections; see Clause 13.5(c), CSA-S16.1-M89.)

r_y = Radius of gyration about the weak, Y–Y, axis, mm.

r_x/r_y = Ratio of radius of gyration of X–X axis to that for Y–Y axis.

M_{rx} = Factored moment resistance for bending about the X–X axis, computed considering $L \le L_u$, using the Class of the section considering bending about the X–X axis only and the value of F_y shown, for Class 1 and 2 section, $\phi Z_x F_y \times 10^{-6}$; for Class 3 sections, $\phi S_x F_y \times 10^{-6}$; and for Class 4 sections Clause 13.5(c) of S16.1-M89, kN·m.

M_{ry} = Factored moment resistance for bending about the Y–Y axis computed using the Class of the section considering bending about the Y–Y axis only and the value of F_y shown, for Class 1 and 2 section, $\phi Z_y F_y \times 10^{-6}$; for Class 3 sections, $\phi S_y F_y \times 10^{-6}$; and for class 4 sections, Clause 13.5(c) of S16.1-M89, kN·m.

For hollow structural sections which have identical X–X and Y–Y axis properties, only one value of the relevant properties is shown.

The following additional constants are tabulated for the tables containing WWF and W shapes:

J = St. Venant torsional constant, 10^3 mm^4

C_w = Warping torsional constant, 10^9 mm^6

L_u = Maximum unsupported length of compression flange for which no reduction in M_r is required, mm.

F_y = Specified minimum yield strength of the section, MPa.

Design of Axially Loaded Columns

The design of axially loaded columns (columns theoretically not subjected to combined bending and compression) involves the determination of the governing effective length and the selection of a section with the required resistance at that effective length. Factored axial compressive resistance tables for columns enable a designer to select a suitable section directly, without following a trial-and-error procedure.

Since the factored axial compressive resistances, C_r, (listed in the tables supplied) have been computed on the basis of the least radius of gyration, r_y, for each section, the tables apply directly only to columns unbraced about the Y–Y axis. In certain cases, however, it is necessary to investigate the capacity of a column with reference to both the X–X axis and the Y–Y axis, or with reference only to the X–X axis. The ratio r_x/r_y included in the table of properties at the bottom of each resistance table provides a convenient means of investigating the strength of a column with respect to the X–X axis.

In general, a column having an effective length $K_x L_x$ with respect to the X–X axis will be able to carry a factored load equal to the tabulated factored axial compressive resistance based upon the effective length $K_y L_y$ with respect to the Y–Y axis if $K_x L_x \le K_y L_y (r_x/r_y)$

Examples

1. Given:

A W310 column is required to carry a factored axial load of 3 600 kN. The effective length $K_y L_y$ along the weak axis is 4 500 mm. The effective length $K_x L_x$ along the strong axis is 7 600 mm. Use CSA G40.21-M 300W steel.

Solution:

With $K_y L_y$ = 4 500, the lightest W310 section with sufficient factored axial compressive resistance is W310x143. C_r = 3 850 kN; r_x/r_y = 1.75.

$K_x L_x$ = 7 600 mm

$K_y L_y (r_x/r_y)$ = 4 500 × 1.75

= 7 880 mm > 7 600 — OK

The W310x143 has a factored compressive resistance of 3 850 kN with an effective length of $K_x L_x$ = 7 880 mm, and hence the section is adequate. Use W310x143.

2. **Given:**

Same as example 1, except $K_x L_x$ = 9 500 mm

Solution:

$K_y L_y$ = 4 500 mm

$K_x L_x$ = 9 500 mm

Equivalent $K_y L_y$, for $K_x L_x$ of 9 500 mm = $K_x L_x / (r_x/r_y)$

Assuming that a heavy W310 section will be adequate, r_x/r_y = 1.75.

Equivalent $K_y L_y$ = 9 500 / 1.75

= 5 430 mm > 4 500

Therefore, $K_x L_x$ governs, and the effective $K_y L_y$ is 5 430 mm.

With $K_y L_y$ = 5 430 mm, a W310x158 is the lightest W310 that has a factored axial compressive resistance greater than the factored axial load of 3 600 kN (C_r for 5 500 mm = 3 790 kN; r_x/r_y = 1.76)

Use W310x158

Other Steel Grades

For columns of steels other than G40.21-M 350W and 300W for WWF shapes, 300W for W shapes, or 350W for HSS, see example 1, page 4–10, for a method of determining the factored axial compressive resistance of the section. For steel grades with higher specified minimum yield points, these tables are conservative.

WWF COLUMNS

Factored Axial Compressive Resistances, C_r, in kN

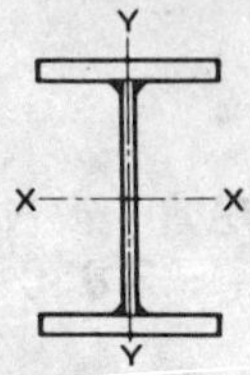

G40.21-M300W

$\phi = 0.90$

Designation	WWF650				
Mass (kg/m)	864 ¤	739 ¤	598	499	400 *
Effective length (KL) in millimetres with respect to least radius of gyration					
0	29 700	25 400	20 600	17 200	13 800
2 000	29 600	25 400	20 600	17 200	13 800
2 250	29 500	25 300	20 500	17 100	13 700
2 500	29 300	25 200	20 400	17 000	13 600
2 750	29 100	25 100	20 300	16 900	13 500
3 000	29 000	24 900	20 200	16 800	13 500
3 250	28 800	24 800	20 100	16 700	13 400
3 500	28 600	24 600	20 000	16 600	13 300
3 750	28 400	24 500	19 900	16 500	13 200
4 000	28 200	24 300	19 700	16 400	13 100
4 250	28 000	24 200	19 600	16 300	13 100
4 500	27 800	24 000	19 500	16 200	13 000
4 750	27 600	23 900	19 400	16 100	12 900
5 000	27 400	23 700	19 200	16 000	12 800
5 250	27 200	23 500	19 100	15 900	12 700
5 500	27 000	23 400	19 000	15 800	12 600
6 000	26 500	23 000	18 700	15 500	12 400
6 500	26 000	22 700	18 400	15 300	12 200
7 000	25 500	22 300	18 100	15 000	11 900
7 500	25 000	21 900	17 800	14 800	11 700
8 000	24 400	21 500	17 400	14 500	11 500
8 500	23 900	21 000	17 100	14 200	11 200
9 000	23 300	20 600	16 700	13 900	11 000
9 500	22 700	20 100	16 400	13 600	10 700
10 000	22 000	19 700	16 000	13 300	10 400
10 500	21 400	19 200	15 600	12 900	10 200
11 000	20 700	18 700	15 200	12 600	9 860
11 500	20 000	18 200	14 800	12 200	9 560
12 000	19 300	17 600	14 400	11 900	9 260
PROPERTIES AND DESIGN DATA					
Area (mm^2)	110 000	94 100	76 200	63 600	51 000
Z_x (10^3 mm^3)	27 300	25 200	21 100	17 500	13 900
S_x (10^3 mm^3)	23 300	22 100	18 900	15 900	12 600
r_x (mm)	262	277	284	285	284
Z_y (10^3 mm^3)	13 200	12 800	10 600	8 510	6 400
S_y (10^3 mm^3)	8 480	8 450	7 040	5 630	4 230
r_y (mm)	158	171	173	170	164
r_x/r_y	1.66	1.62	1.64	1.68	1.73
M_{rx} (kN·m) ($L < L_u$)	7 370	6 800	5 700	4 720	3 400
M_{ry} (kN·m)	3 560	3 460	2 860	2 300	1 140
J (10^3 mm^4)	132 000	98 400	55 600	29 300	13 300
C_w (10^9 mm^6)	240 000	239 000	206 000	170 000	132 000
L_u (mm)	14 700	14 300	12 800	11 600	11 300
F_y (MPa)	300	300	300	300	300
IMPERIAL SIZE AND MASS					
Mass (lb./ft.)	580	497	402	336	269
Nominal Depth and Width (in.)	26 x 26				

¤ Welding does not fully develop web strength.
* Class 3 flanges.

G40.21-M300W

$\phi = 0.90$

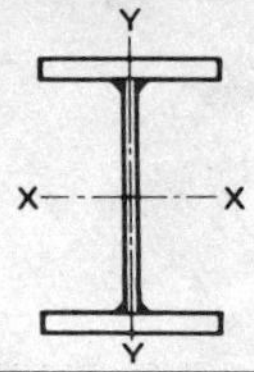

WWF COLUMNS

Factored Axial Compressive Resistances, C_r, in kN

Designation	WWF600				
Mass (kg/m)	793 ¤	680 ¤	551	460	369 *
Effective length (KL) in millimetres with respect to least radius of gyration					
0	27 300	23 400	19 000	15 800	12 700
2 000	27 100	23 300	18 900	15 800	12 600
2 250	27 000	23 200	18 800	15 700	12 600
2 500	26 800	23 100	18 700	15 600	12 500
2 750	26 600	22 900	18 600	15 500	12 400
3 000	26 500	22 800	18 500	15 400	12 300
3 250	26 300	22 700	18 400	15 300	12 300
3 500	26 100	22 500	18 300	15 200	12 200
3 750	25 900	22 400	18 200	15 100	12 100
4 000	25 700	22 200	18 000	15 000	12 000
4 250	25 500	22 100	17 900	14 900	11 900
4 500	25 300	21 900	17 800	14 800	11 800
4 750	25 100	21 700	17 700	14 700	11 700
5 000	24 900	21 600	17 500	14 600	11 600
5 250	24 600	21 400	17 400	14 500	11 500
5 500	24 400	21 200	17 200	14 300	11 400
6 000	23 900	20 900	16 900	14 100	11 200
6 500	23 400	20 500	16 600	13 800	11 000
7 000	22 900	20 100	16 300	13 600	10 800
7 500	22 400	19 700	16 000	13 300	10 500
8 000	21 800	19 200	15 700	13 000	10 300
8 500	21 200	18 800	15 300	12 700	10 000
9 000	20 600	18 300	14 900	12 400	9 760
9 500	20 000	17 800	14 600	12 000	9 490
10 000	19 300	17 400	14 200	11 700	9 200
10 500	18 700	16 800	13 800	11 400	8 910
11 000	18 000	16 300	13 300	11 000	8 600
11 500	17 300	15 800	12 900	10 600	8 290
12 000	16 500	15 200	12 500	10 200	7 970
PROPERTIES AND DESIGN DATA					
Area (mm^2)	101 000	86 600	70 200	58 600	47 000
Z_x (10^3 mm^3)	22 900	21 200	17 800	14 800	11 800
S_x (10^3 mm^3)	19 400	18 500	15 900	13 400	10 700
r_x (mm)	240	253	261	262	261
Z_y (10^3 mm^3)	11 200	10 900	9 050	7 250	5 460
S_y (10^3 mm^3)	7 230	7 200	6 000	4 800	3 600
r_y (mm)	147	158	160	157	152
r_x/r_y	1.63	1.60	1.63	1.67	1.72
M_{rx} (kN·m) ($L < L_u$)	6 180	5 720	4 810	4 000	2 890
M_{ry} (kN·m)	3 020	2 940	2 440	1 960	972
J (10^3 mm^4)	121 000	90 700	51 300	27 000	12 200
C_w (10^9 mm^6)	158 000	158 000	136 000	113 000	87 800
L_u (mm)	14 600	14 000	12 400	11 100	10 600
F_y (MPa)	300	300	300	300	300
IMPERIAL SIZE AND MASS					
Mass (lb./ft.)	531	456	371	309	248
Nominal Depth and Width (in.)	24 x 24				

¤ Welding does not fully develop web strength.

* Class 3 flanges.

WWF COLUMNS

Factored Axial Compressive Resistances, C_r, in kN

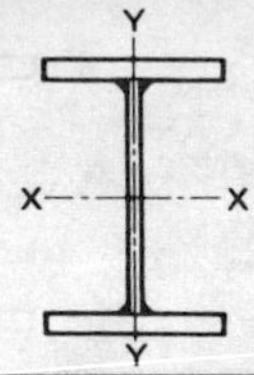

G40.21-M300W

$\phi = 0.90$

Designation		WWF550					WWF500		
Mass (kg/m)		721 ¤	620 ¤	503	420	280 *	651 ¤	561 ¤	456
Effective length (KL) in millimetres with respect to least radius of gyration	0	24 800	21 400	17 300	14 500	9 610	22 400	19 300	15 700
	2 000	24 600	21 200	17 200	14 400	9 540	22 100	19 100	15 600
	2 250	24 400	21 100	17 100	14 300	9 480	21 900	19 000	15 500
	2 500	24 300	21 000	17 000	14 200	9 420	21 700	18 900	15 300
	2 750	24 100	20 800	16 900	14 100	9 350	21 600	18 700	15 200
	3 000	23 900	20 700	16 800	14 000	9 290	21 400	18 600	15 100
	3 250	23 700	20 600	16 700	13 900	9 220	21 200	18 400	15 000
	3 500	23 500	20 400	16 600	13 800	9 150	21 000	18 300	14 900
	3 750	23 300	20 200	16 500	13 700	9 070	20 800	18 100	14 800
	4 000	23 100	20 100	16 300	13 600	9 000	20 600	18 000	14 600
	4 250	22 900	19 900	16 200	13 500	8 920	20 400	17 800	14 500
	4 500	22 700	19 800	16 100	13 400	8 840	20 100	17 600	14 400
	4 750	22 500	19 600	15 900	13 300	8 760	19 900	17 400	14 200
	5 000	22 300	19 400	15 800	13 100	8 680	19 700	17 200	14 100
	5 250	22 000	19 200	15 700	13 000	8 590	19 400	17 100	13 900
	5 500	21 800	19 100	15 500	12 900	8 510	19 200	16 900	13 800
	6 000	21 300	18 700	15 200	12 600	8 330	18 700	16 500	13 400
	6 500	20 800	18 300	14 900	12 400	8 140	18 100	16 100	13 100
	7 000	20 200	17 900	14 600	12 100	7 940	17 600	15 600	12 800
	7 500	19 700	17 400	14 200	11 800	7 740	17 000	15 200	12 400
	8 000	19 100	17 000	13 900	11 500	7 520	16 400	14 700	12 000
	8 500	18 500	16 500	13 500	11 100	7 300	15 700	14 200	11 600
	9 000	17 900	16 000	13 100	10 800	7 070	15 100	13 700	11 200
	9 500	17 200	15 500	12 700	10 500	6 830	14 400	13 200	10 800
	10 000	16 500	15 000	12 300	10 100	6 580	13 700	12 600	10 400
	10 500	15 800	14 500	11 900	9 750	6 330	12 800	12 100	9 940
	11 000	15 100	13 900	11 400	9 380	6 070	12 000	11 400	9 470
	11 500	14 200	13 400	11 000	8 990	5 790	11 300	10 800	8 920
	12 000	13 500	12 700	10 500	8 550	5 470	10 700	10 200	8 430
PROPERTIES AND DESIGN DATA									
Area (mm^2)		92 000	79 100	64 200	53 600	35 600	83 000	71 600	58 200
Z_x (10^3 mm^3)		19 000	17 600	14 800	12 400	8 250	15 400	14 300	12 100
S_x (10^3 mm^3)		16 000	15 200	13 100	11 100	7 530	12 800	12 300	10 600
r_x (mm)		218	230	237	239	241	196	207	214
Z_y (10^3 mm^3)		9 470	9 180	7 610	6 100	3 810	7 850	7 590	6 290
S_y (10^3 mm^3)		6 080	6 050	5 040	4 030	2 520	5 030	5 000	4 170
r_y (mm)		135	145	147	144	140	123	132	134
r_x/r_y		1.61	1.59	1.61	1.66	1.72	1.59	1.57	1.60
M_{rx} (kN·m) ($L < L_u$)		5 130	4 750	4 000	3 350	2 030	4 160	3 860	3 270
M_{ry} (kN·m)		2 560	2 480	2 050	1 650	680	2 120	2 050	1 700
J (10^3 mm^4)		110 000	83 100	47 000	24 700	6 410	99 400	75 400	42 700
C_w (10^9 mm^6)		100 000	99 900	86 700	72 100	47 800	60 800	60 500	52 700
L_u (mm)		14 400	13 800	12 000	10 500	9 480	14 400	13 700	11 700
F_y (MPa)		300	300	300	300	300	300	300	300
IMPERIAL SIZE AND MASS									
Mass (lb./ft.)		484	416	338	282	188	437	377	306
Nominal Depth and Width (in.)		22 x 22					20 x 20		

¤ Welding does not fully develop web strength.

* Class 3 flanges.

G40.21-M300W
$\phi = 0.90$

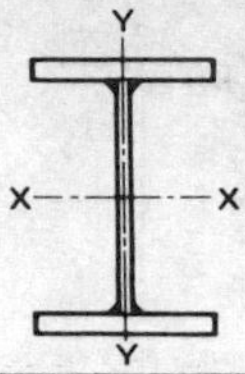

WWF COLUMNS
Factored Axial Compressive Resistances, C_r, in kN

Designation	WWF500						
Mass (kg/m)	381	343	306	276	254 *	223 *	197 **
Effective length (KL) in millimetres with respect to least radius of gyration							
0	13 100	11 800	10 500	9 500	8 720	7 700	6 180
2 000	13 000	11 700	10 400	9 390	8 610	7 600	6 130
2 250	12 900	11 600	10 300	9 330	8 550	7 540	6 100
2 500	12 800	11 500	10 200	9 260	8 480	7 490	6 070
2 750	12 700	11 400	10 200	9 190	8 420	7 430	6 040
3 000	12 600	11 300	10 100	9 110	8 350	7 370	6 000
3 250	12 500	11 200	9 990	9 040	8 280	7 300	5 970
3 500	12 400	11 100	9 910	8 960	8 200	7 240	5 930
3 750	12 300	11 000	9 810	8 880	8 130	7 170	5 900
4 000	12 200	10 900	9 720	8 790	8 050	7 100	5 860
4 250	12 100	10 800	9 620	8 710	7 970	7 030	5 820
4 500	11 900	10 700	9 520	8 620	7 890	6 960	5 770
4 750	11 800	10 600	9 420	8 530	7 800	6 880	5 730
5 000	11 700	10 500	9 320	8 440	7 720	6 810	5 690
5 250	11 600	10 400	9 210	8 340	7 630	6 730	5 640
5 500	11 400	10 300	9 100	8 240	7 530	6 650	5 600
6 000	11 200	10 000	8 870	8 040	7 340	6 480	5 500
6 500	10 900	9 740	8 630	7 830	7 140	6 300	5 390
7 000	10 600	9 470	8 370	7 610	6 940	6 120	5 280
7 500	10 300	9 180	8 110	7 380	6 720	5 930	5 160
8 000	9 930	8 880	7 840	7 140	6 490	5 730	5 030
8 500	9 600	8 570	7 550	6 880	6 250	5 520	4 860
9 000	9 250	8 240	7 260	6 620	6 010	5 300	4 690
9 500	8 880	7 910	6 950	6 350	5 760	5 080	4 510
10 000	8 510	7 560	6 630	6 070	5 490	4 850	4 320
10 500	8 120	7 200	6 280	5 790	5 200	4 590	4 120
11 000	7 670	6 770	5 910	5 440	4 890	4 320	3 880
11 500	7 230	6 380	5 570	5 130	4 610	4 070	3 650
12 000	6 830	6 030	5 260	4 840	4 350	3 840	3 450

PROPERTIES AND DESIGN DATA

Property	381	343	306	276	254 *	223 *	197 **
Area (mm^2)	48 600	43 800	39 000	35 200	32 300	28 500	25 200
Z_x (10^3 mm^3)	10 100	9 100	8 060	7 420	6 780	6 010	5 410
S_x (10^3 mm^3)	9 010	8 140	7 240	6 740	6 160	5 500	4 620
r_x (mm)	215	216	215	218	218	219	223
Z_y (10^3 mm^3)	5 040	4 420	3 800	3 530	3 160	2 770	2 510
S_y (10^3 mm^3)	3 330	2 920	2 500	2 330	2 080	1 830	1 420
r_y (mm)	131	129	127	129	127	127	129
r_x/r_y	1.64	1.67	1.69	1.69	1.72	1.72	1.73
M_{rx} (kN·m) ($L < L_u$)	2 730	2 460	2 180	2 000	1 660	1 480	1 250
M_{ry} (kN·m)	1 360	1 190	1 030	953	562	494	384
J (10^3 mm^4)	22 500	15 400	10 200	7 920	5 820	3 970	2 870
C_w (10^9 mm^6)	44 100	39 400	34 500	32 500	29 400	26 200	24 000
L_u (mm)	10 000	9 330	8 750	8 620	8 830	8 550	8 800
F_y (MPa)	300	300	300	300	300	300	300

IMPERIAL SIZE AND MASS

Mass (lb./ft.)	256	230	205	185	170	150	132
Nominal Depth and Width (in.)	20 x 20						

* Class 3 flanges.
** Class 4 flanges: area and resistance calculated according to CAN/CSA-M16.1-M89 Clause 13.3.3.

WWF COLUMNS

Factored Axial Compressive Resistances, C_r, in kN

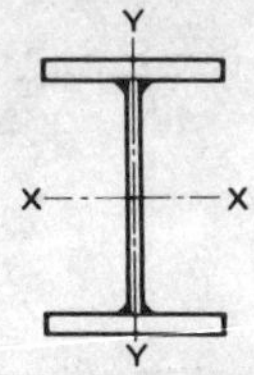

G40.21-M300W

$\phi = 0.90$

Designation	WWF450								
Mass (kg/m)	503 ¤	409	342	308	274	248	228	201 *	177 *
Effective length (KL) in millimetres with respect to least radius of gyration: 0	17 300	14 100	11 800	10 600	9 450	8 530	7 830	6 910	6 100
2 000	17 000	13 900	11 600	10 400	9 270	8 380	7 680	6 780	5 990
2 250	16 900	13 800	11 500	10 300	9 190	8 310	7 620	6 720	5 940
2 500	16 700	13 700	11 400	10 200	9 110	8 240	7 550	6 660	5 890
2 750	16 600	13 500	11 300	10 200	9 030	8 170	7 480	6 600	5 840
3 000	16 500	13 400	11 200	10 100	8 940	8 090	7 410	6 540	5 780
3 250	16 300	13 300	11 100	9 960	8 850	8 010	7 330	6 470	5 730
3 500	16 100	13 200	11 000	9 860	8 760	7 930	7 260	6 410	5 670
3 750	16 000	13 000	10 900	9 750	8 660	7 840	7 180	6 340	5 610
4 000	15 800	12 900	10 700	9 650	8 560	7 760	7 090	6 260	5 550
4 250	15 600	12 800	10 600	9 530	8 460	7 670	7 010	6 190	5 480
4 500	15 400	12 600	10 500	9 420	8 360	7 570	6 920	6 110	5 420
4 750	15 300	12 500	10 400	9 300	8 250	7 480	6 830	6 030	5 350
5 000	15 100	12 300	10 200	9 180	8 140	7 380	6 740	5 950	5 280
5 250	14 900	12 200	10 100	9 050	8 020	7 280	6 650	5 870	5 210
5 500	14 700	12 000	9 950	8 920	7 900	7 180	6 550	5 780	5 130
6 000	14 300	11 700	9 670	8 660	7 660	6 960	6 350	5 600	4 980
6 500	13 800	11 300	9 360	8 380	7 400	6 740	6 130	5 410	4 820
7 000	13 400	11 000	9 050	8 080	7 130	6 500	5 910	5 220	4 650
7 500	12 900	10 600	8 720	7 780	6 850	6 250	5 680	5 010	4 470
8 000	12 400	10 200	8 370	7 460	6 560	6 000	5 430	4 800	4 290
8 500	11 900	9 780	8 010	7 120	6 250	5 730	5 180	4 570	4 100
9 000	11 300	9 350	7 640	6 780	5 940	5 450	4 920	4 340	3 900
9 500	10 800	8 900	7 250	6 410	5 580	5 160	4 620	4 080	3 690
10 000	10 100	8 410	6 800	5 990	5 210	4 820	4 320	3 810	3 440
10 500	9 470	7 890	6 370	5 610	4 880	4 510	4 040	3 570	3 230
11 000	8 900	7 410	5 980	5 270	4 580	4 240	3 800	3 350	3 030
11 500	8 380	6 980	5 630	4 960	4 320	3 990	3 580	3 160	2 850
12 000	7 910	6 590	5 320	4 680	4 070	3 760	3 370	2 980	2 690
PROPERTIES AND DESIGN DATA									
Area (mm^2)	64 100	52 200	43 600	39 300	35 000	31 600	29 000	25 600	22 600
Z_x (10^3 mm^3)	11 400	9 640	8 100	7 290	6 470	5 960	5 450	4 840	4 360
S_x (10^3 mm^3)	9 620	8 380	7 150	6 480	5 770	5 380	4 920	4 400	4 000
r_x (mm)	184	190	192	193	193	196	196	197	200
Z_y (10^3 mm^3)	6 150	5 100	4 090	3 580	3 080	2 860	2 560	2 250	2 040
S_y (10^3 mm^3)	4 050	3 380	2 700	2 360	2 030	1 890	1 690	1 490	1 350
r_y (mm)	119	121	118	116	114	116	114	114	116
r_x/r_y	1.55	1.57	1.63	1.66	1.69	1.69	1.72	1.73	1.72
M_{rx} (kN·m) ($L < L_u$)	3 080	2 600	2 190	1 970	1 750	1 610	1 470	1 190	1 080
M_{ry} (kN·m)	1 660	1 380	1 100	967	832	772	691	402	364
J (10^3 mm^4)	67 800	38 400	20 200	13 900	9 140	7 120	5 230	3 570	2 580
C_w (10^9 mm^6)	34 700	30 400	25 500	22 900	20 100	18 900	17 200	15 300	14 000
L_u (mm)	13 700	11 500	9 620	8 860	8 190	8 020	7 700	7 900	7 730
F_y (MPa)	300	300	300	300	300	300	300	300	300
IMPERIAL SIZE AND MASS									
Mass (lb./ft.)	337	275	229	207	184	166	152	134	119
Nominal Depth and Width (in.)	18 x 18								

¤ Welding does not fully develop web strength.
* Class 3 flanges.

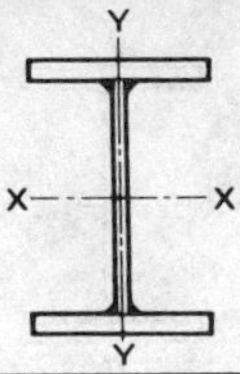

G40.21-M300W

$\phi = 0.90$

WWF COLUMNS

Factored Axial Compressive Resistances, C_r, in kN

Designation	WWF400								
Mass (kg/m)	444 ¤	362	303	273	243	220	202	178	157 *
Effective length (KL) in millimetres with respect to least radius of gyration									
0	15 300	12 500	10 400	9 400	8 370	7 560	6 940	6 130	5 430
2 000	14 900	12 200	10 200	9 160	8 150	7 360	6 750	5 960	5 290
2 250	14 800	12 100	10 100	9 070	8 070	7 290	6 690	5 910	5 230
2 500	14 600	12 000	9 970	8 980	7 980	7 220	6 620	5 850	5 180
2 750	14 500	11 800	9 870	8 880	7 900	7 140	6 550	5 780	5 130
3 000	14 300	11 700	9 760	8 790	7 810	7 060	6 470	5 720	5 070
3 250	14 200	11 600	9 650	8 680	7 710	6 980	6 390	5 650	5 010
3 500	14 000	11 500	9 530	8 580	7 620	6 890	6 310	5 580	4 950
3 750	13 800	11 300	9 410	8 470	7 510	6 800	6 230	5 500	4 880
4 000	13 600	11 200	9 290	8 360	7 410	6 710	6 140	5 430	4 820
4 250	13 500	11 000	9 160	8 240	7 300	6 610	6 060	5 350	4 750
4 500	13 300	10 900	9 030	8 120	7 190	6 520	5 960	5 270	4 680
4 750	13 100	10 700	8 890	7 990	7 080	6 410	5 870	5 180	4 600
5 000	12 900	10 600	8 750	7 870	6 960	6 310	5 770	5 100	4 530
5 250	12 700	10 400	8 610	7 740	6 840	6 200	5 670	5 010	4 450
5 500	12 500	10 200	8 460	7 600	6 720	6 090	5 570	4 920	4 370
6 000	12 000	9 890	8 160	7 320	6 460	5 860	5 360	4 730	4 210
6 500	11 500	9 520	7 830	7 020	6 190	5 620	5 130	4 530	4 030
7 000	11 100	9 130	7 490	6 710	5 900	5 370	4 890	4 320	3 850
7 500	10 500	8 720	7 140	6 390	5 600	5 100	4 650	4 100	3 660
8 000	10 000	8 300	6 770	6 050	5 290	4 820	4 390	3 870	3 460
8 500	9 450	7 860	6 380	5 690	4 940	4 520	4 100	3 620	3 250
9 000	8 800	7 370	5 930	5 280	4 580	4 190	3 800	3 350	3 010
9 500	8 190	6 850	5 510	4 910	4 260	3 900	3 530	3 120	2 800
10 000	7 640	6 400	5 150	4 580	3 970	3 640	3 290	2 910	2 610
10 500	7 150	5 990	4 820	4 290	3 720	3 400	3 080	2 720	2 440
11 000	6 710	5 620	4 520	4 020	3 490	3 190	2 890	2 550	2 290
11 500	6 310	5 290	4 250	3 780	3 280	3 000	2 720	2 400	2 150
12 000	5 950	4 990	4 000	3 560	3 090	2 830	2 560	2 260	2 030
PROPERTIES AND DESIGN DATA									
Area (mm^2)	56 600	46 200	38 600	34 800	31 000	28 000	25 700	22 700	20 100
Z_x (10^3 mm^3)	8 770	7 480	6 300	5 680	5 050	4 660	4 260	3 790	3 420
S_x (10^3 mm^3)	7 300	6 410	5 500	5 000	4 470	4 170	3 830	3 430	3 120
r_x (mm)	161	166	169	170	170	173	173	174	176
Z_y (10^3 mm^3)	4 870	4 030	3 230	2 840	2 440	2 260	2 020	1 780	1 610
S_y (10^3 mm^3)	3 200	2 670	2 130	1 870	1 600	1 490	1 330	1 170	1 070
r_y (mm)	106	108	105	104	102	103	102	102	103
r_x/r_y	1.52	1.54	1.61	1.63	1.67	1.68	1.70	1.71	1.71
M_{rx} (kN·m) ($L < L_u$)	2 370	2 020	1 700	1 530	1 360	1 260	1 150	1 020	842
M_{ry} (kN·m)	1 310	1 090	872	767	659	610	545	481	289
J (10^3 mm^4)	60 100	34 100	17 900	12 300	8 110	6 320	4 640	3 170	2 290
C_w (10^9 mm^6)	18 500	16 300	13 800	12 400	11 000	10 300	9 380	8 390	7 700
L_u (mm)	13 800	11 400	9 310	8 440	7 690	7 470	7 120	6 830	7 070
F_y (MPa)	300	300	300	300	300	300	300	300	300
IMPERIAL SIZE AND MASS									
Mass (lb./ft.)	298	243	203	183	163	147	135	119	105
Nominal Depth and Width (in.)	16 x 16								

¤ Welding does not fully develop web strength.

* Class 3 flanges.

WWF COLUMNS

Factored Axial Compressive Resistances, Cr, in kN

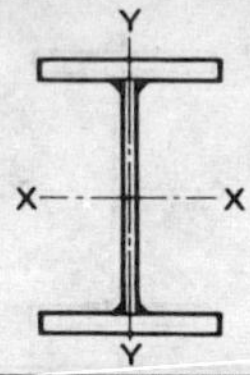

G40.21-M300W

$\phi = 0.90$

Designation	WWF 350							
Mass (kg/m)	315	263	238	212	192	176	155	137
Effective length (KL) in millimetres with respect to least radius of gyration								
0	10 900	9 070	8 180	7 290	6 590	6 050	5 350	4 720
2 000	10 500	8 760	7 880	7 010	6 350	5 820	5 140	4 550
2 250	10 400	8 660	7 790	6 930	6 270	5 750	5 080	4 500
2 500	10 300	8 550	7 700	6 840	6 200	5 680	5 020	4 440
2 750	10 100	8 440	7 600	6 750	6 110	5 600	4 950	4 380
3 000	10 000	8 330	7 490	6 660	6 030	5 520	4 880	4 320
3 250	9 870	8 210	7 380	6 560	5 940	5 440	4 810	4 260
3 500	9 720	8 090	7 270	6 450	5 850	5 360	4 730	4 200
3 750	9 580	7 970	7 160	6 350	5 760	5 270	4 650	4 130
4 000	9 430	7 840	7 030	6 240	5 660	5 180	4 570	4 060
4 250	9 270	7 700	6 910	6 120	5 560	5 080	4 490	3 980
4 500	9 110	7 560	6 780	6 000	5 450	4 980	4 400	3 910
4 750	8 940	7 420	6 650	5 880	5 350	4 880	4 310	3 830
5 000	8 770	7 270	6 510	5 750	5 230	4 780	4 220	3 750
5 250	8 590	7 110	6 370	5 620	5 120	4 670	4 120	3 670
5 500	8 410	6 960	6 220	5 490	5 000	4 560	4 020	3 580
6 000	8 030	6 630	5 920	5 210	4 750	4 330	3 820	3 410
6 500	7 630	6 280	5 590	4 920	4 490	4 090	3 600	3 220
7 000	7 200	5 920	5 260	4 610	4 220	3 830	3 370	3 020
7 500	6 760	5 540	4 890	4 250	3 920	3 540	3 110	2 810
8 000	6 250	5 080	4 490	3 900	3 600	3 250	2 850	2 580
8 500	5 760	4 690	4 130	3 600	3 310	2 990	2 630	2 370
9 000	5 340	4 340	3 830	3 330	3 070	2 770	2 430	2 200
9 500	4 960	4 030	3 550	3 090	2 850	2 570	2 260	2 040
10 000	4 620	3 760	3 310	2 880	2 650	2 390	2 100	1 900
10 500	4 320	3 510	3 090	2 690	2 480	2 240	1 960	1 770
11 000	4 050	3 290	2 890	2 510	2 320	2 090	1 840	1 660
11 500	3 800	3 080	2 720	2 360	2 180	1 960	1 720	1 560
12 000	3 570	2 900	2 550	2 220	2 050	1 850	1 620	1 470
PROPERTIES AND DESIGN DATA								
Area (mm^2)	40 200	33 600	30 300	27 000	24 400	22 400	19 800	17 500
Z_x (10^3 mm^3)	5 580	4 730	4 280	3 810	3 520	3 220	2 870	2 590
S_x (10^3 mm^3)	4 710	4 070	3 720	3 330	3 120	2 870	2 580	2 350
r_x (mm)	143	146	146	147	150	150	151	153
Z_y (10^3 mm^3)	3 090	2 480	2 170	1 870	1 740	1 550	1 360	1 240
S_y (10^3 mm^3)	2 040	1 630	1 430	1 230	1 140	1 020	899	817
r_y (mm)	94.2	92.3	90.8	89.2	90.5	89.4	89.0	90.4
r_x/r_y	1.52	1.58	1.61	1.65	1.66	1.68	1.70	1.69
M_{rx} (kN·m) ($L < L_u$)	1 510	1 280	1 160	1 030	950	869	775	699
M_{ry} (kN·m)	834	670	586	505	470	418	367	335
J (10^3 mm^4)	29 800	15 700	10 800	7 070	5 520	4 060	2 760	2 000
C_w (10^9 mm^6)	8 040	6 870	6 210	5 490	5 190	4 720	4 230	3 890
L_u (mm)	11 500	9 130	8 130	7 270	6 980	6 580	6 230	6 070
F_y (MPa)	300	300	300	300	300	300	300	300
IMPERIAL SIZE AND MASS								
Mass (lb./ft.)	211	177	159	142	128	118	104	92
Nominal Depth and Width (in.)	14 x 14							

G40.21-M 300W
$\phi = 0.90$

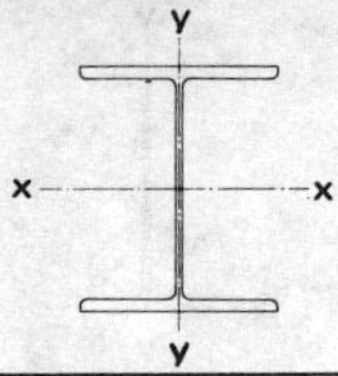

W COLUMNS
Factored Axial Compressive Resistances, C_r, in kN

Designation	W360			W310		
Mass (kg/m)	79	72	64 *	226	202	179
Effective length (KL) in millimetres with respect to least radius of gyration						
0	2 730	2 460	2 060	7 800	6 970	6 160
2 000	2 390	2 150	1 820	7 430	6 630	5 850
2 250	2 320	2 080	1 770	7 330	6 540	5 770
2 500	2 230	2 010	1 710	7 220	6 440	5 680
2 750	2 150	1 930	1 650	7 110	6 340	5 590
3 000	2 060	1 850	1 580	6 990	6 230	5 500
3 250	1 960	1 760	1 520	6 870	6 120	5 400
3 500	1 870	1 670	1 440	6 740	6 010	5 290
3 750	1 760	1 580	1 370	6 610	5 880	5 190
4 000	1 650	1 470	1 270	6 470	5 760	5 070
4 250	1 520	1 360	1 170	6 330	5 630	4 960
4 500	1 410	1 260	1 090	6 180	5 500	4 840
4 750	1 310	1 170	1 010	6 030	5 360	4 710
5 000	1 220	1 090	941	5 870	5 210	4 590
5 250	1 140	1 020	880	5 710	5 070	4 450
5 500	1 070	955	824	5 540	4 910	4 320
6 000	948	845	729	5 190	4 600	4 030
6 500	844	752	649	4 820	4 260	3 730
7 000	756	674	581	4 380	3 870	3 380
7 500	681	606	523	4 000	3 530	3 080
8 000	612	544	467	3 660	3 230	2 820
8 500	545	484	416	3 370	2 970	2 600
9 000	489	434	373	3 110	2 750	2 400
9 500	441	392	337	2 890	2 540	2 220
10 000				2 680	2 370	2 060
10 500				2 500	2 200	1 920
11 000				2 340	2 060	1 800
11 500				2 190	1 930	1 680
12 000				2 050	1 810	1 580
PROPERTIES AND DESIGN DATA						
Area (mm^2)	10 100	9 110	8 140	28 900	25 800	22 800
Z_x (10^3 mm^3)	1 430	1 280	1 140	3 980	3 510	3 050
S_x (10^3 mm^3)	1 280	1 150	1 030	3 420	3 050	2 680
r_x (mm)	150	149	148	144	142	140
Z_y (10^3 mm^3)	362	322	284	1 830	1 610	1 400
S_y (10^3 mm^3)	236	210	186	1 190	1 050	919
r_y (mm)	48.9	48.5	48.1	80.9	80.2	79.5
r_x/r_y	3.07	3.07	3.08	1.78	1.77	1.76
M_{rx} (kN·m) ($L < L_u$)	386	346	308	1 070	948	823
M_{ry} (kN·m)	97.7	86.9	76.7	494	435	378
J (10^3 mm^4)	814	603	438	10 800	7 740	5 380
C_w (10^9 mm^6)	687	600	524	4 620	3 960	3 340
L_u (mm)	3 270	3 190	3 110	7 510	6 970	6 490
F_y (MPa)	300	300	300	300	300	300
IMPERIAL SIZE AND MASS						
Mass (lb./ft.)	53	48	43	152	136	120
Nominal Depth and Width (in.)		14 x 8			12 x 12	

* Resistance for this column calculated according to CAN/CSA-S16.1-M89 Clause 13.3.3.

W COLUMNS
Factored Axial Compressive Resistances, C_r, in kN

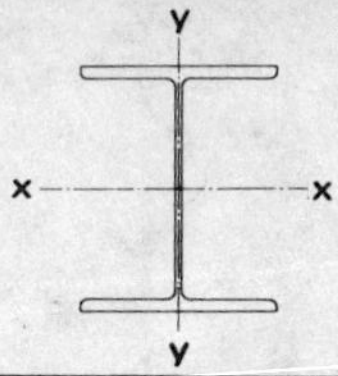

G40.21-M 300W
$\phi = 0.9$

Designation	W310						W310	
Mass (kg/m)	158	143	129	118	107	97†	86	79
Effective length (KL) in millimetres with respect to least radius of gyration								
0	5 430	4 910	4 450	4 050	3 670	3 320	2 970	2 730
2 000	5 160	4 670	4 230	3 840	3 480	3 150	2 740	2 510
2 250	5 080	4 600	4 160	3 780	3 430	3 100	2 690	2 460
2 500	5 000	4 530	4 100	3 720	3 380	3 050	2 630	2 410
2 750	4 920	4 460	4 030	3 660	3 320	3 000	2 570	2 350
3 000	4 840	4 380	3 960	3 600	3 260	2 940	2 500	2 290
3 250	4 750	4 300	3 890	3 530	3 200	2 890	2 430	2 230
3 500	4 660	4 220	3 810	3 460	3 130	2 830	2 360	2 160
3 750	4 560	4 130	3 730	3 380	3 070	2 770	2 290	2 090
4 000	4 460	4 040	3 640	3 310	3 000	2 710	2 210	2 020
4 250	4 360	3 940	3 560	3 230	2 920	2 640	2 130	1 940
4 500	4 250	3 850	3 470	3 150	2 850	2 570	2 050	1 870
4 750	4 140	3 750	3 370	3 060	2 770	2 500	1 960	1 780
5 000	4 030	3 640	3 280	2 970	2 690	2 430	1 870	1 700
5 250	3 910	3 530	3 180	2 880	2 610	2 350	1 780	1 610
5 500	3 790	3 420	3 080	2 790	2 520	2 270	1 670	1 510
6 000	3 530	3 190	2 860	2 590	2 340	2 110	1 490	1 340
6 500	3 250	2 940	2 620	2 370	2 140	1 930	1 330	1 200
7 000	2 950	2 660	2 370	2 150	1 940	1 750	1 200	1 090
7 500	2 690	2 430	2 160	1 960	1 770	1 590	1 090	987
8 000	2 460	2 220	1 980	1 790	1 620	1 460	996	900
8 500	2 260	2 050	1 820	1 650	1 490	1 340	912	824
9 000	2 090	1 890	1 680	1 520	1 370	1 240	838	757
9 500	1 940	1 750	1 560	1 410	1 270	1 140	772	697
10 000	1 800	1 630	1 450	1 310	1 180	1 060	714	644
10 500	1 680	1 520	1 350	1 220	1 100	990	656	589
11 000	1 570	1 410	1 260	1 140	1 030	924	600	539
11 500	1 460	1 320	1 180	1 060	961	864	551	495
12 000	1 370	1 240	1 100	997	900	809	508	457
PROPERTIES AND DESIGN DATA								
Area (mm^2)	20 100	18 200	16 500	15 000	13 600	12 300	11 000	10 100
Z_x ($10^3 mm^3$)	2 670	2 420	2 160	1 950	1 770	1 590	1 420	1 280
S_x ($10^3 mm^3$)	2 360	2 150	1 940	1 750	1 590	1 440	1 280	1 160
r_x (mm)	139	138	137	135	135	134	135	132
Z_y ($10^3 mm^3$)	1 220	1 110	991	893	806	725	533	478
S_y ($10^3 mm^3$)	805	729	652	588	531	478	351	314
r_y (mm)	78.9	78.8	77.8	77.5	77.3	77.0	63.6	62.9
r_x/r_y	1.76	1.75	1.76	1.74	1.75	1.74	2.12	2.10
M_{rx} (kN•m) ($L < L_u$)	721	653	583	526	478	389	383	346
M_{ry} (kN•m)	329	300	268	241	218	129	144	129
J ($10^3 mm^4$)	3 780	2 870	2 130	1 600	1 220	912	877	657
C_w ($10^9 mm^6$)	2 840	2 540	2 220	1 970	1 760	1 560	961	847
L_u (mm)	6 080	5 820	5 580	5 390	5 220	5 410	4 250	4 140
F_y (MPa)	300	300	300	300	300	300	300	300
IMPERIAL SIZE AND MASS								
Mass (lb./ft.)	106	96	87	79	72	65	58	53
Nominal Depth and Width (in.)	12 x 12						12 X 10	

† Class 3 in bending.

Only shapes available from Canadian mills are listed, with those highlighted in white being commonly used sizes which are generally readily available.

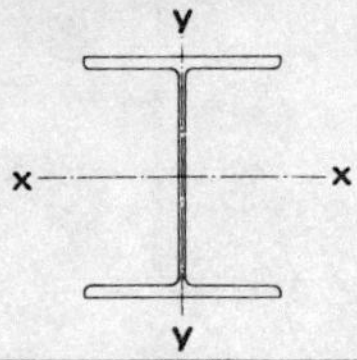

W COLUMNS
Factored Axial Compressive Resistances, C_r, in kN

Designation	W310			W250				
Mass (kg/m)	74	67	60	167	149	131	115	101
Effective length (KL) in millimetres with respect to least radius of gyration								
0	2 560	2 300	2 050	5 750	5 130	4 510	3 940	3 480
2 000	2 260	2 020	1 800	5 360	4 780	4 190	3 660	3 230
2 250	2 190	1 960	1 740	5 270	4 690	4 120	3 590	3 170
2 500	2 110	1 890	1 680	5 170	4 600	4 030	3 520	3 100
2 750	2 030	1 820	1 620	5 060	4 500	3 950	3 440	3 030
3 000	1 950	1 740	1 550	4 940	4 400	3 860	3 360	2 960
3 250	1 870	1 660	1 480	4 830	4 290	3 760	3 280	2 890
3 500	1 770	1 580	1 410	4 700	4 180	3 660	3 190	2 810
3 750	1 680	1 490	1 330	4 580	4 060	3 560	3 100	2 730
4 000	1 580	1 410	1 250	4 440	3 940	3 450	3 000	2 640
4 250	1 460	1 300	1 150	4 300	3 820	3 340	2 900	2 550
4 500	1 350	1 200	1 070	4 160	3 690	3 220	2 800	2 460
4 750	1 260	1 120	991	4 010	3 550	3 100	2 690	2 360
5 000	1 170	1 040	924	3 850	3 410	2 970	2 580	2 260
5 250	1 100	974	864	3 690	3 260	2 840	2 470	2 160
5 500	1 030	913	810	3 530	3 110	2 710	2 340	2 040
6 000	911	808	716	3 150	2 770	2 410	2 090	1 820
6 500	812	720	638	2 830	2 490	2 160	1 870	1 630
7 000	728	645	572	2 560	2 250	1 950	1 690	1 470
7 500	656	581	515	2 330	2 050	1 780	1 540	1 340
8 000	593	524	464	2 130	1 870	1 620	1 400	1 220
8 500	529	467	413	1 950	1 710	1 490	1 290	1 120
9 000	474	419	370	1 800	1 580	1 370	1 180	1 030
9 500	428	378	334	1 660	1 460	1 260	1 090	949
10 000				1 530	1 350	1 170	1 010	878
10 500				1 420	1 250	1 080	937	814
11 000				1 320	1 160	1 000	862	746
11 500				1 220	1 060	918	792	685
12 000				1 120	980	847	730	632
PROPERTIES AND DESIGN DATA								
Area (mm^2)	9 490	8 510	7 590	21 300	19 000	16 700	14 600	12 900
Z_x ($10^3 mm^3$)	1 190	1 060	941	2 430	2 130	1 850	1 600	1 400
S_x ($10^3 mm^3$)	1 060	949	849	2 080	1 840	1 610	1 410	1 240
r_x (mm)	132	131	130	119	117	115	114	113
Z_y ($10^3 mm^3$)	350	310	275	1 140	1 000	870	753	656
S_y ($10^3 mm^3$)	229	203	180	746	656	571	495	432
r_y (mm)	49.7	49.3	49.1	68.1	67.4	66.8	66.3	65.6
r_x/r_y	2.66	2.66	2.65	1.75	1.74	1.72	1.72	1.72
M_{rx} (kN•m) ($L < L_u$)	321	286	254	656	575	499	432	378
M_{ry} (kN•m)	94.5	83.7	74.2	308	270	235	203	177
J ($10^3 mm^4$)	745	545	397	6 310	4 510	3 120	2 130	1 490
C_w ($10^9 mm^6$)	505	439	384	1 630	1 390	1 160	976	829
L_u (mm)	3 380	3 280	3 200	6 670	6 160	5 680	5 270	4 950
F_y (MPa)	300	300	300	300	300	300	300	300
IMPERIAL SIZE AND MASS								
Mass (lb./ft.)	50	45	40	112	100	88	77	68
Nominal Depth and Width (in.)	12 x 8			10 x 10				

Only shapes available from Canadian mills are listed, with those highlighted in white being commonly used sizes which are generally readily available.

W COLUMNS
Factored Axial Compressive Resistances, C_r, in kN

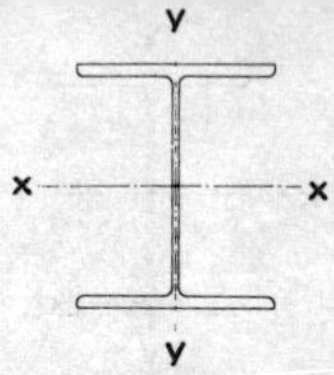

G40.21-M 300W
$\phi = 0.9$

Designation		W250			W250	
Mass (kg/m)	89	80	73	67	58	49
Effective length (KL) in millimetres with respect to least radius of gyration						
0	3 080	2 750	2 510	2 310	2 000	1 690
2 000	2 850	2 550	2 320	2 040	1 770	1 480
2 250	2 800	2 500	2 270	1 980	1 720	1 440
2 500	2 740	2 450	2 230	1 920	1 660	1 390
2 750	2 680	2 390	2 180	1 850	1 600	1 330
3 000	2 610	2 340	2 120	1 780	1 540	1 280
3 250	2 550	2 280	2 070	1 710	1 470	1 220
3 500	2 470	2 210	2 010	1 630	1 400	1 160
3 750	2 400	2 150	1 950	1 550	1 330	1 100
4 000	2 320	2 080	1 880	1 460	1 250	1 030
4 250	2 240	2 000	1 820	1 360	1 160	950
4 500	2 160	1 930	1 750	1 260	1 080	880
4 750	2 080	1 850	1 680	1 170	1 000	818
5 000	1 990	1 770	1 610	1 100	933	763
5 250	1 900	1 690	1 530	1 030	873	713
5 500	1 790	1 590	1 440	962	819	669
6 000	1 590	1 420	1 280	852	725	592
6 500	1 430	1 270	1 150	760	646	527
7 000	1 290	1 150	1 040	682	580	472
7 500	1 170	1 040	944	615	522	425
8 000	1 070	953	861	557	473	383
8 500	980	873	789	500	423	341
9 000	901	803	725	449	379	306
9 500	831	740	669	405	342	276
10 000	769	684	618	367	311	
10 500	712	634	572			
11 000	652	580	523			
11 500	599	532	480			
12 000	552	491	443			
PROPERTIES AND DESIGN DATA						
Area (mm^2)	11 400	10 200	9 280	8 550	7 420	6 250
Z_x ($10^3 mm^3$)	1 230	1 090	985	901	770	633
S_x ($10^3 mm^3$)	1 100	982	891	806	693	572
r_x (mm)	112	111	110	110	108	106
Z_y ($10^3 mm^3$)	574	513	463	332	283	228
S_y ($10^3 mm^3$)	378	338	306	218	186	150
r_y (mm)	65.2	65.0	64.7	51.0	50.3	49.2
r_x/r_y	1.72	1.71	1.70	2.16	2.15	2.15
M_{rx} (kN•m) ($L < L_u$)	332	294	266	243	208	171
M_{ry} (kN•m)	155	139	125	89.6	76.4	61.6
J ($10^3 mm^4$)	1 040	757	575	625	409	241
C_w ($10^9 mm^6$)	713	623	553	324	268	211
L_u (mm)	4 690	4 520	4 390	3 570	3 410	3 240
F_y (MPa)	300	300	300	300	300	300
IMPERIAL SIZE AND MASS						
Mass (lb./ft.)	60	54	49	45	39	33
Nominal Depth and Width (in.)		10 x 10			10 x 8	

Only shapes available from Canadian mills are listed, with those highlighted in white being commonly used sizes which are generally readily available.

G40.21-M 300W
$\phi = 0.90$

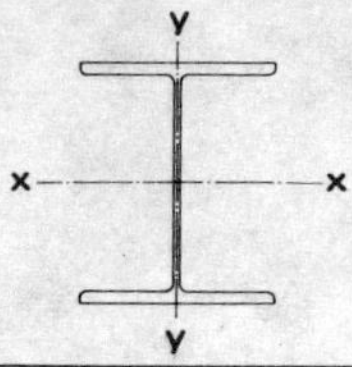

W COLUMNS
Factored Axial Compressive Resistances, C_r, in kN

Designation	W200						W200	
Mass (kg/m)	100	86	71	59	52	46	42	36
Effective length (KL) in millimetres with respect to least radius of gyration								
0	3 430	3 000	2 460	2 040	1 800	1 580	1 430	1 240
2 000	3 070	2 680	2 190	1 810	1 600	1 400	1 200	1 030
2 250	2 990	2 610	2 130	1 760	1 550	1 360	1 140	983
2 500	2 900	2 530	2 070	1 710	1 500	1 320	1 090	935
2 750	2 810	2 450	2 000	1 650	1 450	1 270	1 030	883
3 000	2 710	2 360	1 930	1 590	1 400	1 220	967	828
3 250	2 610	2 270	1 850	1 520	1 340	1 170	901	770
3 500	2 500	2 170	1 780	1 460	1 280	1 120	825	703
3 750	2 390	2 070	1 690	1 390	1 220	1 060	753	641
4 000	2 270	1 970	1 610	1 310	1 150	1 000	690	587
4 250	2 150	1 860	1 520	1 230	1 080	937	635	540
4 500	2 010	1 730	1 410	1 140	1 000	868	587	499
4 750	1 870	1 610	1 310	1 060	932	807	544	463
5 000	1 750	1 510	1 220	992	870	753	506	430
5 250	1 630	1 410	1 150	929	814	705	472	401
5 500	1 530	1 320	1 070	871	763	661	441	375
6 000	1 360	1 170	953	772	676	585	387	329
6 500	1 210	1 050	851	689	603	522	343	291
7 000	1 090	941	764	618	542	469	299	254
7 500	987	850	690	558	489	423	263	222
8 000	895	771	626	506	443	383	232	197
8 500	815	702	570	457	400	344		
9 000	735	631	511	410	359	309		
9 500	663	569	461	370	323	278		
10 000	601	516	418	336	294	253		
10 500	548	471	381					
11 000								
11 500								
12 000								
PROPERTIES AND DESIGN DATA								
Area (mm^2)	12 700	11 100	9 110	7 560	6 660	5 860	5 310	4 580
Z_x ($10^3 mm^3$)	1 150	981	803	653	570	496	446	380
S_x ($10^3 mm^3$)	989	853	709	582	512	448	399	342
r_x (mm)	94.3	92.4	91.7	89.9	89.0	88.1	87.8	86.7
Z_y ($10^3 mm^3$)	533	458	375	303	266	230	165	141
S_y ($10^3 mm^3$)	349	300	246	199	175	151	108	92.6
r_y (mm)	53.7	53.2	52.8	51.9	51.7	51.1	41.2	40.8
r_x/r_y	1.76	1.74	1.74	1.73	1.72	1.72	2.13	2.12
M_{rx} (kN•m) ($L < L_u$)	310	265	217	176	154	134	120	103
M_{ry} (kN•m)	144	124	101	81.8	71.8	62.1	44.5	38.1
J ($10^3 mm^4$)	2 090	1 400	818	465	324	221	223	146
C_w ($10^9 mm^6$)	386	318	250	196	167	141	84.0	69.6
L_u (mm)	5 020	4 620	4 150	3 780	3 620	3 460	2 850	2 730
F_y (MPa)	300	300	300	300	300	300	300	300
IMPERIAL SIZE AND MASS								
Mass (lb./ft.)	67	58	48	40	35	31	28	24
Nominal Depth and Width (in.)	8 x 8						8 x 6½	

Only shapes available from Canadian mills are listed, with those highlighted in white being commonly used sizes which are generally readily available.

W COLUMNS
Factored Axial Compressive Resistances, C_r, in kN

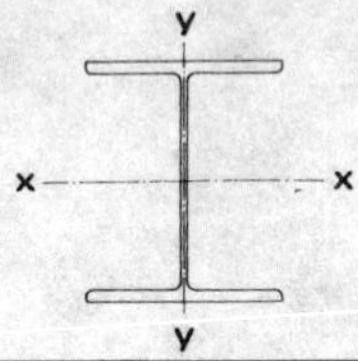

G40.21-M 300W
$\phi = 0.90$

Designation	W200		W150		
Mass (kg/m)	31	27	37	30	22†
Effective length (KL) in millimetres with respect to least radius of gyration					
0	1 080	915	1 280	1 020	769
2 000	807	674	1 040	832	616
2 250	749	622	991	790	582
2 500	685	566	937	746	546
2 750	612	502	878	698	508
3 000	545	446	816	648	468
3 250	489	400	746	590	421
3 500	441	361	676	534	381
3 750	401	328	616	487	347
4 000	365	299	564	446	317
4 250	335	273	519	410	291
4 500	308	251	479	378	269
4 750	284	231	444	350	249
5 000	262	213	412	325	231
5 250	241	195	384	303	214
5 500	221	178	358	282	200
6 000	187	151	314	247	174
6 500			273	214	149
7 000			237	186	130
7 500			208	163	
8 000					
8 500					
9 000					
9 500					
10 000					
10 500					
11 000					
11 500					
12 000					
PROPERTIES AND DESIGN DATA					
Area (mm^2)	4 000	3 390	4 730	3 790	2 850
Z_x ($10^3 mm^3$)	335	279	310	244	176
S_x ($10^3 mm^3$)	299	249	274	219	159
r_x (mm)	88.6	87.2	68.5	67.4	65.2
Z_y ($10^3 mm^3$)	93.8	76.1	140	111	77.6
S_y ($10^3 mm^3$)	61.1	49.6	91.8	72.6	50.9
r_y (mm)	32.0	31.2	38.7	38.3	36.8
r_x/r_y	2.77	2.79	1.77	1.76	1.77
M_{rx} (kN•m) ($L < L_u$)	90.4	75.3	83.7	65.9	42.9
M_{ry} (kN•m)	25.3	20.5	37.8	30.0	13.7
J ($10^3 mm^4$)	119	71.3	193	101	41.8
C_w ($10^9 mm^6$)	40.9	32.5	40.0	30.3	20.4
L_u (mm)	2 150	2 050	2 910	2 680	2 590
F_y (MPa)	300	300	300	300	300
IMPERIAL SIZE AND MASS					
Mass (lb./ft.)	21	18	25	20	15
Nominal Depth and Width (in.)	8 x 5¼		6 x 6		

† Class 3 in bending.

Only shapes available from Canadian mills are listed, with those highlighted in white being commonly used sizes which are generally readily available.

NOTES

WWF COLUMNS

Factored Axial Compressive Resistance, C_r, in kN

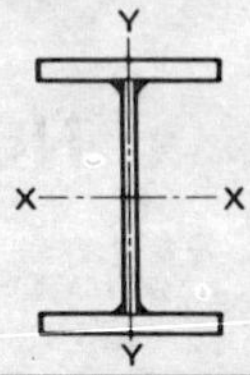

G40.21-M350W

$\phi = 0.9$

Designation	WWF650				
Mass (kg/m)	864 °	739 °	598	499	400 **
Effective length (KL) in millimetres with respect to least radius of gyration					
0	34 600	29 600	24 000	20 000	15 500
2 000	34 500	29 600	24 000	20 000	15 500
2 250	34 300	29 400	23 800	19 900	15 400
2 500	34 000	29 300	23 700	19 800	15 300
2 750	33 800	29 100	23 600	19 700	15 300
3 000	33 600	28 900	23 400	19 500	15 200
3 250	33 400	28 700	23 300	19 400	15 100
3 500	33 100	28 600	23 200	19 300	15 000
3 750	32 900	28 400	23 000	19 200	14 900
4 000	32 600	28 200	22 800	19 000	14 900
4 250	32 400	28 000	22 700	18 900	14 800
4 500	32 100	27 800	22 500	18 800	14 700
4 750	31 800	27 600	22 400	18 600	14 600
5 000	31 500	27 400	22 200	18 500	14 500
5 250	31 300	27 100	22 000	18 300	14 400
5 500	31 000	26 900	21 800	18 200	14 300
6 000	30 400	26 400	21 500	17 900	14 100
6 500	29 700	26 000	21 100	17 500	13 900
7 000	29 100	25 500	20 700	17 200	13 600
7 500	28 400	24 900	20 300	16 800	13 300
8 000	27 600	24 400	19 800	16 500	13 000
8 500	26 900	23 800	19 400	16 100	12 600
9 000	26 100	23 300	18 900	15 700	12 300
9 500	25 300	22 600	18 400	15 300	12 000
10 000	24 500	22 000	18 000	14 800	11 600
10 500	23 600	21 400	17 400	14 400	11 200
11 000	22 800	20 700	16 900	13 900	10 800
11 500	21 900	20 000	16 400	13 500	10 400
12 000	20 900	19 300	15 800	13 000	10 000
PROPERTIES AND DESIGN DATA					
Area (mm^2)	110 000	94 100	76 200	63 600	51 000
Z_x (10^3 mm^3)	27 300	25 200	21 100	17 500	13 900
S_x (10^3 mm^3)	23 300	22 100	18 900	15 900	12 400
r_x (mm)	262	277	284	285	284
Z_y (10^3 mm^3)	13 200	12 800	10 600	8 510	6 400
S_y (10^3 mm^3)	8 480	8 450	7 040	5 630	4 120
r_y (mm)	158	171	173	170	164
r_x/r_y	1.66	1.62	1.64	1.68	1.73
M_{rx} (kN·m) ($L < L_u$)	8 600	7 940	6 650	5 510	3 920
M_{yx} (kN·m)	4 160	4 030	3 340	2 680	1 300
J (10^3 mm^4)	132 000	98 400	55 600	29 300	13 300
C_w (10^9 mm^6)	240 000	239 000	206 000	170 000	132 000
L_u (mm)	13 000	12 600	11 500	10 500	10 300
F_y (MPa)	350	350	350	350	350
IMPERIAL SIZE AND MASS					
Mass (lb./ft.)	580	497	402	336	269
Nominal Depth and Width (in.)	26 x 26				

° Welding does not fully develop web strength.

** Class 4 flanges: area and resistance calculated according to CAN/CSA-M16.1-M89 Clause 13.3.3.

G40.21-M350W

$\phi = 0.9$

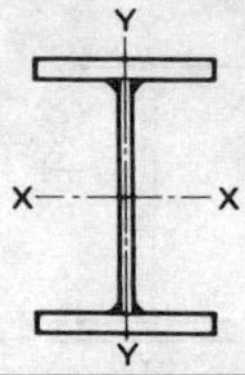

WWF COLUMNS

Factored Axial Compressive Resistance, C_r, in kN

Designation		WWF600				
Mass (kg/m)		793 ª	680 ª	551	460	369 *
Effective length (KL) in millimetres with respect to least radius of gyration	0	31 800	27 300	22 100	18 500	14 800
	2 000	31 500	27 100	22 000	18 400	14 700
	2 250	31 300	27 000	21 900	18 200	14 600
	2 500	31 100	26 800	21 700	18 100	14 500
	2 750	30 900	26 600	21 600	18 000	14 400
	3 000	30 700	26 500	21 500	17 900	14 300
	3 250	30 400	26 300	21 300	17 800	14 200
	3 500	30 200	26 100	21 200	17 600	14 100
	3 750	29 900	25 900	21 000	17 500	14 000
	4 000	29 700	25 700	20 900	17 400	13 900
	4 250	29 400	25 500	20 700	17 200	13 800
	4 500	29 100	25 300	20 500	17 100	13 600
	4 750	28 900	25 100	20 400	16 900	13 500
	5 000	28 600	24 800	20 200	16 800	13 400
	5 250	28 300	24 600	20 000	16 600	13 300
	5 500	28 000	24 400	19 800	16 500	13 100
	6 000	27 300	23 900	19 400	16 100	12 800
	6 500	26 700	23 400	19 000	15 800	12 600
	7 000	26 000	22 900	18 600	15 400	12 300
	7 500	25 300	22 300	18 200	15 100	11 900
	8 000	24 600	21 800	17 700	14 700	11 600
	8 500	23 800	21 200	17 300	14 300	11 300
	9 000	23 000	20 600	16 800	13 900	10 900
	9 500	22 200	19 900	16 300	13 400	10 600
	10 000	21 300	19 300	15 800	13 000	10 200
	10 500	20 400	18 600	15 200	12 500	9 790
	11 000	19 500	17 900	14 700	12 100	9 390
	11 500	18 400	17 200	14 100	11 600	8 970
	12 000	17 400	16 400	13 500	11 000	8 480
PROPERTIES AND DESIGN DATA						
Area (mm^2)		101 000	86 600	70 200	58 600	47 000
Z_x (10^3 mm^3)		22 900	21 200	17 800	14 800	11 800
S_x (10^3 mm^3)		19 400	18 500	15 900	13 400	10 700
r_x (mm)		240	253	261	262	261
Z_y (10^3 mm^3)		11 200	10 900	9 050	7 250	5 460
S_y (10^3 mm^3)		7 230	7 200	6 000	4 800	3 600
r_y (mm)		147	158	160	157	152
r_x/r_y		1.63	1.60	1.63	1.67	1.72
M_{rx} (kN·m) ($L < L_u$)		7 210	6 680	5 610	4 660	3 370
M_{yx} (kN·m)		3 530	3 430	2 850	2 280	1 130
J (10^3 mm^4)		121 000	90 700	51 300	27 000	12 200
C_w (10^9 mm^6)		158 000	158 000	136 000	113 000	87 800
L_u (mm)		12 800	12 300	11 100	9 970	9 630
F_y (MPa)		350	350	350	350	350
IMPERIAL SIZE AND MASS						
Mass (lb./ft.)		531	456	371	309	248
Nominal Depth and Width (in.)		24 x 24				

ª Welding does not fully develop web strength.

* Class 3 flanges.

WWF COLUMNS

Factored Axial Compressive Resistance, C_r, in kN

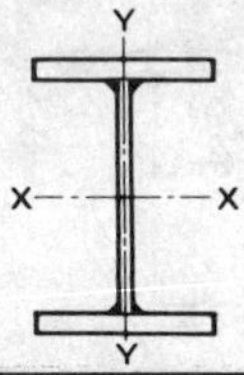

G40.21-M350W

$\phi = 0.9$

Designation	WWF550					WWF500		
Mass (kg/m)	721 □	620 □	503	420	280 **	651 □	561 □	456
Effective length (KL) in millimetres with respect to least radius of gyration								
0	29 000	24 900	20 200	16 900	10 700	26 100	22 600	18 300
2 000	28 600	24 700	20 000	16 700	10 700	25 600	22 200	18 100
2 250	28 400	24 500	19 900	16 600	10 600	25 400	22 100	17 900
2 500	28 200	24 300	19 800	16 500	10 600	25 200	21 900	17 800
2 750	27 900	24 200	19 600	16 400	10 500	25 000	21 700	17 700
3 000	27 700	24 000	19 500	16 200	10 400	24 700	21 500	17 500
3 250	27 500	23 800	19 300	16 100	10 400	24 500	21 300	17 400
3 500	27 200	23 600	19 200	16 000	10 300	24 200	21 100	17 200
3 750	26 900	23 400	19 000	15 800	10 200	24 000	20 900	17 000
4 000	26 700	23 200	18 900	15 700	10 100	23 700	20 700	16 900
4 250	26 400	23 000	18 700	15 600	10 100	23 400	20 500	16 700
4 500	26 100	22 800	18 500	15 400	9 990	23 100	20 200	16 500
4 750	25 800	22 500	18 300	15 300	9 900	22 800	20 000	16 300
5 000	25 500	22 300	18 200	15 100	9 820	22 500	19 800	16 100
5 250	25 200	22 100	18 000	14 900	9 730	22 200	19 500	15 900
5 500	24 900	21 800	17 800	14 800	9 640	21 900	19 300	15 700
6 000	24 300	21 300	17 400	14 400	9 450	21 200	18 800	15 300
6 500	23 600	20 800	17 000	14 100	9 230	20 500	18 200	14 900
7 000	22 900	20 300	16 500	13 700	8 970	19 700	17 600	14 400
7 500	22 100	19 700	16 100	13 300	8 700	18 900	17 000	14 000
8 000	21 400	19 100	15 600	12 900	8 420	18 100	16 400	13 500
8 500	20 600	18 500	15 100	12 500	8 130	17 300	15 800	12 900
9 000	19 700	17 900	14 600	12 000	7 820	16 400	15 100	12 400
9 500	18 900	17 200	14 100	11 600	7 510	15 400	14 400	11 900
10 000	18 000	16 500	13 500	11 100	7 180	14 400	13 600	11 300
10 500	16 900	15 800	13 000	10 600	6 840	13 500	12 800	10 600
11 000	15 900	15 000	12 400	10 100	6 450	12 700	12 000	9 970
11 500	15 000	14 200	11 700	9 520	6 070	11 900	11 300	9 390
12 000	14 200	13 400	11 100	9 000	5 740	11 200	10 700	8 860
PROPERTIES AND DESIGN DATA								
Area (mm^2)	92 000	79 100	64 200	53 600	35 600	83 000	71 600	58 200
Z_x (10^3 mm^3)	19 000	17 600	14 800	12 400	8 250	15 400	14 300	12 100
S_x (10^3 mm^3)	16 000	15 200	13 100	11 100	7 310	12 800	12 300	10 600
r_x (mm)	218	230	237	239	241	196	207	214
Z_y (10^3 mm^3)	9 470	9 180	7 610	6 100	3 810	7 850	7 590	6 290
S_y (10^3 mm^3)	6 080	6 050	5 040	4 030	2 380	5 030	5 000	4 170
r_y (mm)	135	145	147	144	140	123	132	134
r_x/r_y	1.61	1.59	1.61	1.66	1.72	1.59	1.57	1.6
M_{rx} (kN·m) ($L < L_u$)	5 980	5 540	4 660	3 910	2 300	4 850	4 500	3 810
M_{yx} (kN·m)	2 980	2 890	2 400	1 920	750	2 470	2 390	1 980
J (10^3 mm^4)	110 000	83 100	47 000	24 700	6 410	99 400	75 400	42 700
C_w (10^9 mm^6)	100 000	99 900	86 700	72 100	47 800	60 800	60 500	52 700
L_u (mm)	12 500	12 100	10 700	9 410	8 810	12 500	11 900	10 300
F_y (MPa)	350	350	350	350	350	350	350	350
IMPERIAL SIZE AND MASS								
Mass (lb./ft.)	484	416	338	282	188	437	377	306
Nominal Depth and Width (in.)	22 x 22					20 x 20		

□ Welding does not fully develop web strength.

** Class 4 flanges: area and resistance calculated according to CAN/CSA-M16.1-M89 Clause 13.3.3.

G40.21-M350W

$\phi = 0.9$

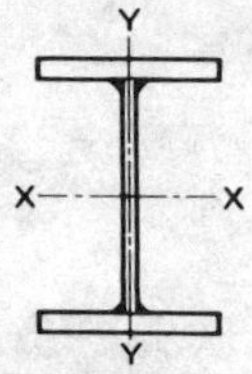

WWF COLUMNS

Factored Axial Compressive Resistance, C_r, in kN

Designation		WWF500						
Mass (kg/m)		381	343	306	276	254 *	223 **	197 **
Effective length (KL) in millimetres with respect to least radius of gyration	0	15 300	13 800	12 300	11 100	10 200	8 420	6 870
	2 000	15 100	13 600	12 100	10 900	10 000	8 320	6 790
	2 250	15 000	13 500	12 000	10 800	9 920	8 270	6 760
	2 500	14 800	13 400	11 900	10 700	9 840	8 220	6 720
	2 750	14 700	13 200	11 800	10 600	9 750	8 170	6 680
	3 000	14 600	13 100	11 700	10 500	9 660	8 120	6 640
	3 250	14 500	13 000	11 600	10 400	9 570	8 060	6 600
	3 500	14 300	12 900	11 400	10 300	9 470	8 010	6 550
	3 750	14 200	12 700	11 300	10 200	9 370	7 950	6 510
	4 000	14 000	12 600	11 200	10 100	9 270	7 880	6 460
	4 250	13 900	12 500	11 100	10 000	9 170	7 820	6 410
	4 500	13 700	12 300	10 900	9 900	9 060	7 750	6 360
	4 750	13 600	12 200	10 800	9 790	8 950	7 680	6 310
	5 000	13 400	12 000	10 700	9 660	8 830	7 600	6 250
	5 250	13 200	11 900	10 500	9 540	8 710	7 520	6 200
	5 500	13 100	11 700	10 400	9 410	8 590	7 440	6 140
	6 000	12 700	11 400	10 100	9 140	8 340	7 270	6 020
	6 500	12 300	11 000	9 760	8 870	8 080	7 080	5 890
	7 000	11 900	10 700	9 420	8 570	7 810	6 860	5 750
	7 500	11 500	10 300	9 080	8 270	7 520	6 610	5 610
	8 000	11 100	9 890	8 710	7 950	7 220	6 340	5 450
	8 500	10 600	9 480	8 340	7 620	6 900	6 070	5 290
	9 000	10 200	9 050	7 940	7 270	6 580	5 780	5 110
	9 500	9 690	8 600	7 540	6 910	6 240	5 490	4 880
	10 000	9 170	8 100	7 060	6 510	5 850	5 140	4 610
	10 500	8 590	7 590	6 620	6 100	5 480	4 810	4 340
	11 000	8 070	7 130	6 210	5 730	5 150	4 520	4 080
	11 500	7 600	6 710	5 850	5 390	4 850	4 260	3 840
	12 000	7 180	6 330	5 520	5 090	4 570	4 020	3 620
PROPERTIES AND DESIGN DATA								
Area (mm^2)		48 600	43 800	39 000	35 200	32 300	28 500	25 200
Z_x (10^3 mm^3)		10 100	9 100	8 060	7 420	6 780	6 010	5 410
S_x (10^3 mm^3)		9 010	8 140	7 240	6 740	6 160	5 170	4 300
r_x (mm)		215	216	215	218	218	219	223
Z_y (10^3 mm^3)		5 040	4 420	3 800	3 530	3 160	2 770	2 510
S_y (10^3 mm^3)		3 330	2 920	2 500	2 330	2 080	1 620	1 220
r_y (mm)		131	129	127	129	127	127	129
r_x/r_y		1.64	1.67	1.69	1.69	1.72	1.72	1.73
M_{rx} (kN·m) ($L < L_u$)		3 180	2 870	2 540	2 340	1 940	1 630	1 350
M_{yx} (kN·m)		1 590	1 390	1 200	1 110	655	511	384
J (10^3 mm^4)		22 500	15 400	10 200	7 920	5 820	3 970	2 870
C_w (10^9 mm^6)		44 100	39 400	34 500	32 500	29 400	26 200	24 000
L_u (mm)		8 970	8 390	7 930	7 830	8 040	8 090	8 390
F_y (MPa)		350	350	350	350	350	350	350
IMPERIAL SIZE AND MASS								
Mass (lb./ft.)		256	230	205	185	170	150	132
Nominal Depth and Width (in.)		20 x 20						

* Class 3 flanges.

** Class 4 flanges: area and resistance calculated according to CAN/CSA-M16.1-M89 Clause 13.3.3.

WWF COLUMNS

Factored Axial Compressive Resistance, Cr, in kN

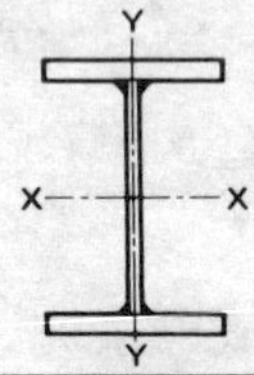

G40.21-M350W

$\phi = 0.9$

Designation	WWF450								
Mass (kg/m)	503 ▫	409	342	308	274	248	228	201 *	177 **
Effective length (KL) in millimetres with respect to least radius of gyration									
0	20 200	16 400	13 700	12 400	11 000	9 950	9 140	8 060	6 600
2 000	19 800	16 100	13 400	12 100	10 800	9 720	8 910	7 870	6 490
2 250	19 600	16 000	13 300	12 000	10 700	9 640	8 830	7 790	6 450
2 500	19 400	15 800	13 200	11 900	10 600	9 540	8 740	7 720	6 410
2 750	19 200	15 700	13 100	11 700	10 400	9 450	8 650	7 640	6 360
3 000	19 000	15 500	12 900	11 600	10 300	9 350	8 560	7 560	6 310
3 250	18 800	15 400	12 800	11 500	10 200	9 240	8 460	7 470	6 260
3 500	18 600	15 200	12 600	11 400	10 100	9 140	8 360	7 380	6 210
3 750	18 400	15 000	12 500	11 200	9 970	9 030	8 260	7 290	6 160
4 000	18 200	14 800	12 300	11 100	9 840	8 910	8 150	7 190	6 100
4 250	17 900	14 700	12 200	10 900	9 700	8 800	8 040	7 100	6 050
4 500	17 700	14 500	12 000	10 800	9 560	8 670	7 920	7 000	5 990
4 750	17 500	14 300	11 900	10 600	9 420	8 550	7 810	6 890	5 920
5 000	17 200	14 100	11 700	10 500	9 280	8 420	7 690	6 780	5 860
5 250	17 000	13 900	11 500	10 300	9 120	8 290	7 560	6 670	5 790
5 500	16 700	13 700	11 300	10 100	8 970	8 150	7 430	6 560	5 720
6 000	16 100	13 200	10 900	9 790	8 650	7 870	7 170	6 330	5 560
6 500	15 600	12 800	10 500	9 420	8 310	7 570	6 880	6 080	5 360
7 000	15 000	12 300	10 100	9 030	7 950	7 260	6 590	5 820	5 150
7 500	14 300	11 800	9 680	8 620	7 580	6 930	6 280	5 550	4 930
8 000	13 700	11 300	9 230	8 200	7 190	6 590	5 960	5 260	4 700
8 500	13 000	10 700	8 750	7 760	6 790	6 240	5 620	4 960	4 440
9 000	12 200	10 100	8 230	7 260	6 310	5 830	5 230	4 620	4 160
9 500	11 400	9 490	7 660	6 750	5 870	5 430	4 870	4 300	3 870
10 000	10 600	8 860	7 150	6 300	5 480	5 070	4 540	4 010	3 610
10 500	9 960	8 300	6 700	5 900	5 130	4 740	4 250	3 750	3 380
11 000	9 350	7 790	6 290	5 540	4 810	4 450	3 990	3 520	3 170
11 500	8 800	7 330	5 920	5 210	4 530	4 190	3 750	3 310	2 980
12 000	8 300	6 910	5 580	4 910	4 270	3 950	3 540	3 120	2 810

PROPERTIES AND DESIGN DATA

Property	503	409	342	308	274	248	228	201	177
Area (mm^2)	64 100	52 200	43 600	39 300	35 000	31 600	29 000	25 600	22 600
Z_x (10^3 mm^3)	11 400	9 640	8 100	7 290	6 470	5 960	5 450	4 840	4 360
S_x (10^3 mm^3)	9 620	8 380	7 150	6 480	5 770	5 380	4 920	4 400	3 800
r_x (mm)	184	190	192	193	193	196	196	197	200
Z_y (10^3 mm^3)	6 150	5 100	4 090	3 580	3 080	2 860	2 560	2 250	2 040
S_y (10^3 mm^3)	4 050	3 380	2 700	2 360	2 030	1 890	1 690	1 490	1 220
r_y (mm)	119	121	118	116	114	116	114	114	116
r_x/r_y	1.55	1.57	1.63	1.66	1.69	1.69	1.72	1.73	1.72
M_{rx} (kN·m) ($L < L_u$)	3 590	3 040	2 550	2 300	2 040	1 880	1 720	1 390	1 200
M_{yx} (kN·m)	1 940	1 610	1 290	1 130	970	901	806	469	384
J (10^3 mm^4)	67 800	38 400	20 200	13 900	9 140	7 120	5 230	3 570	2 580
C_w (10^9 mm^6)	34 700	30 400	25 500	22 900	20 100	18 900	17 200	15 300	14 000
L_u (mm)	11 800	10 100	8 530	7 920	7 380	7 250	7 000	7 200	7 280
F_y (MPa)	350	350	350	350	350	350	350	350	350

IMPERIAL SIZE AND MASS

Mass (lb./ft.)	337	275	229	207	184	166	152	134	119
Nominal Depth and Width (in.)	18 x 18								

▫ Welding does not fully develop web strength.

* Class 3 flanges. ** Class 4 flanges: area and resistance calculated according to CAN/CSA-M16.1-M89 Clause 13.3.3.

G40.21-M350W
$\phi = 0.9$

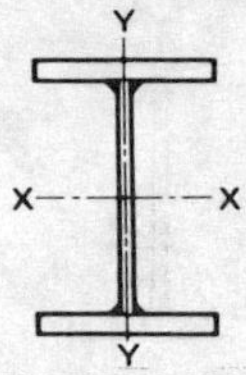

WWF COLUMNS
Factored Axial Compressive Resistance, C_r, in kN

Designation	WWF400								
Mass (kg/m)	444 °	362	303	273	243	220	202	178 *	157 *
Effective length (KL) in millimetres with respect to least radius of gyration									
0	17 800	14 600	12 200	11 000	9 760	8 820	8 100	7 150	6 330
2 000	17 300	14 100	11 800	10 600	9 440	8 540	7 830	6 920	6 130
2 250	17 100	14 000	11 700	10 500	9 340	8 440	7 740	6 840	6 060
2 500	16 900	13 800	11 500	10 400	9 230	8 350	7 650	6 760	5 990
2 750	16 700	13 700	11 400	10 300	9 120	8 250	7 560	6 680	5 920
3 000	16 500	13 500	11 300	10 100	9 000	8 140	7 460	6 590	5 850
3 250	16 300	13 400	11 100	10 000	8 880	8 030	7 360	6 500	5 770
3 500	16 100	13 200	11 000	9 860	8 750	7 920	7 260	6 410	5 690
3 750	15 900	13 000	10 800	9 720	8 620	7 800	7 150	6 310	5 600
4 000	15 600	12 800	10 600	9 570	8 490	7 680	7 030	6 210	5 520
4 250	15 400	12 600	10 500	9 420	8 350	7 560	6 920	6 110	5 430
4 500	15 200	12 400	10 300	9 260	8 200	7 430	6 800	6 000	5 330
4 750	14 900	12 200	10 100	9 100	8 050	7 300	6 670	5 890	5 240
5 000	14 600	12 000	9 940	8 930	7 900	7 160	6 550	5 780	5 140
5 250	14 400	11 800	9 750	8 760	7 740	7 020	6 410	5 670	5 040
5 500	14 100	11 600	9 560	8 580	7 570	6 870	6 280	5 550	4 930
6 000	13 500	11 100	9 150	8 210	7 230	6 570	6 000	5 300	4 720
6 500	12 900	10 600	8 730	7 820	6 870	6 250	5 700	5 030	4 490
7 000	12 200	10 100	8 280	7 410	6 490	5 910	5 380	4 760	4 240
7 500	11 500	9 580	7 810	6 980	6 100	5 560	5 050	4 460	3 990
8 000	10 800	9 020	7 300	6 500	5 640	5 160	4 680	4 130	3 710
8 500	10 000	8 370	6 730	5 990	5 200	4 760	4 310	3 810	3 420
9 000	9 260	7 750	6 240	5 550	4 820	4 410	3 990	3 530	3 160
9 500	8 610	7 210	5 800	5 160	4 480	4 100	3 710	3 280	2 940
10 000	8 030	6 720	5 410	4 810	4 170	3 820	3 460	3 050	2 740
10 500	7 510	6 290	5 050	4 500	3 900	3 570	3 230	2 860	2 560
11 000	7 040	5 900	4 740	4 210	3 650	3 340	3 030	2 670	2 400
11 500	6 610	5 540	4 450	3 960	3 430	3 140	2 840	2 510	2 250
12 000	6 230	5 220	4 190	3 720	3 230	2 960	2 680	2 360	2 120
PROPERTIES AND DESIGN DATA									
Area (mm^2)	56 600	46 200	38 600	34 800	31 000	28 000	25 700	22 700	20 100
Z_x (10^3 mm^3)	8 770	7 480	6 300	5 680	5 050	4 660	4 260	3 790	3 420
S_x (10^3 mm^3)	7 300	6 410	5 500	5 000	4 470	4 170	3 830	3 430	3 120
r_x (mm)	161	166	169	170	170	173	173	174	176
Z_y (10^3 mm^3)	4 870	4 030	3 230	2 840	2 440	2 260	2 020	1 780	1 610
S_y (10^3 mm^3)	3 200	2 670	2 130	1 870	1 600	1 490	1 330	1 170	1 070
r_y (mm)	106	108	105	104	102	103	102	102	103
r_x/r_y	1.52	1.54	1.61	1.63	1.67	1.68	1.70	1.71	1.71
M_{rx} (kN·m) ($L < L_u$)	2 760	2 360	1 980	1 790	1 590	1 470	1 340	1 080	983
M_{yx} (kN·m)	1 530	1 270	1 020	895	769	712	636	369	337
J (10^3 mm^4)	60 100	34 100	17 900	12 300	8 110	6 320	4 640	3 170	2 290
C_w (10^9 mm^6)	18 500	16 300	13 800	12 400	11 000	10 300	9 380	8 390	7 700
L_u (mm)	11 900	9 930	8 190	7 490	6 890	6 720	6 430	6 600	6 440
F_y (MPa)	350	350	350	350	350	350	350	350	350
IMPERIAL SIZE AND MASS									
Mass (lb./ft.)	298	243	203	183	163	147	135	119	105
Nominal Depth and Width (in.)	16 x 16								

° Welding does not fully develop web strength.
* Class 3 flanges.

WWF COLUMNS

Factored Axial Compressive Resistance, C_r, in kN

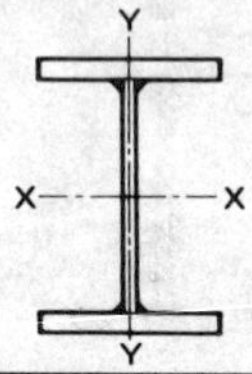

G40.21-M350W

$\phi = 0.9$

Designation		WWF350							
Mass (kg/m)		315	263	238	212	192	176	155	137
Effective length (KL) in millimetres with respect to least radius of gyration	0	12 700	10 600	9 540	8 500	7 690	7 060	6 240	5 510
	2 000	12 200	10 100	9 130	8 120	7 350	6 740	5 950	5 270
	2 250	12 000	10 000	9 010	8 010	7 250	6 650	5 870	5 200
	2 500	11 900	9 880	8 890	7 900	7 150	6 560	5 790	5 130
	2 750	11 700	9 740	8 760	7 780	7 050	6 460	5 700	5 050
	3 000	11 500	9 590	8 620	7 650	6 940	6 350	5 610	4 970
	3 250	11 300	9 440	8 480	7 520	6 820	6 250	5 520	4 890
	3 500	11 200	9 280	8 330	7 390	6 700	6 130	5 420	4 810
	3 750	11 000	9 110	8 180	7 250	6 580	6 020	5 310	4 720
	4 000	10 800	8 940	8 020	7 100	6 450	5 900	5 210	4 620
	4 250	10 600	8 760	7 850	6 950	6 320	5 770	5 090	4 530
	4 500	10 300	8 580	7 680	6 800	6 180	5 640	4 980	4 430
	4 750	10 100	8 390	7 510	6 640	6 040	5 510	4 860	4 330
	5 000	9 890	8 190	7 330	6 470	5 890	5 370	4 740	4 220
	5 250	9 660	7 990	7 140	6 300	5 740	5 230	4 610	4 110
	5 500	9 420	7 780	6 950	6 120	5 580	5 080	4 480	4 000
	6 000	8 910	7 340	6 540	5 750	5 250	4 780	4 210	3 770
	6 500	8 380	6 880	6 120	5 360	4 910	4 450	3 920	3 520
	7 000	7 820	6 390	5 640	4 910	4 520	4 090	3 590	3 240
	7 500	7 170	5 830	5 150	4 480	4 130	3 730	3 270	2 950
	8 000	6 580	5 350	4 720	4 100	3 780	3 420	3 000	2 710
	8 500	6 060	4 930	4 340	3 780	3 480	3 140	2 760	2 490
	9 000	5 610	4 560	4 020	3 490	3 220	2 910	2 550	2 310
	9 500	5 200	4 230	3 730	3 240	2 990	2 690	2 370	2 140
	10 000	4 840	3 930	3 470	3 010	2 780	2 510	2 200	1 990
	10 500	4 520	3 670	3 230	2 810	2 590	2 340	2 050	1 860
	11 000	4 230	3 430	3 020	2 620	2 420	2 180	1 920	1 730
	11 500	3 970	3 220	2 830	2 460	2 270	2 050	1 800	1 620
	12 000	3 730	3 020	2 660	2 310	2 130	1 920	1 690	1 530

PROPERTIES AND DESIGN DATA

Property	315	263	238	212	192	176	155	137
Area (mm^2)	40 200	33 600	30 300	27 000	24 400	22 400	19 800	17 500
Z_x (10^3 mm^3)	5 580	4 730	4 280	3 810	3 520	3 220	2 870	2 590
S_x (10^3 mm^3)	4 710	4 070	3 720	3 330	3 120	2 870	2 580	2 350
r_x (mm)	143	146	146	147	150	150	151	153
Z_y (10^3 mm^3)	3 090	2 480	2 170	1 870	1 740	1 550	1 360	1 240
S_y (10^3 mm^3)	2 040	1 630	1 430	1 230	1 140	1 020	899	817
r_y (mm)	94.2	92.3	90.8	89.2	90.5	89.4	89.0	90.4
r_x/r_y	1.52	1.58	1.61	1.65	1.66	1.68	1.70	1.69
M_{rx} (kN·m) ($L < L_u$)	1 760	1 490	1 350	1 200	1 110	1 010	904	816
M_{yx} (kN·m)	973	781	684	589	548	488	428	391
J (10^3 mm^4)	29 800	15 700	10 800	7 070	5 520	4 060	2 760	2 000
C_w (10^9 mm^6)	8 040	6 870	6 210	5 490	5 190	4 720	4 230	3 890
L_u (mm)	9 920	7 970	7 150	6 450	6 230	5 910	5 630	5 510
F_y (MPa)	350	350	350	350	350	350	350	350

IMPERIAL SIZE AND MASS

Mass (lb./ft.)	211	177	159	142	128	118	104	92
Nominal Depth and Width (in.)	14 x 14							

FACTORED AXIAL COMPRESSIVE RESISTANCES
Hollow Structural Sections

Class C

For rectangular hollow sections, see page 4–52.

For square hollow sections, see page 4–60.

For round hollow sections, see page 4–68.

Class H

For rectangular hollow sections, see page 4–76.

For square hollow sections, see page 4–84.

For round hollow sections, see page 4–92.

RECTANGULAR HOLLOW SECTIONS
Factored Axial Compressive Resistances, C_r, in kN

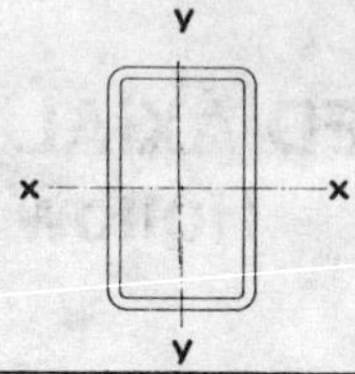

G40.21-M 350W

$\phi = 0.90$

Designation (mm x mm x mm)	HSS 305 x 203					HSS 254 x 152		
	13	11	9.5	††8.0	6.4‡	13	11	9.5
Mass (kg/m)	93.0	82.4	71.3	60.1	48.6	72.7	64.6	56.1
Effective length (KL) in millimetres with respect to least radius of gyration								
0	3 720	3 310	2 860	2 410	1 730	2 920	2 590	2 250
1 000	3 700	3 300	2 850	2 410	1 720	2 860	2 540	2 210
1 250	3 660	3 260	2 820	2 380	1 710	2 810	2 500	2 170
1 500	3 610	3 220	2 790	2 350	1 690	2 750	2 450	2 130
1 750	3 560	3 170	2 750	2 320	1 670	2 690	2 400	2 090
2 000	3 510	3 130	2 710	2 290	1 650	2 630	2 340	2 040
2 250	3 460	3 080	2 670	2 250	1 630	2 570	2 290	1 990
2 500	3 400	3 030	2 630	2 220	1 610	2 500	2 230	1 940
2 750	3 340	2 980	2 580	2 180	1 590	2 420	2 160	1 880
3 000	3 280	2 920	2 540	2 140	1 570	2 340	2 090	1 830
3 250	3 210	2 870	2 490	2 100	1 540	2 260	2 020	1 770
3 500	3 140	2 810	2 440	2 060	1 520	2 170	1 950	1 700
3 750	3 070	2 740	2 380	2 010	1 490	2 080	1 870	1 640
4 000	3 000	2 680	2 330	1 970	1 460	1 990	1 790	1 570
4 250	2 920	2 610	2 270	1 920	1 430	1 890	1 700	1 490
4 500	2 840	2 540	2 210	1 870	1 400	1 790	1 610	1 420
4 750	2 760	2 470	2 150	1 820	1 370	1 670	1 510	1 330
5 000	2 680	2 390	2 090	1 770	1 340	1 560	1 410	1 250
5 250	2 590	2 320	2 020	1 710	1 310	1 460	1 320	1 170
5 500	2 500	2 240	1 950	1 660	1 270	1 370	1 240	1 100
5 750	2 410	2 160	1 880	1 600	1 240	1 290	1 170	1 030
6 000	2 310	2 070	1 810	1 540	1 200	1 220	1 100	973
6 500	2 090	1 880	1 650	1 410	1 120	1 090	986	871
7 000	1 890	1 700	1 490	1 270	1 030	980	888	784
7 500	1 720	1 550	1 360	1 160	948	887	803	710
8 000	1 580	1 420	1 250	1 060	869	806	730	645
8 500	1 450	1 310	1 150	977	799	735	666	589
9 000	1 340	1 200	1 060	902	738	672	610	539
9 500	1 240	1 120	979	835	683	606	552	490
10 000	1 150	1 040	909	776	635	549	500	444
10 500	1 070	964	846	722	591	500	456	405
PROPERTIES AND DESIGN DATA								
Area (mm^2)	11 800	10 500	9 090	7 660	6 190	9 260	8 230	7 150
Z_x ($10^3 mm^3$)	1 190	1 060	925	787	640	746	671	589
S_x ($10^3 mm^3$)	964	867	762	652	535	592	537	475
r_x (mm)	111	112	113	114	115	90.1	91.0	91.9
Z_y ($10^3 mm^3$)	896	802	701	596	486	522	470	413
S_y ($10^3 mm^3$)	769	693	611	524	368	441	401	357
r_y (mm)	81.2	81.9	82.7	83.4	84.1	60.2	61.0	61.7
r_x/r_y	1.37	1.37	1.37	1.37	1.36	1.50	1.49	1.49
M_{rx} (kN•m)	375	334	291	248	202	235	211	186
M_{ry} (kN•m)	282	253	221	165	116	164	148	130
IMPERIAL SIZE AND MASS								
Mass (lb./ft.)	62.5	55.4	47.9	40.4	32.6	48.9	43.4	37.7
Thickness (in.)	.500	.438	.375	.313	.250	.500	.438	.375
Size (in.)	12 x 8					10 x 6		

†† Class 3 in bending about Y-Y axis.

‡ Class 4 in bending about Y-Y axis. Resistances and S_y for this column calculated according to CAN/CSA-S16.1-M89 Clauses 13.3.3 and 13.5(c).

G40.21-M 350W

$\phi = 0.90$

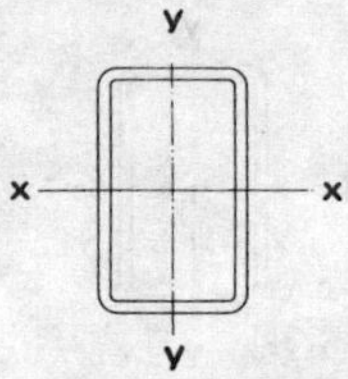

RECTANGULAR HOLLOW SECTIONS
Factored Axial Compressive Resistances, C_r, in kN

Designation (mm x mm x mm)	HSS 254 x 152		HSS 203 x 152					
	8.0	6.4‡	13	11	9.5	8.0	6.4	4.8‡
Mass (kg/m)	47.5	38.4	62.6	55.7	48.5	41.1	33.4	25.5
Effective length (KL) in millimetres with respect to least radius of gyration								
0	1 910	1 460	2 510	2 240	1 950	1 650	1 340	954
1 000	1 870	1 440	2 450	2 190	1 910	1 620	1 310	940
1 250	1 840	1 420	2 410	2 150	1 870	1 590	1 290	927
1 500	1 810	1 400	2 360	2 110	1 840	1 560	1 270	912
1 750	1 770	1 380	2 310	2 060	1 800	1 530	1 240	897
2 000	1 730	1 350	2 250	2 010	1 760	1 490	1 210	880
2 250	1 690	1 330	2 190	1 960	1 710	1 450	1 180	863
2 500	1 650	1 300	2 130	1 900	1 660	1 420	1 150	844
2 750	1 600	1 270	2 060	1 850	1 610	1 370	1 120	824
3 000	1 550	1 240	1 990	1 790	1 560	1 330	1 080	803
3 250	1 500	1 210	1 920	1 720	1 510	1 280	1 050	781
3 500	1 450	1 170	1 840	1 650	1 450	1 240	1 010	757
3 750	1 390	1 130	1 760	1 580	1 390	1 190	970	733
4 000	1 340	1 090	1 670	1 510	1 330	1 130	928	707
4 250	1 280	1 040	1 590	1 430	1 260	1 080	885	679
4 500	1 210	992	1 490	1 350	1 190	1 020	840	650
4 750	1 150	941	1 380	1 260	1 110	956	789	614
5 000	1 070	879	1 290	1 170	1 040	893	737	573
5 250	1 000	824	1 210	1 100	971	837	691	537
5 500	942	774	1 140	1 030	912	785	649	504
5 750	886	729	1 070	970	858	739	610	475
6 000	836	687	1 010	915	809	697	576	448
6 500	748	615	900	818	724	624	515	401
7 000	674	555	809	735	651	561	464	361
7 500	611	502	732	665	589	508	420	327
8 000	555	457	664	604	535	462	382	297
8 500	507	418	605	550	488	421	348	272
9 000	465	383	547	500	446	385	319	249
9 500	424	351	493	451	402	348	290	227
10 000	384	318	448	409	364	316	262	206
10 500	350	289	408	373	332	288	239	187
PROPERTIES AND DESIGN DATA								
Area (mm^2)	6 050	4 900	7 970	7 100	6 180	5 240	4 250	3 250
Z_x (10^3mm^3)	503	411	528	476	419	360	295	228
S_x (10^3mm^3)	409	338	423	385	343	297	246	192
r_x (mm)	92.7	93.6	73.4	74.2	75.1	75.9	76.7	77.5
Z_y (10^3mm^3)	353	290	432	390	344	295	243	188
S_y (10^3mm^3)	309	255	358	327	292	254	211	157
r_y (mm)	62.4	63.1	58.5	59.3	60.0	60.7	61.5	62.2
r_x/r_y	1.49	1.48	1.25	1.25	1.25	1.25	1.25	1.25
M_{rx} (kN•m)	158	129	166	150	132	113	92.9	71.8
M_{ry} (kN•m)	111	80.3	136	123	108	92.9	76.5	49.3
IMPERIAL SIZE AND MASS								
Mass (lb./ft.)	31.9	25.8	42.1	37.5	32.6	27.6	22.4	17.1
Thickness (in.)	.313	.250	.500	.438	.375	.313	.250	.188
Size (in.)	10 x 6		8 x 6					

‡ Class 4 in bending about Y-Y axis. Resistances and S_y for this column calculated according to CAN/CSA-S16.1-M89 Clauses 13.3.3 and 13.5(c).

RECTANGULAR HOLLOW SECTIONS
Factored Axial Compressive Resistances, C_r, in kN

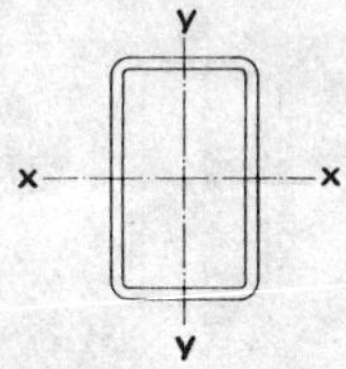

G40.21-M 350W

$\phi = 0.90$

Designation (mm x mm x mm)	HSS 203 x 102						HSS 178 x 127#		
	13	11	9.5	8.0	6.4	4.8 ‡	13	11	9.5
Mass (kg/m)	52.4	46.9	40.9	34.8	28.3	21.7	52.4	46.9	40.9
Effective length (KL) in millimetres with respect to least radius of gyration									
0	2 100	1 880	1 640	1 400	1 140	800	2 100	1 880	1 640
1 000	1 980	1 770	1 550	1 320	1 080	768	2 020	1 810	1 580
1 250	1 910	1 710	1 500	1 280	1 050	750	1 970	1 770	1 550
1 500	1 840	1 650	1 450	1 230	1 010	730	1 920	1 720	1 510
1 750	1 760	1 580	1 390	1 190	972	707	1 860	1 670	1 460
2 000	1 670	1 510	1 320	1 130	930	683	1 800	1 610	1 420
2 250	1 580	1 420	1 250	1 080	885	657	1 730	1 560	1 370
2 500	1 480	1 340	1 180	1 010	836	628	1 660	1 490	1 310
2 750	1 370	1 240	1 100	949	785	598	1 580	1 430	1 260
3 000	1 250	1 150	1 020	880	730	564	1 500	1 360	1 200
3 250	1 120	1 030	920	800	669	523	1 420	1 280	1 130
3 500	1 020	932	833	725	606	473	1 330	1 200	1 070
3 750	927	848	758	660	552	432	1 230	1 120	995
4 000	847	776	694	604	506	395	1 120	1 020	914
4 250	778	713	638	556	465	364	1 040	944	842
4 500	718	658	589	513	429	336	959	874	780
4 750	664	609	545	475	398	311	890	811	724
5 000	615	565	506	441	369	289	829	756	675
5 250	572	525	470	410	344	269	774	706	630
5 500	533	490	439	383	321	252	724	661	590
5 750	498	457	410	358	300	235	679	620	554
6 000	461	426	384	335	281	221	638	583	521
6 500	396	366	330	290	245	193	567	518	463
7 000	344	318	286	252	213	168	506	462	414
7 500	302	279	251	221	187	147	447	411	370
8 000			223	196	165	130	395	363	327
8 500						116	352	324	291
9 000							316	290	261
9 500							286	262	236
10 000									
10 500									
PROPERTIES AND DESIGN DATA									
Area (mm^2)	6 680	5 970	5 210	4 430	3 610	2 760	6 680	5 970	5 210
Z_x ($10^3 mm^3$)	405	367	325	281	232	180	377	342	303
S_x ($10^3 mm^3$)	307	282	253	221	185	145	297	272	244
r_x (mm)	68.4	69.3	70.3	71.2	72.2	73.1	62.8	63.7	64.6
Z_y ($10^3 mm^3$)	246	224	199	172	143	111	298	270	240
S_y ($10^3 mm^3$)	201	186	168	148	125	93.3	243	224	202
r_y (mm)	39.1	39.8	40.5	41.2	42.0	42.7	48.1	48.8	49.6
r_x/r_y	1.75	1.74	1.73	1.73	1.72	1.71	1.31	1.30	1.30
M_{rx} (kN•m)	128	116	102	88.5	73.1	56.7	119	108	95.4
M_{ry} (kN•m)	77.5	70.6	62.7	54.2	45.0	29.4	93.9	85.0	75.6
IMPERIAL SIZE AND MASS									
Mass (lb./ft.)	35.2	31.5	27.5	23.4	19.0	14.6	35.2	31.5	27.5
Thickness (in.)	.500	.438	.375	.313	.250	.188	.500	.438	.375
Size (in.)	8 x 4						7 x 5		

‡ Class 4 in bending about Y-Y axis. Resistances and S_y for this column calculated according to CAN/CSA-S16.1-M89 Clauses 13.3.3 and 13.5(c).

#Check manufacturer for current availability.

G40.21-M 350W

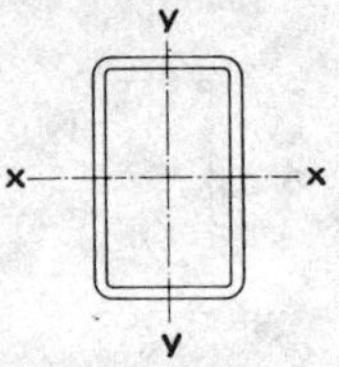

RECTANGULAR HOLLOW SECTIONS
Factored Axial Compressive Resistances, C_r, in kN

$\phi = 0.90$

Designation (mm x mm x mm)	HSS 178 x 127#			HSS 152 x 102				
	8.0	6.4	†† 4.8	11	9.5	8.0	6.4	4.8
Mass (kg/m)	34.8	28.3	21.7	38.0	33.3	28.4	23.2	17.9
Effective length (KL) in millimetres with respect to least radius of gyration								
0	1 400	1 140	869	1 520	1 340	1 140	932	718
1 000	1 350	1 100	842	1 430	1 260	1 080	881	680
1 250	1 320	1 080	823	1 380	1 210	1 040	853	659
1 500	1 280	1 050	803	1 330	1 170	1 000	822	636
1 750	1 250	1 020	782	1 270	1 120	959	789	611
2 000	1 210	989	759	1 200	1 060	914	752	583
2 250	1 170	955	734	1 130	1 000	865	713	554
2 500	1 120	920	707	1 060	938	812	671	523
2 750	1 070	883	679	976	870	756	627	489
3 000	1 030	843	650	885	794	697	579	454
3 250	972	801	618	796	714	626	524	413
3 500	917	757	586	720	646	567	475	374
3 750	859	711	551	655	588	516	432	341
4 000	792	659	513	599	538	472	396	312
4 250	730	607	473	550	494	434	364	287
4 500	676	562	438	507	456	400	336	265
4 750	628	523	407	468	421	370	311	245
5 000	585	487	379	434	391	344	288	228
5 250	547	455	355	403	363	320	268	212
5 500	512	427	332	376	338	298	250	198
5 750	481	401	312	350	316	278	234	185
6 000	452	377	294	323	293	260	219	173
6 500	402	335	262	277	251	223	188	150
7 000	359	300	234	241	218	194	164	130
7 500	323	270	211	211	192	170	144	114
8 000	285	240	188				127	101
8 500	254	214	168					
9 000	228	192	150					
9 500	206	173	136					
10 000	187	157	123					
10 500								
PROPERTIES AND DESIGN DATA								
Area (mm^2)	4 430	3 610	2 760	4 840	4 240	3 620	2 960	2 280
Z_x ($10^3 mm^3$)	261	216	167	230	205	178	148	116
S_x ($10^3 mm^3$)	213	178	140	179	162	143	121	95.6
r_x (mm)	65.4	66.2	67.1	53.1	54.0	54.8	55.7	56.5
Z_y ($10^3 mm^3$)	207	171	133	172	154	134	112	87.8
S_y ($10^3 mm^3$)	177	148	117	140	128	113	96.1	76.6
r_y (mm)	50.3	51.1	51.8	38.4	39.1	39.9	40.6	41.3
r_x/r_y	1.30	1.30	1.29	1.38	1.38	1.38	1.37	1.37
M_{rx} (kN•m)	82.2	68.0	52.6	72.4	64.6	56.1	46.6	36.5
M_{ry} (kN•m)	65.2	53.9	36.9	54.2	48.5	42.2	35.3	27.7
IMPERIAL SIZE AND MASS								
Mass (lb./ft.)	23.4	19.0	14.6	25.5	22.4	19.1	15.6	12.0
Thickness (in.)	.313	.250	.188	.438	.375	.313	.250	.188
Size (in.)	7 x 5			6 x 4				

†† Class 3 in bending about Y-Y axis.
\# Check manufacturer for current availability.

RECTANGULAR HOLLOW SECTIONS
Factored Axial Compressive Resistances, C_r, in kN

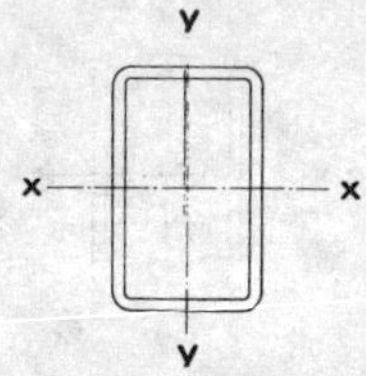

G40.21-M 350W

$\phi = 0.90$

Designation (mm x mm x mm)	HSS 127 x 76				HSS 127 x 64			
	9.5	8.0	6.4	4.8	9.5	8.0	6.4	4.8
Mass (kg/m)	25.7	22.1	18.2	14.1	23.8	20.5	16.9	13.1
Effective length (KL) in millimetres with respect to least radius of gyration								
0	1 030	888	731	564	954	822	677	526
1 000	922	798	659	511	813	706	586	459
1 250	870	755	625	485	748	653	544	428
1 500	812	706	587	457	675	593	497	393
1 750	747	653	545	426	593	526	444	354
2 000	676	595	499	392	500	448	383	310
2 250	594	530	449	354	426	382	327	265
2 500	516	461	391	311	368	330	283	230
2 750	453	405	344	274	322	289	248	201
3 000	402	359	305	243	283	255	219	178
3 250	359	321	273	217	251	226	194	158
3 500	322	289	246	196	224	202	174	142
3 750	291	261	222	177	197	180	156	127
4 000	264	237	202	161	174	159	138	114
4 250	240	216	184	147	155	141	123	101
4 500	216	196	168	135	140	127	110	91
4 750	195	176	152	122		115	100	82
5 000	176	160	138	111			91	75
5 250	161	146	125	101				
5 500	147	134	115	93				
5 750		123	106	85				
6 000			98	79				
6 500								
7 000								
7 500								
8 000								
8 500								
9 000								
9 500								
10 000								
10 500								
PROPERTIES AND DESIGN DATA								
Area (mm^2)	3 280	2 820	2 320	1 790	3 030	2 610	2 150	1 670
Z_x ($10^3 mm^3$)	126	111	93.3	73.8	112	98.9	83.6	66.3
S_x ($10^3 mm^3$)	96.4	86.4	74.0	59.5	83.2	75.1	64.8	52.4
r_x (mm)	43.2	44.1	45.1	45.9	41.7	42.7	43.7	44.6
Z_y ($10^3 mm^3$)	87.5	77.2	65.2	51.7	67.4	60.0	51.0	40.7
S_y ($10^3 mm^3$)	70.5	63.8	55.1	44.7	53.6	49.1	42.9	35.2
r_y (mm)	28.6	29.4	30.1	30.8	23.7	24.4	25.1	25.9
r_x/r_y	1.51	1.50	1.50	1.49	1.76	1.75	1.74	1.73
M_{rx} (kN•m)	39.7	35.0	29.4	23.2	35.3	31.2	26.3	20.9
M_{ry} (kN•m)	27.6	24.3	20.5	16.3	21.2	18.9	16.1	12.8
IMPERIAL SIZE AND MASS								
Mass (lb./ft.)	17.3	14.9	12.2	9.45	16.0	13.8	11.4	8.81
Thickness (in.)	.375	.313	.250	.188	.375	.313	.250	.188
Size (in.)	5 x 3				5 x 2½			

G40.21-M 350W

$\phi = 0.90$

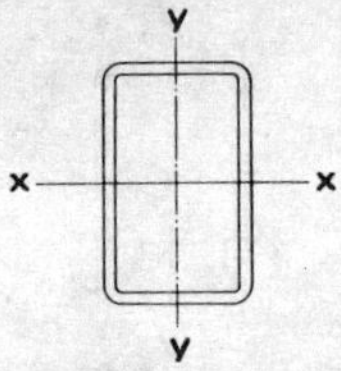

RECTANGULAR HOLLOW SECTIONS
Factored Axial Compressive Resistances, C_r, in kN

	Designation (mm x mm x mm)	HSS 127 x 51				HSS 102 x 76			
		9.5	8.0	6.4	4.8	9.5	8.0	6.4	4.8
	Mass (kg/m)	21.9	18.9	15.6	12.2	21.9	18.9	15.6	12.2
Effective length (KL) in millimetres with respect to least radius of gyration	0	879	759	627	488	879	759	627	488
	1 000	683	600	503	397	779	677	563	440
	1 250	594	528	447	356	732	639	532	417
	1 500	489	444	384	309	680	595	498	392
	1 750	397	361	313	255	622	548	460	363
	2 000	330	300	261	213	559	495	419	332
	2 250	279	254	221	181	485	435	372	299
	2 500	239	218	190	155	420	377	324	260
	2 750	206	189	164	135	369	332	284	229
	3 000	175	162	144	118	327	294	252	203
	3 250	150	139	123	102	292	262	225	181
	3 500	131	121	107	89	262	236	203	163
	3 750		106	94	78	236	213	183	148
	4 000				69	214	193	166	134
	4 250					193	176	151	122
	4 500					173	157	137	112
	4 750					156	142	124	101
	5 000					141	129	112	91
	5 250					129	117	102	83
	5 500					118	108	94	76
	5 750							86	70
	6 000								
	6 500								
	7 000								
	7 500								
	8 000								
	8 500								
	9 000								
	9 500								
	10 000								
	10 500								

PROPERTIES AND DESIGN DATA

	HSS 127 x 51 9.5	8.0	6.4	4.8	HSS 102 x 76 9.5	8.0	6.4	4.8
Area (mm^2)	2 790	2 410	1 990	1 550	2 790	2 410	1 990	1 550
Z_x ($10^3 mm^3$)	97.7	86.9	73.8	58.9	87.6	77.7	65.9	52.5
S_x ($10^3 mm^3$)	70.0	63.8	55.5	45.3	67.2	61.0	52.8	43.0
r_x (mm)	39.9	41.0	42.1	43.1	35.0	35.8	36.7	37.5
Z_y ($10^3 mm^3$)	48.9	44.0	37.9	30.5	71.3	63.5	54.0	43.1
S_y ($10^3 mm^3$)	37.8	35.3	31.4	26.2	56.3	51.4	44.8	36.6
r_y (mm)	18.6	19.3	20.0	20.7	27.7	28.5	29.3	30.0
r_x/r_y	2.15	2.13	2.10	2.08	1.26	1.26	1.25	1.25
M_{rx} (kN•m)	30.8	27.4	23.2	18.6	27.6	24.5	20.8	16.5
M_{ry} (kN•m)	15.4	13.9	11.9	9.61	22.5	20.0	17.0	13.6

IMPERIAL SIZE AND MASS

Mass (lb./ft.)	14.7	12.7	10.5	8.17	14.7	12.7	10.5	8.17
Thickness (in.)	.375	.313	.250	.188	.375	.313	.250	.188
Size (in.)	5 x 2				4 x 3			

RECTANGULAR HOLLOW SECTIONS
Factored Axial Compressive Resistances, C_r, in kN

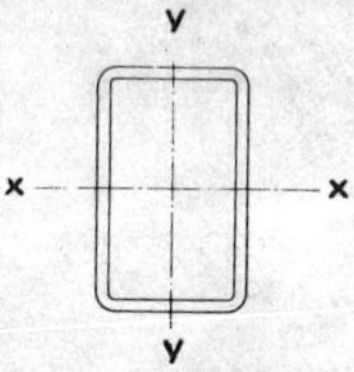

G40.21-M 350W

$\phi = 0.90$

Designation (mm x mm x mm)	HSS 102 x 51					HSS 89 x 64		
	8.0	6.4	4.8	3.8	3.2	8.0	6.4	4.8
Mass (kg/m)	15.8	13.1	10.3	8.37	7.09	15.8	13.1	10.3
Effective length (KL) in millimetres with respect to least radius of gyration								
0	633	526	413	337	284	633	526	413
1 000	494	417	333	274	233	536	450	356
1 250	432	369	297	246	209	492	415	330
1 500	358	312	256	213	183	442	375	301
1 750	290	254	210	176	152	387	331	268
2 000	241	211	175	147	126	324	280	230
2 250	204	179	148	125	107	276	239	196
2 500	175	153	128	107	92	239	207	170
2 750	151	133	111	93	80	208	181	148
3 000	129	115	97	82	70	183	159	131
3 250	110	98	83	71	61	162	141	116
3 500	96	85	72	61	53	144	126	104
3 750	84	75	64	54	47	127	111	93
4 000			56	48	41	112	98	82
4 250						100	88	73
4 500						90	79	66
4 750							71	59
5 000								
5 250								
5 500								
5 750								
6 000								
6 500								
7 000								
7 500								
8 000								
8 500								
9 000								
9 500								
10 000								
10 500								
PROPERTIES AND DESIGN DATA								
Area (mm^2)	2 010	1 670	1 310	1 070	903	2 010	1 670	1 310
Z_x ($10^3 mm^3$)	58.8	50.6	40.8	33.9	29.0	54.9	47.1	38.0
S_x ($10^3 mm^3$)	43.5	38.4	31.8	26.8	23.1	42.1	37.1	30.6
r_x (mm)	33.2	34.2	35.1	35.7	36.1	30.5	31.4	32.3
Z_y ($10^3 mm^3$)	35.4	30.7	24.9	20.8	17.9	43.2	37.2	30.1
S_y ($10^3 mm^3$)	27.9	25.1	21.1	18.0	15.6	34.3	30.4	25.3
r_y (mm)	18.8	19.5	20.3	20.7	21.0	23.3	24.0	24.8
r_x/r_y	1.77	1.75	1.73	1.73	1.72	1.31	1.31	1.30
M_{rx} (kN•m)	18.5	15.9	12.9	10.7	9.13	17.3	14.8	12.0
M_{ry} (kN•m)	11.2	9.67	7.84	6.55	5.64	13.6	11.7	9.48
IMPERIAL SIZE AND MASS								
Mass (lb./ft.)	10.6	8.81	6.89	5.62	4.76	10.6	8.81	6.89
Thickness (in.)	.313	.250	.188	.150	.125	.313	.250	.188
Size (in.)	4 x 2					3½ x 2½		

G40.21-M 350W

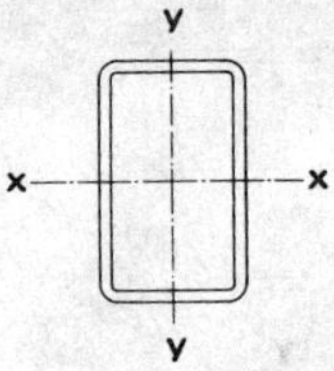

RECTANGULAR HOLLOW SECTIONS
Factored Axial Compressive Resistances, C_r, in kN

$\phi = 0.90$

Designation (mm x mm x mm)		HSS 89 x 64		HSS 76 x 51				HSS 51 x 25	
		3.8	3.2	8.0	6.4	4.8	3.8	3.2	2.5
Mass (kg/m)		8.37	7.09	12.6	10.6	8.35	6.85	3.28	2.71
Effective length (KL) in millimetres with respect to least radius of gyration	0	337	284	504	425	334	275	132	109
	1 000	292	247	385	332	266	220	53	46
	1 250	271	230	332	290	235	196	38	33
	1 500	248	211	269	240	199	168	29	25
	1 750	222	189	218	195	162	137	21	19
	2 000	192	164	181	162	135	114		15
	2 250	164	140	152	137	114	97		
	2 500	142	122	130	117	98	83		
	2 750	124	106	111	101	85	72		
	3 000	109	94	94	86	73	63		
	3 250	97	84	81	74	63	54		
	3 500	87	75	70	65	55	47		
	3 750	78	67		57	48	41		
	4 000	69	60						
	4 250	62	53						
	4 500	55	48						
	4 750	50	43						
	5 000	45	39						
	5 250								
	5 500								
	5 750								
	6 000								
	6 500								
	7 000								
	7 500								
	8 000								
	8 500								
	9 000								
	9 500								
	10 000								
	10 500								
PROPERTIES AND DESIGN DATA									
Area (mm^2)		1 070	903	1 600	1 350	1 060	872	418	345
Z_x ($10^3 mm^3$)		31.5	27.0	35.9	31.4	25.7	21.6	6.33	5.35
S_x ($10^3 mm^3$)		25.8	22.3	26.6	24.1	20.3	17.3	4.81	4.15
r_x (mm)		32.8	33.1	25.1	26.1	27.0	27.5	17.1	17.5
Z_y ($10^3 mm^3$)		25.0	21.4	26.7	23.5	19.3	16.3	3.84	3.27
S_y ($10^3 mm^3$)		21.4	18.5	20.5	18.8	16.0	13.8	3.14	2.75
r_y (mm)		25.2	25.5	18.0	18.8	19.6	20.0	9.77	10.1
r_x/r_y		1.30	1.30	1.39	1.39	1.38	1.37	1.75	1.74
M_{rx} (kN•m)		9.92	8.50	11.3	9.89	8.10	6.80	1.99	1.69
M_{ry} (kN•m)		7.87	6.74	8.41	7.40	6.08	5.13	1.21	1.03
IMPERIAL SIZE AND MASS									
Mass (lb./ft.)		5.62	4.76	8.46	7.11	5.61	4.60	2.21	1.82
Thickness (in.)		.150	.125	.313	.250	.188	.150	.125	.100
Size (in.)		3½ x 2½		3 x 2				2 x 1	

SQUARE HOLLOW SECTIONS
Factored Axial Compressive Resistances, C_r, in kN

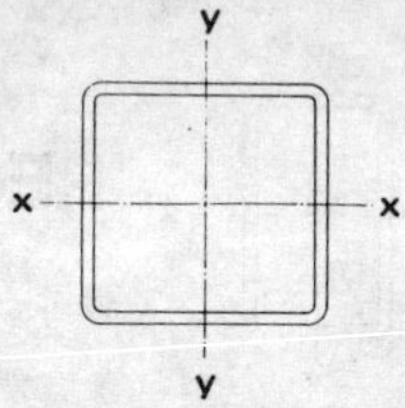

G40.21-M 350W

$\phi = 0.90$

Designation (mm x mm x mm)	HSS 305 x 305					HSS 254 x 254		
	13	11	9.5	††8.0	‡6.4	13	11	9.5
Mass (kg/m)	113	100	86.5	72.8	58.7	93.0	82.4	71.3
Effective length (KL) in millimetres with respect to least radius of gyration								
0	4 540	4 030	3 460	2 920	1 910	3 720	3 310	2 860
1 000	4 540	4 030	3 460	2 920	1 910	3 720	3 310	2 860
1 250	4 540	4 030	3 460	2 920	1 910	3 700	3 290	2 850
1 500	4 510	4 010	3 450	2 910	1 900	3 660	3 260	2 820
1 750	4 470	3 980	3 420	2 890	1 890	3 620	3 220	2 790
2 000	4 440	3 950	3 390	2 860	1 880	3 580	3 190	2 760
2 250	4 400	3 910	3 360	2 840	1 870	3 540	3 150	2 730
2 500	4 360	3 880	3 330	2 810	1 860	3 490	3 110	2 700
2 750	4 310	3 840	3 300	2 790	1 850	3 450	3 070	2 660
3 000	4 270	3 800	3 270	2 760	1 830	3 400	3 030	2 630
3 250	4 220	3 760	3 230	2 730	1 820	3 350	2 990	2 590
3 500	4 180	3 720	3 200	2 700	1 810	3 300	2 940	2 550
3 750	4 130	3 670	3 160	2 670	1 800	3 250	2 900	2 510
4 000	4 080	3 630	3 120	2 640	1 780	3 190	2 850	2 470
4 250	4 020	3 580	3 090	2 610	1 770	3 130	2 800	2 430
4 500	3 970	3 540	3 040	2 570	1 750	3 080	2 740	2 380
4 750	3 910	3 490	3 000	2 540	1 730	3 010	2 690	2 340
5 000	3 860	3 440	2 960	2 500	1 720	2 950	2 640	2 290
5 250	3 800	3 390	2 920	2 470	1 700	2 890	2 580	2 240
5 500	3 740	3 330	2 870	2 430	1 680	2 820	2 520	2 190
5 750	3 680	3 280	2 830	2 390	1 670	2 750	2 460	2 140
6 000	3 610	3 220	2 780	2 350	1 650	2 680	2 400	2 080
6 500	3 480	3 110	2 680	2 270	1 610	2 530	2 270	1 970
7 000	3 340	2 990	2 580	2 190	1 570	2 380	2 130	1 860
7 500	3 200	2 860	2 470	2 100	1 520	2 210	1 980	1 730
8 000	3 050	2 730	2 360	2 000	1 470	2 020	1 820	1 590
8 500	2 890	2 590	2 240	1 910	1 420	1 870	1 680	1 470
9 000	2 720	2 440	2 120	1 800	1 370	1 730	1 550	1 360
9 500	2 530	2 280	1 980	1 690	1 300	1 600	1 440	1 260
10 000	2 360	2 120	1 850	1 570	1 240	1 490	1 340	1 180
10 500	2 210	1 990	1 730	1 470	1 180	1 400	1 260	1 100
PROPERTIES AND DESIGN DATA								
Area (mm^2)	14 400	12 800	11 000	9 280	7 480	11 800	10 500	9 090
Z ($10^3 mm^3$)	1 560	1 390	1 210	1 030	833	1 060	945	825
S ($10^3 mm^3$)	1 330	1 190	1 040	886	625	888	799	703
r (mm)	118	119	120	121	121	97.6	98.4	99.1
M_r (kN•m)	491	438	381	279	196	334	298	260
IMPERIAL SIZE AND MASS								
Mass (lb./ft.)	76.1	67.3	58.1	48.9	39.4	62.5	55.4	47.9
Thickness (in.)	.500	.438	.375	.313	.250	.500	.438	.375
Size (in.)	12 x 12					10 x 10		

†† Class 3 in bending.
‡ Class 4 in bending. Resistances and S for this column calculated according to **CSA-S16.1-M89** Clauses 13.3.3 and 13.5(c).

G40.21-M 350W

$\phi = 0.90$

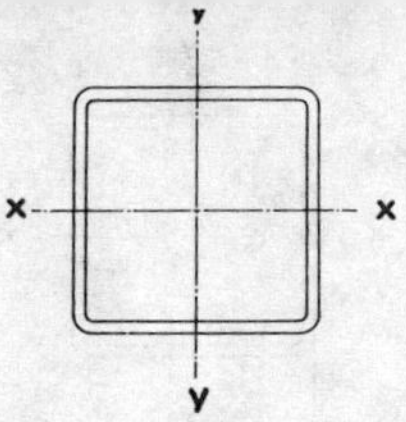

SQUARE HOLLOW SECTIONS
Factored Axial Compressive Resistances, C_r, in kN

Designation (mm x mm x mm)	HSS 254 x 254		HSS 203 x 203				
	8.0	‡6.4	13	11	9.5	8.0	6.4
Mass (kg/m)	60.1	48.6	72.7	64.6	56.1	47.5	38.4
Effective length (KL) in millimetres with respect to least radius of gyration							
0	2 410	1 790	2 920	2 590	2 250	1 910	1 540
1 000	2 410	1 790	2 900	2 580	2 240	1 900	1 540
1 250	2 400	1 790	2 860	2 540	2 210	1 870	1 520
1 500	2 380	1 770	2 820	2 510	2 180	1 850	1 500
1 750	2 350	1 760	2 780	2 470	2 150	1 820	1 480
2 000	2 330	1 750	2 740	2 440	2 120	1 800	1 460
2 250	2 300	1 730	2 690	2 400	2 080	1 770	1 430
2 500	2 280	1 720	2 640	2 350	2 050	1 740	1 410
2 750	2 250	1 700	2 590	2 310	2 010	1 700	1 380
3 000	2 220	1 690	2 540	2 260	1 970	1 670	1 360
3 250	2 190	1 670	2 480	2 210	1 930	1 640	1 330
3 500	2 150	1 650	2 420	2 160	1 880	1 600	1 300
3 750	2 120	1 630	2 360	2 110	1 840	1 560	1 270
4 000	2 090	1 610	2 300	2 050	1 790	1 520	1 240
4 250	2 050	1 590	2 230	2 000	1 740	1 480	1 200
4 500	2 010	1 570	2 170	1 940	1 690	1 440	1 170
4 750	1 970	1 550	2 090	1 870	1 640	1 400	1 140
5 000	1 930	1 530	2 020	1 810	1 580	1 350	1 100
5 250	1 890	1 510	1 950	1 740	1 530	1 300	1 060
5 500	1 850	1 480	1 870	1 680	1 470	1 250	1 020
5 750	1 810	1 460	1 790	1 610	1 410	1 200	984
6 000	1 760	1 430	1 690	1 530	1 340	1 150	943
6 500	1 670	1 360	1 520	1 370	1 210	1 040	849
7 000	1 580	1 280	1 380	1 240	1 090	938	769
7 500	1 470	1 200	1 250	1 130	995	854	700
8 000	1 360	1 110	1 150	1 030	910	781	641
8 500	1 250	1 020	1 050	949	836	718	589
9 000	1 160	945	969	874	771	662	543
9 500	1 070	878	896	809	713	612	502
10 000	1 000	818	831	750	661	568	466
10 500	936	764	772	697	615	528	434
PROPERTIES AND DESIGN DATA							
Area (mm^2)	7 660	6 190	9 260	8 230	7 150	6 050	4 900
Z ($10^3 mm^3$)	702	571	650	584	513	438	359
S ($10^3 mm^3$)	602	492	538	488	432	373	308
r (mm)	99.9	101	76.8	77.6	78.4	79.2	79.9
M_r (kN•m)	221	155	205	184	162	138	113
IMPERIAL SIZE AND MASS							
Mass (lb./ft.)	40.4	32.6	48.9	43.4	37.7	31.9	25.8
Thickness (in.)	.313	.250	.500	.438	.375	.313	.250
Size (in.)	10 x 10		8 x 8				

‡ Class 4 in bending. Resistances and S for this column calculated according to CSA-S16.1-M89 Clauses 13.3.3 and 13.5(c).

SQUARE HOLLOW SECTIONS
Factored Axial Compressive Resistances, C_r, in kN

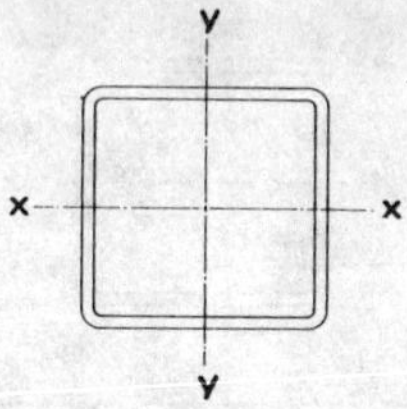

G40.21-M 350W

$\phi = 0.90$

Designation (mm x mm x mm)	HSS 178 x 178						HSS 152 x 152		
	13	11	9.5	8.0	6.4	††4.8	13	11	9.5
Mass (kg/m)	62.6	55.7	48.5	41.1	33.4	25.5	52.4	46.9	40.9
Effective length (KL) in millimetres with respect to least radius of gyration									
0	2 510	2 240	1 950	1 650	1 340	1 020	2 100	1 880	1 640
1 000	2 470	2 210	1 920	1 630	1 320	1 010	2 050	1 830	1 600
1 250	2 440	2 170	1 890	1 610	1 300	998	2 010	1 800	1 570
1 500	2 400	2 140	1 860	1 580	1 280	982	1 970	1 760	1 540
1 750	2 350	2 100	1 830	1 550	1 260	966	1 920	1 720	1 500
2 000	2 310	2 060	1 790	1 520	1 240	949	1 870	1 680	1 470
2 250	2 260	2 010	1 760	1 490	1 210	930	1 820	1 630	1 430
2 500	2 200	1 970	1 720	1 460	1 190	911	1 760	1 580	1 390
2 750	2 150	1 920	1 680	1 430	1 160	890	1 700	1 530	1 340
3 000	2 090	1 870	1 630	1 390	1 130	869	1 640	1 470	1 290
3 250	2 030	1 820	1 590	1 350	1 100	846	1 570	1 410	1 240
3 500	1 970	1 760	1 540	1 310	1 070	823	1 500	1 350	1 190
3 750	1 900	1 700	1 490	1 270	1 040	798	1 430	1 290	1 140
4 000	1 830	1 640	1 440	1 230	1 000	772	1 350	1 220	1 080
4 250	1 760	1 580	1 390	1 190	969	746	1 270	1 150	1 020
4 500	1 690	1 520	1 330	1 140	932	718	1 180	1 070	953
4 750	1 610	1 450	1 280	1 090	894	690	1 090	997	887
5 000	1 530	1 380	1 220	1 040	855	660	1 020	931	828
5 250	1 430	1 300	1 150	989	815	629	956	871	775
5 500	1 350	1 220	1 080	930	766	593	897	817	727
5 750	1 270	1 150	1 020	876	722	559	843	769	684
6 000	1 200	1 090	960	827	682	528	794	724	645
6 500	1 070	973	861	742	611	474	709	647	576
7 000	970	878	777	670	552	428	637	581	517
7 500	879	797	705	608	501	389	574	524	467
8 000	801	726	643	555	457	355	521	475	424
8 500	733	665	589	508	419	325	471	433	386
9 000	673	611	541	467	385	299	422	387	347
9 500	620	562	498	430	355	276	381	349	313
10 000	570	519	460	398	328	255	345	317	284
10 500	519	473	421	365	303	236	315	289	259
PROPERTIES AND DESIGN DATA									
Area (mm^2)	7 970	7 100	6 180	5 240	4 250	3 250	6 680	5 970	5 210
Z ($10^3 mm^3$)	484	436	385	330	271	209	341	310	275
S ($10^3 mm^3$)	396	361	322	279	231	181	275	253	227
r (mm)	66.4	67.2	68.0	68.8	69.6	70.3	56.0	56.8	57.6
M_r (kN•m)	152	137	121	104	85.4	57.0	107	97.6	86.6
IMPERIAL SIZE AND MASS									
Mass (lb./ft.)	42.1	37.5	32.6	27.6	22.4	17.1	35.2	31.5	27.5
Thickness (in.)	.500	.438	.375	.313	.250	.188	.500	.438	.375
Size (in.)	7 x 7						6 x 6		

†† Class 3 in bending.

G40.21-M 350W

$\phi = 0.90$

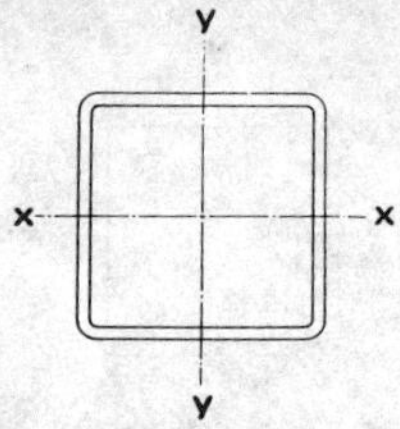

SQUARE HOLLOW SECTIONS
Factored Axial Compressive Resistances, C_r, in kN

Designation (mm x mm x mm)	HSS 152 x 152			HSS 127 x 127				
	8.0	6.4	4.8	11	9.5	8.0	6.4	4.8
Mass (kg/m)	34.8	28.3	21.7	38.0	33.3	28.4	23.2	17.9
Effective length (KL) in millimetres with respect to least radius of gyration								
0	1 400	1 140	869	1 520	1 340	1 140	932	718
1 000	1 360	1 110	851	1 460	1 280	1 100	898	693
1 250	1 340	1 090	836	1 420	1 250	1 070	877	677
1 500	1 310	1 070	820	1 380	1 220	1 040	853	659
1 750	1 280	1 050	802	1 340	1 180	1 010	828	640
2 000	1 250	1 020	784	1 290	1 140	974	801	619
2 250	1 220	996	764	1 240	1 090	938	771	598
2 500	1 180	968	743	1 180	1 040	899	740	574
2 750	1 150	938	720	1 120	995	857	707	549
3 000	1 110	907	697	1 060	942	813	672	523
3 250	1 070	874	672	996	886	767	635	495
3 500	1 020	840	646	928	827	718	596	466
3 750	977	804	619	846	758	662	554	436
4 000	930	766	591	776	696	608	508	400
4 250	880	727	562	715	641	560	468	369
4 500	825	685	532	661	593	518	433	341
4 750	768	637	495	613	550	481	402	317
5 000	717	595	462	571	512	448	375	295
5 250	671	557	433	533	478	418	350	276
5 500	630	523	406	498	447	391	328	258
5 750	593	492	382	467	420	367	307	242
6 000	559	464	361	439	394	345	289	228
6 500	499	415	322	389	350	306	257	203
7 000	449	373	290	345	312	273	229	181
7 500	406	337	262	302	274	241	204	162
8 000	368	306	238	267	242	213	180	143
8 500	335	279	217	238	216	190	160	127
9 000	303	254	198	214	194	171	144	114
9 500	273	229	179			154	130	103
10 000	248	207	162					
10 500	226	189	148					
PROPERTIES AND DESIGN DATA								
Area (mm²)	4 430	3 610	2 760	4 840	4 240	3 620	2 960	2 280
Z ($10^3 mm^3$)	237	195	152	205	183	159	132	103
S ($10^3 mm^3$)	198	166	130	164	149	132	111	88.1
r (mm)	58.4	59.2	59.9	46.4	47.2	48.0	48.8	49.6
M_r (kN•m)	74.7	61.4	47.9	64.6	57.6	50.1	41.6	32.4
IMPERIAL SIZE AND MASS								
Mass (lb./ft.)	23.4	19.0	14.6	25.5	22.4	19.1	15.6	12.0
Thickness (in.)	.313	.250	.188	.438	.375	.313	.250	.188
Size (in.)	6 x 6			5 x 5				

SQUARE HOLLOW SECTIONS
Factored Axial Compressive Resistances, C_r, in kN

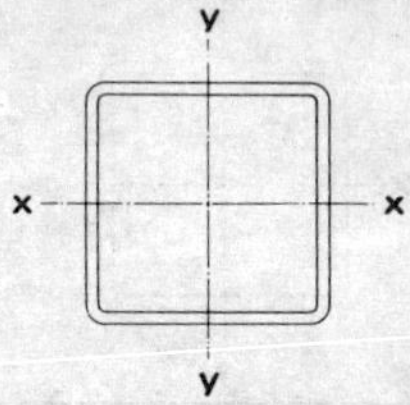

G40.21-M 350W

$\phi = 0.90$

Designation (mm x mm x mm)	HSS 102 x 102				HSS 89 x 89			
	9.5	8.0	6.4	4.8	9.5	8.0	6.4	4.8
Mass (kg/m)	25.7	22.1	18.2	14.1	21.9	18.9	15.6	12.2
Effective length (KL) in millimetres with respect to least radius of gyration								
0	1 030	888	731	564	879	759	627	488
1 000	964	831	686	530	800	694	576	450
1 250	928	801	662	513	762	662	550	431
1 500	889	768	636	493	719	627	522	410
1 750	845	732	607	472	673	588	491	387
2 000	798	693	576	448	621	546	458	362
2 250	747	651	542	423	566	500	421	334
2 500	693	606	506	397	502	448	382	305
2 750	635	558	468	368	442	395	337	271
3 000	567	502	424	337	392	351	299	241
3 250	509	451	381	302	351	314	268	216
3 500	460	407	345	274	317	283	242	195
3 750	418	371	314	249	287	257	220	177
4 000	382	339	287	228	261	234	200	161
4 250	351	311	264	209	238	214	183	148
4 500	323	286	243	193	218	196	168	136
4 750	298	265	224	179	200	180	155	125
5 000	276	245	208	166	182	164	142	115
5 250	256	228	193	154	165	150	130	106
5 500	238	212	180	143	151	137	119	97
5 750	219	196	168	134	139	126	109	89
6 000	202	181	155	124	128	116	101	82
6 500	173	155	133	107			87	70
7 000	151	135	115	93				
7 500		118	101	81				
8 000								
8 500								
9 000								
9 500								
10 000								
10 500								
PROPERTIES AND DESIGN DATA								
Area (mm^2)	3 280	2 820	2 320	1 790	2 790	2 410	1 990	1 550
Z ($10^3 mm^3$)	110	96.6	81.3	64.3	80.2	71.2	60.5	48.2
S ($10^3 mm^3$)	87.4	78.4	67.3	54.2	62.7	57.0	49.5	40.3
r (mm)	36.8	37.6	38.4	39.2	31.6	32.4	33.2	34.0
M_r (kN•m)	34.6	30.4	25.6	20.3	25.3	22.4	19.1	15.2
IMPERIAL SIZE AND MASS								
Mass (lb./ft.)	17.3	14.9	12.2	9.45	14.7	12.7	10.5	8.17
Thickness (in.)	.375	.313	.250	.188	.375	.313	.250	.188
Size (in.)	4 x 4				3½ x 3½			

G40.21-M 350W

$\phi = 0.90$

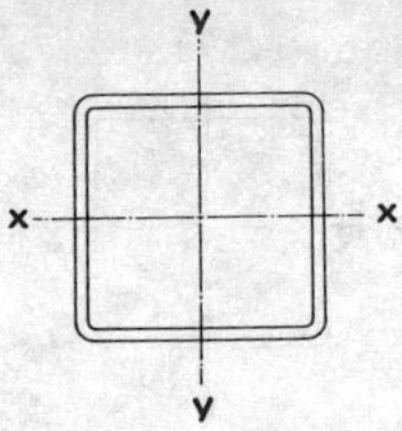

SQUARE HOLLOW SECTIONS
Factored Axial Compressive Resistances, C_r, in kN

Designation (mm x mm x mm)	HSS 76 x 76			HSS 64 x 64			
	8.0	6.4	4.8	6.4	4.8	3.8	3.2
Mass (kg/m)	15.8	13.1	10.3	10.6	8.35	6.85	5.82
Effective length (KL) in millimetres with respect to least radius of gyration							
0	633	526	413	425	334	275	233
1 000	559	468	369	358	284	235	200
1 250	524	440	348	327	261	217	185
1 500	486	409	325	292	235	196	168
1 750	443	375	300	253	207	174	149
2 000	395	338	272	211	174	147	127
2 250	341	294	240	180	148	125	108
2 500	296	255	208	155	128	109	94
2 750	259	224	183	136	112	95	82
3 000	229	199	162	119	99	84	72
3 250	205	177	145	106	87	74	64
3 500	183	159	130	93	78	66	57
3 750	165	144	118	82	68	59	51
4 000	150	130	107	72	60	52	45
4 250	134	118	97	64	54	46	40
4 500	120	105	87	58	48	41	36
4 750	108	95	79			37	33
5 000	98	86	71				
5 250	90	79	65				
5 500		72	60				
5 750			55				
6 000							
6 500							
7 000							
7 500							
8 000							
8 500							
9 000							
9 500							
10 000							
10 500							

PROPERTIES AND DESIGN DATA

Property	8.0	6.4	4.8	6.4	4.8	3.8	3.2
Area (mm^2)	2 010	1 670	1 310	1 350	1 060	872	741
Z ($10^3 mm^3$)	49.7	42.7	34.4	28.0	22.9	19.2	16.5
S ($10^3 mm^3$)	39.0	34.4	28.5	22.1	18.7	15.9	13.9
r (mm)	27.2	28.0	28.8	22.8	23.6	24.1	24.4
M_r (kN•m)	15.7	13.5	10.8	8.82	7.21	6.05	5.20

IMPERIAL SIZE AND MASS

Mass (lb./ft.)	10.6	8.81	6.89	7.11	5.61	4.60	3.91
Thickness (in.)	.313	.250	.188	.250	.188	.150	.125
Size (in.)	3 x 3			2½ x 2½			

SQUARE HOLLOW SECTIONS
Factored Axial Compressive Resistances, C_r, in kN

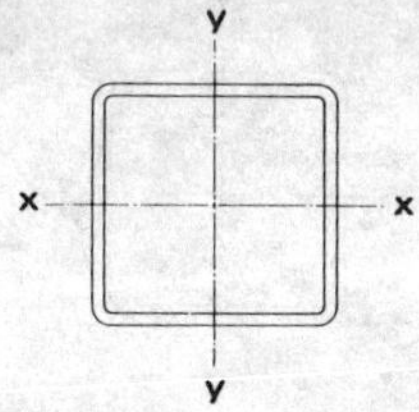

G40.21-M 350W

$\phi = 0.90$

Designation (mm x mm x mm)	HSS 51 x 51					HSS 38 x 38			
	6.4	4.8	3.8	3.2	2.8	4.8	3.8	3.2	2.5
Mass (kg/m)	8.05	6.45	5.33	4.55	4.05	4.54	3.81	3.28	2.71
Effective length (KL) in millimetres with respect to least radius of gyration									
0	324	259	214	183	163	182	153	132	109
1 000	245	200	167	144	129	110	96	85	71
1 250	209	173	146	127	113	82	72	64	54
1 500	168	142	122	106	96	63	56	50	42
1 750	136	115	99	86	78	50	45	40	34
2 000	113	96	82	72	65	41	37	33	28
2 250	95	81	69	61	55	33	29	26	23
2 500	81	69	59	52	47	27	24	22	19
2 750	69	60	51	45	41			18	16
3 000	58	50	44	39	35				
3 250	50	43	38	33	30				
3 500	43	38	33	29	26				
3 750			29	25	23				
4 000									
4 250									
4 500									
4 750									
5 000									
5 250									
5 500									
5 750									
6 000									
6 500									
7 000									
7 500									
8 000									
8 500									
9 000									
9 500									
10 000									
10 500									
PROPERTIES AND DESIGN DATA									
Area (mm^2)	1 030	821	679	580	516	578	485	418	345
Z ($10^3 mm^3$)	16.3	13.8	11.7	10.2	9.15	6.91	6.04	5.34	4.51
S ($10^3 mm^3$)	12.5	10.9	9.54	8.42	7.64	5.27	4.77	4.31	3.71
r (mm)	17.6	18.4	18.9	19.2	19.4	13.2	13.7	14.0	14.3
M_r (kN•m)	5.13	4.35	3.69	3.21	2.88	2.18	1.90	1.68	1.42
IMPERIAL SIZE AND MASS									
Mass (lb./ft.)	5.41	4.33	3.58	3.06	2.72	3.05	2.56	2.21	1.82
Thickness (in.)	.250	.188	.150	.125	.110	.188	.150	.125	.100
Size (in.)	2 x 2					1½ x 1½			

G40.21-M 350W

$\phi = 0.90$

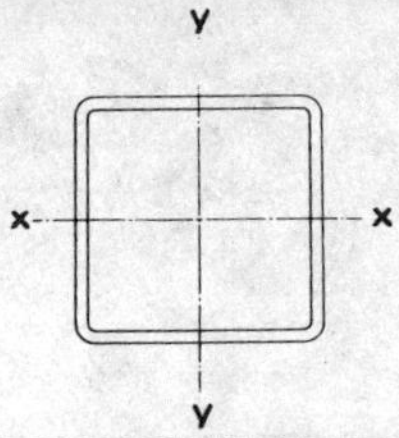

SQUARE HOLLOW SECTIONS
Factored Axial Compressive Resistances, C_r, in kN

Designation (mm x mm x mm)		HSS 32 x 32			HSS 25 x 25	
		3.8	3.2	2.5	3.2	2.5
Mass (kg/m)		3.06	2.65	2.20	2.01	1.69
Effective length (KL) in millimetres with respect to least radius of gyration	0	123	106	89	81	68
	1 000	59	53	46	28	25
	1 250	43	39	34	20	18
	1 500	33	30	26	14	13
	1 750	25	23	20	11	10
	2 000	20	18	16		
	2 250		14	13		
	2 500					
	2 750					
	3 000					
	3 250					
	3 500					
	3 750					
	4 000					
	4 250					
	4 500					
	4 750					
	5 000					
	5 250					
	5 500					
	5 750					
	6 000					
	6 500					
	7 000					
	7 500					
	8 000					
	8 500					
	9 000					
	9 500					
	10 000					
	10 500					

PROPERTIES AND DESIGN DATA

Area (mm^2)	389	338	281	257	216
Z ($10^3 mm^3$)	3.92	3.51	3.01	2.05	1.79
S ($10^3 mm^3$)	3.01	2.77	2.44	1.56	1.41
r (mm)	11.1	11.4	11.7	8.79	9.12
M_r (kN•m)	1.23	1.11	0.948	0.646	0.564

IMPERIAL SIZE AND MASS

Mass (lb./ft.)	2.05	1.78	1.48	1.35	1.14
Thickness (in.)	.150	.125	.100	.125	.100
Size (in.)		1¼ x 1¼		1 x 1	

ROUND HOLLOW SECTIONS
Factored Axial Compressive Resistances, C_r, in kN

G40.21-M 350W

$\phi = 0.90$

Designation (mm x mm)	HSS 406					HSS 356		
	13	11	9.5	8.0	6.4	13	11	9.5
Mass (kg/m)	123	108	93.3	78.1	62.6	107	94.6	81.3
Effective length (KL) in millimetres with respect to least radius of gyration								
0	4 950	4 350	3 750	3 130	2 510	4 320	3 780	3 280
1 000	4 950	4 350	3 750	3 130	2 510	4 320	3 780	3 280
1 250	4 950	4 350	3 750	3 130	2 510	4 320	3 780	3 280
1 500	4 950	4 350	3 750	3 130	2 510	4 300	3 760	3 260
1 750	4 920	4 330	3 730	3 120	2 500	4 260	3 740	3 240
2 000	4 890	4 300	3 710	3 100	2 490	4 230	3 710	3 210
2 250	4 850	4 270	3 680	3 080	2 470	4 190	3 670	3 180
2 500	4 820	4 240	3 650	3 060	2 450	4 150	3 640	3 160
2 750	4 780	4 200	3 620	3 030	2 430	4 110	3 610	3 130
3 000	4 740	4 170	3 600	3 010	2 410	4 070	3 570	3 100
3 250	4 700	4 140	3 570	2 980	2 390	4 030	3 540	3 060
3 500	4 660	4 100	3 540	2 960	2 370	3 990	3 500	3 030
3 750	4 620	4 060	3 500	2 930	2 350	3 940	3 460	3 000
4 000	4 570	4 030	3 470	2 910	2 330	3 900	3 420	2 960
4 250	4 530	3 990	3 440	2 880	2 310	3 850	3 380	2 930
4 500	4 480	3 950	3 400	2 850	2 290	3 800	3 330	2 890
4 750	4 440	3 910	3 370	2 820	2 260	3 750	3 290	2 850
5 000	4 390	3 860	3 330	2 790	2 240	3 700	3 250	2 810
5 250	4 340	3 820	3 290	2 760	2 210	3 640	3 200	2 770
5 500	4 290	3 780	3 260	2 730	2 190	3 590	3 150	2 730
5 750	4 240	3 730	3 220	2 690	2 160	3 530	3 100	2 690
6 000	4 180	3 680	3 180	2 660	2 130	3 470	3 050	2 650
6 500	4 070	3 590	3 090	2 590	2 080	3 350	2 950	2 560
7 000	3 950	3 490	3 010	2 520	2 020	3 230	2 840	2 460
7 500	3 830	3 380	2 920	2 450	1 960	3 090	2 720	2 360
8 000	3 710	3 270	2 820	2 370	1 900	2 960	2 610	2 260
8 500	3 580	3 160	2 720	2 290	1 830	2 810	2 480	2 150
9 000	3 440	3 040	2 620	2 200	1 770	2 660	2 350	2 040
9 500	3 300	2 920	2 520	2 120	1 700	2 490	2 200	1 910
10 000	3 150	2 790	2 410	2 030	1 620	2 330	2 060	1 780
10 500	3 000	2 660	2 290	1 930	1 550	2 180	1 930	1 670
PROPERTIES AND DESIGN DATA								
Area (mm^2)	15 700	13 800	11 900	9 950	7 980	13 700	12 000	10 400
Z ($10^3 mm^3$)	1 970	1 740	1 500	1 260	1 020	1 490	1 320	1 140
S ($10^3 mm^3$)	1 500	1 330	1 150	972	786	1 130	1 010	873
r (mm)	139	140	140	141	141	121	122	122
M_r (kN•m)	621	548	472	397	248	469	416	359
IMPERIAL SIZE AND MASS								
Mass (lb./ft.)	82.9	72.9	62.7	52.5	42.1	72.2	63.5	54.7
Thickness (in.)	.500	.438	.375	.313	.250	.500	.438	.375
Size (in.)	16 OD					14 OD		

ROUND HOLLOW SECTIONS
Factored Axial Compressive Resistances, C_r, in kN

Designation (mm x mm)	HSS 356		HSS 324				
	8.0	6.4	13	11	9.5	8.0	6.4
Mass (kg/m)	68.2	54.7	97.5	85.9	73.9	61.9	49.7
Effective length (KL) in millimetres with respect to least radius of gyration							
0	2 730	2 200	3 910	3 430	2 960	2 490	1 990
1 000	2 730	2 200	3 910	3 430	2 960	2 490	1 990
1 250	2 730	2 200	3 900	3 430	2 960	2 490	1 990
1 500	2 720	2 190	3 870	3 400	2 940	2 470	1 980
1 750	2 700	2 170	3 840	3 370	2 910	2 440	1 960
2 000	2 680	2 150	3 800	3 340	2 890	2 420	1 940
2 250	2 660	2 140	3 760	3 310	2 860	2 400	1 920
2 500	2 640	2 120	3 720	3 280	2 830	2 370	1 900
2 750	2 610	2 100	3 680	3 240	2 800	2 350	1 880
3 000	2 590	2 080	3 640	3 210	2 770	2 320	1 860
3 250	2 560	2 060	3 600	3 170	2 730	2 300	1 840
3 500	2 530	2 030	3 550	3 130	2 700	2 270	1 820
3 750	2 510	2 010	3 510	3 090	2 670	2 240	1 800
4 000	2 480	1 990	3 460	3 050	2 630	2 210	1 770
4 250	2 450	1 970	3 410	3 000	2 590	2 180	1 750
4 500	2 420	1 940	3 360	2 960	2 550	2 150	1 720
4 750	2 390	1 920	3 300	2 910	2 510	2 110	1 700
5 000	2 350	1 890	3 250	2 860	2 470	2 080	1 670
5 250	2 320	1 860	3 190	2 810	2 430	2 040	1 640
5 500	2 290	1 840	3 130	2 760	2 390	2 010	1 610
5 750	2 250	1 810	3 070	2 710	2 340	1 970	1 580
6 000	2 220	1 780	3 010	2 660	2 300	1 930	1 550
6 500	2 140	1 720	2 890	2 550	2 200	1 850	1 490
7 000	2 060	1 660	2 750	2 430	2 100	1 770	1 420
7 500	1 980	1 590	2 610	2 310	2 000	1 690	1 350
8 000	1 900	1 520	2 470	2 190	1 890	1 600	1 280
8 500	1 810	1 450	2 300	2 050	1 770	1 500	1 200
9 000	1 720	1 380	2 130	1 900	1 640	1 390	1 120
9 500	1 610	1 290	1 980	1 770	1 520	1 290	1 040
10 000	1 510	1 210	1 850	1 650	1 420	1 210	968
10 500	1 410	1 130	1 730	1 540	1 330	1 130	906
PROPERTIES AND DESIGN DATA							
Area (mm^2)	8 680	6 970	12 400	10 900	9 410	7 890	6 330
Z ($10^3 mm^3$)	961	775	1 230	1 090	942	794	640
S ($10^3 mm^3$)	738	598	930	827	719	608	493
r (mm)	123	123	110	111	111	112	112
M_r (kN•m)	303	188	387	343	297	250	202
IMPERIAL SIZE AND MASS							
Mass (lb./ft.)	45.8	36.8	65.5	57.7	49.6	41.6	33.4
Thickness (in.)	.313	.250	.500	.438	.375	.313	.250
Size (in.)	14 OD		12.75 OD				

ROUND HOLLOW SECTIONS
Factored Axial Compressive Resistances, C_r, in kN

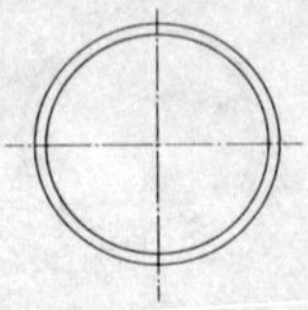

G40.21-M 350W

$\phi = 0.90$

Designation (mm x mm)	HSS 273					HSS 219		
	13	11	9.5	8.0	6.4	13	11	9.5
Mass (kg/m)	81.6	71.9	61.9	52.0	41.8	64.6	57.1	49.3
Effective length (KL) in millimetres with respect to least radius of gyration								
0	3 280	2 890	2 490	2 090	1 680	2 590	2 290	1 980
1 000	3 280	2 890	2 490	2 090	1 680	2 570	2 270	1 960
1 250	3 250	2 860	2 470	2 070	1 660	2 530	2 240	1 930
1 500	3 210	2 830	2 440	2 050	1 650	2 500	2 210	1 910
1 750	3 180	2 800	2 410	2 030	1 630	2 460	2 170	1 880
2 000	3 140	2 770	2 380	2 000	1 610	2 420	2 140	1 840
2 250	3 100	2 730	2 350	1 980	1 590	2 370	2 100	1 810
2 500	3 060	2 690	2 320	1 950	1 570	2 330	2 060	1 780
2 750	3 010	2 660	2 290	1 920	1 550	2 280	2 010	1 740
3 000	2 970	2 620	2 260	1 890	1 520	2 230	1 970	1 700
3 250	2 920	2 570	2 220	1 870	1 500	2 170	1 920	1 660
3 500	2 870	2 530	2 180	1 830	1 480	2 120	1 870	1 620
3 750	2 820	2 490	2 140	1 800	1 450	2 060	1 820	1 580
4 000	2 770	2 440	2 110	1 770	1 420	2 000	1 770	1 530
4 250	2 710	2 390	2 060	1 740	1 400	1 930	1 710	1 480
4 500	2 650	2 340	2 020	1 700	1 370	1 870	1 660	1 440
4 750	2 590	2 290	1 980	1 660	1 340	1 800	1 600	1 390
5 000	2 530	2 240	1 930	1 630	1 310	1 730	1 540	1 330
5 250	2 470	2 180	1 890	1 590	1 280	1 660	1 470	1 280
5 500	2 410	2 130	1 840	1 550	1 250	1 580	1 410	1 220
5 750	2 340	2 070	1 790	1 510	1 210	1 490	1 330	1 160
6 000	2 270	2 010	1 740	1 460	1 180	1 410	1 260	1 100
6 500	2 130	1 880	1 630	1 380	1 110	1 270	1 130	984
7 000	1 980	1 750	1 520	1 280	1 040	1 140	1 020	890
7 500	1 800	1 600	1 390	1 170	950	1 040	927	809
8 000	1 650	1 470	1 270	1 080	872	950	847	739
8 500	1 520	1 350	1 170	992	803	871	777	678
9 000	1 410	1 250	1 080	918	743	802	715	624
9 500	1 310	1 160	1 010	852	690	740	661	577
10 000	1 220	1 080	936	793	642	685	612	534
10 500	1 130	1 010	874	740	599	636	568	496
PROPERTIES AND DESIGN DATA								
Area (mm^2)	10 400	9 160	7 890	6 620	5 320	8 230	7 270	6 270
Z ($10^3 mm^3$)	862	764	662	559	452	542	482	419
S ($10^3 mm^3$)	646	577	502	427	347	402	360	315
r (mm)	92.2	92.7	93.2	93.8	94.3	73.1	73.6	74.2
M_r (kN•m)	272	241	209	176	142	171	152	132
IMPERIAL SIZE AND MASS								
Mass (lb./ft.)	54.8	48.3	41.6	34.9	28.1	43.4	38.4	33.1
Thickness (in.)	.500	.438	.375	.313	.250	.500	.438	.375
Size (in.)	10.75 OD					8.625 OD		

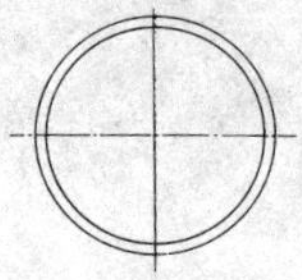

ROUND HOLLOW SECTIONS
Factored Axial Compressive Resistances, C_r, in kN

$\phi = 0.90$

Designation (mm x mm)	HSS 219			HSS 168			
	8.0	6.4	4.8	9.5	8.0	6.4	4.8
Mass (kg/m)	41.4	33.3	25.3	37.3	31.4	25.4	19.3
Effective length (KL) in millimetres with respect to least radius of gyration							
0	1 660	1 340	1 010	1 500	1 260	1 020	775
1 000	1 650	1 330	1 010	1 460	1 230	993	757
1 250	1 630	1 310	994	1 430	1 210	974	743
1 500	1 600	1 290	980	1 400	1 180	954	727
1 750	1 580	1 270	966	1 370	1 150	932	711
2 000	1 550	1 250	950	1 330	1 120	909	693
2 250	1 520	1 230	934	1 290	1 090	884	675
2 500	1 500	1 210	916	1 250	1 060	857	655
2 750	1 470	1 180	898	1 210	1 020	829	634
3 000	1 430	1 160	879	1 170	987	800	612
3 250	1 400	1 130	859	1 120	948	769	588
3 500	1 370	1 100	839	1 070	907	736	564
3 750	1 330	1 070	817	1 020	864	702	538
4 000	1 290	1 040	795	964	819	667	512
4 250	1 250	1 010	771	907	773	630	484
4 500	1 210	980	747	841	718	587	452
4 750	1 170	947	722	782	668	546	421
5 000	1 130	912	696	730	624	510	393
5 250	1 080	876	669	683	584	477	368
5 500	1 040	839	642	641	548	448	345
5 750	983	800	613	603	515	421	324
6 000	929	756	579	568	485	397	306
6 500	835	679	520	507	433	354	273
7 000	755	614	470	455	389	318	246
7 500	686	558	428	411	351	287	222
8 000	627	510	391	372	319	261	201
8 500	575	468	359	337	290	237	183
9 000	530	431	331	302	260	213	165
9 500	490	399	306	272	234	192	149
10 000	453	369	283	247	212	174	135
10 500	421	343	263	225	194	159	123
PROPERTIES AND DESIGN DATA							
Area (mm^2)	5 270	4 240	3 220	4 750	4 000	3 230	2 460
Z ($10^3 mm^3$)	355	288	220	241	205	167	128
S ($10^3 mm^3$)	269	219	169	179	153	126	97.6
r (mm)	74.7	75.3	75.8	56.2	56.8	57.3	57.8
M_r (kN•m)	112	90.7	69.3	75.9	64.6	52.6	40.3
IMPERIAL SIZE AND MASS							
Mass (lb./ft.)	27.8	22.4	17.0	25.1	21.1	17.0	13.0
Thickness (in.)	.313	.250	.188	.375	.313	.250	.188
Size (in.)	8.625 OD			6.625 OD			

ROUND HOLLOW SECTIONS
Factored Axial Compressive Resistances, C_r, in kN

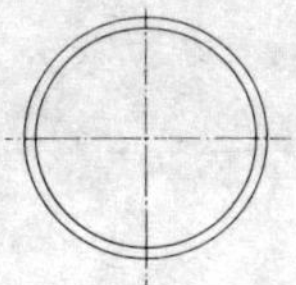

G40.21-M 350W

$\phi = 0.90$

Designation (mm x mm)	HSS 141				HSS 114		
	9.5	8.0	6.4	4.8	8.0	6.4	4.8
Mass (kg/m)	31.0	26.1	21.1	16.1	20.9	16.9	12.9
Effective length (KL) in millimetres with respect to least radius of gyration							
0	1 240	1 050	847	646	838	677	517
1 000	1 190	1 010	815	621	784	635	485
1 250	1 160	982	795	606	756	613	469
1 500	1 130	954	773	590	725	588	451
1 750	1 090	924	749	572	692	562	431
2 000	1 050	892	723	553	655	533	409
2 250	1 010	857	696	532	615	501	386
2 500	968	820	667	510	573	468	361
2 750	921	781	635	487	527	432	334
3 000	871	740	603	462	475	391	304
3 250	818	696	568	436	427	351	273
3 500	762	649	531	409	386	317	247
3 750	697	596	489	378	351	289	225
4 000	639	546	449	347	321	264	206
4 250	589	503	414	320	294	242	189
4 500	544	466	383	296	271	223	174
4 750	505	432	355	275	251	206	161
5 000	470	402	331	256	232	191	149
5 250	439	376	309	239	216	178	139
5 500	411	351	289	224	201	166	129
5 750	385	330	271	210	186	154	121
6 000	361	310	255	197	171	142	111
6 500	320	275	226	175	147	122	96
7 000	286	245	202	156	128	106	83
7 500	250	215	178	138	112	93	73
8 000	221	190	157	122			
8 500	197	169	140	109			
9 000	177	152	126	98			
9 500			114	88			
10 000							
10 500							
PROPERTIES AND DESIGN DATA							
Area (mm^2)	3 950	3 330	2 690	2 050	2 660	2 150	1 640
Z ($10^3 mm^3$)	166	142	116	89.1	90.1	74.1	57.4
S ($10^3 mm^3$)	122	105	86.9	67.7	66.1	55.1	43.2
r (mm)	46.7	47.2	47.8	48.3	37.7	38.2	38.8
M_r (kN•m)	52.3	44.7	36.5	28.1	28.4	23.3	18.1
IMPERIAL SIZE AND MASS							
Mass (lb./ft.)	20.8	17.6	14.2	10.8	14.0	11.4	8.68
Thickness (in.)	.375	.313	.250	.188	.313	.250	.188
Size (in.)	5.562 OD				4.5 OD		

G40.21-M 350W

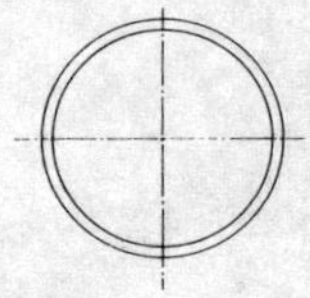

ROUND HOLLOW SECTIONS
Factored Axial Compressive Resistances, C_r, in kN

$\phi = 0.90$

Designation (mm x mm)	HSS 102				HSS 89			
	8.0	6.4	4.8	3.8	8.0	6.4	4.8	3.8
Mass (kg/m)	18.4	14.9	11.4	9.19	15.9	12.9	9.92	8.00
Effective length (KL) in millimetres with respect to least radius of gyration								
0	737	598	457	369	636	520	397	321
500	726	590	451	364	621	508	388	315
750	703	572	437	353	597	489	374	303
1 000	677	551	422	341	569	466	357	290
1 250	647	528	404	327	537	441	339	275
1 500	614	502	385	311	501	413	317	258
1 750	578	473	363	294	462	381	294	240
2 000	538	442	340	276	419	347	269	219
2 250	495	408	315	256	370	309	241	197
2 500	449	371	288	234	321	268	210	172
2 750	396	329	256	209	282	236	184	151
3 000	352	293	228	186	250	209	163	134
3 250	316	263	205	167	223	187	146	120
3 500	285	237	185	151	201	168	131	108
3 750	258	215	168	137	181	152	119	98
4 000	235	196	153	125	164	138	108	89
4 250	215	179	140	114	150	126	98	81
4 500	198	165	129	105	135	114	90	74
4 750	182	152	119	97	121	102	81	67
5 000	167	140	110	90	110	93	73	60
5 250	152	128	101	82	100	85	67	55
5 500	139	117	92	75	92	78	61	50
5 750	128	108	84	69	85	71	56	46
6 000	118	99	78	64				43
6 500	102	85	67	55				
7 000								
7 500								
8 000								
8 500								
9 000								
9 500								
PROPERTIES AND DESIGN DATA								
Area (mm^2)	2 340	1 900	1 450	1 170	2 020	1 650	1 260	1 020
Z ($10^3 mm^3$)	69.9	57.7	44.8	36.5	52.3	43.4	33.9	27.6
S ($10^3 mm^3$)	50.8	42.6	33.6	27.6	37.6	31.7	25.2	20.8
r (mm)	33.2	33.8	34.3	34.6	28.8	29.3	29.8	30.1
M_r (kN•m)	22.0	18.2	14.1	11.5	16.5	13.7	10.7	8.69
IMPERIAL SIZE AND MASS								
Mass (lb./ft.)	12.3	10.0	7.67	6.17	10.7	8.69	6.66	5.37
Thickness (in.)	.313	.250	.188	.150	.313	.250	.188	.150
Size (in.)	4 OD				3.5 OD			

ROUND HOLLOW SECTIONS
Factored Axial Compressive Resistances, C_r, in kN

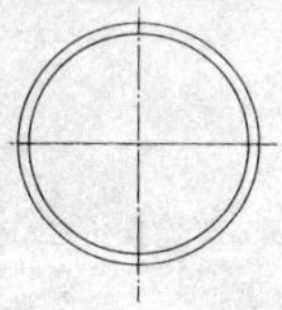

G40.21-M 350W

$\phi = 0.90$

Designation (mm x mm)	HSS 73				HSS 60			
	6.4	4.8	3.8	3.2	6.4	4.8	3.8	3.2
Mass (kg/m)	10.4	8.04	6.50	5.48	8.45	6.54	5.31	4.48
Effective length (KL) in millimetres with respect to least radius of gyration								
0	419	321	261	220	340	263	213	180
500	402	309	251	212	319	247	201	170
750	381	294	239	202	296	230	187	158
1 000	357	275	224	189	268	209	171	145
1 250	328	254	207	175	236	185	152	129
1 500	296	230	189	160	198	158	130	111
1 750	260	204	167	142	161	128	106	91
2 000	219	173	143	122	134	107	89	76
2 250	187	148	122	104	113	91	75	64
2 500	162	128	105	90	97	78	64	55
2 750	141	112	92	79	84	67	56	48
3 000	124	98	81	69	72	58	49	42
3 250	110	87	72	62	62	50	42	36
3 500	98	78	65	55	54	44	36	31
3 750	87	69	57	49	47	38	32	27
4 000	77	61	51	43				24
4 250	68	54	45	39				
4 500	61	49	41	35				
4 750		44	37	31				
5 000								
5 250								
5 500								
5 750								
6 000								
6 500								
7 000								
7 500								
8 000								
8 500								
9 000								
9 500								

PROPERTIES AND DESIGN DATA

	HSS 73 6.4	HSS 73 4.8	HSS 73 3.8	HSS 73 3.2	HSS 60 6.4	HSS 60 4.8	HSS 60 3.8	HSS 60 3.2
Area (mm^2)	1 330	1 020	828	698	1 080	834	676	571
Z ($10^3 mm^3$)	28.3	22.3	18.3	15.5	18.6	14.8	12.2	10.4
S ($10^3 mm^3$)	20.4	16.4	13.6	11.7	13.2	10.7	8.99	7.74
r (mm)	23.7	24.2	24.5	24.7	19.2	19.7	20.0	20.2
M_r (kN•m)	8.91	7.02	5.76	4.88	5.86	4.66	3.84	3.28

IMPERIAL SIZE AND MASS

	HSS 73 6.4	HSS 73 4.8	HSS 73 3.8	HSS 73 3.2	HSS 60 6.4	HSS 60 4.8	HSS 60 3.8	HSS 60 3.2
Mass (lb./ft.)	7.01	5.40	4.37	3.68	5.68	4.40	3.57	3.01
Thickness (in.)	.250	.188	.150	.125	.250	.188	.150	.125
Size (in.)	2.875 OD				2.375 OD			

G40.21-M 350W

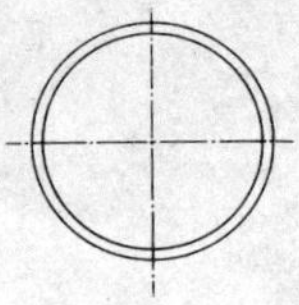

ROUND HOLLOW SECTIONS
Factored Axial Compressive Resistances, C_r, in kN

$\phi = 0.90$

Designation (mm x mm)	HSS 48				HSS 42		HSS 33		HSS 27	
	4.8	3.8	3.2	2.8	3.2	2.5	3.2	2.5	3.2	2.5
Mass (kg/m)	5.13	4.18	3.54	3.13	3.06	2.48	2.37	1.93	1.84	1.51
Effective length (KL) in millimetres with respect to least radius of gyration										
0	206	168	142	126	123	100	95	77	74	61
500	187	153	130	115	109	89	78	64	54	45
750	167	137	117	104	95	78	62	51	36	30
1 000	144	119	101	90	78	64	43	36	24	20
1 250	115	96	83	74	58	49	32	27	17	15
1 500	90	75	65	58	45	38	24	20	12	10
1 750	72	60	52	46	36	30	18	16		
2 000	59	50	43	38	30	25	14	12		
2 250	50	42	36	32	24	20				
2 500	41	35	30	27	20	17				
2 750	34	29	25	22	16	14				
3 000	29	25	21	19						
3 250										
3 500										
3 750										
4 000										
4 250										
4 500										
4 750										
5 000										
5 250										
5 500										
5 750										
6 000										
6 500										
7 000										
7 500										
8 000										
8 500										
9 000										
9 500										
PROPERTIES AND DESIGN DATA										
Area (mm^2)	654	533	451	399	390	316	302	246	235	193
Z ($10^3 mm^3$)	9.09	7.56	6.48	5.79	4.85	4.00	2.91	2.42	1.77	1.49
S ($10^3 mm^3$)	6.48	5.50	4.77	4.29	3.54	2.96	2.09	1.77	1.24	1.07
r (mm)	15.5	15.8	16.0	16.1	13.8	14.1	10.7	10.9	8.39	8.59
M_r (kN•m)	2.86	2.38	2.04	1.82	1.53	1.26	0.917	0.762	0.558	0.469
IMPERIAL SIZE AND MASS										
Mass (lb./ft.)	3.45	2.81	2.38	2.10	2.06	1.67	1.59	1.30	1.24	1.02
Thickness (in.)	.188	.150	.125	.110	.125	.100	.125	.100	.125	.100
Size (in.)	1.9 OD				1.66 OD		1.315 OD		1.05 OD	

RECTANGULAR HOLLOW SECTIONS
Factored Axial Compressive Resistances, C_r, in kN

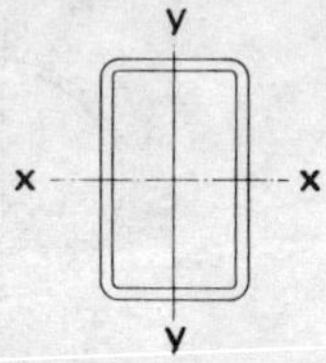

G40.21-M 350W CLASS H $\phi = 0.90$

Designation (mm x mm x mm)	HSS 305 x 203					HSS 254 x 152		
	13	11	9.5	†† 8.0	6.4 ‡	13	11	9.5
Mass (kg/m)	93.0	82.4	71.3	60.1	48.6	72.7	64.6	56.1
Effective length (KL) in millimetres with respect to least radius of gyration								
0	3 720	3 310	2 860	2 410	1 730	2 920	2 590	2 250
1 000	3 720	3 310	2 860	2 410	1 730	2 910	2 590	2 250
1 250	3 720	3 310	2 860	2 410	1 720	2 900	2 580	2 240
1 500	3 710	3 300	2 860	2 410	1 720	2 890	2 570	2 230
1 750	3 700	3 290	2 850	2 400	1 720	2 870	2 550	2 220
2 000	3 680	3 280	2 840	2 390	1 710	2 840	2 520	2 190
2 250	3 660	3 260	2 820	2 380	1 710	2 800	2 490	2 170
2 500	3 640	3 240	2 810	2 370	1 700	2 760	2 460	2 140
2 750	3 610	3 210	2 780	2 350	1 690	2 710	2 410	2 100
3 000	3 570	3 180	2 760	2 330	1 680	2 650	2 370	2 060
3 250	3 530	3 150	2 730	2 300	1 660	2 590	2 310	2 020
3 500	3 490	3 110	2 700	2 280	1 650	2 520	2 250	1 970
3 750	3 440	3 070	2 660	2 250	1 630	2 450	2 190	1 910
4 000	3 390	3 020	2 620	2 220	1 610	2 360	2 120	1 850
4 250	3 330	2 970	2 580	2 180	1 590	2 280	2 040	1 790
4 500	3 270	2 920	2 540	2 140	1 570	2 180	1 960	1 720
4 750	3 210	2 860	2 490	2 100	1 550	2 080	1 870	1 640
5 000	3 130	2 800	2 430	2 060	1 520	1 970	1 780	1 560
5 250	3 060	2 730	2 380	2 010	1 490	1 860	1 680	1 480
5 500	2 980	2 660	2 320	1 960	1 460	1 730	1 570	1 390
5 750	2 890	2 590	2 260	1 910	1 430	1 590	1 450	1 290
6 000	2 810	2 510	2 190	1 860	1 400	1 480	1 340	1 190
6 500	2 610	2 350	2 050	1 740	1 330	1 280	1 160	1 030
7 000	2 400	2 160	1 890	1 610	1 250	1 120	1 020	905
7 500	2 160	1 950	1 720	1 470	1 160	998	907	803
8 000	1 920	1 730	1 530	1 310	1 050	895	813	720
8 500	1 720	1 560	1 370	1 170	960	801	730	648
9 000	1 560	1 410	1 240	1 060	869	717	653	580
9 500	1 420	1 280	1 130	963	790	646	589	523
10 000	1 300	1 170	1 030	881	723	585	533	474
10 500	1 190	1 080	948	811	665	533	486	431
PROPERTIES AND DESIGN DATA								
Area (mm^2)	11 800	10 500	9 090	7 660	6 190	9 260	8 230	7 150
Z_x ($10^3 mm^3$)	1 190	1 060	925	787	640	746	671	589
S_x ($10^3 mm^3$)	964	867	762	652	535	592	537	475
r_x (mm)	111	112	113	114	115	90.1	91.0	91.9
Z_y ($10^3 mm^3$)	896	802	701	596	486	522	470	413
S_y ($10^3 mm^3$)	769	693	611	524	368	441	401	357
r_y (mm)	81.2	81.9	82.7	83.4	84.1	60.2	61.0	61.7
r_x/r_y	1.37	1.37	1.37	1.37	1.36	1.50	1.49	1.49
M_{rx} (kN•m)	375	334	291	248	202	235	211	186
M_{ry} (kN•m)	282	253	221	165	116	164	148	130
IMPERIAL SIZE AND MASS								
Mass (lb./ft.)	62.5	55.4	47.9	40.4	32.6	48.9	43.4	37.7
Thickness (in.)	.500	438	.375	.313	.250	.500	.438	.375
Size (in.)	12 x 8					10 x 6		

†† Class 3 in bending about Y-Y axis.

‡ Class 4 in bending about Y-Y axis. Resistances and S_y for this column calculated according to CAN/CSA-S16.1-M89 Clauses 13.3.3 and 13.5(c).

G40.21-M 350W CLASS H $\phi = 0.90$

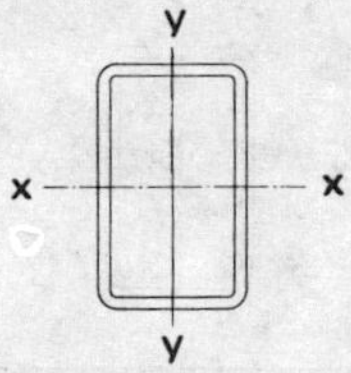

RECTANGULAR HOLLOW SECTIONS
Factored Axial Compressive Resistances, C_r, in kN

Designation (mm x mm x mm)	HSS 254 x 152		HSS 203 x 152					
	8.0	6.4 ‡	13	11	9.5	8.0	6.4	4.8 ‡
Mass (kg/m)	47.5	38.4	62.6	55.7	48.5	41.1	33.4	25.5
Effective length (KL) in millimetres with respect to least radius of gyration								
0	1 910	1 460	2 510	2 240	1 950	1 650	1 340	954
1 000	1 900	1 460	2 510	2 230	1 940	1 650	1 340	954
1 250	1 900	1 460	2 500	2 230	1 940	1 640	1 330	951
1 500	1 890	1 450	2 480	2 210	1 930	1 630	1 330	947
1 750	1 880	1 450	2 460	2 190	1 910	1 620	1 320	942
2 000	1 860	1 440	2 430	2 170	1 890	1 610	1 300	934
2 250	1 840	1 420	2 400	2 140	1 870	1 590	1 290	925
2 500	1 810	1 410	2 360	2 110	1 840	1 560	1 270	915
2 750	1 780	1 390	2 320	2 070	1 810	1 540	1 250	902
3 000	1 750	1 370	2 260	2 030	1 770	1 500	1 220	888
3 250	1 710	1 340	2 210	1 980	1 730	1 470	1 200	872
3 500	1 670	1 320	2 140	1 920	1 680	1 430	1 170	854
3 750	1 620	1 290	2 080	1 860	1 630	1 390	1 130	834
4 000	1 580	1 260	2 000	1 800	1 580	1 340	1 100	813
4 250	1 520	1 220	1 920	1 730	1 520	1 300	1 060	790
4 500	1 470	1 180	1 830	1 650	1 450	1 240	1 020	764
4 750	1 400	1 140	1 740	1 570	1 380	1 190	974	737
5 000	1 340	1 100	1 640	1 490	1 310	1 130	926	707
5 250	1 270	1 040	1 540	1 400	1 230	1 060	876	676
5 500	1 200	983	1 410	1 290	1 150	992	823	641
5 750	1 110	918	1 300	1 190	1 060	915	760	595
6 000	1 030	850	1 210	1 100	979	847	704	550
6 500	891	735	1 050	955	849	734	610	477
7 000	781	645	920	839	745	645	535	418
7 500	693	572	818	746	662	573	475	371
8 000	621	512	733	669	594	513	426	332
8 500	561	463	652	596	531	460	383	300
9 000	502	415	584	534	475	412	343	269
9 500	452	374	526	481	428	371	309	242
10 000	409	339	477	436	388	336	280	219
10 500	373	308	434	397	353	306	255	199
PROPERTIES AND DESIGN DATA								
Area (mm^2)	6 050	4 900	7 970	7 100	6 180	5 240	4 250	3 250
Z_x ($10^3 mm^3$)	503	411	528	476	419	360	295	228
S_x ($10^3 mm^3$)	409	338	423	385	343	297	246	192
r_x (mm)	92.7	93.6	73.4	74.2	75.1	75.9	76.7	77.5
Z_y ($10^3 mm^3$)	353	290	432	390	344	295	243	188
S_y ($10^3 mm^3$)	309	255	358	327	292	254	211	165
r_y (mm)	62.4	63.1	58.5	59.3	60.0	60.7	61.5	62.2
r_x/r_y	1.49	1.48	1.25	1.25	1.25	1.25	1.25	1.25
M_{rx} (kN•m)	158	129	166	150	132	113	92.9	71.8
M_{ry} (kN•m)	111	80.3	136	123	108	92.9	76.5	50.8
IMPERIAL SIZE AND MASS								
Mass (lb./ft.)	31.9	25.8	42.1	37.5	32.6	27.6	22.4	17.1
Thickness (in.)	.313	.250	.500	.438	.375	.313	.250	.188
Size (in.)	10 x 6		8 x 6					

‡ Class 4 in bending about Y-Y axis. Resistances and S_y for this column calculated according to CAN/CSA-S16.1-M89 Clauses 13.3.3 and 13.5(c).

RECTANGULAR HOLLOW SECTIONS
Factored Axial Compressive Resistances, C_r, in kN

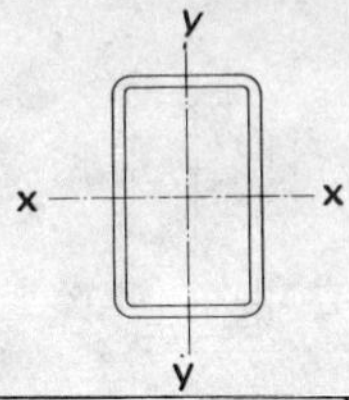

G40.21-M 350W
CLASS H
$\phi = 0.90$

Designation (mm x mm x mm)	HSS 203 x 102						HSS 178 x 127#		
	13	11	9.5	8.0	6.4	4.8 ‡	13	11	9.5
Mass (kg/m)	52.4	46.9	40.9	34.8	28.3	21.7	52.4	46.9	40.9
Effective length (KL) in millimetres with respect to least radius of gyration									
0	2 100	1 880	1 640	1 400	1 140	800	2 100	1 880	1 640
1 000	2 080	1 860	1 630	1 380	1 130	795	2 100	1 870	1 640
1 250	2 050	1 840	1 610	1 370	1 120	788	2 080	1 860	1 620
1 500	2 010	1 800	1 580	1 340	1 100	779	2 060	1 840	1 610
1 750	1 960	1 760	1 540	1 310	1 070	766	2 030	1 810	1 590
2 000	1 900	1 710	1 500	1 280	1 050	750	1 990	1 780	1 560
2 250	1 830	1 640	1 440	1 230	1 010	732	1 940	1 740	1 530
2 500	1 740	1 570	1 380	1 180	974	710	1 890	1 700	1 490
2 750	1 650	1 490	1 310	1 130	929	685	1 830	1 650	1 440
3 000	1 540	1 400	1 240	1 070	880	656	1 760	1 590	1 400
3 250	1 420	1 300	1 150	995	826	624	1 690	1 520	1 340
3 500	1 290	1 180	1 060	919	766	589	1 610	1 450	1 280
3 750	1 140	1 050	948	832	701	548	1 520	1 370	1 220
4 000	1 020	937	844	740	624	492	1 420	1 290	1 150
4 250	912	841	757	664	560	441	1 320	1 200	1 070
4 500	825	760	684	600	505	398	1 190	1 090	984
4 750	751	692	623	545	460	362	1 080	993	892
5 000	689	634	570	499	420	331	987	905	813
5 250	635	584	525	459	387	304	905	830	745
5 500	582	538	486	425	358	281	834	765	687
5 750	534	494	446	392	331	261	772	708	635
6 000	492	455	410	361	305	241	718	658	590
6 500	421	390	352	309	261	206	628	575	515
7 000	366	338	305	268	227	179	545	501	451
7 500	321	296	267	235	199	157	477	438	394
8 000			237	208	176	138	421	387	348
8 500						123	375	344	310
9 000							336	309	278
9 500							303	279	251
10 000									
10 500									
PROPERTIES AND DESIGN DATA									
Area (mm^2)	6 680	5 970	5 210	4 430	3 610	2 760	6 680	5 970	5 210
Z_x ($10^3 mm^3$)	405	367	325	281	232	180	377	342	303
S_x ($10^3 mm^3$)	307	282	253	221	185	145	297	272	244
r_x (mm)	68.4	69.3	70.3	71.2	72.2	73.1	62.8	63.7	64.6
Z_y ($10^3 mm^3$)	246	224	199	172	143	111	298	270	240
S_y ($10^3 mm^3$)	201	186	168	148	125	93.3	243	224	202
r_y (mm)	39.1	39.8	40.5	41.2	42.0	42.7	48.1	48.8	49.6
r_x/r_y	1.75	1.74	1.73	1.73	1.72	1.71	1.31	1.30	1.30
M_{rx} (kN•m)	128	116	102	88.5	73.1	56.7	119	108	95.4
M_{ry} (kN•m)	77.5	70.6	62.7	54.2	45.0	29.4	93.9	85.0	75.6
IMPERIAL SIZE AND MASS									
Mass (lb./ft.)	35.2	31.5	27.5	23.4	19.0	14.6	35.2	31.5	27.5
Thickness (in.)	.500	.438	.375	.313	.250	.188	.500	.438	.375
Size (in.)	8 x 4						7 x 5		

‡ Class 4 in bending about Y-Y axis. Resistances and S_y for this column calculated according to **CAN/CSA-S16.1-M89** Clauses 13.3.3 and 13.5(c).

#Check manufacturer for current availability.

G40.21-M 350W CLASS H $\phi = 0.90$

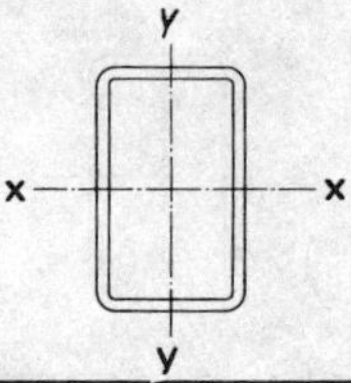

RECTANGULAR HOLLOW SECTIONS
Factored Axial Compressive Resistances, C_r, in kN

Designation (mm x mm x mm)	HSS 178 x 127#			HSS 152 x 102				
	8.0	6.4	††4.8	11	9.5	8.0	6.4	4.8
Mass (kg/m)	34.8	28.3	21.7	38.0	33.3	28.4	23.2	17.9
Effective length (KL) in millimetres with respect to least radius of gyration								
0	1 400	1 140	869	1 520	1 340	1 140	932	718
1 000	1 390	1 130	867	1 510	1 320	1 130	924	712
1 250	1 380	1 130	862	1 480	1 300	1 110	912	704
1 500	1 370	1 120	854	1 450	1 280	1 090	896	692
1 750	1 350	1 100	844	1 420	1 250	1 070	876	677
2 000	1 330	1 080	831	1 370	1 210	1 040	850	658
2 250	1 300	1 060	815	1 310	1 160	997	821	636
2 500	1 270	1 040	797	1 250	1 110	954	786	610
2 750	1 230	1 010	776	1 180	1 040	904	747	581
3 000	1 190	979	753	1 100	977	849	704	549
3 250	1 150	944	727	1 010	902	787	656	513
3 500	1 100	905	698	907	820	720	603	474
3 750	1 050	863	666	800	724	641	541	430
4 000	988	817	632	712	645	571	481	383
4 250	925	768	596	640	579	512	432	343
4 500	858	715	556	579	524	463	390	310
4 750	778	653	511	528	477	422	355	282
5 000	709	595	466	484	437	386	325	258
5 250	650	545	427	446	403	356	299	237
5 500	598	501	393	407	369	328	277	220
5 750	554	464	363	373	339	301	254	203
6 000	514	431	337	344	312	277	234	187
6 500	449	375	294	295	267	237	201	160
7 000	394	331	259	256	232	206	174	139
7 500	345	290	227	225	204	181	153	121
8 000	304	256	201				135	107
8 500	271	227	178					
9 000	243	204	160					
9 500	219	184	144					
10 000	199	167	131					
10 500								
PROPERTIES AND DESIGN DATA								
Area (mm^2)	4 430	3 610	2 760	4 840	4 240	3 620	2 960	2 280
Z_x ($10^3 mm^3$)	261	216	167	230	205	178	148	116
S_x ($10^3 mm^3$)	213	178	140	179	162	143	121	95.6
r_x (mm)	65.4	66.2	67.1	53.1	54.0	54.8	55.7	56.5
Z_y ($10^3 mm^3$)	207	171	133	172	154	134	112	87.8
S_y ($10^3 mm^3$)	177	148	117	140	128	113	96.1	76.6
r_y (mm)	50.3	51.1	51.8	38.4	39.1	39.9	40.6	41.3
r_x/r_y	1.30	1.30	1.29	1.38	1.38	1.38	1.37	1.37
M_{rx} (kN•m)	82.2	68.0	52.6	72.4	64.6	56.1	46.6	36.5
M_{ry} (kN•m)	65.2	53.9	36.9	54.2	48.5	42.2	35.3	27.7
IMPERIAL SIZE AND MASS								
Mass (lb./ft.)	23.4	19.0	14.6	25.5	22.4	19.1	15.6	12.0
Thickness (in.)	.313	.250	.188	.438	.375	.313	.250	.188
Size (in.)	7 x 5			6 x 4				

†† Class 3 in bending about Y-Y axis.
# Check manufacturer for current availability.

RECTANGULAR HOLLOW SECTIONS
Factored Axial Compressive Resistances, C_r, in kN

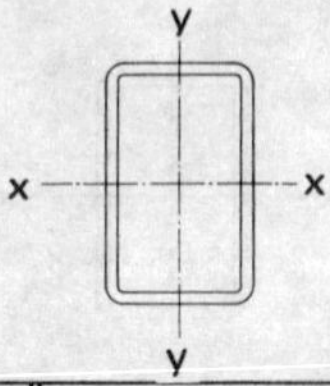

G40.21-M 350W CLASS H $\phi = 0.90$

Designation (mm x mm x mm)	HSS 127 x 76				HSS 127 x 64			
	9.5	8.0	6.4	4.8	9.5	8.0	6.4	4.8
Mass (kg/m)	25.7	22.1	18.2	14.1	23.8	20.5	16.9	13.1
Effective length (KL) in millimetres with respect to least radius of gyration								
0	1 030	888	731	564	954	822	677	526
1 000	999	862	710	549	900	779	644	503
1 250	968	836	691	535	854	742	616	482
1 500	926	803	665	516	794	694	579	455
1 750	874	761	632	492	721	635	533	422
2 000	811	710	592	463	633	564	478	383
2 250	739	651	547	429	527	479	415	337
2 500	656	584	494	391	436	396	343	282
2 750	558	504	433	347	369	334	289	238
3 000	477	431	370	297	318	288	249	204
3 250	414	374	320	258	277	251	217	178
3 500	364	328	281	226	240	219	190	157
3 750	324	292	250	201	210	191	167	138
4 000	289	262	224	180	186	169	147	121
4 250	257	233	201	162	165	151	131	108
4 500	230	209	180	145	148	135	117	97
4 750	207	188	162	130		122	106	87
5 000	188	170	147	118			96	79
5 250	171	155	133	108				
5 500	157	142	122	98				
5 750		130	112	90				
6 000			104	83				
6 500								
7 000								
7 500								
8 000								
8 500								
9 000								
9 500								
10 000								
10 500								
PROPERTIES AND DESIGN DATA								
Area (mm^2)	3 280	2 820	2 320	1 790	3 030	2 610	2 150	1 670
Z_x ($10^3 mm^3$)	126	111	93.3	73.8	112	98.9	83.6	66.3
S_x ($10^3 mm^3$)	96.4	86.4	74.0	59.5	83.2	75.1	64.8	52.4
r_x (mm)	43.2	44.1	45.1	45.9	41.7	42.7	43.7	44.6
Z_y ($10^3 mm^3$)	87.5	77.2	65.2	51.7	67.4	60.0	51.0	40.7
S_y ($10^3 mm^3$)	70.5	63.8	55.1	44.7	53.6	49.1	42.9	35.2
r_y (mm)	28.6	29.4	30.1	30.8	23.7	24.4	25.1	25.9
r_x/r_y	1.51	1.50	1.50	1.49	1.76	1.75	1.74	1.73
M_{rx} (kN•m)	39.7	35.0	29.4	23.2	35.3	31.2	26.3	20.9
M_{ry} (kN•m)	27.6	24.3	20.5	16.3	21.2	18.9	16.1	12.8
IMPERIAL SIZE AND MASS								
Mass (lb./ft.)	17.3	14.9	12.2	9.45	16.0	13.8	11.4	8.81
Thickness (in.)	.375	.313	.250	.188	.375	.313	.250	.188
Size (in.)	5 x 3				5 x 2½			

G40.21-M 350W
CLASS H
$\phi = 0.90$

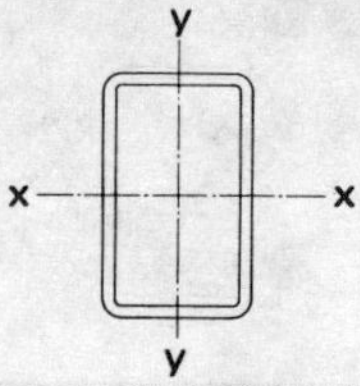

RECTANGULAR HOLLOW SECTIONS
Factored Axial Compressive Resistances, C_r, in kN

Designation (mm x mm x mm)		HSS 127 x 51				HSS 102 x 76			
		9.5	8.0	6.4	4.8	9.5	8.0	6.4	4.8
Mass (kg/m)		21.9	18.9	15.6	12.2	21.9	18.9	15.6	12.2
Effective length (KL) in millimetres with respect to least radius of gyration	0	879	759	627	488	879	759	627	488
	1 000	782	683	570	448	847	734	608	475
	1 250	708	624	525	415	818	711	590	461
	1 500	613	549	468	374	780	680	566	444
	1 750	493	456	397	323	732	641	536	422
	2 000	388	358	315	261	675	595	500	395
	2 250	316	291	256	212	609	541	458	364
	2 500	265	243	213	176	532	480	410	329
	2 750	221	205	181	150	448	407	353	287
	3 000	187	173	153	127	383	348	302	245
	3 250	160	148	131	109	333	302	262	213
	3 500	139	129	114	95	293	266	230	187
	3 750		113	100	83	262	237	205	166
	4 000				73	231	211	184	149
	4 250					205	187	163	133
	4 500					184	168	146	119
	4 750					166	151	132	107
	5 000					150	137	119	97
	5 250					137	125	109	89
	5 500					125	114	100	81
	5 750							91	75
	6 000								
	6 500								
	7 000								
	7 500								
	8 000								
	8 500								
	9 000								
	9 500								
	10 000								
	10 500								
PROPERTIES AND DESIGN DATA									
Area (mm^2)		2 790	2 410	1 990	1 550	2 790	2 410	1 990	1 550
Z_x ($10^3 mm^3$)		97.7	86.9	73.8	58.9	87.6	77.7	65.9	52.5
S_x ($10^3 mm^3$)		70.0	63.8	55.5	45.3	67.2	61.0	52.8	43.0
r_x (mm)		39.9	41.0	42.1	43.1	35.0	35.8	36.7	37.5
Z_y ($10^3 mm^3$)		48.9	44.0	37.9	30.5	71.3	63.5	54.0	43.1
S_y ($10^3 mm^3$)		37.8	35.3	31.4	26.2	56.3	51.4	44.8	36.6
r_y (mm)		18.6	19.3	20.0	20.7	27.7	28.5	29.3	30.0
r_x/r_y		2.15	2.13	2.10	2.08	1.26	1.26	1.25	1.25
M_{rx} (kN•m)		30.8	27.4	23.2	18.6	27.6	24.5	20.8	16.5
M_{ry} (kN•m)		15.4	13.9	11.9	9.61	22.5	20.0	17.0	13.6
IMPERIAL SIZE AND MASS									
Mass (lb./ft.)		14.7	12.7	10.5	8.17	14.7	12.7	10.5	8.17
Thickness (in.)		.375	.313	.250	.188	.375	.313	.250	.188
Size (in.)		5 x 2				4 x 3			

RECTANGULAR HOLLOW SECTIONS
Factored Axial Compressive Resistances, C_r, in kN

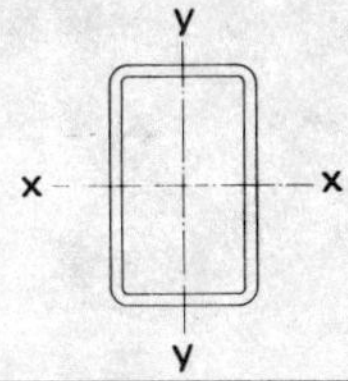

G40.21-M 350W CLASS H $\phi = 0.90$

Designation (mm x mm x mm)	HSS 102 x 51					HSS 89 x 64		
	8.0	6.4	4.8	3.8	3.2	8.0	6.4	4.8
Mass (kg/m)	15.8	13.1	10.3	8.37	7.09	15.8	13.1	10.3
Effective length (KL) in millimetres with respect to least radius of gyration								
0	633	526	413	337	284	633	526	413
1 000	565	475	376	309	262	595	497	392
1 250	513	435	348	287	244	563	472	374
1 500	447	384	311	258	220	522	440	351
1 750	362	322	267	223	192	472	401	322
2 000	285	253	213	180	156	412	354	288
2 250	232	205	173	146	126	339	297	248
2 500	194	171	144	122	105	281	246	204
2 750	162	145	123	103	89	238	208	173
3 000	137	122	104	88	76	205	179	148
3 250	118	105	89	75	65	178	157	130
3 500	102	91	77	65	57	154	136	113
3 750	90	80	68	57	50	135	119	99
4 000			60	51	44	119	105	88
4 250						106	93	78
4 500						95	84	70
4 750							76	63
5 000								
5 250								
5 500								
5 750								
6 000								
6 500								
7 000								
7 500								
8 000								
8 500								
9 000								
9 500								
10 000								
10 500								
PROPERTIES AND DESIGN DATA								
Area (mm^2)	2 010	1 670	1 310	1 070	903	2 010	1 670	1 310
Z_x ($10^3 mm^3$)	58.8	50.6	40.8	33.9	29.0	54.9	47.1	38.0
S_x ($10^3 mm^3$)	43.5	38.4	31.8	26.8	23.1	42.1	37.1	30.6
r_x (mm)	33.2	34.2	35.1	35.7	36.1	30.5	31.4	32.3
Z_y ($10^3 mm^3$)	35.4	30.7	24.9	20.8	17.9	43.2	37.2	30.1
S_y ($10^3 mm^3$)	27.9	25.1	21.1	18.0	15.6	34.3	30.4	25.3
r_y (mm)	18.8	19.5	20.3	20.7	21.0	23.3	24.0	24.8
r_x/r_y	1.77	1.75	1.73	1.73	1.72	1.31	1.31	1.30
M_{rx} (kN•m)	18.5	15.9	12.9	10.7	9.13	17.3	14.8	12.0
M_{ry} (kN•m)	11.2	9.67	7.84	6.55	5.64	13.6	11.7	9.48
IMPERIAL SIZE AND MASS								
Mass (lb./ft.)	10.6	8.81	6.89	5.62	4.76	10.6	8.81	6.89
Thickness (in.)	.313	.250	.188	.150	.125	.313	.250	.188
Size (in.)	4 x 2					3½ x 2½		

G40.21-M 350W
CLASS H
$\phi = 0.90$

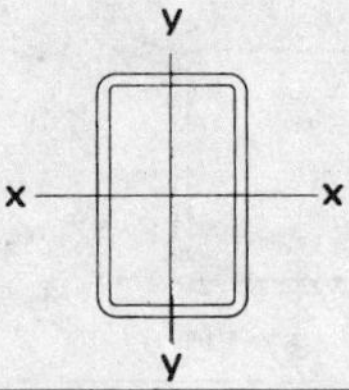

RECTANGULAR HOLLOW SECTIONS
Factored Axial Compressive Resistances, C_r, in kN

Designation (mm x mm x mm)	HSS 89 x 64		HSS 76 x 51				HSS 51 x 25	
	3.8	3.2	8.0	6.4	4.8	3.8	3.2	2.5
Mass (kg/m)	8.37	7.09	12.6	10.6	8.35	6.85	3.28	2.71
Effective length (KL) in millimetres with respect to least radius of gyration								
0	337	284	504	425	334	275	132	109
1 000	321	271	443	379	302	250	63	56
1 250	307	260	398	345	277	230	43	38
1 500	289	245	339	300	245	205	31	27
1 750	266	226	267	243	206	174	23	20
2 000	239	204	210	191	162	138		16
2 250	208	178	171	156	131	112		
2 500	172	148	143	130	110	93		
2 750	145	125	119	109	93	79		
3 000	125	107	100	92	78	67		
3 250	109	94	86	79	67	57		
3 500	96	82	75	69	58	50		
3 750	84	72		60	51	44		
4 000	74	64						
4 250	66	57						
4 500	59	51						
4 750	53	46						
5 000	48	42						
5 250								
5 500								
5 750								
6 000								
6 500								
7 000								
7 500								
8 000								
8 500								
9 000								
9 500								
10 000								
10 500								
PROPERTIES AND DESIGN DATA								
Area (mm^2)	1 070	903	1 600	1 350	1 060	872	418	345
Z_x ($10^3 mm^3$)	31.5	27.0	35.9	31.4	25.7	21.6	6.33	5.35
S_x ($10^3 mm^3$)	25.8	22.3	26.6	24.1	20.3	17.3	4.81	4.15
r_x (mm)	32.8	33.1	25.1	26.1	27.0	27.5	17.1	17.5
Z_y ($10^3 mm^3$)	25.0	21.4	26.7	23.5	19.3	16.3	3.84	3.27
S_y ($10^3 mm^3$)	21.4	18.5	20.5	18.8	16.0	13.8	3.14	2.75
r_y (mm)	25.2	25.5	18.0	18.8	19.6	20.0	9.77	10.1
r_x/r_y	1.30	1.30	1.39	1.39	1.38	1.37	1.75	1.74
M_{rx} (kN•m)	9.92	8.50	11.3	9.89	8.10	6.80	1.99	1.69
M_{ry} (kN•m)	7.87	6.74	8.41	7.40	6.08	5.13	1.21	1.03
IMPERIAL SIZE AND MASS								
Mass (lb./ft.)	5.62	4.76	8.46	7.11	5.61	4.60	2.21	1.82
Thickness (in.)	.150	.125	.313	.250	.188	.150	.125	.100
Size (in.)	3½ x 2½		3 x 2				2 x 1	

SQUARE HOLLOW SECTIONS
Factored Axial Compressive Resistances, C_r, in kN

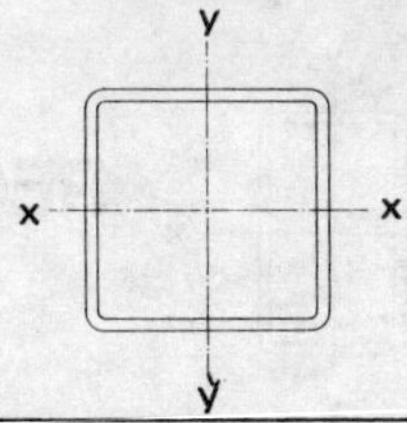

G40.21-M 350W
CLASS H
$\phi = 0.90$

Designation (mm x mm x mm)	HSS 305 x 305					HSS 254 x 254		
	13	11	9.5	††8.0	‡6.4	13	11	9.5
Mass (kg/m)	113	100	86.5	72.8	58.7	93.0	82.4	71.3
Effective length (KL) in millimetres with respect to least radius of gyration								
0	4 540	4 030	3 460	2 920	1 910	3 720	3 310	2 860
1 000	4 540	4 030	3 460	2 920	1 910	3 720	3 310	2 860
1 250	4 540	4 030	3 460	2 920	1 910	3 720	3 310	2 860
1 500	4 540	4 030	3 470	2 920	1 910	3 720	3 310	2 860
1 750	4 530	4 030	3 460	2 920	1 910	3 710	3 300	2 860
2 000	4 530	4 030	3 460	2 920	1 910	3 700	3 290	2 850
2 250	4 520	4 020	3 460	2 920	1 900	3 690	3 280	2 840
2 500	4 510	4 010	3 450	2 910	1 900	3 680	3 270	2 830
2 750	4 500	4 000	3 440	2 900	1 900	3 660	3 260	2 820
3 000	4 490	3 990	3 430	2 890	1 890	3 640	3 240	2 800
3 250	4 470	3 970	3 420	2 880	1 890	3 610	3 220	2 790
3 500	4 450	3 960	3 400	2 870	1 880	3 590	3 190	2 770
3 750	4 430	3 940	3 390	2 860	1 880	3 550	3 170	2 740
4 000	4 400	3 920	3 370	2 840	1 870	3 520	3 140	2 720
4 250	4 370	3 890	3 350	2 830	1 860	3 480	3 100	2 690
4 500	4 340	3 860	3 320	2 810	1 860	3 440	3 070	2 660
4 750	4 310	3 840	3 300	2 790	1 850	3 400	3 030	2 630
5 000	4 270	3 800	3 270	2 770	1 840	3 350	2 990	2 600
5 250	4 230	3 770	3 250	2 740	1 830	3 300	2 950	2 560
5 500	4 190	3 730	3 210	2 720	1 820	3 250	2 900	2 520
5 750	4 150	3 700	3 180	2 690	1 800	3 200	2 850	2 480
6 000	4 100	3 650	3 150	2 660	1 790	3 140	2 800	2 430
6 500	4 000	3 570	3 070	2 600	1 760	3 010	2 690	2 340
7 000	3 890	3 470	2 990	2 530	1 730	2 870	2 570	2 230
7 500	3 770	3 360	2 900	2 460	1 700	2 720	2 430	2 120
8 000	3 630	3 250	2 800	2 380	1 660	2 550	2 290	2 000
8 500	3 490	3 120	2 700	2 290	1 620	2 370	2 130	1 860
9 000	3 340	2 990	2 580	2 190	1 570	2 160	1 950	1 710
9 500	3 170	2 840	2 460	2 090	1 520	1 960	1 770	1 550
10 000	3 000	2 690	2 330	1 990	1 470	1 790	1 620	1 420
10 500	2 810	2 530	2 200	1 870	1 410	1 640	1 480	1 300
PROPERTIES AND DESIGN DATA								
Area (mm^2)	14 400	12 800	11 000	9 280	7 480	11 800	10 500	9 090
Z ($10^3 mm^3$)	1 560	1 390	1 210	1 030	833	1 060	945	825
S ($10^3 mm^3$)	1 330	1 190	1 040	886	625	888	799	703
r (mm)	118	119	120	121	121	97.6	98.4	99.1
M_r (kN•m)	491	438	381	279	197	334	298	260
IMPERIAL SIZE AND MASS								
Mass (lb./ft.)	76.1	67.3	58.1	48.9	39.4	62.5	55.4	47.9
Thickness (in.)	.500	.438	.375	.313	.250	.500	.438	.375
Size (in.)	12 x 12					10 x 10		

†† Class 3 in bending

‡ Class 4 in bending. Pesistances and S for this column calculated according to CAN/CSA-S16.1-M89 Clauses 13.3.3 and 13.5(c).

G40.21-M 350W CLASS H $\phi = 0.90$

SQUARE HOLLOW SECTIONS Factored Axial Compressive Resistances, C_r, in kN

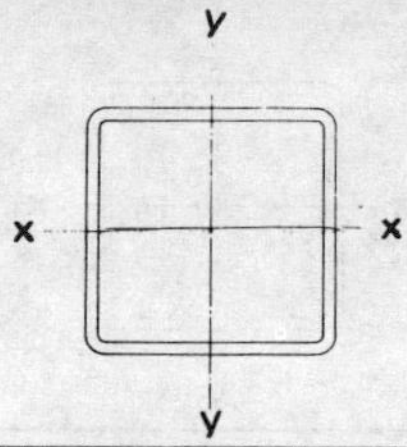

Designation (mm x mm x mm)		HSS 254 x 254		HSS 203 x 203				
		8.0	‡6.4	13	11	9.5	8.0	6.4
Mass (kg/m)		60.1	48.6	72.7	64.6	56.1	47.5	38.4
Effective length (KL) in millimetres with respect to least radius of gyration	0	2 410	1 790	2 920	2 590	2 250	1 910	1 540
	1 000	2 410	1 790	2 920	2 590	2 250	1 910	1 540
	1 250	2 410	1 790	2 910	2 590	2 250	1 900	1 540
	1 500	2 410	1 790	2 910	2 580	2 250	1 900	1 540
	1 750	2 410	1 790	2 900	2 580	2 240	1 890	1 530
	2 000	2 400	1 790	2 880	2 560	2 230	1 890	1 530
	2 250	2 400	1 780	2 860	2 550	2 210	1 870	1 520
	2 500	2 390	1 780	2 840	2 530	2 200	1 860	1 510
	2 750	2 380	1 770	2 810	2 500	2 180	1 840	1 500
	3 000	2 360	1 770	2 780	2 480	2 160	1 830	1 480
	3 250	2 350	1 760	2 750	2 450	2 130	1 800	1 460
	3 500	2 330	1 750	2 710	2 410	2 100	1 780	1 450
	3 750	2 310	1 740	2 670	2 380	2 070	1 760	1 420
	4 000	2 290	1 730	2 620	2 340	2 030	1 730	1 400
	4 250	2 270	1 720	2 570	2 290	2 000	1 700	1 380
	4 500	2 250	1 700	2 510	2 240	1 960	1 660	1 350
	4 750	2 220	1 690	2 450	2 190	1 910	1 630	1 320
	5 000	2 190	1 670	2 390	2 140	1 870	1 590	1 290
	5 250	2 160	1 660	2 320	2 080	1 820	1 550	1 260
	5 500	2 130	1 640	2 250	2 020	1 770	1 500	1 220
	5 750	2 090	1 620	2 180	1 950	1 710	1 460	1 190
	6 000	2 060	1 600	2 100	1 880	1 650	1 410	1 150
	6 500	1 980	1 560	1 930	1 740	1 530	1 310	1 070
	7 000	1 890	1 510	1 730	1 570	1 390	1 190	977
	7 500	1 800	1 450	1 530	1 390	1 230	1 060	870
	8 000	1 700	1 380	1 360	1 230	1 090	941	774
	8 500	1 590	1 290	1 220	1 110	980	845	695
	9 000	1 470	1 200	1 110	1 000	887	764	628
	9 500	1 330	1 090	1 010	914	808	696	572
	10 000	1 210	992	926	837	740	637	524
	10 500	1 110	909	852	773	682	587	482

PROPERTIES AND DESIGN DATA

	HSS 254 x 254 8.0	‡6.4	HSS 203 x 203 13	11	9.5	8.0	6.4
Area (mm^2)	7 660	6 190	9 260	8 230	7 150	6 050	4 900
Z ($10^3 mm^3$)	702	571	650	584	513	438	359
S ($10^3 mm^3$)	602	492	538	488	432	373	308
r (mm)	99.9	101	76.8	77.6	78.4	79.2	79.9
M_r (kN•m)	221	155	205	184	162	138	113

IMPERIAL SIZE AND MASS

Mass (lb./ft.)	40.4	32.6	48.9	43.4	37.7	31.9	25.8
Thickness (in.)	.313	.250	.500	.438	.375	.313	.250
Size (in.)	10 x 10		8 x 8				

‡ Class 4 in bending. Resistances and S for this column calculated according to CAN/CSA-S16.1-M89 Clauses 13.3.3 and 13.5(c).

SQUARE HOLLOW SECTIONS
Factored Axial Compressive Resistances, C_r, in kN

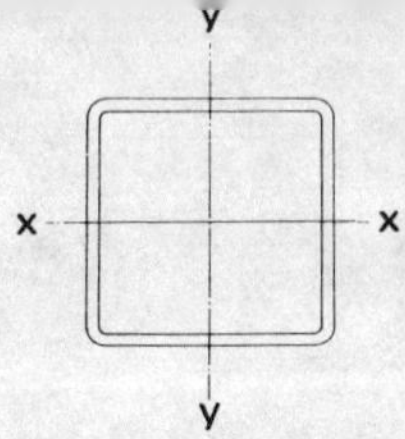

G40.21-M 350W
CLASS H
$\phi = 0.90$

Designation (mm x mm x mm)	HSS 178 x 178						HSS 152 x 152		
	13	11	9.5	8.0	6.4	†† 4.8	13	11	9.5
Mass (kg/m)	62.6	55.7	48.5	41.1	33.4	25.5	52.4	46.9	40.9
Effective length (KL) in millimetres with respect to least radius of gyration									
0	2 510	2 240	1 950	1 650	1 340	1 020	2 100	1 880	1 640
1 000	2 510	2 240	1 950	1 650	1 340	1 020	2 100	1 880	1 640
1 250	2 500	2 230	1 940	1 650	1 340	1 020	2 090	1 870	1 630
1 500	2 490	2 220	1 940	1 640	1 330	1 020	2 080	1 860	1 620
1 750	2 480	2 210	1 920	1 630	1 320	1 010	2 060	1 840	1 610
2 000	2 460	2 190	1 910	1 620	1 320	1 010	2 030	1 820	1 590
2 250	2 440	2 170	1 890	1 610	1 300	998	2 000	1 790	1 570
2 500	2 410	2 150	1 870	1 590	1 290	988	1 960	1 760	1 540
2 750	2 370	2 120	1 850	1 570	1 280	977	1 920	1 720	1 510
3 000	2 340	2 090	1 820	1 550	1 260	963	1 870	1 680	1 470
3 250	2 290	2 050	1 790	1 520	1 240	948	1 820	1 640	1 440
3 500	2 250	2 010	1 750	1 490	1 210	931	1 760	1 590	1 390
3 750	2 190	1 960	1 720	1 460	1 190	913	1 700	1 530	1 350
4 000	2 140	1 910	1 670	1 430	1 160	892	1 630	1 470	1 290
4 250	2 080	1 860	1 630	1 390	1 130	871	1 550	1 410	1 240
4 500	2 010	1 800	1 580	1 350	1 100	847	1 470	1 340	1 180
4 750	1 940	1 740	1 530	1 310	1 070	822	1 390	1 260	1 120
5 000	1 870	1 680	1 470	1 260	1 030	795	1 300	1 180	1 050
5 250	1 790	1 610	1 420	1 210	994	766	1 190	1 090	976
5 500	1 700	1 540	1 350	1 160	953	736	1 090	1 000	897
5 750	1 610	1 460	1 290	1 110	910	704	1 010	925	828
6 000	1 520	1 380	1 220	1 050	865	670	935	857	767
6 500	1 310	1 190	1 060	920	762	593	813	745	666
7 000	1 150	1 050	929	804	666	519	716	655	586
7 500	1 020	925	822	712	589	459	637	583	521
8 000	909	827	735	636	526	409	565	519	465
8 500	820	746	662	573	474	369	502	461	414
9 000	746	677	601	520	430	334	450	413	370
9 500	672	612	545	473	392	305	405	372	334
10 000	608	554	494	428	355	277	367	337	302
10 500	553	505	449	390	323	252	335	307	276
PROPERTIES AND DESIGN DATA									
Area (mm^2)	7 970	7 100	6 180	5 240	4 250	3 250	6 680	5 970	5 210
Z ($10^3 mm^3$)	484	436	385	330	271	209	341	310	275
S ($10^3 mm^3$)	396	361	322	279	231	181	275	253	227
r (mm)	66.4	67.2	68.0	68.8	69.6	70.3	56.0	56.8	57.6
M_r (kN•m)	152	137	121	104	85.4	57.0	107	97.6	86.6
IMPERIAL SIZE AND MASS									
Mass (lb./ft.)	42.1	37.5	32.6	27.6	22.4	17.1	35.2	31.5	27.5
Thickness (in.)	.500	.438	.375	.313	.250	.188	.500	.438	.375
Size (in.)	7 x 7						6 x 6		

†† Class 3 in bending.

G40.21-M 350W
CLASS H
$\phi = 0.90$

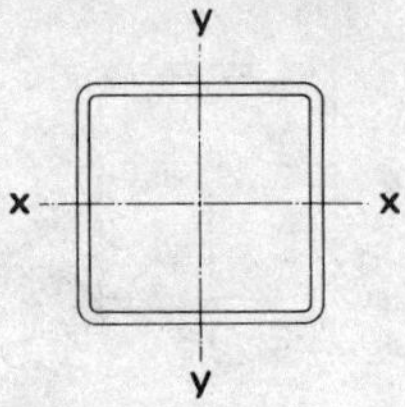

SQUARE HOLLOW SECTIONS
Factored Axial Compressive Resistances, C_r, in kN

Designation (mm x mm x mm)	HSS 152 x 152			HSS 127 x 127				
	8.0	6.4	4.8	11	9.5	8.0	6.4	4.8
Mass (kg/m)	34.8	28.3	21.7	38.0	33.3	28.4	23.2	17.9
Effective length (KL) in millimetres with respect to least radius of gyration								
0	1 400	1 140	869	1 520	1 340	1 140	932	718
1 000	1 390	1 140	869	1 520	1 330	1 140	929	716
1 250	1 390	1 130	866	1 500	1 320	1 130	922	711
1 500	1 380	1 130	861	1 490	1 300	1 110	912	704
1 750	1 370	1 120	854	1 460	1 280	1 100	899	694
2 000	1 350	1 100	845	1 430	1 260	1 080	883	682
2 250	1 330	1 090	834	1 400	1 230	1 050	864	668
2 500	1 310	1 070	821	1 350	1 190	1 020	841	651
2 750	1 290	1 050	806	1 310	1 150	991	816	632
3 000	1 260	1 030	790	1 250	1 110	955	787	611
3 250	1 230	1 000	771	1 200	1 060	914	755	587
3 500	1 190	976	750	1 130	1 010	869	720	561
3 750	1 150	946	727	1 060	946	821	681	532
4 000	1 110	913	703	986	882	768	640	501
4 250	1 070	877	676	899	813	711	595	468
4 500	1 020	839	648	810	732	644	543	431
4 750	965	798	617	735	664	584	492	390
5 000	910	754	584	671	606	533	449	356
5 250	851	708	550	616	556	489	411	326
5 500	782	653	510	568	512	451	379	300
5 750	721	603	471	526	475	417	351	278
6 000	668	558	436	490	441	388	326	258
6 500	580	484	378	425	385	339	285	226
7 000	510	425	332	368	333	294	248	197
7 500	453	378	295	322	292	257	217	173
8 000	406	339	265	285	258	227	192	152
8 500	361	302	236	254	229	202	171	136
9 000	323	270	212	227	206	181	153	122
9 500	291	244	191			164	138	110
10 000	264	221	173					
10 500	240	201	157					
PROPERTIES AND DESIGN DATA								
Area (mm^2)	4 430	3 610	2 760	4 840	4 240	3 620	2 960	2 280
Z ($10^3 mm^3$)	237	195	152	205	183	159	132	103
S ($10^3 mm^3$)	198	166	130	164	149	132	111	88.1
r (mm)	58.4	59.2	59.9	46.4	47.2	48.0	48.8	49.6
M_r (kN•m)	74.7	61.4	47.9	64.6	57.6	50.1	41.6	32.4
IMPERIAL SIZE AND MASS								
Mass (lb./ft.)	23.4	19.0	14.6	25.5	22.4	19.1	15.6	12.0
Thickness (in.)	.313	.250	.188	.438	.375	.313	.250	.188
Size (in.)	6 x 6			5 x 5				

SQUARE HOLLOW SECTIONS
Factored Axial Compressive Resistances, C_r, in kN

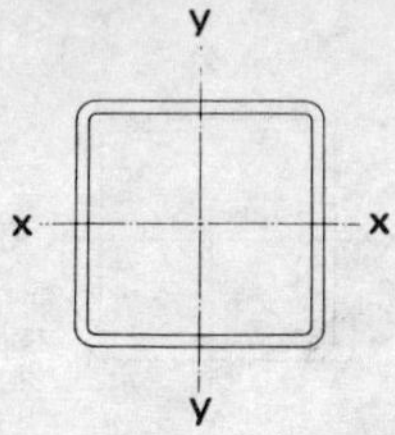

G40.21-M 350W CLASS H $\phi = 0.90$

Designation (mm x mm x mm)	HSS 102 x 102				HSS 89 x 89			
	9.5	8.0	6.4	4.8	9.5	8.0	6.4	4.8
Mass (kg/m)	25.7	22.1	18.2	14.1	21.9	18.9	15.6	12.2
Effective length (KL) in millimetres with respect to least radius of gyration								
0	1 030	888	731	564	879	759	627	488
1 000	1 020	877	722	558	858	743	614	479
1 250	1 000	864	712	550	837	726	601	470
1 500	980	845	697	540	809	703	583	457
1 750	951	821	679	526	774	674	561	440
2 000	915	793	656	509	731	639	534	420
2 250	874	759	630	490	682	599	502	397
2 500	827	720	599	467	625	553	466	370
2 750	773	676	565	442	561	500	425	340
3 000	713	627	526	413	485	439	379	306
3 250	647	572	483	382	420	380	327	266
3 500	569	508	435	347	368	333	287	233
3 750	502	449	383	307	327	295	254	206
4 000	448	400	342	273	293	264	227	184
4 250	403	359	307	245	264	238	205	166
4 500	365	325	278	222	237	215	186	151
4 750	333	297	253	202	214	194	168	137
5 000	306	272	232	185	194	175	152	124
5 250	278	249	214	171	176	160	138	113
5 500	254	228	195	157	161	146	126	103
5 750	233	209	179	144	148	134	116	95
6 000	215	192	165	132	137	124	107	87
6 500	184	165	141	113			92	75
7 000	160	143	123	98				
7 500		126	108	86				
8 000								
8 500								
9 000								
9 500								
10 000								
10 500								
PROPERTIES AND DESIGN DATA								
Area (mm^2)	3 280	2 820	2 320	1 790	2 790	2 410	1 990	1 550
Z ($10^3 mm^3$)	110	96.6	81.3	64.3	80.2	71.2	60.5	48.2
S ($10^3 mm^3$)	87.4	78.4	67.3	54.2	62.7	57.0	49.5	40.3
r (mm)	36.8	37.6	38.4	39.2	31.6	32.4	33.2	34.0
M_r (kN•m)	34.6	30.4	25.6	20.3	25.3	22.4	19.1	15.2
IMPERIAL SIZE AND MASS								
Mass (lb./ft.)	17.3	14.9	12.2	9.45	14.7	12.7	10.5	8.17
Thickness (in.)	.375	.313	.250	.188	.375	.313	.250	.188
Size (in.)	4 x 4				3½ x 3½			

G40.21-M 350W
CLASS H
$\phi = 0.90$

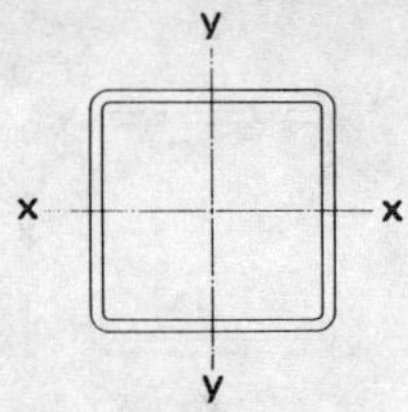

SQUARE HOLLOW SECTIONS
Factored Axial Compressive Resistances, C_r, in kN

Designation (mm x mm x mm)	HSS 76 x 76			HSS 64 x 64			
	8.0	6.4	4.8	6.4	4.8	3.8	3.2
Mass (kg/m)	15.8	13.1	10.3	10.6	8.35	6.85	5.82
Effective length (KL) in millimetres with respect to least radius of gyration							
0	633	526	413	425	334	275	233
1 000	609	508	399	398	315	260	221
1 250	587	491	387	376	298	247	211
1 500	558	468	371	347	277	230	197
1 750	522	440	350	311	251	210	180
2 000	480	407	326	269	220	186	160
2 250	430	368	297	219	183	156	136
2 500	371	324	264	181	151	129	112
2 750	312	273	225	154	128	109	95
3 000	267	234	193	133	110	94	82
3 250	233	203	167	115	96	82	71
3 500	205	179	147	99	83	71	62
3 750	182	159	131	87	73	62	54
4 000	161	141	117	77	64	55	48
4 250	143	126	104	68	57	49	43
4 500	128	112	93	61	51	44	38
4 750	115	101	84			40	35
5 000	105	92	76				
5 250	95	84	69				
5 500		77	63				
5 750			58				
6 000							
6 500							
7 000							
7 500							
8 000							
8 500							
9 000							
9 500							
10 000							
10 500							
PROPERTIES AND DESIGN DATA							
Area (mm^2)	2 010	1 670	1 310	1 350	1 060	872	741
Z ($10^3 mm^3$)	49.7	42.7	34.4	28.0	22.9	19.2	16.5
S ($10^3 mm^3$)	39.0	34.4	28.5	22.1	18.7	15.9	13.9
r (mm)	27.2	28.0	28.8	22.8	23.6	24.1	24.4
M_r (kN•m)	15.7	13.5	10.8	8.82	7.21	6.05	5.20
IMPERIAL SIZE AND MASS							
Mass (lb./ft.)	10.6	8.81	6.89	7.11	5.61	4.60	3.91
Thickness (in.)	.313	.250	.188	.250	.188	.150	.125
Size (in.)	3 x 3			2½ x 2½			

SQUARE HOLLOW SECTIONS
Factored Axial Compressive Resistances, C_r, in kN

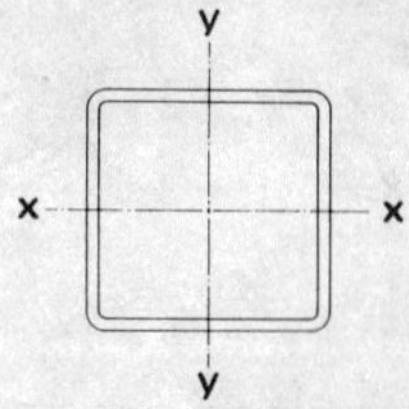

G40.21-M 350W
CLASS H
ϕ = 0.9

Designation (mm x mm x mm)	HSS 51 x 51					HSS 38 x 38			
	6.4	4.8	3.8	3.2	2.8	4.8	3.8	3.2	2.5
Mass (kg/m)	8.05	6.45	5.33	4.55	4.05	4.54	3.81	3.28	2.71
Effective length (KL) in millimetres with respect to least radius of gyration									
0	324	259	214	183	163	182	153	132	109
1 000	283	229	191	164	146	135	116	102	85
1 250	252	207	174	150	134	101	91	81	69
1 500	213	178	152	131	118	73	65	59	50
1 750	165	142	124	109	99	56	50	45	38
2 000	130	112	97	85	77	44	39	35	30
2 250	106	91	79	69	63	35	31	28	24
2 500	88	76	66	58	53	28	26	23	20
2 750	73	64	55	49	44			19	16
3 000	62	54	47	41	37				
3 250	53	46	40	35	32				
3 500	46	40	35	31	28				
3 750			31	27	24				
4 000									
4 250									
4 500									
4 750									
5 000									
5 250									
5 500									
5 750									
6 000									
6 500									
7 000									
7 500									
8 000									
8 500									
9 000									
9 500									
10 000									
10 500									
PROPERTIES AND DESIGN DATA									
Area (mm^2)	1 030	821	679	580	516	578	485	418	345
Z ($10^3 mm^3$)	16.3	13.8	11.7	10.2	9.15	6.91	6.04	5.34	4.51
S ($10^3 mm^3$)	12.5	10.9	9.54	8.42	7.64	5.27	4.77	4.31	3.71
r (mm)	17.6	18.4	18.9	19.2	19.4	13.2	13.7	14.0	14.3
M_r (kN•m)	5.13	4.35	3.69	3.21	2.88	2.18	1.90	1.68	1.42
IMPERIAL SIZE AND MASS									
Mass (lb./ft.)	5.41	4.33	3.58	3.06	2.72	3.05	2.56	2.21	1.82
Thickness (in.)	.250	.188	.150	.125	.110	.188	.150	.125	.100
Size (in.)	2 x 2					1½ x 1½			

G40.21-M 350W
CLASS H
$\phi = 0.90$

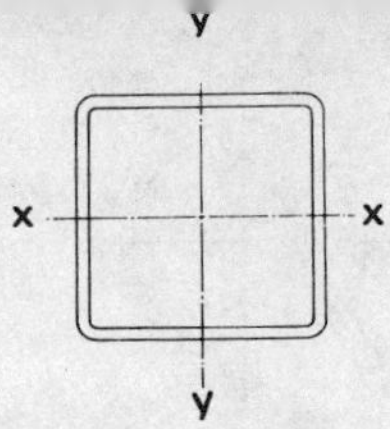

SQUARE HOLLOW SECTIONS
Factored Axial Compressive Resistances, C_r, in kN

Designation (mm x mm x mm)	HSS 32 x 32			HSS 25 x 25	
	3.8	3.2	2.5	3.2	2.5
Mass (kg/m)	3.06	2.65	2.20	2.01	1.69
Effective length (KL) in millimetres with respect to least radius of gyration					
0	123	106	89	81	68
1 000	75	67	58	32	29
1 250	50	45	40	22	20
1 500	37	33	29	15	14
1 750	27	25	22	11	10
2 000	21	19	17		
2 250		15	13		
2 500					
2 750					
3 000					
3 250					
3 500					
3 750					
4 000					
4 250					
4 500					
4 750					
5 000					
5 250					
5 500					
5 750					
6 000					
6 500					
7 000					
7 500					
8 000					
8 500					
9 000					
9 500					
10 000					
10 500					
PROPERTIES AND DESIGN DATA					
Area (mm^2)	389	338	281	257	216
Z ($10^3 mm^3$)	3.92	3.51	3.01	2.05	1.79
S ($10^3 mm^3$)	3.01	2.77	2.44	1.56	1.41
r (mm)	11.1	11.4	11.7	8.79	9.12
M_r (kN•m)	1.23	1.11	0.948	0.646	0.564
IMPERIAL SIZE AND MASS					
Mass (lb./ft.)	2.05	1.78	1.48	1.35	1.14
Thickness (in.)	.150	.125	.100	.125	.100
Size (in.)	1¼ x 1¼			1 x 1	

ROUND HOLLOW SECTIONS
Factored Axial Compressive Resistances, C_r, in kN

G40.21-M 350W
CLASS H
$\phi = 0.90$

Effective length (KL) in millimetres with respect to least radius of gyration	HSS 406					HSS 356		
Designation (mm x mm)	13	11	9.5	8.0	6.4	13	11	9.5
Mass (kg/m)	123	108	93.3	78.1	62.6	107	94.6	81.3
0	4 950	4 350	3 750	3 130	2 510	4 320	3 780	3 280
1 000	4 950	4 350	3 750	3 130	2 510	4 320	3 780	3 280
1 250	4 950	4 350	3 750	3 130	2 510	4 320	3 780	3 280
1 500	4 950	4 350	3 750	3 130	2 510	4 320	3 780	3 280
1 750	4 950	4 350	3 750	3 130	2 510	4 310	3 780	3 280
2 000	4 950	4 350	3 750	3 130	2 510	4 310	3 780	3 270
2 250	4 940	4 340	3 750	3 130	2 510	4 310	3 770	3 270
2 500	4 940	4 340	3 740	3 130	2 510	4 300	3 760	3 260
2 750	4 930	4 330	3 740	3 120	2 510	4 290	3 760	3 250
3 000	4 920	4 320	3 730	3 120	2 500	4 270	3 740	3 250
3 250	4 910	4 320	3 720	3 110	2 500	4 260	3 730	3 230
3 500	4 890	4 300	3 710	3 100	2 490	4 240	3 720	3 220
3 750	4 880	4 290	3 700	3 090	2 480	4 220	3 700	3 210
4 000	4 860	4 270	3 690	3 080	2 470	4 200	3 680	3 190
4 250	4 840	4 260	3 670	3 070	2 460	4 170	3 660	3 170
4 500	4 820	4 240	3 650	3 060	2 450	4 140	3 630	3 150
4 750	4 790	4 220	3 640	3 040	2 440	4 110	3 610	3 130
5 000	4 770	4 190	3 620	3 030	2 430	4 080	3 580	3 100
5 250	4 740	4 170	3 600	3 010	2 410	4 050	3 550	3 080
5 500	4 710	4 140	3 570	2 990	2 400	4 010	3 520	3 050
5 750	4 680	4 120	3 550	2 970	2 380	3 970	3 490	3 020
6 000	4 640	4 090	3 520	2 950	2 370	3 930	3 450	2 990
6 500	4 570	4 020	3 470	2 900	2 330	3 840	3 370	2 920
7 000	4 480	3 950	3 410	2 850	2 290	3 740	3 280	2 850
7 500	4 390	3 870	3 340	2 800	2 240	3 630	3 190	2 760
8 000	4 290	3 780	3 260	2 740	2 190	3 510	3 090	2 680
8 500	4 180	3 690	3 180	2 670	2 140	3 380	2 980	2 580
9 000	4 070	3 590	3 090	2 600	2 080	3 240	2 860	2 480
9 500	3 940	3 480	3 000	2 520	2 020	3 090	2 730	2 370
10 000	3 810	3 360	2 900	2 440	1 960	2 930	2 590	2 250
10 500	3 670	3 240	2 800	2 350	1 890	2 770	2 450	2 120
PROPERTIES AND DESIGN DATA								
Area (mm^2)	15 700	13 800	11 900	9 950	7 980	13 700	12 000	10 400
Z ($10^3 mm^3$)	1 970	1 740	1 500	1 260	1 020	1 490	1 320	1 140
S ($10^3 mm^3$)	1 500	1 330	1 150	972	786	1 130	1 010	873
r (mm)	139	140	140	141	141	121	122	122
M_r (kN•m)	621	548	472	397	248	469	416	359
IMPERIAL SIZE AND MASS								
Mass (lb./ft.)	82.9	72.9	62.7	52.5	42.1	72.2	63.5	54.7
Thickness (in.)	.500	.438	.375	.313	.250	.500	.438	.375
Size (in.)	16 OD					14 OD		

G40.21-M 350W
CLASS H
$\phi = 0.90$

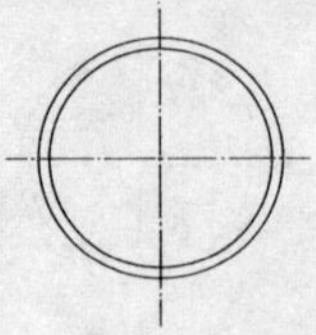

ROUND HOLLOW SECTIONS
Factored Axial Compressive Resistances, C_r, in kN

Designation (mm x mm)	HSS 356		HSS 324				
	8.0	6.4	13	11	9.5	8.0	6.4
Mass (kg/m)	68.2	54.7	97.5	85.9	73.9	61.9	49.7
Effective length (KL) in millimetres with respect to least radius of gyration							
0	2 730	2 200	3 910	3 430	2 960	2 490	1 990
1 000	2 730	2 200	3 910	3 430	2 960	2 490	1 990
1 250	2 730	2 200	3 910	3 430	2 960	2 490	1 990
1 500	2 730	2 200	3 910	3 430	2 960	2 490	1 990
1 750	2 730	2 200	3 900	3 430	2 960	2 480	1 990
2 000	2 730	2 190	3 900	3 430	2 960	2 480	1 990
2 250	2 730	2 190	3 890	3 420	2 950	2 480	1 990
2 500	2 720	2 190	3 880	3 410	2 950	2 470	1 980
2 750	2 720	2 180	3 870	3 400	2 940	2 460	1 980
3 000	2 710	2 180	3 850	3 390	2 920	2 450	1 970
3 250	2 700	2 170	3 830	3 370	2 910	2 440	1 960
3 500	2 690	2 160	3 810	3 350	2 890	2 430	1 950
3 750	2 680	2 150	3 790	3 330	2 880	2 410	1 940
4 000	2 660	2 140	3 760	3 310	2 860	2 400	1 920
4 250	2 650	2 130	3 730	3 290	2 840	2 380	1 910
4 500	2 630	2 110	3 700	3 260	2 810	2 360	1 890
4 750	2 610	2 100	3 670	3 230	2 790	2 340	1 880
5 000	2 590	2 080	3 630	3 200	2 760	2 320	1 860
5 250	2 570	2 070	3 590	3 160	2 730	2 290	1 840
5 500	2 550	2 050	3 550	3 130	2 700	2 270	1 820
5 750	2 530	2 030	3 500	3 090	2 670	2 240	1 800
6 000	2 500	2 010	3 460	3 050	2 630	2 210	1 780
6 500	2 440	1 960	3 350	2 960	2 560	2 150	1 720
7 000	2 380	1 910	3 240	2 860	2 470	2 080	1 670
7 500	2 320	1 860	3 120	2 760	2 380	2 010	1 610
8 000	2 240	1 800	2 980	2 640	2 280	1 920	1 540
8 500	2 160	1 740	2 840	2 520	2 170	1 840	1 470
9 000	2 080	1 670	2 680	2 380	2 060	1 740	1 400
9 500	1 990	1 600	2 520	2 240	1 930	1 640	1 320
10 000	1 890	1 520	2 330	2 090	1 800	1 530	1 230
10 500	1 790	1 440	2 140	1 910	1 650	1 400	1 130
PROPERTIES AND DESIGN DATA							
Area (mm^2)	8 680	6 970	12 400	10 900	9 410	7 890	6 330
Z ($10^3 mm^3$)	961	775	1 230	1 090	942	794	640
S ($10^3 mm^3$)	738	598	930	827	719	608	493
r (mm)	123	123	110	111	111	112	112
M_r (kN•m)	303	188	387	343	297	250	202
IMPERIAL SIZE AND MASS							
Mass (lb./ft.)	45.8	36.8	65.5	57.7	49.6	41.6	33.4
Thickness (in.)	.313	.250	.500	.438	.375	.313	.250
Size (in.)	14 OD		12.75 OD				

ROUND HOLLOW SECTIONS
Factored Axial Compressive Resistances, C_r, in kN

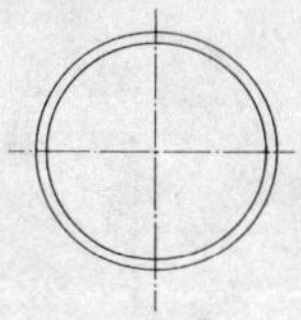

G40.21-M 350W
CLASS
$\phi = 0.9$

Designation (mm x mm)	HSS 273					HSS 219		
	13	11	9.5	8.0	6.4	13	11	9.5
Mass (kg/m)	81.6	71.9	61.9	52.0	41.8	64.6	57.1	49.3
Effective length (KL) in millimetres with respect to least radius of gyration								
0	3 280	2 890	2 490	2 090	1 680	2 590	2 290	1 980
1 000	3 280	2 890	2 490	2 090	1 680	2 590	2 290	1 980
1 250	3 280	2 890	2 490	2 090	1 680	2 590	2 290	1 970
1 500	3 270	2 880	2 480	2 080	1 670	2 580	2 280	1 970
1 750	3 270	2 880	2 480	2 080	1 670	2 570	2 270	1 960
2 000	3 260	2 870	2 470	2 070	1 670	2 560	2 260	1 950
2 250	3 250	2 860	2 460	2 070	1 660	2 540	2 240	1 930
2 500	3 230	2 850	2 450	2 060	1 650	2 510	2 220	1 920
2 750	3 210	2 830	2 440	2 050	1 650	2 490	2 200	1 900
3 000	3 190	2 810	2 420	2 030	1 640	2 460	2 170	1 870
3 250	3 170	2 790	2 400	2 020	1 620	2 420	2 140	1 850
3 500	3 140	2 770	2 380	2 000	1 610	2 380	2 110	1 820
3 750	3 110	2 740	2 360	1 980	1 590	2 340	2 070	1 790
4 000	3 070	2 710	2 340	1 960	1 580	2 290	2 030	1 750
4 250	3 040	2 680	2 310	1 940	1 560	2 240	1 990	1 720
4 500	3 000	2 640	2 280	1 910	1 540	2 190	1 940	1 680
4 750	2 950	2 600	2 250	1 890	1 520	2 130	1 890	1 630
5 000	2 900	2 560	2 210	1 860	1 500	2 070	1 830	1 590
5 250	2 860	2 520	2 170	1 830	1 470	2 000	1 770	1 540
5 500	2 800	2 470	2 140	1 800	1 450	1 930	1 710	1 490
5 750	2 750	2 420	2 090	1 760	1 420	1 850	1 650	1 430
6 000	2 690	2 370	2 050	1 730	1 390	1 780	1 580	1 370
6 500	2 560	2 260	1 960	1 650	1 330	1 610	1 430	1 250
7 000	2 420	2 140	1 850	1 560	1 260	1 410	1 260	1 100
7 500	2 270	2 010	1 740	1 470	1 190	1 240	1 110	974
8 000	2 100	1 860	1 620	1 370	1 110	1 110	992	868
8 500	1 910	1 700	1 480	1 250	1 020	998	892	781
9 000	1 720	1 530	1 330	1 130	917	905	809	707
9 500	1 560	1 390	1 210	1 020	831	826	738	645
10 000	1 430	1 270	1 100	935	759	757	677	592
10 500	1 310	1 160	1 010	858	696	688	616	540
PROPERTIES AND DESIGN DATA								
Area (mm^2)	10 400	9 160	7 890	6 620	5 320	8 230	7 270	6 270
Z ($10^3 mm^3$)	862	764	662	559	452	542	482	419
S ($10^3 mm^3$)	646	577	502	427	347	402	360	315
r (mm)	92.2	92.7	93.2	93.8	94.3	73.1	73.6	74.2
M_r (kN•m)	272	241	209	176	142	171	152	132
IMPERIAL SIZE AND MASS								
Mass (lb./ft.)	54.8	48.3	41.6	34.9	28.1	43.4	38.4	33.1
Thickness (in.)	.500	.438	.375	.313	.250	.500	.438	.375
Size (in.)	10.75 OD					8.625 OD		

G40.21-M 350W
CLASS H
$\phi = 0.90$

ROUND HOLLOW SECTIONS
Factored Axial Compressive Resistances, C_r, in kN

Designation (mm x mm)	HSS 219			HSS 168			
	8.0	6.4	4.8	9.5	8.0	6.4	4.8
Mass (kg/m)	41.4	33.3	25.3	37.3	31.4	25.4	19.3
Effective length (KL) in millimetres with respect to least radius of gyration							
0	1 660	1 340	1 010	1 500	1 260	1 020	775
1 000	1 660	1 340	1 010	1 490	1 260	1 020	774
1 250	1 660	1 330	1 010	1 490	1 250	1 010	771
1 500	1 650	1 330	1 010	1 480	1 240	1 010	766
1 750	1 650	1 330	1 010	1 460	1 230	996	759
2 000	1 640	1 320	1 000	1 440	1 220	984	750
2 250	1 630	1 310	995	1 420	1 200	970	740
2 500	1 610	1 300	987	1 400	1 180	953	727
2 750	1 600	1 290	977	1 370	1 150	934	713
3 000	1 580	1 270	966	1 330	1 130	912	697
3 250	1 560	1 250	953	1 300	1 100	888	679
3 500	1 530	1 240	940	1 260	1 060	861	658
3 750	1 510	1 210	924	1 210	1 030	832	637
4 000	1 480	1 190	907	1 160	985	800	613
4 250	1 450	1 170	889	1 110	942	766	587
4 500	1 410	1 140	869	1 050	895	729	560
4 750	1 380	1 110	848	991	845	689	530
5 000	1 340	1 080	826	927	792	647	499
5 250	1 300	1 050	802	851	731	599	464
5 500	1 260	1 020	776	782	671	551	426
5 750	1 210	981	749	722	620	508	393
6 000	1 160	943	721	669	574	471	364
6 500	1 060	862	660	582	499	409	316
7 000	939	766	589	512	439	360	278
7 500	829	676	520	456	391	320	247
8 000	739	603	463	404	348	285	221
8 500	664	542	416	359	309	254	197
9 000	601	490	377	322	277	227	176
9 500	548	447	343	290	249	205	159
10 000	503	410	315	263	226	186	144
10 500	460	376	289	240	206	169	131

PROPERTIES AND DESIGN DATA

Area (mm^2)	5 270	4 240	3 220	4 750	4 000	3 230	2 460
Z ($10^3 mm^3$)	355	288	220	241	205	167	128
S ($10^3 mm^3$)	269	219	169	179	153	126	97.6
r (mm)	74.7	75.3	75.8	56.2	56.8	57.3	57.8
M_r (kN•m)	112	90.7	69.3	75.9	64.6	52.6	40.3

IMPERIAL SIZE AND MASS

Mass (lb./ft.)	27.8	22.4	17.0	25.1	21.1	17.0	13.0
Thickness (in.)	.313	.250	.188	.375	.313	.250	.188
Size (in.)	8.625 OD			6.625 OD			

ROUND HOLLOW SECTIONS Factored Axial Compressive Resistances, C_r, in kN

G40.21-M 350W CLASS
$\phi = 0.90$

Designation (mm x mm)	HSS 141 9.5	HSS 141 8.0	HSS 141 6.4	HSS 141 4.8	HSS 114 8.0	HSS 114 6.4	HSS 114 4.8
Mass (kg/m)	31.0	26.1	21.1	16.1	20.9	16.9	12.9
Effective length (KL) in millimetres with respect to least radius of gyration							
0	1 240	1 050	847	646	838	677	517
1 000	1 240	1 040	844	643	827	669	511
1 250	1 230	1 040	837	638	815	659	504
1 500	1 210	1 020	828	631	797	646	494
1 750	1 190	1 010	815	622	775	628	481
2 000	1 170	988	800	611	748	607	465
2 250	1 140	965	781	597	717	582	447
2 500	1 110	937	760	581	680	554	426
2 750	1 070	906	736	563	639	521	402
3 000	1 030	871	708	542	593	485	375
3 250	980	832	678	520	542	445	346
3 500	929	789	644	495	482	399	313
3 750	872	743	607	467	425	352	276
4 000	811	693	568	438	379	314	246
4 250	742	638	525	406	341	282	221
4 500	669	575	475	369	308	255	200
4 750	607	521	431	335	281	232	182
5 000	554	476	393	305	258	213	167
5 250	508	437	361	280	236	196	154
5 500	469	402	332	258	216	179	141
5 750	434	373	308	239	198	164	129
6 000	404	347	286	222	182	151	119
6 500	351	302	250	194	156	130	102
7 000	304	262	217	168	136	113	88
7 500	266	229	190	147	119	99	78
8 000	235	202	167	130			
8 500	209	180	149	116			
9 000	188	162	134	104			
9 500			121	94			
10 000							
10 500							

PROPERTIES AND DESIGN DATA

	HSS 141 9.5	HSS 141 8.0	HSS 141 6.4	HSS 141 4.8	HSS 114 8.0	HSS 114 6.4	HSS 114 4.8
Area (mm^2)	3 950	3 330	2 690	2 050	2 660	2 150	1 640
Z (10^3mm^3)	166	142	116	89.1	90.1	74.1	57.4
S (10^3mm^3)	122	105	86.9	67.7	66.1	55.1	43.2
r (mm)	46.7	47.2	47.8	48.3	37.7	38.2	38.8
M_r (kN•m)	52.3	44.7	36.5	28.1	28.4	23.3	18.1

IMPERIAL SIZE AND MASS

Mass (lb./ft.)	20.8	17.6	14.2	10.8	14.0	11.4	8.68
Thickness (in.)	.375	.313	.250	.188	.313	.250	.188
Size (in.)	5.562 OD				4.5 OD		

G40.21-M 350W
CLASS H
$\phi = 0.90$

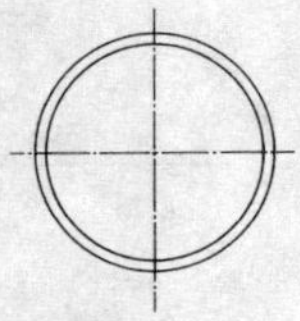

ROUND HOLLOW SECTIONS
Factored Axial Compressive Resistances, C_r, in kN

Designation (mm x mm)	HSS 102				HSS 89			
	8.0	6.4	4.8	3.8	8.0	6.4	4.8	3.8
Mass (kg/m)	18.4	14.9	11.4	9.19	15.9	12.9	9.92	8.00
Effective length (KL) in millimetres with respect to least radius of gyration								
0	737	598	457	369	636	520	397	321
500	737	598	457	369	635	519	396	321
750	732	595	454	367	629	514	393	318
1 000	722	587	449	362	616	504	385	312
1 250	707	575	440	355	597	489	375	304
1 500	686	559	428	346	571	469	360	292
1 750	660	538	413	334	540	444	342	278
2 000	628	514	394	319	502	415	320	260
2 250	591	485	373	302	458	380	294	240
2 500	548	451	348	283	408	340	265	217
2 750	500	414	321	261	348	293	231	190
3 000	445	372	290	236	297	250	197	163
3 250	385	323	253	207	258	217	171	141
3 500	337	283	221	182	227	191	150	124
3 750	299	250	196	161	202	170	133	110
4 000	267	224	175	143	180	152	120	99
4 250	241	202	158	129	160	135	107	88
4 500	219	183	143	117	144	121	96	79
4 750	197	166	130	107	129	109	86	71
5 000	179	150	118	97	117	99	78	64
5 250	162	137	107	88	107	90	71	59
5 500	149	125	98	80	98	83	65	54
5 750	136	115	90	74	90	76	60	49
6 000	126	106	83	68				46
6 500	108	91	71	58				
7 000								
7 500								
8 000								
8 500								
9 000								
9 500								
PROPERTIES AND DESIGN DATA								
Area (mm^2)	2 340	1 900	1 450	1 170	2 020	1 650	1 260	1 020
Z ($10^3 mm^3$)	69.9	57.7	44.8	36.5	52.3	43.4	33.9	27.6
S ($10^3 mm^3$)	50.8	42.6	33.6	27.6	37.6	31.7	25.2	20.8
r (mm)	33.2	33.8	34.3	34.6	28.8	29.3	29.8	30.1
M_r (kN•m)	22.0	18.2	14.1	11.5	16.5	13.7	10.7	8.69
IMPERIAL SIZE AND MASS								
Mass (lb./ft.)	12.3	10.0	7.67	6.17	10.7	8.69	6.66	5.37
Thickness (in.)	.313	.250	.188	.150	.313	.250	.188	.150
Size (in.)	4 OD				3.5 OD			

ROUND HOLLOW SECTIONS
Factored Axial Compressive Resistances, C_r, in kN

G40.21-M 350W
CLASS H
ϕ = 0.90

Designation (mm x mm)	HSS 73				HSS 60			
	6.4	4.8	3.8	3.2	6.4	4.8	3.8	3.2
Mass (kg/m)	10.4	8.04	6.50	5.48	8.45	6.54	5.31	4.48
Effective length (KL) in millimetres with respect to least radius of gyration								
0	419	321	261	220	340	263	213	180
500	417	320	260	219	336	260	211	178
750	409	314	255	215	325	252	204	173
1 000	395	304	247	209	306	238	193	164
1 250	375	289	236	199	279	218	178	151
1 500	349	270	221	187	245	193	159	135
1 750	316	246	202	171	202	163	135	116
2 000	278	218	180	153	159	129	107	92
2 250	231	184	153	131	129	104	87	75
2 500	191	152	126	108	108	87	72	62
2 750	162	129	107	91	91	74	62	53
3 000	139	111	92	79	77	62	52	45
3 250	122	97	80	69	66	53	45	38
3 500	105	84	70	60	57	46	39	33
3 750	92	74	61	52	50	41	34	29
4 000	81	65	54	46				26
4 250	73	58	48	41				
4 500	65	52	43	37				
4 750		47	39	33				
5 000								
5 250								
5 500								
5 750								
6 000								
6 500								
7 000								
7 500								
8 000								
8 500								
9 000								
9 500								
PROPERTIES AND DESIGN DATA								
Area (mm^2)	1 330	1 020	828	698	1 080	834	676	571
Z ($10^3 mm^3$)	28.3	22.3	18.3	15.5	18.6	14.8	12.2	10.4
S ($10^3 mm^3$)	20.4	16.4	13.6	11.7	13.2	10.7	8.99	7.74
r (mm)	23.7	24.2	24.5	24.7	19.2	19.7	20.0	20.2
M_r (kN•m)	8.91	7.02	5.76	4.88	5.86	4.66	3.84	3.28
IMPERIAL SIZE AND MASS								
Mass (lb./ft.)	7.01	5.40	4.37	3.68	5.68	4.40	3.57	3.01
Thickness (in.)	.250	.188	.150	.125	.250	.188	.150	.125
Size (in.)	2.875 OD				2.375 OD			

G40.21-M 350W
CLASS H
$\phi = 0.90$

ROUND HOLLOW SECTIONS
Factored Axial Compressive Resistances, C_r, in kN

Designation (mm x mm)	HSS 48				HSS 42		HSS 33		HSS 27	
	4.8	3.8	3.2	2.8	3.2	2.5	3.2	2.5	3.2	2.5
Mass (kg/m)	5.13	4.18	3.54	3.13	3.06	2.48	2.37	1.93	1.84	1.51
Effective length (KL) in millimetres with respect to least radius of gyration										
0	206	168	142	126	123	100	95	77	74	61
500	201	164	139	123	118	96	88	72	63	53
750	189	155	131	116	109	89	75	62	46	39
1 000	170	140	119	106	94	77	54	46	27	23
1 250	144	119	102	91	74	62	36	31	18	16
1 500	110	93	80	72	53	45	26	22	13	11
1 750	84	70	61	54	41	34	20	17		
2 000	66	56	48	43	32	27	15	13		
2 250	54	45	39	35	26	22				
2 500	44	37	32	29	21	18				
2 750	36	31	27	24	17	15				
3 000	31	26	23	20						
3 250										
3 500										
3 750										
4 000										
4 250										
4 500										
4 750										
5 000										
5 250										
5 500										
5 750										
6 000										
6 500										
7 000										
7 500										
8 000										
8 500										
9 000										
9 500										
PROPERTIES AND DESIGN DATA										
Area (mm^2)	654	533	451	399	390	316	302	246	235	193
Z ($10^3 mm^3$)	9.09	7.56	6.48	5.79	4.85	4.00	2.91	2.42	1.77	1.49
S ($10^3 mm^3$)	6.48	5.50	4.77	4.29	3.54	2.96	2.09	1.77	1.24	1.07
r (mm)	15.5	15.8	16.0	16.1	13.8	14.1	10.7	10.9	8.39	8.59
M_r (kN•m)	2.86	2.38	2.04	1.82	1.53	1.26	0.917	0.762	0.558	0.469
IMPERIAL SIZE AND MASS										
Mass (lb./ft.)	3.45	2.81	2.38	2.10	2.06	1.67	1.59	1.30	1.24	1.02
Thickness (in.)	1.88	.150	.125	.110	.125	.100	.125	.100	.125	.100
Size (in.)	1.9 OD				1.66 OD		1.315 OD		1.05 OD	

DESIGN OF BEAM-COLUMNS

General

For members which are subjected to both axial compression and bending (usually referred to as beam-columns), Clause 13.8 of CAN/CSA-S16.1-M89, *Limit States Design of Steel Structures* is applicable, along with Appendix F of that Standard.

Clause 13.8.1 applies to all sections except Class 1 of I-shaped members; and Clause 13.8.2 applies only to the Class 1 sections of I-shaped members. If a more detailed design of W or HSS shaped Class 1 or 2 sections is required than that given in Clause 13.8, Appendix F is available. The appendix takes advantage of the redistribution of stress after initiation of yielding under specified loads.

In each of Clauses 13.8.1 and 13.8.2, a single comprehensive axial-flexural interaction expression applies for checking cross-sectional strength, overall member strength and lateral torsional buckling strength of the member. The first and last of these three checks correspond to those in S16.1-M84 for strength and stability, while the check for overall member strength applies to members which have sufficient lateral bracing to be torsionally stable, or which are inherently stable like HSS. The three checks involve the use of different values for the variables in the expressions, as outlined below.

For all sections except Class 1 of I-shaped members, the following design expression applies (Clause 13.8.1).

$$\frac{C_f}{C_r} + \frac{U_{1x} M_{fx}}{M_{rx}} + \frac{U_{1y} M_{fy}}{M_{ry}} \leq 1.0 \qquad [1]$$

where

M_{fx}, M_{fy} = maximum factored end moments as defined in Clause 8.6, *Stability Effects*, to include frame sway effects.

C_f = factored axial load.

When checking cross-sectional strength:

C_r = factored axial compressive resistance of the member as defined in Clause 13.3, *Axial Compression*, with $\lambda = 0$.

M_{rx}, M_{ry} = factored moment resistances of the member as defined in Clause 13.5, *Bending — Laterally Supported Members*, (for the appropriate class of section).

U_{1x}, U_{1y} = 1.0.

When checking overall member strength:

C_r = factored axial compressive resistance of the member as defined in Clause 13.3, *Axial Compression*, with K = 1.0, based on the greater slenderness ratio, L_x/r_x or L_y/r_y. For uniaxial strong axis bending, C_{rx} may be used.

M_{rx}, M_{ry} = factored moment resistances of the member as defined in Clause 13.5, *Bending — Laterally Supported Members*, (for the appropriate class of section).

$$U_1 = \frac{\omega_1}{1 - \left(\frac{C_f}{C_e}\right)} \qquad [2]$$

and is a term which includes factors to account for 1) the effects of a varying moment along the member, and 2) the Pδ moment application which accounts for the effects of member curvature.

ω_1 = coefficient used to determine the bending effect which is equivalent to that derived from an ideal uniform bending moment that approximates the effect due to the actual design moments.

C_e = Euler buckling load equal to $\pi^2 EI/(KL)^2$ where KL is the effective column length in the plane of buckling.

When checking lateral torsional buckling strength:

C_r = factored axial compressive resistance of the member as defined in Clause 13.3, *Axial Compression*, with K = 1.0, based on weak-axis buckling.

M_{rx} = factored moment resistance of the member as defined in Clause 13.6, *Bending — Laterally Unsupported Members*, (for the appropriate class of section).

M_{ry} = factored moment resistance of the member as defined in Clause 13.5, *Bending — Laterally Supported Members*, (for the appropriate class of section).

U_{1x} = as defined for the overall member strength check, but not less than 1.0.

U_{1y} = as defined for the overall member strength check.

For Class 1 sections of I-shaped members, the following design expression applies (Clause 13.8.2):

$$\frac{C_f}{C_r} + \frac{0.85\,U_{1x}\,M_{fx}}{M_{rx}} + \frac{0.60\,U_{1y}\,M_{fy}}{M_{ry}} \leq 1.0 \qquad [3]$$

in which the coefficients 0.85 and 0.60 reflect the increased bending resistance of Class 1 sections. Other terms are the same as for [1].

The member must also satisfy the condition:

$$\frac{M_{fx}}{M_{rx}} + \frac{M_{fy}}{M_{ry}} \leq 1.0 \qquad [4]$$

The design of a beam-column may be controlled by any of the three failure modes. Often the requirement guarding against one failure mode governs for one loading condition, while the requirement against another mode governs for a different loading condition.

The value of ω_1, (ω_{1x} or ω_{1y}), depends upon the direction and magnitude of the end moments, and any presence of transverse loads. The value of ω_1 may be determined by analysis, or the following values may be used:

(a) for members not subjected to transverse loads between supports,

$\omega_1 = 0.6 - 0.4\kappa$ but not less than 0.4 (see Table 4–7, page 4–24)

where

$\kappa = M_{f1}/M_{f2}$ for moments at opposite ends of the unbraced length, positive for double curvature and negative for single curvature, in which

M_{f1} = the smaller factored end moment, and

M_{f2} = the larger factored end moment.

(b) for members subjected to distributed loads or a series of point loads between supports,

ω_1 = 1.0.

(c) for members subjected to a concentrated load or moment between supports,

$\omega_1 = 0.85$.

The member may be considered to be divided into two segments at the point of load (or moment) application. Each segment is then treated as a member which depends on its own flexural stiffness to prevent side-sway in the plane of bending considered. In computing the slenderness ratio KL /r for use in [1] or [3], the total length of the member is used.

It is important to remember that the factored axial compressive resistance, C_r, in [1] and [3] is determined as though only compressive loads were present. Therefore, it is always the value associated with the maximum effective slenderness ratio, regardless of the axis about which bending occurs, while C_e is always computed using the slenderness ratio in the plane of bending.

References

Additional information on beam-columns can be obtained from:

PART TWO of this Handbook.

KENNEDY, D.J.L., PICARD, A., and BEAULIEU, D. 1990. New Canadian provisions for the design of steel beam-columns. Canadian Journal of Civil Engineering, **17** (6), December.

KULAK, G.L., ADAMS, P.F., and GILMOR, M.I. 1990. Limit states design in structural steel. Canadian Institute of Steel Construction, Willowdale, Ontario.

PICARD, A., and BEAULIEU, D. 1991. Calcul des charpentes d'acier, 2nd edition. Canadian Institute of Steel Construction, Willowdale, Ontario.

SSRC. 1988. Guide to stability design criteria for metal structures, 4th edition. Structural Stability Research Council. John Wiley, New York, N.Y.

CISC. Column selection program. Canadian Institute of Steel Construction, Willowdale, Ontario.

Material provided in this Handbook should help the designer minimize design time when a computer is not available. Designers with access to IBM-compatible computers should become familiar with the CISC-CSCC Column Selection Program. Other programmable devices can substantially reduce the time required by the designer to select a suitable beam-column. Contact CISC for more information.

Design

The selection of a suitable steel section to use as a beam-column is a trial-and-error procedure. A trial section must be selected and then its suitability checked in accordance with the appropriate design expressions [1],[3] or [4]. The procedure is then repeated if the initial choice is not suitable.

The following terms are a result of the analysis for forces and moments:

C_f = factored axial load,

M_{fx}, M_{fy} = factored bending moments.

The following terms can be determined from tables in this Handbook:

C_r = factored axial compressive resistance, ($\leq \phi A F_y$), or,

= value in tables of Factored Axial Compressive Resistances of columns for the appropriate effective length, page 4–30*,

M_{rx} = factored moment resistance for X–X axis bending,

= tabulated value with properties in tables of Factored Axial Compressive Resistances of columns when the unsupported length of compression flange $L \leq L_u$, page 4–30*, or,

= tabulated in Factored Moment Resistances for Columns table under the appropriate heading for length of compression flange which is unsupported, or may be interpolated from these values, page 4–105*,

M_{ry} = factored moment resistance for Y–Y axis bending

= tabulated value with properties in the Column tables, page 4–30*,

ω_{1x}, ω_{1y} = equivalent uniform bending factors to account for moment gradient

= tabulated in Table 4–7, page 4–24,

U = $\dfrac{1}{1-\left(\dfrac{C_f}{C_e}\right)}$, an amplification factor to account for secondary effects of axial force acting on the deformed member, ($U_1 = \omega_1 U$)

= tabulated in Table 4–9, page 4–26, values for various C_f/C_e ratios,

C_e = Euler buckling load

= tabulated as C_e/A for KL/r ratios from 1 to 200 in Table 4–8, page 4–25.

* First page of the series of like tables.

Trial Selection by Equivalent-Loads Method

Any convenient method can be used to choose a trial section. One method is that of equivalent-loads. Basically, this method involves converting applied loads and bending moments into equivalent total axial loads computed from the "cross-sectional strength" requirement and from the "lateral torsional buckling strength" requirement. The trial section is chosen for the larger equivalent load from the Factored Axial Compressive Resistances column tables (pages 4–30 to 4–91), and is then checked using the appropriate expression [1] or [3]. This method is applicable only to prismatic members governed by [1] or [3].

The "overall member strength" requirement need not be used when making trial selections for I-shaped members because it seldom governs for those members; but that check is still made when confirming the section.

Beam-Columns Without Transverse Loads Between Brace Points

Cross-Sectional Strength Requirement

$$C_f' = C_f + B\,M_{f2} \qquad [5]$$

where

C_f' = equivalent total factored axial load for cross-sectional strength (kN)

C_f = applied factored axial load (kN)

B = appropriate bending factor (m^{-1}) chosen from Table 4–6 (page 4–22)

M_{f2} = larger applied factored end moment (kN·m)

Lateral Torsional Buckling Strength Requirement

$$C_f'' = C_f + (\omega_1 U) M_{f2} \left(B \frac{C_{rL}}{C_{r0}} \right) \qquad [6]$$

where

C_f'' = equivalent total factored axial load for lateral torsional buckling (kN)

ω_1 = equivalent uniform bending factor

C_{rL} = factored axial compressive resistance at the actual effective length for the trial section chosen (kN)

C_{r0} = factored axial compressive resistance at an effective length of zero millimetres for the trial section chosen (kN)

$U = \dfrac{1}{1 - \left(\dfrac{C_f}{C_e} \right)}$

an amplification factor listed in Table 4–9 for various combinations of C_f / C_e in the plane of bending, where $\omega_1 U$ must be ≥ 1.0.

C_f, M_{f2} and B are the same as for [5].

Multiplying the larger end moment, M_{f2}, with the value of ω_1, produces a uniform moment for the unbraced segment, which would have the same effect on the beam-column as the applied end moments so far as lateral torsional buckling strength is concerned.

The expression C_{rL} / C_{r0} in [6] modifies the factor B to account for the difference between the capacity of the beam-column along the unbraced length and that at a brace location. If [6] is used with moments at the brace location, then $C_{rL} = C_{r0}$, and the ratio C_{rL} / C_{r0} can be deleted.

The other refinement which [6] provides (as compared to [5]) is the introduction of the amplification factor, U. This factor takes into account the increased second-order moment produced by the applied load as the curvature of the column under load increases. Accordingly, U increases rapidly as the length of the column (and hence the effective slenderness ratio in the plane of bending increases) and/or as the applied axial load increases.

When bending about two axes is involved, the second term of [5] and [6] should be expanded to two terms using variables appropriate to each axis.

Although [6] looks cumbersome, it is easily used since all the unknowns in the equation can be picked directly from tables.

However, [6] can be simplified with a possible reduction in accuracy, by assuming that the ratio $C_{rL} / C_{r0} = 0.70$. This is an average value for 300 MPa yield steel columns with KL / r from 40 to 80.

Beam-Columns With Transverse Loads Between Brace Points

For beam-columns subject to transverse loading between the brace locations, [5] satisfies the cross-sectional strength requirement if the larger end moment is greater than the maximum bending moment along the unbraced length.

The lateral torsional buckling strength requirement is satisfied by a modified form of Eqn. 6 identified as Eqn. 6a. If the maximum bending moment along the unbraced length is greater than the larger end moment, Eqn. 6a will satisfy both the cross-sectional strength and the lateral torsional buckling strength requirements for the beam-column.

$$C_f'' = C_f + U\,M_{f2}'\left(B\,\frac{C_{rL}}{C_{r0}}\right) \qquad [6a]$$

M_{f2}' = the greater of the maximum bending moment along the unbraced length, or the larger end moment.

The other symbols are the same as for [6].

Suggested Design Procedure

A. *Trial Section Selection*

1. Choose a nominal size group (WWF500, W310, etc.) from which the trial section will be chosen and select the appropriate bending factor or factors from Table 4–6.
2. Use [5] to compute equivalent axial load for cross-sectional strength, C_f'.
3. Use C_f' to select an initial trial section from the tables of Factored Axial Compressive Resistances of columns. C_f' must not be greater than C_r in the tables at zero effective length.
4. Use [6] to compute equivalent axial load for lateral torsional buckling strength, C_f''. The value of ω_1 can be chosen from Table 4–7, and the value of B will be the same value used to establish C_f'. The values of C_{rL} and C_{r0} will be those listed in the table of Factored Axial Compressive Resistances of columns for the initial trial section chosen in step 3. The value of U will be selected from Table 4–9, using the properties of the initial trial section chosen in step 3; the product of $\omega_1 U$ must not be less than 1.0. Alternatively, an assumed value of C_{rL}/C_{r0} can be chosen such as 0.70. If the column is subject to transverse loads, use [6a] for this step.
5. Use the Factored Axial Compressive Resistance tables to verify that the initial trial section is adequate for C_f'' at the appropriate effective length, or choose a new trial section to satisfy C_f'' at the appropriate KL.

B. Suitability of the Trial Section

1. Use the properties and resistances of the trial section to check its suitability in accordance with the requirements of Clause 13.8 of S16.1-M89. Since most of the values used in these equations are tabulated this should be a quick computation.
2. If the trial section is unsuitable, choose a new trial section (which should be the next lighter or heavier section listed) and verify suitability.

Examples

1. Given:

Design a steel column for the third storey of a six storey building for the loading conditions shown. Moments are due to rigidly framed beams and gravity loading, and cause bending about the X–X axis of the column. The PΔ effects have been included in the analysis. Beams framing to the minor axis have flexible connections. Steel is G40.21-M 300W.

Solution:

$L = 3\,700$ mm $\quad M_{f1} = 200$ kN·m

$C_f = 2\,000$ kN $\quad M_{f2} = 300$ kN·m

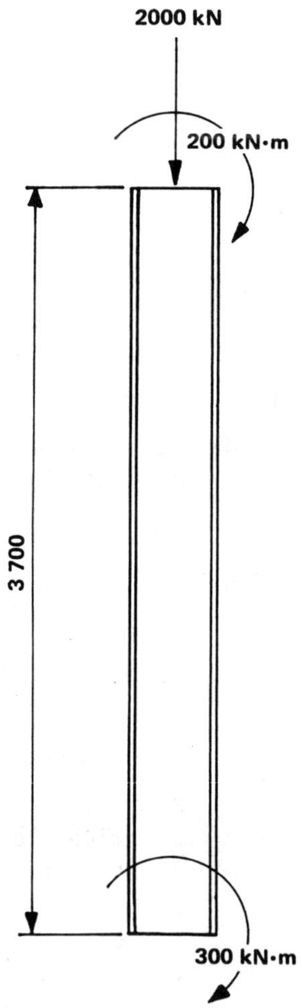

A. *Trial Section Selection*

Assume W310, Class 1 section

From Table 4–6 (page 4–22), for a Class 1 W shape, 310x310

$B_x = 6.34\ m^{-1}$

$C_f' = C_f + B_x M_{f2}$ [5]

$= 2\,000 + 6.34\,(300) = 3\,900$ kN

Select W310x129 from table on page 4–38

C_{r0} (for KL = 0) = 4 450 kN, comfortably > 3 900 kN

(A W310x118 with C_{r0} of 4 050 was tried,

but found inadequate.)

$A = 16\,500\ mm^2$, $r_x = 137$ mm, $r_y = 77.8$ mm, $r_x/r_y = 1.76$

$M_{rx} = 583$ kN·m, $L_u = 5\,580$ mm

$C_{rL} = 3\,750$ kN — by interpolation (KL = 3 700 mm)

$$C_f'' = C_f + (\omega_1 U)\, M_{f2}\left(B_x \frac{C_{rL}}{C_{r0}}\right) \quad [6]$$

$M_{f1} / M_{f2} = 200/300 = 0.67$ (double curvature)

From Table 4–7 (page 4–24), $\omega_1 = 0.4$

$KL_x/r_x = 1.0\,(3\,700/137) = 27.0$

From Table 4–8 (page 4–25), $C_e/A = 2\,702$ MPa

Then, $C_e = 2\,702\ \text{MPa} \times 16\,500\ mm^2 = 44\,600$ kN

and $C_f/C_e = 2\,000/44\,600 = 0.04$

From Table 4–9 (page 4–26), U = 1.04

Where $\omega_1 U < 1.0$, use $\omega_1 U = 1.0$

From Table 4–6 (page 4–22), $B_x = 6.34\ m^{-1}$ (as for C_f')

$C_f'' = 2\,000 + (1.0)\,300\,(6.34 \times 3\,750/4\,450)$

$= 2\,000 + 1600$

$= 3\,600$ kN $< 3\,750$ — OK $\qquad$ Use W310x129 as trial section.

B. Suitability of the Trial Section

Confirm class of section as a beam-column.

Although Table 5-1, Classes of Sections in Bending (page 5–8), lists the W310x129 as Class 1, the addition of axial load might change that class (as per "Webs in axial compression" line, Table 1 in Clause 11.2 of S16.1-M89). However, an examination of Table 4–3, Class of Section for Beam-Columns (page 4–9), shows that the W310x129 is not listed there which means that it is always a Class 1 section, and [3] applies:

$$\frac{C_f}{C_r} + \frac{0.85\,U_{1x}\,M_{fx}}{M_{rx}} \leq 1.0$$

i) Cross-sectional strength check

$C_f = 2\,000$ kN

$M_{fx} = 300$ kN·m

From Table of Factored Axial Compressive Resistances, page 4–38,

$C_r = C_{r0} = 4\,450$ kN, and $M_{rx} = 583$ kN.m

U_1 is taken as 1.0 for cross-sectional strength check

Therefore,

$$\frac{2\,000}{4\,450} + \frac{0.85 \times 1.0 \times 300}{583} = 0.449 + 0.437 = 0.886 \quad \leq 1.0 \quad \text{— OK}$$

ii) Overall member strength check

$KL_x/r_x = 1.0\,(3\,700/137) = 27.0$

From Table 4–4 (page 4–11), $C_{rx}/A = 255$ MPa, for $F_y = 300$ MPa

$C_r = C_{rx} = 255 \times 16\,500 / 10^3 = 4\,210$ kN

$U_1 = \omega_1 U = 0.4 \times 1.04 = 0.42$

Therefore,

$$\frac{2\,000}{4\,210} + \frac{0.85 \times 0.42 \times 300}{583} = 0.475 + 0.184 = 0.659 \quad \leq 1.0 \quad \text{— OK}$$

iii) Lateral torsional buckling strength check

$C_r = C_{ry} = C_{rL} = 3\,750$ kN (by interpolation, table on page 4-38)

$L = 3\,700$ mm $\quad L_u = 5\,580$ mm

Therefore, $M_{rx} = 583$ kN·m

$U_1 = \omega_1 U$ (but ≥ 1.0) $= 0.4 \times 1.04 \quad \therefore U_1 = 1.0$

Therefore,

$$\frac{2\,000}{3\,750} + \frac{0.85 \times 1.0 \times 300}{583} = 0.533 + 0.437 = 0.970 \quad \leq 1.0 \quad \text{— OK}$$

Use the W310x129 column section.

Comments:

1. C_r could more accurately be determined by computing the KL /r values and entering the tables of Unit Factored Compressive Resistances for the larger KL /r and multiplying that value by the area of the column.

2. When $L > L_u$ the tables on pages 4–113 to 4-115, Factored Moment Resistances of Columns, will be more useful. (Caution: if a column section changes from Class 2 to Class 3 on account of high axial loads, M_{rx} or M'_{rx} values need to be adjusted as noted on page 4-111.)

2. **Given:**

Same as example 1, except that beams framing to Y–Y axis have rigid connections and induce moments of 100 kN·m at each end of the column. The direction of the moments is such that double curvature is induced in the column.

Solution:

$L = 3\,700$ mm $M_{fx1} = 200$ kN·m $M_{fy1} = 100$ kN·m

$C_f = 2\,000$ kN $M_{fx2} = 300$ kN·m $M_{fy2} = 100$ kN·m

A. *Trial Section Selection*

Assume W310 section, Class 1

From Table 4–6 (page 4–22), for a Class 1 W shape, 310x310

$B_x = 6.34$ and $B_y = 9.78$

$$C_f' = C_f + B_x M_{fx2} + B_y M_{fy2} \qquad [5]$$

$$= 2\,000 + (6.34 \times 300) + (9.78 \times 100)$$

$$= 4\,880 \text{ kN}$$

Select W310x143 from table on page 4–38, a Class 1 section in G40.21-M 300W.

$C_{r0} = 4\,910$ kN > 4 880 — OK

$A = 18\,200\ \text{mm}^2$ $M_{rx} = 653$ kN·m

$r_x = 138$ mm $M_{ry} = 300$ kN·m

$r_y = 78.8$ mm

$C_{rL} = 4\,150$ kN — by interpolation (KL = 3 700 mm)

$$C_f'' = C_f + (\omega_{1x} U_x) M_{fx2} \left(B_x \frac{C_{rLx}}{C_{r0}}\right) + (\omega_{1y} U_y) M_{fy2} \left(B_y \frac{C_{rLy}}{C_{r0}}\right) \qquad [6]$$

$M_{fx1}/M_{fx2} = 200/300 = 0.67$ (double curvature)

$M_{fy1}/M_{fy2} = 100/100 = 1.00$ (double curvature)

From Table 4–7 (page 4–24), $\omega_{1x} = 0.40$, $\omega_{1y} = 0.40$

$KL_x/r_x = 1.0\,(3\,700/138) = 26.8$

$KL_y/r_y = 1.0\,(3\,700/78.8) = 47.0$

From Table 4–8 (page 4–25), $(C_e/A)_x = 2\,740$ MPa $(KL/r = 26.8)$

$(C_e/A)_y = 892$ MPa $(KL/r = 47.0)$

$$C_{ex} = 2\,740\ \text{MPa} \times 18\,200\ \text{mm}^2 = 49\,900 \times 10^3\ \text{N}$$

$$= 49\,900\ \text{kN}$$

$$C_{ey} = 892\ \text{MPa} \times 18\,200\ \text{mm}^2 = 16\,200 \times 10^3\ \text{N}$$

$$= 16\,200\ \text{kN}$$

$C_f / C_{ex} = 2\ 000/49\ 900 = 0.04$

$C_f / C_{ey} = 2\ 000/16\ 200 = 0.12$

From Table 4–9 (page 4–26), $U_x = 1.04$, $U_y = 1.14$

Since $\omega_{1x} U_x = U_{1x} < 1.0$, use 1.0

but $\omega_{1y} U_y = 0.4(1.14) = 0.456$

From Table 4–6, $B_x = 6.34\ m^{-1}$ and $B_y = 9.78\ m^{-1}$ (as for C_f')

$$C_f'' = 2\ 000 + (1.0)\ 300\ (6.34 \times 4\ 150 / 4\ 910)$$
$$+ (0.456)\ 100\ (9.78 \times 4\ 150 / 4\ 910)$$
$$= 3\ 980\ \text{kN}$$

Since C_r for L = 3 700 mm is 4 150 kN > 3 980 kN, the W310x143 satisfies the lateral torsional buckling strength requirement as well as the cross-sectional strength requirement.

Use W310x143 as a trial section.

B. *Suitability of the Trial Section*

This section is a heavier one of the same series as the W310x129 used in Example 1, and will also be Class 1. Therefore, [3] applies:

$$\frac{C_f}{C_r} + \frac{0.85\ U_{1x}\ M_{fx}}{M_{rx}} + \frac{0.60\ U_{1y}\ M_{fy}}{M_{ry}} \leq 1.0$$

i) *Cross-sectional strength check*

$C_f = 2\ 000$ kN

$M_{fx} = 300$ kN·m

$M_{fy} = 100$ kN·m

From Table of Factored Axial Compressive Resistances, page 4–38,

$C_r = C_{r0} = 4\ 910$ kN, $M_{rx} = 653$ kN·m and $M_{ry} = 300$ kN·m

U_{1x} and U_{1y} are taken as 1.0 for cross-sectional strength check

Therefore,

$$\frac{2\ 000}{4\ 910} + \frac{0.85 \times 1.0 \times 300}{653} + \frac{0.60 \times 1.0 \times 100}{300}$$
$$= 0.407 + 0.391 + 0.200 = 0.998 \quad \leq 1.0 \quad \text{— OK}$$

ii) *Overall member strength check*

$KL_x/r_x = 1.0\ (3\ 700/138) = 26.8$

$KL_y/r_y = 1.0\ (3\ 700/78.8) = 47.0$

From Table 4–4 (page 4–11), $C_r / A = 228$ MPa (use $KL/r = 47.0$)

$C_r = 228 \times 18\ 200 / 10^3 = 4\ 150$ kN

$$U_{1x} = \omega_{1x} U_x = 0.4 \times 1.04 = 0.42$$

$$U_{1y} = \omega_{1y} U_y = 0.4 \times 1.14 = 0.46$$

Therefore,

$$\frac{2\,000}{4\,150} + \frac{0.85 \times 0.42 \times 300}{653} + \frac{0.60 \times 0.46 \times 100}{300}$$

$$= 0.482 + 0.164 + 0.092 = 0.738 \quad \leq 1.0 \quad \text{— OK}$$

iii) Lateral torsional buckling check

$C_r = C_{ry} = C_{rL} = 4\,150$ kN

$L = 3\,700$ mm $\quad < L_u = 5\,820$ mm

Therefore, $M_{rx} = 653$ kN·m

$U_{1x} = \omega_{1x} U_x \quad (\text{but} \geq 1.0) = 0.4 \times 1.04 \quad \therefore U_{1x} = 1.0$

$U_{1y} = \omega_{1y} U_y = 0.4 \times 1.14 = 0.46$

Therefore,

$$\frac{2\,000}{4\,150} + \frac{0.85 \times 1.0 \times 300}{653} + \frac{0.60 \times 0.46 \times 100}{300}$$

$$= 0.482 + 0.391 + 0.092 = 0.965 \quad \leq 1.0 \quad \text{— OK}$$

Use the W310x143 column section.

Shear

Where beams with large end moments are connected to columns with thin webs, a check for shear capacity in the column web will be necessary.

FACTORED MOMENT RESISTANCES OF COLUMNS

The tables on pages 4–113 to 4–115 inclusive list: 1) the factored moment resistance for strong axis bending, M_{rx}, for cases where the unsupported length of compression flange, L, is less than L_u, and 2) the factored moment resistance, M'_{rx}, where L is greater than L_u. The listings in G40.21-M steel are for WWF, grade 350W shapes, and for WWF and W, grade 300W shapes normally used as columns. Sections are ordered as in Part Six of this book, all of the sections of the same nominal dimensions are listed together.

The M_{rx} and M'_{rx} values are based on the class of the section in bending about the X–X axis, without axial load. However, the class of a section used as a beam-column is a function of the ratio of the factored axial load to the axial compressive load at yield stress, C_f/C_y, in accordance with CAN/CSA-S16.1-M89 Clause 11, width-thickness criteria. For example, a W410x39 of G40.21-M grade 300W steel becomes a Class 3 section when C_f/C_y exceeds 0.64, based upon the S16.1-M89 Class 2 limit for h/w of

$$\frac{1700}{\sqrt{F_y}}\left(1-0.61\frac{C_f}{C_y}\right)$$

Table 4–3 on page 4–7 lists the classes for sections when they are used as beam-columns, and the values of C_f/C_y at which the class changes.

Thus, sections whose loading causes a change from Class 2 to Class 3 need to have their tabulated values of M_{rx} and M'_{rx} adjusted. A simplified method is to multiply the listed values by the factor S_x / Z_x.

G40.21-M 350W
$\phi = 0.90$

FACTORED MOMENT RESISTANCES ¤ OF COLUMNS, M_{rx} and M'_{rx} (kN·m)

Section	M_{rx}	M'_{rx} for the Following Unsupported Lengths in Millimetres									
		6 000	7 000	8 000	9 000	10 000	12 000	14 000	16 000	18 000	20 000
WWF650X864	8 600	....	....	....	....	....	....	8 480	8 240	8 000	7 760
WWF650X739	7 940	....	....	....	....	....	....	7 770	7 530	7 300	7 060
WWF650X598	6 650	....	....	....	....	....	6 580	6 340	6 090	5 840	5 600
WWF650X499	5 510	....	....	....	....	....	5 330	5 070	4 810	4 540	4 280
** WWF650X400	3 920	....	....	....	....	....	3 750	3 540	3 320	3 100	2 870
WWF600X793	7 210	....	....	....	....	....	....	7 090	6 890	6 700	6 500
WWF600X680	6 680	....	....	....	....	....	....	6 510	6 320	6 120	5 930
WWF600X551	5 610	....	....	....	....	....	5 510	5 300	5 100	4 890	4 690
WWF600X460	4 660	....	....	....	....	....	4 440	4 220	4 000	3 780	3 560
* WWF600X369	3 370	....	....	....	....	3 340	3 150	2 960	2 770	2 570	2 370
WWF550X721	5 980	....	....	....	....	....	....	5 870	5 710	5 550	5 390
WWF550X620	5 540	....	....	....	....	....	....	5 390	5 230	5 070	4 910
WWF550X503	4 660	....	....	....	....	....	4 550	4 380	4 210	4 050	3 880
WWF550X420	3 910	....	....	....	....	3 850	3 670	3 480	3 300	3 120	2 940
** WWF550X280	2 300	....	....	....	2 290	2 220	2 070	1 920	1 760	1 610	1 420
WWF500X651	4 850	....	....	....	....	....	....	4 760	4 630	4 510	4 380
WWF500X561	4 500	....	....	....	....	....	....	4 370	4 250	4 120	4 000
WWF500X456	3 810	....	....	....	....	....	3 700	3 560	3 430	3 300	3 160
WWF500X381	3 180	....	....	....	....	3 110	2 960	2 810	2 670	2 520	2 380
WWF500X343	2 870	....	....	....	2 820	2 740	2 590	2 430	2 280	2 130	1 980
WWF500X306	2 540	....	....	2 530	2 460	2 380	2 220	2 060	1 900	1 740	1 540
WWF500X276	2 340	....	....	2 320	2 250	2 170	2 020	1 860	1 690	1 520	1 340
* WWF500X254	1 940	....	....	....	1 880	1 820	1 680	1 550	1 410	1 270	1 110
** WWF500X223	1 630	....	....	....	1 580	1 530	1 410	1 290	1 160	1 030	892
** WWF500X197	1 350	....	....	....	1 330	1 280	1 190	1 090	983	869	752
WWF450X503	3 590	....	....	....	....	....	3 580	3 480	3 390	3 290	3 190
WWF450X409	3 040	....	....	....	....	....	2 940	2 830	2 730	2 630	2 530
WWF450X342	2 550	....	....	....	2 520	2 470	2 350	2 240	2 120	2 010	1 900
WWF450X308	2 300	....	....	2 290	2 230	2 170	2 050	1 930	1 810	1 690	1 570
WWF450X274	2 040	....	....	2 000	1 940	1 870	1 740	1 620	1 490	1 360	1 210
WWF450X248	1 880	....	....	1 830	1 770	1 700	1 580	1 450	1 320	1 180	1 040
WWF450X228	1 720	....	....	1 660	1 590	1 530	1 400	1 260	1 130	976	861
* WWF450X201	1 390	....	....	1 350	1 300	1 240	1 130	1 020	905	779	684
** WWF450X177	1 200	....	....	1 160	1 120	1 070	974	874	763	653	570
WWF400X444	2 760	....	....	....	....	....	....	2 690	2 620	2 540	2 470
WWF400X362	2 360	....	....	....	....	2 350	2 280	2 200	2 120	2 040	1 970
WWF400X303	1 980	....	....	....	1 950	1 900	1 820	1 730	1 640	1 560	1 480
WWF400X273	1 790	....	....	1 760	1 720	1 670	1 580	1 490	1 400	1 310	1 220
WWF400X243	1 590	....	1 580	1 540	1 490	1 440	1 340	1 240	1 140	1 040	932
WWF400X220	1 470	....	1 450	1 400	1 360	1 310	1 210	1 110	1 010	900	801
WWF400X202	1 340	....	1 310	1 260	1 210	1 160	1 060	959	846	739	656
* WWF400X178	1 080	....	1 060	1 020	982	940	854	769	673	585	518
* WWF400X157	983	....	961	921	879	836	749	662	560	484	427
WWF350X315	1 760	....	....	....	....	....	1 700	1 640	1 590	1 530	1 480
WWF350X263	1 490	....	....	....	1 460	1 420	1 360	1 300	1 240	1 180	1 110
WWF350X238	1 350	....	....	1 320	1 280	1 250	1 180	1 120	1 050	985	920
WWF350X212	1 200	....	1 180	1 140	1 110	1 070	999	928	858	783	701
WWF350X192	1 110	....	1 080	1 040	1 010	970	898	827	756	671	600
WWF350X176	1 010	....	973	935	897	859	784	710	625	549	490
WWF350X155	904	890	852	813	774	734	656	572	490	430	383
WWF350X137	816	798	760	722	682	643	565	475	405	354	315

¤ Based on class of section for X–X axis bending only, $\omega_2 = 1.0$.
* Class 3 flanges.
** Class 4 flanges.

FACTORED MOMENT RESISTANCES ¤ OF COLUMNS, M_{rx} and M'_{rx} (kN·m)

G40.21-M 300W
$\phi = 0.90$

Section	M_{rx}	M'_{rx} for the Following Unsupported Lengths in Millimetres							
		8 000	9 000	10 000	12 000	14 000	16 000	18 000	20 000
WWF650X864	7 370	....	....	....	....	....	7 260	7 090	6 910
WWF650X739	6 800	....	....	....	....	....	6 650	6 480	6 300
WWF650X598	5 700	....	....	....	....	5 590	5 410	5 230	5 050
WWF650X499	4 720	....	....	....	4 690	4 500	4 310	4 120	3 920
* WWF650X400	3 400	....	....	....	3 340	3 190	3 020	2 850	2 680
WWF600X793	6 180	....	....	....	....	....	6 080	5 940	5 800
WWF600X680	5 720	....	....	....	....	....	5 580	5 440	5 290
WWF600X551	4 810	....	....	....	....	4 680	4 540	4 380	4 240
WWF600X460	4 000	....	....	....	3 920	3 760	3 600	3 440	3 280
* WWF600X369	2 890	....	....	....	2 790	2 650	2 510	2 360	2 220
WWF550X721	5 130	....	....	....	....	....	5 040	4 920	4 800
WWF550X620	4 750	....	....	....	....	4 740	4 620	4 500	4 390
WWF550X503	4 000	....	....	....	....	3 880	3 750	3 630	3 510
WWF550X420	3 350	....	....	....	3 240	3 110	2 980	2 840	2 710
* WWF550X280	2 030	....	....	2 000	1 890	1 770	1 650	1 530	1 400
WWF500X651	4 160	....	....	....	....	....	4 080	3 990	3 900
WWF500X561	3 860	....	....	....	....	3 850	3 750	3 660	3 570
WWF500X456	3 270	....	....	....	3 250	3 150	3 060	2 960	2 860
WWF500X381	2 730	....	....	....	2 620	2 510	2 410	2 300	2 200
WWF500X343	2 460	....	....	2 420	2 300	2 190	2 080	1 970	1 860
WWF500X306	2 180	....	2 160	2 100	1 990	1 870	1 750	1 630	1 520
WWF500X276	2 000	....	1 980	1 930	1 810	1 690	1 570	1 460	1 340
* WWF500X254	1 660	....	....	1 610	1 510	1 410	1 310	1 210	1 110
* WWF500X223	1 480	....	1 460	1 420	1 320	1 220	1 120	1 020	892
** WWF500X197	1 250	....	1 240	1 200	1 120	1 040	947	858	752
WWF450X503	3 080	....	....	....	....	3 070	2 990	2 920	2 850
WWF450X409	2 600	....	....	....	2 580	2 510	2 430	2 360	2 280
WWF450X342	2 190	....	....	2 170	2 090	2 000	1 920	1 840	1 750
WWF450X308	1 970	....	1 960	1 920	1 830	1 740	1 650	1 560	1 480
WWF450X274	1 750	....	1 710	1 660	1 570	1 470	1 380	1 290	1 200
WWF450X248	1 610	....	1 560	1 520	1 420	1 330	1 240	1 140	1 040
WWF450X228	1 470	1 460	1 410	1 360	1 270	1 170	1 070	976	861
* WWF450X201	1 190	1 180	1 150	1 110	1 030	946	864	779	684
* WWF450X177	1 080	1 070	1 030	995	915	833	750	653	570

¤ Based on class of section for X–X axis bending only, $\omega_2 = 1.0$.
* Class 3 flanges.
** Class 4 flanges.

G40.21-M 300W
$\phi = 0.90$

FACTORED MOMENT RESISTANCES[¤] OF COLUMNS, M_{rx} and M'_{rx} (kN·m)

Section	M_{rx}	M'_{rx} for the Following Unsupported Lengths in Millimetres									
		4 000	5 000	6 000	7 000	8 000	9 000	10 000	12 000	14 000	16 000
WWF400X444	2 370	....	....	....	....	....	....	....	....	2 360	2 310
WWF400X362	2 020	....	....	....	....	....	....	....	2 000	1 950	1 890
WWF400X303	1 700	....	....	....	....	....	....	1 680	1 610	1 550	1 490
WWF400X273	1 530	....	....	....	....	....	1 510	1 480	1 410	1 340	1 280
WWF400X243	1 360	....	....	....	....	1 350	1 320	1 280	1 210	1 140	1 060
WWF400X220	1 260	....	....	....	....	1 240	1 200	1 170	1 090	1 020	951
WWF400X202	1 150	....	....	....	....	1 120	1 080	1 040	968	893	819
WWF400X178	1 020	....	....	....	....	981	943	905	828	752	673
* WWF400X157	842	....	....	....	...	815	784	753	689	625	560
WWF350X315	1 510	....	....	....	....	....	....	....	1 500	1 460	1 410
WWF350X263	1 280	....	....	....	....	....	....	1 260	1 210	1 160	1 120
WWF350X238	1 160	....	....	....	....	....	1 130	1 110	1 060	1 010	962
WWF350X212	1 030	....	....	....	....	1 010	982	956	903	851	799
WWF350X192	950	....	....	....	...	923	896	869	816	763	711
WWF350X176	869	....	....	....	858	830	802	774	719	664	610
WWF350X155	775	....	....	....	753	724	696	667	610	553	490
WWF350X137	699	....	....	....	673	645	616	587	530	472	405
W310X226	1 070	....	....	....	....	1 060	1 040	1 010	964	916	868
W310X202	948	....	....	....	947	922	898	874	826	778	731
W310X179	824	....	....	....	811	787	763	739	692	646	600
W310X158	721	....	....	....	699	675	652	628	582	536	490
W310X143	653	...	...	649	626	602	578	555	509	463	412
W310X129	583	...	...	573	550	527	504	481	435	390	336
W310X118	526	...	...	513	490	467	444	421	375	324	279
W310X107	478	...	...	461	438	415	393	370	324	272	234
* W310X97	389	...	...	379	361	342	324	305	268	226	194
W250X167	656	...	...	...	651	634	618	601	569	537	505
W250X149	575	...	...	...	561	545	529	513	482	450	419
W250X131	500	...	...	494	478	462	447	431	400	369	338
W250X115	432	...	...	420	405	389	374	358	328	297	262
W250X101	378	...	377	361	346	330	315	300	270	236	205
W250X89	332	...	327	312	296	281	266	250	220	186	161
W250X80	294	...	287	272	257	242	227	212	179	151	130
W250X73	266	...	257	242	227	212	198	183	149	126	108
W200X100	310	...	...	300	289	279	269	258	238	217	194
W200X86	265	...	261	251	240	230	220	210	190	169	147
W200X71	217	...	208	198	188	178	168	158	137	117	101
W200X59	176	174	164	154	144	134	124	114	93.4	79.3	69.0
W200X52	154	150	140	131	121	111	101	89.7	73.4	62.2	54.0
W200X46	134	129	119	109	99.8	90.1	78.6	69.6	56.7	48.0	41.6

¤ Based on class of section for X–X axis bending only, $\omega_2 = 1.0$.
* Class 3 flanges.

NOTES

G40.21-M
300W
$\phi = 0.90$

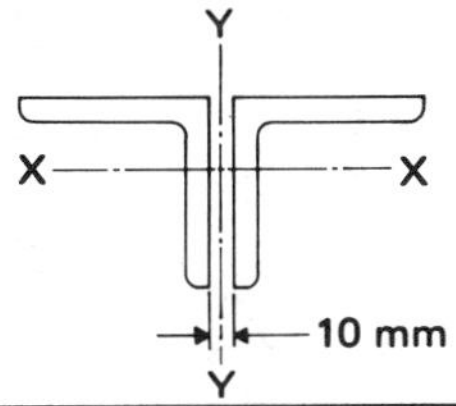

DOUBLE ANGLE STRUTS
Imperial Series – Equal Leg Angles

Factored Axial Compressive Resistances – kN
Legs 10 mm Back to Back *

Designation (mm x mm x mm)		L 152 x 152				L 127 x 127			
		19	16	13**	9.5**	16	13	9.5**	7.9**
Mass (kg/m)		85.4	71.9	58.1	44.1	59.4	48.1	36.6	30.7
Effective length (KL) in millimetres with respect to indicated axis									
X–X Axis	0	2 940	2 470	1 890	1 180	2 040	1 650	1 110	820
	500	2 940	2 470	1 890	1 180	2 040	1 650	1 110	819
	1 000	2 840	2 390	1 850	1 160	1 940	1 570	1 070	795
	1 500	2 700	2 280	1 780	1 120	1 810	1 470	1 030	764
	2 000	2 540	2 140	1 710	1 080	1 670	1 360	973	725
	2 500	2 360	1 990	1 610	1 030	1 500	1 220	906	679
	3 000	2 150	1 820	1 480	976	1 300	1 070	819	623
	3 500	1 920	1 630	1 330	910	1 080	888	685	552
	4 000	1 660	1 410	1 150	829	900	741	572	484
	4 500	1 410	1 200	982	741	764	630	486	412
	5 000	1 220	1 040	851	655	657	542	418	354
	5 500	1 070	911	745	573	571	471	364	308
	6 000	945	804	658	507	500	413	319	271
	6 500	840	715	585	451	434	360	279	238
	7 000	751	640	523	404	377	313	242	206
	7 500	675	575	471	363	331	274	213	181
	8 000	595	509	418	323				
	8 500	530	453	372	288				
	9 000	476	407	334	258				
	9 500				233				
Y–Y Axis	0	2 940	2 470	1 420	817	2 040	1 660	812	566
	500	2 940	2 470	1 420	817	2 040	1 660	812	565
	1 000	2 920	2 450	1 410	816	2 010	1 620	811	564
	1 500	2 840	2 380	1 410	815	1 940	1 570	809	563
	2 000	2 750	2 300	1 410	812	1 850	1 500	806	561
	2 500	2 640	2 220	1 400	809	1 760	1 420	800	557
	3 000	2 530	2 120	1 390	805	1 660	1 340	791	552
	3 500	2 410	2 020	1 380	798	1 550	1 250	776	544
	4 000	2 270	1 900	1 340	789	1 430	1 150	755	533
	4 500	2 130	1 780	1 300	777	1 290	1 040	725	516
	5 000	1 970	1 650	1 240	759	1 140	913	683	493
	5 500	1 810	1 510	1 180	736	1 000	803	622	463
	6 000	1 610	1 340	1 100	706	891	713	554	426
	6 500	1 450	1 210	1 020	669	797	638	482	382
	7 000	1 310	1 090	925	625	718	574	421	337
	7 500	1 190	991	825	574	650	520	370	298
	8 000	1 090	905	732	516	591	473	328	266
	8 500	998	830	653	465	540	431	292	238
	9 000	919	764	585	420	495	395	262	214
	9 500	848	705	528	381	450	359	236	193
	10 000	785	653	478	346	408	325	214	175
PROPERTIES OF 2 ANGLES — 10 mm BACK TO BACK									
Area (mm^2)		10 900	9 160	7 400	5 610	7 570	6 130	4 660	3 910
r_x (mm)		46.3	46.7	47.1	47.6	38.6	39.1	39.5	39.8
r_y (mm)		68.1	67.6	67.1	66.6	57.5	57.0	56.4	56.2
* r_z (mm)		29.7	29.8	30.0	30.2	24.8	25.0	25.1	25.2
* r_z / r_x		0.642	0.638	0.637	0.635	0.642	0.639	0.635	0.634
* r_z / r_y		0.436	0.441	0.447	0.453	0.431	0.439	0.445	0.448
IMPERIAL SIZE AND MASS									
Mass (lb./ft.)		57.4	48.4	39.2	29.8	40.0	32.4	24.6	20.6
Thickness (in.)		3/4	5/8	1/2	3/8	5/8	1/2	3/8	5/16
Size (in.)		6 x 6				5 x 5			

* See page 4–4 for more information regarding these tables.
** Factored axial compressive resistances calculated according to Clause 13.3.3, CAN/CSA-S16.1-M89.

DOUBLE ANGLE STRUTS
Imperial Series – Equal Leg Angles

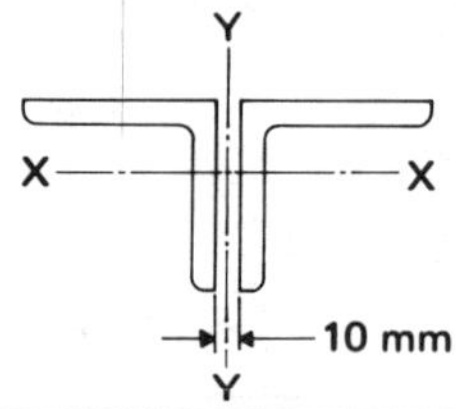

G40.21-M
300W
$\phi = 0.90$

Factored Axial Compressive Resistances – kN
Legs 10 mm Back to Back *

Designation (mm x mm x mm)	L 102 x 102				L 89 x 89			
	13	9.5	7.9**	6.4**	13	9.5	7.9	6.4**
Mass (kg/m)	38.1	29.1	24.4	19.7	32.9	25.2	21.2	17.1
Effective length (KL) in millimetres with respect to indicated axis — X–X Axis								
0	1 310	1 000	761	525	1 130	866	728	502
500	1 290	989	754	521	1 110	848	713	496
1 000	1 210	924	720	500	1 010	779	656	469
1 500	1 100	841	676	471	895	690	582	433
2 000	965	742	619	435	750	582	492	386
2 500	809	626	529	390	577	451	383	312
3 000	637	495	419	332	450	352	299	243
3 500	515	400	339	276	361	283	240	196
4 000	426	331	281	228	296	232	197	161
4 500	358	279	236	192	243	192	164	134
5 000	304	237	201	164	199	157	134	109
5 500	254	199	169	138			112	91.3
6 000	215	168	143	117				
Y–Y Axis								
0	1 310	1 000	560	360	1 130	867	729	361
500	1 310	1 000	559	360	1 130	867	729	360
1 000	1 270	967	558	359	1 080	827	695	359
1 500	1 210	921	556	358	1 020	779	655	357
2 000	1 140	866	552	355	948	723	606	353
2 500	1 060	804	544	351	865	658	551	346
3 000	968	734	529	345	771	584	488	333
3 500	868	656	507	334	662	497	414	313
4 000	751	565	468	318	553	416	346	282
4 500	641	483	418	295	471	354	294	237
5 000	555	418	361	266	406	305	254	197
5 500	486	366	304	230	354	266	221	165
6 000	429	323	259	198	311	233	194	140
6 500	382	287	222	171	276	206	172	120
7 000	342	257	193	150	242	180	149	104
7 500	307	230	169	132	212	158	131	91.4
8 000	272	203	149	117	188	140	116	80.7
8 500	242	181	133	104				
9 000	217	163	119	93.1				
PROPERTIES OF 2 ANGLES — 10 mm BACK TO BACK								
Area (mm^2)	4 860	3 710	3 110	2 510	4 190	3 210	2 700	2 180
r_x (mm)	31.1	31.5	31.7	31.9	26.9	27.3	27.5	27.7
r_y (mm)	46.9	46.4	46.1	45.8	41.7	41.1	40.8	40.5
* r_z (mm)	19.9	20.1	20.2	20.3	17.3	17.4	17.5	17.6
* r_z / r_x	0.641	0.638	0.637	0.636	0.644	0.637	0.636	0.636
* r_z / r_y	0.424	0.433	0.438	0.443	0.415	0.423	0.429	0.435
IMPERIAL SIZE AND MASS								
Mass (lb./ft.)	25.6	19.6	16.4	13.2	22.2	17.0	14.4	11.6
Thickness (in.)	1/2	3/8	5/16	1/4	1/2	3/8	5/16	1/4
Size (in.)	4 x 4				3 1/2 x 3 1/2			

* See page 4–4 for more information regarding these tables.
** Factored axial compressive resistances calculated according to Clause 13.3.3, CAN/CSA-S16.1-M89.

G40.21-M
300W
$\phi = 0.90$

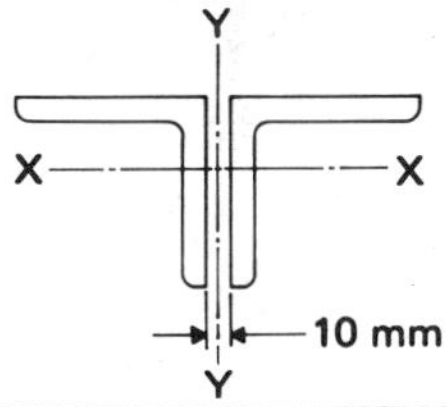

DOUBLE ANGLE STRUTS
Imperial Series – Equal Leg Angles

Factored Axial Compressive Resistances – kN
Legs 10 mm Back to Back *

Designation (mm x mm x mm)		L 76 x 76				L 64 x 64		L 51 x 51				
		13	9.5	7.9	6.4**	6.4	4.8**	9.5	7.9	6.4	4.8	3.2**
Mass (kg/m)		27.9	21.4	18.0	14.6	12.0	9.14	13.8	11.7	9.52	7.24	4.92
Effective length (KL) in millimetres with respect to indicated axis: X–X Axis	0	958	735	619	473	414	277	474	402	327	249	131
	500	924	709	598	462	392	268	433	369	300	229	128
	1 000	825	635	536	429	338	243	341	293	240	184	116
	1 500	695	538	456	372	266	205	220	191	158	122	93.4
	2 000	529	415	353	291	184	143	147	128	106	82.0	58.5
	2 500	390	306	261	215	135	105	104	90.6	75.4	58.6	37.5
	3 000	301	237	202	166	103	79.9	73.3	64.0	53.3	41.6	26.0
	3 500	239	188	160	133	77.7	60.9					
	4 000	188	149	128	106							
	4 500	151	119	102	84.7							
Y–Y Axis	0	958	734	618	352	413	200	472	402	327	249	83.6
	500	953	730	614	351	407	200	461	392	318	242	83.3
	1 000	903	690	580	350	378	199	420	357	289	219	82.4
	1 500	840	641	538	347	342	196	368	311	251	190	80.1
	2 000	765	582	488	338	299	190	304	255	205	154	74.2
	2 500	678	514	430	318	247	177	230	192	153	115	61.7
	3 000	579	434	361	287	194	151	179	149	119	89.3	46.6
	3 500	471	353	294	248	156	118	144	119	95.3	71.4	35.6
	4 000	392	294	244	202	129	92.4	117	97.6	77.8	58.2	27.9
	4 500	332	248	207	162	108	74.0	95.3	78.8	62.5	46.6	22.3
	5 000	285	213	177	133	91.3	60.5	78.0	64.5	51.2	38.2	
	5 500	247	184	153	111	76.1	50.3					
	6 000	214	159	132	93.4	64.5	42.5					
	6 500	184	137	113	79.9							
	7 000	160	119	98.4	69.1							
PROPERTIES OF 2 ANGLES — 10 mm BACK TO BACK												
Area (mm^2)		3 550	2 720	2 290	1 850	1 530	1 160	1 750	1 490	1 210	922	626
r_x (mm)		22.8	23.2	23.4	23.7	19.5	19.8	15.1	15.3	15.5	15.7	15.9
r_y (mm)		36.6	36.0	35.7	35.4	30.3	30.1	26.0	25.6	25.3	25.0	24.7
* r_z (mm)		14.8	14.9	15.0	15.0	12.5	12.6	9.89	9.90	9.93	10.0	10.1
* r_z / r_x		0.648	0.641	0.640	0.633	0.640	0.637	0.656	0.648	0.642	0.637	0.636
* r_z / r_y		0.404	0.414	0.420	0.424	0.413	0.419	0.380	0.387	0.392	0.400	0.409
IMPERIAL SIZE AND MASS												
Mass (lb./ft.)		18.8	14.4	12.2	9.80	8.20	6.14	9.40	7.84	6.38	4.88	3.30
Thickness (in.)		1/2	3/8	5/16	1/4	1/4	3/16	3/8	5/16	1/4	3/16	1/8
Size (in.)		3 x 3				2 1/2 x 2 1/2		2 x 2				

* See page 4–4 for more information regarding these tables.
** Factored axial compressive resistances calculated according to Clause 13.3.3, CAN/CSA-S16.1-M89.

DOUBLE ANGLE STRUTS
Imperial Series – Equal Leg Angles

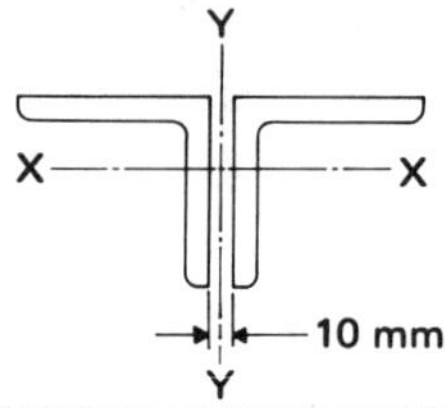

G40.21-M
300W
$\phi = 0.90$

Factored Axial Compressive Resistances – kN
Legs 10 mm Back to Back *

Designation (mm x mm x mm)		L 44 x 44					L 38 x 38		
		9.5	7.9	6.4	4.8	3.2**	6.4	4.8	3.2**
Mass (kg/m)		11.9	10.1	8.24	6.30	4.28	6.96	5.34	3.64
Effective length (KL) in millimetres with respect to indicated axis									
X–X Axis	0	409	348	283	217	126	240	184	119
	500	363	311	254	195	121	206	159	112
	1 000	264	229	189	146	105	132	104	84.3
	1 500	156	136	113	88.3	69.4	75.2	59.2	42.5
	2 000	103	89.6	74.6	58.3	39.0	47.1	37.3	23.9
	2 500	68.3	60.0	50.3	39.5	25.0			
	3 000								
Y–Y Axis	0	408	348	284	217	85.1	239	184	83.4
	500	394	336	273	209	84.8	228	175	83.0
	1 000	354	301	244	186	83.6	199	152	80.4
	1 500	301	255	206	156	79.7	160	121	70.4
	2 000	234	197	156	117	67.7	113	84.8	50.6
	2 500	173	145	115	86.5	48.1	82.8	62.1	33.6
	3 000	134	112	89.1	66.7	34.7	63.3	47.4	23.7
	3 500	106	89.2	70.7	52.9	26.0	48.6	36.2	17.6
	4 000	84.9	70.7	55.7	41.5	20.1	37.7	28.1	
	4 500	67.8	56.6	44.5	33.2				
	5 000								
PROPERTIES OF 2 ANGLES — 10 mm BACK TO BACK									
Area (mm^2)		1 510	1 290	1 050	802	546	887	680	464
r_x (mm)		13.1	13.3	13.5	13.7	13.9	11.4	11.6	11.8
r_y (mm)		23.5	23.2	22.8	22.5	22.2	20.3	20.0	19.6
* r_z (mm)		8.67	8.66	8.68	8.73	8.82	7.42	7.45	7.52
* r_z / r_x		0.663	0.654	0.645	0.638	0.636	0.652	0.642	0.637
* r_z / r_y		0.369	0.373	0.381	0.388	0.397	0.366	0.372	0.384
IMPERIAL SIZE AND MASS									
Mass (lb./ft.)		8.00	6.78	5.54	4.24	2.88	4.68	3.60	2.46
Thickness (in.)		3/8	5/16	1/4	3/16	1/8	1/4	3/16	1/8
Size (in.)		1 3/4 x 1 3/4					1 1/2 x 1 1/2		

* See page 4–4 for more information regarding these tables.
** Factored axial compressive resistances calculated according to Clause 13.3.3, CAN/CSA-S16.1-M89.

G40.21-M
300W
$\phi = 0.90$

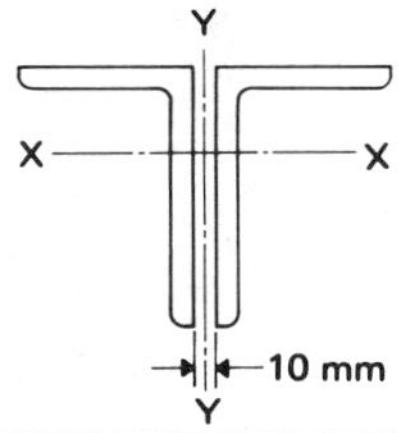

DOUBLE ANGLE STRUTS
Imperial Series – Unequal Leg Angles

Factored Axial Compressive Resistances – kN
Long Legs 10 mm Back to Back *

Designation (mm x mm x mm)		L 178 x 102		L 152 x 102				L 152 x 89			
		13**	9.5**	16	13**	9.5**	7.9**	16	13**	9.5**	7.9**
Mass (kg/m)		53.3	40.5	59.4	48.1	36.6	30.7	56.2	45.5	34.6	29.0
Effective length (KL) in millimetres with respect to indicated axis											
X–X Axis	0	1 660	1 120	2 040	1 600	1 090	810	1 930	1 510	1 020	788
	500	1 660	1 120	2 040	1 600	1 090	810	1 930	1 510	1 020	788
	1 000	1 640	1 100	1 980	1 560	1 060	795	1 870	1 470	999	773
	1 500	1 590	1 070	1 890	1 500	1 030	772	1 790	1 420	964	749
	2 000	1 540	1 040	1 780	1 430	982	745	1 690	1 360	923	718
	2 500	1 480	996	1 660	1 350	930	714	1 580	1 280	875	681
	3 000	1 410	951	1 530	1 240	872	677	1 450	1 180	821	640
	3 500	1 330	901	1 380	1 120	805	630	1 310	1 070	758	593
	4 000	1 240	845	1 210	991	729	573	1 150	939	688	540
	4 500	1 150	784	1 030	846	641	505	981	803	606	478
	5 000	1 020	710	896	734	566	450	850	696	535	426
	5 500	896	641	785	643	496	404	744	610	469	383
	6 000	795	583	694	568	438	365	658	539	415	347
	6 500	712	533	617	506	391	329	586	480	370	314
	7 000	641	489	553	453	350	295	525	430	331	281
	7 500	580	446	497	408	315	266	472	387	298	253
	8 000	528	406	445	366	284	240	422	347	268	229
	8 500	482	371	396	326	253	213	376	309	239	204
	9 000	441	340	355	292	227	191	337	277	214	183
	9 500	401	310	321	264	204	173	305	250	194	165
	10 000	364	281								
Y–Y Axis	0	1 230	745	2 040	1 220	769	548	1 930	1 160	729	520
	500	1 230	742	2 040	1 220	767	547	1 920	1 150	727	519
	1 000	1 220	736	1 960	1 210	763	544	1 820	1 140	720	513
	1 500	1 200	725	1 850	1 200	754	537	1 690	1 120	706	503
	2 000	1 170	708	1 730	1 170	739	527	1 540	1 080	682	487
	2 500	1 120	680	1 580	1 120	715	510	1 370	999	644	460
	3 000	1 060	641	1 420	1 050	680	486	1 170	900	591	422
	3 500	971	589	1 230	957	632	452	949	779	521	371
	4 000	854	522	1 030	850	571	408	791	638	435	320
	4 500	709	456	879	727	497	360	670	514	357	276
	5 000	589	399	759	601	419	317	575	422	297	234
	5 500	496	343	662	503	355	278	498	352	250	199
	6 000	422	296	583	427	304	241	432	298	213	171
	6 500	364	257	516	366	263	209	371	255	183	148
	7 000	316	225	457	318	229	184	322	221	159	129
	7 500	277	198	401	278	201	162				
	8 000	245		354	245	178	144				
	8 500			316							
PROPERTIES OF 2 ANGLES — 10 mm BACK TO BACK											
Area (mm^2)		6 790	5 160	7 570	6 130	4 660	3 910	7 160	5 800	4 410	3 700
r_x (mm)		57.3	57.8	48.1	48.5	49.0	49.2	48.2	48.6	49.0	49.4
r_y (mm)		40.2	39.7	42.7	42.1	41.6	41.3	36.6	36.0	35.4	35.2
* r_z (mm)		22.2	22.4	22.0	22.2	22.4	22.5	19.1	19.3	19.5	19.6
* r_z / r_x		0.387	0.388	0.458	0.458	0.457	0.458	0.397	0.397	0.398	0.397
* r_z / r_y		0.552	0.564	0.515	0.527	0.538	0.545	0.522	0.536	0.551	0.557
IMPERIAL SIZE AND MASS											
Mass (lb./ft.)		35.8	27.2	40.0	32.4	24.6	20.6	37.8	30.6	23.4	19.6
Thickness (in.)		1/2	3/8	5/8	1/2	3/8	5/16	5/8	1/2	3/8	5/16
Size (in.)		7 x 4		6 x 4				6 x 3 1/2			

* See page 4–4 for more information regarding these tables.
** Factored axial compressive resistances calculated according to Clause 13.3.3, CAN/CSA-S16.1-M89.

DOUBLE ANGLE STRUTS
Imperial Series – Unequal Leg Angles

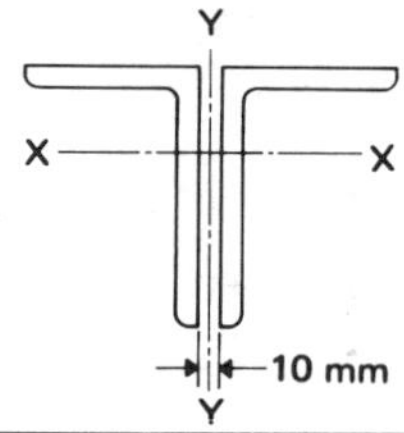

G40.21-M
300W
$\phi = 0.90$

Factored Axial Compressive Resistances – kN
Long Legs 10 mm Back to Back *

Designation (mm x mm x mm)		L 127 x 89						L 127 x 76			
		19	16	13	9.5**	7.9**	6.4**	13	9.5**	7.9**	6.4**
Mass (kg/m)		59.0	49.9	40.5	30.9	25.9	20.9	38.0	29.0	24.3	19.6
Effective length (KL) in millimetres with respect to indicated axis											
X–X Axis	0	2 030	1 720	1 390	989	768	529	1 310	923	720	514
	500	2 030	1 710	1 390	988	768	529	1 310	923	719	514
	1 000	1 930	1 630	1 330	951	744	514	1 240	889	694	499
	1 500	1 810	1 530	1 250	905	710	495	1 170	847	662	480
	2 000	1 670	1 410	1 150	850	668	471	1 080	796	623	457
	2 500	1 500	1 280	1 040	784	618	443	981	737	578	426
	3 000	1 320	1 120	919	708	560	410	866	667	525	388
	3 500	1 100	941	774	599	489	365	730	566	461	342
	4 000	916	785	646	501	422	316	610	473	398	296
	4 500	778	667	549	426	360	277	519	402	340	260
	5 000	670	575	473	367	310	245	447	347	293	231
	5 500	582	500	412	320	270	218	389	302	256	206
	6 000	511	439	361	281	237	193	342	265	225	182
	6 500	447	385	319	248	210	170	302	235	199	161
	7 000	388	334	276	216	183	149	262	204	173	141
	7 500	340	293	242	189	160	130	229	179	152	124
	8 000			215	167	142	115	203	159	134	109
Y–Y Axis	0	2 030	1 720	1 390	733	542	349	1 310	690	510	336
	500	2 030	1 710	1 390	731	541	348	1 290	688	509	334
	1 000	1 930	1 630	1 320	726	537	345	1 210	680	502	330
	1 500	1 810	1 520	1 230	715	529	341	1 100	661	489	321
	2 000	1 670	1 400	1 130	695	515	334	977	626	465	305
	2 500	1 500	1 260	1 010	661	492	322	829	574	428	280
	3 000	1 310	1 090	873	613	458	301	659	495	375	245
	3 500	1 090	903	717	544	412	270	534	395	308	209
	4 000	913	753	598	456	354	237	442	311	246	178
	4 500	775	639	507	370	293	207	372	250	200	149
	5 000	667	550	436	305	244	180	317	205	165	125
	5 500	580	478	379	255	205	154	267	171	138	105
	6 000	509	418	331	216	175	133	226	145	117	90.0
	6 500	444	363	286	186	151	115				
	7 000	386	315	248	161	131	101				
	7 500	338	277	218							
PROPERTIES OF 2 ANGLES — 10 mm BACK TO BACK											
Area (mm^2)		7 520	6 360	5 160	3 930	3 300	2 660	4 840	3 690	3 100	2 500
r_x (mm)		39.3	39.7	40.1	40.6	40.8	41.0	40.3	40.8	41.0	41.2
r_y (mm)		39.2	38.5	37.9	37.4	37.1	36.8	32.0	31.4	31.1	30.8
* r_z (mm)		19.0	19.1	19.2	19.3	19.4	19.6	16.5	16.6	16.7	16.8
* r_z / r_x		0.484	0.482	0.478	0.476	0.475	0.478	0.409	0.407	0.407	0.407
* r_z / r_y		0.485	0.496	0.507	0.516	0.523	0.533	0.516	0.529	0.537	0.545
IMPERIAL SIZE AND MASS											
Mass (lb./ft.)		39.6	33.6	27.2	20.8	17.4	14.0	25.6	19.6	16.4	13.2
Thickness (in.)		3/4	5/8	1/2	3/8	5/16	1/4	1/2	3/8	5/16	1/4
Size (in.)		5 x 3 1/2						5 x 3			

* See page 4–4 for more information regarding these tables.
** Factored axial compressive resistances calculated according to Clause 13.3.3, CAN/CSA-S16.1-M89.

G40.21-M
300W
$\phi = 0.90$

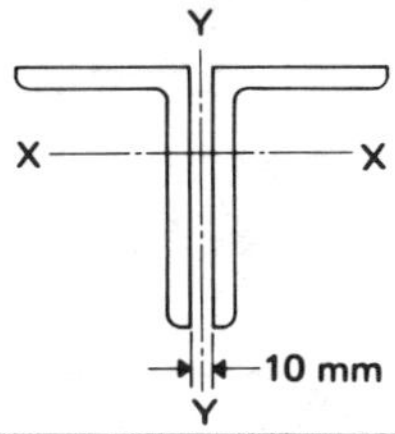

DOUBLE ANGLE STRUTS
Imperial Series – Unequal Leg Angles

Factored Axial Compressive Resistances – kN
Long Legs 10 mm Back to Back *

Designation (mm x mm x mm)		L 102 x 89				L 102 x 76				L 89 x 76
		13	9.5	7.9**	6.4**	13	9.5	7.9**	6.4**	7.9
Mass (kg/m)		35.5	27.1	22.8	18.4	33.0	25.2	21.2	17.1	19.6
Effective length (KL) in millimetres with respect to indicated axis										
X–X Axis	0	1 220	933	739	514	1 130	868	690	499	674
	500	1 210	923	733	510	1 120	856	684	496	661
	1 000	1 130	863	697	489	1 050	802	648	475	609
	1 500	1 030	788	647	461	955	733	602	448	542
	2 000	906	697	584	426	845	650	546	409	461
	2 500	764	591	499	382	715	553	470	360	362
	3 000	604	469	397	318	567	441	376	300	282
	3 500	489	380	322	262	459	357	305	248	227
	4 000	405	315	267	217	380	296	253	206	187
	4 500	340	265	224	183	320	249	213	173	156
	5 000	290	226	191	156	272	212	181	148	127
	5 500	243	190	161	132	229	179	154	126	106
	6 000	206	161	137	112	194	152	130	106	
	6 500							112	91.4	
Y–Y Axis	0	1 220	934	544	361	1 130	867	513	354	675
	500	1 220	933	543	361	1 120	858	511	354	669
	1 000	1 160	888	541	359	1 060	806	507	351	630
	1 500	1 090	834	535	357	975	741	497	344	582
	2 000	1 010	770	525	352	875	663	476	332	523
	2 500	917	695	505	343	759	572	442	310	455
	3 000	809	611	474	328	621	463	390	279	374
	3 500	681	512	428	306	504	376	323	237	303
	4 000	569	427	370	271	418	312	255	192	252
	4 500	484	363	306	230	353	263	205	156	213
	5 000	417	313	252	192	302	225	168	129	182
	5 500	363	272	211	162	261	192	140	108	157
	6 000	318	239	178	138	220	163	118	92.0	133
	6 500	281	210	153	119	189	140	101	79.0	115
	7 000	244	182	133	103					
	7 500	214	159	116	90.7					
	8 000	189								
PROPERTIES OF 2 ANGLES — 10 mm BACK TO BACK										
Area (mm^2)		4 530	3 460	2 910	2 340	4 200	3 210	2 700	2 180	2 500
r_x (mm)		31.5	31.9	32.1	32.3	31.8	32.2	32.5	32.7	27.9
r_y (mm)		40.2	39.7	39.4	39.1	34.0	33.4	33.1	32.8	34.3
* r_z (mm)		18.4	18.5	18.6	18.7	16.2	16.4	16.5	16.6	15.9
* r_z / r_x		0.585	0.580	0.579	0.578	0.509	0.509	0.508	0.508	0.570
* r_z / r_y		0.458	0.466	0.472	0.478	0.476	0.491	0.498	0.506	0.464
IMPERIAL SIZE AND MASS										
Mass (lb./ft.)		23.8	18.2	15.4	12.4	22.2	17.0	14.4	11.6	13.2
Thickness (in.)		1/2	3/8	5/16	1/4	1/2	3/8	5/16	1/4	5/16
Size (in.)		4 x 3 1/2				4 x 3				3 1/2 x 3

* See page 4–4 for more information regarding these tables.
** Factored axial compressive resistances calculated according to Clause 13.3.3, CAN/CSA-S16.1-M89.

DOUBLE ANGLE STRUTS
Imperial Series – Unequal Leg Angles

Factored Axial Compressive Resistances – kN
Long Legs 10 mm Back to Back *

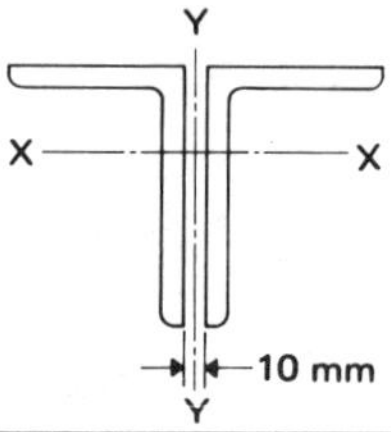

G40.21-M
300W
$\phi = 0.90$

Designation (mm x mm x mm)		L 64 x 51				L 51 x 38		
		9.5	7.9	6.4	4.8**	7.9	6.4	4.8
Mass (kg/m)		15.7	13.3	10.8	8.18	10.1	8.11	6.28
Effective length (KL) in millimetres with respect to indicated axis								
X–X Axis	0	539	456	370	263	347	283	216
X–X Axis	500	512	434	352	253	321	262	200
X–X Axis	1 000	442	375	305	225	257	210	162
X–X Axis	1 500	348	296	243	186	170	141	109
X–X Axis	2 000	241	207	170	131	114	94.2	73.2
X–X Axis	2 500	176	151	124	96.0	81.2	67.4	52.5
X–X Axis	3 000	134	115	94.8	73.3	57.5	47.9	37.4
X–X Axis	3 500	102	87.5	72.3	56.1			
X–X Axis	4 000				43.6			
Y–Y Axis	0	540	456	370	191	348	284	216
Y–Y Axis	500	524	442	358	190	330	268	204
Y–Y Axis	1 000	474	399	322	187	285	230	174
Y–Y Axis	1 500	409	343	276	177	223	178	133
Y–Y Axis	2 000	329	274	217	157	154	122	90.5
Y–Y Axis	2 500	244	203	161	123	113	89.1	65.9
Y–Y Axis	3 000	189	157	124	89.2	85.9	67.7	49.9
Y–Y Axis	3 500	151	125	99.1	66.9	64.9	50.8	37.2
Y–Y Axis	4 000	123	101	79.5	51.8			
Y–Y Axis	4 500	98.0	80.9	63.5	41.3			
PROPERTIES OF 2 ANGLES — 10 mm BACK TO BACK								
Area (mm^2)		2 000	1 690	1 370	1 040	1 290	1 050	801
r_x (mm)		19.5	19.7	19.9	20.1	15.6	15.8	16.0
r_y (mm)		24.6	24.3	23.9	23.6	19.4	19.0	18.6
* r_z (mm)		10.7	10.7	10.8	10.9	8.09	8.12	8.18
* r_z / r_x		0.548	0.543	0.542	0.541	0.518	0.513	0.510
* r_z / r_y		0.435	0.440	0.452	0.462	0.417	0.427	0.440
IMPERIAL SIZE AND MASS								
Mass (lb./ft.)		10.6	9.00	7.24	5.50	6.78	5.54	4.24
Thickness (in.)		3/8	5/16	1/4	3/16	5/16	1/4	3/16
Size (in.)		2 1/2 x 2				2 x 1 1/2		

* See page 4–4 for more information regarding these tables.
** Factored axial compressive resistances calculated according to Clause 13.3.3, CAN/CSA-S16.1-M89.

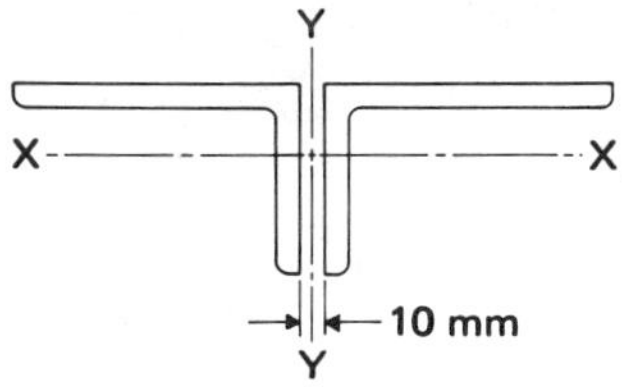

DOUBLE ANGLE STRUTS
Imperial Series – Unequal Leg Angles

Factored Axial Compressive Resistances – kN
Short Legs 10 mm Back to Back *

Designation (mm x mm x mm)		L 178 x 102		L 152 x 102				L 152 x 89			
		13**	9.5**	16	13**	9.5**	7.9**	16	13**	9.5**	7.9**
Mass (kg/m)		53.3	40.5	59.4	48.1	36.6	30.7	56.2	45.5	34.6	29.0
Effective length (KL) in millimetres with respect to indicated axis											
X–X Axis	0	1 660	1 120	2 040	1 600	1 090	810	1 930	1 510	1 020	788
	500	1 640	1 100	2 010	1 580	1 080	802	1 870	1 480	1 000	774
	1 000	1 540	1 040	1 860	1 480	1 010	765	1 690	1 360	927	721
	1 500	1 410	951	1 670	1 360	934	716	1 460	1 190	829	646
	2 000	1 240	845	1 430	1 170	834	654	1 160	958	702	551
	2 500	1 010	710	1 150	951	708	558	859	711	553	438
	3 000	790	583	897	744	576	458	666	552	429	357
	3 500	636	489	724	600	465	384	531	441	343	291
	4 000	524	406	596	495	384	325	429	359	280	238
	4 500	438	340	499	415	322	273	343	286	224	191
	5 000	360	281	412	345	269	229			184	157
	5 500	301	234	344	288	224	191				
	6 000						162				
Y–Y Axis	0	1 210	726	2 040	1 210	757	536	1 930	1 140	715	507
	500	1 210	726	2 040	1 210	757	536	1 930	1 140	715	507
	1 000	1 210	726	2 030	1 210	757	536	1 930	1 140	715	506
	1 500	1 210	726	1 980	1 210	757	536	1 880	1 140	715	506
	2 000	1 210	725	1 930	1 210	756	535	1 830	1 140	714	506
	2 500	1 210	725	1 870	1 200	755	535	1 770	1 140	714	506
	3 000	1 210	725	1 800	1 200	753	534	1 710	1 140	713	505
	3 500	1 210	724	1 720	1 190	751	532	1 640	1 130	712	504
	4 000	1 210	723	1 640	1 180	747	530	1 570	1 120	710	503
	4 500	1 200	722	1 550	1 150	741	527	1 490	1 100	706	501
	5 000	1 200	720	1 460	1 120	731	523	1 400	1 070	699	499
	5 500	1 180	718	1 360	1 070	713	516	1 310	1 030	686	495
	6 000	1 160	714	1 250	1 020	689	505	1 220	982	666	487
	6 500	1 130	707	1 130	956	659	489	1 100	929	639	475
	7 000	1 100	697	1 020	892	623	467	1 000	871	608	457
	7 500	1 050	681	929	822	583	439	911	808	572	434
	8 000	1 010	659	850	747	538	407	834	741	533	405
	8 500	961	631	781	668	490	374	767	670	489	374
	9 000	904	598	721	597	440	344	708	599	443	343
	9 500	835	562	667	537	396	316	655	538	399	316
	10 000	762	523	619	485	359	292	608	486	361	292
PROPERTIES OF 2 ANGLES — 10 mm BACK TO BACK											
Area (mm^2)		6 790	5 160	7 570	6 130	4 660	3 910	7 160	5 800	4 410	3 700
r_x (mm)		28.5	28.9	28.9	29.4	29.8	30.0	24.3	24.7	25.1	25.3
r_y (mm)		87.7	87.1	74.1	73.5	72.9	72.6	76.1	75.5	74.9	74.6
* r_z (mm)		22.2	22.4	22.0	22.2	22.4	22.5	19.1	19.3	19.5	19.6
* r_z / r_x		0.780	0.776	0.760	0.756	0.752	0.751	0.786	0.781	0.777	0.774
* r_z / r_y		0.253	0.257	0.297	0.302	0.307	0.310	0.251	0.256	0.260	0.263
IMPERIAL SIZE AND MASS											
Mass (lb./ft.)		35.8	27.2	40.0	32.4	24.6	20.6	37.8	30.6	23.4	19.6
Thickness (in.)		1/2	3/8	5/8	1/2	3/8	5/16	5/8	1/2	3/8	5/16
Size (in.)		7 x 4		6 x 4				6 x 3 1/2			

* See page 4–4 for more information regarding these tables.
** Factored axial compressive resistances calculated according to Clause 13.3.3, CAN/CSA-S16.1-M89.

DOUBLE ANGLE STRUTS
Imperial Series – Unequal Leg Angles

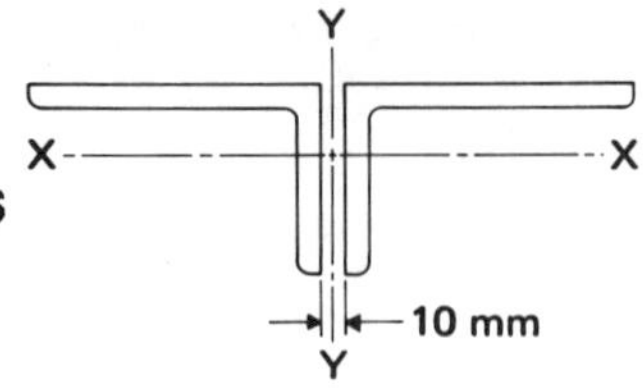

G40.21-M
300W
$\phi = 0.90$

Factored Axial Compressive Resistances – kN
Short Legs 10 mm Back to Back *

Designation (mm x mm x mm)		L 127 x 89						L 127 x 76			
		19	16	13	9.5**	7.9**	6.4**	13	9.5**	7.9**	6.4**
Mass (kg/m)		59.0	49.9	40.5	30.9	25.9	20.9	38.0	29.0	24.3	19.6
Effective length (KL) in millimetres with respect to indicated axis											
X–X Axis	0	2 030	1 720	1 390	989	768	529	1 310	923	720	514
	500	1 970	1 670	1 360	968	755	521	1 250	893	697	501
	1 000	1 790	1 520	1 230	899	705	492	1 100	808	632	463
	1 500	1 550	1 320	1 080	806	635	454	898	689	542	402
	2 000	1 250	1 070	884	684	545	402	649	508	422	317
	2 500	928	802	665	518	435	326	477	374	316	248
	3 000	720	623	517	403	342	266	366	287	243	200
	3 500	575	498	414	323	274	221	286	226	192	158
	4 000	469	407	338	264	224	183	222	175	148	123
	4 500	374	326	273	214	182	149				
	5 000		267	223	175	149	122				
Y–Y Axis	0	2 030	1 720	1 390	725	534	341	1 310	681	500	325
	500	2 030	1 720	1 390	725	534	341	1 310	681	500	324
	1 000	2 000	1 690	1 370	725	533	341	1 290	680	500	324
	1 500	1 940	1 640	1 330	724	533	340	1 250	680	500	324
	2 000	1 870	1 580	1 280	723	532	340	1 210	679	499	324
	2 500	1 790	1 510	1 230	720	531	339	1 160	678	499	323
	3 000	1 710	1 440	1 160	716	528	338	1 100	676	498	323
	3 500	1 610	1 360	1 100	708	525	337	1 040	671	496	322
	4 000	1 510	1 270	1 020	692	518	334	975	659	492	320
	4 500	1 390	1 170	941	667	506	331	904	638	483	318
	5 000	1 270	1 060	855	636	486	325	827	610	468	313
	5 500	1 130	943	755	599	461	315	738	578	446	306
	6 000	1 000	839	672	550	430	298	656	539	419	293
	6 500	901	752	602	495	394	276	589	490	388	275
	7 000	813	679	543	436	354	251	532	437	353	252
	7 500	738	616	493	381	313	228	483	383	316	228
	8 000	673	561	449	336	276	208	440	337	279	208
	8 500	616	514	411	298	246	189	403	299	247	189
	9 000	566	472	377	267	220	172	370	267	221	173
	9 500	522	435	348	240	198	156	341	240	199	157
	10 000	482	401	321	217	179	141	316	217	180	142
PROPERTIES OF 2 ANGLES — 10 mm BACK TO BACK											
Area (mm^2)		7 520	6 360	5 160	3 930	3 300	2 660	4 840	3 690	3 100	2 500
r_x (mm)		24.8	25.2	25.6	26.0	26.2	26.4	21.1	21.5	21.6	21.9
r_y (mm)		63.1	62.5	61.9	61.3	61.0	60.7	63.8	63.2	62.9	62.6
* r_z (mm)		19.0	19.1	19.2	19.3	19.4	19.6	16.5	16.6	16.7	16.8
* r_z / r_x		0.766	0.759	0.751	0.744	0.742	0.741	0.783	0.774	0.772	0.767
* r_z / r_y		0.301	0.306	0.310	0.315	0.318	0.323	0.259	0.263	0.266	0.268
IMPERIAL SIZE AND MASS											
Mass (lb./ft.)		39.6	33.6	27.2	20.8	17.4	14.0	25.6	19.6	16.4	13.2
Thickness (in.)		3/4	5/8	1/2	3/8	5/16	1/4	1/2	3/8	5/16	1/4
Size (in.)		5 x 3 1/2						5 x 3			

* See page 4–4 for more information regarding these tables.
** Factored axial compressive resistances calculated according to Clause 13.3.3, CAN/CSA-S16.1-M89.

G40.21-M
300W
$\phi = 0.90$

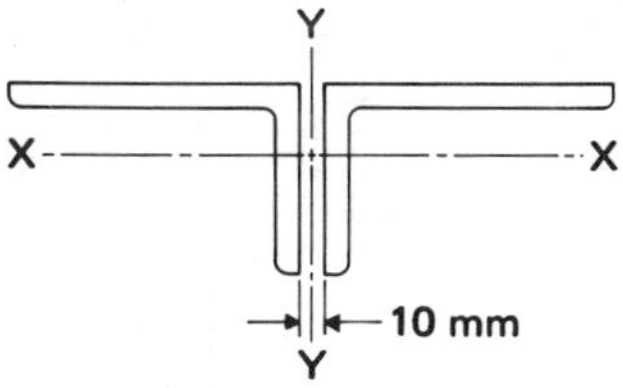

DOUBLE ANGLE STRUTS
Imperial Series – Unequal Leg Angles

Factored Axial Compressive Resistances – kN
Short Legs 10 mm Back to Back *

Designation (mm x mm x mm)			L 102 x 89				L 102 x 76				L 89 x 76
			13	9.5	7.9**	6.4**	13	9.5	7.9**	6.4**	7.9
Mass (kg/m)			35.5	27.1	22.8	18.4	33.0	25.2	21.2	17.1	19.6
Effective length (KL) in millimetres with respect to indicated axis	X–X Axis	0	1 220	933	739	514	1 130	868	690	499	674
		500	1 190	913	727	506	1 090	834	669	487	651
		1 000	1 090	836	680	479	965	741	608	452	582
		1 500	960	738	613	442	801	620	525	395	492
		2 000	798	618	524	394	593	464	396	315	377
		2 500	609	474	404	322	436	342	292	240	278
		3 000	474	370	315	256	336	264	225	185	215
		3 500	380	297	253	205	266	209	178	147	171
		4 000	311	243	208	169	206	163	140	116	135
		4 500	254	200	171	139			112	92.6	108
		5 000	208	163	140	114					
	Y–Y Axis	0	1 220	934	541	359	1 130	867	508	349	675
		500	1 220	934	541	359	1 130	867	508	349	675
		1 000	1 190	905	540	358	1 100	842	507	349	646
		1 500	1 130	863	538	357	1 060	805	506	348	610
		2 000	1 070	815	535	355	1 000	762	504	347	568
		2 500	999	760	528	353	938	714	499	345	520
		3 000	920	698	514	347	868	660	489	341	465
		3 500	831	629	491	339	791	599	470	333	402
		4 000	732	550	458	324	706	533	443	318	337
		4 500	626	470	414	303	608	457	406	297	287
		5 000	542	407	364	272	527	396	361	271	247
		5 500	475	357	308	237	462	347	312	239	216
		6 000	420	315	261	203	409	307	264	206	190
		6 500	374	281	224	175	365	274	226	177	168
		7 000	335	251	194	152	327	246	195	154	148
		7 500	302	226	170	134	295	221	171	135	130
		8 000	270	202	150	118	267	200	150	119	115
		8 500	241	180	133	105	238	178	133	105	
		9 000	216	161	119	94.0	214	160	119	94.3	
		9 500	195	145	107		193	144	107	84.8	
		10 000					175				

PROPERTIES OF 2 ANGLES — 10 mm BACK TO BACK

Property	13	9.5	7.9**	6.4**	13	9.5	7.9**	6.4**	7.9
Area (mm^2)	4 530	3 460	2 910	2 340	4 200	3 210	2 700	2 180	2 500
r_x (mm)	26.4	26.8	27.1	27.2	21.9	22.3	22.5	22.8	23.0
r_y (mm)	48.5	47.9	47.6	47.4	50.2	49.6	49.3	49.0	42.3
* r_z (mm)	18.4	18.5	18.6	18.7	16.2	16.4	16.5	16.6	15.9
* r_z / r_x	0.696	0.689	0.687	0.686	0.739	0.735	0.733	0.730	0.691
* r_z / r_y	0.379	0.386	0.391	0.395	0.323	0.331	0.335	0.339	0.376

IMPERIAL SIZE AND MASS

Mass (lb./ft.)	23.8	18.2	15.4	12.4	22.2	17.0	14.4	11.6	13.2
Thickness (in.)	1/2	3/8	5/16	1/4	1/2	3/8	5/16	1/4	5/16
Size (in.)	4 x 3 1/2				4 x 3				3 1/2 x 3

* See page 4–4 for more information regarding these tables.
** Factored axial compressive resistances calculated according to Clause 13.3.3, CAN/CSA-S16.1-M89.

DOUBLE ANGLE STRUTS
Imperial Series – Unequal Leg Angles

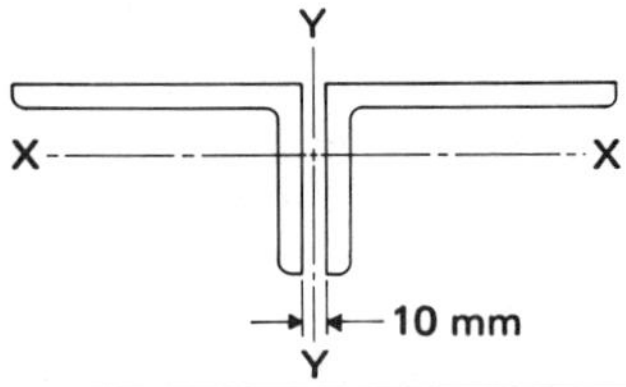

G40.21-M
300W
$\phi = 0.90$

Factored Axial Compressive Resistances – kN
Short Legs 10 mm Back to Back *

Designation (mm x mm x mm)		L 64 x 51				L 51 x 38		
		9.6	7.9	6.4	4.8**	7.9	6.4	4.8
Mass (kg/m)		15.7	13.3	10.8	8.18	10.1	8.22	6.28
Effective length (KL) in millimetres with respect to indicated axis								
X–X Axis	0	539	456	370	263	347	283	216
	500	491	416	338	245	295	242	185
	1 000	381	325	266	201	179	149	116
	1 500	240	207	171	132	101	84.5	66.2
	2 000	160	138	114	88.5	61.7	52.0	41.1
	2 500	111	96.4	80.2	62.6			
	3 000			56.7	44.2			
Y–Y Axis	0	540	456	370	189	348	284	216
	500	534	451	365	189	341	277	211
	1 000	501	422	342	188	313	254	193
	1 500	458	386	312	186	277	225	170
	2 000	408	343	277	181	234	189	142
	2 500	349	292	235	169	181	145	109
	3 000	279	233	187	149	141	113	84.7
	3 500	226	189	151	120	114	91.0	67.9
	4 000	188	156	125	93.5	93.3	74.6	55.7
	4 500	158	132	105	74.4	77.1	61.4	45.6
	5 000	135	112	89.7	60.6	63.1	50.3	37.3
	5 500	114	94.9	75.6	50.3			
	6 000	96.8	80.4	64.0	42.3			
	6 500	83.2						
PROPERTIES OF 2 ANGLES — 10 mm BACK TO BACK								
Area (mm^2)		2 000	1 690	1 370	1 040	1 290	1 050	801
r_x (mm)		14.6	14.8	15.0	15.2	10.8	11.0	11.2
r_y (mm)		32.6	32.3	32.0	31.6	27.3	27.0	26.6
* r_z (mm)		10.7	10.7	10.8	10.9	8.09	8.12	8.18
* r_z / r_x		0.731	0.721	0.718	0.716	0.749	0.741	0.732
* r_z / r_y		0.328	0.331	0.338	0.345	0.296	0.301	0.308
IMPERIAL SIZE AND MASS								
Mass (lb./ft.)		10.6	9.00	7.24	5.50	6.78	5.54	4.24
Thickness (in.)		3/8	5/16	1/4	3/16	5/16	1/4	3/16
Size (in.)		2 1/2 x 2				2 x 1 1/2		

* See page 4–4 for more information regarding these tables.
** Factored axial compressive resistances calculated according to Clause 13.3.3, CAN/CSA-S16.1-M89.

G40.21-M 300W

$\phi = 0.90$

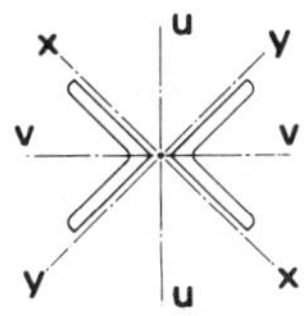

DOUBLE ANGLE STRUTS

Imperial Series – Star Shaped *

Factored Axial Compressive Resistances – kN

Designation (mm x mm x mm)		L 152 x 152		L 127 x 127		L 102 x 102		L 89 x 89		
		19	16	16	13	13	9.5	13	9.5	7.9
Mass (kg/m)		85.4	71.9	59.4	48.1	38.1	29.1	32.9	25.2	21.2
Legs		12 mm apart		12 mm apart		10 mm apart		8 mm apart		
Effective length (KL) in millimetres with respect to indicated axis										
U–U Axis	0	2 940	2 470	2 040	1 660	1 310	1 000	1 130	867	729
	500	2 940	2 470	2 040	1 660	1 310	1 000	1 130	867	729
	1 000	2 940	2 470	2 020	1 640	1 280	976	1 090	835	702
	1 500	2 870	2 410	1 970	1 590	1 230	937	1 040	794	666
	2 000	2 790	2 340	1 900	1 530	1 170	891	982	746	625
	2 500	2 710	2 270	1 830	1 470	1 110	839	913	690	577
	3 000	2 620	2 190	1 750	1 400	1 040	781	836	628	523
	3 500	2 520	2 100	1 660	1 330	956	717	749	558	464
	4 000	2 410	2 010	1 560	1 250	868	646	650	477	394
	4 500	2 290	1 910	1 460	1 160	767	564	555	407	336
	5 000	2 170	1 800	1 340	1 070	666	490	480	352	290
	5 500	2 040	1 690	1 220	958	585	430	420	308	254
	6 000	1 890	1 570	1 090	852	519	381	371	271	224
	6 500	1 740	1 430	974	765	464	340	330	241	199
	7 000	1 570	1 290	880	691	417	305	295	215	177
	7 500	1 430	1 180	800	627	377	275	266	192	157
	8 000	1 310	1 080	730	573	342	250	236	170	139
	8 500	1 210	990	670	525	311	226	210	151	124
	9 000	1 120	914	616	482	280	203	188	136	
	9 500	1 030	846	569	445	253	183			
	10 000	960	786	526	411	229	166			
V–V Axis	0	2 940	2 470	2 040	1 660	1 310	1 000	1 130	867	729
	500	2 940	2 470	2 040	1 660	1 310	1 000	1 120	859	723
	1 000	2 890	2 430	1 980	1 610	1 250	952	1 050	810	682
	1 500	2 790	2 350	1 890	1 540	1 170	894	971	747	630
	2 000	2 680	2 250	1 790	1 450	1 080	825	871	673	568
	2 500	2 550	2 150	1 670	1 360	968	746	754	586	496
	3 000	2 400	2 030	1 540	1 260	847	655	614	482	410
	3 500	2 250	1 900	1 390	1 140	704	549	498	391	333
	4 000	2 070	1 760	1 230	1 010	588	458	414	325	277
	4 500	1 890	1 600	1 050	865	499	389	349	275	234
	5 000	1 670	1 420	911	750	430	335	299	235	200
	5 500	1 470	1 250	799	658	374	292	257	203	173
	6 000	1 300	1 110	706	582	327	256	217	172	147
	6 500	1 170	998	629	518	286	225	187	148	126
	7 000	1 050	899	563	465	248	195			
	7 500	953	815	507	418	218	171			
	8 000	867	742	455	378					
	8 500	792	678	406	336					
	9 000	726	621	364	301					
	9 500	664	571	328	272					
	10 000	602	517							
PROPERTIES OF 2 STARRED ANGLES										
Area (mm^2)		10 900	9 160	7 570	6 130	4 860	3 710	4 190	3 210	2 700
r_x (mm)		78.0	76.6	66.5	65.0	53.6	52.1	47.0	45.5	44.9
r_y (mm)		58.2	58.9	48.7	49.3	39.1	39.7	33.8	34.4	34.7
* r_z (mm)		29.7	29.8	24.8	25.0	19.9	20.1	17.3	17.4	17.5
* r_z / r_x		0.381	0.389	0.373	0.385	0.371	0.386	0.368	0.383	0.390
* r_z / r_y		0.510	0.506	0.509	0.507	0.508	0.506	0.512	0.507	0.504
IMPERIAL SIZE AND MASS										
Mass (lb./ft.)		57.4	48.4	40.0	32.4	25.6	19.6	22.2	17.0	14.4
Thickness (in.)		3/4	5/8	5/8	1/2	1/2	3/8	1/2	3/8	5/16
Size (in.)		6 x 6		5 x 5		4 x 4		3 1/2 x 3 1/2		

* See page 4–4 for more information regarding these tables.
See Clauses 19.1.4 and 19.1.5 CAN/CSA-S16.1-M89 for interconnecting requirements.

DOUBLE ANGLE STRUTS
Imperial Series – Star Shaped *

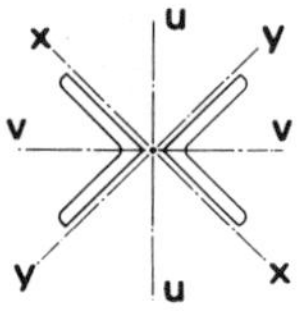

G40.21-M
300W
$\phi = 0.90$

Factored Axial Compressive Resistances – kN

Designaton (mm x mm x mm)	L 76 x 76			L 64 x 64	L 51 x 51			
	13	9.5	7.9	6.4	9.5	7.9	6.4	4.8
Mass (kg/m)	27.9	21.4	18.0	12.0	13.8	11.7	9.52	7.24
Legs	8 mm apart			8 mm apart	6 mm apart			
Effective length (KL) in millimetres with respect to indicated axis — U–U Axis								
0	958	735	619	414	474	402	327	249
500	958	734	617	409	464	394	320	243
1 000	917	699	588	385	429	364	294	223
1 500	865	658	552	354	385	324	260	196
2 000	804	608	509	318	331	276	219	164
2 500	734	551	460	275	265	218	170	126
3 000	655	487	404	224	207	170	133	98.0
3 500	564	410	339	182	167	137	107	78.6
4 000	472	343	283	151	138	113	87.5	64.4
4 500	402	292	240	128	115	93.9	72.3	52.8
5 000	347	251	207	109	95.3	76.9	59.1	43.2
5 500	302	219	180	93.7	79.5	64.2		
6 000	266	192	158	79.4				
6 500	235	170	139	68.2				
7 000	207	147	120					
7 500	181	129	106					
8 000	160	114						
V–V Axis								
0	958	734	618	413	472	402	327	249
500	941	722	608	401	447	381	310	237
1 000	870	668	564	363	382	327	267	205
1 500	779	601	508	314	295	255	210	162
2 000	669	519	441	253	202	176	146	113
2 500	534	418	359	188	147	128	107	83.0
3 000	417	327	281	146	112	97.6	81.2	63.3
3 500	336	264	226	116	83.8	73.5	61.5	48.2
4 000	277	217	187	94.8				
4 500	231	182	156	75.5				
5 000	191	151	131					
5 500	159	126	109					
PROPERTIES OF 2 STARRED ANGLES								
Area (mm^2)	3 550	2 720	2 290	1 530	1 750	1 490	1 210	922
r_x (mm)	41.9	40.3	39.7	33.8	28.9	28.1	27.3	26.7
r_y (mm)	28.7	29.2	29.6	24.7	18.9	19.2	19.5	19.8
* r_z (mm)	14.8	14.9	15.0	12.5	9.89	9.90	9.93	10.0
* r_z / r_x	0.353	0.369	0.378	0.370	0.342	0.352	0.363	0.375
* r_z / r_y	0.516	0.510	0.507	0.507	0.523	0.516	0.510	0.505
IMPERIAL SIZE AND MASS								
Mass (lb./ft.)	18.8	14.4	12.2	8.20	9.40	7.84	6.38	4.88
Thickness (in.)	1/2	3/8	5/16	1/4	3/8	5/16	1/4	3/16
Size (in.)	3 x 3			2 1/2 x 2 1/2	2 x 2			

* See page 4–4 for more information regarding these tables.
See Clauses 19.1.4 and 19.1.5 CAN/CSA-S16.1-M89 for interconnecting requirements.

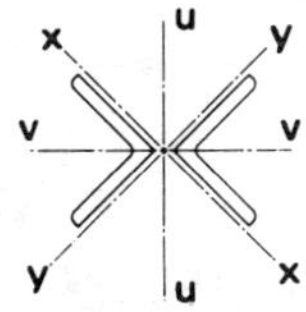

DOUBLE ANGLE STRUTS
Imperial Series – Star Shaped *

Factored Axial Compressive Resistances – kN

Designation (mm x mm x mm)		L 44 x 44				L 38 x 38	
		9.5	7.9	6.4	4.8	6.4	4.8
Mass (kg/m)		11.9	10.1	8.24	6.30	6.96	5.34
Legs		6 mm apart				6 mm apart	
Effective length (KL) in millimetres with respect to indicated axis							
U–U Axis	0	409	348	283	217	240	184
U–U Axis	500	398	339	275	210	230	176
U–U Axis	1 000	364	309	249	189	205	156
U–U Axis	1 500	320	269	216	162	171	128
U–U Axis	2 000	266	221	174	129	127	94.2
U–U Axis	2 500	203	166	130	95.2	93.9	69.3
U–U Axis	3 000	158	129	101	73.7	72.4	53.3
U–U Axis	3 500	127	103	80.3	58.8	57.3	42.1
U–U Axis	4 000	104	84.5	65.4	47.3	44.7	32.5
U–U Axis	4 500	84.6	68.2	52.2	37.8		
U–U Axis	5 000	69.3	55.8				
V–V Axis	0	408	348	284	217	239	184
V–V Axis	500	378	324	264	202	217	167
V–V Axis	1 000	309	266	218	168	167	130
V–V Axis	1 500	213	185	154	120	104	81.8
V–V Axis	2 000	143	124	104	81.2	68.8	54.4
V–V Axis	2 500	103	89.4	74.8	58.7	47.4	37.8
V–V Axis	3 000	74.0	64.7	54.5	43.0		
PROPERTIES OF 2 STARRED ANGLES							
Area (mm^2)		1 510	1 290	1 050	802	887	680
r_x (mm)		26.4	25.6	24.8	24.1	22.2	21.6
r_y (mm)		16.4	16.6	16.9	17.2	14.3	14.6
* r_z (mm)		8.67	8.66	8.68	8.73	7.42	7.45
* r_z / r_x		0.329	0.339	0.351	0.362	0.334	0.346
* r_z / r_y		0.529	0.521	0.514	0.506	0.519	0.509
IMPERIAL SIZE AND MASS							
Mass (lb./ft.)		8.00	6.78	5.54	4.24	4.68	3.60
Thickness (in.)		3/8	5/16	1/4	3/16	1/4	3/16
Size (in.)		1 3/4 x 1 3/4				1 1/2 x 1 1/2	

* See page 4–4 for more information regarding these tables.
See Clauses 19.1.4 and 19.1.5 CAN/CSA-S16.1-M89 for interconnecting requirements.

DOUBLE ANGLE STRUTS
Metric Series – Equal Leg Angles

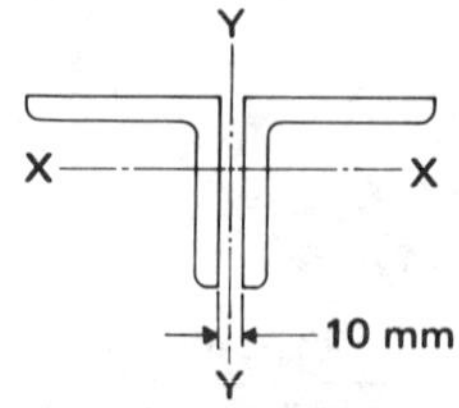

G40.21-M
300W
$\phi = 0.90$

Factored Axial Compressive Resistances – kN
Legs 10 mm Back to Back *

Effective length (KL) in millimetres with respect to indicated axis	Size (mm)	200×200						150×150			
	Thickness (mm)	30	25	20	16**	13**	10**	20	16	13	10**
	Mass (kg/m)	174	147	119	96.5	79.0	61.2	87.9	71.3	58.6	45.5
X–X Axis	0	5 990	5 080	4 100	3 060	2 170	1 380	3 020	2 450	2 010	1 280
	500	5 990	5 080	4 100	3 060	2 170	1 380	3 020	2 450	2 010	1 280
	1 000	5 900	5 000	4 050	3 030	2 150	1 370	2 920	2 370	1 950	1 250
	1 500	5 710	4 840	3 920	2 960	2 110	1 340	2 770	2 250	1 850	1 210
	2 000	5 490	4 650	3 770	2 880	2 060	1 310	2 600	2 120	1 740	1 160
	2 500	5 240	4 450	3 600	2 800	2 000	1 280	2 410	1 970	1 620	1 110
	3 000	4 960	4 220	3 420	2 700	1 940	1 240	2 190	1 790	1 480	1 040
	3 500	4 660	3 960	3 220	2 580	1 870	1 200	1 950	1 600	1 320	968
	4 000	4 320	3 680	3 000	2 440	1 780	1 150	1 660	1 370	1 140	874
	4 500	3 960	3 380	2 760	2 250	1 690	1 090	1 420	1 170	970	763
	5 000	3 560	3 060	2 510	2 040	1 580	1 030	1 230	1 010	840	661
	5 500	3 140	2 690	2 210	1 810	1 460	958	1 070	888	735	579
	6 000	2 790	2 390	1 970	1 610	1 330	888	947	783	649	511
	6 500	2 500	2 150	1 760	1 440	1 190	826	841	696	577	454
	7 000	2 250	1 930	1 590	1 300	1 080	771	752	622	516	406
	7 500	2 040	1 750	1 440	1 180	977	721	669	557	463	365
	8 000	1 860	1 600	1 310	1 070	890	675	592	492	409	323
	8 500	1 700	1 460	1 200	984	815	634	527	439	364	288
	9 000	1 560	1 340	1 100	903	749	586	473	394	327	258
	9 500	1 440	1 230	1 020	832	690	540				
	10 000	1 310	1 130	936	768	637	499				
Y–Y Axis	0	5 990	5 080	4 100	2 270	1 530	846	3 020	2 450	2 010	902
	500	5 990	5 080	4 100	2 270	1 530	846	3 020	2 450	2 010	902
	1 000	5 990	5 080	4 100	2 270	1 530	846	3 000	2 430	1 990	901
	1 500	5 900	4 990	4 030	2 270	1 520	845	2 910	2 360	1 940	900
	2 000	5 770	4 880	3 940	2 270	1 520	843	2 820	2 280	1 870	897
	2 500	5 630	4 760	3 840	2 260	1 520	841	2 710	2 200	1 800	894
	3 000	5 470	4 630	3 740	2 250	1 510	839	2 590	2 100	1 720	888
	3 500	5 310	4 490	3 620	2 250	1 510	836	2 470	1 990	1 630	881
	4 000	5 130	4 330	3 490	2 230	1 500	832	2 330	1 880	1 530	869
	4 500	4 940	4 170	3 360	2 210	1 490	827	2 170	1 750	1 430	853
	5 000	4 730	3 990	3 210	2 190	1 480	820	2 010	1 620	1 320	831
	5 500	4 520	3 810	3 060	2 150	1 460	812	1 840	1 470	1 200	802
	6 000	4 290	3 610	2 900	2 110	1 440	803	1 640	1 310	1 070	766
	6 500	4 050	3 400	2 730	2 040	1 410	791	1 470	1 180	956	723
	7 000	3 790	3 180	2 550	1 960	1 380	776	1 330	1 060	864	666
	7 500	3 500	2 930	2 340	1 870	1 340	757	1 210	967	785	594
	8 000	3 210	2 690	2 140	1 770	1 290	734	1 100	883	717	531
	8 500	2 960	2 480	1 980	1 660	1 230	710	1 010	809	657	477
	9 000	2 740	2 290	1 830	1 540	1 170	683	932	745	604	429
	9 500	2 540	2 130	1 700	1 420	1 090	655	860	687	558	389
	10 000	2 370	1 980	1 580	1 290	1 000	627	796	636	516	353

PROPERTIES OF 2 ANGLES — 10 mm BACK TO BACK

Area (mm2)	22 200	18 800	15 200	12 300	10 100	7 800	11 200	9 090	7 460	5 800
r_x (mm)	60.3	60.8	61.6	62.1	62.6	63.0	45.5	46.1	46.4	46.8
r_y (mm)	89.3	88.5	87.6	87.0	86.6	86.1	67.5	66.8	66.3	65.8
* r_z (mm)	39.0	39.1	39.3	39.5	39.7	39.9	29.3	29.4	29.6	29.8
* r_z / r_x	0.647	0.642	0.638	0.636	0.634	0.633	0.644	0.638	0.637	0.637
* r_z / r_y	0.437	0.442	0.449	0.454	0.458	0.463	0.434	0.440	0.446	0.453

IMPERIAL SIZE AND MASS

Mass (lb./ft.)	117	98.8	80.0	64.8	53.1	41.1	59.1	47.9	39.4	30.6
Thickness (in.)	1.18	0.98	0.79	0.63	0.51	0.39	0.79	0.63	0.51	0.39
Size (in.)	7.87×7.87						5.91×5.91			

* See page 4–4 for more information regarding these tables.
** Factored axial compressive resistances calculated according to Clause 13.3.3 CAN/CSA-S16.1-M89.

G40.21-M
300W
$\phi = 0.90$

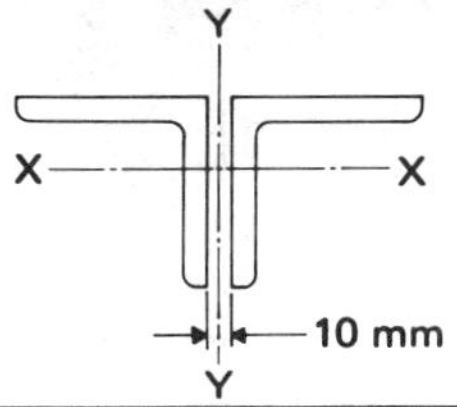

DOUBLE ANGLE STRUTS
Metric Series – Equal Leg Angles

Factored Axial Compressive Resistances – kN
Legs 10 mm Back to Back *

Size (mm)			125 x 125				100 x 100				
Thickness (mm)			16	13	10**	8**	16	13	10	8**	6**
Mass (kg/m)			58.8	48.4	37.7	30.4	46.2	38.2	29.8	24.1	18.3
Effective length (KL) in millimetres with respect to indicated axis	X–X Axis	0	2 020	1 660	1 190	827	1 590	1 310	1 030	764	474
		500	2 020	1 660	1 190	825	1 570	1 290	1 010	757	470
		1 000	1 910	1 580	1 150	800	1 450	1 200	942	721	450
		1 500	1 790	1 470	1 100	767	1 310	1 090	855	674	424
		2 000	1 640	1 350	1 040	727	1 140	952	750	609	391
		2 500	1 470	1 210	952	679	939	789	627	510	349
		3 000	1 270	1 060	830	620	735	617	491	401	297
		3 500	1 040	871	689	545	594	499	397	324	250
		4 000	871	727	574	468	490	412	329	268	207
		4 500	739	617	488	398	411	346	276	225	174
		5 000	635	531	420	342	345	292	234	191	148
		5 500	552	461	365	298	287	243	195	160	124
		6 000	483	404	320	261	244	206	165	135	105
		6 500	417	350	278	228					
		7 000	362	304	241	198					
		7 500	318	267	212	173					
	Y–Y Axis	0	2 020	1 660	886	575	1 590	1 310	1 030	565	320
		500	2 020	1 660	885	575	1 590	1 310	1 030	565	319
		1 000	1 980	1 630	884	574	1 540	1 270	989	564	318
		1 500	1 910	1 570	881	573	1 460	1 210	940	561	317
		2 000	1 830	1 500	877	570	1 380	1 130	883	556	315
		2 500	1 740	1 430	871	567	1 280	1 050	818	547	311
		3 000	1 640	1 340	859	561	1 170	959	744	531	305
		3 500	1 520	1 250	841	552	1 050	856	662	501	295
		4 000	1 400	1 140	812	539	905	736	566	458	281
		4 500	1 260	1 030	764	521	773	629	483	406	260
		5 000	1 110	900	706	496	669	544	418	348	234
		5 500	974	792	640	464	586	476	365	292	204
		6 000	865	703	567	424	517	420	322	248	176
		6 500	774	628	491	372	460	373	286	213	153
		7 000	697	566	427	328	411	334	256	184	134
		7 500	630	512	375	290	370	299	228	161	118
		8 000	573	465	332	258	327	264	202	142	104
		8 500	523	424	295	230	291	236	180	127	92.8
		9 000	479	388	265	207	261	211	161	113	83.2
		9 500	434	351	238	187					
		10 000	393	318	216	169					
PROPERTIES OF 2 ANGLES — 10 mm BACK TO BACK											
Area (mm^2)			7 490	6 160	4 800	3 870	5 890	4 860	3 800	3 070	2 330
r_x (mm)			38.0	38.4	38.8	39.1	30.0	30.4	30.8	31.0	31.4
r_y (mm)			56.7	56.2	55.7	55.4	46.7	46.2	45.6	45.3	45.0
* r_z (mm)			24.4	24.5	24.7	24.8	19.5	19.5	19.7	19.8	19.9
* r_z / r_x			0.642	0.638	0.636	0.634	0.650	0.642	0.640	0.638	0.634
* r_z / r_y			0.430	0.436	0.443	0.448	0.418	0.422	0.432	0.437	0.442
IMPERIAL SIZE AND MASS											
Mass (lb./ft.)			39.5	32.5	25.3	20.4	31.0	25.7	20.0	16.2	12.3.
Thickness (in.)			0.63	0.51	0.39	0.31	0.63	0.51	0.39	0.31	0.24
Size (in.)			4.92 x 4.92				3.94 x 3.94				

* See page 4–4 for more information regarding these tables.
** Factored axial compressive resistances calculated according to Clause 13.3.3 CAN/CSA-S16.1-M89.

DOUBLE ANGLE STRUTS
Metric Series – Equal Leg Angles

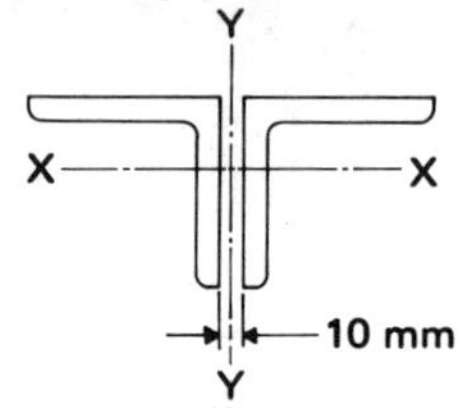

G40.21-M
300W
$\phi = 0.90$

Factored Axial Compressive Resistances – kN
Legs 10 mm Back to Back *

Size (mm)	90 x 90				75 x 75				
Thickness (mm)	13	10	8	6**	13	10	8	6**	5**
Mass (kg/m)	34.1	26.7	21.6	16.4	28.0	22.0	17.8	13.6	11.4
Effective length (KL) in millimetres with respect to indicated axis — X–X Axis									
0	1 170	918	743	459	962	756	613	430	319
500	1 150	898	727	453	925	729	591	420	312
1 000	1 050	826	670	430	823	651	529	389	291
1 500	932	733	596	399	689	548	447	343	261
2 000	784	620	506	358	518	417	342	265	218
2 500	607	483	396	303	382	308	253	196	165
3 000	473	376	309	238	294	237	195	151	128
3 500	380	303	249	192	233	189	155	120	102
4 000	312	249	204	158	183	149	122	95.5	80.8
4 500	258	206	170	132		119	97.9	76.4	64.6
5 000	211	169	139	108					
5 500		141	116	90.0					
Y–Y Axis									
0	1 170	918	742	323	961	756	613	317	223
500	1 170	918	742	322	956	751	608	316	223
1 000	1 120	877	709	321	904	709	574	315	222
1 500	1 060	827	668	320	840	658	532	312	220
2 000	986	769	620	316	764	596	481	306	216
2 500	901	701	565	311	675	524	422	291	209
3 000	806	624	502	301	572	440	351	263	196
3 500	696	535	429	285	465	358	285	226	177
4 000	582	447	358	262	387	297	237	182	145
4 500	496	381	305	226	328	252	201	146	118
5 000	428	329	263	189	281	216	172	120	97.4
5 500	373	286	229	159	244	187	149	99.9	81.5
6 000	328	252	201	136	211	160	127	84.4	69.1
6 500	291	223	178	117	181	138	109	72.2	59.3
7 000	256	195	156	101	157	120	94.9		
7 500	225	171	137	88.9					
8 000	199	151	121	78.5					
PROPERTIES OF 2 ANGLES — 10 mm BACK TO BACK									
Area (mm^2)	4 340	3 400	2 750	2 090	3 560	2 800	2 270	1 730	1 450
r_x (mm)	27.2	27.5	27.8	28.1	22.4	22.8	23.0	23.3	23.4
r_y (mm)	42.2	41.6	41.3	40.9	36.2	35.6	35.2	34.9	34.7
* r_z (mm)	17.6	17.6	17.7	17.9	14.6	14.6	14.7	14.8	14.9
* r_z / r_x	0.647	0.639	0.636	0.637	0.653	0.642	0.640	0.635	0.636
* r_z / r_y	0.417	0.423	0.429	0.438	0.403	0.410	0.418	0.424	0.429
IMPERIAL SIZE AND MASS									
Mass (lb./ft.)	22.0	17.9	14.5	11.0	18.8	14.8	12.0	9.14	7.66
Thickness (in.)	0.51	0.39	0.31	0.24	0.51	0.39	0.31	0.24	0.20
Size (in.)	3.54 x 3.54				2.95 x 2.95				

* See page 4–4 for more information regarding these tables.
** Factored axial compressive resistances calculated according to Clause 13.3.3 CAN/CSA-S16.1-M89.

G40.21-M
300W
$\phi = 0.90$

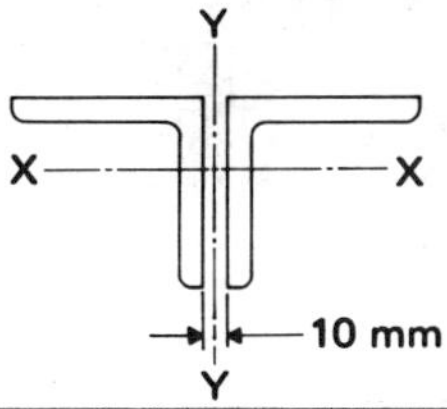

DOUBLE ANGLE STRUTS
Metric Series – Equal Leg Angles

Factored Axial Compressive Resistances – kN

Legs 10 mm Back to Back *

		Size (mm)	65 x 65				55 x 55					
		Thickness (mm)	10	8	6	5**	10	8	6	5	4**	3**
		Mass (kg/m)	18.8	15.3	11.7	9.8	15.7	12.8	9.8	8.2	6.7	5.0
Effective length (KL) in millimetres with respect to indicated axis	X–X Axis	0	648	527	402	303	540	441	337	284	198	121
		500	615	500	383	293	501	409	314	264	193	119
		1 000	530	433	333	267	409	336	260	219	177	110
		1 500	417	344	266	224	282	233	183	155	147	93.5
		2 000	289	240	187	158	189	157	123	105	91.7	68.1
		2 500	211	176	137	116	136	113	89.1	75.5	58.7	45.5
		3 000	161	134	105	88.5	98.0	81.7	64.8	55.1	40.8	31.6
		3 500	122	102	80.2	67.9						
		4 000			62.2	52.7						
	Y–Y Axis	0	648	526	402	221	540	440	338	284	136	69.8
		500	640	519	397	220	529	430	330	277	136	69.7
		1 000	598	485	370	219	487	395	302	254	135	68.9
		1 500	545	441	336	216	433	350	267	224	131	67.3
		2 000	482	389	295	209	367	296	224	187	123	63.8
		2 500	407	327	246	195	287	229	172	143	101	57.6
		3 000	323	258	194	167	224	179	134	112	74.6	49.3
		3 500	261	209	157	132	180	144	108	89.5	56.5	39.1
		4 000	216	173	129	103	148	118	88.4	73.4	44.0	31.2
		4 500	182	145	109	82.8	123	97.4	72.6	60.1	35.1	25.3
		5 000	155	123	92.4	67.6	100	79.7	59.4	49.2	28.6	20.8
		5 500	130	103	76.9	56.2	83.9					
		6 000	110	87.4	65.2	47.5						
PROPERTIES OF 2 ANGLES — 10 mm BACK TO BACK												
		Area (mm^2)	2 400	1 950	1 490	1 250	2 000	1 630	1 250	1 050	848	642
		r_x (mm)	19.5	19.8	20.1	20.2	16.4	16.6	16.9	17.0	17.1	17.3
		r_y (mm)	31.7	31.3	30.9	30.7	27.7	27.3	26.9	26.7	26.5	26.3
		* r_z (mm)	12.7	12.7	12.8	12.9	10.7	10.7	10.8	10.8	10.9	11.0
		* r_z / r_x	0.650	0.641	0.637	0.638	0.654	0.644	0.640	0.636	0.636	0.636
		* r_z / r_y	0.401	0.406	0.414	0.420	0.386	0.392	0.401	0.404	0.411	0.418
IMPERIAL SIZE AND MASS												
		Mass (lb./ft.)	12.6	10.3	7.86	6.59	10.5	8.60	6.59	5.54	4.48	3.39
		Thickness (in.)	0.39	0.31	0.24	0.20	0.39	0.31	0.24	0.20	0.16	0.12
		Size (in.)	2.56 x 2.56				2.17 x 2.17					

* See page 4–4 for more information regarding these tables.

** Factored axial compressive resistances calculated according to Clause 13.3.3 CAN/CSA-S16.1-M89.

DOUBLE ANGLE STRUTS
Metric Series – Equal Leg Angles

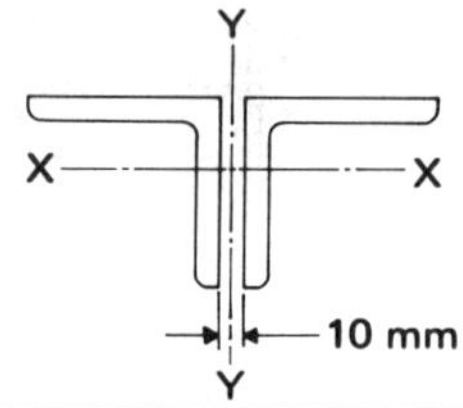

G40.21-M
300W
$\phi = 0.90$

Factored Axial Compressive Resistances – kN
Legs 10 mm Back to Back *

Size (mm)		45 x 45					35 x 35			
Thickness (mm)		8	6	5	4	3**	6	5	4	3**
Mass (kg/m)		10.3	7.9	6.7	5.4	4.1	6.0	5.1	4.1	3.2
Effective length (KL) in millimetres with respect to indicated axis: X–X Axis	0	354	272	230	186	115	207	176	143	104
	500	317	245	207	168	110	174	148	120	96.9
	1 000	234	184	155	127	96.5	102	87.8	72.2	66.1
	1 500	140	111	94.6	77.3	67.3	57.7	49.5	40.8	30.8
	2 000	92.1	73.4	62.5	51.1	37.9	34.8	30.0	24.8	17.3
	2 500	61.8	49.7	42.4	34.8	24.2				
Y–Y Axis	0	354	273	230	186	75.1	207	176	143	72.7
	500	342	263	221	179	74.9	196	166	135	72.3
	1 000	307	235	197	159	73.8	168	142	115	69.5
	1 500	261	198	166	134	70.7	131	109	87.6	58.5
	2 000	202	151	126	101	62.3	90.0	74.5	59.6	39.3
	2 500	149	112	92.9	74.2	45.3	65.6	54.3	43.4	25.9
	3 000	115	86.2	71.6	57.2	33.0	49.9	41.2	32.9	18.2
	3 500	91.8	68.5	56.8	45.4	24.9	37.5	30.8	24.5	13.5
	4 000	73.0	54.0	44.7	35.6	19.3				
	4 500	58.4	43.2	35.8	28.5					
PROPERTIES OF 2 ANGLES — 10 mm BACK TO BACK										
Area (mm^2)		1 310	1 010	850	688	522	768	650	528	402
r_x (mm)		13.4	13.7	13.8	13.9	14.0	10.5	10.6	10.7	10.8
r_y (mm)		23.4	22.9	22.7	22.5	22.3	19.1	18.8	18.6	18.4
* r_z (mm)		8.76	8.79	8.82	8.87	8.93	6.81	6.83	6.86	6.91
* r_z / r_x		0.653	0.644	0.641	0.638	0.636	0.652	0.646	0.641	0.637
* r_z / r_y		0.374	0.384	0.389	0.394	0.400	0.357	0.363	0.369	0.376
IMPERIAL SIZE AND MASS										
Mass (lb./ft.)		6.92	5.32	4.48	3.63	2.75	4.05	3.43	2.78	2.12
Thickness (in.)		0.31	0.24	0.20	0.16	0.12	0.24	0.20	0.16	0.12
Size (in.)		1.77 x 1.77					1.38 x 1.38			

* See page 4–4 for more information regarding these tables.
** Factored axial compressive resistances calculated according to Clause 13.3.3 CAN/CSA-S16.1-M89.

G40.21-M 300W
$\phi = 0.90$

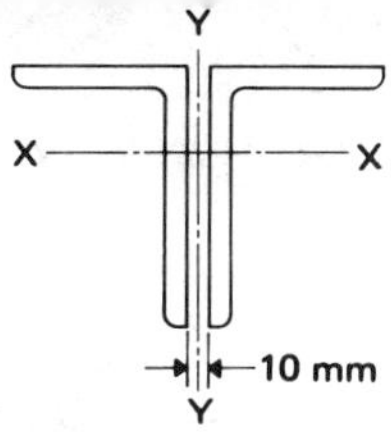

DOUBLE ANGLE STRUTS
Metric Series – Unequal Leg Angles

Factored Axial Compressive Resistances – kN
Long Legs 10 mm Back to Back *

Size (mm)	150 x 100				125 x 90			
Thickness (mm)	16	13	10**	8**	16	13	10**	8**
Mass (kg/m)	58.8	48.4	37.7	30.4	50.0	41.2	32.2	26.0
Effective length (KL) in millimetres with respect to indicated axis — X–X Axis								
0	2 020	1 660	1 150	816	1 720	1 420	1 060	778
500	2 020	1 660	1 150	816	1 720	1 410	1 050	777
1 000	1 960	1 610	1 120	800	1 630	1 350	1 010	753
1 500	1 870	1 540	1 080	777	1 530	1 260	962	717
2 000	1 760	1 450	1 030	749	1 410	1 160	900	673
2 500	1 640	1 350	977	716	1 270	1 050	825	621
3 000	1 500	1 240	913	676	1 110	921	726	560
3 500	1 350	1 120	840	624	920	769	608	486
4 000	1 170	977	756	564	768	642	508	415
4 500	1 000	834	659	496	652	545	432	352
5 000	869	723	571	442	561	469	372	304
5 500	761	633	501	396	488	408	324	264
6 000	672	559	442	358	427	358	284	232
6 500	598	498	394	320	373	313	249	205
7 000	535	446	353	286	324	272	216	177
7 500	481	401	317	258	284	238	190	156
8 000	428	358	284	231				138
8 500	381	318	253	206				
9 000	342	286	227	185				
9 500		258	205	167				
Y–Y Axis								
0	2 020	1 660	829	556	1 720	1 420	793	552
500	2 020	1 660	827	554	1 720	1 410	791	552
1 000	1 930	1 590	822	550	1 630	1 340	786	548
1 500	1 830	1 500	812	544	1 530	1 260	775	541
2 000	1 700	1 390	795	532	1 410	1 160	753	527
2 500	1 550	1 270	767	515	1 270	1 040	717	505
3 000	1 380	1 130	726	489	1 110	906	660	472
3 500	1 190	964	671	453	926	750	586	427
4 000	995	805	602	406	773	626	497	371
4 500	847	685	511	355	657	532	404	306
5 000	731	591	426	311	565	457	333	255
5 500	638	515	360	270	491	397	278	214
6 000	561	453	307	233	431	348	235	183
6 500	496	401	265	202	376	303	202	157
7 000	436	351	230	177	327	263	175	137
7 500	382	307	202	156	287	231	153	120
8 000	338	272	179	139				
PROPERTIES OF 2 ANGLES — 10 mm BACK TO BACK								
Area (mm^2)	7 490	6 160	4 800	3 870	6 370	5 250	4 100	3 310
r_x (mm)	47.4	47.8	48.3	48.5	39.0	39.4	39.8	40.1
r_y (mm)	41.9	41.4	40.9	40.6	39.2	38.7	38.1	37.8
* r_z (mm)	21.6	21.7	21.9	22.0	19.2	19.3	19.5	19.6
* r_z / r_x	0.456	0.454	0.453	0.454	0.492	0.490	0.490	0.489
* r_z / r_y	0.516	0.524	0.535	0.542	0.490	0.499	0.512	0.519
IMPERIAL SIZE AND MASS								
Mass (lb./ft.)	39.5	32.5	25.3	20.4	33.6	27.7	21.6	17.5
Thickness (in.)	0.63	0.51	0.39	0.31	0.63	0.51	0.39	0.31
Size (in.)	5.91 x 3.94				4.92 x 3.54			

* See page 4–4 for more information regarding these tables.
** Factored axial compressive resistances calculated according to Clause 1.3.3.3 CAN/CSA-S16.1-M89.

DOUBLE ANGLE STRUTS
Metric Series – Unequal Leg Angles

G40.21-M
300W
$\phi = 0.90$

Factored Axial Compressive Resistances – kN
Long Legs 10 mm Back to Back *

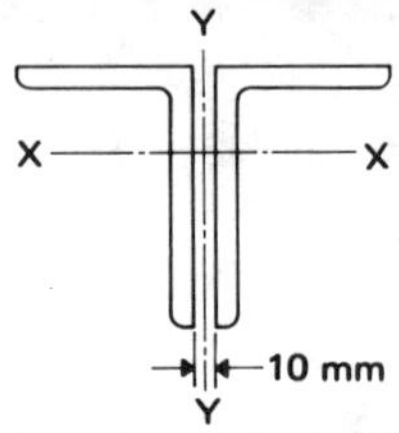

Size (mm)		125 x 75				100 x 90			
Thickness (mm)		13	10**	8**	6**	13	10	8**	6**
Mass (kg/m)		38.2	29.8	24.1	18.3	36.1	28.3	22.9	17.3
Effective length (KL) in millimetres with respect to indicated axis — X–X Axis	0	1 310	975	720	465	1 240	972	747	466
	500	1 310	974	719	465	1 220	959	740	463
	1 000	1 250	937	693	451	1 140	894	703	443
	1 500	1 170	890	660	434	1 030	813	650	418
	2 000	1 080	835	620	413	906	715	581	385
	2 500	975	767	574	387	756	600	490	343
	3 000	856	676	519	352	592	472	387	289
	3 500	716	568	452	308	479	382	313	239
	4 000	598	474	387	267	396	316	259	198
	4 500	508	403	329	235	332	265	218	167
	5 000	438	347	284	208	281	226	185	142
	5 500	381	302	247	186	234	188	155	119
	6 000	334	265	217	166	199	159	131	101
	6 500	293	233	191	147				
	7 000	254	203	166	128				
	7 500	223	178	146	112				
	8 000		157	129	99.1				
Y–Y Axis	0	1 310	739	515	298	1 240	972	552	322
	500	1 300	736	513	297	1 240	971	551	321
	1 000	1 210	727	506	293	1 190	926	549	320
	1 500	1 100	705	493	284	1 120	871	544	318
	2 000	974	665	467	270	1 040	807	534	314
	2 500	822	599	428	247	942	732	515	307
	3 000	651	510	373	215	836	647	484	296
	3 500	527	403	302	184	713	547	439	278
	4 000	436	316	240	157	596	457	383	253
	4 500	367	253	194	133	507	388	320	218
	5 000	312	207	160	112	437	335	263	185
	5 500	262	173	134	94.4	381	292	220	156
	6 000	222	146	114	80.7	334	256	186	134
	6 500					296	226	160	115
	7 000					258	197	138	100
	7 500					226	172	121	88.1
	8 000					200	152	107	
PROPERTIES OF 2 ANGLES — 10 mm BACK TO BACK									
Area (mm^2)		4 860	3 800	3 070	2 330	4 600	3 600	2 910	2 210
r_x (mm)		39.6	40.0	40.3	40.6	30.7	31.1	31.4	31.6
r_y (mm)		31.6	31.0	30.7	30.3	41.1	40.5	40.2	39.8
* r_z (mm)		16.2	16.3	16.4	16.6	18.4	18.5	18.6	18.7
* r_z / r_x		0.409	0.407	0.407	0.409	0.599	0.595	0.592	0.591
* r_z / r_y		0.513	0.526	0.534	0.548	0.448	0.457	0.463	0.470
IMPERIAL SIZE AND MASS									
Mass (lb./ft.)		25.7	20.0	16.2	12.3	24.3	19.0	15.4	11.6
Thickness (in.)		0.51	0.39	0.31	0.24	0.51	0.39	0.31	0.24
Size (in.)		4.92 x 2.95				3.94 x 3.54			

* See page 4–4 for more information regarding these tables.
** Factored axial compressive resistances calculated according to Clause 1.3.3.3 CAN/CSA-S16.1-M89.

G40.21-M
300W
$\phi = 0.90$

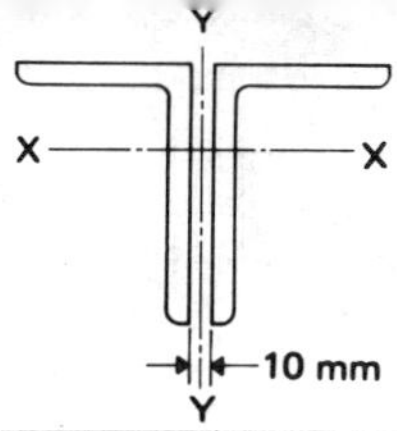

DOUBLE ANGLE STRUTS
Metric Series – Unequal Leg Angles

Factored Axial Compressive Resistances – kN

Long Legs 10 mm Back to Back *

Size (mm)	100 x 75				90 x 75				
Thickness (mm)	13	10	8**	6**	13	10	8	6**	5**
Mass (kg/m)	33.1	25.9	21.0	15.9	31.0	24.3	19.7	15.0	12.6
Effective length (KL) in millimetres with respect to indicated axis — X–X Axis									
0	1 140	891	689	452	1 070	837	678	444	327
500	1 120	879	682	448	1 040	820	664	439	324
1 000	1 050	821	645	429	960	756	613	416	308
1 500	951	748	597	404	853	674	548	386	287
2 000	836	660	537	371	722	573	468	343	260
2 500	701	557	455	325	563	451	370	283	223
3 000	552	440	361	269	439	352	289	223	183
3 500	446	356	292	224	353	283	233	180	151
4 000	369	295	242	185	290	233	192	148	125
4 500	310	248	204	156	241	194	160	124	104
5 000	264	211	173	133	197	159	131	102	86.0
5 500	220	177	146	112	165	133	110	85.0	71.8
6 000	186	150	123	94.8					
Effective length (KL) in millimetres with respect to indicated axis — Y–Y Axis									
0	1 140	891	515	318	1 070	837	678	320	221
500	1 130	882	513	318	1 060	830	671	319	220
1 000	1 060	828	509	315	997	781	631	317	219
1 500	974	760	498	310	921	719	581	313	216
2 000	872	678	476	298	830	646	520	304	211
2 500	754	583	438	279	724	560	449	286	202
3 000	612	470	383	250	597	458	365	259	187
3 500	497	381	315	212	485	372	296	221	162
4 000	412	316	248	173	403	309	246	177	137
4 500	348	267	199	141	341	261	207	143	114
5 000	297	228	163	117	292	223	177	118	94.6
5 500	255	194	135	97.9	252	192	152	98.6	79.7
6 000	216	164	114	83.2	214	163	129	83.6	67.9
6 500	185	141	97.8		184	140	111	71.7	58.4
PROPERTIES OF 2 ANGLES — 10 mm BACK TO BACK									
Area (mm^2)	4 210	3 300	2 670	2 030	3 950	3 100	2 510	1 910	1 600
r_x (mm)	31.1	31.5	31.8	32.0	27.6	28.0	28.3	28.6	28.7
r_y (mm)	33.6	33.1	32.7	32.4	34.6	34.0	33.6	33.3	33.1
* r_z (mm)	16.0	16.1	16.2	16.3	15.6	15.7	15.8	15.9	16.0
* r_z / r_x	0.515	0.511	0.510	0.509	0.564	0.561	0.559	0.556	0.557
* r_z / r_y	0.476	0.486	0.495	0.503	0.451	0.462	0.470	0.477	0.483
IMPERIAL SIZE AND MASS									
Mass (lb./ft.)	22.2	17.4	14.1	10.7	20.8	16.3	13.2	10.1	8.47
Thickness (in.)	0.51	0.39	0.31	0.24	0.51	0.39	0.31	0.24	0.20
Size (in.)	3.94 x 2.95				3.54 x 2.95				

* See page 4–4 for more information regarding these tables.

** Factored axial compressive resistances calculated according to Clause 1.3.3.3 CAN/CSA-S16.1-M89.

DOUBLE ANGLE STRUTS
Metric Series – Unequal Leg Angles

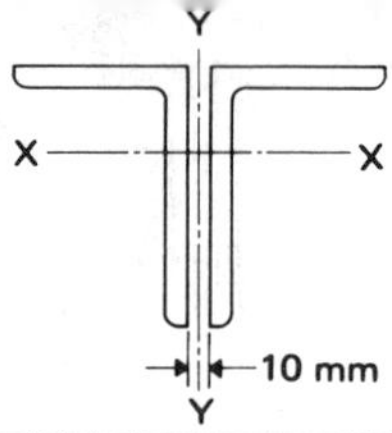

G40.21-M
300W
$\phi = 0.90$

Factored Axial Compressive Resistances – kN
Long Legs 10 mm Back to Back *

Size (mm)		90 x 65				80 x 60			
Thickness (mm)		10	8	6**	5**	10	8	6**	5**
Mass (kg/m)		22.8	18.5	14.0	11.8	20.4	16.6	12.6	10.6
Effective length (KL) in millimetres with respect to indicated axis — X–X Axis	0	783	635	430	319	702	570	405	309
	500	768	623	424	316	682	554	396	304
	1 000	709	575	398	300	619	503	366	285
	1 500	633	515	365	280	536	437	327	258
	2 000	541	442	322	253	435	356	273	219
	2 500	428	352	268	212	324	266	205	173
	3 000	334	275	212	173	252	207	159	135
	3 500	269	221	171	144	201	165	127	108
	4 000	221	182	141	119	164	135	104	88.4
	4 500	185	152	118	99.3	131	108	83.7	71.4
	5 000	152	126	97.4	82.3	108	88.6	68.6	58.4
	5 500	127	105	81.3	68.6				
Y–Y Axis	0	783	634	306	219	702	570	297	219
	500	769	623	305	218	687	557	296	218
	1 000	712	576	301	216	632	512	292	215
	1 500	640	516	292	211	562	453	281	208
	2 000	552	443	274	199	476	381	259	193
	2 500	444	353	246	179	371	295	223	169
	3 000	347	276	204	150	289	230	173	136
	3 500	280	222	157	123	233	185	131	104
	4 000	231	183	124	98.3	191	152	102	82.0
	4 500	193	153	99.4	79.8	159	125	81.2	65.9
	5 000	160	126	81.5	65.9	130	102	66.3	54.1
	5 500	134	105	67.9	55.2	108			
PROPERTIES OF 2 ANGLES — 10 mm BACK TO BACK									
Area (mm^2)		2 900	2 350	1 790	1 500	2 600	2 110	1 610	1 350
r_x (mm)		28.3	28.6	28.9	29.0	25.0	25.2	25.4	25.6
r_y (mm)		29.1	28.7	28.4	28.2	27.6	27.2	26.8	26.6
* r_z (mm)		13.9	14.0	14.2	14.2	12.8	12.9	13.0	13.0
* r_z / r_x		0.491	0.490	0.492	0.490	0.513	0.512	0.511	0.507
* r_z / r_y		0.478	0.488	0.500	0.504	0.464	0.474	0.485	0.489
IMPERIAL SIZE AND MASS									
Mass (lb./ft.)		15.3	12.4	9.41	7.93	13.7	11.2	8.47	7.12
Thickness (in.)		0.39	0.31	0.24	0.20	0.39	0.31	0.24	0.20
Size (in.)		3.54 x 2.56				3.15 x 2.36			

* See page 4–4 for more information regarding these tables.
** Factored axial compressive resistances calculated according to Clause 1.3.3.3 CAN/CSA-S16.1-M89.

G40.21-M
300W
$\phi = 0.90$

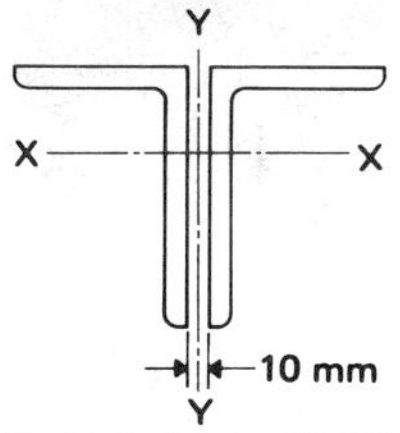

DOUBLE ANGLE STRUTS
Metric Series – Unequal Leg Angles

Factored Axial Compressive Resistances – kN
Long Legs 10 mm Back to Back *

Size (mm)		75 x 50			65 x 50			
Thickness (mm)		8	6**	5**	8	6	5**	4**
Mass (kg/m)		14.7	11.2	9.4	13.4	10.3	8.6	7.0
Effective length (KL) in millimetres with respect to indicated axis — X–X Axis	0	505	367	288	462	353	280	200
	500	488	358	280	440	337	269	197
	1 000	439	328	258	383	295	241	186
	1 500	375	288	228	307	238	200	167
	2 000	293	228	188	216	169	143	133
	2 500	217	168	142	158	124	105	91.0
	3 000	168	130	110	121	94.9	80.3	63.2
	3 500	134	104	87.9	92.9	73.2	62.0	46.4
	4 000	107	83.6	70.8	72.1	56.8	48.1	35.5
	4 500	85.4	66.8	56.5				
Y–Y Axis	0	505	276	206	462	354	205	134
	500	487	274	204	447	342	204	133
	1 000	435	267	199	402	306	200	130
	1 500	367	248	186	343	260	188	122
	2 000	280	209	162	268	201	164	104
	2 500	207	157	123	198	148	126	78.2
	3 000	160	112	89.8	153	115	90.7	57.8
	3 500	127	83.7	67.7	122	91.2	67.8	43.9
	4 000	100	64.8	52.7	97.7	72.4	52.5	34.2
	4 500	80.0	51.5		78.1	57.9	41.8	27.4
PROPERTIES OF 2 ANGLES — 10 mm BACK TO BACK								
Area (mm^2)		1 870	1 430	1 200	1 710	1 310	1 100	888
r_x (mm)		23.7	24.0	24.1	20.2	20.5	20.6	20.8
r_y (mm)		22.9	22.5	22.3	23.7	23.3	23.1	23.0
* r_z (mm)		10.8	10.9	10.9	10.6	10.7	10.8	10.8
* r_z / r_x		0.456	0.455	0.452	0.524	0.522	0.523	0.519
* r_z / r_y		0.472	0.484	0.489	0.447	0.459	0.468	0.470
IMPERIAL SIZE AND MASS								
Mass (lb./ft.)		9.88	7.53	6.33	9.00	6.92	5.80	4.68
Thickness (in.)		0.31	0.24	0.20	0.31	0.24	0.20	0.16
Size (in.)		2.95 x 1.97			2.56 x 1.97			

* See page 4–4 for more information regarding these tables.
** Factored axial compressive resistances calculated according to Clause 1.3.3.3 CAN/CSA-S16.1-M89.

DOUBLE ANGLE STRUTS
Metric Series – Unequal Leg Angles

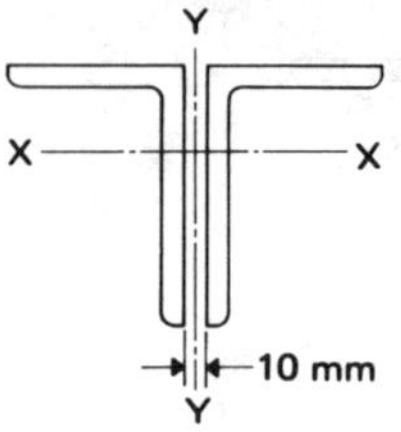

G40.21-M
300W
$\phi = 0.90$

Factored Axial Compressive Resistances – kN
Long Legs 10 mm Back to Back *

Size (mm)	55 x 35				45 x 30			
Thickness (mm)	6	5	4**	3**	6	5	4	3**
Mass (kg/m)	7.9	6.7	5.4	4.1	6.5	5.5	4.5	3.4
Effective length (KL) in millimetres with respect to indicated axis — X–X Axis								
0	272	230	170	113	224	189	153	104
500	255	215	165	110	202	171	139	99.0
1 000	213	180	150	102	153	131	107	85.1
1 500	154	131	123	84.3	94.0	81.0	66.4	59.7
2 000	104	88.2	78.8	59.0	62.2	53.7	44.0	33.6
2 500	75.2	63.8	50.5	39.2	42.5	36.9	30.3	21.5
3 000	55.4	47.1	35.0	27.2				
3 500		35.2	25.7	20.0				
Y–Y Axis								
0	273	230	121	70.4	224	189	153	70.2
500	255	214	119	69.2	206	174	141	69.0
1 000	211	177	110	64.1	165	138	111	62.4
1 500	150	125	85.9	51.0	109	90.5	72.2	43.9
2 000	101	84.0	53.6	35.9	72.9	60.5	48.2	27.0
2 500	73.3	60.6	35.7	24.8	52.1	43.1	34.1	17.9
3 000	53.6	44.1	25.3	17.9	36.9	30.4	24.1	
PROPERTIES OF 2 ANGLES — 10 mm BACK TO BACK								
Area (mm^2)	1 010	850	688	522	828	700	568	432
r_x (mm)	17.4	17.5	17.6	17.8	14.0	14.2	14.3	14.5
r_y (mm)	17.1	16.9	16.7	16.4	15.6	15.4	15.2	14.9
* r_z (mm)	7.55	7.59	7.65	7.72	6.44	6.47	6.51	6.57
* r_z / r_x	0.435	0.433	0.434	0.434	0.459	0.456	0.456	0.454
* r_z / r_y	0.442	0.449	0.458	0.471	0.413	0.420	0.428	0.441
IMPERIAL SIZE AND MASS								
Mass (lb./ft.)	5.32	4.48	3.63	2.80	4.37	3.69	3.00	2.28
Thickness (in.)	0.24	0.20	0.16	0.12	0.24	0.20	0.16	0.12
Size (in.)	2.17 x 1.38				1.77 x 1.18			

* See page 4–4 for more information regarding these tables.
** Factored axial compressive resistances calculated according to Clause 1.3.3.3 CAN/CSA-S16.1-M89.

G40.21-M
300W
$\phi = 0.90$

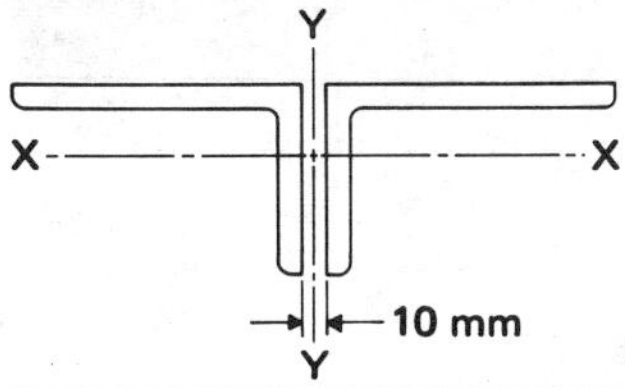

DOUBLE ANGLE STRUTS
Metric Series – Unequal Leg Angles

Factored Axial Compressive Resistances – kN
Short Legs 10 mm Back to Back *

Size (mm)		150 x 100				125 x 90			
Thickness (mm)		16	13	10**	8**	16	13	10**	8**
Mass (kg/m)		58.8	48.4	37.7	30.4	50.0	41.2	32.2	26.0
Effective length (KL) in millimetres with respect to indicated axis — X–X Axis	0	2 020	1 660	1 150	816	1 720	1 420	1 060	778
	500	1 980	1 630	1 130	807	1 670	1 380	1 030	765
	1 000	1 830	1 510	1 070	768	1 520	1 260	961	717
	1 500	1 630	1 350	977	717	1 330	1 100	864	647
	2 000	1 400	1 160	866	645	1 090	913	723	558
	2 500	1 100	926	723	544	821	691	551	447
	3 000	863	723	572	445	638	537	429	350
	3 500	695	583	461	373	511	431	344	281
	4 000	572	480	380	311	417	352	282	230
	4 500	477	401	318	260	337	286	230	188
	5 000	392	331	263	217	276	234	188	154
	5 500	327	276	220	181				
Y–Y Axis	0	2 020	1 660	818	544	1 720	1 420	785	545
	500	2 020	1 660	818	544	1 720	1 420	785	545
	1 000	2 010	1 650	817	543	1 690	1 400	785	544
	1 500	1 960	1 610	817	543	1 640	1 350	784	544
	2 000	1 900	1 560	816	542	1 580	1 300	782	543
	2 500	1 840	1 510	815	542	1 510	1 240	779	541
	3 000	1 770	1 450	813	541	1 430	1 180	773	539
	3 500	1 700	1 390	811	539	1 350	1 100	760	534
	4 000	1 620	1 320	806	537	1 250	1 030	738	525
	4 500	1 530	1 250	797	534	1 150	943	708	510
	5 000	1 430	1 170	782	529	1 040	850	666	488
	5 500	1 330	1 090	760	521	918	749	616	460
	6 000	1 220	995	730	509	817	666	562	427
	6 500	1 100	894	695	490	732	596	502	388
	7 000	995	809	655	466	660	538	438	344
	7 500	906	737	610	437	599	488	383	302
	8 000	829	674	560	403	545	444	337	267
	8 500	761	619	499	369	499	406	299	237
	9 000	702	571	447	339	458	373	268	212
	9 500	649	528	402	311	422	343	240	191
	10 000	602	489	364	286	387	314	217	173
PROPERTIES OF 2 ANGLES — 10 mm BACK TO BACK									
Area (mm^2)		7 490	6 160	4 800	3 870	6 370	5 250	4 100	3 310
r_x (mm)		28.3	28.7	29.0	29.3	25.6	26.0	26.4	26.6
r_y (mm)		73.3	72.7	72.2	71.8	61.2	60.7	60.1	59.8
* r_z (mm)		21.6	21.7	21.9	22.0	19.2	19.3	19.5	19.6
* r_z / r_x		0.764	0.757	0.754	0.750	0.749	0.743	0.740	0.736
* r_z / r_y		0.295	0.298	0.303	0.306	0.314	0.318	0.324	0.328
IMPERIAL SIZE AND MASS									
Mass (lb./ft.)		39.5	32.5	25.3	20.4	33.6	27.7	21.6	17.5
Thickness (in.)		0.63	0.51	0.39	0.31	0.63	0.51	0.39	0.31
Size (in.)		5.91 x 3.94				4.92 x 3.54			

* See page 4–4 for more information regarding these tables.
** Factored axial compressive resistances calculated according to Clause 1.3.3.3 CAN/CSA-S16.1-M89.

DOUBLE ANGLE STRUTS
Metric Series – Unequal Leg Angles

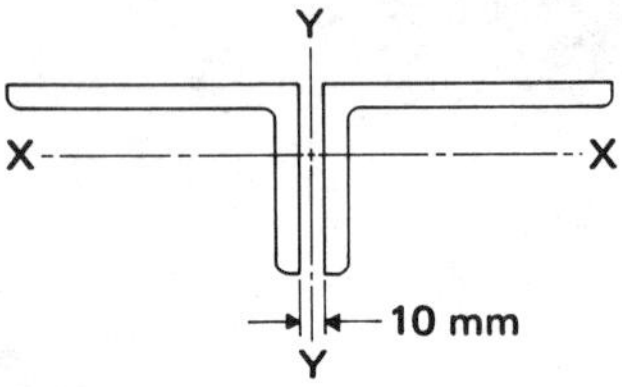

G40.21-M
300W
$\phi = 0.90$

Factored Axial Compressive Resistances – kN
Short Legs 10 mm Back to Back *

Size (mm)		125 x 75				100 x 90			
Thickness (mm)		13	10**	8**	6**	13	10	8**	6**
Mass (kg/m)		38.2	29.8	24.1	18.3	36.1	28.3	22.9	17.3
Effective length (KL) in millimetres with respect to indicated axis									
X–X Axis	0	1 310	975	720	465	1 240	972	747	466
	500	1 250	941	696	453	1 210	950	735	460
	1 000	1 100	846	630	418	1 110	873	689	437
	1 500	889	703	538	364	983	773	623	405
	2 000	635	507	415	284	824	651	531	363
	2 500	466	372	307	222	634	504	413	307
	3 000	357	285	236	179	494	392	322	248
	3 500	277	222	185	143	397	315	259	200
	4 000	215	172	143	110	325	259	213	164
	4 500					267	214	177	137
	5 000					219	175	144	112
	5 500							120	93.3
Y–Y Axis	0	1 310	729	505	287	1 240	972	550	320
	500	1 310	729	505	287	1 240	972	550	320
	1 000	1 300	729	505	287	1 200	939	549	319
	1 500	1 260	729	505	287	1 150	895	546	318
	2 000	1 210	728	504	287	1 080	843	542	316
	2 500	1 160	726	503	286	1 010	783	533	314
	3 000	1 100	723	502	286	922	716	517	309
	3 500	1 040	715	500	285	829	642	490	300
	4 000	972	697	495	283	721	555	451	287
	4 500	899	672	484	281	616	474	404	267
	5 000	820	639	467	278	533	410	351	240
	5 500	728	596	443	271	467	359	295	209
	6 000	648	549	415	260	413	317	250	179
	6 500	581	497	382	244	367	282	214	155
	7 000	524	441	346	223	329	252	186	135
	7 500	476	385	305	203	295	227	162	119
	8 000	434	339	269	185	263	201	143	105
	8 500	397	301	239	169	234	179	127	93.3
	9 000	365	269	213	154	210	160	114	83.5
	9 500	336	241	192	141				
	10 000	311	218	173	128				
PROPERTIES OF 2 ANGLES — 10 mm BACK TO BACK									
Area (mm^2)		4 860	3 800	3 070	2 330	4 600	3 600	2 910	2 210
r_x (mm)		20.7	21.0	21.3	21.5	26.9	27.2	27.5	27.8
r_y (mm)		63.0	62.4	62.0	61.6	47.4	46.8	46.5	46.1
* r_z (mm)		16.2	16.3	16.4	16.6	18.4	18.5	18.6	18.7
* r_z / r_x		0.783	0.775	0.771	0.771	0.685	0.679	0.677	0.672
* r_z / r_y		0.257	0.261	0.265	0.269	0.388	0.395	0.400	0.406
IMPERIAL SIZE AND MASS									
Mass (lb./ft.)		25.7	20.0	16.2	12.3	24.3	19.0	15.4	11.6
Thickness (in.)		0.51	0.39	0.31	0.24	0.51	0.39	0.31	0.24
Size (in.)		4.92 x 2.95				3.94 x 3.54			

* See page 4–4 for more information regarding these tables.
** Factored axial compressive resistances calculated according to Clause 1.3.3.3 CAN/CSA-S16.1-M89.

G40.21-M
300W
$\phi = 0.90$

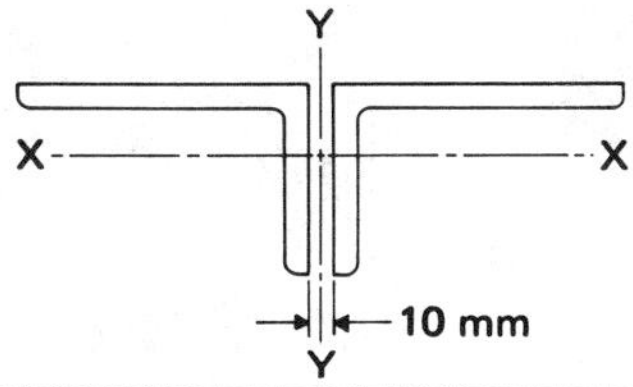

DOUBLE ANGLE STRUTS
Metric Series – Unequal Leg Angles

Factored Axial Compressive Resistances – kN
Short Legs 10 mm Back to Back*

Size (mm)		100 x 75				90 x 75				
Thickness (mm)		13	10	8**	6**	13	10	8	6**	5**
Mass (kg/m)		33.1	25.9	21.0	15.9	31.0	24.3	19.7	15.0	12.6
Effective length (KL) in millimetres with respect to indicated axis										
X–X Axis	0	1 140	891	689	452	1 070	837	678	444	327
	500	1 090	856	667	441	1 020	805	653	434	320
	1 000	962	758	604	408	908	715	581	402	298
	1 500	792	629	513	359	753	597	487	355	267
	2 000	579	466	382	284	557	445	367	282	223
	2 500	426	343	281	217	410	328	271	210	174
	3 000	328	264	217	168	316	253	209	162	137
	3 500	258	209	171	133	250	200	166	128	108
	4 000	200	162	134	104	194	156	130	101	85.6
	4 500							104	80.9	68.4
Y–Y Axis	0	1 140	891	510	314	1 070	837	678	317	218
	500	1 140	891	510	314	1 070	837	678	317	218
	1 000	1 100	864	509	313	1 020	803	650	317	217
	1 500	1 060	826	508	313	971	761	615	316	217
	2 000	999	781	506	312	909	711	574	314	216
	2 500	935	730	500	310	837	653	527	310	214
	3 000	863	672	488	306	755	588	473	301	210
	3 500	783	608	466	299	665	516	414	285	203
	4 000	696	538	435	286	561	433	347	262	192
	4 500	596	460	395	267	478	369	295	232	174
	5 000	517	398	349	242	413	319	255	194	151
	5 500	453	349	298	212	361	279	223	163	132
	6 000	401	309	252	184	318	245	196	138	113
	6 500	357	275	216	158	282	218	174	118	97.1
	7 000	320	246	186	137	252	194	154	103	84.4
	7 500	288	222	163	120	221	170	135	89.7	74.0
	8 000	261	199	143	106	196	150	120	79.0	65.3
	8 500	232	177	127	94.4	175	134	107	70.2	58.1
	9 000	208	159	113	84.4					
	9 500	188	144	102	75.9					
PROPERTIES OF 2 ANGLES — 10 mm BACK TO BACK										
Area (mm^2)		4 210	3 300	2 670	2 030	3 950	3 100	2 510	1 910	1 600
r_x (mm)		21.5	21.9	22.1	22.4	21.9	22.2	22.5	22.8	22.9
r_y (mm)		49.4	48.8	48.4	48.1	44.0	43.5	43.1	42.7	42.6
* r_z (mm)		16.0	16.1	16.2	16.3	15.6	15.7	15.8	15.9	16.0
* r_z / r_x		0.744	0.736	0.732	0.727	0.713	0.707	0.703	0.698	0.698
* r_z / r_y		0.324	0.330	0.335	0.339	0.355	0.361	0.367	0.372	0.376
IMPERIAL SIZE AND MASS										
Mass (lb./ft.)		22.2	17.4	14.1	10.7	20.8	16.3	13.2	10.1	8.47
Thickness (in.)		0.51	0.39	0.31	0.24	0.51	0.39	0.31	0.24	0.20
Size (in.)		3.94 x 2.95				3.54 x 2.95				

* See page 4–4 for more information regarding these tables.
** Factored axial compressive resistances calculated according to Clause 1.3.3.3 CAN/CSA-S16.1-M89.

DOUBLE ANGLE STRUTS
Metric Series – Unequal Leg Angles

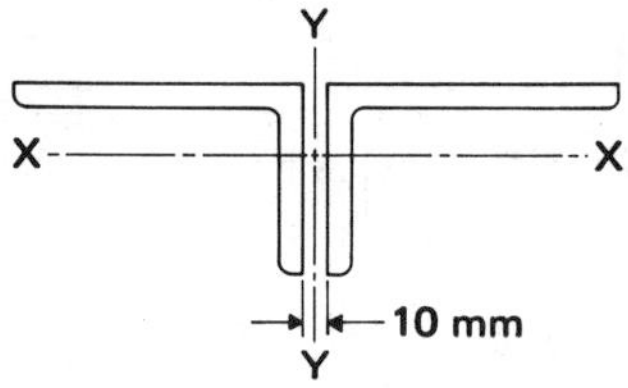

G40.21-M
300W
$\phi = 0.90$

Factored Axial Compressive Resistances – kN
Short Legs 10 mm Back to Back *

Size (mm)	90 x 65				80 x 60			
Thickness (mm)	10	8	6**	5**	10	8	6**	5**
Mass (kg/m)	22.8	18.5	14.0	11.8	20.4	16.6	12.6	10.6
Effective length (KL) in millimetres with respect to indicated axis — X–X Axis								
0	783	635	430	319	702	570	405	309
500	739	600	412	309	656	533	385	297
1 000	631	513	364	280	546	446	333	263
1 500	484	396	297	235	393	324	253	205
2 000	330	271	211	174	265	219	171	145
2 500	240	198	154	131	192	158	124	105
3 000	182	150	117	99.9	141	117	92.1	78.2
3 500	136	112	88.2	75.4		87.3	68.7	58.3
Y–Y Axis								
0	783	634	301	215	702	570	293	215
500	783	634	301	215	701	569	293	215
1 000	754	610	301	214	668	542	293	215
1 500	716	579	300	214	628	509	292	214
2 000	671	542	299	213	581	470	289	212
2 500	620	500	296	212	526	425	282	209
3 000	562	453	289	209	464	373	267	202
3 500	498	400	276	203	391	313	245	188
4 000	423	339	256	191	327	261	214	168
4 500	361	289	230	174	278	222	177	143
5 000	312	250	198	152	239	191	144	118
5 500	273	218	165	132	208	167	120	98.5
6 000	240	192	140	114	183	146	101	83.3
6 500	214	171	119	98.1	161	128	86.4	71.3
7 000	191	152	103	85.1	140	111	74.6	61.6
7 500	169	135	90.2	74.4	123	97.7	65.1	53.8
8 000	149	119	79.5	65.6	109			
8 500	133	106	70.5	58.3				
PROPERTIES OF 2 ANGLES — 10 mm BACK TO BACK								
Area (mm^2)	2 900	2 350	1 790	1 500	2 600	2 110	1 610	1 350
r_x (mm)	18.7	18.9	19.2	19.4	17.3	17.5	17.8	17.9
r_y (mm)	44.9	44.5	44.1	43.9	40.2	39.8	39.4	39.2
* r_z (mm)	13.9	14.0	14.2	14.2	12.8	12.9	13.0	13.0
* r_z / r_x	0.745	0.739	0.739	0.734	0.741	0.736	0.731	0.725
* r_z / r_y	0.310	0.315	0.322	0.323	0.318	0.324	0.330	0.332
IMPERIAL SIZE AND MASS								
Mass (lb./ft.)	15.3	12.4	9.41	7.93	13.7	11.2	8.47	7.12
Thickness (in.)	0.39	0.31	0.24	0.20	0.39	0.31	0.24	0.20
Size (in.)	3.54 x 2.56				3.15 x 2.36			

* See page 4–4 for more information regarding these tables.
** Factored axial compressive resistances calculated according to Clause 1.3.3.3 CAN/CSA-S16.1-M89.

G40.21-M
300W
$\phi = 0.90$

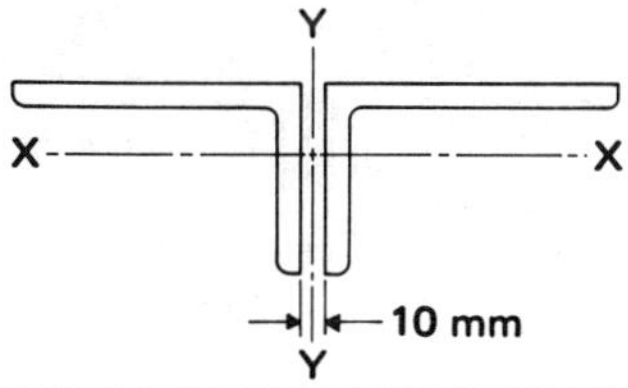

DOUBLE ANGLE STRUTS
Metric Series – Unequal Leg Angles

Factored Axial Compressive Resistances – kN
Short Legs 10 mm Back to Back *

Effective length (KL) in millimetres with respect to indicated axis		75 x 50			65 x 50			
Size (mm)		75 x 50			65 x 50			
Thickness (mm)		8	6**	5**	8	6	5**	4**
Mass (kg/m)		14.7	11.2	9.4	13.4	10.3	8.6	7.0
X–X Axis	0	505	367	288	462	353	280	200
X–X Axis	500	457	339	266	420	322	260	193
X–X Axis	1 000	348	270	216	324	251	212	172
X–X Axis	1 500	214	168	143	204	159	136	126
X–X Axis	2 000	142	112	95.0	135	106	90.6	73.9
X–X Axis	2 500	97.2	77.3	65.8	93.8	73.7	63.6	47.3
X–X Axis	3 000							32.9
Y–Y Axis1	0	505	272	201	462	354	203	131
Y–Y Axis1	500	503	271	201	457	350	203	131
Y–Y Axis1	1 000	479	271	201	429	328	202	130
Y–Y Axis1	1 500	448	270	200	394	301	201	129
Y–Y Axis1	2 000	412	268	199	352	268	196	127
Y–Y Axis1	2 500	370	259	196	303	230	183	120
Y–Y Axis1	3 000	322	242	186	245	184	163	107
Y–Y Axis1	3 500	266	215	171	198	150	135	86.8
Y–Y Axis1	4 000	222	184	150	165	124	105	68.3
Y–Y Axis1	4 500	189	150	123	139	104	83.8	54.7
Y–Y Axis1	5 000	162	122	101	119	89.2	68.2	44.7
Y–Y Axis1	5 500	141	101	83.5	101	75.8	56.5	37.2
Y–Y Axis1	6 000	123	85.0	70.3	85.7	64.2	47.6	31.4
Y–Y Axis1	6 500	107	72.5	60.1	73.7	55.2	40.6	
Y–Y Axis1	7 000	93.1	62.6	51.9				
Y–Y Axis1	7 500	81.7	54.6	45.3				

PROPERTIES OF 2 ANGLES — 10 mm BACK TO BACK

Property	75 x 50 × 8	75 x 50 × 6**	75 x 50 × 5**	65 x 50 × 8	65 x 50 × 6	65 x 50 × 5**	65 x 50 × 4**
Area (mm^2)	1 870	1 430	1 200	1 710	1 310	1 100	888
r_x (mm)	14.1	14.4	14.5	14.5	14.7	14.9	15.0
r_y (mm)	38.6	38.2	38.0	33.2	32.8	32.6	32.4
* r_z (mm)	10.8	10.9	10.9	10.6	10.7	10.8	10.8
* r_z / r_x	0.764	0.757	0.751	0.732	0.726	0.727	0.720
* r_z / r_y	0.280	0.285	0.287	0.319	0.326	0.331	0.333

IMPERIAL SIZE AND MASS

Mass (lb./ft.)	9.88	7.53	6.33	9.00	6.92	5.80	4.68
Thickness (in.)	0.31	0.24	0.20	0.31	0.24	0.20	0.16
Size (in.)	2.95 x 1.97			2.56 x 1.97			

* See page 4–4 for more information regarding these tables.
** Factored axial compressive resistances calculated according to Clause 1.3.3.3 CAN/CSA-S16.1-M89.

DOUBLE ANGLE STRUTS
Metric Series – Unequal Leg Angles

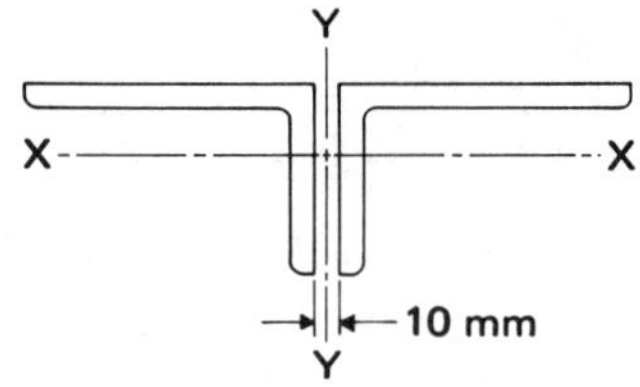

G40.21-M
300W
$\phi = 0.90$

Factored Axial Compressive Resistances – kN
Short Legs 10 mm Back to Back *

Size (mm)		55 x 35				45 x 30			
Thickness (mm)		6	5	4**	3**	6	5	4	3**
Mass (kg/m)		7.9	6.7	5.4	4.1	6.5	5.5	4.5	3.4
Effective length (KL) in millimetres with respect to indicated axis									
X–X Axis	0	272	230	170	113	224	189	153	104
	500	223	189	155	105	171	146	119	90.9
	1 000	122	104	101	72.7	80.4	69.3	57.4	48.5
	1 500	68.0	58.3	45.3	35.7	42.0	36.4	30.4	21.6
	2 000			25.5	20.1				
Y–Y Axis	0	273	230	118	66.2	224	189	153	67.4
	500	268	226	118	66.2	217	184	149	67.4
	1 000	249	209	117	66.0	197	166	135	67.1
	1 500	224	188	116	65.6	171	144	116	66.1
	2 000	195	163	112	64.6	139	116	93.6	61.0
	2 500	159	132	99.5	61.3	103	86.4	69.3	48.4
	3 000	124	103	78.1	52.2	80.2	67.0	53.8	34.3
	3 500	100	83.2	58.0	42.0	64.1	53.5	42.9	25.5
	4 000	82.8	68.6	44.6	32.9	52.3	43.6	34.9	19.6
	4 500	69.4	57.5	35.4	26.2	41.8	34.8	27.8	15.5
	5 000	58.0	47.9	28.7	21.3	34.3			
	5 500	48.3	39.9	23.8	17.7				
PROPERTIES OF 2 ANGLES — 10 mm BACK TO BACK									
Area (mm^2)		1 010	850	688	522	828	700	568	432
r_x (mm)		9.77	9.89	10.0	10.2	8.35	8.46	8.59	8.71
r_y (mm)		29.7	29.4	29.2	29.0	25.0	24.8	24.6	24.4
* r_z (mm)		7.55	7.59	7.65	7.72	6.44	6.47	6.51	6.57
* r_z / r_x		0.773	0.768	0.764	0.760	0.771	0.765	0.758	0.754
* r_z / r_y		0.254	0.258	0.262	0.266	0.258	0.261	0.265	0.269
IMPERIAL SIZE AND MASS									
Mass (lb./ft.)		5.32	4.48	3.63	2.80	4.37	3.69	3.00	2.28
Thickness (in.)		0.24	0.20	0.16	0.12	0.24	0.20	0.16	0.12
Size (in.)		2.17 x 1.38				1.77 x 1.18			

* See page 4–4 for more information regarding these tables.
** Factored axial compressive resistances calculated according to Clause 1.3.3.3 CAN/CSA-S16.1-M89.

G40.21-M
300W
$\phi = 0.90$

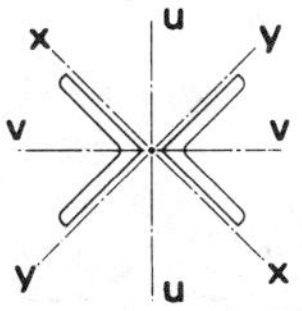

DOUBLE ANGLE STRUTS
Metric Series – Star Shaped *

Factored Axial Compressive Resistances – kN

Size (mm)		200 x 200			150 x 150			125 x 125	
Legs		16 mm apart			12 mm apart			12 mm apart	
Thickness (mm)		30	25	20	20	16	13	16	13
Mass (kg/m)		174	147	119	87.9	71.3	58.6	58.8	48.4
Effective length (KL) in millimetres with respect to indicated axis: U–U Axis	0	5 990	5 080	4 100	3 020	2 450	2 010	2 020	1 660
	500	5 990	5 080	4 100	3 020	2 450	2 010	2 020	1 660
	1 000	5 990	5 080	4 100	3 020	2 450	2 010	2 000	1 640
	1 500	5 950	5 030	4 060	2 950	2 390	1 960	1 940	1 590
	2 000	5 850	4 940	3 990	2 870	2 320	1 900	1 880	1 540
	2 500	5 730	4 850	3 910	2 780	2 250	1 840	1 800	1 480
	3 000	5 610	4 740	3 820	2 690	2 170	1 770	1 720	1 410
	3 500	5 480	4 630	3 720	2 580	2 080	1 700	1 630	1 330
	4 000	5 340	4 500	3 620	2 470	1 990	1 620	1 530	1 250
	4 500	5 190	4 370	3 510	2 350	1 890	1 530	1 430	1 160
	5 000	5 030	4 240	3 390	2 220	1 780	1 440	1 310	1 060
	5 500	4 870	4 090	3 270	2 080	1 660	1 340	1 190	949
	6 000	4 690	3 940	3 140	1 940	1 540	1 240	1 060	844
	6 500	4 500	3 770	3 000	1 770	1 400	1 120	948	757
	7 000	4 310	3 600	2 850	1 610	1 260	1 010	857	684
	7 500	4 110	3 430	2 700	1 460	1 150	921	778	621
	8 000	3 890	3 240	2 540	1 340	1 050	843	710	567
	8 500	3 670	3 040	2 360	1 230	968	774	651	519
	9 000	3 410	2 810	2 190	1 140	894	714	599	477
	9 500	3 170	2 620	2 030	1 050	827	661	552	440
	10 000	2 960	2 440	1 900	979	768	613	511	406
V–V Axis	0	5 990	5 080	4 100	3 020	2 450	2 010	2 020	1 660
	500	5 990	5 080	4 100	3 020	2 450	2 010	2 020	1 660
	1 000	5 970	5 060	4 090	2 970	2 410	1 980	1 960	1 610
	1 500	5 830	4 940	4 000	2 860	2 330	1 910	1 870	1 540
	2 000	5 670	4 810	3 890	2 740	2 230	1 830	1 760	1 450
	2 500	5 490	4 660	3 780	2 610	2 120	1 750	1 640	1 360
	3 000	5 290	4 500	3 650	2 460	2 000	1 650	1 510	1 250
	3 500	5 080	4 320	3 510	2 290	1 870	1 540	1 360	1 130
	4 000	4 850	4 130	3 350	2 110	1 730	1 430	1 190	993
	4 500	4 600	3 930	3 190	1 910	1 570	1 300	1 020	848
	5 000	4 340	3 700	3 020	1 680	1 390	1 150	881	735
	5 500	4 050	3 470	2 830	1 480	1 220	1 010	772	644
	6 000	3 750	3 220	2 630	1 310	1 090	901	682	569
	6 500	3 400	2 940	2 410	1 170	972	807	607	507
	7 000	3 080	2 660	2 180	1 060	875	727	544	454
	7 500	2 810	2 420	1 990	957	793	659	489	409
	8 000	2 570	2 220	1 820	871	721	599	437	366
	8 500	2 360	2 040	1 680	795	659	548	389	326
	9 000	2 180	1 880	1 550	728	603	502	349	293
	9 500	2 020	1 740	1 430	662	552	461	315	264
	10 000	1 870	1 620	1 330	600	500	417		
PROPERTIES OF 2 STARRED ANGLES									
Area (mm^2)		22 200	18 800	15 200	11 200	9 090	7 460	7 490	6 160
r_x (mm)		105	103	100	77.6	75.8	74.4	65.7	64.3
r_y (mm)		75.7	76.8	77.7	57.3	58.1	58.6	47.9	48.4
* r_z (mm)		39.0	39.1	39.3	29.3	29.4	29.6	24.4	24.5
* r_z / r_x		0.372	0.380	0.391	0.378	0.388	0.398	0.372	0.381
* r_z / r_y		0.515	0.509	0.506	0.511	0.506	0.505	0.510	0.506
IMPERIAL SIZE AND MASS									
Mass (lb./ft.)		117	98.8	80.0	59.1	47.9	39.4	39.5	32.5
Thickness (in.)		1.18	0.98	0.79	0.79	0.63	0.51	0.63	0.51
Size (in.)		7.87 x 7.87			5.91 x 5.91			4.92 x 4.92	

* See page 4–4 for more information regarding these tables.
See Clauses 19.1.4 and 19.1.5 CAN/CSA-S16.1-M89 for interconnecting requirements.

DOUBLE ANGLE STRUTS

Metric Series – Star Shaped *

Factored Axial Compressive Resistances – kN

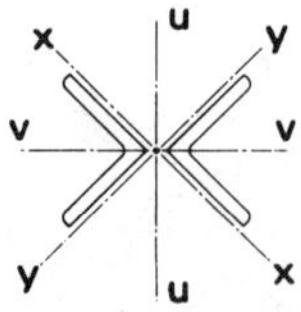

G40.21-M

300W

$\phi = 0.90$

Size (mm)		100 x 100			90 x 90			75 x 75		
Legs		10 mm apart			10 mm apart			8 mm apart		
Thickness (mm)		16	13	10	13	10	8	13	10	8
Mass (kg/m)		46.2	38.2	29.8	34.1	26.7	21.6	28.0	22.0	17.8
Effective length (KL) in millimetres with respect to indicated axis										
U–U Axis	0	1 590	1 310	1 030	1 170	918	743	962	756	613
	500	1 590	1 310	1 030	1 170	918	742	961	755	612
	1 000	1 550	1 280	999	1 140	888	717	918	720	582
	1 500	1 500	1 230	958	1 090	847	683	866	676	546
	2 000	1 430	1 170	911	1 030	799	643	804	625	503
	2 500	1 350	1 100	857	960	744	597	733	566	453
	3 000	1 260	1 030	796	884	682	546	652	499	396
	3 500	1 170	948	729	800	614	488	559	420	330
	4 000	1 060	858	656	707	535	422	467	351	276
	4 500	946	754	570	604	457	360	397	298	234
	5 000	822	655	495	524	395	312	343	257	201
	5 500	722	575	434	459	346	273	299	223	175
	6 000	641	510	385	406	306	241	263	196	154
	6 500	572	455	343	362	272	214	232	173	134
	7 000	515	409	308	324	244	191	204	150	116
	7 500	465	369	278	292	219	172	178	132	102
	8 000	422	335	252	262	195	152	158	116	
	8 500	385	305	227	233	174	135			
	9 000	348	273	204	209	156	122			
	9 500	314	247	184	189	141				
	10 000	285	224	167						
V–V Axis	0	1 590	1 310	1 030	1 170	918	742	961	756	613
	500	1 580	1 310	1 020	1 160	911	737	942	742	602
	1 000	1 500	1 240	973	1 090	859	696	869	685	557
	1 500	1 400	1 160	911	1 010	794	644	775	614	500
	2 000	1 280	1 070	838	906	716	583	660	526	431
	2 500	1 150	956	754	786	626	511	520	419	346
	3 000	991	831	657	644	518	426	406	327	270
	3 500	813	685	545	523	421	346	327	264	218
	4 000	678	571	455	434	350	288	269	217	180
	4 500	575	485	386	367	296	243	224	181	150
	5 000	494	417	332	313	253	208	184	150	124
	5 500	429	362	289	271	219	180	153	125	104
	6 000	375	317	253	229	186	154			
	6 500	323	275	220	197	160	132			
	7 000	280	238	191			115			
	7 500	246	209	168						
PROPERTIES OF 2 STARRED ANGLES										
Area (mm^2)		5 890	4 860	3 800	4 340	3 400	2 750	3 560	2 800	2 270
r_x (mm)		54.3	52.9	51.6	48.8	47.5	46.6	41.5	40.1	39.2
r_y (mm)		37.7	38.3	38.8	34.1	34.8	35.2	28.1	28.6	29.0
* r_z (mm)		19.5	19.5	19.7	17.6	17.6	17.7	14.6	14.6	14.7
* r_z / r_x		0.359	0.368	0.382	0.361	0.370	0.380	0.351	0.364	0.375
* r_z / r_y		0.518	0.509	0.507	0.516	0.506	0.503	0.519	0.510	0.506
IMPERIAL SIZE AND MASS										
Mass (lb./ft.)		31.0	25.7	20.0	22.9	17.9	14.5	18.8	14.8	12.0
Thickness (in.)		0.63	0.51	0.39	0.51	0.39	0.31	0.51	0.39	0.31
Size (in.)		3.94 x 3.94			3.54 x 3.54			2.95 x 2.95		

* See page 4–4 for more information regarding these tables.

See Clauses 19.1.4 and 19.1.5 CAN/CSA-S16.1-M89 for interconnecting requirements.

G40.21-M 300W
$\phi = 0.90$

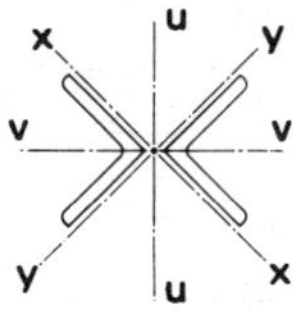

DOUBLE ANGLE STRUTS
Metric Series – Star Shaped *
Factored Axial Compressive Resistances – kN

Size (mm)		65 x 65			55 x 55			
Legs		8 mm apart			8 mm apart			
Thickness (mm)		10	8	6	10	8	6	5
Mass (kg/m)		18.8	15.3	11.7	15.7	12.8	9.8	8.2
Effective length (KL) in millimetres with respect to indicated axis								
U–U Axis	0	648	527	402	540	441	337	284
U–U Axis	500	644	523	399	534	434	332	279
U–U Axis	1 000	609	493	376	499	405	309	259
U–U Axis	1 500	566	457	347	456	368	280	233
U–U Axis	2 000	514	413	312	404	324	244	203
U–U Axis	2 500	454	362	271	344	272	202	165
U–U Axis	3 000	384	302	223	274	215	158	129
U–U Axis	3 500	312	245	181	222	174	128	104
U–U Axis	4 000	260	204	150	184	144	106	86.1
U–U Axis	4 500	220	172	127	155	121	88.5	72.1
U–U Axis	5 000	189	148	109	132	103	74.6	60.3
U–U Axis	5 500	163	128	93.7	111	85.7	62.1	50.3
U–U Axis	6 000	141	109	79.5	94.0	72.6	52.7	
U–U Axis	6 500	121	93.8	68.3				
U–U Axis	7 000	105	81.6					
V–V Axis	0	648	526	402	540	440	338	284
V–V Axis	500	629	512	391	515	421	323	272
V–V Axis	1 000	569	464	356	450	369	285	240
V–V Axis	1 500	491	402	310	363	300	234	198
V–V Axis	2 000	395	326	253	258	216	170	145
V–V Axis	2 500	293	243	190	189	159	125	106
V–V Axis	3 000	227	189	148	145	122	95.9	81.7
V–V Axis	3 500	181	151	118	112	94.5	75.2	64.3
V–V Axis	4 000	147	123	96.4	86.7	73.3	58.3	49.8
V–V Axis	4 500	118	98.5	77.6				
V–V Axis	5 000		80.7	63.5				
PROPERTIES OF 2 STARRED ANGLES								
Area (mm^2)		2 400	1 950	1 490	2 000	1 630	1 250	1 050
r_x (mm)		36.1	35.2	34.3	32.1	31.2	30.3	29.7
r_y (mm)		24.6	25.0	25.4	20.5	20.9	21.3	21.5
* r_z (mm)		12.7	12.7	12.8	10.7	10.7	10.8	10.8
* r_z / r_x		0.352	0.361	0.373	0.333	0.343	0.357	0.363
* r_z / r_y		0.517	0.509	0.505	0.521	0.512	0.507	0.503
IMPERIAL SIZE AND MASS								
Mass (lb./ft.)		12.6	10.3	7.86	10.5	8.60	6.59	5.54
Thickness (in.)		0.39	0.31	0.24	0.39	0.31	0.24	0.20
Size (in.)		2.56 x 2.56			2.17 x 2.17			

* See page 4–4 for more information regarding these tables.
See Clauses 19.1.4 and 19.1.5 CAN/CSA-S16.1-M89 for interconnecting requirements.

DOUBLE ANGLE STRUTS
Metric Series – Star Shaped *

Factored Axial Compressive Resistances – kN

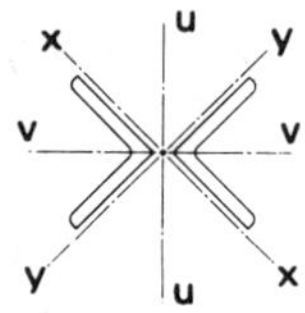

G40.21-M
300W
$\phi = 0.90$

Size (mm)		45 x 45				35 x 35		
Legs		6 mm apart				6 mm apart		
Thickness (mm)		8	6	5	4	6	5	4
Mass (kg/m)		10.3	7.9	6.7	5.4	6.0	5.1	4.1
Effective length (KL) in millimetres with respect to indicated axis	U–U Axis							
	0	354	272	230	186	207	176	143
	500	345	265	223	180	198	167	136
	1 000	314	240	201	162	174	146	118
	1 500	275	208	173	139	141	117	93.5
	2 000	227	168	139	109	101	83.3	65.4
	2 500	172	125	103	80.7	74.2	61.1	47.9
	3 000	133	96.7	79.5	62.5	56.9	46.7	36.6
	3 500	107	77.3	63.5	49.8	44.1	36.0	27.9
	4 000	87.4	63.0	51.3	39.9	34.2	27.9	
	4 500	70.8	50.2	41.0	31.9			
	5 000	57.9						
	V–V Axis							
	0	354	273	230	186	207	176	143
	500	330	255	215	174	185	157	128
	1 000	272	212	179	146	134	115	94.9
	1 500	192	152	130	106	79.5	68.7	57.0
	2 000	129	102	87.5	71.9	52.1	45.2	37.5
	2 500	93.3	73.9	63.3	52.1	34.7	30.2	25.3
	3 000	68.0	54.2	46.6	38.6			
	3 500				28.8			
PROPERTIES OF 2 STARRED ANGLES								
Area (mm^2)		1 310	1 010	850	688	768	650	528
r_x (mm)		25.9	24.8	24.4	23.9	20.8	20.4	19.9
r_y (mm)		16.9	17.2	17.4	17.6	13.1	13.3	13.5
* r_z (mm)		8.76	8.79	8.82	8.87	6.81	6.83	6.86
* r_z / r_x		0.339	0.354	0.361	0.371	0.327	0.335	0.345
* r_z / r_y		0.520	0.512	0.507	0.505	0.520	0.512	0.509
IMPERIAL SIZE AND MASS								
Mass (lb./ft.)		6.92	5.32	4.48	3.63	4.05	3.43	2.78
Thickness (in.)		0.31	0.24	0.20	0.16	0.24	0.20	0.16
Size (in.)		1.77 x 1.77				1.38 x 1.38		

* See page 4–4 for more information regarding these tables.
See Clauses 19.1.4 and 19.1.5 CAN/CSA-S16.1-M89 for interconnecting requirements.

NOTES

COLUMN BASE PLATES

When steel columns bear on concrete footings, steel base plates are required to distribute the column load to the footing without exceeding the bearing resistance of the concrete. In general the ends of columns are saw-cut or milled to a plane surface so as to bear evenly on the base plate. Connection of the column to the base plate and then to the footing depends on the loading conditions. For columns carrying vertical gravity loads only, this connection is required only to hold the parts in line, and two anchor bolts are generally sufficient.

For base plates subjected to vertical gravity loads only, the following assumptions and design method are recommended:

1. The factored gravity load is assumed uniformly distributed over the base plate within a rectangle of $0.95d \times 0.80b$ (see diagram).
2. The base plate exerts a uniform pressure over the footing.
3. The base plate projecting beyond the area of $0.95d \times 0.80b$ acts as a cantilever subject to the uniform bearing pressure.

C_f = total factored column load, kN

A = $B \times C$ = Area of plate, mm^2

t_p = plate thickness, mm^2

F_y = specified minimum yield strength of base plate steel, MPa

f'_c = specified 28-day strength of concrete, MPa

ϕ = 0.90 for steel

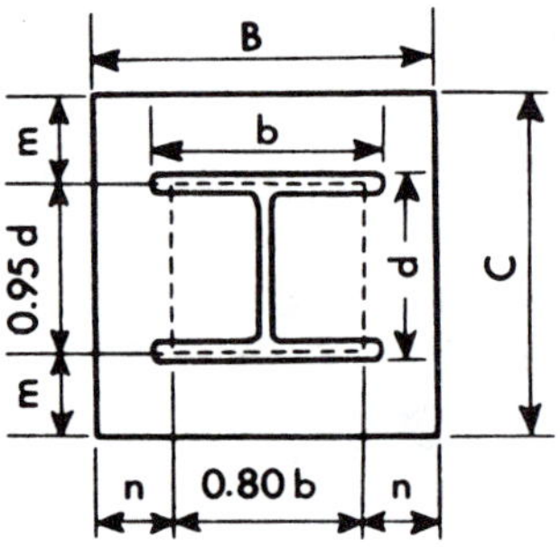

1. Determine the required area $A = C_f / B_r$ where B_r is the factored bearing resistance per unit of bearing area. For concrete, B_r is assumed to be $0.85\ \phi_c\ f'_c$ where $\phi_c = 0.60$ in bearing. (Clause 10.15 of CAN3-A23.3-M84 states when B_r may be increased.)
2. Determine B and C so that the dimensions *m* and *n* (the projections of the plate beyond the area $0.95d \times 0.80b$) are approximately equal.
3. Determine *m* and *n* and solve for t_p, where:

$$t_p = \sqrt{\frac{2\,C_f\,m^2}{BC\,\phi\,F_y}} \quad \text{or} \quad \sqrt{\frac{2\,C_f\,n^2}{BC\,\phi\,F_y}} \quad \text{whichever is greater.}$$

These formulae were derived by equating the factored moment acting on the portion of the plate taken as a cantilever to the factored moment resistance of the plate ($M_r = \phi\ Z\ F_y$) and solving for the plate thickness, t_p.

To minimize deflection of the base plate, the thickness should be generally not less than about 1/5 of the overhang, *m* or *n*.

$\phi Z F_y = \phi \frac{bt^2}{4} F_y$ = PLATE BEARING STRESS × CANTILEVER × MOM ARM

$= \frac{C_f}{C} \times m \times \frac{m}{2}$

Examples

1. **Given:**

A W310x118 column subjected to a factored axial load of 2 500 kN is supported by a concrete foundation whose 28-day specified strength is 20 MPa. Design the base plate assuming 300 MPa steel.

Solution:

For W310x118, b = 307 mm, d = 314 mm.

$$\text{Area of plate required} = \frac{2\,500}{0.85 \times 0.60 \times 20/10^3} = 245\,000 \text{ mm}^2$$

Try B = C = 500 mm; A = 250 000 mm^2

Determine m and n

$0.95\text{d} = 0.95 \times 314 = 298$ mm

Therefore, $\text{m} = (500 - 298)/2 = 101$ mm

$0.80\text{b} = 0.80 \times 307 = 246$ mm

Therefore, $\text{n} = (500 - 246)/2 = 127$ mm

Use n for design

$$\text{Plate thickness required} = \sqrt{\frac{2 \times 2\,500 \times 127^2}{500 \times 500 \times 0.90 \times 300/10^3}} = 34.5 \text{ mm}$$

$$\frac{\text{n}}{5} = \frac{127}{5} = 25 \text{ mm} \quad < 34.5 \text{ mm} \quad \text{— OK}$$

Use 35 mm (first preference plate thickness).

Since the plate thickness of 35 mm is less than 65 mm, F_y = 300 MPa for G40.21-M grade 300W steel. For plates greater than 65 mm in thickness, F_y = 280 MPa for 300W steel (see Table 6–3).

Therefore, use PL 35x500 x 500 for the base plate.

2. **Given:**

An HSS 203x203x9.5 column supports a factored axial load of 1 470 kN.

Select a base plate assuming f'_c = 20 MPa and F_y = 300 MPa.

Solution:

$$\text{Area required is } \frac{1\,470}{0.85 \times 0.60 \times 20/10^3} = 144\,000 \text{ mm}^2$$

$$\text{B} = \text{C} = \sqrt{\text{A}} = \sqrt{144 \times 10^3} = 380 \text{ mm}$$

$$\text{n} = \frac{380 - (203 - 9)}{2} = 93 \text{ mm}$$

$$\text{Therefore, } t_p = \sqrt{\frac{2 \times 1\,470 \times 93^2}{380 \times 380 \times 0.9 \times 300/10^3}} = 25.5 \text{ mm}$$

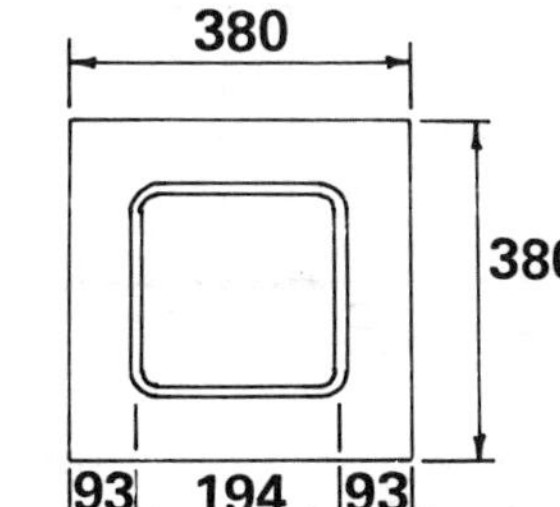

Use 30 mm (first preference plate thickness).

Therefore, use PL 30x380 x 380 for the base plate.

Design Chart

As an alternative to computing the plate thickness, Figure 4–1 provides a means of selecting t_p knowing the length of cantilever *m* or *n* and the unit factored bearing resistance.

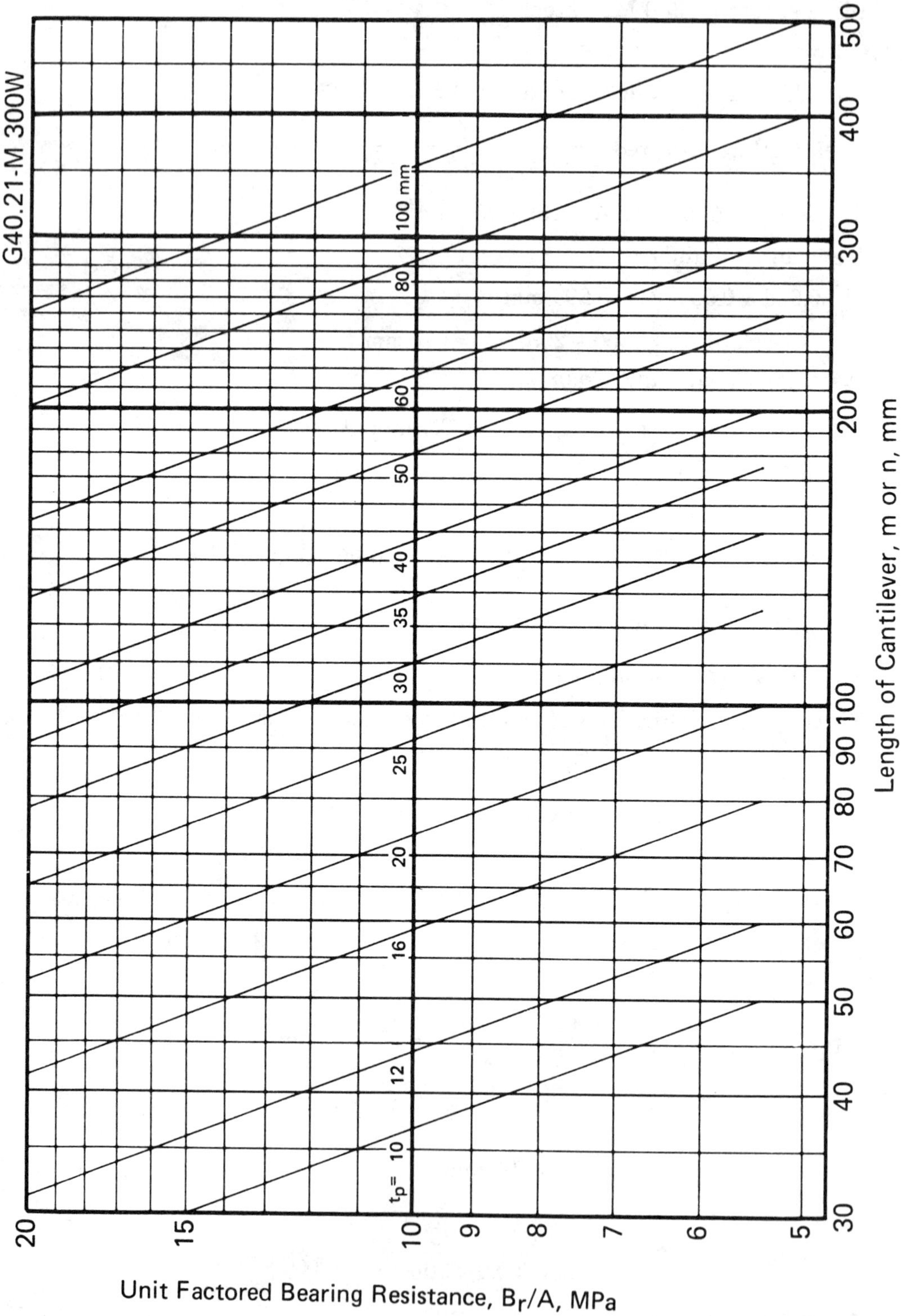

FIGURE 4-1
COLUMN BASE PLATE THICKNESS, t_p

Example

Given:

Same as example 1

Solution:

Unit factored bearing resistance is $0.85 \times 0.60 \times 20 = 10.2$ MPa

From Figure 4–1 for 10.2 MPa and n = 112, select t_p = 35 mm.

Base plate assemblies (including anchor bolts) which are subjected to applied bending moments, uplift tension, and shear forces must be designed to resist all such forces.

Lightly Loaded Base Plates

For lightly loaded base plates where the required bearing area is less than or about equal to the area bounded by the column dimensions *b* and *d*, the above method does not give realistic results for the base plate thickness, and other methods have been proposed in the literature. Fling (1970) uses a yield line theory to derive an equation for plate thickness. When modified for limit states design the equation becomes:

$$t_p = 0.43\, b\, \beta \sqrt{\frac{B_r}{\phi\, F_y\, (1 - \beta^2)}}$$

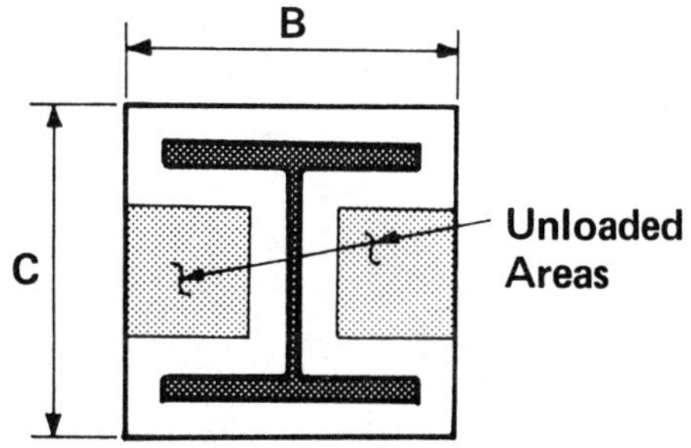

where

$B_r = 0.85\, \phi_c\, f'_c$

$\beta = \sqrt{0.75 + \frac{1}{4\lambda^2}} - \frac{1}{2\lambda}$

$\lambda = 2d / b$

b = column width, mm

d = column depth, mm

Stockwell (1975) assumes an effective bearing area where only an H-shaped pattern under a W column is loaded, and that the remainder of the base plate is unloaded. The assumed width of flange strips can be derived from the required bearing area, and the thickness of plate can be determined by the expression:

$$t_p = \sqrt{\frac{2\, C_f\, m^2}{A\, \phi\, F_y}}$$

where

A = the effective bearing area

m = half the width of the bearing strips

References

FLING, R.S. 1970. Design of steel bearing plates. Engineering Journal, AISC, **7**(2), April.

STOCKWELL, F.J.Jr. 1975. Preliminary base plate selection. Engineering Journal, AISC, **12**(3), third quarter.

ANCHOR BOLTS

Theoretically, anchor bolts are not required at base plates for concentrically loaded columns carrying gravity loads only since neither end moments nor horizontal forces are present. In practice, however, anchor bolts are provided to locate the column base, to provide a means for levelling the base plate, and to resist nominal end moments and horizontal forces which may occur.

Clauses 25 and 26 cover requirements for the design of column bases and anchor bolts for cases where anchor bolts are required to transfer end moments and horizontal forces due to lateral loads.

Fabricators normally supply anchor bolts manufactured from round bar stock. The bars are threaded at one end to receive a washer and nut and may be bent at the other end to form a hook, or both ends may be threaded. The material used for most common applications is usually produced to CSA G40.21-M Grade 300W (F_y = 300 MPa) or to ASTM A36 (F_y = 248 MPa). However, ASTM A36 round bar stock is generally more readily available, and should be assumed for anchor bolt design in most situations. For specialized applications, fastener suppliers or fabricators should be consulted.

The diameter of anchor bolt holes in base plates should provide for possible horizontal adjustments for alignment purposes. The following table of Suggested Anchor Bolt Hole Sizes can be used as a guide, although actual sizes used by fabricators may vary depending on shop and field practices.

References

CHIEN, E.Y.L. 1991. Low Rise Building Design Aid. Canadian Institute of Steel Construction, Willowdale, Ont.

CHIEN, E.Y.L. 1985. Single Storey Building Design Aid. Canadian Institute of Steel Construction, Willowdale, Ont.

K.S.M. 1974. Structural engineering aspects of headed concrete anchors and deformed bar anchors in the concrete construction industry. K.S.M. Welding Systems Division, Omark Industries.

NIXON, D. and ADAMS, P. F. 1978. Design of light industrial buildings. Proceedings, Canadian Structural Engineering Conference, C.I.S.C., Willowdale, Ontario.

STELCO Inc. 1983. Structural steels: selection and uses, part 8.

SUGGESTED ANCHOR BOLT HOLE SIZES*

Bolt Diameter	Suggested Hole Diameter
mm	mm
20	26
22	28
24	30
27	33
30	42
36	48
42	54
48	60
56	68
64	76

* Anchor bolt holes in base plates which will receive anchor bolts that are grouted may be flame cut.

* Anchor bolt hole sizes will vary with individual fabricators depending on shop and field practices,

PART FIVE
FLEXURAL MEMBERS

GENERAL INFORMATION

Class of Sections in Bending

See page 5–4

Factored Ultimate Shear Stress in Girder Webs

The tables on pages 5–9 to 5–11 list the factored ultimate shear stress, ϕF_s, in a girder web, computed in accordance with the requirements of Clause 13.4.1.1 of CAN/CSA S16.1-M89, and the required gross area of pairs of intermediate stiffeners, computed in accordance with the requirements of Clause 15.7.3 of S16.1-M89. Values are provided for minimum specified yield strength levels, F_y, of 248, 300 and 350 MPa, for aspect ratios (a/h) from 0.50 to 3.00, and for web slenderness ratios (h/w) varying between 60 and 320 for $F_y = 248$, between 50 and 260 for $F_y = 300$, and between 50 and 220 for $F_y = 350$. The required gross area of stiffeners is provided as a percentage of web area (hw) and is shown in italics.

Plate Girders

For design information and illustrative example, see page 5–12.

Composite Beams

Tables for Shear Resistance of Shear Studs in solid slabs and in deck-slabs are given on pages 5–23 to 5–25, and Trial Selection Tables for composite beams with various combinations of cover slab and cellular steel deck (hollow composite construction), and with solid slabs are given on pages 5–28 to 5–83. See page 5–20 for explanatory text.

Deflection of Flexural Members

See page 5–84 for design chart, table, and illustrative examples.

Factored Resistance of Beams

Beam Selection Tables, which list the factored moment resistance of beams under various conditions of lateral support, are provided on pages 5–94 to 5–100 to facilitate the design of flexural members. See page 5–88 for explanatory text.

Beam Load Tables which list total uniformly distributed factored loads for laterally supported beams of various spans are provided on pages 5–102 to 5–131. For explanatory text, see page 5–89.

Beam Diagrams and Formulae

Pages 5–132 to 5–150 contain diagrams and formulae to facilitate the design of flexural members in accordance with "elastic theory".

Beam Bearing Plates

See page 5–151 for design information, design chart and illustrative example.

Beams with Web Holes

See page 5–154 for design information, design tables and illustrative example.

Guide for Floor Vibration

See Appendix G of CAN/CSA S16.1-M89, page 1–127.

CLASS OF SECTIONS IN BENDING

Table 5-1, pages 5-5 to 5-8, lists the Class of section in bending of WWF and W shapes, for four common grades of steel. Listed are Canadian sizes of WWF and W shapes, and non-Canadian sizes of W shapes provided in Part-6 of this Handbook. For WWF shapes values are given for CAN/CSA-G40.21-M grades 300W and 350W; for Canadian W shapes values are given for G40.21-M grades 300W and 350W, and for ASTM A36 and ASTM A572 grade 50; for non-Canadian W shapes values are given for ASTM A36 and ASTM A572 grade 50. For these four steel grades all S shapes are Class 1, and all C and MC shapes are Class 3.

Table 5-1 also lists for each section size the ratios b/t and h/w, where; b = one-half the flange width, t = flange thickness, h = clear distance between flanges and w = web thickness. See also "Limits on Width-Thickness Ratios", page 4-5.

CLASS OF SECTIONS IN BENDING

TABLE 5-1

Section	CSA		ASTM		$\frac{b}{t}$	$\frac{h}{w}$
	G40.21		A36	A572		
	300W	350W		50		
WWF2000X732	2	3			5.50	95.0
X648	2	3			6.88	96.0
X607	2	3			7.86	96.5
X542	2	3			8.33	97.0
WWF1800X700	2	2			5.50	85.0
X659	2	2			6.11	85.5
X617	2	2			6.88	86.0
X575	2	2			7.86	86.5
X510	2	2			8.33	87.0
WWF1600X622	2	3			5.50	93.7
X580	2	3			6.11	94.3
X538	2	3			6.88	95.0
X496	2	3			7.86	95.6
X431	2	3			8.33	96.2
WWF1400X597	2	2			5.50	81.2
X513	2	2			6.88	82.5
X471	2	2			7.86	83.1
X405	2	2			8.33	83.7
X358	2	2			6.67	83.7
WWF1200X487	2	2			6.88	70.0
X418	2	2			7.14	70.6
X380	2	2			8.33	71.2
X333	2	2			6.67	71.2
X302	2	2			8.00	71.8
X263	2	2			6.00	71.8
WWF1100X458	2	2			6.88	72.8
X388	2	2			7.14	73.5
X351	2	2			8.33	74.2
X304	2	2			6.67	74.2
X273	2	2			8.00	75.0
X234	2	2			6.00	75.0
WWF1000X447	2	2			6.88	65.7
X377	2	2			7.14	66.4
X340	2	2			8.33	67.1
X293	2	2			6.67	67.1
X262	2	2			8.00	67.8
X223	2	2			6.00	67.8
X200	2	2			7.50	68.5
WWF900X417	2	2			6.88	74.5
X347	2	2			7.14	75.4
X309	2	2			8.33	76.3
X262	2	2			6.67	76.3
X231	2	2			8.00	77.2
X192	2	2			6.00	77.2
X169	2	2			7.50	78.1
WWF800X339	2	2			7.14	66.3
X300	2	2			8.33	67.2
X253	2	2			6.67	67.2
X223	2	2			8.00	68.1
X184	2	2			6.00	68.1
X161	2	2			7.50	69.0
WWF700X245	1	1			6.67	58.1
X214	1	2			8.00	59.0
X196	2	3			9.09	59.6
X175	1	2			6.00	59.0
X152	1	2			7.50	60.0
WWF650X864	1	1			5.42	8.83
X739	1	1			5.42	17.6
X598	1	1			6.50	27.5
X499	1	2			8.13	28.5
X400	3	4			10.8	29.5
WWF600X793	1	1			5.00	8.00
X680	1	1			5.00	16.0
X551	1	1			6.00	25.0
X460	1	1			7.50	26.0
X369	3	3			10.0	27.0
WWF550X721	1	1			4.58	7.17
X620	1	1			4.58	14.3
X503	1	1			5.50	22.5
X420	1	1			6.88	23.5
X280	3	4			11.0	31.2
WWF500X651	1	1			4.17	6.33
X561	1	1			4.17	12.6
X456	1	1			5.00	20.0
X381	1	1			6.25	21.0
X343	1	1			7.14	21.5
X306	1	2			8.33	22.0
X276	2	2			8.93	27.7
X254	3	3			10.0	28.1
X223	3	4			11.3	32.5
X197	4	4			12.5	41.8
WWF450X503	1	1			3.75	11.0
X409	1	1			4.50	17.5
X342	1	1			5.62	18.5
X308	1	1			6.43	19.0
X274	1	1			7.50	19.5
X248	1	2			8.04	24.6
X228	2	2			9.00	25.0
X201	3	3			10.2	29.0
X177	3	4			11.2	37.2
WWF400X444	1	1			3.33	9.33
X362	1	1			4.00	15.0
X303	1	1			5.00	16.0
X273	1	1			5.71	16.5
X243	1	1			6.67	17.0
X220	1	1			7.14	21.5
X202	1	2			8.00	21.8
X178	2	3			9.09	25.4
X157	3	3			10.0	32.7
WWF350X315	1	1			3.50	12.5
X263	1	1			4.38	13.5
X238	1	1			5.00	14.0

* Not available from Canadian mills.
Check with manufacturer for availability in grades other than CSA G40.21–300W.

TABLE 5-1 CLASS OF SECTIONS IN BENDING

Section	CSA G40.21 300W	CSA G40.21 350W	ASTM A36	ASTM A572 50	b/t	h/w
WWF350X212	1	1			5.83	14.5
X192	1	1			6.25	18.3
X176	1	1			7.00	18.7
X155	1	2			7.95	21.8
X137	2	2			8.75	28.1
W1000X488*			1	1	5.18	40.1
X444*			1	1	5.66	43.9
X399*			1	1	6.28	48.8
X363*			1	1	7.03	51.5
X328*			1	2	8.30	51.5
X286*			2	3	10.6	51.5
W1000X976*			1	1	2.38	18.5
X883*			1	1	2.59	20.4
X790*			1	1	2.83	22.6
X714*			1	1	3.10	25.0
X649*			1	1	3.38	27.2
X591*			1	1	3.66	29.9
X539*			1	1	3.98	32.6
X483*			1	1	4.39	36.5
X443*			1	1	4.80	39.3
X412*			1	1	5.03	43.9
X371*			1	1	5.54	48.8
X321*			1	1	6.45	56.2
X296*			1	1	7.38	56.2
W1000X272*			1	1	4.84	56.2
X249*			1	1	5.77	56.2
X222*			1	1	7.11	57.9
W920X1262*			1	1	2.00	13.2
X1188*			1	1	2.10	14.0
X1072*			1	1	2.28	15.4
X967*			1	1	2.48	16.9
X876*			1	1	2.70	18.6
X784*			1	1	2.96	20.7
X722*			1	1	3.19	22.2
X653*			1	1	3.48	24.5
X585*			1	1	3.82	27.3
X534*			1	1	4.16	29.8
X488*			1	1	4.49	32.7
X446*			1	1	4.95	35.3
X417*			1	1	5.29	37.7
X387*			1	1	5.74	39.8
X365*			1	1	6.11	41.7
X342*			1	1	6.53	43.9
W920X313*			1	1	4.48	40.9
X289*			1	1	4.81	44.5
X271*			1	1	5.12	46.9
X253*			1	1	5.48	49.9
X238*			1	1	5.89	52.3
X223*			1	1	6.36	54.3
X201*			1	1	7.56	56.8
W840X922*			1	1	2.39	15.9
X845*			1	1	2.56	17.3
X767*			1	1	2.77	19.0

Section	CSA G40.21 300W	CSA G40.21 350W	ASTM A36	ASTM A572 50	b/t	h/w
W840X697*			1	1	3.02	20.6
X631*			1	1	3.29	22.7
X577*			1	1	3.55	24.9
X527*			1	1	3.85	27.0
X473*			1	1	4.23	30.2
X433*			1	1	4.60	32.7
X392*			1	1	5.03	36.1
X359*			1	1	5.66	37.7
X329*			1	1	6.19	40.4
X299*			1	1	6.85	43.8
W840X226*			1	1	5.49	49.5
X210*			1	1	6.00	51.7
X193*			1	1	6.73	54.1
X176*			1	1	7.77	57.0
W760X865*			1	1	2.29	14.4
X783*			1	1	2.48	15.8
X710*			1	1	2.69	17.4
X644*			1	1	2.93	18.8
X582*			1	1	3.19	20.8
X531*			1	1	3.45	22.8
X484*			1	1	3.74	24.8
X434*			1	1	4.12	27.8
X389*			1	1	4.59	30.5
X350*			1	1	5.01	34.1
X314*			1	1	5.75	36.5
X284*			1	1	6.35	39.9
X257*			1	1	7.03	43.3
W760X196*			1	1	5.28	46.1
X185*			1	1	5.66	48.2
X173*			1	1	6.18	49.9
X161*			1	1	6.89	52.1
X147*			1	1	7.79	54.5
X134*			1	2	8.52	60.4
W690X802*			1	1	2.15	12.9
X735*			1	1	2.30	14.0
X667*			1	1	2.50	15.4
X605*			1	1	2.72	16.7
X548*			1	1	2.95	18.4
X500*			1	1	3.19	20.1
X457*			1	1	3.46	21.8
X419*			1	1	3.71	24.0
X384*			1	1	4.02	25.9
X350*			1	1	4.40	27.9
X323*			1	1	4.71	30.6
X289*			1	1	5.24	34.0
X265*			1	1	5.93	35.0
X240*			1	1	6.50	38.5
X217*			1	1	7.16	41.9
W690X170*			1	1	5.42	44.6
X152*			1	1	6.02	49.3
X140*			1	1	6.72	52.1
X125*			1	1	7.76	55.1

* Not available from Canadian mills.
Check with manufacturer for availability in grades other than CSA G40.21–300W.

CLASS OF SECTIONS IN BENDING

TABLE 5-1

Section	CSA G40.21 300W	CSA G40.21 350W	ASTM A36	ASTM A572 50	b/t	h/w
W610X732*			1	1	2.00	11.4
X670*			1	1	2.13	12.4
X608*			1	1	2.31	13.7
X551*			1	1	2.51	14.8
X498*			1	1	2.72	16.3
X455*			1	1	2.94	17.9
X415*			1	1	3.18	19.4
X372*			1	1	3.49	21.7
X341*			1	1	3.79	23.5
X307*			1	1	4.14	25.9
X285*			1	1	4.43	27.8
X262*			1	1	4.81	30.2
X241	1	1	1	1	5.31	32.0
X217	1	1	1	1	5.92	34.7
X195	1	1	1	1	6.70	37.2
X174	1	1	1	1	7.52	40.9
X155	2	2	1	2	8.53	45.2
W610X140	1	1	1	1	5.18	43.7
X125	1	1	1	1	5.84	48.1
X113	1	1	1	1	6.59	51.2
X101	1	1	1	1	7.65	54.6
X91	2	2	1	2	8.94	59.0
X84	2	3	2	3	9.66	63.6
W610X92*			1	1	5.97	52.6
X82*			1	1	6.95	57.4
W530X599*			1	1	2.14	11.4
X543*			1	1	2.33	12.4
X496*			1	1	2.51	13.5
X447*			1	1	2.73	15.0
X409*			1	1	2.94	16.2
X370*			1	1	3.21	19.9
X331*			1	1	3.54	19.8
X300*			1	1	3.85	21.7
X272*			1	1	4.22	23.8
X248*			1	1	4.57	26.4
X219*			1	1	5.45	27.4
X196*			1	1	6.01	30.3
X182*			1	1	6.45	33.1
X165*			1	1	7.05	35.8
X150*			1	1	7.68	39.6
W530X138	1	1	1	1	4.53	34.2
X123	1	1	1	1	5.00	38.2
X109	1	1	1	1	5.61	43.3
X101	1	1	1	1	6.03	46.1
X92	1	1	1	1	6.70	49.2
X82	1	2	1	2	7.86	52.7
X72	2	3	2	3	9.50	56.4
W530X85*			1	1	5.03	48.8
X74*			1	1	6.10	51.8
X66*			1	1	7.24	56.5
W460X464*			1	1	2.19	11.1
X421*			1	1	2.38	12.0
X384*			1	1	2.56	13.2
W460X349*			1	1	2.76	14.5
X315*			1	1	3.02	15.9
X286*			1	1	3.28	17.5
X260*			1	1	3.58	18.9
X235*			1	1	3.92	20.8
X213*			1	1	4.25	23.1
X193*			1	1	4.64	25.2
X177*			1	1	5.32	25.8
X158*			1	1	5.94	28.5
X144*			1	1	6.40	31.4
X128*			1	1	7.19	35.1
X113*			1	2	8.09	39.7
W460X106	1	1	1	1	4.71	34.0
X97	1	1	1	1	5.08	37.5
X89	1	1	1	1	5.42	40.8
X82	1	1	1	1	5.97	43.2
X74	1	1	1	1	6.55	47.6
X67	1	1	1	1	7.48	50.4
X61	2	2	1	2	8.75	52.8
W460X68*			1	1	5.00	47.1
X60*			1	1	5.75	53.6
X52*			1	1	7.04	56.3
W410X149*			1	1	5.30	25.6
X132*			1	1	5.92	28.6
X114*			1	1	6.76	32.8
X100*			1	1	7.69	38.2
W410X85	1	1	1	1	4.97	35.0
X74	1	1	1	1	5.62	39.3
X67	1	1	1	1	6.22	43.3
X60	1	1	1	1	6.95	49.6
X54	1	2	1	2	8.12	50.9
W410X46	1	1	1	1	6.25	54.5
X39	1	2	1	2	7.95	59.7
W360X1086*			1	1	1.82	4.10
X990*			1	1	1.95	4.45
X900*			1	1	2.08	4.86
X818*			1	1	2.25	5.29
X744*			1	1	2.43	5.76
X677*			1	1	2.63	6.27
W360X634*			1	1	2.75	6.72
X592*			1	1	2.91	7.14
X551*			1	1	3.09	7.64
X509*			1	1	3.32	8.20
X463*			1	1	3.59	8.97
X421*			1	1	3.89	9.78
X382*			1	1	4.23	10.7
X347*			1	1	4.62	11.7
X314*			1	1	5.06	12.8
X287*			1	1	5.45	14.1
X262*			1	1	5.98	15.2
X237*			1	1	6.54	16.9
X216*			1	1	7.11	18.5

* Not available from Canadian mills.
Check with manufacturer for availability in grades other than CSA G40.21–300W.

TABLE 5-1 CLASS OF SECTIONS IN BENDING

Section	CSA G40.21 300W	CSA G40.21 350W	ASTM A36	ASTM A572 50	b/t	h/w
W360X196*			1	1	7.14	19.4
X179*			1	1	7.80	21.3
X162*			1	2	8.51	24.0
X147*			2	3	9.34	26.0
X134*			2	3	10.2	28.5
W360X122*			1	1	5.92	24.6
X110*			1	1	6.43	28.0
X101*			1	1	6.97	30.6
X91*			1	1	7.74	33.8
W360X79	1	1	1	1	6.10	34.0
X72	1	1	1	1	6.75	37.1
X64	1	1	1	1	7.52	41.6
W360X57	1	1	1	1	6.56	42.0
X51	1	1	1	1	7.37	46.2
X45	2	2	1	2	8.72	48.1
W360X39	1	1	1	1	5.98	51.1
X33	1	1	1	1	7.47	57.4
W310X500*			1	1	2.26	6.16
X454*			1	1	2.45	6.75
X415*			1	1	2.66	7.16
X375*			1	1	2.89	7.85
X342*			1	1	3.12	8.49
X313*			1	1	3.36	9.25
W310X283*			1	1	3.65	10.3
X253*			1	1	4.03	11.3
X226	1		1	1	4.45	12.5
X202	1		1	1	4.95	13.8
X179	1		1	1	5.57	15.4
X158	1	1	1	1	6.18	17.9
X143	1	1	1	1	6.75	19.8
X129	1	1	1	1	7.48	21.1
X118	1	2	1	2	8.21	23.2
X107	2	2	1	2	9.00	25.5
X97	3	3	2	3	9.90	28.0
W310X86	1	2	1	1	7.79	30.4
X79	2	2	1	2	8.70	31.4
W310X74	1	1	1	1	6.29	29.5
X67	1	1	1	1	6.99	32.5
X60	1	1	1	1	7.75	37.0
W310X52	1	1	1	1	6.33	38.3
X45	1	1	1	1	7.41	44.1
X39	2	2	1	2	8.51	50.1
X31	3	4	3	4	11.0	58.2
W310X33*			1	1	4.72	44.3
X28*			1	1	5.73	48.7

Section	CSA G40.21 300W	CSA G40.21 350W	ASTM A36	ASTM A572 50	b/t	h/w
W310X24*			1	1	7.54	52.2
X21*			1	2	8.86	57.3
W250X167	1	1	1	1	4.17	11.7
X149	1	1	1	1	4.63	13.0
X131	1	1	1	1	5.20	14.6
X115	1	1	1	1	5.86	16.7
X101	1	1	1	1	6.56	18.8
X89	1	1	1	1	7.40	21.0
X80	1	2	1	2	8.17	23.9
X73	2	2	1	2	8.94	26.2
W250X67	1	1	1	1	6.50	25.4
X58	1	1	1	1	7.52	28.1
X49	2	3	1	3	9.18	30.5
W250X45	1	1	1	1	5.69	31.5
X39	1	1	1	1	6.56	36.3
X33	1	2	1	2	8.02	39.3
X24	3	4	3	4	11.3	48.2
W250X28*			1	1	5.10	37.5
X25*			1	1	6.07	39.5
X22*			1	1	7.39	41.4
X18*			2	3	9.53	50.2
W200X100	1	1	1	1	4.43	12.5
X86	1	1	1	1	5.07	13.9
X71	1	1	1	1	5.92	17.7
X59	1	1	1	1	7.22	19.9
X52	1	2	1	2	8.10	22.8
X46	2	3	2	3	9.23	25.2
W200X42	1	1	1	1	7.03	25.3
X36	1	2	1	2	8.09	29.2
W200X31	1	1	1	1	6.57	29.6
X27	1	2	1	2	7.92	32.9
X21	3	3	2	3	10.3	38.2
W200X22*			1	1	6.38	30.6
X19*			1	2	7.85	32.9
X15*			2	3	9.62	44.0
W150X37	1	1	1	1	6.64	17.1
X30	1	2	1	2	8.23	21.1
X22	3	4	3	4	11.5	23.9
W150X24*			1	1	4.95	21.1
X18*			1	1	7.18	24.1
X14*			1	2	9.09	32.3
W130X28*			1	1	5.87	15.9
X24*			1	1	6.98	18.0
W100X19*			1	1	5.85	12.4

* Not available from Canadian mills.
Check with manufacturer for availability in grades other than CSA G40.21–300W.

F_y = 248 MPa
ϕ = 0.90

FACTORED ULTIMATE SHEAR STRESS, ϕF_s GIRDER WEBS (Elastic Analysis), (MPa)#

Required Gross Area of Pairs of Intermediate Stiffeners (Percent of Web Area, hw) * +

Web Slender-ness Ratio $\frac{h}{w}$	Aspect Raito $\frac{a}{h}$: Stiffener Spacing to Web Depth										No Inter-mediate Stiffeners
	0.50	0.67	0.75	1.00	1.25	1.50	1.75	2.00	2.50	3.00	
60										147 *0.77*	147
70								147 *1.06*	144 *0.89*	141 *0.77*	135
80				147 *1.46*	144 *1.37*	137 *1.26*	132 *1.15*	129 *1.06*	127 *0.89*	125 *0.77*	118
90			147 *1.50*	140 *1.46*	129 *1.37*	125 *1.26*	123 *1.15*	120 *1.06*	117 *0.89*	115 *0.83*	105
100			147 *1.50*	128 *1.46*	123 *1.37*	119 *1.39*	116 *1.95*	113 *2.19*	107 *2.26*	103 *2.13*	86.5
110		147 *1.49*	137 *1.50*	123 *1.46*	118 *2.52*	111 *3.33*	106 *3.61*	102 *3.64*	95.8 *3.42*	91.5 *3.10*	71.4
120	147 *1.38*	137 *1.49*	128 *1.50*	119 *2.77*	111 *4.30*	104 *4.81*	98.0 *4.88*	93.6 *4.75*	87.1 *4.30*	82.5 *3.83*	60.0
130	147 *1.38*	128 *1.49*	125 *1.50*	114 *4.53*	104 *5.69*	97.4 *5.96*	91.7 *5.86*	87.1 *5.61*	80.3 *4.99*	75.6 *4.40*	51.1
140	147 *1.38*	126 *1.49*	122 *2.09*	109 *5.92*	99.7 *6.80*	92.5 *6.88*	86.7 *6.64*	82.0 *6.29*	75.0 *5.53*	70.0 *4.86*	44.1
150	138 *1.38*	123 *1.74*	119 *3.76*	105 *7.05*	95.8 *7.68*	88.5 *7.62*	82.7 *7.27*	77.9 *6.84*	70.7 *5.97*	65.5 *5.22*	38.4
160	129 *1.38*	121 *3.32*	116 *5.12*	102 *7.97*	92.7 *8.41*	85.3 *8.22*	79.4 *7.79*	74.5 *7.29*	67.1 *6.33*		33.7
170	127 *1.38*	118 *4.64*	113 *6.25*	99.2 *8.73*	90.0 *9.02*	82.6 *8.72*	76.6 *8.21*	71.7 *7.66*			29.9
180	126 *1.38*	115 *5.74*	110 *7.19*	97.0 *9.37*	87.8 *9.52*	80.4 *9.14*	74.3 *8.57*	69.3 *7.97*			26.7
190	124 *1.68*	113 *6.68*	108 *7.99*	95.2 *9.91*	86.0 *9.95*	78.5 *9.49*	72.4 *8.88*				23.9
200	123 *2.87*	111 *7.48*	106 *8.67*	93.6 *10.4*	84.4 *10.3*	76.9 *9.79*					21.6
220	119 *4.77*	108 *8.76*	103 *9.77*	91.0 *11.1*	81.9 *10.9*						17.8
240	116 *6.21*	105 *9.73*	101 *10.6*	89.1 *11.7*							15.0
260	114 *7.34*	103 *10.5*	99.2 *11.3*								12.8
280	112 *8.23*	102 *11.1*	97.8 *11.8*								11.0
300	110 *8.95*	101 *11.6*	96.7 *12.2*								9.61
320	109 *9.54*										8.45

 # Clause 13.4.1.1, CAN/CSA S16.1-M89

* For single stiffeners on one side of web only, multiply percentages shown by 1.8 for angle stiffeners and by 2.4 for plate stiffeners.

\+ When stiffener F_y is not the same as the web F_y, multiply gross area by ratio (F_y web / F_y stiffener)

FACTORED ULTIMATE SHEAR STRESS, ϕF_s GIRDER WEBS (Elastic Analysis), (MPa)#

F_y = 300 MPa
ϕ = 0.90

Required Gross Area of Pairs of Intermediate Stiffeners
(Percent of Web Area, hw) * +

Web Slender-ness Ratio $\frac{h}{w}$	Aspect Ratio $\frac{a}{h}$: Stiffener Spacing to Web Depth										No Inter-mediate Stiffeners
	0.50	0.67	0.75	1.00	1.25	1.50	1.75	2.00	2.50	3.00	
50											178
60										178 *0.77*	174
70					178 *1.37*	172 *1.26*	166 *1.15*	163 *1.06*	158 *0.89*	155 *0.77*	149
80			178 *1.50*	173 *1.46*	159 *1.37*	153 *1.26*	150 *1.15*	148 *1.06*	144 *0.89*	141 *0.77*	130
90			178 *1.50*	155 *1.46*	149 *1.37*	145 *1.26*	141 *1.75*	138 *2.02*	131 *2.12*	127 *2.02*	106
100		178 *1.49*	166 *1.50*	149 *1.46*	143 *2.52*	135 *3.33*	128 *3.61*	123 *3.64*	116 *3.42*	111 *3.10*	86.5
110	178 *1.38*	164 *1.49*	154 *1.50*	144 *2.96*	133 *4.46*	124 *4.94*	118 *4.99*	112 *4.84*	104 *4.37*	98.9 *3.90*	71.4
120	178 *1.38*	154 *1.49*	151 *1.50*	136 *4.83*	125 *5.93*	116 *6.16*	110 *6.03*	104 *5.75*	95.8 *5.10*	89.9 *4.50*	60.0
130	175 *1.38*	151 *1.49*	147 *2.62*	130 *6.28*	119 *7.08*	110 *7.11*	103 *6.84*	97.6 *6.46*	89.0 *5.67*	83.0 *4.98*	51.1
140	163 *1.38*	148 *2.41*	142 *4.33*	125 *7.43*	114 *7.99*	105 *7.87*	98.3 *7.49*	92.5 *7.03*	83.7 *6.12*	77.4 *5.35*	44.1
150	155 *1.38*	144 *4.01*	138 *5.70*	122 *8.36*	110 *8.73*	102 *8.48*	94.3 *8.01*	88.3 *7.48*	79.3 *6.49*	72.9 *5.65*	38.4
160	153 *1.38*	140 *5.32*	134 *6.83*	118 *9.12*	107 *9.33*	98.3 *8.98*	91.0 *8.44*	85.0 *7.86*	75.8 *6.78*		33.7
170	151 *1.38*	137 *6.41*	131 *7.76*	116 *9.75*	105 *9.83*	95.6 *9.39*	88.2 *8.79*	82.1 *8.16*			29.9
180	149 *2.64*	134 *7.32*	129 *8.54*	114 *10.3*	102 *10.2*	93.4 *9.74*	85.9 *9.09*	79.8 *8.42*			26.7
190	146 *3.79*	132 *8.10*	127 *9.21*	112 *10.7*	101 *10.6*	91.5 *10.0*	84.0 *9.34*				23.9
200	144 *4.77*	130 *8.75*	125 *9.77*	110 *11.1*	99.0 *10.9*	89.9 *10.3*					21.6
220	140 *6.34*	127 *9.81*	122 *10.7*	108 *11.7*	96.5 *11.4*						17.8
240	137 *7.53*	125 *10.6*	120 *11.4*	106 *12.2*							15.0
260	134 *8.46*	123 *11.2*	118 *11.9*								12.8

 # Clause 13.4.1.1, CAN/CSA S16.1-M89

* For single stiffeners on one side of web only, multiply percentages shown by 1.8 for angle stiffeners and by 2.4 for plate stiffeners.

\+ When stiffener F_y is not the same as the web F_y, multiply gross area by ratio (F_y web / F_y stiffener)

F_y = 350 MPa
ϕ = 0.90

FACTORED ULTIMATE SHEAR STRESS, ϕF_s GIRDER WEBS (Elastic Analysis), (MPa)#

Required Gross Area of Pairs of Intermediate Stiffeners
(Percent of Web Area, hw) * +

Web Slenderness Ratio $\frac{h}{w}$	Aspect Ratio $\frac{a}{h}$: Stiffener Spacing to Web Depth										No Intermediate Stiffeners
	0.50	0.67	0.75	1.00	1.25	1.50	1.75	2.00	2.50	3.00	
50										208 *0.77*	207
60							208 *1.15*	205 *1.06*	199 *0.89*	196 *0.77*	188
70				208 *1.46*	196 *1.37*	186 *1.26*	181 *1.15*	178 *1.06*	174 *0.89*	172 *0.77*	161
80			208 *1.50*	187 *1.46*	177 *1.37*	172 *1.26*	168 *1.15*	165 *1.29*	160 *1.54*	156 *1.54*	135
90		208 *1.49*	199 *1.50*	176 *1.46*	168 *1.86*	161 *2.79*	154 *3.15*	148 *3.24*	140 *3.09*	134 *2.83*	106
100	208 *1.38*	195 *1.49*	181 *1.50*	169 *2.53*	157 *4.11*	147 *4.66*	140 *4.74*	133 *4.63*	124 *4.21*	118 *3.75*	86.5
110	208 *1.38*	180 *1.49*	176 *1.50*	160 *4.63*	147 *5.78*	137 *6.03*	129 *5.92*	122 *5.66*	113 *5.03*	106 *4.44*	71.4
120	205 *1.38*	176 *1.49*	172 *2.55*	152 *6.23*	139 *7.04*	129 *7.08*	121 *6.82*	114 *6.44*	104 *5.65*	97.1 *4.96*	60.0
130	189 *1.38*	173 *2.48*	166 *4.39*	146 *7.48*	133 *8.03*	123 *7.90*	114 *7.51*	108 *7.05*	97.4 *6.14*	90.1 *5.36*	51.1
140	180 *1.38*	168 *4.18*	160 *5.85*	141 *8.46*	128 *8.81*	118 *8.54*	109 *8.07*	103 *7.53*	92.0 *6.52*	84.5 *5.69*	44.1
150	178 *1.38*	163 *5.56*	156 *7.03*	137 *9.26*	124 *9.44*	114 *9.07*	105 *8.51*	98.4 *7.92*	87.7 *6.84*	80.0 *5.94*	38.4
160	176 *1.69*	159 *6.68*	152 *8.00*	134 *9.91*	121 *9.95*	111 *9.49*	102 *8.88*	95.0 *8.24*	84.2 *7.09*		33.7
170	173 *3.08*	156 *7.62*	149 *8.80*	132 *10.5*	119 *10.4*	108 *9.85*	99.4 *9.18*	92.2 *8.51*			29.9
180	169 *4.24*	153 *8.40*	147 *9.47*	129 *10.9*	117 *10.7*	106 *10.1*	97.1 *9.43*	89.9 *8.73*			26.7
190	167 *5.22*	151 *9.06*	145 *10.0*	128 *11.3*	115 *11.0*	104 *10.4*	95.2 *9.65*				23.9
200	164 *6.06*	149 *9.63*	143 *10.5*	126 *11.6*	113 *11.3*	102 *10.6*					21.6
220	160 *7.41*	146 *10.5*	140 *11.3*	123 *12.1*	111 *11.7*						17.8

 # Clause 13.4.1.1, CAN/CSA S16.1-M89

* For single stiffeners on one side of web only, multiply percentages shown by 1.8 for angle stiffeners and by 2.4 for plate stiffeners.

\+ When stiffener F_y is not the same as the web F_y, multiply gross area by ratio (F_y web / F_y stiffener)

PLATE GIRDERS

General

When the required capacity of a steel flexural member exceeds that which can be provided by the rolled or welded wide flange sections listed in Part Six in this book, a plate girder can usually provide an economical solution. A plate girder basically consists of flange plates connected to a relatively thin web plate. Transverse or longitudinal stiffeners (usually consisting of plates or small shapes) may be used to increase the strength of the web. Heavy concentrated loads and reactions are supported directly on bearing stiffeners.

Clause 15 of CAN/CSA S16.1-M89, "Limit States Design of Steel Structures", contains requirements for the design of plate girders. Clause 15.1 specifies that girders shall be proportioned on the basis of the section properties, i.e. by the moment of inertia method, where $M_r = \phi\,(I/y)\,F_y$. To obtain a trial section for calculating section properties an approximate method is frequently used. One such method assumes the applied bending moment resisted by the flanges alone, so that the required area of one flange, A_f, is given approximately by $A_f = \dfrac{M_f}{\phi\,F_y\,d_{eff}}$, where M_f = bending moment in the girder under factored load, ϕ = the performance factor, 0.90, F_y = specified minimum yield stress and d_{eff} = an assumed effective girder depth. The moment of resistance of the trial section should then be checked using the expression $M_r = \phi\,(I/y)\,F_y > M_f$, where I = the moment of inertia of the girder (gross or net) in accordance with Clause 15.1 of S16.1-M89, and y = the distance from the neutral axis to the extreme fibre of a plate girder flange. Formulae in Part Seven can be used to compute I.

Flanges

The compression flange may fail either by buckling or by yielding, and will generally govern the flange design of a symmetrical plate girder. S16.1-M89 clauses intended to minimize the possibility of buckling are:

1. Local flange buckling
   - Clause 11.2, Table 1
   - $b/2t \leq 200/\sqrt{F_y}$
   - Limits on Width-Thickness Ratios, page 4-5
2. Lateral buckling
   - Clause 13.6
3. Vertical buckling
   - Clause 13.4.1.3
   - $\dfrac{h}{w} \leq \dfrac{83\,000}{F_y}$

where: F_y = specified minimum yield point of compression flange steel, (MPa)

h = clear depth of web between flanges, (mm)

w = web thickness, (mm)

b = flange width, (mm)

t = flange thickness, (mm)

The portion of a girder web subject to compressive stress tends to deflect laterally, causing a stress redistribution within the girder which results in an increase to the compression flange stress. For relatively thick webs this increase is insignificant, but when the web slenderness ratio (h/w) exceeds $1900/\sqrt{M_f/\phi S}$, Clause 15.4 of S16.1-M89 limits the factored moment resistance to M'_r as follows:

$$M'_r = M_r [\, 1.0 - (\, 0.0005 A_w / A_f \,)\, (\, h / w - 1900 / \sqrt{M_f / \phi S} \,)]$$

where M'_r = reduced factored moment resistance

M_r = factored moment resistance determined by Clause 13.5 or Clause 13.6 but not exceeding $\phi\, M_y$

S = elastic section modulus, I/y, (mm^3)

ϕ = performance factor, 0.90

A_w = web area, (mm^2)

A_f = compression flange area, (mm^2)

and the other symbols are as previously defined.

When designing plate girders with long spans, economy may be achieved by reducing the area of flange plates where the bending moment is substantially less than the maximum. The cost of making flange splices must, however, be balanced against the weight savings achieved.

Webs

Plate girders are subjected to a complex combination of bending and shear stresses. The shear capacity of unstiffened plate girder webs decreases rapidly as the web slenderness ratio increases. Transverse web stiffeners increase the factored shear resistance of plate girder webs, with the amount of increase depending on the aspect ratio, a/h, where a = stiffener spacing and h = clear web depth.

Clause 13.4.1.1 of CAN/CSA S16.1-M89 specifies factored shear resistances for both stiffened and unstiffened webs (which have been proportioned elastically) for different values of the web slenderness ratio. The tables on pages 5-9, 5-10 and 5-11 list the factored ultimate shear stress, $\phi\, F_s$, for plate girder webs, from which the factored shear resistance can be computed according to Clause 13.4.1.1. The values are tabulated for various combinations of web slenderness ratio, h/w, stiffener aspect ratio, a/h, and for specified minimum yield strengths, F_y, of 248, 300 and 350 MPa.

The formulae in Clause 13.4.1.1 consider the post buckling strength of transversely stiffened plate girder webs in establishing the factored shear resistance. In addition to shear resistance, web crippling and yielding of thin webs subject to loads acting on the compression edge of the web plate, Clause 15.9, S16.1-M89 must be investigated.

Stiffeners

Bearing stiffeners, intermediate transverse stiffeners and longitudinal stiffeners are used with thin web plate girders. Longitudinal stiffeners are seldom used in building construction but are common for bridge construction. CAN/CSA S16.1-M89 provides detailed requirements for bearing stiffeners, Clause 15.6 , and for intermediate transverse stiffeners, Clause 15.7. The tables on pages 5-9, 5-10 and 5-11 list also the required gross area of pairs of intermediate transverse stiffeners as a percentage of the web area.

When intermediate transverse stiffeners are required, the maximum spacing must not exceed 3h for web slenderness ratios, h/w, less than or equal to 150. For $h / w \geq 150$, the maximum spacing, a_{max} , is established by the formula:

$$a_{max} = \frac{67\,500\,h}{(h/w)^2}$$

Thin-web stiffened plate girders may be less economical than plate girders with thicker webs which do not require intermediate transverse stiffeners, since possible saving in weight can sometimes be more than overcome by the additional fabrication costs. Framing angles at the ends of plate girders can be assumed to provide web stability at girder ends, provided they comprise at least 2/3 of the girder depth.

References

Additional information on plate girders can be obtained from the following references:

1. Part 2 of the Handbook.
2. Kulak, Adams and Gilmor, "Limit States Design in Structural Steel", CISC. 1990
3. Picard, A., Beaulieu, D., "Calcul des charpentes d'acier", CISC, 1991.
4. Balser and Thürlimann, "Strength of Plate Girders in Bending", Proc. ASCE. J. Struc. Div., Vol. 87, No. ST6, Aug. 1961.
5. Basler, "Strength of Plate Girders in Shear", Proc. ASCE. J. Struc. Div., Vol. 87, No. ST7, Oct. 1961.
6. Basler, "Strength of Plate Girders Under Combined Bending and Shear", Proc. ASCE. J. Struc. Div., Vol. 87, Nos. ST7, Oct. 1961.

Example

Given:

Design a simply supported welded plate girder spanning 22 metres and loaded as shown. The compression flange is laterally supported along the entire length of the member. The uniform load includes an estimate of 3.50 kN/m for the girder dead load. Total depth is limited to 1800 mm. Use G40.21-M 300W Steel.

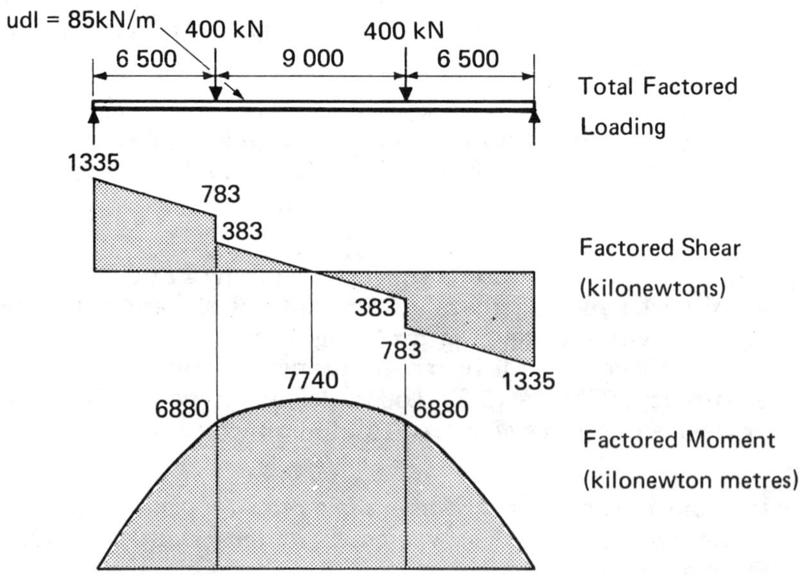

Solution:

Trial Section - Flanges

Approximate flange area, $A_f = \dfrac{M_f}{\phi F_y d_{eff}}$

Assume d_{eff} = 1770 mm

Maximum ultimate stress = ϕF_y = 0.90 x 300 = 270 MPa

Required $A_f = \dfrac{7\,740 \times 10^6}{250 \times 1770} \times 90\,\% = 14\,600\ \text{mm}^2$

where the 90% factor allows for the contribution of the web to flexural capacity

Flange slenderness ratio, $b/2t \leq 200/\sqrt{F_y} = 11.5$ (page 4-5)

Try 30 x 500 flange plate, A_f = 500 x 30 = 15 000 mm^2

$b/2t = 500/(2 \times 30) = 8.33 < 11.5$ - OK (Class 1)

Trial Section - Web

Maximum h/w permitted = 83 000 / F_y (Clause 13.4.1.3, S16.1-M89)

= 83 000 / 300 = 277

Maximum h/w permitted before the factored moment resistance is reduced

$= 1900/\sqrt{M_f/\phi S}$ (Clause 15.4, S16.1-M89)

For $M_f = \phi S F_y$, $h/w = 1900/\sqrt{F_y} = 1900/\sqrt{300} = 110$

Try 10 x 1740 Web, (10 mm is a first preference thickness)

h / w = 1 740 / 10 = 174

Factored ultimate shear stress in girder web

$= 1\,335 \times 10^3 / (1\,740 \times 10) = 76.7$ MPa

From table on page 5-10, for h/w = 180, a/h = 2.00 (stiffened web)

ϕF_s = 79.8 MPa > 76.7 - OK

A_w = 10 x 1740 = 17 400 mm^2

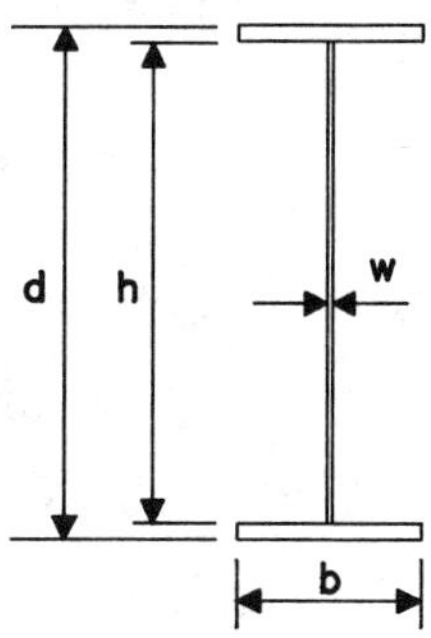

Check Trial Section

(a) Compute Girder Properties

Moment of inertia $I = \dfrac{b\,d^3 - (b - w)\,h^3}{12}$

$= \dfrac{500 \times 1800^3 - (500 - 10)\,1740^3}{12} = 27\,900 \times 10^6\ \text{mm}^4$

Section modulus $S = I/y = 27\,900 \times 10^6 / 900 = 31\,000 \times 10^3\ \text{mm}^3$

Flange area A_f = 15 000 mm^2

Web area A_w = 17 400 mm^2

(b) Calculate moment resistance,

Since h/w exceeds 110, check reduced moment resistance

$M'_r = M_r [\, 1.0 - (\, 0.0005\, A_w / A_f\,)\,(\, h/w - 1900 / \sqrt{M_f / \phi S}\,)]$

$M_r = \phi\, M_y = \phi\, S\, F_y = 0.9 \times 31\,000 \times 10^3 \times 300 \times 10^{-6} = 8\,370$ kN·m (Clause 15.4)

$A_w / A_f = 17\,400 / 15\,000 = 1.16$

$1\,900 / \sqrt{M_f / \phi S} = 1\,900 / \sqrt{(\,7\,740 \times 10^6\,) / (\,0.9 \times 31\,000 \times 10^3\,)} = 114$

$M'_r = 8\,370\, [\, 1.0 - 0.0005 \times 1.16\, (\, 174 - 114\,)\,]$ kN·m

$= 8080$ kN·m $> 7\,740$ - OK

Locate Web Stiffeners

(a) Establish end panel length for no tension field. (Clause 15.7.1)

By Clause 13.4.1.1(d), for h/w = 174 and $F_t = 0$

$$V_r = 0.9 \times 17\,400 \times \frac{180\,000\, k_v}{174^2 \times 1000} = 93.1\, k_v$$

and equating V_r to V_f = 1335 kN, gives $k_v = 14.34$

but $k_v = 4 + \dfrac{5.34}{(a/h)^2}$, therefore $a/h = \sqrt{5.34 / 10.34} = 0.719$

therefore maximum end panel length = 0.719 x 1740 = 1250 mm

(b) Check shear resistance of unstiffened web.

Using table on page 5-10 and interpolating for h/w = 174

$\phi F_s = 29.9 - 0.4\,(\,29.9 - 26.7\,) = 28.6$ MPa

Therefore V_r (unstiffened web) = 28.6 x 17 400 / 1000 = 498 kN > 383

Therefore stiffeners are not required between the two concentrated loads, but are required at the concentrated load locations.

Using an end panel length of 1000 mm, and 2 equal panels between the end stiffeners and the stiffeners at the concentrated loads gives an intermediate stiffener spacing of

(6 500 – 1 000) / 2 = 2 750 mm ,

and an aspect ratio a / h = 2 750 / 1 740 = 1.58

From table on page 5-10 for a/h = 1.75 and interpolating for h/w = 174

$\phi F_s = 88.2 - 0.4\,(\,88.2 - 85.9) = 87.3$ MPa (Consertive)

V_r (stiffened web) = 87.3 x 17 400 / 1000 = 1 520 kN > V_f

Stiffener Size

(a) Intermediate Stiffeners

From table on page 5-10, with h/w = 174 and a/h = 1.58 , the required total area of a pair of intermediate stiffeners is approximately 9.5 % of the web area,

$= 0.095 \times 17\ 400 = 1\ 650\ \text{mm}^2$

Required I $= (h / 50)^4 = (1\ 740 / 50)^4 = 1.47 \times 10^6 \text{mm}^4$ (Claue 15.7.3)

Maximum $b / t = 200 / \sqrt{F_y} = 11.5$

Use 2 - 10 x 110 stiffeners

$A_s = 2 \times 10 \times 110 = 2\ 200\ \text{mm}^2 > 1\ 650$

$I = 10 \times 230^3 / 12 = 10.14 \times 10^6\ \text{mm}^4 > 1.47 \times 10^6\ \text{mm}^4$

$b / t = 110 / 10 = 11.0 < 11.5$

(b) Stiffeners at concentrated loads

Assuming the concentrated load is applied to the top flange of the girder, the web must be checked for web crippling and web yielding. (Clause 15.9(a))

Assume the load is applied through a column base plate (N = 300mm) and that k = flange thickness plus 8mm (assumed weld size),

B_r (web yielding)

$= 1.10 \times 0.9 \times 10\ [\ 300 + 5\ (\ 30 + 8\ )\ ]\ 300 / 10^3 = 1\ 460\ \text{kN} > 400$

B_r (web crippling)

$= 300 \times 0.9 \times 10^2\ [\ 1 + 3\ (\ 300 / 1\ 740\ )\ (\ 10 / 30\ )^{1.5}\ ]\ (\sqrt{300 \times (\ 30 / 10\ )}) / 10^3$

$= 891\ \text{kN} > 400$

Therefore stiffeners are not required to bear at the top flange.

Use 2 - 10 x 110 intermediate stiffeners as above.

Combined Shear and Moment

Combined shear and moment in girder webs is governed by Clause 13.4.1.4 of S16.1-M89. The worst effect will occur under the concentrated load; $V_f = 783$ kN, $M_f = 6\ 880$ kN·m

Check whether $h / w = 174$ exceeds $502 \sqrt{k_v / F_y}$

$k_v = 5.34 + 4 / (\ a / h\ )^2 = 5.34 + 4 / 1.58^2 = 6.94$

Therefore $502 \sqrt{k_v / F_y} = 502 \times \sqrt{6.94 / 300} = 76.4 < 174$

Therefore Clause 13.4.1.4 must be checked

From table on page 5-10, for $h / w = 174$, and $a / h = 1.58$, $\phi F_s = 87.3$ MPa

Therefore $V_r = 87.3 \times 17\ 400 / 10^3 = 1\ 520$ kN

$$\frac{V_f}{V_r} = \frac{783}{1\ 520} = 0.515 < 1$$

$$\frac{M_f}{M_r} = \frac{6\,880}{8\,080} = 0.851 < 1$$

$$0.727\frac{M_f}{M_r} + 0.455\frac{V_f}{V_r} = (0.727 \times 0.851) + (0.455 \times 0.515) = 0.853 < 1 \text{ - OK}$$

Final Selection

Use —— 30 x 500 flange plates

10 x 1 740 web plate

10 x 110 intermediate stiffeners in pairs spaced as shown

Girder UDL excluding stiffeners

Flanges — 2 x 30 x 500 = 30 000

Web — 10 x 1 740 = 17 400

Total area — = 47 400 mm^2

Dead Load = 47 400 x 7 850 kg / m^3 x 9.81/ (10^6 x 10^3) = 3.65 kg / m

Assumed dead load = 3.50 kg/m

Since there is reserve moment and shear capacity, this slight increase in UDL has no effect on the design. If desired the flange area could be reduced, and a thinner web with additional stiffeners might be used to obtain a lighter girder, provided the saving in weight is not offset by possible additional fabrication costs.

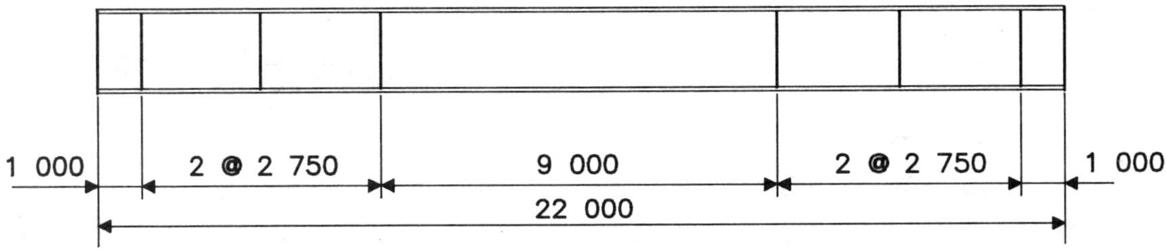

Girder-to-Column Connection

Assuming the girder ends bear on top of supporting columns, then bearing stiffeners designed according to Clause 15.6 of S16.1-M89 must be investigated.

$1100 / \sqrt{F_y} = 1100 / \sqrt{300} = 63.5 < h / w = 174$

therefore bearing stiffeners are required.

Try 2 – 16 x 175 stiffeners

$b / t = 175 / 16 = 10.9 < 200 / \sqrt{F_y} = 11.5$

Area = $(2 \times 175 \times 16) + (12 \times 10 \times 10) = 5\,600 + 1\,200 = 6\,800$ mm^2

$I = 16 \times 360^3 / 12 = 62.2 \times 10^6$ mm^4

$r = \sqrt{62.2 \times 10^6 / 6\,800} = 95.6$ mm

$$\frac{KL}{r} = 0.75 \times 1\,740 / 95.6 = 13.7 \text{ (for top flange laterally supported at the reaction)}$$

Using Table 4–4 on page 4–11, $C_r / A = 268$ MPa

Therefore $C_r = 268 \times 6\,800 / 1000 = 1\,820$ kN > 1 335 kN

Bearing on contact area (Clause 13.10.1(a)

Assume 150 mm bearing (stiffener clipped 25 mm to clear fillet welds)

Bearing area = $2 \times 150 \times 16 = 4\,800$ mm^2

Therefore $B_r = 1.50 \times 0.9 \times 300$ MPa $\times 4\,800$ mm^2 / 1000 = 1 940 kN > 1 335 kN

If the girder ends are framed to the supporting columns, the girder-to-column connections may consist of pairs of header angles bolted or welded to the web plate and connected to the supporting column flange. These connection angles must also serve as end web stiffeners, therefore their length must be almost equal to the clear depth of the girder web.

Note: Although not considered in this example, a deep header connection as described above can develop significant bending moment at the column, and the effects of this moment upon the members must be considered in design. A method for assessing this moment is described in "Advanced Design in Structural Steel" by J. E. Lothers, Prentice Hall, 1960.

COMPOSITE BEAMS

General

A composite beam, in general, consists of a steel beam and concrete slab so interconnected that both the steel beam and the slab act jointly to resist bending. Several combinations which effectively act as composite beams occur in practice. These include a steel beam or girder with a concrete slab inter-connected with mechanical shear connectors, a steel beam or girder with a ribbed concrete slab formed by steel deck inter-connected by mechanical shear connectors, and a steel beam or girder fully encased by the concrete in such a way that the encased beam and the concrete slab behave monolithically. Clause 17 of CAN/CSA S16.1-M89, "Limit States Design of Steel Structures", contains requirements for composite beams.

Some advantages of composite construction are:

- Reduced weight of steel members.
- Reduced depth of steel members.
- Reduced deflections under superimposed load.
- Simplified changes to electrical services when steel deck is used.

Composite construction is most advantageous when heavy loads and long spans are involved. For this reason composite construction is widely used for bridges. For building construction, composite beams consisting of steel beams with steel deck and concrete cover slab utilizing steel stud shear connectors welded to the beam top flange are most frequently used. Other types of composite construction used in buildings include composite trusses and joists, and stub-girders.

A comprehensive examination of the design and behaviour of steel-concrete composite floor systems and their components is contained in the CISC publication "Design and Construction of Composite Floor Systems", (Reference 2). This book contains design aids and examples for composite beams, girders, trusses, and stub-girders. Other design aids include CISC / CSCC developed computer programs for both main frame and personal computers.

Tables

The Composite Beam Trial Selection Tables, pages 5-28 to 5-83 are based on the use of CAN/CSA G40.21-M 300W steel and list composite members for the practical range of Canadian rolled W-shapes from 150 mm to 610 mm nominal depth, and for three WWF-section sizes. W-shapes highlighted in blue are those which are highlighted in the tables of Properites and Dimension in Part 6 of this Handbook. (See also "Principal Sources of Structural Steel sections", page 6-18, and "Rolled Structural Shapes" page 6-42 for comments regarding availability). Tables are provided for the following combinations of deck-slab concrete strength and concrete density:

- 75 mm steel deck with 65 mm cover slab with f'_c of 20 MPa, 2300 kg / m^3 concrete.
- 75 mm steel deck with 85 mm cover slab with f'_c of 20 MPa, 2300 kg / m^3 concrete.
- 75 mm steel deck with 75 mm cover slab with f'_c of 25 MPa, 2300 kg / m^3 concrete.
- 75 mm steel deck with 85 mm cover slab with f'_c of 25 MPa, 2000 kg / m^3 concrete.
- 75 mm steel deck with 85 mm cover slab with f'_c of 25 MPa, 1850 kg / m^3 concrete.
- 38 mm steel deck with 65 mm cover slab with f'_c of 20 MPa, 2300 kg / m^3 concrete.
- 130 mm solid slab with f'_c of 20 MPa, 2300 kg / m^3 concrete.

The tables show steel shapes listed in descending order of nominal depth and mass, and include the following properties, design data and resistances:

b = flange width of steel shape, (mm)

t = flange thickness of steel shape, (mm)

d = overall depth of steel shape, (mm)

b_1 = effective width of slab used in computing values of M_{rc}, Q_r, I_t, and S_t, (mm). (Refer to Clause 17.4 of CAN/CSA S16.1-M89 for appropriate design effective width)

M_{rc} = factored moment resistance of composite beam for percentage of full shear connection equal to 100%, 70% and 40%, (kN·m)

Q_r = required sum of factored shear resistances between adjacent points of maximum and zero moment for 100 per cent shear connection, (kN). Q_r = lesser of $\phi A_s F_y$ or $0.85 \phi_c b_1 t_c f'_c$, where t_c = effective slab thickness or effective cover slab thickness.

I_t = moment of inertia of the composite section, transformed into steel properties, computed using mass density as shown on each table, (10^6 mm^4)

S_t = section modulus of the composite section related to the extreme fibre of the bottom flange of the steel beam based on the value of I_t, (10^3 mm^3)

M_r = factored moment resistance of laterally supported bare steel section, (kN·m)

V_r = factored shear resistance of the bare steel beam, (kN)

L_u = maximum unsupported length of compression flange of the steel beam alone for which no reduction in M_r is required, (mm)

I_x = moment of inertia about the x-x axis of the bare steel beam, (10^6 mm^4)

S_x = section modulus of the bare steel beam, (10^3 mm^3)

M'_r = factored moment resistance of the bare steel beam for an unsupported length L′, (kN·m)

Since the concrete slab and / or the steel deck prevent movement of the top flange, lateral buckling is not a consideration at composite action. During construction, however, the unsupported length of the compression flange may be greater than L_u, and the moment resistance for the non-composite shape for the appropriate unsupported length of compression flange must be used.

The tabulated factored shear resistance, V_r, is computed according to Clause 13.4.1.1, S16.1-M89 for the appropriate h/w ratio.

Shear Connectors

Clauses 17.9.5 and 17.9.6, S16.1-M89 stipulate the amount of total factored horizontal shear force that must be resisted by shear connectors.

For full (i.e. 100%) shear connection the total factored horizontal shear force, V_h, to be transferred between the point of maximum positive moment and adjacent points of zero moment is either

- $\phi A_s F_y$ when the plastic neutral axis is in the slab, or
- $0.85 \phi_c b_1 t_c f'_c$ when the plastic neutral axis is in the steel section.

For partial shear connection the total factored horizontal shear force, V_h, is the sum of the factored resistances of all the shear connectors between the point of maximum positive moment and each adjacent point of zero moment. Clause 17.9.4,

S16.1-M89, limits the minimum amount of partial shear connection to 40 percent of either $\phi A_s F_y$ or $0.85 \phi_c b_1 t_c f'_c$, whichever is the lesser, when computing flexural strength.

Generally, shear connectors may be uniformly spaced in regions of positive or negative bending. However, when a concentrated load occurs within a region of positive bending, the number of shear connectors and the shear connector spacing is determined by Clause 17.9.8, S16.1-M89.

Tables 5-2, 5-3 and 5-4 provide values of factored shear resistance, q_r , for the most common sizes of end-welded shear studs according to the requirements of Clause 17.7, S16.1-M89 when the stud height is at least four stud diameters, and when the stud projection in a ribbed slab is at least two stud diameters above the top surface of the steel deck.

Table 5-2 on page 5-23 gives values of q_r for stud diameters of 3/4 inch (19 mm), 5/8 inch (15.9 mm), and 1/2 inch (12.7 mm) in solid slabs, or in deck-slabs with ribs parallel to the beam, based on three concrete strength levels, f'_c, of 20 MPa, 25 MPa, and 30 MPa for both normal density (2300 kg / m^3) and semi-low density (1850 kg / m^3) concrete. Values are calculated according to Clause 17.7.2.1 and Clause 17.7.2.2, S16.1-M89.

Table 5-3 on page 5-24 and Table 5-4 on page 5-25 give values of q_r for 3/4 inch (19 mm) and 5/8 inch (15.9 mm) diameter studs respectively in ribbed slabs for 75 mm or 38 mm deck, with ribs perpendicular to the beam, calculated according to Clause 17.7.2.3, S16.1-M89. Values are given for three concrete strength levels, f'_c, of 20 MPa, 25 MPa, and 30 MPa for both normal density (2300 kg / m^3) and semi-low density (1850 kg / m^3) concrete.

Deflections

Composite beams are stiffer than similar non-composite beams, and deflections are reduced when composite construction is used. Due to creep of the concrete slab over time, maximum deflections may increase, especially if the full load is sustained. Appendix L of CAN/CSA Sl6.l-M89 provides guidance for estimating deflections caused by shrinkage of the concrete slab. Beam deflection during construction, due to loads supported prior to hardening of the concrete while the steel beam alone supports the loads, should be checked. Cambering or the use of temporary shores will reduce the total final deflection.

For steel beams unshored during construction, Clause 17.11, Sl6.l-M89 limits the total stress in the tension flange to $0.90 F_y$, due to the specified loads applied before the concrete strength reaches $0.75 f'_c$, plus at the same location, the remaining specified loads acting on the composite section.

Other Composite Members

Other composite members suitable for floor construction include composite trusses, composite open web steel joists, and stub-girders. Optimum spans for performance and economy depend on overall building considerations such as storey height restrictions, and integration of building services.

For composite trusses and joists, Clause 17.9.2, S16.1-M89 stipulates that the area of the top chord shall be neglected in determining the properties of the composite section, and that the factored moment resistance of the composite truss or joist shall be computed on the basis of full shear connection with the plastic neutral axis in the slab.

Composite stub-girders use wide flange column shapes with short W shape stubs shop welded to the top of the girders and interconnected with the deck-slab by shear connectors to provide vierendeel girder action. Deck-slabs usually consist of 75 mm composite steel deck with 75 mm or 85 mm cover slabs.

Examples and design information for composite trusses, composite joists and stub girders, including design tables for stub giders, are contained in the CISC publication Design and Construction of Composite Floor Systems.

References

Additional information on composite beams can be obtained from the following references:

1. Commentary on Clause 17, Part 2 of the Handbook
2. Chien, E.Y.L., and Ritchie, J.K., Design and Construction of Composite Floor Systems, CISC, 1984.
3. Kulak, Adams and Gilmor, Limit States Design in Structural Steel, CISC, 1990.
4. Picard, A., Beaulieu, D., Calcul des charpentes d'acier, CISC, 1991.

Factored Shear Resistance of Shear Studs, q_r (kN), in **Solid Slabs** or in **Deck-Slabs with Ribs Parallel to Beam** — Table 5-2

Stud Diameter	Stud Location	Factored Shear Resistance of a Stud, q_r (kN) for listed f'_c as					
		w_c = 2 300 kg/m³			w_c = 1 850 kg/m³		
		20 MPa	25 MPa	30 MPa	20 MPa	25 MPa	30 MPa
3/4" (19 mm)	Solid Slab (or) Deck-Slab with $w_d/h_d \geq 1.5$	73.9	87.3	94.1	62.7	74.2	85.0
5/8" (15.9 mm)		51.7	61.2	65.9	43.9	51.9	59.6
1/2" (12.7 mm)		33.0	39.0	42.1	28.0	33.1	38.0
3/4" (19 mm)	Deck-Slab with $w_d/h_d \approx 1.4$, $h/h_d \approx 2.0$	62.0	73.4	79.1	52.7	62.3	71.4
5/8" (15.9 mm)		43.5	51.4	55.4	36.9	43.6	50.0
1/2" (12.7 mm)		27.7	32.8	35.3	23.5	27.8	31.9
3/4" (19 mm)	Deck-Slab with $w_d/h_d \approx 1.4$, $h/h_d \geq 2.2$	73.9	87.3	94.1	62.7	74.2	85.0
5/8" (15.9 mm)		51.7	61.2	65.9	43.9	51.9	59.6
1/2" (12.7 mm)		33.0	39.0	42.1	28.0	33.1	38.0

Notes: q_r = factored shear resistance per stud connection, (kN)
= least of a) $0.5\,\phi_{sc} A_{sc} \sqrt{f'_c E_c}$ b) $415\,\phi_{sc} A_{sc}$ and
c) $0.6\,(w_d/h_d)(h/h_d - 1)$ times lesser of a) and b)
where, $\phi_{sc} = 0.8$, A_{sc} = stud cross sectional area, $E_c = w_c^{1.5}\, 0.043 \sqrt{f'_c}$

Table 5-3

Factored Shear Resistance of **3/4" Dia.** Studs, q_r (kN), with Shear Transverse to Concrete Ribs of Deck-Slabs Formed by 75 mm or 38 mm Deep Steel Deck

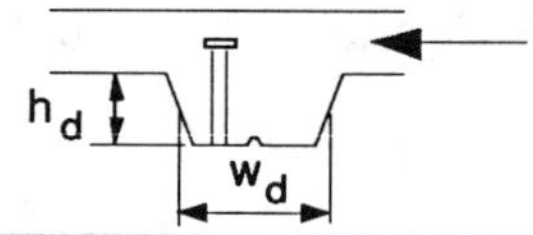

Deck		Stud Connector(s)				Pullout Cone Area	Factored Shear Resistance (kN) of Stud(s), $q_{r(rib)}$, for listed f'_c as					
							N.D. w_c = 2 300 kg/m³			S.L.D. w_c = 1 850 kg/m³		
Depth h_d mm	$\frac{w_d}{h_d}$	Diameter in. (mm)	Length mm	value of n	Edge Distance mm	A_p x 10^3 mm²	20 MPa	25 MPa	30 MPa	20 MPa	25 MPa	30 MPa
75	2.4	3/4" (19)	115	1	Int.	52.2	65.4	73.1	80.1	55.6	62.1	68.0
				1	65	41.3	51.7	57.8	63.3	44.0	49.1	53.8
				1	35	34.3	43.0	48.0	52.6	36.5	40.8	44.7
				2	Int.	70.0	87.7	98.0	107	74.5	83.3	91.3
			150	1	Int.	80.9	73.9	87.3	94.1	62.7	74.2	85.0
				1	65	59.4	73.9	83.2	91.1	62.7	70.7	77.4
				1	35	50.7	63.5	71.0	77.8	54.0	60.3	66.1
				2	Int.	103	129	144	158	110	123	134
75	2.0	3/4" (19)	115	1	Int.	48.4	60.6	67.8	74.2	51.5	57.6	63.1
				1	65	38.3	48.0	53.6	58.7	40.8	45.6	49.9
				1	35	31.8	39.8	44.5	48.8	33.8	37.8	41.5
				2	Int.	64.9	81.3	90.9	99.5	69.1	77.2	84.6
			150	1	Int.	71.3	73.9	87.3	94.1	62.7	74.2	85.0
				1	65	52.6	65.9	73.6	80.7	56.0	62.6	68.6
				1	35	44.8	56.1	62.7	68.7	47.7	53.3	58.4
				2	Int.	91.2	114	128	140	97.1	109	119
38	2.5	3/4" (19)	75	1	Int.	20.5	44.7	50.0	54.8	38.0	42.5	46.6
				1	65	19.3	42.1	47.1	51.6	35.8	40.0	43.8
				1	35	15.2	33.2	37.1	40.6	28.2	31.5	34.5
				2	Int.	31.2	68.1	76.1	83.4	57.9	64.7	70.9
			100	1	Int.	32.0	69.8	78.1	85.5	59.4	66.4	72.7
				1	65	27.3	59.6	66.6	73.0	50.6	56.6	62.0
				1	35	22.1	48.2	53.9	59.1	41.0	45.8	50.2
				2	Int.	45.3	98.9	111	121	84.0	94.0	103
38	1.4	3/4" (19)	75	1	Int.	13.6	29.7	33.2	36.4	25.2	28.2	30.9
				1	65	13.0	28.4	31.7	34.7	24.1	27.0	29.5
				1	35	10.4	22.7	25.4	27.8	19.3	21.6	23.6
				2	Int.	21.6	47.1	52.7	57.7	40.1	44.8	49.1
			100	1	Int.	27.6	60.2	67.3	73.8	51.2	57.2	62.7
				1	65	24.9	54.3	60.8	66.6	46.2	51.6	56.6
				1	35	19.9	43.4	48.6	53.2	36.9	41.3	45.2
				2	Int.	40.9	89.3	99.8	109	75.9	84.8	92.9

$q_{r(rib)}$ = factored shear resistance (kN) per rib connection (Total for 1- or 2-stud connection as noted)
= least of a) $0.5\,n\,\phi_{sc}A_{sc}\sqrt{f'_c E_c}$, b) $415\,n\,\phi_{sc}A_{sc}$ and c) $C\phi_{sc}\rho A_p\sqrt{f'_c}$
where, n = number of studs per rib, A_{sc} = stud cross sectional area, A_p = pullout cone area,
ρ = 1.0 for N.D. concrete or = 0.85 for S.L.D. concrete,
C = 0.35 for 75 mm deck or 0.61 for 38 mm deck, ϕ_{sc} = 0.8, $E_c = w_c^{1.5}\,0.043\sqrt{f'_c}$

Notes: 1. Stud length listed = length after welding.
Minimum length prior to welding ≈ stud length listed + 10 mm as fusion allowance.
2. Double studs transversely spaced at minimum 4 stud diameters.
3. Int. = interior condition.
4. Studs placed off-centre in ribs of 75 mm deck and on-centre in ribs of 38 mm deck.

Table 5-4

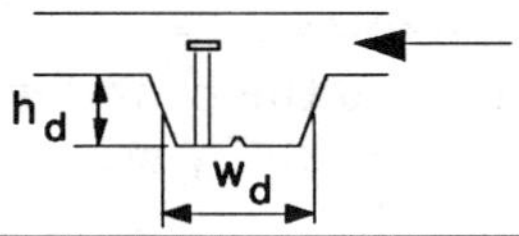

Factored Shear Resistance of **5/8" Dia.** Studs, q_r (kN), with Shear Transverse to Concrete Ribs of Deck-Slabs Formed by 75 mm or 38 mm Deep Steel Deck

Deck		Stud Connector(s)				Pullout Cone Area	Factored Shear Resistance (kN) of Stud(s), $q_{r(rib)}$, for listed f'_c as					
							N.D. w_c = 2 300 kg/m³			S.L.D. w_c = 1 850 kg/m³		
Depth h_d mm	$\frac{w_d}{h_d}$	Diameter in. (mm)	Length mm	value of n	Edge Distance mm	A_p x 10³ mm²	20 MPa	25 MPa	30 MPa	20 MPa	25 MPa	30 MPa
75	2.4	5/8"	115	1	Int.	52.2	51.7	61.2	65.9	43.9	51.9	59.6
		(15.9)		1	65	41.3	51.7	57.8	63.3	43.9	49.1	53.8
				1	35	34.3	43.0	48.0	52.6	36.5	40.8	44.7
				2	Int.	67.1	84.0	93.9	103	71.4	79.8	87.5
			150	1	Int.	80.9	51.7	61.2	65.9	43.9	51.9	59.6
				1	65	59.4	51.7	61.2	65.9	43.9	51.9	59.6
				1	35	50.7	51.7	61.2	65.9	43.9	51.9	59.6
				2	Int.	99.5	103	122	132	87.9	104	119
75	2.0	5/8"	115	1	Int.	48.4	51.7	61.2	65.9	43.9	51.9	59.6
		(15.9)		1	65	38.3	48.0	53.6	58.7	40.8	45.6	49.9
				1	35	31.8	39.8	44.5	48.8	33.8	37.8	41.5
				2	Int.	62.3	78.0	87.2	95.5	66.3	74.1	81.2
			150	1	Int.	71.3	51.7	61.2	65.9	43.9	51.9	59.6
				1	65	52.6	51.7	61.2	65.9	43.9	51.9	59.6
				1	35	44.8	51.7	61.2	65.9	43.9	51.9	58.4
				2	Int.	87.9	103	122	132	87.9	104	115
38	2.5	5/8"	75	1	Int.	20.5	44.7	50.0	54.8	38.0	42.5	46.6
		(15.9)		1	65	19.3	42.1	47.1	51.6	35.8	40.0	43.8
				1	35	15.2	33.2	37.1	40.6	28.2	31.5	34.5
				2	Int.	29.4	64.2	71.7	78.6	54.5	61.0	66.8
			100	1	Int.	32.0	51.7	61.2	65.9	43.9	51.9	59.6
				1	65	27.3	51.7	61.2	65.9	43.9	51.9	59.6
				1	35	22.1	48.2	53.9	59.1	41.0	45.8	50.2
				2	Int.	43.2	94.3	105	115	80.1	89.6	98.1
38	1.4	5/8"	75	1	Int.	13.6	29.7	33.2	36.4	25.2	28.2	30.9
		(15.9)		1	65	13.0	28.4	31.7	34.7	24.1	27.0	29.5
				1	35	10.4	22.7	25.4	27.8	19.3	21.6	23.6
				2	Int.	20.3	44.3	49.5	54.3	37.7	42.1	46.1
			100	1	Int.	27.6	51.7	61.2	65.9	43.9	51.9	59.6
				1	65	24.9	51.7	60.8	65.9	43.9	51.6	56.6
				1	35	19.9	43.4	48.6	53.2	36.9	41.3	45.2
				2	Int.	38.8	84.7	94.7	104	72.0	80.5	88.2

$q_{r(rib)}$ = factored shear resistance (kN) per rib connection (Total for 1- or 2-stud connection as noted)
= least of a) $0.5\, n\, \phi_{sc} A_{sc} \sqrt{f'_c E_c}$, b) $415\, n\, \phi_{sc} A_{sc}$ and c) $C \phi_{sc} \rho A_p \sqrt{f'_c}$
where, n = number of studs per rib, A_{sc} = stud cross sectional area, A_p = pullout cone area,
ρ = 1.0 for N.D. concrete or = 0.85 for S.L.D. concrete,
C = 0.35 for 75 mm deck or 0.61 for 38 mm deck, ϕ_{sc} = 0.8, $E_c = w_c^{1.5}\, 0.043 \sqrt{f'_c}$

Notes: 1. Stud length listed = length after welding.
Minimum length prior to welding ≈ stud length listed + 10 mm as fusion allowance.
2. Double studs transversely spaced at minimum 4 stud diameters.
3. Int. = interior condition.
4. Studs placed off-centre in ribs of 75 mm deck and on-centre in ribs of 38 mm deck.

Example

Given

Select a simply supported composite beam to span 12 m and carry a uniformly distributed specified load of 15 kN/m live load and 9 kN/m dead load. Beams are spaced at 3 m on centre and support a 75 mm steel deck (ribs perpendicular to the beam) with a 65 mm cover slab of 20 MPa normal density concrete. Use G40.21-M 300W steel and limit live load deflections to L / 300.

Solution

Total factored load $= (1.25 \times 9) + (1.50 \times 15) = 33.8\,\text{kN/m}$

Therefore $M_f = 33.8 \times 12^2 / 8 = 608\,\text{kN·m}$

and $V_f = 33.8 \times 12 / 2 = 203\,\text{kN}$

Compute minimum I_{reqd} for deflection limit L / 300 using Figure 5–1 and Table 5–4 on pages 5–86 and 5–87.

Total specified live load, $W = 15 \times 12 = 180\,\text{kN}$

$B_d = 1.0$ simple span udl (Table 5–4)

$C_d = 2.8$ for 12 m span and $L / \Delta = 300$ (Figure 5.1)

$I_{reqd} = W \times C_d \times B_d$

$= (180 \times 2.8 \times 1.0)\, 1.15 = 580 \times 10^6\,\text{mm}^4$ (with 15% allowance for creep)

Effective Width (Clause 17.4.1)

a) $0.25\,L = 0.25 \times 12\,000\,\text{mm} = 3\,000\,\text{mm}$

b) beam spacing = 3 m = 3 000 mm

Therefore maximum effective width = 3 000 mm

Beam Selection

From composite beam selection tables for 75 mm steel deck with 65 mm cover slab and $b_1 = 3000$ mm, page 5-30, the first suitable shape is a W460 x 67 with M_{rc} for 40 per cent shear connection = 615 kN·m > 608 kN·m

$V_r = 688\,\text{kN} > 203\,\text{kN}$

$I_t = 991 \times 10^6\,\text{mm}^4$

For 40% shear connection, $I_e = I_s + 0.85\,(p)^{0.25}\,(I_t - I_s)$ Clause 17.3.1(a)

$= 300 + 0.85\,(0.4)^{0.25}\,(991 - 300) = 767 \times 10^6\,\text{mm}^4 > 580 \times 10^6\,\text{mm}^4$

$Q_r = 1\,990\,\text{kN}$; $S_t = 2\,140 \times 10^3\,\text{mm}^3$; $M_r = 405\,\text{kN·m}$; $L_u = 2\,660\,\text{mm}$

Check Steel Beam Under Construction Loads (Clause 17.12)

Clause 17.12 requires that the steel section alone must be capable of supporting all factored loads applied before concrete hardens. In this case the steel deck will provide lateral support to the compression flange of the beam, thus $M_r = 405$ kN·m applies.

Assuming dead load due to deck-slab and steel beam as 7 kN/m and construction live load as 2.5 kN/m, the total factored load applied before the concrete hardens is $(1.25 \times 7) + (1.5 \times 2.5) = 12.5 \text{ kN / m}$

$$M_f = 12.5 \times 12^2 / 8 = 225 \text{ kN·m} < 405 \text{ kN·m}$$

Check Unshored Beam Tension Flange (Clause 17.11)

Assume that load applied before concrete strength reaches 0.75 f'_c is the specified dead load (7 kN/m), and that the remaining dead load (9 – 7 = 2 kN/m) and the specified live load act on the composite section.

Stress in tension flange due to specified load acting on steel beam alone,

S_x of steel beam $= 1\,320 \times 10^3 \text{ mm}^3$

$$f_1 = \frac{M_1}{S_x} = \frac{7 \times 12\,000^2}{8 \times 1\,320 \times 10^3} = 95.5 \text{ MPa}$$

Stress in tension flange due to specified live and superimposed dead loads acting on composite section,

$$f_2 = \frac{M_2}{S_t} = \frac{(15 + 2) \times 12\,000^2}{8 \times 2\,140 \times 10^3} = 143 \text{ MPa}$$

$$0.9\,F_y = 0.9 \times 300 = 270 \text{ MPa}$$

$$f_1 + f_2 = 95.5 + 143 = 239 \text{ MPa} < 270 \text{ MPa}$$

Shear Connectors

Q_r (100 % connection) $= 1\,990$ kN

Assume 3/4 inch (19 mm) diameter studs,

Minimum flange thickness $= 19 / 2.5 = 7.6 \text{ mm} < 12.7 \text{ mm}$ (Clause 17.6.5)

From Table 5–3 for 3/4 inch (19 mm) diameter studs; $h_d = 75$, $w_d / h_d = 2.0$, 20 MPa 2 300 kg/m³ concrete; factored shear resistance per stud, $q_r = 60.6$ kN

$$\text{Number of studs required} = \frac{2 \times Q_r \times (\text{ \% shear connection}/ 100)}{q_r}$$

$$= \frac{2 \times 1\,990 \times 40 / 100}{60.6} = 26.2 \quad \text{Use 28 studs}$$

Since there are no concentrated loads, the studs can be spaced uniformly along the full length of the beam as permitted by the deck flutes.

COMPOSITE MEMBERS
Trial Selection Table

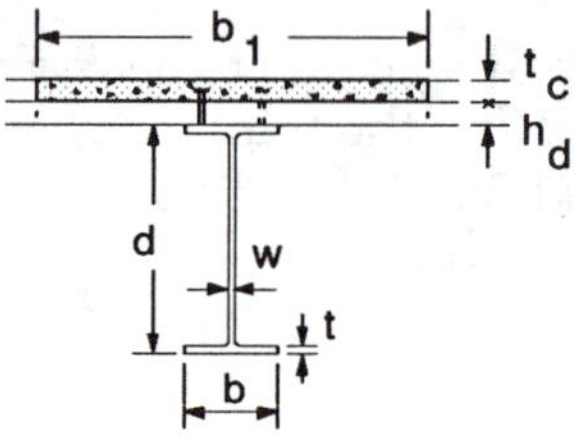

75 mm Deck with 65 mm Slab
$\phi = 0.9$, $\phi_c = 0.6$, $\phi_{sc} = 0.8$

300W
20 MPa

Steel Shape	b_1 mm	Composite Beam* Factored Resistances M_{rc} (kN·m) for Shear Connections = 100%	70%	40%	Q_r (kN) for 100%	I_t x10⁶ mm⁴	S_t x10³ mm³	Steel Shape Data	Non-Composite Shape Unbraced Condition L' mm	M'r kN·m	L' mm	M'r kN·m
WWF900X169	5 000	2 960	2 850	2 600	3 320	7 070	8 920	Mr 1 990	5 000	1 860	12 000	625
	4 130	2 890	2 760	2 510	2 740	6 760	8 810	Vr 1 690	6 000	1 680	14 000	491
b=300	3 250	2 800	2 650	2 410	2 150	6 360	8 660	Lu 4 130	7 000	1 480	16 000	402
t=20	2 380	2 650	2 510	2 310	1 580	5 830	8 450	Ix 2930	8 000	1 250	18 000	339
d=900	1 500	2 450	2 330	2 190	995	5 110	8 100	Sx 6 510	10 000	845	20 000	293
WWF800X161	5 000	2 560	2 450	2 250	3 320	5 570	7 760	Mr 1 710	5 000	1 610	12 000	581
	4 130	2 480	2 390	2 180	2 740	5 330	7 670	Vr 1 490	6 000	1 470	14 000	460
b=300	3 250	2 410	2 290	2 090	2 150	5 020	7 540	Lu 4 230	7 000	1 310	16 000	379
t=20	2 380	2 290	2 170	2 000	1 580	4 600	7 360	Ix 2250	8 000	1 130	18 000	322
d=800	1 500	2 120	2 020	1 890	995	4 020	7 060	Sx 5 610	10 000	778	20 000	280
WWF700X152	4 500	2 140	2 050	1 880	2 980	4 170	6 600	Mr 1 440	5 000	1 370	10 000	711
	3 710	2 070	1 990	1 820	2 460	3 980	6 520	Vr 1 290	6 000	1 260	12 000	537
b=300	2 930	2 010	1 910	1 750	1 940	3 740	6 410	Lu 4 330	7 000	1 140	14 000	429
t=20	2 140	1 910	1 810	1 670	1 420	3 420	6 240	Ix 1 660	8 000	1 000	16 000	356
d=700	1 350	1 770	1 690	1 590	895	2 980	5 980	Sx 4 760	9 000	842	18 000	305
W610X174	4 500	2 140	2 050	1 870	2 980	3 710	6 750	Mr 1 450	5 000	1 430	10 000	924
W24X117	3 710	2 070	1 980	1 810	2 460	3 530	6 660	Vr 1 540	6 000	1 350	12 000	709
b=325	2 930	2 000	1 900	1 740	1 940	3 300	6 530	Lu 4 830	7 000	1 250	14 000	574
t=21.6	2 140	1 900	1 800	1 670	1 420	3 010	6 340	Ix 1 470	8 000	1 150	16 000	482
d=616	1 350	1 760	1 680	1 590	895	2 610	6 050	Sx 4 780	9 000	1 040	18 000	415
W610X155	4 000	1 890	1 810	1 650	2 650	3 260	5 970	Mr 1 280	5 000	1 260	10 000	762
W24X104	3 300	1 830	1 750	1 590	2 190	3 100	5 880	Vr 1 390	6 000	1 180	11 000	659
b=324	2 600	1 770	1 680	1 530	1 720	2 900	5 770	Lu 4 740	7 000	1 080	12 000	579
t=19	1 900	1 680	1 590	1 470	1 260	2 640	5 600	Ix 1 290	8 000	988	14 000	465
d=612	1 200	1 550	1 480	1 400	796	2 290	5 340	Sx 4 220	9 000	886	16 000	388
W610X140	4 000	1 760	1 670	1 500	2 650	3 010	5 360	Mr 1 120	4 000	1 050	9 000	486
W24X94	3 300	1 700	1 610	1 450	2 190	2 870	5 290	Vr 1 440	5 000	946	10 000	422
b=230	2 600	1 630	1 530	1 380	1 720	2 680	5 180	Lu 3 320	6 000	829	12 000	334
t=22.2	1 900	1 530	1 440	1 320	1 260	2 440	5 020	Ix 1 120	7 000	695	14 000	277
d=618	1 200	1 400	1 330	1 250	796	2 100	4 780	Sx 3 630	8 000	573	16 000	237
W610X125	4 000	1 590	1 510	1 360	2 650	2 720	4 780	Mr 991	4 000	923	9 000	396
W24X84	3 300	1 530	1 450	1 300	2 190	2 590	4 720	Vr 1 300	5 000	821	10 000	342
b=229	2 600	1 470	1 380	1 250	1 720	2 430	4 630	Lu 3 250	6 000	708	12 000	269
t=19.6	1 900	1 380	1 300	1 180	1 260	2 220	4 500	Ix 985	7 000	575	14 000	222
d=612	1 200	1 260	1 200	1 110	796	1 920	4 290	Sx 3 220	8 000	470	16 000	189
W610X113	4 000	1 460	1 380	1 250	2 650	2 490	4 330	Mr 888	4 000	818	9 000	328
W24X76	3 300	1 410	1 330	1 190	2 190	2 380	4 270	Vr 1 210	5 000	719	10 000	282
b=228	2 600	1 350	1 270	1 140	1 720	2 230	4 200	Lu 3 180	6 000	610	12 000	220
t=17.3	1 900	1 270	1 190	1 080	1 260	2 040	4 080	Ix 875	7 000	481	14 000	180
d=608	1 200	1 160	1 090	1 010	796	1 770	3 890	Sx 2 880	8 000	391	16 000	153

Note : * 20 MPa, 2 300 kg/m³ Concrete. G40.21-M 300W Steel.

Units : d, b, t, L_u (mm) $I_x = 10^6$ mm⁴ $S_x = 10^3$ mm³ M_r = kN·m V_r = kN

COMPOSITE MEMBERS
Trial Selection Table

300W
20 MPa

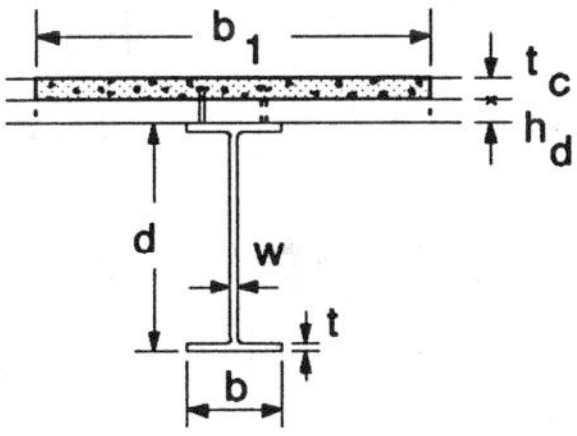

75 mm Deck with 65 mm Slab
$\phi = 0.9,\ \phi_c = 0.6,\ \phi_{sc} = 0.8$

Steel Shape	b_1 mm	Composite Beam*: Factored Resistances: M_{rc} (kN·m) for Shear Connections = 100%	70%	40%	Q_r (kN) for 100%	I_t x10⁶ mm⁴	S_t x10³ mm³	Steel Shape Data		Non-Composite Shape: Unbraced Condition: L' mm	M'r kN·m	L' mm	M'r kN·m
W610X101	3 500	1 300	1 240	1 100	2 320	2 190	3 850	Mr	783	4 000	713	9 000	267
W24X68	2 890	1 260	1 190	1 060	1 920	2 090	3 800	Vr	1 130	5 000	619	10 000	228
b=228	2 280	1 200	1 130	1 010	1 510	1 960	3 730	Lu	3 110	6 000	512	11 000	199
t=14.9	1 660	1 130	1 050	949	1 100	1 790	3 620	Ix	764	7 000	396	12 000	176
d=604	1 050	1 020	961	889	696	1 550	3 450	Sx	2 530	8 000	320	14 000	144
W610X91	3 500	1 180	1 120	996	2 320	1 970	3 430	Mr	691	4 000	619	9 000	217
W24X61	2 890	1 140	1 080	951	1 920	1 890	3 390	Vr	1 030	5 000	530	10 000	185
b=227	2 280	1 090	1 020	901	1 510	1 780	3 330	Lu	3 020	6 000	421	11 000	161
t=12.7	1 660	1 020	945	845	1 100	1 620	3 240	Ix	667	7 000	325	12 000	142
d=598	1 050	916	857	785	696	1 410	3 090	Sx	2 230	8 000	261	14 000	115
W610X84	3 500	1 110	1 040	935	2 320	1 840	3 180	Mr	637	3 000	636	8 000	232
W24X56	2 890	1 060	1 010	892	1 920	1 760	3 140	Vr	956	4 000	566	9 000	192
b=226	2 280	1 020	955	844	1 510	1 670	3 090	Lu	2 990	5 000	482	10 000	163
t=11.7	1 660	953	886	789	1 100	1 530	3 010	Ix	613	6 000	377	12 000	125
d=596	1 050	858	801	730	696	1 330	2 870	Sx	2 060	7 000	289	14 000	101
W530X123	3 500	1 390	1 310	1 170	2 320	2 140	4 260	Mr	867	4 000	794	9 000	361
W21X83	2 890	1 340	1 260	1 130	1 920	2 030	4 200	Vr	1 270	5 000	706	10 000	316
b=212	2 280	1 280	1 200	1 080	1 510	1 900	4 110	Lu	3 100	6 000	613	11 000	281
t=21.2	1 660	1 190	1 120	1 020	1 100	1 720	3 980	Ix	761	7 000	505	12 000	253
d=544	1 050	1 090	1 030	964	696	1 480	3 770	Sx	2 800	8 000	421	14 000	211
W530X109	3 500	1 250	1 180	1 060	2 320	1 930	3 790	Mr	764	4 000	692	9 000	291
W21X73	2 890	1 200	1 140	1 020	1 920	1 840	3 740	Vr	1 120	5 000	608	10 000	254
b=211	2 280	1 150	1 080	970	1 510	1 720	3 660	Lu	3 040	6 000	517	11 000	225
t=18.8	1 660	1 080	1 010	918	1 100	1 570	3 550	Ix	667	7 000	413	12 000	202
d=540	1 050	984	928	862	696	1 350	3 380	Sx	2 480	8 000	342	14 000	168
W530X101	3 500	1 180	1 110	999	2 320	1 820	3 530	Mr	707	3 000	707	8 000	301
W21X68	2 890	1 130	1 070	957	1 920	1 730	3 490	Vr	1 050	4 000	635	9 000	255
b=210	2 280	1 080	1 020	911	1 510	1 630	3 420	Lu	2 990	5 000	553	10 000	222
t=17.4	1 660	1 020	951	860	1 100	1 480	3 320	Ix	617	6 000	462	12 000	176
d=538	1 050	924	870	805	696	1 280	3 160	Sx	2 300	7 000	365	14 000	146
W530X92	3 500	1 100	1 030	925	2 320	1 670	3 220	Mr	637	3 000	633	8 000	253
W21X62	2 890	1 050	992	885	1 920	1 600	3 180	Vr	971	4 000	565	9 000	214
b=209	2 280	1 000	943	840	1 510	1 500	3 120	Lu	2 930	5 000	486	10 000	185
t=15.6	1 660	941	878	789	1 100	1 370	3 040	Ix	552	6 000	393	12 000	146
d=534	1 050	852	800	735	696	1 180	2 890	Sx	2 070	7 000	309	14 000	120
W530X82	3 500	997	932	835	2 320	1 500	2 860	Mr	559	3 000	551	8 000	204
W21X55	2 890	951	896	797	1 920	1 440	2 830	Vr	894	4 000	487	9 000	172
b=209	2 280	903	851	754	1 510	1 350	2 780	Lu	2 860	5 000	412	10 000	148
t=13.3	1 660	848	790	705	1 100	1 240	2 710	Ix	479	6 000	321	12 000	115
d=528	1 050	765	714	651	696	1 070	2 580	Sx	1 810	7 000	251	14 000	94.8

Note : * 20 MPa, 2 300 kg/m³ Concrete. G40.21-M 300W Steel.

Units : d, b, t, L_u (mm) $I_x = 10^6$ mm⁴ $S_x = 10^3$ mm³ M_r = kN·m V_r = kN

COMPOSITE MEMBERS
Trial Selection Table

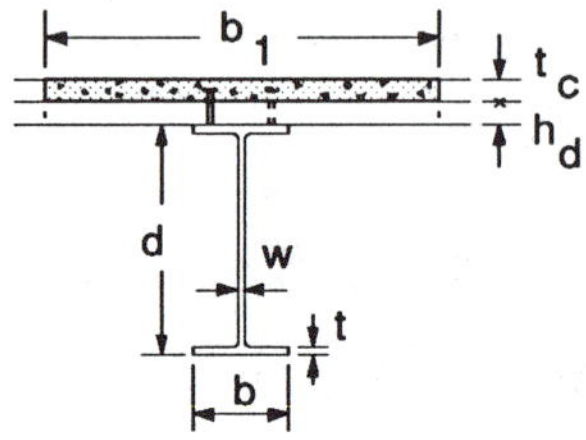

75 mm Deck with 65 mm Slab
$\phi = 0.9$, $\phi_c = 0.6$, $\phi_{sc} = 0.8$

300W
20 MPa

Steel Shape	b_1	Composite Beam*							Non-Composite Shape			
		Factored Resistances							Unbraced Condition			
		M_{rc} (kN·m) for Shear Connections =			Q_r (kN) for	I_t $\times 10^6$	S_t $\times 10^3$	Steel Shape Data	L'	M'r	L'	M'r
	mm	100%	70%	40%	100%	mm^4	mm^3		mm	kN·m	mm	kN·m
W530X72	3 500	897	835	745	2 320	1 320	2 490	Mr 475	3 000	462	8 000	155
W21X48	2 890	853	800	709	1 920	1 270	2 460	Vr 831	4 000	402	9 000	129
b=207	2 280	806	759	667	1 510	1 200	2 420	Lu 2 750	5 000	331	10 000	111
t=10.9	1 660	755	702	619	1 100	1 100	2 360	Ix 402	6 000	247	12 000	85.9
d=524	1 050	678	629	567	696	957	2 250	Sx 1 530	7 000	191	14 000	70.1
W460X106	3 500	1 100	1 030	918	2 320	1 510	3 330	Mr 645	3 000	640	8 000	308
W18X71	2 890	1 050	987	879	1 920	1 440	3 280	Vr 1 060	4 000	579	9 000	266
b=194	2 280	997	936	836	1 510	1 350	3 210	Lu 2 910	5 000	512	10 000	235
t=20.6	1 660	934	872	788	1 100	1 220	3 110	Ix 488	6 000	444	12 000	190
d=470	1 050	847	797	737	696	1 040	2 940	Sx 2 080	7 000	366	14 000	160
W460X97	3 500	1 020	950	851	2 320	1 400	3 050	Mr 589	3 000	581	8 000	264
W18X65	2 890	970	912	815	1 920	1 330	3 010	Vr 947	4 000	522	9 000	227
b=193	2 280	921	867	774	1 510	1 250	2 950	Lu 2 870	5 000	457	10 000	200
t=19	1 660	864	808	728	1 100	1 140	2 860	Ix 445	6 000	389	12 000	161
d=466	1 050	784	737	678	696	976	2 720	Sx 1 910	7 000	314	14 000	135
W460X89	3 500	961	894	801	2 320	1 310	2 830	Mr 543	3 000	534	8 000	231
W18X60	2 890	914	857	767	1 920	1 250	2 800	Vr 868	4 000	477	9 000	198
b=192	2 280	865	816	727	1 510	1 180	2 750	Lu 2 830	5 000	414	10 000	174
t=17.7	1 660	812	760	683	1 100	1 070	2 670	Ix 410	6 000	343	12 000	140
d=464	1 050	737	691	634	696	923	2 540	Sx 1 770	7 000	276	14 000	117
W460X82	3 500	894	830	742	2 320	1 210	2 590	Mr 494	3 000	482	8 000	195
W18X55	2 890	848	793	709	1 920	1 160	2 560	Vr 812	4 000	427	9 000	167
b=191	2 280	800	754	672	1 510	1 090	2 510	Lu 2 770	5 000	365	10 000	146
t=16	1 660	750	703	629	1 100	995	2 440	Ix 370	6 000	292	12 000	117
d=460	1 050	680	637	582	696	859	2 330	Sx 1 610	7 000	234	14 000	97.3
W460X74	3 500	833	771	688	2 320	1 110	2 360	Mr 446	3 000	433	8 000	164
W18X50	2 890	788	734	657	1 920	1 070	2 330	Vr 735	4 000	380	9 000	140
b=190	2 280	742	697	622	1 510	1 010	2 290	Lu 2 730	5 000	320	10 000	122
t=14.5	1 660	692	651	580	1 100	923	2 230	Ix 333	6 000	249	12 000	96.9
d=458	1 050	629	588	534	696	800	2 130	Sx 1 460	7 000	198	14 000	80.6
W460X67	3 000	745	691	615	1 990	991	2 140	Mr 405	3 000	390	8 000	140
W18X45	2 480	706	660	586	1 640	950	2 120	Vr 688	4 000	339	9 000	119
b=190	1 950	666	626	553	1 290	895	2 080	Lu 2 660	5 000	281	10 000	103
t=12.7	1 430	624	581	516	948	820	2 030	Ix 300	6 000	214	11 000	91.4
d=454	900	561	523	474	597	707	1 930	Sx 1 320	7 000	169	12 000	82
W460X61	3 000	685	632	560	1 990	890	1 910	Mr 354	3 000	336	8 000	111
W18X41	2 480	647	602	533	1 640	855	1 880	Vr 650	4 000	288	9 000	93.9
b=189	1 950	607	570	500	1 290	808	1 850	Lu 2 580	5 000	231	10 000	81.3
t=10.8	1 430	567	527	464	948	743	1 810	Ix 259	6 000	172	11 000	71.7
d=450	900	507	470	423	597	642	1 720	Sx 1 150	7 000	135	12 000	64.1

Note : * 20 MPa, 2 300 kg/m³ Concrete. G40.21-M 300W Steel.

Units : d, b, t, L_u (mm) $I_x = 10^6\ mm^4$ $S_x = 10^3\ mm^3$ M_r = kN·m V_r = kN

300W
20 MPa

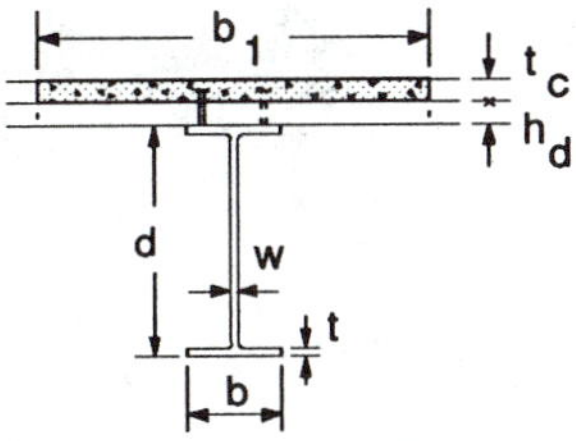

COMPOSITE MEMBERS
Trial Selection Table

75 mm Deck with 65 mm Slab
$\phi = 0.9$, $\phi_c = 0.6$, $\phi_{sc} = 0.8$

Steel Shape	b_1 mm	Composite Beam*: Factored Resistances: M_{rc} (kN·m) for Shear Connections = 100%	70%	40%	Q_r (kN) for 100%	I_t x10⁶ mm⁴	S_t x10³ mm³	Steel Shape Data	Non-Composite Shape: Unbraced Condition: L' mm	M'r kN·m	L' mm	M'r kN·m
W410X85	3 500	857	791	704	2 320	1 070	2 500	Mr 467	3 000	455	8 000	205
W16X57	2 890	810	754	672	1 920	1 020	2 470	Vr 812	4 000	406	9 000	178
b=181	2 280	762	715	636	1 510	962	2 430	Lu 2 730	5 000	354	10 000	157
t=18.2	1 660	711	665	595	1 100	876	2 350	Ix 315	6 000	297	12 000	127
d=418	1 050	644	603	550	696	752	2 240	Sx 1 510	7 000	243	14 000	107
W410X74	3 500	783	720	637	2 320	960	2 220	Mr 408	3 000	394	8 000	163
W16X50	2 890	737	683	608	1 920	921	2 190	Vr 716	4 000	348	9 000	140
b=180	2 280	690	646	575	1 510	868	2 160	Lu 2 670	5 000	297	10 000	124
t=16	1 660	641	601	535	1 100	794	2 100	Ix 275	6 000	239	12 000	99.8
d=414	1 050	581	542	492	696	685	2 000	Sx 1 330	7 000	194	14 000	83.8
W410X67	3 500	725	664	584	2 320	874	2 010	Mr 367	3 000	352	8 000	135
W16X45	2 890	681	628	558	1 920	840	1 980	Vr 643	4 000	307	9 000	116
b=179	2 280	635	592	527	1 510	794	1 950	Lu 2 610	5 000	258	10 000	102
t=14.4	1 660	587	551	490	1 100	730	1 900	Ix 246	6 000	201	12 000	81.7
d=410	1 050	532	496	447	696	632	1 820	Sx 1 200	7 000	161	14 000	68.4
W410X60	3 000	631	579	510	1 990	762	1 760	Mr 321	3 000	306	8 000	109
W16X40	2 480	593	548	488	1 640	732	1 740	Vr 560	4 000	264	9 000	93.2
b=178	1 950	554	517	461	1 290	691	1 710	Lu 2 580	5 000	217	10 000	81.4
t=12.8	1 430	513	483	429	948	635	1 670	Ix 216	6 000	165	11 000	72.2
d=408	900	465	434	392	597	549	1 600	Sx 1 060	7 000	131	12 000	65
W410X54	3 000	574	523	459	1 840	684	1 570	Mr 284	3 000	266	8 000	86.1
W16X36	2 480	548	504	445	1 640	659	1 560	Vr 540	4 000	225	9 000	73.2
b=177	1 950	509	473	419	1 290	624	1 530	Lu 2 480	5 000	176	10 000	63.6
t=10.9	1 430	469	440	388	948	575	1 490	Ix 186	6 000	132	11 000	56.3
d=404	900	423	393	352	597	499	1 430	Sx 924	7 000	104	12 000	50.5
W410X46	3 000	503	455	396	1 590	602	1 360	Mr 239	2 000	236	7 000	61.7
W16X31	2 480	494	451	395	1 590	582	1 350	Vr 504	3 000	195	8 000	51.8
b=140	1 950	460	424	373	1 290	553	1 330	Lu 1 930	4 000	142	9 000	44.6
t=11.2	1 430	420	393	343	948	513	1 300	Ix 156	5 000	99.9	10 000	39.2
d=404	900	376	348	307	597	448	1 240	Sx 773	6 000	76.4	12 000	31.6
W410X39	3 000	428	386	333	1 350	514	1 150	Mr 197	2 000	193	7 000	44.1
W16X26	2 480	422	383	332	1 350	498	1 140	Vr 456	3 000	155	8 000	36.6
b=140	1 950	408	374	327	1 290	476	1 120	Lu 1 860	4 000	105	9 000	31.3
t=8.8	1 430	370	344	299	948	444	1 100	Ix 127	5 000	73.1	10 000	27.4
d=400	900	329	303	265	597	391	1 050	Sx 634	6 000	55.2	12 000	21.9
W360X79	3 000	694	638	566	1 990	784	2 130	Mr 386	4 000	364	8 000	225
W14X53	2 480	654	606	543	1 640	747	2 100	Vr 593	4 500	348	9 000	195
b=205	1 950	612	574	515	1 290	699	2 060	Lu 3 270	5 000	331	10 000	172
t=16.8	1 430	570	537	484	948	634	2 000	Ix 227	6 000	298	11 000	154
d=354	900	519	489	448	597	540	1 890	Sx 1 280	7 000	264	12 000	139

Note : * 20 MPa, 2 300 kg/m³ Concrete. G40.21-M 300W Steel.

Units : d, b, t, L_u (mm) $I_x = 10^6$ mm⁴ $S_x = 10^3$ mm³ M_r = kN·m V_r = kN

COMPOSITE MEMBERS
Trial Selection Table

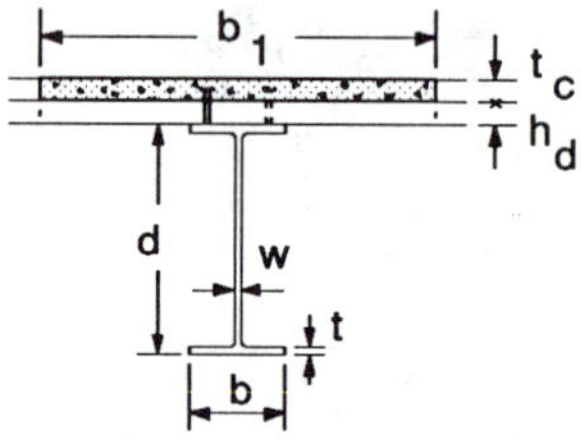

75 mm Deck with 65 mm Slab
$\phi = 0.9, \phi_c = 0.6, \phi_{sc} = 0.8$

300W
20 MPa

Steel Shape	b_1 mm	Composite Beam* — Factored Resistances — M_{rc} (kN·m) for Shear Connections = 100%	70%	40%	Q_r (kN) for 100%	I_t x10⁶ mm⁴	S_t x10³ mm³	Steel Shape Data	Non-Composite Shape — Unbraced Condition — L' mm	M'r kN·m	L' mm	M'r kN·m
W360X72	3 000	643	589	519	1 990	713	1 920	Mr 346	4 000	322	8 000	186
W14X48	2 480	604	558	497	1 640	681	1 900	Vr 536	4 500	306	9 000	161
b=204	1 950	563	526	471	1 290	639	1 860	Lu 3 190	5 000	290	10 000	141
t=15.1	1 430	522	492	441	948	582	1 810	Ix 201	6 000	257	11 000	126
d=350	900	475	446	406	597	498	1 720	Sx 1 150	7 000	222	12 000	114
W360X64	3 000	596	543	474	1 990	648	1 720	Mr 308	4 000	283	8 000	153
W14X43	2 480	558	512	455	1 640	621	1 700	Vr 478	4 500	268	9 000	131
b=203	1 950	518	481	431	1 290	584	1 670	Lu 3 110	5 000	252	10 000	115
t=13.5	1 430	477	449	402	948	535	1 630	Ix 178	6 000	220	11 000	102
d=348	900	433	407	369	597	459	1 550	Sx 1 030	7 000	183	12 000	92.4
W360X57	2 500	527	482	424	1 660	585	1 540	Mr 273	3 000	259	6 000	147
W14X38	2 060	494	456	404	1 370	559	1 510	Vr 504	3 500	243	7 000	119
b=172	1 630	461	430	381	1 080	526	1 490	Lu 2 550	4 000	225	8 000	99.8
t=13.1	1 190	427	400	355	789	479	1 450	Ix 161	4 500	207	9 000	86
d=358	750	385	359	326	497	410	1 370	Sx 897	5 000	189	10 000	75.7
W360X51	2 500	488	444	388	1 660	530	1 370	Mr 241	3 000	227	6 000	121
W14X34	2 060	456	419	370	1 370	508	1 360	Vr 457	3 500	212	7 000	97
b=171	1 630	424	393	349	1 080	479	1 330	Lu 2 500	4 000	195	8 000	81
t=11.6	1 190	390	365	324	789	438	1 300	Ix 141	4 500	178	9 000	69.5
d=356	750	351	327	295	497	377	1 240	Sx 796	5 000	159	10 000	60.9
W360X45	2 500	442	400	347	1 550	474	1 220	Mr 210	3 000	195	6 000	96.1
W14X30	2 060	419	382	335	1 370	455	1 200	Vr 433	3 500	181	7 000	76.5
b=171	1 630	387	358	315	1 080	431	1 180	Lu 2 430	4 000	165	8 000	63.4
t=9.8	1 190	354	331	290	789	396	1 150	Ix 122	4 500	148	9 000	54.1
d=352	750	317	294	262	497	342	1 100	Sx 691	5 000	128	10 000	47.2
W360X39	2 500	391	351	302	1 340	421	1 060	Mr 179	2 000	173	6 000	54.2
W14X26	2 060	383	348	301	1 340	405	1 050	Vr 410	2 500	157	7 000	44.3
b=128	1 630	354	324	283	1 080	385	1 030	Lu 1 790	3 000	139	8 000	37.5
t=10.7	1 190	321	298	259	789	356	1 000	Ix 102	4 000	97.2	9 000	32.5
d=354	750	285	263	231	497	309	960	Sx 580	5 000	69.8	10 000	28.8
W360X33	2 500	330	294	251	1 130	357	886	Mr 146	2 000	139	6 000	38.1
W14X22	2 060	324	292	251	1 130	345	877	Vr 362	2 500	124	7 000	30.8
b=127	1 630	313	285	247	1 080	329	864	Lu 1 720	3 000	108	8 000	25.9
t=8.5	1 190	281	259	224	789	306	845	Ix 82.7	4 000	70.3	9 000	22.3
d=350	750	247	228	198	497	269	811	Sx 474	5 000	49.7	10 000	19.6
W310X86	3 500	708	644	563	2 320	728	2 170	Mr 383	5 000	367	10 000	248
W12X58	2 890	662	608	540	1 920	694	2 140	Vr 503	6 000	344	11 000	221
b=254	2 280	615	571	515	1 510	650	2 100	Lu 4 250	7 000	320	12 000	200
t=16.3	1 660	566	533	484	1 100	590	2 040	Ix 199	8 000	297	13 000	182
d=310	1 050	516	489	449	696	502	1 940	Sx 1 280	9 000	273	14 000	167

Note : * 20 MPa, 2 300 kg/m³ Concrete. G40.21-M 300W Steel.

Units : d, b, t, L_u (mm) $I_x = 10^6$ mm⁴ $S_x = 10^3$ mm³ M_r = kN·m V_r = kN

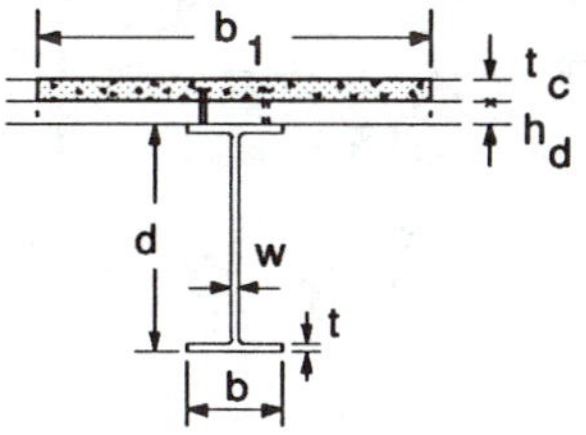

300W
20 MPa

COMPOSITE MEMBERS
Trial Selection Table

75 mm Deck with 65 mm Slab
$\phi = 0.9$, $\phi_c = 0.6$, $\phi_{sc} = 0.8$

Steel Shape	b_1 mm	Composite Beam* Factored Resistances M_{rc} (kN·m) for Shear Connections = 100%	70%	40%	Q_r (kN) for 100%	I_t x10⁶ mm⁴	S_t x10³ mm³	Steel Shape Data	Non-Composite Shape Unbraced Condition L' mm	M'r kN·m	L' mm	M'r kN·m
W310X79	3 000	629	574	505	1 990	645	1 970	Mr 346	5 000	327	8 000	258
W12X53	2 480	590	543	485	1 640	614	1 940	Vr 480	5 500	316	9 000	235
b=254	1 950	549	511	461	1 290	573	1 900	Lu 4 140	6 000	305	10 000	207
t=14.6	1 430	508	479	433	948	519	1 840	Ix 177	6 500	293	11 000	184
d=306	900	463	437	401	597	440	1 750	Sx 1 160	7 000	282	12 000	166
W310X74	3 000	609	554	484	1 990	621	1 850	Mr 321	4 000	307	8 000	206
W12X50	2 480	570	523	463	1 640	592	1 830	Vr 519	4 500	295	9 000	179
b=205	1 950	529	491	438	1 290	553	1 790	Lu 3 380	5 000	282	10 000	159
t=16.3	1 430	487	458	410	948	502	1 740	Ix 165	6 000	258	11 000	142
d=310	900	441	414	378	597	426	1 650	Sx 1 060	7 000	233	12 000	129
W310X67	3 000	565	511	442	1 990	561	1 660	Mr 286	4 000	270	8 000	169
W12X45	2 480	526	481	423	1 640	537	1 640	Vr 463	4 500	258	9 000	147
b=204	1 950	486	449	400	1 290	503	1 610	Lu 3 280	5 000	246	10 000	129
t=14.6	1 430	445	417	373	948	458	1 570	Ix 145	6 000	222	11 000	116
d=306	900	402	377	341	597	391	1 490	Sx 949	7 000	198	12 000	105
W310X60	2 500	489	444	387	1 660	490	1 470	Mr 254	4 000	237	6 500	179
W12X40	2 060	456	418	370	1 370	467	1 450	Vr 406	4 500	226	7 000	166
b=203	1 630	423	393	351	1 080	438	1 430	Lu 3 200	5 000	214	8 000	139
t=13.1	1 190	389	366	327	789	397	1 390	Ix 129	5 500	203	9 000	120
d=304	750	352	331	301	497	338	1 310	Sx 849	6 000	191	10 000	106
W310X52	2 500	464	420	363	1 660	466	1 330	Mr 226	3 000	216	6 000	130
W12X35	2 060	432	394	347	1 370	445	1 310	Vr 431	3 500	203	7 000	106
b=167	1 630	400	369	327	1 080	419	1 290	Lu 2 570	4 000	189	8 000	89.4
t=13.2	1 190	365	342	304	789	382	1 250	Ix 118	4 500	176	9 000	77.4
d=318	750	329	307	278	497	327	1 190	Sx 747	5 000	162	10 000	68.4
W310X45	2 500	410	368	315	1 540	405	1 140	Mr 191	3 000	180	6 000	98.2
W12X30	2 060	388	351	305	1 370	388	1 120	Vr 369	3 500	168	7 000	79.3
b=166	1 630	356	327	288	1 080	367	1 100	Lu 2 490	4 000	155	8 000	66.5
t=11.2	1 190	323	301	267	789	337	1 080	Ix 99.2	4 500	142	9 000	57.3
d=314	750	288	270	242	497	291	1 030	Sx 634	5 000	128	10 000	50.4
W310X39	2 500	359	319	272	1 330	356	991	Mr 165	3 000	153	6 000	77.7
W12X26	2 060	351	316	271	1 330	343	980	Vr 320	3 500	142	7 000	62.2
b=165	1 630	323	294	257	1 080	326	965	Lu 2 440	4 000	130	8 000	51.8
t=9.7	1 190	290	268	237	789	300	942	Ix 85.1	4 500	117	9 000	44.3
d=310	750	256	240	213	497	261	902	Sx 549	5 000	103	10 000	38.8
W250X80	3 000	564	509	440	1 990	512	1 790	Mr 294	5 000	287	8 000	242
W10X54	2 480	525	478	420	1 640	486	1 770	Vr 429	5 500	280	9 000	227
b=255	1 950	484	446	399	1 290	452	1 730	Lu 4 520	6 000	272	10 000	212
t=15.6	1 430	443	414	375	948	407	1 670	Ix 126	6 500	265	11 000	197
d=256	900	400	378	347	597	342	1 580	Sx 982	7 000	257	12 000	179

Note : * 20 MPa, 2 300 kg/m³ Concrete. G40.21-M 300W Steel.

Units : d, b, t, L_u (mm) $I_x = 10^6$ mm⁴ $S_x = 10^3$ mm³ M_r = kN·m V_r = kN

COMPOSITE MEMBERS
Trial Selection Table

75 mm Deck with 65 mm Slab
$\phi = 0.9, \phi_c = 0.6, \phi_{sc} = 0.8$

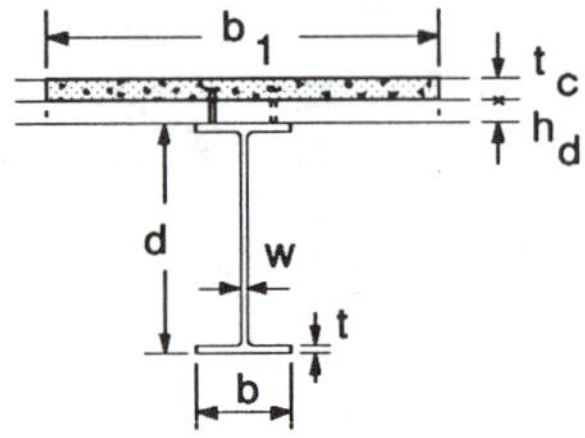

300W
20 MPa

Steel Shape	b_1 mm	Composite Beam* — Factored Resistances — M_{rc} (kN·m) for Shear Connections = 100%	70%	40%	Q_r (kN) for 100%	I_t x10⁶ mm⁴	S_t x10³ mm³	Steel Shape Data	Non-Composite Shape — Unbraced Condition — L' mm	M'r kN·m	L' mm	M'r kN·m
W250X73	3 000	531	477	409	1 990	473	1 640	Mr 266	5 000	257	8 000	212
W10X49	2 480	492	446	389	1 640	450	1 610	Vr 389	5 500	250	9 000	198
b=254	1 950	452	415	369	1 290	420	1 580	Lu 4 390	6 000	242	10 000	183
t=14.2	1 430	411	384	347	948	379	1 530	Ix 113	6 500	235	11 000	165
d=254	900	369	349	319	597	320	1 450	Sx 891	7 000	227	12 000	149
W250X67	3 000	511	457	388	1 990	450	1 520	Mr 243	4 000	237	8 000	174
W10X45	2 480	473	427	369	1 640	429	1 500	Vr 409	4 500	229	9 000	157
b=204	1 950	432	395	349	1 290	401	1 470	Lu 3 570	5 000	221	10 000	139
t=15.7	1 430	391	363	325	948	364	1 420	Ix 104	6 000	205	11 000	125
d=258	900	349	328	297	597	307	1 350	Sx 806	7 000	189	12 000	114
W250X58	2 500	430	385	328	1 660	376	1 300	Mr 208	4 000	199	6 500	160
W10X39	2 060	397	360	312	1 370	358	1 280	Vr 359	4 500	191	7 000	153
b=203	1 630	365	334	295	1 080	335	1 260	Lu 3 410	5 000	184	8 000	137
t=13.5	1 190	331	308	275	789	303	1 220	Ix 87.3	5 500	176	9 000	119
d=252	750	295	277	251	497	255	1 150	Sx 693	6 000	168	10 000	105
W250X49	2 500	387	344	289	1 660	322	1 100	Mr 171	4 000	160	6 500	123
W10X33	2 060	356	319	273	1 370	308	1 080	Vr 327	4 500	153	7 000	115
b=202	1 630	324	294	257	1 080	290	1 060	Lu 3 240	5 000	146	8 000	97.2
t=11	1 190	290	268	237	789	264	1 030	Ix 70.6	5 500	138	9 000	84
d=248	750	256	239	214	497	224	979	Sx 572	6 000	130	10 000	74.1
W250X45	2 500	375	332	279	1 540	326	1 040	Mr 163	3 000	151	6 000	90.6
W10X30	2 060	352	315	269	1 370	312	1 020	Vr 360	3 500	142	7 000	75.2
b=148	1 630	320	290	252	1 080	294	1 010	Lu 2 360	4 000	132	8 000	64.4
t=13	1 190	287	264	232	789	269	979	Ix 71.1	4 500	122	9 000	56.3
d=266	750	252	234	208	497	230	930	Sx 534	5 000	112	10 000	50.1
W250X39	2 500	325	286	239	1 330	284	897	Mr 139	3 000	126	6 000	69.2
W10X26	2 060	318	283	238	1 330	273	886	Vr 308	3 500	117	7 000	57
b=147	1 630	290	261	224	1 080	259	871	Lu 2 280	4 000	108	8 000	48.6
t=11.2	1 190	257	235	206	789	238	849	Ix 60.1	4 500	98.6	9 000	42.3
d=262	750	223	208	183	497	205	811	Sx 459	5 000	88	10 000	37.6
W250X33	2 000	272	240	200	1 130	233	750	Mr 114	3 000	102	5 500	55.6
W10X22	1 650	263	234	198	1 090	224	740	Vr 280	3 500	93.1	6 000	49.4
b=146	1 300	237	214	185	862	211	727	Lu 2 180	4 000	84	6 500	44.4
t=9.1	950	211	194	169	630	194	708	Ix 48.9	4 500	74.1	7 000	40.3
d=258	600	185	171	150	398	167	675	Sx 379	5 000	63.6	8 000	34.1
W200X71	3 000	478	424	354	1 990	378	1 470	Mr 217	5 000	208	8 000	178
W8X48	2 480	439	393	335	1 640	359	1 440	Vr 393	5 500	203	9 000	168
b=206	1 950	399	361	314	1 290	334	1 410	Lu 4 150	6 000	198	10 000	158
t=17.4	1 430	357	329	293	948	300	1 360	Ix 76.6	6 500	193	11 000	148
d=216	900	314	295	267	597	250	1 280	Sx 709	7 000	188	12 000	137

Note : * 20 MPa, 2 300 kg/m³ Concrete. G40.21-M 300W Steel.

Units : d, b, t, L_u (mm) $I_x = 10^6$ mm⁴ $S_x = 10^3$ mm³ M_r = kN·m V_r = kN

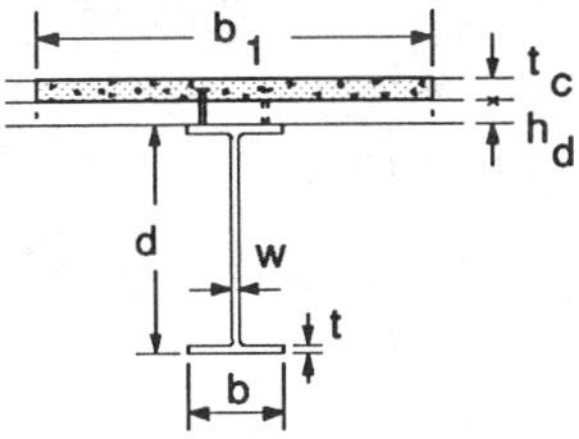

COMPOSITE MEMBERS
Trial Selection Table

300W
20 MPa

75 mm Deck with 65 mm Slab
$\phi = 0.9, \phi_c = 0.6, \phi_{sc} = 0.8$

Steel Shape	b_1 mm	Composite Beam*: Factored Resistances M_{rc} (kN·m) for Shear Connections = 100%	70%	40%	Q_r (kN) for 100%	I_t $\times10^6$ mm⁴	S_t $\times10^3$ mm³	Steel Shape Data		Non-Composite Shape, Unbraced Condition: L' mm	M'r kN·m	L' mm	M'r kN·m
W200X59	3 000	428	376	308	1 990	318	1 220	Mr	176	4 000	174	8 000	134
W8X40	2 480	390	346	289	1 640	304	1 200	Vr	341	4 500	169	9 000	124
b=205	1 950	351	315	270	1 290	284	1 180	Lu	3 780	5 000	164	10 000	114
t=14.2	1 430	311	284	249	948	258	1 140	Ix	61.1	6 000	154	11 000	103
d=210	900	269	251	225	597	217	1 080	Sx	582	7 000	144	12 000	93.4
W200X52	2 500	363	319	264	1 660	272	1 070	Mr	154	4 000	150	6 500	126
W8X35	2 060	331	294	247	1 370	259	1 050	Vr	290	4 500	145	7 000	121
b=204	1 630	299	269	232	1 080	242	1 030	Lu	3 620	5 000	140	8 000	111
t=12.6	1 190	265	243	214	789	218	997	Ix	52.7	5 500	135	9 000	101
d=206	750	231	216	194	497	183	941	Sx	512	6 000	131	10 000	89.7
W200X46	2 500	334	291	238	1 580	244	943	Mr	134	4 000	129	6 500	105
W8X31	2 060	308	272	226	1 370	233	929	Vr	262	4 500	124	7 000	99.8
b=203	1 630	276	247	210	1 080	218	912	Lu	3 460	5 000	119	8 000	90.1
t=11	1 190	243	221	194	789	198	885	Ix	45.5	5 500	114	9 000	78.6
d=204	750	209	195	174	497	168	839	Sx	448	6 000	109	10 000	69.6
W200X42	2 000	290	255	210	1 330	216	846	Mr	120	3 000	119	5 500	93.5
W8X28	1 650	265	235	198	1 090	205	833	Vr	264	3 500	114	6 000	88.4
b=166	1 300	239	215	185	862	192	816	Lu	2 850	4 000	109	6 500	83.4
t=11.8	950	212	194	170	630	173	790	Ix	40.9	4 500	104	7 000	77.7
d=206	600	184	172	153	398	145	745	Sx	399	5 000	98.6	8 000	66.7
W200X36	2 000	261	227	185	1 240	189	733	Mr	103	3 000	100	5 500	75.5
W8X24	1 650	242	213	177	1 090	180	722	Vr	223	3 500	95.3	6 000	70.6
b=165	1 300	217	193	164	862	169	709	Lu	2 730	4 000	90.4	6 500	64.7
t=10.2	950	191	173	151	630	154	688	Ix	34.4	4 500	85.5	7 000	59.2
d=202	600	164	152	135	398	130	652	Sx	342	5 000	80.5	8 000	50.6
W200X31	2 000	236	205	166	1 080	176	654	Mr	90.5	3 000	81.1	5 500	50.6
W8X21	1 650	230	202	165	1 080	169	645	Vr	240	3 500	75.3	6 000	45.6
b=134	1 300	206	183	153	862	159	633	Lu	2 150	4 000	69.5	6 500	41.5
t=10.2	950	180	162	140	630	146	616	Ix	31.4	4 500	63.6	7 000	38
d=210	600	153	141	123	398	125	585	Sx	299	5 000	57	8 000	32.7
W200X27	2 000	203	174	141	915	152	557	Mr	75.3	3 000	65.4	5 500	36.4
W8X18	1 650	198	172	140	915	147	549	Vr	215	3 500	59.7	6 000	32.6
b=133	1 300	188	165	137	862	139	540	Lu	2 050	4 000	54	6 500	29.6
t=8.4	950	162	145	124	630	128	526	Ix	25.8	4 500	47.5	7 000	27.1
d=208	600	136	125	108	398	111	502	Sx	249	5 000	41.2	8 000	23.1
W150X30	2 500	204	170	133	1 020	131	566	Mr	65.9	3 000	64.1	6 000	46.8
W6X20	2 060	199	168	132	1 020	126	558	Vr	186	3 500	61.2	7 000	40.4
b=153	1 630	193	165	131	1 020	120	548	Lu	2 680	4 000	58.3	8 000	34.8
t=9.3	1 190	165	144	118	789	110	534	Ix	17.2	4 500	55.4	9 000	30.6
d=158	750	133	119	102	497	95.1	510	Sx	219	5 000	52.5	10 000	27.3

Note : * 20 MPa, 2 300 kg/m³ Concrete. G40.21-M 300W Steel.

Units : d, b, t, L_u (mm) $I_x = 10^6$ mm⁴ $S_x = 10^3$ mm³ M_r = kN·m V_r = kN

COMPOSITE MEMBERS
Trial Selection Table

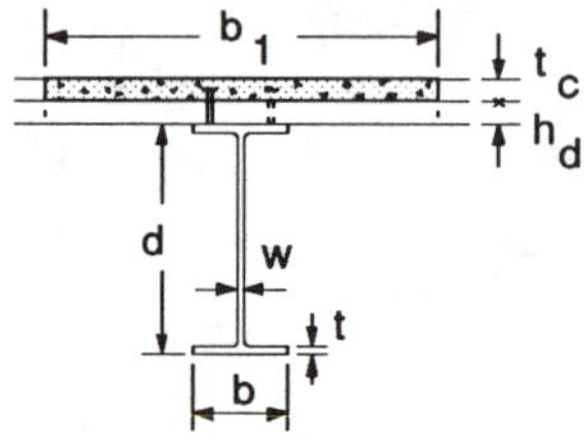

75 mm Deck with 85 mm Slab
$\phi = 0.9$, $\phi_c = 0.6$, $\phi_{sc} = 0.8$

300W
20 MPa

Steel Shape	b_1 mm	Composite Beam* Factored Resistances M_{rc} (kN·m) for Shear Connections = 100%	70%	40%	Q_r (kN) for 100%	I_t x10⁶ mm⁴	S_t x10³ mm³	Steel Shape Data	Non-Composite Shape Unbraced Condition L' mm	M'r kN·m	L' mm	M'r kN·m
WWF900X169	5 000	3 130	3 000	2 760	4 340	7 660	9 190	Mr 1 990	5 000	1 860	12 000	625
	4 130	3 030	2 920	2 660	3 580	7 350	9 090	Vr 1 690	6 000	1 680	14 000	491
b=300	3 250	2 930	2 800	2 540	2 820	6 950	8 950	Lu 4 130	7 000	1 480	16 000	402
t=20	2 380	2 800	2 650	2 410	2 060	6 410	8 750	Ix 2930	8 000	1 250	18 000	339
d=900	1 500	2 580	2 440	2 260	1 300	5 620	8 400	Sx 6 510	10 000	845	20 000	293
WWF800X161	5 000	2 720	2 590	2 390	4 340	6 050	8 010	Mr 1 710	5 000	1 610	12 000	581
	4 130	2 620	2 510	2 310	3 580	5 810	7 920	Vr 1 490	6 000	1 470	14 000	460
b=300	3 250	2 520	2 420	2 200	2 820	5 500	7 800	Lu 4 230	7 000	1 310	16 000	379
t=20	2 380	2 410	2 290	2 090	2 060	5 070	7 630	Ix 2250	8 000	1 130	18 000	322
d=800	1 500	2 230	2 110	1 960	1 300	4 440	7 330	Sx 5 610	10 000	778	20 000	280
WWF700X152	4 500	2 290	2 170	2 000	3 900	4 550	6 830	Mr 1 440	5 000	1 370	10 000	711
	3 710	2 200	2 100	1 930	3 220	4 360	6 760	Vr 1 290	6 000	1 260	12 000	537
b=300	2 930	2 110	2 020	1 850	2 540	4 120	6 650	Lu 4 330	7 000	1 140	14 000	429
t=20	2 140	2 010	1 910	1 750	1 860	3 790	6 490	Ix 1 660	8 000	1 000	16 000	356
d=700	1 350	1 860	1 770	1 640	1 170	3 300	6 230	Sx 4 760	9 000	842	18 000	305
W610X174	4 500	2 290	2 170	1 990	3 900	4 090	7 020	Mr 1 450	5 000	1 430	10 000	924
W24X117	3 710	2 200	2 100	1 910	3 220	3 910	6 930	Vr 1 540	6 000	1 350	12 000	709
b=325	2 930	2 110	2 020	1 830	2 540	3 670	6 810	Lu 4 830	7 000	1 250	14 000	574
t=21.6	2 140	2 010	1 900	1 740	1 860	3 350	6 620	Ix 1 470	8 000	1 150	16 000	482
d=616	1 350	1 850	1 760	1 640	1 170	2 900	6 320	Sx 4 780	9 000	1 040	18 000	415
W610X155	4 000	2 030	1 920	1 760	3 470	3 600	6 200	Mr 1 280	5 000	1 260	10 000	762
W24X104	3 300	1 950	1 860	1 690	2 860	3 430	6 120	Vr 1 390	6 000	1 180	11 000	659
b=324	2 600	1 870	1 780	1 620	2 250	3 220	6 010	Lu 4 740	7 000	1 080	12 000	579
t=19	1 900	1 770	1 680	1 530	1 650	2 940	5 850	Ix 1 290	8 000	988	14 000	465
d=612	1 200	1 630	1 550	1 440	1 040	2 550	5 590	Sx 4 220	9 000	886	16 000	388
W610X140	4 000	1 890	1 790	1 610	3 470	3 320	5 590	Mr 1 120	4 000	1 050	9 000	486
W24X94	3 300	1 810	1 720	1 540	2 860	3 170	5 510	Vr 1 440	5 000	946	10 000	422
b=230	2 600	1 730	1 640	1 470	2 250	2 980	5 410	Lu 3 320	6 000	829	12 000	334
t=22.2	1 900	1 630	1 530	1 380	1 650	2 720	5 260	Ix 1 120	7 000	695	14 000	277
d=618	1 200	1 490	1 400	1 290	1 040	2 350	5 010	Sx 3 630	8 000	573	16 000	237
W610X125	4 000	1 720	1 620	1 460	3 470	2 990	4 980	Mr 991	4 000	923	9 000	396
W24X84	3 300	1 640	1 550	1 400	2 860	2 860	4 920	Vr 1 300	5 000	821	10 000	342
b=229	2 600	1 560	1 480	1 330	2 250	2 700	4 830	Lu 3 250	6 000	708	12 000	269
t=19.6	1 900	1 480	1 390	1 250	1 650	2 470	4 710	Ix 985	7 000	575	14 000	222
d=612	1 200	1 340	1 260	1 160	1 040	2 150	4 490	Sx 3 220	8 000	470	16 000	189
W610X113	4 000	1 590	1 490	1 340	3 470	2 720	4 500	Mr 888	4 000	818	9 000	328
W24X76	3 300	1 510	1 430	1 290	2 860	2 620	4 450	Vr 1 210	5 000	719	10 000	282
b=228	2 600	1 440	1 360	1 220	2 250	2 470	4 380	Lu 3 180	6 000	610	12 000	220
t=17.3	1 900	1 350	1 270	1 140	1 650	2 280	4 270	Ix 875	7 000	481	14 000	180
d=608	1 200	1 230	1 150	1 050	1 040	1 980	4 080	Sx 2 880	8 000	391	16 000	153

Note : * 20 MPa, 2 300 kg/m³ Concrete. G40.21-M 300W Steel.

Units : d, b, t, L_u (mm) $I_x = 10^6$ mm⁴ $S_x = 10^3$ mm³ M_r = kN·m V_r = kN

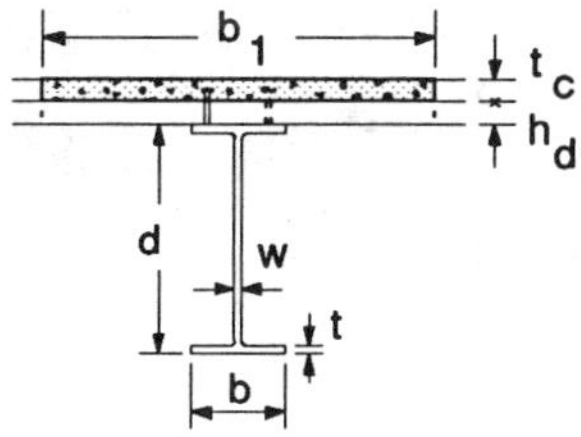

300W
20 MPa

COMPOSITE MEMBERS
Trial Selection Table

75 mm Deck with 85 mm Slab
$\phi = 0.9, \phi_c = 0.6, \phi_{sc} = 0.8$

Steel Shape	b_1 mm	Composite Beam*: Factored Resistances: M_{rc} (kN·m) for Shear Connections = 100%	70%	40%	Q_r (kN) for 100%	I_t $\times 10^6$ mm^4	S_t $\times 10^3$ mm^3	Steel Shape Data	Non-Composite Shape: Unbraced Condition: L' mm	M'r kN·m	L' mm	M'r kN·m
W610X101	3 500	1 420	1 330	1 190	3 030	2 400	4 010	Mr 783	4 000	713	9 000	267
W24X68	2 890	1 350	1 280	1 140	2 510	2 310	3 960	Vr 1 130	5 000	619	10 000	228
b=228	2 280	1 280	1 210	1 080	1 980	2 180	3 900	Lu 3 110	6 000	512	11 000	199
t=14.9	1 660	1 210	1 130	1 010	1 440	2 000	3 790	Ix 764	7 000	396	12 000	176
d=604	1 050	1 090	1 020	927	910	1 730	3 620	Sx 2 530	8 000	320	14 000	144
W610X91	3 500	1 290	1 210	1 080	3 030	2 160	3 570	Mr 691	4 000	619	9 000	217
W24X61	2 890	1 230	1 160	1 030	2 510	2 080	3 530	Vr 1 030	5 000	530	10 000	185
b=227	2 280	1 160	1 100	971	1 980	1 970	3 480	Lu 3 020	6 000	421	11 000	161
t=12.7	1 660	1 090	1 020	901	1 440	1 810	3 390	Ix 667	7 000	325	12 000	142
d=598	1 050	983	914	823	910	1 580	3 240	Sx 2 230	8 000	261	14 000	115
W610X84	3 500	1 210	1 120	1 010	2 890	2 010	3 310	Mr 637	3 000	636	8 000	232
W24X56	2 890	1 150	1 080	969	2 510	1 940	3 270	Vr 956	4 000	566	9 000	192
b=226	2 280	1 090	1 030	912	1 980	1 840	3 220	Lu 2 990	5 000	482	10 000	163
t=11.7	1 660	1 020	957	844	1 440	1 700	3 140	Ix 613	6 000	377	12 000	125
d=596	1 050	922	856	768	910	1 490	3 010	Sx 2 060	7 000	289	14 000	101
W530X123	3 500	1 500	1 410	1 260	3 030	2 370	4 460	Mr 867	4 000	794	9 000	361
W21X83	2 890	1 430	1 350	1 210	2 510	2 260	4 390	Vr 1 270	5 000	706	10 000	316
b=212	2 280	1 360	1 290	1 140	1 980	2 120	4 310	Lu 3 100	6 000	613	11 000	281
t=21.2	1 660	1 280	1 200	1 080	1 440	1 930	4 180	Ix 761	7 000	505	12 000	253
d=544	1 050	1 160	1 090	1 000	910	1 660	3 970	Sx 2 800	8 000	421	14 000	211
W530X109	3 500	1 370	1 280	1 150	3 030	2 130	3 960	Mr 764	4 000	692	9 000	291
W21X73	2 890	1 300	1 220	1 100	2 510	2 040	3 910	Vr 1 120	5 000	608	10 000	254
b=211	2 280	1 230	1 170	1 040	1 980	1 920	3 840	Lu 3 040	6 000	517	11 000	225
t=18.8	1 660	1 160	1 080	970	1 440	1 760	3 730	Ix 667	7 000	413	12 000	202
d=540	1 050	1 050	982	898	910	1 520	3 560	Sx 2 480	8 000	342	14 000	168
W530X101	3 500	1 290	1 210	1 080	3 030	2 000	3 690	Mr 707	3 000	707	8 000	301
W21X68	2 890	1 230	1 150	1 030	2 510	1 920	3 640	Vr 1 050	4 000	635	9 000	255
b=210	2 280	1 160	1 100	976	1 980	1 810	3 580	Lu 2 990	5 000	553	10 000	222
t=17.4	1 660	1 090	1 020	912	1 440	1 660	3 490	Ix 617	6 000	462	12 000	176
d=538	1 050	986	923	840	910	1 440	3 320	Sx 2 300	7 000	365	14 000	146
W530X92	3 500	1 210	1 120	1 000	3 030	1 840	3 370	Mr 637	3 000	633	8 000	253
W21X62	2 890	1 140	1 070	957	2 510	1 770	3 330	Vr 971	4 000	565	9 000	214
b=209	2 280	1 080	1 020	904	1 980	1 670	3 270	Lu 2 930	5 000	486	10 000	185
t=15.6	1 660	1 010	946	840	1 440	1 540	3 190	Ix 552	6 000	393	12 000	146
d=534	1 050	912	851	770	910	1 330	3 050	Sx 2 070	7 000	309	14 000	120
W530X82	3 500	1 090	1 010	895	2 840	1 640	2 990	Mr 559	3 000	551	8 000	204
W21X55	2 890	1 040	972	867	2 510	1 580	2 960	Vr 894	4 000	487	9 000	172
b=209	2 280	977	919	816	1 980	1 500	2 910	Lu 2 860	5 000	412	10 000	148
t=13.3	1 660	909	855	755	1 440	1 390	2 840	Ix 479	6 000	321	12 000	115
d=528	1 050	822	765	686	910	1 210	2 720	Sx 1 810	7 000	251	14 000	94.8

Note : * 20 MPa, 2 300 kg/m³ Concrete. G40.21-M 300W Steel.

Units : d, b, t, L_u (mm) $I_x = 10^6$ mm⁴ $S_x = 10^3$ mm³ M_r = kN·m V_r = kN

COMPOSITE MEMBERS
Trial Selection Table

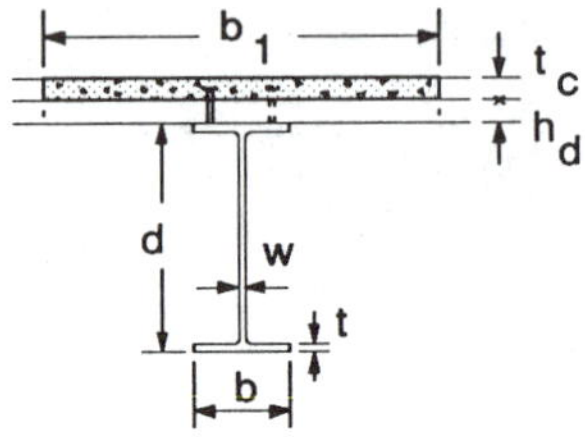

75 mm Deck with 85 mm Slab
$\phi = 0.9$, $\phi_c = 0.6$, $\phi_{sc} = 0.8$

300W
20 MPa

Steel Shape	b_1 mm	Composite Beam*						Steel Shape Data	Non-Composite Shape			
		Factored Resistances				I_t x10⁶ mm⁴	S_t x10³ mm³		Unbraced Condition			
		M_{rc} (kN·m) for Shear Connections =			Q_r (kN) for 100%				L' mm	M'r kN·m	L' mm	M'r kN·m
		100%	70%	40%								
W530X72	3 500	958	881	776	2 470	1 440	2 600	M r 475	3 000	462	8 000	155
W21X48	2 890	940	872	773	2 470	1 390	2 570	V r 831	4 000	402	9 000	129
b=207	2 280	879	823	727	1 980	1 330	2 530	L u 2 750	5 000	331	10 000	111
t=10.9	1 660	812	763	668	1 440	1 230	2 470	I x 402	6 000	247	12 000	85.9
d=524	1 050	732	678	601	910	1 080	2 370	S x 1 530	7 000	191	14 000	70.1
W460X106	3 500	1 210	1 120	997	3 030	1 680	3 490	M r 645	3 000	640	8 000	308
W18X71	2 890	1 140	1 070	950	2 510	1 610	3 440	V r 1 060	4 000	579	9 000	266
b=194	2 280	1 080	1 010	897	1 980	1 510	3 380	L u 2 910	5 000	512	10 000	235
t=20.6	1 660	1 000	939	837	1 440	1 380	3 280	I x 488	6 000	444	12 000	190
d=470	1 050	905	847	771	910	1 190	3 120	S x 2 080	7 000	366	14 000	160
W460X97	3 500	1 130	1 040	925	3 030	1 550	3 200	M r 589	3 000	581	8 000	264
W18X65	2 890	1 070	990	883	2 510	1 490	3 160	V r 947	4 000	522	9 000	227
b=193	2 280	997	936	833	1 980	1 400	3 100	L u 2 870	5 000	457	10 000	200
t=19	1 660	926	871	775	1 440	1 280	3 020	I x 445	6 000	389	12 000	161
d=466	1 050	839	784	711	910	1 110	2 870	S x 1 910	7 000	314	14 000	135
W460X89	3 500	1 070	986	871	3 030	1 450	2 970	M r 543	3 000	534	8 000	231
W18X60	2 890	1 010	934	832	2 510	1 390	2 930	V r 868	4 000	477	9 000	198
b=192	2 280	941	880	785	1 980	1 320	2 890	L u 2 830	5 000	414	10 000	174
t=17.7	1 660	870	820	729	1 440	1 210	2 810	I x 410	6 000	343	12 000	140
d=464	1 050	789	738	666	910	1 050	2 680	S x 1 770	7 000	276	14 000	117
W460X82	3 500	985	903	794	2 810	1 330	2 710	M r 494	3 000	482	8 000	195
W18X55	2 890	940	869	771	2 510	1 280	2 680	V r 812	4 000	427	9 000	167
b=191	2 280	875	816	727	1 980	1 220	2 640	L u 2 770	5 000	365	10 000	146
t=16	1 660	806	759	674	1 440	1 120	2 570	I x 370	6 000	292	12 000	117
d=460	1 050	730	682	613	910	974	2 460	S x 1 610	7 000	234	14 000	97.3
W460X74	3 500	901	823	721	2 550	1 230	2 470	M r 446	3 000	433	8 000	164
W18X50	2 890	879	810	716	2 510	1 180	2 440	V r 735	4 000	380	9 000	140
b=190	2 280	815	758	675	1 980	1 120	2 410	L u 2 730	5 000	320	10 000	122
t=14.5	1 660	747	704	624	1 440	1 040	2 350	I x 333	6 000	249	12 000	96.9
d=458	1 050	676	632	565	910	906	2 250	S x 1 460	7 000	198	14 000	80.6
W460X67	3 000	817	748	657	2 340	1 100	2 250	M r 405	3 000	390	8 000	140
W18X45	2 480	784	725	641	2 150	1 060	2 220	V r 688	4 000	339	9 000	119
b=190	1 950	729	680	601	1 690	1 000	2 190	L u 2 660	5 000	281	10 000	103
t=12.7	1 430	672	631	555	1 240	923	2 130	I x 300	6 000	214	11 000	91.4
d=454	900	605	562	501	780	803	2 040	S x 1 320	7 000	169	12 000	82
W460X61	3 000	735	669	584	2 100	983	2 000	M r 354	3 000	336	8 000	111
W18X41	2 480	720	662	581	2 100	948	1 980	V r 650	4 000	288	9 000	93.9
b=189	1 950	669	621	547	1 690	901	1 950	L u 2 580	5 000	231	10 000	81.3
t=10.8	1 430	614	575	503	1 240	835	1 900	I x 259	6 000	172	11 000	71.7
d=450	900	550	509	450	780	729	1 820	S x 1 150	7 000	135	12 000	64.1

Note : * 20 MPa, 2 300 kg/m³ Concrete. G40.21-M 300W Steel.

Units : d, b, t, L_u (mm) $I_x = 10^6$ mm⁴ $S_x = 10^3$ mm³ M_r = kN·m V_r = kN

300W
20 MPa

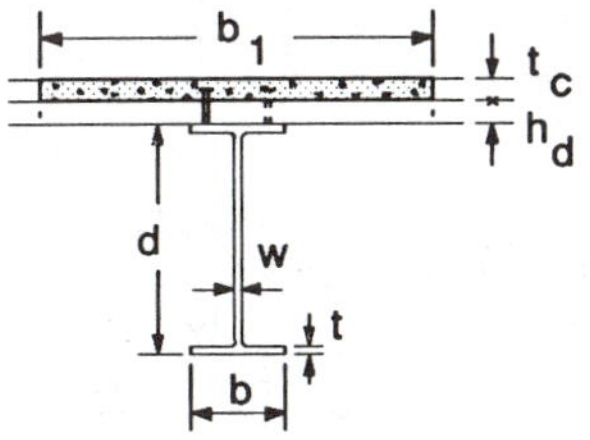

COMPOSITE MEMBERS
Trial Selection Table

75 mm Deck with 85 mm Slab
$\phi = 0.9, \phi_c = 0.6, \phi_{sc} = 0.8$

Steel Shape	b_1 mm	Composite Beam*: Factored Resistances: M_{rc} (kN·m) for Shear Connections = 100%	70%	40%	Q_r (kN) for 100%	I_t x10⁶ mm⁴	S_t x10³ mm³	Steel Shape Data	Non-Composite Shape: Unbraced Condition: L' mm	M'r kN·m	L' mm	M'r kN·m
W410X85	3 500	957	874	761	2 920	1 190	2 640	Mr 467	3 000	455	8 000	205
W16X57	2 890	903	831	733	2 510	1 140	2 600	Vr 812	4 000	406	9 000	178
b=181	2 280	837	778	690	1 980	1 080	2 560	Lu 2 730	5 000	354	10 000	157
t=18.2	1 660	767	721	639	1 440	993	2 490	Ix 315	6 000	297	12 000	127
d=418	1 050	692	646	581	910	858	2 370	Sx 1 510	7 000	243	14 000	107
W410X74	3 500	853	774	672	2 580	1 070	2 340	Mr 408	3 000	394	8 000	163
W16X50	2 890	828	759	664	2 510	1 030	2 310	Vr 716	4 000	348	9 000	140
b=180	2 280	764	707	625	1 980	974	2 270	Lu 2 670	5 000	297	10 000	124
t=16	1 660	696	652	577	1 440	898	2 220	Ix 275	6 000	239	12 000	99.8
d=414	1 050	625	584	521	910	781	2 120	Sx 1 330	7 000	194	14 000	83.8
W410X67	3 500	772	697	602	2 320	968	2 110	Mr 367	3 000	352	8 000	135
W16X45	2 890	756	689	600	2 320	934	2 090	Vr 643	4 000	307	9 000	116
b=179	2 280	708	652	575	1 980	889	2 060	Lu 2 610	5 000	258	10 000	102
t=14.4	1 660	641	598	530	1 440	823	2 010	Ix 246	6 000	201	12 000	81.7
d=410	1 050	573	536	476	910	720	1 930	Sx 1 200	7 000	161	14 000	68.4
W410X60	3 000	677	611	530	2 050	844	1 860	Mr 321	3 000	306	8 000	109
W16X40	2 480	662	604	527	2 050	815	1 830	Vr 560	4 000	264	9 000	93.2
b=178	1 950	615	568	502	1 690	774	1 810	Lu 2 580	5 000	217	10 000	81.4
t=12.8	1 430	560	523	464	1 240	717	1 760	Ix 216	6 000	165	11 000	72.2
d=408	900	501	469	417	780	626	1 690	Sx 1 060	7 000	131	12 000	65
W410X54	3 000	610	549	473	1 840	758	1 660	Mr 284	3 000	266	8 000	86.1
W16X36	2 480	599	543	472	1 840	732	1 640	Vr 540	4 000	225	9 000	73.2
b=177	1 950	570	523	459	1 690	698	1 610	Lu 2 480	5 000	176	10 000	63.6
t=10.9	1 430	515	480	422	1 240	649	1 580	Ix 186	6 000	132	11 000	56.3
d=404	900	457	426	376	780	569	1 510	Sx 924	7 000	104	12 000	50.5
W410X46	3 000	534	478	409	1 590	666	1 430	Mr 239	2 000	236	7 000	61.7
W16X31	2 480	526	473	408	1 590	645	1 420	Vr 504	3 000	195	8 000	51.8
b=140	1 950	512	467	405	1 590	617	1 400	Lu 1 930	4 000	142	9 000	44.6
t=11.2	1 430	466	431	377	1 240	576	1 370	Ix 156	5 000	99.9	10 000	39.2
d=404	900	409	380	332	780	509	1 320	Sx 773	6 000	76.4	12 000	31.6
W410X39	3 000	455	405	344	1 350	568	1 210	Mr 197	2 000	193	7 000	44.1
W16X26	2 480	449	402	343	1 350	552	1 200	Vr 456	3 000	155	8 000	36.6
b=140	1 950	439	397	341	1 350	529	1 180	Lu 1 860	4 000	105	9 000	31.3
t=8.8	1 430	415	381	331	1 240	497	1 160	Ix 127	5 000	73.1	10 000	27.4
d=400	900	359	334	288	780	443	1 120	Sx 634	6 000	55.2	12 000	21.9
W360X79	3 000	788	716	619	2 600	881	2 250	Mr 386	4 000	364	8 000	225
W14X53	2 480	734	672	590	2 150	844	2 220	Vr 593	4 500	348	9 000	195
b=205	1 950	676	626	557	1 690	794	2 180	Lu 3 270	5 000	331	10 000	172
t=16.8	1 430	618	580	518	1 240	726	2 120	Ix 227	6 000	298	11 000	154
d=354	900	557	523	472	780	622	2 020	Sx 1 280	7 000	264	12 000	139

Note : * 20 MPa, 2 300 kg/m³ Concrete. G40.21-M 300W Steel.

Units : d, b, t, L_u (mm) $I_x = 10^6$ mm⁴ $S_x = 10^3$ mm³ M_r = kN·m V_r = kN

COMPOSITE MEMBERS
Trial Selection Table

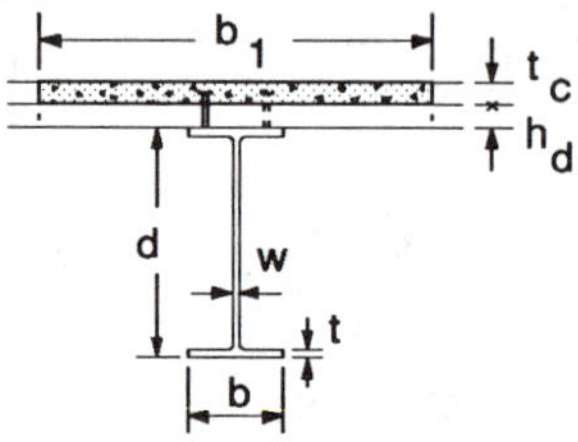

75 mm Deck with 85 mm Slab
$\phi = 0.9, \phi_c = 0.6, \phi_{sc} = 0.8$

300W
20 MPa

Steel Shape	b_1	Composite Beam*						Steel Shape Data	Non-Composite Shape			
		Factored Resistances				I_t	S_t		Unbraced Condition			
		M_{rc} (kN·m) for Shear Connections =			Q_r (kN) for	x10⁶	x10³		L'	M'r	L'	M'r
	mm	100%	70%	40%	100%	mm⁴	mm³		mm	kN·m	mm	kN·m
W360X72	3 000	725	655	562	2 460	800	2 030	M r 346	4 000	322	8 000	186
W14X48	2 480	683	622	542	2 150	768	2 010	V r 536	4 500	306	9 000	161
b=204	1 950	626	577	511	1 690	725	1 970	L u 3 190	5 000	290	10 000	141
t=15.1	1 430	569	532	475	1 240	665	1 920	I x 201	6 000	257	11 000	126
d=350	900	509	479	430	780	573	1 830	S x 1 150	7 000	222	12 000	114
W360X64	3 000	655	588	503	2 200	726	1 820	M r 308	4 000	283	8 000	153
W14X43	2 480	635	576	497	2 150	699	1 800	V r 478	4 500	268	9 000	131
b=203	1 950	580	532	468	1 690	661	1 770	L u 3 110	5 000	252	10 000	115
t=13.5	1 430	524	487	434	1 240	610	1 730	I x 178	6 000	220	11 000	102
d=348	900	465	438	392	780	528	1 650	S x 1 030	7 000	183	12 000	92.4
W360X57	2 500	586	529	454	1 950	655	1 620	M r 273	3 000	259	6 000	147
W14X38	2 060	559	509	443	1 790	629	1 600	V r 504	3 500	243	7 000	119
b=172	1 630	513	473	416	1 410	595	1 580	L u 2 550	4 000	225	8 000	99.8
t=13.1	1 190	466	435	384	1 030	547	1 540	I x 161	4 500	207	9 000	86
d=358	750	416	388	346	650	472	1 460	S x 897	5 000	189	10 000	75.7
W360X51	2 500	529	474	406	1 740	592	1 450	M r 241	3 000	227	6 000	121
W14X34	2 060	516	468	404	1 740	570	1 440	V r 457	3 500	212	7 000	97
b=171	1 630	475	436	382	1 410	541	1 410	L u 2 500	4 000	195	8 000	81
t=11.6	1 190	428	398	351	1 030	499	1 380	I x 141	4 500	178	9 000	69.5
d=356	750	380	355	315	650	433	1 320	S x 796	5 000	159	10 000	60.9
W360X45	2 500	473	421	359	1 550	529	1 290	M r 210	3 000	195	6 000	96.1
W14X30	2 060	463	416	357	1 550	510	1 280	V r 433	3 500	181	7 000	76.5
b=171	1 630	438	399	347	1 410	486	1 260	L u 2 430	4 000	165	8 000	63.4
t=9.8	1 190	392	363	318	1 030	450	1 230	I x 122	4 500	148	9 000	54.1
d=352	750	344	321	282	650	393	1 170	S x 691	5 000	128	10 000	47.2
W360X39	2 500	418	370	313	1 340	469	1 120	M r 179	2 000	173	6 000	54.2
W14X26	2 060	410	366	312	1 340	454	1 110	V r 410	2 500	157	7 000	44.3
b=128	1 630	399	361	310	1 340	433	1 090	L u 1 790	3 000	139	8 000	37.5
t=10.7	1 190	359	329	286	1 030	403	1 070	I x 102	4 000	97.2	9 000	32.5
d=354	750	311	289	251	650	355	1 020	S x 580	5 000	69.8	10 000	28.8
W360X33	2 500	352	310	260	1 130	397	942	M r 146	2 000	139	6 000	38.1
W14X22	2 060	347	308	260	1 130	385	931	V r 362	2 500	124	7 000	30.8
b=127	1 630	339	304	258	1 130	369	917	L u 1 720	3 000	108	8 000	25.9
t=8.5	1 190	318	290	250	1 030	346	897	I x 82.7	4 000	70.3	9 000	22.3
d=350	750	272	252	217	650	307	864	S x 474	5 000	49.7	10 000	19.6
W310X86	3 500	812	730	619	2 970	824	2 310	M r 383	5 000	367	10 000	248
W12X58	2 890	754	683	590	2 510	789	2 280	V r 503	6 000	344	11 000	221
b=254	2 280	689	631	556	1 980	743	2 240	L u 4 250	7 000	320	12 000	200
t=16.3	1 660	621	578	519	1 440	679	2 180	I x 199	8 000	297	13 000	182
d=310	1 050	552	523	474	910	582	2 070	S x 1 280	9 000	273	14 000	167

Note : * 20 MPa, 2 300 kg/m³ Concrete. G40.21-M 300W Steel.

Units : d, b, t, L_u (mm) $I_x = 10^6$ mm⁴ $S_x = 10^3$ mm³ M_r = kN·m V_r = kN

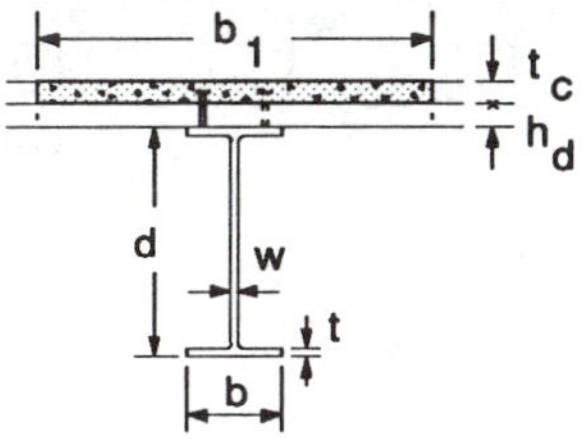

300W
20 MPa

COMPOSITE MEMBERS
Trial Selection Table
75 mm Deck with 85 mm Slab
$\phi = 0.9, \phi_c = 0.6, \phi_{sc} = 0.8$

Steel Shape	b_1 mm	Composite Beam*							Non-Composite Shape			
		Factored Resistances				I_t x10⁶ mm⁴	S_t x10³ mm³	Steel Shape Data	Unbraced Condition			
		M_{rc} (kN·m) for Shear Connections =			Q_r (kN) for 100%				L' mm	M'r kN·m	L' mm	M'r kN·m
		100%	70%	40%								
W310X79	3 000	723	651	556	2 600	731	2 090	Mr 346	5 000	327	8 000	258
W12X53	2 480	669	608	528	2 150	700	2 060	Vr 480	5 500	316	9 000	235
b=254	1 950	612	563	498	1 690	657	2 020	Lu 4 140	6 000	305	10 000	207
t=14.6	1 430	555	518	465	1 240	599	1 970	Ix 177	6 500	293	11 000	184
d=306	900	495	468	423	780	511	1 870	Sx 1 160	7 000	282	12 000	166
W310X74	3 000	700	629	533	2 560	703	1 970	Mr 321	4 000	307	8 000	206
W12X50	2 480	649	588	507	2 150	673	1 940	Vr 519	4 500	295	9 000	179
b=205	1 950	592	543	477	1 690	633	1 910	Lu 3 380	5 000	282	10 000	159
t=16.3	1 430	535	497	442	1 240	579	1 850	Ix 165	6 000	258	11 000	142
d=310	900	474	446	400	780	495	1 760	Sx 1 060	7 000	233	12 000	129
W310X67	3 000	633	564	476	2 300	635	1 770	Mr 286	4 000	270	8 000	169
W12X45	2 480	604	545	465	2 150	609	1 750	Vr 463	4 500	258	9 000	147
b=204	1 950	549	500	436	1 690	575	1 720	Lu 3 280	5 000	246	10 000	129
t=14.6	1 430	492	455	404	1 240	528	1 670	Ix 145	6 000	222	11 000	116
d=306	900	433	407	363	780	454	1 590	Sx 949	7 000	198	12 000	105
W310X60	2 500	557	499	423	2 050	555	1 570	Mr 254	4 000	237	6 500	179
W12X40	2 060	521	471	406	1 790	532	1 550	Vr 406	4 500	226	7 000	166
b=203	1 630	476	435	382	1 410	501	1 520	Lu 3 200	5 000	214	8 000	139
t=13.1	1 190	428	398	354	1 030	458	1 480	Ix 129	5 500	203	9 000	120
d=304	750	379	356	319	650	393	1 410	Sx 849	6 000	191	10 000	106
W310X52	2 500	511	455	385	1 800	525	1 410	Mr 226	3 000	216	6 000	130
W12X35	2 060	496	448	382	1 790	504	1 390	Vr 431	3 500	203	7 000	106
b=167	1 630	452	412	358	1 410	478	1 370	Lu 2 570	4 000	189	8 000	89.4
t=13.2	1 190	404	374	330	1 030	439	1 330	Ix 118	4 500	176	9 000	77.4
d=318	750	355	333	296	650	379	1 270	Sx 747	5 000	162	10 000	68.4
W310X45	2 500	441	389	327	1 540	455	1 210	Mr 191	3 000	180	6 000	98.2
W12X30	2 060	431	385	326	1 540	439	1 190	Vr 369	3 500	168	7 000	79.3
b=166	1 630	407	368	317	1 410	417	1 180	Lu 2 490	4 000	155	8 000	66.5
t=11.2	1 190	361	332	291	1 030	386	1 150	Ix 99.2	4 500	142	9 000	57.3
d=314	750	313	293	259	650	336	1 100	Sx 634	5 000	128	10 000	50.4
W310X39	2 500	385	338	283	1 330	400	1 060	Mr 165	3 000	153	6 000	77.7
W12X26	2 060	378	334	282	1 330	386	1 040	Vr 320	3 500	142	7 000	62.2
b=165	1 630	367	329	280	1 330	368	1 030	Lu 2 440	4 000	130	8 000	51.8
t=9.7	1 190	327	299	260	1 030	343	1 000	Ix 85.1	4 500	117	9 000	44.3
d=310	750	281	262	231	650	301	964	Sx 549	5 000	103	10 000	38.8
W250X80	3 000	658	586	491	2 600	589	1 920	Mr 294	5 000	287	8 000	242
W10X54	2 480	604	543	462	2 150	562	1 890	Vr 429	5 500	280	9 000	227
b=255	1 950	547	498	434	1 690	526	1 850	Lu 4 520	6 000	272	10 000	212
t=15.6	1 430	490	453	404	1 240	477	1 800	Ix 126	6 500	265	11 000	197
d=256	900	430	406	367	780	403	1 700	Sx 982	7 000	257	12 000	179

Note : * 20 MPa, 2 300 kg/m³ Concrete. G40.21-M 300W Steel.

Units : d, b, t, L_u (mm) $I_x = 10^6$ mm⁴ $S_x = 10^3$ mm³ M_r = kN·m V_r = kN

COMPOSITE MEMBERS
Trial Selection Table

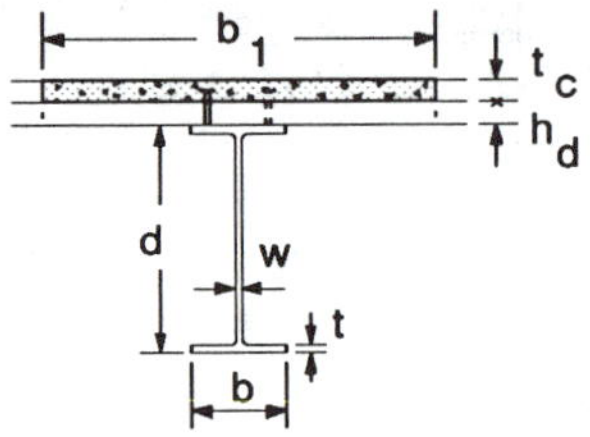

75 mm Deck with 85 mm Slab
$\phi = 0.9$, $\phi_c = 0.6$, $\phi_{sc} = 0.8$

300W
20 MPa

Steel Shape	b_1	Composite Beam*						Steel Shape Data	Non-Composite Shape			
		Factored Resistances				I_t	S_t		Unbraced Condition			
		M_{rc} (kN·m) for Shear Connections =			Q_r (kN) for	x10⁶	x10³		L'	M'r	L'	M'r
	mm	100%	70%	40%	100%	mm⁴	mm³		mm	kN·m	mm	kN·m
W250X73	3 000	617	547	454	2 510	542	1 750	Mr 266	5 000	257	8 000	212
W10X49	2 480	570	511	431	2 150	518	1 730	Vr 389	5 500	250	9 000	198
b=254	1 950	514	466	403	1 690	487	1 700	Lu 4 390	6 000	242	10 000	183
t=14.2	1 430	458	421	374	1 240	443	1 650	Ix 113	6 500	235	11 000	165
d=254	900	399	375	339	780	377	1 560	Sx 891	7 000	227	12 000	149
W250X67	3 000	580	512	423	2 310	515	1 630	Mr 243	4 000	237	8 000	174
W10X45	2 480	550	491	411	2 150	493	1 610	Vr 409	4 500	229	9 000	157
b=204	1 950	495	446	382	1 690	464	1 570	Lu 3 570	5 000	221	10 000	139
t=15.7	1 430	438	401	353	1 240	424	1 530	Ix 104	6 000	205	11 000	125
d=258	900	379	355	317	780	362	1 450	Sx 806	7 000	189	12 000	114
W250X58	2 500	494	437	361	2 000	432	1 400	Mr 208	4 000	199	6 500	160
W10X39	2 060	462	413	347	1 790	413	1 380	Vr 359	4 500	191	7 000	153
b=203	1 630	417	377	324	1 410	388	1 350	Lu 3 410	5 000	184	8 000	137
t=13.5	1 190	369	339	299	1 030	354	1 310	Ix 87.3	5 500	176	9 000	119
d=252	750	320	300	268	650	301	1 240	Sx 693	6 000	168	10 000	105
W250X49	2 500	423	370	304	1 690	370	1 180	Mr 171	4 000	160	6 500	123
W10X33	2 060	411	364	302	1 690	355	1 160	Vr 327	4 500	153	7 000	115
b=202	1 630	375	336	284	1 410	335	1 140	Lu 3 240	5 000	146	8 000	97.2
t=11	1 190	329	299	260	1 030	307	1 110	Ix 70.6	5 500	138	9 000	84
d=248	750	281	261	231	650	264	1 060	Sx 572	6 000	130	10 000	74.1
W250X45	2 500	406	354	291	1 540	371	1 110	Mr 163	3 000	151	6 000	90.6
W10X30	2 060	396	349	290	1 540	357	1 100	Vr 360	3 500	142	7 000	75.2
b=148	1 630	371	332	280	1 410	338	1 080	Lu 2 360	4 000	132	8 000	64.4
t=13	1 190	325	295	255	1 030	312	1 050	Ix 71.1	4 500	122	9 000	56.3
d=266	750	277	257	225	650	270	1 000	Sx 534	5 000	112	10 000	50.1
W250X39	2 500	352	305	250	1 330	323	964	Mr 139	3 000	126	6 000	69.2
W10X26	2 060	345	301	248	1 330	311	951	Vr 308	3 500	117	7 000	57
b=147	1 630	333	296	247	1 330	297	935	Lu 2 280	4 000	108	8 000	48.6
t=11.2	1 190	295	266	228	1 030	275	912	Ix 60.1	4 500	98.6	9 000	42.3
d=262	750	248	229	200	650	240	873	Sx 459	5 000	88	10 000	37.6
W250X33	2 000	294	255	209	1 130	266	807	Mr 114	3 000	102	5 500	55.6
W10X22	1 650	288	252	208	1 130	256	795	Vr 280	3 500	93.1	6 000	49.4
b=146	1 300	278	247	207	1 130	243	781	Lu 2 180	4 000	84	6 500	44.4
t=9.1	950	241	218	187	824	224	761	Ix 48.9	4 500	74.1	7 000	40.3
d=258	600	204	189	164	520	195	728	Sx 379	5 000	63.6	8 000	34.1
W200X71	3 000	560	490	397	2 460	439	1 590	Mr 217	5 000	208	8 000	178
W8X48	2 480	518	458	377	2 150	419	1 560	Vr 393	5 500	203	9 000	168
b=206	1 950	462	412	348	1 690	392	1 530	Lu 4 150	6 000	198	10 000	158
t=17.4	1 430	405	367	319	1 240	356	1 480	Ix 76.6	6 500	193	11 000	148
d=216	900	345	320	286	780	300	1 400	Sx 709	7 000	188	12 000	137

Note : * 20 MPa, 2 300 kg/m³ Concrete. G40.21-M 300W Steel.

Units : d, b, t, L_u (mm) $I_x = 10^6$ mm⁴ $S_x = 10^3$ mm³ M_r = kN·m V_r = kN

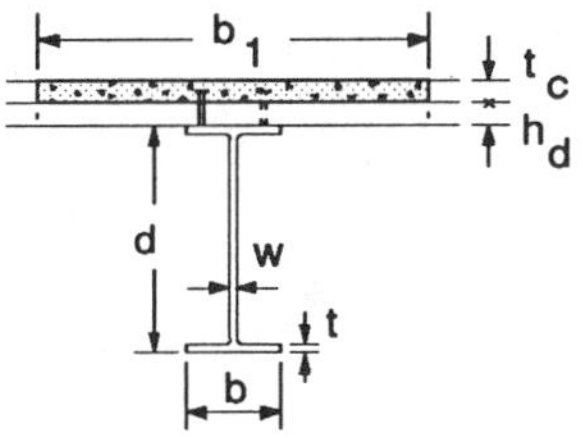

COMPOSITE MEMBERS
Trial Selection Table

300W
20 MPa

75 mm Deck with 85 mm Slab
$\phi = 0.9,\ \phi_c = 0.6,\ \phi_{sc} = 0.8$

Steel Shape	b_1	Composite Beam*						Steel Shape Data	Non-Composite Shape			
		Factored Resistances				I_t	S_t		Unbraced Condition			
		M_{rc} (kN·m) for Shear Connections =			Q_r (kN) for	$\times10^6$	$\times10^3$		L'	M'r	L'	M'r
	mm	100%	70%	40%	100%	mm⁴	mm³		mm	kN·m	mm	kN·m
W200X59	3 000	473	408	327	2 040	370	1 330	Mr 176	4 000	174	8 000	134
W8X40	2 480	459	401	325	2 040	354	1 310	Vr 341	4 500	169	9 000	124
b=205	1 950	412	365	303	1 690	333	1 280	Lu 3 780	5 000	164	10 000	114
t=14.2	1 430	357	321	274	1 240	305	1 240	Ix 61.1	6 000	154	11 000	103
d=210	900	299	275	243	780	260	1 180	Sx 582	7 000	144	12 000	93.4
W200X52	2 500	410	354	285	1 800	316	1 160	Mr 154	4 000	150	6 500	126
W8X35	2 060	395	347	282	1 790	302	1 140	Vr 290	4 500	145	7 000	121
b=204	1 630	351	311	259	1 410	284	1 120	Lu 3 620	5 000	140	8 000	111
t=12.6	1 190	304	274	236	1 030	259	1 080	Ix 52.7	5 500	135	9 000	101
d=206	750	256	236	209	650	220	1 030	Sx 512	6 000	131	10 000	89.7
W200X46	2 500	365	314	251	1 580	283	1 030	Mr 134	4 000	129	6 500	105
W8X31	2 060	355	308	249	1 580	271	1 010	Vr 262	4 500	124	7 000	99.8
b=203	1 630	327	289	238	1 410	256	989	Lu 3 460	5 000	119	8 000	90.1
t=11	1 190	281	252	214	1 030	235	961	Ix 45.5	5 500	114	9 000	78.6
d=204	750	234	215	189	650	201	915	Sx 448	6 000	109	10 000	69.6
W200X42	2 000	327	283	227	1 430	251	919	Mr 120	3 000	119	5 500	93.5
W8X28	1 650	316	277	225	1 430	240	904	Vr 264	3 500	114	6 000	88.4
b=166	1 300	280	248	207	1 130	226	886	Lu 2 850	4 000	109	6 500	83.4
t=11.8	950	242	219	188	824	205	859	Ix 40.9	4 500	104	7 000	77.7
d=206	600	204	189	166	520	174	814	Sx 399	5 000	98.6	8 000	66.7
W200X36	2 000	285	244	195	1 240	220	798	Mr 103	3 000	100	5 500	75.5
W8X24	1 650	277	240	194	1 240	211	785	Vr 223	3 500	95.3	6 000	70.6
b=165	1 300	257	227	186	1 130	199	770	Lu 2 730	4 000	90.4	6 500	64.7
t=10.2	950	221	198	167	824	182	748	Ix 34.4	4 500	85.5	7 000	59.2
d=202	600	183	168	147	520	156	712	Sx 342	5 000	80.5	8 000	50.6
W200X31	2 000	258	220	175	1 080	204	712	Mr 90.5	3 000	81.1	5 500	50.6
W8X21	1 650	252	217	174	1 080	196	701	Vr 240	3 500	75.3	6 000	45.6
b=134	1 300	242	212	173	1 080	186	687	Lu 2 150	4 000	69.5	6 500	41.5
t=10.2	950	210	187	157	824	172	668	Ix 31.4	4 500	63.6	7 000	38
d=210	600	172	157	135	520	148	638	Sx 299	5 000	57	8 000	32.7
W200X27	2 000	221	187	148	915	177	608	Mr 75.3	3 000	65.4	5 500	36.4
W8X18	1 650	217	185	148	915	170	599	Vr 215	3 500	59.7	6 000	32.6
b=133	1 300	210	182	147	915	162	587	Lu 2 050	4 000	54	6 500	29.6
t=8.4	950	192	169	140	824	150	571	Ix 25.8	4 500	47.5	7 000	27.1
d=208	600	155	141	120	520	131	547	Sx 249	5 000	41.2	8 000	23.1
W150X30	2 500	224	185	141	1 020	156	631	Mr 65.9	3 000	64.1	6 000	46.8
W6X20	2 060	220	183	140	1 020	150	619	Vr 186	3 500	61.2	7 000	40.4
b=153	1 630	213	179	139	1 020	142	605	Lu 2 680	4 000	58.3	8 000	34.8
t=9.3	1 190	201	174	137	1 020	132	588	Ix 17.2	4 500	55.4	9 000	30.6
d=158	750	156	138	115	650	115	561	Sx 219	5 000	52.5	10 000	27.3

Note : * 20 MPa, 2 300 kg/m³ Concrete. G40.21-M 300W Steel.

Units : d, b, t, L_u (mm) $I_x = 10^6$ mm⁴ $S_x = 10^3$ mm³ M_r = kN·m V_r = kN

COMPOSITE MEMBERS
Trial Selection Table

75 mm Deck with 75 mm Slab
$\phi = 0.9, \phi_c = 0.6, \phi_{sc} = 0.8$

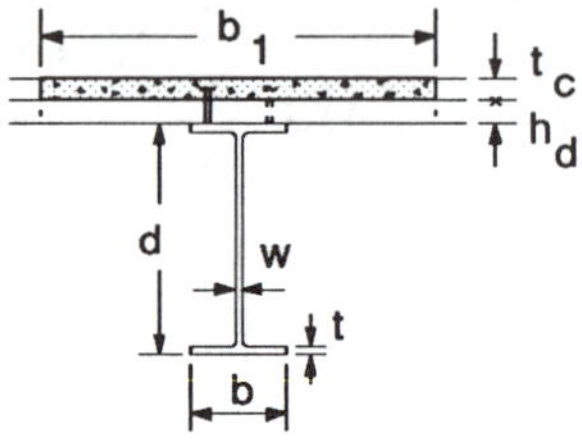

300W
25 MPa

Steel Shape	b_1 mm	Composite Beam* Factored Resistances M_{rc} (kN·m) for Shear Connections = 100%	70%	40%	Q_r (kN) for 100%	I_t x10⁶ mm⁴	S_t x10³ mm³	Steel Shape Data	Non-Composite Shape Unbraced Condition L' mm	M'r kN·m	L' mm	M'r kN·m
WWF900X169	5 000	3 160	3 020	2 800	4 780	7 550	9 110	Mr 1 990	5 000	1 860	12 000	625
	4 130	3 060	2 940	2 700	3 950	7 250	9 020	Vr 1 690	6 000	1 680	14 000	491
b=300	3 250	2 950	2 840	2 580	3 110	6 850	8 880	Lu 4 130	7 000	1 480	16 000	402
t=20	2 380	2 830	2 690	2 440	2 280	6 320	8 680	Ix 2930	8 000	1 250	18 000	339
d=900	1 500	2 620	2 470	2 290	1 430	5 550	8 340	Sx 6 510	10 000	845	20 000	293
WWF800X161	5 000	2 750	2 610	2 420	4 780	5 960	7 930	Mr 1 710	5 000	1 610	12 000	581
	4 130	2 650	2 530	2 340	3 950	5 720	7 850	Vr 1 490	6 000	1 470	14 000	460
b=300	3 250	2 550	2 450	2 240	3 110	5 410	7 740	Lu 4 230	7 000	1 310	16 000	379
t=20	2 380	2 440	2 320	2 120	2 280	5 000	7 560	Ix 2250	8 000	1 130	18 000	322
d=800	1 500	2 260	2 140	1 980	1 430	4 380	7 270	Sx 5 610	10 000	778	20 000	280
WWF700X152	4 500	2 310	2 190	2 020	4 300	4 470	6 760	Mr 1 440	5 000	1 370	10 000	711
	3 710	2 220	2 120	1 950	3 550	4 290	6 690	Vr 1 290	6 000	1 260	12 000	537
b=300	2 930	2 130	2 040	1 870	2 800	4 050	6 590	Lu 4 330	7 000	1 140	14 000	429
t=20	2 140	2 030	1 940	1 770	2 050	3 720	6 430	Ix 1 660	8 000	1 000	16 000	356
d=700	1 350	1 890	1 790	1 660	1 290	3 250	6 170	Sx 4 760	9 000	842	18 000	305
W610X174	4 500	2 320	2 190	2 010	4 300	4 010	6 940	Mr 1 450	5 000	1 430	10 000	924
W24X117	3 710	2 230	2 120	1 940	3 550	3 830	6 850	Vr 1 540	6 000	1 350	12 000	709
b=325	2 930	2 130	2 040	1 860	2 800	3 600	6 730	Lu 4 830	7 000	1 250	14 000	574
t=21.6	2 140	2 030	1 930	1 760	2 050	3 290	6 550	Ix 1 470	8 000	1 150	16 000	482
d=616	1 350	1 880	1 780	1 650	1 290	2 850	6 260	Sx 4 780	9 000	1 040	18 000	415
W610X155	4 000	2 050	1 940	1 780	3 830	3 530	6 130	Mr 1 280	5 000	1 260	10 000	762
W24X104	3 300	1 970	1 870	1 710	3 160	3 370	6 050	Vr 1 390	6 000	1 180	11 000	659
b=324	2 600	1 880	1 800	1 640	2 490	3 160	5 950	Lu 4 740	7 000	1 080	12 000	579
t=19	1 900	1 790	1 700	1 550	1 820	2 890	5 790	Ix 1 290	8 000	988	14 000	465
d=612	1 200	1 660	1 570	1 460	1 150	2 510	5 530	Sx 4 220	9 000	886	16 000	388
W610X140	4 000	1 920	1 810	1 640	3 830	3 250	5 520	Mr 1 120	4 000	1 050	9 000	486
W24X94	3 300	1 840	1 740	1 570	3 160	3 110	5 450	Vr 1 440	5 000	946	10 000	422
b=230	2 600	1 750	1 660	1 490	2 490	2 920	5 350	Lu 3 320	6 000	829	12 000	334
t=22.2	1 900	1 660	1 560	1 400	1 820	2 670	5 200	Ix 1 120	7 000	695	14 000	277
d=618	1 200	1 510	1 420	1 300	1 150	2 310	4 960	Sx 3 630	8 000	573	16 000	237
W610X125	4 000	1 740	1 630	1 480	3 830	2 930	4 920	Mr 991	4 000	923	9 000	396
W24X84	3 300	1 660	1 570	1 420	3 160	2 800	4 860	Vr 1 300	5 000	821	10 000	342
b=229	2 600	1 580	1 500	1 350	2 490	2 650	4 780	Lu 3 250	6 000	708	12 000	269
t=19.6	1 900	1 490	1 410	1 260	1 820	2 430	4 650	Ix 985	7 000	575	14 000	222
d=612	1 200	1 360	1 280	1 170	1 150	2 110	4 450	Sx 3 220	8 000	470	16 000	189
W610X113	4 000	1 610	1 510	1 360	3 830	2 670	4 450	Mr 888	4 000	818	9 000	328
W24X76	3 300	1 530	1 440	1 310	3 160	2 560	4 400	Vr 1 210	5 000	719	10 000	282
b=228	2 600	1 450	1 380	1 240	2 490	2 430	4 330	Lu 3 180	6 000	610	12 000	220
t=17.3	1 900	1 370	1 290	1 160	1 820	2 230	4 220	Ix 875	7 000	481	14 000	180
d=608	1 200	1 250	1 170	1 060	1 150	1 940	4 040	Sx 2 880	8 000	391	16 000	153

Note : * 25 MPa, 2 300 kg/m³ Concrete. G40.21-M 300W Steel.

Units : d, b, t, L_u (mm) $I_x = 10^6$ mm⁴ $S_x = 10^3$ mm³ M_r = kN·m V_r = kN

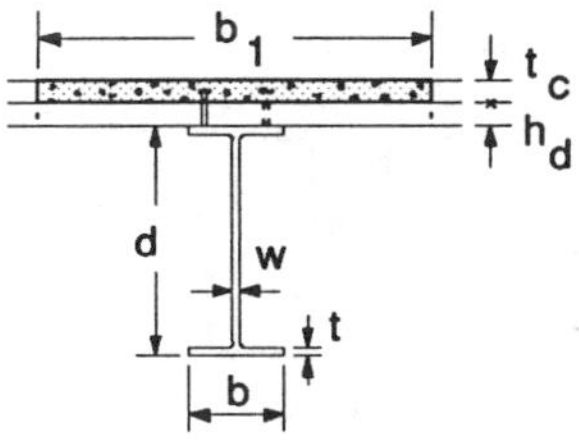

300W
25 MPa

COMPOSITE MEMBERS
Trial Selection Table
75 mm Deck with 75 mm Slab
$\phi = 0.9, \phi_c = 0.6, \phi_{sc} = 0.8$

Steel Shape	b_1	Composite Beam*						Non-Composite Shape					
		Factored Resistances				I_t	S_t	Steel Shape Data		Unbraced Condition			
		M_{rc} (kN·m) for Shear Connections =			Q_r (kN) for	$\times 10^6$	$\times 10^3$			L'	M'r	L'	M'r
	mm	100%	70%	40%	100%	mm^4	mm^3			mm	kN·m	mm	kN·m
W610X101	3 500	1 440	1 340	1 210	3 350	2 350	3 960	Mr	783	4 000	713	9 000	267
W24X68	2 890	1 370	1 290	1 160	2 760	2 260	3 910	Vr	1 130	5 000	619	10 000	228
b=228	2 280	1 300	1 230	1 100	2 180	2 130	3 850	Lu	3 110	6 000	512	11 000	199
t=14.9	1 660	1 220	1 150	1 020	1 590	1 960	3 750	Ix	764	7 000	396	12 000	176
d=604	1 050	1 110	1 040	938	1 000	1 700	3 580	Sx	2 530	8 000	320	14 000	144
W610X91	3 500	1 300	1 210	1 080	3 130	2 110	3 530	Mr	691	4 000	619	9 000	217
W24X61	2 890	1 250	1 170	1 050	2 760	2 030	3 490	Vr	1 030	5 000	530	10 000	185
b=227	2 280	1 180	1 110	988	2 180	1 930	3 430	Lu	3 020	6 000	421	11 000	161
t=12.7	1 660	1 110	1 040	916	1 590	1 770	3 350	Ix	667	7 000	325	12 000	142
d=598	1 050	1 000	930	834	1 000	1 550	3 210	Sx	2 230	8 000	261	14 000	115
W610X84	3 500	1 200	1 120	998	2 890	1 970	3 260	Mr	637	3 000	636	8 000	232
W24X56	2 890	1 170	1 100	985	2 760	1 900	3 230	Vr	956	4 000	566	9 000	192
b=226	2 280	1 100	1 040	928	2 180	1 800	3 180	Lu	2 990	5 000	482	10 000	163
t=11.7	1 660	1 030	973	858	1 590	1 670	3 110	Ix	613	6 000	377	12 000	125
d=596	1 050	939	872	779	1 000	1 460	2 980	Sx	2 060	7 000	289	14 000	101
W530X123	3 500	1 530	1 430	1 280	3 350	2 320	4 390	Mr	867	4 000	794	9 000	361
W21X83	2 890	1 450	1 370	1 230	2 760	2 210	4 340	Vr	1 270	5 000	706	10 000	316
b=212	2 280	1 380	1 300	1 160	2 180	2 080	4 250	Lu	3 100	6 000	613	11 000	281
t=21.2	1 660	1 300	1 220	1 090	1 590	1 890	4 130	Ix	761	7 000	505	12 000	253
d=544	1 050	1 180	1 100	1 010	1 000	1 630	3 930	Sx	2 800	8 000	421	14 000	211
W530X109	3 500	1 390	1 290	1 160	3 350	2 090	3 910	Mr	764	4 000	692	9 000	291
W21X73	2 890	1 320	1 240	1 110	2 760	2 000	3 860	Vr	1 120	5 000	608	10 000	254
b=211	2 280	1 250	1 180	1 050	2 180	1 880	3 790	Lu	3 040	6 000	517	11 000	225
t=18.8	1 660	1 170	1 100	985	1 590	1 720	3 690	Ix	667	7 000	413	12 000	202
d=540	1 050	1 070	997	908	1 000	1 490	3 510	Sx	2 480	8 000	342	14 000	168
W530X101	3 500	1 310	1 220	1 100	3 350	1 960	3 640	Mr	707	3 000	707	8 000	301
W21X68	2 890	1 250	1 170	1 050	2 760	1 880	3 590	Vr	1 050	4 000	635	9 000	255
b=210	2 280	1 170	1 110	992	2 180	1 770	3 530	Lu	2 990	5 000	553	10 000	222
t=17.4	1 660	1 100	1 040	925	1 590	1 630	3 440	Ix	617	6 000	462	12 000	176
d=538	1 050	1 000	937	850	1 000	1 410	3 280	Sx	2 300	7 000	365	14 000	146
W530X92	3 500	1 210	1 130	1 010	3 190	1 800	3 320	Mr	637	3 000	633	8 000	253
W21X62	2 890	1 160	1 080	972	2 760	1 730	3 280	Vr	971	4 000	565	9 000	214
b=209	2 280	1 090	1 030	918	2 180	1 640	3 230	Lu	2 930	5 000	486	10 000	185
t=15.6	1 660	1 020	961	854	1 590	1 500	3 150	Ix	552	6 000	393	12 000	146
d=534	1 050	927	865	780	1 000	1 310	3 010	Sx	2 070	7 000	309	14 000	120
W530X82	3 500	1 080	999	888	2 840	1 610	2 940	Mr	559	3 000	551	8 000	204
W21X55	2 890	1 060	984	880	2 760	1 550	2 910	Vr	894	4 000	487	9 000	172
b=209	2 280	992	930	829	2 180	1 470	2 870	Lu	2 860	5 000	412	10 000	148
t=13.3	1 660	920	868	767	1 590	1 360	2 800	Ix	479	6 000	321	12 000	115
d=528	1 050	836	778	696	1 000	1 180	2 680	Sx	1 810	7 000	251	14 000	94.8

Note : * 25 MPa, 2 300 kg/m³ Concrete. G40.21-M 300W Steel.

Units : d, b, t, L_u (mm) $I_x = 10^6$ mm⁴ $S_x = 10^3$ mm³ M_r = kN·m V_r = kN

COMPOSITE MEMBERS
Trial Selection Table

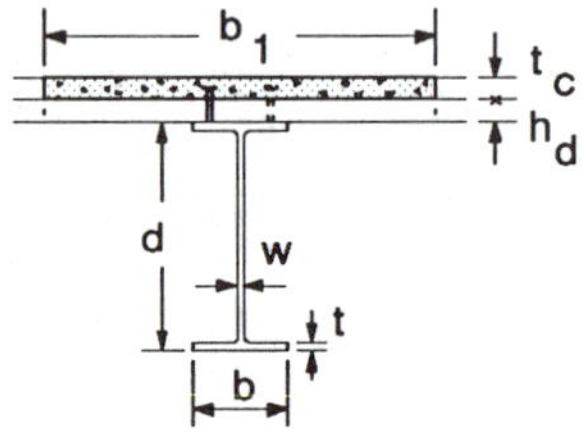

75 mm Deck with 75 mm Slab
$\phi = 0.9$, $\phi_c = 0.6$, $\phi_{sc} = 0.8$

300W
25 MPa

Steel Shape	b_1 mm	Composite Beam* Factored Resistances M_{rc} (kN·m) for Shear Connections = 100%	70%	40%	Q_r (kN) for 100%	I_t x10⁶ mm⁴	S_t x10³ mm³	Steel Shape Data	Non-Composite Shape Unbraced Condition L' mm	M'r kN·m	L' mm	M'r kN·m
W530X72	3 500	950	872	769	2 470	1 410	2 560	Mr 475	3 000	462	8 000	155
W21X48	2 890	936	865	766	2 470	1 360	2 530	Vr 831	4 000	402	9 000	129
b=207	2 280	893	833	740	2 180	1 300	2 490	Lu 2 750	5 000	331	10 000	111
t=10.9	1 660	823	775	680	1 590	1 200	2 440	Ix 402	6 000	247	12 000	85.9
d=524	1 050	745	691	611	1 000	1 050	2 340	Sx 1 530	7 000	191	14 000	70.1
W460X106	3 500	1 230	1 140	1 010	3 350	1 640	3 430	Mr 645	3 000	640	8 000	308
W18X71	2 890	1 160	1 080	965	2 760	1 570	3 390	Vr 1 060	4 000	579	9 000	266
b=194	2 280	1 090	1 020	912	2 180	1 480	3 330	Lu 2 910	5 000	512	10 000	235
t=20.6	1 660	1 010	954	849	1 590	1 350	3 230	Ix 488	6 000	444	12 000	190
d=470	1 050	920	860	780	1 000	1 160	3 070	Sx 2 080	7 000	366	14 000	160
W460X97	3 500	1 150	1 060	934	3 320	1 510	3 150	Mr 589	3 000	581	8 000	264
W18X65	2 890	1 080	1 000	895	2 760	1 450	3 110	Vr 947	4 000	522	9 000	227
b=193	2 280	1 010	947	846	2 180	1 370	3 050	Lu 2 870	5 000	457	10 000	200
t=19	1 660	938	884	787	1 590	1 250	2 970	Ix 445	6 000	389	12 000	161
d=466	1 050	852	797	720	1 000	1 080	2 830	Sx 1 910	7 000	314	14 000	135
W460X89	3 500	1 070	981	865	3 080	1 410	2 920	Mr 543	3 000	534	8 000	231
W18X60	2 890	1 020	947	843	2 760	1 360	2 890	Vr 868	4 000	477	9 000	198
b=192	2 280	955	891	797	2 180	1 290	2 840	Lu 2 830	5 000	414	10 000	174
t=17.7	1 660	882	832	740	1 590	1 180	2 770	Ix 410	6 000	343	12 000	140
d=464	1 050	802	750	675	1 000	1 020	2 640	Sx 1 770	7 000	276	14 000	117
W460X82	3 500	979	894	786	2 810	1 300	2 670	Mr 494	3 000	482	8 000	195
W18X55	2 890	957	882	781	2 760	1 250	2 640	Vr 812	4 000	427	9 000	167
b=191	2 280	889	827	738	2 180	1 190	2 600	Lu 2 770	5 000	365	10 000	146
t=16	1 660	817	769	684	1 590	1 090	2 540	Ix 370	6 000	292	12 000	117
d=460	1 050	741	693	621	1 000	952	2 430	Sx 1 610	7 000	234	14 000	97.3
W460X74	3 500	894	814	714	2 550	1 190	2 430	Mr 446	3 000	433	8 000	164
W18X50	2 890	879	806	712	2 550	1 150	2 400	Vr 735	4 000	380	9 000	140
b=190	2 280	829	768	685	2 180	1 090	2 370	Lu 2 730	5 000	320	10 000	122
t=14.5	1 660	758	712	634	1 590	1 010	2 310	Ix 333	6 000	249	12 000	96.9
d=458	1 050	685	642	573	1 000	885	2 220	Sx 1 460	7 000	198	14 000	80.6
W460X67	3 000	812	740	650	2 340	1 070	2 210	Mr 405	3 000	390	8 000	140
W18X45	2 480	797	733	648	2 340	1 030	2 180	Vr 688	4 000	339	9 000	119
b=190	1 950	741	688	611	1 860	975	2 150	Lu 2 660	5 000	281	10 000	103
t=12.7	1 430	681	640	565	1 370	901	2 100	Ix 300	6 000	214	11 000	91.4
d=454	900	615	572	509	861	784	2 010	Sx 1 320	7 000	169	12 000	82
W460X61	3 000	728	661	578	2 100	956	1 970	Mr 354	3 000	336	8 000	111
W18X41	2 480	716	655	576	2 100	923	1 940	Vr 650	4 000	288	9 000	93.9
b=189	1 950	681	630	557	1 860	878	1 920	Lu 2 580	5 000	231	10 000	81.3
t=10.8	1 430	623	583	512	1 370	814	1 870	Ix 259	6 000	172	11 000	71.7
d=450	900	559	518	457	861	712	1 800	Sx 1 150	7 000	135	12 000	64.1

Note : * 25 MPa, 2 300 kg/m³ Concrete. G40.21-M 300W Steel.

Units : d, b, t, L_u (mm) $I_x = 10^6$ mm⁴ $S_x = 10^3$ mm³ M_r = kN·m V_r = kN

300W
25 MPa

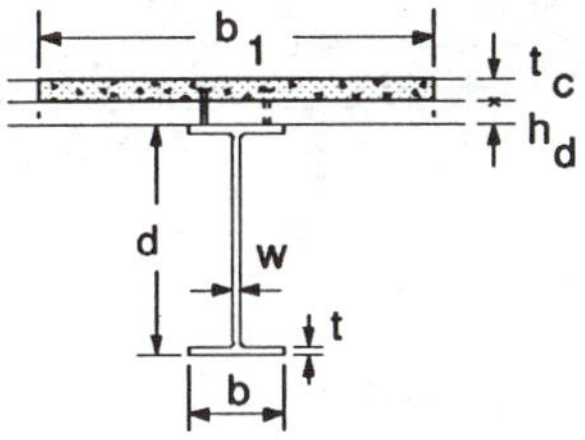

75 mm Deck with 75 mm Slab

$\phi = 0.9, \phi_c = 0.6, \phi_{sc} = 0.8$

Steel Shape	b_1 mm	Composite Beam* Factored Resistances M_{rc} (kN·m) for Shear Connections = 100%	70%	40%	Q_r (kN) for 100%	I_t x10^6 mm^4	S_t x10^3 mm^3	Steel Shape Data	Non-Composite Shape Unbraced Condition L' mm	M'r kN·m	L' mm	M'r kN·m
W410X85	3 500	952	865	754	2 920	1 160	2 590	Mr 467	3 000	455	8 000	205
W16X57	2 890	920	844	742	2 760	1 110	2 560	Vr 812	4 000	406	9 000	178
b=181	2 280	852	788	700	2 180	1 050	2 510	Lu 2 730	5 000	354	10 000	157
t=18.2	1 660	779	730	648	1 590	967	2 450	Ix 315	6 000	297	12 000	127
d=418	1 050	702	656	588	1 000	837	2 340	Sx 1 510	7 000	243	14 000	107
W410X74	3 500	846	765	664	2 580	1 040	2 290	Mr 408	3 000	394	8 000	163
W16X50	2 890	830	757	662	2 580	998	2 270	Vr 716	4 000	348	9 000	140
b=180	2 280	778	717	634	2 180	947	2 230	Lu 2 670	5 000	297	10 000	124
t=16	1 660	707	660	586	1 590	874	2 180	Ix 275	6 000	239	12 000	99.8
d=414	1 050	634	593	529	1 000	761	2 090	Sx 1 330	7 000	194	14 000	83.8
W410X67	3 500	764	688	596	2 320	940	2 070	Mr 367	3 000	352	8 000	135
W16X45	2 890	751	681	594	2 320	908	2 050	Vr 643	4 000	307	9 000	116
b=179	2 280	721	662	583	2 180	864	2 020	Lu 2 610	5 000	258	10 000	102
t=14.4	1 660	652	606	538	1 590	801	1 970	Ix 246	6 000	201	12 000	81.7
d=410	1 050	580	544	483	1 000	702	1 890	Sx 1 200	7 000	161	14 000	68.4
W410X60	3 000	670	604	524	2 050	820	1 820	Mr 321	3 000	306	8 000	109
W16X40	2 480	658	598	522	2 050	791	1 800	Vr 560	4 000	264	9 000	93.2
b=178	1 950	627	576	509	1 860	752	1 770	Lu 2 580	5 000	217	10 000	81.4
t=12.8	1 430	569	530	471	1 370	698	1 730	Ix 216	6 000	165	11 000	72.2
d=408	900	507	476	423	861	610	1 660	Sx 1 060	7 000	131	12 000	65
W410X54	3 000	603	541	468	1 840	735	1 620	Mr 284	3 000	266	8 000	86.1
W16X36	2 480	594	537	466	1 840	711	1 610	Vr 540	4 000	225	9 000	73.2
b=177	1 950	579	530	464	1 840	678	1 580	Lu 2 480	5 000	176	10 000	63.6
t=10.9	1 430	524	486	429	1 370	631	1 550	Ix 186	6 000	132	11 000	56.3
d=404	900	463	434	382	861	554	1 490	Sx 924	7 000	104	12 000	50.5
W410X46	3 000	527	471	404	1 590	646	1 400	Mr 239	2 000	236	7 000	61.7
W16X31	2 480	520	467	403	1 590	626	1 390	Vr 504	3 000	195	8 000	51.8
b=140	1 950	509	462	401	1 590	599	1 370	Lu 1 930	4 000	142	9 000	44.6
t=11.2	1 430	475	437	383	1 370	560	1 340	Ix 156	5 000	99.9	10 000	39.2
d=404	900	415	387	337	861	495	1 290	Sx 773	6 000	76.4	12 000	31.6
W410X39	3 000	448	398	339	1 350	550	1 190	Mr 197	2 000	193	7 000	44.1
W16X26	2 480	443	396	339	1 350	535	1 180	Vr 456	3 000	155	8 000	36.6
b=140	1 950	435	392	337	1 350	514	1 160	Lu 1 860	4 000	105	9 000	31.3
t=8.8	1 430	422	385	335	1 350	483	1 140	Ix 127	5 000	73.1	10 000	27.4
d=400	900	365	339	294	861	431	1 100	Sx 634	6 000	55.2	12 000	21.9
W360X79	3 000	795	718	619	2 730	854	2 210	Mr 386	4 000	364	8 000	225
W14X53	2 480	749	683	596	2 370	818	2 180	Vr 593	4 500	348	9 000	195
b=205	1 950	689	635	564	1 860	770	2 140	Lu 3 270	5 000	331	10 000	172
t=16.8	1 430	628	587	525	1 370	705	2 080	Ix 227	6 000	298	11 000	154
d=354	900	564	530	478	861	605	1 980	Sx 1 280	7 000	264	12 000	139

Note : * 25 MPa, 2 300 kg/m³ Concrete. G40.21-M 300W Steel.

Units : d, b, t, L_u (mm) $I_x = 10^6$ mm⁴ $S_x = 10^3$ mm³ M_r = kN·m V_r = kN

COMPOSITE MEMBERS
Trial Selection Table

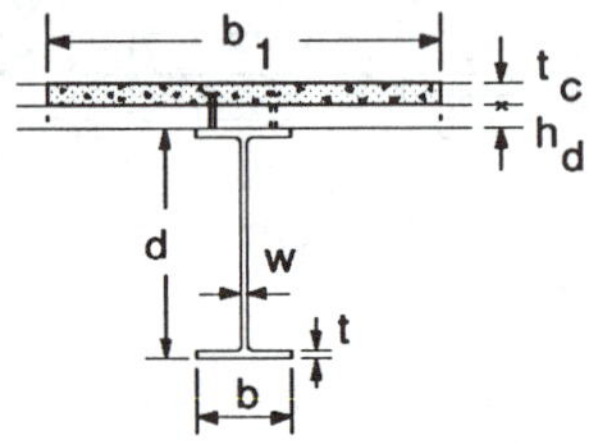

75 mm Deck with 75 mm Slab
$\phi = 0.9$, $\phi_c = 0.6$, $\phi_{sc} = 0.8$

300W
25 MPa

Steel Shape	b_1 mm	Composite Beam* Factored Resistances M_{rc} (kN·m) for Shear Connections = 100%	70%	40%	Q_r (kN) for 100%	I_t x10⁶ mm⁴	S_t x10³ mm³	Steel Shape Data	Non-Composite Shape Unbraced Condition L' mm	M'r kN·m	L' mm	M'r kN·m
W360X72	3 000	720	647	555	2 460	775	1 990	Mr 346	4 000	322	8 000	186
W14X48	2 480	697	633	548	2 370	744	1 970	Vr 536	4 500	306	9 000	161
b=204	1 950	639	586	518	1 860	703	1 930	Lu 3 190	5 000	290	10 000	141
t=15.1	1 430	579	539	481	1 370	646	1 890	Ix 201	6 000	257	11 000	126
d=350	900	516	485	436	861	557	1 800	Sx 1 150	7 000	222	12 000	114
W360X64	3 000	649	580	496	2 200	702	1 790	Mr 308	4 000	283	8 000	153
W14X43	2 480	636	574	494	2 200	676	1 760	Vr 478	4 500	268	9 000	131
b=203	1 950	592	540	473	1 860	641	1 740	Lu 3 110	5 000	252	10 000	115
t=13.5	1 430	533	494	440	1 370	591	1 700	Ix 178	6 000	220	11 000	102
d=348	900	471	444	397	861	513	1 620	Sx 1 030	7 000	183	12 000	92.4
W360X57	2 500	582	523	449	1 950	635	1 590	Mr 273	3 000	259	6 000	147
W14X38	2 060	569	516	447	1 950	610	1 570	Vr 504	3 500	243	7 000	119
b=172	1 630	523	480	422	1 560	577	1 550	Lu 2 550	4 000	225	8 000	99.8
t=13.1	1 190	473	440	390	1 140	531	1 510	Ix 161	4 500	207	9 000	86
d=358	750	421	394	351	717	459	1 440	Sx 897	5 000	189	10 000	75.7
W360X51	2 500	524	468	401	1 740	573	1 420	Mr 241	3 000	227	6 000	121
W14X34	2 060	513	463	399	1 740	552	1 410	Vr 457	3 500	212	7 000	97
b=171	1 630	485	443	387	1 560	524	1 380	Lu 2 500	4 000	195	8 000	81
t=11.6	1 190	436	404	357	1 140	484	1 350	Ix 141	4 500	178	9 000	69.5
d=356	750	385	360	319	717	421	1 290	Sx 796	5 000	159	10 000	60.9
W360X45	2 500	467	415	354	1 550	512	1 260	Mr 210	3 000	195	6 000	96.1
W14X30	2 060	459	411	353	1 550	494	1 250	Vr 433	3 500	181	7 000	76.5
b=171	1 630	447	405	351	1 550	471	1 230	Lu 2 430	4 000	165	8 000	63.4
t=9.8	1 190	399	368	323	1 140	436	1 200	Ix 122	4 500	148	9 000	54.1
d=352	750	349	326	286	717	382	1 150	Sx 691	5 000	128	10 000	47.2
W360X39	2 500	411	364	309	1 340	453	1 100	Mr 179	2 000	173	6 000	54.2
W14X26	2 060	405	361	308	1 340	439	1 080	Vr 410	2 500	157	7 000	44.3
b=128	1 630	396	357	306	1 340	419	1 070	Lu 1 790	3 000	139	8 000	37.5
t=10.7	1 190	366	334	291	1 140	391	1 050	Ix 102	4 000	97.2	9 000	32.5
d=354	750	316	294	255	717	344	1 000	Sx 580	5 000	69.8	10 000	28.8
W360X33	2 500	346	305	257	1 130	383	919	Mr 146	2 000	139	6 000	38.1
W14X22	2 060	342	303	256	1 130	372	909	Vr 362	2 500	124	7 000	30.8
b=127	1 630	335	299	255	1 130	357	897	Lu 1 720	3 000	108	8 000	25.9
t=8.5	1 190	324	294	253	1 130	335	878	Ix 82.7	4 000	70.3	9 000	22.3
d=350	750	276	256	221	717	298	847	Sx 474	5 000	49.7	10 000	19.6
W310X86	3 500	807	721	611	2 970	795	2 250	Mr 383	5 000	367	10 000	248
W12X58	2 890	771	696	597	2 760	762	2 230	Vr 503	6 000	344	11 000	221
b=254	2 280	703	642	562	2 180	719	2 190	Lu 4 250	7 000	320	12 000	200
t=16.3	1 660	632	585	525	1 590	657	2 130	Ix 199	8 000	297	13 000	182
d=310	1 050	559	528	480	1 000	565	2 030	Sx 1 280	9 000	273	14 000	167

Note : * 25 MPa, 2 300 kg/m³ Concrete. G40.21-M 300W Steel.

Units : d, b, t, L_u (mm) $I_x = 10^6$ mm⁴ $S_x = 10^3$ mm³ M_r = kN·m V_r = kN

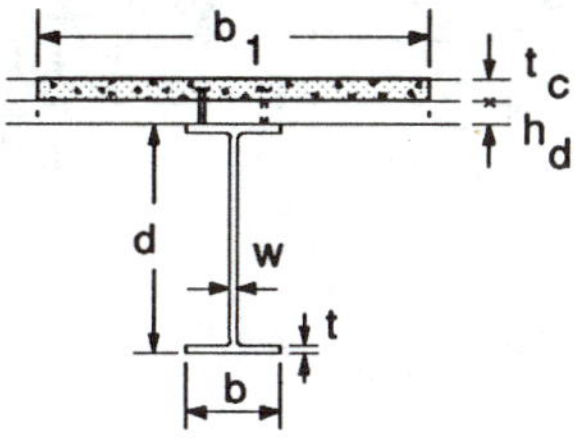

COMPOSITE MEMBERS
Trial Selection Table

300W
25 MPa

75 mm Deck with 75 mm Slab
$\phi = 0.9, \phi_c = 0.6, \phi_{sc} = 0.8$

Steel Shape	b_1 mm	Composite Beam*: Factored Resistances: M_{rc} (kN·m) for Shear Connections = 100%	70%	40%	Q_r (kN) for 100%	I_t x10⁶ mm⁴	S_t x10³ mm³	Steel Shape Data	Non-Composite Shape: Unbraced Condition: L' mm	M'r kN·m	L' mm	M'r kN·m
W310X79	3 000	729	653	556	2 730	706	2 040	Mr 346	5 000	327	8 000	258
W12X53	2 480	684	619	534	2 370	676	2 020	Vr 480	5 500	316	9 000	235
b=254	1 950	624	571	504	1 860	635	1 980	Lu 4 140	6 000	305	10 000	207
t=14.6	1 430	564	524	471	1 370	580	1 930	Ix 177	6 500	293	11 000	184
d=306	900	501	474	428	861	496	1 830	Sx 1 160	7 000	282	12 000	166
W310X74	3 000	696	621	526	2 560	678	1 930	Mr 321	4 000	307	8 000	206
W12X50	2 480	664	599	513	2 370	650	1 900	Vr 519	4 500	295	9 000	179
b=205	1 950	605	552	483	1 860	612	1 870	Lu 3 380	5 000	282	10 000	159
t=16.3	1 430	545	504	448	1 370	560	1 820	Ix 165	6 000	258	11 000	142
d=310	900	481	452	405	861	479	1 730	Sx 1 060	7 000	233	12 000	129
W310X67	3 000	627	557	470	2 300	612	1 730	Mr 286	4 000	270	8 000	169
W12X45	2 480	613	550	467	2 300	588	1 710	Vr 463	4 500	258	9 000	147
b=204	1 950	560	509	441	1 860	556	1 680	Lu 3 280	5 000	246	10 000	129
t=14.6	1 430	501	462	409	1 370	510	1 640	Ix 145	6 000	222	11 000	116
d=306	900	439	412	368	861	439	1 560	Sx 949	7 000	198	12 000	105
W310X60	2 500	553	493	417	2 050	535	1 530	Mr 254	4 000	237	6 500	179
W12X40	2 060	533	480	411	1 970	513	1 510	Vr 406	4 500	226	7 000	166
b=203	1 630	486	442	386	1 560	484	1 490	Lu 3 200	5 000	214	8 000	139
t=13.1	1 190	436	403	358	1 140	443	1 450	Ix 129	5 500	203	9 000	120
d=304	750	384	361	323	717	380	1 380	Sx 849	6 000	191	10 000	106
W310X52	2 500	506	449	380	1 800	507	1 380	Mr 226	3 000	216	6 000	130
W12X35	2 060	495	444	378	1 800	487	1 360	Vr 431	3 500	203	7 000	106
b=167	1 630	461	419	363	1 560	462	1 340	Lu 2 570	4 000	189	8 000	89.4
t=13.2	1 190	412	379	335	1 140	425	1 300	Ix 118	4 500	176	9 000	77.4
d=318	750	361	338	300	717	367	1 240	Sx 747	5 000	162	10 000	68.4
W310X45	2 500	435	383	323	1 540	439	1 180	Mr 191	3 000	180	6 000	98.2
W12X30	2 060	427	379	321	1 540	423	1 170	Vr 369	3 500	168	7 000	79.3
b=166	1 630	415	374	320	1 540	403	1 150	Lu 2 490	4 000	155	8 000	66.5
t=11.2	1 190	368	337	295	1 140	373	1 120	Ix 99.2	4 500	142	9 000	57.3
d=314	750	318	297	263	717	325	1 080	Sx 634	5 000	128	10 000	50.4
W310X39	2 500	379	332	279	1 330	385	1 030	Mr 165	3 000	153	6 000	77.7
W12X26	2 060	373	329	278	1 330	372	1 020	Vr 320	3 500	142	7 000	62.2
b=165	1 630	364	325	276	1 330	356	1 000	Lu 2 440	4 000	130	8 000	51.8
t=9.7	1 190	335	304	264	1 140	331	981	Ix 85.1	4 500	117	9 000	44.3
d=310	750	285	265	234	717	291	944	Sx 549	5 000	103	10 000	38.8
W250X80	3 000	666	591	492	2 750	565	1 870	Mr 294	5 000	287	8 000	242
W10X54	2 480	619	554	469	2 370	540	1 850	Vr 429	5 500	280	9 000	227
b=255	1 950	559	506	439	1 860	506	1 810	Lu 4 520	6 000	272	10 000	212
t=15.6	1 430	499	459	408	1 370	459	1 760	Ix 126	6 500	265	11 000	197
d=256	900	436	410	371	861	389	1 660	Sx 982	7 000	257	12 000	179

Note : * 25 MPa, 2 300 kg/m³ Concrete. G40.21-M 300W Steel.

Units : d, b, t, L_u (mm) $I_x = 10^6$ mm⁴ $S_x = 10^3$ mm³ M_r = kN·m V_r = kN

COMPOSITE MEMBERS
Trial Selection Table

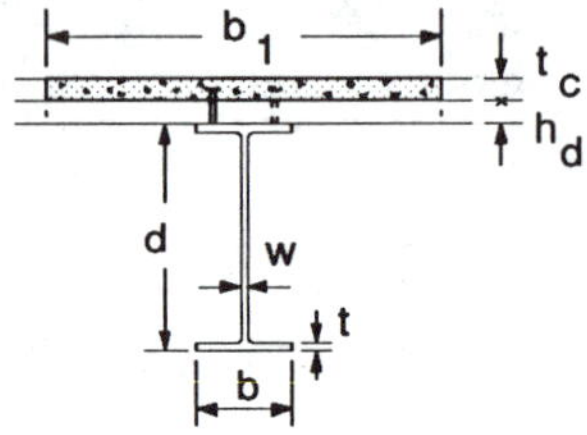

75 mm Deck with 75 mm Slab
$\phi = 0.9$, $\phi_c = 0.6$, $\phi_{sc} = 0.8$

300W
25 MPa

Steel Shape	b_1 mm	Composite Beam*: Factored Resistances: M_{rc} (kN·m) for Shear Connections = 100%	70%	40%	Q_r (kN) for 100%	I_t x10⁶ mm⁴	S_t x10³ mm³	Steel Shape Data	Non-Composite Shape: Unbraced Condition: L' mm	M'r kN·m	L' mm	M'r kN·m
W250X73	3 000	612	539	447	2 510	520	1 710	Mr 266	5 000	257	8 000	212
W10X49	2 480	585	521	437	2 370	498	1 680	Vr 389	5 500	250	9 000	198
b=254	1 950	526	474	408	1 860	468	1 650	Lu 4 390	6 000	242	10 000	183
t=14.2	1 430	467	428	378	1 370	427	1 610	Ix 113	6 500	235	11 000	165
d=254	900	405	380	343	861	363	1 530	Sx 891	7 000	227	12 000	149
W250X67	3 000	574	504	416	2 310	494	1 580	Mr 243	4 000	237	8 000	174
W10X45	2 480	560	497	414	2 310	474	1 560	Vr 409	4 500	229	9 000	157
b=204	1 950	507	455	387	1 860	446	1 530	Lu 3 570	5 000	221	10 000	139
t=15.7	1 430	448	408	357	1 370	408	1 490	Ix 104	6 000	205	11 000	125
d=258	900	385	359	321	861	349	1 420	Sx 806	7 000	189	12 000	114
W250X58	2 500	490	430	356	2 000	414	1 360	Mr 208	4 000	199	6 500	160
W10X39	2 060	474	421	352	1 970	396	1 340	Vr 359	4 500	191	7 000	153
b=203	1 630	427	384	328	1 560	373	1 320	Lu 3 410	5 000	184	8 000	137
t=13.5	1 190	377	344	302	1 140	340	1 280	Ix 87.3	5 500	176	9 000	119
d=252	750	326	304	271	717	290	1 210	Sx 693	6 000	168	10 000	105
W250X49	2 500	418	363	299	1 690	354	1 150	Mr 171	4 000	160	6 500	123
W10X33	2 060	408	359	297	1 690	340	1 130	Vr 327	4 500	153	7 000	115
b=202	1 630	385	343	288	1 560	322	1 110	Lu 3 240	5 000	146	8 000	97.2
t=11	1 190	336	304	264	1 140	295	1 080	Ix 70.6	5 500	138	9 000	84
d=248	750	286	265	234	717	254	1 030	Sx 572	6 000	130	10 000	74.1
W250X45	2 500	400	348	287	1 540	356	1 080	Mr 163	3 000	151	6 000	90.6
W10X30	2 060	392	344	285	1 540	342	1 070	Vr 360	3 500	142	7 000	75.2
b=148	1 630	380	338	284	1 540	325	1 050	Lu 2 360	4 000	132	8 000	64.4
t=13	1 190	332	300	259	1 140	300	1 030	Ix 71.1	4 500	122	9 000	56.3
d=266	750	282	261	229	717	260	980	Sx 534	5 000	112	10 000	50.1
W250X39	2 500	346	299	245	1 330	309	936	Mr 139	3 000	126	6 000	69.2
W10X26	2 060	340	296	244	1 330	299	925	Vr 308	3 500	117	7 000	57
b=147	1 630	331	292	243	1 330	285	910	Lu 2 280	4 000	108	8 000	48.6
t=11.2	1 190	302	271	231	1 140	264	889	Ix 60.1	4 500	98.6	9 000	42.3
d=262	750	252	232	203	717	231	852	Sx 459	5 000	88	10 000	37.6
W250X33	2 000	289	251	206	1 130	254	783	Mr 114	3 000	102	5 500	55.6
W10X22	1 650	284	248	205	1 130	245	773	Vr 280	3 500	93.1	6 000	49.4
b=146	1 300	276	244	204	1 130	233	760	Lu 2 180	4 000	84	6 500	44.4
t=9.1	950	247	222	190	908	216	742	Ix 48.9	4 500	74.1	7 000	40.3
d=258	600	208	192	166	574	188	710	Sx 379	5 000	63.6	8 000	34.1
W200X71	3 000	556	483	391	2 460	419	1 540	Mr 217	5 000	208	8 000	178
W8X48	2 480	532	468	383	2 370	400	1 520	Vr 393	5 500	203	9 000	168
b=206	1 950	474	421	353	1 860	375	1 490	Lu 4 150	6 000	198	10 000	158
t=17.4	1 430	414	374	323	1 370	341	1 440	Ix 76.6	6 500	193	11 000	148
d=216	900	351	325	289	861	288	1 360	Sx 709	7 000	188	12 000	137

Note : * 25 MPa, 2 300 kg/m³ Concrete. G40.21-M 300W Steel.

Units : d, b, t, L_u (mm) $I_x = 10^6$ mm⁴ $S_x = 10^3$ mm³ M_r = kN·m V_r = kN

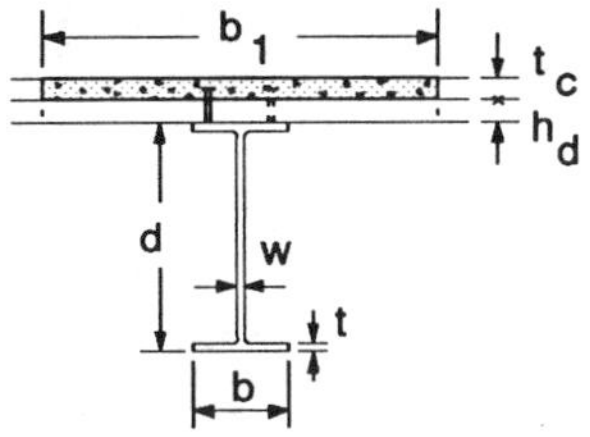

300W
25 MPa

COMPOSITE MEMBERS
Trial Selection Table
75 mm Deck with 75 mm Slab
$\phi = 0.9, \phi_c = 0.6, \phi_{sc} = 0.8$

Steel Shape	b_1 mm	Composite Beam*						Steel Shape Data	Non-Composite Shape: Unbraced Condition			
		Factored Resistances: M_{rc} (kN·m) for Shear Connections =			Q_r (kN) for	I_t x10⁶	S_t x10³		L' mm	M'r kN·m	L' mm	M'r kN·m
		100%	70%	40%	100%	mm⁴	mm³					
W200X59	3 000	466	400	321	2 040	352	1 280	Mr 176	4 000	174	8 000	134
W8X40	2 480	455	395	319	2 040	338	1 270	Vr 341	4 500	169	9 000	124
b=205	1 950	424	373	307	1 860	318	1 240	Lu 3 780	5 000	164	10 000	114
t=14.2	1 430	366	327	278	1 370	291	1 210	Ix 61.1	6 000	154	11 000	103
d=210	900	305	280	246	861	249	1 150	Sx 582	7 000	144	12 000	93.4
W200X52	2 500	404	348	280	1 800	301	1 120	Mr 154	4 000	150	6 500	126
W8X35	2 060	393	343	278	1 800	288	1 100	Vr 290	4 500	145	7 000	121
b=204	1 630	360	318	263	1 560	271	1 080	Lu 3 620	5 000	140	8 000	111
t=12.6	1 190	311	279	238	1 140	247	1 050	Ix 52.7	5 500	135	9 000	101
d=206	750	261	240	212	717	211	999	Sx 512	6 000	131	10 000	89.7
W200X46	2 500	359	307	246	1 580	269	991	Mr 134	4 000	129	6 500	105
W8X31	2 060	351	303	245	1 580	258	977	Vr 262	4 500	124	7 000	99.8
b=203	1 630	337	295	241	1 560	244	959	Lu 3 460	5 000	119	8 000	90.1
t=11	1 190	289	257	217	1 140	224	933	Ix 45.5	5 500	114	9 000	78.6
d=204	750	239	218	191	717	192	889	Sx 448	6 000	109	10 000	69.6
W200X42	2 000	322	277	223	1 430	239	889	Mr 120	3 000	119	5 500	93.5
W8X28	1 650	314	273	222	1 430	229	876	Vr 264	3 500	114	6 000	88.4
b=166	1 300	287	254	210	1 240	215	859	Lu 2 850	4 000	109	6 500	83.4
t=11.8	950	248	223	190	908	196	834	Ix 40.9	4 500	104	7 000	77.7
d=206	600	208	191	168	574	167	791	Sx 399	5 000	98.6	8 000	66.7
W200X36	2 000	280	239	191	1 240	209	771	Mr 103	3 000	100	5 500	75.5
W8X24	1 650	274	236	190	1 240	200	760	Vr 223	3 500	95.3	6 000	70.6
b=165	1 300	264	231	189	1 240	189	746	Lu 2 730	4 000	90.4	6 500	64.7
t=10.2	950	226	202	170	908	174	726	Ix 34.4	4 500	85.5	7 000	59.2
d=202	600	187	171	149	574	149	691	Sx 342	5 000	80.5	8 000	50.6
W200X31	2 000	253	215	172	1 080	194	687	Mr 90.5	3 000	81.1	5 500	50.6
W8X21	1 650	248	212	171	1 080	187	678	Vr 240	3 500	75.3	6 000	45.6
b=134	1 300	240	209	170	1 080	177	665	Lu 2 150	4 000	69.5	6 500	41.5
t=10.2	950	215	191	159	908	164	648	Ix 31.4	4 500	63.6	7 000	38
d=210	600	176	160	138	574	142	619	Sx 299	5 000	57	8 000	32.7
W200X27	2 000	216	183	145	915	168	586	Mr 75.3	3 000	65.4	5 500	36.4
W8X18	1 650	213	181	145	915	162	578	Vr 215	3 500	59.7	6 000	32.6
b=133	1 300	207	178	144	915	154	568	Lu 2 050	4 000	54	6 500	29.6
t=8.4	950	197	173	142	908	143	554	Ix 25.8	4 500	47.5	7 000	27.1
d=208	600	159	143	122	574	125	531	Sx 249	5 000	41.2	8 000	23.1
W150X30	2 500	218	180	137	1 020	146	603	Mr 65.9	3 000	64.1	6 000	46.8
W6X20	2 060	214	178	137	1 020	141	593	Vr 186	3 500	61.2	7 000	40.4
b=153	1 630	209	175	136	1 020	134	581	Lu 2 680	4 000	58.3	8 000	34.8
t=9.3	1 190	200	171	134	1 020	124	566	Ix 17.2	4 500	55.4	9 000	30.6
d=158	750	161	141	116	717	109	542	Sx 219	5 000	52.5	10 000	27.3

Note : * 25 MPa, 2 300 kg/m³ Concrete. G40.21-M 300W Steel.

Units : d, b, t, L_u (mm) $I_x = 10^6$ mm⁴ $S_x = 10^3$ mm³ M_r = kN·m V_r = kN

COMPOSITE MEMBERS
Trial Selection Table

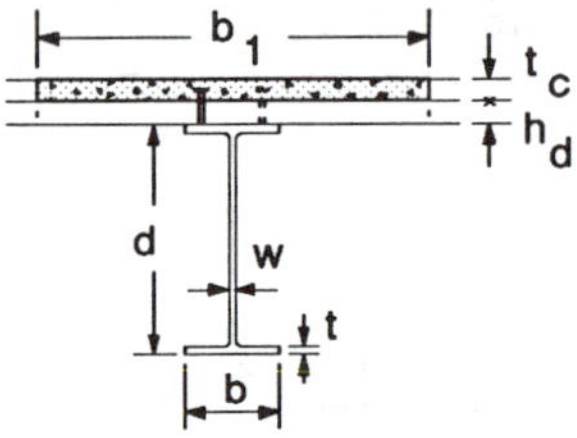

75 mm Deck with 85 mm Slab

$\phi = 0.9$, $\phi_c = 0.6$, $\phi_{sc} = 0.8$

300W
25 MPa

Steel Shape	b_1 mm	Composite Beam* – Factored Resistances – M_{rc} (kN·m) for Shear Connections = 100%	70%	40%	Q_r (kN) for 100%	I_t x10⁶ mm⁴	S_t x10³ mm³	Steel Shape Data	Non-Composite Shape – Unbraced Condition – L' mm	M'r kN·m	L' mm	M'r kN·m
WWF900X169	5 000	3 260	3 110	2 880	5 420	7 510	9 140	Mr 1 990	5 000	1 860	12 000	625
	4 130	3 140	3 010	2 780	4 480	7 190	9 030	Vr 1 690	6 000	1 680	14 000	491
b=300	3 250	3 020	2 910	2 650	3 520	6 780	8 890	Lu 4 130	7 000	1 480	16 000	402
t=20	2 380	2 890	2 760	2 500	2 580	6 240	8 680	Ix 2930	8 000	1 250	18 000	339
d=900	1 500	2 690	2 530	2 330	1 630	5 460	8 320	Sx 6 510	10 000	845	20 000	293
WWF800X161	5 000	2 850	2 700	2 490	5 420	5 930	7 960	Mr 1 710	5 000	1 610	12 000	581
	4 130	2 740	2 600	2 410	4 480	5 690	7 880	Vr 1 490	6 000	1 470	14 000	460
b=300	3 250	2 620	2 510	2 300	3 520	5 360	7 750	Lu 4 230	7 000	1 310	16 000	379
t=20	2 380	2 490	2 390	2 170	2 580	4 930	7 570	Ix 2250	8 000	1 130	18 000	322
d=800	1 500	2 320	2 190	2 010	1 630	4 310	7 260	Sx 5 610	10 000	778	20 000	280
WWF700X152	4 500	2 410	2 270	2 080	4 880	4 460	6 800	Mr 1 440	5 000	1 370	10 000	711
	3 710	2 300	2 180	2 010	4 020	4 260	6 710	Vr 1 290	6 000	1 260	12 000	537
b=300	2 930	2 190	2 090	1 920	3 180	4 020	6 600	Lu 4 330	7 000	1 140	14 000	429
t=20	2 140	2 080	1 990	1 820	2 320	3 680	6 440	Ix 1 660	8 000	1 000	16 000	356
d=700	1 350	1 940	1 830	1 690	1 460	3 200	6 170	Sx 4 760	9 000	842	18 000	305
W610X174	4 500	2 420	2 270	2 070	4 880	4 000	6 970	Mr 1 450	5 000	1 430	10 000	924
W24X117	3 710	2 310	2 180	2 000	4 020	3 810	6 880	Vr 1 540	6 000	1 350	12 000	709
b=325	2 930	2 200	2 100	1 910	3 180	3 570	6 750	Lu 4 830	7 000	1 250	14 000	574
t=21.6	2 140	2 080	1 980	1 800	2 320	3 250	6 560	Ix 1 470	8 000	1 150	16 000	482
d=616	1 350	1 920	1 820	1 680	1 460	2 810	6 260	Sx 4 780	9 000	1 040	18 000	415
W610X155	4 000	2 130	2 010	1 830	4 340	3 510	6 160	Mr 1 280	5 000	1 260	10 000	762
W24X104	3 300	2 040	1 930	1 770	3 580	3 350	6 080	Vr 1 390	6 000	1 180	11 000	659
b=324	2 600	1 940	1 850	1 690	2 820	3 140	5 970	Lu 4 740	7 000	1 080	12 000	579
t=19	1 900	1 840	1 750	1 590	2 060	2 860	5 800	Ix 1 290	8 000	988	14 000	465
d=612	1 200	1 700	1 610	1 480	1 300	2 470	5 530	Sx 4 220	9 000	886	16 000	388
W610X140	4 000	2 000	1 880	1 690	4 340	3 240	5 550	Mr 1 120	4 000	1 050	9 000	486
W24X94	3 300	1 910	1 800	1 620	3 580	3 090	5 470	Vr 1 440	5 000	946	10 000	422
b=230	2 600	1 810	1 720	1 540	2 820	2 900	5 370	Lu 3 320	6 000	829	12 000	334
t=22.2	1 900	1 700	1 610	1 440	2 060	2 640	5 210	Ix 1 120	7 000	695	14 000	277
d=618	1 200	1 560	1 460	1 330	1 300	2 280	4 950	Sx 3 630	8 000	573	16 000	237
W610X125	4 000	1 820	1 700	1 530	4 290	2 920	4 950	Mr 991	4 000	923	9 000	396
W24X84	3 300	1 730	1 630	1 470	3 580	2 800	4 880	Vr 1 300	5 000	821	10 000	342
b=229	2 600	1 640	1 550	1 390	2 820	2 630	4 800	Lu 3 250	6 000	708	12 000	269
t=19.6	1 900	1 540	1 460	1 300	2 060	2 400	4 660	Ix 985	7 000	575	14 000	222
d=612	1 200	1 410	1 320	1 200	1 300	2 080	4 450	Sx 3 220	8 000	470	16 000	189
W610X113	4 000	1 660	1 540	1 380	3 890	2 670	4 480	Mr 888	4 000	818	9 000	328
W24X76	3 300	1 600	1 500	1 350	3 580	2 560	4 420	Vr 1 210	5 000	719	10 000	282
b=228	2 600	1 510	1 420	1 280	2 820	2 410	4 350	Lu 3 180	6 000	610	12 000	220
t=17.3	1 900	1 410	1 340	1 190	2 060	2 210	4 230	Ix 875	7 000	481	14 000	180
d=608	1 200	1 290	1 210	1 090	1 300	1 920	4 040	Sx 2 880	8 000	391	16 000	153

Note : * 25 MPa, 2 000 kg/m³ Concrete. G40.21-M 300W Steel.

Units : d, b, t, L_u (mm) $I_x = 10^6$ mm⁴ $S_x = 10^3$ mm³ M_r = kN·m V_r = kN

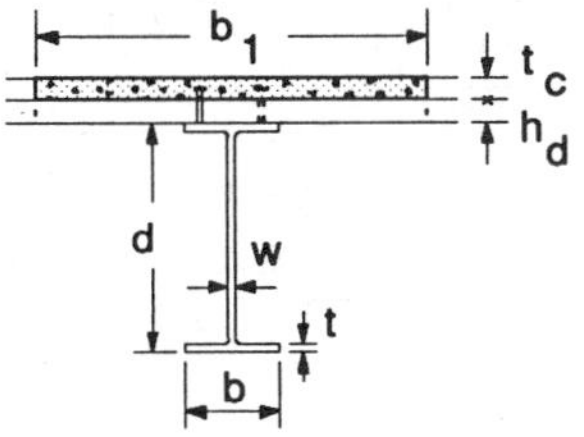

300W
25 MPa

COMPOSITE MEMBERS
Trial Selection Table

75 mm Deck with 85 mm Slab
$\phi = 0.9, \phi_c = 0.6, \phi_{sc} = 0.8$

Steel Shape	b_1 mm	Composite Beam*						Steel Shape Data	Non-Composite Shape			
		Factored Resistances							Unbraced Condition			
		M_{rc} (kN·m) for Shear Connections =			Q_r (kN) for	I_t x10⁶	S_t x10³		L'	M'r	L'	M'r
		100%	70%	40%	100%	mm⁴	mm³		mm	kN·m	mm	kN·m
W610X101	3 500	1 480	1 380	1 240	3 510	2 350	3 990	Mr 783	4 000	713	9 000	267
W24X68	2 890	1 430	1 340	1 200	3 130	2 250	3 940	Vr 1 130	5 000	619	10 000	228
b=228	2 280	1 350	1 270	1 140	2 470	2 120	3 870	Lu 3 110	6 000	512	11 000	199
t=14.9	1 660	1 260	1 190	1 050	1 800	1 940	3 760	Ix 764	7 000	396	12 000	176
d=604	1 050	1 150	1 070	962	1 140	1 680	3 580	Sx 2 530	8 000	320	14 000	144
W610X91	3 500	1 330	1 230	1 100	3 130	2 120	3 550	Mr 691	4 000	619	9 000	217
W24X61	2 890	1 300	1 220	1 090	3 130	2 030	3 510	Vr 1 030	5 000	530	10 000	185
b=227	2 280	1 230	1 150	1 030	2 470	1 920	3 450	Lu 3 020	6 000	421	11 000	161
t=12.7	1 660	1 140	1 080	949	1 800	1 760	3 360	Ix 667	7 000	325	12 000	142
d=598	1 050	1 040	963	858	1 140	1 530	3 210	Sx 2 230	8 000	261	14 000	115
W610X84	3 500	1 230	1 140	1 010	2 890	1 970	3 290	Mr 637	3 000	636	8 000	232
W24X56	2 890	1 210	1 130	1 010	2 890	1 900	3 250	Vr 956	4 000	566	9 000	192
b=226	2 280	1 150	1 080	965	2 470	1 800	3 200	Lu 2 990	5 000	482	10 000	163
t=11.7	1 660	1 070	1 010	890	1 800	1 650	3 120	Ix 613	6 000	377	12 000	125
d=596	1 050	972	903	802	1 140	1 440	2 980	Sx 2 060	7 000	289	14 000	101
W530X123	3 500	1 600	1 490	1 330	3 790	2 320	4 420	Mr 867	4 000	794	9 000	361
W21X83	2 890	1 520	1 420	1 270	3 130	2 210	4 360	Vr 1 270	5 000	706	10 000	316
b=212	2 280	1 430	1 350	1 200	2 470	2 060	4 270	Lu 3 100	6 000	613	11 000	281
t=21.2	1 660	1 340	1 260	1 120	1 800	1 870	4 140	Ix 761	7 000	505	12 000	253
d=544	1 050	1 210	1 140	1 030	1 140	1 610	3 930	Sx 2 800	8 000	421	14 000	211
W530X109	3 500	1 460	1 350	1 210	3 750	2 090	3 940	Mr 764	4 000	692	9 000	291
W21X73	2 890	1 380	1 290	1 160	3 130	2 000	3 880	Vr 1 120	5 000	608	10 000	254
b=211	2 280	1 300	1 220	1 090	2 470	1 870	3 810	Lu 3 040	6 000	517	11 000	225
t=18.8	1 660	1 210	1 140	1 020	1 800	1 710	3 700	Ix 667	7 000	413	12 000	202
d=540	1 050	1 100	1 030	930	1 140	1 470	3 510	Sx 2 480	8 000	342	14 000	168
W530X101	3 500	1 360	1 260	1 120	3 480	1 960	3 670	Mr 707	3 000	707	8 000	301
W21X68	2 890	1 300	1 220	1 090	3 130	1 880	3 620	Vr 1 050	4 000	635	9 000	255
b=210	2 280	1 220	1 150	1 030	2 470	1 770	3 550	Lu 2 990	5 000	553	10 000	222
t=17.4	1 660	1 140	1 070	956	1 800	1 610	3 450	Ix 617	6 000	462	12 000	176
d=538	1 050	1 040	968	872	1 140	1 390	3 290	Sx 2 300	7 000	365	14 000	146
W530X92	3 500	1 250	1 150	1 020	3 190	1 800	3 350	Mr 637	3 000	633	8 000	253
W21X62	2 890	1 220	1 130	1 010	3 130	1 730	3 300	Vr 971	4 000	565	9 000	214
b=209	2 280	1 140	1 070	954	2 470	1 630	3 250	Lu 2 930	5 000	486	10 000	185
t=15.6	1 660	1 050	996	884	1 800	1 490	3 160	Ix 552	6 000	393	12 000	146
d=534	1 050	959	895	801	1 140	1 290	3 010	Sx 2 070	7 000	309	14 000	120
W530X82	3 500	1 110	1 020	899	2 840	1 610	2 970	Mr 559	3 000	551	8 000	204
W21X55	2 890	1 090	1 010	896	2 840	1 550	2 940	Vr 894	4 000	487	9 000	172
b=209	2 280	1 040	968	864	2 470	1 470	2 890	Lu 2 860	5 000	412	10 000	148
t=13.3	1 660	955	901	796	1 800	1 350	2 820	Ix 479	6 000	321	12 000	115
d=528	1 050	866	807	717	1 140	1 170	2 690	Sx 1 810	7 000	251	14 000	94.8

Note : * 25 MPa, 2 000 kg/m³ Concrete. G40.21-M 300W Steel.

Units : d, b, t, L_u (mm) $I_x = 10^6$ mm⁴ $S_x = 10^3$ mm³ M_r = kN·m V_r = kN

COMPOSITE MEMBERS
Trial Selection Table

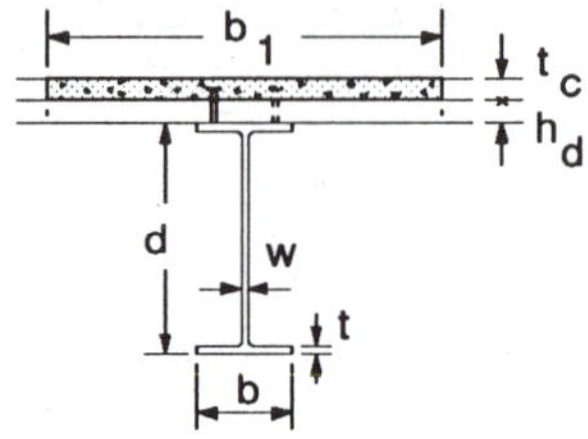

75 mm Deck with 85 mm Slab
$\phi = 0.9$, $\phi_c = 0.6$, $\phi_{sc} = 0.8$

300W
25 MPa

Steel Shape	b_1	Composite Beam*							Non-Composite Shape			
		Factored Resistances				I_t	S_t	Steel Shape Data	Unbraced Condition			
		M_{rc} (kN·m) for Shear Connections =			Q_r (kN) for	x10⁶	x10³		L'	M'r	L'	M'r
	mm	100%	70%	40%	100%	mm⁴	mm³		mm	kN·m	mm	kN·m
W530X72	3 500	975	889	778	2 470	1 420	2 580	Mr 475	3 000	462	8 000	155
W21X48	2 890	961	882	776	2 470	1 370	2 550	Vr 831	4 000	402	9 000	129
b=207	2 280	938	871	772	2 470	1 300	2 510	Lu 2 750	5 000	331	10 000	111
t=10.9	1 660	857	805	709	1 800	1 200	2 450	Ix 402	6 000	247	12 000	85.9
d=524	1 050	772	718	631	1 140	1 050	2 350	Sx 1 530	7 000	191	14 000	70.1
W460X106	3 500	1 290	1 190	1 040	3 650	1 640	3 470	Mr 645	3 000	640	8 000	308
W18X71	2 890	1 220	1 130	1 000	3 130	1 570	3 420	Vr 1 060	4 000	579	9 000	266
b=194	2 280	1 140	1 060	947	2 470	1 470	3 350	Lu 2 910	5 000	512	10 000	235
t=20.6	1 660	1 050	990	878	1 800	1 340	3 250	Ix 488	6 000	444	12 000	190
d=470	1 050	952	889	800	1 140	1 150	3 080	Sx 2 080	7 000	366	14 000	160
W460X97	3 500	1 180	1 080	948	3 320	1 520	3 180	Mr 589	3 000	581	8 000	264
W18X65	2 890	1 140	1 050	932	3 130	1 450	3 130	Vr 947	4 000	522	9 000	227
b=193	2 280	1 060	987	880	2 470	1 370	3 080	Lu 2 870	5 000	457	10 000	200
t=19	1 660	974	917	815	1 800	1 250	2 990	Ix 445	6 000	389	12 000	161
d=466	1 050	881	824	740	1 140	1 070	2 840	Sx 1 910	7 000	314	14 000	135
W460X89	3 500	1 100	1 000	878	3 080	1 420	2 950	Mr 543	3 000	534	8 000	231
W18X60	2 890	1 080	992	874	3 080	1 360	2 920	Vr 868	4 000	477	9 000	198
b=192	2 280	1 000	931	829	2 470	1 290	2 860	Lu 2 830	5 000	414	10 000	174
t=17.7	1 660	918	862	767	1 800	1 170	2 780	Ix 410	6 000	343	12 000	140
d=464	1 050	829	776	694	1 140	1 010	2 650	Sx 1 770	7 000	276	14 000	117
W460X82	3 500	1 010	914	798	2 810	1 310	2 700	Mr 494	3 000	482	8 000	195
W18X55	2 890	988	904	795	2 810	1 260	2 670	Vr 812	4 000	427	9 000	167
b=191	2 280	936	865	769	2 470	1 190	2 620	Lu 2 770	5 000	365	10 000	146
t=16	1 660	852	798	711	1 800	1 090	2 550	Ix 370	6 000	292	12 000	117
d=460	1 050	766	718	640	1 140	943	2 440	Sx 1 610	7 000	234	14 000	97.3
W460X74	3 500	920	831	724	2 550	1 200	2 460	Mr 446	3 000	433	8 000	164
W18X50	2 890	904	824	722	2 550	1 160	2 430	Vr 735	4 000	380	9 000	140
b=190	2 280	875	806	713	2 470	1 100	2 390	Lu 2 730	5 000	320	10 000	122
t=14.5	1 660	793	740	659	1 800	1 010	2 330	Ix 333	6 000	249	12 000	96.9
d=458	1 050	708	666	592	1 140	878	2 230	Sx 1 460	7 000	198	14 000	80.6
W460X67	3 000	835	757	660	2 340	1 080	2 230	Mr 405	3 000	390	8 000	140
W18X45	2 480	820	750	657	2 340	1 030	2 210	Vr 688	4 000	339	9 000	119
b=190	1 950	780	721	638	2 110	977	2 170	Lu 2 660	5 000	281	10 000	103
t=12.7	1 430	711	666	588	1 550	898	2 110	Ix 300	6 000	214	11 000	91.4
d=454	900	637	594	525	975	777	2 020	Sx 1 320	7 000	169	12 000	82
W460X61	3 000	749	676	586	2 100	965	1 990	Mr 354	3 000	336	8 000	111
W18X41	2 480	737	670	584	2 100	929	1 970	Vr 650	4 000	288	9 000	93.9
b=189	1 950	718	661	581	2 100	881	1 930	Lu 2 580	5 000	231	10 000	81.3
t=10.8	1 430	652	608	534	1 550	813	1 890	Ix 259	6 000	172	11 000	71.7
d=450	900	580	540	473	975	706	1 800	Sx 1 150	7 000	135	12 000	64.1

Note : * 25 MPa, 2 000 kg/m³ Concrete. G40.21-M 300W Steel.

Units : d, b, t, L_u (mm) $I_x = 10^6$ mm⁴ $S_x = 10^3$ mm³ M_r = kN·m V_r = kN

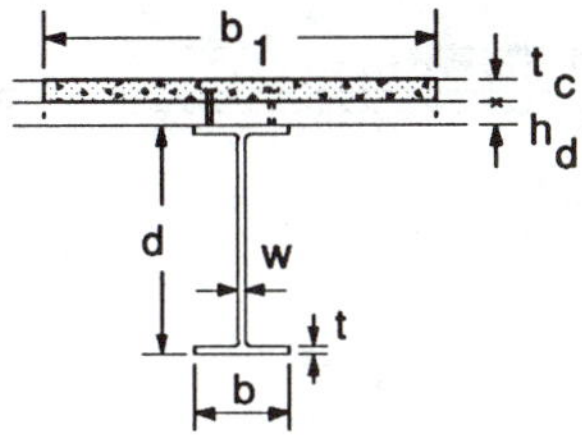

300W
25 MPa

COMPOSITE MEMBERS
Trial Selection Table

75 mm Deck with 85 mm Slab
$\phi = 0.9, \phi_c = 0.6, \phi_{sc} = 0.8$

Steel Shape	b_1 mm	Composite Beam*						Steel Shape Data	Non-Composite Shape			
		Factored Resistances				I_t	S_t		Unbraced Condition			
		M_{rc} (kN·m) for Shear Connections =			Q_r (kN) for	x10⁶	x10³		L'	M'r	L'	M'r
		100%	70%	40%	100%	mm⁴	mm³		mm	kN·m	mm	kN·m
W410X85	3 500	981	885	765	2 920	1 170	2 620	Mr 467	3 000	455	8 000	205
W16X57	2 890	961	876	762	2 920	1 120	2 580	Vr 812	4 000	406	9 000	178
b=181	2 280	899	828	730	2 470	1 060	2 540	Lu 2 730	5 000	354	10 000	157
t=18.2	1 660	814	759	674	1 800	964	2 470	Ix 315	6 000	297	12 000	127
d=418	1 050	727	681	606	1 140	830	2 350	Sx 1 510	7 000	243	14 000	107
W410X74	3 500	872	783	675	2 580	1 050	2 320	Mr 408	3 000	394	8 000	163
W16X50	2 890	856	775	672	2 580	1 010	2 290	Vr 716	4 000	348	9 000	140
b=180	2 280	824	755	662	2 470	951	2 260	Lu 2 670	5 000	297	10 000	124
t=16	1 660	742	689	611	1 800	873	2 200	Ix 275	6 000	239	12 000	99.8
d=414	1 050	657	617	546	1 140	756	2 100	Sx 1 330	7 000	194	14 000	83.8
W410X67	3 500	787	704	605	2 320	951	2 100	Mr 367	3 000	352	8 000	135
W16X45	2 890	774	698	603	2 320	916	2 070	Vr 643	4 000	307	9 000	116
b=179	2 280	755	688	600	2 320	869	2 040	Lu 2 610	5 000	258	10 000	102
t=14.4	1 660	686	634	561	1 800	802	1 990	Ix 246	6 000	201	12 000	81.7
d=410	1 050	602	566	501	1 140	698	1 910	Sx 1 200	7 000	161	14 000	68.4
W410X60	3 000	690	618	532	2 050	829	1 840	Mr 321	3 000	306	8 000	109
W16X40	2 480	679	612	530	2 050	798	1 820	Vr 560	4 000	264	9 000	93.2
b=178	1 950	661	603	527	2 050	757	1 790	Lu 2 580	5 000	217	10 000	81.4
t=12.8	1 430	598	554	491	1 550	698	1 750	Ix 216	6 000	165	11 000	72.2
d=408	900	526	495	438	975	606	1 670	Sx 1 060	7 000	131	12 000	65
W410X54	3 000	621	554	475	1 840	745	1 650	Mr 284	3 000	266	8 000	86.1
W16X36	2 480	612	550	474	1 840	719	1 630	Vr 540	4 000	225	9 000	73.2
b=177	1 950	598	542	471	1 840	683	1 600	Lu 2 480	5 000	176	10 000	63.6
t=10.9	1 430	553	510	449	1 550	632	1 570	Ix 186	6 000	132	11 000	56.3
d=404	900	482	452	397	975	551	1 500	Sx 924	7 000	104	12 000	50.5
W410X46	3 000	543	482	410	1 590	656	1 430	Mr 239	2 000	236	7 000	61.7
W16X31	2 480	536	478	409	1 590	634	1 410	Vr 504	3 000	195	8 000	51.8
b=140	1 950	525	473	407	1 590	605	1 390	Lu 1 930	4 000	142	9 000	44.6
t=11.2	1 430	503	461	402	1 550	563	1 360	Ix 156	5 000	99.9	10 000	39.2
d=404	900	433	404	352	975	494	1 300	Sx 773	6 000	76.4	12 000	31.6
W410X39	3 000	461	408	345	1 350	560	1 210	Mr 197	2 000	193	7 000	44.1
W16X26	2 480	456	405	344	1 350	543	1 190	Vr 456	3 000	155	8 000	36.6
b=140	1 950	449	401	343	1 350	520	1 180	Lu 1 860	4 000	105	9 000	31.3
t=8.8	1 430	435	395	341	1 350	487	1 150	Ix 127	5 000	73.1	10 000	27.4
d=400	900	383	355	307	975	431	1 110	Sx 634	6 000	55.2	12 000	21.9
W360X79	3 000	822	737	630	2 730	862	2 240	Mr 386	4 000	364	8 000	225
W14X53	2 480	798	725	624	2 690	824	2 210	Vr 593	4 500	348	9 000	195
b=205	1 950	729	668	588	2 110	773	2 160	Lu 3 270	5 000	331	10 000	172
t=16.8	1 430	659	612	546	1 550	704	2 100	Ix 227	6 000	298	11 000	154
d=354	900	583	550	492	975	601	1 990	Sx 1 280	7 000	264	12 000	139

Note : * 25 MPa, 2 000 kg/m³ Concrete. G40.21-M 300W Steel.

Units : d, b, t, L_u (mm) $I_x = 10^6$ mm⁴ $S_x = 10^3$ mm³ M_r = kN·m V_r = kN

COMPOSITE MEMBERS
Trial Selection Table

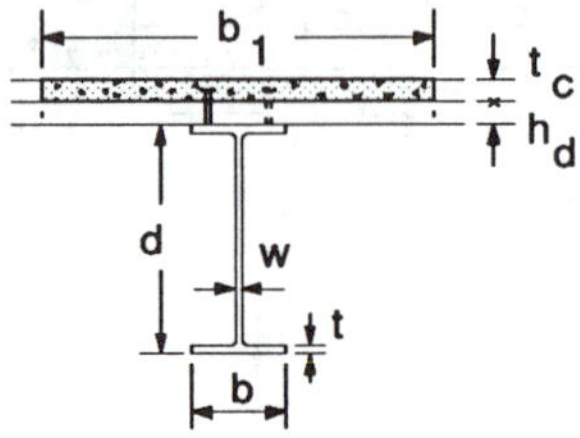

75 mm Deck with 85 mm Slab
$\phi = 0.9, \phi_c = 0.6, \phi_{sc} = 0.8$

300W
25 MPa

Steel Shape	b_1	Composite Beam*							Non-Composite Shape			
		Factored Resistances				I_t	S_t	Steel Shape Data	Unbraced Condition			
		M_{rc} (kN·m) for Shear Connections =			Q_r (kN) for	x10⁶	x10³		L'	M'r	L'	M'r
	mm	100%	70%	40%	100%	mm⁴	mm³		mm	kN·m	mm	kN·m
W360X72	3 000	745	665	565	2 460	784	2 020	Mr 346	4 000	322	8 000	186
W14X48	2 480	728	657	563	2 460	751	1 990	Vr 536	4 500	306	9 000	161
b=204	1 950	678	619	539	2 110	707	1 960	Lu 3 190	5 000	290	10 000	141
t=15.1	1 430	609	563	501	1 550	646	1 900	Ix 201	6 000	257	11 000	126
d=350	900	535	504	450	975	554	1 810	Sx 1 150	7 000	222	12 000	114
W360X64	3 000	671	596	505	2 200	712	1 810	Mr 308	4 000	283	8 000	153
W14X43	2 480	658	589	503	2 200	684	1 790	Vr 478	4 500	268	9 000	131
b=203	1 950	631	573	495	2 110	645	1 760	Lu 3 110	5 000	252	10 000	115
t=13.5	1 430	563	518	459	1 550	593	1 710	Ix 178	6 000	220	11 000	102
d=348	900	490	461	411	975	511	1 640	Sx 1 030	7 000	183	12 000	92.4
W360X57	2 500	601	536	457	1 950	642	1 610	Mr 273	3 000	259	6 000	147
W14X38	2 060	589	530	455	1 950	615	1 590	Vr 504	3 500	243	7 000	119
b=172	1 630	556	507	442	1 770	581	1 560	Lu 2 550	4 000	225	8 000	99.8
t=13.1	1 190	498	461	406	1 290	531	1 520	Ix 161	4 500	207	9 000	86
d=358	750	437	410	363	813	456	1 450	Sx 897	5 000	189	10 000	75.7
W360X51	2 500	541	480	408	1 740	581	1 440	Mr 241	3 000	227	6 000	121
W14X34	2 060	531	475	406	1 740	558	1 430	Vr 457	3 500	212	7 000	97
b=171	1 630	516	468	404	1 740	529	1 400	Lu 2 500	4 000	195	8 000	81
t=11.6	1 190	460	424	373	1 290	485	1 370	Ix 141	4 500	178	9 000	69.5
d=356	750	401	375	331	813	419	1 300	Sx 796	5 000	159	10 000	60.9
W360X45	2 500	482	426	361	1 550	520	1 280	Mr 210	3 000	195	6 000	96.1
W14X30	2 060	474	422	359	1 550	500	1 270	Vr 433	3 500	181	7 000	76.5
b=171	1 630	462	416	357	1 550	475	1 250	Lu 2 430	4 000	165	8 000	63.4
t=9.8	1 190	423	388	338	1 290	438	1 220	Ix 122	4 500	148	9 000	54.1
d=352	750	365	341	298	813	381	1 160	Sx 691	5 000	128	10 000	47.2
W360X39	2 500	425	374	314	1 340	462	1 120	Mr 179	2 000	173	6 000	54.2
W14X26	2 060	419	371	313	1 340	445	1 100	Vr 410	2 500	157	7 000	44.3
b=128	1 630	410	366	312	1 340	424	1 080	Lu 1 790	3 000	139	8 000	37.5
t=10.7	1 190	390	354	306	1 290	394	1 060	Ix 102	4 000	97.2	9 000	32.5
d=354	750	331	308	266	813	344	1 010	Sx 580	5 000	69.8	10 000	28.8
W360X33	2 500	357	313	261	1 130	391	936	Mr 146	2 000	139	6 000	38.1
W14X22	2 060	353	310	261	1 130	379	925	Vr 362	2 500	124	7 000	30.8
b=127	1 630	347	307	260	1 130	362	911	Lu 1 720	3 000	108	8 000	25.9
t=8.5	1 190	335	302	258	1 130	338	891	Ix 82.7	4 000	70.3	9 000	22.3
d=350	750	292	269	232	813	299	856	Sx 474	5 000	49.7	10 000	19.6
W310X86	3 500	837	742	623	2 970	806	2 290	Mr 383	5 000	367	10 000	248
W12X58	2 890	816	731	620	2 970	771	2 260	Vr 503	6 000	344	11 000	221
b=254	2 280	750	680	587	2 470	724	2 220	Lu 4 250	7 000	320	12 000	200
t=16.3	1 660	667	614	545	1 800	658	2 150	Ix 199	8 000	297	13 000	182
d=310	1 050	582	547	495	1 140	562	2 050	Sx 1 280	9 000	273	14 000	167

Note : * 25 MPa, 2 000 kg/m³ Concrete. G40.21-M 300W Steel.

Units : d, b, t, L_u (mm) $I_x = 10^6$ mm⁴ $S_x = 10^3$ mm³ M_r = kN·m V_r = kN

300W
25 MPa

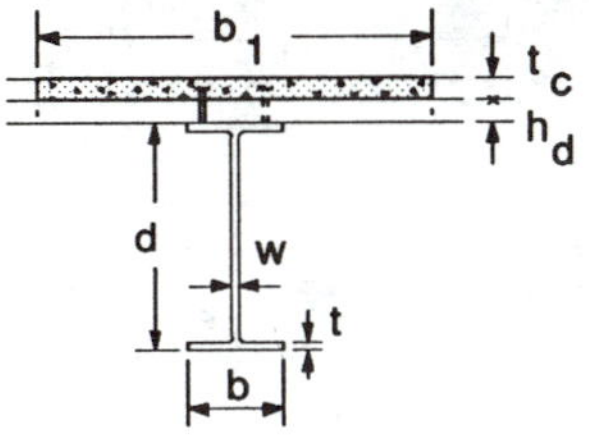

COMPOSITE MEMBERS
Trial Selection Table

75 mm Deck with 85 mm Slab
$\phi = 0.9, \phi_c = 0.6, \phi_{sc} = 0.8$

Steel Shape	b_1 mm	Composite Beam* Factored Resistances M_{rc} (kN·m) for Shear Connections = 100%	70%	40%	Q_r (kN) for 100%	I_t x10⁶ mm⁴	S_t x10³ mm³	Steel Shape Data	Non-Composite Shape Unbraced Condition L' mm	M'r kN·m	L' mm	M'r kN·m
W310X79	3 000	756	673	566	2 730	715	2 080	Mr 346	5 000	327	8 000	258
W12X53	2 480	733	660	561	2 690	682	2 050	Vr 480	5 500	316	9 000	235
b=254	1 950	664	604	525	2 110	639	2 010	Lu 4 140	6 000	305	10 000	207
t=14.6	1 430	594	549	489	1 550	580	1 950	Ix 177	6 500	293	11 000	184
d=306	900	521	491	442	975	493	1 850	Sx 1 160	7 000	282	12 000	166
W310X74	3 000	721	639	537	2 560	688	1 960	Mr 321	4 000	307	8 000	206
W12X50	2 480	703	631	534	2 560	657	1 930	Vr 519	4 500	295	9 000	179
b=205	1 950	645	585	505	2 110	616	1 890	Lu 3 380	5 000	282	10 000	159
t=16.3	1 430	575	529	467	1 550	561	1 840	Ix 165	6 000	258	11 000	142
d=310	900	500	470	419	975	477	1 740	Sx 1 060	7 000	233	12 000	129
W310X67	3 000	650	573	479	2 300	622	1 760	Mr 286	4 000	270	8 000	169
W12X45	2 480	636	566	477	2 300	596	1 730	Vr 463	4 500	258	9 000	147
b=204	1 950	600	541	463	2 110	561	1 700	Lu 3 280	5 000	246	10 000	129
t=14.6	1 430	531	486	427	1 550	512	1 660	Ix 145	6 000	222	11 000	116
d=306	900	458	429	381	975	438	1 570	Sx 949	7 000	198	12 000	105
W310X60	2 500	574	507	425	2 050	543	1 560	Mr 254	4 000	237	6 500	179
W12X40	2 060	559	500	423	2 050	519	1 540	Vr 406	4 500	226	7 000	166
b=203	1 630	519	470	404	1 770	488	1 510	Lu 3 200	5 000	214	8 000	139
t=13.1	1 190	460	423	373	1 290	444	1 470	Ix 129	5 500	203	9 000	120
d=304	750	400	375	335	813	379	1 390	Sx 849	6 000	191	10 000	106
W310X52	2 500	524	461	387	1 800	515	1 400	Mr 226	3 000	216	6 000	130
W12X35	2 060	513	456	385	1 800	493	1 380	Vr 431	3 500	203	7 000	106
b=167	1 630	494	446	381	1 770	466	1 360	Lu 2 570	4 000	189	8 000	89.4
t=13.2	1 190	436	399	350	1 290	426	1 320	Ix 118	4 500	176	9 000	77.4
d=318	750	376	352	311	813	366	1 260	Sx 747	5 000	162	10 000	68.4
W310X45	2 500	450	394	329	1 540	447	1 200	Mr 191	3 000	180	6 000	98.2
W12X30	2 060	442	390	328	1 540	430	1 190	Vr 369	3 500	168	7 000	79.3
b=166	1 630	430	384	326	1 540	408	1 170	Lu 2 490	4 000	155	8 000	66.5
t=11.2	1 190	392	357	309	1 290	376	1 140	Ix 99.2	4 500	142	9 000	57.3
d=314	750	334	310	274	813	325	1 090	Sx 634	5 000	128	10 000	50.4
W310X39	2 500	392	342	284	1 330	393	1 050	Mr 165	3 000	153	6 000	77.7
W12X26	2 060	386	339	283	1 330	379	1 040	Vr 320	3 500	142	7 000	62.2
b=165	1 630	377	334	282	1 330	361	1 020	Lu 2 440	4 000	130	8 000	51.8
t=9.7	1 190	358	323	277	1 290	334	996	Ix 85.1	4 500	117	9 000	44.3
d=310	750	301	278	244	813	292	954	Sx 549	5 000	103	10 000	38.8
W250X80	3 000	694	610	503	2 750	575	1 910	Mr 294	5 000	287	8 000	242
W10X54	2 480	668	595	496	2 690	547	1 880	Vr 429	5 500	280	9 000	227
b=255	1 950	599	539	460	2 110	510	1 840	Lu 4 520	6 000	272	10 000	212
t=15.6	1 430	529	484	425	1 550	461	1 780	Ix 126	6 500	265	11 000	197
d=256	900	456	426	384	975	388	1 680	Sx 982	7 000	257	12 000	179

Note : * 25 MPa, 2 000 kg/m³ Concrete. G40.21-M 300W Steel.

Units : d, b, t, L_u (mm) $I_x = 10^6$ mm⁴ $S_x = 10^3$ mm³ M_r = kN·m V_r = kN

COMPOSITE MEMBERS
Trial Selection Table

75 mm Deck with 85 mm Slab
$\phi = 0.9$, $\phi_c = 0.6$, $\phi_{sc} = 0.8$

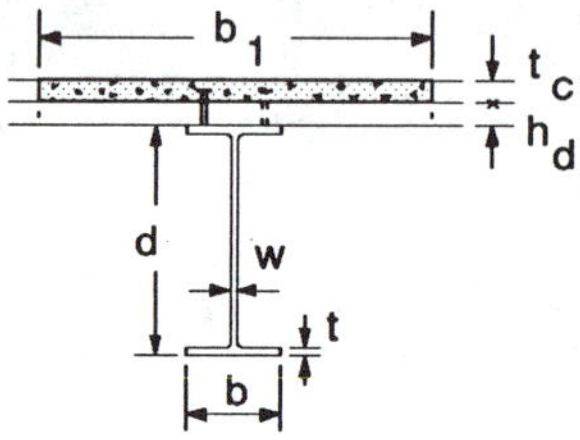

300W
25 MPa

Steel Shape	b_1 (mm)	Composite Beam*: Factored Resistances: M_{rc} (kN·m) for Shear Connections = 100%	70%	40%	Q_r (kN) for 100%	I_t x10⁶ mm⁴	S_t x10³ mm³	Steel Shape Data	Non-Composite Shape: Unbraced Condition: L' mm	M'r kN·m	L' mm	M'r kN·m
W250X73	3 000	637	557	457	2 510	530	1 740	Mr 266	5 000	257	8 000	212
W10X49	2 480	620	548	454	2 510	506	1 720	Vr 389	5 500	250	9 000	198
b=254	1 950	566	507	429	2 110	473	1 680	Lu 4 390	6 000	242	10 000	183
t=14.2	1 430	497	452	394	1 550	429	1 630	Ix 113	6 500	235	11 000	165
d=254	900	424	395	355	975	363	1 540	Sx 891	7 000	227	12 000	149
W250X67	3 000	597	520	426	2 310	504	1 620	Mr 243	4 000	237	8 000	174
W10X45	2 480	583	513	423	2 310	482	1 590	Vr 409	4 500	229	9 000	157
b=204	1 950	546	487	409	2 110	452	1 560	Lu 3 570	5 000	221	10 000	139
t=15.7	1 430	477	432	374	1 550	411	1 510	Ix 104	6 000	205	11 000	125
d=258	900	404	375	333	975	349	1 430	Sx 806	7 000	189	12 000	114
W250X58	2 500	510	444	364	2 000	422	1 390	Mr 208	4 000	199	6 500	160
W10X39	2 060	497	438	362	2 000	403	1 370	Vr 359	4 500	191	7 000	153
b=203	1 630	460	411	346	1 770	378	1 340	Lu 3 410	5 000	184	8 000	137
t=13.5	1 190	402	364	316	1 290	343	1 300	Ix 87.3	5 500	176	9 000	119
d=252	750	341	317	282	813	290	1 230	Sx 693	6 000	168	10 000	105
W250X49	2 500	435	375	305	1 690	362	1 170	Mr 171	4 000	160	6 500	123
W10X33	2 060	425	371	304	1 690	347	1 150	Vr 327	4 500	153	7 000	115
b=202	1 630	411	364	302	1 690	327	1 130	Lu 3 240	5 000	146	8 000	97.2
t=11	1 190	360	324	277	1 290	298	1 100	Ix 70.6	5 500	138	9 000	84
d=248	750	301	278	244	813	255	1 050	Sx 572	6 000	130	10 000	74.1
W250X45	2 500	415	359	293	1 540	364	1 110	Mr 163	3 000	151	6 000	90.6
W10X30	2 060	407	355	292	1 540	349	1 090	Vr 360	3 500	142	7 000	75.2
b=148	1 630	395	349	290	1 540	330	1 070	Lu 2 360	4 000	132	8 000	64.4
t=13	1 190	357	320	273	1 290	303	1 040	Ix 71.1	4 500	122	9 000	56.3
d=266	750	298	274	239	813	260	992	Sx 534	5 000	112	10 000	50.1
W250X39	2 500	359	308	251	1 330	317	958	Mr 139	3 000	126	6 000	69.2
W10X26	2 060	353	305	250	1 330	305	944	Vr 308	3 500	117	7 000	57
b=147	1 630	344	301	248	1 330	290	928	Lu 2 280	4 000	108	8 000	48.6
t=11.2	1 190	326	290	244	1 290	268	904	Ix 60.1	4 500	98.6	9 000	42.3
d=262	750	268	244	212	813	232	864	Sx 459	5 000	88	10 000	37.6
W250X33	2 000	301	258	210	1 130	261	801	Mr 114	3 000	102	5 500	55.6
W10X22	1 650	295	256	210	1 130	251	790	Vr 280	3 500	93.1	6 000	49.4
b=146	1 300	287	252	208	1 130	237	775	Lu 2 180	4 000	84	6 500	44.4
t=9.1	950	266	238	201	1 030	218	755	Ix 48.9	4 500	74.1	7 000	40.3
d=258	600	220	202	174	650	189	720	Sx 379	5 000	63.6	8 000	34.1
W200X71	3 000	580	500	401	2 460	429	1 580	Mr 217	5 000	208	8 000	178
W8X48	2 480	564	492	398	2 460	408	1 550	Vr 393	5 500	203	9 000	168
b=206	1 950	513	454	375	2 110	381	1 510	Lu 4 150	6 000	198	10 000	158
t=17.4	1 430	444	398	339	1 550	344	1 470	Ix 76.6	6 500	193	11 000	148
d=216	900	370	340	301	975	288	1 380	Sx 709	7 000	188	12 000	137

Note : * 25 MPa, 2 000 kg/m³ Concrete. G40.21-M 300W Steel.

Units : d, b, t, L_u (mm) $I_x = 10^6$ mm⁴ $S_x = 10^3$ mm³ M_r = kN·m V_r = kN

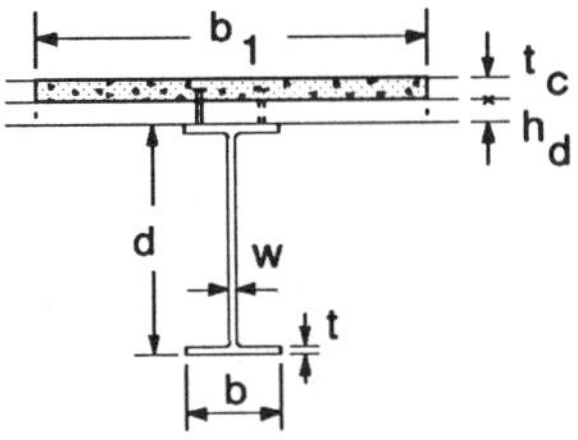

300W
25 MPa

COMPOSITE MEMBERS
Trial Selection Table

75 mm Deck with 85 mm Slab
$\phi = 0.9, \phi_c = 0.6, \phi_{sc} = 0.8$

Steel Shape	b_1	Composite Beam*							Non-Composite Shape			
		Factored Resistances				I_t	S_t	Steel Shape Data	Unbraced Condition			
		M_{rc} (kN·m) for Shear Connections =			Q_r (kN) for	$\times 10^6$	$\times 10^3$		L'	M'r	L'	M'r
	mm	100%	70%	40%	100%	mm^4	mm^3		mm	kN·m	mm	kN·m
W200X59	3 000	486	415	329	2 040	362	1 320	Mr 176	4 000	174	8 000	134
W8X40	2 480	475	409	328	2 040	346	1 300	Vr 341	4 500	169	9 000	124
b=205	1 950	457	400	325	2 040	325	1 270	Lu 3 780	5 000	164	10 000	114
t=14.2	1 430	395	351	294	1 550	295	1 230	Ix 61.1	6 000	154	11 000	103
d=210	900	324	295	257	975	250	1 160	Sx 582	7 000	144	12 000	93.4
W200X52	2 500	422	360	287	1 800	309	1 150	Mr 154	4 000	150	6 500	126
W8X35	2 060	411	355	285	1 800	295	1 130	Vr 290	4 500	145	7 000	121
b=204	1 630	393	345	281	1 770	277	1 110	Lu 3 620	5 000	140	8 000	111
t=12.6	1 190	336	299	252	1 290	251	1 070	Ix 52.7	5 500	135	9 000	101
d=206	750	276	252	221	813	212	1 010	Sx 512	6 000	131	10 000	89.7
W200X46	2 500	375	318	252	1 580	277	1 020	Mr 134	4 000	129	6 500	105
W8X31	2 060	367	314	251	1 580	265	1 000	Vr 262	4 500	124	7 000	99.8
b=203	1 630	354	308	249	1 580	250	981	Lu 3 460	5 000	119	8 000	90.1
t=11	1 190	313	277	230	1 290	228	952	Ix 45.5	5 500	114	9 000	78.6
d=204	750	254	231	200	813	194	903	Sx 448	6 000	109	10 000	69.6
W200X42	2 000	337	287	229	1 430	246	912	Mr 120	3 000	119	5 500	93.5
W8X28	1 650	328	283	227	1 430	234	897	Vr 264	3 500	114	6 000	88.4
b=166	1 300	313	275	224	1 410	219	878	Lu 2 850	4 000	109	6 500	83.4
t=11.8	950	268	239	201	1 030	199	850	Ix 40.9	4 500	104	7 000	77.7
d=206	600	221	202	176	650	168	803	Sx 399	5 000	98.6	8 000	66.7
W200X36	2 000	293	248	196	1 240	215	792	Mr 103	3 000	100	5 500	75.5
W8X24	1 650	286	245	195	1 240	206	779	Vr 223	3 500	95.3	6 000	70.6
b=165	1 300	277	240	194	1 240	194	763	Lu 2 730	4 000	90.4	6 500	64.7
t=10.2	950	246	217	180	1 030	177	741	Ix 34.4	4 500	85.5	7 000	59.2
d=202	600	199	181	156	650	150	703	Sx 342	5 000	80.5	8 000	50.6
W200X31	2 000	263	222	176	1 080	200	706	Mr 90.5	3 000	81.1	5 500	50.6
W8X21	1 650	258	220	175	1 080	192	695	Vr 240	3 500	75.3	6 000	45.6
b=134	1 300	251	216	174	1 080	182	681	Lu 2 150	4 000	69.5	6 500	41.5
t=10.2	950	234	206	169	1 030	167	662	Ix 31.4	4 500	63.6	7 000	38
d=210	600	189	170	145	650	143	630	Sx 299	5 000	57	8 000	32.7
W200X27	2 000	225	189	149	915	173	603	Mr 75.3	3 000	65.4	5 500	36.4
W8X18	1 650	222	187	148	915	167	594	Vr 215	3 500	59.7	6 000	32.6
b=133	1 300	216	185	148	915	158	582	Lu 2 050	4 000	54	6 500	29.6
t=8.4	950	207	180	146	915	146	566	Ix 25.8	4 500	47.5	7 000	27.1
d=208	600	171	153	129	650	127	541	Sx 249	5 000	41.2	8 000	23.1
W150X30	2 500	228	187	141	1 020	153	625	Mr 65.9	3 000	64.1	6 000	46.8
W6X20	2 060	225	185	141	1 020	147	613	Vr 186	3 500	61.2	7 000	40.4
b=153	1 630	219	183	140	1 020	139	600	Lu 2 680	4 000	58.3	8 000	34.8
t=9.3	1 190	210	178	139	1 020	128	582	Ix 17.2	4 500	55.4	9 000	30.6
d=158	750	176	154	124	813	111	555	Sx 219	5 000	52.5	10 000	27.3

Note : * 25 MPa, 2 000 kg/m³ Concrete. G40.21-M 300W Steel.

Units : d, b, t, L_u (mm) $I_x = 10^6\ mm^4$ $S_x = 10^3\ mm^3$ M_r = kN·m V_r = kN

COMPOSITE MEMBERS
Trial Selection Table

75 mm Deck with 85 mm Slab
$\phi = 0.9$, $\phi_c = 0.6$, $\phi_{sc} = 0.8$

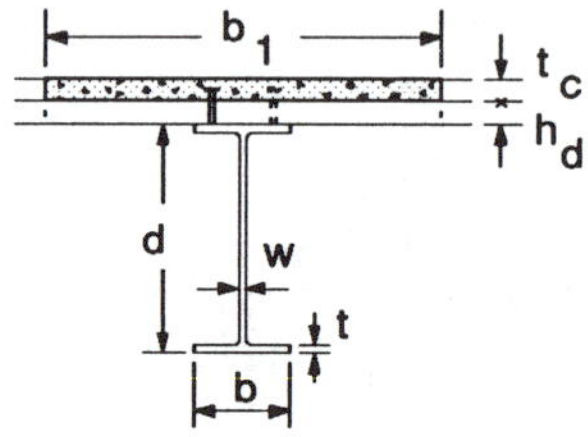

300W
25 MPa

Steel Shape	b_1	Composite Beam*						Steel Shape Data	Non-Composite Shape			
		Factored Resistances				I_t	S_t		Unbraced Condition			
		M_{rc} (kN·m) for Shear Connections =			Q_r (kN) for	x10⁶	x10³		L'	M'r	L'	M'r
	mm	100%	70%	40%	100%	mm⁴	mm³		mm	kN·m	mm	kN·m
WWF900X169	5 000	3 260	3 110	2 880	5 420	7 320	9 080	Mr 1 990	5 000	1 860	12 000	625
	4 130	3 140	3 010	2 780	4 480	6 990	8 970	Vr 1 690	6 000	1 680	14 000	491
b=300	3 250	3 020	2 910	2 650	3 520	6 580	8 810	Lu 4 130	7 000	1 480	16 000	402
t=20	2 380	2 890	2 760	2 500	2 580	6 030	8 590	Ix 2930	8 000	1 250	18 000	339
d=900	1 500	2 690	2 530	2 330	1 630	5 270	8 230	Sx 6 510	10 000	845	20 000	293
WWF800X161	5 000	2 850	2 700	2 490	5 420	5 780	7 910	Mr 1 710	5 000	1 610	12 000	581
	4 130	2 740	2 600	2 410	4 480	5 530	7 820	Vr 1 490	6 000	1 470	14 000	460
b=300	3 250	2 620	2 510	2 300	3 520	5 200	7 690	Lu 4 230	7 000	1 310	16 000	379
t=20	2 380	2 490	2 390	2 170	2 580	4 770	7 500	Ix 2250	8 000	1 130	18 000	322
d=800	1 500	2 320	2 190	2 010	1 630	4 160	7 180	Sx 5 610	10 000	778	20 000	280
WWF700X152	4 500	2 410	2 270	2 080	4 880	4 340	6 750	Mr 1 440	5 000	1 370	10 000	711
	3 710	2 300	2 180	2 010	4 020	4 140	6 660	Vr 1 290	6 000	1 260	12 000	537
b=300	2 930	2 190	2 090	1 920	3 180	3 890	6 540	Lu 4 330	7 000	1 140	14 000	429
t=20	2 140	2 080	1 990	1 820	2 320	3 560	6 370	Ix 1 660	8 000	1 000	16 000	356
d=700	1 350	1 940	1 830	1 690	1 460	3 090	6 090	Sx 4 760	9 000	842	18 000	305
W610X174	4 500	2 420	2 270	2 070	4 880	3 880	6 920	Mr 1 450	5 000	1 430	10 000	924
W24X117	3 710	2 310	2 180	2 000	4 020	3 690	6 820	Vr 1 540	6 000	1 350	12 000	709
b=325	2 930	2 200	2 100	1 910	3 180	3 450	6 680	Lu 4 830	7 000	1 250	14 000	574
t=21.6	2 140	2 080	1 980	1 800	2 320	3 140	6 490	Ix 1 470	8 000	1 150	16 000	482
d=616	1 350	1 920	1 820	1 680	1 460	2 710	6 180	Sx 4 780	9 000	1 040	18 000	415
W610X155	4 000	2 130	2 010	1 830	4 340	3 410	6 110	Mr 1 280	5 000	1 260	10 000	762
W24X104	3 300	2 040	1 930	1 770	3 580	3 240	6 020	Vr 1 390	6 000	1 180	11 000	659
b=324	2 600	1 940	1 850	1 690	2 820	3 030	5 910	Lu 4 740	7 000	1 080	12 000	579
t=19	1 900	1 840	1 750	1 590	2 060	2 750	5 730	Ix 1 290	8 000	988	14 000	465
d=612	1 200	1 700	1 610	1 480	1 300	2 380	5 460	Sx 4 220	9 000	886	16 000	388
W610X140	4 000	2 000	1 880	1 690	4 340	3 150	5 500	Mr 1 120	4 000	1 050	9 000	486
W24X94	3 300	1 910	1 800	1 620	3 580	3 000	5 420	Vr 1 440	5 000	946	10 000	422
b=230	2 600	1 810	1 720	1 540	2 820	2 800	5 310	Lu 3 320	6 000	829	12 000	334
t=22.2	1 900	1 700	1 610	1 440	2 060	2 550	5 150	Ix 1 120	7 000	695	14 000	277
d=618	1 200	1 560	1 460	1 330	1 300	2 190	4 890	Sx 3 630	8 000	573	16 000	237
W610X125	4 000	1 820	1 700	1 530	4 290	2 850	4 910	Mr 991	4 000	923	9 000	396
W24X84	3 300	1 730	1 630	1 470	3 580	2 710	4 840	Vr 1 300	5 000	821	10 000	342
b=229	2 600	1 640	1 550	1 390	2 820	2 550	4 750	Lu 3 250	6 000	708	12 000	269
t=19.6	1 900	1 540	1 460	1 300	2 060	2 320	4 610	Ix 985	7 000	575	14 000	222
d=612	1 200	1 410	1 320	1 200	1 300	2 000	4 390	Sx 3 220	8 000	470	16 000	189
W610X113	4 000	1 660	1 540	1 380	3 890	2 600	4 440	Mr 888	4 000	818	9 000	328
W24X76	3 300	1 600	1 500	1 350	3 580	2 490	4 390	Vr 1 210	5 000	719	10 000	282
b=228	2 600	1 510	1 420	1 280	2 820	2 340	4 300	Lu 3 180	6 000	610	12 000	220
t=17.3	1 900	1 410	1 340	1 190	2 060	2 140	4 190	Ix 875	7 000	481	14 000	180
d=608	1 200	1 290	1 210	1 090	1 300	1 850	3 990	Sx 2 880	8 000	391	16 000	153

Note : * 25 MPa, 1 850 kg/m³ Concrete. G40.21-M 300W Steel.

Units : d, b, t, L_u (mm) $I_x = 10^6$ mm⁴ $S_x = 10^3$ mm³ M_r = kN·m V_r = kN

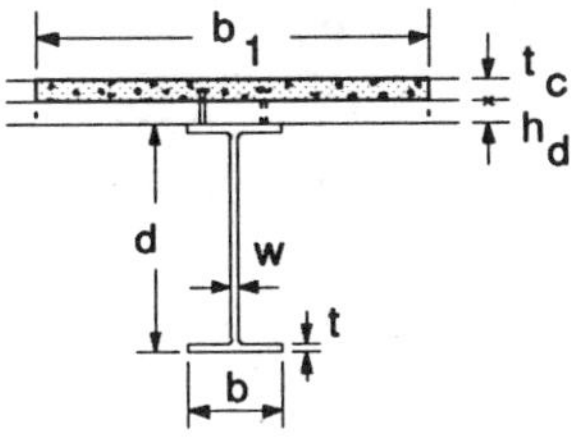

COMPOSITE MEMBERS
Trial Selection Table

300W
25 MPa

75 mm Deck with 85 mm Slab
$\phi = 0.9, \phi_c = 0.6, \phi_{sc} = 0.8$

Steel Shape	b_1	Composite Beam*						Steel Shape Data	Non-Composite Shape			
		Factored Resistances				I_t	S_t		Unbraced Condition			
		M_{rc} (kN·m) for Shear Connections =			Q_r (kN) for	$\times 10^6$	$\times 10^3$		L'	M'r	L'	M'r
	mm	100%	70%	40%	100%	mm^4	mm^3		mm	kN·m	mm	kN·m
W610X101	3 500	1 480	1 380	1 240	3 510	2 290	3 960	Mr 783	4 000	713	9 000	267
W24X68	2 890	1 430	1 340	1 200	3 130	2 190	3 900	Vr 1 130	5 000	619	10 000	228
b=228	2 280	1 350	1 270	1 140	2 470	2 060	3 830	Lu 3 110	6 000	512	11 000	199
t=14.9	1 660	1 260	1 190	1 050	1 800	1 870	3 720	Ix 764	7 000	396	12 000	176
d=604	1 050	1 150	1 070	962	1 140	1 620	3 530	Sx 2 530	8 000	320	14 000	144
W610X91	3 500	1 330	1 230	1 100	3 130	2 060	3 530	Mr 691	4 000	619	9 000	217
W24X61	2 890	1 300	1 220	1 090	3 130	1 980	3 480	Vr 1 030	5 000	530	10 000	185
b=227	2 280	1 230	1 150	1 030	2 470	1 860	3 420	Lu 3 020	6 000	421	11 000	161
t=12.7	1 660	1 140	1 080	949	1 800	1 700	3 330	Ix 667	7 000	325	12 000	142
d=598	1 050	1 040	963	858	1 140	1 470	3 170	Sx 2 230	8 000	261	14 000	115
W610X84	3 500	1 230	1 140	1 010	2 890	1 930	3 270	Mr 637	3 000	636	8 000	232
W24X56	2 890	1 210	1 130	1 010	2 890	1 850	3 230	Vr 956	4 000	566	9 000	192
b=226	2 280	1 150	1 080	965	2 470	1 750	3 170	Lu 2 990	5 000	482	10 000	163
t=11.7	1 660	1 070	1 010	890	1 800	1 600	3 090	Ix 613	6 000	377	12 000	125
d=596	1 050	972	903	802	1 140	1 390	2 940	Sx 2 060	7 000	289	14 000	101
W530X123	3 500	1 600	1 490	1 330	3 790	2 250	4 390	Mr 867	4 000	794	9 000	361
W21X83	2 890	1 520	1 420	1 270	3 130	2 140	4 320	Vr 1 270	5 000	706	10 000	316
b=212	2 280	1 430	1 350	1 200	2 470	1 990	4 220	Lu 3 100	6 000	613	11 000	281
t=21.2	1 660	1 340	1 260	1 120	1 800	1 800	4 090	Ix 761	7 000	505	12 000	253
d=544	1 050	1 210	1 140	1 030	1 140	1 540	3 870	Sx 2 800	8 000	421	14 000	211
W530X109	3 500	1 460	1 350	1 210	3 750	2 030	3 900	Mr 764	4 000	692	9 000	291
W21X73	2 890	1 380	1 290	1 160	3 130	1 940	3 850	Vr 1 120	5 000	608	10 000	254
b=211	2 280	1 300	1 220	1 090	2 470	1 810	3 770	Lu 3 040	6 000	517	11 000	225
t=18.8	1 660	1 210	1 140	1 020	1 800	1 640	3 650	Ix 667	7 000	413	12 000	202
d=540	1 050	1 100	1 030	930	1 140	1 410	3 460	Sx 2 480	8 000	342	14 000	168
W530X101	3 500	1 360	1 260	1 120	3 480	1 910	3 640	Mr 707	3 000	707	8 000	301
W21X68	2 890	1 300	1 220	1 090	3 130	1 820	3 590	Vr 1 050	4 000	635	9 000	255
b=210	2 280	1 220	1 150	1 030	2 470	1 710	3 520	Lu 2 990	5 000	553	10 000	222
t=17.4	1 660	1 140	1 070	956	1 800	1 560	3 410	Ix 617	6 000	462	12 000	176
d=538	1 050	1 040	968	872	1 140	1 340	3 240	Sx 2 300	7 000	365	14 000	146
W530X92	3 500	1 250	1 150	1 020	3 190	1 760	3 320	Mr 637	3 000	633	8 000	253
W21X62	2 890	1 220	1 130	1 010	3 130	1 680	3 280	Vr 971	4 000	565	9 000	214
b=209	2 280	1 140	1 070	954	2 470	1 580	3 220	Lu 2 930	5 000	486	10 000	185
t=15.6	1 660	1 050	996	884	1 800	1 440	3 120	Ix 552	6 000	393	12 000	146
d=534	1 050	959	895	801	1 140	1 240	2 970	Sx 2 070	7 000	309	14 000	120
W530X82	3 500	1 110	1 020	899	2 840	1 580	2 950	Mr 559	3 000	551	8 000	204
W21X55	2 890	1 090	1 010	896	2 840	1 510	2 910	Vr 894	4 000	487	9 000	172
b=209	2 280	1 040	968	864	2 470	1 420	2 860	Lu 2 860	5 000	412	10 000	148
t=13.3	1 660	955	901	796	1 800	1 300	2 790	Ix 479	6 000	321	12 000	115
d=528	1 050	866	807	717	1 140	1 130	2 650	Sx 1 810	7 000	251	14 000	94.8

Note : * 25 MPa, 1 850 kg/m³ Concrete. G40.21-M 300W Steel.

Units : d, b, t, L_u (mm) $I_x = 10^6\ mm^4$ $S_x = 10^3\ mm^3$ M_r = kN·m V_r = kN

COMPOSITE MEMBERS
Trial Selection Table

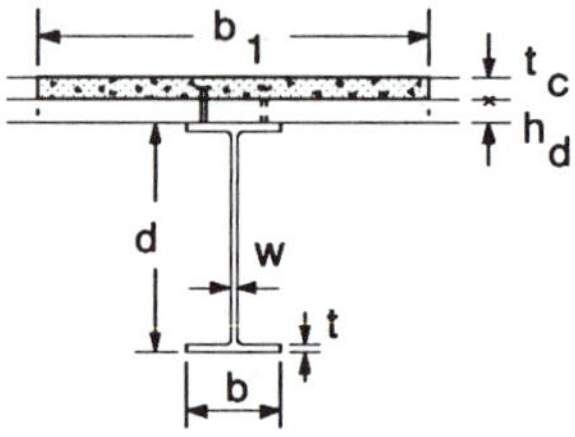

75 mm Deck with 85 mm Slab
$\phi = 0.9$, $\phi_c = 0.6$, $\phi_{sc} = 0.8$

300W
25 MPa

Steel Shape	b_1 mm	Composite Beam* Factored Resistances M_{rc} (kN·m) for Shear Connections = 100%	70%	40%	Q_r (kN) for 100%	I_t x10⁶ mm⁴	S_t x10³ mm³	Steel Shape Data	Non-Composite Shape Unbraced Condition L' mm	M'r kN·m	L' mm	M'r kN·m
W530X72	3 500	975	889	778	2 470	1 390	2 560	Mr 475	3 000	462	8 000	155
W21X48	2 890	961	882	776	2 470	1 330	2 530	Vr 831	4 000	402	9 000	129
b=207	2 280	938	871	772	2 470	1 260	2 490	Lu 2 750	5 000	331	10 000	111
t=10.9	1 660	857	805	709	1 800	1 160	2 430	Ix 402	6 000	247	12 000	85.9
d=524	1 050	772	718	631	1 140	1 010	2 320	Sx 1 530	7 000	191	14 000	70.1
W460X106	3 500	1 290	1 190	1 040	3 650	1 600	3 440	Mr 645	3 000	640	8 000	308
W18X71	2 890	1 220	1 130	1 000	3 130	1 520	3 390	Vr 1 060	4 000	579	9 000	266
b=194	2 280	1 140	1 060	947	2 470	1 420	3 310	Lu 2 910	5 000	512	10 000	235
t=20.6	1 660	1 050	990	878	1 800	1 290	3 210	Ix 488	6 000	444	12 000	190
d=470	1 050	952	889	800	1 140	1 100	3 030	Sx 2 080	7 000	366	14 000	160
W460X97	3 500	1 180	1 080	948	3 320	1 480	3 150	Mr 589	3 000	581	8 000	264
W18X65	2 890	1 140	1 050	932	3 130	1 410	3 110	Vr 947	4 000	522	9 000	227
b=193	2 280	1 060	987	880	2 470	1 320	3 050	Lu 2 870	5 000	457	10 000	200
t=19	1 660	974	917	815	1 800	1 200	2 950	Ix 445	6 000	389	12 000	161
d=466	1 050	881	824	740	1 140	1 030	2 800	Sx 1 910	7 000	314	14 000	135
W460X89	3 500	1 100	1 000	878	3 080	1 390	2 930	Mr 543	3 000	534	8 000	231
W18X60	2 890	1 080	992	874	3 080	1 330	2 890	Vr 868	4 000	477	9 000	198
b=192	2 280	1 000	931	829	2 470	1 250	2 840	Lu 2 830	5 000	414	10 000	174
t=17.7	1 660	918	862	767	1 800	1 130	2 750	Ix 410	6 000	343	12 000	140
d=464	1 050	829	776	694	1 140	973	2 610	Sx 1 770	7 000	276	14 000	117
W460X82	3 500	1 010	914	798	2 810	1 280	2 680	Mr 494	3 000	482	8 000	195
W18X55	2 890	988	904	795	2 810	1 220	2 640	Vr 812	4 000	427	9 000	167
b=191	2 280	936	865	769	2 470	1 150	2 600	Lu 2 770	5 000	365	10 000	146
t=16	1 660	852	798	711	1 800	1 050	2 520	Ix 370	6 000	292	12 000	117
d=460	1 050	766	718	640	1 140	906	2 400	Sx 1 610	7 000	234	14 000	97.3
W460X74	3 500	920	831	724	2 550	1 180	2 440	Mr 446	3 000	433	8 000	164
W18X50	2 890	904	824	722	2 550	1 130	2 410	Vr 735	4 000	380	9 000	140
b=190	2 280	875	806	713	2 470	1 070	2 370	Lu 2 730	5 000	320	10 000	122
t=14.5	1 660	793	740	659	1 800	977	2 310	Ix 333	6 000	249	12 000	96.9
d=458	1 050	708	666	592	1 140	845	2 200	Sx 1 460	7 000	198	14 000	80.6
W460X67	3 000	835	757	660	2 340	1 050	2 220	Mr 405	3 000	390	8 000	140
W18X45	2 480	820	750	657	2 340	1 010	2 190	Vr 688	4 000	339	9 000	119
b=190	1 950	780	721	638	2 110	948	2 150	Lu 2 660	5 000	281	10 000	103
t=12.7	1 430	711	666	588	1 550	868	2 090	Ix 300	6 000	214	11 000	91.4
d=454	900	637	594	525	975	747	1 990	Sx 1 320	7 000	169	12 000	82
W460X61	3 000	749	676	586	2 100	944	1 980	Mr 354	3 000	336	8 000	111
W18X41	2 480	737	670	584	2 100	906	1 950	Vr 650	4 000	288	9 000	93.9
b=189	1 950	718	661	581	2 100	856	1 920	Lu 2 580	5 000	231	10 000	81.3
t=10.8	1 430	652	608	534	1 550	787	1 870	Ix 259	6 000	172	11 000	71.7
d=450	900	580	540	473	975	679	1 780	Sx 1 150	7 000	135	12 000	64.1

Note : * 25 MPa, 1 850 kg/m³ Concrete. G40.21-M 300W Steel.

Units : d, b, t, L_u (mm) $I_x = 10^6$ mm⁴ $S_x = 10^3$ mm³ M_r = kN·m V_r = kN

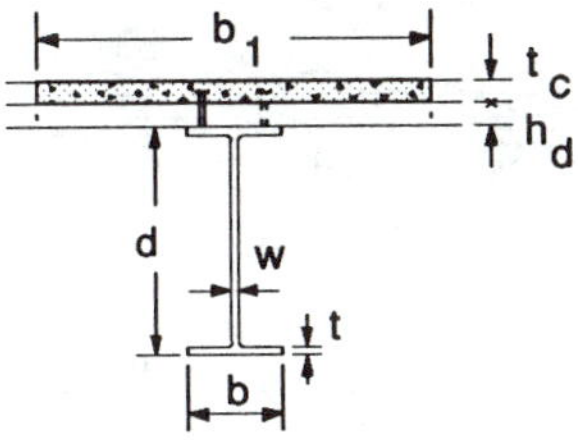

300W
25 MPa

COMPOSITE MEMBERS
Trial Selection Table

75 mm Deck with 85 mm Slab
$\phi = 0.9, \phi_c = 0.6, \phi_{sc} = 0.8$

Steel Shape	b_1 mm	Composite Beam* — Factored Resistances — M_{rc} (kN·m) for Shear Connections = 100%	70%	40%	Q_r (kN) for 100%	I_t $\times 10^6$ mm^4	S_t $\times 10^3$ mm^3	Steel Shape Data	Non-Composite Shape — Unbraced Condition — L' mm	M'r kN·m	L' mm	M'r kN·m
W410X85	3 500	981	885	765	2 920	1 140	2 600	Mr 467	3 000	455	8 000	205
W16X57	2 890	961	876	762	2 920	1 090	2 560	Vr 812	4 000	406	9 000	178
b=181	2 280	899	828	730	2 470	1 020	2 510	Lu 2 730	5 000	354	10 000	157
t=18.2	1 660	814	759	674	1 800	930	2 440	Ix 315	6 000	297	12 000	127
d=418	1 050	727	681	606	1 140	796	2 310	Sx 1 510	7 000	243	14 000	107
W410X74	3 500	872	783	675	2 580	1 020	2 310	Mr 408	3 000	394	8 000	163
W16X50	2 890	856	775	672	2 580	979	2 280	Vr 716	4 000	348	9 000	140
b=180	2 280	824	755	662	2 470	923	2 240	Lu 2 670	5 000	297	10 000	124
t=16	1 660	742	689	611	1 800	844	2 170	Ix 275	6 000	239	12 000	99.8
d=414	1 050	657	617	546	1 140	726	2 070	Sx 1 330	7 000	194	14 000	83.8
W410X67	3 500	787	704	605	2 320	930	2 080	Mr 367	3 000	352	8 000	135
W16X45	2 890	774	698	603	2 320	893	2 060	Vr 643	4 000	307	9 000	116
b=179	2 280	755	688	600	2 320	845	2 020	Lu 2 610	5 000	258	10 000	102
t=14.4	1 660	686	634	561	1 800	776	1 970	Ix 246	6 000	201	12 000	81.7
d=410	1 050	602	566	501	1 140	671	1 880	Sx 1 200	7 000	161	14 000	68.4
W410X60	3 000	690	618	532	2 050	811	1 830	Mr 321	3 000	306	8 000	109
W16X40	2 480	679	612	530	2 050	778	1 810	Vr 560	4 000	264	9 000	93.2
b=178	1 950	661	603	527	2 050	735	1 780	Lu 2 580	5 000	217	10 000	81.4
t=12.8	1 430	598	554	491	1 550	675	1 730	Ix 216	6 000	165	11 000	72.2
d=408	900	526	495	438	975	583	1 650	Sx 1 060	7 000	131	12 000	65
W410X54	3 000	621	554	475	1 840	729	1 640	Mr 284	3 000	266	8 000	86.1
W16X36	2 480	612	550	474	1 840	701	1 620	Vr 540	4 000	225	9 000	73.2
b=177	1 950	598	542	471	1 840	664	1 590	Lu 2 480	5 000	176	10 000	63.6
t=10.9	1 430	553	510	449	1 550	612	1 550	Ix 186	6 000	132	11 000	56.3
d=404	900	482	452	397	975	531	1 480	Sx 924	7 000	104	12 000	50.5
W410X46	3 000	543	482	410	1 590	643	1 420	Mr 239	2 000	236	7 000	61.7
W16X31	2 480	536	478	409	1 590	620	1 400	Vr 504	3 000	195	8 000	51.8
b=140	1 950	525	473	407	1 590	589	1 380	Lu 1 930	4 000	142	9 000	44.6
t=11.2	1 430	503	461	402	1 550	546	1 350	Ix 156	5 000	99.9	10 000	39.2
d=404	900	433	404	352	975	476	1 290	Sx 773	6 000	76.4	12 000	31.6
W410X39	3 000	461	408	345	1 350	550	1 200	Mr 197	2 000	193	7 000	44.1
W16X26	2 480	456	405	344	1 350	532	1 190	Vr 456	3 000	155	8 000	36.6
b=140	1 950	449	401	343	1 350	508	1 170	Lu 1 860	4 000	105	9 000	31.3
t=8.8	1 430	435	395	341	1 350	473	1 140	Ix 127	5 000	73.1	10 000	27.4
d=400	900	383	355	307	975	416	1 100	Sx 634	6 000	55.2	12 000	21.9
W360X79	3 000	822	737	630	2 730	839	2 220	Mr 386	4 000	364	8 000	225
W14X53	2 480	798	725	624	2 690	799	2 190	Vr 593	4 500	348	9 000	195
b=205	1 950	729	668	588	2 110	747	2 140	Lu 3 270	5 000	331	10 000	172
t=16.8	1 430	659	612	546	1 550	678	2 080	Ix 227	6 000	298	11 000	154
d=354	900	583	550	492	975	575	1 970	Sx 1 280	7 000	264	12 000	139

Note : * 25 MPa, 1 850 kg/m³ Concrete. G40.21-M 300W Steel.

Units : d, b, t, L_u (mm) $I_x = 10^6\ mm^4$ $S_x = 10^3\ mm^3$ M_r = kN·m V_r = kN

COMPOSITE MEMBERS
Trial Selection Table

75 mm Deck with 85 mm Slab
$\phi = 0.9$, $\phi_c = 0.6$, $\phi_{sc} = 0.8$

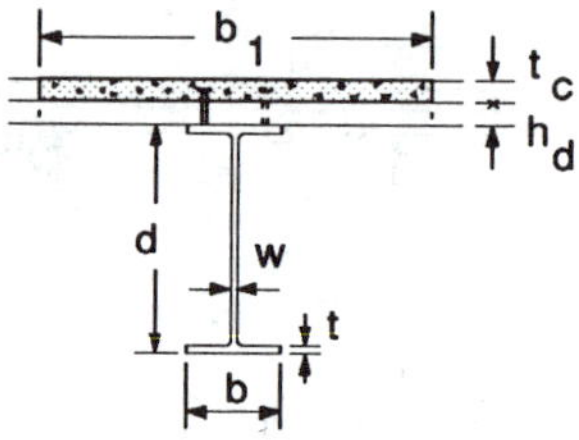

300W
25 MPa

Steel Shape	b_1 mm	Composite Beam* — Factored Resistances — M_{rc} (kN·m) for Shear Connections = 100%	70%	40%	Q_r (kN) for 100%	I_t x10⁶ mm⁴	S_t x10³ mm³	Steel Shape Data	Non-Composite Shape — Unbraced Condition — L' mm	M'r kN·m	L' mm	M'r kN·m
W360X72	3 000	745	665	565	2 460	764	2 000	Mr 346	4 000	322	8 000	186
W14X48	2 480	728	657	563	2 460	730	1 980	Vr 536	4 500	306	9 000	161
b=204	1 950	678	619	539	2 110	684	1 940	Lu 3 190	5 000	290	10 000	141
t=15.1	1 430	609	563	501	1 550	623	1 880	Ix 201	6 000	257	11 000	126
d=350	900	535	504	450	975	531	1 790	Sx 1 150	7 000	222	12 000	114
W360X64	3 000	671	596	505	2 200	695	1 800	Mr 308	4 000	283	8 000	153
W14X43	2 480	658	589	503	2 200	665	1 780	Vr 478	4 500	268	9 000	131
b=203	1 950	631	573	495	2 110	626	1 740	Lu 3 110	5 000	252	10 000	115
t=13.5	1 430	563	518	459	1 550	572	1 700	Ix 178	6 000	220	11 000	102
d=348	900	490	461	411	975	490	1 610	Sx 1 030	7 000	183	12 000	92.4
W360X57	2 500	601	536	457	1 950	626	1 600	Mr 273	3 000	259	6 000	147
W14X38	2 060	589	530	455	1 950	598	1 580	Vr 504	3 500	243	7 000	119
b=172	1 630	556	507	442	1 770	563	1 550	Lu 2 550	4 000	225	8 000	99.8
t=13.1	1 190	498	461	406	1 290	512	1 500	Ix 161	4 500	207	9 000	86
d=358	750	437	410	363	813	437	1 430	Sx 897	5 000	189	10 000	75.7
W360X51	2 500	541	480	408	1 740	568	1 430	Mr 241	3 000	227	6 000	121
W14X34	2 060	531	475	406	1 740	544	1 410	Vr 457	3 500	212	7 000	97
b=171	1 630	516	468	404	1 740	513	1 390	Lu 2 500	4 000	195	8 000	81
t=11.6	1 190	460	424	373	1 290	469	1 350	Ix 141	4 500	178	9 000	69.5
d=356	750	401	375	331	813	402	1 280	Sx 796	5 000	159	10 000	60.9
W360X45	2 500	482	426	361	1 550	508	1 270	Mr 210	3 000	195	6 000	96.1
W14X30	2 060	474	422	359	1 550	488	1 260	Vr 433	3 500	181	7 000	76.5
b=171	1 630	462	416	357	1 550	462	1 240	Lu 2 430	4 000	165	8 000	63.4
t=9.8	1 190	423	388	338	1 290	424	1 200	Ix 122	4 500	148	9 000	54.1
d=352	750	365	341	298	813	366	1 150	Sx 691	5 000	128	10 000	47.2
W360X39	2 500	425	374	314	1 340	452	1 110	Mr 179	2 000	173	6 000	54.2
W14X26	2 060	419	371	313	1 340	435	1 090	Vr 410	2 500	157	7 000	44.3
b=128	1 630	410	366	312	1 340	413	1 080	Lu 1 790	3 000	139	8 000	37.5
t=10.7	1 190	390	354	306	1 290	381	1 050	Ix 102	4 000	97.2	9 000	32.5
d=354	750	331	308	266	813	331	1 000	Sx 580	5 000	69.8	10 000	28.8
W360X33	2 500	357	313	261	1 130	384	930	Mr 146	2 000	139	6 000	38.1
W14X22	2 060	353	310	261	1 130	371	918	Vr 362	2 500	124	7 000	30.8
b=127	1 630	347	307	260	1 130	354	904	Lu 1 720	3 000	108	8 000	25.9
t=8.5	1 190	335	302	258	1 130	329	883	Ix 82.7	4 000	70.3	9 000	22.3
d=350	750	292	269	232	813	288	846	Sx 474	5 000	49.7	10 000	19.6
W310X86	3 500	837	742	623	2 970	785	2 270	Mr 383	5 000	367	10 000	248
W12X58	2 890	816	731	620	2 970	748	2 240	Vr 503	6 000	344	11 000	221
b=254	2 280	750	680	587	2 470	700	2 200	Lu 4 250	7 000	320	12 000	200
t=16.3	1 660	667	614	545	1 800	634	2 130	Ix 199	8 000	297	13 000	182
d=310	1 050	582	547	495	1 140	538	2 020	Sx 1 280	9 000	273	14 000	167

Note : * 25 MPa, 1 850 kg/m³ Concrete. G40.21-M 300W Steel.

Units : d, b, t, L_u (mm) $I_x = 10^6$ mm⁴ $S_x = 10^3$ mm³ M_r = kN·m V_r = kN

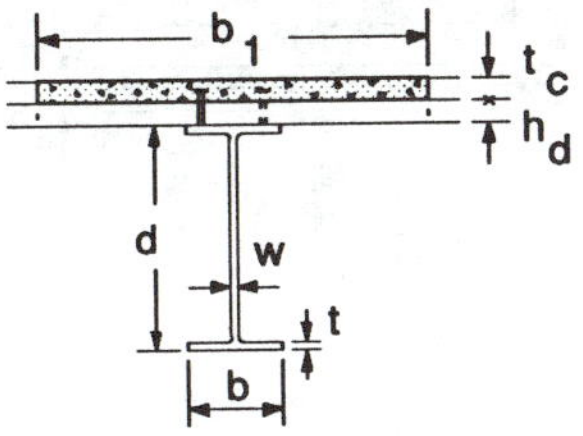

300W
25 MPa

COMPOSITE MEMBERS
Trial Selection Table

75 mm Deck with 85 mm Slab
$\phi = 0.9,\ \phi_c = 0.6,\ \phi_{sc} = 0.8$

Steel Shape	b_1 mm	Composite Beam* — Factored Resistances — M_{rc} (kN·m) for Shear Connections = 100%	70%	40%	Q_r (kN) for 100%	I_t x10⁶ mm⁴	S_t x10³ mm³	Steel Shape Data	Non-Composite Shape — Unbraced Condition — L' mm	M'r kN·m	L' mm	M'r kN·m
W310X79	3 000	756	673	566	2 730	695	2 060	Mr 346	5 000	327	8 000	258
W12X53	2 480	733	660	561	2 690	662	2 030	Vr 480	5 500	316	9 000	235
b=254	1 950	664	604	525	2 110	617	1 980	Lu 4 140	6 000	305	10 000	207
t=14.6	1 430	594	549	489	1 550	558	1 920	Ix 177	6 500	293	11 000	184
d=306	900	521	491	442	975	471	1 820	Sx 1 160	7 000	282	12 000	166
W310X74	3 000	721	639	537	2 560	669	1 940	Mr 321	4 000	307	8 000	206
W12X50	2 480	703	631	534	2 560	638	1 910	Vr 519	4 500	295	9 000	179
b=205	1 950	645	585	505	2 110	596	1 870	Lu 3 380	5 000	282	10 000	159
t=16.3	1 430	575	529	467	1 550	540	1 810	Ix 165	6 000	258	11 000	142
d=310	900	500	470	419	975	457	1 710	Sx 1 060	7 000	233	12 000	129
W310X67	3 000	650	573	479	2 300	606	1 740	Mr 286	4 000	270	8 000	169
W12X45	2 480	636	566	477	2 300	579	1 720	Vr 463	4 500	258	9 000	147
b=204	1 950	600	541	463	2 110	543	1 690	Lu 3 280	5 000	246	10 000	129
t=14.6	1 430	531	486	427	1 550	494	1 640	Ix 145	6 000	222	11 000	116
d=306	900	458	429	381	975	420	1 550	Sx 949	7 000	198	12 000	105
W310X60	2 500	574	507	425	2 050	529	1 550	Mr 254	4 000	237	6 500	179
W12X40	2 060	559	500	423	2 050	504	1 520	Vr 406	4 500	226	7 000	166
b=203	1 630	519	470	404	1 770	472	1 490	Lu 3 200	5 000	214	8 000	139
t=13.1	1 190	460	423	373	1 290	428	1 450	Ix 129	5 500	203	9 000	120
d=304	750	400	375	335	813	363	1 370	Sx 849	6 000	191	10 000	106
W310X52	2 500	524	461	387	1 800	502	1 390	Mr 226	3 000	216	6 000	130
W12X35	2 060	513	456	385	1 800	480	1 370	Vr 431	3 500	203	7 000	106
b=167	1 630	494	446	381	1 770	451	1 340	Lu 2 570	4 000	189	8 000	89.4
t=13.2	1 190	436	399	350	1 290	411	1 300	Ix 118	4 500	176	9 000	77.4
d=318	750	376	352	311	813	351	1 240	Sx 747	5 000	162	10 000	68.4
W310X45	2 500	450	394	329	1 540	437	1 190	Mr 191	3 000	180	6 000	98.2
W12X30	2 060	442	390	328	1 540	419	1 180	Vr 369	3 500	168	7 000	79.3
b=166	1 630	430	384	326	1 540	396	1 160	Lu 2 490	4 000	155	8 000	66.5
t=11.2	1 190	392	357	309	1 290	363	1 130	Ix 99.2	4 500	142	9 000	57.3
d=314	750	334	310	274	813	312	1 070	Sx 634	5 000	128	10 000	50.4
W310X39	2 500	392	342	284	1 330	384	1 040	Mr 165	3 000	153	6 000	77.7
W12X26	2 060	386	339	283	1 330	370	1 030	Vr 320	3 500	142	7 000	62.2
b=165	1 630	377	334	282	1 330	351	1 010	Lu 2 440	4 000	130	8 000	51.8
t=9.7	1 190	358	323	277	1 290	324	986	Ix 85.1	4 500	117	9 000	44.3
d=310	750	301	278	244	813	281	943	Sx 549	5 000	103	10 000	38.8
W250X80	3 000	694	610	503	2 750	558	1 890	Mr 294	5 000	287	8 000	242
W10X54	2 480	668	595	496	2 690	530	1 860	Vr 429	5 500	280	9 000	227
b=255	1 950	599	539	460	2 110	492	1 820	Lu 4 520	6 000	272	10 000	212
t=15.6	1 430	529	484	425	1 550	443	1 760	Ix 126	6 500	265	11 000	197
d=256	900	456	426	384	975	370	1 650	Sx 982	7 000	257	12 000	179

Note : * 25 MPa, 1 850 kg/m³ Concrete. G40.21-M 300W Steel.

Units : d, b, t, L_u (mm) $I_x = 10^6$ mm⁴ $S_x = 10^3$ mm³ M_r = kN·m V_r = kN

COMPOSITE MEMBERS
Trial Selection Table

75 mm Deck with 85 mm Slab
$\phi = 0.9,\ \phi_c = 0.6,\ \phi_{sc} = 0.8$

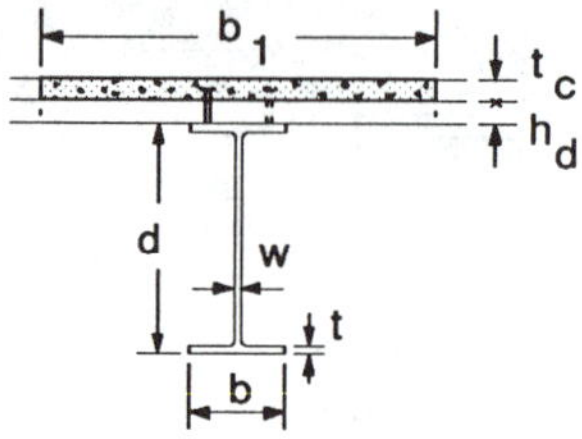

300W
25 MPa

Steel Shape	b_1	Composite Beam*						Steel Shape Data	Non-Composite Shape			
		Factored Resistances				I_t	S_t		Unbraced Condition			
		M_{rc} (kN·m) for Shear Connections =			Q_r (kN) for	x10⁶	x10³		L'	M'r	L'	M'r
	mm	100%	70%	40%	100%	mm⁴	mm³		mm	kN·m	mm	kN·m
W250X73	3 000	637	557	457	2 510	515	1 730	Mr 266	5 000	257	8 000	212
W10X49	2 480	620	548	454	2 510	490	1 700	Vr 389	5 500	250	9 000	198
b=254	1 950	566	507	429	2 110	457	1 660	Lu 4 390	6 000	242	10 000	183
t=14.2	1 430	497	452	394	1 550	412	1 610	Ix 113	6 500	235	11 000	165
d=254	900	424	395	355	975	346	1 520	Sx 891	7 000	227	12 000	149
W250X67	3 000	597	520	426	2 310	490	1 600	Mr 243	4 000	237	8 000	174
W10X45	2 480	583	513	423	2 310	467	1 580	Vr 409	4 500	229	9 000	157
b=204	1 950	546	487	409	2 110	437	1 540	Lu 3 570	5 000	221	10 000	139
t=15.7	1 430	477	432	374	1 550	395	1 500	Ix 104	6 000	205	11 000	125
d=258	900	404	375	333	975	333	1 410	Sx 806	7 000	189	12 000	114
W250X58	2 500	510	444	364	2 000	411	1 370	Mr 208	4 000	199	6 500	160
W10X39	2 060	497	438	362	2 000	391	1 350	Vr 359	4 500	191	7 000	153
b=203	1 630	460	411	346	1 770	365	1 320	Lu 3 410	5 000	184	8 000	137
t=13.5	1 190	402	364	316	1 290	329	1 280	Ix 87.3	5 500	176	9 000	119
d=252	750	341	317	282	813	277	1 210	Sx 693	6 000	168	10 000	105
W250X49	2 500	435	375	305	1 690	353	1 160	Mr 171	4 000	160	6 500	123
W10X33	2 060	425	371	304	1 690	337	1 140	Vr 327	4 500	153	7 000	115
b=202	1 630	411	364	302	1 690	316	1 120	Lu 3 240	5 000	146	8 000	97.2
t=11	1 190	360	324	277	1 290	287	1 090	Ix 70.6	5 500	138	9 000	84
d=248	750	301	278	244	813	244	1 030	Sx 572	6 000	130	10 000	74.1
W250X45	2 500	415	359	293	1 540	355	1 100	Mr 163	3 000	151	6 000	90.6
W10X30	2 060	407	355	292	1 540	340	1 080	Vr 360	3 500	142	7 000	75.2
b=148	1 630	395	349	290	1 540	320	1 060	Lu 2 360	4 000	132	8 000	64.4
t=13	1 190	357	320	273	1 290	292	1 030	Ix 71.1	4 500	122	9 000	56.3
d=266	750	298	274	239	813	250	978	Sx 534	5 000	112	10 000	50.1
W250X39	2 500	359	308	251	1 330	310	950	Mr 139	3 000	126	6 000	69.2
W10X26	2 060	353	305	250	1 330	298	936	Vr 308	3 500	117	7 000	57
b=147	1 630	344	301	248	1 330	282	920	Lu 2 280	4 000	108	8 000	48.6
t=11.2	1 190	326	290	244	1 290	259	895	Ix 60.1	4 500	98.6	9 000	42.3
d=262	750	268	244	212	813	223	853	Sx 459	5 000	88	10 000	37.6
W250X33	2 000	301	258	210	1 130	255	794	Mr 114	3 000	102	5 500	55.6
W10X22	1 650	295	256	210	1 130	244	783	Vr 280	3 500	93.1	6 000	49.4
b=146	1 300	287	252	208	1 130	230	768	Lu 2 180	4 000	84	6 500	44.4
t=9.1	950	266	238	201	1 030	211	747	Ix 48.9	4 500	74.1	7 000	40.3
d=258	600	220	202	174	650	181	710	Sx 379	5 000	63.6	8 000	34.1
W200X71	3 000	580	500	401	2 460	417	1 560	Mr 217	5 000	208	8 000	178
W8X48	2 480	564	492	398	2 460	395	1 530	Vr 393	5 500	203	9 000	168
b=206	1 950	513	454	375	2 110	367	1 500	Lu 4 150	6 000	198	10 000	158
t=17.4	1 430	444	398	339	1 550	330	1 450	Ix 76.6	6 500	193	11 000	148
d=216	900	370	340	301	975	274	1 360	Sx 709	7 000	188	12 000	137

Note : * 25 MPa, 1 850 kg/m³ Concrete. G40.21-M 300W Steel.

Units : d, b, t, L_u (mm) $I_x = 10^6$ mm⁴ $S_x = 10^3$ mm³ M_r = kN·m V_r = kN

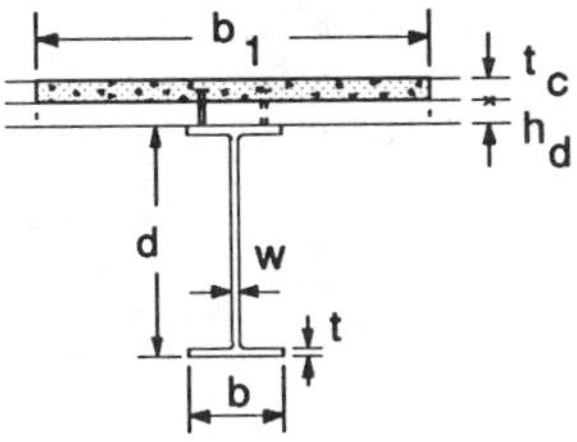

COMPOSITE MEMBERS
Trial Selection Table

300W
25 MPa

75 mm Deck with 85 mm Slab
$\phi = 0.9, \phi_c = 0.6, \phi_{sc} = 0.8$

Steel Shape	b_1	Composite Beam*						Steel Shape Data	Non-Composite Shape			
		Factored Resistances				I_t	S_t		Unbraced Condition			
		M_{rc} (kN·m) for Shear Connections =			Q_r (kN) for	$\times10^6$	$\times10^3$		L'	M'r	L'	M'r
	mm	100%	70%	40%	100%	mm^4	mm^3		mm	kN·m	mm	kN·m
W200X59	3 000	486	415	329	2 040	352	1 300	Mr 176	4 000	174	8 000	134
W8X40	2 480	475	409	328	2 040	336	1 280	Vr 341	4 500	169	9 000	124
b=205	1 950	457	400	325	2 040	314	1 250	Lu 3 780	5 000	164	10 000	114
t=14.2	1 430	395	351	294	1 550	284	1 210	Ix 61.1	6 000	154	11 000	103
d=210	900	324	295	257	975	239	1 150	Sx 582	7 000	144	12 000	93.4
W200X52	2 500	422	360	287	1 800	301	1 140	Mr 154	4 000	150	6 500	126
W8X35	2 060	411	355	285	1 800	286	1 120	Vr 290	4 500	145	7 000	121
b=204	1 630	393	345	281	1 770	267	1 090	Lu 3 620	5 000	140	8 000	111
t=12.6	1 190	336	299	252	1 290	241	1 060	Ix 52.7	5 500	135	9 000	101
d=206	750	276	252	221	813	202	997	Sx 512	6 000	131	10 000	89.7
W200X46	2 500	375	318	252	1 580	270	1 010	Mr 134	4 000	129	6 500	105
W8X31	2 060	367	314	251	1 580	257	991	Vr 262	4 500	124	7 000	99.8
b=203	1 630	354	308	249	1 580	242	971	Lu 3 460	5 000	119	8 000	90.1
t=11	1 190	313	277	230	1 290	219	941	Ix 45.5	5 500	114	9 000	78.6
d=204	750	254	231	200	813	185	890	Sx 448	6 000	109	10 000	69.6
W200X42	2 000	337	287	229	1 430	239	903	Mr 120	3 000	119	5 500	93.5
W8X28	1 650	328	283	227	1 430	227	888	Vr 264	3 500	114	6 000	88.4
b=166	1 300	313	275	224	1 410	212	868	Lu 2 850	4 000	109	6 500	83.4
t=11.8	950	268	239	201	1 030	191	839	Ix 40.9	4 500	104	7 000	77.7
d=206	600	221	202	176	650	160	790	Sx 399	5 000	98.6	8 000	66.7
W200X36	2 000	293	248	196	1 240	209	784	Mr 103	3 000	100	5 500	75.5
W8X24	1 650	286	245	195	1 240	200	771	Vr 223	3 500	95.3	6 000	70.6
b=165	1 300	277	240	194	1 240	187	755	Lu 2 730	4 000	90.4	6 500	64.7
t=10.2	950	246	217	180	1 030	170	732	Ix 34.4	4 500	85.5	7 000	59.2
d=202	600	199	181	156	650	144	692	Sx 342	5 000	80.5	8 000	50.6
W200X31	2 000	263	222	176	1 080	195	699	Mr 90.5	3 000	81.1	5 500	50.6
W8X21	1 650	258	220	175	1 080	187	688	Vr 240	3 500	75.3	6 000	45.6
b=134	1 300	251	216	174	1 080	176	674	Lu 2 150	4 000	69.5	6 500	41.5
t=10.2	950	234	206	169	1 030	161	655	Ix 31.4	4 500	63.6	7 000	38
d=210	600	189	170	145	650	137	621	Sx 299	5 000	57	8 000	32.7
W200X27	2 000	225	189	149	915	169	597	Mr 75.3	3 000	65.4	5 500	36.4
W8X18	1 650	222	187	148	915	163	588	Vr 215	3 500	59.7	6 000	32.6
b=133	1 300	216	185	148	915	154	576	Lu 2 050	4 000	54	6 500	29.6
t=8.4	950	207	180	146	915	142	560	Ix 25.8	4 500	47.5	7 000	27.1
d=208	600	171	153	129	650	122	534	Sx 249	5 000	41.2	8 000	23.1
W150X30	2 500	228	187	141	1 020	149	618	Mr 65.9	3 000	64.1	6 000	46.8
W6X20	2 060	225	185	141	1 020	143	606	Vr 186	3 500	61.2	7 000	40.4
b=153	1 630	219	183	140	1 020	135	593	Lu 2 680	4 000	58.3	8 000	34.8
t=9.3	1 190	210	178	139	1 020	124	576	Ix 17.2	4 500	55.4	9 000	30.6
d=158	750	176	154	124	813	107	548	Sx 219	5 000	52.5	10 000	27.3

Note : * 25 MPa, 1 850 kg/m³ Concrete. G40.21-M 300W Steel.

Units : d, b, t, L_u (mm) $I_x = 10^6\ mm^4$ $S_x = 10^3\ mm^3$ M_r = kN·m V_r = kN

COMPOSITE MEMBERS
Trial Selection Table

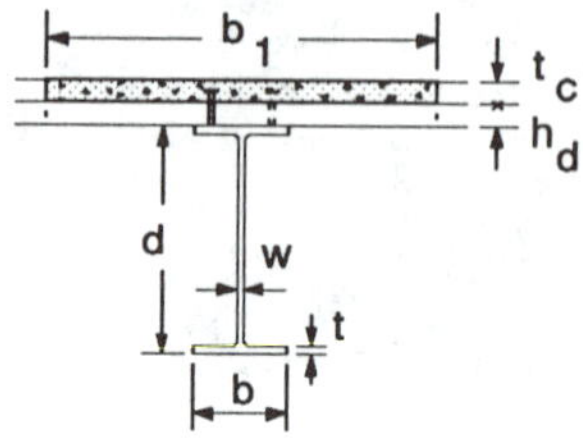

38 mm Deck with 65 mm Slab
$\phi = 0.9$, $\phi_c = 0.6$, $\phi_{sc} = 0.8$

300W
20 MPa

Steel Shape	b_1	Composite Beam*						Steel Shape Data	Non-Composite Shape			
		Factored Resistances				I_t	S_t		Unbraced Condition			
		M_{rc} (kN·m) for Shear Connections =			Q_r (kN) for	$\times 10^6$	$\times 10^3$		L'	M'r	L'	M'r
	mm	100%	70%	40%	100%	mm^4	mm^3		mm	kN·m	mm	kN·m
WWF900X169	5 000	2 840	2 770	2 550	3 320	6 540	8 490	Mr 1 990	5 000	1 860	12 000	625
	4 130	2 790	2 690	2 470	2 740	6 270	8 400	Vr 1 690	6 000	1 680	14 000	491
b=300	3 250	2 720	2 590	2 380	2 150	5 920	8 280	Lu 4 130	7 000	1 480	16 000	402
t=20	2 380	2 600	2 460	2 290	1 580	5 460	8 100	Ix 2930	8 000	1 250	18 000	339
d=900	1 500	2 420	2 310	2 180	995	4 830	7 810	Sx 6 510	10 000	845	20 000	293
WWF800X161	5 000	2 430	2 370	2 200	3 320	5 110	7 350	Mr 1 710	5 000	1 610	12 000	581
	4 130	2 380	2 320	2 140	2 740	4 900	7 270	Vr 1 490	6 000	1 470	14 000	460
b=300	3 250	2 330	2 230	2 060	2 150	4 630	7 170	Lu 4 230	7 000	1 310	16 000	379
t=20	2 380	2 240	2 130	1 970	1 580	4 270	7 010	Ix 2250	8 000	1 130	18 000	322
d=800	1 500	2 080	1 990	1 880	995	3 780	6 770	Sx 5 610	10 000	778	20 000	280
WWF700X152	4 500	2 030	1 970	1 840	2 980	3 780	6 210	Mr 1 440	5 000	1 370	10 000	711
	3 710	1 980	1 930	1 780	2 460	3 620	6 140	Vr 1 290	6 000	1 260	12 000	537
b=300	2 930	1 940	1 860	1 720	1 940	3 420	6 050	Lu 4 330	7 000	1 140	14 000	429
t=20	2 140	1 860	1 780	1 650	1 420	3 150	5 920	Ix 1 660	8 000	1 000	16 000	356
d=700	1 350	1 740	1 670	1 570	895	2 780	5 710	Sx 4 760	9 000	842	18 000	305
W610X174	4 500	2 030	1 970	1 820	2 980	3 330	6 310	Mr 1 450	5 000	1 430	10 000	924
W24X117	3 710	1 980	1 920	1 770	2 460	3 180	6 230	Vr 1 540	6 000	1 350	12 000	709
b=325	2 930	1 930	1 850	1 710	1 940	2 990	6 130	Lu 4 830	7 000	1 250	14 000	574
t=21.6	2 140	1 850	1 760	1 640	1 420	2 740	5 980	Ix 1 470	8 000	1 150	16 000	482
d=616	1 350	1 730	1 660	1 570	895	2 410	5 750	Sx 4 780	9 000	1 040	18 000	415
W610X155	4 000	1 790	1 740	1 610	2 650	2 930	5 570	Mr 1 280	5 000	1 260	10 000	762
W24X104	3 300	1 750	1 690	1 560	2 190	2 790	5 500	Vr 1 390	6 000	1 180	11 000	659
b=324	2 600	1 710	1 630	1 510	1 720	2 630	5 410	Lu 4 740	7 000	1 080	12 000	579
t=19	1 900	1 630	1 560	1 450	1 260	2 410	5 280	Ix 1 290	8 000	988	14 000	465
d=612	1 200	1 520	1 460	1 390	796	2 120	5 080	Sx 4 220	9 000	886	16 000	388
W610X140	4 000	1 660	1 600	1 460	2 650	2 690	5 000	Mr 1 120	4 000	1 050	9 000	486
W24X94	3 300	1 620	1 550	1 410	2 190	2 570	4 930	Vr 1 440	5 000	946	10 000	422
b=230	2 600	1 570	1 490	1 360	1 720	2 420	4 840	Lu 3 320	6 000	829	12 000	334
t=22.2	1 900	1 490	1 410	1 300	1 260	2 210	4 720	Ix 1 120	7 000	695	14 000	277
d=618	1 200	1 370	1 310	1 230	796	1 940	4 520	Sx 3 630	8 000	573	16 000	237
W610X125	4 000	1 490	1 440	1 320	2 650	2 420	4 450	Mr 991	4 000	923	9 000	396
W24X84	3 300	1 450	1 400	1 270	2 190	2 320	4 400	Vr 1 300	5 000	821	10 000	342
b=229	2 600	1 410	1 340	1 220	1 720	2 190	4 320	Lu 3 250	6 000	708	12 000	269
t=19.6	1 900	1 340	1 270	1 160	1 260	2 010	4 220	Ix 985	7 000	575	14 000	222
d=612	1 200	1 240	1 170	1 100	796	1 760	4 050	Sx 3 220	8 000	470	16 000	189
W610X113	4 000	1 360	1 310	1 210	2 650	2 210	4 020	Mr 888	4 000	818	9 000	328
W24X76	3 300	1 320	1 280	1 160	2 190	2 120	3 970	Vr 1 210	5 000	719	10 000	282
b=228	2 600	1 280	1 230	1 110	1 720	2 000	3 910	Lu 3 180	6 000	610	12 000	220
t=17.3	1 900	1 220	1 160	1 060	1 260	1 840	3 820	Ix 875	7 000	481	14 000	180
d=608	1 200	1 130	1 070	996	796	1 610	3 660	Sx 2 880	8 000	391	16 000	153

Note : * 20 MPa, 2 300 kg/m³ Concrete. G40.21-M 300W Steel.

Units : d, b, t, L_u (mm) $I_x = 10^6\ mm^4$ $S_x = 10^3\ mm^3$ M_r = kN·m V_r = kN

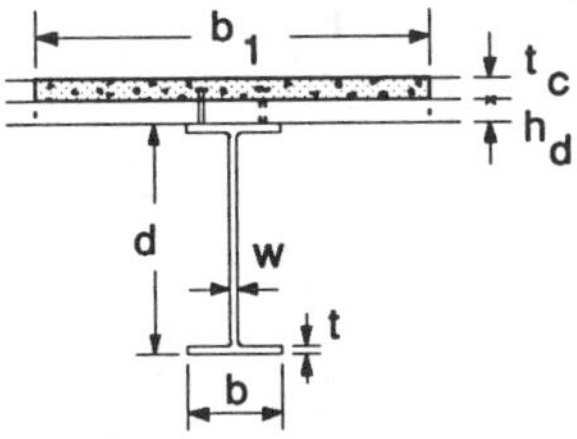

300W
20 MPa

COMPOSITE MEMBERS
Trial Selection Table

38 mm Deck with 65 mm Slab
$\phi = 0.9$, $\phi_c = 0.6$, $\phi_{sc} = 0.8$

Steel Shape	b_1 (mm)	Composite Beam* — Factored Resistances — M_{rc} (kN·m) for Shear Connections = 100%	70%	40%	Q_r (kN) for 100%	I_t $\times 10^6$ mm^4	S_t $\times 10^3$ mm^3	Steel Shape Data	Non-Composite Shape — Unbraced Condition — L' mm	M'r kN·m	L' mm	M'r kN·m
W610X101	3 500	1 220	1 180	1 070	2 320	1 940	3 570	Mr 783	4 000	713	9 000	267
W24X68	2 890	1 180	1 140	1 030	1 920	1 860	3 530	Vr 1 130	5 000	619	10 000	228
b=228	2 280	1 150	1 090	984	1 510	1 760	3 470	Lu 3 110	6 000	512	11 000	199
t=14.9	1 660	1 090	1 020	933	1 100	1 610	3 380	Ix 764	7 000	396	12 000	176
d=604	1 050	996	943	878	696	1 410	3 240	Sx 2 530	8 000	320	14 000	144
W610X91	3 500	1 100	1 060	962	2 320	1 750	3 180	Mr 691	4 000	619	9 000	217
W24X61	2 890	1 070	1 030	923	1 920	1 670	3 150	Vr 1 030	5 000	530	10 000	185
b=227	2 280	1 030	979	879	1 510	1 580	3 100	Lu 3 020	6 000	421	11 000	161
t=12.7	1 660	977	916	829	1 100	1 460	3 020	Ix 667	7 000	325	12 000	142
d=598	1 050	891	839	775	696	1 280	2 900	Sx 2 230	8 000	261	14 000	115
W610X84	3 500	1 020	985	901	2 320	1 630	2 940	Mr 637	3 000	636	8 000	232
W24X56	2 890	992	957	864	1 920	1 570	2 910	Vr 956	4 000	566	9 000	192
b=226	2 280	960	916	821	1 510	1 480	2 870	Lu 2 990	5 000	482	10 000	163
t=11.7	1 660	913	857	773	1 100	1 370	2 800	Ix 613	6 000	377	12 000	125
d=596	1 050	832	783	720	696	1 200	2 690	Sx 2 060	7 000	289	14 000	101
W530X123	3 500	1 300	1 250	1 140	2 320	1 890	3 930	Mr 867	4 000	794	9 000	361
W21X83	2 890	1 260	1 210	1 100	1 920	1 800	3 880	Vr 1 270	5 000	706	10 000	316
b=212	2 280	1 220	1 160	1 050	1 510	1 690	3 810	Lu 3 100	6 000	613	11 000	281
t=21.2	1 660	1 150	1 090	1 000	1 100	1 540	3 700	Ix 761	7 000	505	12 000	253
d=544	1 050	1 060	1 010	954	696	1 340	3 540	Sx 2 800	8 000	421	14 000	211
W530X109	3 500	1 170	1 120	1 030	2 320	1 700	3 490	Mr 764	4 000	692	9 000	291
W21X73	2 890	1 130	1 090	989	1 920	1 620	3 450	Vr 1 120	5 000	608	10 000	254
b=211	2 280	1 100	1 040	948	1 510	1 530	3 390	Lu 3 040	6 000	517	11 000	225
t=18.8	1 660	1 040	983	901	1 100	1 400	3 300	Ix 667	7 000	413	12 000	202
d=540	1 050	958	910	851	696	1 220	3 160	Sx 2 480	8 000	342	14 000	168
W530X101	3 500	1 090	1 050	964	2 320	1 590	3 250	Mr 707	3 000	707	8 000	301
W21X68	2 890	1 060	1 020	929	1 920	1 530	3 210	Vr 1 050	4 000	635	9 000	255
b=210	2 280	1 030	979	889	1 510	1 440	3 160	Lu 2 990	5 000	553	10 000	222
t=17.4	1 660	976	922	844	1 100	1 320	3 080	Ix 617	6 000	462	12 000	176
d=538	1 050	898	852	795	696	1 150	2 950	Sx 2 300	7 000	365	14 000	146
W530X92	3 500	1 010	970	890	2 320	1 460	2 960	Mr 637	3 000	633	8 000	253
W21X62	2 890	979	943	856	1 920	1 400	2 930	Vr 971	4 000	565	9 000	214
b=209	2 280	945	904	817	1 510	1 330	2 880	Lu 2 930	5 000	486	10 000	185
t=15.6	1 660	900	850	773	1 100	1 220	2 810	Ix 552	6 000	393	12 000	146
d=534	1 050	826	782	725	696	1 070	2 700	Sx 2 070	7 000	309	14 000	120
W530X82	3 500	911	872	801	2 320	1 310	2 630	Mr 559	3 000	551	8 000	204
W21X55	2 890	880	846	768	1 920	1 260	2 600	Vr 894	4 000	487	9 000	172
b=209	2 280	847	812	731	1 510	1 190	2 560	Lu 2 860	5 000	412	10 000	148
t=13.3	1 660	808	762	688	1 100	1 100	2 500	Ix 479	6 000	321	12 000	115
d=528	1 050	739	696	641	696	961	2 400	Sx 1 810	7 000	251	14 000	94.8

Note : * 20 MPa, 2 300 kg/m³ Concrete. G40.21-M 300W Steel.

Units : d, b, t, L_u (mm) $I_x = 10^6$ mm⁴ $S_x = 10^3$ mm³ M_r = kN·m V_r = kN

COMPOSITE MEMBERS
Trial Selection Table

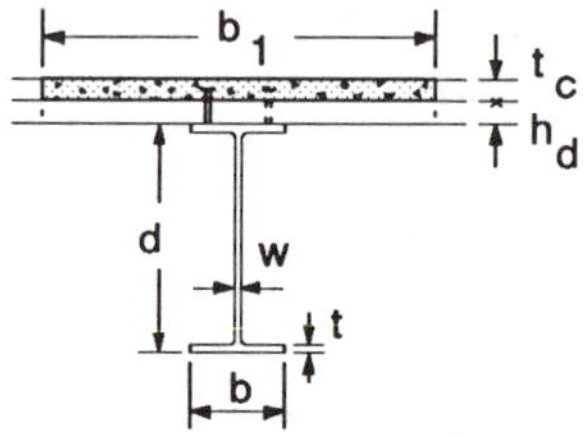

38 mm Deck with 65 mm Slab
$\phi = 0.9$, $\phi_c = 0.6$, $\phi_{sc} = 0.8$

300W
20 MPa

Steel Shape	b_1 mm	Composite Beam* — Factored Resistances — M_{rc} (kN·m) for Shear Connections = 100%	70%	40%	Q_r (kN) for 100%	I_t x10⁶ mm⁴	S_t x10³ mm³	Steel Shape Data	Non-Composite Shape — Unbraced Condition — L' mm	M'r kN·m	L' mm	M'r kN·m
W530X72	3 500	811	775	710	2 320	1 140	2 280	Mr 475	3 000	462	8 000	155
W21X48	2 890	782	750	680	1 920	1 100	2 250	Vr 831	4 000	402	9 000	129
b=207	2 280	750	720	645	1 510	1 050	2 220	Lu 2 750	5 000	331	10 000	111
t=10.9	1 660	715	673	603	1 100	969	2 170	Ix 402	6 000	247	12 000	85.9
d=524	1 050	652	611	557	696	852	2 080	Sx 1 530	7 000	191	14 000	70.1
W460X106	3 500	1 010	967	883	2 320	1 300	3 030	Mr 645	3 000	640	8 000	308
W18X71	2 890	977	937	850	1 920	1 250	2 990	Vr 1 060	4 000	579	9 000	266
b=194	2 280	941	897	813	1 510	1 170	2 930	Lu 2 910	5 000	512	10 000	235
t=20.6	1 660	893	844	772	1 100	1 070	2 850	Ix 488	6 000	444	12 000	190
d=470	1 050	821	779	727	696	930	2 720	Sx 2 080	7 000	366	14 000	160
W460X97	3 500	933	890	817	2 320	1 200	2 770	Mr 589	3 000	581	8 000	264
W18X65	2 890	899	863	786	1 920	1 150	2 740	Vr 947	4 000	522	9 000	227
b=193	2 280	865	828	751	1 510	1 090	2 690	Lu 2 870	5 000	457	10 000	200
t=19	1 660	823	780	712	1 100	995	2 620	Ix 445	6 000	389	12 000	161
d=466	1 050	758	719	668	696	867	2 510	Sx 1 910	7 000	314	14 000	135
W460X89	3 500	875	834	767	2 320	1 130	2 580	Mr 543	3 000	534	8 000	231
W18X60	2 890	843	807	738	1 920	1 080	2 550	Vr 868	4 000	477	9 000	198
b=192	2 280	809	776	705	1 510	1 020	2 500	Lu 2 830	5 000	414	10 000	174
t=17.7	1 660	771	732	666	1 100	937	2 440	Ix 410	6 000	343	12 000	140
d=464	1 050	711	673	624	696	818	2 340	Sx 1 770	7 000	276	14 000	117
W460X82	3 500	808	769	708	2 320	1 030	2 350	Mr 494	3 000	482	8 000	195
W18X55	2 890	777	743	681	1 920	994	2 330	Vr 812	4 000	427	9 000	167
b=191	2 280	744	715	650	1 510	941	2 290	Lu 2 770	5 000	365	10 000	146
t=16	1 660	709	674	613	1 100	866	2 240	Ix 370	6 000	292	12 000	117
d=460	1 050	654	619	571	696	758	2 150	Sx 1 610	7 000	234	14 000	97.3
W460X74	3 500	748	710	653	2 320	951	2 140	Mr 446	3 000	433	8 000	164
W18X50	2 890	717	685	629	1 920	915	2 120	Vr 735	4 000	380	9 000	140
b=190	2 280	686	658	600	1 510	868	2 090	Lu 2 730	5 000	320	10 000	122
t=14.5	1 660	652	622	564	1 100	801	2 040	Ix 333	6 000	249	12 000	96.9
d=458	1 050	603	570	524	696	703	1 960	Sx 1 460	7 000	198	14 000	80.6
W460X67	3 000	672	639	586	1 990	848	1 940	Mr 405	3 000	390	8 000	140
W18X45	2 480	646	617	562	1 640	816	1 920	Vr 688	4 000	339	9 000	119
b=190	1 950	618	593	534	1 290	772	1 890	Lu 2 660	5 000	281	10 000	103
t=12.7	1 430	588	557	502	948	712	1 850	Ix 300	6 000	214	11 000	91.4
d=454	900	539	507	465	597	622	1 770	Sx 1 320	7 000	169	12 000	82
W460X61	3 000	612	581	530	1 990	759	1 730	Mr 354	3 000	336	8 000	111
W18X41	2 480	586	559	508	1 640	731	1 710	Vr 650	4 000	288	9 000	93.9
b=189	1 950	559	536	481	1 290	693	1 680	Lu 2 580	5 000	231	10 000	81.3
t=10.8	1 430	532	503	450	948	642	1 640	Ix 259	6 000	172	11 000	71.7
d=450	900	485	455	414	597	562	1 580	Sx 1 150	7 000	135	12 000	64.1

Note : * 20 MPa, 2 300 kg/m³ Concrete. G40.21-M 300W Steel.

Units : d, b, t, L_u (mm) $I_x = 10^6$ mm⁴ $S_x = 10^3$ mm³ M_r = kN·m V_r = kN

300W
20 MPa

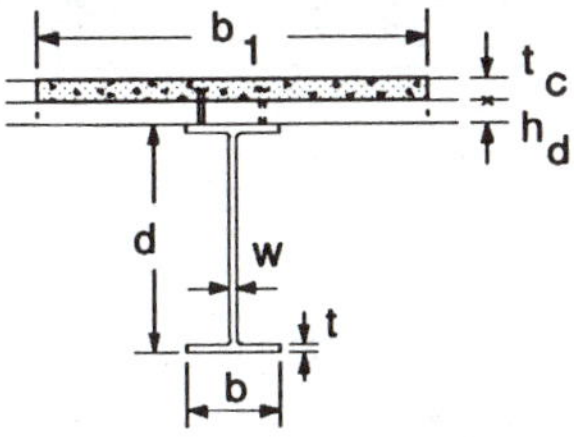

COMPOSITE MEMBERS
Trial Selection Table
38 mm Deck with 65 mm Slab
$\phi = 0.9, \phi_c = 0.6, \phi_{sc} = 0.8$

Steel Shape	b_1 mm	Composite Beam* — Factored Resistances — M_{rc} (kN·m) for Shear Connections = 100%	70%	40%	Q_r (kN) for 100%	I_t x10⁶ mm⁴	S_t x10³ mm³	Steel Shape Data	Non-Composite Shape — Unbraced Condition — L' mm	M'r kN·m	L' mm	M'r kN·m
W410X85	3 500	771	731	669	2 320	906	2 250	Mr 467	3 000	455	8 000	205
W16X57	2 890	739	704	644	1 920	869	2 230	Vr 812	4 000	406	9 000	178
b=181	2 280	706	676	614	1 510	821	2 190	Lu 2 730	5 000	354	10 000	157
t=18.2	1 660	670	637	579	1 100	753	2 130	Ix 315	6 000	297	12 000	127
d=418	1 050	618	585	540	696	656	2 040	Sx 1 510	7 000	243	14 000	107
W410X74	3 500	697	659	602	2 320	810	2 000	Mr 408	3 000	394	8 000	163
W16X50	2 890	667	634	580	1 920	779	1 970	Vr 716	4 000	348	9 000	140
b=180	2 280	634	607	552	1 510	738	1 940	Lu 2 670	5 000	297	10 000	124
t=16	1 660	600	573	519	1 100	680	1 900	Ix 275	6 000	239	12 000	99.8
d=414	1 050	555	524	481	696	595	1 820	Sx 1 330	7 000	194	14 000	83.8
W410X67	3 500	640	604	549	2 320	736	1 800	Mr 367	3 000	352	8 000	135
W16X45	2 890	610	579	530	1 920	709	1 780	Vr 643	4 000	307	9 000	116
b=179	2 280	579	553	505	1 510	673	1 760	Lu 2 610	5 000	258	10 000	102
t=14.4	1 660	546	523	473	1 100	623	1 720	Ix 246	6 000	201	12 000	81.7
d=410	1 050	506	478	437	696	547	1 650	Sx 1 200	7 000	161	14 000	68.4
W410X60	3 000	558	527	481	1 990	641	1 580	Mr 321	3 000	306	8 000	109
W16X40	2 480	533	506	464	1 640	618	1 570	Vr 560	4 000	264	9 000	93.2
b=178	1 950	506	483	442	1 290	586	1 540	Lu 2 580	5 000	217	10 000	81.4
t=12.8	1 430	478	458	415	948	542	1 510	Ix 216	6 000	165	11 000	72.2
d=408	900	443	419	384	597	475	1 450	Sx 1 060	7 000	131	12 000	65
W410X54	3 000	506	475	432	1 840	574	1 410	Mr 284	3 000	266	8 000	86.1
W16X36	2 480	487	461	421	1 640	554	1 400	Vr 540	4 000	225	9 000	73.2
b=177	1 950	461	439	400	1 290	526	1 370	Lu 2 480	5 000	176	10 000	63.6
t=10.9	1 430	434	415	374	948	489	1 350	Ix 186	6 000	132	11 000	56.3
d=404	900	401	377	343	597	429	1 290	Sx 924	7 000	104	12 000	50.5
W410X46	3 000	444	414	373	1 590	504	1 220	Mr 239	2 000	236	7 000	61.7
W16X31	2 480	435	410	371	1 590	487	1 200	Vr 504	3 000	195	8 000	51.8
b=140	1 950	412	391	354	1 290	465	1 190	Lu 1 930	4 000	142	9 000	44.6
t=11.2	1 430	385	368	329	948	433	1 160	Ix 156	5 000	99.9	10 000	39.2
d=404	900	354	332	299	597	383	1 120	Sx 773	6 000	76.4	12 000	31.6
W410X39	3 000	379	351	313	1 350	428	1 030	Mr 197	2 000	193	7 000	44.1
W16X26	2 480	372	348	312	1 350	416	1 020	Vr 456	3 000	155	8 000	36.6
b=140	1 950	361	341	308	1 290	398	1 000	Lu 1 860	4 000	105	9 000	31.3
t=8.8	1 430	335	320	285	948	373	982	Ix 127	5 000	73.1	10 000	27.4
d=400	900	307	288	256	597	332	947	Sx 634	6 000	55.2	12 000	21.9
W360X79	3 000	620	586	537	1 990	650	1 890	Mr 386	4 000	364	8 000	225
W14X53	2 480	593	564	518	1 640	622	1 870	Vr 593	4 500	348	9 000	195
b=205	1 950	565	540	496	1 290	585	1 840	Lu 3 270	5 000	331	10 000	172
t=16.8	1 430	535	513	470	948	536	1 790	Ix 227	6 000	298	11 000	154
d=354	900	497	473	439	597	465	1 710	Sx 1 280	7 000	264	12 000	139

Note : * 20 MPa, 2 300 kg/m³ Concrete. G40.21-M 300W Steel.

Units : d, b, t, L_u (mm) $I_x = 10^6\ mm^4$ $S_x = 10^3\ mm^3$ M_r = kN·m V_r = kN

COMPOSITE MEMBERS
Trial Selection Table

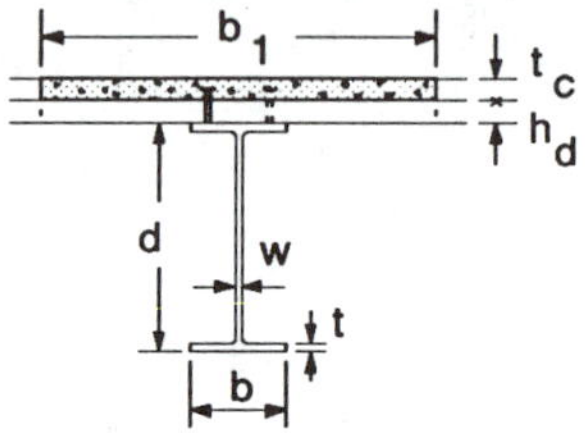

38 mm Deck with 65 mm Slab
$\phi = 0.9, \phi_c = 0.6, \phi_{sc} = 0.8$

300W
20 MPa

Steel Shape	b_1 mm	Composite Beam* — Factored Resistances — M_{rc} (kN·m) for Shear Connections = 100%	70%	40%	Q_r (kN) for 100%	I_t x10⁶ mm⁴	S_t x10³ mm³	Steel Shape Data	Non-Composite Shape — Unbraced Condition — L' mm	M'r kN·m	L' mm	M'r kN·m
W360X72	3 000	570	537	489	1 990	589	1 710	Mr 346	4 000	322	8 000	186
W14X48	2 480	543	515	473	1 640	565	1 690	Vr 536	4 500	306	9 000	161
b=204	1 950	515	492	452	1 290	533	1 660	Lu 3 190	5 000	290	10 000	141
t=15.1	1 430	487	467	427	948	490	1 620	Ix 201	6 000	257	11 000	126
d=350	900	453	430	397	597	426	1 550	Sx 1 150	7 000	222	12 000	114
W360X64	3 000	522	491	445	1 990	534	1 530	Mr 308	4 000	283	8 000	153
W14X43	2 480	497	470	430	1 640	514	1 510	Vr 478	4 500	268	9 000	131
b=203	1 950	470	447	411	1 290	486	1 490	Lu 3 110	5 000	252	10 000	115
t=13.5	1 430	442	425	388	948	448	1 450	Ix 178	6 000	220	11 000	102
d=348	900	411	391	360	597	391	1 400	Sx 1 030	7 000	183	12 000	92.4
W360X57	2 500	465	439	399	1 660	484	1 360	Mr 273	3 000	259	6 000	147
W14X38	2 060	443	420	384	1 370	464	1 350	Vr 504	3 500	243	7 000	119
b=172	1 630	421	402	365	1 080	439	1 330	Lu 2 550	4 000	225	8 000	99.8
t=13.1	1 190	397	379	344	789	403	1 290	Ix 161	4 500	207	9 000	86
d=358	750	366	347	318	497	350	1 240	Sx 897	5 000	189	10 000	75.7
W360X51	2 500	427	401	363	1 660	437	1 220	Mr 241	3 000	227	6 000	121
W14X34	2 060	406	383	350	1 370	420	1 200	Vr 457	3 500	212	7 000	97
b=171	1 630	384	365	333	1 080	398	1 180	Lu 2 500	4 000	195	8 000	81
t=11.6	1 190	361	345	312	789	367	1 160	Ix 141	4 500	178	9 000	69.5
d=356	750	333	315	288	497	320	1 110	Sx 796	5 000	159	10 000	60.9
W360X45	2 500	385	360	324	1 550	389	1 080	Mr 210	3 000	195	6 000	96.1
W14X30	2 060	368	347	315	1 370	375	1 070	Vr 433	3 500	181	7 000	76.5
b=171	1 630	347	330	299	1 080	357	1 050	Lu 2 430	4 000	165	8 000	63.4
t=9.8	1 190	325	310	279	789	330	1 030	Ix 122	4 500	148	9 000	54.1
d=352	750	299	281	255	497	289	986	Sx 691	5 000	128	10 000	47.2
W360X39	2 500	341	316	282	1 340	345	934	Mr 179	2 000	173	6 000	54.2
W14X26	2 060	333	313	281	1 340	333	924	Vr 410	2 500	157	7 000	44.3
b=128	1 630	314	296	267	1 080	317	911	Lu 1 790	3 000	139	8 000	37.5
t=10.7	1 190	291	277	247	789	295	890	Ix 102	4 000	97.2	9 000	32.5
d=354	750	266	250	224	497	259	855	Sx 580	5 000	69.8	10 000	28.8
W360X33	2 500	288	265	235	1 130	291	781	Mr 146	2 000	139	6 000	38.1
W14X22	2 060	283	263	234	1 130	282	773	Vr 362	2 500	124	7 000	30.8
b=127	1 630	273	257	231	1 080	270	762	Lu 1 720	3 000	108	8 000	25.9
t=8.5	1 190	252	239	213	789	252	747	Ix 82.7	4 000	70.3	9 000	22.3
d=350	750	229	215	191	497	224	719	Sx 474	5 000	49.7	10 000	19.6
W310X86	3 500	622	584	529	2 320	592	1 910	Mr 383	5 000	367	10 000	248
W12X58	2 890	591	558	512	1 920	566	1 890	Vr 503	6 000	344	11 000	221
b=254	2 280	559	532	493	1 510	534	1 850	Lu 4 250	7 000	320	12 000	200
t=16.3	1 660	525	505	468	1 100	488	1 810	Ix 199	8 000	297	13 000	182
d=310	1 050	490	471	439	696	424	1 730	Sx 1 280	9 000	273	14 000	167

Note : * 20 MPa, 2 300 kg/m³ Concrete. G40.21-M 300W Steel.

Units : d, b, t, L_u (mm) $I_x = 10^6$ mm⁴ $S_x = 10^3$ mm³ M_r = kN·m V_r = kN

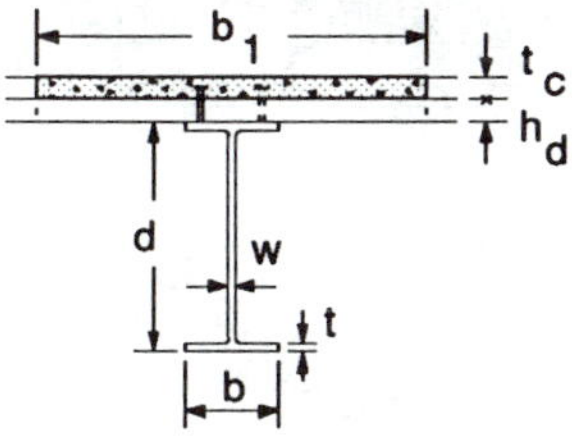

300W
20 MPa

COMPOSITE MEMBERS
Trial Selection Table

38 mm Deck with 65 mm Slab
$\phi = 0.9$, $\phi_c = 0.6$, $\phi_{sc} = 0.8$

Steel Shape	b_1	Composite Beam*						Steel Shape Data	Non-Composite Shape			
		Factored Resistances				I_t	S_t		Unbraced Condition			
		M_{rc} (kN·m) for Shear Connections =			Q_r (kN) for	$\times 10^6$	$\times 10^3$		L'	M'r	L'	M'r
	mm	100%	70%	40%	100%	mm^4	mm^3		mm	kN·m	mm	kN·m
W310X79	3 000	555	522	475	1 990	523	1 730	Mr 346	5 000	327	8 000	258
W12X53	2 480	529	501	460	1 640	500	1 700	Vr 480	5 500	316	9 000	235
b=254	1 950	501	478	442	1 290	470	1 670	Lu 4 140	6 000	305	10 000	207
t=14.6	1 430	473	455	419	948	430	1 630	Ix 177	6 500	293	11 000	184
d=306	900	441	422	392	597	371	1 560	Sx 1 160	7 000	282	12 000	166
W310X74	3 000	536	503	455	1 990	503	1 630	Mr 321	4 000	307	8 000	206
W12X50	2 480	509	481	439	1 640	481	1 600	Vr 519	4 500	295	9 000	179
b=205	1 950	481	457	419	1 290	453	1 580	Lu 3 380	5 000	282	10 000	159
t=16.3	1 430	452	433	396	948	414	1 540	Ix 165	6 000	258	11 000	142
d=310	900	419	399	369	597	358	1 470	Sx 1 060	7 000	233	12 000	129
W310X67	3 000	491	460	413	1 990	453	1 460	Mr 286	4 000	270	8 000	169
W12X45	2 480	466	438	398	1 640	435	1 440	Vr 463	4 500	258	9 000	147
b=204	1 950	438	415	381	1 290	410	1 410	Lu 3 280	5 000	246	10 000	129
t=14.6	1 430	410	393	359	948	377	1 380	Ix 145	6 000	222	11 000	116
d=306	900	380	361	333	597	326	1 320	Sx 949	7 000	198	12 000	105
W310X60	2 500	428	401	362	1 660	396	1 290	Mr 254	4 000	237	6 500	179
W12X40	2 060	406	383	350	1 370	379	1 280	Vr 406	4 500	226	7 000	166
b=203	1 630	383	365	335	1 080	357	1 250	Lu 3 200	5 000	214	8 000	139
t=13.1	1 190	360	345	316	789	327	1 220	Ix 129	5 500	203	9 000	120
d=304	750	334	318	293	497	283	1 170	Sx 849	6 000	191	10 000	106
W310X52	2 500	403	377	339	1 660	377	1 160	Mr 226	3 000	216	6 000	130
W12X35	2 060	382	359	326	1 370	362	1 150	Vr 431	3 500	203	7 000	106
b=167	1 630	360	341	311	1 080	342	1 130	Lu 2 570	4 000	189	8 000	89.4
t=13.2	1 190	336	322	292	789	315	1 100	Ix 118	4 500	176	9 000	77.4
d=318	750	310	295	270	497	273	1 050	Sx 747	5 000	162	10 000	68.4
W310X45	2 500	353	328	292	1 540	327	996	Mr 191	3 000	180	6 000	98.2
W12X30	2 060	337	316	285	1 370	314	984	Vr 369	3 500	168	7 000	79.3
b=166	1 630	316	299	272	1 080	299	970	Lu 2 490	4 000	155	8 000	66.5
t=11.2	1 190	294	280	255	789	276	947	Ix 99.2	4 500	142	9 000	57.3
d=314	750	270	257	234	497	241	909	Sx 634	5 000	128	10 000	50.4
W310X39	2 500	309	285	253	1 330	286	867	Mr 165	3 000	153	6 000	77.7
W12X26	2 060	302	281	251	1 330	276	858	Vr 320	3 500	142	7 000	62.2
b=165	1 630	283	266	241	1 080	264	846	Lu 2 440	4 000	130	8 000	51.8
t=9.7	1 190	261	248	226	789	245	828	Ix 85.1	4 500	117	9 000	44.3
d=310	750	238	227	206	497	215	797	Sx 549	5 000	103	10 000	38.8
W250X80	3 000	491	458	410	1 990	402	1 540	Mr 294	5 000	287	8 000	242
W10X54	2 480	464	436	396	1 640	384	1 520	Vr 429	5 500	280	9 000	227
b=255	1 950	436	413	380	1 290	359	1 490	Lu 4 520	6 000	272	10 000	212
t=15.6	1 430	408	390	361	948	327	1 450	Ix 126	6 500	265	11 000	197
d=256	900	378	363	338	597	280	1 380	Sx 982	7 000	257	12 000	179

Note : * 20 MPa, 2 300 kg/m³ Concrete. G40.21-M 300W Steel.

Units : d, b, t, L_u (mm) $I_x = 10^6\ mm^4$ $S_x = 10^3\ mm^3$ M_r = kN·m V_r = kN

COMPOSITE MEMBERS
Trial Selection Table

38 mm Deck with 65 mm Slab
$\phi = 0.9$, $\phi_c = 0.6$, $\phi_{sc} = 0.8$

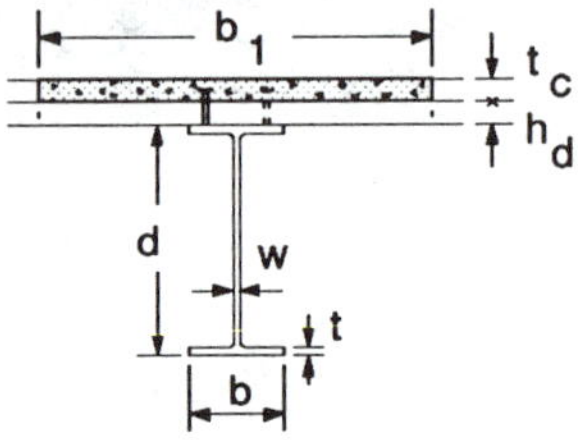

300W
20 MPa

Steel Shape	b_1	Composite Beam*: Factored Resistances: M_{rc} (kN·m) for Shear Connections = 100%	70%	40%	Q_r (kN) for 100%	I_t x10⁶ mm^4	S_t x10³ mm^3	Steel Shape Data	Non-Composite Shape: Unbraced Condition: L'	M'r	L'	M'r
	mm								mm	kN·m	mm	kN·m
W250X73	3 000	457	425	379	1 990	370	1 410	Mr 266	5 000	257	8 000	212
W10X49	2 480	431	404	365	1 640	354	1 390	Vr 389	5 500	250	9 000	198
b=254	1 950	404	381	350	1 290	332	1 360	Lu 4 390	6 000	242	10 000	183
t=14.2	1 430	376	359	333	948	303	1 320	Ix 113	6 500	235	11 000	165
d=254	900	347	334	310	597	260	1 260	Sx 891	7 000	227	12 000	149
W250X67	3 000	438	406	359	1 990	352	1 300	Mr 243	4 000	237	8 000	174
W10X45	2 480	412	384	345	1 640	337	1 290	Vr 409	4 500	229	9 000	157
b=204	1 950	384	361	329	1 290	317	1 260	Lu 3 570	5 000	221	10 000	139
t=15.7	1 430	356	339	311	948	290	1 230	Ix 104	6 000	205	11 000	125
d=258	900	327	313	288	597	249	1 170	Sx 806	7 000	189	12 000	114
W250X58	2 500	369	342	304	1 660	294	1 110	Mr 208	4 000	199	6 500	160
W10X39	2 060	347	324	292	1 370	280	1 100	Vr 359	4 500	191	7 000	153
b=203	1 630	325	306	279	1 080	264	1 080	Lu 3 410	5 000	184	8 000	137
t=13.5	1 190	301	287	263	789	241	1 050	Ix 87.3	5 500	176	9 000	119
d=252	750	277	264	244	497	207	1 000	Sx 693	6 000	168	10 000	105
W250X49	2 500	326	301	264	1 660	250	938	Mr 171	4 000	160	6 500	123
W10X33	2 060	305	284	252	1 370	240	926	Vr 327	4 500	153	7 000	115
b=202	1 630	284	266	241	1 080	226	910	Lu 3 240	5 000	146	8 000	97.2
t=11	1 190	261	248	225	789	208	886	Ix 70.6	5 500	138	9 000	84
d=248	750	238	227	207	497	179	846	Sx 572	6 000	130	10 000	74.1
W250X45	2 500	318	292	256	1 540	255	892	Mr 163	3 000	151	6 000	90.6
W10X30	2 060	301	280	248	1 370	245	881	Vr 360	3 500	142	7 000	75.2
b=148	1 630	280	262	236	1 080	232	866	Lu 2 360	4 000	132	8 000	64.4
t=13	1 190	257	244	220	789	214	845	Ix 71.1	4 500	122	9 000	56.3
d=266	750	234	222	201	497	185	808	Sx 534	5 000	112	10 000	50.1
W250X39	2 500	276	252	219	1 330	222	770	Mr 139	3 000	126	6 000	69.2
W10X26	2 060	269	248	218	1 330	213	761	Vr 308	3 500	117	7 000	57
b=147	1 630	250	233	208	1 080	203	749	Lu 2 280	4 000	108	8 000	48.6
t=11.2	1 190	228	215	194	789	188	731	Ix 60.1	4 500	98.6	9 000	42.3
d=262	750	205	195	176	497	164	701	Sx 459	5 000	88	10 000	37.6
W250X33	2 000	230	210	184	1 130	181	642	Mr 114	3 000	102	5 500	55.6
W10X22	1 650	222	206	182	1 090	174	634	Vr 280	3 500	93.1	6 000	49.4
b=146	1 300	206	192	172	862	165	623	Lu 2 180	4 000	84	6 500	44.4
t=9.1	950	188	178	160	630	153	608	Ix 48.9	4 500	74.1	7 000	40.3
d=258	600	170	161	145	398	133	582	Sx 379	5 000	63.6	8 000	34.1
W200X71	3 000	405	372	325	1 990	285	1 230	Mr 217	5 000	208	8 000	178
W8X48	2 480	379	350	310	1 640	272	1 210	Vr 393	5 500	203	9 000	168
b=206	1 950	351	327	295	1 290	254	1 190	Lu 4 150	6 000	198	10 000	158
t=17.4	1 430	322	304	279	948	231	1 150	Ix 76.6	6 500	193	11 000	148
d=216	900	292	279	258	597	196	1 090	Sx 709	7 000	188	12 000	137

Note : * 20 MPa, 2 300 kg/m³ Concrete. G40.21-M 300W Steel.

Units : d, b, t, L_u (mm) $I_x = 10^6$ mm⁴ $S_x = 10^3$ mm³ M_r = kN·m V_r = kN

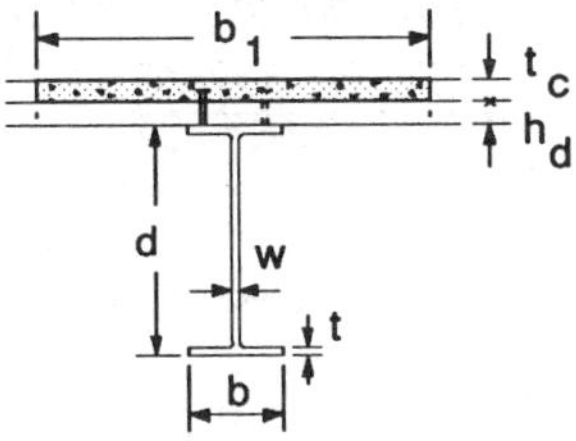

300W
20 MPa

COMPOSITE MEMBERS
Trial Selection Table

38 mm Deck with 65 mm Slab
$\phi = 0.9,\ \phi_c = 0.6,\ \phi_{sc} = 0.8$

Steel Shape	b_1	Composite Beam*							Non-Composite Shape			
		Factored Resistances							Unbraced Condition			
		M_{rc} (kN·m) for Shear Connections =			Q_r (kN) for	I_t x10⁶	S_t x10³	Steel Shape Data	L'	M'r	L'	M'r
	mm	100%	70%	40%	100%	mm^4	mm^3		mm	kN·m	mm	kN·m
W200X59	3 000	355	324	279	1 990	239	1 020	Mr 176	4 000	174	8 000	134
W8X40	2 480	330	303	265	1 640	229	1 010	Vr 341	4 500	169	9 000	124
b=205	1 950	303	281	250	1 290	215	989	Lu 3 780	5 000	164	10 000	114
t=14.2	1 430	276	259	235	948	196	961	Ix 61.1	6 000	154	11 000	103
d=210	900	247	236	216	597	168	914	Sx 582	7 000	144	12 000	93.4
W200X52	2 500	302	276	239	1 660	203	891	Mr 154	4 000	150	6 500	126
W8X35	2 060	281	259	227	1 370	194	878	Vr 290	4 500	145	7 000	121
b=204	1 630	259	241	216	1 080	183	861	Lu 3 620	5 000	140	8 000	111
t=12.6	1 190	236	222	203	789	166	836	Ix 52.7	5 500	135	9 000	101
d=206	750	213	203	186	497	142	795	Sx 512	6 000	131	10 000	89.7
W200X46	2 500	275	250	215	1 580	182	787	Mr 134	4 000	129	6 500	105
W8X31	2 060	257	236	206	1 370	174	776	Vr 262	4 500	124	7 000	99.8
b=203	1 630	236	219	194	1 080	164	762	Lu 3 460	5 000	119	8 000	90.1
t=11	1 190	214	201	182	789	150	741	Ix 45.5	5 500	114	9 000	78.6
d=204	750	191	182	167	497	129	706	Sx 448	6 000	109	10 000	69.6
W200X42	2 000	241	221	191	1 330	161	706	Mr 120	3 000	119	5 500	93.5
W8X28	1 650	224	207	182	1 090	154	695	Vr 264	3 500	114	6 000	88.4
b=166	1 300	207	192	172	862	144	681	Lu 2 850	4 000	109	6 500	83.4
t=11.8	950	188	178	161	630	131	661	Ix 40.9	4 500	104	7 000	77.7
d=206	600	170	161	147	398	112	628	Sx 399	5 000	98.6	8 000	66.7
W200X36	2 000	215	195	167	1 240	140	611	Mr 103	3 000	100	5 500	75.5
W8X24	1 650	202	185	161	1 090	134	602	Vr 223	3 500	95.3	6 000	70.6
b=165	1 300	185	171	151	862	127	591	Lu 2 730	4 000	90.4	6 500	64.7
t=10.2	950	167	157	141	630	116	574	Ix 34.4	4 500	85.5	7 000	59.2
d=202	600	149	142	129	398	99.8	547	Sx 342	5 000	80.5	8 000	50.6
W200X31	2 000	196	177	150	1 080	132	547	Mr 90.5	3 000	81.1	5 500	50.6
W8X21	1 650	190	174	149	1 080	127	539	Vr 240	3 500	75.3	6 000	45.6
b=134	1 300	174	160	141	862	120	529	Lu 2 150	4 000	69.5	6 500	41.5
t=10.2	950	156	146	130	630	110	515	Ix 31.4	4 500	63.6	7 000	38
d=210	600	138	131	117	398	95.5	492	Sx 299	5 000	57	8 000	32.7
W200X27	2 000	169	151	128	915	114	465	Mr 75.3	3 000	65.4	5 500	36.4
W8X18	1 650	165	148	127	915	109	459	Vr 215	3 500	59.7	6 000	32.6
b=133	1 300	156	143	124	862	104	450	Lu 2 050	4 000	54	6 500	29.6
t=8.4	950	139	129	114	630	96.1	439	Ix 25.8	4 500	47.5	7 000	27.1
d=208	600	121	115	102	398	84	420	Sx 249	5 000	41.2	8 000	23.1
W150X30	2 500	166	144	117	1 020	92.7	460	Mr 65.9	3 000	64.1	6 000	46.8
W6X20	2 060	161	142	117	1 020	89.1	452	Vr 186	3 500	61.2	7 000	40.4
b=153	1 630	155	139	116	1 020	84.5	443	Lu 2 680	4 000	58.3	8 000	34.8
t=9.3	1 190	136	124	106	789	78	430	Ix 17.2	4 500	55.4	9 000	30.6
d=158	750	114	106	94.2	497	67.9	411	Sx 219	5 000	52.5	10 000	27.3

Note : * 20 MPa, 2 300 kg/m³ Concrete. G40.21-M 300W Steel.

Units : d, b, t, L_u (mm) $I_x = 10^6\ mm^4$ $S_x = 10^3\ mm^3$ M_r = kN·m V_r = kN

COMPOSITE MEMBERS
Trial Selection Table

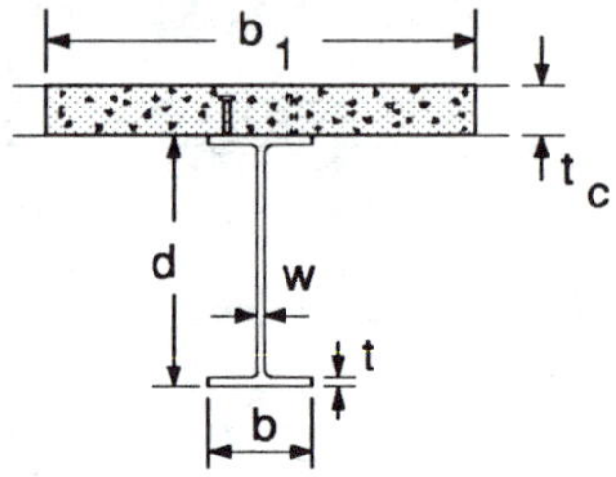

130 mm Slab
$\phi = 0.9, \phi_c = 0.6, \phi_{sc} = 0.8$

300W
20 MPa

Steel Shape	b_1	Composite Beam*						Steel Shape Data	Non-Composite Shape			
		Factored Resistances				I_t	S_t		Unbraced Condition			
		M_{rc} (kN·m) for Shear Connections =			Q_r (kN) for	x10⁶	x10³		L '	M 'r	L '	M 'r
	mm	100%	70%	40%	100%	mm⁴	mm³		mm	kN·m	mm	kN·m
WWF900X169	5 000	3 050	2 980	2 830	5 830	7 390	8 770	Mr 1 990	5 000	1 860	12 000	625
	4 130	2 980	2 940	2 800	5 480	7 160	8 700	Vr 1 690	6 000	1 680	14 000	491
b=300	3 250	2 900	2 850	2 690	4 310	6 860	8 600	Lu 4 130	7 000	1 480	16 000	402
t=20	2 380	2 810	2 760	2 550	3 160	6 430	8 460	Ix 2930	8 000	1 250	18 000	339
d=900	1 500	2 680	2 560	2 370	1 990	5 760	8 210	Sx 6 510	10 000	845	20 000	293
WWF800X161	5 000	2 630	2 560	2 420	5 540	5 760	7 600	Mr 1 710	5 000	1 610	12 000	581
	4 130	2 570	2 530	2 410	5 480	5 590	7 530	Vr 1 490	6 000	1 470	14 000	460
b=300	3 250	2 490	2 450	2 320	4 310	5 350	7 450	Lu 4 230	7 000	1 310	16 000	379
t=20	2 380	2 400	2 370	2 200	3 160	5 020	7 320	Ix 2250	8 000	1 130	18 000	322
d=800	1 500	2 300	2 210	2 050	1 990	4 510	7 110	Sx 5 610	10 000	778	20 000	280
WWF700X152	4 500	2 220	2 160	2 030	5 240	4 290	6 440	Mr 1 440	5 000	1 370	10 000	711
	3 710	2 150	2 110	2 000	4 920	4 150	6 380	Vr 1 290	6 000	1 260	12 000	537
b=300	2 930	2 080	2 040	1 940	3 890	3 970	6 300	Lu 4 330	7 000	1 140	14 000	429
t=20	2 140	2 000	1 970	1 840	2 840	3 720	6 190	Ix 1 660	8 000	1 000	16 000	356
d=700	1 350	1 910	1 850	1 710	1 790	3 320	6 010	Sx 4 760	9 000	842	18 000	305
W610X174	4 500	2 230	2 190	2 060	5 970	3 830	6 580	Mr 1 450	5 000	1 430	10 000	924
W24X117	3 710	2 160	2 120	2 000	4 920	3 700	6 510	Vr 1 540	6 000	1 350	12 000	709
b=325	2 930	2 090	2 050	1 920	3 890	3 520	6 420	Lu 4 830	7 000	1 250	14 000	574
t=21.6	2 140	2 000	1 970	1 820	2 840	3 270	6 290	Ix 1 470	8 000	1 150	16 000	482
d=616	1 350	1 900	1 830	1 700	1 790	2 900	6 080	Sx 4 780	9 000	1 040	18 000	415
W610X155	4 000	1 970	1 930	1 820	5 300	3 370	5 810	Mr 1 280	5 000	1 260	10 000	762
W24X104	3 300	1 910	1 870	1 770	4 380	3 250	5 750	Vr 1 390	6 000	1 180	11 000	659
b=324	2 600	1 840	1 810	1 700	3 450	3 090	5 670	Lu 4 740	7 000	1 080	12 000	579
t=19	1 900	1 770	1 740	1 610	2 520	2 870	5 550	Ix 1 290	8 000	988	14 000	465
d=612	1 200	1 680	1 620	1 500	1 590	2 550	5 370	Sx 4 220	9 000	886	16 000	388
W610X140	4 000	1 840	1 780	1 660	4 830	3 090	5 220	Mr 1 120	4 000	1 050	9 000	486
W24X94	3 300	1 780	1 740	1 630	4 380	2 980	5 160	Vr 1 440	5 000	946	10 000	422
b=230	2 600	1 710	1 670	1 560	3 450	2 840	5 090	Lu 3 320	6 000	829	12 000	334
t=22.2	1 900	1 640	1 600	1 460	2 520	2 640	4 980	Ix 1 120	7 000	695	14 000	277
d=618	1 200	1 540	1 470	1 350	1 590	2 350	4 800	Sx 3 630	8 000	573	16 000	237
W610X125	4 000	1 650	1 590	1 470	4 290	2 760	4 650	Mr 991	4 000	923	9 000	396
W24X84	3 300	1 600	1 560	1 460	4 290	2 670	4 600	Vr 1 300	5 000	821	10 000	342
b=229	2 600	1 530	1 500	1 410	3 450	2 550	4 530	Lu 3 250	6 000	708	12 000	269
t=19.6	1 900	1 460	1 440	1 320	2 520	2 380	4 440	Ix 985	7 000	575	14 000	222
d=612	1 200	1 380	1 320	1 210	1 590	2 120	4 290	Sx 3 220	8 000	470	16 000	189
W610X113	4 000	1 500	1 440	1 330	3 890	2 510	4 200	Mr 888	4 000	818	9 000	328
W24X76	3 300	1 460	1 420	1 320	3 890	2 430	4 160	Vr 1 210	5 000	719	10 000	282
b=228	2 600	1 410	1 380	1 290	3 450	2 320	4 100	Lu 3 180	6 000	610	12 000	220
t=17.3	1 900	1 340	1 310	1 210	2 520	2 180	4 020	Ix 875	7 000	481	14 000	180
d=608	1 200	1 260	1 210	1 110	1 590	1 950	3 880	Sx 2 880	8 000	391	16 000	153

Note : * 20 MPa, 2 300 kg/m³ Concrete. G40.21-M 300W Steel.

Units : d, b, t, L_u (mm) $I_x = 10^6$ mm⁴ $S_x = 10^3$ mm³ M_r = kN·m V_r = kN

300W
20 MPa

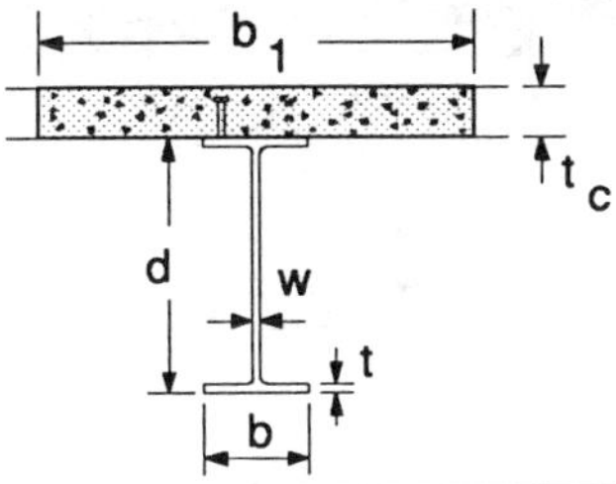

COMPOSITE MEMBERS
Trial Selection Table
130 mm Slab
$\phi = 0.9, \phi_c = 0.6, \phi_{sc} = 0.8$

Steel Shape	b_1	Composite Beam*						Steel Shape Data	Non-Composite Shape			
		Factored Resistances				I_t	S_t		Unbraced Condition			
		M_{rc} (kN·m) for Shear Connections =			Q_r (kN) for	$\times10^6$	$\times10^3$		L'	M'r	L'	M'r
	mm	100%	70%	40%	100%	mm^4	mm^3		mm	kN·m	mm	kN·m
W610X101	3 500	1 340	1 290	1 190	3 510	2 210	3 740	Mr 783	4 000	713	9 000	267
W24X68	2 890	1 310	1 270	1 180	3 510	2 140	3 700	Vr 1 130	5 000	619	10 000	228
b=228	2 280	1 260	1 230	1 150	3 020	2 050	3 640	Lu 3 110	6 000	512	11 000	199
t=14.9	1 660	1 200	1 170	1 070	2 200	1 910	3 570	Ix 764	7 000	396	12 000	176
d=604	1 050	1 130	1 070	977	1 390	1 710	3 450	Sx 2 530	8 000	320	14 000	144
W610X91	3 500	1 210	1 150	1 050	3 130	1 970	3 330	Mr 691	4 000	619	9 000	217
W24X61	2 890	1 180	1 140	1 050	3 130	1 910	3 290	Vr 1 030	5 000	530	10 000	185
b=227	2 280	1 130	1 110	1 030	3 020	1 830	3 250	Lu 3 020	6 000	421	11 000	161
t=12.7	1 660	1 080	1 060	963	2 200	1 720	3 180	Ix 667	7 000	325	12 000	142
d=598	1 050	1 010	967	873	1 390	1 540	3 070	Sx 2 230	8 000	261	14 000	115
W610X84	3 500	1 120	1 060	971	2 890	1 840	3 080	Mr 637	3 000	636	8 000	232
W24X56	2 890	1 090	1 050	967	2 890	1 780	3 050	Vr 956	4 000	566	9 000	192
b=226	2 280	1 060	1 030	961	2 890	1 710	3 000	Lu 2 990	5 000	482	10 000	163
t=11.7	1 660	1 000	984	903	2 200	1 610	2 940	Ix 613	6 000	377	12 000	125
d=596	1 050	942	905	816	1 390	1 440	2 850	Sx 2 060	7 000	289	14 000	101
W530X123	3 500	1 450	1 410	1 300	4 240	2 170	4 130	Mr 867	4 000	794	9 000	361
W21X83	2 890	1 400	1 370	1 280	3 830	2 100	4 080	Vr 1 270	5 000	706	10 000	316
b=212	2 280	1 340	1 310	1 210	3 020	1 990	4 010	Lu 3 100	6 000	613	11 000	281
t=21.2	1 660	1 280	1 250	1 140	2 200	1 850	3 920	Ix 761	7 000	505	12 000	253
d=544	1 050	1 200	1 140	1 050	1 390	1 640	3 770	Sx 2 800	8 000	421	14 000	211
W530X109	3 500	1 300	1 250	1 150	3 750	1 950	3 670	Mr 764	4 000	692	9 000	291
W21X73	2 890	1 260	1 230	1 150	3 750	1 880	3 630	Vr 1 120	5 000	608	10 000	254
b=211	2 280	1 210	1 180	1 100	3 020	1 790	3 570	Lu 3 040	6 000	517	11 000	225
t=18.8	1 660	1 150	1 120	1 030	2 200	1 670	3 490	Ix 667	7 000	413	12 000	202
d=540	1 050	1 080	1 030	943	1 390	1 480	3 360	Sx 2 480	8 000	342	14 000	168
W530X101	3 500	1 220	1 170	1 070	3 480	1 820	3 420	Mr 707	3 000	707	8 000	301
W21X68	2 890	1 180	1 150	1 060	3 480	1 760	3 380	Vr 1 050	4 000	635	9 000	255
b=210	2 280	1 130	1 110	1 030	3 020	1 680	3 320	Lu 2 990	5 000	553	10 000	222
t=17.4	1 660	1 070	1 050	967	2 200	1 570	3 250	Ix 617	6 000	462	12 000	176
d=538	1 050	1 010	968	884	1 390	1 400	3 140	Sx 2 300	7 000	365	14 000	146
W530X92	3 500	1 120	1 070	975	3 190	1 670	3 120	Mr 637	3 000	633	8 000	253
W21X62	2 890	1 090	1 050	971	3 190	1 610	3 080	Vr 971	4 000	565	9 000	214
b=209	2 280	1 050	1 020	955	3 020	1 540	3 030	Lu 2 930	5 000	486	10 000	185
t=15.6	1 660	989	969	893	2 200	1 440	2 970	Ix 552	6 000	393	12 000	146
d=534	1 050	927	894	813	1 390	1 290	2 860	Sx 2 070	7 000	309	14 000	120
W530X82	3 500	1 000	948	861	2 840	1 480	2 770	Mr 559	3 000	551	8 000	204
W21X55	2 890	981	936	858	2 840	1 430	2 740	Vr 894	4 000	487	9 000	172
b=209	2 280	944	919	852	2 840	1 380	2 690	Lu 2 860	5 000	412	10 000	148
t=13.3	1 660	890	871	804	2 200	1 290	2 640	Ix 479	6 000	321	12 000	115
d=528	1 050	830	803	727	1 390	1 160	2 550	Sx 1 810	7 000	251	14 000	94.8

Note : * 20 MPa, 2 300 kg/m³ Concrete. G40.21-M 300W Steel.

Units : d, b, t, L_u (mm) $I_x = 10^6\ mm^4$ $S_x = 10^3\ mm^3$ M_r = kN·m V_r = kN

COMPOSITE MEMBERS
Trial Selection Table

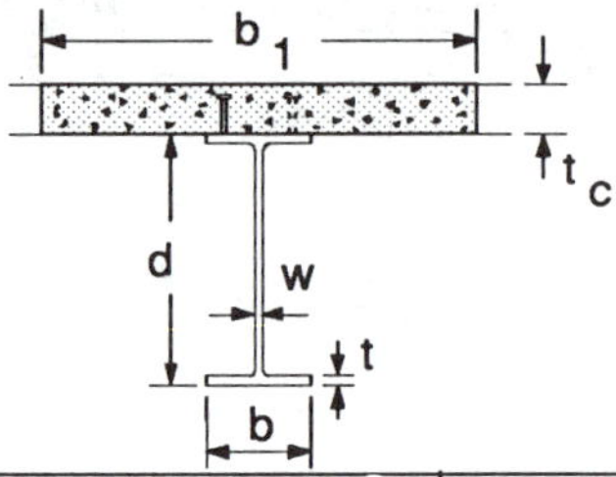

130 mm Slab
$\phi = 0.9, \phi_c = 0.6, \phi_{sc} = 0.8$

300W
20 MPa

Steel Shape	b_1 mm	Composite Beam* Factored Resistances M_{rc} (kN·m) for Shear Connections = 100%	70%	40%	Q_r (kN) for 100%	I_t $\times 10^6$ mm^4	S_t $\times 10^3$ mm^3	Steel Shape Data	Non-Composite Shape Unbraced Condition L' mm	M'r kN·m	L' mm	M'r kN·m
W530X72	3 500	884	829	746	2 470	1 290	2 410	Mr 475	3 000	462	8 000	155
W21X48	2 890	866	820	743	2 470	1 250	2 380	Vr 831	4 000	402	9 000	129
b=207	2 280	838	806	739	2 470	1 200	2 340	Lu 2 750	5 000	331	10 000	111
t=10.9	1 660	791	774	714	2 200	1 130	2 290	Ix 402	6 000	247	12 000	85.9
d=524	1 050	733	712	641	1 390	1 020	2 210	Sx 1 530	7 000	191	14 000	70.1
W460X106	3 500	1 140	1 090	993	3 650	1 510	3 210	Mr 645	3 000	640	8 000	308
W18X71	2 890	1 110	1 070	987	3 650	1 450	3 160	Vr 1 060	4 000	579	9 000	266
b=194	2 280	1 050	1 020	949	3 020	1 380	3 110	Lu 2 910	5 000	512	10 000	235
t=20.6	1 660	990	966	886	2 200	1 280	3 030	Ix 488	6 000	444	12 000	190
d=470	1 050	922	887	810	1 390	1 140	2 910	Sx 2 080	7 000	366	14 000	160
W460X97	3 500	1 050	996	903	3 320	1 380	2 940	Mr 589	3 000	581	8 000	264
W18X65	2 890	1 020	980	897	3 320	1 340	2 900	Vr 947	4 000	522	9 000	227
b=193	2 280	970	946	877	3 020	1 270	2 850	Lu 2 870	5 000	457	10 000	200
t=19	1 660	911	889	821	2 200	1 190	2 780	Ix 445	6 000	389	12 000	161
d=466	1 050	846	819	748	1 390	1 060	2 680	Sx 1 910	7 000	314	14 000	135
W460X89	3 500	982	925	836	3 080	1 290	2 730	Mr 543	3 000	534	8 000	231
W18X60	2 890	954	911	832	3 080	1 250	2 700	Vr 868	4 000	477	9 000	198
b=192	2 280	911	888	823	3 020	1 190	2 650	Lu 2 830	5 000	414	10 000	174
t=17.7	1 660	853	833	771	2 200	1 110	2 590	Ix 410	6 000	343	12 000	140
d=464	1 050	791	769	702	1 390	993	2 490	Sx 1 770	7 000	276	14 000	117
W460X82	3 500	900	844	760	2 810	1 180	2 500	Mr 494	3 000	482	8 000	195
W18X55	2 890	877	832	757	2 810	1 140	2 470	Vr 812	4 000	427	9 000	167
b=191	2 280	841	815	751	2 810	1 090	2 420	Lu 2 770	5 000	365	10 000	146
t=16	1 660	787	768	713	2 200	1 020	2 370	Ix 370	6 000	292	12 000	117
d=460	1 050	727	709	647	1 390	917	2 280	Sx 1 610	7 000	234	14 000	97.3
W460X74	3 500	825	769	691	2 550	1 080	2 280	Mr 446	3 000	433	8 000	164
W18X50	2 890	806	760	688	2 550	1 050	2 250	Vr 735	4 000	380	9 000	140
b=190	2 280	776	745	683	2 550	1 010	2 210	Lu 2 730	5 000	320	10 000	122
t=14.5	1 660	727	710	659	2 200	943	2 160	Ix 333	6 000	249	12 000	96.9
d=458	1 050	668	654	598	1 390	848	2 080	Sx 1 460	7 000	198	14 000	80.6
W460X67	3 000	747	699	629	2 340	970	2 070	Mr 405	3 000	390	8 000	140
W18X45	2 480	728	690	626	2 340	938	2 040	Vr 688	4 000	339	9 000	119
b=190	1 950	699	675	621	2 340	897	2 000	Lu 2 660	5 000	281	10 000	103
t=12.7	1 430	654	639	590	1 900	842	1 960	Ix 300	6 000	214	11 000	91.4
d=454	900	603	587	532	1 190	753	1 890	Sx 1 320	7 000	169	12 000	82
W460X61	3 000	672	625	559	2 100	867	1 850	Mr 354	3 000	336	8 000	111
W18X41	2 480	657	618	556	2 100	839	1 820	Vr 650	4 000	288	9 000	93.9
b=189	1 950	633	606	552	2 100	803	1 780	Lu 2 580	5 000	231	10 000	81.3
t=10.8	1 430	594	581	536	1 900	755	1 740	Ix 259	6 000	172	11 000	71.7
d=450	900	545	532	479	1 190	678	1 680	Sx 1 150	7 000	135	12 000	64.1

Note : * 20 MPa, 2 300 kg/m³ Concrete. G40.21-M 300W Steel.

Units : d, b, t, L_u (mm) $I_x = 10^6\ mm^4$ $S_x = 10^3\ mm^3$ M_r = kN·m V_r = kN

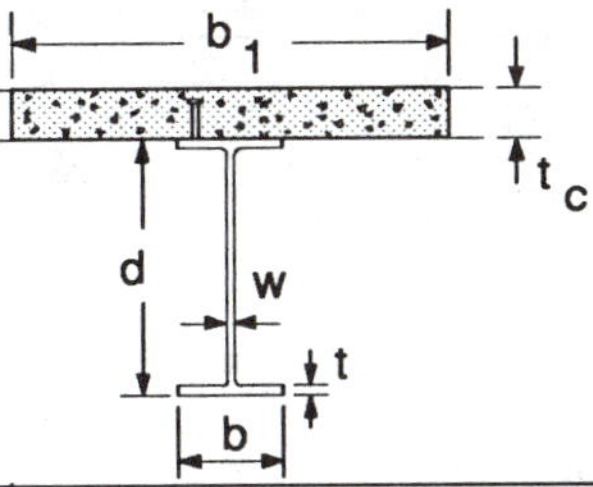

COMPOSITE MEMBERS
Trial Selection Table

300W
20 MPa

130 mm Slab
$\phi = 0.9, \phi_c = 0.6, \phi_{sc} = 0.8$

Steel Shape	b_1	Composite Beam*						Steel Shape Data	Non-Composite Shape			
		Factored Resistances							Unbraced Condition			
		M_{rc} (kN·m) for Shear Connections =			Q_r (kN) for	I_t x10^6	S_t x10^3		L'	M'r	L'	M'r
	mm	100%	70%	40%	100%	mm^4	mm^3		mm	kN·m	mm	kN·m
W410X85	3 500	869	813	726	2 920	1 050	2 420	Mr 467	3 000	455	8 000	205
W16X57	2 890	844	800	722	2 920	1 010	2 380	Vr 812	4 000	406	9 000	178
b=181	2 280	806	781	716	2 920	963	2 330	Lu 2 730	5 000	354	10 000	157
t=18.2	1 660	750	730	675	2 200	898	2 270	Ix 315	6 000	297	12 000	127
d=418	1 050	688	670	612	1 390	800	2 190	Sx 1 510	7 000	243	14 000	107
W410X74	3 500	776	720	641	2 580	933	2 150	Mr 408	3 000	394	8 000	163
W16X50	2 890	756	710	637	2 580	901	2 110	Vr 716	4 000	348	9 000	140
b=180	2 280	726	695	633	2 580	861	2 070	Lu 2 670	5 000	297	10 000	124
t=16	1 660	676	658	609	2 200	806	2 020	Ix 275	6 000	239	12 000	99.8
d=414	1 050	617	603	551	1 390	721	1 940	Sx 1 330	7 000	194	14 000	83.8
W410X67	3 500	702	648	575	2 320	846	1 950	Mr 367	3 000	352	8 000	135
W16X45	2 890	686	640	572	2 320	818	1 910	Vr 643	4 000	307	9 000	116
b=179	2 280	662	628	568	2 320	783	1 880	Lu 2 610	5 000	258	10 000	102
t=14.4	1 660	619	603	557	2 200	734	1 830	Ix 246	6 000	201	12 000	81.7
d=410	1 050	562	549	504	1 390	660	1 760	Sx 1 200	7 000	161	14 000	68.4
W410X60	3 000	615	568	505	2 050	738	1 710	Mr 321	3 000	306	8 000	109
W16X40	2 480	601	561	503	2 050	713	1 680	Vr 560	4 000	264	9 000	93.2
b=178	1 950	578	550	499	2 050	682	1 650	Lu 2 580	5 000	217	10 000	81.4
t=12.8	1 430	541	527	487	1 900	640	1 610	Ix 216	6 000	165	11 000	72.2
d=408	900	491	480	441	1 190	574	1 550	Sx 1 060	7 000	131	12 000	65
W410X54	3 000	555	510	451	1 840	661	1 530	Mr 284	3 000	266	8 000	86.1
W16X36	2 480	544	504	450	1 840	639	1 500	Vr 540	4 000	225	9 000	73.2
b=177	1 950	525	496	447	1 840	611	1 470	Lu 2 480	5 000	176	10 000	63.6
t=10.9	1 430	495	480	442	1 840	575	1 430	Ix 186	6 000	132	11 000	56.3
d=404	900	447	437	399	1 190	517	1 380	Sx 924	7 000	104	12 000	50.5
W410X46	3 000	487	444	390	1 590	582	1 340	Mr 239	2 000	236	7 000	61.7
W16X31	2 480	478	440	388	1 590	562	1 310	Vr 504	3 000	195	8 000	51.8
b=140	1 950	464	433	386	1 590	538	1 280	Lu 1 930	4 000	142	9 000	44.6
t=11.2	1 430	441	422	383	1 590	507	1 240	Ix 156	5 000	99.9	10 000	39.2
d=404	900	398	388	354	1 190	458	1 200	Sx 773	6 000	76.4	12 000	31.6
W410X39	3 000	415	376	328	1 350	498	1 140	Mr 197	2 000	193	7 000	44.1
W16X26	2 480	409	373	327	1 350	481	1 120	Vr 456	3 000	155	8 000	36.6
b=140	1 950	399	369	325	1 350	460	1 090	Lu 1 860	4 000	105	9 000	31.3
t=8.8	1 430	382	360	322	1 350	434	1 060	Ix 127	5 000	73.1	10 000	27.4
d=400	900	347	338	308	1 190	394	1 010	Sx 634	6 000	55.2	12 000	21.9
W360X79	3 000	716	668	593	2 730	761	2 050	Mr 386	4 000	364	8 000	225
W14X53	2 480	690	656	589	2 730	732	2 010	Vr 593	4 500	348	9 000	195
b=205	1 950	651	632	577	2 590	695	1 970	Lu 3 270	5 000	331	10 000	172
t=16.8	1 430	603	586	543	1 900	646	1 920	Ix 227	6 000	298	11 000	154
d=354	900	550	537	495	1 190	571	1 840	Sx 1 280	7 000	264	12 000	139

Note : * 20 MPa, 2 300 kg/m³ Concrete. G40.21-M 300W Steel.

Units : d, b, t, L_u (mm) $I_x = 10^6\ mm^4$ $S_x = 10^3\ mm^3$ M_r = kN·m V_r = kN

COMPOSITE MEMBERS
Trial Selection Table

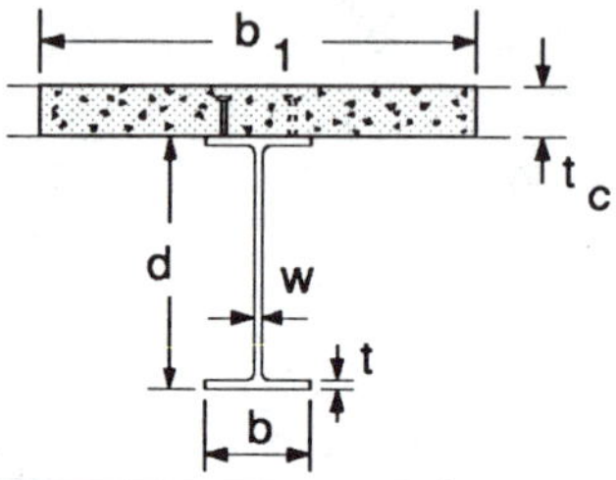

130 mm Slab
$\phi = 0.9$, $\phi_c = 0.6$, $\phi_{sc} = 0.8$

300W
20 MPa

Steel Shape	b_1 mm	Composite Beam* — Factored Resistances — M_{rc} (kN·m) for Shear Connections = 100%	70%	40%	Q_r (kN) for 100%	I_t x10⁶ mm⁴	S_t x10³ mm³	Steel Shape Data		Non-Composite Shape — Unbraced Condition — L' mm	M'r kN·m	L' mm	M'r kN·m
W360X72	3 000	651	603	533	2 460	689	1 850	Mr	346	4 000	322	8 000	186
W14X48	2 480	631	593	529	2 460	663	1 820	Vr	536	4 500	306	9 000	161
b=204	1 950	598	577	524	2 460	631	1 780	Lu	3 190	5 000	290	10 000	141
t=15.1	1 430	552	537	496	1 900	588	1 730	Ix	201	6 000	257	11 000	126
d=350	900	501	489	452	1 190	521	1 660	Sx	1 150	7 000	222	12 000	114
W360X64	3 000	589	542	476	2 200	625	1 670	Mr	308	4 000	283	8 000	153
W14X43	2 480	573	534	473	2 200	602	1 640	Vr	478	4 500	268	9 000	131
b=203	1 950	547	521	469	2 200	573	1 600	Lu	3 110	5 000	252	10 000	115
t=13.5	1 430	505	491	452	1 900	535	1 560	Ix	178	6 000	220	11 000	102
d=348	900	455	445	412	1 190	476	1 500	Sx	1 030	7 000	183	12 000	92.4
W360X57	2 500	528	488	431	1 950	565	1 480	Mr	273	3 000	259	6 000	147
W14X38	2 060	512	480	428	1 950	544	1 460	Vr	504	3 500	243	7 000	119
b=172	1 630	488	468	425	1 950	518	1 420	Lu	2 550	4 000	225	8 000	99.8
t=13.1	1 190	451	438	404	1 580	483	1 390	Ix	161	4 500	207	9 000	86
d=358	750	409	399	365	995	429	1 330	Sx	897	5 000	189	10 000	75.7
W360X51	2 500	477	438	385	1 740	510	1 330	Mr	241	3 000	227	6 000	121
W14X34	2 060	464	432	383	1 740	491	1 310	Vr	457	3 500	212	7 000	97
b=171	1 630	445	422	380	1 740	469	1 280	Lu	2 500	4 000	195	8 000	81
t=11.6	1 190	412	401	369	1 580	437	1 240	Ix	141	4 500	178	9 000	69.5
d=356	750	372	363	332	995	390	1 190	Sx	796	5 000	159	10 000	60.9
W360X45	2 500	426	389	340	1 550	456	1 190	Mr	210	3 000	195	6 000	96.1
W14X30	2 060	416	384	339	1 550	439	1 170	Vr	433	3 500	181	7 000	76.5
b=171	1 630	401	377	336	1 550	419	1 140	Lu	2 430	4 000	165	8 000	63.4
t=9.8	1 190	375	364	332	1 550	392	1 100	Ix	122	4 500	148	9 000	54.1
d=352	750	335	327	299	995	351	1 060	Sx	691	5 000	128	10 000	47.2
W360X39	2 500	377	342	297	1 340	405	1 040	Mr	179	2 000	173	6 000	54.2
W14X26	2 060	370	338	296	1 340	390	1 020	Vr	410	2 500	157	7 000	44.3
b=128	1 630	358	333	294	1 340	372	993	Lu	1 790	3 000	139	8 000	37.5
t=10.7	1 190	338	323	291	1 340	349	962	Ix	102	4 000	97.2	9 000	32.5
d=354	750	302	294	267	995	313	921	Sx	580	5 000	69.8	10 000	28.8
W360X33	2 500	319	286	247	1 130	346	888	Mr	146	2 000	139	6 000	38.1
W14X22	2 060	313	284	246	1 130	332	864	Vr	362	2 500	124	7 000	30.8
b=127	1 630	305	280	245	1 130	317	840	Lu	1 720	3 000	108	8 000	25.9
t=8.5	1 190	291	273	243	1 130	297	812	Ix	82.7	4 000	70.3	9 000	22.3
d=350	750	262	255	231	995	268	776	Sx	474	5 000	49.7	10 000	19.6
W310X86	3 500	723	667	583	2 970	701	2 100	Mr	383	5 000	367	10 000	248
W12X58	2 890	697	654	579	2 970	672	2 050	Vr	503	6 000	344	11 000	221
b=254	2 280	657	635	573	2 970	638	2 010	Lu	4 250	7 000	320	12 000	200
t=16.3	1 660	601	583	536	2 200	591	1 950	Ix	199	8 000	297	13 000	182
d=310	1 050	542	528	494	1 390	522	1 860	Sx	1 280	9 000	273	14 000	167

Note : * 20 MPa, 2 300 kg/m³ Concrete. G40.21-M 300W Steel.

Units : d, b, t, L_u (mm) $I_x = 10^6$ mm⁴ $S_x = 10^3$ mm³ M_r = kN·m V_r = kN

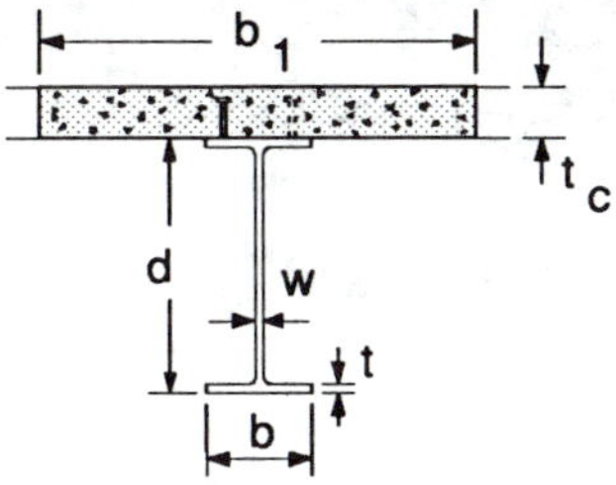

00W
0 MPa

COMPOSITE MEMBERS
Trial Selection Table
130 mm Slab
$\phi = 0.9, \phi_c = 0.6, \phi_{sc} = 0.8$

Steel Shape	b_1	Composite Beam*						Steel Shape Data	Non-Composite Shape			
		Factored Resistances				I_t	S_t		Unbraced Condition			
		M_{rc} (kN·m) for Shear Connections =			Q_r (kN) for	$\times 10^6$	$\times 10^3$		L'	M'r	L'	M'r
	mm	100%	70%	40%	100%	mm^4	mm^3		mm	kN·m	mm	kN·m
W310X79	3 000	650	603	530	2 730	621	1 890	Mr 346	5 000	327	8 000	258
W12X53	2 480	625	591	526	2 730	596	1 850	Vr 480	5 500	316	9 000	235
b=254	1 950	585	567	514	2 590	564	1 810	Lu 4 140	6 000	305	10 000	207
t=14.6	1 430	538	522	482	1 900	522	1 760	Ix 177	6 500	293	11 000	184
d=306	900	486	475	442	1 190	459	1 680	Sx 1 160	7 000	282	12 000	166
W310X74	3 000	623	575	503	2 560	597	1 790	Mr 321	4 000	307	8 000	206
W12X50	2 480	600	564	499	2 560	573	1 750	Vr 519	4 500	295	9 000	179
b=205	1 950	565	547	493	2 560	542	1 710	Lu 3 380	5 000	282	10 000	159
t=16.3	1 430	518	502	461	1 900	503	1 660	Ix 165	6 000	258	11 000	142
d=310	900	466	454	420	1 190	443	1 580	Sx 1 060	7 000	233	12 000	129
W310X67	3 000	564	516	449	2 300	539	1 610	Mr 286	4 000	270	8 000	169
W12X45	2 480	546	507	446	2 300	517	1 580	Vr 463	4 500	258	9 000	147
b=204	1 950	518	493	441	2 300	490	1 540	Lu 3 280	5 000	246	10 000	129
t=14.6	1 430	474	459	420	1 900	455	1 490	Ix 145	6 000	222	11 000	116
d=306	900	424	412	381	1 190	402	1 430	Sx 949	7 000	198	12 000	105
W310X60	2 500	496	456	398	2 050	470	1 430	Mr 254	4 000	237	6 500	179
W12X40	2 060	478	447	395	2 050	451	1 400	Vr 406	4 500	226	7 000	166
b=203	1 630	452	434	391	2 050	428	1 360	Lu 3 200	5 000	214	8 000	139
t=13.1	1 190	413	401	368	1 580	396	1 320	Ix 129	5 500	203	9 000	120
d=304	750	371	362	335	995	350	1 260	Sx 849	6 000	191	10 000	106
W310X52	2 500	457	417	363	1 800	447	1 290	Mr 226	3 000	216	6 000	130
W12X35	2 060	443	411	361	1 800	429	1 260	Vr 431	3 500	203	7 000	106
b=167	1 630	423	401	358	1 800	408	1 230	Lu 2 570	4 000	189	8 000	89.4
t=13.2	1 190	389	377	344	1 580	379	1 190	Ix 118	4 500	176	9 000	77.4
d=318	750	347	338	312	995	336	1 140	Sx 747	5 000	162	10 000	68.4
W310X45	2 500	395	357	309	1 540	388	1 120	Mr 191	3 000	180	6 000	98.2
W12X30	2 060	385	352	307	1 540	373	1 090	Vr 369	3 500	168	7 000	79.3
b=166	1 630	370	345	305	1 540	355	1 060	Lu 2 490	4 000	155	8 000	66.5
t=11.2	1 190	344	332	301	1 540	331	1 030	Ix 99.2	4 500	142	9 000	57.3
d=314	750	304	296	273	995	295	982	Sx 634	5 000	128	10 000	50.4
W310X39	2 500	345	310	267	1 330	343	987	Mr 165	3 000	153	6 000	77.7
W12X26	2 060	338	306	266	1 330	329	960	Vr 320	3 500	142	7 000	62.2
b=165	1 630	327	301	264	1 330	313	933	Lu 2 440	4 000	130	8 000	51.8
t=9.7	1 190	307	291	261	1 330	292	901	Ix 85.1	4 500	117	9 000	44.3
d=310	750	271	263	242	995	261	860	Sx 549	5 000	103	10 000	38.8
W250X80	3 000	587	540	466	2 750	489	1 730	Mr 294	5 000	287	8 000	242
W10X54	2 480	561	527	462	2 750	467	1 690	Vr 429	5 500	280	9 000	227
b=255	1 950	520	502	449	2 590	439	1 640	Lu 4 520	6 000	272	10 000	212
t=15.6	1 430	473	457	417	1 900	404	1 580	Ix 126	6 500	265	11 000	197
d=256	900	421	410	382	1 190	352	1 500	Sx 982	7 000	257	12 000	179

Note : * 20 MPa, 2 300 kg/m³ Concrete. G40.21-M 300W Steel.

Units : d, b, t, L_u (mm) $I_x = 10^6\ mm^4$ $S_x = 10^3\ mm^3$ M_r = kN·m V_r = kN

COMPOSITE MEMBERS
Trial Selection Table

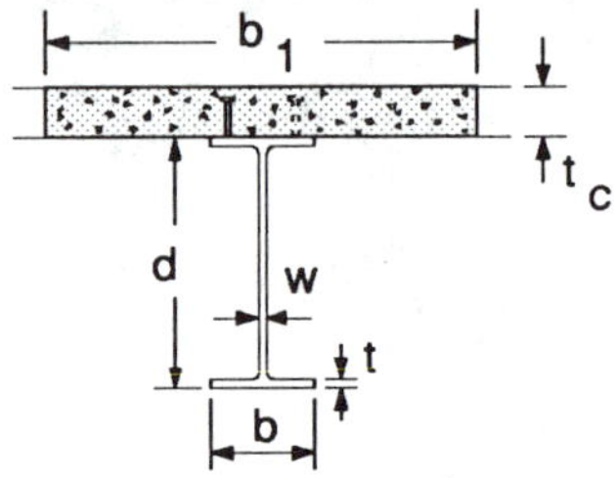

130 mm Slab
$\phi = 0.9, \phi_c = 0.6, \phi_{sc} = 0.8$

300W
20 MPa

Steel Shape	b_1 mm	Composite Beam* — Factored Resistances — M_{rc} (kN·m) for Shear Connections = 100%	70%	40%	Q_r (kN) for 100%	I_t x10⁶ mm⁴	S_t x10³ mm³	Steel Shape Data	Non-Composite Shape — Unbraced Condition — L' mm	M'r kN·m	L' mm	M'r kN·m
W250X73	3 000	541	494	424	2 510	451	1 590	Mr 266	5 000	257	8 000	212
W10X49	2 480	520	483	420	2 510	430	1 550	Vr 389	5 500	250	9 000	198
b=254	1 950	486	467	415	2 510	405	1 500	Lu 4 390	6 000	242	10 000	183
t=14.2	1 430	440	425	386	1 900	373	1 450	Ix 113	6 500	235	11 000	165
d=254	900	390	379	352	1 190	326	1 380	Sx 891	7 000	227	12 000	149
W250X67	3 000	511	463	395	2 310	429	1 480	Mr 243	4 000	237	8 000	174
W10X45	2 480	493	454	392	2 310	410	1 440	Vr 409	4 500	229	9 000	157
b=204	1 950	464	440	388	2 310	386	1 400	Lu 3 570	5 000	221	10 000	139
t=15.7	1 430	420	406	366	1 900	356	1 350	Ix 104	6 000	205	11 000	125
d=258	900	370	359	331	1 190	312	1 280	Sx 806	7 000	189	12 000	114
W250X58	2 500	434	395	337	2 000	359	1 270	Mr 208	4 000	199	6 500	160
W10X39	2 060	417	386	335	2 000	342	1 230	Vr 359	4 500	191	7 000	153
b=203	1 630	392	374	331	2 000	322	1 200	Lu 3 410	5 000	184	8 000	137
t=13.5	1 190	354	342	309	1 580	297	1 150	Ix 87.3	5 500	176	9 000	119
d=252	750	312	303	280	995	259	1 090	Sx 693	6 000	168	10 000	105
W250X49	2 500	373	334	283	1 690	308	1 090	Mr 171	4 000	160	6 500	123
W10X33	2 060	361	328	281	1 690	294	1 050	Vr 327	4 500	153	7 000	115
b=202	1 630	343	320	279	1 690	277	1 020	Lu 3 240	5 000	146	8 000	97.2
t=11	1 190	312	301	270	1 580	255	979	Ix 70.6	5 500	138	9 000	84
d=248	750	272	264	242	995	224	928	Sx 572	6 000	130	10 000	74.1
W250X45	2 500	359	322	273	1 540	312	1 030	Mr 163	3 000	151	6 000	90.6
W10X30	2 060	349	317	271	1 540	298	1 000	Vr 360	3 500	142	7 000	75.2
b=148	1 630	334	309	269	1 540	282	968	Lu 2 360	4 000	132	8 000	64.4
t=13	1 190	308	296	265	1 540	260	931	Ix 71.1	4 500	122	9 000	56.3
d=266	750	268	260	237	995	230	883	Sx 534	5 000	112	10 000	50.1
W250X39	2 500	312	277	234	1 330	274	906	Mr 139	3 000	126	6 000	69.2
W10X26	2 060	305	273	232	1 330	261	876	Vr 308	3 500	117	7 000	57
b=147	1 630	294	268	231	1 330	247	846	Lu 2 280	4 000	108	8 000	48.6
t=11.2	1 190	274	258	227	1 330	229	811	Ix 60.1	4 500	98.6	9 000	42.3
d=262	750	238	230	210	995	203	768	Sx 459	5 000	88	10 000	37.6
W250X33	2 000	261	232	196	1 130	224	753	Mr 114	3 000	102	5 500	55.6
W10X22	1 650	254	229	195	1 130	214	729	Vr 280	3 500	93.1	6 000	49.4
b=146	1 300	244	224	193	1 130	201	703	Lu 2 180	4 000	84	6 500	44.4
t=9.1	950	226	215	190	1 130	186	675	Ix 48.9	4 500	74.1	7 000	40.3
d=258	600	196	190	173	796	165	639	Sx 379	5 000	63.6	8 000	34.1
W200X71	3 000	487	439	368	2 460	358	1 430	Mr 217	5 000	208	8 000	178
W8X48	2 480	466	428	365	2 460	340	1 390	Vr 393	5 500	203	9 000	168
b=206	1 950	433	413	359	2 460	318	1 340	Lu 4 150	6 000	198	10 000	158
t=17.4	1 430	387	372	332	1 900	291	1 290	Ix 76.6	6 500	193	11 000	148
d=216	900	336	324	297	1 190	251	1 210	Sx 709	7 000	188	12 000	137

Note : * 20 MPa, 2 300 kg/m³ Concrete. G40.21-M 300W Steel.

Units : d, b, t, L_u (mm) $I_x = 10^6$ mm⁴ $S_x = 10^3$ mm³ M_r = kN·m V_r = kN

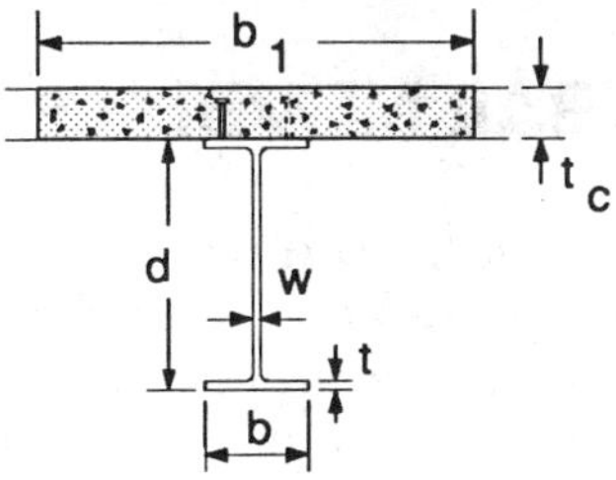

300W
20 MPa

130 mm Slab

$\phi = 0.9, \phi_c = 0.6, \phi_{sc} = 0.8$

Steel Shape	b_1 (mm)	Composite Beam*: Factored Resistances: M_{rc} (kN·m) for Shear Connections = 100%	70%	40%	Q_r (kN) for 100%	I_t x10⁶ mm⁴	S_t x10³ mm³	Steel Shape Data	Non-Composite Shape: Unbraced Condition: L' mm	M'r kN·m	L' mm	M'r kN·m
W200X59	3 000	412	365	303	2 040	304	1 220	Mr 176	4 000	174	8 000	134
W8X40	2 480	397	358	301	2 040	288	1 180	Vr 341	4 500	169	9 000	124
b=205	1 950	375	347	297	2 040	269	1 130	Lu 3 780	5 000	164	10 000	114
t=14.2	1 430	337	324	286	1 900	247	1 080	Ix 61.1	6 000	154	11 000	103
d=210	900	289	278	253	1 190	214	1 020	Sx 582	7 000	144	12 000	93.4
W200X52	2 500	356	316	263	1 800	259	1 060	Mr 154	4 000	150	6 500	126
W8X35	2 060	342	310	261	1 800	245	1 020	Vr 290	4 500	145	7 000	121
b=204	1 630	322	300	258	1 800	229	986	Lu 3 620	5 000	140	8 000	111
t=12.6	1 190	288	276	245	1 580	210	942	Ix 52.7	5 500	135	9 000	101
d=206	750	247	239	218	995	182	885	Sx 512	6 000	131	10 000	89.7
W200X46	2 500	318	280	232	1 580	234	955	Mr 134	4 000	129	6 500	105
W8X31	2 060	308	275	230	1 580	221	919	Vr 262	4 500	124	7 000	99.8
b=203	1 630	292	267	228	1 580	207	882	Lu 3 460	5 000	119	8 000	90.1
t=11	1 190	264	254	223	1 580	189	840	Ix 45.5	5 500	114	9 000	78.6
d=204	750	224	217	196	995	164	788	Sx 448	6 000	109	10 000	69.6
W200X42	2 000	284	252	210	1 430	205	842	Mr 120	3 000	119	5 500	93.5
W8X28	1 650	273	247	208	1 430	194	813	Vr 264	3 500	114	6 000	88.4
b=166	1 300	257	239	206	1 430	182	781	Lu 2 850	4 000	109	6 500	83.4
t=11.8	950	229	220	195	1 260	166	746	Ix 40.9	4 500	104	7 000	77.7
d=206	600	197	190	173	796	144	701	Sx 399	5 000	98.6	8 000	66.7
W200X36	2 000	248	218	180	1 240	181	746	Mr 103	3 000	100	5 500	75.5
W8X24	1 650	240	214	179	1 240	171	717	Vr 223	3 500	95.3	6 000	70.6
b=165	1 300	228	208	177	1 240	160	687	Lu 2 730	4 000	90.4	6 500	64.7
t=10.2	950	207	198	173	1 240	146	653	Ix 34.4	4 500	85.5	7 000	59.2
d=202	600	176	169	153	796	127	612	Sx 342	5 000	80.5	8 000	50.6
W200X31	2 000	225	197	162	1 080	171	675	Mr 90.5	3 000	81.1	5 500	50.6
W8X21	1 650	219	194	161	1 080	162	648	Vr 240	3 500	75.3	6 000	45.6
b=134	1 300	210	189	160	1 080	151	620	Lu 2 150	4 000	69.5	6 500	41.5
t=10.2	950	194	181	157	1 080	138	588	Ix 31.4	4 500	63.6	7 000	38
d=210	600	165	159	142	796	121	551	Sx 299	5 000	57	8 000	32.7
W200X27	2 000	194	168	137	915	151	593	Mr 75.3	3 000	65.4	5 500	36.4
W8X18	1 650	189	166	137	915	142	566	Vr 215	3 500	59.7	6 000	32.6
b=133	1 300	183	162	136	915	132	538	Lu 2 050	4 000	54	6 500	29.6
t=8.4	950	171	157	134	915	121	509	Ix 25.8	4 500	47.5	7 000	27.1
d=208	600	147	141	126	796	106	474	Sx 249	5 000	41.2	8 000	23.1
W150X30	2 500	193	163	128	1 020	137	654	Mr 65.9	3 000	64.1	6 000	46.8
W6X20	2 060	189	161	128	1 020	127	614	Vr 186	3 500	61.2	7 000	40.4
b=153	1 630	182	158	127	1 020	116	574	Lu 2 680	4 000	58.3	8 000	34.8
t=9.3	1 190	171	152	125	1 020	104	531	Ix 17.2	4 500	55.4	9 000	30.6
d=158	750	145	139	120	995	89.3	484	Sx 219	5 000	52.5	10 000	27.3

Note : * 20 MPa, 2 300 kg/m³ Concrete. G40.21-M 300W Steel.

Units : d, b, t, L_u (mm) $I_x = 10^6$ mm⁴ $S_x = 10^3$ mm³ M_r = kN·m V_r = kN

DEFLECTION OF FLEXURAL MEMBERS

CAN/CSA S16.1-M89, *Limit States Design of Steel Structures*, considers deflection to be a serviceability limit state which must be accounted for in the design of flexural members. Appendix I of the S16.1-M89 Standard, *Recommended Maximum Values for Deflections for Specified Live and Wind Loads*, provides some guidance to designers. Deflections tend to be more significant with longer clear spans, shallower members and with the use of high strength steels. Deflection calculations are based on specified loads.

Three methods for dealing with deflection of prismatic beams are summarized below:

1. Compute the required minimum moment of inertia to satisfy the deflection constraint, prior to selection of the beam size.

   $I_{reqd} = W \cdot C_d \cdot B_d$

   where

   I_{reqd} = required value of moment of inertia, (10^6 mm^4)

   W = specified load value as described in Table 5–5, (kN)

   C_d = value of deflection constant obtained from Figure 5–1 for the appropriate span, L , and span/deflection limit, L/Δ . ($10^6 mm^4/kN$)

   B_d = a number to relate the actual load and support condition to a uniformly distributed load (udl) on a simply supported beam, Table 5–5. Values of B_d are computed for the maximum deflection within the span. For a uniformly distributed load (udl), $B_d = 1.0$

The actual deflection of a beam can be computed as:

   $\Delta = (I_{reqd}/I)\,\Delta_m$

   where

   Δ = actual deflection, (mm)

   I = moment of inertia of beam, ($10^6 mm^4$)

   Δ_m = maximum deflection permitted, (mm)

   I_{reqd} = moment of inertia required to meet Δ_m, ($10^6 mm^4$)

2. Compute deflections using the formulae for deflection of beams included with the Beam Diagrams and Formulae provided on pages 5–132 to 5–144.

3. The beam load tables for WWF shapes, W shapes and rectangular HSS on pages 5–102 through 5–131 list approximate deflections for the various steel sections and spans, based on the tabulated uniformly distributed total factored loads at an assumed stress of 215 MPa for 300W steel, and 240 MPa for 350W steel. (See "Vertical Deflection" on page 5–90). Deflections, for live load only or for total load, which result in a stress different from the assumed stress level, can be determined by multiplying the tabulated deflection by the ratio of actual stress to the assumed stress of 215 MPa or 240 MPa.

Examples

Given:

A W410x85 section has been chosen for a simply supported non-composite beam spanning 10 m and subjected to a uniformly distributed specified load of 15 kN/m live and 7 kN/m dead. Check for live load deflection assuming the beam is laterally supported, 300W steel and deflection is limited to = L/300 = 33 mm.

Solutions:

Method 1

Using Figure 5–1 and Table 5–5 on pages 5–86 and 5–87.

$B_d = 1.0$ (simple span udl, Figure 5–5)

$C_d = 1.95 \times 10^6 \text{ mm}^4 / \text{kN}$ (for $L / \Delta = 300$ and $L = 10$ m, Figure 5–1)

$I_{reqd} = W \cdot C_d \cdot B_d$

$= (15 \times 10) \times 1.95 \times 10^6 \times 1.0$

$= 293 \times 10^6 \text{ mm}^4$

For W410x85, $I_x = 315 \times 10^6 \text{ mm}^4$

Actual deflection, $\Delta = (I_{reqd} / I)\, \Delta_m$

$= (293 / 315)\ 33 = 31$ mm

Method 2

From Beam Diagrams and Formulae, page 5–132

$$\Delta = \frac{5\, w\, L^4}{384\, E\, I}$$

$I = 315 \times 10^6 \text{ mm}^4$

$E = 200\,000$ MPa

$$\text{Therefore } \Delta = \frac{5 \times 15 \times (10 \times 10^3)^4}{384 \times 200\,000 \times 315 \times 10^6} = 31 \text{ mm}$$

Method 3

From beam load tables, page 5–113, approximate deflection for W410x85 beam, span 10 m, loaded to a stress of 215 MPa = 55 mm

$$\text{Stress due to live load is } \frac{M}{S} = \frac{W\,L}{8\,S}$$

$$= \frac{(15 \times 10^3)(10 \times 10^3)}{8 \times 1\,510 \times 10^3}$$

$= 124$ MPa

Live load deflection is $(124 / 215)\ 55 = 32$ mm

DEFLECTION CONSTANT C_d

Figure 5-1

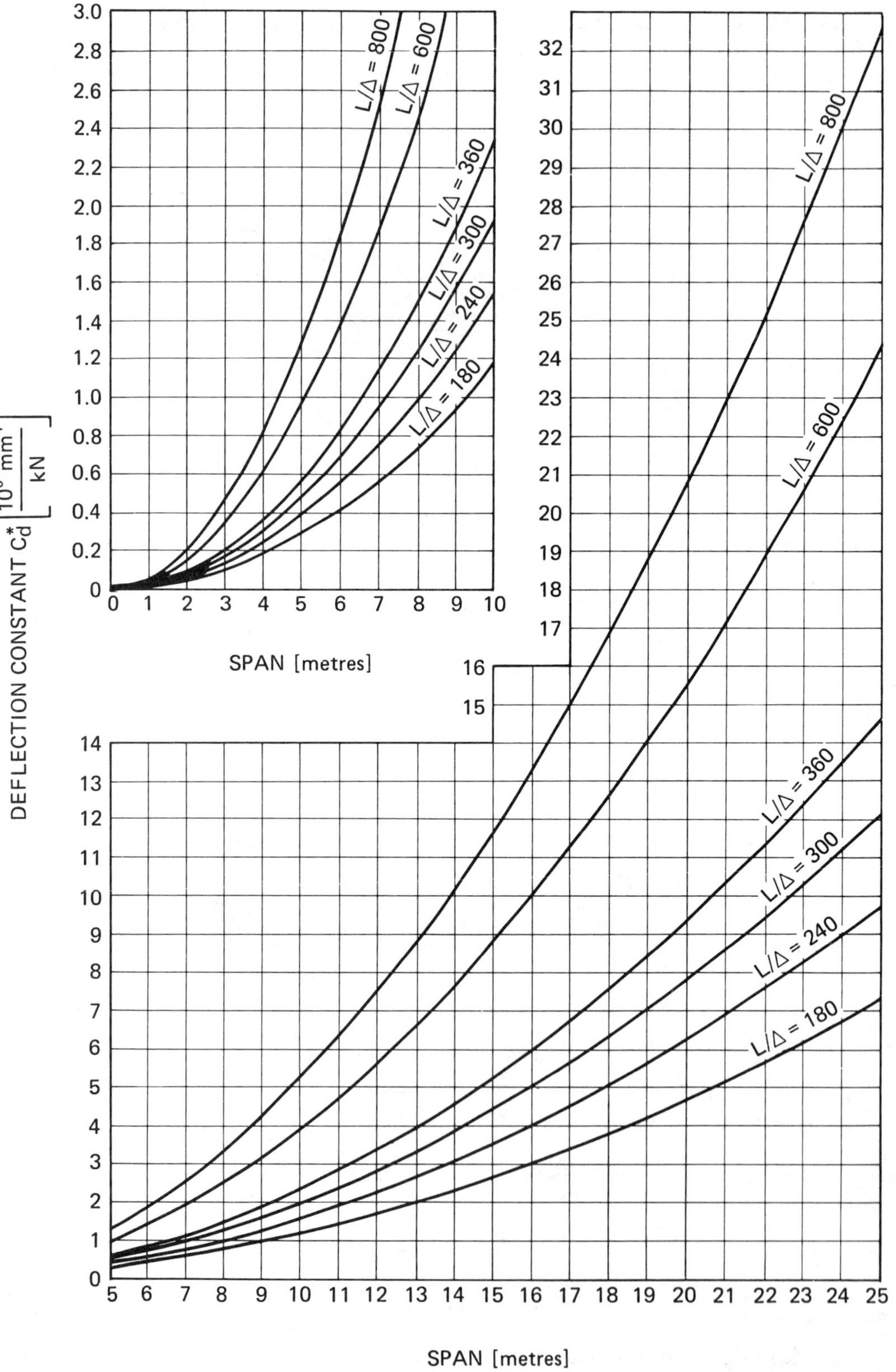

*C_d computed for simple span udl. For other conditions see Table 5-5 for values of B_d.

Table 5-5

Values of B_d for Various Loadings & Support Conditions

LOADING CONDITION	a/L	B_d	LOADING CONDITION	a/L	B_d
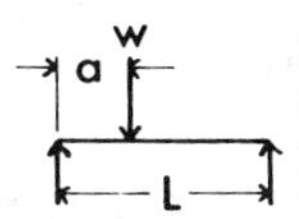	1. .8 .6 .5 .4 .2	0.0 .91 1.5 1.6 1.5 .91	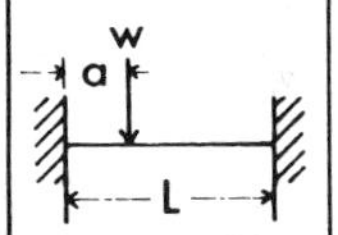	1. .8 .6 .5 .4 .2	0.0 .16 .37 .40 .37 .16
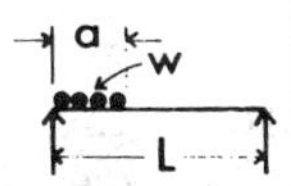	1. .8 .6 .4 .2	1.0 1.13 1.10 .86 .48	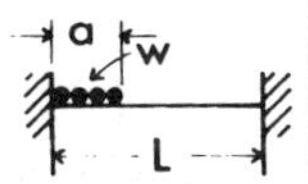	1. .8 .6 .4 .2	.20 .24 .23 .16 .057
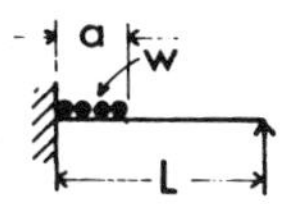	1. .8 .6 .4 .2	.415 .452 .390 .242 .079	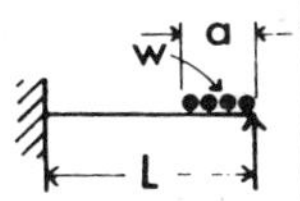	1. .8 .6 .4 .2	.415 .500 .539 .463 .268
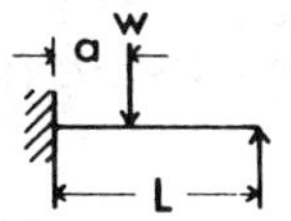	1. .8 .6 .5 .4 .2	0.0 .51 .75 .72 .59 .22	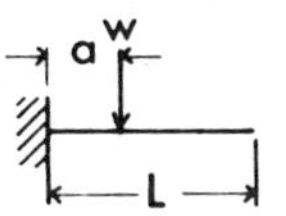	1. .8 .6 .4 .2	25.6 17.8 10.9 5.27 1.41

LOADING CONDITION	B_d	LOADING CONDITION	B_d
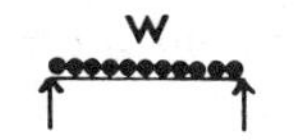	1.0	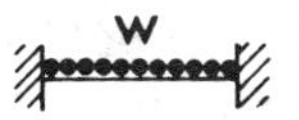	.20
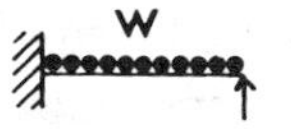	.42	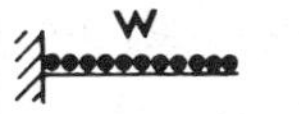	9.6
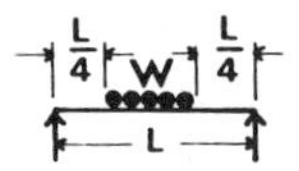	1.4	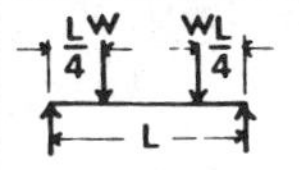	2.2
	1.0		1.3
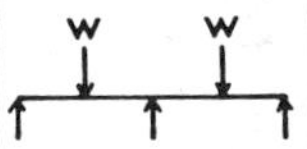	.72	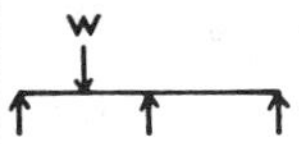	1.2
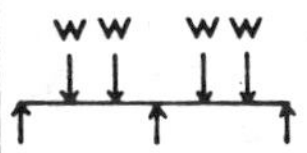	.93	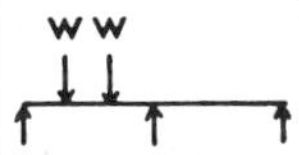	1.9
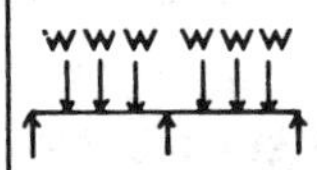	1.6	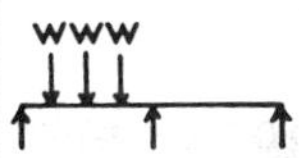	2.7
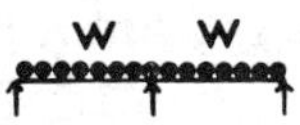	.42	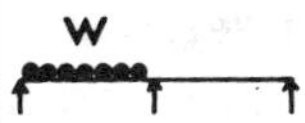	.71
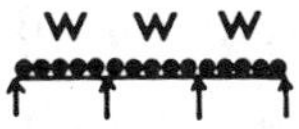	.53		.77
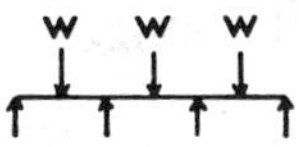	.89	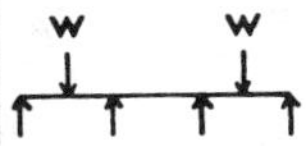	1.24
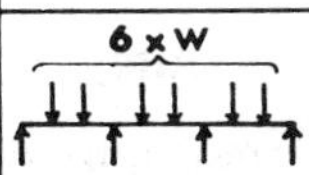	1.44	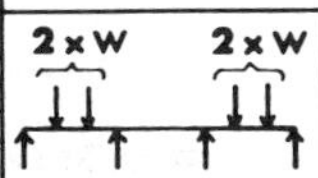	2.1
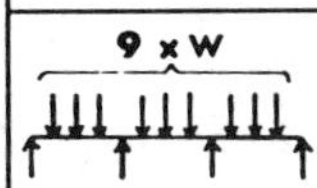	2.1	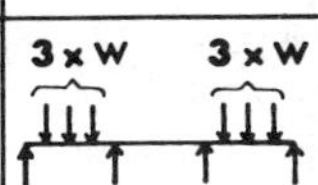	2.9

$$I_{REQUIRED} = W.C_d.B_d$$

Where I 10^6 mm^4
W kN
C_d from graph
B_d from this table
= 1.0 for simple span, udl.

LOADING CONDITION	n	B_d	LOADING CONDITION	n	B_d
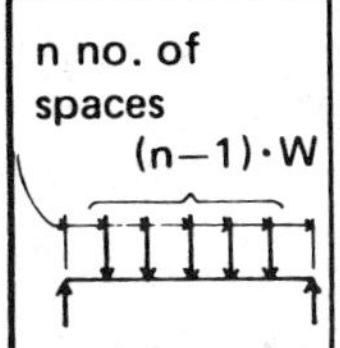	2 3 4 5 6 7	1.60 2.72 3.80 4.84 6.14 6.87	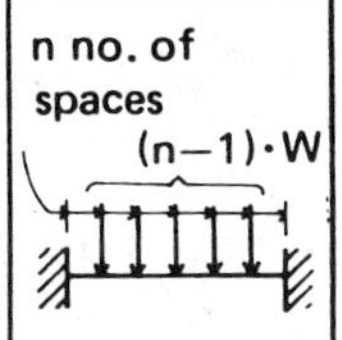	2 3 4 5 6 7	.40 .55 .80 1.0 1.2 1.4

FACTORED RESISTANCE OF BEAMS

General

The blue pages contain tables which can be used to select flexural members and tables for estimating maximum beam reactions. These tables include Beam Selection Tables on pages 5–94 to 5–100 and Beam Load Tables on pages 5–102 to 5–131. The beam selection tables facilitate proportioning of flexural members subject to forces and moments determined by elastic analysis. Some values in these tables may also be used for selection of Class 1 flexural members for which the moments and forces have been determined by plastic analysis. The beam load tables list total uniformly distributed factored loads for laterally supported beams.

When using these tables, factored moments or forces must be equal to or less than the appropriate factored resistances, V_r, M_r or M_r' given in the tables.

Beam Selection Tables

Table

The beam selection tables list beam sizes in descending order of their factored moment resistance, M_r (shown in bold), based on full lateral support (Clause 13.5, CAN/CSA S16.1-M89). Listed beams include all WWF 2000 to WWF 700 beam sizes, Canadian W shapes normally used as beams, and all Canadian C shapes. Two sets of tables are included: 1) based on G40.21-M grade 300W material for the WWF shapes, W shapes and C shapes, with the C shapes listed separately; 2) based on G40.21-M 350W material for WWF shapes only.

Shapes shown in bold type are generally the economy sections, based on both M_r and mass. Other shapes listed below a bold-type section are heavier sections, but may be economy sections when depth limitations require a shallower beam, or when shear resistance of a coped beam influences the beam selection. (See comments on availability of rolled shapes under "Principal Sources of Structural Steel Sections" on page 6–17). Guidance as to factored shear resistance of coped beams is provided in the Beam Load Tables. (See comments on page 5–90.)

For each beam size, the tables list the maximum unsupported length, L_u, for which the factored moment resistance, M_r, is applicable. In addition the tables list the factored moment resistance, M_r', for laterally unsupported beams (Clause 13.6, S16.1-M89) for selected values of the unbraced beam length greater than L_u. For other values of unbraced length greater than L_u, M_r' can be interpolated.

The following items are included in the table:

V_r = factored shear resistance, (kN)

= $\phi A_w F_s$ (Clause 13.4.1.1, S16.1-M89)

I_x = moment of inertia about X-X axis, (10^6 mm^4)

b = flange width, (mm)

L_u = maximum unsupported beam length for which $\mathbf{M_r}$ is applicable, (mm)

$\mathbf{M_r}$ = factored moment resistance for laterally supported member, (kN·m)

= $\phi Z F_y$ for Class 1 and Class 2 sections

= $\phi S F_y$ for Class 3 sections (Clause 13.5, S16.1-M89)

M_r' = factored moment resistance for tabulated unbraced beam length when greater than L_u (kN·m)

(M_r' is computed according to Clause 13.6 of S16.1-M89 using $\omega_2 = 1.0$ in the expression for M_u.)

Design

Compute the maximum bending moment in the beam under factored loads, M_f, and the required moment of inertia, I_{reqd}, to meet the deflection limit using the specified loads. (For I_{reqd}, see "Deflection of Flexural Members", page 5–84)

For a laterally supported beam, proceed up the M_r column until a value of $M_r > M_f$ is obtained. Any beam above will satisfy the factored moment requirement. Check to ensure that $V_r > V_f$, the maximum factored beam shear, that $I_x > I_{reqd}$, and that L_u is greater than the maximum unsupported beam length.

For a laterally unsupported beam, proceed up the M_r column until a value of $M_r > M_f$ is reached, then move to the right across the table to the column headed by the unsupported length (or the first listed unsupported length greater than that required) to obtain a value of M_r'. Proceed up this column comparing a few beams that have an $M_r' > M_f$ and choose the lightest section. Check $V_r > V_f$ and $I_x > I_{reqd}$ if deflection check is necessary.

Other Steel Grades

To use the Beam Selection Tables for beams with a minimum specified yield strength less than 300 MPa, multiply the tabulated values by the ratio of $F_y / 300$ where F_y is the minimum specified yield strength in MPa.

For beams of a steel grade with F_y greater than 300 MPa, use of the tabulated values is conservative.

Plastic Analysis

For beams analysed plastically (Clause 8.5, S16.1-M89), the Beam Selection Tables may be used to facilitate selection of a beam size as follows:

Proceed up the M_r column until a value of $M_r > M_f$ is obtained.

Any beam above will satisfy the factored moment requirement, provided the beam is a Class 1 section. (See Table 5–1 on pages 5–5 to 5–8 for the Class of Sections in Bending for various steel grades).

Check to ensure that $0.83\,V_r > V_f$, the maximum factored beam shear. (Clause 13.4.1.2, S16.1-M89)

Provide suitable lateral bracing. (Clause 13.7, S16.1-M89)

Beam Load Tables — WWF Beams and W Shapes

Loads

The beam load tables list total uniformly distributed factored loads for simply supported beams with the top flange fully supported (ie. the unsupported length of beam is less than or equal to L_u), based on G40.21-M 300W material. Beam sizes listed include the same sections listed in the beam selection tables. To obtain the net supported load (factored), the beam factored dead load should be deducted from the total tabulated load.

For laterally supported beams the beam load tables may also be used to estimate loads for other loading conditions by dividing the tabulated values by the coefficient of the "Equivalent Tabular Load" value for the particular loading condition. (See "Beam Diagrams and Formulae", pages 5–132 to 5–144.) Thus, for a simple beam, laterally supported and carrying equal concentrated loads at third points (loading condition 9), each factored concentrated load is 3/8 of the tabulated uniformly distributed factored load, and the total load is 3/4 of the tabulated load for the same span.

For steel grades where F_y is less than 300 MPa, multiply the tabulated values by the ratio of $F_y/300$, where F_y is in magapascals. For steel grades with F_y greater than 300 MPa, the tabluated values are conservative.

The beam load tables (sometimes referred to as book loads) are frequently used to obtain maximum reactions for design of connections, when beam reactions are not provided on the structural design drawings. It should be noted that reactions computed from beam load tables are generally lower than beam reactions for compositely designed beams.

Vertical Deflection

The column headed "Approximate Deflection" lists the approximate theoretical mid-span deflection, at assumed bending stress levels of 215 MPa for 300W steel and 240 MPa for 350W steel, for beams of various spans designed to support the tabulated factored loads.

The listed deflections are based on the nominal depth of the beam, and are calculated using the formula $\Delta = \frac{5}{384}\frac{W\,L^3}{EI}$.

For E = 200 000 MPa and an assumed bending unit stress of 215 MPa this formula reduces to $\Delta = \frac{224 \times 10^{-6} \times L^2}{d}$ where:

Δ = deflection, (mm)

W = total uniform load including the dead load of the beam, (kN)

L = beam span, (mm)

E = modulus of elasticity, (MPa)

I = moment of inertia of beam, (mm^4)

d = depth of beam, (mm)

More accurate deflections can be determined by multiplying the approximate deflection values listed by the ratio of actual bending stress to the assumed unit bending stress of 215 MPa for 300W steel or 240 MPa for 350W steel. (See also "Deflection of Flexural Members", page 5–84).

Web Shear

For beams with very short spans, with high end shear, and with flanges coped at the supports, the loads for beams may be limited by the shear capacity of the web rather than the bending capacity of the section. The designer should consider the effect of copes on the load carrying capacity of beams when selecting appropriate member sizes.

Both the depth and length of copes can vary considerably depending on the relative size and elevations of intersecting beams. For rolled shapes and for welded wide flange sections, the beam load tables list the factored shear resistance, V_r, for uncoped beams (Clause 13.4.1.1, S16.1-M89). Also listed for rolled shapes is the factored shear resistance for singly coped beams, V_{r1}, and for doubly coped beams, V_{r2}, for cope depths of 1.5 times the beam k-distance, where $V_{r1} = \phi\,(d - 1.5k)\,w\,F_s$ and $V_{r2} = \phi\,(d - 3k)\,w\,F_s$. V_{r1} and V_{r2} should be adjusted as necessary for the actual depth and length of copes required, to account for possible non-uniform shear distribution across the effective web depth and to account for local web instability.

Web Crippling & Yielding

Bearing stiffeners are required when the factored compressive resistance (Clause 15.9, S16.1-M89) of the web is exceeded. For most common beam sizes the bearing resistance is governed by web crippling, except for very short bearing lengths when web yielding may govern.

The beam load tables for W shapes list values of, R (kN), the maximum factored end reaction for 100 mm of bearing per Clause 15.9(b)(ii), and values of the increment in bearing resistance, G (kN), per 10 millimetres of bearing length. For steels other than G40.21-M 300W, values of R and G can be computed by multiplying the listed values by the ratio $\sqrt{F_y / 300}$.

For WWF shapes, values of R and G are not listed, since the web slenderness of WWF shapes generally exceeds $1100/\sqrt{F_y}$, and bearing stiffeners would be required (Clause 15.6.1, S16.1-M89).

Proper lateral support must be provided for the top flanges of beams at the reaction point to ensure that the web crippling strength is not decreased.

Properties & Dimensions

The properties and dimensions listed on the beam load tables for rolled shapes include the beam depth d, the flange width width b, the flange thickness t, the web thickness w, and the beam k-distance, (all in millimetres). Dimensions d, t, w and k are required for calculating the compressive resistance of the web (yielding or crippling) per Clause 15.9, S16.1-M89.

Beam Load Tables — Rectangular HSS

The beam load tables for rectangular Hollow Structural Sections (HSS) on pages 5–126 to 5–131 are based on G40.21-M 350W steel and list total uniformly distributed factored loads for laterally supported rectangular hollow sections in strong axis bending. As with the beam load tables for W shapes, the "Equivalent Tabular Load" values for other loading conditions may be used.

Approximate deflections listed are based on an assumed bending stress of 240 MPa.

Examples

1. Given:

Design a simply supported beam spanning 8 metres to carry a uniformly distributed load of 15 kN/m specified live load and 7 kN/m specified dead load. The dead load includes an assumed beam dead load of 0.7 kN/m. Live load deflection is limited to L/300. Assume the beam frames into supporting members and that the beam is laterally supported. Use G40.21-M 300W steel.

Solutions:

(a) Using Beam Selection Tables — Elastic Analysis:

$$\text{Factored load} = \alpha_D D + \alpha_L L = (1.25 \times 7) + (1.50 \times 15) = 31.3 \text{ kN/m}$$

$$M_f \text{ (factored load moment)} = \frac{w L^2}{8} = \frac{31.3 \times 8^2}{8} = 250 \text{ kN·m}$$

$$V_f \text{ (factored end shear)} = \frac{w L}{2} = \frac{31.3 \times 8}{2} = 125 \text{ kN}$$

Compute I_{reqd} to meet deflection limit, see page 5–84

For udl, $B_d = 1.0$; — from Figure 5–1, for $L / \Delta = 300$, $C_d = 1.25 \times 10^6$

$I_{reqd} = W \cdot C_d \cdot B_d = (15 \times 8) \times 1.25 \times 10^6 \times 1.0 = 150 \times 10^6 \text{ mm}^4$

From table page 5–97, select a W410x54

$M_r = 284 \quad > 250$ kN·m — OK

$V_r = 540 \quad > 125$ kN — OK

$I_x = 186 \times 10^6 \text{ mm}^4 \quad > 150 \times 10^6 \text{ mm}^4$ — OK

Use W410x54

(b) Using Beam Load Tables:

Total factored load, $W_f = 31.3 \times 8 = 250$ kN

End reaction $= 250 / 2 = 125$ kN

From tables page 5–113, select a W410x54

W_r (8 000) $= 284 \quad > 250$ kN — OK

$V_r = 540 \quad > 125$ kN — (uncoped)

$V_{r1} = 484 \quad > 125$ kN — (top flange coped to 1.5 k)

(If the beam were load bearing rather than framing into a girder, it would be necessary to check that R listed in the tables is greater than or equal to the factored end reaction.)

Approximate deflection listed at assumed stress of 215 MPa = 35 mm

$$\text{Stress at specified live load} = \frac{M_{Live}}{S_x} = \frac{15}{31.3} \times \frac{250 \text{ kN·m} \times 10^6}{924 \times 10^3} = 130 \text{ MPa}$$

$$\text{Live load deflection} = \frac{130}{215} \times 35 = 21 \text{ mm}$$

$$\frac{L}{300} = \frac{8\,000}{300} = 27 \text{ mm} \quad > 21 \text{ mm} \quad \text{— OK}$$

Use W410x54

2. Given:

Same as in Example (1), except assume that beam is laterally supported at quarter points, mid span and ends of beam.

Solution:

As in (1), select a W410x54

$L_u = 2\,480$ mm $\quad > 8\,000 / 4 = 2\,000$ mm — OK

Use W410x54

3. Given:

Same as in Example (1), except assume that beam is laterally supported at mid-point and ends of beam only.

Solution:

As in (1), select a W410x54 on basis of M_r and check for M_r' for unbraced length of 8 000 / 2 = 4 000 mm

M_r' (4 000) = 225 kN·m < 250 kN·m

Thus the W410x54 is not adequate, therefore check further up the table for lightest section with M_r' > 250 kN·m for unbraced length of 4 000 mm.

For W410x60, M_r' (4 000) = 264 kN·m > 250 kN·m — OK

V_r = 560 kN > 125 kN — (uncoped)

$I_x = 216 \times 10^6\ mm^4 > 187 \times 10^6\ mm^4$

Use W410x60

BEAM SELECTION TABLE
WWF and W Shapes

G40.21-M 300W
$\phi = 0.90$

Designation*	V_r	I_x	b	L_u	Factored Moment Resistance (kN·m)						
					M_r	M'_r					
						Unbraced length (mm)					
	kN	$10^6 mm^4$	mm	mm	$\leq L_u$	8 000	10 000	12 000	14 000	16 000	20 000
WWF2000X732	3 640	63 900	550	7 770	**19 400**	19 200	17 600	15 700	13 600	11 100	7 610
WWF1800X700	4 070	50 400	550	7 920	**16 900**	....	15 500	14 000	12 300	10 200	7 120
WWF2000X648	3 600	54 200	550	7 460	**16 700**	16 300	14 800	13 000	10 800	8 510	5 760
WWF1800X659	4 050	46 600	550	7 770	15 700	15 600	14 300	12 800	11 100	8 990	6 190
WWF2000X607	3 590	49 300	550	7 270	**15 300**	14 800	13 300	11 600	9 360	7 330	4 920
WWF1800X617	4 020	42 700	550	7 600	14 500	14 300	13 000	11 500	9 840	7 810	5 330
WWF1600X622	2 360	37 600	550	8 270	14 000	....	13 100	12 000	10 700	9 360	6 580
WWF1800X575	4 000	38 800	550	7 420	**13 300**	12 900	11 700	10 300	8 540	6 700	4 540
WWF2000X542	3 570	41 400	500	6 300	**13 100**	11 900	10 200	8 100	6 070	4 750	3 200
WWF1600X580	2 350	34 600	550	8 130	12 900	....	12 000	10 800	9 610	8 170	5 680
WWF1400X597	2 670	28 100	550	8 480	11 900	....	11 300	10 400	9 400	8 380	6 140
WWF1600X538	2 330	31 600	550	7 970	**11 800**	....	10 800	9 760	8 540	7 050	4 850
WWF1800X510	3 980	32 400	500	6 440	**11 300**	10 400	9 020	7 350	5 530	4 340	2 940
WWF1600X496	2 320	28 400	550	7 800	**10 700**	10 600	9 720	8 660	7 480	6 010	4 090
WWF1400X513	2 670	23 500	550	8 150	9 990	....	9 290	8 430	7 480	6 400	4 460
WWF1400X471	2 660	21 100	550	7 980	**9 020**	9 010	8 300	7 470	6 540	5 410	3 730
WWF1600X431	2 300	23 400	500	6 830	**8 940**	8 460	7 480	6 340	4 920	3 870	2 620
WWF1200X487	2 670	16 700	550	8 370	8 260	....	7 790	7 140	6 440	5 690	4 090
WWF1400X405	2 640	17 300	500	6 990	**7 510**	7 170	6 400	5 500	4 390	3 470	2 380
WWF1100X458	2 050	13 700	550	8 580	7 320	....	6 970	6 440	5 860	5 250	3 900
WWF1200X418	2 670	13 800	500	7 370	6 910	6 740	6 130	5 430	4 680	3 780	2 670
WWF1000X447	2 050	11 100	550	8 750	6 530	....	6 270	5 830	5 350	4 850	3 730
WWF1400X358	2 640	14 500	400	5 410	**6 400**	5 340	4 310	3 130	2 400	1 920	1 360
WWF1200X380	2 670	12 300	500	7 180	6 160	5 940	5 360	4 690	3 890	3 100	2 160
WWF1100X388	2 050	11 200	500	7 570	6 050	5 950	5 450	4 890	4 280	3 550	2 530
WWF900X417	1 260	8 680	550	9 080	5 620	....	5 460	5 120	4 750	4 370	3 550
WWF1000X377	2 050	9 120	500	7 700	5 400	5 340	4 920	4 440	3 940	3 360	2 420
WWF1100X351	2 050	9 940	500	7 380	**5 370**	5 240	4 760	4 210	3 610	2 900	2 030

*Nominal depth in millimetres and mass in kilograms per metre. For imperial designations, see page 6-19.

BEAM SELECTION TABLE
WWF and W Shapes

Designation*	V_r	I_x	b	L_u	Factored Moment Resistance (kN·m)						
					M_r	M'_r					
						Unbraced length (mm)					
	kN	$10^6 mm^4$	mm	mm	$\leq L_u$	5 000	6 000	8 000	10 000	12 000	16 000
WWF1200X333	2 670	10 200	400	5 590	**5 210**	....	5 100	4 460	3 700	2 780	1 740
WWF1000X340	2 050	8 060	500	7 500	4 780	....	....	4 690	4 280	3 820	2 720
WWF900X347	1 260	7 100	500	7 970	4 620	....	....	4 610	4 280	3 920	3 130
WWF1200X302	2 670	8 970	400	5 390	**4 620**	....	4 460	3 840	3 100	2 250	1 390
WWF1100X304	2 050	8 220	400	5 760	4 510	....	4 460	3 940	3 330	2 590	1 630
WWF900X309	1 260	6 240	500	7 780	4 050	....	....	4 020	3 700	3 350	2 520
WWF800X339	1 260	5 500	500	8 180	4 020	....	....	....	3 780	3 490	2 880
WWF1000X293	2 050	6 640	400	5 860	**4 000**	....	3 970	3 540	3 030	2 420	1 550
WWF1100X273	2 050	7 160	400	5 580	**3 970**	....	3 880	3 380	2 790	2 080	1 280
WWF1200X263	2 670	7 250	300	3 850	**3 830**	3 460	3 070	2 100	1 430	1 060	684
WWF800X300	1 260	4 840	500	7 930	3 540	....	....	3 530	3 270	2 980	2 360
WWF1000X262	2 050	5 780	400	5 670	**3 510**	....	3 450	3 030	2 540	1 930	1 210
WWF900X262	1 260	5 110	400	6 120	3 350	....	....	3 040	2 650	2 240	1 450
WWF1100X234	2 050	5 720	300	4 010	**3 240**	2 990	2 680	1 930	1 320	978	632
WWF800X253	1 260	3 950	400	6 240	2 920	....	....	2 680	2 370	2 040	1 380
WWF900X231	1 260	4 410	400	5 930	**2 920**	....	2 910	2 590	2 220	1 760	1 110
WWF1000X223	2 050	4 590	300	4 100	**2 840**	2 640	2 390	1 790	1 230	920	602
WWF800X223	1 260	3 410	400	6 050	**2 520**	....	....	2 270	1 970	1 630	1 050
WWF700X245	1 250	2 950	400	6 420	2 490	....	....	2 320	2 080	1 840	1 310
WWF1000X200	2 050	3 940	300	3 910	**2 480**	2 250	2 010	1 390	945	699	449
WWF900X192	1 260	3 460	300	4 310	**2 320**	2 210	2 030	1 600	1 120	842	555
WWF700X214	1 260	2 540	400	6 190	2 150	....	....	1 960	1 730	1 480	982
W610X241	2 030	2 150	329	5 200	**2 070**	....	1 990	1 770	1 540	1 300	911
WWF800X184	1 260	2 660	300	4 410	2 000	1 920	1 770	1 440	1 040	791	529
WWF900X169	1 260	2 930	300	4 130	1 990	1 860	1 680	1 240	845	625	402
WWF700X196	1 260	2 300	400	6 060	1 940	....	....	1 750	1 520	1 260	812
W610X217	1 850	1 910	328	5 080	**1 850**	....	1 760	1 550	1 320	1 070	745
WWF800X161	1 260	2 250	300	4 230	1 710	1 610	1 470	1 140	778	581	379
WWF700X175	1 260	1 970	300	4 520	1 690	1 640	1 530	1 270	965	741	505
W610X195	1 710	1 680	327	4 940	**1 640**	1 630	1 540	1 340	1 120	870	598
W610X174	1 540	1 470	325	4 830	**1 450**	1 430	1 350	1 150	924	709	482
WWF700X152	1 260	1 660	300	4 330	1 440	1 370	1 260	1 000	711	537	356
W610X155	1 390	1 290	324	4 740	**1 280**	1 260	1 180	988	762	579	388

*Nominal depth in millimetres and mass in kilograms per metre. For imperial designations, see page 6-19.

BEAM SELECTION TABLE
WWF and W Shapes

G40.21-M 300W
$\phi = 0.90$

Designation*	V_r	I_x	b	L_u	Factored Moment Resistance (kN·m)						
					M_r	M_r'					
						Unbraced length (mm)					
	kN	$10^6 mm^4$	mm	mm	$\le L_u$	3 000	4 000	5 000	6 000	8 000	10 000
W610X140	1 440	1 120	230	3 320	**1 120**	....	1 050	946	829	573	422
W610X125	1 300	985	229	3 250	**991**	...	923	821	708	470	342
W530X138	1 440	861	214	3 180	975	...	906	815	720	515	390
W610X113	1 210	875	228	3 180	**888**	...	818	719	610	391	282
W530X123	1 270	761	212	3 100	867	...	794	706	613	421	316
W610X101	1 130	764	228	3 110	**783**	...	713	619	512	320	228
W530X109	1 120	667	211	3 040	764	...	692	608	517	342	254
W530X101	1 050	617	210	2 990	**707**	...	635	553	462	301	222
W610X91	1 030	667	227	3 020	**691**	...	619	530	421	261	185
W460X106	1 060	488	194	2 910	645	640	579	512	444	308	235
W530X92	971	552	209	2 930	637	633	565	486	393	253	185
W610X84	881	613	226	2 990	**637**	636	566	482	377	232	163
W460X97	947	445	193	2 870	589	581	522	457	389	264	200
W530X82	894	479	209	2 860	**559**	551	487	412	321	204	148
W460X89	868	410	192	2 830	543	534	477	414	343	231	174
W460X82	812	370	191	2 770	**494**	482	427	365	292	195	146
W530X72	831	402	207	2 750	**475**	462	402	331	247	155	111
W410X85	812	315	181	2 730	467	455	406	354	297	205	157
W460X74	735	333	190	2 730	446	433	380	320	249	164	122
W410X74	716	275	180	2 670	408	394	348	297	239	163	124
W460X67	688	300	190	2 660	**405**	390	339	281	214	140	103
W360X79	593	227	205	3 270	386	...	364	331	298	225	172
W310X86	503	199	254	4 250	383	...	...	367	344	297	248
W410X67	643	246	179	2 610	**367**	352	307	258	201	135	102
W460X61	650	259	189	2 580	**354**	336	288	231	172	111	81.3
W310X79	480	177	254	4 140	346	...	...	327	305	258	207
W360X72	536	201	204	3 190	346	...	322	290	257	186	141
W310X74	519	165	205	3 380	321	...	307	282	258	206	159
W410X60	560	216	178	2 580	**321**	306	264	217	165	109	81.4
W360X64	478	178	203	3 110	308	...	283	252	220	153	115
W310X67	463	145	204	3 280	286	...	270	246	222	169	129

*Nominal depth in millimetres and mass in kilograms per metre. For imperial designations, see page 6-19.

G40.21-M 300W
$\phi = 0.90$

BEAM SELECTION TABLE
WWF and W Shapes

Designation*	V_r	I_x	b	L_u	Factored Moment Resistance (kN·m)						
					M_r	M'_r					
						Unbraced length (mm)					
	kN	$10^6 mm^4$	mm	mm	$\leq L_u$	2 000	3 000	4 000	5 000	6 000	8 000
W410X54	540	186	177	2 480	**284**	...	266	225	176	132	86.1
W360X57	504	161	172	2 550	273	...	259	225	189	147	99.8
W310X60	406	129	203	3 200	254	...	...	237	214	191	139
W250X67	409	104	204	3 570	243	...	...	237	221	205	174
W360X51	457	141	171	2 500	**241**	...	227	195	159	121	81.0
W410X46	504	156	140	1 930	**239**	236	195	142	99.9	76.4	51.8
W310X52	431	118	167	2 570	226	...	216	189	162	130	89.4
W360X45	433	122	171	2 430	**210**	...	195	165	128	96.1	63.4
W250X58	359	87.3	203	3 410	208	...	...	199	184	168	137
W410X39	447	127	140	1 860	**197**	193	155	105	73.1	55.2	36.6
W310X45	369	99.2	166	2 490	191	...	180	155	128	98.2	66.5
W360X39	410	102	128	1 790	**179**	173	139	97.2	69.8	54.2	37.5
W250X49	327	70.6	202	3 240	171	...	...	160	146	130	97.2
W310X39	320	85.1	165	2 440	**165**	...	153	130	103	77.7	51.8
W250X45	360	71.1	148	2 360	163	...	151	132	112	90.6	64.4
W360X33	362	82.7	127	1 720	**146**	139	108	70.3	49.7	38.1	25.9
W250X39	308	60.1	147	2 280	139	...	126	108	88.0	69.2	48.6
W200X42	264	40.9	166	2 850	120	...	119	109	98.6	88.4	66.7
+W310X31	264	67.0	164	2 460	**118**	...	110	93.4	72.6	54.3	35.6
W250X33	280	48.9	146	2 180	114	...	102	84.0	63.6	49.4	34.1
W200X36	223	34.4	165	2 730	103	...	100	90.4	80.5	70.6	50.6
W200X31	240	31.4	134	2 150	**90.4**	...	81.1	69.5	57.0	45.6	32.7
W150X37	234	22.2	154	2 910	83.7	...	83.2	77.2	71.3	65.4	53.3
W200X27	215	25.8	133	2 050	**75.3**	...	65.4	54.0	41.2	32.6	23.1
+W250X24	226	34.7	145	2 190	**74.2**	...	65.6	53.0	38.2	29.0	19.4
W150X30	186	17.2	153	2 680	65.9	...	64.1	58.3	52.5	46.8	34.8
+W200X21	182	19.8	133	2 080	52.6	...	45.7	37.1	27.4	21.4	14.8
+W150X22	157	12.1	152	2 590	42.9	...	41.2	36.8	32.2	27.3	19.3

+Class 3 Section.

*Nominal depth in millimetres and mass in kilograms per metre. For imperial designations, see page 6-19.

BEAM SELECTION TABLE
C Shapes

G40.21-M 300W
$\phi = 0.90$

Designation*	V_r	I_x	b	L_u	Factored Moment Resistance (kN·m)						
					M_r	M'_r					
						Unbraced length (mm)					
	kN	$10^6 mm^4$	mm	mm	$\leq L_u$	1 500	2 000	2 500	3 000	4 000	5 000
+C250X37	607	37.9	73	1 630	**80.7**	...	77.0	71.9	66.9	57.0	46.1
+C250X30	435	32.7	69	1 460	**69.4**	69.0	63.4	57.7	52.0	39.9	31.1
+C230X30	465	25.5	67	1 500	**59.9**	...	55.7	51.6	47.4	39.1	30.8
+C250X23	276	27.8	65	1 340	**59.1**	57.3	51.4	45.2	38.9	27.4	21.2
+C230X22	294	21.3	63	1 330	**50.2**	48.7	44.0	39.2	34.5	24.9	19.5
+C200X28	449	18.2	64	1 540	48.6	...	45.8	42.9	40.0	34.2	27.8
+C230X20	241	19.8	61	1 280	**46.7**	44.7	39.9	35.0	29.8	21.2	16.5
+C200X21	279	14.9	59	1 290	39.7	38.2	34.6	31.1	27.6	20.4	16.0
+C200X17	203	13.5	57	1 210	**35.9**	33.8	30.1	26.3	22.3	16.0	12.5
+C180X18	254	10.0	55	1 250	30.5	29.2	26.6	24.1	21.6	16.2	12.8
+C180X15	168	8.86	53	1 150	**26.9**	25.0	22.2	19.5	16.4	11.9	9.4
+C150X19	301	7.12	54	1 470	25.3	25.2	23.7	22.2	20.8	17.9	14.6
+C150X16	217	6.22	51	1 240	22.1	21.2	19.4	17.7	16.0	12.4	9.8
+C150X12	138	5.36	48	1 080	**19.1**	17.4	15.5	13.6	11.5	8.4	6.7
+C130X17	272	4.36	52	1 690	18.5	...	18.0	17.1	16.2	14.5	12.8
+C130X13	188	3.66	47	1 270	15.6	15.0	13.9	12.9	11.8	9.5	7.6
+C130X10	109	3.09	44	1 050	**13.1**	12.0	10.7	9.4	8.1	6.0	4.7
+C100X11	149	1.91	43	1 340	10.1	9.9	9.3	8.6	8.0	6.8	5.5
+C100X9	115	1.77	42	1 180	**9.34**	8.9	8.2	7.5	6.9	5.4	4.3
+C100X8	85.4	1.61	40	1 050	**8.53**	7.8	7.1	6.4	5.6	4.2	3.3
+C75X9	122	.846	40	1 680	6.02	...	5.8	5.6	5.3	4.7	4.2
+C75X7	89.4	.749	37	1 320	**5.32**	5.2	4.9	4.6	4.2	3.6	2.9
+C75X6	58.2	.670	35	1 080	**4.75**	4.4	4.1	3.7	3.4	2.6	2.1

+Class 3 Section.
*Nominal depth in millimetres and mass in kilograms per metre. For imperial designations, see page 6-19.

G40.21-M 350W
$\phi = 0.90$

BEAM SELECTION TABLE
WWF Shapes

Designation*	V_r	I_x	b	L_u	M_r	M'_r (Factored Moment Resistance (kN·m), Unbraced length (mm))					
	kN	$10^6 mm^4$	mm	mm	$\leq L_u$	8 000	10 000	12 000	14 000	16 000	20 000
+WWF2000X732	3 640	63 900	550	7 620	**20 100**	19 800	18 100	16 100	13 800	11 100	7 610
WWF1800X700	4 070	50 400	550	7 300	**19 800**	19 200	17 400	15 300	12 800	10 200	7 120
WWF1800X659	4 050	46 600	550	7 170	**18 300**	17 700	15 900	13 800	11 300	8 990	6 190
+WWF2000X648	3 600	54 200	550	7 370	**17 100**	16 600	15 000	13 100	10 800	8 510	5 760
WWF1800X617	4 020	42 700	550	7 020	**16 900**	16 200	14 500	12 400	9 900	7 810	5 330
+WWF2000X607	3 590	49 300	550	7 210	**15 500**	15 000	13 500	11 600	9 360	7 330	4 920
WWF1800X575	4 000	38 800	550	6 860	**15 500**	14 700	13 000	11 000	8 540	6 700	4 540
+WWF1600X622	2 360	37 600	550	8 010	14 800	....	13 700	12 400	11 000	9 370	6 580
WWF1400X597	2 730	28 100	550	7 790	13 900	13 800	12 700	11 500	10 100	8 610	6 140
+WWF1600X580	2 350	34 600	550	7 890	13 600	....	12 500	11 200	9 850	8 170	5 680
WWF1800X510	3 980	32 400	500	5 960	**13 200**	11 600	9 750	7 350	5 530	4 340	2 940
+WWF2000X542	3 570	41 400	500	6 310	13 000	11 900	10 200	8 100	6 070	4 750	3 200
+WWF1600X538	2 330	31 600	550	7 750	12 400	12 300	11 300	10 100	8 710	7 050	4 850
WWF1400X513	2 680	23 500	550	7 510	11 700	11 400	10 400	9 250	7 950	6 400	4 460
+WWF1600X496	2 320	28 400	550	7 620	**11 200**	11 000	10 000	8 890	7 590	6 010	4 090
WWF1400X471	2 660	21 100	550	7 360	**10 500**	10 200	9 280	8 150	6 840	5 410	3 730
WWF1200X487	2 890	16 700	550	7 710	9 640	9 540	8 760	7 870	6 910	5 770	4 090
+WWF1600X431	2 300	23 400	500	6 720	**9 230**	8 670	7 630	6 410	4 920	3 870	2 620
WWF1400X405	2 640	17 300	500	6 450	**8 760**	8 080	7 030	5 800	4 390	3 470	2 380
WWF1100X458	2 210	13 700	550	7 890	8 540	8 500	7 850	7 120	6 340	5 450	3 900
WWF1200X418	2 890	13 800	500	6 790	8 060	7 630	6 790	5 850	4 700	3 780	2 670
WWF1000X447	2 210	11 100	550	8 030	7 620	....	7 080	6 470	5 820	5 140	3 730
WWF1400X358	2 640	14 500	400	5 000	**7 470**	5 830	4 320	3 130	2 400	1 920	1 360
WWF1200X380	2 890	12 300	500	6 630	7 180	6 720	5 920	5 000	3 890	3 100	2 160
WWF1100X388	2 210	11 200	500	6 970	7 060	6 750	6 070	5 300	4 400	3 550	2 530
WWF900X417	1 370	8 680	550	8 310	6 550	....	6 180	5 710	5 210	4 690	3 550
WWF1000X377	2 210	9 120	500	7 080	6 300	6 060	5 480	4 840	4 140	3 360	2 420
WWF1100X351	2 210	9 940	500	6 810	**6 270**	5 930	5 280	4 530	3 620	2 900	2 030

+ Class 3 Section

*Nominal depth in millimetres and mass in kilograms per metre. For imperial designations, see page 6-19.

BEAM SELECTION TABLE
WWF Shapes

G40.21-M 350W
$\phi = 0.90$

Designation*	V_r	I_x	b	L_u	Factored Moment Resistance (kN·m)						
					M_r	M_r'					
						Unbraced length (mm)					
	kN	$10^6 mm^4$	mm	mm	$\le L_u$	5 000	6 000	8 000	10 000	12 000	16 000
WWF1200X333	2 890	10 200	400	5 150	**6 080**	....	5 770	4 900	3 820	2 780	1 740
WWF1000X340	2 210	8 060	500	6 910	5 580	....	....	5 310	4 760	4 130	2 720
WWF900X347	1 370	7 100	500	7 320	5 390	....	....	5 250	4 790	4 300	3 150
WWF1200X302	2 890	8 970	400	4 980	**5 390**	5 380	5 040	4 200	3 110	2 250	1 390
WWF1100X304	2 210	8 220	400	5 310	5 260	....	5 060	4 350	3 530	2 590	1 630
WWF900X309	1 370	6 240	500	7 160	4 720	....	....	4 560	4 130	3 650	2 520
WWF800X339	1 370	5 500	500	7 490	4 690	....	....	4 610	4 240	3 850	2 980
WWF1000X293	2 210	6 640	400	5 400	**4 660**	....	4 510	3 920	3 230	2 420	1 550
WWF1100X273	2 210	7 160	400	5 150	**4 630**	....	4 390	3 720	2 860	2 080	1 280
WWF1200X263	2 890	7 250	300	3 560	**4 470**	3 850	3 320	2 100	1 430	1 060	684
WWF800X300	1 370	4 840	500	7 280	4 130	....	....	4 010	3 660	3 270	2 360
WWF1000X262	2 210	5 780	400	5 230	**4 100**	....	3 910	3 340	2 650	1 930	1 210
WWF900X262	1 370	5 110	400	5 630	**3 910**	....	3 830	3 380	2 860	2 240	1 450
WWF1100X234	2 210	5 720	300	3 710	**3 780**	3 340	2 930	1 930	1 320	978	632
WWF800X253	1 370	3 950	400	5 730	3 400	....	3 360	2 990	2 570	2 090	1 380
WWF900X231	1 350	4 410	400	5 460	**3 400**	....	3 300	2 870	2 360	1 760	1 110
WWF1000X223	2 210	4 590	300	3 790	**3 310**	2 960	2 620	1 790	1 230	920	602
WWF800X223	1 370	3 410	400	5 570	**2 940**	....	2 880	2 530	2 120	1 630	1 050
WWF700X245	1 370	2 950	400	5 880	2 900	....	2 890	2 600	2 280	1 940	1 310
WWF1000X200	2 210	3 940	300	3 610	**2 890**	2 510	2 180	1 390	945	699	449
WWF900X192	1 350	3 460	300	3 980	**2 710**	2 490	2 240	1 620	1 120	842	555
WWF700X214	1 370	2 540	400	5 690	2 500	....	2 470	2 190	1 870	1 500	982
WWF800X184	1 370	2 660	300	4 060	**2 330**	2 170	1 970	1 490	1 040	791	529
WWF900X169	1 340	2 930	300	3 820	**2 320**	2 080	1 840	1 240	845	625	402
+WWF700X196	1 370	2 300	400	5 860	2 070	....	2 050	1 830	1 570	1 260	812
WWF800X161	1 370	2 250	300	3 900	**1 990**	1 810	1 620	1 140	778	581	379
WWF700X175	1 370	1 970	300	4 160	1 980	1 860	1 700	1 350	965	741	505
WWF700X152	1 370	1 660	300	3 990	1 680	1 540	1 390	1 020	711	537	356

*Nominal depth in millimetres and mass in kilograms per metre. For imperial designations, see page 6-19.

NOTES

BEAM LOAD TABLES
WWF Shapes

G40.21-M
300W

Total Uniformly Distributed Factored Loads
for Laterally Supported Beams – kN

$\phi = 0.90$

Designation	WWF2000				Approx. Deflect. (mm)	WWF1800					Approx. Deflect. (mm)
Mass (kg/m)	732	648	607	542		700	659	617	575	510	
Span in Millimetres											
11 000					14					7 960	15
11 500					15					7 850	16
12 000					16					7 520	18
12 500					18					7 220	19
13 000					19				8 000	6 940	21
13 500					20				7 860	6 690	23
14 000					22			8 050	7 580	6 450	24
14 500				7 140	24			8 000	7 310	6 230	26
15 000				6 980	25			7 730	7 070	6 020	28
15 500				6 760	27		8 090	7 480	6 840	5 820	30
16 000				6 550	29		7 860	7 250	6 630	5 640	32
16 500				6 350	30	8 140	7 620	7 030	6 430	5 470	34
17 000			7 170	6 160	32	7 970	7 400	6 820	6 240	5 310	36
17 500			6 990	5 990	34	7 740	7 180	6 630	6 060	5 160	38
18 000		7 210	6 790	5 820	36	7 520	6 980	6 440	5 890	5 020	40
18 500		7 200	6 610	5 660	38	7 320	6 800	6 270	5 730	4 880	43
19 000		7 010	6 440	5 510	40	7 130	6 620	6 100	5 580	4 750	45
19 500		6 830	6 270	5 370	43	6 940	6 450	5 950	5 440	4 630	47
20 000		6 660	6 110	5 240	45	6 770	6 290	5 800	5 300	4 510	50
20 500		6 500	5 960	5 110	47	6 610	6 130	5 660	5 170	4 400	52
21 000	7 280	6 350	5 820	4 990	49	6 450	5 990	5 520	5 050	4 300	55
21 500	7 210	6 200	5 690	4 870	52	6 300	5 850	5 400	4 930	4 200	58
22 000	7 050	6 060	5 560	4 760	54	6 160	5 710	5 270	4 820	4 100	60
22 500	6 890	5 920	5 430	4 660	57	6 020	5 590	5 160	4 710	4 010	63
23 000	6 740	5 790	5 320	4 560	59	5 890	5 470	5 040	4 610	3 930	66
23 500	6 600	5 670	5 200	4 460	62	5 760	5 350	4 940	4 510	3 840	69
24 000	6 460	5 550	5 090	4 360	65	5 640	5 240	4 830	4 420	3 760	72
24 500	6 330	5 440	4 990	4 280	67	5 530	5 130	4 730	4 330	3 680	75
25 000	6 200	5 330	4 890	4 190	70	5 420	5 030	4 640	4 240	3 610	78
25 500	6 080	5 230	4 790	4 110	73	5 310	4 930	4 550	4 160	3 540	81
26 000	5 960	5 130	4 700	4 030	76	5 210	4 840	4 460	4 080	3 470	84
26 500	5 850	5 030	4 610	3 950	79	5 110	4 740	4 380	4 000	3 410	87
27 000	5 740	4 940	4 530	3 880	82	5 020	4 660	4 300	3 930	3 340	91
27 500	5 640	4 850	4 450	3 810	85	4 920	4 570	4 220	3 860	3 280	94
28 000	5 540	4 760	4 370	3 740	88	4 840	4 490	4 140	3 790	3 220	98
DESIGN DATA AND PROPERTIES											
V_r (kN)	3 640	3 600	3 590	3 570		4 070	4 050	4 020	4 000	3 980	
L_u (mm)	7 770	7 460	7 270	6 300		7 920	7 770	7 600	7 420	6 440	
d (mm)	2 000	2 000	2 000	2 000		1 800	1 800	1 800	1 800	1 800	
b (mm)	550	550	550	500		550	550	550	550	500	
t (mm)	50.0	40.0	35.0	30.0		50.0	45.0	40.0	35.0	30.0	
w (mm)	20.0	20.0	20.0	20.0		20.0	20.0	20.0	20.0	20.0	
k (mm)	61	51	46	41		61	56	51	46	41	
IMPERIAL SIZE AND MASS											
Mass (lb./ft.)	490	436	408	364		470	442	415	388	344	
Nominal Depth (in.)	79					71					

BEAM LOAD TABLES
WWF Shapes

Total Uniformly Distributed Factored Loads
for Laterally Supported Beams – kN

Designation	WWF1600					Approx. Deflect. (mm)	WWF1400					Approx. Deflect. (mm)
Mass (kg/m)	622	580	538	496	431		597	513	471	405	358	
Span in Millimetres												
9 000						11						13
9 500						13					5 290	14
10 000						14					5 120	16
10 500						15					4 880	18
11 000						17				5 290	4 650	19
11 500						19				5 220	4 450	21
12 000						20				5 000	4 270	23
12 500						22				4 800	4 100	25
13 000						24				4 620	3 940	27
13 500						26			5 330	4 450	3 790	29
14 000						27			5 150	4 290	3 660	31
14 500						29		5 350	4 980	4 140	3 530	34
15 000						32		5 330	4 810	4 000	3 410	36
15 500					4 600	34		5 160	4 650	3 870	3 300	38
16 000					4 470	36		5 000	4 510	3 750	3 200	41
16 500					4 330	38		4 840	4 370	3 640	3 100	44
17 000					4 210	40		4 700	4 240	3 530	3 010	46
17 500					4 080	43	5 350	4 570	4 120	3 430	2 920	49
18 000				4 630	3 970	45	5 280	4 440	4 010	3 340	2 840	52
18 500				4 620	3 860	48	5 140	4 320	3 900	3 250	2 770	55
19 000				4 500	3 760	51	5 000	4 210	3 800	3 160	2 690	58
19 500				4 390	3 670	53	4 870	4 100	3 700	3 080	2 620	61
20 000			4 660	4 280	3 580	56	4 750	4 000	3 610	3 000	2 560	64
20 500			4 600	4 170	3 490	59	4 640	3 900	3 520	2 930	2 500	67
21 000			4 500	4 070	3 400	62	4 530	3 810	3 440	2 860	2 440	71
21 500		4 690	4 390	3 980	3 320	65	4 420	3 720	3 360	2 790	2 380	74
22 000		4 680	4 290	3 890	3 250	68	4 320	3 630	3 280	2 730	2 330	77
22 500		4 580	4 200	3 800	3 180	71	4 220	3 550	3 210	2 670	2 280	81
23 000		4 480	4 100	3 720	3 110	74	4 130	3 480	3 140	2 610	2 230	85
23 500	4 720	4 380	4 020	3 640	3 040	77	4 040	3 400	3 070	2 560	2 180	88
24 000	4 660	4 290	3 930	3 560	2 980	81	3 960	3 330	3 010	2 500	2 130	92
24 500	4 570	4 200	3 850	3 490	2 920	84	3 880	3 260	2 940	2 450	2 090	96
25 000	4 480	4 120	3 780	3 420	2 860	88	3 800	3 200	2 890	2 400	2 050	100
25 500	4 390	4 040	3 700	3 350	2 800	91	3 730	3 130	2 830	2 360	2 010	104
26 000	4 300	3 960	3 630	3 290	2 750	95	3 660	3 070	2 780	2 310	1 970	108
DESIGN DATA AND PROPERTIES												
V_r (kN)	2 360	2 350	2 330	2 320	2 300		2 670	2 670	2 660	2 640	2 640	
L_u (mm)	8 270	8 130	7 970	7 800	6 830		8 480	8 150	7 980	6 990	5 410	
d (mm)	1 600	1 600	1 600	1 600	1 600		1 400	1 400	1 400	1 400	1 400	
b (mm)	550	550	550	550	500		550	550	550	500	400	
t (mm)	50.0	45.0	40.0	35.0	30.0		50.0	40.0	35.0	30.0	30.0	
w (mm)	16.0	16.0	16.0	16.0	16.0		16.0	16.0	16.0	16.0	16.0	
k (mm)	61	56	51	44	39		61	51	44	39	39	
IMPERIAL SIZE AND MASS												
Mass (lb./ft.)	419	388	361	333	289		402	344	316	272	240	
Nominal Depth (in.)	63						55					

G40.21-M
300W
$\phi = 0.90$

Total Uniformly Distributed Factored Loads
for Laterally Supported Beams – kN

Designation	WWF1200						Approximate Deflection (mm)
Mass (kg/m)	487	418	380	333	302	263	
Span in Millimetres							
5 000							5
5 500						5 350	6
6 000						5 110	7
6 500					5 350	4 720	8
7 000					5 280	4 380	9
7 500				5 350	4 920	4 090	11
8 000				5 210	4 620	3 830	12
8 500				4 900	4 340	3 610	13
9 000			5 350	4 630	4 100	3 410	15
9 500			5 180	4 390	3 890	3 230	17
10 000		5 350	4 920	4 170	3 690	3 070	19
10 500		5 270	4 690	3 970	3 520	2 920	21
11 000		5 030	4 480	3 790	3 360	2 790	23
11 500		4 810	4 280	3 620	3 210	2 670	25
12 000	5 350	4 610	4 100	3 470	3 080	2 560	27
12 500	5 290	4 420	3 940	3 340	2 960	2 450	29
13 000	5 080	4 250	3 790	3 210	2 840	2 360	32
13 500	4 900	4 100	3 650	3 090	2 740	2 270	34
14 000	4 720	3 950	3 520	2 980	2 640	2 190	37
14 500	4 560	3 810	3 400	2 880	2 550	2 120	39
15 000	4 410	3 690	3 280	2 780	2 460	2 040	42
15 500	4 260	3 570	3 180	2 690	2 380	1 980	45
16 000	4 130	3 460	3 080	2 610	2 310	1 920	48
16 500	4 010	3 350	2 980	2 530	2 240	1 860	51
17 000	3 890	3 250	2 900	2 450	2 170	1 800	54
17 500	3 780	3 160	2 810	2 380	2 110	1 750	57
18 000	3 670	3 070	2 740	2 320	2 050	1 700	60
18 500	3 570	2 990	2 660	2 250	2 000	1 660	64
19 000	3 480	2 910	2 590	2 190	1 940	1 610	67
19 500	3 390	2 840	2 530	2 140	1 890	1 570	71
20 000	3 300	2 760	2 460	2 080	1 850	1 530	75
20 500	3 220	2 700	2 400	2 030	1 800	1 500	78
21 000	3 150	2 630	2 340	1 980	1 760	1 460	82
21 500	3 070	2 570	2 290	1 940	1 720	1 430	86
22 000	3 000	2 510	2 240	1 900	1 680	1 390	90
DESIGN DATA AND PROPERTIES							
V_r (kN)	2 670	2 670	2 670	2 670	2 670	2 670	
L_u (mm)	8 370	7 370	7 180	5 590	5 390	3 850	
d (mm)	1 200	1 200	1 200	1 200	1 200	1 200	
b (mm)	550	500	500	400	400	300	
t (mm)	40.0	35.0	30.0	30.0	25.0	25.0	
w (mm)	16.0	16.0	16.0	16.0	16.0	16.0	
k (mm)	51	44	39	39	34	34	
IMPERIAL SIZE AND MASS							
Mass (lb./ft.)	326	281	255	224	203	176	
Nominal Depth (in.)	47						

BEAM LOAD TABLES
WWF Shapes

Total Uniformly Distributed Factored Loads
for Laterally Supported Beams – kN

Designation	WWF1100						Approximate Deflection (mm)
Mass (kg/m)	458	388	351	304	273	234	
Span in Millimetres							
6 000						4 090	7
6 500						3 990	9
7 000						3 700	10
7 500					4 090	3 460	11
8 000					3 970	3 240	13
8 500				4 090	3 740	3 050	15
9 000				4 010	3 530	2 880	16
9 500				3 800	3 340	2 730	18
10 000			4 090	3 610	3 180	2 590	20
10 500			4 090	3 440	3 020	2 470	22
11 000			3 910	3 280	2 890	2 360	25
11 500		4 090	3 740	3 140	2 760	2 250	27
12 000		4 030	3 580	3 010	2 650	2 160	29
12 500		3 870	3 440	2 890	2 540	2 070	32
13 000		3 720	3 310	2 780	2 440	1 990	34
13 500		3 580	3 180	2 670	2 350	1 920	37
14 000	4 090	3 460	3 070	2 580	2 270	1 850	40
14 500	4 040	3 340	2 960	2 490	2 190	1 790	43
15 000	3 900	3 230	2 870	2 400	2 120	1 730	46
15 500	3 780	3 120	2 770	2 330	2 050	1 670	49
16 000	3 660	3 020	2 690	2 250	1 980	1 620	52
16 500	3 550	2 930	2 600	2 190	1 920	1 570	55
17 000	3 440	2 850	2 530	2 120	1 870	1 520	59
17 500	3 340	2 760	2 460	2 060	1 810	1 480	62
18 000	3 250	2 690	2 390	2 000	1 760	1 440	66
18 500	3 160	2 620	2 320	1 950	1 720	1 400	70
19 000	3 080	2 550	2 260	1 900	1 670	1 360	74
19 500	3 000	2 480	2 200	1 850	1 630	1 330	77
20 000	2 930	2 420	2 150	1 800	1 590	1 300	81
20 500	2 860	2 360	2 100	1 760	1 550	1 260	86
21 000	2 790	2 300	2 050	1 720	1 510	1 230	90
21 500	2 720	2 250	2 000	1 680	1 480	1 210	94
22 000	2 660	2 200	1 950	1 640	1 440	1 180	99
22 500	2 600	2 150	1 910	1 600	1 410	1 150	103
23 000	2 540	2 100	1 870	1 570	1 380	1 130	108
DESIGN DATA AND PROPERTIES							
V_r (kN)	2 050	2 050	2 050	2 050	2 050	2 050	
L_u (mm)	8 580	7 570	7 380	5 760	5 580	4 010	
d (mm)	1 100	1 100	1 100	1 100	1 100	1 100	
b (mm)	550	500	500	400	400	300	
t (mm)	40.0	35.0	30.0	30.0	25.0	25.0	
w (mm)	14.0	14.0	14.0	14.0	14.0	14.0	
k (mm)	51	44	39	39	34	34	
IMPERIAL SIZE AND MASS							
Mass (lb./ft.)	307	260	236	204	184	157	
Nominal Depth (in.)	43						

BEAM LOAD TABLES
WWF Shapes

G40.21-M
300W

Total Uniformly Distributed Factored Loads
for Laterally Supported Beams – kN

$\phi = 0.90$

Designation		WWF1000							Approx. Deflect. (mm)
Mass (kg/m)		447	377	340	293	262	223	200	
Span in Millimetres	4 000								4
	4 500							4 090	5
	5 000							3 960	6
	5 500						4 090	3 600	7
	6 000						3 780	3 300	8
	6 500					4 090	3 490	3 050	9
	7 000					4 010	3 240	2 830	11
	7 500				4 090	3 740	3 020	2 640	13
	8 000				4 000	3 510	2 840	2 480	14
	8 500				3 760	3 300	2 670	2 330	16
	9 000			4 090	3 550	3 120	2 520	2 200	18
	9 500			4 020	3 360	2 960	2 390	2 080	20
	10 000			3 820	3 200	2 810	2 270	1 980	22
	10 500		4 090	3 640	3 040	2 670	2 160	1 890	25
	11 000		3 930	3 480	2 910	2 550	2 060	1 800	27
	11 500		3 760	3 320	2 780	2 440	1 970	1 720	30
	12 000		3 600	3 190	2 660	2 340	1 890	1 650	32
	12 500	4 090	3 460	3 060	2 560	2 250	1 810	1 580	35
	13 000	4 020	3 320	2 940	2 460	2 160	1 740	1 520	38
	13 500	3 870	3 200	2 830	2 370	2 080	1 680	1 470	41
	14 000	3 730	3 090	2 730	2 280	2 010	1 620	1 420	44
	14 500	3 600	2 980	2 640	2 200	1 940	1 560	1 370	47
	15 000	3 480	2 880	2 550	2 130	1 870	1 510	1 320	50
	15 500	3 370	2 790	2 470	2 060	1 810	1 460	1 280	54
	16 000	3 270	2 700	2 390	2 000	1 760	1 420	1 240	57
	16 500	3 170	2 620	2 320	1 940	1 700	1 380	1 200	61
	17 000	3 080	2 540	2 250	1 880	1 650	1 330	1 160	65
	17 500	2 990	2 470	2 180	1 830	1 600	1 300	1 130	69
	18 000	2 900	2 400	2 120	1 780	1 560	1 260	1 100	73
	18 500	2 830	2 340	2 070	1 730	1 520	1 230	1 070	77
	19 000	2 750	2 270	2 010	1 680	1 480	1 190	1 040	81
	19 500	2 680	2 220	1 960	1 640	1 440	1 160	1 020	85
	20 000	2 610	2 160	1 910	1 600	1 400	1 130	990	90
	20 500	2 550	2 110	1 860	1 560	1 370	1 110	966	94
	21 000	2 490	2 060	1 820	1 520	1 340	1 080	943	99
DESIGN DATA AND PROPERTIES									
V_r (kN)		2 050	2 050	2 050	2 050	2 050	2 050	2 050	
L_u (mm)		8 750	7 700	7 500	5 860	5 670	4 100	3 910	
d (mm)		1 000	1 000	1 000	1 000	1 000	1 000	1 000	
b (mm)		550	500	500	400	400	300	300	
t (mm)		40.0	35.0	30.0	30.0	25.0	25.0	20.0	
w (mm)		14.0	14.0	14.0	14.0	14.0	14.0	14.0	
k (mm)		51	44	39	39	34	34	29	
IMPERIAL SIZE AND MASS									
Mass (lb./ft.)		300	253	228	197	176	150	134	
Nominal Depth (in.)		39							

BEAM LOAD TABLES
WWF Shapes

Total Uniformly Distributed Factored Loads for Laterally Supported Beams – kN

Designation	WWF900							Approx. Deflect. (mm)
Mass (kg/m)	417	347	309	262	231	192	169	
Span in Millimetres 6 000							2 530	9
6 500							2 450	11
7 000						2 530	2 270	12
7 500						2 480	2 120	14
8 000						2 320	1 990	16
8 500						2 180	1 870	18
9 000					2 530	2 060	1 770	20
9 500					2 460	1 960	1 680	22
10 000					2 330	1 860	1 590	25
10 500				2 530	2 220	1 770	1 520	27
11 000				2 440	2 120	1 690	1 450	30
11 500				2 330	2 030	1 620	1 380	33
12 000				2 230	1 940	1 550	1 330	36
12 500			2 530	2 140	1 870	1 490	1 270	39
13 000			2 490	2 060	1 790	1 430	1 220	42
13 500			2 400	1 980	1 730	1 380	1 180	45
14 000			2 310	1 910	1 670	1 330	1 140	49
14 500		2 530	2 230	1 850	1 610	1 280	1 100	52
15 000		2 460	2 160	1 790	1 560	1 240	1 060	56
15 500		2 380	2 090	1 730	1 500	1 200	1 030	60
16 000		2 310	2 020	1 670	1 460	1 160	995	64
16 500		2 240	1 960	1 620	1 410	1 130	965	68
17 000		2 170	1 910	1 580	1 370	1 090	936	72
17 500	2 530	2 110	1 850	1 530	1 330	1 060	910	76
18 000	2 500	2 050	1 800	1 490	1 300	1 030	884	81
18 500	2 430	2 000	1 750	1 450	1 260	1 000	860	85
19 000	2 360	1 940	1 700	1 410	1 230	978	838	90
19 500	2 300	1 890	1 660	1 370	1 200	953	816	95
20 000	2 250	1 850	1 620	1 340	1 170	929	796	100
20 500	2 190	1 800	1 580	1 310	1 140	906	777	105
21 000	2 140	1 760	1 540	1 280	1 110	885	758	110
21 500	2 090	1 720	1 510	1 250	1 080	864	740	115
22 000	2 040	1 680	1 470	1 220	1 060	844	724	120
22 500	2 000	1 640	1 440	1 190	1 040	826	708	126
23 000	1 950	1 610	1 410	1 160	1 010	808	692	132
DESIGN DATA AND PROPERTIES								
V_r (kN)	1 260	1 260	1 260	1 260	1 260	1 260	1 260	
L_u (mm)	9 080	7 970	7 780	6 120	5 930	4 310	4 130	
d (mm)	900	900	900	900	900	900	900	
b (mm)	550	500	500	400	400	300	300	
t (mm)	40.0	35.0	30.0	30.0	25.0	25.0	20.0	
w (mm)	11.0	11.0	11.0	11.0	11.0	11.0	11.0	
k (mm)	51	44	39	39	34	34	29	
IMPERIAL SIZE AND MASS								
Mass (lb./ft.)	279	233	208	176	156	128	113	
Nominal Depth (in.)	35							

BEAM LOAD TABLES
WWF Shapes

G40.21-M
300W
$\phi = 0.90$

Total Uniformly Distributed Factored Loads
for Laterally Supported Beams – kN

Designation	WWF800						Approximate Deflection (mm)
Mass (kg/m)	339	300	253	223	184	161	
Span in Millimetres							
5 000						2 530	7
5 500						2 480	8
6 000					2 530	2 280	10
6 500					2 460	2 100	12
7 000					2 290	1 950	14
7 500				2 530	2 130	1 820	16
8 000				2 520	2 000	1 710	18
8 500				2 370	1880	1 610	20
9 000			2 530	2 240	1 780	1 520	23
9 500			2 460	2 120	1 680	1 440	25
10 000			2 330	2 020	1 600	1 360	28
10 500			2 220	1 920	1 520	1 300	31
11 000		2 530	2 120	1 830	1 460	1 240	34
11 500		2 460	2 030	1 750	1 390	1 190	37
12 000		2 360	1 940	1 680	1 330	1 140	40
12 500	2 530	2 260	1 870	1 610	1 280	1 090	44
13 000	2 480	2 180	1 790	1 550	1 230	1 050	47
13 500	2 380	2 100	1 730	1 490	1 190	1 010	51
14 000	2 300	2 020	1 670	1 440	1 140	975	55
14 500	2 220	1 950	1 610	1 390	1 100	941	59
15 000	2 150	1 890	1 560	1 340	1 070	910	63
15 500	2 080	1 830	1 500	1 300	1 030	881	67
16 000	2 010	1 770	1 460	1 260	1 000	853	72
16 500	1 950	1 720	1 410	1 220	970	827	76
17 000	1 890	1 660	1 370	1 190	942	803	81
17 500	1 840	1 620	1 330	1 150	915	780	86
18 000	1 790	1 570	1 300	1 120	889	758	91
18 500	1 740	1 530	1 260	1 090	865	738	96
19 000	1 690	1 490	1 230	1 060	842	718	101
19 500	1 650	1 450	1 200	1 040	821	700	106
20 000	1 610	1 420	1 170	1 010	800	683	112
20 500	1 570	1 380	1 140	984	781	666	118
21 000	1 530	1 350	1 110	961	762	650	123
21 500	1 500	1 320	1 080	938	744	635	129
22 000	1 460	1 290	1 060	917	728	621	136
DESIGN DATA AND PROPERTIES							
V_r (kN)	1 260	1 260	1 260	1 260	1 260	1 260	
L_u (mm)	8 180	7 930	6 240	6 050	4 410	4 230	
d (mm)	800	800	800	800	800	800	
b (mm)	500	500	400	400	300	300	
t (mm)	35.0	30.0	30.0	25.0	25.0	20.0	
w (mm)	11.0	11.0	11.0	11.0	11.0	11.0	
k (mm)	44	39	39	34	34	29	
IMPERIAL SIZE AND MASS							
Mass (lb./ft.)	228	202	170	150	123	108	
Nominal Depth (in.)	31						

BEAM LOAD TABLES
WWF Shapes

Total Uniformly Distributed Factored Loads
for Laterally Supported Beams – kN

Designation	WWF700					Approximate Deflection (mm)
Mass (kg/m)	245	214	196	175	152	
Span in Millimetres						
4 000						5
4 500					2 530	6
5 000				2 530	2 300	8
5 500				2 460	2 090	10
6 000			2 530	2 260	1 920	12
6 500		2 530	2 390	2 080	1 770	14
7 000		2 450	2 220	1 940	1 640	16
7 500	2 510	2 290	2 070	1 810	1 530	18
8 000	2 490	2 150	1 940	1 690	1 440	20
8 500	2 340	2 020	1 830	1 590	1 350	23
9 000	2 210	1 910	1 730	1 500	1 280	26
9 500	2 090	1 810	1 640	1 430	1 210	29
10 000	1 990	1 720	1 550	1 350	1 150	32
10 500	1 900	1 640	1 480	1 290	1 090	35
11 000	1 810	1 560	1 410	1 230	1 040	39
11 500	1 730	1 490	1 350	1 180	999	42
12 000	1 660	1 430	1 290	1 130	958	46
12 500	1 590	1 370	1 240	1 080	919	50
13 000	1 530	1 320	1 200	1 040	884	54
13 500	1 470	1 270	1 150	1 000	851	58
14 000	1 420	1 230	1 110	967	821	63
14 500	1 370	1 180	1 070	934	792	67
15 000	1 330	1 140	1 040	903	766	72
15 500	1 280	1 110	1 000	874	741	77
16 000	1 240	1 070	971	846	718	82
16 500	1 210	1 040	941	821	696	87
17 000	1 170	1 010	914	797	676	92
17 500	1 140	981	887	774	657	98
18 000	1 100	954	863	752	638	104
18 500	1 080	928	839	732	621	110
19 000	1 050	904	817	713	605	116
19 500	1 020	881	796	695	589	122
20 000	995	859	777	677	575	128
20 500	970	838	758	661	561	134
21 000	947	818	740	645	547	141
DESIGN DATA AND PROPERTIES						
V_r (kN)	1 250	1 260	1 260	1 260	1 260	
L_u (mm)	6 420	6 190	6 060	4 520	4 330	
d (mm)	700	700	700	700	700	
b (mm)	400	400	400	300	300	
t (mm)	30.0	25.0	22.0	25.0	20.0	
w (mm)	11.0	11.0	11.0	11.0	11.0	
k (mm)	39	34	31	34	29	
IMPERIAL SIZE AND MASS						
Mass (lb./ft.)	164	144	132	117	102	
Nominal Depth (in.)	28					

BEAM LOAD TABLES
W Shapes

G40.21-M
300W
$\phi = 0.90$

Total Uniformly Distributed Factored Loads
for Laterally Supported Beams – kN

Designation	W610											Approx. Deflect. (mm)
Mass (kg/m)	241	217	195	174	155	140	125	113	101	91	84	
Span in Millimetres												
2 000												1
2 500								2 430	2 260	2 050	1 760	2
3 000						2 880	2 600	2 370	2 090	1 840	1 700	3
3 500			3 410	3 070	2 770	2 560	2 260	2 030	1 790	1 580	1 460	4
4 000	4 060	3 690	3 280	2 890	2 550	2 240	1 980	1 780	1 570	1 380	1 270	6
4 500	3 680	3 290	2 910	2 570	2 270	1 990	1 760	1 580	1 390	1 230	1 130	7
5 000	3 310	2 960	2 620	2 320	2 040	1 790	1 580	1 420	1 250	1 110	1 020	9
5 500	3 010	2 690	2 380	2 100	1 860	1 630	1 440	1 290	1 140	1 000	927	11
6 000	2 760	2 470	2 180	1 930	1 700	1 490	1 320	1 180	1 040	922	850	13
6 500	2 550	2 280	2 020	1 780	1 570	1 380	1 220	1 090	964	851	784	16
7 000	2 370	2 110	1 870	1 650	1 460	1 280	1 130	1 020	895	790	728	18
7 500	2 210	1 970	1 750	1 540	1 360	1 200	1 060	948	835	737	680	21
8 000	2 070	1 850	1 640	1 450	1 280	1 120	991	888	783	691	637	24
8 500	1 950	1 740	1 540	1 360	1 200	1 060	933	836	737	651	600	27
9 000	1 840	1 640	1 460	1 290	1 140	996	881	790	696	614	566	30
9 500	1 740	1 560	1 380	1 220	1 080	944	834	748	659	582	537	33
10 000	1 660	1 480	1 310	1 160	1 020	896	793	711	626	553	510	37
10 500	1 580	1 410	1 250	1 100	973	854	755	677	597	527	485	40
11 000	1 510	1 340	1 190	1 050	929	815	721	646	569	503	463	44
11 500	1 440	1 290	1 140	1 010	888	779	689	618	545	481	443	49
12 000	1 380	1 230	1 090	965	851	747	661	592	522	461	425	53
12 500	1 320	1 180	1 050	926	817	717	634	569	501	442	408	57
13 000	1 270	1 140	1 010	891	786	690	610	547	482	425	392	62
13 500	1 230	1 100	971	858	757	664	587	526	464	410	378	67
14 000	1 180	1 060	937	827	730	640	566	508	447	395	364	72
14 500	1 140	1 020	904	798	705	618	547	490	432	381	352	77
15 000	1 100	986	874	772	681	598	528	474	418	369	340	83
15 500	1 070	955	846	747	659	578	511	458	404	357	329	88
16 000	1 040	925	819	724	639	560	495	444	392	346	319	94
16 500	1 000	897	795	702	619	543	480	431	380	335	309	100
17 000	975	870	771	681	601	527	466	418	368	325	300	106
17 500	947	845	749	662	584	512	453	406	358	316	291	112
18 000	920	822	728	643	568	498	440	395	348	307	283	119
18 500	896	800	709	626	552	485	428	384	339	299	276	126
19 000	872	779	690	609	538	472	417	374	330	291	268	133
DESIGN DATA AND PROPERTIES												
V_r (kN)	2 030	1 850	1 710	1 540	1 390	1 440	1 300	1 210	1 130	1 030	881	
V_{r1} (kN)	1 790	1 640	1 530	1 380	1 260	1 300	1 170	1 100	1 030	944	812	
V_{r2} (kN)	1 550	1 430	1 340	1 230	1 130	1 150	1 050	992	939	862	743	
R (kN)	1 190	1 010	867	714	585	637	524	458	397	336	289	
G (kN)	20.4	18.1	16.9	14.5	12.4	11.5	9.85	9.37	9.02	8.43	7.33	
L_u (mm)	5 200	5 080	4 940	4 830	4 740	3 320	3 250	3 180	3 110	3 020	2 990	
d (mm)	635	628	622	616	611	617	612	608	603	598	596	
b (mm)	329	328	327	325	324	230	229	228	228	227	226	
t (mm)	31.0	27.7	24.4	21.6	19.0	22.2	19.6	17.3	14.9	12.7	11.7	
w (mm)	17.9	16.5	15.4	14.0	12.7	13.1	11.9	11.2	10.5	9.7	9.0	
k (mm)	50	47	44	41	38	41	39	36	34	32	31	
IMPERIAL SIZE AND MASS												
Mass (lb./ft.)	162	146	131	117	104	94	84	76	68	62	55	
Nominal Depth (in.)	24											

Sections highlighted in white are commonly used sizes which are generally readily available.

Total Uniformly Distributed Factored Loads
for Laterally Supported Beams – kN

Designation	W530							Approx. Deflect. (mm)
Mass (kg/m)	138	123	109	101	92	82	72	
Span in Millimetres								
2 000						1 790	1 660	2
2 500	2 880	2 540	2 230	2 090	1 940	1 790	1 520	3
3 000	2 600	2 310	2 040	1 890	1 700	1 490	1 270	4
3 500	2 230	1 980	1 750	1 620	1 460	1 280	1 090	5
4 000	1 950	1 730	1 530	1 420	1 270	1 120	950	7
4 500	1 730	1 540	1 360	1 260	1 130	994	845	9
5 000	1 560	1 390	1 220	1 130	1 020	894	760	11
5 500	1 420	1 260	1 110	1 030	927	813	691	13
6 000	1 300	1 160	1 020	943	850	745	634	15
6 500	1 200	1 070	940	871	784	688	585	18
7 000	1 110	991	873	808	728	639	543	21
7 500	1 040	924	815	755	680	596	507	24
8 000	975	867	764	707	637	559	475	27
8 500	917	816	719	666	600	526	447	31
9 000	866	770	679	629	566	497	422	34
9 500	821	730	643	596	537	471	400	38
10 000	780	693	611	566	510	447	380	42
10 500	743	660	582	539	485	426	362	47
11 000	709	630	556	514	463	406	346	51
11 500	678	603	532	492	443	389	331	56
12 000	650	578	509	472	425	373	317	61
12 500	624	555	489	453	408	358	304	66
13 000	600	533	470	435	392	344	292	71
13 500	578	514	453	419	378	331	282	77
14 000	557	495	437	404	364	319	272	83
14 500	538	478	422	390	352	308	262	89
15 000	520	462	408	377	340	298	253	95
15 500	503	447	394	365	329	288	245	102
16 000	487	433	382	354	319	279	238	108
DESIGN DATA AND PROPERTIES								
V_r (kN)	1 440	1 270	1 120	1 050	971	894	831	
V_{r1} (kN)	1 280	1 130	1 000	943	878	820	767	
V_{r2} (kN)	1 110	997	887	841	785	747	703	
R (kN)	812	647	508	448	390	335	292	
G (kN)	17.2	13.7	10.8	9.70	8.94	8.56	8.66	
L_u (mm)	3 180	3 100	3 040	2 990	2 930	2 860	2 750	
d (mm)	549	544	539	537	533	528	524	
b (mm)	214	212	211	210	209	209	207	
t (mm)	23.6	21.2	18.8	17.4	15.6	13.3	10.9	
w (mm)	14.7	13.1	11.6	10.9	10.2	9.5	8.9	
k (mm)	42	39	37	35	34	29	27	
IMPERIAL SIZE AND MASS								
Mass (lb./ft.)	93	83	73	68	62	55	48	
Nominal Depth (in.)	21							

Sections highlighted in white are commonly used sizes which are generally readily available.

BEAM LOAD TABLES
W Shapes

G40.21-M 300W

$\phi = 0.90$

Total Uniformly Distributed Factored Loads
for Laterally Supported Beams – kN

Designation	W460							Approx. Deflect. (mm)
Mass (kg/m)	106	97	89	82	74	67	61	
Span in Millimetres								
2 000	2 110	1 890		1 620	1 470	1 380	1 300	2
2 500	2 060	1 880	1 740	1 580	1 430	1 300	1 130	3
3 000	1 720	1 570	1 450	1 320	1 190	1 080	943	4
3 500	1 480	1 340	1 240	1 130	1 020	926	808	6
4 000	1 290	1 180	1 080	988	891	810	707	8
4 500	1 150	1 050	965	878	792	720	629	10
5 000	1 030	942	868	791	713	648	566	12
5 500	939	856	789	719	648	589	514	15
6 000	860	785	724	659	594	540	472	18
6 500	794	724	668	608	548	498	435	21
7 000	737	673	620	565	509	463	404	24
7 500	688	628	579	527	475	432	377	27
8 000	645	589	543	494	446	405	354	31
8 500	607	554	511	465	419	381	333	35
9 000	574	523	482	439	396	360	314	39
9 500	543	496	457	416	375	341	298	44
10 000	516	471	434	395	356	324	283	49
10 500	492	448	413	376	339	309	269	54
11 000	469	428	395	359	324	295	257	59
11 500	449	409	378	344	310	282	246	64
12 000	430	392	362	329	297	270	236	70
12 500	413	377	347	316	285	259	226	76
13 000	397	362	334	304	274	249	218	82
13 500	382	349	322	293	264	240	210	89
14 000	369	336	310	282	255	231	202	95
DESIGN DATA AND PROPERTIES								
V_r (kN)	1 060	947	868	812	735	688	650	
V_{r1} (kN)	927	837	770	724	658	620	589	
V_{r2} (kN)	799	727	672	637	581	551	528	
R (kN)	620	510	434	384	317	281	254	
G (kN)	14.5	11.7	9.89	9.25	7.70	7.47	7.67	
L_u (mm)	2 910	2 870	2 830	2 770	2 730	2 660	2 580	
d (mm)	469	466	463	460	457	454	450	
b (mm)	194	193	192	191	190	190	189	
t (mm)	20.6	19.0	17.7	16.0	14.5	12.7	10.8	
w (mm)	12.6	11.4	10.5	9.9	9.0	8.5	8.1	
k (mm)	38	36	35	33	31	30	28	
IMPERIAL SIZE AND MASS								
Mass (lb./ft.)	71	65	60	55	50	45	41	
Nominal Depth (in.)	18							

Sections highlighted in white are commonly used sizes which are generally readily available.

BEAM LOAD TABLES
W Shapes

Total Uniformly Distributed Factored Loads
for Laterally Supported Beams – kN

Designation	W410							Approx. Deflect. (mm)
Mass (kg/m)	85	74	67	60	54	46	39	
Span in Millimetres								
1 000								1
1 500						1 010	895	1
2 000	1 620	1 430	1 290	1 120	1 080	956	788	2
2 500	1 500	1 300	1 180	1 030	907	765	631	3
3 000	1 250	1 090	979	857	756	637	526	5
3 500	1 070	932	839	734	648	546	451	7
4 000	934	815	734	643	567	478	394	9
4 500	830	725	653	571	504	425	350	11
5 000	747	652	588	514	454	382	315	14
5 500	679	593	534	467	412	348	287	17
6 000	623	544	490	428	378	319	263	20
6 500	575	502	452	395	349	294	243	23
7 000	534	466	420	367	324	273	225	27
7 500	498	435	392	343	302	255	210	31
8 000	467	408	367	321	284	239	197	35
8 500	440	384	346	302	267	225	186	39
9 000	415	362	326	286	252	212	175	44
9 500	393	343	309	271	239	201	166	49
10 000	374	326	294	257	227	191	158	55
10 500	356	311	280	245	216	182	150	60
11 000	340	297	267	234	206	174	143	66
11 500	325	284	255	224	197	166	137	72
12 000	311	272	245	214	189	159	131	79
12 500	299	261	235	206	181	153	126	85
13 000	287	251	226	198	174	147	121	92
DESIGN DATA AND PROPERTIES								
V_r (kN)	812	716	643	560	540	504	447	
V_{r1} (kN)	710	630	568	498	484	452	404	
V_{r2} (kN)	608	544	492	436	428	399	360	
R (kN)	478	379	313	240	226	198	165	
G (kN)	11.9	9.67	8.10	6.13	6.72	5.32	5.22	
L_u (mm)	2 730	2 670	2 610	2 580	2 480	1 930	1 860	
d (mm)	417	413	410	407	403	403	399	
b (mm)	181	180	179	178	177	140	140	
t (mm)	18.2	16.0	14.4	12.8	10.9	11.2	8.8	
w (mm)	10.9	9.7	8.8	7.7	7.5	7.0	6.4	
k (mm)	35	33	31	30	28	28	26	
IMPERIAL SIZE AND MASS								
Mass (lb./ft.)	57	50	45	40	36	31	26	
Nominal Depth (in.)	16							

Sections highlighted in white are commonly used sizes which are generally readily available.

BEAM LOAD TABLES
W Shapes

G40.21-M
300W
$\phi = 0.90$

Total Uniformly Distributed Factored Loads
for Laterally Supported Beams – kN

Designation	W360								Approx. Deflect. (mm)
Mass (kg/m)	79	72	64	57	51	45	39	33	
Span in Millimetres									
1 000									1
1 500						866	820	723	1
2 000				1 010	914	841	715	585	2
2 500	1 190	1 070	955	873	772	673	572	468	4
3 000	1 030	922	821	727	644	561	477	390	6
3 500	883	790	704	623	552	481	409	334	8
4 000	772	691	616	545	483	421	357	293	10
4 500	686	614	547	485	429	374	318	260	13
5 000	618	553	492	436	386	337	286	234	16
5 500	562	503	448	397	351	306	260	213	19
6 000	515	461	410	364	322	280	238	195	22
6 500	475	425	379	336	297	259	220	180	26
7 000	441	395	352	312	276	240	204	167	30
7 500	412	369	328	291	257	224	191	156	35
8 000	386	346	308	273	241	210	179	146	40
8 500	363	325	290	257	227	198	168	138	45
9 000	343	307	274	242	215	187	159	130	50
9 500	325	291	259	230	203	177	151	123	56
10 000	309	276	246	218	193	168	143	117	62
10 500	294	263	235	208	184	160	136	111	69
11 000	281	251	224	198	176	153	130	106	75
DESIGN DATA AND PROPERTIES									
V_r (kN)	593	536	478	504	457	433	410	362	
V_{r1} (kN)	503	458	410	441	401	383	361	321	
V_{r2} (kN)	412	380	342	377	345	333	313	281	
R (kN)	374	314	252	262	217	199	178	141	
G (kN)	9.80	8.44	6.82	7.37	6.34	6.68	5.09	4.60	
L_u (mm)	3 270	3 190	3 110	2 550	2 500	2 430	1 790	1 720	
d (mm)	354	350	347	358	355	352	353	349	
b (mm)	205	204	203	172	171	171	128	127	
t (mm)	16.8	15.1	13.5	13.1	11.6	9.8	10.7	8.5	
w (mm)	9.4	8.6	7.7	7.9	7.2	6.9	6.5	5.8	
k (mm)	36	34	33	30	29	27	28	25	
IMPERIAL SIZE AND MASS									
Mass (lb./ft.)	53	48	43	38	34	30	26	22	
Nominal Depth (in.)	14								

Sections highlighted in white are commonly used sizes which are generally readily available.

BEAM LOAD TABLES
W Shapes

Total Uniformly Distributed Factored Loads
for Laterally Supported Beams – kN

Designation	W310									Approx. Deflect. (mm)
Mass (kg/m)	86	79	74	67	60	52	45	39	31*	
Span in Millimetres										
1 000										1
1 500									527	2
2 000			1 040	927	813	861	739	641	473	3
2 500		960	1 030	916	813	723	612	527	378	5
3 000	1 000	922	857	763	678	603	510	439	315	7
3 500	876	790	734	654	581	517	437	376	270	9
4 000	767	691	643	572	508	452	382	329	237	12
4 500	682	614	571	509	452	402	340	293	210	15
5 000	613	553	514	458	407	362	306	264	189	18
5 500	558	503	467	416	370	329	278	240	172	22
6 000	511	461	428	382	339	301	255	220	158	26
6 500	472	425	395	352	313	278	235	203	146	31
7 000	438	395	367	327	290	258	218	188	135	35
7 500	409	369	343	305	271	241	204	176	126	41
8 000	383	346	321	286	254	226	191	165	118	46
8 500	361	325	302	269	239	213	180	155	111	52
9 000	341	307	286	254	226	201	170	146	105	59
9 500	323	291	271	241	214	190	161	139	100	65
10 000	307	276	257	229	203	181	153	132	95	72
DESIGN DATA AND PROPERTIES										
V_r (kN)	503	480	519	463	406	431	369	320	264	
V_{r1} (kN)	420	405	434	391	346	370	320	279	231	
V_{r2} (kN)	337	329	348	318	286	309	271	237	199	
R (kN)	364	340	387	318	248	251	190	147	105	
G (kN)	10.5	10.7	11.5	9.64	7.43	7.34	5.73	4.55	3.64	
L_u (mm)	4 250	4 140	3 380	3 280	3 200	2 570	2 490	2 440	2 460	
d (mm)	310	306	310	306	303	317	313	310	306	
b (mm)	254	254	205	204	203	167	166	165	164	
t (mm)	16.3	14.6	16.3	14.6	13.1	13.2	11.2	9.7	7.4	
w (mm)	9.1	8.8	9.4	8.5	7.5	7.6	6.6	5.8	4.9	
k (mm)	33	32	33	32	30	30	28	27	25	
IMPERIAL SIZE AND MASS										
Mass (lb./ft.)	58	53	50	45	40	35	30	26	21	
Nominal Depth (in.)	12									

* Class 3 section.
Sections highlighted in white are commonly used sizes which are generally readily available.

BEAM LOAD TABLES
W Shapes

G40.21-M 300W

Total Uniformly Distributed Factored Loads
for Laterally Supported Beams – kN

$\phi = 0.90$

Designation	W250							Approx. Deflect. (mm)
Mass (kg/m)	67	58	49	45	39	33	24*	
Span in Millimetres								
1 000							453	1
1 500				720	616	561	396	2
2 000	818	719	654	650	554	458	297	4
2 500	778	665	547	520	443	366	238	6
3 000	649	554	456	433	369	305	198	8
3 500	556	475	391	372	317	262	170	11
4 000	487	416	342	325	277	229	149	14
4 500	432	370	304	289	246	204	132	18
5 000	389	333	273	260	222	183	119	22
5 500	354	302	249	236	201	167	108	27
6 000	324	277	228	217	185	153	99	32
6 500	299	256	210	200	170	141	91	38
7 000	278	238	195	186	158	131	85	44
7 500	259	222	182	173	148	122	79	50
8 000	243	208	171	163	139	114	74	57
DESIGN DATA AND PROPERTIES								
V_r (kN)	409	359	327	360	308	280	226	
V_{r1} (kN)	331	293	272	314	271	249	205	
V_{r2} (kN)	252	227	216	267	234	218	184	
R (kN)	368	300	260	266	201	174	120	
G (kN)	12.2	10.6	10.4	8.90	6.87	6.78	5.39	
L_u (mm)	3 570	3 410	3 240	2 360	2 280	2 180	2 190	
d (mm)	257	252	247	266	262	258	253	
b (mm)	204	203	202	148	147	146	145	
t (mm)	15.7	13.5	11.0	13.0	11.2	9.1	6.4	
w (mm)	8.9	8.0	7.4	7.6	6.6	6.1	5.0	
k (mm)	33	31	28	23	21	19	17	
IMPERIAL SIZE AND MASS								
Mass (lb./ft.)	45	39	33	30	26	22	16	
Nominal Depth (in.)	10							

* Class 3 section.
Sections highlighted in white are commonly used sizes which are generally readily available.

Total Uniformly Distributed Factored Loads
for Laterally Supported Beams – kN

Designation	W200					Approximate Deflection (mm)
Mass (kg/m)	42	36	31	27	21*	
Span in Millimetres						
1 000				430	364	1
1 500	529	446	479	402	281	3
2 000	482	410	362	301	211	4
2 500	385	328	289	241	168	7
3 000	321	274	241	201	140	10
3 500	275	235	207	172	120	14
4 000	241	205	181	151	105	18
4 500	214	182	161	134	94	23
5 000	193	164	145	121	84	28
5 500	175	149	132	110	77	34
6 000	161	137	121	100	70	40
6 500	148	126	111	93	65	47
7 000	138	117	103	86	60	55
DESIGN DATA AND PROPERTIES						
V_r (kN)	264	223	240	215	182	
V_{r1} (kN)	216	185	205	187	160	
V_{r2} (kN)	168	147	171	159	139	
R (kN)	263	196	207	173	133	
G (kN)	10.8	8.11	8.58	7.83	6.72	
L_u (mm)	2 850	2 730	2 150	2 050	2 080	
d (mm)	205	201	210	207	203	
b (mm)	166	165	134	133	133	
t (mm)	11.8	10.2	10.2	8.4	6.4	
w (mm)	7.2	6.2	6.4	5.8	5.0	
k (mm)	25	23	20	18	16	
IMPERIAL SIZE AND MASS						
Mass (lb./ft.)	28	24	21	18	14	
Nominal Depth (in.)	8					

* Class 3 section.
Sections highlighted in white are commonly used sizes which are generally readily available.

Total Uniformly Distributed Factored Loads
for Laterally Supported Beams – kN

$\phi = 0.90$

Designation	WWF2000				Approx. Deflect. (mm)	WWF1800					Approx. Deflect. (mm)
Mass (kg/m)	732*	648*	607*	542*		700	659	617	575	510	
Span in Millimetres											
13 000					22					7 960	25
13 500					24					7 800	26
14 000					26					7 520	28
14 500				7 140	27					7 260	31
15 000				6 960	29				8 000	7 020	33
15 500				6 730	31				7 980	6 800	35
16 000				6 520	33				7 730	6 580	37
16 500				6 320	36			8 050	7 500	6 380	40
17 000			7 170	6 140	38			7 960	7 280	6 200	42
17 500			7 100	5 960	40			7 730	7 070	6 020	44
18 000			6 900	5 800	42		8 090	7 520	6 870	5 850	47
18 500		7 210	6 720	5 640	45		7 930	7 320	6 690	5 690	50
19 000		7 190	6 540	5 490	47	8 140	7 720	7 120	6 510	5 540	52
19 500		7 000	6 370	5 350	50	8 100	7 520	6 940	6 340	5 400	55
20 000		6 830	6 210	5 220	52	7 900	7 330	6 770	6 190	5 270	58
20 500		6 660	6 060	5 090	55	7 710	7 150	6 600	6 040	5 140	61
21 000		6 500	5 920	4 970	58	7 520	6 980	6 440	5 890	5 020	64
21 500		6 350	5 780	4 850	60	7 350	6 820	6 290	5 760	4 900	67
22 000	7 280	6 210	5 650	4 740	63	7 180	6 670	6 150	5 620	4 790	70
22 500	7 160	6 070	5 520	4 640	66	7 020	6 520	6 010	5 500	4 680	74
23 000	7 000	5 940	5 400	4 540	69	6 870	6 380	5 880	5 380	4 580	77
23 500	6 850	5 810	5 290	4 440	72	6 720	6 240	5 760	5 260	4 480	80
24 000	6 710	5 690	5 180	4 350	75	6 580	6 110	5 640	5 160	4 390	84
24 500	6 570	5 580	5 070	4 260	78	6 450	5 990	5 520	5 050	4 300	87
25 000	6 440	5 460	4 970	4 170	82	6 320	5 870	5 410	4 950	4 210	91
25 500	6 320	5 360	4 870	4 090	85	6 200	5 750	5 310	4 850	4 130	94
26 000	6 190	5 250	4 780	4 010	88	6 080	5 640	5 200	4 760	4 050	98
26 500	6 080	5 150	4 690	3 940	92	5 960	5 530	5 110	4 670	3 980	102
27 000	5 960	5 060	4 600	3 860	95	5 850	5 430	5 010	4 580	3 900	106
27 500	5 860	4 970	4 520	3 790	99	5 750	5 330	4 920	4 500	3 830	110
28 000	5 750	4 880	4 440	3 730	102	5 640	5 240	4 830	4 420	3 760	114
28 500	5 650	4 790	4 360	3 660	106	5 540	5 150	4 750	4 340	3 700	118
29 000	5 550	4 710	4 280	3 600	110	5 450	5 060	4 670	4 270	3 630	122
29 500	5 460	4 630	4 210	3 540	114	5 360	4 970	4 590	4 190	3 570	126
30 000	5 370	4 550	4 140	3 480	118	5 270	4 890	4 510	4 120	3 510	131
DESIGN DATA AND PROPERTIES											
V_r (kN)	3 640	3 600	3 590	3 570		4 070	4 050	4 020	4 000	3 980	
L_u (mm)	7 620	7 370	7 210	6 310		7 300	7 170	7 020	6 860	5 960	
d (mm)	2 000	2 000	2 000	2 000		1 800	1 800	1 800	1 800	1 800	
b (mm)	550	550	550	500		550	550	550	550	500	
t (mm)	50.0	40.0	35.0	30.0		50.0	45.0	40.0	35.0	30.0	
w (mm)	20.0	20.0	20.0	20.0		20.0	20.0	20.0	20.0	20.0	
k (mm)	61	51	46	41		61	56	51	46	41	
IMPERIAL SIZE AND MASS											
Mass (lb./ft.)	490	436	408	364		470	442	415	388	344	
Nominal Depth (in.)	79					71					

* Class 3 section

BEAM LOAD TABLES
WWF Shapes

Total Uniformly Distributed Factored Loads
for Laterally Supported Beams – kN

Designation	WWF1600					Approx. Deflect. (mm)	WWF1400					Approx. Deflect. (mm)
Mass (kg/m)	622*	580*	538*	496*	431*		597	513	471	405	358	
Span in Millimetres												
11 000						20					5 290	23
11 500						22					5 190	25
12 000						24					4 980	27
12 500						26					4 780	29
13 000						28				5 290	4 590	32
13 500						30				5 190	4 420	34
14 000						32				5 000	4 270	37
14 500						34				4 830	4 120	39
15 000						37				4 670	3 980	42
15 500						39			5 330	4 520	3 850	45
16 000					4 600	42			5 260	4 380	3 730	48
16 500					4 480	44			5 100	4 250	3 620	51
17 000					4 340	47		5 370	4 950	4 120	3 510	54
17 500					4 220	50		5 330	4 810	4 000	3 410	57
18 000					4 100	53		5 180	4 680	3 890	3 320	60
18 500					3 990	56		5 040	4 550	3 790	3 230	64
19 000				4 630	3 890	59		4 910	4 430	3 690	3 140	67
19 500				4 590	3 790	62		4 780	4 320	3 590	3 060	71
20 000				4 470	3 690	65	5 450	4 660	4 210	3 500	2 990	75
20 500				4 360	3 600	69	5 410	4 550	4 110	3 420	2 910	78
21 000			4 660	4 260	3 520	72	5 280	4 440	4 010	3 340	2 840	82
21 500			4 630	4 160	3 430	76	5 160	4 340	3 920	3 260	2 780	86
22 000			4 520	4 070	3 360	79	5 040	4 240	3 830	3 180	2 720	90
22 500			4 420	3 980	3 280	83	4 930	4 140	3 740	3 110	2 650	95
23 000		4 690	4 330	3 890	3 210	86	4 820	4 050	3 660	3 050	2 600	99
23 500		4 640	4 240	3 810	3 140	90	4 720	3 970	3 580	2 980	2 540	103
24 000		4 550	4 150	3 730	3 080	94	4 620	3 880	3 510	2 920	2 490	108
24 500		4 450	4 060	3 650	3 010	98	4 530	3 810	3 440	2 860	2 440	112
25 000	4 720	4 360	3 980	3 580	2 950	102	4 440	3 730	3 370	2 800	2 390	117
25 500	4 660	4 280	3 900	3 510	2 900	106	4 350	3 660	3 300	2 750	2 340	121
26 000	4 560	4 200	3 830	3 440	2 840	110	4 260	3 590	3 240	2 690	2 300	126
26 500	4 480	4 120	3 760	3 380	2 790	115	4 180	3 520	3 180	2 640	2 250	131
27 000	4 400	4 040	3 690	3 310	2 740	119	4 110	3 450	3 120	2 600	2 210	136
27 500	4 320	3 970	3 620	3 250	2 680	124	4 030	3 390	3 060	2 550	2 170	141
28 000	4 240	3 900	3 560	3 200	2 640	128	3 960	3 330	3 010	2 500	2 130	146
DESIGN DATA AND PROPERTIES												
V_r (kN)	2 360	2 350	2 330	2 320	2 300		2 730	2 680	2 660	2 640	2 640	
L_u (mm)	8 010	7 890	7 750	7 620	6 720		7 790	7 510	7 360	6 450	5 000	
d (mm)	1 600	1 600	1 600	1 600	1 600		1 400	1 400	1 400	1 400	1 400	
b (mm)	550	550	550	550	500		550	550	550	500	400	
t (mm)	50.0	45.0	40.0	35.0	30.0		50.0	40.0	35.0	30.0	30.0	
w (mm)	16.0	16.0	16.0	16.0	16.0		16.0	16.0	16.0	16.0	16.0	
k (mm)	61	56	51	44	39		61	51	44	39	39	
IMPERIAL SIZE AND MASS												
Mass (lb./ft.)	419	388	361	333	289		402	344	316	272	240	
Nominal Depth (in.)	63						55					

* Class 3 section

BEAM LOAD TABLES
WWF Shapes

G40.21-M
350W
$\phi = 0.90$

Total Uniformly Distributed Factored Loads
for Laterally Supported Beams – kN

Designation	WWF1200						Approximate Deflection (mm)
Mass (kg/m)	487	418	380	333	302	263	
Span in Millimetres							
6 000						5 780	8
6 500						5 500	9
7 000					5 780	5 110	11
7 500					5 750	4 770	12
8 000				5 780	5 390	4 470	14
8 500				5 720	5 070	4 210	16
9 000				5 400	4 790	3 980	18
9 500			5 780	5 120	4 540	3 770	20
10 000			5 750	4 860	4 310	3 580	22
10 500			5 470	4 630	4 100	3 410	24
11 000		5 780	5 220	4 420	3 920	3 250	26
11 500		5 610	5 000	4 230	3 750	3 110	29
12 000		5 380	4 790	4 050	3 590	2 980	31
12 500		5 160	4 600	3 890	3 450	2 860	34
13 000	5 780	4 960	4 420	3 740	3 320	2 750	37
13 500	5 710	4 780	4 260	3 600	3 190	2 650	40
14 000	5 510	4 610	4 100	3 470	3 080	2 560	43
14 500	5 320	4 450	3 960	3 350	2 970	2 470	46
15 000	5 140	4 300	3 830	3 240	2 870	2 390	49
15 500	4 980	4 160	3 710	3 140	2 780	2 310	52
16 000	4 820	4 030	3 590	3 040	2 690	2 240	56
16 500	4 670	3 910	3 480	2 950	2 610	2 170	59
17 000	4 540	3 800	3 380	2 860	2 540	2 100	63
17 500	4 410	3 690	3 280	2 780	2 460	2 040	67
18 000	4 280	3 580	3 190	2 700	2 390	1 990	71
18 500	4 170	3 490	3 110	2 630	2 330	1 930	75
19 000	4 060	3 400	3 020	2 560	2 270	1 880	79
19 500	3 950	3 310	2 950	2 490	2 210	1 840	83
20 000	3 860	3 230	2 870	2 430	2 160	1 790	87
20 500	3 760	3 150	2 800	2 370	2 100	1 750	92
21 000	3 670	3 070	2 740	2 320	2 050	1 700	96
21 500	3 590	3 000	2 670	2 260	2 000	1 660	101
22 000	3 500	2 930	2 610	2 210	1 960	1 630	105
22 500	3 430	2 870	2 550	2 160	1 920	1 590	110
23 000	3 350	2 800	2 500	2 120	1 870	1 560	115
DESIGN DATA AND PROPERTIES							
V_r (kN)	2 890	2 890	2 890	2 890	2 890	2 890	
L_u (mm)	7 710	6 790	6 630	5 150	4 980	3 560	
d (mm)	1 200	1 200	1 200	1 200	1 200	1 200	
b (mm)	550	500	500	400	400	300	
t (mm)	40.0	35.0	30.0	30.0	25.0	25.0	
w (mm)	16.0	16.0	16.0	16.0	16.0	16.0	
k (mm)	51	44	39	39	34	34	
IMPERIAL SIZE AND MASS							
Mass (lb./ft.)	326	281	255	224	203	176	
Nominal Depth (in.)	47						

G40.21-M
350W
$\phi = 0.90$

BEAM LOAD TABLES
WWF Shapes

Total Uniformly Distributed Factored Loads
for Laterally Supported Beams – kN

Designation	WWF1100						Approximate Deflection (mm)
Mass (kg/m)	458	388	351	304	273	234	
Span in Millimetres							
6 000							9
6 500						4 420	10
7 000						4 320	12
7 500						4 030	13
8 000					4 420	3 780	15
8 500					4 360	3 560	17
9 000					4 120	3 360	19
9 500				4 420	3 900	3 180	21
10 000				4 210	3 700	3 020	24
10 500				4 010	3 530	2 880	26
11 000			4 420	3 830	3 370	2 750	29
11 500			4 360	3 660	3 220	2 630	31
12 000			4 180	3 510	3 090	2 520	34
12 500		4 420	4 010	3 370	2 960	2 420	37
13 000		4 340	3 860	3 240	2 850	2 330	40
13 500		4 180	3 720	3 120	2 740	2 240	43
14 000		4 030	3 580	3 010	2 650	2 160	47
14 500		3 890	3 460	2 900	2 560	2 090	50
15 000	4 420	3 760	3 340	2 810	2 470	2 020	53
15 500	4 410	3 640	3 240	2 720	2 390	1 950	57
16 000	4 270	3 530	3 130	2 630	2 320	1 890	61
16 500	4 140	3 420	3 040	2 550	2 240	1 830	65
17 000	4 020	3 320	2 950	2 480	2 180	1 780	69
17 500	3 900	3 230	2 870	2 400	2 120	1 730	73
18 000	3 790	3 140	2 790	2 340	2 060	1 680	77
18 500	3 690	3 050	2 710	2 280	2 000	1 640	81
19 000	3 590	2 970	2 640	2 220	1 950	1 590	86
19 500	3 500	2 900	2 570	2 160	1 900	1 550	90
20 000	3 420	2 820	2 510	2 100	1 850	1 510	95
20 500	3 330	2 750	2 450	2 050	1 810	1 480	100
21 000	3 250	2 690	2 390	2 000	1 760	1 440	105
21 500	3 180	2 620	2 330	1 960	1 720	1 410	110
22 000	3 100	2 570	2 280	1 910	1 680	1 380	115
22 500	3 040	2 510	2 230	1 870	1 650	1 340	120
23 000	2 970	2 450	2 180	1 830	1 610	1 320	126
DESIGN DATA AND PROPERTIES							
V_r (kN)	2 210	2 210	2 210	2 210	2 210	2 210	
L_u (mm)	7 890	6 970	6 810	5 310	5 150	3 710	
d (mm)	1 100	1 100	1 100	1 100	1 100	1 100	
b (mm)	550	500	500	400	400	300	
t (mm)	40.0	35.0	30.0	30.0	25.0	25.0	
w (mm)	14.0	14.0	14.0	14.0	14.0	14.0	
k (mm)	51	44	39	39	34	34	
IMPERIAL SIZE AND MASS							
Mass (lb./ft.)	307	260	236	204	184	157	
Nominal Depth (in.)	43						

Total Uniformly Distributed Factored Loads
for Laterally Supported Beams – kN

Designation	WWF1000							Approx. Deflect. (mm)
Mass (kg/m)	447	377	340	293	262	223	200	
Span in Millimetres								
5 000							4 420	7
5 500						4 420	4 200	8
6 000						4 410	3 850	9
6 500						4 070	3 560	11
7 000					4 420	3 780	3 300	13
7 500					4 370	3 530	3 080	15
8 000				4 420	4 100	3 310	2 890	17
8 500				4 390	3 850	3 110	2 720	19
9 000				4 140	3 640	2 940	2 570	21
9 500				3 930	3 450	2 780	2 430	24
10 000			4 420	3 730	3 280	2 650	2 310	26
10 500			4 250	3 550	3 120	2 520	2 200	29
11 000		4 420	4 060	3 390	2 980	2 400	2 100	32
11 500		4 380	3 880	3 240	2 850	2 300	2 010	35
12 000		4 200	3 720	3 110	2 730	2 200	1 930	38
12 500		4 030	3 570	2 980	2 620	2 120	1 850	41
13 000		3 880	3 430	2 870	2 520	2 040	1 780	44
13 500	4 420	3 730	3 300	2 760	2 430	1 960	1 710	48
14 000	4 360	3 600	3 190	2 660	2 340	1 890	1 650	51
14 500	4 210	3 480	3 080	2 570	2 260	1 820	1 590	55
15 000	4 070	3 360	2 970	2 490	2 180	1 760	1 540	59
15 500	3 930	3 250	2 880	2 410	2 110	1 710	1 490	63
16 000	3 810	3 150	2 790	2 330	2 050	1 650	1 440	67
16 500	3 700	3 060	2 700	2 260	1 980	1 600	1 400	71
17 000	3 590	2 960	2 620	2 190	1 930	1 560	1 360	76
17 500	3 480	2 880	2 550	2 130	1 870	1 510	1 320	80
18 000	3 390	2 800	2 480	2 070	1 820	1 470	1 280	85
18 500	3 300	2 720	2 410	2 020	1 770	1 430	1 250	89
19 000	3 210	2 650	2 350	1 960	1 720	1 390	1 220	94
19 500	3 130	2 580	2 290	1 910	1 680	1 360	1 180	99
20 000	3 050	2 520	2 230	1 860	1 640	1 320	1 160	105
20 500	2 980	2 460	2 180	1 820	1 600	1 290	1 130	110
21 000	2 900	2 400	2 120	1 780	1 560	1 260	1 100	115
21 500	2 840	2 340	2 080	1 740	1 520	1 230	1 080	121
22 000	2 770	2 290	2 030	1 700	1 490	1 200	1 050	126
DESIGN DATA AND PROPERTIES								
V_r (kN)	2 210	2 210	2 210	2 210	2 210	2 210	2 210	
L_u (mm)	8 030	7 080	6 910	5 400	5 230	3 790	3 610	
d (mm)	1 000	1 000	1 000	1 000	1 000	1 000	1 000	
b (mm)	550	500	500	400	400	300	300	
t (mm)	40.0	35.0	30.0	30.0	25.0	25.0	20.0	
w (mm)	14.0	14.0	14.0	14.0	14.0	14.0	14.0	
k (mm)	51	44	39	39	34	34	29	
IMPERIAL SIZE AND MASS								
Mass (lb./ft.)	300	253	228	197	176	150	134	
Nominal Depth (in.)	39							

BEAM LOAD TABLES
WWF Shapes

Total Uniformly Distributed Factored Loads
for Laterally Supported Beams – kN

Designation	WWF900							Approx. Deflect. (mm)
Mass (kg/m)	417	347	309	262	231	192	169	
Span in Millimetres								
6 000								10
6 500							2 680	12
7 000							2 650	14
7 500						2 710	2 480	16
8 000						2 710	2 320	19
8 500						2 550	2 180	21
9 000						2 410	2 060	24
9 500						2 280	1 960	26
10 000					2 710	2 170	1 860	29
10 500					2 590	2 060	1 770	32
11 000				2 730	2 470	1 970	1 690	35
11 500				2 720	2 370	1 880	1 620	38
12 000				2 600	2 270	1 810	1 550	42
12 500				2 500	2 180	1 730	1 490	45
13 000				2 400	2 090	1 670	1 430	49
13 500			2 730	2 320	2 020	1 600	1 380	53
14 000			2 700	2 230	1 940	1 550	1 330	57
14 500			2 610	2 160	1 880	1 500	1 280	61
15 000			2 520	2 080	1 810	1 440	1 240	65
15 500		2 730	2 440	2 020	1 760	1 400	1 200	70
16 000		2 690	2 360	1 950	1 700	1 360	1 160	74
16 500		2 610	2 290	1 890	1 650	1 310	1 130	79
17 000		2 540	2 220	1 840	1 600	1 280	1 090	84
17 500		2 460	2 160	1 790	1 560	1 240	1 060	89
18 000		2 390	2 100	1 740	1 510	1 200	1 030	94
18 500		2 330	2 040	1 690	1 470	1 170	1 000	99
19 000	2 730	2 270	1 990	1 640	1 430	1 140	977	105
19 500	2 690	2 210	1 940	1 600	1 400	1 110	952	110
20 000	2 620	2 160	1 890	1 560	1 360	1 080	929	116
20 500	2 560	2 100	1 840	1 520	1 330	1 060	906	122
21 000	2 500	2 050	1 800	1 490	1 300	1 030	884	128
21 500	2 440	2 000	1 760	1 450	1 270	1 010	864	134
22 000	2 380	1 960	1 720	1 420	1 240	985	844	141
22 500	2 330	1 920	1 680	1 390	1 210	963	825	147
23 000	2 280	1 870	1 640	1 360	1 180	942	807	154
DESIGN DATA AND PROPERTIES								
V_r (kN)	1 370	1 370	1 370	1 370	1 350	1 350	1 340	
L_u (mm)	8 310	7 320	7 160	5 630	5 460	3 980	3 820	
d (mm)	900	900	900	900	900	900	900	
b (mm)	550	500	500	400	400	300	300	
t (mm)	40.0	35.0	30.0	30.0	25.0	25.0	20.0	
w (mm)	11.0	11.0	11.0	11.0	11.0	11.0	11.0	
k (mm)	51	44	39	39	34	34	29	
IMPERIAL SIZE AND MASS								
Mass (lb./ft.)	279	233	208	176	156	128	113	
Nominal Depth (in.)	35							

BEAM LOAD TABLES
WWF Shapes

G40.21-M
350W
$\phi = 0.90$

Total Uniformly Distributed Factored Loads
for Laterally Supported Beams – kN

Designation	WWF800						Approximate Deflection (mm)
Mass (kg/m)	339	300	253	223	184	161	
Span in Millimetres							
5 000							8
5 500						2 730	10
6 000						2 650	12
6 500					2 730	2 450	14
7 000					2 670	2 280	16
7 500					2 490	2 120	18
8 000					2 330	1 990	21
8 500				2 730	2 200	1 870	24
9 000				2 620	2 080	1 770	26
9 500			2 730	2 480	1 970	1 680	29
10 000			2 720	2 350	1 870	1 590	33
10 500			2 590	2 240	1 780	1 520	36
11 000			2 470	2 140	1 700	1 450	40
11 500			2 370	2 050	1 620	1 380	43
12 000		2 730	2 270	1 960	1 560	1 330	47
12 500		2 640	2 180	1 880	1 490	1 270	51
13 000		2 540	2 090	1 810	1 440	1 220	55
13 500	2 730	2 440	2 020	1 740	1 380	1 180	60
14 000	2 680	2 360	1 940	1 680	1 330	1 140	64
14 500	2 590	2 280	1 880	1 620	1 290	1 100	69
15 000	2 500	2 200	1 810	1 570	1 240	1 060	74
15 500	2 420	2 130	1 760	1 520	1 200	1 030	78
16 000	2 350	2 060	1 700	1 470	1 170	995	84
16 500	2 280	2 000	1 650	1 430	1 130	965	89
17 000	2 210	1 940	1 600	1 380	1 100	937	94
17 500	2 150	1 890	1 560	1 340	1 070	910	100
18 000	2 090	1 830	1 510	1 310	1 040	885	106
18 500	2 030	1 780	1 470	1 270	1 010	861	112
19 000	1 980	1 740	1 430	1 240	983	838	118
19 500	1 930	1 690	1 400	1 210	958	817	124
20 000	1 880	1 650	1 360	1 180	934	796	131
20 500	1 830	1 610	1 330	1 150	911	777	137
21 000	1 790	1 570	1 300	1 120	889	758	144
21 500	1 750	1 540	1 270	1 100	869	741	151
22 000	1 710	1 500	1 240	1 070	849	724	158
DESIGN DATA AND PROPERTIES							
V_r (kN)	1 370	1 370	1 370	1 370	1 370	1 370	
L_u (mm)	7 490	7 280	5 730	5 570	4 060	3 900	
d (mm)	800	800	800	800	800	800	
b (mm)	500	500	400	400	300	300	
t (mm)	35.0	30.0	30.0	25.0	25.0	20.0	
w (mm)	11.0	11.0	11.0	11.0	11.0	11.0	
k (mm)	44	39	39	34	34	29	
IMPERIAL SIZE AND MASS							
Mass (lb./ft.)	228	202	170	150	123	108	
Nominal Depth (in.)	31						

BEAM LOAD TABLES
WWF Shapes

Total Uniformly Distributed Factored Loads for Laterally Supported Beams – kN

Designation	WWF700					Approximate Deflection (mm)
Mass (kg/m)	245	214	196	175	152	
Span in Millimetres 4 000						6
4 500					2 730	8
5 000					2 680	9
5 500				2 730	2 440	11
6 000			2 730	2 630	2 230	13
6 500			2 540	2 430	2 060	16
7 000		2 730	2 360	2 260	1 920	18
7 500		2 670	2 200	2 110	1 790	21
8 000	2 730	2 500	2 070	1 980	1 680	24
8 500	2 730	2 360	1 940	1 860	1 580	27
9 000	2 580	2 230	1 840	1 760	1 490	30
9 500	2 440	2 110	1 740	1 660	1 410	34
10 000	2 320	2 000	1 650	1 580	1 340	37
10 500	2 210	1 910	1 570	1 500	1 280	41
11 000	2 110	1 820	1 500	1 440	1 220	45
11 500	2 020	1 740	1 440	1 370	1 170	49
12 000	1 930	1 670	1 380	1 320	1 120	54
12 500	1 860	1 600	1 320	1 260	1 070	58
13 000	1 780	1 540	1 270	1 220	1 030	63
13 500	1 720	1 480	1 220	1 170	993	68
14 000	1 660	1 430	1 180	1 130	958	73
14 500	1 600	1 380	1 140	1 090	925	78
15 000	1 550	1 340	1 100	1 050	894	84
15 500	1 500	1 290	1 070	1 020	865	90
16 000	1 450	1 250	1 030	988	838	96
16 500	1 410	1 210	1 000	958	813	102
17 000	1 360	1 180	972	929	789	108
17 500	1 330	1 140	945	903	766	114
18 000	1 290	1 110	918	878	745	121
18 500	1 260	1 080	894	854	725	128
19 000	1 220	1 050	870	832	706	135
19 500	1 190	1 030	848	810	688	142
20 000	1 160	1 000	827	790	670	149
20 500	1 130	977	806	771	654	157
21 000	1 100	954	787	752	638	165
DESIGN DATA AND PROPERTIES						
V_r (kN)	1 370	1 370	1 370	1 370	1 370	
L_u (mm)	5 880	5 690	5 860	4 160	3 990	
d (mm)	700	700	700	700	700	
b (mm)	400	400	400	300	300	
t (mm)	30.0	25.0	22.0	25.0	20.0	
w (mm)	11.0	11.0	11.0	11.0	11.0	
k (mm)	39	34	31	34	29	
IMPERIAL SIZE AND MASS						
Mass (lb./ft.)	164	144	132	117	102	
Nominal Depth (in.)	28					

350W

BEAM LOAD TABLES[1]
Rectangular HSS

G40.21-M 350W
∅ = 0.90

Total Uniformly Distributed Factored
Loads for Laterally Supported Beams — kN

Designation mm x mm mm	HSS 305 x 203					Approx. Deflection (mm)	HSS 254 x 152					Approx. Deflection (mm)
	13	11	9.5	8.0	6.4		13	11	9.5	8.0	6.4	
Mass (kg/m)	93	82.4	71.3	60.1	48.6		72.7	64.6	56.1	47.5	38.4	
Span in millimetres												
500							2 150	1 940	1 710	1 470	1 210	
1 000	2 680	2 410	2 110	1 800	1 480	1	1 880	1 690	1 480	1 270	1 040	1
1 500	2 000	1 780	1 550	1 320	1 080	2	1 250	1 130	990	845	690	2
2 000	1 500	1 340	1 160	992	806	3	940	845	742	634	518	4
2 500	1 200	1 070	932	793	645	5	752	676	594	507	414	6
3 000	1 000	890	777	661	538	8	627	564	495	423	345	9
3 500	857	763	666	567	461	10	537	483	424	362	296	13
4 000	750	668	583	496	403	14	470	423	371	317	259	16
4 500	666	594	518	441	358	17	418	376	330	282	230	21
5 000	600	534	466	397	323	21	376	338	297	254	207	26
5 500	545	486	424	361	293	26	342	307	270	230	188	31
6 000	500	445	388	331	269	31	313	282	247	211	173	37
6 500	461	411	359	305	248	36	289	260	228	195	159	43
7 000	428	382	333	283	230	42	269	242	212	181	148	50
7 500	400	356	311	264	215	48	251	225	198	169	138	58
8 000	375	334	291	248	202	55	235	211	186	158	129	66
8 500	353	314	274	233	190	62	221	199	175	149	122	74
9 000	333	297	259	220	179	69	209	188	165	141	115	83
9 500	316	281	245	209	170	77	198	178	156	133	109	93
10 000	300	267	233	198	161	86	188	169	148	127	104	103
PROPERTIES AND DESIGN DATA												
S_x $10^3 mm^3$	964	867	762	652	535		592	537	475	409	338	
Z_x $10^3 mm^3$	1 190	1 060	925	787	640		746	671	589	503	411	
I_x $10^6 mm^4$	147	132	116	99.4	81.5		75.2	68.2	60.3	52.0	42.9	
C mm^2	6 450	5 790	5 080	4 340	3 550		5 160	4 660	4 110	3 530	2 900	
V_r kN	1 340	1 200	1 060	902	738		1 070	969	855	735	604	
IMPERIAL SIZE AND MASS												
Mass (lb./ft.)	62.5	55.4	47.9	40.4	32.6		48.9	43.4	37.7	31.9	25.8	
Thickness (in.)	.500	.438	.375	.313	.250		.500	.438	.375	.313	.250	
Size (in.)	12 x 8						10 x 6					

[1] For strong axis bending only.

BEAM LOAD TABLES[1]
Rectangular HSS

Total Uniformly Distributed Factored Loads for Laterally Supported Beams – kN

Designation mm x mm mm	HSS 203 x 152						HSS 203 x 102						Approx. Deflection (mm)
	13	11	9.5	8.0	6.5	4.8	13	11	9.5	8.0	6.4	4.8	
Mass (kg/m)	62.6	55.7	48.5	41.1	33.4	25.5	52.4	46.9	40.9	34.8	28.3	21.7	
Span in millimetres													
500	1 610	1 470	1 310	1 130	939	732	1 610	1 470	1 310	1 130	939	732	
1 000	1 330	1 200	1 060	907	743	575	1 020	925	819	708	585	454	1
1 500	887	800	704	605	496	383	680	617	546	472	390	302	3
2 000	665	600	528	454	372	287	510	462	409	354	292	227	5
2 500	532	480	422	363	297	230	408	370	328	283	234	181	8
3 000	444	400	352	302	248	192	340	308	273	236	195	151	12
3 500	380	343	302	259	212	164	292	264	234	202	167	130	16
4 000	333	300	264	227	186	144	255	231	205	177	146	113	21
4 500	296	267	235	202	165	128	227	206	182	157	130	101	26
5 000	266	240	211	181	149	115	204	185	164	142	117	91	32
5 500	242	218	192	165	135	104	186	168	149	129	106	82	39
6 000	222	200	176	151	124	96	170	154	136	118	97	76	46
6 500	205	185	162	140	114	88	157	142	126	109	90	70	54
7 000	190	171	151	130	106	82	146	132	117	101	84	65	63

PROPERTIES AND DESIGN DATA

S_x $10^3 mm^3$	423	385	343	297	246	192	307	282	253	221	185	145
Z_x $10^3 mm^3$	528	476	419	360	295	228	405	367	325	281	232	180
I_x $10^6 mm^4$	43.0	39.1	34.8	30.2	25.0	19.5	31.2	28.7	25.8	22.5	18.8	14.7
C mm^2	3 870	3 530	3 150	2 730	2 260	1 760	3 870	3 530	3 150	2 730	2 260	1 760
V_r kN	805	734	654	567	469	366	805	734	654	567	469	366

IMPERIAL SIZE AND MASS

Mass (lb./ft.)	42.1	37.5	32.6	27.6	22.4	17.1	35.2	31.5	27.5	23.4	19.0	14.6
Thickness (in.)	.500	.438	.375	.313	.250	.188	.500	.438	.375	.313	.250	.188
Size (in.)	8 x 6						8 x 4					

[1] For strong axis bending only.

BEAM LOAD TABLES[1]
Rectangular HSS

G40.21-M 350W
$\phi = 0.90$

Total Uniformly Distributed Factored
Loads for Laterally Supported Beams – kN

Designation mm x mm mm		HSS 178 x 127 #						Approx. Deflection (mm)
		13	11	9.5	8.0	6.4	4.8	
Mass (kg/m)		52.4	46.9	40.9	34.8	28.3	21.7	
Span in millimetres	500	1 340	1 230	1 110	965	805	631	
	1 000	950	862	764	658	544	421	1
	1 500	633	575	509	438	363	281	3
	2 000	475	431	382	329	272	210	6
	2 500	380	345	305	263	218	168	9
	3 000	317	287	255	219	181	140	13
	3 500	271	246	218	188	156	120	18
	4 000	238	215	191	164	136	105	23
	4 500	211	192	170	146	121	94	30
	5 000	190	172	153	132	109	84	37
	5 500	173	157	139	120	99	77	44
	6 000	158	144	127	110	91	70	53

PROPERTIES AND DESIGN DATA

	13	11	9.5	8.0	6.4	4.8
S_x $10^3 mm^3$	297	272	244	213	178	140
Z_x $10^3 mm^3$	377	342	303	261	216	167
I_x $10^6 mm^4$	26.4	24.2	21.7	18.9	15.8	12.4
C mm^2	3 230	2 970	2 660	2 320	1 940	1 520
V_r kN	671	617	553	483	402	315

IMPERIAL SIZE AND MASS

Mass (lb./ft.)	35.2	31.5	27.5	23.4	19.0	14.6
Thickness (in.)	.500	.438	.375	.313	.250	.188
Size (in.)	7 x 5 #					

Check manufacturer for current availability.
[1] For strong axis bending only.

Total Uniformly Distributed Factored Loads for Laterally Supported Beams – kN

Designation mm x mm mm	HSS 152 x 102					Approx. Deflection (mm)	HSS 127 x 76				Approx. Deflection (mm)
	11	9.5	8.0	6.4	4.8		9.5	8.0	6.4	4.8	
Mass (kg/m)	38	33.3	28.4	23.2	17.9		25.7	22.1	18.2	14.1	
Span in millimetres											
500	999	906	797	671	530		635	559	470	372	1
1 000	580	517	449	373	292	2	318	280	235	186	2
1 500	386	344	299	249	195	4	212	186	157	124	5
2 000	290	258	224	186	146	7	159	140	118	93	8
2 500	232	207	179	149	117	11	127	112	94	74	13
3 000	193	172	150	124	97	15	106	93	78	62	19
3 500	166	148	128	107	84	21	91	80	67	53	25
4 000	145	129	112	93	73	28	79	70	59	46	33
4 500	129	115	100	83	65	35	71	62	52	41	42
5 000	116	103	90	75	58	43	64	56	47	37	51
PROPERTIES AND DESIGN DATA											
S_x $10^3 mm^3$	179	162	143	121	95.6		96.4	86.4	74.0	59.5	
Z_x $10^3 mm^3$	230	205	178	148	116		126	111	93.3	73.8	
I_x $10^6 mm^4$	13.6	12.4	10.9	9.18	7.28		6.12	5.49	4.70	3.78	
C mm^2	2 400	2 180	1 920	1 610	1 270		1 690	1 510	1 290	1 030	
V_r kN	499	453	399	335	265		352	315	268	214	
IMPERIAL SIZE AND MASS											
Mass (lb./ft.)	25.5	22.4	19.1	15.6	12.0		17.3	14.9	12.2	9.45	
Thickness (in.)	.438	.375	.313	.250	.188		.375	.313	.250	.188	
Size (in.)	6 x 4						5 x 3				

[1] For strong axis bending only.

BEAM LOAD TABLES[1]
Rectangular HSS

Total Uniformly Distributed Factored
Loads for Laterally Supported Beams — kN

Designation mm x mm mm	HSS 127 x 64				HSS 127 x 51				Approx. Deflection (mm)
	9.5	8.0	6.4	4.8	9.5	8.0	6.4	4.8	
Mass (kg/m)	23.8	20.5	16.9	13.1	21.9	18.9	15.6	12.2	
Span in millimetres 500	564	498	421	334	492	438	372	297	1
1 000	282	249	211	167	246	219	186	148	2
1 500	188	166	140	111	164	146	124	99	5
2 000	141	125	105	84	123	109	93	74	8
2 500	113	100	84	67	98	88	74	59	13
3 000	94	83	70	56	82	73	62	49	19
3 500	81	71	60	48	70	63	53	42	25
4 000	71	62	53	42	62	55	46	37	33
PROPERTIES AND DESIGN DATA									
S_x $10^3 mm^3$	83.2	75.1	64.8	52.4	70.0	63.8	55.5	45.3	
Z_x $10^3 mm^3$	112	98.9	83.6	66.3	97.7	86.9	73.8	58.9	
I_x $10^6 mm^4$	5.28	4.77	4.11	3.33	4.45	4.05	3.53	2.87	
C mm^2	1 690	1 510	1 290	1 030	1 690	1 510	1 290	1 030	
V_r kN	352	315	268	214	352	315	268	214	
IMPERIAL SIZE AND MASS									
Mass (lb./ft.)	16.0	13.8	11.4	8.81	14.7	12.7	10.5	8.17	
Thickness (in.)	.375	.313	.250	.188	.375	.313	.250	.188	
Size (in.)	5 x 2½				5 x 2				

[1] For strong axis bending only.

BEAM LOAD TABLES[1]
Rectangular HSS

Total Uniformly Distributed Factored Loads for Laterally Supported Beams – kN

Designation mm x mm mm	HSS 102 x 76				HSS 102 x 51					Approx. Deflection (mm)
	9.5	8.0	6.4	4.8	8.0	6.4	4.8	3.8	3.2	
Mass (kg/m)	21.9	18.9	15.6	12.2	15.8	13.1	10.3	8.37	7.09	
Span in millimetres 500	442	392	332	265	296	255	206	171	146	1
1 000	221	196	166	132	148	128	103	85	73	3
1 500	147	131	111	88	99	85	69	57	49	6
2 000	110	98	83	66	74	64	51	43	37	10
2 500	88	78	66	53	59	51	41	34	29	16
3 000	74	65	55	44	49	43	34	28	24	23
3 500	63	56	47	38	42	36	29	24	21	31
4 000	55	49	42	33	37	32	26	21	18	41
PROPERTIES AND DESIGN DATA										
S_x $10^3 mm^3$	67.2	61.0	52.8	43.0	43.5	38.4	31.8	26.8	23.1	
Z_x $10^3 mm^3$	87.6	77.7	65.9	52.5	58.8	50.6	40.8	33.9	29.0	
I_x $10^6 mm^4$	3.41	3.10	2.68	2.18	2.21	1.95	1.61	1.36	1.17	
C mm^2	1 210	1 110	968	789	1 110	968	789	658	565	
V_r kN	252	231	201	164	231	201	164	137	118	
IMPERIAL SIZE AND MASS										
Mass (lb./ft.)	14.7	12.7	10.5	8.17	10.6	8.81	6.89	5.62	4.76	
Thickness (in.)	.375	.313	.250	.188	.313	.250	.188	.150	.125	
Size (in.)	4 x 3				4 x 2					

[1] For strong axis bending only.

BEAM DIAGRAMS AND FORMULAE

Equivalent Tabular Load is the uniformly distributed factored load given in the Beam Load Tables

1 SIMPLE BEAM—UNIFORMLY DISTRIBUTED LOAD

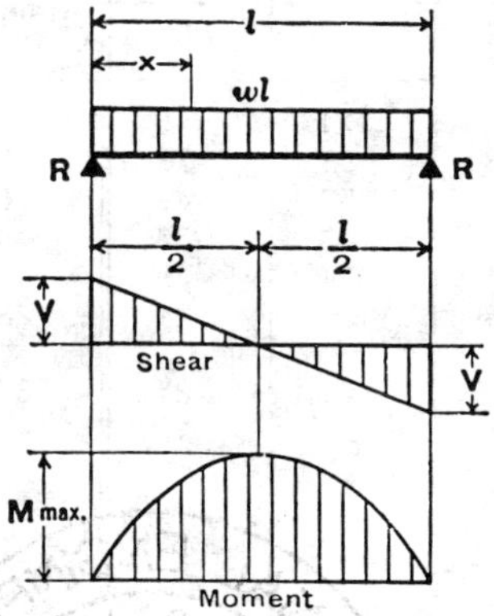

Equivalent Tabular Load . . . $= wl$

$R = V$ . . . $= \frac{wl}{2}$

V_x . . . $= w\left(\frac{l}{2} - x\right)$

M max. (at center) . . . $= \frac{wl^2}{8}$

M_x . . . $= \frac{wx}{2}(l - x)$

Δmax. (at center) . . . $= \frac{5\,wl^4}{384\,EI}$

Δx . . . $= \frac{wx}{24EI}(l^3 - 2lx^2 + x^3)$

2 SIMPLE BEAM—LOAD INCREASING UNIFORMLY TO ONE END

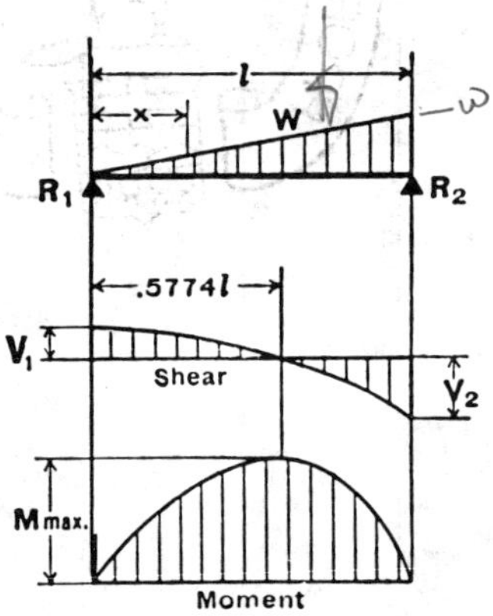

Equivalent Tabular Load . . . $= \frac{16W}{9\sqrt{3}} = 1.0264W$

$R_1 = V_1$ . . . $= \frac{W}{3}$

$R_2 = V_2$ max. . . . $= \frac{2W}{3}$

V_x . . . $= \frac{W}{3} - \frac{Wx^2}{l^2}$

M max. $\left(\text{at } x = \frac{l}{\sqrt{3}} = .5774l\right)$ . . . $= \frac{2Wl}{9\sqrt{3}} = .1283\,Wl$

M_x . . . $= \frac{Wx}{3l^2}(l^2 - x^2)$

Δmax. $\left(\text{at } x = l\sqrt{1 - \sqrt{\frac{8}{15}}} = .5193l\right) = .01304\,\frac{Wl^3}{EI}$

Δx . . . $= \frac{Wx}{180EI\,l^2}(3x^4 - 10l^2x^2 + 7l^4)$

3 SIMPLE BEAM—LOAD INCREASING UNIFORMLY TO CENTER

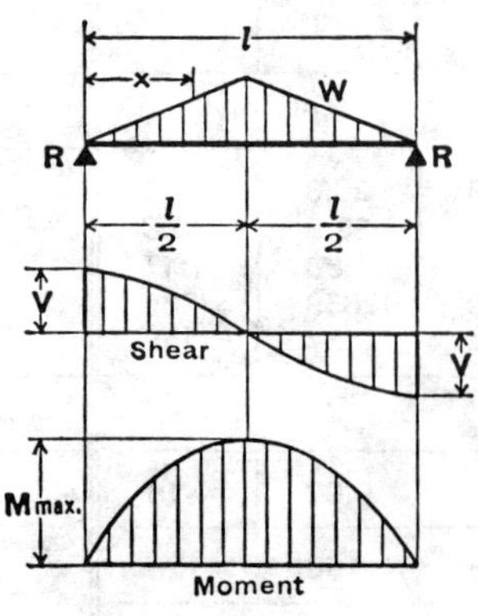

Equivalent Tabular Load . . . $= \frac{4W}{3}$

$R = V$ . . . $= \frac{W}{2}$

V_x $\left(\text{when } x < \frac{l}{2}\right)$ . . . $= \frac{W}{2l^2}(l^2 - 4x^2)$

M max. (at center) . . . $= \frac{Wl}{6}$

M_x $\left(\text{when } x < \frac{l}{2}\right)$ . . . $= Wx\left(\frac{1}{2} - \frac{2x^2}{3l^2}\right)$

Δmax. (at center) . . . $= \frac{Wl^3}{60EI}$

Δx . . . $= \frac{Wx}{480\,EI\,l^2}(5l^2 - 4x^2)^2$

Note: For deflection calculations, use specified loads.

4 SIMPLE BEAM—UNIFORM LOAD PARTIALLY DISTRIBUTED

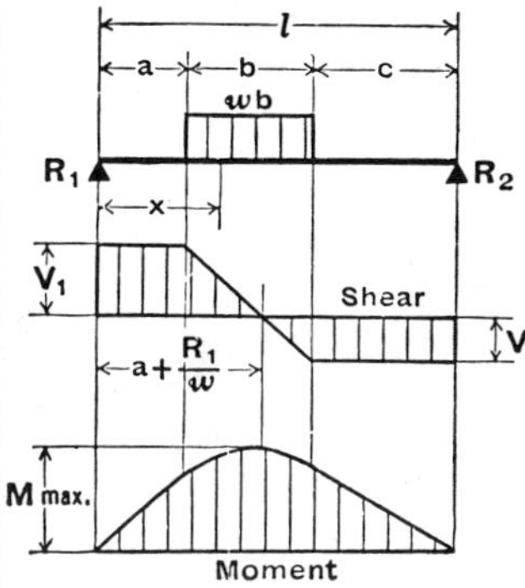

$R_1 = V_1$ $\left(\text{max. when } a < c\right)$ . . . $= \dfrac{wb}{2l}(2c + b)$

$R_2 = V_2$ $\left(\text{max. when } a > c\right)$ . . . $= \dfrac{wb}{2l}(2a + b)$

V_x $\left(\text{when } x > a \text{ and } < (a+b)\right)$. $= R_1 - w(x - a)$

M max. $\left(\text{at } x = a + \dfrac{R_1}{w}\right)$ . . . . $= R_1\left(a + \dfrac{R_1}{2w}\right)$

M_x $\left(\text{when } x < a\right)$ . . . . . $= R_1 x$

M_x $\left(\text{when } x > a \text{ and } < (a+b)\right)$. $= R_1 x - \dfrac{w}{2}(x-a)^2$

M_x $\left(\text{when } x > (a+b)\right)$ . . . . $= R_2(l - x)$

5 SIMPLE BEAM—UNIFORM LOAD PARTIALLY DISTRIBUTED AT ONE END

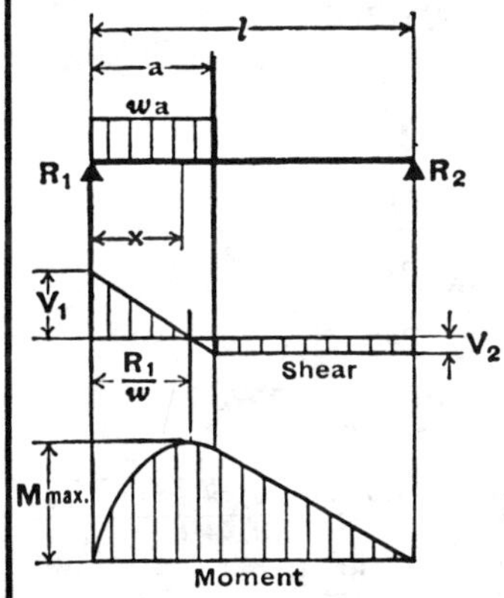

$R_1 = V_1$ max. . . . . $= \dfrac{wa}{2l}(2l - a)$

$R_2 = V_2$ . . . . $= \dfrac{wa^2}{2l}$

V $\left(\text{when } x < a\right)$ . . . . . $= R_1 - wx$

M max. $\left(\text{at } x = \dfrac{R_1}{w}\right)$ . . . . . $= \dfrac{R_1^2}{2w}$

M_x $\left(\text{when } x < a\right)$ . . . . . $= R_1 x - \dfrac{wx^2}{2}$

M_x $\left(\text{when } x > a\right)$ . . . . . $= R_2(l - x)$

Δx $\left(\text{when } x < a\right)$ . . . . . $= \dfrac{wx}{24EIl}\left(a^2(2l-a)^2 - 2ax^2(2l-a) + lx^3\right)$

Δx $\left(\text{when } x > a\right)$ . . . . . $= \dfrac{wa^2(l-x)}{24EIl}(4xl - 2x^2 - a^2)$

6 SIMPLE BEAM—UNIFORM LOADS PARTIALLY DISTRIBUTED AT EACH END

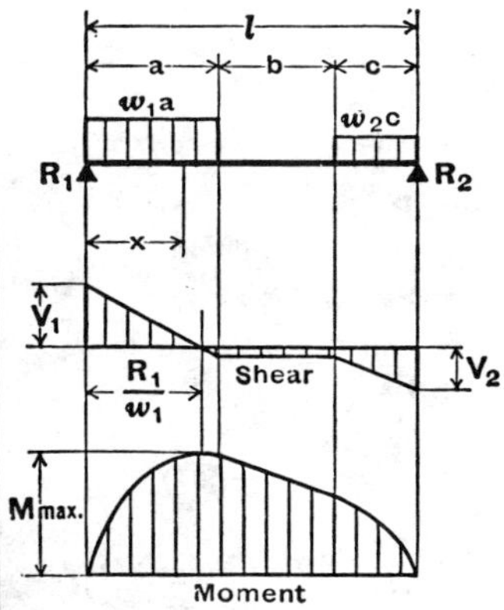

$R_1 = V_1$ . . . . $= \dfrac{w_1 a(2l - a) + w_2 c^2}{2l}$

$R_2 = V_2$ . . . . $= \dfrac{w_2 c(2l - c) + w_1 a^2}{2l}$

V_x $\left(\text{when } x < a\right)$ . . . . . $= R_1 - w_1 x$

V_x $\left(\text{when } x > a \text{ and } < (a+b)\right)$. $= R_1 - w_1 a$

V_x $\left(\text{when } x > (a+b)\right)$ . . . . $= R_2 - w_2(l - x)$

M max. $\left(\text{at } x = \dfrac{R_1}{w_1} \text{ when } R_1 < w_1 a\right)$. $= \dfrac{R_1^2}{2w_1}$

M max. $\left(\text{at } x = l - \dfrac{R_2}{w_2} \text{ when } R_2 < w_2 c\right)$ $= \dfrac{R_2^2}{2w_2}$

M_x $\left(\text{when } x < a\right)$ . . . . . $= R_1 x - \dfrac{w_1 x^2}{2}$

M_x $\left(\text{when } x > a \text{ and } < (a+b)\right)$. $= R_1 x - \dfrac{w_1 a}{2}(2x - a)$

M_x $\left(\text{when } x > (a+b)\right)$ . . . . $= R_2(l - x) - \dfrac{w_2(l - x)^2}{2}$

Note: For deflection calculations, use specified loads.

BEAM DIAGRAMS AND FORMULAE

Equivalent Tabular Load is the uniformly
distributed factored load given in the Beam Load Tables

7 SIMPLE BEAM—CONCENTRATED LOAD AT CENTER

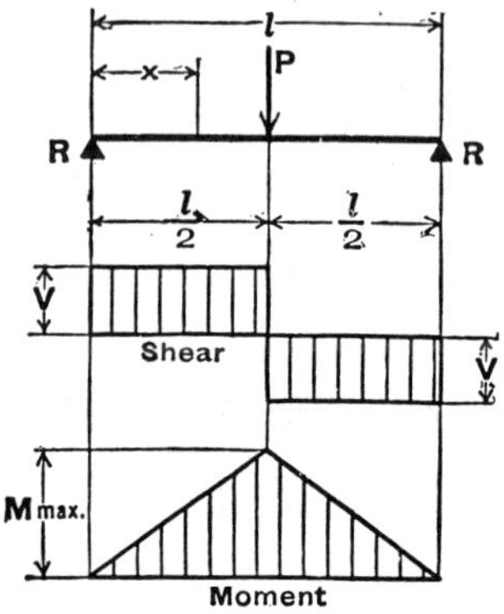

- Equivalent Tabular Load . . . . . . $= 2P$
- $R = V$ . . . . $= \frac{P}{2}$
- M max. (at point of load) . . . . . . $= \frac{Pl}{4}$
- M_x $\left(\text{when } x < \frac{l}{2}\right)$ . . . . . . $= \frac{Px}{2}$
- Δmax. (at point of load) . . . . . . $= \frac{Pl^3}{48EI}$
- Δx $\left(\text{when } x < \frac{l}{2}\right)$ . . . . . . $= \frac{Px}{48EI}(3l^2 - 4x^2)$

8 SIMPLE BEAM—CONCENTRATED LOAD AT ANY POINT

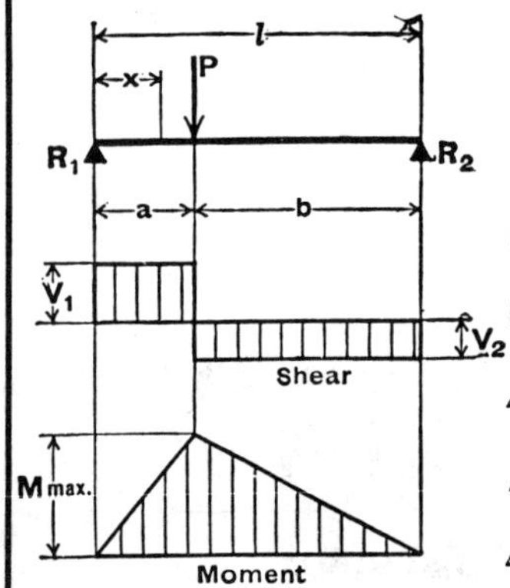

- Equivalent Tabular Load . . . . . . $= \frac{8\,Pab}{l^2}$
- $R_1 = V_1$ (max. when $a < b$) . . . . $= \frac{Pb}{l}$
- $R_2 = V_2$ (max. when $a > b$) . . . . $= \frac{Pa}{l}$
- M max. (at point of load) . . . . $= \frac{Pab}{l}$
- M_x (when $x < a$) . . . . $= \frac{Pbx}{l}$
- Δmax. $\left(\text{at } x = \sqrt{\frac{a(a+2b)}{3}} \text{ when } a > b\right)$ $= \frac{Pab(a+2b)\sqrt{3a(a+2b)}}{27\,EI\,l}$
- Δa (at point of load) . . . . $= \frac{Pa^2b^2}{3EI\,l}$
- Δx (when $x < a$) . . . . $= \frac{Pbx}{6EI\,l}(l^2 - b^2 - x^2)$

9 SIMPLE BEAM—TWO EQUAL CONCENTRATED LOADS SYMMETRICALLY PLACED

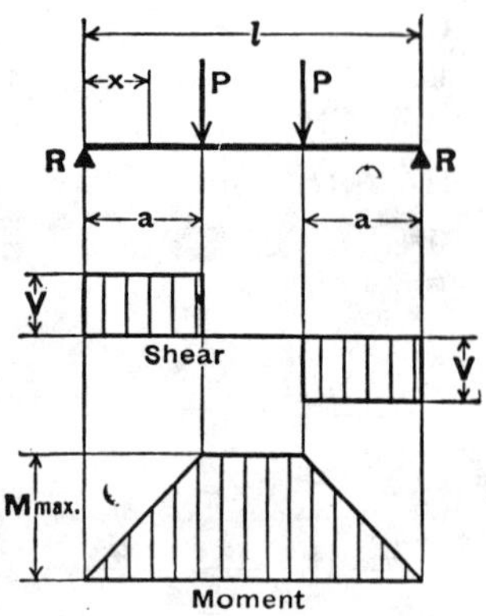

- Equivalent Tabular Load . . . . . . $= \frac{8\,Pa}{l}$
- $R = V$ . . . . $= P$
- M max. (between loads) . . . . . . $= Pa$
- M_x (when $x < a$) . . . . . . $= Px$
- Δmax. (at center) . . . . $= \frac{Pa}{24EI}(3l^2 - 4a^2)$
- Δx (when $x < a$) . . . . $= \frac{Px}{6EI}(3la - 3a^2 - x^2)$
- Δx (when $x > a$ and $< (l - a)$) . . $= \frac{Pa}{6EI}(3lx - 3x^2 - a^2)$

Note: For deflection calculations, use specified loads.

Equivalent Tabular Load is the uniformly distributed factored load given in the Beam Load Tables

10 SIMPLE BEAM—TWO EQUAL CONCENTRATED LOADS UNSYMMETRICALLY PLACED

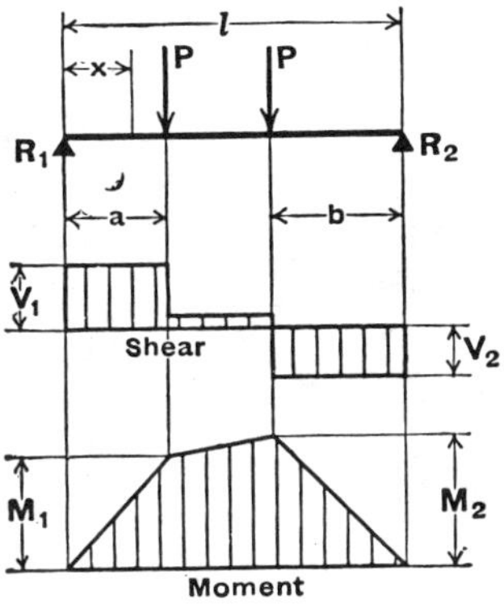

$R_1 = V_1$ (max. when $a < b$) . . . . $= \frac{P}{l}(l - a + b)$

$R_2 = V_2$ (max. when $a > b$) . . . . $= \frac{P}{l}(l - b + a)$

V_x (when $x > a$ and $< (l - b)$) . . $= \frac{P}{l}(b - a)$

M_1 (max. when $a > b$) . . . . $= R_1 a$

M_2 (max. when $a < b$) . . . . $= R_2 b$

M_x (when $x < a$) . . . . . . $= R_1 x$

M_x (when $x > a$ and $< (l - b)$) . . $= R_1 x - P(x - a)$

11 SIMPLE BEAM—TWO UNEQUAL CONCENTRATED LOADS UNSYMMETRICALLY PLACED

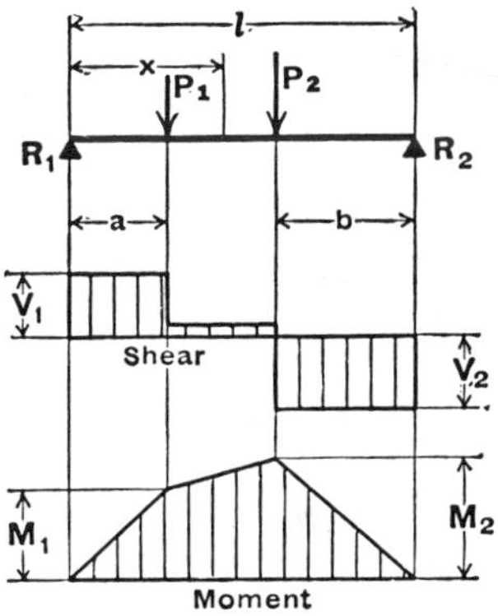

$R_1 = V_1$ . . . . $= \frac{P_1(l - a) + P_2 b}{l}$

$R_2 = V_2$ . . . . $= \frac{P_1 a + P_2(l - b)}{l}$

V_x (when $x > a$ and $< (l - b)$) . . $= R_1 - P_1$

M_1 (max. when $R_1 < P_1$) . . . $= R_1 a$

M_2 (max. when $R_2 < P_2$) . . . $= R_2 b$

M_x (when $x < a$) . . . . . . $= R_1 x$

M_x (when $x > a$ and $< (l - b)$) . . $= R_1 x - P_1(x - a)$

12 BEAM FIXED AT ONE END, SUPPORTED AT OTHER—UNIFORMLY DISTRIBUTED LOAD

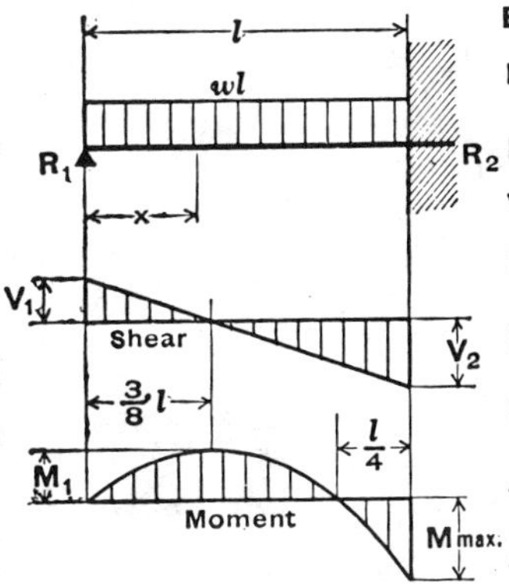

Equivalent Tabular Load . . . . . $= wl$

$R_1 = V_1$ . . . . $= \frac{3wl}{8}$

$R_2 = V_2$ max. . . . . $= \frac{5wl}{8}$

V_x . . . . $= R_1 - wx$

M max. . . . . $= \frac{wl^2}{8}$

M_1 $\left(\text{at } x = \frac{3}{8}l\right)$ . . . . $= \frac{9}{128}wl^2$

M_x . . . . $= R_1 x - \frac{wx^2}{2}$

Δmax. $\left(\text{at } x = \frac{l}{16}(1 + \sqrt{33}) = .4215l\right)$. $= \frac{wl^4}{185EI}$

Δx . . . . $= \frac{wx}{48EI}(l^3 - 3lx^2 + 2x^3)$

Note: For deflection calculations, use specified loads.

BEAM DIAGRAMS AND FORMULAE

Equivalent Tabular Load is the uniformly distributed factored load given in the Beam Load Tables

13 BEAM FIXED AT ONE END, SUPPORTED AT OTHER— CONCENTRATED LOAD AT CENTER

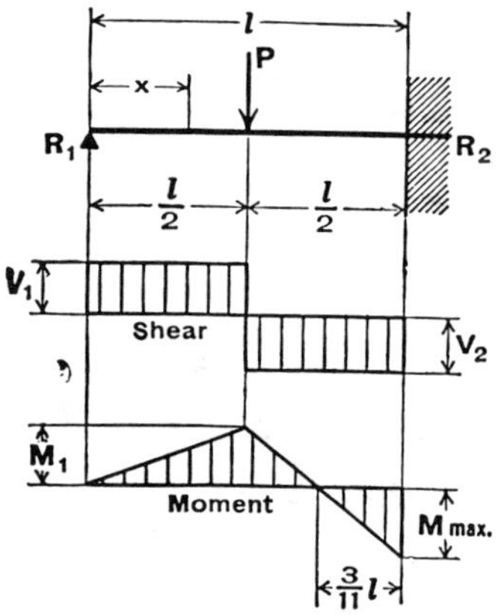

Equivalent Tabular Load . . . . . $= \frac{3P}{2}$

$R_1 = V_1$ . . . . $= \frac{5P}{16}$

$R_2 = V_2$ max. . . . . $= \frac{11P}{16}$

M max. (at fixed end) . . . . $= \frac{3Pl}{16}$

M_1 (at point of load) . . . . . $= \frac{5Pl}{32}$

M_x $\left(\text{when } x < \frac{l}{2}\right)$ . . . . . . $= \frac{5Px}{16}$

M_x $\left(\text{when } x > \frac{l}{2}\right)$ . . . . . . $= P\left(\frac{l}{2} - \frac{11x}{16}\right)$

Δmax. $\left(\text{at } x = l\sqrt{\frac{1}{5}} = .4472l\right)$ . . $= \frac{Pl^3}{48EI\sqrt{5}} = .009317\,\frac{Pl^3}{EI}$

Δx (at point of load) . . . . . $= \frac{7Pl^3}{768EI}$

Δx $\left(\text{when } x < \frac{l}{2}\right)$ . . . . . . $= \frac{Px}{96EI}(3l^2 - 5x^2)$

Δx $\left(\text{when } x > \frac{l}{2}\right)$ . . . . . . $= \frac{P}{96EI}(x - l)^2(11x - 2l)$

14 BEAM FIXED AT ONE END, SUPPORTED AT OTHER— CONCENTRATED LOAD AT ANY POINT

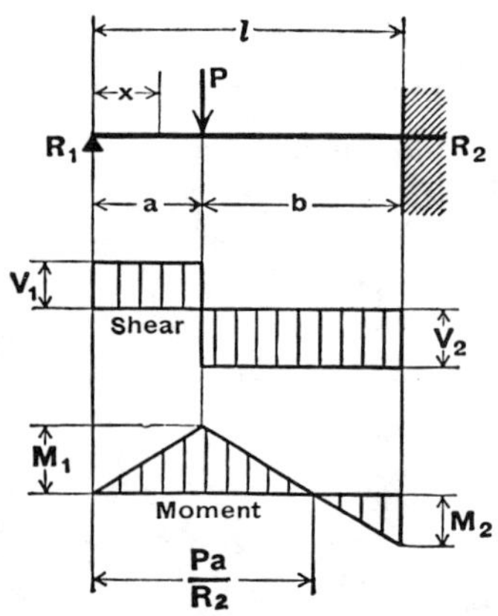

$R_1 = V_1$ . . . . $= \frac{Pb^2}{2l^3}(a + 2l)$

$R_2 = V_2$ . . . . $= \frac{Pa}{2l^3}(3l^2 - a^2)$

M_1 (at point of load) . . . . . $= R_1 a$

M_2 (at fixed end) . . . . . . $= \frac{Pab}{2l^2}(a + l)$

M_x (when $x < a$) . . . . . . $= R_1 x$

M_x (when $x > a$) . . . . . . $= R_1 x - P(x - a)$

Δmax. $\left(\text{when } a < .414l \text{ at } x = l\,\frac{l^2+a^2}{3l^2-a^2}\right) = \frac{Pa}{3EI}\,\frac{(l^2 - a^2)^3}{(3l^2 - a^2)^2}$

Δmax. $\left(\text{when } a > .414l \text{ at } x = l\sqrt{\frac{a}{2l+a}}\right) = \frac{Pab^2}{6EI}\sqrt{\frac{a}{2l + a}}$

Δa (at point of load) . . . . . $= \frac{Pa^2b^3}{12EIl^3}(3l + a)$

Δx (when $x < a$) . . . . . . $= \frac{Pb^2x}{12EIl^3}(3al^2 - 2lx^2 - ax^2)$

Δx (when $x > a$) . . . . . . $= \frac{Pa}{12EIl^3}(l - x)^2(3l^2x - a^2x - 2a^2l)$

Note: For deflection calculations, use specified loads.

Equivalent Tabular Load is the uniformly distributed factored load given in the Beam Load Tables

15 BEAM FIXED AT BOTH ENDS—UNIFORMLY DISTRIBUTED LOADS

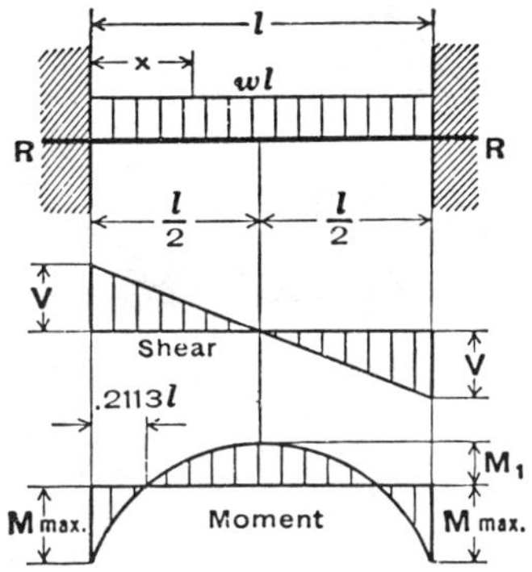

Equivalent Tabular Load . . . . $= \frac{2wl}{3}$

R = V . . . . $= \frac{wl}{2}$

V_x . . . . $= w\left(\frac{l}{2} - x\right)$

M max. (at ends) . . . . $= \frac{wl^2}{12}$

M_1 (at center) . . . . $= \frac{wl^2}{24}$

M_x . . . . $= \frac{w}{12}(6lx - l^2 - 6x^2)$

Δmax. (at center) . . . . $= \frac{wl^4}{384EI}$

Δx . . . . $= \frac{wx^2}{24EI}(l - x)^2$

16 BEAM FIXED AT BOTH ENDS—CONCENTRATED LOAD AT CENTER

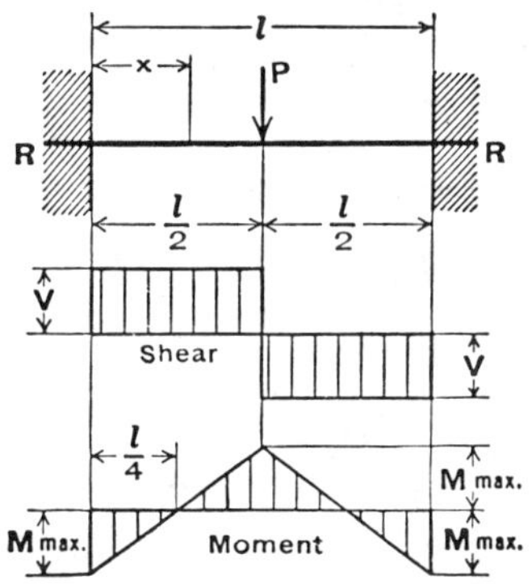

Equivalent Tabular Load . . . . = P

R = V . . . . $= \frac{P}{2}$

M max. (at center and ends) . . . $= \frac{Pl}{8}$

M_x $\left(\text{when } x < \frac{l}{2}\right)$ . . . . . . $= \frac{P}{8}(4x - l)$

Δmax. (at center) . . . . $= \frac{Pl^3}{192EI}$

Δx . . . . $= \frac{Px^2}{48EI}(3l - 4x)$

17 BEAM FIXED AT BOTH ENDS—CONCENTRATED LOAD AT ANY POINT

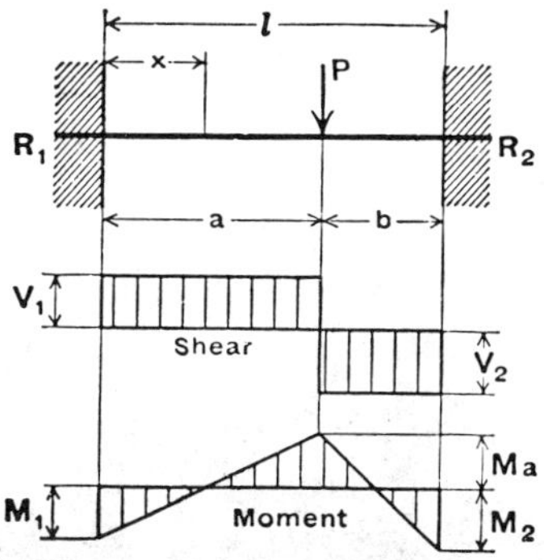

$R_1 = V_1$ (max. when a < b) . . . . $= \frac{Pb^2}{l^3}(3a + b)$

$R_2 = V_2$ (max. when a > b) . . . . $= \frac{Pa^2}{l^3}(a + 3b)$

M_1 (max. when a < b) . . . . $= \frac{Pab^2}{l^2}$

M_2 (max. when a > b) . . . . $= \frac{Pa^2b}{l^2}$

M_a (at point of load) . . . . $= \frac{2Pa^2b^2}{l^3}$

M_x (when x < a) . . . . . . $= R_1x - \frac{Pab^2}{l^2}$

Δmax. $\left(\text{when } a > b \text{ at } x = \frac{2al}{3a + b}\right)$. $= \frac{2Pa^3b^2}{3EI\,(3a + b)^2}$

Δa (at point of load) . . . . $= \frac{Pa^3b^3}{3EIl^3}$

Δx (when x < a) . . . . . . $= \frac{Pb^2x^2}{6EIl^3}(3al - 3ax - bx)$

Note: For deflection calculations, use specified loads.

BEAM DIAGRAMS AND FORMULAE

Equivalent Tabular Load is the uniformly
distributed factored load given in the Beam Load Tables

18 CANTILEVER BEAM—LOAD INCREASING UNIFORMLY TO FIXED END

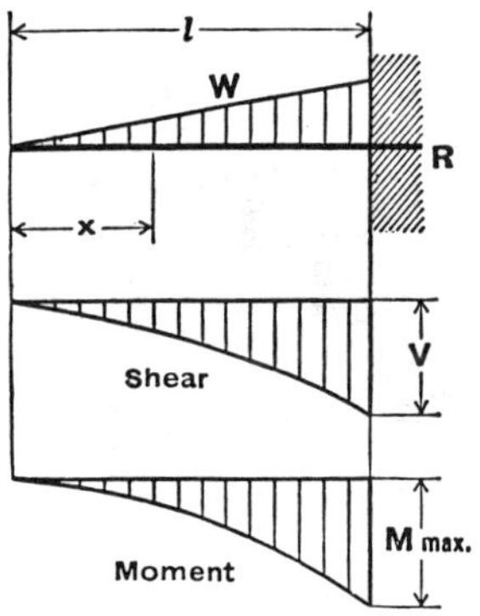

Equivalent Tabular Load . . . . $= \frac{8}{3} W$

R = V . . . . $= W$

V_x . . . . $= W \frac{x^2}{l^2}$

M max. (at fixed end) . . . . . . $= \frac{Wl}{3}$

M_x . . . . $= \frac{Wx^3}{3l^2}$

Δmax. (at free end) . . . . . . $= \frac{Wl^3}{15EI}$

Δx . . . . $= \frac{W}{60EIl^2}(x^5 - 5l^4x + 4l^5)$

19 CANTILEVER BEAM—UNIFORMLY DISTRIBUTED LOAD

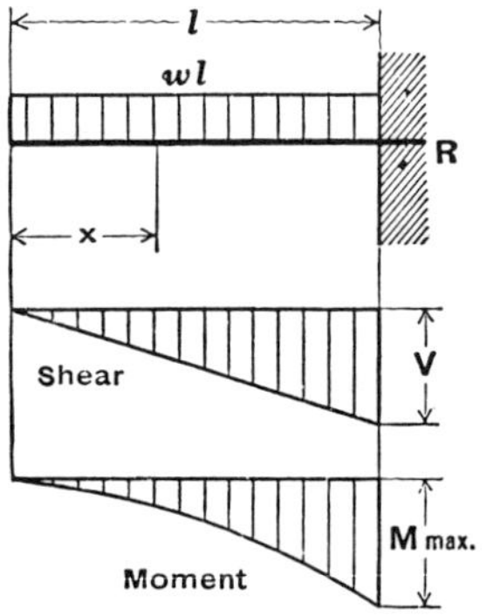

Equivalent Tabular Load . . . . $= 4wl$

R = V . . . . $= wl$

V_x . . . . $= wx$

M max. (at fixed end) . . . . . . $= \frac{wl^2}{2}$

M_x . . . . $= \frac{wx^2}{2}$

Δmax. (at free end) . . . . . . $= \frac{wl^4}{8EI}$

Δx . . . . $= \frac{w}{24EI}(x^4 - 4l^3x + 3l^4)$

20 BEAM FIXED AT ONE END, FREE BUT GUIDED AT OTHER—UNIFORMLY DISTRIBUTED LOAD

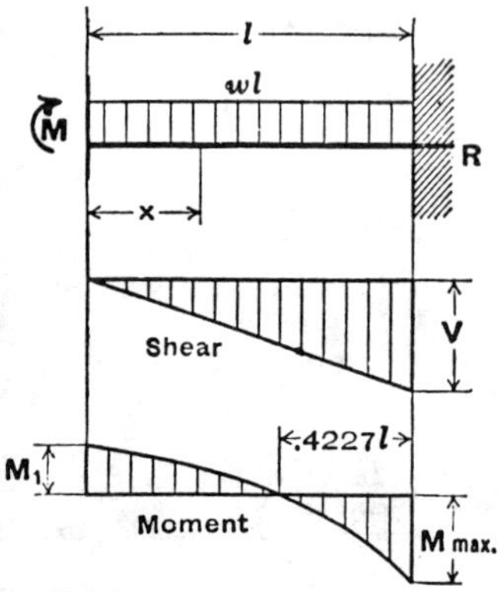

Equivalent Tabular Load . . . . $= \frac{8}{3} wl$

R = V . . . . $= wl$

V_x . . . . $= wx$

M max. (at fixed end) . . . . . . $= \frac{wl^2}{3}$

M_1 (at deflected end) . . . . . $= \frac{wl^2}{6}$

M_x . . . . $= \frac{w}{6}(l^2 - 3x^2)$

Δmax. (at deflected end) . . . . . $= \frac{wl^4}{24EI}$

Δx . . . . $= \frac{w(l^2 - x^2)^2}{24EI}$

Note: For deflection calculations, use specified loads.

Equivalent Tabular Load is the uniformly distributed factored load given in the Beam Load Tables

21 CANTILEVER BEAM—CONCENTRATED LOAD AT ANY POINT

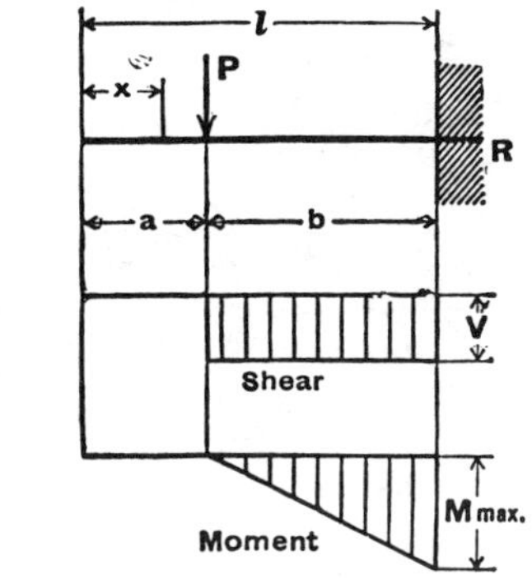

Equivalent Tabular Load . . . . $= \frac{8Pb}{l}$

R = V $\left(\text{when } x < a\right)$ . . . . . . $= P$

M max. $\left(\text{at fixed end}\right)$ . . . . . . $= Pb$

M_x $\left(\text{when } x > a\right)$ . . . . . . $= P(x - a)$

Δmax. $\left(\text{at free end}\right)$ . . . . . . $= \frac{Pb^2}{6EI}(3l - b)$

Δa $\left(\text{at point of load}\right)$ . . . . . $= \frac{Pb^3}{3EI}$

Δx $\left(\text{when } x < a\right)$ . . . . . . $= \frac{Pb^2}{6EI}(3l - 3x - b)$

Δx $\left(\text{when } x > a\right)$ . . . . . . $= \frac{P(l - x)^2}{6EI}(3b - l + x)$

22 CANTILEVER BEAM—CONCENTRATED LOAD AT FREE END

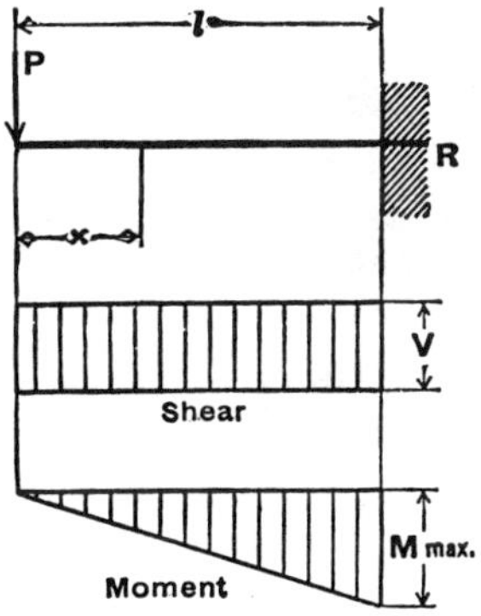

Equivalent Tabular Load . . . . $= 8P$

R = V . . . . $= P$

M max. $\left(\text{at fixed end}\right)$ . . . . . . $= Pl$

M_x . . . . $= Px$

Δmax. $\left(\text{at free end}\right)$ . . . . . $= \frac{Pl^3}{3EI}$

Δx . . . . $= \frac{P}{6EI}(2l^3 - 3l^2x + x^3)$

23 BEAM FIXED AT ONE END, FREE BUT GUIDED AT OTHER—CONCENTRATED LOAD AT GUIDED END

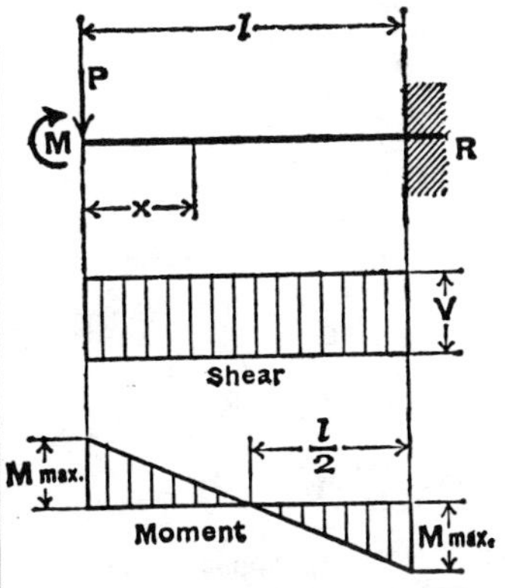

Equivalent Tabular Load . . . . $= 4P$

R = V . . . . $= P$

M max. $\left(\text{at both ends}\right)$ . . . . $= \frac{Pl}{2}$

M_x . . . . $= P\left(\frac{l}{2} - x\right)$

Δmax. $\left(\text{at deflected end}\right)$ . . . . $= \frac{Pl^3}{12EI}$

Δx . . . . $= \frac{P(l - x)^2}{12EI}(l + 2x)$

Note: For deflection calculations, use specified loads.

24 BEAM OVERHANGING ONE SUPPORT—UNIFORMLY DISTRIBUTED LOAD

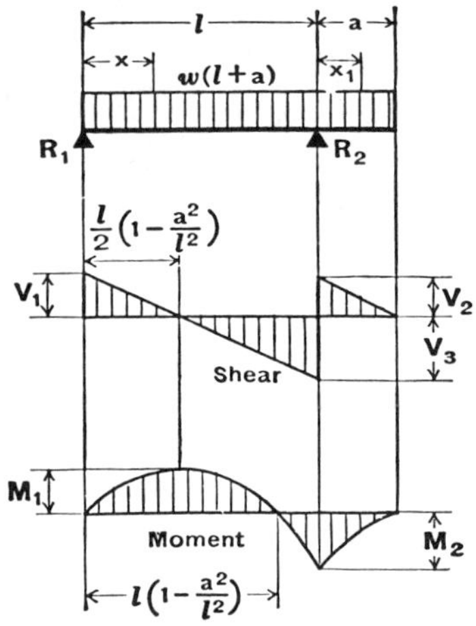

$R_1 = V_1 \quad . \; . \; . \; . \; . \; . \; . \; . \; . \; . \quad = \dfrac{w}{2l}(l^2 - a^2)$

$R_2 = V_2 + V_3 \quad . \; . \; . \; . \; . \; . \; . \; . \quad = \dfrac{w}{2l}(l + a)^2$

$V_2 \quad . \; . \; . \; . \; . \; . \; . \; . \; . \; . \; . \; . \; . \quad = wa$

$V_3 \quad . \; . \; . \; . \; . \; . \; . \; . \; . \; . \; . \; . \; . \quad = \dfrac{w}{2l}(l^2 + a^2)$

$V_x \quad \left(\text{between supports}\right) \quad . \; . \quad = R_1 - wx$

$V_{x_1} \quad \left(\text{for overhang}\right) \quad . \; . \; . \; . \quad = w(a - x_1)$

$M_1 \quad \left(\text{at } x = \dfrac{l}{2}\left[1 - \dfrac{a^2}{l^2}\right]\right) \quad . \; . \quad = \dfrac{w}{8l^2}(l + a)^2(l - a)^2$

$M_2 \quad \left(\text{at } R_2\right) \quad . \; . \; . \; . \; . \; . \; . \quad = \dfrac{wa^2}{2}$

$M_x \quad \left(\text{between supports}\right) \quad . \; . \quad = \dfrac{wx}{2l}(l^2 - a^2 - xl)$

$M_{x_1} \quad \left(\text{for overhang}\right) \quad . \; . \; . \; . \quad = \dfrac{w}{2}(a - x_1)^2$

$\Delta x \quad \left(\text{between supports}\right) \quad . \; . \quad = \dfrac{wx}{24EIl}(l^4 - 2l^2x^2 + lx^3 - 2a^2l^2 + 2a^2x^2)$

$\Delta x_1 \quad \left(\text{for overhang}\right) \quad . \; . \; . \; . \quad = \dfrac{wx_1}{24EI}(4a^2l - l^3 + 6a^2x_1 - 4ax_1^2 + x_1^3)$

25 BEAM OVERHANGING ONE SUPPORT—UNIFORMLY DISTRIBUTED LOAD ON OVERHANG

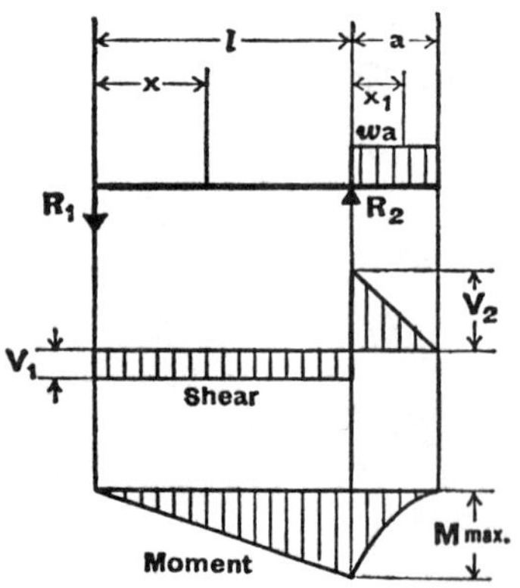

$R_1 = V_1 \quad . \; . \; . \; . \; . \; . \; . \; . \; . \; . \quad = \dfrac{wa^2}{2l}$

$R_2 = V_1 + V_2 \quad . \; . \; . \; . \; . \; . \; . \; . \quad = \dfrac{wa}{2l}(2l + a)$

$V_2 \quad . \; . \; . \; . \; . \; . \; . \; . \; . \; . \; . \; . \; . \quad = wa$

$V_{x_1} \quad \left(\text{for overhang}\right) \quad . \; . \; . \; . \quad = w(a - x_1)$

$M\ max. \left(\text{at } R_2\right) \quad . \; . \; . \; . \; . \; . \; . \quad = \dfrac{wa^2}{2}$

$M_x \quad \left(\text{between supports}\right) \quad . \; . \quad = \dfrac{wa^2x}{2l}$

$M_{x_1} \quad \left(\text{for overhang}\right) \quad . \; . \; . \; . \quad = \dfrac{w}{2}(a - x_1)^2$

$\Delta max. \quad \left(\text{between supports at } x = \dfrac{l}{\sqrt{3}}\right) = \dfrac{wa^2l^2}{18\sqrt{3}EI} = .03208\,\dfrac{wa^2l^2}{EI}$

$\Delta max. \quad \left(\text{for overhang at } x_1 = a\right) \quad . \quad = \dfrac{wa^3}{24EI}(4l + 3a)$

$\Delta x \quad \left(\text{between supports}\right) \quad . \; . \quad = \dfrac{wa^2x}{12EIl}(l^2 - x^2)$

$\Delta x_1 \quad \left(\text{for overhang}\right) \quad . \; . \; . \; . \quad = \dfrac{wx_1}{24EI}(4a^2l + 6a^2x_1 - 4ax_1^2 + x_1^3)$

Note: For deflection calculations, use specified loads.

Equivalent Tabular Load is the uniformly distributed factored load given in the Beam Load Tables

26 BEAM OVERHANGING ONE SUPPORT—CONCENTRATED LOAD AT END OF OVERHANG

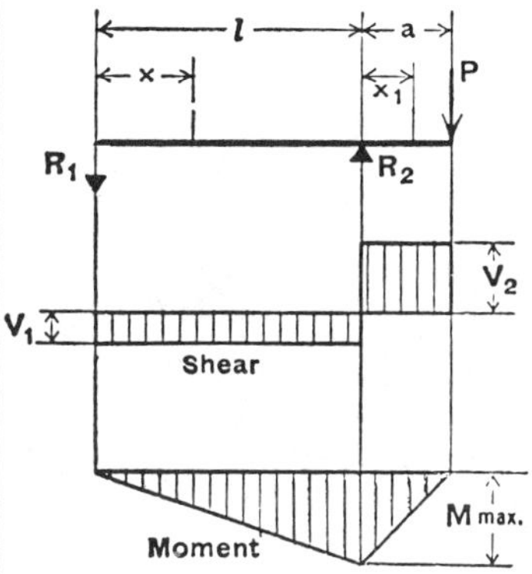

$R_1 = V_1$ . . . . $= \frac{Pa}{l}$

$R_2 = V_1 + V_2$ . . . . $= \frac{P}{l}(l + a)$

V_2 . . . . $= P$

M max. (at R_2) . . . . $= Pa$

M_x (between supports) . . $= \frac{Pax}{l}$

M_{x_1} (for overhang) . . . . . $= P(a - x_1)$

Δmax. (between supports at $x = \frac{l}{\sqrt{3}}$) $= \frac{Pal^2}{9\sqrt{3}EI} = .06415 \frac{Pal^2}{EI}$

Δmax. (for overhang at $x_1 = a$) . $= \frac{Pa^2}{3EI}(l + a)$

Δx (between supports) . . $= \frac{Pax}{6EIl}(l^2 - x^2)$

Δx_1 (for overhang) . . . . . $= \frac{Px_1}{6EI}(2al + 3ax_1 - x_1^2)$

27 BEAM OVERHANGING ONE SUPPORT—UNIFORMLY DISTRIBUTED LOAD BETWEEN SUPPORTS

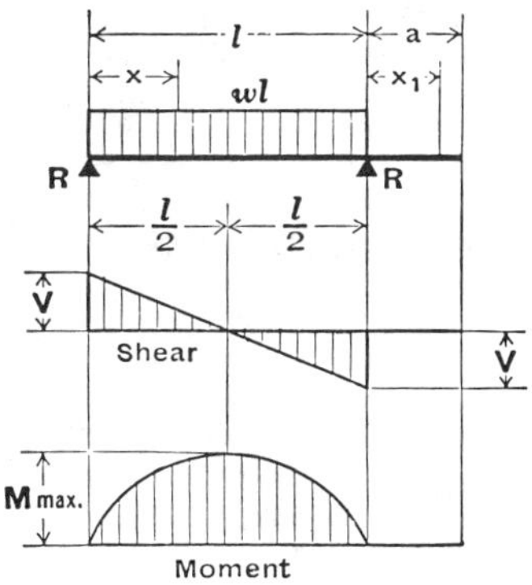

Equivalent Tabular Load . . . . $= wl$

$R = V$ . . . . $= \frac{wl}{2}$

V_x . . . . $= w\left(\frac{l}{2} - x\right)$

M max. (at center) . . . . . . $= \frac{wl^2}{8}$

M_x . . . . $= \frac{wx}{2}(l - x)$

Δmax. (at center) . . . . . . $= \frac{5wl^4}{384EI}$

Δx . . . . $= \frac{wx}{24EI}(l^3 - 2lx^2 + x^3)$

Δx_1 . . . . $= \frac{wl^3x_1}{24EI}$

28 BEAM OVERHANGING ONE SUPPORT—CONCENTRATED LOAD ANY POINT BETWEEN SUPPORTS

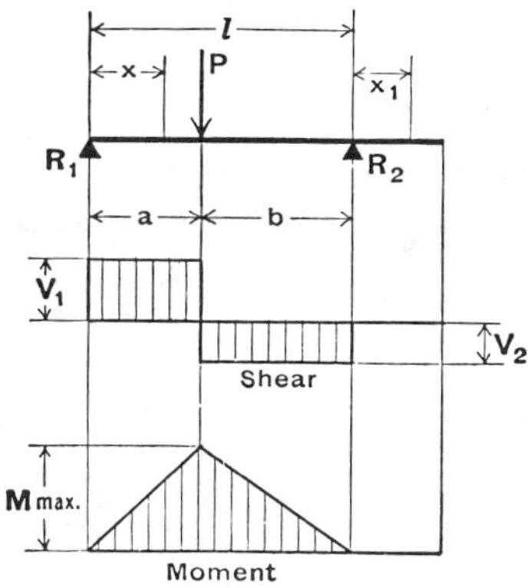

Equivalent Tabular Load . . . . $= \frac{8Pab}{l^2}$

$R_1 = V_1$ (max. when $a < b$) . . . $= \frac{Pb}{l}$

$R_2 = V_2$ (max. when $a > b$) . . . $= \frac{Pa}{l}$

M max. (at point of load) . . . $= \frac{Pab}{l}$

M_x (when $x < a$) . . . . $= \frac{Pbx}{l}$

Δmax. (at $x = \sqrt{\frac{a(a+2b)}{3}}$ when $a > b$) $= \frac{Pab\,(a + 2b)\sqrt{3a\,(a + 2b)}}{27EIl}$

Δa (at point of load) . . . $= \frac{Pa^2b^2}{3EIl}$

Δx (when $x < a$) . . . . $= \frac{Pbx}{6EIl}(l^2 - b^2 - x^2)$

Δx (when $x > a$) . . . . $= \frac{Pa\,(l - x)}{6EIl}(2lx - x^2 - a^2)$

Δx_1 . . . . $= \frac{Pabx_1}{6EIl}(l + a)$

Note: For deflection calculations, use specified loads.

29 BEAM—UNIFORMLY DISTRIBUTED LOAD AND VARIABLE END MOMENTS

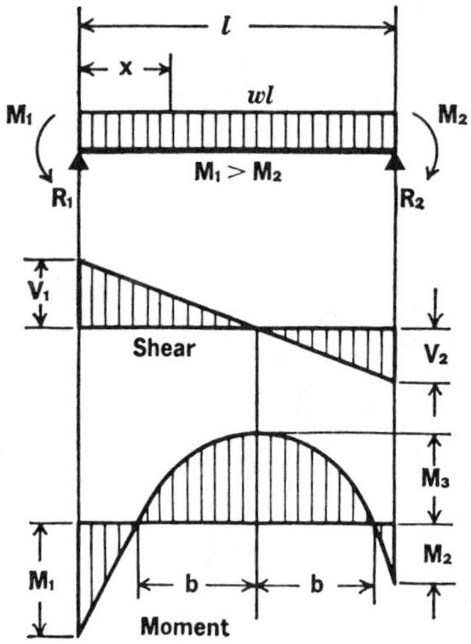

$$R_1 = V_1 = \frac{wl}{2} + \frac{M_1 - M_2}{l}$$

$$R_2 = V_2 = \frac{wl}{2} - \frac{M_1 - M_2}{l}$$

$$V_x = w\left(\frac{l}{2} - x\right) + \frac{M_1 - M_2}{l}$$

$$M_3 \left(\text{at } x = \frac{l}{2} + \frac{M_1 - M_2}{wl}\right)$$

$$= \frac{wl^2}{8} - \frac{M_1 + M_2}{2} + \frac{(M_1 - M_2)^2}{2wl^2}$$

$$M_x = \frac{wx}{2}(l - x) + \left(\frac{M_1 - M_2}{l}\right) x - M_1$$

$$b\left(\begin{matrix}\text{To locate}\\ \text{inflection points}\end{matrix}\right) = \sqrt{\frac{l^2}{4} - \left(\frac{M_1 + M_2}{w}\right) + \left(\frac{M_1 - M_2}{wl}\right)^2}$$

$$\Delta_x = \frac{wx}{24EI}\left[x^3 - \left(2l + \frac{4M_1}{wl} - \frac{4M_2}{wl}\right)x^2 + \frac{12M_1}{w}x + l^3 - \frac{8M_1 l}{w} - \frac{4M_2 l}{w}\right]$$

30 BEAM—CONCENTRATED LOAD AT CENTER AND VARIABLE END MOMENTS

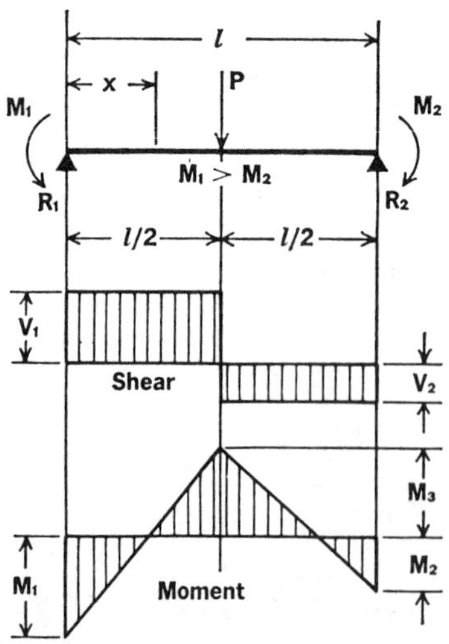

$$R_1 = V_1 = \frac{P}{2} + \frac{M_1 - M_2}{l}$$

$$R_2 = V_2 = \frac{P}{2} - \frac{M_1 - M_2}{l}$$

$$M_3 \text{ (At center)} = \frac{Pl}{4} - \frac{M_1 + M_2}{2}$$

$$M_x \left(\text{When } x < \frac{l}{2}\right) = \left(\frac{P}{2} + \frac{M_1 - M_2}{l}\right) x - M_1$$

$$M_x \left(\text{When } x > \frac{l}{2}\right) = \frac{P}{2}(l - x) + \frac{(M_1 - M_2)x}{l} - M_1$$

$$\Delta_x \left(\text{When } x < \frac{l}{2}\right) = \frac{Px}{48EI}\left(3l^2 - 4x^2 - \frac{8(l - x)}{Pl}[M_1(2l - x) + M_2(l + x)]\right)$$

Note: For deflection calculations, use specified loads.

31 SIMPLE BEAM—ONE CONCENTRATED MOVING LOAD

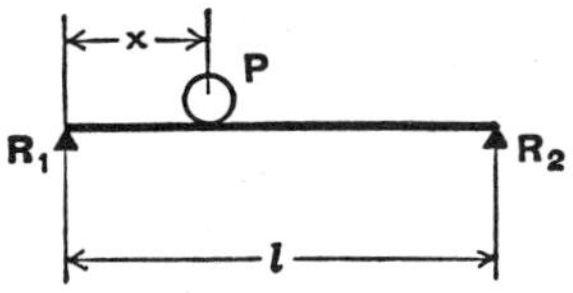

$$R_1 \text{ max.} = V_1 \text{ max.} \left(\text{at } x = o\right) \quad \ldots\ldots \quad = P$$

$$M \text{ max.} \left(\text{at point of load, when } x = \frac{l}{2}\right) \quad \ldots \quad = \frac{Pl}{4}$$

32 SIMPLE BEAM—TWO EQUAL CONCENTRATED MOVING LOADS

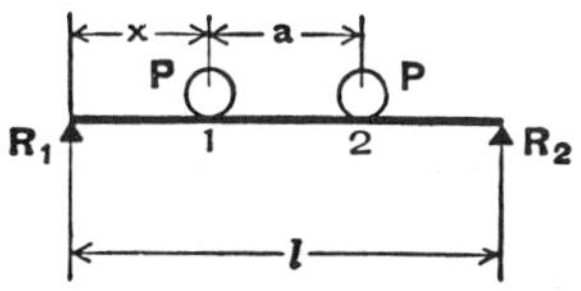

$$R_1 \text{ max.} = V_1 \text{ max.} \left(\text{at } x = o\right) \quad \ldots\ldots \quad = P\left(2 - \frac{a}{l}\right)$$

M max.

- [when $a < (2-\sqrt{2})\, l = .586l$ under load 1 at $x = \frac{1}{2}\left(l - \frac{a}{2}\right)$] $= \frac{P}{2l}\left(l - \frac{a}{2}\right)^2$
- [when $a > (2-\sqrt{2})\, l = .586l$ with one load at center of span (case 31)] $= \frac{Pl}{4}$

33 SIMPLE BEAM—TWO UNEQUAL CONCENTRATED MOVING LOADS

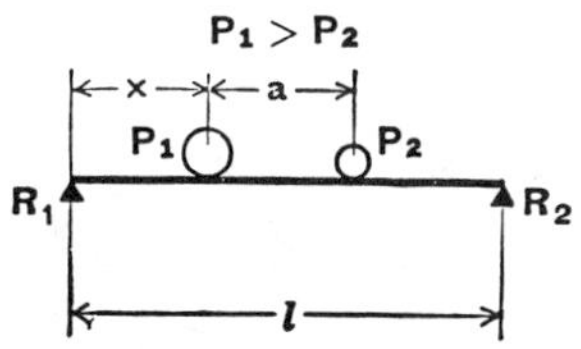

$$R_1 \text{ max.} = V_1 \text{ max.} \left(\text{at } x = o\right) \quad \ldots\ldots \quad = P_1 + P_2 \frac{l - a}{l}$$

M max.

- $\left[\text{under } P_1, \text{ at } x = \frac{1}{2}\left(l - \frac{P_2 a}{P_1 + P_2}\right)\right] = \left(P_1 + P_2\right)\frac{x^2}{l}$
- [M max. may occur with larger load at center of span and other load off span (case 31)] $= \frac{P_1 l}{4}$

GENERAL RULES FOR SIMPLE BEAMS CARRYING MOVING CONCENTRATED LOADS

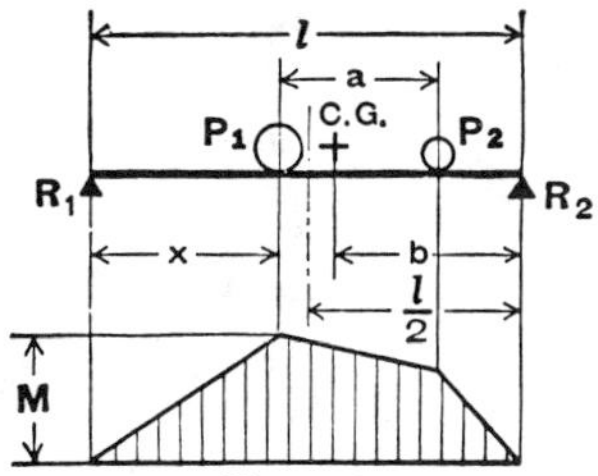

Moment

The maximum shear due to moving concentrated loads occurs at one support when one of the loads is at that support. With several moving loads, the location that will produce maximum shear must be determined by trial.

The maximum bending moment produced by moving concentrated loads occurs under one of the loads when that load is as far from one support as the center of gravity of all the moving loads on the beam is from the other support.

In the accompanying diagram, the maximum bending moment occurs under load P_1 when $x = b$. It should also be noted that this condition occurs when the center line of the span is midway between the center of gravity of loads and the nearest concentrated load.

Note: For deflection calculations, use specified loads.

BEAM DIAGRAMS AND FORMULAE

Equivalent Tabular Load is the uniformly
distributed factored load given in the Beam Load Tables

34 CONTINUOUS BEAM—TWO EQUAL SPANS—UNIFORM LOAD ON ONE SPAN

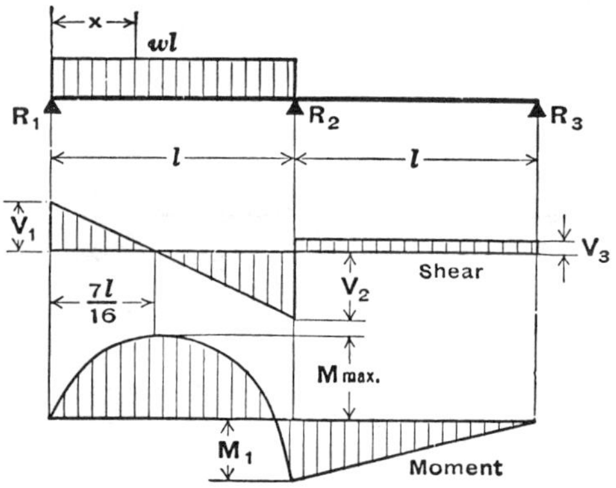

Equivalent Tabular Load . = $\frac{49}{64}wl$

$R_1 = V_1$ . . . . = $\frac{7}{16}wl$

$R_2 = V_2 + V_3$ . . . . . = $\frac{5}{8}wl$

$R_3 = V_3$ . . . . = $-\frac{1}{16}wl$

V_2 . . . . = $\frac{9}{16}wl$

M max. $\left(\text{at } x = \frac{7}{16}l\right)$ . . = $\frac{49}{512}wl^2$

M_1 (at support R_2) . = $\frac{1}{16}wl^2$

M_x (when $x < l$) . . = $\frac{wx}{16}(7l - 8x)$

Δ Max. (0.472 l from R_1) = 0.0092 wl^4/EI

35 CONTINUOUS BEAM—TWO EQUAL SPANS—CONCENTRATED LOAD AT CENTER OF ONE SPAN

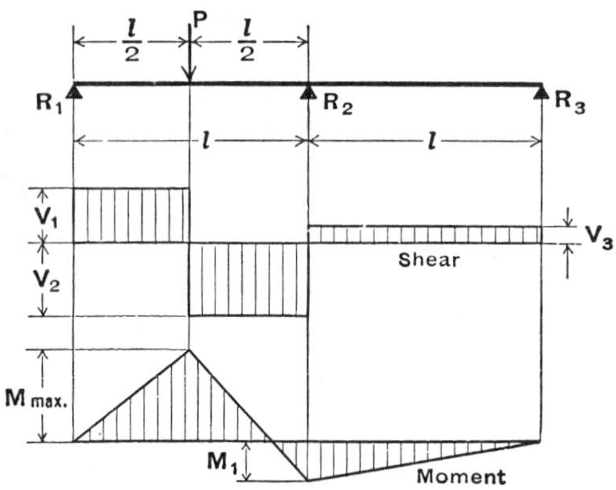

Equivalent Tabular Load . = $\frac{13}{8}P$

$R_1 = V_1$ . . . . = $\frac{13}{32}P$

$R_2 = V_2 + V_3$ . . . . . = $\frac{11}{16}P$

$R_3 = V_3$ . . . . = $-\frac{3}{32}P$

V_2 . . . . = $\frac{19}{32}P$

M max. (at point of load) . = $\frac{13}{64}Pl$

M_1 (at support R_2) . = $\frac{3}{32}Pl$

Δ Max. (0.480 l from R_1) = 0.015 Pl^3/EI

36 CONTINUOUS BEAM—TWO EQUAL SPANS—CONCENTRATED LOAD AT ANY POINT

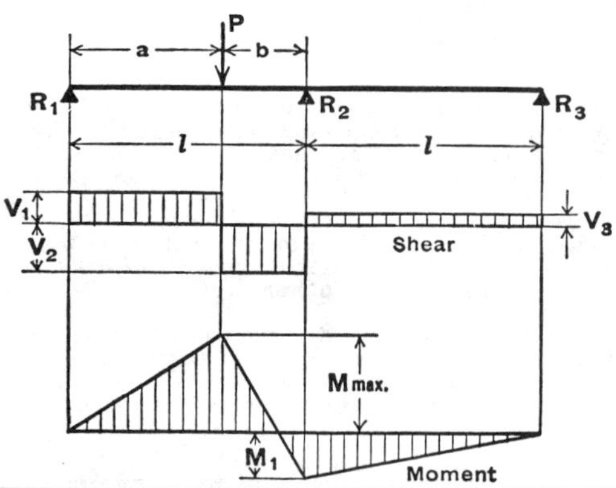

$R_1 = V_1$ . . . . = $\frac{Pb}{4l^3}\left(4l^2 - a(l+a)\right)$

$R_2 = V_2 + V_3$ . . . . . = $\frac{Pa}{2l^3}\left(2l^2 + b(l+a)\right)$

$R_3 = V_3$ . . . . = $-\frac{Pab}{4l^3}(l + a)$

V_2 . . . . = $\frac{Pa}{4l^3}\left(4l^2 + b(l+a)\right)$

M max. (at point of load) . = $\frac{Pab}{4l^3}\left(4l^2 - a(l+a)\right)$

M_1 (at support R_2) . = $\frac{Pab}{4l^2}(l + a)$

Note: For deflection calculations, use specified loads.

Moment = Coefficient x W x L
Reaction = Coefficient x W
Where: W = Total uniformly distributed load on one span
L = Length of one span

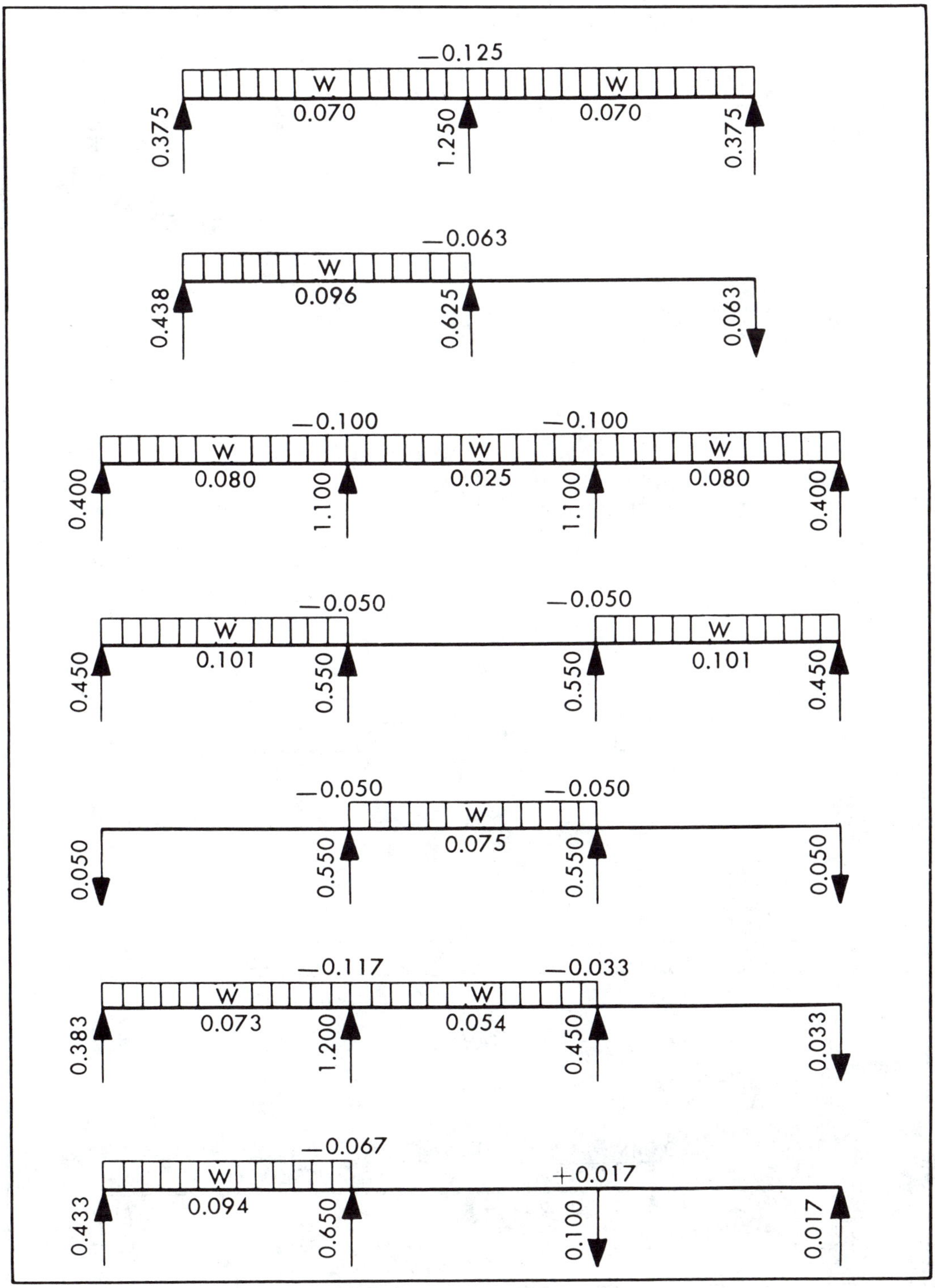

MOMENTS, REACTIONS
Equal Span Continuous Beams

UNIFORMLY DISTRIBUTED LOADS

Moment = Coefficient x W x L
Reaction = Coefficient x W
Where: W = Total uniformly distributed load on one span
L = Length of one span

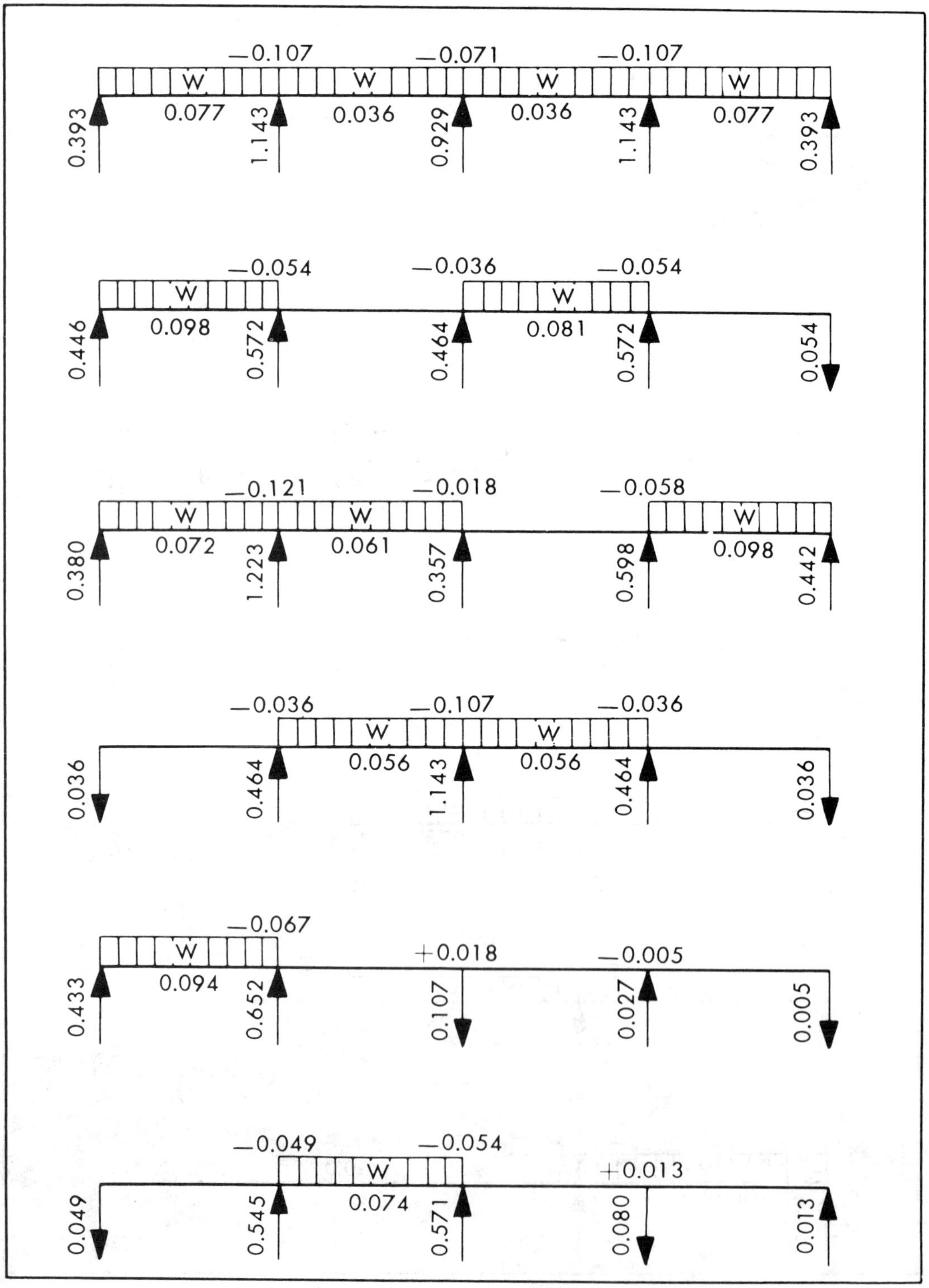

MOMENTS, REACTIONS
Equal Span Continuous Beams
CENTRAL POINT LOADS

Moment = Coefficient x W x L
Reaction = Coefficient x W
Where: W = The concentrated load on one span
L = Length of one span

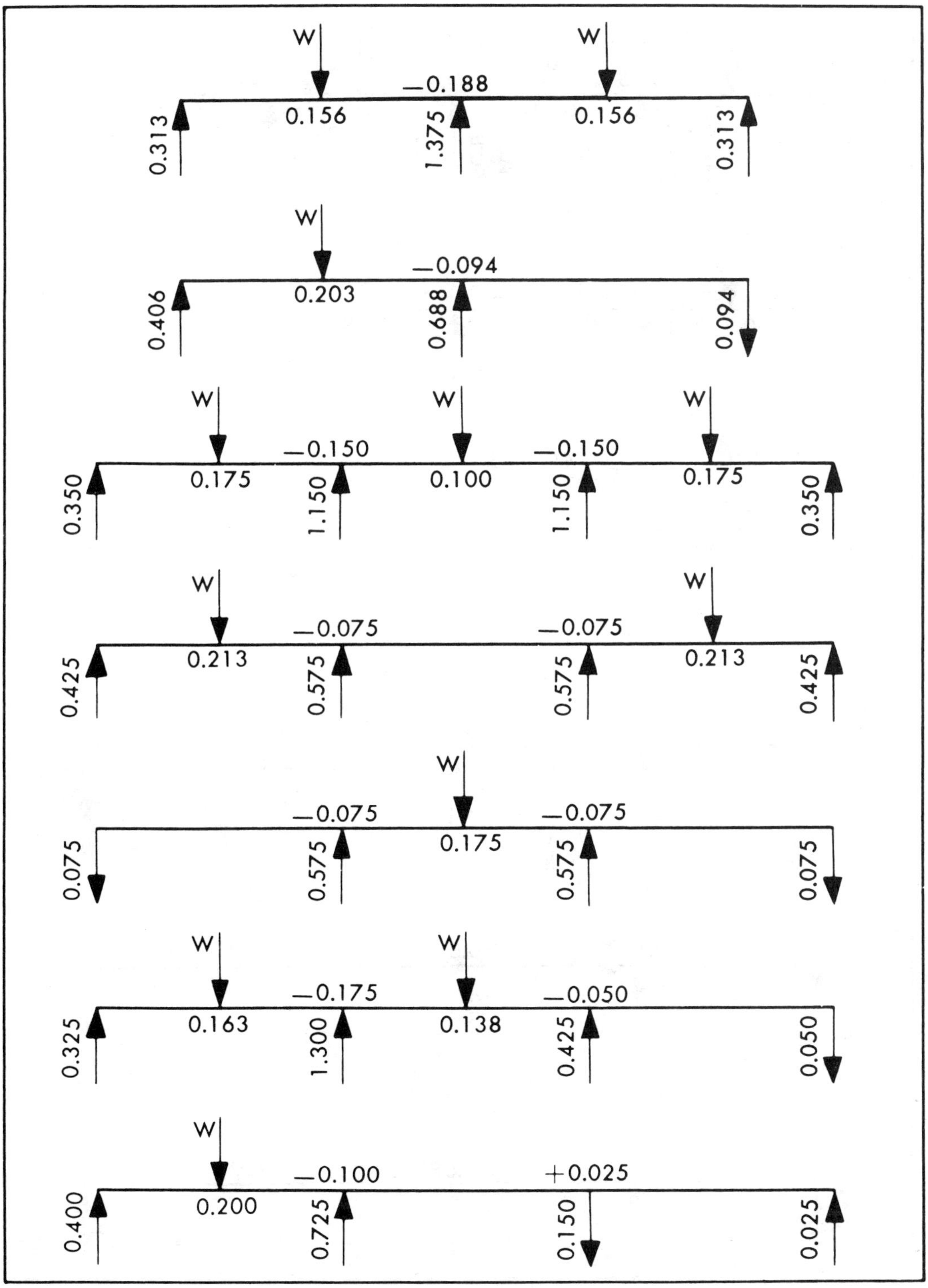

MOMENTS, REACTIONS
Equal Span Continuous Beams

CENTRAL POINT LOADS

Moment = Coefficient x W x L
Reaction = Coefficient x W
Where: W = The concentrated load on one span
L = Length of one span

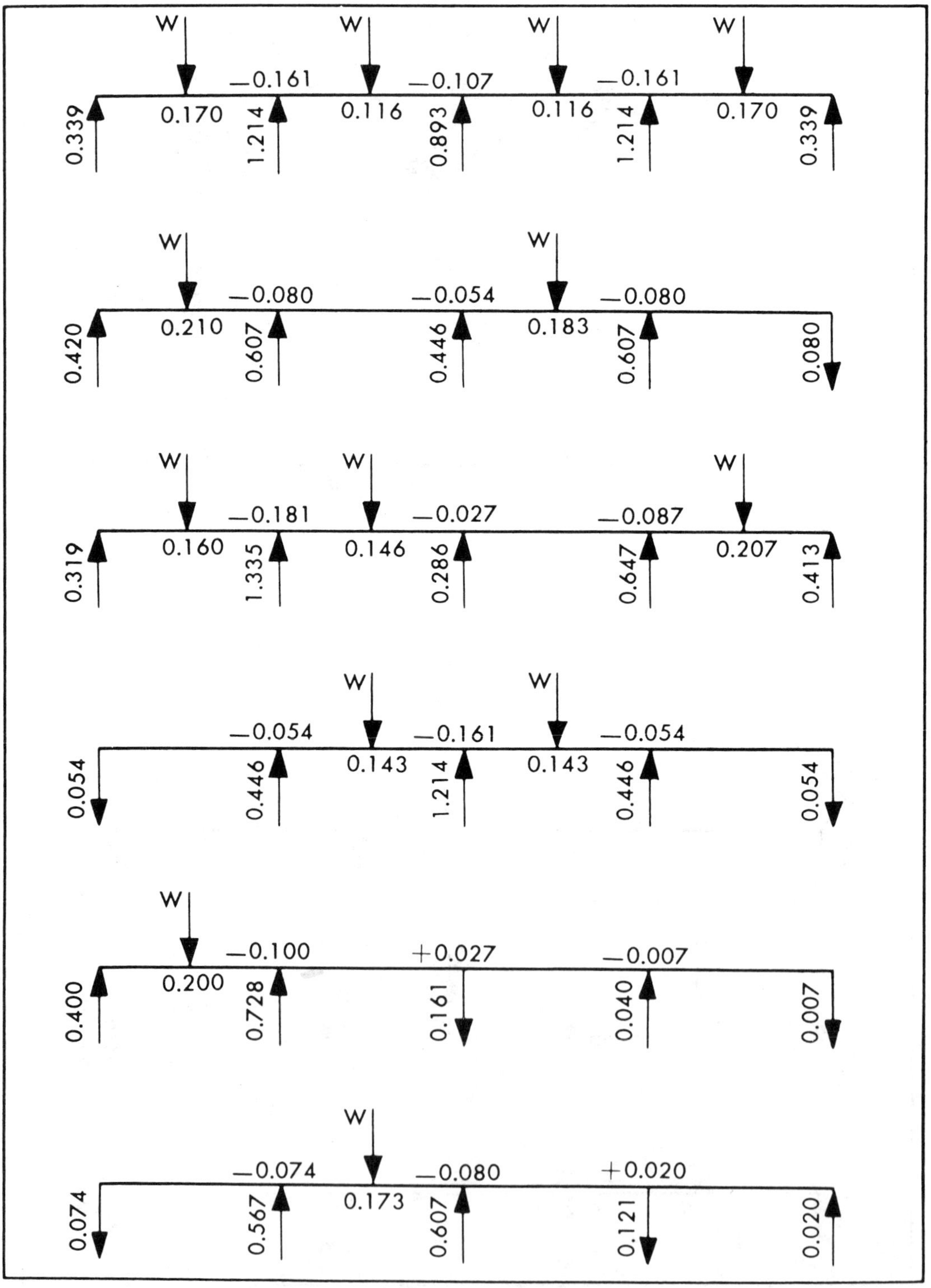

MOMENTS, REACTIONS
Equal Span Continuous Beams

POINT LOADS AT THIRD POINTS OF SPAN

Moment = Coefficient x W x L
Reaction = Coefficient x W
Where: W = The total load on one span
L = Length of one span

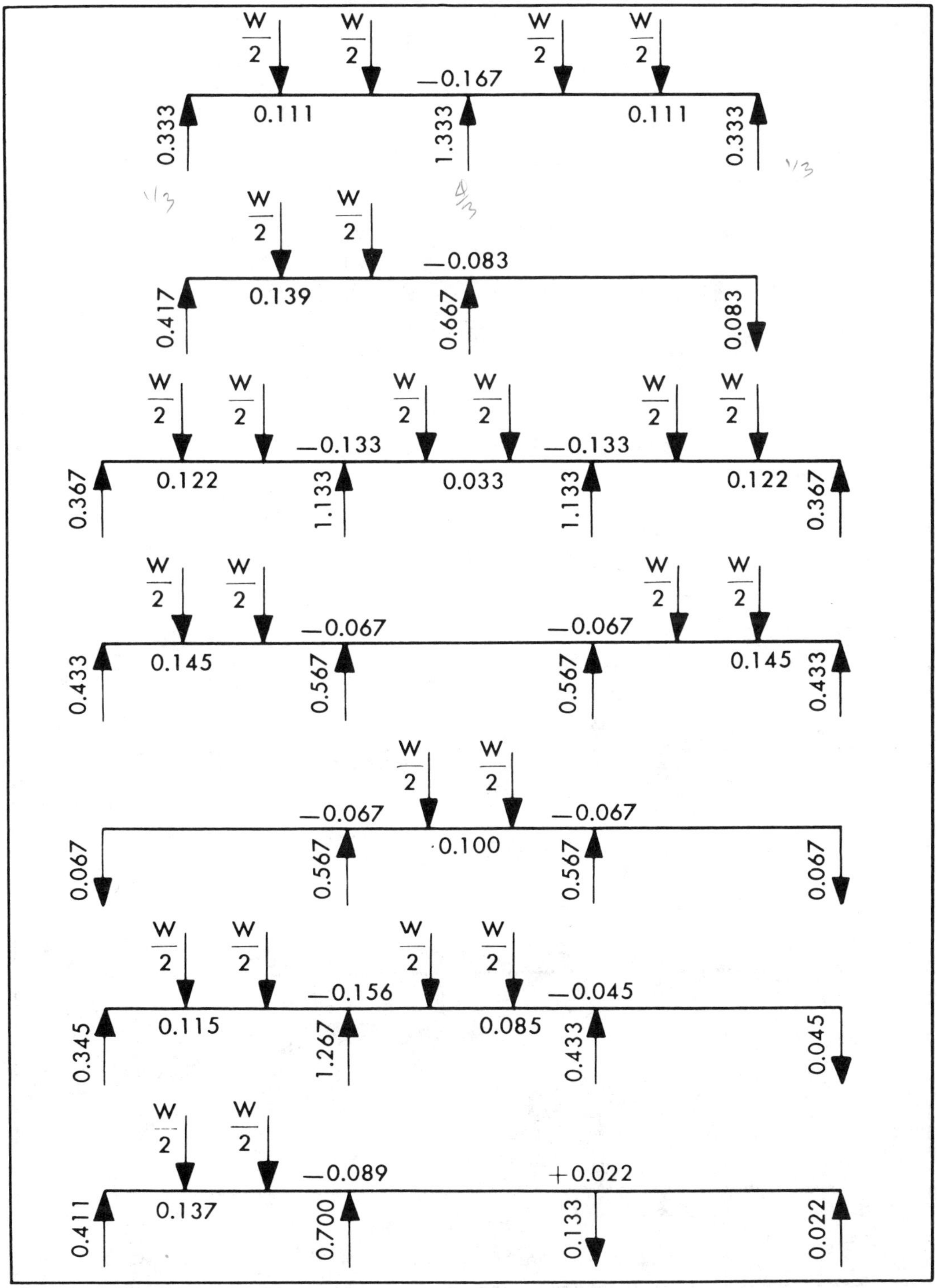

MOMENTS, REACTIONS
Equal Span Continuous Beams

POINT LOADS AT THIRD POINTS OF SPAN

Moment = Coefficient x W x L
Reaction = Coefficient x W
Where: W = The total load on one span
L = Length of one span

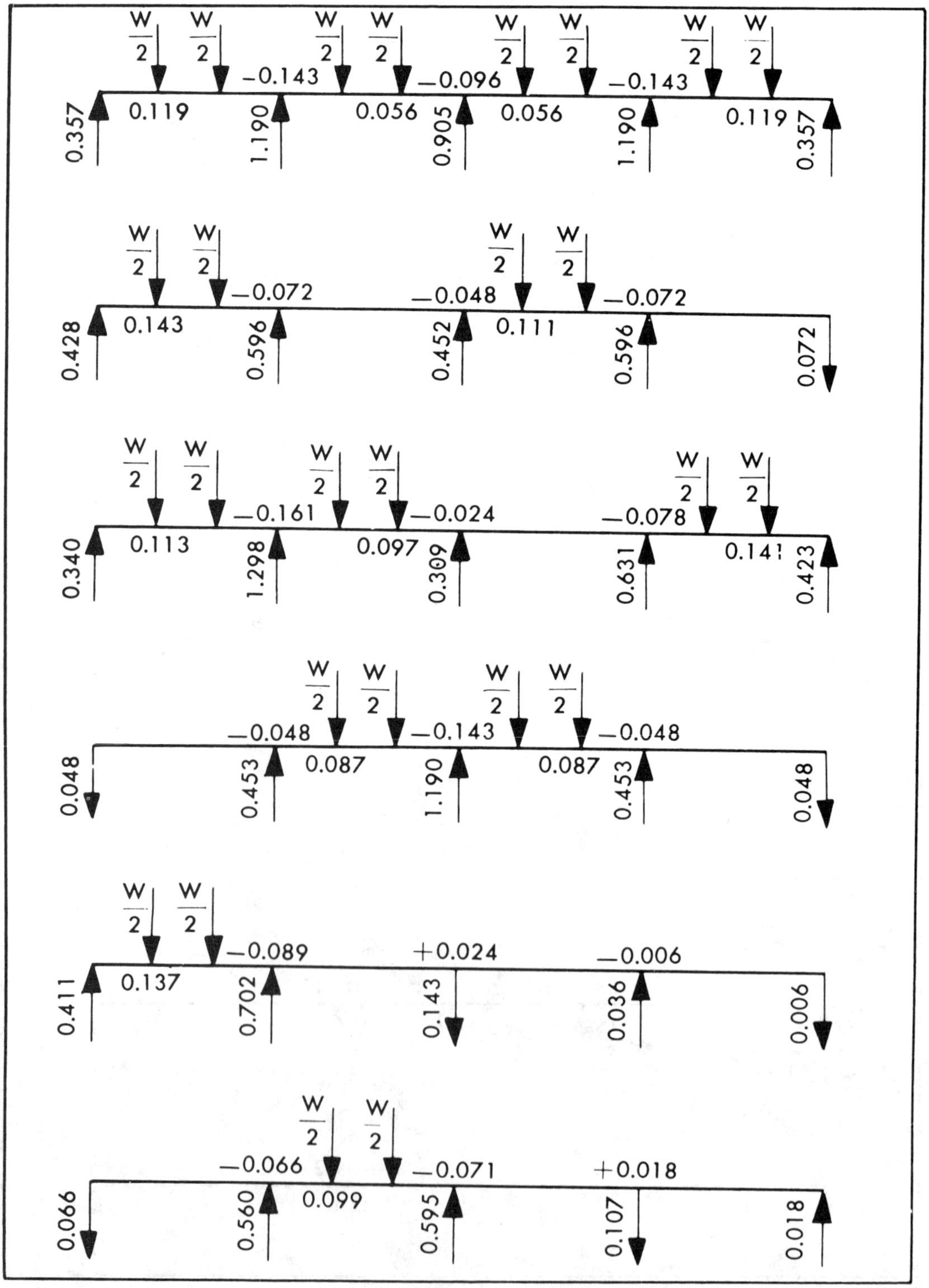

BEAM BEARING PLATES

General

When a flexural member is supported by a masonry wall or pier, the beam reaction must be distributed over sufficient area to avoid exceeding the bearing capacity of the masonry or concrete. Steel bearing plates may be used for this purpose.

Bearing plates are usually set in place and grouted level at the required elevation before positioning the beam. Thus, even though the beam flange may be able to distribute the reaction to supporting masonry or concrete, a bearing plate can be useful to facilitate erection. Some form of anchorage is required to ensure that the beam is connected to the pier or wall either longitudinally or for uplift forces.

Design Chart

Figure 5–2 on page 5–153 provides a graph to determine the thickness of bearing plates using G40.21-M 300W steel, for beams without bearing stiffeners, based on the following assumptions:

- The beam reaction R is uniformly distributed to the bearing plate over an effective area of width 2k and length C;
- The bearing pressure between the effective area of the bearing plate and the concrete or masonry support is uniform over the area of the plate;
- The bearing pressure under the portion of plate projecting beyond the k-distance from the centre line of the beam is ignored, since in practice the flange may be slightly "curled".

Equating the factored moment acting on the portion of the bearing plate taken as a cantilever, to the factored moment resistance of the plate, ($M_r = \phi Z F_y$), the bearing plate thickness is calculated as:

$$t_p = \sqrt{\frac{2 B_r n^2}{A \phi F_y}}$$

where

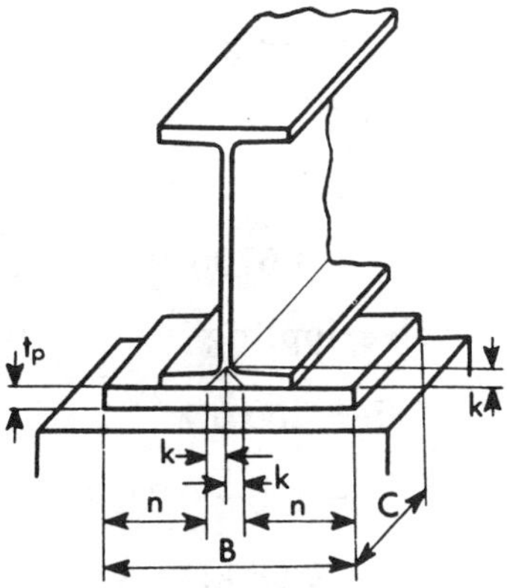

F_y = specified minimum yield strength of the bearing plate steel, (MPa)

A = $B \times C$ = area of plate, (mm^2)

t_p = required thickness of bearing plate, (mm)

k = beam k-distance = distance from web toe of fillet to outer face of flange, (mm)

n = $B/2 - k$, (mm)

b = width of beam flange, (mm)

To minimize deflection of the bearing plate, the thickness generally should not be less than about one fifth of the overhang, ie. $t_p \geq (B - b)/10$.

Use of chart

1. Required area, A = beam reaction due to factored loads divided by the unit factored concrete bearing resistance, ($0.85 \phi_c f'_c$), where $\phi_c = 0.60$;

2. Determine C and solve for B. (C, the length of bearing, is usually governed by the available wall thickness or other structural considerations;

3. Determine n and enter Figure 5–2 to determine t_p.

Example

Given:

A W610x140 beam has a factored end reaction of 850 kN and is supported on a concrete pier with 28 day compressive strength of 20 MPa. Design the bearing plate assuming G40.21-M 300W steel and a concrete bearing length of 300 mm.

Solution:

Unit factored bearing resistance of concrete is

$0.85 \times 0.60 \times 20 = 10.2$ MPa

Area required is $(850 \times 10^3) / 10.2 = 83\,300$ mm^2

Therefore required B is $83\,300 / 300 = 278$ mm

For W610x140,

$b = 230$ mm, $t = 22.2$ mm, $k = 41$ mm, $w = 13.1$ mm

Select B = 280 mm (greater than flange width, b = 230 mm

$n = (B / 2) - k = (280 / 2) - 41 = 99$ mm

From Figure 5–2, for unit factored bearing resistance of 10.2 MPa and n of 94 mm,

minimum t_p = 26 mm — select 30 mm

Use plate 30x300 x 280

Check for web crippling and web yielding (Clause 15.9(b), S16.1-M89)

Web yielding

$$B_r = 1.10\,\phi\, w\,(N + 2.5k)\,F_y$$

$$= 1.10 \times 0.9 \times 13.1\,(300 + (2.5 \times 41))\,300 / 10^3 \text{ kN}$$

$$= 1\,570 \text{ kN}$$

Web crippling

$$B_r = 150\,\phi\, w^2\,[\,1 + 3\,(N / d)(w / t)^{1.5}\,]\,(F_y\, t / w)^{0.5}$$

$$= 150 \times 0.9 \times 13.1^2\,[1 + 3\,(300 / 617)\,(13.1 / 22.2)^{1.5}\,]\,(300 \times 22.2 / 13.1)^{0.5}$$

$$= 868 \text{ kN}$$

Therefore web crippling, Clause 15.9(b)(ii), governs and B_r = 868 kN

Alternatively, the bearing resistance can be established using the values of R and G from the beam load tables for a W610x140, on page 5–110.

R = 637 kN, G = 11.5 kN

Bearing capacity for bearing length of 300 mm is

$$R + (200 \times G / 10) = 637 + (200 \times 11.5 / 10) = 867 \text{ kN}$$

FIGURE 5-2
BEAM BEARING PLATE THICKNESS, t_p

G40.21-M 300W

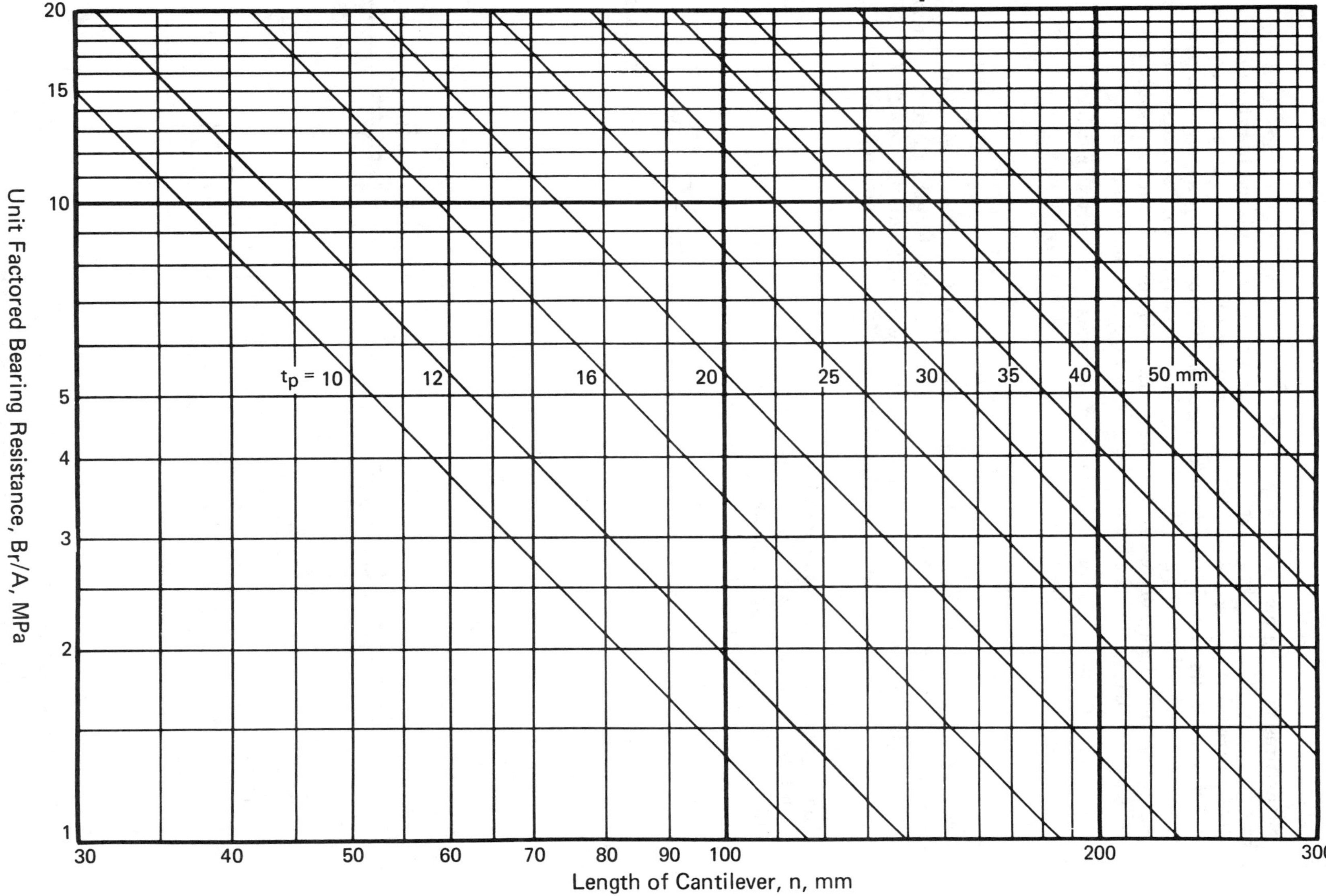

BEAMS WITH WEB HOLES

General

Structures may support a variety of pipes, ducts, conduit and other services, and efforts to reduce floor heights have led to these items being placed in the same plane as the structural floor members. Structural systems using stub girders, trusses and open web steel joists provide openings for structural/mechanical integration; however, when beams with solid webs are used it may be necessary to cut openings through the webs. This section, based on research summarized by Redwood and Shrivastava (1980), describes a method to account for web holes during design of the member.

Special precautions may be required if it becomes necessary to cut holes in beam webs after construction is complete.

Design

The formulas are applicable for beams of Class 1 and Class 2 sections with openings between 0.3 and 0.7 times the depth of the beam, and hole lengths up to three times the hole height. The steel should meet the requirements of Clause 8.5(a) of CAN/CSA-S16.1-M89 and exhibit the characteristics necessary to achieve moment redistribution, such as G40.21-M grade 300W.

The hole corner should have a radius at least equal to the larger of 16 mm or twice the web thickness. Fatigue loading considerations have not been accounted for in the formulae of this section and if holes are necessary in a member subjected to fatigue, some guidance is available from Frost & Leffler (1971).

Special design considerations are required if concentrated loads are to be located within the hole length or within one beam depth from either end of a hole.

The width-to-thickness ratio of outstanding reinforcing plates should meet Class 1 requirements.

Nomenclature

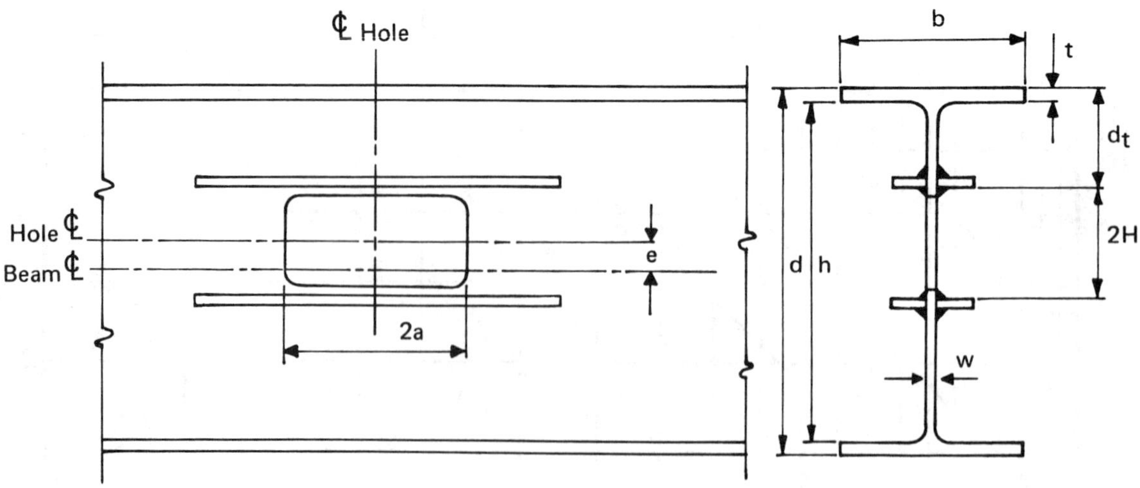

The following nomenclature is used throughout this section:

A_f area of one flange (b·t)

A_r area of reinforcement along top or bottom edge of the hole

A_w area of web (d·w)

e eccentricity of centreline of the hole above or below beam centreline — *always positive*

M_f bending moment due to factored loads at centreline of hole

M_r factored moment resistance of an unperforated beam

M_o, M_l values of moment resistance defined in web hole formulae

R radius of a circular hole

s length of web between adjacent holes

V_f shear force at centreline of hole due to factored loads

V_r' factored shear resistance based on plastic analysis of an unperforated beam $= 0.55\ \phi\ w\ d\ F_y$. This can be obtained by multiplying tabulated values of V_r (the factored shear resistance based on elastic analysis) given in the "Beam Selection Tables" or the "Beam Load Tables" by 0.833

V_o, V_l values of shear resistance defined in web hole formulae

Web Stability

This section of the Handbook is valid for the following range of values:

For Class One Sections	**For Class Two Sections**
$V_f \leq 0.67\ V_r'$ and in addition, for rectangular holes $a / H \leq 3.0$ $(a / H) + 6\ (2\ H / d) \leq 5.6$	$V_f \leq 0.45\ V_r'$ and in addition, for rectangular holes $a / H \leq 2.2$ $(a / H) + 6\ (2\ H / d) \leq 5.6$

If these values are exceeded, refer to Redwood and Shrivastava (1980).

Deflections

One or two small circular holes normally result in negligible additional deflections; however, deflections of beams with large holes will increase because of local deformations caused by:

(a) effect of rotation produced by change in length of the tee sections above and below the hole;

(b) local bending over the length of the hole;

(c) shear deformations.

Multiple Holes

To avoid effects of interaction between two adjacent holes which may occur with high shear, the length of the web between the holes should satisfy the following, where s = clear length of solid web between the holes:

Rectangular holes

$$s \geq 2H$$

$$s \geq 2a\left(\frac{V_f / V_r'}{1-(V_f / V_r')}\right)$$

Circular holes

$$s \geq 3R$$

$$s \geq 2R\left(\frac{V_f / V_r'}{1-(V_f / V_r')}\right)$$

where in each case the length, height or radius refers to that of the larger of the two holes.

Lateral Stability

The presence of a web hole has only a minor effect on the lateral stability of a beam, when the strength of the beam is governed by the resistance of a section remote from the hole. For members that may be susceptible to lateral buckling, refer to the paper by Redwood and Shrivastava.

Unreinforced Holes

According to Clause 15.10.2 of S16.1-M89 (see Part Two), unreinforced circular openings may be used under stipulated conditions. Round holes that are not covered by Clause 15.10.2 may be checked using the unreinforced hole formulas below by equating a and H to hole radius R as follows:

$2a = 0.9R$ and $2H = 1.8R$

Beam Resistance — Unreinforced Holes

Web stability must always be confirmed (page 5–155), and compression zone stability of the tee section must be checked when $2a > 4\,d_t$.

The factored shear force V_f and factored moment M_f applied at the web hole centreline must satisfy:

$$V_f \leq V_l \qquad [1]$$

$$M_f \leq M_o - (M_o - M_l)\ V_f / V_l \qquad [2]$$

in which

$$\frac{M_o}{M_r} = 1 - \frac{\dfrac{A_w}{4\,A_f}\left[\left(\dfrac{2H}{d}\right)^2 + \left(\dfrac{4e}{d}\right)\left(\dfrac{2H}{d}\right)\right]}{1+\dfrac{A_w}{4\,A_f}} \qquad [3]$$

$$\frac{M_l}{M_r} = \frac{1-\dfrac{2}{\sqrt{3}}\left(\dfrac{A_w}{A_f}\right)\left(\dfrac{a}{d}\right)\sqrt{\dfrac{\alpha_2}{1+\alpha_2}}}{1+\dfrac{A_w}{4\,A_f}} \qquad [4]$$

$$\frac{V_l}{V_r'} = \frac{2}{\sqrt{3}}\left(\frac{a}{d}\right)\left(\frac{\alpha_1}{\sqrt{1+\alpha_1}} + \frac{\alpha_2}{\sqrt{1+\alpha_2}}\right) \qquad [5]$$

where

$$\alpha_1 = \frac{3}{16}\left(\frac{d}{a}\right)^2\left(1-\frac{2H}{d}-\frac{2e}{d}\right)^2 \quad [6]$$

$$\alpha_2 = \frac{3}{16}\left(\frac{d}{a}\right)^2\left(1-\frac{2H}{d}+\frac{2e}{d}\right)^2 \quad [7]$$

Table 5–6, page 5–164, and Table 5–7, page 5–165, provide a means of evaluating [1] and [2]. For further explanation of these tables, see page 5–160.

Reinforced Holes

Horizontal Bars Only

Equal areas of reinforcement should be placed above and below the opening, with the reinforcement as close as possible to the edges of the hole. Welds attaching the reinforcement to the beam web should be continuous and may be placed on only one side of the reinforcing bar (with a short weld at each end on the opposite side of the bar to maintain alignment). Within the length of the hole, the welds should develop twice the factored tensile resistance of the reinforcement except that the weld capacity need not exceed 1.15 a w F_y. The reinforcement should extend past the hole far enough for the weld to develop the factored tensile resistance of the reinforcement but not less than a distance of a/2.

Reinforcement may be placed on only one side of the web of Class 1 sections (for economy) providing the following conditions are satisfied:

$A_r \leq 0.333\, A_f$

$M_f \leq 20\, V_f\, d$ — at the hole centreline

$a / H \leq 2.5$

$d_t / w \leq 370 / \sqrt{F_y}$

Round holes may be checked using the reinforced hole formulae by relating *a* and *H* to *R* as follows:

$2a = 0.9R$ and $2H = 2R$

Once it is established that hole reinforcement is required, Table 5–8 provides a means of checking the resistance of a beam with a reinforced hole for an assumed area of reinforcement. To determine minimum reinforcement requirements, a flow chart for writing programs for a programmable calculator is shown on page 5–159.

Vertical Bars

The compression zone stability of the reinforced tee should be checked by treating it as an axially loaded column with effective length equal to 2a.

If it is determined that web instability could be a problem, vertical reinforcing at the ends of the hole will be required. Attachment of both vertical and horizontal bars is generally more economical when the horizontal bars are placed on one side of the web with the vertical bars on the other side.

Beam Resistance — Holes with Horizontal Reinforcing Bars

Web stability and compression zone stability must be checked in addition to the following strength criteria. The factored shear force V_f and factored moment M_f at the web hole centreline must satisfy the following, where A_r is less than A_f :

$V_f \leq V_l$ [8a]

$$V_f / V_r' \le 1 - \frac{2H}{d} \tag{8b}$$

$$M_f \le M_o - (M_o - M_l)\, V_f / V_l \tag{9a}$$

$$M_f \le M_r \tag{9b}$$

in which

$$\left(\frac{M_o}{M_r}\right)_a = 1 + \frac{\frac{A_r}{A_f}\left(\frac{2H}{d}\right) - \frac{A_w}{4A_f}\left[\left(\frac{2H}{d}\right)^2 + 4\left(\frac{2H}{d}\right)\left(\frac{e}{d}\right) - 4\left(\frac{e}{d}\right)^2\right]}{1 + \frac{A_w}{4\,A_f}} \quad \text{— for } \frac{e}{d} \le \frac{A_r}{A_w} \tag{10a}$$

$$\text{or} \quad \left(\frac{M_o}{M_r}\right)_b = \left(\frac{M_o}{M_r}\right)_a - \frac{\frac{A_w}{A_f}\left(\frac{e}{d} - \frac{A_r}{A_w}\right)^2}{1 + \frac{A_w}{4\,A_f}} \quad \text{— for } \frac{e}{d} > \frac{A_r}{A_w} \tag{10b}$$

$$\frac{M_l}{M_r} = \frac{1 - \frac{A_r}{A_f}}{1 + \frac{A_w}{4\,A_f}} \tag{11}$$

$$\frac{V_l}{V_r'} = \sqrt{3}\left(\frac{d}{a}\right)\left(\frac{A_r}{A_w}\right)\left(1 - \frac{2H}{d}\right) \tag{12}$$

Flow Chart

The flow chart on page 5–159 is provided as a guide in developing programs for programmable calculators and computers. The logic provided determines the minimum A_r which will satisfy [9a]. It is anticipated that the individuals implementing the program will modify it to their own needs.

References

FROST, R.W., and LEFFLER, R.E. 1971. Fatigue tests of beams with rectangular web holes. Journal of the Structural Division, ASCE, **97**(ST2): 509-527.

PART TWO of this Handbook.

REDWOOD, R.G. 1974. The influence of web holes on the design of steel beams. Proceedings, Canadian Structural Engineering Conference, Canadian Steel Construction Council, Willowdale, Ontario.

REDWOOD, R.G. 1971. Simplified plastic analysis for reinforced web holes. Engineering Journal, AISC, **8**(3): 128-131.

REDWOOD, R.G., and SHRIVASTAVA, S.C. 1980. Design recommendations for steel beams with web holes. Canadian Journal of Civil Engineering, **7**(4), December.

REDWOOD, R.G., and WONG, P. 1982. Web holes in composite beams with steel deck. Procedings, Canadian Structrual Engineering Conference, Canadian Steel Construction Council, Willowdale, Ontario.

REINFORCED HOLE PROGRAM FLOWCHART

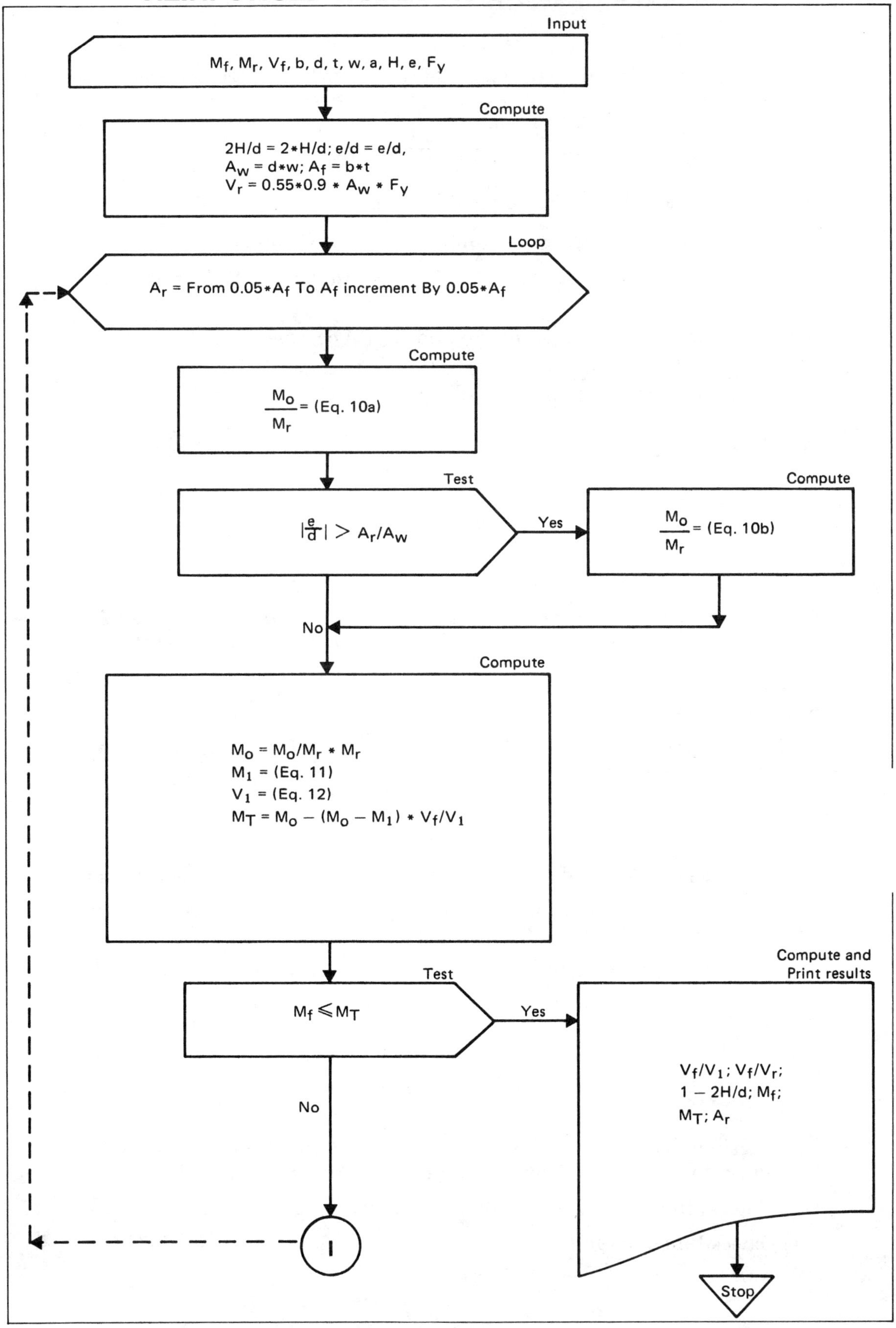

Tables

(a) *Unreinforced Holes*

Table 5–6, page 5–164 gives the values of constants C_1 and C_2 for unreinforced holes where,

$$C_1 = \frac{M_o}{M_r} \quad \text{and} \quad C_2 = \frac{M_o / M_r - M_l / M_r}{V_l / V_r'}$$

where M_o, M_l, and V_l are defined in [3] to [5].

Table 5–7, page 5–165, gives the value of constant C_3 taken as V_l / V_r'.

A_w/A_f varies from 0.5 to 2.25, $2H/d$ from 0.3 to 0.6, a/H from 0.50 to 2.2, and $e/d = 0$.

Written in terms of the constants C_1 and C_2, [2] becomes

$$\frac{M_f}{M_r} \le C_1 - C_2\left(\frac{V_f}{V_r'}\right) \qquad [13]$$

and [1] becomes

$$\frac{V_f}{V_r'} \le C_3 \qquad [14]$$

Use

For concentric ($e/d = 0$) unreinforced holes, compute A_w/A_f, $2H/d$ and a/H. Determine C_1, C_2 and C_3 with the aid of Tables 5–6 and 5–7 for use in [13] and [14].

(b) *Reinforced Holes*

Table 5–8, pages 5–166 to 5–168, gives the values of the constants C_4 and C_5 for reinforced holes where

$$C_4 = \frac{M_o}{M_r} \quad \text{and} \quad C_5 = \frac{M_o / M_r - M_l / M_r}{V_l / V_r'}$$

where M_o, M_l and V_l are defined in [10] to [12] for concentric holes ($e/d = 0$).

A_w/A_f varies from 0.5 to 2.25, $2H/d$ from 0.3 to 0.6 and a/H from 0.45 to 2.2, for the three values of A_r/A_f, 0.33, 0.67 and 1.0.

Written in terms of the constants C_4 and C_5, [9a] becomes

$$\frac{M_f}{M_r} \le C_4 - C_5\left(\frac{V_f}{V_r'}\right) \qquad [15]$$

Use

For concentric ($e/d = 0$) reinforced holes, compute A_w/A_f, $2H/d$, and a/H. Determine C_4 and C_5, for one of the assumed values of A_r/A_f, for use in [15].

Calculate V_l from [12] for use in [8a].

Then check [8b] and [9b].

Example

Given:

A simple span W610x101 beam spanning 12 m supports a factored total uniformly distributed load of 420 kN (35 kN/m). Check the adequacy of the section for 2 rectangular holes located as shown. Steel is G40.21-M 300W. Lateral support to compression flange is provided.

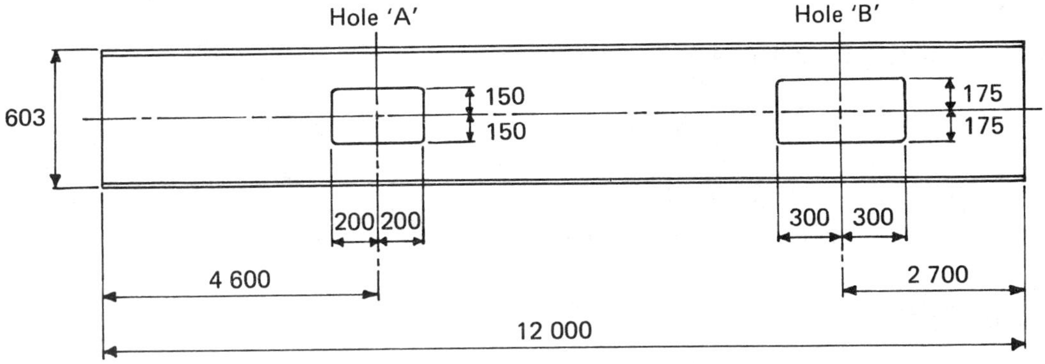

Solution for Hole 'A'

Class of beam: From page 5–7, for G40.21-M 300W steel, W610x101 is Class 1

From page 5–96, M_r = 783 kN·m

V_r = 1 130 kN

For use in formulas for holes in beam webs, V_r must be reduced to V_r' used for plastically analysed beams.

Therefore, $V_r' = 0.833 \times 1\,130 = 941$ kN

At centreline of hole,

$M_f = 35\text{ kN/m} \times 4.6\text{ m} \times (12 - 4.6)/2 = 596\text{ kN·m}$

$V_f = 35\text{ kN/m} \times ((12/2) - 4.6) = 49\text{ kN}$

Therefore, $\frac{M_f}{M_r} = \frac{596}{783} = 0.761$, and $\frac{V_f}{V_r'} = \frac{49}{941} = 0.052$

Check web stability

Refer to page 5–155.

$\frac{V_f}{V_r'} = 0.052 \quad \leq 0.67$ — OK

$a/H = 1.33 \quad \leq 3.0$ (limit for Class 1 beam) — OK

$a/H + 6(2H/d) = 200/150 + 6(300/603) = 1.33 + 6(0.498) = 4.3 \quad \leq 5.6$ — OK

Check compression zone stability

OK, if $2a \leq 4\, d_t$

$\leq 4\,((603/2) - 150) = 606$ mm

$2a = 400$ mm < 606 — OK

Check for unreinforced hole

$$\frac{A_w}{A_f} = \frac{10.5 \times 603}{228 \times 14.9} = 1.86 \quad \text{and} \quad \frac{2H}{d} = 0.50,$$

$C_1 = 0.92$, from Table 5–6, page 5–164

For $a / H = 1.33$ — use 1.4

$C_2 = 1.9$, from Table 5–6

$C_3 = 0.263$, from Table 5–7, page 5–165

$$\frac{M_f}{M_r} \leq C_1 - C_2 \left(\frac{V_f}{V_r'} \right) \quad [13]$$

$$\leq 0.92 - 1.9\,(0.052) = 0.82$$

$M_f / M_r = 0.76 \quad < 0.82$ — OK

$$\frac{V_f}{V_r'} \leq C_3 \quad [14]$$

$0.052 < 0.263$ — OK

Therefore, reinforcement not required.

Solution for Hole 'B'

At centreline of hole,

$M_f = 35\text{ kN/m} \times 2.7\text{ m} \times (12 - 2.7) / 2 = 440\text{ kN·m}$

$V_f = 35\text{ kN/m} \times ((12 / 2) - 2.7) = 116\text{ kN}$

Therefore, $\dfrac{M_f}{M_r} = \dfrac{440}{783} = 0.561$, and $\dfrac{V_f}{V_r'} = \dfrac{116}{941} = 0.123$

Check spacing between holes

Use 2H of larger hole.

OK, if $s \geq 2H$

≥ 350

$s = 12\,000 - (2\,700 + 4\,600) = 4\,700 \quad > 350$ — OK

Check web stability

Refer to page 5–155.

$\dfrac{V_f}{V_r'} = 0.12 \quad \leq 0.67$ — OK

$a / H = 1.7 \quad \leq 3.0$ (limit for Class 1 beam) — OK

$a / H + 6(2\,H / d) = 300/175 + 6(350/603) = 1.71 + 6(0.580) = 5.2 \quad \leq 5.6$ — OK

Check compression zone stability

OK, if $2a \leq 4\ d_t$ (unreinforced tee)

$\leq 4\ ((603/2) - 175) = 506$ mm

$2a = 600$ mm > 506 — not adequate

Check for unreinforced hole

From Table 5–6, page 5–164, for $A_w/A_f = 1.86$ (use 2.0)

and $2H/d = 0.58$ (use 0.60),

$C_1 = 0.88$

For $a/H = 1.71$ (use 1.8),

$C_2 = 3.83$

$$\frac{M_f}{M_r} \leq C_1 - C_2\left(\frac{V_f}{V_r'}\right) \quad [13]$$

$\leq 0.88 - 3.83\ (0.123) = 0.409$

$M_f/M_r = 0.561 > 0.409$ — reinforcement required

Reinforcement

Assume $A_r/A_f = 0.33$ (maximum permitted for one-sided reinforcement)

From Table 5–8, page 5–166,

for $\frac{A_r}{A_f} = 0.33,\ \frac{A_w}{A_f} = 2.0,\ \frac{2H}{d} = 0.60,$

$C_4 = 1.013$

For $\frac{a}{H} = 1.7$, $C_5 = 2.51$ (by interpolation)

$$\frac{M_f}{M_r} \leq C_4 - C_5\left(\frac{V_f}{V_r'}\right) \quad [15]$$

$\leq 1.01 - 2.51\,(0.123) = 0.70$

$M_f/M_r = 0.56 < 0.70$ — OK

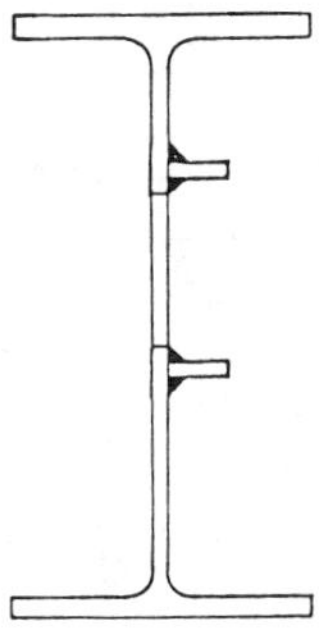

Section through beam at Hole 'B'

Further refinement of A_r/A_f can be accomplished by using the expressions previously given.

Check one-sided reinforcement

$M_f \leq 20\ V_f\ d$ at hole centreline (see page 5–157)

$\leq 20 \times 116 \times 603 = 1\,400\,000$ kN·mm $= 1\,400$kN·m

$M_f = 440$ kN·m $< 1\,400$ kN·m — OK

$a/H = 1.7 \leq 2.5$ — OK

$$\frac{d_t}{w} \leq \frac{370}{\sqrt{F_y}} \leq 21.4$$

VALUES OF C_1 AND C_2

Table 5-6

For Unreinforced Concentric Holes in Beam Webs

$\frac{A_w}{A_f}$	$\frac{2H}{d}$	C_1	C_2 For following $\frac{a}{H}$ values							
			.50	1.0	1.2	1.4	1.6	1.8	2.0	2.2
0.50	.30	.990	0.204	0.271	0.300	0.330	0.360	0.391	0.423	0.455
	.35	.986	0.226	0.315	0.353	0.392	0.433	0.474	0.516	0.558
	.40	.982	0.252	0.367	0.417	0.468	0.520	0.574	0.628	0.682
	.45	.978	0.283	0.432	0.495	0.561	0.628	0.696	0.764	0.833
	.50	.972	0.321	0.511	0.593	0.676	0.761	0.846	0.933	1.020
	.55	.966	0.368	0.612	0.715	0.820	0.927	1.035	1.143	1.252
	.60	.960	0.428	0.740	0.872	1.005	1.140	1.275	1.411	*
0.75	.30	.986	0.290	0.385	0.426	0.468	0.512	0.556	0.601	0.647
	.35	.981	0.321	0.447	0.502	0.557	0.615	0.673	0.733	0.793
	.40	.975	0.358	0.522	0.593	0.665	0.740	0.815	0.892	0.970
	.45	.968	0.402	0.613	0.704	0.797	0.892	0.989	1.086	1.184
	.50	.961	0.456	0.726	0.842	0.961	1.081	1.203	1.325	1.449
	.55	.952	0.522	0.869	1.016	1.166	1.317	1.470	1.624	1.779
	.60	.943	0.608	1.052	1.239	1.428	1.619	1.812	2.005	*
1.00	.30	.982	0.367	0.488	0.540	0.593	0.648	0.705	0.762	0.820
	.35	.975	0.407	0.567	0.635	0.706	0.779	0.853	0.928	1.005
	.40	.968	0.454	0.661	0.751	0.843	0.937	1.033	1.130	1.228
	.45	.960	0.510	0.777	0.892	1.010	1.130	1.252	1.376	1.500
	.50	.950	0.577	0.920	1.067	1.217	1.369	1.523	1.679	1.835
	.55	.939	0.662	1.101	1.287	1.477	1.669	1.862	2.057	2.253
	.60	.928	0.770	1.333	1.569	1.809	2.051	2.295	2.539	*
1.25	.30	.979	0.437	0.581	0.643	0.706	0.772	0.839	0.907	0.976
	.35	.971	0.485	0.675	0.756	0.841	0.927	1.015	1.105	1.196
	.40	.962	0.540	0.787	0.894	1.003	1.115	1.229	1.345	1.462
	.45	.952	0.607	0.925	1.062	1.202	1.346	1.491	1.638	1.786
	.50	.940	0.687	1.095	1.270	1.449	1.630	1.813	1.999	2.185
	.55	.928	0.788	1.310	1.532	1.758	1.987	2.217	2.449	2.682
	.60	.914	0.916	1.587	1.868	2.154	2.442	2.732	3.023	*
1.50	.30	.975	0.500	0.666	0.736	0.809	0.884	0.961	1.039	1.118
	.35	.967	0.555	0.773	0.866	0.963	1.062	1.163	1.266	1.370
	.40	.956	0.619	0.902	1.024	1.149	1.278	1.408	1.541	1.675
	.45	.945	0.695	1.059	1.216	1.377	1.541	1.708	1.876	2.046
	.50	.932	0.787	1.255	1.455	1.659	1.867	2.077	2.289	2.502
	.55	.917	0.902	1.501	1.755	2.014	2.276	2.540	2.805	3.072
	.60	.902	1.050	1.817	2.140	2.467	2.797	3.129	3.463	*
1.75	.30	.973	0.558	0.743	0.822	0.903	0.987	1.072	1.159	1.248
	.35	.963	0.619	0.862	0.967	1.075	1.185	1.298	1.413	1.529
	.40	.951	0.691	1.006	1.142	1.282	1.426	1.572	1.719	1.869
	.45	.938	0.776	1.182	1.357	1.537	1.720	1.906	2.094	2.283
	.50	.924	0.879	1.400	1.624	1.852	2.084	2.318	2.555	2.793
	.55	.908	1.007	1.675	1.959	2.247	2.539	2.834	3.131	3.429
	.60	.890	1.171	2.028	2.388	2.753	3.122	3.492	3.864	*
2.00	.30	.970	0.611	0.813	0.900	0.989	1.081	1.174	1.270	1.366
	.35	.959	0.678	0.944	1.059	1.177	1.298	1.421	1.547	1.674
	.40	.947	0.757	1.102	1.251	1.405	1.561	1.721	1.883	2.047
	.45	.933	0.849	1.295	1.486	1.683	1.884	2.087	2.293	2.500
	.50	.917	0.962	1.534	1.778	2.028	2.282	2.539	2.798	3.059
	.55	.899	1.103	1.835	2.145	2.461	2.781	3.104	3.429	3.755
	.60	.880	1.283	2.221	2.616	3.016	3.419	3.825	4.232	*
2.25	.30	.968	0.660	0.878	0.972	1.068	1.167	1.268	1.371	1.476
	.35	.956	0.733	1.020	1.144	1.271	1.402	1.535	1.671	1.808
	.40	.942	0.817	1.190	1.351	1.517	1.686	1.859	2.034	2.211
	.45	.927	0.917	1.398	1.605	1.818	2.034	2.254	2.476	2.700
	.50	.910	1.039	1.656	1.920	2.190	2.465	2.742	3.022	3.303
	.55	.891	1.191	1.981	2.317	2.658	3.004	3.352	3.703	4.056
	.60	.870	1.386	2.399	2.825	3.257	3.692	4.131	4.571	*

*a/H plus 6(2H/d) exceeds 5.6.

$d_t / w = ((603 / 2) - 175) / 10.5 = 12.0 \quad < 21.4 \quad$ — OK

Therefore, one-sided reinforcement OK

$A_r = 0.33, \quad A_f = 0.33(228 \times 14.9) = 1\,120 \text{ mm}^2$

Check shear

$$V_l = \sqrt{3}\left(\frac{d}{a}\right)\left(\frac{A_r}{A_w}\right)\left(1 - \frac{2H}{d}\right)V_r' \qquad [12]$$

$$= \sqrt{3}\left(\frac{603}{300}\right)\left(\frac{1\,120}{10.5 \times 603}\right)(1 - 0.58)\,941 = 243$$

$V_f \le V_l \quad$ [8a]

$116 < 243 \quad$ — OK

$$V_f / V_r' \le 1 - \frac{2H}{d} \qquad [8b]$$

$\le 1 - 0.58 = 0.42$

$V_f / V_r' = 0.123 \quad < 0.42 \quad$ — OK

Try 16 x 70 reinforcement

$$\frac{b}{t} \le \frac{145}{\sqrt{F_y}} \quad \text{(for Class 1)}$$

≤ 8.37

$b / t = 70 / 16 = 4.3 \quad < 8.37 \quad$ — OK

Therefore, use 16 x 70 one-sided reinforcement.

VALUES OF C_3

For Unreinforced Concentric Holes in Beam Webs

Table 5–7

$\frac{2H}{d}$	a/H							
	0.5	1.0	1.2	1.4	1.6	1.8	2.0	2.2
.30	.680	.627	.602	.575	.549	.523	.498	.474
.35	.621	.552	.521	.490	.461	.433	.407	.384
.40	.560	.475	.441	.408	.378	.351	.327	.305
.45	.497	.400	.364	.332	.303	.279	.257	.238
.50	.433	.327	.293	.263	.238	.217	.199	.183
.55	.368	.260	.229	.203	.182	.165	.150	.138
.60	.302	.200	.173	.152	.136	.122	.111	.102

VALUES OF C_4 AND C_5

For Reinforced Concentric Holes in Beam Webs

Table 5-8

$A_r/A_f = 0.33$

$\frac{A_w}{A_f}$	$\frac{2H}{d}$	C_4	C_5 For following $\frac{a}{H}$ values							
			.45	1.0	1.2	1.4	1.6	1.8	2.0	2.2
0.50	.30	1.079	0.041	0.090	0.108	0.126	0.144	0.162	0.180	0.199
	.35	1.090	0.052	0.116	0.139	0.162	0.186	0.209	0.232	0.255
	.40	1.101	0.066	0.147	0.176	0.205	0.235	0.264	0.293	0.323
	.45	1.111	0.083	0.184	0.220	0.257	0.294	0.330	0.367	0.404
	.50	1.120	0.103	0.229	0.274	0.320	0.366	0.411	0.457	0.503
	.55	1.129	0.128	0.284	0.341	0.398	0.455	0.511	0.568	0.625
	.60	1.138	0.159	0.354	0.425	0.496	0.567	0.637	0.708	*
0.75	.30	1.070	0.064	0.142	0.170	0.198	0.227	0.255	0.283	0.311
	.35	1.079	0.081	0.181	0.217	0.253	0.290	0.326	0.362	0.398
	.40	1.087	0.102	0.228	0.273	0.319	0.364	0.410	0.455	0.501
	.45	1.094	0.127	0.283	0.340	0.397	0.453	0.510	0.566	0.623
	.50	1.101	0.158	0.350	0.420	0.491	0.561	0.631	0.701	0.771
	.55	1.107	0.195	0.433	0.519	0.606	0.693	0.779	0.866	0.952
	.60	1.112	0.241	0.536	0.643	0.750	0.858	0.965	1.072	*
1.00	.30	1.062	0.088	0.196	0.235	0.275	0.314	0.353	0.392	0.432
	.35	1.069	0.112	0.250	0.300	0.350	0.400	0.449	0.499	0.549
	.40	1.075	0.141	0.313	0.375	0.438	0.500	0.563	0.625	0.688
	.45	1.079	0.174	0.387	0.464	0.542	0.619	0.697	0.774	0.851
	.50	1.083	0.214	0.476	0.572	0.667	0.762	0.857	0.953	1.048
	.55	1.086	0.263	0.585	0.702	0.819	0.936	1.053	1.170	1.287
	.60	1.088	0.324	0.721	0.865	1.009	1.153	1.297	1.441	*
1.25	.30	1.055	0.114	0.254	0.304	0.355	0.406	0.457	0.507	0.558
	.35	1.060	0.145	0.322	0.386	0.450	0.515	0.579	0.643	0.708
	.40	1.063	0.180	0.401	0.481	0.561	0.642	0.722	0.802	0.882
	.45	1.066	0.222	0.494	0.593	0.692	0.791	0.890	0.989	1.088
	.50	1.067	0.273	0.606	0.727	0.848	0.969	1.090	1.211	1.333
	.55	1.068	0.333	0.741	0.889	1.037	1.185	1.333	1.481	1.629
	.60	1.067	0.408	0.907	1.089	1.270	1.452	1.633	1.815	*
1.50	.30	1.048	0.141	0.314	0.376	0.439	0.502	0.565	0.627	0.690
	.35	1.051	0.178	0.396	0.476	0.555	0.634	0.713	0.793	0.872
	.40	1.053	0.222	0.492	0.591	0.689	0.788	0.886	0.985	1.083
	.45	1.054	0.272	0.605	0.726	0.847	0.968	1.089	1.210	1.331
	.50	1.053	0.332	0.738	0.886	1.033	1.181	1.329	1.476	1.624
	.55	1.051	0.404	0.899	1.078	1.258	1.438	1.618	1.797	1.977
	.60	1.047	0.493	1.096	1.315	1.534	1.753	1.973	2.192	*
1.75	.30	1.042	0.169	0.376	0.451	0.526	0.601	0.676	0.751	0.827
	.35	1.044	0.213	0.473	0.568	0.663	0.757	0.852	0.947	1.041
	.40	1.044	0.264	0.586	0.704	0.821	0.938	1.055	1.173	1.290
	.45	1.043	0.323	0.718	0.861	1.005	1.149	1.292	1.436	1.579
	.50	1.040	0.393	0.873	1.048	1.222	1.397	1.572	1.746	1.921
	.55	1.035	0.477	1.059	1.271	1.483	1.694	1.906	2.118	2.330
	.60	1.030	0.579	1.286	1.543	1.801	2.058	2.315	2.572	*
2.00	.30	1.037	0.198	0.440	0.528	0.615	0.703	0.791	0.879	0.967
	.35	1.037	0.249	0.553	0.663	0.774	0.884	0.995	1.105	1.216
	.40	1.036	0.307	0.683	0.819	0.956	1.092	1.229	1.365	1.502
	.45	1.033	0.375	0.833	1.000	1.167	1.333	1.500	1.667	1.833
	.50	1.028	0.455	1.010	1.212	1.415	1.617	1.819	2.021	2.223
	.55	1.021	0.550	1.221	1.466	1.710	1.954	2.198	2.443	2.687
	.60	1.013	0.665	1.478	1.774	2.069	2.365	2.660	2.956	*
2.25	.30	1.032	0.227	0.505	0.606	0.707	0.808	0.909	1.010	1.111
	.35	1.031	0.285	0.634	0.760	0.887	1.014	1.141	1.267	1.394
	.40	1.028	0.351	0.781	0.937	1.093	1.249	1.405	1.562	1.718
	.45	1.023	0.428	0.951	1.141	1.331	1.521	1.712	1.902	2.092
	.50	1.017	0.517	1.150	1.380	1.610	1.839	2.069	2.299	2.529
	.55	1.008	0.623	1.386	1.663	1.940	2.217	2.494	2.771	3.048
	.60	0.998	0.752	1.671	2.005	2.340	2.674	3.008	3.342	*

*a/H plus 6(2H/d) exceeds 5.6.

VALUES OF C_4 AND C_5

For Reinforced Concentric Holes in Beam Webs

Table 5-8

Ar/Af = 0.67

$\frac{A_w}{A_f}$	$\frac{2H}{d}$	C_4	C_5							
			For following $\frac{a}{H}$ values							
			.45	1.0	1.2	1.4	1.6	1.8	2.0	2.2
0.50	.30	1.168	0.036	0.081	0.097	0.113	0.129	0.146	0.162	0.178
	.35	1.194	0.047	0.105	0.126	0.146	0.167	0.188	0.209	0.230
	.40	1.219	0.060	0.133	0.160	0.187	0.213	0.240	0.266	0.293
	.45	1.244	0.076	0.168	0.201	0.235	0.269	0.302	0.336	0.369
	.50	1.269	0.095	0.210	0.253	0.295	0.337	0.379	0.421	0.463
	.55	1.292	0.119	0.264	0.316	0.369	0.422	0.474	0.527	0.580
	.60	1.316	0.149	0.331	0.397	0.463	0.530	0.596	0.662	*
0.75	.30	1.154	0.055	0.122	0.146	0.170	0.195	0.219	0.243	0.267
	.35	1.177	0.071	0.157	0.188	0.219	0.251	0.282	0.314	0.345
	.40	1.199	0.089	0.199	0.239	0.278	0.318	0.358	0.398	0.438
	.45	1.221	0.112	0.250	0.300	0.350	0.400	0.450	0.500	0.549
	.50	1.241	0.140	0.312	0.374	0.437	0.499	0.561	0.624	0.686
	.55	1.261	0.175	0.389	0.467	0.545	0.623	0.700	0.778	0.856
	.60	1.280	0.219	0.487	0.584	0.682	0.779	0.876	0.974	*
1.00	.30	1.142	0.073	0.162	0.195	0.227	0.260	0.292	0.325	0.357
	.35	1.162	0.094	0.209	0.251	0.292	0.334	0.376	0.418	0.459
	.40	1.181	0.119	0.264	0.317	0.370	0.422	0.475	0.528	0.581
	.45	1.199	0.149	0.330	0.397	0.463	0.529	0.595	0.661	0.727
	.50	1.217	0.185	0.411	0.494	0.576	0.658	0.740	0.823	0.905
	.55	1.233	0.230	0.511	0.614	0.716	0.818	0.920	1.023	1.125
	.60	1.248	0.287	0.637	0.765	0.892	1.020	1.147	1.275	*
1.25	.30	1.131	0.092	0.203	0.244	0.285	0.325	0.366	0.407	0.448
	.35	1.149	0.117	0.261	0.313	0.365	0.417	0.469	0.521	0.574
	.40	1.165	0.148	0.329	0.395	0.460	0.526	0.592	0.658	0.723
	.45	1.180	0.185	0.410	0.492	0.574	0.656	0.738	0.821	0.903
	.50	1.194	0.229	0.509	0.611	0.713	0.814	0.916	1.018	1.120
	.55	1.207	0.284	0.631	0.757	0.883	1.009	1.135	1.261	1.388
	.60	1.219	0.353	0.784	0.940	1.097	1.254	1.410	1.567	*
1.50	.30	1.121	0.110	0.245	0.293	0.342	0.391	0.440	0.489	0.538
	.35	1.136	0.141	0.313	0.375	0.438	0.500	0.563	0.625	0.688
	.40	1.150	0.177	0.393	0.472	0.550	0.629	0.708	0.786	0.865
	.45	1.163	0.220	0.489	0.587	0.685	0.783	0.881	0.978	1.076
	.50	1.174	0.272	0.605	0.726	0.847	0.968	1.089	1.210	1.332
	.55	1.184	0.336	0.748	0.897	1.047	1.196	1.346	1.495	1.645
	.60	1.193	0.417	0.926	1.111	1.296	1.481	1.667	1.852	*
1.75	.30	1.112	0.129	0.286	0.343	0.400	0.457	0.514	0.571	0.629
	.35	1.125	0.164	0.364	0.437	0.510	0.583	0.656	0.729	0.802
	.40	1.137	0.206	0.457	0.549	0.640	0.731	0.823	0.914	1.006
	.45	1.147	0.255	0.567	0.681	0.794	0.908	1.021	1.135	1.248
	.50	1.156	0.315	0.700	0.840	0.980	1.120	1.260	1.400	1.540
	.55	1.163	0.388	0.862	1.035	1.207	1.380	1.552	1.725	1.897
	.60	1.169	0.479	1.065	1.278	1.491	1.704	1.917	2.130	*
2.00	.30	1.103	0.147	0.327	0.392	0.458	0.523	0.589	0.654	0.719
	.35	1.115	0.187	0.416	0.499	0.583	0.666	0.749	0.832	0.916
	.40	1.124	0.234	0.521	0.625	0.729	0.833	0.938	1.042	1.146
	.45	1.133	0.290	0.645	0.774	0.903	1.032	1.161	1.290	1.419
	.50	1.139	0.357	0.794	0.953	1.111	1.270	1.429	1.588	1.746
	.55	1.144	0.439	0.975	1.170	1.365	1.560	1.755	1.951	2.146
	.60	1.147	0.540	1.201	1.441	1.681	1.921	2.162	2.402	*
2.25	.30	1.096	0.166	0.368	0.442	0.516	0.589	0.663	0.737	0.810
	.35	1.105	0.211	0.468	0.561	0.655	0.749	0.842	0.936	1.029
	.40	1.113	0.263	0.584	0.701	0.818	0.935	1.052	1.169	1.286
	.45	1.119	0.325	0.722	0.866	1.011	1.155	1.300	1.444	1.588
	.50	1.123	0.399	0.887	1.064	1.241	1.419	1.596	1.773	1.951
	.55	1.126	0.489	1.087	1.304	1.521	1.738	1.956	2.173	2.390
	.60	1.126	0.600	1.334	1.601	1.868	2.135	2.402	2.669	*

*a/H plus 6(2H/d) exceeds 5.6.

VALUES OF C_4 AND C_5

For Reinforced Concentric Holes in Beam Webs

Table 5-8

Ar/Af = 1.00

$\frac{A_w}{A_f}$	$\frac{2H}{d}$	C_4	C_5							
			For following $\frac{a}{H}$ values							
			.45	1.0	1.2	1.4	1.6	1.8	2.0	2.2
0.50	.30	1.257	0.035	0.078	0.093	0.109	0.124	0.140	0.155	0.171
	.35	1.298	0.045	0.101	0.121	0.141	0.161	0.182	0.202	0.222
	.40	1.338	0.058	0.129	0.154	0.180	0.206	0.232	0.257	0.283
	.45	1.378	0.073	0.163	0.195	0.228	0.260	0.293	0.325	0.358
	.50	1.417	0.092	0.204	0.245	0.286	0.327	0.368	0.409	0.450
	.55	1.455	0.116	0.257	0.308	0.359	0.411	0.462	0.513	0.565
	.60	1.493	0.145	0.323	0.388	0.453	0.517	0.582	0.647	*
0.75	.30	1.238	0.052	0.115	0.138	0.161	0.184	0.207	0.230	0.253
	.35	1.275	0.067	0.149	0.178	0.208	0.238	0.268	0.297	0.327
	.40	1.312	0.085	0.189	0.227	0.265	0.303	0.341	0.379	0.416
	.45	1.347	0.107	0.239	0.286	0.334	0.382	0.429	0.477	0.525
	.50	1.382	0.135	0.299	0.359	0.419	0.479	0.538	0.598	0.658
	.55	1.415	0.169	0.375	0.449	0.524	0.599	0.674	0.749	0.824
	.60	1.448	0.212	0.470	0.564	0.659	0.753	0.847	0.941	*
1.00	.30	1.222	0.068	0.151	0.181	0.212	0.242	0.272	0.302	0.333
	.35	1.255	0.088	0.195	0.234	0.273	0.312	0.351	0.390	0.429
	.40	1.288	0.112	0.248	0.297	0.347	0.397	0.446	0.496	0.545
	.45	1.320	0.140	0.312	0.374	0.436	0.499	0.561	0.623	0.686
	.50	1.350	0.175	0.390	0.468	0.546	0.624	0.701	0.779	0.857
	.55	1.380	0.219	0.487	0.584	0.681	0.779	0.876	0.973	1.071
	.60	1.408	0.274	0.610	0.732	0.854	0.975	1.097	1.219	*
1.25	.30	1.207	0.084	0.187	0.224	0.261	0.299	0.336	0.373	0.411
	.35	1.238	0.108	0.240	0.289	0.337	0.385	0.433	0.481	0.529
	.40	1.267	0.137	0.305	0.366	0.427	0.488	0.548	0.609	0.670
	.45	1.295	0.172	0.382	0.459	0.535	0.612	0.688	0.764	0.841
	.50	1.321	0.215	0.477	0.572	0.668	0.763	0.858	0.954	1.049
	.55	1.347	0.267	0.594	0.713	0.832	0.951	1.069	1.188	1.307
	.60	1.371	0.334	0.742	0.891	1.039	1.188	1.336	1.485	*
1.50	.30	1.194	0.100	0.222	0.266	0.310	0.354	0.399	0.443	0.487
	.35	1.221	0.128	0.285	0.342	0.399	0.456	0.512	0.569	0.626
	.40	1.247	0.162	0.360	0.432	0.504	0.576	0.648	0.720	0.792
	.45	1.272	0.203	0.451	0.541	0.631	0.721	0.811	0.901	0.991
	.50	1.295	0.252	0.561	0.673	0.785	0.898	1.010	1.122	1.234
	.55	1.317	0.314	0.697	0.837	0.976	1.116	1.255	1.395	1.534
	.60	1.338	0.391	0.869	1.043	1.217	1.391	1.565	1.738	*
1.75	.30	1.181	0.115	0.256	0.307	0.358	0.409	0.460	0.512	0.563
	.35	1.206	0.148	0.328	0.394	0.459	0.525	0.591	0.656	0.722
	.40	1.230	0.186	0.414	0.497	0.580	0.663	0.745	0.828	0.911
	.45	1.251	0.233	0.517	0.621	0.724	0.828	0.931	1.034	1.138
	.50	1.272	0.289	0.642	0.771	0.899	1.028	1.156	1.285	1.413
	.55	1.291	0.359	0.797	0.956	1.116	1.275	1.434	1.594	1.753
	.60	1.308	0.446	0.991	1.189	1.387	1.586	1.784	1.982	*
2.00	.30	1.170	0.130	0.289	0.347	0.405	0.463	0.521	0.579	0.637
	.35	1.192	0.167	0.371	0.445	0.519	0.593	0.667	0.741	0.816
	.40	1.213	0.210	0.467	0.560	0.654	0.747	0.841	0.934	1.027
	.45	1.233	0.262	0.582	0.699	0.815	0.932	1.048	1.164	1.281
	.50	1.250	0.325	0.722	0.866	1.010	1.155	1.299	1.443	1.588
	.55	1.266	0.402	0.893	1.072	1.251	1.429	1.608	1.786	1.965
	.60	1.280	0.499	1.109	1.330	1.552	1.774	1.995	2.217	*
2.25	.30	1.160	0.145	0.323	0.387	0.452	0.516	0.581	0.646	0.710
	.35	1.180	0.186	0.413	0.495	0.578	0.660	0.743	0.825	0.908
	.40	1.198	0.234	0.519	0.623	0.726	0.830	0.934	1.038	1.142
	.45	1.215	0.291	0.646	0.775	0.904	1.033	1.162	1.291	1.421
	.50	1.230	0.360	0.799	0.959	1.118	1.278	1.438	1.598	1.758
	.55	1.243	0.444	0.987	1.184	1.382	1.579	1.776	1.974	2.171
	.60	1.254	0.550	1.222	1.467	1.711	1.955	2.200	2.444	*

*a/H plus 6(2H/d) exceeds 5.6.

PART SIX
PROPERTIES AND DIMENSIONS

STRUCTURAL STEELS

General

Canadian structural steels are covered by two sets of standards prepared by the the Canadian Standards Association Technical Committee on Structural Steel, G40. Currently these are the metric (SI) version, CAN/CSA G40.20-M87 and CAN/CSA G40.21-M87, and the imperial version, CAN/CSA G40.20-87 and CAN/CSA G40.21-87. At the time of printing this Handbook, numerous changes to the 1987 edition of these standards had been ballotted and approved by the Technical Committee in preparation for a new edition. Information provided in this section includes these approved changes, and is generally based on the metric version in keeping with Canadian design standards.

CAN/CSA G40.20-M, "General requirements for Rolled or Welded Structural Quality Steel" sets out the general requirements governing the delivery of structural quality steels. These requirements include: Definitions, Chemical Composition, Variations in dimensions, Methods of Testing, Frequency of Testing, Heat Treatment, Repairs of defects, Marking, etc.

CAN/CSA G40.21-M, "Structural Quality Steels" governs the chemical and mechanical properties of 8 types and 8 strength levels of structural steels for general construction and engineering purposes. All strength levels are not available in all types, and selection of the proper grade (type and strength level) is important for a particular application. CAN/CSA G40.21-M 300W is the standard weldable steel normally specified for building construction, while CAN/CSA G40.21-M 350A and CAN/CSA G40.21-M 350AT are atmospheric corosion resistant steels normally used in bridge construction. CAN/CSA G40.21-M 350W is the normal steel used for HSS sections.

The 8 types covered in CAN/CSA G40.21-M are:

(a) **Type G – General Construction Steel.** Steels of this type meet specified strength requirements; however, the chemical control is not such that all of these steels may be welded satisfactorily under normal field conditions. These steels are primarily designed for applications involving bolted connections or for welding under carefully controlled shop conditions.

(b) **Type W – Weldable Steel.** Steels of this type meet specified strength requirements and are suitable for general welded construction where notch toughness at low temperature is not a design requirement. Applications may include buildings, compression members of bridges, etc.

(c) **Type WT – Weldable Notch Tough Steel.** Steels of this type meet specified strength and Charpy V-Notch impact requirements and are suitable for welded construction where notch toughness at low temperature is a design requirement. The purchaser, in addition to specifying the grade, must specify the category of steel required that establishes the Charpy V-Notch test temperature and energy level. Applications may include primary tension members in bridges and similar elements.

(d) **Type R – Atmospheric Corrosion-Resistant Steel.** Steels of this type meet specified strength requirements and display an atmospheric corrosion-resistance approximately four times that of plain carbon steels*. These steels may be readily welded up to the maximum thickness covered by this standard. Applications include unpainted siding, unpainted light structural members, etc, where notch toughness at low temperature is not a design requirement.

* Copper content not exceeding 0.02%

(e) **Type A – Atmospheric Corrosion-Resistant Weldable Steel.** Steels of this type meet specified strength requirements and display an atmospheric corrosion-resistance approximately four times that of plain carbon steels*. These steels are suitable for welded construction where notch toughness at low temperature is not a design requirememt and are often used in structures in the unpainted condition. Applications are similar to those for type W.

* Copper content not exceeding 0.02%

(f) **Type AT – Atmospheric Corrosion-Resistant Weldable Notch Tough Steel.** Steels of this type meet specified strength and Charpy V-Notch impact requirements and display an atmospheric corrosion-resistance of approximately four times that of plain carbon steels*. These steels are suitable for welded construction where notch toughness at low temperature is a design requirement and are often used in structures in the unpainted condition. The purchaser, in addition to specifying the grade, must specify the category of steel required that establishes the Charpy V-Notch test temperature and energy level. Applications may include primary tension members in bridges and similar elements.

* Copper content not exceeding 0.02%

(g) **Type Q – Quenched and Tempered Low Alloy Steel Plate.** Steels of this type meet specified strength requirements. While these steels may be readily welded, the welding and fabrication techniques are of fundamental importance and must not adversely affect the properties of the plate, especially the heat-affected zone. Applications may include bridges and similar structures.

(h) **Type QT – Quenched and Tempered Low Alloy Notch Tough Steel Plate.** Steels of this type meet specified strength and Charpy V-Notch impact requirements. They provide good resistance to brittle fracture and are suitable for structures where notch toughness at low temperature is a design requirement. The purchaser, in addition to specifying the grade, must specify the category of steel required that establishes the Charpy V-Notch test temperature and energy level. While these steels may be readily welded, the welding and fabrication techniques are of fundamental importance and must not adversely affect the properties of the plate, especially the heat-affected zone. Applications may include primary tension members in bridges and similar elements.

Tables

Table 6-1, "Grades, Types, Strength Levels", page 6-6 gives the grade designation of the various types and strength levels of structural steels according to the requirements of CAN/CSA G40.21-M.

Availability of any grade and shape combination should be kept in mind when designing to ensure overall economy, since a specified product may not always be available in the tonnage and time frame contemplated. Local availability should always be checked.

Table 6-3, "Mechanical Properties Summary", page 6-7 provides a summary of the various grades and strength levels, tensile strength and yield strength, and the usual maximum thickness and size available for plates, bars and welded shapes, rolled shapes and sheet piling, and hollow structural sections based on CAN/CSA G40.21-M. Table 6-2, "Shape Size Grouping for Tensile Property Classification", page 6-6 summarizes the size groupings for rolled shapes.

Table 6-4, "Chemical Composition", page 6-8 summarizes the chemical requirements of various grades of steel covered by CAN/CSA G40.21-M.

Steel is identified at the mill as to type and grade according to the requirements of the G40.21-M standard. Table 6-5, "Steel Marking Colour code", page 6-9 summarizes the colour code according to the standard. Normally one end of each piece is marked with the appropriate colour code, however, where products are bundled or are shipped as secured lifts only the top or an outside piece may be marked, or a substantial tag may be used.

The particular standards, CAN/CSA G40.20-M and CAN/CSA G40.21-M should be consulted for more detail. Similar information about steel covered by ASTM standards should be consulted when appropriate.

GRADES, TYPES, STRENGTH LEVELS **

Table 6-1

Type	Yield Strength, MPa							
	230	260	300	350	380	400	480	700
G	230G	–	–	350G	–	400G	–	–
W	–	260W	300W	350W	380W*	400W	480W	–
WT	–	260WT	300WT	350WT	380WT*	400WT	480WT	–
R	–	–	–	350R	–	–	–	–
A	–	–	–	350A	–	400A	480A	–
AT	–	–	–	350AT	–	400AT	480AT	–
Q	–	–	–	–	–	–	–	700Q
QT	–	–	–	–	–	–	–	700QT

*These grades are available in hollow structural sections and angles only.
** See CAN/CSA-G40.21-M

SHAPE SIZE GROUPINGS FOR TENSILE PROPERTY CLASSIFICATION*

Table 6–2

Shape Type	Group 1	Group 2	Group 3	Group 4	Group 5
W Shapes	W610x82 –92 W530x66–85 W460x52–106 W410x39–85 W360x33–79 W310x21–86 W250x18–67 W200x15–71 W150x13–37 W130x23 & 28 W100x19	W1000x222–399 W920x201–313 W840x176–226 W760x134–314 W690x125–265 W610x101–241 W530x92–219 W460x113–213 W410x100–149 W360x91–196 W310x97–158 W250x73–167 W200x86 & 100	W1000x412–488 W920x342–446 W840x299–433 W760x350–389 W690x289–384 W610x262–341 W530x248–331 W460x235–286 W360x216–314 W310x179–283	W1000x539–976 W920x488–1188 W840x473–922 W760x434–865 W690x419–802 W610x372–732 W530x370–599 W460x315–464 W360x347–818 W310x313–500	W920x1262 W360x900–1086
M Shapes	To 56 kg/m				
S Shapes	To 52 kg/m	Over 52 kg/m			
HP Shapes		To 150 kg/m	Over 152 kg/m		
C Shapes	To 30.8 kg/m	Over 30.8 kg/m			
MC Shapes	To 42.4 kg/m	Over 42.4 kg/m			
L Shapes	To 13 mm	Over 13 to 20 mm	Over 20 mm		

Note: Tees cut from W, M, and S shapes fall in the same group as the shape from which they are cut.
* See CAN/CSA-G40.20-M

Table 6-3 MECHANICAL PROPERTIES SUMMARY

CSA G40.21-M*		Tensile Strength F_u (MPa)	Plates, Bars and Welded Shapes			Rolled Shapes and Sheet Piling				Hollow Structural Sections	
			Nominal Maximum Thickness t (mm)	F_y (MPa) min.		Usual Maximum Shape Size Group	F_y (MPa) min.			Usual Maximum Wall Thickness (mm)	F_y (MPa) min.
Type	Grade			t ≤ 65	t > 65		Groups 1 to 3	Group 4	Group 5		
G	230G	380-500[1]	300	230	230	5	230	230	230		
	350G	480-690	30	350[3]		3	350				
	400G	550-720	30	400[3]		3	400				
W	260W	410-590	200	260	250	4	260	260			
	300W	450-620[4]	200	300	280	3	300			16	300
	350W	450-650[2]	150	350	320	2	350			16	350
	380W	480-650								16	380
	400W	520-690	20	400		1	400				
	480W	590-790	20	480		1	480				
WT	260WT	410-590	150	260	250	5	260	260	250		
	300WT	450-620	150	300	280	5	300	290	280		
	350WT	480-650	150	350	320	4	350	330		16	350
	380WT	480-650								16	380
	400WT	520-690	20	400		2	400				
	480WT	590-790	20	480		1	480				
R	350R	480-650	14	350		1	350				
A	350A	480-650	100	350	350	5	350	350	320	16	350
	400A	520-690	40	400		2	400				
	480A	590-790	20	480							
AT	350AT	480-650	100	350	350	5	350	350	320	16	350
	400AT	520-690	40	400		2	400				
	480AT	590-790	20	480							
Q	700Q	800-950	65	700							
QT	700QT	800-950	65	700							

[1] Upper bound F_u = 520 MPa for rolled shapes and sheet piling.
[2] Upper bound F_u = 620 MPa for HSS.
[3] Bar size shapes only.
[4] 410-590 MPa for HSS only.
* See CAN/CSA-G40.21-M.

CHEMICAL COMPOSITION[1] — Table 6-4

CSA G40.21-M* Grade	Chemical Composition (Heat Analysis) Per Cent[14] — All percentages are maxima unless otherwise indicated							
	C	Mn	P	S	Si[12,13]	Other[2]	Cr	Ni	Cu[9]
230G	0.26	1.20[3]	0.05	0.05	0.40	0.10	–	–	–
350G, 400G	0.28[4]	1.65	0.04	0.05	0.40	0.10	–	–	–
260W	0.20[15]	0.50/1.50	0.04	0.05	0.40	0.10	–	–	–
300W[5]	0.22[15]	0.50/1.50	0.04	0.05	0.40	0.10	–	–	–
350W	0.23	0.50/1.50	0.04	0.05	0.40	0.10	–	–	–
380W[6]	0.23	0.50/1.50	0.04	0.05	0.40	0.10	–	–	–
400W	0.23	0.50/1.50	0.04	0.05	0.40	0.10	–	–	–
480W	0.26	0.50/1.50	0.04	0.05	0.40	0.10	–	–	–
260WT	0.20[15]	0.80/1.50	0.03	0.04	0.15/0.40	0.10	–	–	–
300WT	0.22[15]	0.80/1.50	0.03	0.04	0.15/0.40	0.10	–	–	–
350WT	0.22[15]	0.80/1.50[7]	0.03	0.04	0.15/0.40	0.10[8]	–	–	–
380WT[6]	0.22	0.80/1.50	0.03	0.04	0.15/0.40	0.10	–	–	–
400WT	0.22	0.80/1.50[7]	0.03	0.04	0.15/0.40	0.10[8]	–	–	–
480WT	0.26	0.80/1.50[7]	0.03	0.04	0.15/0.40	0.10[8]	–	–	–
350R	0.16	0.75	0.05/0.15	0.04	0.75	0.10	0.30/1.25[10]	0.90[10]	0.20/0.60[10]
350A	0.20	0.75/1.35	0.03	0.04	0.15/0.40	0.10	0.70[11]	0.90[11]	0.20/0.60
400A	0.20	0.75/1.35	0.03	0.04	0.15/0.40	0.10	0.70[11]	0.90[11]	0.20/0.60
480A	0.20	1.00/1.60	0.025	0.035	0.15/0.40	0.12	0.70[11]	0.25/0.50[11]	0.20/0.60
350AT	0.20	0.75/1.35[7]	0.03	0.04	0.15/0.40	0.10	0.70[11]	0.90[11]	0.20/0.60
400AT	0.20	0.75/1.35[7]	0.03	0.04	0.15/0.40	0.10	0.70[11]	0.90[11]	0.20/0.60
480AT	0.20	1.00/1.60[7]	0.025	0.035	0.15/0.40	0.12	0.70[11]	0.25/0.50[11]	0.20/0.60
700Q	0.20	1.50	0.03	0.04	0.15/0.35	–	Boron 0.0005/0.005		–
700QT	0.20	1.50	0.03	0.04	0.15/0.35	–	Boron 0.0005/0.005		–

*See CAN/CSA-G40.21-M

Notes:

1. Actual standard should be consulted for full details.
2. Elements (C_b, V) may be used singly or in combination. Al, when used, is not included in summation. See G40.21-M.
3. May have 1.50% Mn.
4. May have 0.32% C for thicknesses over 20 mm.
5. For HSS 0.26% C and 0.30/1.20% Mn.
6. HSS only.
7. With prior agreement Mn may be increased. See CSA G40.21-M.
8. 0.01/0.02% N may be used but N $\not>$ ¼V.
9. Copper content of 0.20% minimum may be specified.
10. Cr + Ni + Cu $\not<$ 1.00%.
11. Cr + Ni $\not<$ 0.40%.
12. Si content of 0.15% to 0.40% is required for W type steel over 40 mm thickness or bar diameter except as noted in footnote 13.
13. By purchaser's request or producer's option no minimum Si content is required so long as 0.02% of aluminum content is used.
14. Additional alloying elements may be used when approved.
15. For thickness over 100 mm, C may be 0.22% for 260W and 260WT grades and 0.23% for 300W, 300WT, and 350WT grades.

Table 6-5

Steel Grade	Primary Colour	Secondary Colour
230G 350G 400G	Orange Blue Black	Orange Orange Orange
260W 300W 350W 380W 400W 480W	White Green Blue Brown Black Yellow	Green Green Green Green Green Green
260WT 300WT 350WT 380WT 400WT 480WT	White Green Blue Brown Black Yellow	White White White White White White
350R	Blue	Blue
350A 400A 480A	Blue Black Yellow	Yellow Yellow Yellow
350AT 400AT 480AT	Blue Black Yellow	Brown Brown Brown
700Q	Red	Red
700QT	Red	Purple

In this Code the following colour system applies:

Strength Level	Primary Colour	Type	Secondary Colour
260 300 350 380 400 480 700	White Green Blue Brown Black Yellow Red	G W WT R A AT Q QT	Orange Green White Blue Yellow Brown Red Purple

Note: See also page 6—4.

NOTES

STANDARD MILL PRACTICE

General

Rolled structural shapes are produced by passing hot blooms, billets or slabs of steel through a series of grooved rolls. Wear on the rolls can cause the dimensions of the finished product to vary slightly from the theorectical, published dimensions. Standard rolling tolerances have been established to make allowance for roll wear, and other factors. These tolerances are contained in CAN/CSA G40.20-M for shapes supplied according to CSA material standards, and in ASTM Standard A6 for shapes supplied according to ASTM material standards.

Letter symbols for dimensions on sketches shown in this section are in accordance with CAN/CSA G40.20-M, ASTM A6, and mill catalogs.

Methods of increasing area and mass by spreading rolls

Most nominal size groups of rolled shapes contain several specific shapes, each of which is slightly different in mass, area and properties from other shapes in the same size group. Methods used to increase the area and mass, from the minimum nominal size, by spreading the rolls are described below:

For W Shapes (Fig. 1), the thickness of both flange and web is increased, resulting in an increase to the overall beam depth and flange width, with the distance between inside faces of flanges being unchanged.

For S Shapes and Channels (Fig. 2 and 3), the web thickness and flange width are increased by equal amounts, all other dimensions remaining unchanged.

For angles (Fig. 4) the thickness of each leg is increased an equal amount, resulting in a corresponding increase in leg length.

Fig. 1

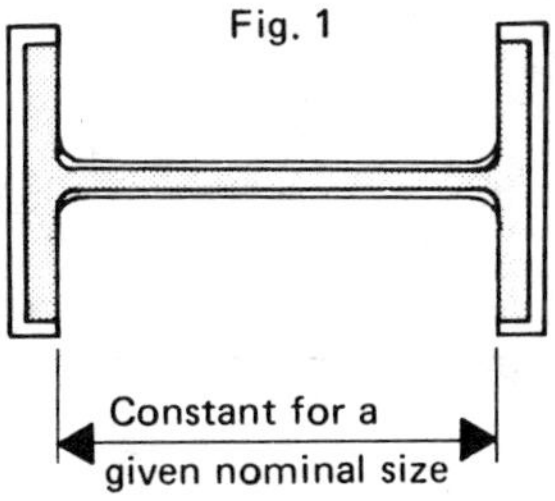

Fig. 2

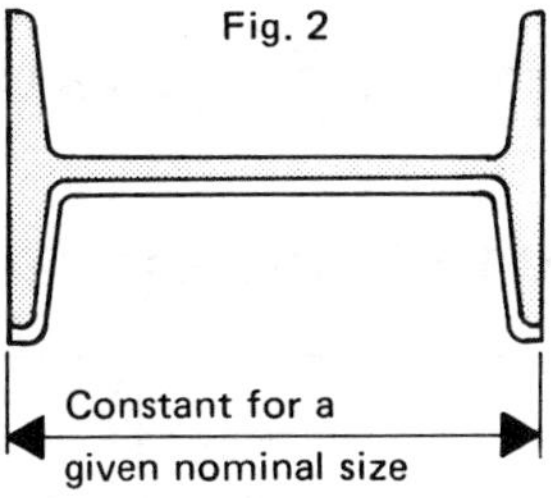

(except S610 and S510)

Fig. 3

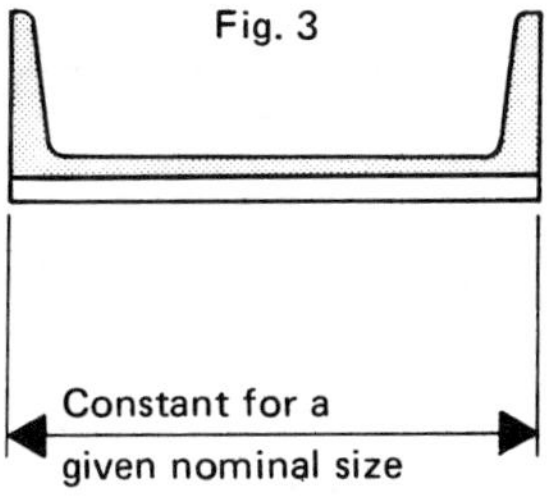

Fig. 4

Tolerances – rolled sections

After a section is rolled, it is cold straightened to meet the specified sweep and camber tolerances. The following table lists Permissible Variations in Straightness for Standard and Wide-Flange Shapes, Bars, and Bar-Size Shapes.

Shapes	Maximum Permissible Variation in Straightness mm	Shape	Maximum Permissible Variation in Straightness, mm
Standard shapes (camber)	$\frac{\text{Length in mm}}{500}$	Wide-flange beams when ordered as columns (camber or sweep):	
Wide-flange beams (camber or sweep)	$\frac{\text{Length in mm}}{1000}$		
Bars and bar-size shapes*	$\frac{\text{Length in mm}}{250}$	Lengths of 14 000 and under	$\frac{\text{Length in mm}}{1000}$, but not more than 10 mm
Steel sheet piling	$\frac{\text{Length in mm}}{1000}$	Lengths over 14 000 mm	$10\text{ mm} + \frac{\text{length} - 14\,000}{1000}$

**Permissible variations do not apply to hot rolled bars if any subsequent heating operation has been performed.*

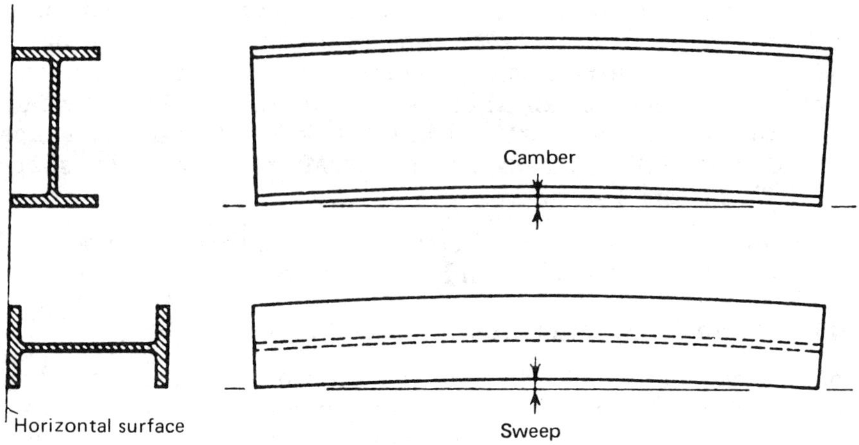

Positions for measuring camber and sweep

Camber is a deflection, approximating a simple regular curve, measured along the depth of a section. It is usually measured half way between two specified points. The length for purposes of determining the "maximum permissible variation" is the distance between the two specified points.

Sweep is a deflection, similar to camber, measured along the width of the section.

WELDED WIDE FLANGE SHAPES

WWF MANUFACTURING TOLERANCES

Nominal Depth mm	Depth A mm		Width of Flange B mm		Combined Warpage and Tilt* mm	Web Off-Centre E mm	Web Flatness** mm	Diagram
	Over	Under	Over	Under	Maximum	Maximum	Maximum	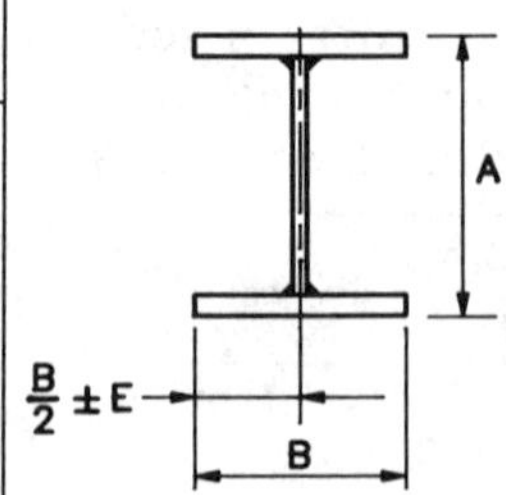
900 mm and under	4	3	6	5	Greater of B/100 or 6	6	A/150	
Over 900 to 2000 mm	5	5	6	5		6		

*The combined warpage and tilt of the flange is measured from the toe of the flange to a line normal to the plane of the web through the intersection of the centreline of the web with the outside surface of the flange plate.

**The deviation from flatness of the web is measured in any length of the web equal to the total depth of the beam.

W SHAPES†

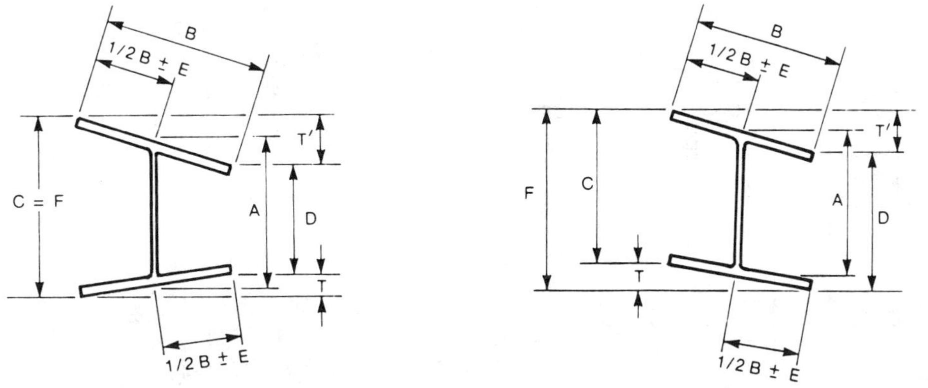

NOTE:
"A" is measured at centreline of web.
"B" is the actual flange width and is measured parallel to the flange.
"F" is measured parallel to the web.

Nominal Depth mm	A Depth mm		B Flange Width mm		T+T′ Out-of-Square mm	C - D Out-of-Parallel mm	E Web Off-Centre mm	F Max. Overall Depth at any Cross-Section mm
	Over	Under	Over	Under	Not Over	Not Over	Not Over	Over Nominal
300 and under*	4	3	6	5	5	5	5	6
Over 300	4	3	6	5	6	6	5	6

*Includes all H-beams rolled on mills having vertical rolls.

PERMISSIBLE VARIATIONS IN LENGTH FOR STANDARD AND WIDE FLANGE SHAPES

Shapes	Variations from Specified Length for Lengths Given, mm									
	To 9 000 incl.		Over 9000 to 12 000 incl.		Over 12 000 to 15 000 incl.		Over 15 000 to 20 000 incl.		Over 20 000	
	Over	Under	Over	Under	Over	Under	Over	Under	Over	Under
All standard sections	13	6	19	6	25	6	29	6	32	6
Wide-flange shapes:										
Beams 610 mm and under in nominal depth.	10	10	13	10	16	10	21	10	24	10
Beams over 610 mm in nominal depth and all columns.	13	13	16	13	19	13	24	13	27	13

When wide flange sections are used as bearing piles, the length tolerance is plus 120 mm, minus 0 mm. See page 6-14 for other tolerances.
†includes HP sections.

S and C SHAPES

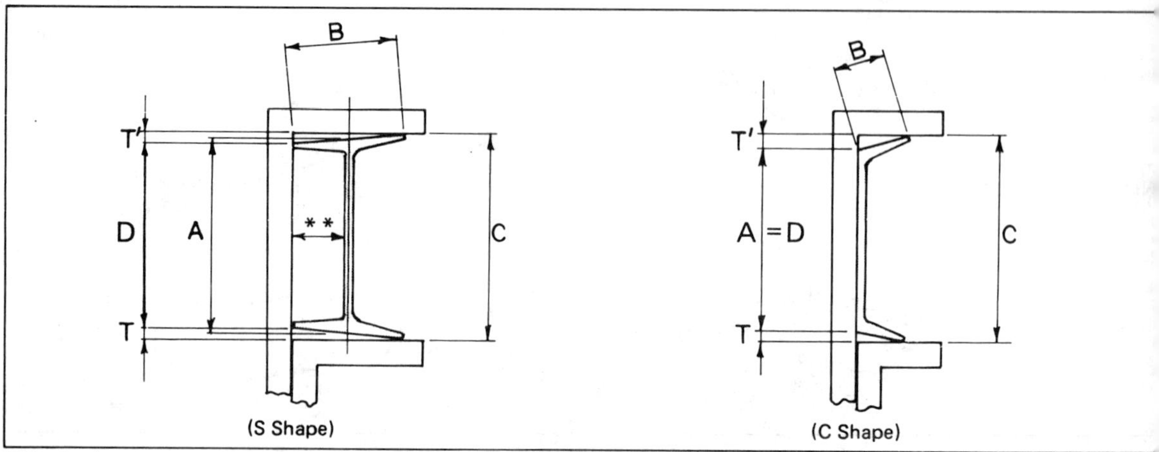

* *Back of square and centreline of web to be parallel when measuring out-of-square

Shape	Nominal Specified Size, mm	Depth, A* mm		Flange Width B mm		Out-of-Square $\frac{T + T'}{B}$	Out-of Parallel $\frac{C\text{-}D}{B}$
		Over	Under	Over	Under		
S Shapes	75 to 180 incl.	2	2	4	3	0.03	0.03
	over 180 to 360 incl.	4	2	4	4	0.03	0.03
	over 360 to 610 incl.	5	3	5	5	0.03	0.03
Channels	75 to 180 incl.	3	2	3	3	0.03	0.03
	over 180 to 360 incl.	3	3	3	4	0.03	0.03
	over 360	5	4	3	5	0.03	0.03

*A is measured at centreline of web for beams and at back of web for channels.

CUTTING TOLERANCES

Section	Variations from Specified Length for Lengths Giving mm									
	to 9 000 mm inclusive		Over 9 000 to 12 000 mm inclusive		Over 12 000 to 15 000 mm inclusive		Over 15 000 to 20 000 mm inclusive		Over 20 000 mm	
	Over	Under	Over	Under	Over	Under	Over	Under	Over	Under
S Shapes and American Standard Channels	13	6	19	6	25	6	29	6	32	6

OTHER TOLERANCES FOR W, S & C SHAPES

Area and Mass Variation: ± 2.5% theoretical or specified amount.
Ends Out-of-Square: S shapes and channels 0.016 mm per mm of depth.

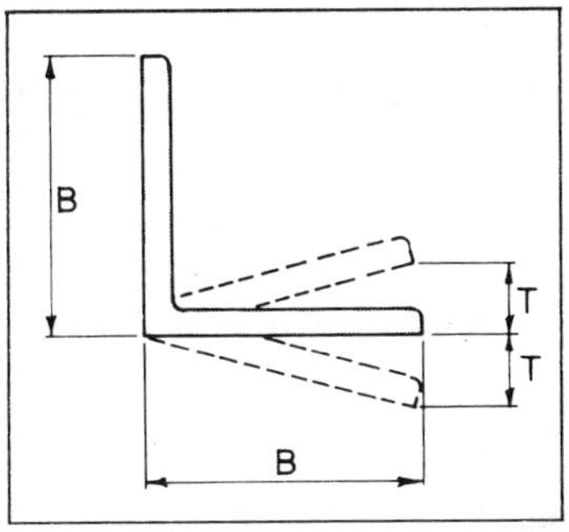

ANGLES

Angle Tolerances (See Tables below)

Area and Mass Variation :
± 2.5% theoretical or specified amount

Ends Out-of-Square:
0.024 mm per mm of leg length or 1½ degrees based on longer leg dimension for unegual leg angle.

+Bar size means rolled angle have max. cross-sect. dimension < 75 mm.

BAR SIZE+

Specified Length of Leg mm	Variations from Thickness for Thicknesses Given over and under mm			Variations from Length of Leg Over and Under mm
	5 and under	Over 5 to 10 incl.	Over 10	
50 and under	0.2	0.2	0.3	1
Over 50	0.3	0.4	0.4	2

Notes:
1. The longer leg of an unequal angle determines the size for permissible variation.
2. Permissible out-of-square in either direction, 1-1/2 degrees.

STRUCTURAL SIZE

Shape	Specified size mm	Length of Leg B, mm		Out-of-Square T/B
		Over	Under	
Angles*	75 to 100 incl.	3	2	0.026
	Over 100 to 150 incl.	3	3	0.026
	Over 150	5	3	0.026

Notes:
*For unequal leg angles, longer leg determines classification.

HOLLOW STRUCTURAL SECTIONS

Class of Section

Class H means hollow sections made by:

(i) A seamless or furnace-buttwelded (continuous-welded) or automatic electric-welded process hot formed to final shape; or

(ii) A seamless or automatic electric-welding process producing a continuous weld, and cold formed to final shape, subsequently stress-relieved by heating to a temperature of 450°C or higher, followed by cooling in air.

Class C means hollow sections that are cold formed from a section produced by a seamless process or by an automatic electric-welding process producing a continuous weld.

Cross-Sectional Dimensions

Outside dimensions measured across the flats or diameter at least 50 mm from either end of a piece, including an allowance for convexity or concavity, shall not vary from the specified dimensions of the section by more than the prescribed tolerances.

Largest Outside Dimension Across Flats or Diameter mm	Tolerance* mm
To 65	±0.5
Over 65 – 90 incl.	±0.8
Over 90 – 140 incl.	±1.0
Over 140	±1 per cent

** Tolerance includes allowance for convexity or concavity. Tolerance may be increased by 50 per cent when applied to the smaller dimension of rectangular sections whose ratio of cross sectional dimensions is between 1.5 and 3, and by 100 per cent when this ratio exceeds 3.*

Mass Variation

Based on a mass density of 7850 kg/m^3, the actual mass shall not deviate from the published mass by more than minus 3.5 or plus 10 per cent.

Wall Thickness

Not more than plus or minus 10 per cent from the nominal specified wall thickness, except for the weld seam.

Maximum Outside Corner radius

Wall Thickness mm	Maximum Outside Corner Radii, mm	
	Perimeter to 700 mm Incl.	Perimeter Over 700 mm
To 3 incl. Over 3 – 4 incl.	6 8	– –
Over 4 – 5 incl. Over 5 – 6 incl. Over 6 – 8 incl.	15 18 21	– 18 24
Over 8 – 10 incl. Over 10 – 13 incl. Over 13	27 36 –	30 39 3 x wall thickness

Corner squareness

For rectangular sections, corners shall be square (90°) within plus or minus 1° for hot formed sections and plus or minus 2° for cold formed sections, with the average slope of the sides being the basis for determination.

Straightness Variation

Deviation from straightness in millimetres shall not exceed total length in millimetres divided by 500.

Permissible Twist

Twist of a rectangular section measured by holding down the side of one end of the section on a flat surface and noting the height above the surface of either corner at the opposite end of that side shall not exceed the prescribed tolerances.

Largest Outside Dimension mm	Maximum Twist per 1000 mm of Length mm
To 65 incl.	0.8
Over 65 – 105 incl.	1.0
Over 105 – 155 incl.	1.1
Over 155 – 205 incl.	1.2
Over 205	1.4

Cutting Tolerances

Tolerances on ordered cold cut lengths are:
- plus 12 and minus 6 millimetres for lengths 7500 mm and under;
- plus 18 and minus 6 millimetres for lenghts over 7500 mm.

Tolerances on ordered hot cut lengths of hot rolled sections are:
- plus or minus 25 millimetres for lengths 7500 mm and under;
- plus or minus 50 millimetres for lengths over 7500 mm.

PRINCIPAL SOURCES OF STRUCTURAL STEEL SECTIONS

General

Standard Canadian and North American sections can be supplied by a number of steel mills in Canada and elsewhere. Principal sources for the various section sizes listed in this Handbook are indicated below. Canadian steel mills produce a wide range of structural sections for the Canadian market; however, most of the large rolled wide flange sections, large angles, and the lighter rolled sections are imported.

Canadian Sizes

Structural sections available from Canadian mills are listed below. Canadian sizes should be specified to the CAN/CSA G40.21 material standards, and generally are the most readily available and provide the greatest overall economy. When Canadian sizes are obtained from non-Canadian sources, the material should be supplied to the CAN/CSA G40.21 Standard. If material substitutions are offered, the sections must be designed using the minimum specified strength levels of the substitute material.

STRUCTURAL SECTIONS AVAILABLE FROM CANADIAN MILLS

W610 x 241—84	WRF — All sizes listed	ANGLES — All sizes less than 200mm in leg size
W530 x 138—72[1]	WWF — All sizes listed	
W460 x 106—61[2]		
W410 x 85—39	C250 x 37—23	HSS — All sizes listed
W360 x 79—33	C230 x 30—20	
W310 x 283—31	C200 x 28—17	Pipe — All Standard sizes
W250 x 167—24[3]	C180 x 18—15	
W200 x 100—21[4]	C150 x 19—12	
W150 x 37—22[5]	C130 x 17—10	
	C100 x 11—8	
	C75 x 9—6	
HP310 x 174—79[6]		
HP250 x 85—62		
HP200 x 54		

[1] Except W530 x 85 and W530 x 74.
[2] Except W460 x 68 and W460 x 60.
[3] Except W250 x 28 and W250 x 25.
[4] Except W200 x 22.
[5] Except W150 x 24
[6] Except HP310 x 125

Non-Canadian Sizes

Structural sections not available from Canadian mills are identified by an asterisk(*) in the Designation Tables on pages 6-21 to 6-25, and in the Tables of Section Properties on pages 6-44 to 6-74. When non-Canadian sizes are required, generally the material should also be specified to the CAN/CSA G40.21 Standards, and if material substitutions are offered the sections must be designed using the minimum specified strength levels applicable to the substitute material.

Principal Sources

Some of the more common sources (for Canada) of structural sections are listed below. Producers' catalogues should be consulted for more information and details about particular sections produced.

The Algoma Steel Corporation Limited
Bethlehem Steel Corp.*
Chaparral Steel*
Sidbec-DOSCO
British Steel Canada Inc.*
Trade ARBED Canada Inc.*
LASCO
Manitoba Rolling Mills (Canada) Limited
Northwestern Steel and Wire*
Sonco
Prudential Steel
Ipsco Inc.
Stelco, Inc.
Standard Tube Canada Limited
USS Division of USX Corp.*
LTV Steel Co.*
Western Canada Steel Limited
Nucor-Yamata Steel Co.*

* *non-Canadian sources*

Availability of Rolled Shapes

Section sizes are generally produced by steel mills according to rolling schedules. For W-shapes produced in Canada, the rollings for a nominal size range typically cycle every four to eight weeks depending on demand. Steel mills and warehouses carry various inventories, usually of the more commonly used sections, and serve as a buffer between rolling cycles to provide ready availability of material. The designer should consider material availability when specifying section sizes, particularly for the heavier mass per metre sizes in a nominal size range and for small quantities of the less commonly used sizes.

The tables of properties and dimensions highlight those section sizes which are more commonly used, and which are generally readily available. Other section sizes are available from mill rollings. Guidance as to the availability of particular sizes can be obtained from local steel fabricators and from steel mill and warehouse sources.

Availability of Angles

Tables of properties and dimensions for imperial series angles on pages 6-76 to 6-80, and for metric series angles on pages 6-88 to 6-91 provide guidance as to the availability of angle sections, based on a rating scale of 1 to 5, where:

1 = angles frequently rolled in Canada,

2 = angles less frequently rolled in Canada,

3 = angles infrequently rolled (i.e. very occasionally) in Canada,

4 = angles not rolled in Canada.

5 = contact mills for information.

Metric series angles are less readily available than imperial series angles, and the designer should check whether they can be obtained, or specify equivalent imperial series angles.

METRIC AND IMPERIAL DESIGNATIONS

General

In Canada, the official designation for structural steel sections for purposes of design, detailing and ordering material, is the metric (SI) designation. For W and HP shapes, angles, and hollow structural sections this is described in the CAN3-G312.3-M78 Standard. Approved (1990) revisions to this Standard with respect to designations for angles and hollow structural sections are included in this handbook. Canadian and North American sections may also be defined using imperial designations. However, all tables of properties and dimensions, and all design tables included elsewhere in this handbook provide only Canadian metric designations, metric properties and metric design information. Tables on pages 6-21 to 6-32 list Canadian (SI) Designations and corresponding Imperial Designations. Shapes identified with an asterisk (*) in these tables are not produced by Canadian mills.

WWF Shapes

Canadian WWF shapes, page 6-21 are hard metric shapes and the Canadian (SI) designation is the total depth in millimetres times the mass in kilograms per metre. The corresponding imperial designation is expressed as inches x lb./ft.

W Shapes

Canadian (SI) designations and imperial designations for W shapes are given on pages 6-22 and 6-23. The Canadian (SI) designation is the nominal depth in millimetres times the mass in kilograms per metre. The corresponding imperial designation is generally the ASTM A6 designation and is expressed as inches x lb./ft.

HP Shapes, S Shapes, M shapes

Canadian (SI) designations and imperial designations for HP, S and M shapes are given on page 6-24. The Canadian (SI) designation is the nominal depth in millimetres times the mass in kilograms per metre and is generally soft converted from the corresponding imperial designation expressed as inches x lb./ft.

C Shapes, MC Shapes

Canadian (SI) designations and imperial designations for C and MC shapes are given on page 6-25. The Canadian (SI) designation is the nominal depth in millimetres times the mass in kilograms per metre and is soft converted from the corresponding imperial designation expressed as inches x lb./ft.

Angles (L)

Revisions to the G312.3 Standard include a metric angle series and an imperial angle series. The metric series is hard metric with the designation being the leg lengths and thickness in millimetres. The designation for the imperial series is a soft conversion of the imperial size, expressed as leg lengths in millimetres and thickness in millimetres to two significant figures. Designations for equal leg angles are given on pages 6-26 and 6-27, and for unequal leg angles on pages 6-28 and 6-29.

Hollow Structural Sections (HSS)

Revisions to the G312.3 Standard for square, rectangular and round hollow structural sections provide for Canadian (SI) designations to be expressed as the nominal outside dimensions in millimetres times the thickness in millimetres to two significant figures. The imperial designation (size) is the outside dimensions times the thickness, both in inches. Designations for rectangular HSS, square HSS and round HSS are given on pages 6-30, 6-31 and 6-32.

NOTES

DESIGNATION TABLE FOR WWF SHAPES

Canadian (SI) Designation (mm x kg/m)	Imperial Designation (in. × lb./ft.)
WWF2000X732	WWF79X490
X648	X436
X607	X408
X542	X364
WWF1800X700	WWF71X470
X659	X442
X617	X415
X575	X388
X510	X344
WWF1600X622	WWF63X419
X580	X388
X538	X361
X496	X333
X431	X289
WWF1400X597	WWF55X402
X513	X344
X471	X316
X405	X272
X358	X240
WWF1200X487	WWF47X326
X418	X281
X380	X255
X333	X224
X302	X203
X263	X176
WWF1100X458	WWF43X307
X388	X260
X351	X236
X304	X204
X273	X184
X234	X157
WWF1000X447	WWF39X300
X377	X253
X340	X228
X293	X197
X262	X176
X223	X150
X200	X134
WWF900X417	WWF35X279
X347	X233
X309	X208
X262	X176
X231	X156
X192	X128
X169	X113
WWF800X339	WWF31X228
X300	X202
X253	X170
X223	X150
X184	X123
X161	X108
WWF700X245	WWF28X164
X214	X144
X196	X132
X175	X117
X152	X102
WWF650X864	WWF26X580
X739	X497
X598	X402
X499	X336
X400	X269
WWF600X793	WWF24X531
X680	X456
X551	X371
X460	X309
X369	X248
WWF550X721	WWF22X484
X620	X416
X503	X338
X420	X282
X280	X188
WWF500X651	WWF20X437
X561	X377
X456	X306
X381	X256
X343	X230
X306	X205
X276	X185
X254	X170
X223	X150
X197	X132
WWF450X503	WWF18X337
X409	X275
X342	X229
X308	X207
X274	X184
X248	X166
X228	X152
X201	X134
X177	X119
WWF400X444	WWF16X298
X362	X243
X303	X203
X273	X183
X243	X163
X220	X147
X202	X135
X178	X119
X157	X105
WWF350X315	WWF14X211
X263	X177
X238	X159
X212	X142
X192	X128
X176	X118
X155	X104
X137	X92

DESIGNATION TABLE FOR W SHAPES

Canadian (SI) Designation (mm x kg/m)	Imperial Designation (in. × lb./ft.)
W1000X488*	W40X328
X444*	X298
X399*	X268
X363*	X244
X328*	X221
X286*	X192
W1000X976*	W40X655
X883*	X593
X790*	X531
X714*	X480
X649*	X436
X591*	X397
X539*	X362
X483*	X324
X443*	X297
X412*	X277
X371*	X249
X321*	X215
X296*	X199
W1000X272*	W40X183
X249*	X167
X222*	X149
W920X1262*	W36X848
X1188*	X798
X1072*	X720
X967*	X650
X876*	X588
X784*	X527
X722*	X485
X653*	X439
X585*	X393
X534*	X359
X488*	X328
X446*	X300
X417*	X280
X387*	X260
X365*	X245
X342*	X230
W920X313*	W36X210
X289*	X194
X271*	X182
X253*	X170
X238*	X160
X223*	X150
X201*	X135
W840X922*	W33X619
X845*	X567
X767*	X515
X697*	X468
X631*	X424
X577*	X387
X527*	X354
X473*	X318
X433*	X291
X392*	X263
X359*	X241
X329*	X221
X299*	X201

Canadian (SI) Designation (mm x kg/m)	Imperial Designation (in. × lb./ft.)
W840X226*	W33X152
X210*	X141
X193*	X130
X176*	X118
W760X865*	W30X581
X783*	X526
X710*	X477
X644*	X433
X582*	X391
X531*	X357
X484*	X326
X434*	X292
X389*	X261
X350*	X235
X314*	X211
X284*	X191
X257*	X173
W760X196*	W30X132
X185*	X124
X173*	X116
X161*	X108
X147*	X99
X134*	X90
W690X802*	W27X539
X735*	X494
X667*	X448
X605*	X407
X548*	X368
X500*	X336
X457*	X307
X419*	X281
X384*	X258
X350*	X235
X323*	X217
X289*	X194
X265*	X178
X240*	X161
X217*	X146
W690X170*	W27X114
X152*	X102
X140*	X94
X125*	X84
W610X732*	W24X492
X670*	X450
X608*	X408
X551*	X370
X498*	X335
X455*	X306
X415*	X279
X372*	X250
X341*	X229
X307*	X207
X285*	X192
X262*	X176
X241	X162
X217	X146
X195	X131
X174	X117
X155	X104

Canadian (SI) Designation (mm x kg/m)	Imperial Designation (in. × lb./ft.)
W610X140	W24X94
X125	X84
X113	X76
X101	X68
X91	X61
X84	X56
W610X92*	W24X62
X82*	X55
W530X599*	W21X402
X543*	X364
X496*	X333
X447*	X300
X409*	X275
X370*	X248
X331*	X223
X300*	X201
X272*	X182
X248*	X166
X219*	X147
X196*	X132
X182*	X122
X165*	X111
X150*	X101
W530X138	W21X93
X123	X83
X109	X73
X101	X68
X92	X62
X82	X55
X72	X48
W530X85*	W21X57
X74*	X50
X66*	X44
W460X464*	W18X311
X421*	X283
X384*	X258
X349*	X234
X315*	X211
X286*	X192
X260*	X175
X235*	X158
X213*	X143
X193*	X130
X177*	X119
X158*	X106
X144*	X97
X128*	X86
X113*	X76
W460X106	W18X71
X97	X65
X89	X60
X82	X55
X74	X50
X67	X45
X61	X41

DESIGNATION TABLE FOR W SHAPES

Canadian (SI) Designation (mm x kg/m)	Imperial Designation (in. × lb./ft.)
W460X68*	W18X46
X60*	X40
X52*	X35
W410X149*	W16X100
X132*	X89
X114*	X77
X100*	X67
W410X85	W16X57
X74	X50
X67	X45
X60	X40
X54	X36
W410X46	W16X31
X39	X26
W360X1086*	W14X730
X990*	X665
X900*	X605
X818*	X550
X744*	X500
X677*	X455
W360X634*	W14X426
X592*	X398
X551*	X370
X509*	X342
X463*	X311
X421*	X283
X382*	X257
X347*	X233
X314*	X211
X287*	X193
X262*	X176
X237*	X159
X216*	X145
W360X196*	W14X132
X179*	X120
X162*	X109
X147*	X99
X134*	X90
W360X122*	W14X82
X110*	X74
X101*	X68
X91*	X61
W360X79	W14X53
X72	X48
X64	X43
W360X57	W14X38
X51	X34
X45	X30
W360X39	W14X26
X33	X22
W310X500*	W12X336
X454*	X305
X415*	X279
X375*	X252
X342*	X230
X313*	X210
W310X283*	W12X190
X253*	X170
X226	X152
X202	X136
X179	X120
X158	X106
X143	X96
X129	X87
X118	X79
X107	X72
X97	X65
W310X86	W12X58
X79	X53
W310X74	W12X50
X67	X45
X60	X40
W310X52	W12X35
X45	X30
X39	X26
X31	X21
W310X33*	W12X22
X28*	X19
X24*	X16
X21*	X14
W250X167	W10X112
X149	X100
X131	X88
X115	X77
X101	X68
X89	X60
X80	X54
X73	X49
W250X67	W10X45
X58	X39
X49	X33
W250X45	W10X30
X39	X26
X33	X22
X24	X16
W250X28*	W10X19
X25*	X17
X22*	X15
X18*	X12
W200X100	W8X67
X86	X58
X71	X48
X59	X40
X52	X35
X46	X31
W200X42	W8X28
X36	X24
W200X31	W8X21
X27	X18
X21	X14
W200X22*	W8X15
X19*	X13
X15*	X10
W150X37	W6X25
X30	X20
X22	X15
W150X24*	W6X16
X18*	X12
X14*	X9
W130X28*	W5X19
X24*	X16

DESIGNATION TABLE FOR HP SHAPES

Old HP Series #	NEW HP SERIES	
Designation (in. x lb./ft.)	Canadian Metric Designation† (mm x kg/m)	ASTM Imperial Designation (in. x lb./ft.)
HP14x117	**HP360x174***	HP14x117
x102	**x152***	x102
x89	**x132***	x89
x73	**x108***	x73
–	**HP330x149***	HP13x100
–	**x129***	x87
–	**x109***	x73
–	**x89***	x60
HP12x117	**HP310x174**[1]	–
x102	**x152**[1]	–
x89	**x132**[1]	–
–	**x125***	HP12x84
x74	**x110**	x74
–	**x94**	x63
x53	**x79**	x53
HP10x57	**HP250x85**	HP10x57
x42	**x62**	x42
HP8x36	**HP200x54**	HP8x36

†Nominal depth in millimetres and mass in kilograms per metre.
#North American mills discontinued this HP series on 1978 09 01.
*Not available from Canadian mills.
[1] Produced only by Algoma.

DESIGNATION TABLE FOR M SHAPES

Old Designation (in. x lb./ft.)	Metric Designation† (mm x kg/m)
M12x11.8	**M310x17.6***
M10x9	**M250x13.4***
M8x6.5	**M200x9.7***
M6x20	**M150x29.8***
M6x4.4	**M150x6.5***
M5x18.9	**M130x28.1***
M4x13	**M100x19***

†Nominal depth in millimetres and mass in kilograms per metre.
*Not available from Canadian Mills.

Note: These shapes have been soft converted from Imperial to SI dimensions.

DESIGNATION TABLE FOR S SHAPES

Old Designation (in. x lb./ft.)	Metric Designation† (mm x kg/m)
S24x121	**S610x180***
x106	**x158***
S24x100	**S610x149***
x90	**x134***
x80	**x119***
S20x96	**S510x143***
x86	**x128***
S20x75	**S510x112***
x66	**x98.2***
S18x70	**S460x104***
x54.7	**x81.4***
S15x50	**S380x74***
x42.9	**x64***
S12x50	**S310x74***
x40.8	**x60.7***
S12x35	**S310x52***
x31.8	**x47***
S10x35	**S250x52***
x25.4	**x38***
S8x23	**S200x34***
x18.4	**x27***
S7x20	**S180x30***
x15.3	**x22.8***
S6x17.25	**S150x26***
x12.5	**x19***
S5x14.75	**S130x22***
x10	**x15***
S4x9.5	**S100x14.1***
x7.7	**x11***
S3x7.5	**S75x11***
x5.7	**x8***

†Nominal depth in millimetres and mass in kilograms per metre.
*Not available from Canadian Mills.

Note: These shapes have been soft converted from Imperial to SI dimensions.

DESIGNATION TABLE FOR C SHAPES

Old Designation (in. x lb./ft.)	Metric Designation† (mm x kg/m)
C15x50	C380x74*
x40	x60*
x33.9	x50*
C12x30	C310x45*
x25	x37*
x20.7	x31*
C10x30	C250x45*
x25	x37
x20	x30
x15.3	x23
C9x20	C230x30
x15	x22
x13.4	x20
C8x18.75	C200x28
x13.75	x21
x11.5	x17
C7x14.75	C180x22
x12.25	x18
x9.8	x15
C6x13	C150x19
x10.5	x16
x8.2	x12
C5x9	C130x13
x6.7	x10
C4x7.25	C100x11
x5.4	x8
C3x6	C75x9
x5	x7
x4.1	x6

†Nominal depth in millimetres and mass in kilograms per metre.

*Not available from Canadian Mills.

Note: These shapes have been soft converted from Imperial to SI dimensions.

DESIGNATION TABLE FOR MC SHAPES

Old Designation (in. x lb./ft.)	Metric Designation† (mm x kg/m)
MC18x58	MC460x86 *
x51.9	x77.2 *
x45.8	x68.2 *
x42.7	x63.5 *
MC13x50	MC330x74 *
x40	x60 *
x35	x52 *
x31.8	x47.3 *
MC12x50	MC310x74 *
x45	x67 *
x40	x60 *
x35	x52 *
x30.9	x46 *
MC12x10.6	MC310x15.8 *
MC10x41.1	MC250x61.2 *
x33.6	x50 *
x28.5	x42.4 *
MC10x25	MC250x37*
x22	x33
MC10x8.4	MC250x12.5 *
MC10x6.5	MC250x9.7 *
MC9x25.4	MC230x37.8 *
x23.9	x35.6 *
MC8x22.8	MC200x33.9 *
x21.4	x31.8 *
MC8x20	MC200x29.8 *
x18.7	x27.8 *
MC8x8.5	MC200x12.6 *
MC7x22.7	MC180x33.8 *
x19.1	x28.4 *
MC7x17.6	MC180x26.2 *
MC6x18	MC150x26.8 *
x15.3	x22.8
MC6x16.3	MC150x24.3 *
x15.1	x22.5 *
MC6x12	MC150x17.9 *

†Nominal depth in millimetres and mass in kilograms per metre.

*Not available from Canadian Mills.

Note: These shapes have been soft converted from Imperial to SI dimensions.

DESIGNATION TABLE FOR EQUAL LEG ANGLES

Metric Series	Imperial Series	
Designation	Designation	Size
mm x mm x mm	mm x mm x mm	in. x in. x in.
L 200 x 200	L 203 x 203	8 x 8
x 30		
	x 29	x 1⅛
	x 25	x 1
x 25		
	x 22	x ⅞
x 20		
	x 19	x ¾
x 16		
	x 16	x ⅝
	x 14	x 9/16
x 13		
	x 13	x ½
x 10		
L 150 x 150	L 152 x 152	6 x 6
	x 25	x 1
	x 22	x ⅞
x 20		
	x 19	x ¾
x 16		
	x 16	x ⅝
	x 14	x 9/16
x 13		
	x 13	x ½
	x 11	x 7/16
x 10		
	x 9.5	x ⅜
	x 7.9	x 5/16
L 125 x 125	L 127 x 127	5 x 5
	x 22	x ⅞
	x 19	x ¾
x 16		
	x 16	x ⅝
x 13		
	x 13	x ½
	x 11	x 7/16
x 10		
	x 9.5	x ⅜
x 8		
	x 7.9	x 5/16
	x 6.4	x ¼
	x 4.8	x 3/16

Metric Series	Imperial Series	
Designation	Designation	Size
mm x mm x mm	mm x mm x mm	in. x in. x in.
L 100 x 100	L 102 x 102	4 x 4
	x 19	x ¾
x 16		
	x 16	x ⅝
x 13		
	x 13	x ½
	x 11	x 7/16
x 10		
	x 9.5	x ⅜
x 8		
	x 7.9	x 5/16
	x 6.4	x ¼
x 6		
	x 4.8	x 3/16
L 90 x 90	L 89 x 89	3½ x 3½
	x 16	x ⅝
x 13		
	x 13	x ½
	x 11	x 7/16
x 10		
	x 9.5	x ⅜
x 8		
	x 7.9	x 5/16
	x 6.4	x ¼
x 6		
	x 4.8	x 3/16
L 75 x 75	L 76 x 76	3 x 3
	x 16	x ⅝
x 13		
	x 13	x ½
	x 11	x 7/16
x 10		
	x 9.5	x ⅜
x 8		
	x 7.9	x 5/16
	x 6.4	x ¼
x 6		
x 5		
	x 4.8	x 3/16
L 65 x 65	L 64 x 64	2½ x 2½
	x 13	x ½
x 10		
	x 9.5	x ⅜
x 8		
	x 7.9	x 5/16
	x 6.4	x ¼
x 6		
x 5		
	x 4.8	x 3/16
	x 3.2	x ⅛

DESIGNATION TABLE FOR EQUAL LEG ANGLES

Metric Series	Imperial Series	
Designation	Designation	Size
mm x mm x mm	mm x mm x mm	in. x in. x in.
L 55 x 55	L 51 x 51	2 x 2
x 10		
	x 9.5	x ⅜
x 8		
	x 7.9	x 5/16
	x 6.4	x ¼
x 6		
x 5		
	x 4.8	x 3/16
x 4		
	x 3.2	x ⅛
x 3		
L 45 x 45	L 44 x 44	1¾ x 1¾
	x 9.5	x ⅜
x 8		
	x 7.9	x 5/16
	x 6.4	x ¼
x 6		
x 5		
	x 4.8	x 3/16
x 4		
	x 3.2	x ⅛
x 3		

Metric Series	Imperial Series	
Designation	Designation	Size
mm x mm x mm	mm x mm x mm	in. x in. x in.
	L 38 x 38	1½ x 1½
	x 9.5	x ⅜
	x 7.9	x 5/16
	x 6.4	x ¼
	x 4.8	x 3/16
	x 4.0	x 5/32
	x 3.2	x ⅛
L 35 x 35		
x 6		
x 5		
x 4		
x 3		
	L 32 x 32	1¼ x 1¼
	x 9.5	x ⅜
	x 7.9	x 5/16
	x 6.4	x ¼
	x 4.8	x 3/16
	x 3.2	x ⅛
L 25 x 25	L 25 x 25	1 x 1
	x 9.5	x ⅜
x 5	x 7.9	x 5/16
	x 6.4	x ¼
x 4		
	x 4.8	x 3/16
x 3		
	x 3.2	x ⅛

DESIGNATION TABLE FOR UNEQUAL LEG ANGLES

Metric Series	Imperial Series	
Designation	Designation	Size
mm x mm x mm	mm x mm x mm	in. x in. x in.
L 200 x 150	L 203 x 152	8 x 6
	x 25	x 1
x 25		
	x 22	x ⅞
x 20		
	x 19	
x 16		x ¾
	x 16	x ⅝
	x 14	x 9/16
x 13		
	x 13	x ½
	x11	x 7/16
L 200 x 100	L 203 x 102	8 x 4
	x 25	x 1
x 20		
	x 19	x ¾
x 16		
	x 14	x 9/16
x 13		
	x 13	x ½
x 10		
	L 178 x 102	7 x 4
	x 19	x ¾
	x 16	x ⅝
	x 13	x ½
	x 9.5	x ⅜
L 150 x 100	L 152 x 102	6 x 4
	x 22	x ⅞
	x 19	x ¾
x 16		
	x 16	x ⅝
	x 14	x 9/16
x 13		
	x 13	x ½
	x 11	x 7/16
x 10		
	x 9.5	x ⅜
x 8		
	x 7.9	x 5/16
	x 6.4	x ¼
	x 4.8	x 3/16
	L 152 x 89	6 x 3½
	x 16	x ⅝
	x 13	x ½
	x 9.5	x ⅜
	x 7.9	x 5/16
	x 6.4	x ¼
L 125 x 90	L 127 x 89	5 x 3½
	x 19	x ¾
x 16		
	x 16	x ⅝
x 13		
	x 13	x ½
	x 11	x 7/16
x 10		
	x 9.5	x ⅜
x 8		
	x 7.9	x 5/16
	x 6.4	x ¼
	x 4.8	x 3/16
L 125 x 75	L 127 x 76	5 x 3
	x 16	x ⅝
x 13		
	x 13	x ½
	x 11	x 7/16
x 10		
	x 9.5	x ⅜
x 8		
	x 7.9	x 5/16
	x 6.4	x ¼
x 6		
	x 4.8	x 3/16
L 100 x 90	L 102 x 89	4 x 3½
	x 16	x ⅝
x 13		
	x 13	x ½
	x 11	x 7/16
x 10		
	x 9.5	x ⅜
x 8		
	x 7.9	x 5/16
	x 6.4	x ¼
x 6		
	x 4.8	x 3/16
L 100 x 75	L 102 x 76	4 x 3
	x 16	x ⅝
x 13		
	x 13	x ½
	x 11	x 7/16
x 10		
	x 9.5	x ⅜
x 8		
	x 7.9	x 5/16
	x 6.4	x ¼
x 6		
	x 4.8	x 3/16

DESIGNATION TABLE FOR UNEQUAL LEG ANGLES

Metric Series	Imperial Series	
Designation	Designation	Size
mm x mm x mm	mm x mm x mm	in. x in. x in.
L 90 x 75	L 89 x 76	3½ x 3
	x 16	x 5/8
x 13		
	x 13	x 1/2
	x 11	x 7/16
x 10		
	x 9.5	x 3/8
x 8		
	x 7.9	x 5/16
	x 6.4	x 1/4
x 6		
x 5		
	x 4.8	x 3/16
L 90 x 65	L 89 x 64	3½ x 2½
	x 16	x 5/8
	x 13	x 1/2
	x 11	x 7/16
x 10		
	x 9.5	x 3/8
x 8		
	x 7.9	x 5/16
	x 6.4	x 1/4
x 6		
x 5		
	x 4.8	x 3/16
L 80 x 60		
x 10		
x 8		
x 6		
x 5		
	L 76 x 64	3 x 2½
	x 16	x 5/8
	x 13	x 1/2
	x 11	x 7/16
	x 9.5	x 3/8
	x 7.9	x 5/16
	x 6.4	x 1/4
	x 4.8	x 3/16
L 75 x 50	L 76 x 51	3 x 2
	x 16	x 5/8
	x 13	x 1/2
	x 11	x 7/16
	x 9.5	x 3/8
x 8		
	x 7.9	x 5/16
	x 6.4	x 1/4
x 6		
x 5		
	x 4.8	x 3/16

Metric Series	Imperial Series	
Designation	Designation	Size
mm x mm x mm	mm x mm x mm	in. x in. x in.
L 65 x 50	L 64 x 51	2½ x 2
	x 9.5	x 3/8
x 8		
	x 7.9	x 5/16
x 6	x 6.4	x 1/4
x 5		
	x 4.8	x 3/16
x 4		
	x 3.2	x1/8
	L 64 x 38	2½ x 1½
	x 7.9	x 5/16
	x 6.4	x 1/4
	x 4.8	x 3/16
L 55 x 35		
x 6		
x 5		
x 4		
x 3		
	L 51 x 38	2 x 1½
	x 9.5	x 3/8
	x 7.9	x 5/16
	x 6.4	x 1/4
	x 4.8	x 3/16
	x 3.2	x1/8
L 45 x 30		
x 6		
x 5		
x 4		
x 3		

DESIGNATION TABLE FOR RECTANGULAR HSS

Canadian (SI) Designation	Imperial Size	
	Size	Thickness
mm x mm x mm	in. x in.	in.
HSS 51 x 25	2 x 1	
x 2.5		.110
x 3.2		.125
HSS 76 x 51	3 x 2	
x 3.8		.150
x 4.8		.188
x 6.4		.250
x 8.0		.313
HSS 89 x 64	3½ x 2½	
x 3.2		.125
x 3.8		.150
x 4.8		.188
x 6.4		.250
x 8.0		.313
HSS 102 x 51	4 x 2	
x 3.2		.125
x 3.8		.150
x 4.8		.188
x 6.4		.250
x 8.0		.313
HSS 102 x 76	4 x 3	
x 4.8		.188
x 6.4		.250
x 8.0		.313
x 9.5		.375
HSS 127 x 51	5 x 2	
x 4.8		.188
x 6.4		.250
x 8.0		.313
x 9.5		.375
HSS 127 x 64	5 x 2½	
x 4.8		.188
x 6.4		.250
x 8.0		.313
x 9.5		.375
HSS 127 x 76	5 x 3	
x 4.8		.188
x 6.4		.250
x 8.0		.313
x 9.5		.375

Canadian (SI) Designation	Imperial Size	
	Size	Thickness
mm x mm x mm	in. x in.	in.
HSS 152 x 102	6 x 4	
x 4.8		.188
x 6.4		.250
x 8.0		.313
x 9.5		.375
x 11		.438
HSS 178 x 127	7 x 5	
x 4.8		.188
x 6.4		.250
x 8.0		.313
x 9.5		.375
x 11		.438
x 13		.500
HSS 203 x 102	8 x 4	
x 4.8		.188
x 6.4		.250
x 8.0		.313
x 9.5		.375
x 11		.438
x 13		.500
HSS 203 x 152	8 x 6	
x 4.8		.188
x 6.4		.250
x 8.0		.313
x 9.5		.375
x 11		.438
x 13		.500
HSS 254 x 152	10 x 6	
x 6.4		.250
x 8.0		.313
x 9.5		.375
x 11		.438
x 13		.500
HSS 305 x 203	12 x 8	
x 6.4		.250
x 8.0		.313
x 9.5		.375
x 11		.438
x 13		.500

DESIGNATION TABLE FOR SQUARE HSS

Canadian (SI) Designation	Imperial Size	
	Size	Thickness
mm x mm x mm	in. x in.	in.
HSS 25 x 25	1 x 1	
x 2.5		.110
x 3.2		.125
HSS 32 x 32	1¼ x 1¼	
x 2.5		.100
x 3.2		.125
x 3.8		.150
HSS 38 x 38	1½ x 1½	
x 2.5		.100
x 3.2		.125
x 3.8		.150
x 4.8		.188
HSS 51 x 51	2 x 2	
x 2.8		.110
x 3.2		.125
x 3.8		.150
x 4.8		.188
x 6.4		.250
HSS 64 x 64	2½ x 2½	
x 3.2		.125
x 3.8		.150
x 4.8		.188
x 6.4		.250
HSS 76 x 76	3 x 3	
x 4.8		.188
x 6.4		.250
x 8.0		.313
HSS 89 x 89	3½ x 3½	
x 4.8		.188
x 6.4		.250
x 8.0		.313
x 9.5		.375
HSS 102 x 102	4 x 4	
x 4.8		.188
x 6.4		.250
x 8.0		.313
x 9.5		.375

Canadian (SI) Designation	Imperial Size	
	Size	Thickness
mm x mm x mm	in. x in.	in.
HSS 127 x 127	5 x 5	
x 4.8		.188
x 6.4		.250
x 8.0		.313
x 9.5		.375
x 11		.438
HSS 152 x 152	6 x 6	
x 4.8		.188
x 6.4		.250
x 8.0		.313
x 9.5		.375
x 11		.438
x 13		.500
HSS 178 x 178	7 x 7	
x 4.8		.188
x 6.4		.250
x 8.0		.313
x 9.5		.375
x 11		.438
x 13		.500
HSS 203 x 203	8 x 8	
x 6.4		.250
x 8.0		.313
x 9.5		.375
x 11		.438
x 13		.500
HSS 254 x 254	10 x 10	
x 6.4		.250
x 8.0		.313
x 9.5		.375
x 11		.438
x 13		.500
HSS 305 x 305	12 x 12	
x 6.4		.250
x 8.0		.313
x 9.5		.375
x 11		.438
x 13		.500

DESIGNATION TABLE FOR ROUND HSS

Canadian (SI) Designation	Imperial Size	
	Diameter	Thickness
mm x mm	in.	in.
HSS 27	1.05	
x 2.5		.110
x 3.2		.125
HSS 33	1.315	
x 2.5		.100
x 3.2		.125
HSS 42	1.66	
x 2.5		.100
x 3.2		.125
HSS 48	1.9	
x 2.8		.110
x 3.2		.125
x 3.8		.150
x 4.8		.188
HSS 60	2.375	
x 3.2		.125
x 3.8		.150
x 4.8		.188
x 6.4		.250
HSS 73	2.875	
x 3.2		.125
x 3.8		.150
x 4.8		.188
x 6.4		.250
HSS 89	3.5	
x 3.8		.150
x 4.8		.188
x 6.4		.250
x 8.0		.313
HSS 102	4	
x 3.8		.150
x 4.8		.188
x 6.4		.250
x 8.0		.313
HSS 114	4.5	
x 4.8		.188
x 6.4		.250
x 8.0		.313

Canadian (SI) Designation	Imperial Size	
	Diameter	Thickness
mm x mm	in.	in.
HSS 141	5.562	
x 4.8		.188
x 6.4		.250
x 8.0		.313
x 9.5		.375
HSS 168	6.625	
x 4.8		.188
x 6.4		.250
x 8.0		.313
x 9.5		.375
HSS 219	8.625	
x 4.8		.188
x 6.4		.250
x 8.0		.313
x 9.5		.375
x 11		.438
x 13		.500
HSS 273	10.75	
x 6.4		.250
x 8.0		.313
x 9.5		.375
x 11		.438
x 13		.500
HSS 324	12.75	
x 6.4		.250
x 8.0		.313
x 9.5		.375
x 11		.438
x 13		.500
HSS 356	14	
x 6.4		.250
x 8.0		.313
x 9.5		.375
x 11		.438
x 13		.500
HSS 406	16	
x 6.4		.250
x 8.0		.313
x 9.5		.375
x 11		.438
x 13		.500

WELDED SHAPES

General

Welded shapes are produced to the requirements of CAN/CSA-G40.20-M Standard, using plate meeting the requirements of CAN/CSA-G40.21-M material Standard. Canadian welded shapes are produced by the Algoma Steel Corporation, Limited and are available in 300W and 350W grade material. If the use of other grades of steel is contemplated, the manufacturer should be consulted. Properties and dimensions of WWF (welded wide flange) shapes and WRF (welded reduced flange) shapes are provided on pages 6-34 to 6-41.

Manufacture

The G40.20 Standard defines welded shapes as I-type sections produced by automatic welding processes from three individual components. Hot rolled plates are flame-cut to required width, then assembled together on a special jig where pressure rolls hold the flange and web plates in contact for automatic submerged arc welding. Welds are made simultaneously along both flange-to-web joints in the horizontal fillet position.

For web thicknesses up to and including 20 mm, the flange-to-web welds develop the full capacity of the flange-to-web joint. For web thicknesses greater than 20 mm, which occurs in the heavier WWF column shapes, the full strength of the web is not developed by the web-to-flange welds, so that additional welding by the purchaser may be required at major connection points. Column WWF shapes with partial strength web welds are identified by an asterisk (*) and the note "welding does not fully develop the web strength for these sections".

Products

WWF beams range in depth from 700 mm to 2000 mm and have equal width flanges which range in width between 300 mm and 550 mm. Flange thickness varies while the overall depth of beams in each size range is constant. The Class in bending for WWF beams for G40.21 grades 300W and 350W steels is given in Table 5-1.

WWF column shapes range in depth and width from 350 mm to 650 mm in increments of 50 mm. Overall depth and flange width in each size range is the same.

WRF shapes are asymmetrical beams in which one flange is narrower than the other. Overall depths vary from 1000 mm to 1800 mm and are constant for all beams in each size range. Flange widths are 300 mm and 550 mm.

Availability of WWF and WRF Shapes

WWF beam and column shapes, and WRF shapes are generally not carried in inventory and are produced on demand. Typical lead time to source welded shapes is eight to ten weeks.

WELDED REDUCED FLANGE SHAPES
WRF1800-WRF1000

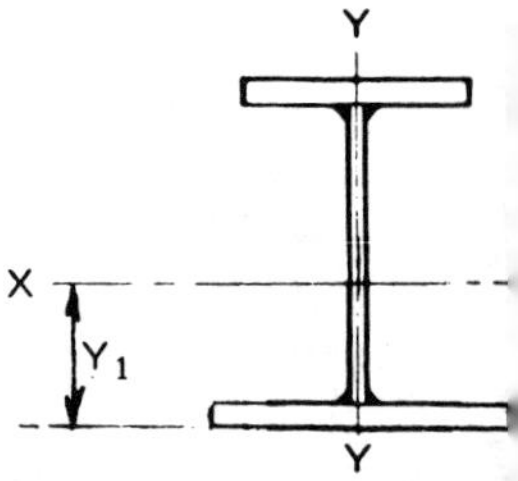

PROPERTIES

Designation ‡	Dead Load	Total Area	Axis X-X					Axis Y-Y			Torsional Constant	Wa Cor
			I_x	S_{xt}	S_{xb}	r_x	Y_1	I_y	S_y	r_y	J	(
	kN/m	mm^2	$10^6 mm^4$	$10^3 mm^3$	$10^3 mm^3$	mm	mm	$10^6 mm^4$	$10^3 mm^3$	mm	$10^3 mm^4$	10^9
WRF1800												
X543	5.33	69 200	35 700	34 200	47 100	718	757	726	2 640	102	29 100	268
X480	4.70	61 100	30 100	29 300	38 900	702	774	565	2 050	96	15 500	211
X416	4.08	53 000	24 300	24 200	30 600	678	795	404	1 470	87	7 830	152
WRF1600												
X491	4.82	62 500	26 600	28 300	40 200	652	660	726	2 640	108	27 900	211
X427	4.19	54 400	22 200	24 000	32 900	639	674	565	2 050	102	14 200	166
X362	3.56	46 200	17 700	19 500	25 500	619	693	403	1 470	93	6 540	120
WRF1400												
X413	4.05	52 600	17 600	21 200	30 800	578	571	645	2 350	111	19 300	143
X348	3.42	44 400	14 200	17 500	24 400	566	584	484	1 760	104	8 880	109
X284	2.78	36 200	10 800	13 500	17 800	545	605	323	1 170	94	3 510	73
WRF1200												
X373	3.66	47 600	12 200	16 900	25 500	506	478	645	2 350	116	18 800	104
X309	3.03	39 300	9 760	13 700	20 000	498	488	484	1 760	111	8 310	79
X244	2.39	31 000	7 240	10 400	14 300	483	505	322	1 170	102	2 940	53
WRF1000												
X340	3.34	43 300	7 980	13 100	20 500	429	389	645	2 340	122	18 400	71
X275	2.70	35 000	6 340	10 500	16 000	426	396	484	1 760	117	7 960	54
X210	2.06	26 700	4 620	7 820	11 300	416	408	322	1 170	110	2 590	37

‡Nominal depth in millimetres and mass in kilograms per metre.

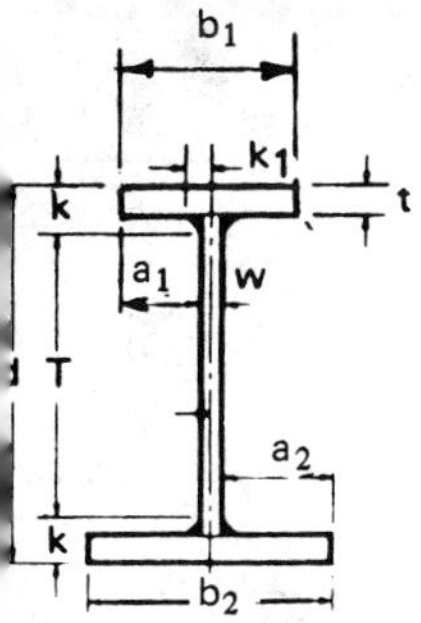

WELDED REDUCED FLANGE SHAPES
WRF1800-WRF1000

DIMENSIONS AND SURFACE AREAS

Nominal Mass	Depth	Flange			Web Thickness	Distances						Surface Area (m^2) per metre of length	
	d	Width b_1	Width b_2	Thickness t	w	a_1	a_2	T	k	k_1	d-2t	Total	Minus Top of Top Flange
kg/m	mm	mm	mm	mm	mm	mm	mm	mm	mm	mm	mm		
543	1 800	300	550	45.0	18.0	141	266	1 688	56	19	1 710	5.26	4.96
480	1 800	300	550	35.0	18.0	141	266	1 708	46	19	1 730	5.26	4.96
416	1 800	300	550	25.0	18.0	141	266	1 728	36	19	1 750	5.26	4.96
491	1 600	300	550	45.0	16.0	142	267	1 492	54	16	1 510	4.87	4.57
427	1 600	300	550	35.0	16.0	142	267	1 512	44	16	1 530	4.87	4.57
362	1 600	300	550	25.0	16.0	142	267	1 532	34	16	1 550	4.87	4.57
413	1 400	300	550	40.0	14.0	143	268	1 302	49	15	1 320	4.47	4.17
348	1 400	300	550	30.0	14.0	143	268	1 322	39	15	1 340	4.47	4.17
284	1 400	300	550	20.0	14.0	143	268	1 342	29	15	1 360	4.47	4.17
373	1 200	300	550	40.0	12.0	144	269	1 102	49	14	1 120	4.08	3.78
309	1 200	300	550	30.0	12.0	144	269	1 122	39	14	1 140	4.08	3.78
244	1 200	300	550	20.0	12.0	144	269	1 142	29	14	1 160	4.08	3.78
340	1 000	300	550	40.0	10.0	145	270	902	49	13	920	3.68	3.38
275	1 000	300	550	30.0	10.0	145	270	922	39	13	940	3.68	3.38
210	1 000	300	550	20.0	10.0	145	270	942	29	13	960	3.68	3.38

WELDED WIDE FLANGE SHAPES
WWF2000-WWF1100

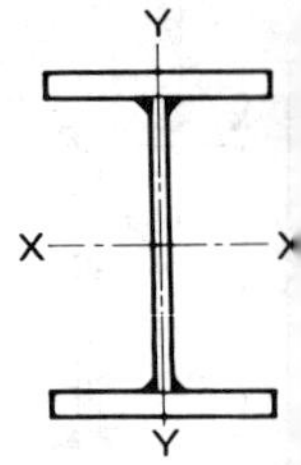

PROPERTIES

Designation‡	Dead Load	Total Area	Axis X-X				Axis Y-Y				Torsional Constant	Warping Constant
			I_x	S_x	r_x	Z_x	I_y	S_y	r_y	Z_y	J	C_w
	kN/m	mm^2	10^6mm^4	10^3mm^3	mm	10^3mm^3	10^6mm^4	10^3mm^3	mm	10^3mm^3	10^3mm^4	10^9mm^6
WWF2000												
•X732	7.17	93 200	63 900	63 900	828	71 800	1 390	5 050	122	7 750	50 900	1 320 000
•X648	6.35	82 600	54 200	54 200	810	61 700	1 110	4 040	116	6 240	28 600	1 070 000
•X607	5.94	77 300	49 300	49 300	799	56 600	972	3 530	112	5 490	20 900	938 000
X542	5.31	69 000	41 400	41 400	775	48 500	626	2 510	95.2	3 950	14 200	608 000
WWF1800												
•X700	6.86	89 200	50 400	56 000	752	62 700	1 390	5 050	125	7 730	50 400	1 060 000
•X659	6.45	83 900	46 600	51 800	745	58 200	1 250	4 540	122	6 980	38 000	962 000
•X617	6.04	78 600	42 700	47 400	737	53 700	1 110	4 040	119	6 220	28 100	860 000
•X575	5.64	73 300	38 800	43 100	728	49 100	972	3 530	115	5 470	20 300	757 000
X510	5.00	65 000	32 400	36 000	706	41 800	626	2 500	98.1	3 930	13 600	490 000
WWF1600												
•X622	6.09	79 200	37 600	47 100	689	51 800	1 390	5 040	132	7 660	47 900	833 000
•X580	5.68	73 800	34 600	43 300	685	47 700	1 250	4 540	130	6 900	35 500	755 000
X538	5.27	68 500	31 600	39 500	679	43 700	1 110	4 040	127	6 150	25 500	675 000
X496	4.85	63 100	28 400	35 500	671	39 600	971	3 530	124	5 390	17 800	595 000
X431	4.21	54 800	23 400	29 300	653	33 100	626	2 500	107	3 850	11 100	385 000
WWF1400												
•X597	5.84	76 000	28 100	40 100	608	44 000	1 390	5 040	135	7 650	47 600	632 000
X513	5.02	65 300	23 500	33 600	600	37 000	1 110	4 040	130	6 140	25 300	513 000
X471	4.61	59 900	21 100	30 200	594	33 400	971	3 530	127	5 380	17 500	452 000
X405	3.97	51 600	17 300	24 800	579	27 800	625	2 500	110	3 840	10 800	293 000
X358	3.50	45 600	14 500	20 800	564	23 700	320	1 600	83.8	2 490	9 030	150 000
WWF1200												
X487	4.78	62 100	16 700	27 900	519	30 600	1 110	4 030	134	6 120	25 000	373 000
X418	4.09	53 200	13 800	23 100	509	25 600	730	2 920	117	4 450	15 800	248 000
X380	3.72	48 400	12 300	20 500	504	22 800	625	2 500	114	3 820	10 600	214 000
X333	3.26	42 400	10 200	17 100	490	19 300	320	1 600	86.9	2 470	8 760	110 000
X302	2.96	38 500	8 970	15 000	483	17 100	267	1 340	83.3	2 070	5 740	92 200
X263	2.58	33 500	7 250	12 100	465	14 200	113	753	58.1	1 200	4 700	39 000
WWF1100												
X458	4.50	58 500	13 700	24 800	484	27 100	1 110	4 030	138	6 100	24 400	312 000
X388	3.81	49 500	11 200	20 400	476	22 400	729	2 920	121	4 430	15 200	207 000
X351	3.44	44 700	9 940	18 100	472	19 900	625	2 500	118	3 800	9 950	179 000
X304	2.97	38 700	8 220	14 900	461	16 700	320	1 600	90.9	2 450	8 150	91 700
X273	2.68	34 800	7 160	13 000	454	14 700	267	1 330	87.6	2 050	5 130	77 100
X234	2.29	29 800	5 720	10 400	438	12 000	113	752	61.6	1 180	4 090	32 600

‡Nominal depth in millimetres and mass in kilograms per metre.

•Maximum piece weight may limit length.

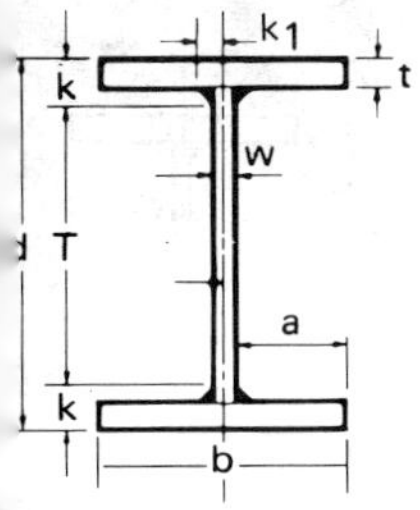

WELDED WIDE FLANGE SHAPES
WWF2000-WWF1100

DIMENSIONS AND SURFACE AREAS

Nominal Mass	Depth d	Flange		Web Thickness w	Distances					Surface Area (m^2) per metre of length	
		Width b	Thickness t		a	T	k	k_1	d-2t	Total	Minus Top of Top flange
kg/m	mm	mm	mm	mm	mm	mm	mm	mm	mm		
732	2 000	550	50.0	20.0	265	1 878	61	20	1 900	6.16	5.61
648	2 000	550	40.0	20.0	265	1 898	51	20	1 920	6.16	5.61
607	2 000	550	35.0	20.0	265	1 908	46	20	1 930	6.16	5.61
542	2 000	500	30.0	20.0	240	1 918	41	20	1 940	5.96	5.46
700	1 800	550	50.0	20.0	265	1 678	61	20	1 700	5.76	5.21
659	1 800	550	45.0	20.0	265	1 688	56	20	1 710	5.76	5.21
617	1 800	550	40.0	20.0	265	1 698	51	20	1 720	5.76	5.21
575	1 800	550	35.0	20.0	265	1 708	46	20	1 730	5.76	5.21
510	1 800	500	30.0	20.0	240	1 718	41	20	1 740	5.56	5.06
622	1 600	550	50.0	16.0	267	1 478	61	18	1 500	5.37	4.82
580	1 600	550	45.0	16.0	267	1 488	56	18	1 510	5.37	4.82
538	1 600	550	40.0	16.0	267	1 498	51	18	1 520	5.37	4.82
496	1 600	550	35.0	16.0	267	1 512	44	16	1 530	5.37	4.82
431	1 600	500	30.0	16.0	242	1 522	39	16	1 540	5.17	4.67
597	1 400	550	50.0	16.0	267	1 278	61	18	1 300	4.97	4.42
513	1 400	550	40.0	16.0	267	1 298	51	18	1 320	4.97	4.42
471	1 400	550	35.0	16.0	267	1 312	44	16	1 330	4.97	4.42
405	1 400	500	30.0	16.0	242	1 322	39	16	1 340	4.77	4.27
358	1 400	400	30.0	16.0	192	1 322	39	16	1 340	4.37	3.97
487	1 200	550	40.0	16.0	267	1 098	51	18	1 120	4.57	4.02
418	1 200	500	35.0	16.0	242	1 112	44	16	1 130	4.37	3.87
380	1 200	500	30.0	16.0	242	1 122	39	16	1 140	4.37	3.87
333	1 200	400	30.0	16.0	192	1 122	39	16	1 140	3.97	3.57
302	1 200	400	25.0	16.0	192	1 132	34	16	1 150	3.97	3.57
263	1 200	300	25.0	16.0	142	1 132	34	16	1 150	3.57	3.27
458	1 100	550	40.0	14.0	268	998	51	17	1 020	4.37	3.82
388	1 100	500	35.0	14.0	243	1 012	44	15	1 030	4.17	3.67
351	1 100	500	30.0	14.0	243	1 022	39	15	1 040	4.17	3.67
304	1 100	400	30.0	14.0	193	1 022	39	15	1 040	3.77	3.37
273	1 100	400	25.0	14.0	193	1 032	34	15	1 050	3.77	3.37
234	1 100	300	25.0	14.0	143	1 032	34	15	1 050	3.37	3.07

WELDED WIDE FLANGE SHAPES
WWF1000-WWF600

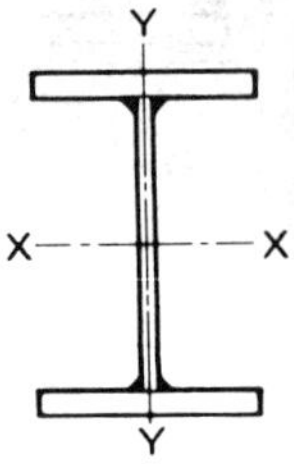

PROPERTIES

Designation‡	Dead Load	Total Area	Axis X-X				Axis Y-Y				Torsional Constant	Warping Constant
			I_x	S_x	r_x	Z_x	I_y	S_y	r_y	Z_y	J	C_w
	kN/m	mm^2	$10^6 mm^4$	$10^3 mm^3$	mm	$10^3 mm^3$	$10^6 mm^4$	$10^3 mm^3$	mm	$10^3 mm^3$	$10^3 mm^4$	$10^9 mm^6$
WWF1000												
X447	4.39	57 100	11 100	22 200	441	24 200	1 110	4 030	139	6 100	24 300	256 000
X377	3.70	48 100	9 120	18 200	435	20 000	729	2 920	123	4 420	15 100	170 000
X340	3.33	43 300	8 060	16 100	431	17 700	625	2 500	120	3 800	9 860	147 000
X293	2.87	37 300	6 640	13 300	422	14 800	320	1 600	92.6	2 450	8 060	75 300
X262	2.57	33 400	5 780	11 600	416	13 000	267	1 330	89.4	2 050	5 040	63 400
X223	2.18	28 400	4 590	9 190	402	10 500	113	752	63.1	1 170	3 990	26 800
X200	1.96	25 600	3 940	7 890	392	9 170	90.2	602	59.4	948	2 480	21 700
WWF900												
X417	4.09	53 200	8 680	19 300	404	20 800	1 110	4 030	144	6 080	23 800	205 000
X347	3.40	44 300	7 100	15 800	400	17 100	729	2 920	128	4 400	14 700	136 000
X309	3.03	39 400	6 240	13 900	398	15 000	625	2 500	126	3 780	9 370	118 000
X262	2.56	33 400	5 110	11 400	391	12 400	320	1 600	97.9	2 430	7 570	60 600
X231	2.26	29 500	4 410	9 810	387	10 800	267	1 330	95.1	2 030	4 540	51 100
X192	1.88	24 500	3 460	7 680	376	8 600	113	751	67.9	1 150	3 500	21 600
X169	1.66	21 600	2 930	6 510	368	7 370	90.1	601	64.6	927	1 980	17 400
WWF800												
X339	3.32	43 200	5 500	13 700	357	14 900	729	2 920	130	4 400	14 600	107 000
X300	2.94	38 300	4 840	12 100	355	13 100	625	2 500	128	3 770	9 330	92 700
X253	2.48	32 300	3 950	9 870	350	10 800	320	1 600	99.5	2 420	7 530	47 400
X223	2.18	28 400	3 410	8 520	347	9 340	267	1 330	97.0	2 020	4 500	40 100
X184	1.80	23 400	2 660	6 640	337	7 410	113	751	69.5	1 150	3 460	16 900
X161	1.57	20 500	2 250	5 610	331	6 320	90.1	601	66.3	924	1 940	13 700
WWF700												
X245	2.39	31 200	2 950	8 420	307	9 210	320	1 600	101	2 420	7 480	35 900
X214	2.10	27 300	2 540	7 270	305	7 950	267	1 330	98.9	2 020	4 460	30 400
X196	1.92	24 900	2 300	6 560	304	7 190	235	1 170	97.1	1 780	3 130	27 000
X175	1.71	22 300	1 970	5 640	297	6 270	113	751	71.2	1 150	3 410	12 800
X152	1.49	19 400	1 660	4 760	293	5 320	90.1	601	68.1	921	1 890	10 400
WWF650												
•X864**	8.46	110 000	7 570	23 300	262	27 300	2 760	8 480	158	13 200	132 000	240 000
•X739**	7.24	94 100	7 200	22 100	277	25 200	2 750	8 450	171	12 800	98 400	239 [illegible]
•X598	5.86	76 200	6 150	18 900	284	21 100	2 290	7 040	173	10 600	55 600	206 000
X499	4.89	63 600	5 170	15 900	285	17 500	1 830	5 630	170	8 510	29 300	170 000
X400	3.92	51 000	4 110	12 600	284	13 900	1 370	4 230	164	6 400	13 300	132 000
WWF600												
•X793**	7.77	101 000	5 830	19 400	240	22 900	2 170	7 230	147	11 200	121 000	158 000
•X680**	6.66	86 600	5 560	18 500	253	21 200	2 160	7 200	158	10 900	90 700	158 000
X551	5.40	70 200	4 770	15 900	261	17 800	1 800	6 000	160	9 050	51 300	136 000
X460	4.51	58 600	4 020	13 400	262	14 800	1 440	4 800	157	7 250	27 000	113 000
X369	3.61	47 000	3 200	10 700	261	11 800	1 080	3 600	152	5 460	12 200	87 800

‡Nominal depth in millimetres and mass in kilograms per metre.

**Welding does not fully develop the web strength for these sections.

•Maximum piece weight may limit length.

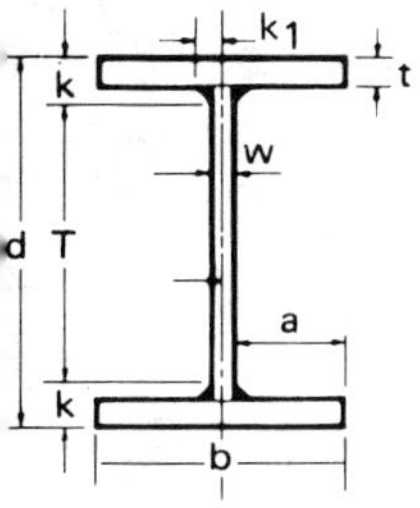

WELDED WIDE FLANGE SHAPES
WWF1000-WWF600

DIMENSIONS AND SURFACE AREAS

Nominal Mass	Depth d	Flange		Web Thickness w	Distances					Surface Area (m^2) per metre of length	
		Width b	Thickness t		a	T	k	k_1	d-2t	Total	Minus Top of Top flange
kg/m	mm	mm	mm	mm	mm	mm	mm	mm	mm		
447	1 000	550	40.0	14.0	268	898	51	17	920	4.17	3.62
377	1 000	500	35.0	14.0	243	912	44	15	930	3.97	3.47
340	1 000	500	30.0	14.0	243	922	39	15	940	3.97	3.47
293	1 000	400	30.0	14.0	193	922	39	15	940	3.57	3.17
262	1 000	400	25.0	14.0	193	932	34	15	950	3.57	3.17
223	1 000	300	25.0	14.0	143	932	34	15	950	3.17	2.87
200	1 000	300	20.0	14.0	143	942	29	15	960	3.17	2.87
417	900	550	40.0	11.0	270	798	51	16	820	3.98	3.43
347	900	500	35.0	11.0	245	812	44	14	830	3.78	3.28
309	900	500	30.0	11.0	245	822	39	14	840	3.78	3.28
262	900	400	30.0	11.0	195	822	39	14	840	3.38	2.98
231	900	400	25.0	11.0	195	832	34	14	850	3.38	2.98
192	900	300	25.0	11.0	145	832	34	14	850	2.98	2.68
169	900	300	20.0	11.0	145	842	29	14	860	2.98	2.68
339	800	500	35.0	11.0	245	712	44	14	730	3.58	3.08
300	800	500	30.0	11.0	245	722	39	14	740	3.58	3.08
253	800	400	30.0	11.0	195	722	39	14	740	3.18	2.78
223	800	400	25.0	11.0	195	732	34	14	750	3.18	2.78
184	800	300	25.0	11.0	145	732	34	14	750	2.78	2.48
161	800	300	20.0	11.0	145	742	29	14	760	2.78	2.48
245	700	400	30.0	11.0	195	622	39	14	640	2.98	2.58
214	700	400	25.0	11.0	195	632	34	14	650	2.98	2.58
196	700	400	22.0	11.0	195	638	31	14	656	2.98	2.58
175	700	300	25.0	11.0	145	632	34	14	650	2.58	2.28
152	700	300	20.0	11.0	145	642	29	14	660	2.58	2.28
864	650	650	60.0	60.0	295	508	71	40	530	3.78	3.13
739	650	650	60.0	30.0	310	508	71	25	530	3.84	3.19
598	650	650	50.0	20.0	315	528	61	20	550	3.86	3.21
499	650	650	40.0	20.0	315	548	51	20	570	3.86	3.21
400	650	650	30.0	20.0	315	568	41	20	590	3.86	3.21
793	600	600	60.0	60.0	270	458	71	40	480	3.48	2.88
680	600	600	60.0	30.0	285	458	71	25	480	3.54	2.94
551	600	600	50.0	20.0	290	478	61	20	500	3.56	2.96
460	600	600	40.0	20.0	290	498	51	20	520	3.56	2.96
369	600	600	30.0	20.0	290	518	41	20	540	3.56	2.96

WELDED WIDE FLANGE SHAPES
WWF550-WWF350

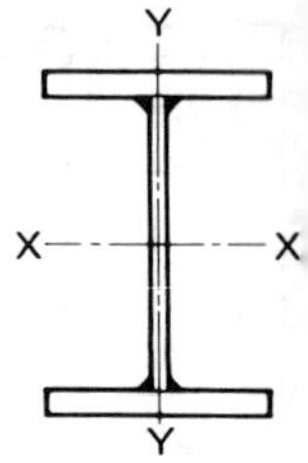

PROPERTIES

Designation‡	Dead Load	Total Area	Axis X-X				Axis Y-Y				Torsional Constant	Warping Constant
			I_x	S_x	r_x	Z_x	I_y	S_y	r_y	Z_y	J	C_w
	kN/m	mm^2	10^6mm^4	10^3mm^3	mm	10^3mm^3	10^6mm^4	10^3mm^3	mm	10^3mm^3	10^3mm^4	10^9mm^6
WWF550												
•X721**	7.08	92 000	4 390	16 000	218	19 000	1 670	6 080	135	9 470	110 000	100 000
•X620**	6.08	79 100	4 190	15 200	230	17 600	1 660	6 050	145	9 180	83 100	99 900
X503	4.94	64 200	3 610	13 100	237	14 800	1 390	5 040	147	7 610	47 000	86 700
X420	4.12	53 600	3 050	11 100	239	12 400	1 110	4 030	144	6 100	24 700	72 100
X280	2.74	35 600	2 070	7 530	241	8 250	693	2 520	140	3 810	6 410	47 800
WWF500												
•X651**	6.38	83 000	3 200	12 800	196	15 400	1 260	5 030	123	7 850	99 400	60 800
X561**	5.51	71 600	3 070	12 300	207	14 300	1 250	5 000	132	7 590	75 400	60 500
X456	4.47	58 200	2 660	10 600	214	12 100	1 040	4 170	134	6 290	42 700	52 700
X381	3.74	48 600	2 250	9 010	215	10 100	834	3 330	131	5 040	22 500	44 100
X343	3.37	43 800	2 040	8 140	216	9 100	729	2 920	129	4 420	15 400	39 400
X306	3.00	39 000	1 810	7 240	215	8 060	625	2 500	127	3 800	10 200	34 500
X276	2.71	35 200	1 680	6 740	218	7 420	583	2 330	129	3 530	7 920	32 500
X254	2.48	32 300	1 540	6 160	218	6 780	521	2 080	127	3 160	5 820	29 400
X223	2.19	28 500	1 370	5 500	219	6 010	458	1 830	127	2 770	3 970	26 200
X197	1.93	25 200	1 250	4 990	223	5 410	417	1 670	129	2 510	2 870	24 000
WWF450												
X503**	4.93	64 100	2 160	9 620	184	11 400	912	4 050	119	6 150	67 800	34 700
X409	4.01	52 200	1 890	8 380	190	9 640	760	3 380	121	5 100	38 400	30 400
X342	3.35	43 600	1 610	7 150	192	8 100	608	2 700	118	4 090	20 200	25 500
X308	3.02	39 300	1 460	6 480	193	7 290	532	2 360	116	3 580	13 900	22 900
X274	2.69	35 000	1 300	5 770	193	6 470	456	2 030	114	3 080	9 140	20 100
X248	2.43	31 600	1 210	5 380	196	5 960	425	1 890	116	2 860	7 120	18 900
X228	2.23	29 000	1 110	4 920	196	5 450	380	1 690	114	2 560	5 230	17 200
X201	1.97	25 600	991	4 400	197	4 840	334	1 490	114	2 250	3 570	15 300
X177	1.74	22 600	901	4 000	200	4 360	304	1 350	116	2 040	2 580	14 000
WWF400												
X444**	4.35	56 600	1 460	7 300	161	8 770	641	3 200	106	4 870	60 100	18 500
X362	3.55	46 200	1 280	6 410	166	7 480	534	2 670	108	4 030	34 100	16 300
X303	2.97	38 600	1 100	5 500	169	6 300	427	2 130	105	3 230	17 900	13 800
X273	2.67	34 800	1 000	5 000	170	5 680	374	1 870	104	2 840	12 300	12 400
X243	2.38	31 000	894	4 470	170	5 050	320	1 600	102	2 440	8 110	11 000
X220	2.15	28 000	834	4 170	173	4 660	299	1 490	103	2 260	6 320	10 300
X202	1.98	25 700	765	3 830	173	4 260	267	1 330	102	2 020	4 640	9 380
X178	1.74	22 700	686	3 430	174	3 790	235	1 170	102	1 780	3 170	8 390
X157	1.54	20 100	625	3 120	176	3 420	213	1 070	103	1 610	2 290	7 700
WWF350												
X315	3.09	40 200	824	4 710	143	5 580	357	2 040	94.2	3 090	29 800	8 040
X263	2.58	33 600	712	4 070	146	4 730	286	1 630	92.3	2 480	15 700	6 870
X238	2.33	30 300	650	3 720	146	4 280	250	1 430	90.8	2 170	10 800	6 210
X212	2.07	27 000	583	3 330	147	3 810	215	1 230	89.2	1 870	7 070	5 490
X192	1.88	24 400	546	3 120	150	3 520	200	1 140	90.5	1 740	5 520	5 190
X176	1.72	22 400	502	2 870	150	3 220	179	1 020	89.4	1 550	4 060	4 720
X155	1.52	19 800	451	2 580	151	2 870	157	899	89.0	1 360	2 760	4 230
X137	1.35	17 500	412	2 350	153	2 590	143	817	90.4	1 240	2 000	3 890

‡ Nominal depth in millimetres and mass in kilograms per metre.

** Welding does not fully develop the web strength for these sections.

•Maximum piece weight may limit length.

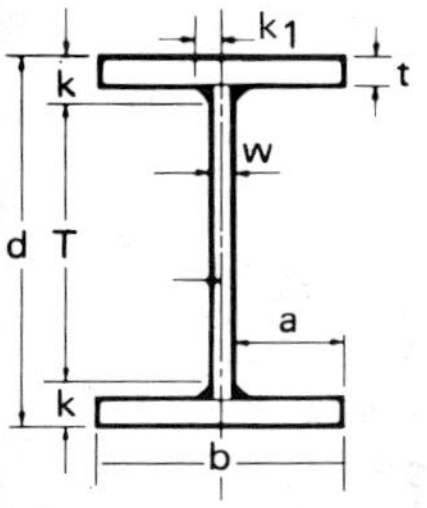

WELDED WIDE FLANGE SHAPES WWF550-WWF350

DIMENSIONS AND SURFACE AREAS

Nominal Mass	Depth d	Flange Width b	Flange Thickness t	Web Thickness w	Distances a	Distances T	Distances k	Distances k_1	Distances d-2t	Surface Area (m^2) per metre of length: Total	Surface Area (m^2) per metre of length: Minus Top of Top flange
kg/m	mm	mm	mm	mm	mm	mm	mm	mm	mm		
721	550	550	60.0	60.0	245	408	71	40	430	3.18	2.63
620	550	550	60.0	30.0	260	408	71	25	430	3.24	2.69
503	550	550	50.0	20.0	265	428	61	20	450	3.26	2.71
420	550	550	40.0	20.0	265	448	51	20	470	3.26	2.71
280	550	550	25.0	16.0	267	482	34	16	500	3.27	2.72
651	500	500	60.0	60.0	220	358	71	40	380	2.88	2.38
561	500	500	60.0	30.0	235	358	71	25	380	2.94	2.44
456	500	500	50.0	20.0	240	378	61	20	400	2.96	2.46
381	500	500	40.0	20.0	240	398	51	20	420	2.96	2.46
343	500	500	35.0	20.0	240	408	46	20	430	2.96	2.46
306	500	500	30.0	20.0	240	418	41	20	440	2.96	2.46
276	500	500	28.0	16.0	242	426	37	16	444	2.97	2.47
254	500	500	25.0	16.0	242	432	34	16	450	2.97	2.47
223	500	500	22.0	14.0	243	438	31	15	456	2.97	2.47
197	500	500	20.0	11.0	245	442	29	14	460	2.98	2.48
503	450	450	60.0	30.0	210	308	71	25	330	2.64	2.19
409	450	450	50.0	20.0	215	328	61	20	350	2.66	2.21
342	450	450	40.0	20.0	215	348	51	20	370	2.66	2.21
308	450	450	35.0	20.0	215	358	46	20	380	2.66	2.21
274	450	450	30.0	20.0	215	368	41	20	390	2.66	2.21
248	450	450	28.0	16.0	217	376	37	16	394	2.67	2.22
228	450	450	25.0	16.0	217	382	34	16	400	2.67	2.22
201	450	450	22.0	14.0	218	388	31	15	406	2.67	2.22
177	450	450	20.0	11.0	220	392	29	14	410	2.68	2.23
444	400	400	60.0	30.0	185	258	71	25	280	2.34	1.94
362	400	400	50.0	20.0	190	278	61	20	300	2.36	1.96
303	400	400	40.0	20.0	190	298	51	20	320	2.36	1.96
273	400	400	35.0	20.0	190	308	46	20	330	2.36	1.96
243	400	400	30.0	20.0	190	318	41	20	340	2.36	1.96
220	400	400	28.0	16.0	192	326	37	16	344	2.37	1.97
202	400	400	25.0	16.0	192	332	34	16	350	2.37	1.97
178	400	400	22.0	14.0	193	338	31	15	356	2.37	1.97
157	400	400	20.0	11.0	195	342	29	14	360	2.38	1.98
315	350	350	50.0	20.0	165	228	61	20	250	2.06	1.71
263	350	350	40.0	20.0	165	248	51	20	270	2.06	1.71
238	350	350	35.0	20.0	165	258	46	20	280	2.06	1.71
212	350	350	30.0	20.0	165	268	41	20	290	2.06	1.71
192	350	350	28.0	16.0	167	276	37	16	294	2.07	1.72
176	350	350	25.0	16.0	167	282	34	16	300	2.07	1.72
155	350	350	22.0	14.0	168	288	31	15	306	2.07	1.72
137	350	350	20.0	11.0	170	292	29	14	310	2.08	1.73

ROLLED STRUCTURAL SHAPES

General

In Canada rolled shapes are produced to requirements of the CAN/CSA-G40.20-M Standard in grades of steel meeting requirements of the CAN/CSA-G40.21-M material Standard. The normal grade of steel used is G40.21-M 300W. See pages 6-3 to 6-9 for more information on steel grades, and pages 6-11 to 6-16 for information on tolerances and mill practice.

The tables of properties and dimensions on pages 6-44 to 6-96 include most of the rolled shapes used in construction. Shapes not rolled in Canada are identified in the tables by an asterisk (*) and the note "not available from Canadian mills". (See "Principal sources of Structural Sections", page 6-17 for comments re Canadian sizes and Non-Canadian sizes).

Special shapes, such as rolled Tees, Zees, Bulb Angles, Carbuilding and Shipbuilding Channels are produced by some mills. These shapes are generally rolled only at irregular intervals and usually by special arrangement. Their use should, therefore, be avoided unless the quantity of any one size can warrant a rolling. Properties and dimensions of these shapes may be obtained from the appropriate mill catalogues.

Properties and Dimensions

The basic metric dimensions used to compute properties of the rolled steel shapes are taken from the recently (1990) ballotted and approved revision of the CAN3-G312.3-M78 Standard "Preferred Metric Dimensions for Structural Steel W and HP Shapes, Angles and Hollow Structural Sections". For shapes not listed in this Standard they are taken from the ASTM A6 Standard.

W shapes rolled in Canada, and by most mills supplying the Canadian market, have essentially parallel flanges, although some non-Canadian mills offer W shapes with slightly tapered (approximately 3°) flanges. Also the web-to-flange fillet radius may vary slightly for different mills. Properties for W shapes are calculated using the smallest theoretical fillet radius while dimensions for detailing are adjusted for the largest theoretical fillet radius.

HP shapes are essentially square (equal flange width and overall depth) with parallel flange surfaces and equal thickness flanges and web.

S shapes and Standard channels (C shapes) have tapered flanges with the inside face sloping at approximately 16⅔% (2 in 12). The tabulated thickness is the mean thickness.

M and MC shapes are essentially shapes that cannot be classified as W, HP, S or C shapes. They are not rolled in Canada, and are only produced by a few mills. Availability should be checked before specifying their use. These shapes may be produced with parallel flanges or with tapered flanges of various slopes. Dimensions and properties provided in this Handbook should be suitable for general use, in spite of possible variations in actual dimensions.

Two values for flange and web thickness are tabulated for W, HP, S, C, M and MC shapes. The first is the theoretical dimension used for calculating properties of the section, and the second is a rounded dimension used for detailing.

Availability of W Shapes

The tables for W shapes highlight in yellow those section sizes which are commonly used and which are generally readily available from inventory sources or from mill rollings. These sizes should be selected for design when small quantities and fast delivery are required. For larger projects requiring significant tonnage of various section sizes, availability will depend on mill rolling cycles, and the most appropriate section size for the design condition should be selected. (See comments on availability of rolled shapes under "Principal sources of Structural Sections" on page 6-17).

Angles

Properties and dimensions in metric units are provided for both metric series hot-rolled angles and imperial series hot-rolled angles for equal-leg angles and for unequal-leg angles. The tables include properties and dimensions for single angles and for two equal-leg angles back to back, two unequal-leg angles with short legs back to back, and two unequal-leg angles with long legs back to back.

In addition, tables are provided which combine both metric series angles and imperial series angles listed in decreasing order of size, and arranged so that the properties and dimensions of single metric and imperial series angles can be readily compared. These tables cover a limited range of angle sizes, and are intended to facilitate design and detailing when substitutions may be required from one series to the other.

The properties of angles produced by cold-forming may be up to 7 percent less than the properties of hot-rolled angles of similar size due to the absence of a heel. Designers encountering cold-formed angles should consult the manufacturer's catalogue for the exact dimensions and properties.

As discussed under "Principal Sources of Structural Steel Sections", metric series angles may be less readily available than imperial series angles. Under the heading "Rolling", the angle tables provide a rating scale of 1 to 5, where 1 = angles frequently rolled in Canada, 2 = angles less frequently rolled in Canada, 3 = angles infrequently rolled (i.e very occasionally) in Canada, 4 = angles not rolled in Canada, and 5 = contact mills for information.

Tees cut from W shapes

Properties and dimensions of Tees are based on W shapes assuming the depth of the Tee equal to one-half the depth of the W shape. Tees are not rolled in Canada, and are usually fabricated from W shapes by splitting the web either using rotary shears or flame cutting and subsequently straightening to meet published tolerances.

W SHAPES
W1000-W920

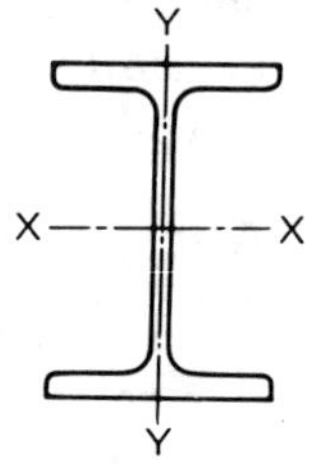

PROPERTIES

Designation‡	Dead Load	Total Area	Axis X-X				Axis Y-Y				Torsional Constant	Warping Constant
			I_x	S_x	r_x	Z_x	I_y	S_y	r_y	Z_y	J	C_w
	kN/m	mm^2	10^6mm^4	10^3mm^3	mm	10^3mm^3	10^6mm^4	10^3mm^3	mm	10^3mm^3	10^3mm^4	10^9mm^6
W1000												
X488*	4.78	62 200	11 100	21 900	422	24 700	690	3 030	105	4 680	31 000	163 000
X444*	4.35	56 600	10 100	20 000	422	22 400	621	2 740	105	4 220	23 600	145 000
X399*	3.91	50 800	8 960	17 900	420	20 100	550	2 440	104	3 750	17 300	128 000
X363*	3.56	46 300	8 000	16 100	416	18 100	487	2 160	103	3 330	12 800	112 000
X328*	3.22	41 900	6 920	14 100	406	15 900	412	1 830	99.2	2 830	8 800	94 000
X286*	2.80	36 500	5 630	11 600	393	13 200	321	1 430	93.8	2 220	5 450	72 300
W1000												
X976*	9.55	124 000	23 500	42 400	435	50 300	1 190	5 540	98.0	8 840	244 000	307 000
X883*	8.66	113 000	21 000	38 400	431	45 300	1 050	4 950	96.4	7 870	185 000	268 000
X790*	7.74	101 000	18 400	34 300	427	40 100	912	4 350	95.0	6 900	135 000	228 000
X714*	7.00	91 000	16 500	31 000	426	36 100	810	3 890	94.3	6 140	101 000	200 000
X649*	6.35	82 600	14 700	28 100	422	32 500	714	3 470	93.0	5 460	76 700	175 000
X591*	5.79	75 300	13 300	25 600	420	29 500	640	3 130	92.2	4 920	59 000	155 000
X539*	5.29	68 700	12 000	23 400	418	26 800	576	2 830	91.6	4 440	45 300	138 000
X483*	4.73	61 500	10 700	20 900	417	23 900	507	2 510	90.8	3 920	33 100	120 000
X443*	4.33	56 400	9 670	19 100	414	21 800	455	2 260	89.8	3 530	25 400	107 000
X412*	4.04	52 500	9 100	18 100	416	20 500	434	2 160	90.9	3 350	21 400	102 000
X371*	3.64	47 300	8 140	16 300	415	18 400	386	1 930	90.3	2 980	15 900	89 600
X321*	3.14	40 900	6 960	14 100	413	15 800	331	1 660	90.0	2 550	10 300	76 100
X296*	2.90	37 800	6 200	12 600	405	14 300	290	1 450	87.6	2 240	7 640	66 000
W1000												
X272*	2.67	34 700	5 540	11 200	400	12 800	140	933	63.5	1 470	8 350	32 200
X249*	2.43	31 700	4 810	9 820	390	11 300	118	783	61.0	1 240	5 820	26 700
X222*	2.17	28 300	4 080	8 410	380	9 800	95.4	636	58.1	1 020	3 900	21 500
W920												
X1262*	12.3	161 000	28 100	52 000	418	62 800	1 900	8 240	109	13 100	519 000	440 000
X1188*	11.6	151 000	26 100	48 900	416	58 800	1 750	7 660	108	12 200	439 000	401 000
X1072*	10.5	137 000	23 000	44 000	410	52 400	1 530	6 770	106	10 700	329 000	342 000
X967*	9.47	123 000	20 300	39 500	406	46 800	1 340	6 000	104	9 490	246 000	294 000
X876*	8.58	112 000	18 100	35 800	402	42 100	1 190	5 370	103	8 460	186 000	257 000
X784*	7.68	99 800	15 900	32 000	399	37 300	1 030	4 730	102	7 420	136 000	220 000
X722*	7.07	91 900	14 400	29 400	396	34 100	932	4 300	101	6 730	107 000	195 000
X653*	6.40	83 200	12 900	26 600	394	30 700	830	3 850	99.9	6 020	80 500	172 000
X585*	5.73	74 500	11 400	23 800	391	27 400	728	3 410	98.9	5 310	58 900	149 000
X534*	5.23	68 000	10 300	21 700	389	24 800	656	3 090	98.2	4 800	45 100	132 000
X488*	4.78	62 100	9 350	19 900	388	22 600	590	2 800	97.5	4 340	35 000	118 000
X446*	4.38	57 000	8 470	18 200	385	20 600	540	2 550	97.3	3 950	26 800	107 000
X417*	4.10	53 300	7 880	17 000	385	19 200	501	2 370	97.0	3 670	21 900	98 700
X387*	3.79	49 300	7 180	15 600	382	17 600	453	2 160	95.9	3 330	17 300	88 500
X365*	3.57	46 400	6 710	14 600	380	16 500	421	2 010	95.3	3 110	14 400	81 900
X342*	3.35	43 600	6 250	13 700	379	15 400	390	1 870	94.6	2 880	11 900	75 500

‡ Nominal depth in millimetres and mass in kilograms per metre.

* Not available from Canadian mills.

W SHAPES
W1000-W920

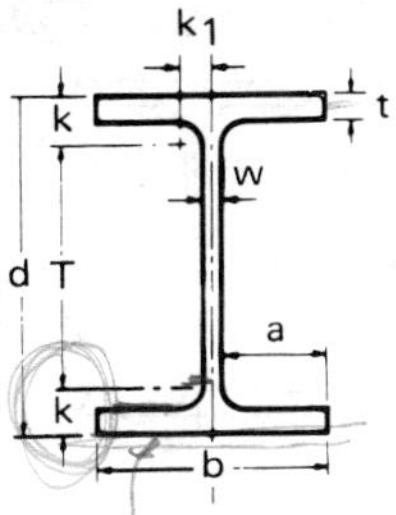

DIMENSIONS AND SURFACE AREAS

Nominal Mass	Theo-retical Mass	Depth d	Flange			Web Thickness w		Distances					Surface Area (m²) per metre of length	
			Width b	Mean Thickness t				a	T	k	k_1	d-2t	Total	Minus Top of Top Flange
kg/m	kg/m	mm	mm	mm	mm	mm	mm	mm	mm	mm	mm	mm		
488	488.0	1 016	455	43.9	44	23.1	23	216	866	75	42	928	3.81	3.35
444	444.3	1 008	453	40.0	40	21.1	21	216	866	71	41	928	3.79	3.33
399	398.7	1 000	451	35.9	36	19.0	19	216	866	67	40	928	3.77	3.32
363	363.3	992	450	32.0	32	18.0	18	216	866	63	39	928	3.75	3.30
328	328.6	982	450	27.1	27	18.0	18	216	866	58	39	928	3.73	3.28
286	286.2	970	450	21.1	21	18.0	18	216	866	52	39	928	3.70	3.25
976	974.5	1 108	428	89.9	90	50.0	50	189	866	121	55	928	3.83	3.40
883	883.4	1 092	424	82.0	82	45.5	46	189	866	113	53	928	3.79	3.36
790	789.9	1 075	419	73.9	74	40.9	41	189	865	105	51	928	3.75	3.33
714	714.5	1 062	416	67.1	67	37.1	37	190	866	98	49	928	3.71	3.30
649	648.3	1 050	412	61.0	61	34.0	34	189	866	92	47	928	3.68	3.27
591	590.9	1 040	409	55.9	56	31.0	31	189	866	87	46	928	3.65	3.24
539	539.4	1 030	407	51.1	51	28.4	28	190	866	82	44	928	3.63	3.22
483	482.9	1 020	404	46.0	46	25.4	25	190	866	77	43	928	3.61	3.20
443	442.5	1 012	402	41.9	42	23.6	24	189	866	73	42	928	3.58	3.18
412	412.2	1 008	402	40.0	40	21.1	21	191	866	71	41	928	3.58	3.18
371	371.2	1 000	400	36.1	36	19.0	19	191	866	67	40	928	3.56	3.16
321	320.9	990	400	31.0	31	16.5	17	192	866	62	38	928	3.55	3.15
296	296.4	982	400	27.1	27	16.5	17	192	866	58	38	928	3.53	3.13
272	272.3	990	300	31.0	31	16.5	17	142	866	62	38	928	3.15	2.85
249	248.7	980	300	26.0	26	16.5	17	142	866	57	38	928	3.13	2.83
222	222.0	970	300	21.1	21	16.0	16	142	866	52	38	928	3.11	2.81
1 262	1 262.9	1 078	461	115	115	64.0	64	199	797	141	56	848	3.87	3.41
1 188	1 188.7	1 066	457	109	109	60.5	61	199	797	135	54	848	3.84	3.38
1 072	1 071.6	1 046	451	99.1	99	55.0	55	198	797	125	52	848	3.79	3.34
967	966.3	1 028	446	89.9	90	50.0	50	198	797	116	49	848	3.74	3.29
876	875.8	1 012	442	82.0	82	45.5	46	198	797	108	47	848	3.70	3.26
784	783.2	996	437	73.9	74	40.9	41	198	797	100	45	848	3.66	3.22
722	721.5	984	434	68.1	68	38.1	38	198	797	94	43	848	3.63	3.19
653	653.1	972	431	62.0	62	34.5	35	199	797	88	41	848	3.60	3.17
585	585.0	960	427	55.9	56	31.0	31	198	797	82	40	848	3.57	3.14
534	533.9	950	425	51.1	51	28.4	28	199	797	77	38	848	3.54	3.12
488	487.7	942	422	47.0	47	25.9	26	198	797	73	37	848	3.52	3.10
446	447.2	933	423	42.7	43	24.0	24	200	792	70	38	848	3.51	3.09
417	418.1	928	422	39.9	40	22.5	23	200	793	67	37	848	3.50	3.08
387	387.0	921	420	36.6	37	21.3	21	200	792	64	36	848	3.48	3.06
365	364.6	916	419	34.3	34	20.3	20	200	793	61	36	848	3.47	3.05
342	342.4	912	418	32.0	32	19.3	19	200	793	59	35	848	3.46	3.04

Sections highlighted in yellow are commonly used sizes and are generally readily available.

W SHAPES
W920-W760

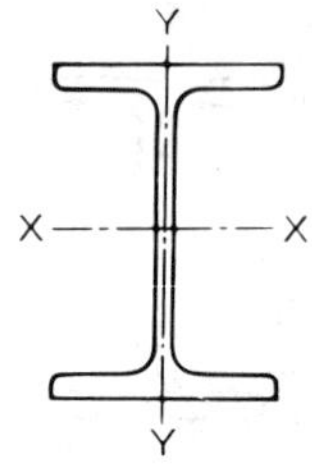

PROPERTIES

Designation‡	Dead Load	Total Area	Axis X-X				Axis Y-Y				Torsional Constant	Warping Constant
			I_x	S_x	r_x	Z_x	I_y	S_y	r_y	Z_y	J	C_w
	kN/m	mm^2	10^6mm^4	10^3mm^3	mm	10^3mm^3	10^6mm^4	10^3mm^3	mm	10^3mm^3	10^3mm^4	10^9mm^6
W920												
X313*	3.06	39 800	5 480	11 800	371	13 600	170	1 100	65.4	1 750	11 600	34 300
X289*	2.83	36 800	5 040	10 900	370	12 600	156	1 020	65.1	1 600	9 230	31 300
X271*	2.66	34 600	4 720	10 200	369	11 800	145	946	64.7	1 490	7 690	28 900
X253*	2.48	32 300	4 370	9 520	368	11 000	134	874	64.4	1 370	6 260	26 500
X238*	2.33	30 400	4 060	8 880	365	10 200	123	806	63.6	1 270	5 140	24 300
X223*	2.19	28 600	3 770	8 270	363	9 540	112	738	62.6	1 160	4 220	22 100
X201*	1.97	25 600	3 250	7 200	356	8 360	94.4	621	60.7	982	2 910	18 400
W840												
X922*	9.04	117 000	17 400	35 700	386	42 300	1 200	5 580	101	8 820	232 000	236 000
X845*	8.28	108 000	15 700	32 600	381	38 500	1 070	5 030	99.5	7 930	182 000	207 000
X767*	7.51	97 600	14 000	29 600	379	34 700	949	4 510	98.6	7 080	138 000	181 000
X697*	6.83	88 800	12 500	26 800	375	31 200	845	4 040	97.5	6 340	105 000	158 000
X631*	6.19	80 400	11 200	24 200	373	28 100	748	3 610	96.5	5 650	79 400	138 000
X577*	5.64	73 400	10 100	22 100	371	25 500	672	3 270	95.7	5 100	61 300	123 000
X527*	5.17	67 200	9 140	20 200	369	23 300	607	2 970	95.0	4 620	47 500	110 000
X473*	4.64	60 300	8 120	18 200	367	20 800	537	2 640	94.4	4 100	34 900	95 800
X433*	4.24	55 200	7 350	16 600	365	18 900	483	2 390	93.5	3 710	26 800	85 500
X392*	3.84	49 900	6 590	15 000	363	17 000	430	2 140	92.8	3 310	20 100	75 300
X359*	3.52	45 800	5 910	13 600	359	15 400	389	1 930	92.2	2 980	15 000	67 400
X329*	3.23	42 000	5 350	12 400	357	14 000	349	1 740	91.2	2 690	11 500	60 000
X299*	2.93	38 100	4 790	11 200	355	12 600	312	1 560	90.5	2 410	8 560	53 200
W840												
X226*	2.22	28 900	3 400	7 990	343	9 160	114	774	62.8	1 210	5 140	19 300
X210*	2.06	26 800	3 110	7 340	341	8 430	103	700	62.0	1 100	4 050	17 300
X193*	1.89	24 700	2 780	6 630	335	7 620	90.3	618	60.5	971	3 050	15 100
X176*	1.72	22 400	2 460	5 900	331	6 810	78.2	536	59.1	844	2 220	13 000
W760												
X865*	8.47	110 000	13 700	30 500	353	36 400	1 050	5 100	97.7	8 050	219 000	172 000
X783*	7.67	99 700	12 200	27 600	350	32 700	927	4 560	96.4	7 170	166 000	149 000
X710*	6.95	90 400	10 900	25 000	347	29 400	821	4 080	95.3	6 400	126 000	129 000
X644*	6.31	82 000	9 640	22 600	343	26 400	724	3 630	94.0	5 690	95 100	112 000
X582*	5.70	74 200	8 600	20 400	340	23 700	644	3 250	93.2	5 080	71 500	98 200
X531*	5.20	67 600	7 750	18 600	339	21 500	578	2 940	92.5	4 580	55 100	87 000
X484*	4.75	61 700	6 980	17 000	336	19 500	517	2 650	91.5	4 120	42 400	76 800
X434*	4.25	55 200	6 180	15 200	335	17 400	455	2 350	90.8	3 640	30 900	66 800
X389*	3.80	49 500	5 440	13 500	332	15 400	399	2 070	89.8	3 210	22 200	57 800
X350*	3.42	44 500	4 860	12 200	330	13 800	355	1 860	89.3	2 860	16 500	50 800
X314*	3.08	40 100	4 270	10 900	326	12 300	316	1 640	88.8	2 540	11 600	44 700
X284*	2.78	36 200	3 810	9 790	324	11 000	280	1 470	87.9	2 260	8 580	39 300
X257*	2.52	32 800	3 420	8 840	323	9 930	250	1 310	87.3	2 020	6 380	34 800

‡ Nominal depth in millimetres and mass in kilograms per metre.

* Not available from Canadian mills.

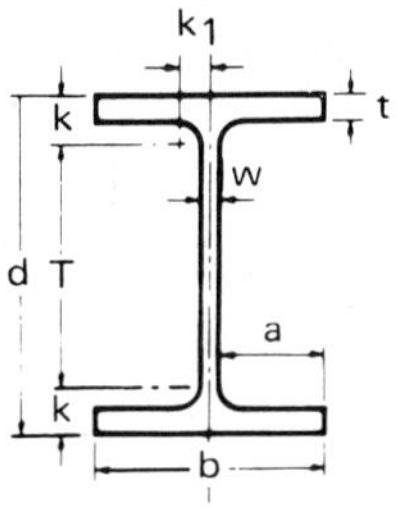

DIMENSIONS AND SURFACE AREAS

Nominal Mass	Theo-retical Mass	Depth d	Flange Width b	Flange Mean Thickness t		Web Thickness w		Distances a	T	k	k_1	d-2t	Surface Area (m^2) per metre of length: Total	Minus Top of Top Flange
kg/m	kg/m	mm	mm	mm	mm	mm	mm	mm	mm	mm	mm	mm		
313	312.7	932	309	34.5	35	21.1	21	144	820	56	31	863	3.06	2.75
289	288.6	927	308	32.0	32	19.4	19	145	819	54	30	863	3.05	2.74
271	271.7	923	307	30.0	30	18.4	18	145	819	52	29	863	3.04	2.73
253	253.7	919	306	27.9	28	17.3	17	145	819	50	29	863	3.03	2.72
238	238.3	915	305	25.9	26	16.5	17	145	819	48	28	863	3.02	2.71
223	224.2	911	304	23.9	24	15.9	16	144	819	46	28	863	3.01	2.70
201	201.3	903	304	20.1	20	15.2	15	145	819	42	28	863	2.99	2.69
922	922.0	977	430	89.9	90	50.0	50	190	758	110	43	797	3.57	3.14
845	844.4	963	425	83.1	83	46.0	46	190	758	103	41	797	3.53	3.11
767	766.1	949	421	75.9	76	41.9	42	190	758	96	39	797	3.50	3.08
697	697.1	935	418	69.1	69	38.6	39	190	758	89	38	797	3.46	3.05
631	631.3	923	414	63.0	63	35.1	35	190	758	83	36	797	3.43	3.02
577	576.1	913	411	57.9	56	32.0	32	190	758	78	34	797	3.41	3.00
527	527.7	903	409	53.1	53	29.5	30	190	758	73	33	797	3.38	2.97
473	473.3	893	406	48.0	48	26.4	26	190	758	68	31	797	3.36	2.95
433	433.3	885	404	43.9	44	24.4	24	190	758	64	30	797	3.34	2.93
392	391.7	877	401	39.9	40	22.1	22	190	758	60	29	797	3.31	2.91
359	359.4	868	403	35.6	36	21.1	21	191	755	57	30	797	3.31	2.90
329	329.4	862	401	32.4	32	19.7	20	191	757	53	29	797	3.29	2.88
299	299.3	855	400	29.2	29	18.2	18	191	756	50	28	797	3.27	2.87
226	226.6	851	294	26.8	27	16.1	16	139	756	48	27	797	2.85	2.55
210	210.8	846	293	24.4	24	15.4	15	139	757	45	27	797	2.83	2.54
193	193.5	840	292	21.7	22	14.7	15	139	755	43	27	797	2.82	2.53
176	176.0	835	292	18.8	19	14.0	14	139	756	40	26	797	2.81	2.52
865	864.3	899	411	89.9	90	50.0	50	181	682	109	42	719	3.34	2.93
783	782.7	883	407	82.0	82	45.5	46	181	682	101	40	719	3.30	2.90
710	709.6	869	403	74.9	75	41.4	41	181	682	94	38	719	3.27	2.86
644	643.5	855	399	68.1	68	38.1	38	181	682	87	36	719	3.23	2.83
582	582.1	843	396	62.0	62	34.5	35	181	682	81	34	719	3.20	2.81
531	530.9	833	393	56.9	57	31.5	32	181	682	76	33	719	3.17	2.78
484	484.6	823	390	52.1	52	29.0	29	181	682	71	32	719	3.15	2.76
434	433.7	813	387	47.0	47	25.9	26	181	682	66	30	719	3.12	2.74
389	388.5	803	385	41.9	42	23.6	24	181	682	61	29	719	3.10	2.71
350	349.5	795	382	38.1	38	21.1	21	181	682	57	28	719	3.08	2.69
314	314.4	786	384	33.4	33	19.7	20	182	681	52	28	719	3.07	2.68
284	283.9	779	382	30.1	30	18.0	18	182	680	49	27	719	3.05	2.67
257	257.6	773	381	27.1	27	16.6	17	182	680	46	26	719	3.04	2.66

Sections highlighted in yellow are commonly used sizes and are generally readily available.

W SHAPES
W760-W610

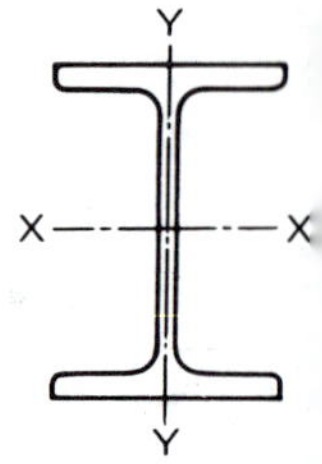

PROPERTIES

Designation‡	Dead Load	Total Area	Axis X-X				Axis Y-Y				Torsional Constant	Warping Constant
			I_x	S_x	r_x	Z_x	I_y	S_y	r_y	Z_y	J	C_w
	kN/m	mm^2	$10^6 mm^4$	$10^3 mm^3$	mm	$10^3 mm^3$	$10^6 mm^4$	$10^3 mm^3$	mm	$10^3 mm^3$	$10^3 mm^4$	$10^9 mm^6$
W760												
X196*	1.93	25 100	2 400	6 240	309	7 170	81.7	610	57.1	959	4 040	11 300
X185*	1.81	23 500	2 230	5 820	308	6 690	75.1	563	56.5	884	3 330	10 300
X173*	1.70	22 100	2 060	5 400	305	6 210	68.7	515	55.8	810	2 690	9 420
X161*	1.57	20 400	1 860	4 900	302	5 660	60.7	457	54.5	720	2 070	8 280
X147*	1.44	18 700	1 660	4 410	298	5 100	52.9	399	53.2	631	1 560	7 160
X134*	1.30	17 000	1 500	4 010	297	4 630	47.7	361	53.0	568	1 180	6 430
W690												
X802*	7.85	102 000	10 600	25 700	322	30 900	875	4 520	92.6	7 140	203 000	119 000
X735*	7.20	93 600	9 540	23 500	319	28 100	783	4 090	91.5	6 440	160 000	104 000
X667*	6.52	84 800	8 490	21 300	316	25 200	693	3 660	90.4	5 740	121 000	90 300
X605*	5.93	77 100	7 550	19 300	313	22 700	615	3 270	89.3	5 130	92 200	78 600
X548*	5.36	69 700	6 710	17 400	310	20 300	543	2 920	88.3	4 560	69 600	68 200
X500*	4.89	63 600	6 050	15 900	308	18 400	487	2 640	87.5	4 110	53 700	60 300
X457*	4.48	58 200	5 450	14 500	306	16 800	439	2 390	86.9	3 720	41 600	53 600
X419*	4.09	53 200	4 940	13 300	305	15 300	395	2 170	86.2	3 370	32 400	47 700
X384*	3.76	48 900	4 470	12 200	302	13 900	357	1 970	85.4	3 050	25 200	42 600
X350*	3.43	44 600	4 020	11 000	300	12 600	319	1 770	84.6	2 740	19 100	37 600
X323*	3.17	41 200	3 700	10 200	300	11 600	294	1 640	84.5	2 530	15 300	34 400
X289*	2.82	36 700	3 250	9 100	298	10 300	256	1 440	83.5	2 220	10 900	29 600
X265*	2.59	33 700	2 900	8 220	293	9 290	231	1 290	82.8	1 990	8 110	26 400
X240*	2.35	30 600	2 610	7 450	292	8 390	206	1 160	82.0	1 780	6 090	23 400
X217*	2.13	27 700	2 340	6 740	291	7 570	185	1 040	81.7	1 600	4 560	20 800
W690												
X170*	1.66	21 600	1 700	4 910	281	5 620	66.2	517	55.4	809	3 050	7 410
X152*	1.49	19 400	1 510	4 380	279	5 000	57.8	455	54.6	710	2 200	6 420
X140*	1.37	17 800	1 360	3 980	276	4 550	51.7	407	53.9	636	1 670	5 720
X125*	1.23	16 000	1 190	3 500	273	4 010	44.1	349	52.5	546	1 180	4 830
W610												
X732*	7.18	93 400	7 940	21 100	292	25 500	699	3 900	86.5	6 160	186 000	76 900
X670*	6.56	85 300	7 090	19 200	288	23 100	619	3 500	85.2	5 510	146 000	66 600
X608*	5.96	77 400	6 310	17 400	286	20 800	551	3 140	84.4	4 930	111 000	58 000
X551*	5.40	70 200	5 580	15 700	282	18 600	484	2 790	83.0	4 380	83 900	49 900
X498*	4.88	63 500	4 950	14 200	279	16 700	426	2 480	81.9	3 890	63 300	43 100
X455*	4.45	57 900	4 450	12 900	277	15 100	381	2 240	81.1	3 500	48 800	37 900
X415*	4.07	52 900	4 000	11 800	275	13 700	343	2 030	80.5	3 160	37 800	33 600
X372*	3.65	47 400	3 530	10 600	273	12 200	302	1 800	79.8	2 800	27 700	29 100
X341*	3.33	43 400	3 180	9 630	271	11 100	271	1 630	79.0	2 520	21 300	25 800
X307*	3.01	39 100	2 840	8 690	270	9 930	240	1 450	78.3	2 240	15 900	22 500
X285*	2.79	36 400	2 610	8 060	268	9 180	221	1 340	77.9	2 070	12 800	20 500
X262*	2.56	33 300	2 360	7 360	266	8 350	198	1 210	77.1	1 870	9 910	18 300
X241	2.37	30 800	2 150	6 780	264	7 670	184	1 120	77.3	1 730	7 700	16 800
X217	2.13	27 800	1 910	6 070	262	6 850	163	995	76.6	1 530	5 600	14 700
X195	1.91	24 900	1 680	5 400	260	6 070	142	871	75.5	1 340	3 970	12 700
X174	1.70	22 200	1 470	4 780	257	5 360	124	761	74.7	1 170	2 800	10 900
X155	1.51	19 700	1 290	4 220	256	4 730	108	666	74.0	1 020	1 950	9 450

‡ Nominal depth in millimetres and mass in kilograms per metre.

* Not available from Canadian mills.

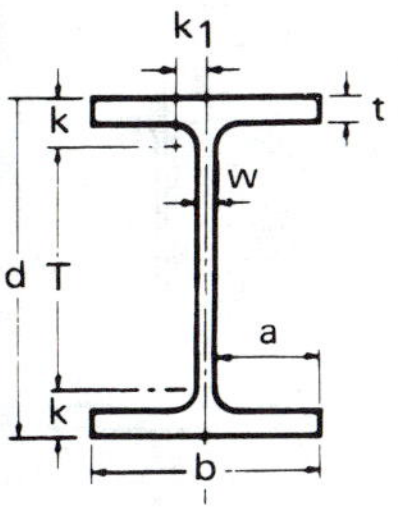

W SHAPES
W760-W610

DIMENSIONS AND SURFACE AREAS

Nominal Mass	Theo-retical Mass	Depth d	Flange Width b	Flange Mean Thickness t		Web Thickness w		Distances a	T	k	k_1	d-2t	Surface Area (m^2) per metre of length Total	Minus Top of Top Flange
kg/m	kg/m	mm	mm	mm	mm	mm	mm	mm	mm	mm	mm	mm		
196	196.8	770	268	25.4	25	15.6	16	126	681	44	26	719	2.58	2.31
185	184.8	766	267	23.6	24	14.9	15	126	679	43	25	719	2.57	2.31
173	173.6	762	267	21.6	22	14.4	14	127	679	41	25	719	2.57	2.30
161	160.4	758	266	19.3	19	13.8	14	126	681	38	25	719	2.55	2.28
147	147.1	753	265	17.0	17	13.2	13	126	680	36	24	719	2.54	2.28
134	133.2	750	264	15.5	16	11.9	12	126	682	34	23	719	2.53	2.27
802	801.4	826	387	89.9	90	50.0	50	169	613	107	40	646	3.10	2.71
735	734.4	812	383	83.1	83	46.0	46	169	613	100	38	646	3.06	2.68
667	665.7	798	379	75.9	76	41.9	42	169	613	93	36	646	3.03	2.65
605	605.1	784	376	69.1	69	38.6	39	169	613	86	35	646	2.99	2.62
548	547.5	772	372	63.0	63	35.1	35	169	613	80	33	646	2.96	2.59
500	499.3	762	369	57.9	58	32.0	32	169	613	75	31	646	2.94	2.57
457	457.0	752	367	53.1	53	29.5	30	169	613	70	30	646	2.91	2.55
419	418.0	744	364	49.0	49	26.9	27	169	613	66	29	646	2.89	2.53
384	383.5	736	362	45.0	45	24.9	25	169	613	62	28	646	2.87	2.51
350	349.9	728	360	40.9	41	23.1	23	169	613	58	27	646	2.85	2.49
323	323.2	722	359	38.1	38	21.1	21	169	613	55	26	646	2.84	2.48
289	287.9	714	356	34.0	34	19.0	19	169	613	51	25	646	2.81	2.46
265	264.5	706	358	30.2	30	18.4	18	170	610	48	26	646	2.81	2.45
240	239.9	701	356	27.4	27	16.8	17	170	611	45	25	646	2.79	2.43
217	217.8	695	355	24.8	25	15.4	15	170	609	43	24	646	2.78	2.43
170	169.9	693	256	23.6	24	14.5	15	121	609	42	24	646	2.38	2.13
152	152.1	688	254	21.1	21	13.1	13	121	610	39	23	646	2.37	2.11
140	139.8	684	254	18.9	19	12.4	12	121	610	37	23	646	2.36	2.11
125	125.6	678	253	16.3	16	11.7	12	121	610	34	23	646	2.34	2.09
732	732.8	753	359	89.9	90	50.0	50	155	544	105	38	573	2.84	2.48
670	669.8	739	354	83.1	83	46.0	46	154	544	98	36	573	2.80	2.45
608	607.9	725	351	75.9	76	41.9	42	155	544	91	34	573	2.77	2.42
551	551.2	711	347	69.1	69	38.6	39	154	544	84	33	573	2.73	2.38
498	498.3	699	343	63.0	63	35.1	35	154	544	78	31	573	2.70	2.36
455	454.2	689	340	57.9	58	32.0	32	154	544	73	29	573	2.67	2.33
415	415.6	679	338	53.1	53	29.5	30	154	544	68	28	573	2.65	2.31
372	372.3	669	335	48.0	48	26.4	26	155	544	63	26	573	2.63	2.29
341	340.4	661	333	43.9	44	24.4	24	155	544	59	25	573	2.61	2.27
307	307.3	653	330	39.9	40	22.1	22	154	544	55	24	573	2.58	2.25
285	285.4	647	329	37.1	37	20.6	21	154	544	52	24	573	2.57	2.24
262	261.2	641	327	34.0	34	19.0	19	154	544	49	23	573	2.55	2.22
241	241.7	635	329	31.0	31	17.9	18	156	534	50	27	573	2.55	2.22
217	217.9	628	328	27.7	28	16.5	17	156	533	47	26	573	2.54	2.21
195	195.6	622	327	24.4	24	15.4	15	156	535	43	25	573	2.52	2.19
174	174.3	616	325	21.6	22	14.0	14	156	533	41	25	573	2.51	2.18
155	154.9	611	324	19.0	19	12.7	13	156	534	38	24	573	2.49	2.17

Sections highlighted in yellow are commonly used sizes and are generally readily available.

W SHAPES
W610-W530

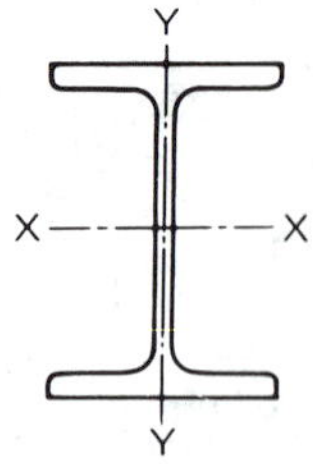

PROPERTIES

Designation‡	Dead Load	Total Area	Axis X-X				Axis Y-Y				Torsional Constant	Warping Constant
			I_x	S_x	r_x	Z_x	I_y	S_y	r_y	Z_y	J	C_w
	kN/m	mm²	10^6mm^4	10^3mm^3	mm	10^3mm^3	10^6mm^4	10^3mm^3	mm	10^3mm^3	10^3mm^4	10^9mm^6
W610												
X140	1.37	17 900	1 120	3 630	250	4 150	45.1	392	50.2	613	2 180	3 990
X125	1.22	15 900	985	3 220	249	3 670	39.3	343	49.7	535	1 540	3 450
X113	1.11	14 400	875	2 880	247	3 290	34.3	300	48.8	469	1 120	2 990
X101	.997	13 000	764	2 530	242	2 900	29.5	259	47.6	404	781	2 550
X91 †	.892	11 600	667	2 230	240	2 560	24.8	219	46.2	343	577	2 130
X84 †	.824	10 700	613	2 060	239	2 360	22.6	200	46.0	313	462	1 930
W610												
X92*	.905	11 800	646	2 140	234	2 510	14.4	161	34.9	258	710	1 250
X82*	.803	10 400	560	1 870	232	2 200	12.1	136	34.1	218	488	1 040
W530												
X599*	5.87	76 300	5 070	15 300	258	18 500	524	3 080	82.9	4 840	122 000	44 300
X543*	5.33	69 300	4 490	13 900	255	16 600	465	2 760	81.9	4 320	92 300	38 400
X496*	4.86	63 200	4 010	12 600	252	15 000	415	2 490	81.0	3 890	71 400	33 500
X447*	4.38	56 900	3 530	11 300	249	13 400	364	2 210	80.0	3 440	53 400	28 800
X409*	4.01	52 100	3 170	10 300	247	12 100	325	1 990	79.0	3 100	41 500	25 300
X370*	3.50	45 500	2 780	9 220	247	10 700	287	1 770	79.4	2 730	29 700	21 900
X331*	3.25	42 200	2 480	8 360	242	9 660	254	1 580	77.6	2 440	22 700	19 000
X300*	2.93	38 200	2 210	7 550	241	8 680	225	1 410	76.7	2 180	17 100	16 600
X272*	2.66	34 600	1 970	6 830	239	7 800	200	1 260	76.0	1 950	12 800	14 600
X248*	2.42	31 400	1 780	6 220	238	7 070	180	1 140	75.7	1 760	9 830	13 000
X219*	2.14	27 900	1 510	5 390	233	6 110	157	986	75.0	1 520	6 420	11 000
X196*	1.92	25 000	1 340	4 840	232	5 460	139	877	74.6	1 350	4 700	9 640
X182*	1.78	23 100	1 240	4 480	232	5 040	127	808	74.1	1 240	3 740	8 820
X165*	1.62	21 100	1 110	4 060	229	4 550	114	726	73.5	1 110	2 830	7 790
X150*	1.47	19 200	1 010	3 710	229	4 150	103	659	73.2	1 010	2 160	7 030
W530												
X138	1.35	17 600	861	3 140	221	3 610	38.7	362	46.9	569	2 500	2 670
X123	1.20	15 700	761	2 800	220	3 210	33.8	319	46.4	499	1 800	2 310
X109	1.06	13 900	667	2 480	219	2 830	29.5	280	46.1	437	1 260	2 000
X101	.995	12 900	617	2 300	219	2 620	26.9	256	45.7	400	1 020	1 820
X92	.907	11 800	552	2 070	216	2 360	23.8	228	44.9	355	762	1 590
X82 †	.808	10 500	479	1 810	214	2 070	20.3	194	44.0	303	530	1 340
X72†	.705	9 160	402	1 530	209	1 760	16.2	156	42.1	245	344	1 060
W530												
X85*	.830	10 800	485	1 810	212	2 100	12.6	152	34.2	242	737	849
X74*	.733	9 520	411	1 550	208	1 810	10.4	125	33.1	200	480	692
X66*	.644	8 370	351	1 340	205	1 560	8.57	104	32.0	166	320	565

‡ Nominal depth in millimetres and mass in kilograms per metre.
* Not available from Canadian mills.
† Produced exclusively by Algoma Steel

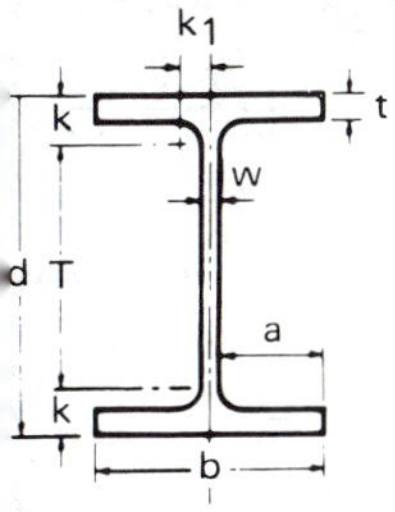

W SHAPES
W610-W530

DIMENSIONS AND SURFACE AREAS

Nominal Mass	Theo-retical Mass	Depth d	Flange Width b	Flange Mean Thickness t		Web Thickness w		Distances a	T	k	k_1	d-2t	Surface Area (m^2) per metre of length Total	Minus Top of Top Flange
kg/m	kg/m	mm	mm	mm	mm	mm	mm	mm	mm	mm	mm	mm		
140	140.1	617	230	22.2	22	13.1	13	109	534	41	24	573	2.13	1.90
125	125.1	612	229	19.6	20	11.9	12	109	533	39	24	573	2.12	1.89
113	113.4	608	228	17.3	17	11.2	11	109	535	36	23	573	2.10	1.88
101	101.7	603	228	14.9	15	10.5	11	109	534	34	23	573	2.10	1.87
91	90.9	598	227	12.7	13	9.7	10	109	534	32	23	573	2.09	1.86
84	84.0	596	226	11.7	12	9.0	9	109	534	31	22	573	2.08	1.85
92	92.3	603	179	15.0	15	10.9	11	84	534	34	23	573	1.90	1.72
82	81.9	599	178	12.8	13	10.0	10	84	534	32	23	573	1.89	1.71
599	598.7	661	340	79.5	80	43.9	44	148	470	96	36	502	2.60	2.26
543	543.6	647	337	72.4	72	40.4	40	149	472	88	34	502	2.56	2.22
496	496.2	635	334	66.5	67	37.1	37	149	472	82	33	502	2.53	2.20
447	446.8	623	330	60.5	61	33.5	34	148	472	76	31	502	2.50	2.17
409	408.9	613	327	55.6	56	31.0	31	148	470	72	30	502	2.47	2.15
370	357.5	603	324	50.5	51	25.2	25	150	472	66	27	502	2.45	2.13
331	331.4	593	322	45.5	46	25.4	25	149	470	62	27	502	2.43	2.10
300	299.7	585	319	41.4	41	23.1	23	148	472	57	26	502	2.40	2.08
272	271.6	577	317	37.6	38	21.1	21	148	470	54	25	502	2.38	2.07
248	246.8	571	315	34.5	35	19.0	19	148	472	50	24	502	2.36	2.05
219	218.9	560	318	29.2	29	18.3	18	150	466	47	26	502	2.36	2.04
196	196.5	554	316	26.3	26	16.5	17	150	466	44	25	502	2.34	2.02
182	181.7	551	315	24.4	24	15.2	15	150	467	42	24	502	2.33	2.02
165	165.3	546	313	22.2	22	14.0	14	150	466	40	24	502	2.32	2.00
150	150.6	543	312	20.3	20	12.7	13	150	467	38	23	502	2.31	1.99
138	138.3	549	214	23.6	24	14.7	15	100	465	42	24	502	1.93	1.71
123	123.3	544	212	21.2	21	13.1	13	100	466	39	23	502	1.91	1.70
109	109.0	539	211	18.8	19	11.6	12	100	465	37	23	502	1.90	1.69
101	101.4	537	210	17.4	17	10.9	11	100	467	35	22	502	1.89	1.68
92	92.5	533	209	15.6	16	10.2	10	100	465	34	22	502	1.88	1.68
82	82.4	528	209	13.3	13	9.5	10	100	471	29	19	502	1.87	1.66
72	71.9	524	207	10.9	11	8.9	9	99	471	27	19	502	1.86	1.65
85	84.7	535	166	16.5	17	10.3	10	78	467	34	22	502	1.71	1.55
74	74.7	529	166	13.6	14	9.7	10	78	465	32	22	502	1.70	1.54
66	65.7	525	165	11.4	11	8.9	9	78	467	29	21	502	1.69	1.52

Sections highlighted in yellow are commonly used sizes and are generally readily available.

W SHAPES
W460-W410

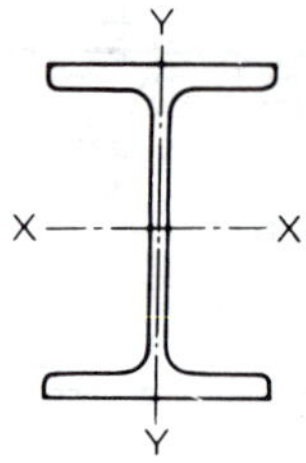

PROPERTIES

Designation‡	Dead Load	Total Area	Axis X-X				Axis Y-Y				Torsional Constant	Warping Constant
			I_x	S_x	r_x	Z_x	I_y	S_y	r_y	Z_y	J	C_w
	kN/m	mm²	10⁶mm⁴	10³mm³	mm	10³mm³	10⁶mm⁴	10³mm³	mm	10³mm³	10³mm⁴	10⁹mm⁶
W460												
X464*	4.54	59 100	2 900	10 200	222	12 300	331	2 170	74.8	3 400	72 600	20 500
X421*	4.13	53 700	2 570	9 250	219	11 100	293	1 940	73.9	3 030	55 400	17 700
X384*	3.76	48 900	2 290	8 420	216	10 000	261	1 750	73.1	2 730	42 800	15 500
X349*	3.42	44 500	2 040	7 640	214	9 010	233	1 570	72.4	2 440	32 900	13 500
X315*	3.08	40 000	1 800	6 850	212	8 020	204	1 390	71.4	2 160	24 400	11 600
X286*	2.80	36 400	1 610	6 230	210	7 250	183	1 260	70.9	1 950	18 600	10 200
X260*	2.55	33 100	1 440	5 650	209	6 530	163	1 130	70.2	1 740	14 100	8 950
X235*	2.30	29 900	1 270	5 090	206	5 840	145	1 010	69.6	1 550	10 500	7 790
X213*	2.08	27 100	1 140	4 620	205	5 270	129	909	69.0	1 400	8 000	6 890
X193*	1.89	24 600	1 020	4 190	204	4 760	115	816	68.4	1 250	6 050	6 070
X177*	1.73	22 600	910	3 780	201	4 280	105	735	68.2	1 130	4 410	5 440
X158*	1.54	20 100	796	3 350	199	3 780	91.4	643	67.4	989	3 120	4 670
X144*	1.41	18 400	726	3 080	199	3 450	83.6	591	67.4	906	2 440	4 230
X128*	1.26	16 400	637	2 730	197	3 050	73.3	520	66.9	796	1 720	3 670
X113*	1.10	14 400	556	2 400	196	2 670	63.3	452	66.3	691	1 180	3 150
W460												
X106	1.03	13 500	488	2 080	190	2 390	25.1	259	43.1	405	1 460	1 260
X97	.947	12 300	445	1 910	190	2 180	22.8	237	43.1	368	1 130	1 140
X89	.876	11 400	410	1 770	190	2 010	20.9	218	42.8	339	907	1 040
X82	.804	10 400	370	1 610	189	1 830	18.6	195	42.3	303	691	918
X74	.728	9 450	333	1 460	188	1 650	16.6	175	41.9	271	517	813
X67 †	.668	8 680	300	1 320	186	1 500	14.6	153	41.0	239	412	709
X61 †	.598	7 760	259	1 150	183	1 310	12.2	129	39.7	202	289	588
W460												
X68*	.672	8 730	297	1 290	184	1 490	9.41	122	32.8	192	509	463
X60*	.584	7 590	255	1 120	183	1 280	7.96	104	32.4	163	335	388
X52*	.510	6 630	212	943	179	1 090	6.34	83.4	30.9	131	210	306
W410												
X149*	1.46	19 000	619	2 870	180	3 250	77.7	586	63.9	900	3 220	3 200
X132*	1.29	16 800	538	2 530	179	2 850	67.4	512	63.3	785	2 260	2 730
X114*	1.12	14 600	462	2 200	178	2 460	57.2	439	62.6	671	1 490	2 300
X100*	.977	12 700	398	1 920	177	2 130	49.5	381	62.4	581	994	1 960
W410												
X85	.833	10 800	315	1 510	171	1 730	18.0	199	40.8	310	926	717
X74	.735	9 550	275	1 330	170	1 510	15.6	173	40.4	269	637	614
X67	.662	8 600	246	1 200	169	1 360	13.8	154	40.1	239	469	540
X60	.584	7 580	216	1 060	169	1 190	12.0	135	39.8	209	328	468
X54	.524	6 810	186	924	165	1 050	10.1	114	38.5	177	226	388
W410												
X46	.453	5 890	156	773	163	885	5.14	73.4	29.5	115	192	197
X39	.384	4 990	127	634	160	730	4.04	57.7	28.5	90.6	111	154

‡ Nominal depth in millimetres and mass in kilograms per metre.
* Not available from Canadian mills.
† Produced exclusively by Algoma Steel

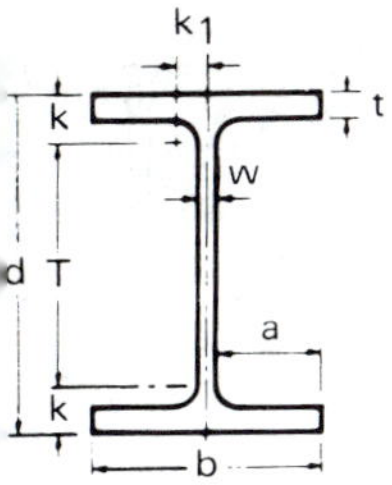

DIMENSIONS AND SURFACE AREAS

Nominal Mass	Theo-retical Mass	Depth d	Flange Width b	Flange Mean Thickness t		Web Thickness w		Distances a	Distances T	Distances k	Distances k1	Distances d-2t	Surface Area (m²) per metre of length: Total	Surface Area (m²) per metre of length: Minus Top of Top Flange
kg/m	kg/m	mm	mm	mm	mm	mm	mm	mm	mm	mm	mm	mm		
464	463.7	567	305	69.6	70	38.6	39	133	402	82	30	428	2.28	1.97
421	421.5	555	302	63.5	64	35.6	36	133	402	76	29	428	2.25	1.95
384	384.2	545	299	58.4	58	32.5	33	134	404	70	27	428	2.22	1.92
349	349.0	535	296	53.6	54	29.5	30	133	402	66	26	428	2.20	1.90
315	314.3	525	293	48.5	49	26.9	27	133	404	60	24	428	2.17	1.87
286	285.7	517	291	44.4	44	24.4	24	134	404	56	23	428	2.15	1.86
260	260.1	509	289	40.4	40	22.6	23	133	404	52	22	428	2.13	1.84
235	234.9	501	287	36.6	37	20.6	21	133	402	49	21	428	2.11	1.82
213	212.9	495	285	33.5	34	18.5	19	134	402	46	20	428	2.10	1.81
193	193.4	489	283	30.5	31	17.0	17	133	404	42	19	428	2.07	1.79
177	177.3	482	286	26.9	27	16.6	17	135	395	44	24	428	2.07	1.79
158	157.7	476	284	23.9	24	15.0	15	135	395	41	23	428	2.06	1.77
144	144.6	472	283	22.1	22	13.6	14	135	395	39	22	428	2.05	1.76
128	128.4	467	282	19.6	20	12.2	13	135	394	37	21	428	2.04	1.76
113	113.1	463	280	17.3	17	10.8	11	135	396	34	21	428	2.02	1.74
106	105.8	469	194	20.6	21	12.6	13	91	393	38	22	428	1.69	1.50
97	96.6	466	193	19.0	19	11.4	11	91	394	36	21	428	1.68	1.49
89	89.3	463	192	17.7	18	10.5	11	91	393	35	21	428	1.68	1.48
82	81.9	460	191	16.0	16	9.9	10	91	394	33	21	428	1.66	1.47
74	74.2	457	190	14.5	15	9.0	9	91	395	31	20	428	1.65	1.46
67	68.1	454	190	12.7	13	8.5	9	91	394	30	20	428	1.65	1.46
61	60.9	450	189	10.8	11	8.1	8	91	394	28	20	428	1.64	1.45
68	68.5	459	154	15.4	15	9.1	9	73	396	32	20	428	1.51	1.36
60	59.6	455	153	13.3	13	8.0	8	73	396	30	19	428	1.50	1.35
52	52.0	450	152	10.8	11	7.6	8	72	395	28	19	428	1.49	1.34
149	149.3	431	265	25.0	25	14.9	15	125	348	42	23	381	1.89	1.63
132	132.1	425	263	22.2	22	13.3	13	125	348	39	22	381	1.88	1.61
114	114.5	420	261	19.3	19	11.6	12	125	349	36	21	381	1.86	1.60
100	99.6	415	260	16.9	17	10.0	10	125	348	34	20	381	1.85	1.59
85	85.0	417	181	18.2	18	10.9	11	85	347	35	21	381	1.54	1.36
74	74.9	413	180	16.0	16	9.7	10	85	347	33	21	381	1.53	1.35
67	67.5	410	179	14.4	14	8.8	9	85	348	31	20	381	1.52	1.34
60	59.5	407	178	12.8	13	7.7	8	85	347	30	20	381	1.51	1.33
54	53.4	403	177	10.9	11	7.5	8	85	347	28	20	381	1.50	1.32
46	46.2	403	140	11.2	11	7.0	7	67	347	28	19	381	1.35	1.21
39	39.2	399	140	8.8	9	6.4	6	67	347	26	19	381	1.35	1.21

Sections highlighted in yellow are commonly used sizes and are generally readily available.

W SHAPES
W360

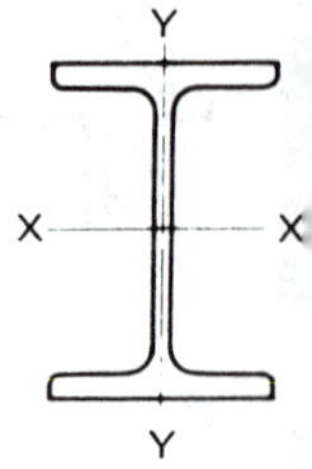

PROPERTIES

Designation‡	Dead Load	Total Area	Axis X-X				Axis Y-Y				Torsional Constant	Warping Constant
			I_x	S_x	r_x	Z_x	I_y	S_y	r_y	Z_y	J	C_w
	kN/m	mm^2	10^6mm^4	10^3mm^3	mm	10^3mm^3	10^6mm^4	10^3mm^3	mm	10^3mm^3	10^3mm^4	10^9mm^6
W360												
X1086*	10.6	139 000	5 960	20 900	207	27 200	1 960	8 650	119	13 400	605 000	96 700
X990*	9.71	126 000	5 190	18 900	203	24 300	1 730	7 740	117	12 000	469 000	82 000
X900*	8.84	115 000	4 500	17 000	198	21 600	1 530	6 940	115	10 700	364 000	69 200
X818*	8.03	104 000	3 920	15 300	194	19 300	1 360	6 200	114	9 560	279 000	58 900
X744*	7.29	94 800	3 420	13 700	190	17 200	1 200	5 550	113	8 550	214 000	50 200
X677*	6.64	86 300	2 990	12 400	186	15 300	1 070	4 990	111	7 680	165 000	43 100
W360												
X634*	6.22	80 800	2 740	11 600	184	14 200	983	4 630	110	7 120	138 000	38 700
X592*	5.81	75 500	2 500	10 800	182	13 100	902	4 280	109	6 570	114 000	34 800
X551*	5.40	70 100	2 260	9 940	180	12 100	825	3 950	108	6 050	92 500	31 000
X509*	4.99	64 900	2 050	9 170	178	11 000	754	3 630	108	5 550	74 000	27 700
X463*	4.53	59 000	1 800	8 280	175	9 880	670	3 250	107	4 980	56 500	23 900
X421*	4.13	53 700	1 600	7 510	173	8 880	601	2 940	106	4 490	43 400	20 800
X382*	3.75	48 700	1 410	6 790	170	7 970	536	2 640	105	4 030	32 900	18 200
X347*	3.40	44 200	1 250	6 140	168	7 140	481	2 380	104	3 630	24 800	15 900
X314*	3.07	39 900	1 100	5 530	166	6 370	426	2 120	103	3 240	18 500	13 800
X287*	2.82	36 600	997	5 070	165	5 810	388	1 940	103	2 960	14 500	12 300
X262*	2.57	33 500	894	4 620	163	5 260	350	1 760	102	2 680	11 100	11 000
X237*	2.31	30 100	788	4 150	162	4 690	310	1 570	101	2 390	8 190	9 500
X216*	2.12	27 600	712	3 790	161	4 260	283	1 430	101	2 180	6 330	8 520
W360												
X196*	1.92	25 000	636	3 420	159	3 840	229	1 220	95.7	1 860	5 140	6 830
X179*	1.75	22 800	575	3 120	159	3 480	207	1 110	95.3	1 680	3 910	6 120
X162*	1.58	20 600	516	2 830	158	3 140	186	1 000	95.0	1 520	2 940	5 430
X147*	1.44	18 800	463	2 570	157	2 840	167	904	94.2	1 370	2 230	4 840
X134*	1.31	17 100	415	2 330	156	2 560	151	817	94.0	1 240	1 680	4 310
W360												
X122*	1.19	15 500	365	2 010	153	2 270	61.5	478	63.0	732	2 110	1 790
X110*	1.08	14 000	331	1 840	154	2 060	55.7	435	63.1	664	1 610	1 610
X101*	0.993	12 900	302	1 690	153	1 880	50.6	397	62.6	606	1 260	1 450
X91*	0.891	11 600	267	1 510	152	1 680	44.8	353	62.1	538	916	1 270
W360												
X79	0.777	10 100	227	1 280	150	1 430	24.2	236	48.9	362	814	687
X72	0.701	9 110	201	1 150	149	1 280	21.4	210	48.5	322	603	600
X64	0.627	8 140	178	1 030	148	1 140	18.8	186	48.1	284	438	524
W360												
X57	0.556	7 220	161	897	149	1 010	11.1	129	39.2	200	334	331
X51	0.496	6 450	141	796	148	894	9.68	113	38.7	174	238	285
X45	0.441	5 730	122	691	146	779	8.18	95.7	37.8	148	160	239
W360												
X39	0.384	4 980	102	580	143	662	3.75	58.6	27.4	91.7	151	110
X33	0.321	4 170	82.7	474	141	542	2.91	45.8	26.4	71.8	85.9	84.3

‡ Nominal depth in millimetres and mass in kilograms per metre.
* Not available from Canadian mills.

When subject to tension, bolted connections are preferred for these sections.

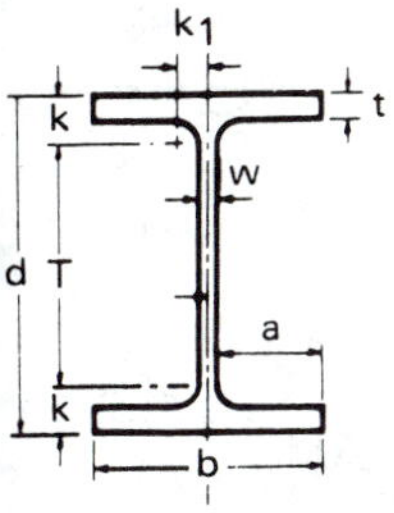

W SHAPES
W360

DIMENSIONS AND SURFACE AREAS

Nominal Mass	Theo-retical Mass	Depth d	Flange Width b	Flange Mean Thickness t		Web Thickness w		Distances a	Distances T	Distances k	Distances k_1	Distances d-2t	Surface Area (m^2) per metre of length: Total	Surface Area (m^2) per metre of length: Minus Top of Top Flange
kg/m	kg/m	mm	mm	mm	mm	mm	mm	mm	mm	mm	mm	mm		
1086	1087.9	569	454	125	125	78.0	78	188	286	142	54	320	2.80	2.35
990	991.0	550	448	115	115	71.9	72	188	287	132	51	320	2.75	2.30
900	902.2	531	442	106	106	65.9	66	188	286	123	48	320	2.70	2.26
818	819.0	514	437	97.0	97	60.5	60	189	287	114	45	320	2.66	2.22
744	744.3	498	432	88.9	89	55.6	56	188	287	106	43	320	2.61	2.18
677	677.8	483	428	81.5	82	51.2	51	189	286	99	41	320	2.58	2.15
634	634.3	474	424	77.1	77	47.6	48	188	287	94	39	320	2.55	2.12
592	592.6	465	421	72.3	72	45.0	45	188	288	89	38	320	2.52	2.10
551	550.6	455	418	67.6	68	42.0	42	188	286	85	36	320	2.50	2.08
509	509.5	446	416	62.7	63	39.1	39	189	287	80	35	320	2.48	2.06
463	462.8	435	412	57.4	57	35.8	36	188	288	74	33	320	2.44	2.03
421	421.7	425	409	52.6	53	32.8	33	188	286	70	32	320	2.42	2.01
382	382.4	416	406	48.0	48	29.8	30	188	287	65	30	320	2.40	1.99
347	347.0	407	404	43.7	44	27.2	27	189	286	61	29	320	2.38	1.97
314	313.4	399	401	39.6	40	24.9	25	188	286	57	28	320	2.35	1.95
287	287.6	393	399	36.6	37	22.6	23	188	286	54	27	320	2.34	1.94
262	262.7	387	398	33.3	33	21.1	21	189	288	50	26	320	2.32	1.92
237	236.3	380	395	30.2	30	18.9	19	188	287	47	25	320	2.30	1.91
216	216.3	375	394	27.7	28	17.3	17	189	286	45	24	320	2.29	1.90
196	196.5	372	374	26.2	26	16.4	16	179	287	43	23	320	2.21	1.83
179	179.2	368	373	23.9	24	15.0	15	179	287	41	23	320	2.20	1.82
162	162.0	364	371	21.8	22	13.3	13	179	287	39	22	320	2.19	1.81
147	147.5	360	370	19.8	20	12.3	12	179	287	37	21	320	2.18	1.81
134	134.0	356	369	18.0	18	11.2	11	179	287	35	21	320	2.17	1.80
122	121.7	363	257	21.7	22	13.0	13	122	280	41	24	320	1.73	1.47
110	110.2	360	256	19.9	20	11.4	11	123	281	39	23	320	1.72	1.47
101	101.2	357	255	18.3	18	10.5	10	123	282	37	23	320	1.71	1.46
91	90.8	353	254	16.4	16	9.5	10	122	282	35	23	320	1.70	1.45
79	79.3	354	205	16.8	17	9.4	9	98	281	36	22	320	1.51	1.30
72	71.5	350	204	15.1	15	8.6	9	98	281	34	22	320	1.50	1.29
64	63.9	347	203	13.5	14	7.7	8	98	280	33	22	320	1.49	1.29
57	56.7	358	172	13.1	13	7.9	8	82	298	30	20	332	1.39	1.22
51	50.6	355	171	11.6	12	7.2	7	82	297	29	19	332	1.38	1.21
45	45.0	352	171	9.8	10	6.9	7	82	298	27	19	332	1.37	1.20
39	39.1	353	128	10.7	11	6.5	6	61	297	28	19	332	1.21	1.08
33	32.8	349	127	8.5	8	5.8	6	61	299	25	19	332	1.19	1.06

Sections highlighted in yellow are commonly used sizes and are generally readily available.

W SHAPES
W310

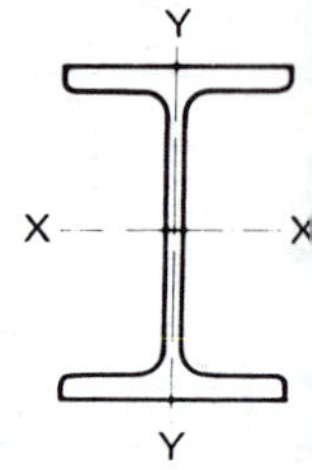

PROPERTIES

Designation ‡	Dead Load	Total Area	Axis X-X				Axis Y-Y				Torsional Constant	Warping Constant
			I_x	S_x	r_x	Z_x	I_y	S_y	r_y	Z_y	J	C_w
	kN/m	mm^2	$10^6 mm^4$	$10^3 mm^3$	mm	$10^3 mm^3$	$10^6 mm^4$	$10^3 mm^3$	mm	$10^3 mm^3$	$10^3 mm^4$	$10^9 mm^6$
W310												
X500*	4.90	63 700	1 690	7 910	163	9 880	494	2 910	88.1	4 490	101 000	15 300
X454*	4.45	57 800	1 480	7 130	160	8 820	436	2 600	86.9	4 000	77 200	13 100
X415*	4.07	52 900	1 300	6 450	157	7 900	391	2 340	86.0	3 610	59 500	11 300
X375*	3.67	47 700	1 130	5 760	154	7 000	343	2 080	84.8	3 200	44 700	9 560
X342*	3.36	43 700	1 010	5 260	152	6 330	310	1 890	84.2	2 910	34 900	8 420
X313*	3.07	39 900	896	4 790	150	5 720	277	1 700	83.3	2 620	27 000	7 350
W310												
X283*	2.77	36 000	787	4 310	148	5 100	246	1 530	82.7	2 340	20 400	6 330
X253*	2.48	32 200	682	3 830	146	4 490	215	1 350	81.7	2 060	14 800	5 370
X226	2.22	28 900	596	3 420	144	3 980	189	1 190	80.9	1 830	10 800	4 620
X202	1.98	25 800	520	3 050	142	3 510	166	1 050	80.2	1 610	7 740	3 960
X179	1.75	22 800	445	2 680	140	3 050	144	919	79.5	1 400	5 380	3 340
X158	1.54	20 100	386	2 360	139	2 670	125	805	78.9	1 220	3 780	2 840
X143	1.40	18 200	348	2 150	138	2 420	113	729	78.8	1 110	2 870	2 540
X129	1.27	16 500	308	1 940	137	2 160	100	652	77.8	991	2 130	2 220
X118	1.15	15 000	275	1 750	135	1 950	90.2	588	77.5	893	1 600	1 970
X107	1.04	13 600	248	1 590	135	1 770	81.2	531	77.3	806	1 220	1 760
X97	0.950	12 300	222	1 440	134	1 590	72.9	478	77.0	725	912	1 560
W310												
X86	0.847	11 000	199	1 280	135	1 420	44.5	351	63.6	533	877	961
X79	0.774	10 100	177	1 160	132	1 280	39.9	314	62.9	478	657	847
W310												
X74	0.730	9 490	165	1 060	132	1 190	23.4	229	49.7	350	745	505
X67	0.655	8 510	145	949	131	1 060	20.7	203	49.3	310	545	439
X60	0.585	7 590	129	849	130	941	18.3	180	49.1	275	397	384
W310												
X52	0.513	6 670	118	747	133	837	10.3	123	39.3	189	308	237
X45	0.438	5 690	99.2	634	132	708	8.55	103	38.8	158	191	195
X39	0.380	4 940	85.1	549	131	610	7.27	88.1	38.4	135	126	164
X31 †	0.310	4 020	67.2	439	129	488	5.45	66.4	36.8	102	73.1	121
W310												
X33*	0.321	4 180	65.0	415	125	480	1.92	37.6	21.4	59.6	122	43.8
X28*	0.278	3 610	54.3	351	123	407	1.58	31.0	20.9	49.2	75.7	35.6
X24*	0.234	3 040	42.7	280	119	328	1.16	22.9	19.5	36.7	42.5	25.7
X21*	0.207	2 690	37.0	244	117	287	0.983	19.5	19.1	31.2	29.4	21.7

‡ Nominal depth in millimetres and mass in kilograms per metre.
* Not available from Canadian mills.
† Produced exclusively by Algoma Steel

☐ When subject to tension, bolted connections are preferred for these sections.

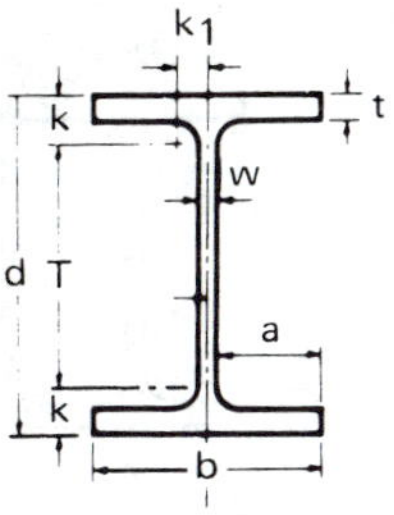

W SHAPES
W310

DIMENSIONS AND SURFACE AREAS

Nominal Mass	Theo-retical Mass	Depth d	Flange Width b	Flange Mean Thickness t		Web Thickness w		Distances a	Distances T	Distances k	Distances k_1	Distances d-2t	Surface Area (m^2) per metre of length: Total	Surface Area (m^2) per metre of length: Minus Top of Top Flange
kg/m	kg/m	mm	mm	mm	mm	mm	mm	mm	mm	mm	mm	mm		
500	500.4	427	340	75.1	75	45.1	45	148	244	92	38	277	2.12	1.78
454	454.0	415	336	68.7	69	41.3	41	148	244	86	36	277	2.09	1.76
415	415.1	403	334	62.7	63	38.9	39	148	244	80	35	277	2.06	1.73
375	374.3	391	330	57.1	57	35.4	35	148	244	74	33	277	2.03	1.70
342	343.3	382	328	52.6	53	32.6	33	148	243	70	32	277	2.01	1.68
313	313.3	374	325	48.3	48	30.0	30	148	245	65	30	277	1.99	1.66
283	283.0	365	322	44.1	44	26.9	27	148	243	61	29	277	1.96	1.64
253	252.9	356	319	39.6	40	24.4	24	148	242	57	28	277	1.94	1.62
226	226.8	348	317	35.6	36	22.1	22	148	242	53	27	277	1.92	1.60
202	202.6	341	315	31.8	32	20.1	20	148	243	49	26	277	1.90	1.59
179	178.8	333	313	28.1	28	18.0	18	148	243	45	25	277	1.88	1.57
158	157.4	327	310	25.1	25	15.5	16	147	243	42	24	277	1.86	1.55
143	143.1	323	309	22.9	23	14.0	14	148	243	40	23	277	1.85	1.54
129	129.6	318	308	20.6	21	13.1	13	148	242	38	22	277	1.84	1.54
118	117.5	314	307	18.7	19	11.9	12	148	242	36	22	277	1.83	1.53
107	106.9	311	306	17.0	17	10.9	11	148	243	34	21	277	1.82	1.52
97	96.8	308	305	15.4	15	9.9	10	148	244	32	21	277	1.81	1.51
86	86.4	310	254	16.3	16	9.1	9	123	244	33	20	277	1.62	1.36
79	78.9	306	254	14.6	15	8.8	9	123	242	32	20	277	1.61	1.36
74	74.5	310	205	16.3	16	9.4	9	98	244	33	20	277	1.42	1.21
67	66.8	306	204	14.6	15	8.5	8	98	242	32	20	277	1.41	1.21
60	59.6	303	203	13.1	13	7.5	8	98	243	30	20	277	1.40	1.20
52	52.3	317	167	13.2	13	7.6	8	80	257	30	20	291	1.29	1.12
45	44.6	313	166	11.2	11	6.6	7	80	257	28	19	291	1.28	1.11
39	38.7	310	165	9.7	10	5.8	6	80	256	27	19	291	1.27	1.10
31	31.6	306	164	7.4	7	5.0	5	80	256	25	18	291	1.26	1.09
33	32.8	313	102	10.8	11	6.6	7	48	269	22	13	291	1.02	0.918
28	28.4	309	102	8.9	9	6.0	6	48	269	20	12	291	1.01	0.912
24	23.8	305	101	6.7	7	5.6	6	48	269	18	12	291	1.00	0.901
21	21.1	303	101	5.7	6	5.1	5	48	269	17	12	291	1.00	0.899

Sections highlighted in yellow are commonly used sizes and are generally readily available.

W SHAPES
W250 - W200

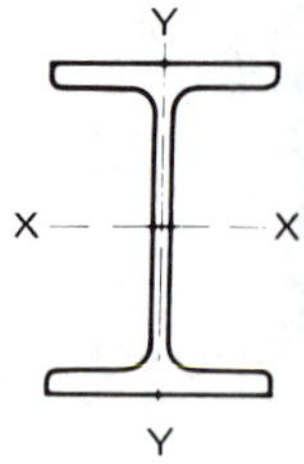

PROPERTIES

Designation‡	Dead Load	Total Area	Axis X-X				Axis Y-Y				Torsional Constant	Warping Constant
			I_x	S_x	r_x	Z_x	I_y	S_y	r_y	Z_y	J	C_w
	kN/m	mm^2	$10^6 mm^4$	$10^3 mm^3$	mm	$10^3 mm^3$	$10^6 mm^4$	$10^3 mm^3$	mm	$10^3 mm^3$	$10^3 mm^4$	$10^9 mm^6$
W250												
X167	1.64	21 300	300	2 080	119	2 430	98.8	746	68.1	1 140	6 310	1 630
X149	1.46	19 000	259	1 840	117	2 130	86.2	656	67.4	1 000	4 510	1 390
X131	1.28	16 700	221	1 610	115	1 850	74.5	571	66.8	870	3 120	1 160
X115	1.12	14 600	189	1 410	114	1 600	64.1	495	66.3	753	2 130	976
X101	0.992	12 900	164	1 240	113	1 400	55.5	432	65.6	656	1 490	829
X89	0.878	11 400	143	1 100	112	1 230	48.4	378	65.2	574	1 040	713
X80	0.786	10 200	126	982	111	1 090	43.1	338	65.0	513	757	623
X73	0.715	9 280	113	891	110	985	38.8	306	64.7	463	575	553
W250												
X67	0.658	8 550	104	806	110	901	22.2	218	51.0	332	625	324
X58	0.571	7 420	87.3	693	108	770	18.8	186	50.3	283	409	268
X49	0.481	6 250	70.6	572	106	633	15.1	150	49.2	228	241	211
W250												
X45	0.440	5 720	71.1	534	111	602	7.03	95.1	35.1	146	261	113
X39	0.379	4 920	60.1	459	111	513	5.94	80.8	34.7	124	169	93.4
X33	0.321	4 170	48.9	379	108	424	4.73	64.7	33.7	99.5	98.5	73.2
X24†	0.239	3 110	34.7	275	106	307	3.26	44.9	32.4	69.0	40.1	49.5
W250												
X28*	0.279	3 630	40.0	307	105	353	1.78	34.8	22.1	54.7	96.7	27.7
X25*	0.249	3 230	34.2	266	103	307	1.49	29.2	21.5	46.2	65.2	23.0
X22*	0.219	2 850	28.9	227	101	263	1.23	24.0	20.8	38.1	43.4	18.7
X18*	0.175	2 270	22.4	179	99.3	207	0.913	18.1	20.1	28.6	22.4	13.8
W200												
X100	0.976	12 700	113	989	94.3	1 150	36.6	349	53.7	533	2 090	386
X86	0.851	11 100	94.7	853	92.4	981	31.4	300	53.2	458	1 400	318
X71	0.701	9 110	76.6	709	91.7	803	25.4	246	52.8	375	818	250
X59	0.582	7 560	61.1	582	89.9	653	20.4	199	51.9	303	465	196
X52	0.513	6 660	52.7	512	89.0	570	17.8	175	51.7	266	324	167
X46	0.451	5 860	45.5	448	88.1	496	15.3	151	51.1	230	221	141
W200												
X42	0.409	5 310	40.9	399	87.8	446	9.00	108	41.2	165	223	84.0
X36	0.352	4 580	34.4	342	86.7	380	7.64	92.6	40.8	141	146	69.6
W200												
X31	0.308	4 000	31.4	299	88.6	335	4.10	61.1	32.0	93.8	119	40.9
X27	0.261	3 390	25.8	249	87.2	279	3.30	49.6	31.2	76.1	71.3	32.5
X21†	0.208	2 710	19.8	195	85.5	218	2.51	37.8	30.4	58.0	35.9	24.3
W200												
X22*	0.220	2 860	20.0	194	83.6	222	1.42	27.8	22.3	43.7	56.6	13.9
X19*	0.191	2 480	16.6	163	81.8	187	1.15	22.6	21.5	35.6	36.2	11.1
X15*	0.147	1 900	12.7	127	81.8	145	0.869	17.4	21.4	27.1	17.6	8.24

‡ Nominal depth in millimetres and mass in kilograms per metre.
* Not available from Canadian mills.
† Produced exclusively by Algoma Steel

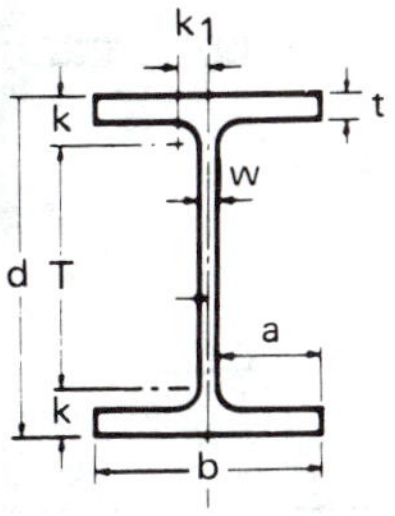

DIMENSIONS AND SURFACE AREAS

Nominal Mass	Theo-retical Mass	Depth d	Flange Width b	Flange Mean Thickness t		Web Thickness w		Distances a	T	k	k1	d-2t	Surface Area (m²) per metre of length Total	Minus Top of Top Flange
kg/m	kg/m	mm	mm	mm	mm	mm	mm	mm	mm	mm	mm	mm		
167	167.4	289	265	31.8	32	19.2	19	123	191	49	25	225	1.60	1.33
149	148.9	282	263	28.4	28	17.3	17	123	192	45	24	225	1.58	1.32
131	131.1	275	261	25.1	25	15.4	15	123	191	42	23	225	1.56	1.30
115	114.8	269	259	22.1	22	13.5	14	123	191	39	23	225	1.55	1.29
101	101.2	264	257	19.6	20	11.9	12	123	190	37	22	225	1.53	1.28
89	89.6	260	256	17.3	17	10.7	11	123	192	34	21	225	1.52	1.26
80	80.1	256	255	15.6	16	9.4	9	123	190	33	20	225	1.52	1.26
73	72.9	253	254	14.2	14	8.6	9	123	191	31	20	225	1.50	1.25
67	67.1	257	204	15.7	16	8.9	9	98	191	33	20	225	1.31	1.11
58	58.2	252	203	13.5	14	8.0	8	98	190	31	20	225	1.30	1.10
49	49.0	247	202	11.0	11	7.4	7	98	191	28	19	225	1.29	1.09
45	44.9	266	148	13.0	13	7.6	8	70	220	23	12	240	1.11	0.960
39	38.7	262	147	11.2	11	6.6	7	70	220	21	12	240	1.10	0.951
33	32.7	258	146	9.1	9	6.1	6	70	220	19	11	240	1.09	0.942
24	24.4	253	145	6.4	6	5.0	5	70	220	17	12	240	1.08	0.930
28	28.5	260	102	10.0	10	6.4	6	48	221	20	11	240	0.916	0.814
25	25.3	257	102	8.4	8	6.1	6	48	222	18	11	240	0.908	0.806
22	22.4	254	102	6.9	7	5.8	6	48	221	17	11	240	0.904	0.802
18	17.9	251	101	5.3	5	4.8	5	48	222	15	11	240	0.894	0.793
100	99.5	229	210	23.7	24	14.5	14	98	156	37	18	181	1.27	1.06
86	86.7	222	209	20.6	21	13.0	13	98	155	34	18	181	1.26	1.05
71	71.5	216	206	17.4	17	10.2	10	98	157	30	16	181	1.23	1.03
59	59.4	210	205	14.2	14	9.1	9	98	157	27	16	181	1.22	1.01
52	52.3	206	204	12.6	13	7.9	8	98	155	26	15	181	1.21	1.01
46	46.0	203	203	11.0	11	7.2	7	98	156	24	15	181	1.20	1.00
42	41.7	205	166	11.8	12	7.2	7	80	156	25	15	181	1.06	0.894
36	35.9	201	165	10.2	10	6.2	6	80	156	23	14	181	1.05	0.885
31	31.4	210	134	10.2	10	6.4	6	64	171	20	11	190	0.944	0.810
27	26.6	207	133	8.4	8	5.8	6	64	172	18	11	190	0.932	0.799
21	21.2	203	133	6.4	6	5.0	5	64	171	16	10	190	0.927	0.794
22	22.4	206	102	8.0	8	6.2	6	48	171	18	11	190	0.808	0.706
19	19.4	203	102	6.5	6	5.8	6	48	172	16	11	190	0.800	0.698
15	15.0	200	100	5.2	5	4.3	4	48	171	15	10	190	0.792	0.692

Sections highlighted in yellow are commonly used sizes and are generally readily available.

W SHAPES
W150 - W100

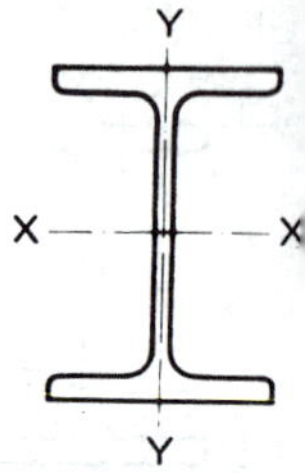

PROPERTIES

Designation ‡	Dead Load	Total Area	Axis X-X				Axis Y-Y				Torsional Constant	Warping Constant
			I_x	S_x	r_x	Z_x	I_y	S_y	r_y	Z_y	J	C_w
	kN/m	mm^2	$10^6 mm^4$	$10^3 mm^3$	mm	$10^3 mm^3$	$10^6 mm^4$	$10^3 mm^3$	mm	$10^3 mm^3$	$10^3 mm^4$	$10^9 mm^6$
W150												
X37	0.364	4 730	22.2	274	68.5	310	7.07	91.8	38.7	140	193	40.0
X30	0.292	3 790	17.2	219	67.4	244	5.56	72.6	38.3	111	101	30.3
X22	0.219	2 850	12.1	159	65.2	176	3.87	50.9	36.8	77.6	41.8	20.4
W150												
X24*	0.235	3 060	13.4	168	66.2	192	1.83	35.8	24.5	55.3	92.9	10.2
X18*	0.176	2 290	9.16	120	63.2	136	1.26	24.7	23.5	38.2	37.3	6.70
X14*	0.133	1 730	6.87	91.5	63.0	103	0.918	18.4	23.0	28.3	17.0	4.79
W130												
X28*	0.275	3 580	10.9	167	55.2	190	3.81	59.6	32.6	90.8	128	13.8
X24*	0.232	3 010	8.80	139	54.1	156	3.11	49.0	32.1	74.6	76.7	10.8
W100												
X19*	0.190	2 470	4.76	89.9	43.9	103	1.61	31.2	25.5	48.0	63.4	3.79

‡ Nominal depth in millimetres and mass in kilograms per metre.
* Not available from Canadian mills.

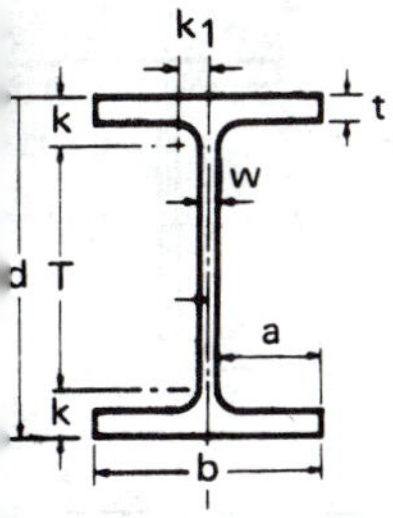

W SHAPES
W150 - W100

DIMENSIONS AND SURFACE AREAS

Nominal Mass	Theo-retical Mass	Depth d	Flange			Web Thickness w		Distances					Surface Area (m^2) per metre of length	
			Width b	Mean Thickness t				a	T	k	k_1	d-2t	Total	Minus Top of Top Flange
kg/m	kg/m	mm	mm	mm	mm	mm	mm	mm	mm	mm	mm	mm		
37	37.1	162	154	11.6	12	8.1	8	73	122	20	10	139	0.926	0.772
30	29.8	157	153	9.3	9	6.6	7	73	123	17	10	139	0.912	0.759
22	22.3	152	152	6.6	7	5.8	6	73	122	15	9	139	0.902	0.750
24	24.0	160	102	10.3	10	6.6	7	48	124	18	10	139	0.712	0.610
18	18.0	153	102	7.1	7	5.8	6	48	123	15	9	139	0.702	0.600
14	13.6	150	100	5.5	6	4.3	4	48	122	14	8	139	0.694	0.594
28	28.1	131	128	10.9	11	6.9	7	61	93	19	10	109	0.760	0.632
24	23.6	127	127	9.1	9	6.1	6	61	93	17	9	109	0.750	0.623
19	19.4	106	103	8.8	9	7.1	7	48	72	17	10	88	0.610	0.507

Sections highlighted in yellow are commonly used sizes and are generally readily available.

HP SHAPES

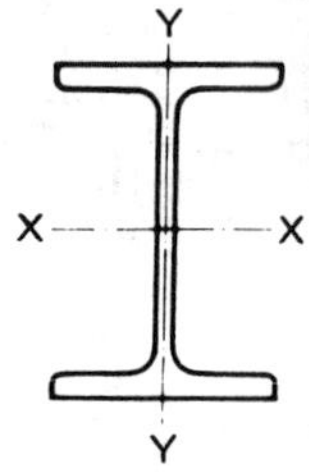

PROPERTIES

Designation‡	Dead Load	Total Area	Axis X-X				Axis Y-Y				Torsional Constant	Warping Constant
			I_x	S_x	r_x	Z_x	I_y	S_y	r_y	Z_y	J	C_w
	kN/m	mm^2	$10^6 mm^4$	$10^3 mm^3$	mm	$10^3 mm^3$	$10^6 mm^4$	$10^3 mm^3$	mm	$10^3 mm^3$	$10^3 mm^4$	$10^9 mm^6$
HP360												
X174*	1.70	22 200	508	2 820	151	3 180	184	973	91.0	1 490	3 310	5 330
X152*	1.49	19 400	439	2 470	150	2 770	159	845	90.5	1 290	2 240	4 540
X132*	1.29	16 800	375	2 140	149	2 380	135	724	89.6	1 110	1 490	3 800
X108*	1.06	13 800	303	1 750	148	1 940	108	585	88.5	891	832	3 000
HP330												
X149*	1.45	18 900	368	2 200	140	2 500	122	727	80.3	1 120	2 590	3 010
X129*	1.26	16 400	315	1 910	139	2 150	104	626	79.6	960	1 720	2 540
X109*	1.07	14 000	263	1 620	137	1 810	86.3	523	78.5	801	1 070	2 070
X89*	0.872	11 300	211	1 320	137	1 460	68.9	420	78.1	641	583	1 630
HP310												
X174†	1.70	22 200	394	2 430	133	2 800	138	843	78.8	1 300	4 470	3 110
X152†	1.48	19 300	338	2 120	132	2 420	115	716	77.2	1 100	3 040	2 550
X132†	1.28	16 700	287	1 830	131	2 070	93.7	599	74.9	922	2 050	2 050
X125*	1.22	15 900	270	1 730	130	1 960	88.2	566	74.5	870	1 760	1 910
X110	1.08	14 100	237	1 540	130	1 730	77.1	497	73.9	763	1 240	1 650
X94	0.916	11 900	196	1 300	128	1 450	63.9	415	73.3	635	764	1 340
X79	0.768	9 980	163	1 090	128	1 210	52.6	344	72.6	525	460	1 090
HP250												
X85	0.837	10 900	123	968	106	1 090	42.3	325	62.3	500	829	606
X62	0.614	7 980	87.5	711	105	792	30.0	234	61.3	358	339	415
HP200												
X54	0.525	6 820	49.8	488	85.5	552	16.7	162	49.5	249	321	155

‡ Nominal depth in millimetres and mass in kilograms per metre.
* Not available from Canadian mills.
† Produced exclusively by Algoma Steel.

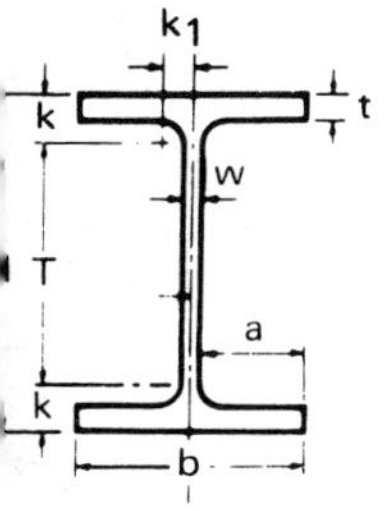

HP SHAPES

DIMENSIONS AND SURFACE AREAS

Nominal Mass	Theo-retical Mass	Depth d	Flange Width b	Flange Mean Thickness t		Web Thickness w		Distances a	Distances T	Distances k	Distances k_1	Distances d-2t	Surface Area (m^2) per metre of length Total	Surface Area (m^2) per metre of length Minus Top of Top Flange
kg/m	kg/m	mm	mm	mm	mm	mm	mm	mm	mm	mm	mm	mm		
174	173.9	361	378	20.4	20	20.4	20	179	289	36	25	320	2.19	1.81
152	152.2	356	376	17.9	18	17.9	18	179	288	34	24	320	2.18	1.80
132	132.1	351	373	15.6	16	15.6	16	179	287	32	23	320	2.16	1.79
108	108.1	346	370	12.8	13	12.8	13	179	288	29	22	320	2.15	1.78
149	148.5	334	335	19.4	19	19.4	19	158	263	36	25	295	1.97	1.63
129	129.1	329	333	16.9	17	16.9	17	158	262	34	24	295	1.96	1.62
109	109.5	324	330	14.4	14	14.4	14	158	263	31	23	295	1.94	1.61
89	89.0	319	328	11.7	12	11.7	12	158	262	29	22	295	1.93	1.60
174	174.0	324	327	23.6	24	23.6	24	152	243	41	28	277	1.91	1.58
152	151.7	319	321	20.8	21	20.8	21	150	244	38	26	277	1.88	1.56
132	131.3	314	313	18.3	18	18.3	18	148	245	35	25	277	1.84	1.53
125	124.7	312	312	17.4	17	17.4	17	148	245	34	24	277	1.84	1.52
110	110.5	308	310	15.5	16	15.4	15	148	243	33	23	277	1.83	1.52
94	93.4	303	308	13.1	13	13.1	13	148	244	30	22	277	1.81	1.50
79	78.3	299	306	11.0	11	11.0	11	148	244	28	21	277	1.80	1.49
85	85.3	254	260	14.4	14	14.4	14	123	193	31	23	225	1.52	1.26
62	62.6	246	256	10.7	11	10.5	10	123	191	28	21	225	1.50	1.24
54	53.5	204	207	11.3	11	11.3	11	98	158	23	17	181	1.21	1.00

M SHAPES

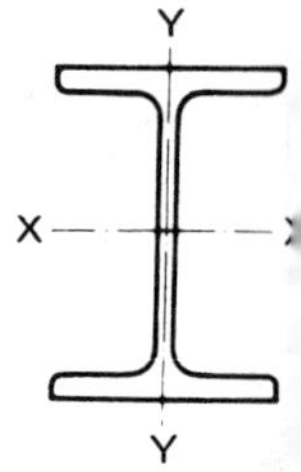

PROPERTIES

Designation ‡	Dead Load	Total Area	Axis X-X				Axis Y-Y				Torsional Constant	Warping Constant
			I_x	S_x	r_x	Z_x	I_y	S_y	r_y	Z_y	J	C_w
	kN/m	mm^2	$10^6 mm^4$	$10^3 mm^3$	mm	$10^3 mm^3$	$10^6 mm^4$	$10^3 mm^3$	mm	$10^3 mm^3$	$10^3 mm^4$	$10^9 mm^6$
M310 X17.6*	0.171	2 230	29.7	195	115	232	0.394	10.2	13.3	17.3	21.8	8.82
M250 X13.4*	0.131	1 700	16.1	126	97.3	150	0.248	7.30	12.1	12.3	13.7	3.84
M200 X9.7*	0.094	1 220	7.56	74.5	78.7	87.5	0.136	4.78	10.6	7.93	7.92	1.34
M150 X29.8*	0.292	3 790	16.1	212	65.2	237	4.87	63.5	35.8	103	121	24.7
M150 X6.5*	0.063	821	2.92	38.4	59.6	44.8	0.065	2.81	8.88	4.63	4.17	0.353
M130 X28.1*	0.276	3 580	10.0	158	52.9	181	3.31	51.2	30.4	82.6	142	11.2
M100 X19*	0.190	2 460	4.41	86.4	42.3	99.8	1.41	28.1	23.9	44.8	78.5	3.02

‡ Nominal depth in millimetres and mass in kilograms per metre.
* Not available from Canadian mills.
Note: These shapes have been soft converted from Imperial to SI dimensions.

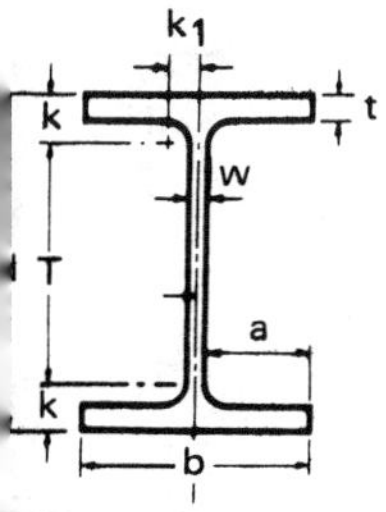

M SHAPES

DIMENSIONS AND SURFACE AREAS

Nominal Mass	Theoretical Mass	Depth d	Flange Width b	Flange Mean Thickness t		Web Thickness w		Distances a	Distances T	Distances k	Distances k_1	Surface Area (m^2) per metre of length: Total	Surface Area (m^2) per metre of length: Minus Top of Top Flange
kg/m	kg/m	mm	mm	mm	mm	mm	mm	mm	mm	mm	mm		
17.6	17.5	305	77	5.7	6	4.5	4	37	277	14	8	0.910	0.833
13.4	13.4	254	68	5.2	5	4.0	4	32	230	12	7	0.772	0.704
9.7	9.6	203	57	4.8	5	3.4	3	27	181	11	6	0.628	0.571
29.8	29.8	152	151	9.6	10	6.4	6	73	111	21	11	0.896	0.745
6.5	6.5	152	46	4.3	4	2.9	3	22	134	9	5	0.482	0.436
28.1	28.1	127	127	10.6	11	8.0	8	60	84	22	12	0.746	0.619
19	19.3	102	100	9.4	9	6.5	6	47	63	19	11	0.592	0.492

S SHAPES
S610 - S180

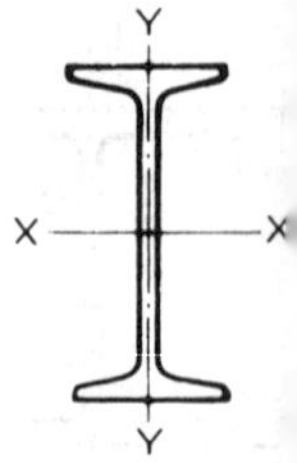

PROPERTIES

Designation #	Dead Load	Total Area	Axis X-X				Axis Y-Y				Torsional Constant	Warping Constant
			I_x	S_x	r_x	Z_x	I_y	S_y	r_y	Z_y	J	C_w
	kN/m	mm^2	$10^6 mm^4$	$10^3 mm^3$	mm	$10^3 mm^3$	$10^6 mm^4$	$10^3 mm^3$	mm	$10^3 mm^3$	$10^3 mm^4$	$10^9 mm^6$
S610												
X180*	1.76	22 900	1 310	4 220	239	5 020	34.7	340	38.9	591	5 330	3 070
X158*	1.54	20 100	1 220	3 940	246	4 580	32.4	324	40.1	544	4 210	2 860
S610												
X149*	1.45	18 900	996	3 270	230	3 930	20.1	218	32.6	392	3 150	1 730
X134*	1.31	17 100	939	3 080	234	3 650	18.9	209	33.2	366	2 520	1 640
X119*	1.16	15 200	879	2 880	240	3 360	17.9	201	34.3	341	2 030	1 540
S510												
X143*	1.40	18 300	702	2 720	196	3 260	21.1	231	34.0	409	3 570	1 280
X128*	1.26	16 400	660	2 560	201	3 020	19.6	219	34.6	377	2 830	1 190
S510												
X112*	1.09	14 200	532	2 090	194	2 500	12.5	155	29.7	273	1 910	745
X98.2*	0.965	12 500	497	1 960	199	2 290	11.7	148	30.6	252	1 490	698
S460												
X104*	1.02	13 300	387	1 690	171	2 050	10.3	129	27.8	237	1 740	495
X81.4*	0.800	10 400	335	1 470	179	1 720	8.77	115	29.0	198	986	423
S380												
X74*	0.731	9 500	203	1 060	146	1 270	6.60	92.3	26.4	163	884	220
X64*	0.627	8 150	187	980	151	1 140	6.11	87.3	27.4	149	641	204
S310												
X74*	0.729	9 470	127	833	116	1 000	6.60	94.9	26.4	168	1 160	137
X60.7*	0.595	7 730	113	744	121	869	5.67	85.3	27.1	144	723	118
S310												
X52*	0.512	6 650	95.8	629	120	736	4.16	64.5	25.0	111	450	88.2
X47*	0.465	6 040	91.1	597	123	690	3.94	62.1	25.5	105	374	83.6
S250												
X52*	0.513	6 660	61.6	485	96.2	584	3.56	56.5	23.1	102	541	51.9
X38*	0.371	4 820	51.4	405	103	465	2.84	48.2	24.3	81.1	251	41.5
S200												
X34*	0.336	4 370	27.0	266	78.6	316	1.81	34.2	20.4	60.2	229	16.8
X27*	0.270	3 500	24.0	237	82.8	272	1.59	31.1	21.3	52.2	140	14.7
S180												
X30*	0.293	3 800	17.8	200	68.4	238	1.34	27.3	18.8	48.4	188	9.44
X22.8*	0.224	2 910	15.4	173	72.7	199	1.12	24.0	19.6	40.2	101	7.88

*Not available from Canadian mills.

For sections available from Canadian mills, the nominal mass has been rounded to the nearest kg/m. Designation consists of nominal depth in millimetres by nominal mass in kilograms per metre.

Note: These shapes have been soft converted from Imperial to SI dimension.

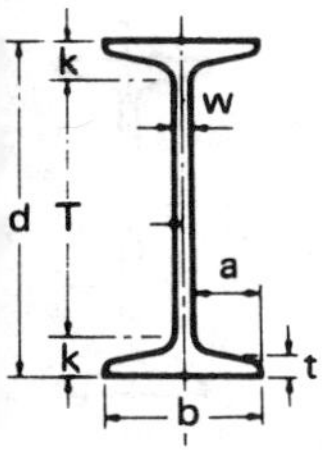

S SHAPES
S610 - S180

DIMENSIONS AND SURFACE AREAS

Nominal Mass	Theoretical Mass	Depth d	Flange			Web Thickness w		Distances			Surface Area (m²) per metre of length	
			Width b	Mean Thickness t				a	T	k	Total	Minus Top of Top Flange
kg/m	kg/m	mm	mm	mm	mm	mm	mm	mm	mm	mm		
180	180.0	622	204	27.7	28	20.3	20	92	522	50	2.02	1.82
158	157.8	622	200	27.7	28	15.7	16	92	522	50	2.01	1.81
149	148.7	610	184	22.1	22	18.9	19	83	524	43	1.92	1.73
134	134.4	610	181	22.1	22	15.9	16	83	524	43	1.91	1.73
119	119.1	610	178	22.1	22	12.7	13	83	524	43	1.91	1.73
143	143.3	516	183	23.4	23	20.3	20	82	424	46	1.72	1.54
128	128.9	516	179	23.4	23	16.8	17	81	424	46	1.71	1.53
112	111.4	508	162	20.2	20	16.1	16	73	428	40	1.63	1.47
98.2	98.4	508	159	20.2	20	12.8	13	73	428	40	1.63	1.47
104	104.7	457	159	17.6	18	18.1	18	71	383	37	1.51	1.35
81.4	81.6	457	152	17.6	18	11.7	12	70	383	37	1.50	1.35
74	74.6	381	143	15.8	16	14.0	14	65	314	34	1.31	1.16
64	63.9	381	140	15.8	16	10.4	10	65	314	34	1.30	1.16
74	74.4	305	139	16.7	17	17.4	17	61	235	35	1.13	0.993
60.7	60.6	305	133	16.7	17	11.7	12	61	235	35	1.12	0.985
52	52.2	305	129	13.8	14	10.9	11	59	246	29	1.10	0.975
47	47.4	305	127	13.8	14	8.9	9	59	246	29	1.10	0.973
52	52.3	254	126	12.5	12	15.1	15	56	201	26	0.982	0.856
38	37.8	254	118	12.5	12	7.9	8	55	202	26	0.964	0.846
34	34.3	203	106	10.8	11	11.2	11	48	155	24	0.808	0.702
27	27.5	203	102	10.8	11	6.9	7	48	155	24	0.800	0.698
30	29.9	178	98	10.0	10	11.4	11	44	134	22	0.726	0.628
22.8	22.9	178	93	10.0	10	6.4	6	44	134	22	0.716	0.623

S SHAPES
S150 - S75

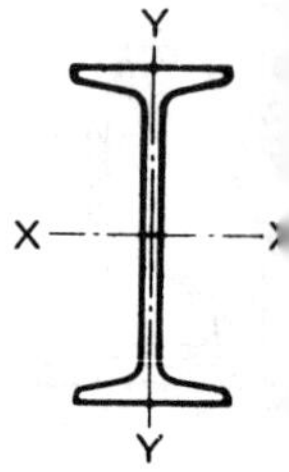

PROPERTIES

Designation#	Dead Load	Total Area	Axis X-X				Axis Y-Y				Torsional Constant	Warping Constant
			I_x	S_x	r_x	Z_x	I_y	S_y	r_y	Z_y	J	C_w
	kN/m	mm^2	$10^6 mm^4$	$10^3 mm^3$	mm	$10^3 mm^3$	$10^6 mm^4$	$10^3 mm^3$	mm	$10^3 mm^3$	$10^3 mm^4$	$10^9 mm^6$
S150												
X26*	0.252	3 270	10.9	144	57.7	173	0.981	21.6	17.3	38.7	155	5.01
X19*	0.183	2 370	9.19	121	62.3	139	0.776	18.2	18.1	30.5	70.1	3.96
S130												
X22*	0.215	2 790	6.33	99.6	47.6	121	0.690	16.6	15.7	30.3	133	2.43
X15*	0.146	1 890	5.12	80.6	52.0	92.8	0.508	13.4	16.4	22.2	47.4	1.79
S100												
X14.1*	0.139	1 800	2.85	55.8	39.8	66.5	0.376	10.6	14.5	18.3	49.9	0.842
X11*	0.112	1 450	2.55	50.1	41.9	57.7	0.324	9.52	14.9	15.8	29.9	0.724
S75												
X11*	0.110	1 430	1.22	32.0	29.2	38.7	0.249	7.77	13.2	13.5	38.2	0.299
X8*	0.083	1 070	1.04	27.4	31.2	31.9	0.190	6.43	13.3	10.6	18.3	0.228

*Not available from Canadian mills.

For sections available from Canadian mills, the nominal mass has been rounded to the nearest kg/m. Designation consists of nominal depth in millimetres by nominal mass in kilograms per metre.

Note: These shapes have been soft converted from Imperial to SI dimension.

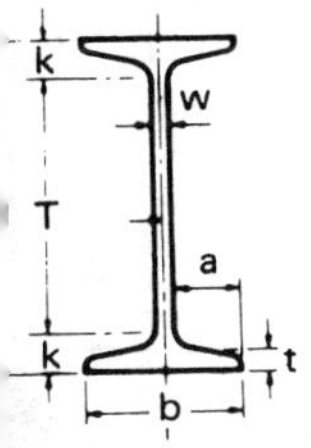

S SHAPES
S150 - S75

DIMENSIONS AND SURFACE AREAS

Nominal Mass	Theo-retical Mass	Depth d	Flange Width b	Flange Mean Thickness t		Web Thickness w		Distances a	Distances T	Distances k	Surface Area (m²) per metre of length: Total	Surface Area (m²) per metre of length: Minus Top of Top Flange
kg/m	kg/m	mm	mm	mm	mm	mm	mm	mm	mm	mm		
26	25.7	152	91	9.1	9	11.8	12	40	111	20	0.644	0.553
19	18.6	152	85	9.1	9	5.9	6	40	111	20	0.632	0.547
22	21.9	127	83	8.3	8	12.5	12	36	90	19	0.562	0.479
15	14.8	127	76	8.3	8	5.4	5	36	90	19	0.548	0.472
14.1	14.2	102	71	7.4	7	8.3	8	32	68	17	0.472	0.401
11	11.4	102	68	7.4	7	4.8	5	32	68	17	0.466	0.398
11	11.2	76	64	6.6	7	8.9	9	28	44	16	0.390	0.326
8	8.4	76	59	6.6	7	4.3	4	28	44	16	0.380	0.321

STANDARD CHANNELS (C SHAPES)

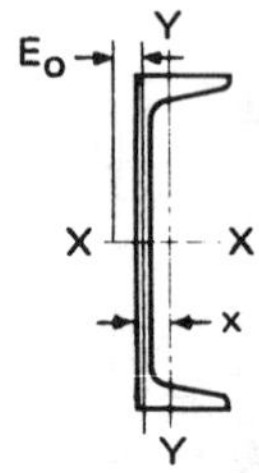

PROPERTIES

Designation#	Dead Load	Total Area	Axis X-X			Axis Y-Y				Shear Centre	Torsional Constant	Warping Constant
			I_x	S_x	r_x	I_y	S_y	r_y	x	E_o	J	C_w
	kN/m	mm^2	10^6 mm^4	10^3 mm^3	mm	10^6 mm^4	10^3 mm^3	mm	mm	mm	10^3 mm^4	10^9 mm^6
C380												
X74*	0.730	9 480	168	881	133	4.60	62.4	22.0	20.3	23.7	1 110	131
X60*	0.583	7 570	145	760	138	3.84	55.5	22.5	19.7	25.9	607	109
X50*	0.495	6 430	131	687	143	3.39	51.4	23.0	20.0	27.6	424	95.2
C310												
X45*	0.438	5 690	67.3	442	109	2.12	33.6	19.3	17.0	21.9	363	39.9
X37*	0.363	4 720	59.9	393	113	1.85	30.9	19.8	17.1	23.7	224	34.6
X31*	0.302	3 920	53.5	351	117	1.59	28.2	20.1	17.5	25.3	153	29.3
C250												
X45*	0.437	5 670	42.8	337	86.9	1.60	26.8	16.8	16.3	17.5	511	20.5
X37	0.365	4 750	37.9	299	89.3	1.40	24.3	17.2	15.7	19.1	290	18.2
X30	0.291	3 780	32.7	257	93.0	1.16	21.5	17.5	15.3	20.8	154	15.0
X23	0.221	2 880	27.8	219	98.2	0.922	18.8	17.9	15.9	22.8	86.8	11.7
C230												
X30	0.292	3 800	25.5	222	81.9	1.01	19.3	16.3	14.8	18.6	180	10.5
X22	0.219	2 840	21.3	186	86.6	0.806	16.8	16.8	14.9	20.9	86.9	8.33
X20	0.195	2 530	19.8	173	88.5	0.716	15.6	16.8	15.1	21.5	69.7	7.35
C200												
X28	0.274	3 560	18.2	180	71.5	0.825	16.6	15.2	14.4	17.0	183	6.67
X21	0.200	2 600	14.9	147	75.7	0.627	13.9	15.5	14.0	19.0	77.8	5.04
X17	0.167	2 170	13.5	133	78.9	0.544	12.8	15.8	14.5	20.3	54.2	4.34
C180												
X22*	0.214	2 780	11.3	127	63.8	0.568	12.8	14.3	13.5	16.4	111	3.47
X18	0.178	2 310	10.0	113	65.8	0.476	11.4	14.4	13.2	17.3	67.3	2.90
X15	0.142	1 850	8.86	99.6	69.2	0.405	10.3	14.8	13.8	19.1	41.6	2.46
C150												
X19	0.189	2 450	7.12	93.7	53.9	0.425	10.3	13.2	12.9	14.9	100	1.84
X16	0.152	1 980	6.22	81.9	56.0	0.351	9.13	13.3	12.6	16.0	54.3	1.53
X12	0.118	1 540	5.36	70.6	59.0	0.279	7.93	13.5	12.8	17.4	31.0	1.21
C130												
X17	0.168	2 190	4.36	68.7	44.6	0.346	8.85	12.6	12.9	14.0	97.2	1.02
X13	0.131	1 700	3.66	57.6	46.4	0.252	7.20	12.2	11.9	14.5	45.7	0.746
X10	0.097	1 260	3.09	48.6	49.5	0.195	6.14	12.4	12.2	16.2	22.8	0.580
C100												
X11	0.106	1 370	1.91	37.4	37.3	0.174	5.52	11.3	11.5	13.5	34.6	0.320
X9	0.092	1 190	1.77	34.6	38.6	0.158	5.18	11.5	11.6	14.5	23.4	0.293
X8	0.079	1 020	1.61	31.6	39.7	0.132	4.65	11.4	11.6	14.9	16.8	0.246
C75												
X9	0.087	1 120	0.85	22.3	27.5	0.123	4.31	10.5	11.4	12.5	30.1	0.118
X7	0.072	933	0.75	19.7	28.3	0.096	3.67	10.1	10.8	12.7	17.7	0.093
X6	0.059	763	0.67	17.6	29.6	0.077	3.21	10.1	10.9	13.5	11.0	0.077

* Not available from Canadian mills.

For sections available from Canadian mills, the nominal mass has been rounded to the nearest kg/m. Designation consists of nominal depth in millimetres by nominal mass in kilograms per metre.

Note: These shapes have been soft converted from Imperial to SI dimensions.

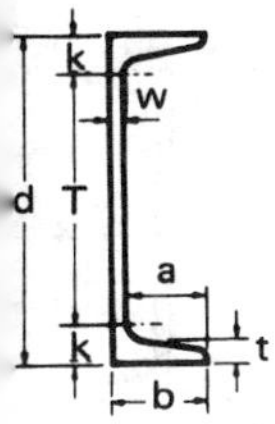

STANDARD CHANNELS (C SHAPES)

DIMENSIONS AND SURFACE AREAS

Nominal Mass	Theoretical Mass	Depth d	Flange: Width b	Flange: Mean Thickness t		Web Thickness w		Distances: a	Distances: T	Distances: k	Surface Area (m^2) per metre of length: Total	Surface Area (m^2) per metre of length: Minus Top of Top Flange
kg/m	kg/m	mm	mm	mm	mm	mm	mm	mm	mm	mm		
74	74.4	381	94	16.5	16	18.2	18	76	312	34	1.10	1.01
60	59.4	381	89	16.5	16	13.2	13	76	312	34	1.09	1.00
50	50.5	381	86	16.5	16	10.2	10	76	312	34	1.09	1.00
45	44.7	305	80	12.7	13	13.0	13	67	250	28	0.904	0.824
37	37.1	305	77	12.7	13	9.8	10	67	250	28	0.898	0.821
31	30.8	305	74	12.7	13	7.2	7	67	250	28	0.892	0.818
45	44.5	254	76	11.1	11	17.1	17	59	206	24	0.778	0.702
37	37.3	254	73	11.1	11	13.4	13	60	206	24	0.774	0.701
30	29.6	254	69	11.1	11	9.6	10	59	206	24	0.764	0.695
23	22.6	254	65	11.1	11	6.1	6	59	206	24	0.756	0.691
30	29.8	229	67	10.5	10	11.4	11	56	184	23	0.704	0.637
22	22.3	229	63	10.5	10	7.2	7	56	184	23	0.696	0.633
20	19.8	229	61	10.5	10	5.9	6	55	184	23	0.690	0.629
28	27.9	203	64	9.9	10	12.4	12	52	159	22	0.638	0.574
21	20.4	203	59	9.9	10	7.7	8	51	159	22	0.626	0.567
17	17.0	203	57	9.9	10	5.6	6	51	159	22	0.622	0.565
22	21.9	178	58	9.3	9	10.6	11	47	137	20	0.566	0.508
18	18.2	178	55	9.3	9	8.0	8	47	137	20	0.560	0.505
15	14.5	178	53	9.3	9	5.3	5	48	137	21	0.558	0.505
19	19.2	152	54	8.7	9	11.1	11	43	112	20	0.498	0.444
16	15.5	152	51	8.7	9	8.0	8	43	112	20	0.492	0.441
12	12.1	152	48	8.7	9	5.1	5	43	112	20	0.486	0.438
17	17.2	127	52	8.1	8	12.0	12	40	90	18	0.438	0.386
13	13.3	127	47	8.1	8	8.3	8	39	90	18	0.426	0.379
10	9.9	127	44	8.1	8	4.8	5	39	90	18	0.420	0.376
11	10.8	102	43	7.5	8	8.2	8	35	66	18	0.360	0.317
9	9.4	102	42	7.5	8	6.3	6	36	66	18	0.360	0.318
8	8.0	102	40	7.5	8	4.7	5	35	66	18	0.354	0.314
9	8.8	76	40	6.9	7	9.0	9	31	44	16	0.294	0.254
7	7.3	76	37	6.9	7	6.6	7	30	44	16	0.286	0.249
6	6.0	76	35	6.9	7	4.3	4	31	44	16	0.284	0.249

MISCELLANEOUS CHANNELS
MC460 - MC 200

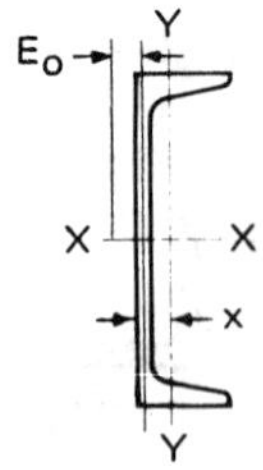

PROPERTIES

Designation#	Dead Load	Total Area	Axis X-X			Axis Y-Y				Shear Centre	Torsional Constant	Warping Constant
			I_x	S_x	r_x	I_y	S_y	r_y	x	E_o	J	C_w
	kN/m	mm²	10^6 mm⁴	10^3 mm³	mm	10^6 mm⁴	10^3 mm³	mm	mm	mm	10^3 mm⁴	10^9 mm⁶
MC460												
X86*	0.851	11 100	283	1 240	160	7.45	87.6	25.9	22.0	26.7	1 190	291
X77.2*	0.759	9 860	262	1 150	163	6.81	82.8	26.3	21.8	27.9	856	265
X68.2*	0.672	8 730	243	1 060	167	6.35	79.5	27.0	22.1	29.6	621	245
X63.5*	0.625	8 120	231	1 010	169	5.94	76.3	27.0	22.2	30.2	524	228
MC330												
X74*	0.731	9 490	131	793	117	6.92	79.4	27.0	24.8	30.7	1 240	150
X60*	0.583	7 570	113	687	122	5.73	70.4	27.5	24.5	33.3	649	123
X52*	0.511	6 640	105	635	126	5.14	65.8	27.8	24.9	34.8	475	110
X47.3*	0.465	6 040	99.8	605	129	4.86	63.7	28.4	25.7	36.6	394	103
MC310												
X74*	0.732	9 510	113	739	109	7.25	92.4	27.6	26.6	29.4	1 360	111
X67*	0.658	8 540	105	690	111	6.60	87.2	27.8	26.3	30.6	982	101
X60*	0.584	7 590	97.5	639	113	5.80	80.5	27.6	26.0	31.3	716	88.7
X52*	0.512	6 660	90.8	595	117	5.31	76.8	28.2	26.8	33.3	527	81.2
X46*	0.455	5 900	84.7	555	120	4.69	71.5	28.2	27.4	34.3	426	71.8
MC310												
X15.8*	0.153	1 990	23.0	151	108	0.159	5.11	8.94	6.8	9.6	24.6	3.11
MC250												
X61.2*	0.603	7 830	66.1	520	91.9	6.73	81.9	29.3	27.8	32.2	952	73.1
X50*	0.493	6 400	58.3	459	95.4	5.57	73.0	29.5	27.7	34.2	509	60.5
X42.4*	0.418	5 430	53.0	417	98.8	4.79	66.9	29.7	28.4	36.1	335	52.0
MC250												
X37*	0.366	4 750	45.8	361	98.2	3.03	48.9	25.3	24.0	30.6	267	33.1
X33*	0.321	4 170	42.8	337	101	2.71	45.9	25.5	25.0	32.1	214	29.7
MC250												
X12.5*	0.122	1 580	13.3	105	91.7	0.138	4.49	9.35	7.2	10.6	17.2	1.87
MC250												
X9.7*	0.095	1 240	9.16	72.2	85.9	0.043	1.82	5.88	4.4	5.9	7.86	0.617
MC230												
X37.8*	0.371	4 830	36.8	322	87.3	3.10	48.7	25.3	24.3	30.4	296	27.5
X35.6*	0.351	4 560	35.7	312	88.5	2.96	47.4	25.5	24.6	31.1	259	26.3
MC200												
X33.9*	0.332	4 310	26.4	260	78.3	2.85	45.4	25.7	25.2	31.5	242	19.7
X31.8*	0.312	4 050	25.6	252	79.5	2.71	44.1	25.9	25.7	32.3	210	18.8
MC200												
X29.8*	0.293	3 800	22.7	223	77.3	1.82	33.1	21.9	21.1	26.1	189	12.6
X27.8*	0.274	3 560	21.9	215	78.4	1.72	32.1	22.0	21.3	26.7	163	11.9
MC200												
X12.6*	0.123	1 600	9.58	94.4	77.4	0.256	7.07	12.6	10.8	15.8	24.1	2.13

* Not available from Canadian mills.

\# For sections available from Canadian mills, the nominal mass has been rounded to the nearest kg/m. Designation consists of nominal depth in millimetres by nominal mass in kilograms per metre.

Note: These shapes have been soft converted from Imperial to SI dimensions.

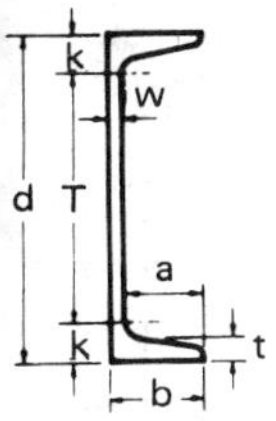

MISCELLANEOUS CHANNELS
MC460 - MC 200

DIMENSIONS AND SURFACE AREAS

Nominal Mass	Theo-retical Mass	Depth d	Flange Width b	Flange Mean Thickness t		Web Thickness w		Distances a	Distances T	Distances k	Surface Area (m^2) per metre of length: Total	Surface Area (m^2) per metre of length: Minus Top of Top Flange
kg/m	kg/m	mm	mm	mm	mm	mm	mm	mm	mm	mm		
86	86.8	457	107	15.9	16	17.8	18	89	387	35	1.31	1.20
77.2	77.4	457	104	15.9	16	15.2	15	89	387	35	1.30	1.20
68.2	68.5	457	102	15.9	16	12.7	13	89	387	35	1.30	1.19
63.5	63.7	457	100	15.9	16	11.4	11	89	387	35	1.29	1.19
74	74.5	330	112	15.5	16	20.0	20	92	261	34	1.07	0.956
60	59.5	330	106	15.5	16	14.2	14	92	261	34	1.06	0.950
52	52.1	330	103	15.5	16	11.4	11	92	261	34	1.05	0.947
47.3	47.4	330	102	15.5	16	9.5	10	92	261	34	1.05	0.946
74	74.7	305	105	17.8	18	21.2	21	84	239	33	0.988	0.883
67	67.0	305	102	17.8	18	18.0	18	84	239	33	0.982	0.880
60	59.6	305	98	17.8	18	15.0	15	83	239	33	0.972	0.874
52	52.3	305	96	17.8	18	11.8	12	84	239	33	0.970	0.874
46	46.3	305	93	17.8	18	9.4	9	84	239	33	0.964	0.871
15.8	15.6	305	38	7.8	8	4.8	5	33	271	17	0.752	0.714
61.2	61.5	254	110	14.6	15	20.2	20	90	189	32	0.908	0.798
50	50.2	254	104	14.6	15	14.6	15	89	189	32	0.894	0.790
42.4	42.6	254	100	14.6	15	10.8	11	89	189	32	0.886	0.786
37	37.3	254	86	14.6	15	9.7	10	76	193	30	0.832	0.746
33	32.8	254	84	14.6	15	7.4	7	77	193	31	0.830	0.746
12.5	12.4	254	38	7.1	7	4.3	4	34	222	16	0.652	0.614
9.7	9.7	254	28	5.1	5	3.9	4	24	233	11	0.612	0.584
37.8	37.9	229	88	14.0	14	11.4	11	77	168	31	0.788	0.700
35.6	35.8	229	87	14.0	14	10.2	10	77	168	31	0.786	0.699
33.9	33.8	203	88	13.3	13	10.8	11	77	145	29	0.736	0.648
31.8	31.8	203	87	13.3	13	9.5	10	77	145	29	0.734	0.647
29.8	29.8	203	76	12.7	13	10.2	10	66	147	28	0.690	0.614
27.8	28.0	203	75	12.7	13	9.0	9	66	147	28	0.688	0.613
12.6	12.5	203	47	7.9	8	4.5	4	43	167	18	0.586	0.539

MISCELLANEOUS CHANNELS
MC180 - MC 150

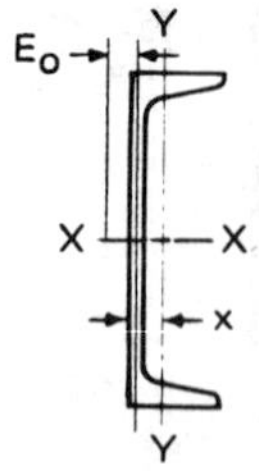

PROPERTIES

Designation#	Dead Load	Total Area	Axis X-X			Axis Y-Y				Shear Centre	Torsional Constant	Warping Constant
			I_x	S_x	r_x	I_y	S_y	r_y	x	E_o	J	C_w
	kN/m	mm^2	10^6mm^4	10^3mm^3	mm	10^6mm^4	10^3mm^3	mm	mm	mm	10^3mm^4	10^9mm^6
MC180												
X33.8*	0.333	4 320	19.9	223	67.9	2.97	45.8	26.2	26.1	31.8	270	15.6
X28.4*	0.279	3 630	18.0	203	70.4	2.48	41.1	26.1	26.8	33.5	175	13.2
MC180												
X26.2*	0.258	3 350	15.8	178	68.7	1.66	30.9	22.3	22.1	27.3	153	8.79
MC150												
X26.8*	0.263	3 410	12.3	162	60.1	2.43	40.6	26.7	28.2	34.0	160	9.08
X22.8*	0.223	2 900	10.5	138	60.2	2.03	33.0	26.5	26.5	33.5	94.7	7.92
MC150												
X24.3*	0.239	3 110	10.9	143	59.2	1.59	30.2	22.6	23.5	28.3	145	5.97
X22.5*	0.221	2 870	10.3	136	59.9	1.43	28.3	22.3	23.6	28.7	122	5.41
MC150												
X17.9*	0.176	2 280	7.76	102	58.3	0.761	16.8	18.3	17.6	22.1	66.4	2.98

* Not available from Canadian mills.

\# For sections available from Canadian mills, the nominal mass has been rounded to the nearest kg/m. Designation consists of nominal depth in millimetres by nominal mass in kilograms per metre.

Note: These shapes have been soft converted from Imperial to SI dimensions.

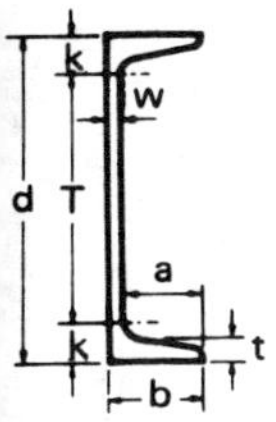

MISCELLANEOUS CHANNELS
MC180 - MC150

DIMENSIONS AND SURFACE AREAS

Nominal Mass	Theoretical Mass	Depth d	Flange			Web Thickness w		Distances			Surface Area (m^2) per metre of length	
			Width b	Mean Thickness t				a	T	k	Total	Minus Top of Top Flange
kg/m	kg/m	mm	mm	mm	mm	mm	mm	mm	mm	mm		
33.8	33.9	178	91	12.7	13	12.8	13	78	121	29	0.694	0.603
28.4	28.5	178	87	12.7	13	8.9	9	78	121	29	0.686	0.599
26.2	26.3	178	76	12.1	12	9.5	10	66	125	27	0.640	0.564
26.8	26.8	152	88	12.1	12	9.6	10	78	99	27	0.636	0.548
22.8	22.8	152	88	9.8	10	8.6	9	79	107	22	0.638	0.550
24.3	24.4	152	76	12.1	12	9.5	10	66	99	26	0.588	0.512
22.5	22.5	152	74	12.1	12	8.0	8	66	99	27	0.584	0.510
17.9	17.9	152	63	9.5	10	7.9	8	55	108	22	0.540	0.477

ANGLES - Imperial Series
Equal Legs

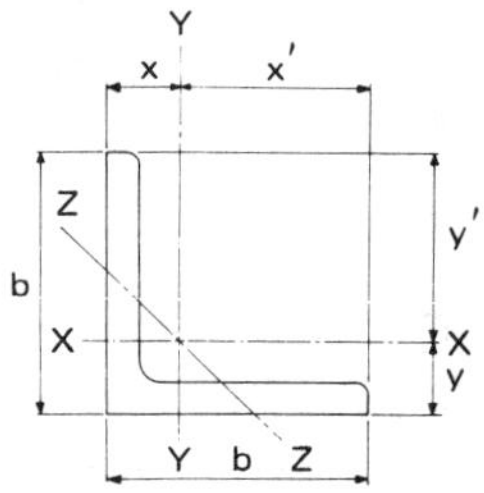

PROPERTIES AND DIMENSIONS

Designation	Size d x b x t	Mass	Dead Load	Area	Axis X-X and Axis Y-Y I	Axis X-X and Axis Y-Y S	Axis X-X and Axis Y-Y r	Axis X-X and Axis Y-Y x or y	Axis Z-Z r	Rolling
	mm x mm x mm	kg/m	kN/m	mm²	10⁶mm⁴	10³mm³	mm	mm	mm	
L203x203	203 x 203									
x29	28.6	84.7	0.831	10800	40.7	287	61.4	61.2	39.6	4
x25	25.4	75.9	0.744	9670	36.9	258	61.8	60.1	39.7	4
x22	22.2	66.9	0.656	8520	33.0	229	62.2	58.9	39.8	4
x19	19.1	58.0	0.569	7390	29.0	200	62.6	57.8	40.0	4
x16	15.9	48.7	0.478	6200	24.7	169	63.1	56.6	40.1	4
x14	14.3	44.0	0.431	5600	22.5	153	63.3	56.0	40.2	4
x13	12.7	39.2	0.385	4990	20.2	137	63.6	55.5	40.3	4
L152x152	152 x 152									
x25	25.4	55.6	0.545	7080	14.6	140	45.5	47.2	29.6	4
x22	22.2	49.1	0.482	6260	13.2	124	45.9	46.1	29.6	4
x19	19.1	42.7	0.419	5440	11.6	109	46.3	45.0	29.7	1
x16	15.9	36.0	0.353	4580	9.99	92.3	46.7	43.9	29.8	1
x14	14.3	32.5	0.319	4140	9.12	83.8	46.9	43.3	29.9	4
x13	12.7	29.0	0.285	3700	8.22	75.2	47.1	42.7	30.0	1
x11	11.1	25.5	0.250	3250	7.29	66.4	47.4	42.1	30.1	4
x9.5	9.53	22.0	0.216	2810	6.36	57.5	47.6	41.5	30.2	2
x7.9	7.94	18.5	0.181	2350	5.38	48.4	47.8	41.0	30.3	4
L127x127	127 x 127									
x22	22.2	40.4	0.396	5150	7.39	84.7	37.9	39.8	24.7	4
x19	19.1	35.2	0.345	4490	6.57	74.4	38.3	38.7	24.8	3
x16	15.9	29.7	0.292	3790	5.66	63.3	38.7	37.6	24.8	1
x13	12.7	24.1	0.236	3060	4.68	51.7	39.1	36.4	25.0	1
x11	11.1	21.2	0.208	2700	4.17	45.7	39.3	35.8	25.0	4
x9.5	9.53	18.3	0.179	2330	3.64	39.7	39.5	35.3	25.1	2
x7.9	7.94	15.3	0.150	1950	3.09	33.5	39.8	34.7	25.2	2
x6.4	6.35	12.3	0.121	1570	2.52	27.1	40.0	34.1	25.4	3
x4.8	4.76	9.31	0.0914	1190	1.92	20.6	40.2	33.5	25.5	3
L102x102	102 x 102									
x19	19.1	27.7	0.272	3530	3.24	46.5	30.3	32.4	19.8	3
x16	15.9	23.5	0.230	2990	2.81	39.8	30.7	31.3	19.9	3
x13	12.7	19.1	0.187	2430	2.34	32.6	31.1	30.2	19.9	2
x11	11.1	16.8	0.165	2140	2.09	28.9	31.3	29.6	20.0	4
x9.5	9.53	14.5	0.143	1850	1.84	25.2	31.5	29.0	20.1	1
x7.9	7.94	12.2	0.120	1560	1.57	21.3	31.7	28.4	20.2	2
x6.4	6.35	9.85	0.0966	1260	1.28	17.3	31.9	27.9	20.3	1
x4.8	4.76	7.44	0.0730	948	0.982	13.1	32.2	27.3	20.4	3
L89x89	88.9 x 88.9									
x16	15.9	20.2	0.198	2570	1.80	29.6	26.5	28.0	17.3	3
x13	12.7	16.5	0.161	2100	1.51	24.4	26.9	26.9	17.3	1
x11	11.1	14.5	0.142	1850	1.36	21.7	27.1	26.3	17.4	4
x9.5	9.53	12.6	0.123	1600	1.19	18.9	27.3	25.7	17.4	1
x7.9	7.94	10.6	0.104	1350	1.02	16.0	27.5	25.2	17.5	2
x6.4	6.35	8.55	0.0838	1090	0.837	13.0	27.7	24.6	17.6	1
x4.8	4.76	6.47	0.0634	824	0.644	9.92	28.0	24.0	17.7	3

See page 6-43 for description of "Rolling" and for qualification of properties of angles produced by cold-forming.

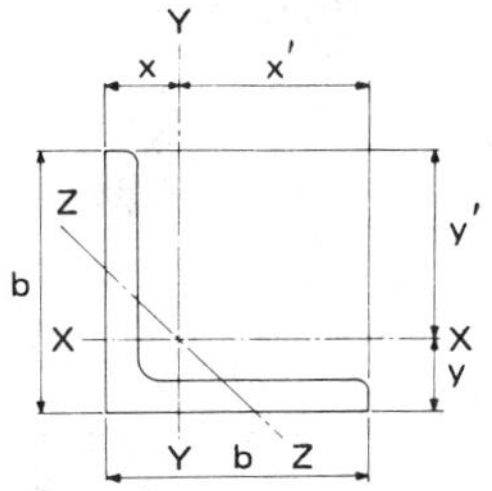

ANGLES - Imperial Series
Equal Legs

PROPERTIES AND DIMENSIONS

Designation	Size d x b x t	Mass	Dead Load	Area	Axis X-X and Axis Y-Y I	Axis X-X and Axis Y-Y S	Axis X-X and Axis Y-Y r	Axis X-X and Axis Y-Y x or y	Axis Z-Z r	Rolling
	mm x mm x mm	kg/m	kN/m	mm^2	10^6mm^4	10^3mm^3	mm	mm	mm	
L76x76	76.2 x 76.2									
x16	15.9	17.0	0.167	2170	1.09	21.3	22.4	24.8	14.8	3
x13	12.7	13.9	0.137	1770	0.923	17.6	22.8	23.7	14.8	1
x11	11.1	12.3	0.121	1570	0.830	15.6	23.0	23.1	14.9	4
x9.5	9.53	10.7	0.105	1360	0.733	13.7	23.2	22.5	14.9	1
x7.9	7.94	9.00	0.0883	1150	0.629	11.6	23.4	22.0	15.0	1
x6.4	6.35	7.28	0.0714	927	0.518	9.45	23.6	21.4	15.0	2
x4.8	4.76	5.52	0.0541	703	0.400	7.22	23.9	20.8	15.1	3
L64x64	63.5 x 63.5									
x13	12.7	11.4	0.112	1450	0.511	11.9	18.8	20.5	12.4	3
x9.5	9.53	8.79	0.0862	1120	0.410	9.28	19.1	19.4	12.4	3
x7.9	7.94	7.42	0.0728	945	0.353	7.90	19.3	18.8	12.4	3
x6.4	6.35	6.01	0.0590	766	0.293	6.46	19.5	18.2	12.5	1
x4.8	4.76	4.57	0.0448	582	0.227	4.96	19.8	17.6	12.6	1
x3.2	3.18	3.09	0.0303	394	0.157	3.39	20.0	17.1	12.7	3
L51x51	50.8 x 50.8									
x9.5	9.53	6.89	0.0676	877	0.199	5.76	15.1	16.2	9.89	2
x7.9	7.94	5.84	0.0573	744	0.173	4.92	15.3	15.6	9.90	2
x6.4	6.35	4.75	0.0466	605	0.145	4.04	15.5	15.0	9.93	1
x4.8	4.76	3.62	0.0355	461	0.113	3.12	15.7	14.5	10.0	1
x3.2	3.18	2.46	0.0241	313	0.0792	2.14	15.9	13.9	10.1	1
L44x44	44.5 x 44.5									
x9.5	9.53	5.95	0.0583	757	0.130	4.33	13.1	14.6	8.67	2
x7.9	7.94	5.05	0.0496	644	0.113	3.71	13.3	14.0	8.66	2
x6.4	6.35	4.12	0.0404	525	0.0949	3.06	13.4	13.4	8.68	2
x4.8	4.76	3.15	0.0309	401	0.0748	2.36	13.7	12.9	8.73	2
x3.2	3.18	2.14	0.0210	273	0.0525	1.63	13.9	12.3	8.82	2
L38x38	38.1 x 38.1									
x9.5	9.53	4.99	0.0489	635	0.0777	3.09	11.1	12.9	7.46	3
x7.9	7.94	4.25	0.0417	542	0.0682	2.65	11.2	12.4	7.42	3
x6.4	6.35	3.48	0.0342	444	0.0577	2.20	11.4	11.8	7.42	2
x4.8	4.76	2.67	0.0262	340	0.0458	1.71	11.6	11.3	7.45	2
x4.0	3.97	2.25	0.0221	287	0.0393	1.45	11.7	11.0	7.48	4
x3.2	3.18	1.82	0.0179	232	0.0324	1.18	11.8	10.7	7.52	2
L32x32	31.8 x 31.8									
x9.5	9.53	4.05	0.0397	515	0.0426	2.08	9.09	11.3	6.31	3
x7.9	7.94	3.47	0.0340	442	0.0377	1.79	9.23	10.8	6.23	3
x6.4	6.35	2.85	0.0280	364	0.0321	1.49	9.40	10.2	6.19	3
x4.8	4.76	2.20	0.0216	280	0.0257	1.16	9.58	9.69	6.20	3
x3.2	3.18	1.51	0.0148	192	0.0184	0.812	9.79	9.12	6.25	3
L25x25	25.4 x 25.4									
x9.5	9.53	3.09	0.0303	393	0.0200	1.27	7.13	9.65	5.20	3
x7.9	7.94	2.67	0.0262	340	0.0178	1.10	7.24	9.14	5.06	3
x6.4	6.35	2.22	0.0217	282	0.0153	0.915	7.37	8.62	4.98	3
x4.8	4.76	1.72	0.0169	219	0.0125	0.719	7.54	8.07	4.94	3
x3.2	3.18	1.19	0.0117	151	0.00905	0.506	7.73	7.52	4.97	3

See page 6-43 for description of "Rolling" and for qualification of properties of angles produced by cold-forming.

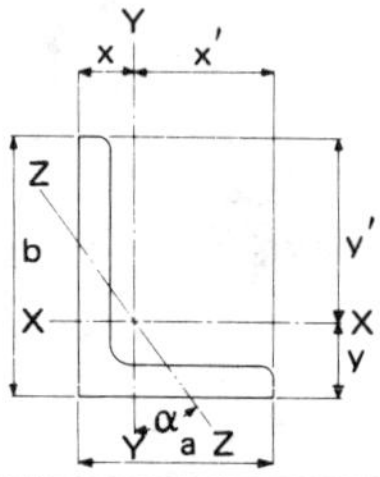

ANGLES – Imperial Series
Unequal Legs

PROPERTIES AND DIMENSIONS

Designation	Size d x b x t	Mass	Dead Load	Area	Axis X-X I	Axis X-X S	Axis X-X r	Axis X-X y	Axis Y-Y I	Axis Y-Y S	Axis Y-Y r	Axis Y-Y x	Axis Z-Z r	tan α	Rolling
	mm x mm x mm	kg/m	kN/m	mm^2	10^6mm^4	10^3mm^3	mm	mm	10^6mm^4	10^3mm^3	mm	mm	mm		
L203x152	203 x 152														
x25	25.4	65.7	0.645	8370	33.5	247	63.3	67.4	16.0	145	43.7	41.9	32.4	0.541	4
x22	22.2	58.0	0.569	7390	30.0	219	63.7	66.2	14.4	129	44.1	40.7	32.5	0.545	4
x19	19.1	50.4	0.494	6420	26.4	191	64.1	65.1	12.7	113	44.5	39.6	32.6	0.549	4
x16	15.9	42.3	0.415	5390	22.5	162	64.6	64.0	10.9	95.9	44.9	38.5	32.8	0.553	4
x14	14.3	38.2	0.375	4870	20.4	146	64.8	63.4	9.94	87.1	45.2	37.9	32.9	0.554	4
x13	12.7	34.1	0.335	4350	18.4	131	65.0	62.8	8.96	78.1	45.4	37.3	33.0	0.556	4
x11	11.1	30.0	0.294	3820	16.3	115	65.3	62.2	7.94	68.9	45.6	36.7	33.1	0.558	4
L203x102	203 x 102														
x25	25.4	55.7	0.547	7100	29.0	230	63.8	77.2	4.90	65.1	26.3	26.7	21.6	0.249	4
x19	19.1	42.9	0.420	5460	22.9	178	64.7	74.8	3.95	50.8	26.9	24.3	21.7	0.260	4
x14	14.3	32.6	0.320	4160	17.8	137	65.4	73.0	3.13	39.4	27.4	22.5	22.0	0.267	4
x13	12.7	29.1	0.286	3710	16.0	123	65.7	72.4	2.84	35.4	27.6	21.9	22.1	0.269	4
L178x102	178 x 102														
x19	19.1	39.1	0.384	4980	15.8	139	56.4	63.8	3.82	50.1	27.7	25.8	21.9	0.325	4
x16	15.9	33.0	0.323	4200	13.6	118	56.8	62.6	3.31	42.7	28.1	24.6	22.1	0.331	4
x13	12.7	26.6	0.261	3390	11.1	95.6	57.3	61.4	2.75	35.0	28.5	23.4	22.2	0.336	2
x9.5	9.53	20.2	0.198	2580	8.60	73.0	57.8	60.2	2.15	26.9	28.9	22.2	22.4	0.341	2
L152x102	152 x 102														
x22	22.2	40.4	0.396	5150	11.5	117	47.2	53.7	4.10	55.9	28.2	28.7	21.9	0.427	4
x19	19.1	35.2	0.345	4490	10.2	102	47.6	52.5	3.66	49.2	28.6	27.5	21.9	0.434	3
x16	15.9	29.7	0.292	3790	8.73	86.8	48.0	51.4	3.17	41.9	28.9	26.4	22.0	0.440	2
x14	14.3	26.9	0.264	3430	7.98	78.8	48.2	50.8	2.91	38.2	29.1	25.8	22.1	0.443	4
x13	12.7	24.1	0.236	3060	7.20	70.7	48.5	50.2	2.64	34.4	29.3	25.2	22.2	0.446	1
x11	11.1	21.2	0.208	2700	6.39	62.4	48.7	49.6	2.35	30.4	29.6	24.6	22.3	0.449	4
x9.5	9.53	18.3	0.179	2330	5.58	54.2	48.9	49.1	2.06	26.5	29.8	24.1	22.4	0.451	2
x7.9	7.94	15.3	0.150	1950	4.72	45.6	49.2	48.5	1.76	22.4	30.0	23.5	22.5	0.454	2
x6.4	6.35	12.3	0.121	1570	3.84	36.9	49.4	47.9	1.44	18.1	30.2	22.9	22.6	0.456	3
x4.8	4.76	9.31	0.0914	1190	2.92	27.9	49.6	47.3	1.10	13.8	30.5	22.3	22.7	0.459	3
L152x89	152 x 88.9														
x16	15.9	28.1	0.275	3580	8.31	84.7	48.2	53.9	2.12	31.8	24.3	22.4	19.1	0.339	2
x13	12.7	22.8	0.223	2900	6.86	69.1	48.6	52.7	1.77	26.1	24.7	21.2	19.3	0.345	1
x9.5	9.53	17.3	0.170	2200	5.32	52.9	49.1	51.6	1.39	20.2	25.1	20.0	19.5	0.351	2
x7.9	7.94	14.5	0.142	1850	4.50	44.6	49.3	51.0	1.19	17.1	25.3	19.4	19.6	0.354	2
x6.4	6.35	11.7	0.115	1490	3.66	36.0	49.6	50.4	0.972	13.9	25.5	18.8	19.7	0.357	3
L127x89	127 x 88.9														
x19	19.1	29.5	0.289	3760	5.80	70.2	39.3	44.4	2.31	36.4	24.8	25.3	19.0	0.464	1
x16	15.9	25.0	0.245	3180	5.01	59.8	39.7	43.2	2.01	31.1	25.2	24.2	19.1	0.472	2
x13	12.7	20.3	0.199	2580	4.16	48.9	40.1	42.1	1.68	25.6	25.6	23.0	19.2	0.479	1
x11	11.1	17.8	0.175	2270	3.70	43.3	40.4	41.5	1.51	22.7	25.8	22.4	19.3	0.482	4
x9.5	9.53	15.4	0.151	1970	3.24	37.6	40.6	40.9	1.33	19.8	26.0	21.9	19.3	0.486	2
x7.9	7.94	13.0	0.127	1650	2.75	31.7	40.8	40.3	1.13	16.7	26.2	21.3	19.4	0.489	2
x6.4	6.35	10.4	0.102	1330	2.24	25.7	41.0	39.7	0.928	13.6	26.4	20.7	19.6	0.492	2
x4.8	4.76	7.89	0.0774	1010	1.71	19.5	41.3	39.1	0.713	10.4	26.6	20.1	19.7	0.495	3

See page 6-43 for description of "Rolling" and for qualification of properties of angles produced by cold-forming.

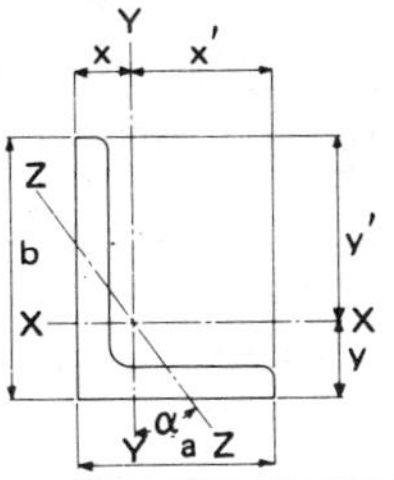

ANGLES – Imperial Series
Unequal Legs

PROPERTIES AND DIMENSIONS

Designation	Size d x b x t	Mass	Dead Load	Area	Axis X-X I	Axis X-X S	Axis X-X r	Axis X-X y	Axis Y-Y I	Axis Y-Y S	Axis Y-Y r	Axis Y-Y x	Axis Z-Z r	Axis Z-Z tan α	Rolling
	mm x mm x mm	kg/m	kN/m	mm²	10⁶mm⁴	10³mm³	mm	mm	10⁶mm⁴	10³mm³	mm	mm	mm		
L127x76	127 x 76.2														
x16	15.9	23.4	0.229	2980	4.74	58.2	39.9	45.6	1.28	22.8	20.7	20.2	16.4	0.349	3
x13	12.7	19.0	0.186	2420	3.93	47.7	40.3	44.5	1.07	18.8	21.1	19.1	16.5	0.357	1
x11	11.1	16.7	0.164	2130	3.51	42.2	40.6	43.9	0.963	16.7	21.3	18.5	16.5	0.361	4
x9.5	9.53	14.5	0.142	1850	3.07	36.7	40.8	43.3	0.849	14.6	21.5	17.9	16.6	0.364	1
x7.9	7.94	12.2	0.119	1550	2.61	30.9	41.0	42.7	0.727	12.3	21.7	17.3	16.7	0.368	2
x6.4	6.35	9.81	0.0962	1250	2.13	25.0	41.2	42.1	0.598	10.1	21.9	16.7	16.8	0.371	2
x4.8	4.76	7.41	0.0727	945	1.63	19.0	41.5	41.5	0.462	7.68	22.1	16.1	17.0	0.374	3
L102x89	102 x 88.9														
x16	15.9	21.8	0.214	2780	2.68	38.9	31.1	33.0	1.89	30.2	26.0	26.5	18.3	0.738	3
x13	12.7	17.8	0.174	2260	2.24	32.0	31.5	31.9	1.58	24.9	26.4	25.4	18.4	0.744	2
x11	11.1	15.7	0.154	2000	2.00	28.3	31.7	31.3	1.42	22.1	26.6	24.8	18.4	0.746	4
x9.5	9.53	13.6	0.133	1730	1.76	24.7	31.9	30.8	1.24	19.2	26.8	24.2	18.5	0.749	2
x7.9	7.94	11.4	0.112	1450	1.50	20.9	32.1	30.2	1.06	16.3	27.1	23.6	18.6	0.751	1
x6.4	6.35	9.20	0.0902	1170	1.23	16.9	32.3	29.6	0.872	13.2	27.3	23.1	18.7	0.753	2
x4.8	4.76	6.96	0.0682	886	0.941	12.9	32.6	29.0	0.671	10.1	27.5	22.5	18.8	0.755	3
L102x76	102 x 76.2														
x16	15.9	20.3	0.199	2580	2.54	38.0	31.4	35.0	1.20	22.2	21.6	22.1	16.2	0.529	3
x13	12.7	16.5	0.162	2100	2.12	31.2	31.8	33.9	1.01	18.3	21.9	21.0	16.2	0.538	2
x11	11.1	14.6	0.143	1850	1.90	27.7	32.0	33.3	0.907	16.3	22.1	20.4	16.3	0.543	4
x9.5	9.53	12.6	0.124	1610	1.67	24.1	32.2	32.7	0.800	14.2	22.3	19.8	16.4	0.547	1
x7.9	7.94	10.6	0.104	1350	1.42	20.4	32.4	32.1	0.686	12.0	22.5	19.2	16.5	0.550	2
x6.4	6.35	8.57	0.0840	1090	1.17	16.5	32.7	31.6	0.565	9.81	22.7	18.7	16.6	0.554	2
x4.8	4.76	6.48	0.0636	826	0.894	12.6	32.9	31.0	0.436	7.50	23.0	18.1	16.7	0.557	3
L89x76	88.9 x 76.2														
x16	15.9	18.6	0.183	2370	1.71	28.9	26.9	29.7	1.15	21.8	22.0	23.3	15.7	0.707	3
x13	12.7	15.2	0.149	1940	1.44	23.8	27.3	28.6	0.969	18.0	22.4	22.2	15.8	0.714	3
x11	11.1	13.4	0.132	1710	1.29	21.2	27.5	28.0	0.871	16.0	22.6	21.7	15.8	0.718	4
x9.5	9.53	11.6	0.114	1480	1.13	18.5	27.7	27.4	0.769	14.0	22.8	21.1	15.9	0.721	3
x7.9	7.94	9.80	0.0961	1250	0.970	15.6	27.9	26.9	0.659	11.8	23.0	20.5	15.9	0.724	2
x6.4	6.35	7.91	0.0776	1010	0.796	12.7	28.1	26.3	0.543	9.65	23.2	19.9	16.0	0.727	3
x4.8	4.76	5.99	0.0588	763	0.613	9.70	28.3	25.7	0.419	7.37	23.4	19.4	16.1	0.729	3
L89x64	88.9 x 63.5														
x16	15.9	17.0	0.167	2170	1.60	28.0	27.2	31.7	0.670	15.1	17.6	19.0	13.6	0.474	3
x13	12.7	13.9	0.137	1770	1.35	23.1	27.6	30.6	0.568	12.5	17.9	17.9	13.6	0.486	3
x11	11.1	12.3	0.121	1570	1.21	20.5	27.8	30.0	0.512	11.1	18.1	17.3	13.6	0.491	4
x9.5	9.53	10.7	0.105	1360	1.07	17.9	28.0	29.5	0.454	9.71	18.3	16.8	13.6	0.496	3
x7.9	7.94	9.00	0.0883	1150	0.912	15.2	28.2	28.9	0.391	8.26	18.5	16.2	13.7	0.501	3
x6.4	6.35	7.28	0.0714	927	0.749	12.4	28.4	28.3	0.323	6.75	18.7	15.6	13.8	0.506	3
x4.8	4.76	5.52	0.0541	703	0.577	9.43	28.7	27.7	0.251	5.17	18.9	15.0	13.9	0.510	3

See page 6-43 for description of "Rolling" and for qualification of properties of angles produced by cold-forming.

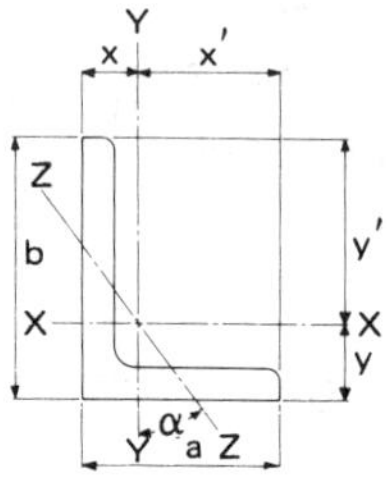

ANGLES – Imperial Series Unequal Legs

PROPERTIES AND DIMENSIONS

Designation	Size d x b x t	Mass	Dead Load	Area	Axis X-X I	Axis X-X S	Axis X-X r	Axis X-X y	Axis Y-Y I	Axis Y-Y S	Axis Y-Y r	Axis Y-Y x	Axis Z-Z r	tan α	Rolling
	mm x mm x mm	kg/m	kN/m	mm^2	10^6mm^4	10^3mm^3	mm	mm	10^6mm^4	10^3mm^3	mm	mm	mm		
L76x64	76.2 x 63.5														
x16	15.9	15.5	0.152	1970	1.03	20.6	22.8	26.5	0.638	14.7	18.0	20.2	13.3	0.656	3
x13	12.7	12.7	0.124	1610	0.867	17.1	23.2	25.4	0.542	12.2	18.3	19.1	13.2	0.667	3
x11	11.1	11.2	0.110	1430	0.780	15.2	23.4	24.8	0.489	10.9	18.5	18.5	13.2	0.672	4
x9.5	9.53	9.74	0.0955	1240	0.690	13.3	23.6	24.3	0.434	9.52	18.7	17.9	13.3	0.676	3
x7.9	7.94	8.21	0.0806	1050	0.592	11.3	23.8	23.7	0.374	8.10	18.9	17.4	13.3	0.680	3
x6.4	6.35	6.65	0.0652	847	0.488	9.20	24.0	23.1	0.309	6.62	19.1	16.8	13.4	0.684	3
x4.8	4.76	5.04	0.0495	642	0.377	7.04	24.2	22.6	0.240	5.08	19.3	16.2	13.5	0.688	3
L76x51	76.2 x 50.8														
x16	15.9	13.9	0.136	1770	0.944	19.8	23.1	28.6	0.327	9.39	13.6	15.9	10.9	0.398	3
x13	12.7	11.4	0.112	1450	0.800	16.4	23.5	27.5	0.280	7.77	13.9	14.8	10.9	0.414	3
x11	11.1	10.1	0.0991	1290	0.721	14.6	23.7	27.0	0.253	6.94	14.0	14.3	10.9	0.422	4
x9.5	9.53	8.79	0.0862	1120	0.638	12.8	23.9	26.4	0.226	6.09	14.2	13.7	10.9	0.428	3
x7.9	7.94	7.42	0.0728	945	0.548	10.9	24.1	25.8	0.196	5.20	14.4	13.1	11.0	0.435	3
x6.4	6.35	6.01	0.0590	766	0.453	8.88	24.3	25.2	0.163	4.26	14.6	12.5	11.0	0.440	3
x4.8	4.76	4.57	0.0448	582	0.350	6.79	24.5	24.6	0.128	3.28	14.8	11.9	11.1	0.446	3
L64x51	63.5 x 50.8														
x9.5	9.53	7.84	0.0769	998	0.380	8.96	19.5	21.1	0.214	5.94	14.6	14.8	10.7	0.614	2
x7.9	7.94	6.63	0.0650	844	0.328	7.64	19.7	20.6	0.186	5.08	14.8	14.2	10.7	0.620	2
x6.4	6.35	5.38	0.0528	685	0.272	6.25	19.9	20.0	0.155	4.17	15.0	13.6	10.8	0.626	2
x4.8	4.76	4.09	0.0401	521	0.212	4.80	20.1	19.4	0.121	3.21	15.2	13.1	10.9	0.631	2
x3.2	3.18	2.77	0.0272	353	0.147	3.28	20.4	18.8	0.0846	2.21	15.5	12.5	11.0	0.636	3
L64x38	63.5 x 38.1														
x7.9	7.94	5.84	0.0573	744	0.296	7.27	19.9	22.8	0.0797	2.85	10.4	10.1	8.18	0.349	4
x6.4	6.35	4.75	0.0466	605	0.246	5.96	20.2	22.2	0.0671	2.35	10.5	9.53	8.23	0.357	4
x4.8	4.76	3.62	0.0355	461	0.192	4.58	20.4	21.6	0.0530	1.82	10.7	8.94	8.31	0.364	4
L51x38	50.8 x 38.1														
x9.5	9.53	5.94	0.0582	756	0.180	5.49	15.4	18.0	0.0854	3.23	10.6	11.6	8.09	0.524	3
x7.9	7.94	5.05	0.0495	643	0.157	4.70	15.6	17.4	0.0748	2.77	10.8	11.1	8.09	0.534	2
x6.4	6.35	4.11	0.0404	524	0.131	3.87	15.8	16.9	0.0630	2.28	11.0	10.5	8.12	0.543	1
x4.8	4.76	3.14	0.0308	401	0.103	2.99	16.0	16.3	0.0499	1.77	11.2	9.93	8.18	0.551	2
x3.2	3.18	2.14	0.0210	273	0.0721	2.06	16.3	15.7	0.0353	1.23	11.4	9.35	8.27	0.558	3

See page 6-43 for description of "Rolling" and for qualification of properties of angles produced by cold-forming.

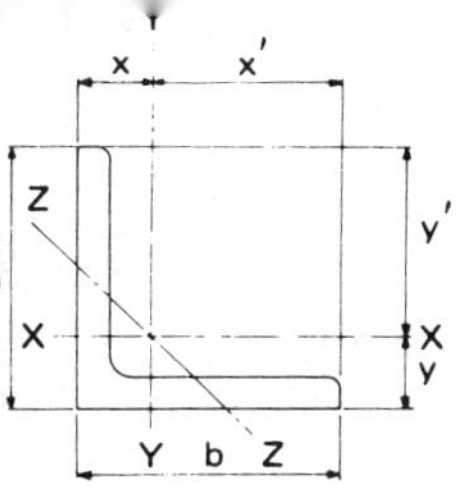

METRIC AND IMPERIAL SIZE ANGLES
Equal Legs

PROPERTIES AND DIMENSIONS

ANGLE SIZE**		Thickness t	Mass	Dead Load	Area	Axis X-X & Axis Y-Y				Axis Z-Z
Metric	Imperial					I	S	r	x or y	r
mm x mm x mm	in x in x in	mm	kg/m	kN/m	mm^2	$10^6 mm^4$	$10^3 mm^3$	mm	mm	mm
200 X 200	8 x 8 (203 x 203)									
200X200X30		**30**	**87.1**	**0.855**	**11 100**	**40.3**	**290**	**60.3**	**60.9**	**39.0**
	8 x 8 x 1 1/8	28.6	84.7	0.831	10 800	40.7	287	61.4	61.2	39.6
	8 x 8 x 1	25.4	75.9	0.744	9 670	36.9	258	61.8	60.1	39.7
200X200X25		**25**	**73.6**	**0.722**	**9 380**	**34.8**	**247**	**60.9**	**59.2**	**39.1**
	8 x 8 x 7/8	22.2	67.0	0.657	8 520	33.0	229	62.2	58.9	39.8
200X200X20		**20**	**59.7**	**0.585**	**7 600**	**28.8**	**202**	**61.6**	**57.4**	**39.3**
	8 x 8 x 3/4	19.0	57.9	0.568	7 350	28.9	199	62.7	57.8	40.0
200X200X16		**16**	**48.2**	**0.473**	**6 140**	**23.7**	**165**	**62.1**	**55.9**	**39.5**
	8 x 8 x 5/8	15.9	48.7	0.478	6 200	24.7	169	63.1	56.6	40.1
	8 x 8 x 9/16	14.3	44.0	0.431	5 600	22.5	153	63.3	56.0	40.2
200X200X13		**13**	**39.5**	**0.387**	**5 030**	**19.7**	**136**	**62.6**	**54.8**	**39.7**
	8 x 8 x 1/2	12.7	39.3	0.385	4 990	20.2	137	63.6	55.5	40.3
200X200X10		**10**	**30.6**	**0.300**	**3 900**	**15.5**	**106**	**63.0**	**53.7**	**39.3**
150 X 150	6 x 6 (152 x 152)									
	6 x 6 x 1	25.4	55.7	0.546	7 080	14.6	140	45.5	47.2	29.6
	6 x 6 x 7/8	22.2	49.3	0.483	6 260	13.2	124	45.9	46.1	29.6
150X150X20		**20**	**44.0**	**0.431**	**5 600**	**11.6**	**110**	**45.5**	**44.8**	**29.3**
	6 x 6 x 3/4	19.0	42.7	0.419	5 420	11.6	108	46.3	45.0	29.7
150X150X16		**16**	**35.7**	**0.350**	**4 540**	**9.63**	**90.3**	**46.0**	**43.4**	**29.4**
	6 x 6 x 5/8	15.9	36.0	0.353	4 580	9.99	92.3	46.7	43.9	29.8
	6 x 6 x 9/16	14.3	32.6	0.320	4 140	9.12	83.8	46.9	43.3	29.9
150X150X13		**13**	**29.3**	**0.287**	**3 730**	**8.05**	**74.7**	**46.4**	**42.3**	**29.6**
	6 x 6 x 1/2	12.7	29.2	0.286	3 700	8.22	75.2	47.1	42.7	30.0
	6 x 6 x 7/16	11.1	25.6	0.251	3 250	7.29	66.4	47.4	42.1	30.1
150X150X10		**10**	**22.8**	**0.223**	**2 900**	**6.37**	**58.6**	**46.9**	**41.2**	**29.8**
	6 x 6 x 3/8	9.5	22.2	0.218	2 800	6.34	57.4	47.6	41.5	30.2
	6 x 6 x 5/16	7.9	18.5	0.181	2 340	5.35	48.2	47.8	40.9	30.3
125 X 125	5 x 5 (127 x 127)									
	5 x 5 x 7/8	22.2	40.5	0.397	5 150	7.39	84.7	37.9	39.8	24.7
	5 x 5 x 3/4	19.0	35.1	0.344	4 460	6.54	74.0	38.3	38.7	24.8
125X125X16		**16**	**29.4**	**0.288**	**3 740**	**5.41**	**61.5**	**38.0**	**37.1**	**24.4**
	5 x 5 x 5/8	15.9	29.8	0.292	3 790	5.66	63.3	38.7	37.6	24.8
125X125X13		**13**	**24.2**	**0.237**	**3 080**	**4.54**	**51.1**	**38.4**	**36.0**	**24.5**
	5 x 5 x 1/2	12.7	24.1	0.236	3 060	4.68	51.7	39.1	36.4	25.0
	5 x 5 x 7/16	11.1	21.3	0.204	2 700	4.17	45.7	39.3	35.8	25.0
125X125X10		**10**	**18.8**	**0.185**	**2 400**	**3.62**	**40.2**	**38.8**	**34.9**	**24.7**
	5 x 5 x 3/8	9.5	18.3	0.179	2 320	3.63	39.6	39.5	35.3	25.1
125X125X8		**8**	**15.2**	**0.149**	**1 940**	**2.96**	**32.6**	**39.1**	**34.2**	**24.8**
	5 x 5 x 5/16	7.9	15.3	0.150	1 940	3.08	33.3	39.8	34.7	25.2

**For angle designations, see page 6-19. See page 6-43 for more information on these tables.

METRIC AND IMPERIAL SIZE ANGLES
Equal Legs

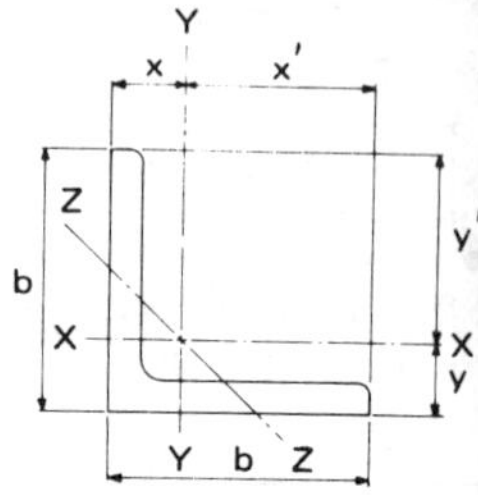

PROPERTIES AND DIMENSIONS

ANGLE SIZE**		Thickness t	Mass	Dead Load	Area	Axis X-X & Axis Y-Y				Axis Z-Z
Metric	Imperial					I	S	r	x or y	r
mm x mm x mm	in x in x in	mm	kg/m	kN/m	mm^2	$10^6 mm^4$	$10^3 mm^3$	mm	mm	mm
100X100	4 x 4 (102 x 102)									
	4 x 4 x 3/4	19.0	27.5	0.270	3 520	3.23	46.3	30.3	32.4	19.8
100X100X16		**16**	**23.1**	**0.227**	**2 940**	**2.65**	**38.3**	**30.0**	**30.8**	**19.5**
	4 x 4 x 5/8	15.9	23.4	0.229	2 990	2.81	39.8	30.7	31.3	19.9
100X100X13		**13**	**19.1**	**0.187**	**2 430**	**2.24**	**31.9**	**30.4**	**29.8**	**19.5**
	4 x 4 x 1/2	12.7	19.0	0.186	2 430	2.34	32.6	31.1	30.2	19.9
	4 x 4 x 7/16	11.1	16.8	0.165	2 140	2.09	28.9	31.3	29.6	20.0
100X100X10		**10**	**14.9**	**0.146**	**1 900**	**1.80**	**25.2**	**30.8**	**28.7**	**19.7**
	4 x 4 x 3/8	9.5	14.6	0.143	1 850	1.83	25.1	31.5	29.0	20.1
100X100X8		**8**	**12.1**	**0.118**	**1 540**	**1.48**	**20.6**	**31.1**	**28.0**	**19.8**
	4 x 4 x 5/16	7.9	12.2	0.120	1 550	1.56	21.2	31.7	28.4	20.2
	4 x 4 x 1/4	6.4	9.80	0.096	1 260	1.29	17.4	31.9	27.9	20.3
100X100X6		**6**	**9.14**	**0.090**	**1 160**	**1.14**	**15.7**	**31.3**	**27.2**	**19.9**
90X90	3 1/2 x 3 1/2 (89 x 89)									
90X90X13		**13**	**17.0**	**0.167**	**2 170**	**1.60**	**25.6**	**27.2**	**27.2**	**17.6**
	3 1/2 x 3 1/2 x 1/2	12.7	16.5	0.162	2 100	1.52	24.4	26.9	26.9	17.3
	3 1/2 x 3 1/2 x 7/16	11.1	14.6	0.143	1 850	1.36	21.7	27.1	26.3	17.4
90X90X10		**10**	**13.3**	**0.131**	**1 700**	**1.29**	**20.2**	**27.6**	**26.2**	**17.6**
	3 1/2 x 3 1/2 x 3/8	9.5	12.6	0.124	1 600	1.19	18.9	27.3	25.7	17.4
90X90X8		**8**	**10.8**	**0.106**	**1 380**	**1.07**	**16.5**	**27.8**	**25.5**	**17.7**
	3 1/2 x 3 1/2 x 5/16	7.9	10.7	0.105	1 340	1.02	16.0	27.5	25.2	17.5
	3 1/2 x 3 1/2 x 1/4	6.4	8.60	0.084	1 100	0.845	13.1	27.7	24.6	17.6
90X90X6		**6**	**8.20**	**0.080**	**1 040**	**0.826**	**12.7**	**28.1**	**24.7**	**17.9**
75X75	3 x 3 (76 x 76)									
75X75X13		**13**	**14.0**	**0.137**	**1 780**	**0.892**	**17.3**	**22.4**	**23.5**	**14.6**
	3 x 3 x 1/2	12.7	14.0	0.137	1 770	0.915	17.5	22.7	23.6	14.8
	3 x 3 x 7/16	11.1	12.4	0.122	1 560	0.823	15.5	22.9	23.1	14.9
75X75X10		**10**	**11.0**	**0.108**	**1 400**	**0.725**	**13.8**	**22.8**	**22.4**	**14.6**
	3 x 3 x 3/8	9.5	10.7	0.105	1 350	0.725	13.5	23.1	22.5	14.9
75X75X8		**8**	**8.92**	**0.087**	**1 140**	**0.602**	**11.3**	**23.0**	**21.7**	**14.7**
	3 x 3 x 5/16	7.9	9.10	0.089	1 140	0.621	11.5	23.4	21.9	15.0
	3 x 3 x 1/4	6.4	7.30	0.072	932	0.517	9.47	23.6	21.4	15.0
75X75X6		**6**	**6.78**	**0.066**	**864**	**0.469**	**8.68**	**23.3**	**21.0**	**14.8**
75X75X5		**5**	**5.69**	**0.056**	**725**	**0.398**	**7.32**	**23.4**	**20.6**	**14.9**
	3 x 3 x 3/16	4.8	5.50	0.054	707	0.400	7.24	23.8	20.8	15.1
65X65	2 1/2 x 2 1/2 (64 x 64)									
	2 1/2 x 2 1/2 x 1/2	12.7	11.4	0.112	1 460	0.524	12.1	18.9	20.6	12.4
65X65X10		**10**	**9.42**	**0.092**	**1 200**	**0.459**	**10.2**	**19.6**	**19.9**	**12.7**
	2 1/2 x 2 1/2 x 3/8	9.5	8.70	0.085	1 130	0.419	9.41	19.3	19.5	12.4
65X65X8		**8**	**7.66**	**0.075**	**976**	**0.383**	**8.36**	**19.8**	**19.2**	**12.7**
	2 1/2 x 2 1/2 x 5/6	7.9	7.40	0.073	949	0.361	8.00	19.5	18.9	12.4
	2 1/2 x 2 1/2 x 1/4	6.4	6.10	0.060	778	0.302	6.62	19.7	18.4	12.5
65X65X6		**6**	**5.84**	**0.057**	**744**	**0.300**	**6.44**	**20.1**	**18.5**	**12.8**
65X65X5		**5**	**4.91**	**0.048**	**625**	**0.255**	**5.45**	**20.2**	**18.1**	**12.9**
	2 1/2 x 2 1/2 x 3/16	4.8	4.60	0.045	591	0.235	5.08	19.9	17.8	12.6

**For angle designations, see page 6-19. See page 6-43 for more information on these tables.

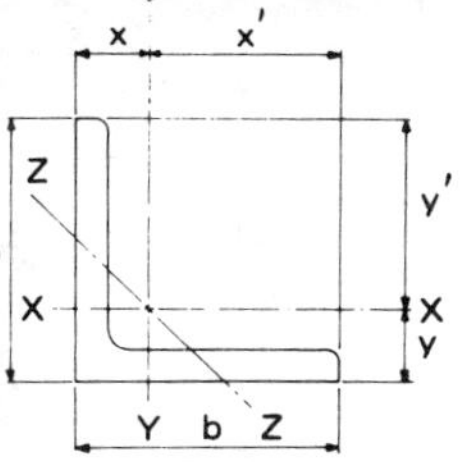

METRIC AND IMPERIAL SIZE ANGLES
Equal Legs

PROPERTIES AND DIMENSIONS

ANGLE SIZE**		Thickness t	Mass	Dead Load	Area	Axis X-X & Axis Y-Y				Axis Z-Z
Metric	Imperial					I	S	r	x or y	r
mm x mm x mm	in x in x in	mm	kg/m	kN/m	mm^2	$10^6 mm^4$	$10^3 mm^3$	mm	mm	mm
55X55	2 x 2 (51 x 51)									
55X55X10		**10**	**7.85**	**0.077**	**1 000**	**0.268**	**7.11**	**16.4**	**17.4**	**10.7**
	2 x 2 x 3/8	9.5	7.00	0.069	879	0.202	5.79	15.1	16.2	9.89
55X55X8		**8**	**6.41**	**0.063**	**816**	**0.225**	**5.87**	**16.6**	**16.7**	**10.7**
	2 x 2 x 5/16	7.9	5.80	0.057	743	0.175	4.94	15.3	15.6	9.90
	2 x 2 x 1/4	6.4	4.70	0.046	612	0.147	4.11	15.5	15.1	9.93
55X55X6		**6**	**4.90**	**0.048**	**624**	**0.177**	**4.54**	**16.9**	**16.0**	**10.8**
55X55X5		**5**	**4.12**	**0.040**	**525**	**0.152**	**3.85**	**17.0**	**15.6**	**10.8**
	2 x 2 x 3/16	4.8	3.60	0.035	467	0.116	3.17	15.7	14.5	10.0
55X55X4		**4**	**3.33**	**0.033**	**424**	**0.125**	**3.13**	**17.1**	**15.2**	**10.9**
	2 x 2 x 1/8	3.2	2.40	0.024	316	0.081	2.17	16.0	13.9	10.1
55X55X3		**3**	**2.52**	**0.025**	**321**	**0.096**	**2.39**	**17.3**	**14.9**	**11.0**
45X45	1 3/4 x 1 3/4 (44 x 44)									
45X45X8		**8**	**5.15**	**0.050**	**656**	**0.118**	**3.02**	**13.4**	**14.2**	**8.76**
	1 3/4 x 1 3/4 x 1/4	6.4	4.10	0.040	522	0.092	3.00	13.3	13.3	8.68
45X45X6		**6**	**3.96**	**0.039**	**504**	**0.094**	**2.98**	**13.7**	**13.4**	**8.79**
45X45X5		**5**	**3.34**	**0.033**	**425**	**0.081**	**2.53**	**13.8**	**13.1**	**8.82**
	1 3/4 x 1 3/4 x 3/16	4.8	3.10	0.030	399	0.073	2.33	13.5	12.8	8.73
45X45X4		**4**	**2.70**	**0.026**	**344**	**0.067**	**2.07**	**13.9**	**12.7**	**8.87**
	1 3/4 x 1 3/4 x 1/8	3.2	2.10	0.021	271	0.051	1.60	13.7	12.2	8.90
45X45X3		**3**	**2.05**	**0.020**	**261**	**0.052**	**1.58**	**14.1**	**12.4**	**8.82**
	1 1/2 x 1 1/2 (38 x 38)									
	1 1/2 x 1 1/2 x 1/4	6.4	3.40	0.033	445	0.058	2.20	11.4	11.8	7.42
	1 1/2 x 1 1/2 x 3/16	4.8	2.70	0.027	342	0.046	1.71	11.6	11.3	7.45
	1 1/2 x 1 1/2 x 5/32	4.0	2.20	0.022	288	0.039	1.45	11.7	11.0	7.48
	1 1/2 x 1 1/2 x 1/8	3.2	1.80	0.018	233	0.032	1.18	11.8	10.7	7.52
35X35										
35X35X6		**6**	**3.01**	**0.030**	**384**	**0.042**	**1.74**	**10.5**	**10.9**	**6.81**
35X35X5		**5**	**2.55**	**0.025**	**325**	**0.036**	**1.49**	**10.6**	**10.6**	**6.83**
35X35X4		**4**	**2.07**	**0.020**	**264**	**0.030**	**1.22**	**10.7**	**10.2**	**6.86**
35X35X3		**3**	**1.58**	**0.015**	**201**	**0.024**	**0.940**	**10.8**	**9.86**	**6.91**
	1 1/4 x 1 1/4 (32 x 32)									
	1 1/4 x 1 1/4 x 1/4	6.4	2.80	0.028	369	0.033	1.52	9.45	10.3	6.19
	1 1/4 x 1 1/4 x 3/16	4.8	2.20	0.022	284	0.026	1.19	9.64	9.75	6.20
	1 1/4 x 1 1/4 x 1/8	3.2	1.50	0.015	195	0.019	0.827	9.85	9.18	6.25
25X25	1 x 1 (25 x 25)									
	1 x 1 x 1/4	6.4	2.20	0.022	279	0.015	0.889	7.24	8.53	4.98
25X25X5		**5**	**1.77**	**0.017**	**225**	**0.012**	**0.724**	**7.39**	**8.06**	**4.87**
	1 x 1 x 3/16	4.8	1.80	0.018	217	0.012	0.700	7.41	7.99	4.94
25X25X4		**4**	**1.44**	**0.014**	**184**	**0.010**	**0.599**	**7.50**	**7.71**	**4.87**
	1 x 1 x 1/8	3.2	1.20	0.012	150	0.009	0.492	7.60	7.42	4.97
25X25X3		**3**	**1.11**	**0.011**	**141**	**0.008**	**0.465**	**7.63**	**7.35**	**4.89**

*For angle designations, see page 6-19. See page 6-43 for more information on these tables.

METRIC AND IMPERIAL SIZE ANGLES
Unequal Legs

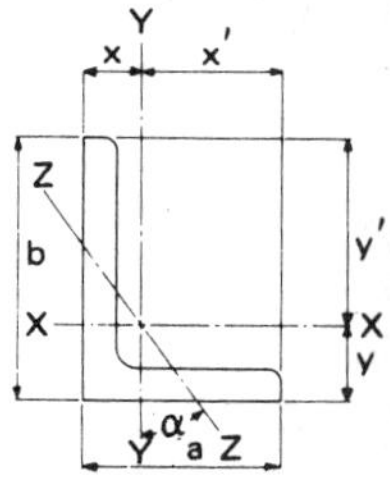

PROPERTIES AND DIMENSIONS

ANGLE SIZE**		Thickness t	Mass	Dead Load	Area	Axis X-X				Axis Y-Y				Axis Z-Z	
Metric	Imperial					I_x	S_x	r_x	y	I_y	S_y	r_y	x	r_z	Tan α
mm × mm × mm	in × in × in	mm	kg/m	kN/m	mm^2	10^6 mm^4	10^3 mm^3	mm	mm	10^6 mm^4	10^3 mm^3	mm	mm	mm	
200 x 150	8 x 6 (203 x 152)														
	8 x 6 x 1	25.4	65.5	0.642	8 370	33.5	247	63.3	67.4	16.0	145	43.7	41.9	32.4	0.541
200 x 150 x 25		**25**	**63.8**	**0.625**	**8 120**	**31.6**	**236**	**62.3**	**66.3**	**15.1**	**139**	**43.2**	**41.3**	**32.0**	**0.543**
	8 x 6 x 7/8	22.2	57.9	0.568	7 390	30.0	219	63.7	66.2	14.4	129	44.1	40.7	32.5	0.545
200 x 150 x 20		**20**	**51.8**	**0.508**	**6 600**	**26.2**	**193**	**63.0**	**64.5**	**12.7**	**115**	**43.8**	**39.5**	**32.1**	**0.540**
	8 x 6 x 3/4	19.0	50.1	0.491	6 380	26.2	190	64.1	65.1	12.7	113	44.5	39.6	32.6	0.549
200 x 150 x 16		**16**	**42.0**	**0.411**	**5 340**	**21.6**	**158**	**63.5**	**63.1**	**10.5**	**93.8**	**44.3**	**38.1**	**32.3**	**0.554**
	8 x 6 x 5/8	15.9	42.2	0.414	5 390	22.5	162	64.6	64.0	10.9	95.9	44.9	38.5	32.8	0.553
	8 x 6 x 9/16	14.3	38.1	0.374	4 870	20.4	146	64.8	63.4	9.94	87.1	45.2	37.9	32.9	0.554
200 x 150 x 13		**13**	**34.4**	**0.337**	**4 380**	**17.9**	**130**	**64.0**	**62.0**	**8.77**	**77.6**	**44.7**	**37.0**	**32.5**	**0.557**
	8 x 6 x 1/2	12.7	34.1	0.334	4 350	18.4	131	65.0	62.8	8.96	78.1	45.4	37.3	33.0	0.556
	8 x 6 x 7/16	11.1	29.9	0.293	3 820	16.3	115	65.3	62.2	7.94	68.9	45.6	36.7	33.1	0.558
200 x 100	8 x 4 (203 x 102)														
	8 x 4 x 1	25.4	55.4	0.543	7 100	29.0	230	63.8	77.2	4.90	65.1	26.3	26.7	21.6	0.249
200 x 100 x 20		**20**	**44.0**	**0.431**	**5 600**	**22.6**	**180**	**63.6**	**74.3**	**3.84**	**50.8**	**26.2**	**24.3**	**21.3**	**0.256**
	8 x 4 x 3/4	19.0	42.5	0.417	5 430	22.8	178	64.7	74.8	3.93	50.6	26.9	24.3	21.7	0.260
200 x 100 x 16		**16**	**35.7**	**0.350**	**4 540**	**18.7**	**147**	**64.2**	**72.8**	**3.22**	**41.8**	**26.6**	**22.8**	**21.4**	**0.262**
	8 x 4 x 9/16	14.3	32.4	0.318	4 160	17.8	137	65.4	73.0	3.13	39.4	27.4	22.5	22.0	0.267
200 x 100 x 13		**13**	**29.3**	**0.287**	**3 730**	**15.6**	**121**	**64.6**	**71.7**	**2.72**	**34.7**	**27.0**	**21.7**	**21.6**	**0.266**
	8 x 4 x 1/2	12.7	29.0	0.284	3 710	16.0	123	65.7	72.4	2.84	35.4	27.6	21.9	22.1	0.269
200 x 100 x 10		**10**	**22.8**	**0.223**	**2 900**	**12.3**	**94.8**	**65.1**	**70.5**	**2.18**	**27.4**	**27.4**	**20.5**	**21.8**	**0.271**
	7 x 4 (178 x 102)														
	7 x 4 x 3/4	19.0	38.8	0.380	4 960	15.8	138	56.4	63.7	3.80	49.4	27.7	25.7	21.9	0.326
	7 x 4 x 5/8	15.9	32.7	0.321	4 200	13.6	118	56.8	62.6	3.31	42.7	28.1	24.6	22.1	0.331
	7 x 4 x 1/2	12.7	26.5	0.260	3 390	11.1	95.6	57.3	61.4	2.75	35.0	28.5	23.4	22.2	0.336
	7 x 4 x 3/8	9.5	20.2	0.198	2 570	8.58	72.8	57.8	60.2	2.14	26.9	28.9	22.2	22.5	0.341
150 x 100	6 x 4 (152 x 102)														
	6 x 4 x 7/8	22.2	40.3	0.395	5 150	11.5	117	47.2	55.7	4.10	55.9	28.2	28.7	21.9	0.427
	6 x 4 x 3/4	19.0	35.0	0.343	4 460	10.1	102	47.6	52.5	3.65	48.9	28.6	27.5	21.9	0.434
150 x 100 x 16		**16**	**29.4**	**0.288**	**3 740**	**8.40**	**84.8**	**47.4**	**50.9**	**3.00**	**40.4**	**28.3**	**25.9**	**21.6**	**0.434**
	6 x 4 x 5/8	15.9	29.6	0.290	3 790	8.73	86.8	48.0	51.4	3.17	41.9	28.9	26.4	22.0	0.440
	6 x 4 x 9/16	14.3	26.8	0.263	3 430	7.98	78.8	48.2	50.8	2.91	38.2	29.1	25.8	22.1	0.443
150 x 100 x 13		**13**	**24.2**	**0.237**	**3 080**	**7.03**	**70.2**	**47.8**	**49.9**	**2.53**	**33.7**	**28.7**	**24.9**	**21.7**	**0.440**
	6 x 4 x 1/2	12.7	24.0	0.235	3 060	7.20	70.7	48.5	50.2	2.64	34.4	29.3	25.2	22.2	0.446
	6 x 4 x 7/16	11.1	21.2	0.208	2 700	6.39	62.4	48.7	49.6	2.35	30.4	29.6	24.6	22.3	0.449
150 x 100 x 10		**10**	**18.8**	**0.185**	**2 400**	**5.58**	**55.1**	**48.2**	**48.8**	**2.03**	**26.6**	**29.1**	**23.8**	**21.9**	**0.445**
	6 x 4 x 3/8	9.5	18.2	0.178	2 320	5.56	54.0	48.9	49.0	2.06	26.4	29.8	24.0	22.4	0.451
150 x 100 x 8		**8**	**15.2**	**0.149**	**1 940**	**4.45**	**44.6**	**48.5**	**48.0**	**1.67**	**21.6**	**29.3**	**23.0**	**22.0**	**0.448**
	6 x 4 x 5/16	7.9	15.3	0.150	1 940	4.70	45.4	49.2	48.5	1.75	22.3	30.0	23.5	22.5	0.454
	6 x 3 1/2 (152 x 89)														
	6 x 3 1/2 x 1/2	12.7	22.7	0.223	2 900	6.86	69.1	48.6	52.7	1.77	26.2	24.7	21.2	19.3	0.346
	6 x 3 1/2 x 3/8	9.5	17.3	0.170	2 200	5.30	52.8	49.1	51.5	1.39	20.2	25.1	20.0	19.5	0.352
	6 x 3 1/2 x 5/16	7.9	14.5	0.142	1 840	4.48	44.4	49.3	50.9	1.18	17.0	25.4	19.4	19.6	0.355

**For angle designations, see page 6-19. See page 6-43 for more information on these tables.

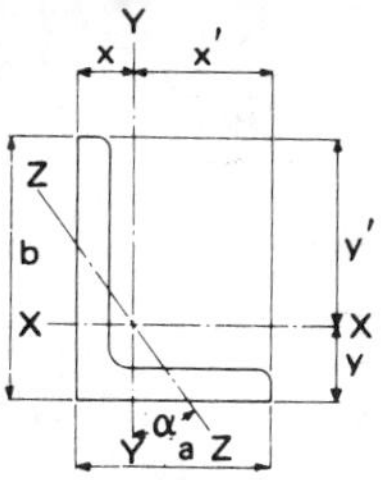

METRIC AND IMPERIAL SIZE ANGLES
Unequal Legs

PROPERTIES AND DIMENSIONS

ANGLE SIZE**		Thickness t	Mass	Dead Load	Area	Axis X-X				Axis Y-Y				Axis Z-Z	
Metric	Imperial					I_x	S_x	r_x	y	I_y	S_y	r_y	x	r_z	Tan α
mm × mm × mm	in × in × in	mm	kg/m	kN/m	mm^2	10^6 mm^4	10^3 mm^3	mm	mm	10^6 mm^4	10^3 mm^3	mm	mm	mm	
125 x 90	5 x 3 1/2 (127 x 89)														
	5 x 3 1/2 x 3/4	19.0	29.3	0.287	3 740	5.78	69.9	39.3	44.3	2.31	36.3	24.9	25.3	19.0	0.465
125 x 90 x 16		**16**	**25.0**	**0.245**	**3 180**	**4.84**	**58.5**	**39.0**	**42.2**	**2.09**	**32.0**	**25.6**	**24.7**	**19.2**	**0.499**
	5 x 3 1/2 x 5/8	15.9	24.9	0.244	3 180	5.01	59.8	39.7	43.2	2.02	31.2	25.2	24.2	19.1	0.473
125 x 90 x 13		**13**	**20.6**	**0.202**	**2 630**	**4.07**	**48.6**	**39.4**	**41.2**	**1.77**	**26.7**	**26.0**	**23.7**	**19.3**	**0.505**
	5 x 3 1/2 x 1/2	12.7	20.2	0.198	2 580	4.16	48.9	40.1	42.1	1.69	25.6	25.6	23.1	19.2	0.480
	5 x 3 1/2 x 7/16	11.1	17.8	0.175	2 270	3.70	43.3	40.4	41.5	1.51	22.7	25.8	22.5	19.3	0.483
125 x 90 x 10		**10**	**16.1**	**0.158**	**2 050**	**3.25**	**38.2**	**39.8**	**40.1**	**1.42**	**21.1**	**26.4**	**22.6**	**19.5**	**0.511**
	5 x 3 1/2 x 3/8	9.5	15.4	0.151	1 960	3.23	37.5	40.6	40.9	1.33	19.8	26.0	21.9	19.4	0.487
125 x 90 x 8		**8**	**13.0**	**0.127**	**1 660**	**2.66**	**31.1**	**40.1**	**39.3**	**1.18**	**17.2**	**26.6**	**21.8**	**19.6**	**0.515**
	5 x 3 1/2 x 5/16	7.9	12.9	0.127	1 640	2.74	31.6	40.8	40.3	1.13	16.7	26.2	21.3	19.5	0.490
	5 x 3 1/2 x 1/4	6.4	10.4	0.102	1 340	2.26	25.9	41.0	39.7	0.938	13.7	26.4	20.7	19.6	0.493
125 x 75	5 x 3 (127 x 76)														
	5 x 3 x 5/8	15.9	23.3	0.228	2 970	4.74	58.2	39.9	45.7	1.27	22.7	20.6	20.2	16.3	0.347
125 x 75 x 13		**13**	**19.1**	**0.187**	**2 430**	**3.82**	**47.1**	**39.6**	**43.9**	**1.04**	**18.5**	**20.7**	**18.9**	**16.2**	**0.356**
	5 x 3 x 1/2	12.7	19.0	0.186	2 420	3.93	47.6	40.3	44.5	1.06	18.7	21.0	19.0	16.4	0.355
	5 x 3 x 7/16	11.1	16.7	0.164	2 130	3.50	42.2	40.6	43.9	0.956	16.6	21.2	18.4	16.5	0.359
125 x 75 x 10		**10**	**14.9**	**0.146**	**1 900**	**3.05**	**37.1**	**40.0**	**42.8**	**0.841**	**14.7**	**21.0**	**17.8**	**16.3**	**0.363**
	5 x 3 x 3/8	9.5	14.5	0.142	1 840	3.06	36.5	40.8	43.3	0.841	14.4	21.4	17.8	16.6	0.363
125 x 75 x 8		**8**	**12.1**	**0.118**	**1 540**	**2.50**	**30.1**	**40.3**	**42.1**	**0.697**	**12.0**	**21.3**	**17.1**	**16.4**	**0.367**
	5 x 3 x 5/16	7.9	12.1	0.119	1 540	2.59	30.8	41.0	42.7	0.719	12.2	21.6	17.2	16.7	0.366
	5 x 3 x 1/4	6.4	9.80	0.096	1 260	2.14	25.2	41.2	42.2	0.598	10.1	21.8	16.7	16.8	0.369
125 x 75 x 6		**6**	**9.14**	**0.090**	**1 160**	**1.92**	**23.0**	**40.6**	**41.3**	**0.542**	**9.23**	**21.6**	**16.3**	**16.6**	**0.372**
100 x 90	4 x 3 1/2 (102 x 89)														
	4 x 3 1/2 x 5/8	15.9	21.8	0.214	2 780	2.69	38.9	31.1	33.0	1.89	30.3	26.1	26.5	18.3	0.740
100 x 90 x 13		**13**	**18.1**	**0.177**	**2 300**	**2.17**	**31.4**	**30.7**	**31.1**	**1.66**	**25.9**	**26.8**	**26.1**	**18.4**	**0.796**
	4 x 3 1/2 x 1/2	12.7	17.6	0.173	2 260	2.24	32.0	31.5	31.9	1.59	24.9	26.5	25.4	18.4	0.746
	4 x 3 1/2 x 7/16	11.1	15.7	0.154	2 000	2.00	28.3	31.7	31.3	1.42	22.1	26.7	24.8	18.4	0.748
100 x 90 x 10		**10**	**14.1**	**0.139**	**1 800**	**1.74**	**24.9**	**31.1**	**30.0**	**1.33**	**20.5**	**27.2**	**25.0**	**18.5**	**0.800**
	4 x 3 1/2 x 3/8	9.5	13.5	0.132	1 720	1.75	24.6	31.9	30.7	1.25	19.2	26.9	24.2	18.5	0.751
100 x 90 x 8		**8**	**11.4**	**0.112**	**1 460**	**1.43**	**20.3**	**31.4**	**29.3**	**1.10**	**16.8**	**27.5**	**24.3**	**18.6**	**0.802**
	4 x 3 1/2 x 5/16	7.9	11.4	0.112	1 450	1.49	20.8	32.1	30.7	1.06	16.3	27.1	23.7	18.6	0.753
	4 x 3 1/2 x 1/4	6.4	9.20	0.090	1 180	1.24	17.1	32.3	29.6	0.881	13.4	27.3	23.1	18.7	0.755
100 x 90 x 6		**6**	**8.67**	**0.085**	**1 100**	**1.11**	**15.5**	**31.7**	**28.5**	**0.853**	**12.8**	**27.8**	**23.5**	**18.7**	**0.805**
100 x 75	4 x 3 (102 x 76)														
	4 x 3 x 5/8	15.9	20.2	0.198	2 580	2.54	37.9	31.4	35.0	1.19	22.1	21.5	22.0	16.1	0.526
100 x 75 x 13		**13**	**16.5**	**0.162**	**2 110**	**2.04**	**30.6**	**31.1**	**33.4**	**0.976**	**18.0**	**21.5**	**20.9**	**16.0**	**0.541**
	4 x 3 x 1/2	12.7	16.4	0.161	2 100	2.12	31.2	31.8	33.9	1.00	18.2	21.8	20.9	16.2	0.536
	4 x 3 x 7/16	11.1	14.5	0.142	1 850	1.90	27.6	32.0	33.3	0.900	16.2	22.0	20.3	16.3	0.540
100 x 75 x 10		**10**	**13.0**	**0.127**	**1 650**	**1.64**	**24.2**	**31.5**	**32.3**	**0.791**	**14.3**	**21.9**	**19.8**	**16.1**	**0.549**
	4 x 3 x 3/8	9.5	12.6	0.124	1 600	1.66	24.0	32.2	32.7	0.792	14.1	22.2	19.7	16.3	0.544
100 x 75 x 8		**8**	**10.5**	**0.103**	**1 340**	**1.35**	**19.7**	**31.8**	**31.5**	**0.656**	**11.7**	**22.2**	**19.0**	**16.2**	**0.554**
	4 x 3 x 5/16	7.9	10.7	0.105	1 340	1.42	20.3	32.5	32.2	0.678	11.9	22.5	19.2	16.4	0.548
	4 x 3 x 1/4	6.4	8.60	0.084	1 100	1.17	16.7	32.7	31.6	0.564	9.83	22.7	18.6	16.5	0.551
100 x 75 x 6		**6**	**7.96**	**0.078**	**1 010**	**1.04**	**15.1**	**32.1**	**30.8**	**0.511**	**9.01**	**22.4**	**18.3**	**16.3**	**0.559**

**For angle designations, see page 6-19. See page 6-43 for more information on these tables.

METRIC AND IMPERIAL SIZE ANGLES
Unequal Legs

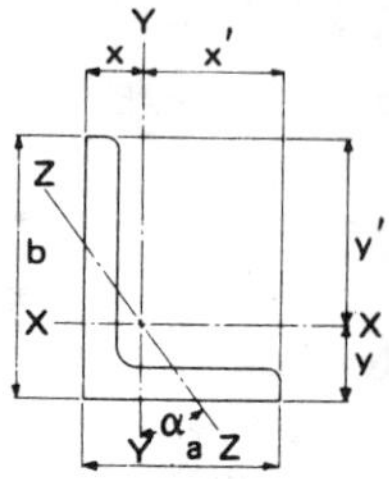

PROPERTIES AND DIMENSIONS

ANGLE SIZE**		Thickness t	Mass	Dead Load	Area	Axis X-X				Axis Y-Y				Axis Z-Z	
Metric	Imperial					I_x	S_x	r_x	y	I_y	S_y	r_y	x	r_z	Tan α
mm × mm × mm	in × in × in	mm	kg/m	kN/m	mm^2	$10^6\ mm^4$	$10^3\ mm^3$	mm	mm	$10^6\ mm^4$	$10^3\ mm^3$	mm	mm	mm	
90 x 75	3½ x 3 (89 x 76)														
90 x 75 x 13		**13**	**15.5**	**0.152**	**1 980**	**1.51**	**24.8**	**27.6**	**29.3**	**0.946**	**17.8**	**21.9**	**21.8**	**15.6**	**0.672**
	3½ x 3 x ½	12.7	15.1	0.148	1 930	1.44	23.9	27.3	28.6	0.962	17.9	22.3	22.1	15.7	0.709
	3½ x 3 x 7/16	11.1	13.5	0.132	1 710	1.29	21.2	27.5	28.1	0.865	15.9	22.5	21.6	15.8	0.712
90 x 75 x 10		**10**	**12.2**	**0.119**	**1 550**	**1.22**	**19.7**	**28.0**	**28.2**	**0.767**	**14.1**	**22.2**	**20.7**	**15.7**	**0.679**
	3½ x 3 x 3/8	9.5	11.7	0.115	1 480	1.13	18.4	27.7	27.5	0.761	13.8	22.7	21.0	15.8	0.715
90 x 75 x 8		**8**	**9.86**	**0.097**	**1 260**	**1.01**	**16.1**	**28.3**	**27.5**	**0.636**	**11.6**	**22.5**	**20.0**	**15.8**	**0.683**
	3½ x 3 x 5/16	7.9	9.80	0.095	1 240	0.968	15.6	27.9	26.9	0.652	11.7	22.9	20.4	15.9	0.718
	3½ x 3 x ¼	6.4	8.00	0.079	1 020	0.804	12.8	28.1	26.4	0.543	9.67	23.1	19.4	16.0	0.721
90 x 75 x 6		**6**	**7.49**	**0.073**	**954**	**0.779**	**12.3**	**28.6**	**26.8**	**0.495**	**8.89**	**22.8**	**19.3**	**15.9**	**0.687**
90 x 75 x 5		**5**	**6.28**	**0.062**	**800**	**0.660**	**10.4**	**28.7**	**26.4**	**0.421**	**7.50**	**22.9**	**18.9**	**16.0**	**0.689**
90 x 65	3½ x 2½ (89 x 64)														
	3½ x 2½ x ½	12.7	13.9	0.136	1 780	1.36	23.2	27.6	30.6	0.581	12.7	18.1	18.1	13.7	0.493
	3½ x 2½ x 7/16	11.1	12.3	0.121	1 580	1.22	20.6	27.8	30.0	0.524	11.3	18.2	17.5	13.7	0.498
90 x 65 x 10		**10**	**11.4**	**0.112**	**1 450**	**1.16**	**19.2**	**28.3**	**29.8**	**0.507**	**10.6**	**18.7**	**17.3**	**13.9**	**0.506**
	3½ x 2½ x 3/8	9.5	10.7	0.105	1 360	1.07	17.9	28.0	29.4	0.463	9.84	18.4	169	13.8	0.503
90 x 65 x 8		**8**	**9.23**	**0.090**	**1 180**	**0.958**	**15.7**	**28.5**	**29.1**	**0.422**	**8.72**	**18.9**	**16.6**	**14.0**	**0.512**
	3½ x 2½ x 5/16	7.9	9.00	0.088	1 150	0.913	15.2	28.0	28.8	0.398	8.35	18.6	16.3	13.8	0.508
	3½ x 2½ x ¼	6.4	7.30	0.072	938	0.759	12.5	28.2	28.3	0.333	6.91	18.8	15.8	13.4	0.512
90 x 65 x 6		**6**	**7.02**	**0.069**	**894**	**0.743**	**12.1**	**28.8**	**28.4**	**0.330**	**6.72**	**19.2**	**15.9**	**14.2**	**0.518**
90 x 65 x 5		**5**	**5.89**	**0.058**	**750**	**0.629**	**10.2**	**29.0**	**28.0**	**0.281**	**5.68**	**19.4**	**15.5**	**14.2**	**0.520**
80 x 60															
80 x 60 x 10		**10**	**10.2**	**0.100**	**1 300**	**0.808**	**15.1**	**24.9**	**26.5**	**0.388**	**8.92**	**17.3**	**16.5**	**12.8**	**0.543**
80 x 60 x 8		**8**	**8.29**	**0.081**	**1 060**	**0.670**	**12.4**	**25.2**	**25.8**	**0.324**	**7.33**	**17.5**	**15.8**	**12.9**	**0.549**
80 x 60 x 6		**6**	**6.31**	**0.062**	**804**	**0.522**	**9.50**	**25.5**	**25.1**	**0.254**	**5.66**	**17.8**	**15.1**	**13.0**	**0.555**
80 x 60 x 5		**5**	**5.30**	**0.052**	**675**	**0.443**	**8.02**	**25.6**	**24.7**	**0.217**	**4.79**	**17.9**	**14.7**	**13.0**	**0.558**
	3 x 2½ (76 x 64)														
	3 x 2½ x ½	12.7	12.6	0.124	1 620	0.863	17.0	23.1	25.2	0.554	12.4	18.5	19.2	13.3	0.682
	3 x 2½ x 7/16	11.1	11.3	0.111	1 430	0.777	15.1	23.3	24.7	0.500	11.0	18.7	18.7	13.3	0.687
	3 x 2½ x 3/8	9.5	9.80	0.096	1 240	0.685	13.2	23.5	24.1	0.442	9.64	18.9	18.1	13.3	0.691
	3 x 2½ x 5/16	7.9	8.30	0.081	1 040	0.587	11.2	23.7	23.5	0.380	8.19	19.1	17.5	13.4	0.695
	3 x 2½ x ¼	6.4	6.70	0.066	855	0.489	9.23	23.9	23.0	0.318	6.77	19.3	17.0	13.5	0.699
	3 x 2½ x 3/16	4.8	5.10	0.050	649	0.379	7.06	24.2	22.4	0.247	5.20	19.5	16.4	13.6	0.702
75 x 50	3 x 2 (76 x 51)														
	3 x 2 x ½	12.7	11.5	0.113	1 450	0.795	16.4	23.4	27.4	0.283	7.83	14.0	14.9	10.9	0.420
	3 x 2 x 7/16	11.1	10.1	0.099	1 290	0.716	14.6	23.6	26.8	0.256	6.99	14.1	14.3	10.9	0.427
	3 x 2 x 3/8	9.5	8.80	0.086	1 120	0.632	12.7	23.8	26.3	0.228	6.12	14.3	13.8	11.0	0.434
75 x 50 x 8		**8**	**7.35**	**0.072**	**936**	**0.525**	**10.6**	**23.7**	**25.5**	**0.187**	**5.06**	**14.1**	**13.0**	**10.8**	**0.434**
	3 x 2 x 5/16	7.9	7.40	0.073	941	0.543	10.8	24.0	25.7	0.197	5.21	14.5	13.2	11.0	0.440
	3 x 2 x ¼	6.4	6.10	0.060	772	0.453	8.90	24.2	25.1	0.166	4.32	14.7	12.6	11.1	0.446
75 x 50 x 6		**6**	**5.60**	**0.055**	**714**	**0.410**	**8.15**	**24.0**	**24.7**	**0.148**	**3.92**	**14.4**	**12.2**	**10.9**	**0.441**
75 x 50 x 5		**5**	**4.71**	**0.046**	**600**	**0.349**	**6.88**	**24.1**	**24.4**	**0.127**	**3.32**	**14.5**	**11.9**	**10.9**	**0.445**
	3 x 2 x 3/16	4.8	4.60	0.045	587	0.351	6.82	24.5	24.5	0.130	3.33	14.9	12.0	11.2	0.451

**For angle designations, see page 6-19. See page 6-43 for more information on these tables.

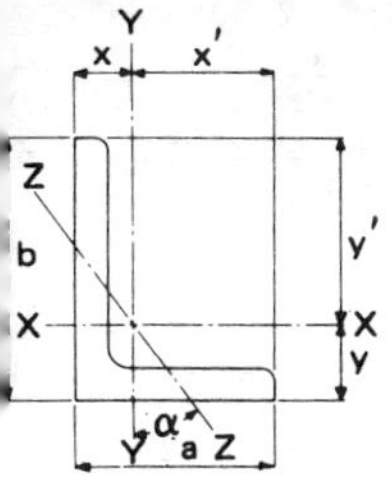

METRIC AND IMPERIAL SIZE ANGLES
Unequal Legs

PROPERTIES AND DIMENSIONS

ANGLE SIZE**		Thickness t	Mass	Dead Load	Area	Axis X-X				Axis Y-Y				Axis Z-Z	
Metric	Imperial					I_x	S_x	r_x	y	I_y	S_y	r_y	x	r_z	Tan α
mm × mm × mm	in × in × in	mm	kg/m	kN/m	mm^2	10^6 mm^4	10^3 mm^3	mm	mm	$10^6 \cdot mm^4$	10^3 mm^3	mm	mm	mm	
65 x 50	2½ x 2 (64 x 51)														
	2½ x 2 x 3/8	9.5	7.90	0.078	1 000	0.388	9.08	19.7	21.3	0.217	5.98	14.7	14.8	10.7	0.609
65 x 50 x 8		**8**	**6.72**	**0.066**	**856**	**0.351**	**8.03**	**20.2**	**21.3**	**0.180**	**4.97**	**14.5**	**13.8**	**10.6**	**0.572**
	2½ x 2 x 5/16	7.9	6.70	0.066	846	0.334	7.73	19.9	20.7	0.188	5.10	14.9	14.2	10.8	0.616
	2½ x 2 x 1/4	6.4	5.40	0.053	695	0.280	6.40	20.1	20.2	0.158	4.24	15.1	13.7	10.8	0.621
65 x 50 x 6		**6**	**5.13**	**0.050**	**654**	**0.275**	**6.19**	**20.5**	**20.6**	**0.142**	**3.85**	**14.7**	**13.1**	**10.7**	**0.580**
65 x 50 x 5		**5**	**4.32**	**0.042**	**550**	**0.235**	**5.24**	**20.7**	**20.2**	**0.122**	**3.27**	**14.9**	**12.7**	**10.8**	**0.583**
	2½ x 2 x 3/16	4.8	4.20	0.041	529	0.218	4.91	20.3	19.6	0.124	3.27	15.3	13.1	10.9	0.626
65 x 50 x 4		**4**	**3.49**	**0.034**	**444**	**0.192**	**4.25**	**20.8**	**19.9**	**0.100**	**2.66**	**15.0**	**12.4**	**10.8**	**0.587**
	2½ x 1½ (64 x 38)														
	2½ x 1½ x 5/16	7.9	5.80	0.057	743	0.301	7.35	20.1	20.3	0.079	2.82	10.3	10.0	8.16	0.343
	2½ x 1½ x 1/4	6.4	4.80	0.047	612	0.253	6.09	20.3	22.5	0.067	2.35	10.5	9.48	8.20	0.350
	2½ x 1½ x 3/16	4.8	3.60	0.035	467	0.197	4.68	20.1	21.9	0.053	1.82	10.7	8.84	8.28	0.357
55 x 35															
55 x 35 x 6		**6**	**3.96**	**0.039**	**504**	**0.152**	**4.23**	**17.4**	**19.0**	**0.048**	**1.85**	**9.77**	**9.04**	**7.55**	**0.396**
55 x 35 x 5		**5**	**3.34**	**0.033**	**425**	**0.130**	**3.59**	**17.5**	**18.7**	**0.041**	**1.58**	**9.89**	**8.68**	**7.59**	**0.401**
55 x 35 x 4		**4**	**2.70**	**0.026**	**344**	**0.107**	**2.92**	**17.7**	**18.3**	**0.034**	**1.29**	**10.0**	**8.31**	**7.65**	**0.406**
55 x 35 x 3		**3**	**2.05**	**0.020**	**261**	**0.083**	**2.23**	**17.8**	**17.9**	**0.027**	**0.994**	**10.2**	**7.94**	**7.72**	**0.411**
	2 x 1½ (51 x 38)														
	2 x 1½ x 1/4	6.4	4.20	0.041	529	0.134	3.92	15.9	17.0	0.063	2.29	10.9	10.5	8.10	0.535
	2 x 1½ x 3/16	4.8	3.10	0.030	404	0.105	3.03	16.1	16.4	0.050	1.78	11.1	9.89	8.16	0.544
	2 x 1½ x 1/8	3.2	2.10	0.021	275	0.073	2.08	16.3	15.8	0.035	1.23	11.3	9.31	8.26	0.551
45 x 30															
45 x 30 x 6		**6**	**3.25**	**0.032**	**414**	**0.082**	**2.79**	**14.0**	**15.7**	**0.029**	**1.32**	**8.35**	**8.22**	**6.44**	**0.426**
45 x 30 x 5		**5**	**2.75**	**0.027**	**350**	**0.070**	**2.37**	**14.2**	**15.4**	**0.025**	**1.13**	**8.46**	**7.86**	**6.47**	**0.433**
45 x 30 x 4		**4**	**2.23**	**0.022**	**284**	**0.058**	**1.94**	**14.3**	**15.0**	**0.021**	**0.930**	**8.58**	**7.49**	**6.51**	**0.439**
45 x 30 x 3		**3**	**1.70**	**0.017**	**216**	**0.045**	**1.49**	**14.5**	**14.6**	**0.016**	**0.717**	**8.72**	**7.12**	**6.57**	**0.445**

**For angle designations, see page 6-19. See page 6-43 for more information on these tables.

ANGLES — Metric Series**
Equal Legs

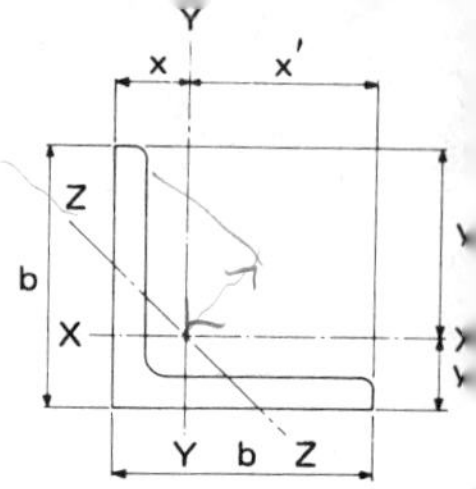

PROPERTIES AND DIMENSIONS

+Size**	Thickness t	Mass	Dead Load	Area	Axis X-X and Axis Y-Y				Axis Z-Z
					I	S	r	x or y	r
mmxmm	mm	kg/m	kN/m	mm²	$10^6 mm^4$	$10^3 mm^3$	mm	mm	mm
200X200	30	87.1	0.855	11 100	40.3	290	60.3	60.9	39.0
	25	73.6	0.722	9 380	34.8	247	60.9	59.2	39.1
	20	59.7	0.585	7 600	28.8	202	61.6	57.4	39.3
	16	48.2	0.473	6 140	23.7	165	62.1	55.9	39.5
	13	39.5	0.387	5 030	19.7	136	62.6	54.8	39.7
	10	30.6	0.300	3 900	15.5	106	63.0	53.7	39.9
150X150	20	44.0	0.431	5 600	11.6	110	45.5	44.8	29.3
	16	35.7	0.350	4 540	9.63	90.3	46.0	43.4	29.4
	13	29.3	0.287	3 730	8.05	74.7	46.4	42.3	29.6
	10	22.8	0.223	2 900	6.37	58.6	46.9	41.2	29.8
125X125	16	29.4	0.288	3 740	5.41	61.5	38.0	37.1	24.4
	13	24.2	0.237	3 080	4.54	51.1	38.4	36.0	24.5
	10	18.8	0.185	2 400	3.62	40.2	38.8	34.9	24.7
	8	15.2	0.149	1 940	2.96	32.6	39.1	34.2	24.8
100X100	16	23.1	0.227	2 940	2.65	38.3	30.0	30.8	19.5
	13	19.1	0.187	2 430	2.24	31.9	30.4	29.8	19.5
	10	14.9	0.146	1 900	1.80	25.2	30.8	28.7	19.7
	8	12.1	0.118	1 540	1.48	20.6	31.1	28.0	19.8
	6	9.14	0.090	1 160	1.14	15.7	31.3	27.2	19.9
90X90	13	17.0	0.167	2 170	1.60	25.6	27.2	27.2	17.6
	10	13.3	0.131	1 700	1.29	20.2	27.6	26.2	17.6
	8	10.8	0.106	1 380	1.07	16.5	27.8	25.5	17.7
	6	8.20	0.080	1 040	0.826	12.7	28.1	24.7	17.9
75X75	13	14.0	0.137	1 780	0.892	17.3	22.4	23.5	14.6
	10	11.0	0.108	1 400	0.725	13.8	22.8	22.4	14.6
	8	8.92	0.087	1 140	0.602	11.3	23.0	21.7	14.7
	6	6.78	0.066	864	0.469	8.68	23.3	21.0	14.8
	5	5.69	0.056	725	0.398	7.32	23.4	20.6	14.9

+Designation consists of the Letter L, the size, and the thickness, eg. L125x90x10. See pages 6-19 for more information.
**Rolling = 5 for all sizes of the Metric Series angles.

Note: The properties of angles currently produced by cold forming are up to 7 percent less than the properties shown in the above tables. Check manufacturers' catalogue for the exact properties and dimensions. **See also page 6-19.**

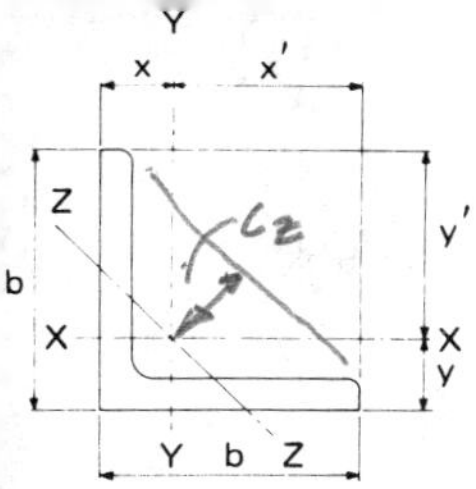

ANGLES — Metric Series**
Equal Legs

PROPERTIES AND DIMENSIONS

+Size**	Thickness t	Mass	Dead Load	Area	Axis X-X and Axis Y-Y				Axis Z-Z
					I	S	r	x or y	r
mmxmm	mm	kg/m	kN/m	mm²	10⁶mm⁴	10³mm³	mm	mm	mm
65X65	10	9.42	0.092	1 200	0.459	10.2	19.6	19.9	12.7
	8	7.66	0.075	976	0.383	8.36	19.8	19.2	12.7
	6	5.84	0.057	744	0.300	6.44	20.1	18.5	12.8
	5	4.91	0.048	625	0.255	5.45	20.2	18.1	12.9
55X55	10	7.85	0.077	1 000	0.268	7.11	16.4	17.4	10.7
	8	6.41	0.063	816	0.225	5.87	16.6	16.7	10.7
	6	4.90	0.048	624	0.177	4.54	16.9	16.0	10.8
	5	4.12	0.040	525	0.152	3.85	17.0	15.6	10.8
	4	3.33	0.033	424	0.125	3.13	17.1	15.2	10.9
	3	2.52	0.025	321	0.096	2.39	17.3	14.9	11.0
45X45	8	5.15	0.050	656	0.118	3.82	13.4	14.2	8.76
	6	3.96	0.039	504	0.094	2.98	13.7	13.4	8.79
	5	3.34	0.033	425	0.081	2.53	13.8	13.1	8.82
	4	2.70	0.026	344	0.067	2.07	13.9	12.7	8.87
	3	2.05	0.020	261	0.052	1.58	14.1	12.4	8.93
35X35	6	3.01	0.030	384	0.042	1.74	10.5	10.9	6.81
	5	2.55	0.025	325	0.036	1.49	10.6	10.6	6.83
	4	2.07	0.020	264	0.030	1.22	10.7	10.2	6.86
	3	1.58	0.015	201	0.024	0.940	10.8	9.86	6.91
25X25	5	1.77	0.017	225	0.012	0.724	7.39	8.06	4.87
	4	1.44	0.014	184	0.010	0.599	7.50	7.71	4.87
	3	1.11	0.011	141	0.008	0.465	7.63	7.35	4.89

+Designation consists of the Letter L, the size, and the thickness, eg. L125x90x10. See pages 6-19 for more information.
**Rolling = 5 for all sizes of the Metric Series angles.

Note: The properties of angles currently produced by cold forming are up to 7 percent less than the properties shown in the above tables. Check manufacturers' catalogue for the exact properties and dimensions. See also page 6-19.

ANGLES — Metric Series**
Unequal Legs

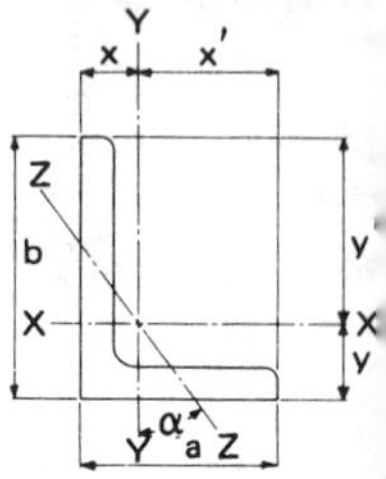

PROPERTIES AND DIMENSIONS

+Size**	Thickness t	Mass	Dead Load	Area	Axis X-X				Axis Y-Y				Axis Z-Z	
					I_x	S_x	r_x	y	I_y	S_y	r_y	x	r_z	Tan α
mmxmm	mm	kg/m	kN/m	mm^2	$10^6 mm^4$	$10^3 mm^3$	mm	mm	$10^6 mm^4$	$10^3 mm^3$	mm	mm	mm	
200X150*	25	63.8	0.625	8 120	31.6	236	62.3	66.3	15.1	139	43.2	41.3	32.0	0.543
	20	51.8	0.508	6 600	26.2	193	63.0	64.5	12.7	115	43.8	39.5	32.1	0.549
	16	42.0	0.411	5 340	21.6	158	63.5	63.1	10.5	93.8	44.3	38.1	32.3	0.554
	13	34.4	0.337	4 380	17.9	130	64.0	62.0	8.77	77.6	44.7	37.0	32.5	0.557
200X100*	20	44.0	0.431	5 600	22.6	180	63.6	74.3	3.84	50.8	26.2	24.3	21.3	0.256
	16	35.7	0.350	4 540	18.7	147	64.2	72.8	3.22	41.8	26.6	22.8	21.4	0.262
	13	29.3	0.287	3 730	15.6	121	64.6	71.7	2.72	34.7	27.0	21.7	21.6	0.266
	10	22.8	0.223	2 900	12.3	94.8	65.1	70.5	2.18	27.4	27.4	20.5	21.8	0.271
150X100	16	29.4	0.288	3 740	8.40	84.8	47.4	50.9	3.00	40.4	28.3	25.9	21.6	0.434
	13	24.2	0.237	3 080	7.03	70.2	47.8	49.9	2.53	33.7	28.7	24.9	21.7	0.440
	10	18.8	0.185	2 400	5.58	55.1	48.2	48.8	2.03	26.6	29.1	23.8	21.9	0.445
	8	15.2	0.149	1 940	4.55	44.6	48.5	48.0	1.67	21.6	29.3	23.0	22.0	0.448
125X90	16	25.0	0.245	3 180	4.84	58.5	39.0	42.2	2.09	32.0	25.6	24.7	19.2	0.499
	13	20.6	0.202	2 630	4.07	48.6	39.4	41.2	1.77	26.7	26.0	23.7	19.3	0.505
	10	16.1	0.158	2 050	3.25	38.2	39.8	40.1	1.42	21.1	26.4	22.6	19.5	0.511
	8	13.0	0.127	1 660	2.66	31.1	40.1	39.3	1.18	17.2	26.6	21.8	19.6	0.515
125X75	13	19.1	0.187	2 430	3.82	47.1	39.6	43.9	1.04	18.5	20.7	18.9	16.2	0.356
	10	14.9	0.146	1 900	3.05	37.1	40.0	42.8	0.841	14.7	21.0	17.8	16.3	0.363
	8	12.1	0.118	1 540	2.50	30.1	40.3	42.1	0.697	12.0	21.3	17.1	16.4	0.367
	6	9.14	0.090	1 160	1.92	23.0	40.6	41.3	0.542	9.23	21.6	16.3	16.6	0.372
100X90	13	18.1	0.177	2 300	2.17	31.4	30.7	31.1	1.66	25.9	26.8	26.1	18.4	0.796
	10	14.1	0.139	1 800	1.74	24.9	31.1	30.0	1.33	20.5	27.2	25.0	18.5	0.800
	8	11.4	0.112	1 460	1.43	20.3	31.4	29.3	1.10	16.8	27.5	24.3	18.6	0.802
	6	8.67	0.085	1 100	1.11	15.5	31.7	28.5	0.853	12.8	27.8	23.5	18.7	0.805

+Designation consists of the Letter L, the size, and the thickness, eg. L125x90x10. See pages 6-19 for more information.
**Rolling = 5 for all sizes of the Metric Series angles.

Note: The properties of angles currently produced by cold forming are up to 7 percent less than the properties shown in th above tables. Check manufacturers' catalogue for the exact properties and dimensions. See also page 6-19.

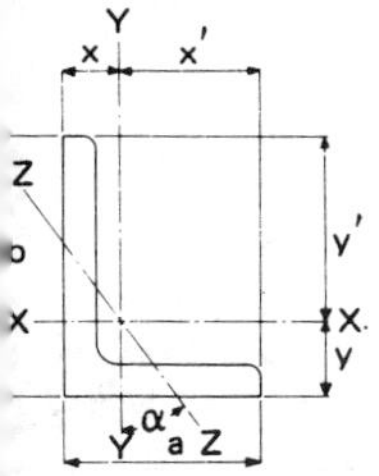

ANGLES — Metric Series** Unequal Legs

PROPERTIES AND DIMENSIONS

+Size**	Thick-ness t	Mass	Dead Load	Area	Axis X-X				Axis Y-Y				Axis Z-Z	
					I_x	S_x	r_x	y	I_y	S_y	r_y	x	r_z	Tan α
mmxmm	mm	kg/m	kN/m	mm^2	10^6mm^4	10^3mm^3	mm	mm	10^6mm^4	10^3mm^3	mm	mm	mm	
100X75	13	16.5	0.162	2 110	2.04	30.6	31.1	33.4	0.976	18.0	21.5	20.9	16.0	0.541
	10	13.0	0.127	1 650	1.64	24.2	31.5	32.3	0.791	14.3	21.9	19.8	16.1	0.549
	8	10.5	0.103	1 340	1.35	19.7	31.8	31.5	0.656	11.7	22.2	19.0	16.2	0.554
	6	7.96	0.078	1 010	1.04	15.1	32.1	30.8	0.511	9.01	22.4	18.3	16.3	0.559
90X75	13	15.5	0.152	1 980	1.51	24.8	27.6	29.3	0.946	17.8	21.9	21.8	15.6	0.672
	10	12.2	0.119	1 550	1.22	19.7	28.0	28.2	0.767	14.1	22.2	20.7	15.7	0.679
	8	9.86	0.097	1 260	1.01	16.1	28.3	27.5	0.636	11.6	22.5	20.0	15.8	0.683
	6	7.49	0.073	954	0.779	12.3	28.6	26.8	0.495	8.89	22.8	19.3	15.9	0.687
	5	6.28	0.062	800	0.660	10.4	28.7	26.4	0.421	7.50	22.9	18.9	16.0	0.689
90X65	10	11.4	0.112	1 450	1.16	19.2	28.3	29.8	0.507	10.6	18.7	17.3	13.9	0.506
	8	9.23	0.090	1 180	0.958	15.7	28.5	29.1	0.422	8.72	18.9	16.6	14.0	0.512
	6	7.02	0.069	894	0.743	12.1	28.8	28.4	0.330	6.72	19.2	15.9	14.2	0.518
	5	5.89	0.058	750	0.629	10.2	29.0	28.0	0.281	5.68	19.4	15.5	14.2	0.520
80X60	10	10.2	0.100	1 300	0.808	15.1	24.9	26.5	0.388	8.92	17.3	16.5	12.8	0.543
	8	8.29	0.081	1 060	0.670	12.4	25.2	25.8	0.324	7.33	17.5	15.8	12.9	0.549
	6	6.31	0.062	804	0.522	9.50	25.5	25.1	0.254	5.66	17.8	15.1	13.0	0.555
	5	5.30	0.052	675	0.443	8.02	25.6	24.7	0.217	4.79	17.9	14.7	13.0	0.558
75X50	8	7.35	0.072	936	0.525	10.6	23.7	25.5	0.187	5.06	14.1	13.0	10.8	0.434
	6	5.60	0.055	714	0.410	8.15	24.0	24.7	0.148	3.92	14.4	12.2	10.9	0.441
	5	4.71	0.046	600	0.349	6.88	24.1	24.4	0.127	3.32	14.5	11.9	10.9	0.445
65X50	8	6.72	0.066	856	0.351	8.03	20.2	21.3	0.180	4.97	14.5	13.8	10.6	0.572
	6	5.13	0.050	654	0.275	6.19	20.5	20.6	0.142	3.85	14.7	13.1	10.7	0.580
	5	4.32	0.042	550	0.235	5.24	20.7	20.2	0.122	3.27	14.9	12.7	10.8	0.583
	4	3.49	0.034	444	0.192	4.25	20.8	19.9	0.100	2.66	15.0	12.4	10.8	0.587
55X35	6	3.96	0.039	504	0.152	4.23	17.4	19.0	0.048	1.85	9.77	9.04	7.55	0.396
	5	3.34	0.033	425	0.130	3.59	17.5	18.7	0.041	1.58	9.89	8.68	7.59	0.401
	4	2.70	0.026	344	0.107	2.92	17.7	18.3	0.034	1.29	10.0	8.31	7.65	0.406
	3	2.05	0.020	261	0.083	2.23	17.8	17.9	0.027	0.994	10.2	7.94	7.72	0.411
45X30	6	3.25	0.032	414	0.082	2.79	14.0	15.7	0.029	1.32	8.35	8.22	6.44	0.426
	5	2.75	0.027	350	0.070	2.37	14.2	15.4	0.025	1.13	8.46	7.86	6.47	0.433
	4	2.23	0.022	284	0.058	1.94	14.3	15.0	0.021	0.930	8.58	7.49	6.51	0.439
	3	1.70	0.017	216	0.045	1.49	14.5	14.6	0.016	0.717	8.72	7.12	6.57	0.445

+Designation consists of the Letter L, the size, and the thickness, eg. L125x90x10. See pages 6-19 for more information.
**Rolling = 5 for all sizes of the Metric Series angles.

Note: The properties of angles currently produced by cold forming are up to 7 percent less than the properties shown in the above tables. Check manufacturers' catalogue for the exact properties and dimensions. See also page 6-19.

STRUCTURAL TEES
Cut from W Shapes

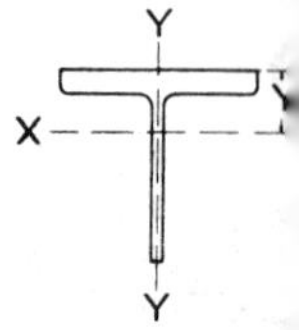

PROPERTIES AND DIMENSIONS

Designation	Dead Load	Total Area	Depth of Tee "d"	Flange		Stem Thick-ness "w"	Axis X-X				Axis Y-Y		
				Width	Aver. Thick-ness		I_x	S_x	r_x	y	I_y	S_y	r_y
	kN/m	mm^2	mm	mm	mm	mm	$10^6 mm^4$	$10^3 mm^3$	mm	mm	$10^6 mm^4$	$10^3 mm^3$	mm
WT460													
X223*	2.18	28 500	466	423	42.7	24.0	510	1 410	134	105	270	1 280	97.3
X208.5*	2.04	26 600	464	422	39.9	22.5	474	1 310	133	103	250	1 190	96.9
X193.5*	1.89	24 700	460	420	36.6	21.3	439	1 230	133	103	226	1 080	95.7
X182.5*	1.79	23 200	458	419	34.3	20.3	413	1 160	133	102	211	1 010	95.4
X171*	1.67	21 800	456	418	32.0	19.3	389	1 100	134	102	195	933	94.6
WT460													
X156.5*	1.53	19 900	466	309	34.5	21.1	410	1 200	144	124	85.2	551	65.4
X144.5*	1.41	18 400	464	308	32.0	19.4	375	1 100	143	122	78.2	508	65.2
X135.5*	1.32	17 300	462	307	30.0	18.4	352	1 030	143	121	72.6	473	64.8
X126.5*	1.24	16 200	460	306	27.9	17.3	328	967	142	120	66.8	437	64.2
X119*	1.16	15 200	458	305	25.9	16.5	308	915	142	120	61.4	403	63.6
X111.5*	1.09	14 300	456	304	23.9	15.9	291	872	143	122	56.1	369	62.6
X100.5*	0.986	12 800	452	304	20.1	15.2	265	812	144	126	47.2	311	60.7
WT420													
X179.5*	1.76	22 900	434	403	35.6	21.1	363	1 080	126	97.6	195	965	92.3
X164.5*	1.61	21 000	431	401	32.4	19.7	333	997	126	97.0	174	870	91.0
X149.5*	1.46	19 100	428	400	29.2	18.2	302	910	126	96.0	156	780	90.4
WT420													
X113*	1.10	14 400	426	294	26.8	16.1	246	777	131	108	56.9	387	62.9
X105*	1.03	13 400	423	293	24.4	15.4	230	733	131	109	51.3	350	61.9
X96.5*	0.946	12 300	420	292	21.7	14.7	213	688	132	111	45.1	309	60.6
X88*	0.863	11 200	418	292	18.8	14.0	196	644	132	114	39.1	268	59.1
WT380													
X157*	1.54	20 000	393	384	33.4	19.7	254	828	113	86.3	158	822	88.9
X142*	1.39	18 100	390	382	30.1	18.0	228	748	112	84.8	140	733	87.9
X128.5*	1.26	16 400	386	381	27.1	16.6	206	682	112	83.9	125	656	87.3
WT380													
X98*	0.961	12 500	385	268	25.4	15.6	175	613	118	99.0	40.9	305	57.2
X92.5*	0.907	11 800	383	267	23.6	14.9	165	580	118	99.1	37.5	281	56.4
X86.5*	0.848	11 100	381	267	21.6	14.4	156	554	119	100	34.4	257	55.7
X80.5*	0.789	10 200	379	266	19.3	13.8	145	523	119	102	30.4	228	54.6
X73.5*	0.721	9 370	376	265	17.0	13.2	134	492	120	104	26.4	200	53.1
WT345													
X132.5*	1.29	16 900	353	358	30.2	18.4	172	624	101	77.4	116	646	82.8
X120*	1.17	15 300	350	356	27.4	16.8	155	565	101	76.1	103	580	82.0
X108.5*	1.06	13 900	348	355	24.8	15.4	140	512	100	74.8	92.6	522	81.6
WT345													
X85*	0.834	10 800	346	256	23.6	14.5	120	464	105	86.9	33.1	259	55.4
X76*	0.745	9 690	344	254	21.1	13.1	107	415	105	85.8	28.9	227	54.6
X70*	0.686	8 910	342	254	18.9	12.4	99.3	389	106	86.5	25.9	204	53.9
X62.5*	0.613	8 000	339	253	16.3	11.7	90.0	359	106	88.3	22.0	174	52.4
WT305													
X120.5	1.18	15 400	318	329	31.0	17.9	122	490	89.0	68.5	92.1	560	77.3
X108.5	1.06	13 900	314	328	27.7	16.5	110	444	89.0	67.4	81.6	497	76.6
X97.5	0.956	12 500	311	327	24.4	15.4	99.4	408	89.2	67.4	71.2	435	75.5
X87	0.853	11 100	308	325	21.6	14.0	88.3	366	89.2	66.5	61.9	381	74.7
X77.5	0.760	9 860	306	324	19.0	12.7	78.6	328	89.3	65.9	53.9	333	73.9

*Not available from Canadian mills.

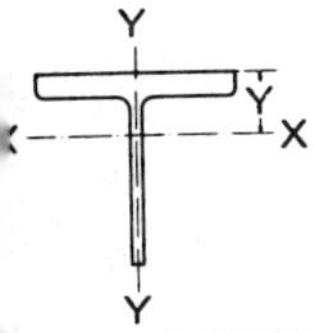

STRUCTURAL TEES
Cut from W Shapes

PROPERTIES AND DIMENSIONS

Designation	Dead Load	Total Area	Depth of Tee "d"	Flange		Stem Thick-ness "w"	Axis X-X				Axis Y-Y		
				Width	Aver. Thick-ness		I_x	S_x	r_x	y	I_y	S_y	r_y
	kN/m	mm²	mm	mm	mm	mm	10^6mm⁴	10^3mm³	mm	mm	10^6mm⁴	10^3mm³	mm
WT305													
X70	0.686	8 930	308	230	22.2	13.1	77.5	333	93.2	76.0	22.6	196	50.3
X62.5	0.613	7 970	306	229	19.6	11.9	69.0	299	93.0	75.4	19.7	172	49.7
X56.5	0.554	7 220	304	228	17.3	11.2	63.2	278	93.6	76.3	17.1	150	48.7
X50.5	0.495	6 480	302	228	14.9	10.5	57.2	255	94.0	77.6	14.7	129	47.6
WT305													
X46*	0.451	5 880	302	179	15.0	10.9	54.6	255	96.4	87.7	7.20	80.5	35.0
X41*	0.402	5 210	300	178	12.8	10.0	48.5	230	96.5	88.9	6.04	67.9	34.0
WT265													
X109.5*	1.07	13 900	280	318	29.2	18.3	85.0	388	78.2	60.8	78.4	493	75.1
X98*	0.961	12 500	277	316	26.3	16.5	75.2	345	77.6	59.0	69.3	438	74.5
X91*	0.892	11 600	276	315	24.4	15.2	68.9	316	77.1	57.7	63.6	404	74.0
X82.5*	0.809	10 500	273	313	22.2	14.0	62.2	288	77.0	56.7	56.8	363	73.5
X75*	0.735	9 590	272	312	20.3	12.7	56.2	260	76.6	55.4	51.4	330	73.2
WT265													
X69	0.677	8 810	274	214	23.6	14.7	59.9	292	82.5	69.4	19.3	181	46.8
X61.5	0.603	7 850	272	212	21.2	13.1	52.6	258	81.9	67.6	16.9	159	46.4
X54.5	0.534	6 940	270	211	18.8	11.6	46.0	226	81.4	66.0	14.8	140	46.2
X50.5	0.495	6 460	268	210	17.4	10.9	42.8	211	81.4	65.7	13.5	128	45.7
X46	0.451	5 890	266	209	15.6	10.2	39.1	195	81.5	65.8	11.9	114	44.9
X41	0.402	5 250	264	209	13.3	9.50	35.1	178	81.8	66.7	10.1	97.0	43.9
WT265													
X42.5*	0.417	5 390	268	166	16.5	10.3	37.6	193	83.5	72.5	6.32	76.1	34.2
X37*	0.363	4 760	264	166	13.6	9.70	33.5	176	83.9	74.5	5.21	62.7	33.1
X33*	0.324	4 190	262	165	11.4	8.90	29.6	159	84.1	75.9	4.29	52.0	32.0
WT230													
X88.5*	0.868	11 300	241	286	26.9	16.6	49.4	260	66.1	51.4	52.5	367	68.2
X79*	0.775	10 000	238	284	23.9	15.0	43.5	231	66.0	50.1	45.7	322	67.6
X72*	0.706	9 210	236	283	22.1	13.6	39.0	208	65.1	48.4	41.8	295	67.4
X64*	0.628	8 180	234	282	19.6	12.2	34.3	184	64.8	47.1	36.7	260	67.0
X56.5*	0.554	7 200	232	280	17.3	10.8	30.0	161	64.5	45.9	31.7	226	66.4
WT230													
X53	0.520	6 740	234	194	20.6	12.6	32.6	184	69.5	57.3	12.6	130	43.2
X48.5	0.476	6 150	233	193	19.0	11.4	29.4	166	69.1	55.8	11.4	118	43.1
X44.5	0.436	5 690	232	192	17.7	10.5	26.8	152	68.6	54.6	10.5	109	43.0
X41	0.402	5 220	230	191	16.0	9.90	24.8	141	68.9	54.8	9.31	97.5	42.2
X37	0.363	4 730	228	190	14.5	9.00	22.3	127	68.7	53.9	8.30	87.4	41.9
X33.5	0.329	4 340	227	190	12.7	8.50	20.5	119	68.7	54.2	7.28	76.6	41.0
X30.5	0.299	3 880	225	189	10.8	8.10	18.7	111	69.4	55.9	6.09	64.5	39.6
WT230													
X34*	0.333	4 360	230	154	15.4	9.10	21.7	127	70.5	59.0	4.70	61.1	32.8
X30*	0.294	3 790	228	153	13.3	8.00	18.7	110	70.2	58.1	3.98	52.0	32.4
X26*	0.255	3 310	225	152	10.8	7.60	16.7	102	71.0	60.8	3.17	41.7	30.9
WT205													
X74.5*	0.731	9 510	216	265	25.0	14.9	32.0	187	58.0	44.7	38.8	293	63.9
X66*	0.647	8 410	212	263	22.2	13.3	27.8	164	57.5	43.1	33.7	256	63.3
X57*	0.559	7 290	210	261	19.3	11.6	23.8	141	57.1	41.6	28.6	219	62.6
X50*	0.490	6 340	208	260	16.9	10.0	20.2	120	56.4	39.7	24.8	191	62.5

*Not available from Canadian mills.

STRUCTURAL TEES
Cut from W Shapes

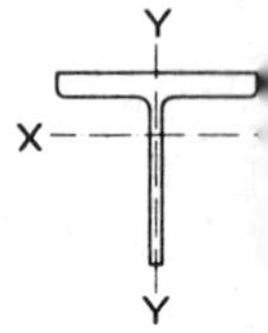

PROPERTIES AND DIMENSIONS

Designation	Dead Load	Total Area	Depth of Tee "d"	Flange		Stem Thickness "w"	Axis X-X				Axis Y-Y		
				Width	Aver. Thickness		I_x	S_x	r_x	y	I_y	S_y	r_y
	kN/m	mm²	mm	mm	mm	mm	$10^6 mm^4$	$10^3 mm^3$	mm	mm	$10^6 mm^4$	$10^3 mm^3$	mm
WT205													
X42.5	0.417	5 410	208	181	18.2	10.9	20.2	127	61.1	49.1	9.02	99.6	40.8
X37	0.363	4 770	206	180	16.0	9.70	17.7	112	60.9	48.1	7.79	86.6	40.4
X33.5	0.329	4 300	205	179	14.4	8.80	15.8	100	60.6	47.3	6.90	77.0	40.1
X30	0.294	3 790	204	178	12.8	7.70	13.8	87.4	60.3	45.9	6.02	67.7	39.9
X27	0.265	3 400	202	177	10.9	7.50	12.7	82.9	61.1	47.9	5.05	57.0	38.5
WT205													
X23	0.226	2 940	202	140	11.2	7.00	11.4	76.0	62.3	51.3	2.57	36.7	29.6
X19.5	0.191	2 500	200	140	8.80	6.40	9.89	67.6	62.9	53.3	2.02	28.8	28.4
WT180													
X543*	5.32	69 300	284	454	125	78.0	307	1 560	66.6	88.1	981	4 320	119
X495*	4.85	63 100	275	448	115	71.9	259	1 350	64.1	82.7	867	3 870	117
X450*	4.41	57 500	266	442	106	65.9	218	1 160	61.6	77.4	767	3 470	115
X409*	4.01	52 200	257	437	97.0	60.5	184	997	59.4	72.4	678	3 100	114
X372*	3.64	47 400	249	432	88.9	55.6	156	864	57.4	67.9	600	2 780	113
X338.5*	3.32	43 200	242	428	81.5	51.2	134	751	55.7	63.8	534	2 500	111
WT180													
X317*	3.10	40 400	237	424	77.1	47.6	119	677	54.3	61.0	491	2 320	110
X296*	2.90	37 700	232	421	72.3	45.0	107	618	53.3	58.5	451	2 140	109
X275.5*	2.70	35 100	228	418	67.6	42.0	95.3	555	52.1	55.7	412	1 970	108
X254.5*	2.49	32 500	223	416	62.7	39.1	84.8	499	51.1	53.0	377	1 810	108
X231.5*	2.27	29 500	218	412	57.4	35.8	73.3	438	49.8	50.0	335	1 630	107
X210.5*	2.06	26 900	212	409	52.6	32.8	63.7	386	48.7	47.2	300	1 470	106
X191*	1.87	24 400	208	406	48.0	29.8	55.4	338	47.6	44.5	268	1 320	105
X173.5*	1.70	22 100	204	404	43.7	27.2	48.2	298	46.7	42.0	240	1 190	104
X157*	1.54	20 000	200	401	39.6	24.9	42.2	264	45.9	39.8	213	1 060	103
X143.5*	1.40	18 300	196	399	36.6	22.6	37.3	235	45.1	37.8	194	972	103
X131*	1.28	16 700	194	398	33.3	21.1	33.7	214	44.9	36.3	175	880	102
X118.5*	1.16	15 000	190	395	30.2	18.9	29.1	187	44.0	34.3	155	786	102
X108*	1.05	13 800	188	394	27.7	17.3	26.0	168	43.4	32.8	141	717	101
WT180													
X98*	0.961	12 500	186	374	26.2	16.4	24.0	157	43.8	32.7	114	611	95.5
X89.5*	0.878	11 400	184	373	23.9	15.0	21.6	141	43.5	31.4	103	555	95.1
X81*	0.794	10 300	182	371	21.8	13.3	18.8	124	42.7	29.8	92.8	500	94.9
X73.5*	0.721	9 400	180	370	19.8	12.3	17.0	113	42.5	28.9	83.6	452	94.3
X67*	0.657	8 530	178	369	18.0	11.2	15.2	101	42.2	27.8	75.4	409	94.0
WT180													
X61*	0.598	7 750	182	257	21.7	13.0	17.1	117	47.0	35.3	30.7	239	62.9
X55*	0.539	7 020	180	256	19.9	11.4	15.0	102	46.2	33.5	27.9	218	63.0
X50.5*	0.495	6 450	178	255	18.3	10.5	13.6	93.2	45.9	32.6	25.3	199	62.6
X45.5*	0.446	5 790	176	254	16.4	9.50	12.0	83.0	45.5	31.6	22.4	176	62.2
WT180													
X39.5	0.387	5 050	177	205	16.8	9.40	11.5	81.2	47.7	35.0	12.1	118	48.9
X36	0.353	4 550	175	204	15.1	8.60	10.3	73.1	47.6	34.2	10.7	105	48.5
X32	0.314	4 070	174	203	13.5	7.70	9.09	64.8	47.3	33.2	9.42	92.8	48.1
WT180													
X28.5	0.279	3 610	179	172	13.1	7.90	9.70	69.4	51.8	39.2	5.56	64.7	39.2
X25.5	0.250	3 220	178	171	11.6	7.20	8.66	62.5	51.9	38.8	4.84	56.6	38.8
X22.5	0.221	2 870	176	171	9.80	6.90	7.96	58.6	52.7	40.2	4.09	47.8	37.8

*Not available from Canadian mills.

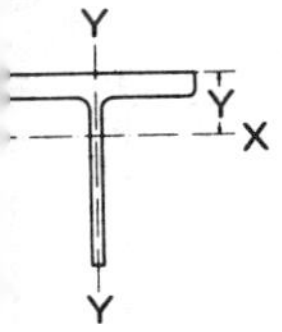

STRUCTURAL TEES
Cut from W Shapes

PROPERTIES AND DIMENSIONS

esignation	Dead Load	Total Area	Depth of Tee "d"	Flange		Stem Thick-ness "w"	Axis X-X				Axis Y-Y		
				Width	Aver. Thick-ness		I_x	S_x	r_x	y	I_y	S_y	r_y
	kN/m	mm²	mm	mm	mm	mm	10⁶mm⁴	10³mm³	mm	mm	10⁶mm⁴	10³mm³	mm
VT180													
X19.5	0.191	2 490	176	128	10.7	6.50	7.22	54.4	53.8	43.7	1.88	29.3	27.5
X16.5	0.162	2 090	174	127	8.50	5.80	6.14	47.3	54.2	44.6	1.45	22.9	26.3
VT155													
X250*	2.45	31 900	214	340	75.1	45.1	79.2	511	49.8	58.6	247	1 450	88.0
X227*	2.22	28 900	208	336	68.7	41.3	67.7	444	48.4	55.0	218	1 300	86.9
X207.5*	2.03	26 400	202	334	62.7	38.9	59.1	396	47.3	52.1	195	1 170	85.9
X187.5*	1.83	23 800	196	330	57.1	35.4	50.1	341	45.9	48.8	172	1 040	85.0
X171*	1.67	21 900	191	328	52.6	32.6	43.8	302	44.7	46.1	155	946	84.1
X156.5*	1.53	20 000	187	325	48.3	30.0	38.5	269	43.9	43.8	139	852	83.4
WT155													
X141.5*	1.38	18 000	182	322	44.1	26.9	32.8	232	42.7	41.0	123	764	82.7
X126.5*	1.24	16 100	178	319	39.6	24.4	28.1	202	41.8	38.6	107	673	81.5
X113	1.10	14 400	174	317	35.6	22.1	24.3	176	41.1	36.4	94.6	597	81.1
X101	0.990	12 900	170	315	31.8	20.1	21.2	156	40.5	34.5	82.9	527	80.2
X89.5	0.878	11 400	166	313	28.1	18.0	18.0	134	39.7	32.4	71.9	459	79.4
X79	0.775	10 000	164	310	25.1	15.5	15.1	113	38.9	30.2	62.4	402	79.0
X71.5	0.701	9 120	162	309	22.9	14.0	13.3	101	38.2	28.8	56.3	365	78.6
X64.5	0.633	8 260	159	308	20.6	13.1	12.0	91.8	38.1	27.9	50.2	326	78.0
X59	0.579	7 490	157	307	18.7	11.9	10.7	82.0	37.8	26.8	45.1	294	77.6
X53.5	0.525	6 810	156	306	17.0	10.9	9.61	74.2	37.6	25.9	40.6	265	77.2
X48.5	0.476	6 170	154	305	15.4	9.90	8.59	66.6	37.3	25.0	36.4	239	76.8
WT155													
X43	0.422	5 500	155	254	16.3	9.10	7.93	61.5	38.0	26.1	22.3	175	63.7
X39.5	0.387	5 030	153	254	14.6	8.80	7.38	58.1	38.3	26.1	20.0	157	63.1
WT155													
X37	0.363	4 740	155	205	16.3	9.40	7.81	62.3	40.6	29.7	11.7	114	49.7
X33.5	0.329	4 250	153	204	14.6	8.50	6.88	55.3	40.2	28.7	10.3	101	49.2
X30	0.294	3 800	152	203	13.1	7.50	5.99	48.3	39.7	27.5	9.14	90.1	49.0
WT155													
X26	0.255	3 330	158	167	13.2	7.60	6.60	52.6	44.5	32.9	5.13	61.4	39.2
X22.5	0.221	2 840	156	166	11.2	6.60	5.59	44.9	44.4	32.1	4.27	51.5	38.8
X19.5	0.191	2 470	155	165	9.70	5.80	4.82	39.0	44.2	31.4	3.63	44.0	38.3
WT155													
X16.5*	0.162	2 090	156	102	10.8	6.60	4.88	42.4	48.3	41.5	0.959	18.8	21.4
X14*	0.137	1 810	154	102	8.90	6.00	4.23	37.6	48.3	41.9	0.790	15.5	20.9
X12*	0.118	1 520	152	101	6.70	5.60	3.63	33.6	48.9	44.4	0.578	11.4	19.5
X10.5*	0.103	1 340	152	101	5.70	5.10	3.22	30.2	49.0	44.8	0.491	9.7	19.1
WT125													
X83.5	0.819	10 700	144	265	31.8	19.2	12.0	105	33.5	30.7	49.4	373	67.9
X74.5	0.731	9 490	141	263	28.4	17.3	10.2	91.3	32.8	28.8	43.1	328	67.4
X65.5	0.642	8 350	138	261	25.1	15.4	8.64	78.1	32.2	26.9	37.2	285	66.7
X57.5	0.564	7 310	134	259	22.1	13.5	7.26	66.3	31.5	25.1	32.0	247	66.2
X50.5	0.495	6 440	132	257	19.6	11.9	6.17	57.0	31.0	23.6	27.7	216	65.6
X44.5	0.436	5 700	130	256	17.3	10.7	5.39	50.2	30.8	22.5	24.2	189	65.2
X40	0.392	5 100	128	255	15.6	9.40	4.61	43.2	30.1	21.2	21.6	169	65.1
X36.5	0.358	4 640	126	254	14.2	8.60	4.12	38.9	29.8	20.4	19.4	153	64.7

*Not available from Canadian mills.

STRUCTURAL TEES
Cut from W Shapes

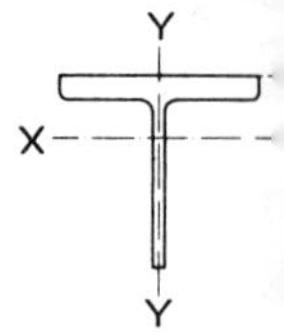

PROPERTIES AND DIMENSIONS

Designation	Dead Load	Total Area	Depth of Tee "d"	Flange		Stem Thick-ness "w"	Axis X-X				Axis Y-Y		
				Width	Aver. Thick-ness		I_x	S_x	r_x	y	I_y	S_y	r_y
	kN/m	mm^2	mm	mm	mm	mm	$10^6 mm^4$	$10^3 mm^3$	mm	mm	$10^6 mm^4$	$10^3 mm^3$	mm
WT125													
X33.5	0.329	4 280	128	204	15.7	8.90	4.29	40.7	31.7	23.1	11.1	109	50.9
X29	0.284	3 710	126	203	13.5	8.00	3.68	35.5	31.5	22.2	9.42	92.8	50.4
X24.5	0.240	3 120	124	202	11.0	7.40	3.21	31.7	32.1	22.1	7.56	74.9	49.2
WT125													
X22.5	0.221	2 860	133	148	13.0	7.60	3.86	36.7	36.7	27.8	3.52	47.5	35.1
X19.5	0.191	2 460	131	147	11.2	6.60	3.26	31.3	36.4	26.7	2.97	40.4	34.7
X16.5	0.162	2 080	129	146	9.10	6.10	2.85	28.1	37.0	27.3	2.36	32.4	33.7
WT125													
X14*	0.137	1 810	130	102	10.0	6.40	2.79	28.7	39.3	32.6	0.888	17.4	22.1
X12.5*	0.123	1 610	128	102	8.40	6.10	2.53	26.6	39.6	33.5	0.746	14.6	21.5
X11*	0.108	1 430	127	102	6.90	5.80	2.27	24.5	39.8	34.6	0.613	12.0	20.7
X9*	0.088	1 140	126	101	5.30	4.80	1.81	19.9	39.8	34.6	0.457	9.0	20.0
WT100													
X50	0.490	6 340	114	210	23.7	14.5	4.55	50.2	26.8	23.8	18.3	174	53.7
X43	0.422	5 530	111	209	20.6	13.0	3.80	42.8	26.2	22.2	15.7	150	53.3
X35.5	0.348	4 550	108	206	17.4	10.2	2.86	32.5	25.1	19.8	12.7	123	52.8
X29.5	0.289	3 780	105	205	14.2	9.10	2.39	27.7	25.1	18.7	10.2	99.5	51.9
X26	0.255	3 330	103	204	12.6	7.90	2.00	23.4	24.5	17.5	8.92	87.4	51.8
X23	0.226	2 930	102	203	11.0	7.20	1.77	20.9	24.6	16.9	7.67	75.6	51.2
WT100													
X21	0.206	2 660	102	166	11.8	7.20	1.76	21.0	25.7	18.6	4.50	54.2	41.1
X18	0.177	2 290	100	165	10.2	6.20	1.46	17.6	25.2	17.5	3.82	46.3	40.8
WT100													
X15.5	0.152	2 000	105	134	10.2	6.40	1.63	19.4	28.5	21.1	2.05	30.6	32.0
X13.5	0.132	1 690	104	133	8.40	5.80	1.41	17.2	28.9	21.1	1.65	24.8	31.2
WT100													
X11*	0.108	1 430	103	102	8.00	6.20	1.36	17.5	30.8	25.3	0.710	13.9	22.3
X9.5*	0.093	1 240	102	102	6.50	5.80	1.20	15.9	31.1	25.9	0.577	11.3	21.6
X7.5*	0.074	952	100	100	5.20	4.30	0.885	11.7	30.5	24.1	0.434	8.7	21.4
WT75													
X18.5	0.181	2 370	81.0	154	11.6	8.10	0.947	14.5	20.0	15.5	3.53	45.9	38.6
X15	0.147	1 900	78.5	153	9.30	6.60	0.725	11.3	19.5	14.2	2.78	36.3	38.3
X11	0.108	1 420	76.0	152	6.60	5.80	0.581	9.4	20.2	14.1	1.93	25.4	36.9
WT75													
X12*	0.118	1 530	80.0	102	10.3	6.60	0.708	11.3	21.5	17.3	0.913	17.9	24.4
X9*	0.088	1 140	76.5	102	7.10	5.80	0.544	9.2	21.8	17.1	0.629	12.3	23.5
X7*	0.069	866	75.0	100	5.50	4.30	0.395	6.7	21.4	15.8	0.459	9.2	23.0
WT65													
X14*	0.137	1 790	65.5	128	10.9	6.90	0.426	8.0	15.4	12.4	1.91	29.8	32.7
X12*	0.118	1 500	63.5	127	9.10	6.10	0.350	6.7	15.3	11.6	1.55	24.5	32.1
WT50													
X9.5*	0.093	1 240	53.0	103	8.80	7.10	0.221	5.3	13.4	11.2	0.803	15.6	25.4

*Not available from Canadian mills.

HOLLOW STRUCTURAL SECTIONS

General

Hollow structural sections are produced in Canada to requirements of the CAN/CSA-G40.20-M Standard to either Class C or Class H, from steel meeting the requirements of the CAN/CSA-G40.21-M material Standard. The normal grade of steel used is G40.21-M 350W. See pages 6-15 and 6-16 for descriptions of the class of hollow section and for data on manufacturing tolerances. Round sections produced in accordance with common pipe specifications may sometimes be used as structural members, but are not classified as hollow structural sections.

Properties and Dimensions

The tables of properties and dimensions on pages 6-98 to 6-103 include the standard Canadian rectangular, square and round hollow structural sections covered by the CAN3-G312.3-M78 Standard and the recently (1990) ballotted and approved revisions to this Standard. The revisions simplify the metric (SI) designation describing Canadian hollow structural sections (see pages 6-30 to 6-32). The new designations (e.g. HSS 127 x 76 x 6.4 class C) are used throughout this Handbook, and should be used by designers and detailers when calling up hollow structural sections, as non-Canadian hollow sections may differ in name, size, designation and material grade from Canadian standard sizes.

Manufacture

Hollow structural sections produced to the CAN/CSA-G40.20-M Standard may be manufactured using either a seamless or welding process. Seamless products are produced by piercing solid material to form a tube or by an extrusion type process. Welded products are manufactured from flat-rolled steel which is formed and joined by various welding processes into a tubular shape. The tubular shape is then either cold-formed or hot-formed to the final shape and, if cold-formed, may be subsequently stress relieved. Class H sections are either hot-formed to final shape, or are cold-formed to final shape and then stress relieved. Class C sections are generally more readily available than class H sections, though Class H sections have greater resistance in direct compression.

Outside dimensions for hollow structural sections are constant for all sizes in the same size range, the inside dimensions changing with material thickness.

Pipe

Properties and dimension for pipe provided in the table on page 6-104 have been soft converted to metric from the ASTM A53 Standard for Welded and Seamless Steel Pipe. Although not a normal structural quality steel, pipe produced in accordance with the ASTM A53 Standard is available in three strength levels, with specified minimum yield strengths of 172, 207, or 241 MPa. These are considerably lower than 350 MPa generally used for hollow structural sections.

Size and wall thickness of seamless and welded mill pipe up to 323.9 mm outside diameter permit threading the ends for joining lengths with couplings or other connectors. The wall thickness of mill pipe has been expressed in terms of "standard wall" (STD), "extra strong" (XS), "double extra strong" (XXS), and in terms of "schedule numbers" (Sch). STD is the same as Sch 40 for all sizes up to and including 273.1 mm outside diameter, XS is the same as Sch 80 for all sizes up to and including 219.1 mm outside diameter, and XXS is the same as Sch 160 for all sizes up to and including 168.3 mm outside diameter. When specifying pipe, it is recommended that outside (external) diameter and wall thickness be used. (e.g. Pipe 114.3 O.D. 8.56).

HOLLOW STRUCTURAL SECTIONS
Rectangular

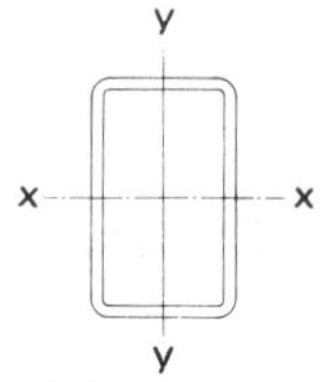

PROPERTIES AND DIMENSIONS

Designation*	Wall Thick-ness	Mass	Dead Load	Area	Axis X-X				Axis Y-Y				Torsional Constant J	Shear Constant C_{rt}
					I_x	S_x	r_x	Z_x	I_y	S_y	r_y	Z_y		
mm x mm x mm	mm	kg/m	kN/m	mm^2	$10^6 mm^4$	$10^3 mm^3$	mm	$10^3 mm^3$	$10^6 mm^4$	$10^3 mm^3$	mm	$10^3 mm^3$	$10^3 mm^4$	mm^2
HSS 305 x 203														
x 13	12.70	93.0	0.912	11 800	147	964	111	1 190	78.1	769	81.2	896	167 000	6 450
x 11	11.13	82.4	0.808	10 500	132	867	112	1 060	70.5	693	81.9	802	149 000	5 790
x 9.5	9.53	71.3	0.699	9 090	116	762	113	925	62.1	611	82.7	701	130 000	5 080
x 8.0	7.95	60.1	0.590	7 660	99.4	652	114	787	53.3	524	83.4	596	111 000	4 340
x 6.4	6.35	48.6	0.476	6 190	81.5	535	115	640	43.8	431	84.1	486	89 800	3 550
HSS 254 x 152														
x 13	12.70	72.7	0.713	9 260	75.2	592	90.1	746	33.6	441	60.2	522	78 200	5 160
x 11	11.13	64.6	0.634	8 230	68.2	537	91.0	671	30.6	401	61.0	470	70 200	4 660
x 9.5	9.53	56.1	0.550	7 150	60.3	475	91.9	589	27.2	357	61.7	413	61 600	4 110
x 8.0	7.95	47.5	0.465	6 050	52.0	409	92.7	503	23.5	309	62.4	353	52 600	3 530
x 6.4	6.35	38.4	0.377	4 900	42.9	338	93.6	411	19.5	256	63.1	290	43 000	2 900
HSS 203 x 152														
x 13	12.70	62.6	0.614	7 970	43.0	423	73.4	528	27.3	358	58.5	432	56 400	3 870
x 11	11.13	55.7	0.547	7 100	39.1	385	74.2	476	24.9	327	59.3	390	50 800	3 530
x 9.5	9.53	48.5	0.476	6 180	34.8	343	75.1	419	22.3	292	60.0	344	44 600	3 150
x 8.0	7.95	41.1	0.403	5 240	30.2	297	75.9	360	19.3	254	60.7	295	38 200	2 730
x 6.4	6.35	33.4	0.327	4 250	25.0	246	76.7	295	16.1	211	61.5	243	31 200	2 260
x 4.8	4.78	25.5	0.250	3 250	19.5	192	77.5	228	12.6	165	62.2	188	24 100	1 760
HSS 203 x 102														
x 13	12.70	52.4	0.514	6 680	31.2	307	68.4	405	10.2	201	39.1	246	27 000	3 870
x 11	11.13	46.9	0.460	5 970	28.7	282	69.3	367	9.46	186	39.8	224	24 600	3 530
x 9.5	9.53	40.9	0.401	5 210	25.8	253	70.3	325	8.56	168	40.5	199	21 900	3 150
x 8.0	7.95	34.8	0.341	4 430	22.5	221	71.2	281	7.54	148	41.2	172	18 900	2 730
x 6.4	6.35	28.3	0.278	3 610	18.8	185	72.2	232	6.35	125	42.0	143	15 600	2 260
x 4.8	4.78	21.7	0.213	2 760	14.7	145	73.1	180	5.03	99.0	42.7	111	12 200	1 760
HSS 178 x 127														
x 13	12.70	52.4	0.514	6 680	26.4	297	62.8	377	15.5	243	48.1	298	33 600	3 230
x 11	11.13	46.9	0.460	5 970	24.2	272	63.7	342	14.2	224	48.8	270	30 400	2 970
x 9.5	9.53	40.9	0.401	5 210	21.7	244	64.6	303	12.8	202	49.6	240	26 900	2 660
x 8.0	7.95	34.8	0.341	4 430	18.9	213	65.4	261	11.2	177	50.3	207	23 100	2 320
x 6.4	6.35	28.3	0.278	3 610	15.8	178	66.2	216	9.40	148	51.1	171	19 000	1 940
x 4.8	4.78	21.7	0.213	2 760	12.4	140	67.1	167	7.41	117	51.8	133	14 700	1 520
HSS 152 x 102														
x 11	11.13	38.0	0.373	4 840	13.6	179	53.1	230	7.13	140	38.4	172	16 300	2 400
x 9.5	9.53	33.3	0.327	4 240	12.4	162	54.0	205	6.50	128	39.1	154	14 500	2 180
x 8.0	7.95	28.4	0.279	3 620	10.9	143	54.8	178	5.76	113	39.9	134	12 600	1 920
x 6.4	6.35	23.2	0.228	2 960	9.18	121	55.7	148	4.88	96.1	40.6	112	10 500	1 610
x 4.8	4.78	17.9	0.175	2 280	7.28	95.6	56.5	116	3.89	76.6	41.3	87.8	8 160	1 270

*Depth x width x thickness, see page 6-19.

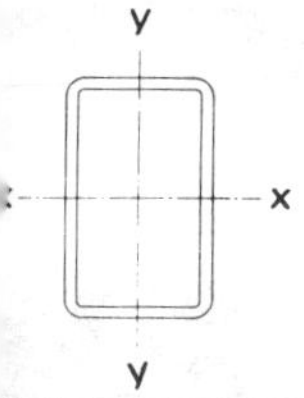

HOLLOW STRUCTURAL SECTIONS
Rectangular

PROPERTIES AND DIMENSIONS

Designation*	Wall Thickness	Mass	Dead Load	Area	Axis X-X				Axis Y-Y				Torsional Constant J	Shear Constant C_{rt}
					I_x	S_x	r_x	Z_x	I_y	S_y	r_y	Z_y		
mm x mm x mm	mm	kg/m	kN/m	mm^2	$10^6 mm^4$	$10^3 mm^3$	mm	$10^3 mm^3$	$10^6 mm^4$	$10^3 mm^3$	mm	$10^3 mm^3$	$10^3 mm^4$	mm^2
HSS 127 x 76														
x 9.5	9.53	25.7	0.252	3 280	6.12	96.4	43.2	126	2.69	70.5	28.6	87.5	6 600	1 690
x 8.0	7.95	22.1	0.217	2 820	5.49	86.4	44.1	111	2.43	63.8	29.4	77.2	5 810	1 510
x 6.4	6.35	18.2	0.178	2 320	4.70	74.0	45.1	93.3	2.10	55.1	30.1	65.2	4 890	1 290
x 4.8	4.78	14.1	0.138	1 790	3.78	59.5	45.9	73.8	1.70	44.7	30.8	51.7	3 860	1 030
HSS 127 x 64														
x 9.5	9.53	23.8	0.234	3 030	5.28	83.2	41.7	112	1.70	53.6	23.7	67.4	4 640	1 690
x 8.0	7.95	20.5	0.201	2 610	4.77	75.1	42.7	98.9	1.56	49.1	24.4	60.0	4 130	1 510
x 6.4	6.35	16.9	0.166	2 150	4.11	64.8	43.7	83.6	1.36	42.9	25.1	51.0	3 510	1 290
x 4.8	4.78	13.1	0.129	1 670	3.33	52.4	44.6	66.3	1.12	35.2	25.9	40.7	2 800	1 030
HSS 127 x 51														
x 9.5	9.53	21.9	0.215	2 790	4.45	70.0	39.9	97.7	0.961	37.8	18.6	48.9	2 930	1 690
x 8.0	7.95	18.9	0.186	2 410	4.05	63.8	41.0	86.9	0.897	35.3	19.3	44.0	2 650	1 510
x 6.4	6.35	15.6	0.153	1 990	3.53	55.5	42.1	73.8	0.798	31.4	20.0	37.9	2 290	1 290
x 4.8	4.78	12.2	0.119	1 550	2.87	45.3	43.1	58.9	0.665	26.2	20.7	30.5	1 850	1 030
HSS 102 x 76														
x 9.5	9.53	21.9	0.215	2 790	3.41	67.2	35.0	87.6	2.15	56.3	27.7	71.3	4 710	1 210
x 8.0	7.95	18.9	0.186	2 410	3.10	61.0	35.8	77.7	1.96	51.4	28.5	63.5	4 170	1 110
x 6.4	6.35	15.6	0.153	1 990	2.68	52.8	36.7	65.9	1.71	44.8	29.3	54.0	3 530	968
x 4.8	4.78	12.2	0.119	1 550	2.18	43.0	37.5	52.5	1.39	36.6	30.0	43.1	2 800	789
HSS 102 x 51														
x 8.0	7.95	15.8	0.155	2 010	2.21	43.5	33.2	58.8	0.709	27.9	18.8	35.4	1 950	1 110
x 6.4	6.35	13.1	0.129	1 670	1.95	38.4	34.2	50.6	0.638	25.1	19.5	30.7	1 690	968
x 4.8	4.78	10.3	0.101	1 310	1.61	31.8	35.1	40.8	0.536	21.1	20.3	24.9	1 370	789
x 3.8	3.81	8.37	0.082	1 070	1.36	26.8	35.7	33.9	0.456	18.0	20.7	20.8	1 140	658
x 3.2	3.18	7.09	0.069	903	1.17	23.1	36.1	29.0	0.397	15.6	21.0	17.9	979	565
HSS 89 x 64														
x 8.0	7.95	15.8	0.155	2 010	1.87	42.1	30.5	54.9	1.09	34.3	23.3	43.2	2 450	908
x 6.4	6.35	13.1	0.129	1 670	1.65	37.1	31.4	47.1	0.966	30.4	24.0	37.2	2 100	806
x 4.8	4.78	10.3	0.101	1 310	1.36	30.6	32.3	38.0	0.803	25.3	24.8	30.1	1 690	667
x 3.8	3.81	8.37	0.082	1 070	1.15	25.8	32.8	31.5	0.679	21.4	25.2	25.0	1 400	561
x 3.2	3.18	7.09	0.069	903	0.990	22.3	33.1	27.0	0.588	18.5	25.5	21.4	1 190	485
HSS 76 x 51														
x 8.0	7.95	12.6	0.123	1 600	1.01	26.6	25.1	35.9	0.522	20.5	18.0	26.7	1 270	706
x 6.4	6.35	10.6	0.104	1 350	0.917	24.1	26.1	31.4	0.477	18.8	18.8	23.5	1 110	645
x 4.8	4.78	8.35	0.082	1 060	0.774	20.3	27.0	25.7	0.407	16.0	19.6	19.3	911	546
x 3.8	3.81	6.85	0.067	872	0.660	17.3	27.5	21.6	0.349	13.8	20.0	16.3	762	465
HSS 51 x 25														
x 3.2	3.18	3.28	0.032	418	0.122	4.81	17.1	6.33	0.040	3.14	9.77	3.84	106	242
x 2.5	2.54	2.71	0.026	345	0.105	4.15	17.5	5.35	0.035	2.75	10.1	3.27	89.8	206

*Depth x width x thickness, see page 6-19

HOLLOW STRUCTURAL SECTIONS
Square

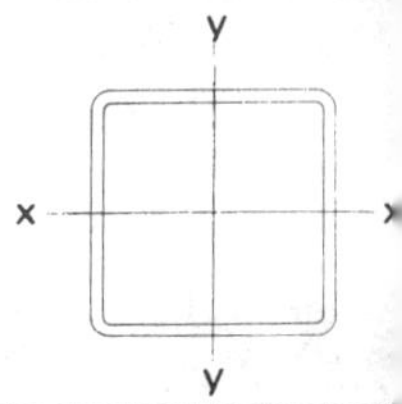

PROPERTIES AND DIMENSIONS

Designation*	Wall Thickness	Mass	Dead Load	Area	I	S	r	Z	Torsional Constant J	Surface Area	Shear Constant C_{rt}
mm x mm x mm	mm	kg/m	kN/m	mm^2	10^6mm^4	10^3mm^3	mm	10^3mm^3	10^3mm^4	m^2/m	mm^2
HSS 305 x 305											
x 13	12.70	113	1.11	14 400	202	1 330	118	1 560	324 000	1.18	6 450
x 11	11.13	100	0.982	12 800	181	1 190	119	1 390	288 000	1.18	5 790
x 9.5	9.53	86.5	0.848	11 000	158	1 040	120	1 210	250 000	1.19	5 080
x 8.0	7.95	72.8	0.714	9 280	135	886	121	1 030	211 000	1.19	4 340
x 6.4	6.35	58.7	0.576	7 480	110	723	121	833	171 000	1.20	3 550
HSS 254 x 254											
x 13	12.70	93.0	0.912	11 800	113	888	97.6	1 060	183 000	0.972	5 160
x 11	11.13	82.4	0.808	10 500	102	799	98.4	945	163 000	0.978	4 660
x 9.5	9.53	71.3	0.699	9 090	89.3	703	99.1	825	142 000	0.983	4 110
x 8.0	7.95	60.1	0.590	7 660	76.5	602	99.9	702	121 000	0.989	3 530
x 6.4	6.35	48.6	0.476	6 190	62.7	494	101	571	97 900	0.994	2 900
HSS 203 x 203											
x 13	12.70	72.7	0.713	9 260	54.7	538	76.8	650	90 700	0.769	3 870
x 11	11.13	64.6	0.634	8 230	49.6	488	77.6	584	81 200	0.775	3 530
x 9.5	9.53	56.1	0.550	7 150	43.9	432	78.4	513	71 000	0.780	3 150
x 8.0	7.95	47.5	0.465	6 050	37.9	373	79.2	438	60 500	0.786	2 730
x 6.4	6.35	38.4	0.377	4 900	31.3	308	79.9	359	49 300	0.791	2 260
HSS 178 x 178											
x 13	12.70	62.6	0.614	7 970	35.2	396	66.4	484	59 200	0.668	3 230
x 11	11.13	55.7	0.547	7 100	32.1	361	67.2	436	53 200	0.673	2 970
x 9.5	9.53	48.5	0.476	6 180	28.6	322	68.0	385	46 700	0.678	2 660
x 8.0	7.95	41.1	0.403	5 240	24.8	279	68.8	330	39 900	0.684	2 320
x 6.4	6.35	33.4	0.327	4 250	20.6	231	69.6	271	32 700	0.689	1 940
x 4.8	4.78	25.5	0.250	3 250	16.1	181	70.3	209	25 200	0.695	1 520
HSS 152 x 152											
x 13	12.70	52.4	0.514	6 680	21.0	275	56.0	341	36 000	0.566	2 580
x 11	11.13	46.9	0.460	5 970	19.3	253	56.8	310	32 500	0.571	2 400
x 9.5	9.53	40.9	0.401	5 210	17.3	227	57.6	275	28 700	0.577	2 180
x 8.0	7.95	34.8	0.341	4 430	15.1	198	58.4	237	24 600	0.582	1 920
x 6.4	6.35	28.3	0.278	3 610	12.6	166	59.2	195	20 300	0.588	1 610
x 4.8	4.78	21.7	0.213	2 760	9.93	130	59.9	152	15 700	0.593	1 270
HSS 127 x 127											
x 11	11.13	38.0	0.373	4 840	10.4	164	46.4	205	18 000	0.470	1 840
x 9.5	9.53	33.3	0.327	4 240	9.47	149	47.2	183	16 000	0.475	1 690
x 8.0	7.95	28.4	0.279	3 620	8.35	132	48.0	159	13 900	0.481	1 510
x 6.4	6.35	23.2	0.228	2 960	7.05	111	48.8	132	11 500	0.486	1 290
x 4.8	4.78	17.9	0.175	2 280	5.60	88.1	49.6	103	8 920	0.492	1 030

*Depth x width x thickness, see page 6-19.

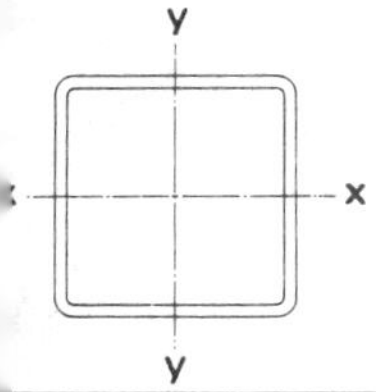

PROPERTIES AND DIMENSIONS

Designation*	Wall Thick-ness	Mass	Dead Load	Area	I	S	r	Z	Torsional Constant J	Surface Area	Shear Constant C_{rt}
mm x mm x mm	mm	kg/m	kN/m	mm^2	10^6mm^4	10^3mm^3	mm	10^3mm^3	10^3mm^4	m^2/m	mm^2
HSS 102 x 102											
x 9.5	9.53	25.7	0.252	3 280	4.44	87.4	36.8	110	7 740	0.374	1 210
x 8.0	7.95	22.1	0.217	2 820	3.98	78.4	37.6	96.6	6 780	0.379	1 110
x 6.4	6.35	18.2	0.178	2 320	3.42	67.3	38.4	81.3	5 670	0.385	968
x 4.8	4.78	14.1	0.138	1 790	2.75	54.2	39.2	64.3	4 450	0.390	789
HSS 89 x 89											
x 9.5	9.53	21.9	0.215	2 790	2.79	62.7	31.6	80.2	4 970	0.323	968
x 8.0	7.95	18.9	0.186	2 410	2.53	57.0	32.4	71.2	4 390	0.328	908
x 6.4	6.35	15.6	0.153	1 990	2.20	49.5	33.2	60.5	3 700	0.334	806
x 4.8	4.78	12.2	0.119	1 550	1.79	40.3	34.0	48.2	2 930	0.339	667
HSS 76 x 76											
x 8.0	7.95	15.8	0.155	2 010	1.49	39.0	27.2	49.7	2 630	0.278	706
x 6.4	6.35	13.1	0.129	1 670	1.31	34.4	28.0	42.7	2 250	0.283	645
x 4.8	4.78	10.3	0.101	1 310	1.08	28.5	28.8	34.4	1 800	0.288	546
HSS 64 x 64											
x 6.4	6.35	10.6	0.104	1 350	0.701	22.1	22.8	28.0	1 230	0.232	484
x 4.8	4.78	8.35	0.082	1 060	0.593	18.7	23.6	22.9	1 000	0.238	424
x 3.8	3.81	6.85	0.067	872	0.506	15.9	24.1	19.2	836	0.241	368
x 3.2	3.18	5.82	0.057	741	0.441	13.9	24.4	16.5	717	0.243	323
HSS 51 x 51											
x 6.4	6.35	8.05	0.079	1 030	0.317	12.5	17.6	16.3	580	0.181	323
x 4.8	4.78	6.45	0.063	821	0.278	10.9	18.4	13.8	485	0.187	303
x 3.8	3.81	5.33	0.052	679	0.242	9.54	18.9	11.7	410	0.190	271
x 3.2	3.18	4.55	0.045	580	0.214	8.42	19.2	10.2	355	0.192	242
x 2.5	2.79	4.05	0.040	516	0.194	7.64	19.4	9.15	318	0.194	221
HSS 38 x 38											
x 4.8	4.78	4.54	0.044	578	0.100	5.27	13.2	6.91	184	0.136	181
x 3.8	3.81	3.81	0.037	485	0.091	4.77	13.7	6.04	160	0.139	174
x 3.2	3.18	3.28	0.032	418	0.082	4.31	14.0	5.34	141	0.141	161
x 2.5	2.54	2.71	0.026	345	0.071	3.71	14.3	4.51	118	0.144	142
HSS 32 x 32											
x 3.8	3.81	3.06	0.030	389	0.048	3.01	11.1	3.92	87.0	0.114	126
x 3.2	3.18	2.65	0.026	338	0.044	2.77	11.4	3.51	77.6	0.116	121
x 2.5	2.54	2.20	0.022	281	0.039	2.44	11.7	3.01	66.1	0.118	110
HSS 25 x 25											
x 3.2	3.18	2.01	0.020	257	0.020	1.56	8.79	2.05	36.3	0.091	80.6
x 2.5	2.54	1.69	0.017	216	0.018	1.41	9.12	1.79	31.6	0.093	77.4

*Depth x width x thickness, see page 6-19.

HOLLOW STRUCTURAL SECTIONS
Round

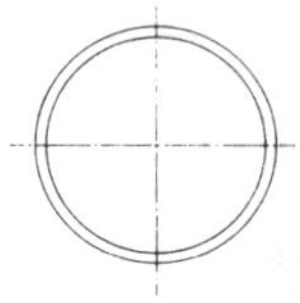

PROPERTIES AND DIMENSIONS

Designation*	Wall Thick-ness	Mass	Dead Load	Area	I	S	r	Z	J	Surface Area	Shear Constant C_{rt}
mm x mm	mm	kg/m	kN/m	mm²	10⁶mm⁴	10³mm³	mm	10³mm³	10³mm⁴	m²/m	mm²
HSS 406											
x 13	12.70	123	1.21	15 700	305	1 500	139	1 970	609 000	1.28	7 860
x 11	11.13	108	1.06	13 800	270	1 330	140	1 740	540 000	1.28	6 910
x 9.5	9.53	93.3	0.915	11 900	234	1 150	140	1 500	468 000	1.28	5 940
x 8.0	7.95	78.1	0.766	9 950	198	972	141	1 260	395 000	1.28	4 970
x 6.4	6.35	62.6	0.614	7 980	160	786	141	1 020	319 000	1.28	3 990
HSS 356											
x 13	12.70	107	1.05	13 700	201	1 130	121	1 490	403 000	1.12	6 850
x 11	11.13	94.6	0.927	12 000	179	1 010	122	1 320	358 000	1.12	6 030
x 9.5	9.53	81.3	0.798	10 400	155	873	122	1 140	310 000	1.12	5 180
x 8.0	7.95	68.2	0.668	8 680	131	738	123	961	262 000	1.12	4 340
x 6.4	6.35	54.7	0.536	6 970	106	598	123	775	213 000	1.12	3 480
HSS 324											
x 13	12.70	97.5	0.956	12 400	151	930	110	1 230	301 000	1.02	6 210
x 11	11.13	85.9	0.842	10 900	134	827	111	1 090	268 000	1.02	5 470
x 9.5	9.53	73.9	0.725	9 410	116	719	111	942	233 000	1.02	4 710
x 8.0	7.95	61.9	0.607	7 890	98.5	608	112	794	197 000	1.02	3 950
x 6.4	6.35	49.7	0.488	6 330	79.9	493	112	640	160 000	1.02	3 170
HSS 273											
x 13	12.70	81.6	0.800	10 400	88.3	646	92.2	862	177 000	0.858	5 200
x 11	11.13	71.9	0.705	9 160	78.7	577	92.7	764	157 000	0.858	4 590
x 9.5	9.53	61.9	0.607	7 890	68.6	502	93.2	662	137 000	0.858	3 950
x 8.0	7.95	52.0	0.510	6 620	58.2	427	93.8	559	116 000	0.858	3 310
x 6.4	6.35	41.8	0.410	5 320	47.4	347	94.3	452	94 700	0.858	2 660
HSS 219											
x 13	12.70	64.6	0.634	8 230	44.0	402	73.1	542	88 000	0.688	4 130
x 11	11.13	57.1	0.560	7 270	39.4	360	73.6	482	78 900	0.688	3 640
x 9.5	9.53	49.3	0.483	6 270	34.5	315	74.2	419	69 000	0.688	3 140
x 8.0	7.95	41.4	0.406	5 270	29.4	269	74.7	355	58 900	0.688	2 640
x 6.4	6.35	33.3	0.327	4 240	24.0	219	75.3	288	48 100	0.688	2 120
x 4.8	4.78	25.3	0.248	3 220	18.5	169	75.8	220	37 000	0.688	1 610
HSS 168											
x 9.5	9.53	37.3	0.366	4 750	15.0	179	56.2	241	30 100	0.529	2 380
x 8.0	7.95	31.4	0.308	4 000	12.9	153	56.8	205	25 800	0.529	2 010
x 6.4	6.35	25.4	0.249	3 230	10.6	126	57.3	167	21 200	0.529	1 620
x 4.8	4.78	19.3	0.189	2 460	8.21	97.6	57.8	128	16 400	0.529	1 230
HSS 141											
x 9.5	9.53	31.0	0.304	3 950	8.61	122	46.7	166	17 200	0.444	1 980
x 8.0	7.95	26.1	0.256	3 330	7.43	105	47.2	142	14 900	0.444	1 670
x 6.4	6.35	21.1	0.207	2 690	6.14	86.9	47.8	116	12 300	0.444	1 350
x 4.8	4.78	16.1	0.158	2 050	4.78	67.7	48.3	89.1	9 560	0.444	1 030

*Diameter x thickness, see page 6-19.

Designation*	Wall Thick-ness	Mass	Dead Load	Area	I	S	r	Z	J	Surface Area	Shear Constant C_{rt}
mm x mm	mm	kg/m	kN/m	mm²	10⁶mm⁴	10³mm³	mm	10³mm³	10³mm⁴	m²/m	mm²
HSS 114											
x 8.0	7.95	20.9	0.204	2 660	3.78	66.1	37.7	90.1	7 550	0.359	1 330
x 6.4	6.35	16.9	0.166	2 150	3.15	55.1	38.2	74.1	6 300	0.359	1 080
x 4.8	4.78	12.9	0.127	1 640	2.47	43.2	38.8	57.4	4 940	0.359	823
HSS 102											
x 8.0	7.95	18.4	0.180	2 340	2.58	50.8	33.2	69.9	5 170	0.319	1 180
x 6.4	6.35	14.9	0.146	1 900	2.16	42.6	33.8	57.7	4 330	0.319	954
x 4.8	4.78	11.4	0.112	1 450	1.71	33.6	34.3	44.8	3 420	0.319	728
x 3.8	3.81	9.19	0.090	1 170	1.40	27.6	34.6	36.5	2 800	0.319	586
HSS 89											
x 8.0	7.95	15.9	0.156	2 020	1.67	37.6	28.8	52.3	3 340	0.279	1 020
x 6.4	6.35	12.9	0.127	1 650	1.41	31.7	29.3	43.4	2 820	0.279	828
x 4.8	4.78	9.92	0.097	1 260	1.12	25.2	29.8	33.9	2 240	0.279	633
x 3.8	3.81	8.00	0.078	1 020	0.924	20.8	30.1	27.6	1 850	0.279	510
HSS 73											
x 6.4	6.35	10.4	0.102	1 330	0.745	20.4	23.7	28.3	1 490	0.229	670
x 4.8	4.78	8.04	0.079	1 020	0.599	16.4	24.2	22.3	1 200	0.229	514
x 3.8	3.81	6.50	0.064	828	0.497	13.6	24.5	18.3	994	0.229	415
x 3.2	3.18	5.48	0.054	698	0.426	11.7	24.7	15.5	852	0.229	349
HSS 60											
x 6.4	6.35	8.45	0.083	1 080	0.397	13.2	19.2	18.6	794	0.189	545
x 4.8	4.78	6.54	0.064	834	0.324	10.7	19.7	14.8	647	0.189	420
x 3.8	3.81	5.31	0.052	676	0.271	8.99	20.0	12.2	542	0.189	339
x 3.2	3.18	4.48	0.044	571	0.233	7.74	20.2	10.4	467	0.189	286
HSS 48											
x 4.8	4.78	5.13	0.050	654	0.157	6.48	15.5	9.09	313	0.152	331
x 3.8	3.81	4.18	0.041	533	0.133	5.50	15.8	7.56	265	0.152	268
x 3.2	3.18	3.54	0.035	451	0.115	4.77	16.0	6.48	231	0.152	226
x 2.8	2.79	3.13	0.031	399	0.104	4.29	16.1	5.79	207	0.152	200
HSS 42											
x 3.2	3.18	3.06	0.030	390	0.075	3.54	13.8	4.85	149	0.133	196
x 2.5	2.54	2.48	0.024	316	0.062	2.96	14.1	4.00	125	0.133	159
HSS 33											
x 3.2	3.18	2.37	0.023	302	0.035	2.09	10.7	2.91	69.7	0.105	153
x 2.5	2.54	1.93	0.019	246	0.029	1.77	10.9	2.42	59.0	0.105	124
HSS 27											
x 3.2	3.18	1.84	0.018	235	0.016	1.24	8.39	1.77	33.1	0.084	120
x 2.5	2.54	1.51	0.015	193	0.014	1.07	8.59	1.49	28.4	0.084	97.4

*Diameter x thickness, see page 6-19.

PIPE
PROPERTIES AND DIMENSIONS[1]

Outside Diameter	Thickness	Weight Class[2]	Mass	Dead Load	Area	I	S	r	Z	Shear Constant	J	Surface Area
mm	mm		kg/m	kN/m	mm²	10⁶mm⁴	10³mm³	mm	10³mm³	mm²	10⁶mm⁴	m²/m
21.3	2.77	STD	1.26	0.012 4	161	0.007 11	0.667	6.64	0.962	81.9	0.0142	0.067
	3.73	XS	1.62	0.015 9	206	0.008 36	0.783	6.36	1.17	106	0.0167	0.067
	7.47	XXS	2.54	0.025 0	325	0.010 1	0.946	5.57	1.58	191	0.0202	0.067
26.7	2.87	STD	1.68	0.016 5	215	0.015 4	1.16	8.48	1.63	108	0.0308	0.083
	3.91	XS	2.19	0.021 5	280	0.018 6	1.40	8.16	2.05	143	0.0373	0.083
	7.82	XXS	3.63	0.035 6	463	0.024 1	1.81	7.21	2.94	257	0.0482	0.083
33.4	3.38	STD	2.50	0.024 5	319	0.036 4	2.18	10.7	3.06	161	0.0727	0.105
	4.55	XS	3.23	0.031 7	412	0.044 0	2.63	10.3	3.82	209	0.0879	0.105
	9.09	XXS	5.45	0.053 4	694	0.058 5	3.50	9.18	5.62	378	0.117	0.105
42.2	3.56	STD	3.38	0.033 1	431	0.081 0	3.84	13.7	5.32	217	0.162	0.132
	4.85	XS	4.46	0.043 8	569	0.101	4.77	13.3	6.79	288	0.201	0.132
	9.70	XXS	7.75	0.076 0	969	0.142	6.73	12.0	10.5	523	0.284	0.132
48.3	3.68	STD	4.05	0.039 7	516	0.129	5.35	15.8	7.34	259	0.258	0.152
	5.08	XS	5.40	0.053 0	689	0.163	6.75	15.4	9.52	348	0.326	0.152
	10.16	XXS	9.54	0.093 5	1 220	0.236	9.80	13.9	15.1	636	0.473	0.152
60.3	3.91	STD	5.43	0.053 3	693	0.277	9.19	20.0	12.5	348	0.554	0.190
	5.54	XS	7.47	0.073 3	953	0.361	12.0	19.5	16.7	480	0.723	0.190
	11.07	XXS	13.4	0.132	1 710	0.546	18.1	17.8	27.3	885	1.09	0.190
73.0	5.16	STD	8.62	0.084 5	1 100	0.637	17.4	24.1	23.8	552	1.27	0.229
	7.01	XS	11.4	0.112	1 450	0.801	21.9	23.5	30.7	732	1.60	0.229
	14.02	XXS	20.4	0.200	2 600	1.19	32.7	21.4	49.7	1350	2.39	0.229
88.9	5.49	STD	11.3	0.111	1 440	1.26	28.3	29.6	38.2	721	2.51	0.279
	7.62	XS	15.3	0.150	1 950	1.62	36.5	28.9	50.5	979	3.24	0.279
	15.24	XXS	27.7	0.271	3 530	2.49	56.1	26.6	83.9	1810	4.99	0.279
101.6	5.74	STD	13.6	0.133	1 730	1.99	39.2	34.0	52.8	866	3.99	0.319
	8.08	XS	18.6	0.183	2 370	2.61	51.5	33.2	70.8	1 190	5.23	0.319
114.3	6.02	STD	16.1	0.157	2 050	3.01	52.7	38.3	70.7	1 030	6.02	0.359
	8.56	XS	22.3	0.219	2 840	4.00	70.0	37.5	95.9	1 430	8.00	0.359
	17.12	XXS	40.9	0.402	5 230	6.36	111	34.9	163	2 670	12.7	0.359
141.3	6.55	STD	21.9	0.213	2 770	6.31	89.3	47.7	119	1 390	12.6	0.444
	9.53	XS	30.9	0.303	3 940	8.60	122	46.7	166	1 980	17.2	0.444
	19.05	XXS	57.4	0.563	7 320	14.0	198	43.7	287	3 720	28.0	0.444
168.3	7.11	STD	28.2	0.277	3 600	11.7	139	57.0	185	1 800	23.4	0.529
	10.97	XS	42.5	0.417	5 420	16.9	200	55.7	272	2 720	33.7	0.529
	21.95	XXS	79.1	0.776	10 100	27.6	328	52.3	473	5 120	55.2	0.529
219.1	8.18	STD	42.5	0.417	5 420	30.2	275	74.6	364	2 710	60.3	0.688
	12.70	XS	64.6	0.633	8 230	44.0	402	73.1	542	4 130	88.0	0.688
	22.23	XXS	108.	1.06	13 700	67.4	616	70.0	865	6 930	135	0.688
273.1	9.27	STD	60.2	0.591	7 680	66.9	490	93.3	645	3 840	134	0.858
	12.70	XS	81.5	0.799	10 400	88.2	646	92.2	862	5 200	176	0.858
	25.40	XXS	155.	1.52	19 800	153	1120	88.0	1560	9 950	306	0.858
323.9	9.53	STD	73.8	0.723	9 410	116	718	111	941	4 710	233	1.02
	12.70	XS	97.4	0.955	12 400	150	929	110	1230	6 210	301	1.02
	25.40	XXS	187.	1.83	23 800	267	1650	106	2270	12 000	534	1.02
355.6	9.53	STD	81.2	0.796	10 400	155	873	122	1140	5 940	310	1.12
	12.70	XS	107.	1.05	13 700	201	1130	121	1490	7 860	403	1.12
406.4	9.53	STD	93.1	0.913	11 900	234	1150	140	1500	5 180	468	1.28
	12.70	XS	123.	1.21	15 700	305	1500	139	1970	6 850	609	1.28
457.2	9.53	STD	105.	1.03	13 400	336	1470	158	1910	6 700	671	1.44
	12.70	XS	139.	1.36	17 700	438	1920	157	2510	8 870	877	1.44

1. These dimensions have been soft converted from those specified in ASTM A53.
2. Class refers to: Standard Weight – STD., Extra Strong – XS, Double Extra Strong – XXS.

COLD FORMED CHANNELS

General

While various proprietary cold formed channels (CFC) are available from Canadian roll formers, the CFC series listed on pages 6-106 to 6-109 is that approved for inclusion in CSA Standard G312.3. CFC sections listed have elements with width-thickness ratios exceeding the Class 3 limits of CAN/CSA-S16.1-M89. Therefore, their factored resistances are determined according to the requirements of CSA Standard CAN/CSA-S136-M89. Many of the resistances are determined using effective cross-sectional properties computed on the basis of some portion of the element(s) in compression notionally removed. The tables on pages 6-106 to 6-109 list both the gross cross-sectional and the effective cross-sectional properties — all computed assuming the inside bend radius to be twice the thickness. For deflection calculations, the effective moment of inertia, I_{xe}, can be conservatively used.

Material

The values in the table are based on CFC sections roll formed from steel meeting the requirement of ASTM A570M Grade 50 (F_y = 345 MPa). Where appropriate, the enhanced yield stress, F'_y, due to the effects of cold working is listed. Factored resistances of a member at a welded zone should be based on F_y and not on F'_y since welding can anneal the steel eliminating the enhancing effect of cold working.

Tables

The designation used is that approved for CSA Standard G312.3 and consists of the letters CFC, the nominal depth, the flange width and sheet steel thickness, all in millimetres. The following symbols are used for the CFC sections in addition to those used for hot rolled shapes:

F'_y = specified minimum yield strength including effects of cold work and forming, (MPa)

I_{xe} = effective moment of inertia about X-X axis, (10^6 mm^4)

S_{xe} = effective section modulus of the tension flange, (10^3 mm^3)

y_e = distance from the neutral axis to the extreme compressive fibre, (mm)

I_{ye} = effective moment of inertia about Y-Y axis assuming the stiffener lips are in tension, (10^6 mm^4)

S_{ye} = effective section modulus of the tension flange (i.e. stiffener lips), (10^3 mm^3)

x_e = distance from the neutral axis to the extreme compressive fibre, (mm)

M_{rxe} = effective factored moment resistance about the X-X axis based on F'_y, (kN·m)

d_i = depth of stiffener lip, (mm)

x_o = distance from shear centre to Y-Y axis, (mm)

r_o = polar radius of gyration, (mm)

J = St. Venant's torsional constant, (10^3 mm^4)

j = torsional-flexural buckling parameter, (mm)

C_w = warping torsional constant, (10^9 mm^6)

These tables have been prepared by Dr. R. Schuster, University of Waterloo.

COLD FORMED CHANNELS
Properties and Dimensions
A570M Grade 50 (F_y = 345 MPa)

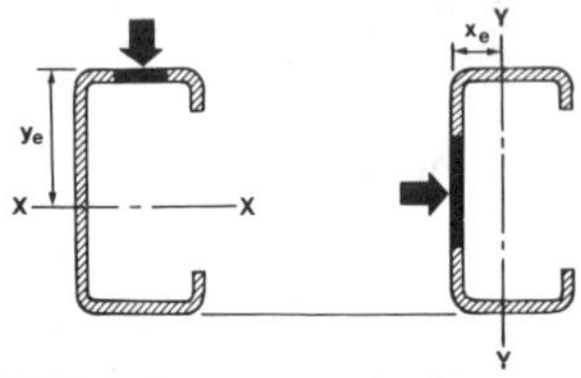

Designation	Mass	Area	F'_y	Effective Section Properties						M_{rxe}	Surface Area	
				X–X Axis			Y–Y Axis				Total	Less top of top flg.
				I_{xe}	S_{xe}	y_e	I_{ye}	S_{ye}	x_e			
	kg/m	mm²	MPa	10⁶mm⁴	10³mm³	mm	10⁶mm⁴	10³mm³	mm	kN·m	m²/m	m²/m
CFC 460 x 127												
x 5.69	33.6	4280	388	127	554	229	7.94	89.7	38.6	194	1.52	1.42
x 4.93	29.4	3740	383	110	478	231	6.84	79.3	40.7	165	1.53	1.43
x 4.17	25.0	3190	345	93.9	401	234	5.70	68.1	43.4	125	1.54	1.44
CFC 380 x 127												
x 5.69	30.2	3850	388	81.9	430	191	7.69	88.8	40.5	150	1.37	1.27
x 4.93	26.4	3370	383	72.2	379	191	6.63	78.5	42.5	131	1.38	1.28
x 4.17	22.5	2870	345	61.9	323	192	5.53	67.5	45.1	100	1.39	1.29
CFC 380 x 89												
x 5.69	25.7	3270	405	63.3	332	191	2.75	41.5	22.7	121	1.16	1.11
x 4.93	22.5	2870	398	56.0	294	191	2.43	37.3	23.9	105	1.17	1.11
x 4.17	19.2	2450	390	47.9	248	193	2.07	32.6	25.5	87.2	1.18	1.12
x 3.43	16.0	2040	383	39.5	201	197	1.69	27.6	27.6	69.2	1.20	1.13
x 2.67	12.6	1600	345	30.4	150	203	1.28	22.0	30.6	46.7	1.21	1.13
CFC 310 x 89												
x 4.93	19.5	2490	398	32.6	214	152	2.33	36.9	25.6	76.5	1.02	.961
x 4.17	16.7	2130	390	28.2	185	152	1.99	32.3	27.1	65.0	1.03	.968
x 3.43	13.9	1780	383	23.6	154	154	1.63	27.3	29.1	53.0	1.04	.974
x 2.67	11.0	1400	345	18.4	117	158	1.24	21.7	31.9	36.2	1.05	.980
CFC 310 x 70												
x 4.93	17.6	2240	411	27.4	180	152	1.15	21.8	17.2	66.4	.919	.878
x 4.17	15.1	1920	401	23.8	156	152	1.00	19.4	18.3	56.5	.930	.885
x 3.43	12.6	1600	392	20.0	130	154	.834	16.6	19.7	45.8	.941	.891
x 2.67	9.91	1260	382	15.6	98.6	158	.644	13.4	21.9	33.9	.952	.898
CFC 250 x 89												
x 4.93	17.6	2240	398	21.0	166	127	2.25	36.4	27.1	59.3	.919	.859
x 4.17	15.1	1920	390	18.2	144	127	1.93	31.9	28.5	50.5	.930	.866
x 3.43	12.6	1600	383	15.4	121	127	1.58	27.0	30.4	41.8	.941	.872
x 2.67	9.91	1260	345	12.1	94.1	129	1.20	21.5	33.1	29.2	.952	.879
x 2.29	8.55	1090	345	10.1	75.3	134	1.00	18.5	34.8	23.4	.957	.882
CFC 250 x 70												
x 4.17	13.4	1710	401	15.3	120	127	.967	19.2	19.4	43.4	.828	.783
x 3.43	11.2	1430	392	12.9	102	127	.809	16.5	20.7	36.0	.839	.790
x 2.67	8.84	1130	382	10.2	79.5	129	.626	13.3	22.8	27.4	.850	.796
x 2.29	7.64	973	345	8.82	67.7	130	.528	11.6	24.2	21.0	.856	.800
x 1.90	6.41	817	345	7.08	52.1	136	.428	9.73	25.8	16.2	.861	.803
CFC 200 x 89												
x 4.17	13.4	1710	390	10.8	106	102	1.84	31.4	30.3	37.2	.828	.764
x 3.43	11.2	1430	383	9.11	89.6	102	1.52	26.6	32.0	30.9	.839	.771
x 2.67	8.84	1130	345	7.24	70.9	102	1.15	21.2	34.5	22.0	.850	.777
x 2.29	7.64	973	345	6.07	57.9	105	.966	18.3	36.1	18.0	.856	.780
x 1.90	6.41	817	345	4.89	45.0	109	.777	15.3	38.0	14.0	.861	.784

Note: For an explanation of this table, see page 6–105.

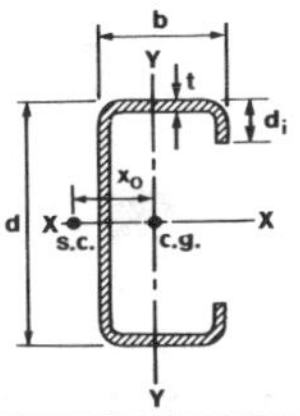

COLD FORMED CHANNELS
Properties and Dimensions

Depth	Flange Width	Stiff'r depth	Thick-ness	Gross Section Properties										
				X–X Axis			Y–Y Axis							
d	b	d_i	t	I_x	S_x	r_x	I_y	S_y	r_y	x_o	r_o	J	j	C_w
mm	mm	mm	mm	10^6mm^4	10^3mm^3	mm	10^6mm^4	10^3mm^3	mm	mm	mm	10^3mm^4	mm	10^9mm^6
457	127	44.4	5.69	127	554	172	8.57	91.9	44.7	87.5	198	46.2	262	377
457	127	44.4	4.93	112	488	173	7.65	82.1	45.2	88.0	199	30.3	259	336
457	127	44.4	4.17	95.8	419	173	6.66	71.5	45.7	88.5	200	18.4	256	292
381	127	44.4	5.69	81.9	430	146	8.10	90.4	45.9	94.8	180	41.5	214	255
381	127	44.4	4.93	72.2	379	147	7.23	80.7	46.4	95.2	181	27.2	212	227
381	127	44.4	4.17	62.1	326	147	6.30	70.4	46.8	95.7	182	16.6	210	197
381	89	31.7	5.69	63.3	332	139	2.82	41.8	29.4	55.7	153	35.3	244	86.8
381	89	31.7	4.93	56.0	294	140	2.56	37.9	29.9	56.2	154	23.2	238	78.1
381	89	31.7	4.17	48.4	254	141	2.26	33.5	30.4	56.6	155	14.2	232	68.5
381	89	31.7	3.43	40.7	214	141	1.94	28.8	30.8	57.1	156	7.98	228	58.4
381	89	31.7	2.67	32.3	170	142	1.57	23.3	31.3	57.6	157	3.80	223	47.1
305	89	31.7	4.93	32.6	214	114	2.40	37.2	31.0	62.4	134	20.2	177	47.7
305	89	31.7	4.17	28.2	185	115	2.12	32.9	31.5	62.9	135	12.3	173	42.0
305	89	31.7	3.43	23.8	156	116	1.82	28.2	32.0	63.3	136	6.96	171	35.9
305	89	31.7	2.67	18.9	124	116	1.47	22.9	32.5	63.8	137	3.31	168	29.0
305	70	25.4	4.93	27.4	180	111	1.16	21.9	22.8	43.3	121	18.1	200	23.0
305	70	25.4	4.17	23.8	156	111	1.04	19.6	23.3	43.8	122	11.1	194	20.4
305	70	25.4	3.43	20.1	132	112	.904	17.1	23.8	44.3	123	6.27	188	17.6
305	70	25.4	2.67	16.1	106	113	.742	14.0	24.2	44.8	124	2.99	183	14.3
254	89	31.7	4.93	21.0	166	96.9	2.26	36.5	31.8	67.6	122	18.1	146	32.3
254	89	31.7	4.17	18.2	144	97.5	2.00	32.3	32.3	68.0	123	11.1	144	28.4
254	89	31.7	3.43	15.4	121	98.1	1.72	27.8	32.8	68.4	124	6.27	142	24.4
254	89	31.7	2.67	12.3	96.8	98.7	1.39	22.5	33.2	68.9	125	2.99	140	19.7
254	89	31.7	2.29	10.7	84.0	99.0	1.22	19.7	33.5	69.1	125	1.90	139	17.3
254	70	25.4	4.17	15.3	120	94.5	.988	19.3	24.1	47.8	109	9.88	150	13.6
254	70	25.4	3.43	12.9	102	95.2	.859	16.8	24.5	48.3	110	5.59	147	11.8
254	70	25.4	2.67	10.4	81.6	95.9	.705	13.8	25.0	48.8	111	2.67	143	9.62
254	70	25.4	2.29	9.01	70.9	96.2	.621	12.2	25.3	49.0	111	1.69	142	8.45
254	70	25.4	1.90	7.62	60.0	96.6	.531	10.4	25.5	49.2	111	.988	140	7.21
203	89	31.7	4.17	10.8	106	79.4	1.85	31.5	33.0	74.2	114	9.88	121	18.0
203	89	31.7	3.43	9.11	89.6	79.9	1.59	27.1	33.4	74.5	114	5.59	120	15.5
203	89	31.7	2.67	7.29	71.7	80.4	1.29	22.0	33.9	74.9	115	2.67	119	12.6
203	89	31.7	2.29	6.34	62.4	80.7	1.13	19.3	34.1	75.1	115	1.69	118	11.0
203	89	31.7	1.90	5.35	52.7	81.0	.962	16.4	34.3	75.3	116	.988	118	9.39

COLD FORMED CHANNELS
Properties and Dimensions
A570M Grade 50 (F_y = 345 MPa)

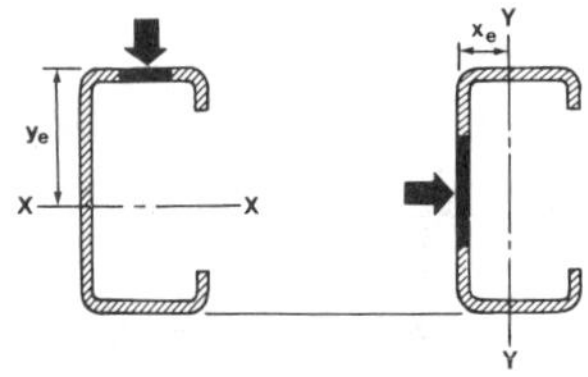

Designation	Mass	Area	F'_y	Effective Section Properties						M_{rxe}	Surface Area	
				X–X Axis			Y–Y Axis				Total	Less top of top flg.
				I_{xe}	S_{xe}	y_e	I_{ye}	S_{ye}	x_e			
	kg/m	mm^2	MPa	10^{6}mm^4	10^{3}mm^3	mm	10^{6}mm^4	10^{3}mm^3	mm	kN·m	m^2/m	m^2/m
CFC 200 x 70												
x 3.43	9.83	1250	392	7.57	74.5	102	.777	16.3	22.1	26.3	.737	.688
x 2.67	7.78	991	382	6.09	59.9	102	.602	13.1	24.0	20.6	.749	.695
x 2.29	6.72	857	345	5.29	51.9	102	.509	11.4	25.3	16.1	.754	.698
x 1.90	5.65	720	345	4.28	40.4	106	.413	9.62	26.9	12.5	.760	.701
CFC 150 x 70												
x 3.43	8.46	1080	392	3.83	50.2	76.2	.726	15.9	24.2	17.7	.636	.587
x 2.67	6.72	856	382	3.09	40.6	76.2	.571	12.9	25.6	14.0	.647	.593
x 2.29	5.81	741	345	2.70	35.4	76.2	.484	11.2	26.8	11.0	.652	.596
x 1.90	4.89	623	345	2.22	28.5	77.9	.394	9.46	28.2	8.85	.658	.600
x 1.52	3.95	503	345	1.72	21.2	81.0	.302	7.60	30.1	6.59	.664	.603
CFC 150 x 50												
x 2.67	5.71	727	395	2.45	32.1	76.2	.243	7.02	16.2	11.4	.550	.516
x 2.29	4.95	630	389	2.14	28.1	76.2	.209	6.18	17.0	9.84	.556	.519
x 1.90	4.17	531	382	1.83	24.0	76.2	.173	5.27	18.1	8.23	.561	.522
x 1.52	3.37	430	345	1.47	18.9	77.5	.134	4.29	19.5	5.88	.567	.525
x 1.22	2.72	347	345	1.11	13.6	81.9	.103	3.45	20.9	4.22	.571	.528

Note: For an explanation of this table, see page 6–105.

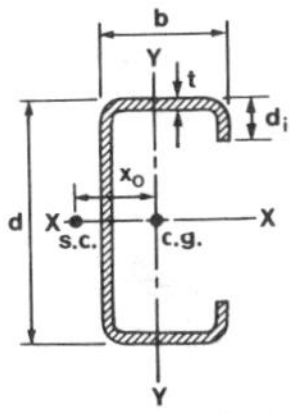

COLD FORMED CHANNELS
Properties and Dimensions

Depth	Flange Width	Stiff'r depth	Thick-ness	Gross Section Properties										
				X–X Axis			Y–Y Axis							
d	b	d_i	t	I_x	S_x	r_x	I_y	S_y	r_y	x_o	r_o	J	j	C_w
mm	mm	mm	mm	$10^6 mm^4$	$10^3 mm^3$	mm	$10^6 mm^4$	$10^3 mm^3$	mm	mm	mm	$10^3 mm^4$	mm	$10^9 mm^6$
203	70	25.4	3.43	7.57	74.5	77.8	.802	16.4	25.3	53.2	97.6	4.91	115	7.31
203	70	25.4	2.67	6.09	59.9	78.4	.658	13.5	25.8	53.7	98.4	2.35	113	5.99
203	70	25.4	2.29	5.30	52.2	78.7	.580	11.9	26.0	53.9	98.8	1.49	112	5.27
203	70	25.4	1.90	4.49	44.2	79.0	.496	10.2	26.2	54.1	99.3	0.87	111	4.50
152	70	25.4	3.43	3.83	50.2	59.6	.726	15.9	25.9	59.5	88.1	4.23	92.7	4.08
152	70	25.4	2.67	3.09	40.6	60.1	.597	13.1	26.4	59.9	88.9	2.03	91.7	3.36
152	70	25.4	2.29	2.70	35.4	60.4	.526	11.6	26.6	60.1	89.2	1.29	91.2	2.96
152	70	25.4	1.90	2.29	30.1	60.6	.450	9.89	26.9	60.2	89.6	0.75	90.8	2.54
152	70	25.4	1.52	1.87	24.5	60.9	.369	8.13	27.1	60.4	90.0	0.39	90.4	2.09
152	51	20.3	2.67	2.45	32.1	58.0	.250	7.07	18.5	39.3	72.5	1.72	85.3	1.31
152	51	20.3	2.29	2.14	28.1	58.3	.222	6.30	18.8	39.5	72.9	1.10	84.2	1.16
152	51	20.3	1.90	1.83	24.0	58.6	.192	5.45	19.0	39.7	73.3	0.64	83.2	1.00
152	51	20.3	1.52	1.49	19.6	58.9	.159	4.52	19.2	39.9	73.7	0.33	82.3	0.83
152	51	20.3	1.22	1.21	15.9	59.2	.131	3.72	19.4	40.1	74.1	0.17	81.7	0.68

NOTES

BUILT-UP SECTIONS

Built-up sections may be fabricated from plate and shapes in various configurations to produce efficient and economical structural sections. Generally the components are joined by welding, though bolting may also be used for some combinations. Frequently used built-up sections include double angles back-to-back, two channels back-to-back or toe-to-toe, and a channel or C shape in combination with a W shape.

Properties and dimensions of double angles are provided for imperial series angles on pages 6-112 to 6-119, and for metric series angles on pages 6-120 to 6-125. Included are equal leg angles and unequal leg angles with long legs back-to-back and with short legs back-to-back. Properites and dimensions of double channels and of C shapes combined with W shapes for some representative combinations are provided in tables on pages 6-126 to 6-128.

Many other combinations of built-up members are possible. Diagrams with expressions and formulae for computing properties of some possible combinations are provided on pages 6-129 to 6-132.

TWO ANGLES EQUAL LEGS - Imperial Series
Back to Back

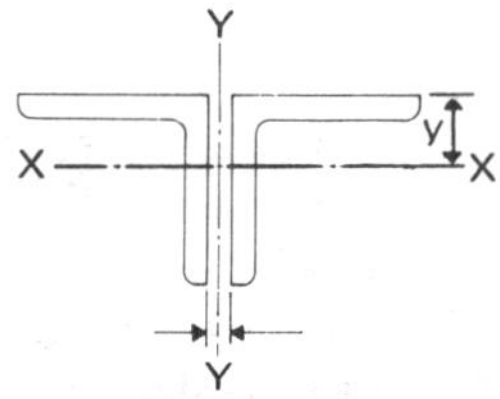

PROPERTIES OF SECTIONS

Designation*	Mass of 2 Angles	Dead Load	Area of 2 Angles	Axis X-X				Radii of Gyration about Axis Y-Y					
				I	S	r	y	back to back of angles, millimetres					
	kg/m	kN/m	mm	10^6 mm^4	10^3 mm^3	mm	mm	0	8	10	12	16	20
L203x203													
x29	169	1.66	21 600	81.4	574	61.4	61.2	86.7	89.6	90.3	91.0	92.5	94.0
x25	152	1.49	19 300	73.8	517	61.8	60.1	86.2	89.0	89.7	90.5	91.9	93.4
x22	134	1.31	17 000	66.0	458	62.2	58.9	85.7	88.5	89.2	89.9	91.4	92.8
x19	116	1.14	14 800	58.0	399	62.6	57.8	85.2	88.0	88.7	89.4	90.9	92.3
x16	97.4	0.955	12 400	49.4	337	63.1	56.6	84.8	87.5	88.2	88.9	90.3	91.8
x14	87.9	0.863	11 200	44.9	306	63.3	56.0	84.6	87.3	88.0	88.7	90.1	91.5
x13	78.4	0.769	9 990	40.4	274	63.6	55.5	84.4	87.0	87.7	88.4	89.8	91.2
L152x152													
x25	111	1.09	14 200	29.3	279	45.5	47.2	65.6	68.5	69.3	70.0	71.5	73.1
x22	98.2	0.963	12 500	26.3	249	45.9	46.1	65.0	67.9	68.7	69.4	70.9	72.5
x19	85.4	0.838	10 900	23.3	218	46.3	45.0	64.5	67.4	68.1	68.9	70.4	71.9
x16	71.9	0.705	9 160	20.0	185	46.7	43.9	64.1	66.9	67.6	68.3	69.8	71.3
x14	65.0	0.638	8 290	18.2	168	46.9	43.3	63.8	66.6	67.3	68.0	69.5	71.0
x13	58.1	0.570	7 400	16.4	150	47.1	42.7	63.6	66.3	67.1	67.8	69.2	70.7
x11	51.0	0.501	6 500	14.6	133	47.4	42.1	63.4	66.1	66.8	67.5	69.0	70.4
x9.5	44.1	0.432	5 610	12.7	115	47.6	41.5	63.2	65.9	66.6	67.3	68.7	70.1
x7.9	36.9	0.362	4 700	10.8	96.8	47.8	41.0	63.0	65.6	66.3	67.0	68.4	69.9
L127x127													
x22	80.8	0.792	10 300	14.8	169	37.9	39.8	55.0	57.9	58.7	59.4	61.0	62.6
x19	70.4	0.691	8 970	13.1	149	38.3	38.7	54.4	57.3	58.1	58.8	60.4	61.9
x16	59.4	0.583	7 570	11.3	127	38.7	37.6	53.9	56.8	57.5	58.3	59.8	61.3
x13	48.1	0.472	6 130	9.37	103	39.1	36.4	53.4	56.2	57.0	57.7	59.2	60.7
x11	42.3	0.415	5 390	8.33	91.4	39.3	35.8	53.2	56.0	56.7	57.4	58.9	60.4
x9.5	36.6	0.359	4 660	7.28	79.4	39.5	35.3	53.0	55.7	56.4	57.2	58.6	60.1
x7.9	30.7	0.301	3 910	6.18	66.9	39.8	34.7	52.8	55.5	56.2	56.9	58.3	59.8
x6.4	24.7	0.242	3 150	5.03	54.2	40.0	34.1	52.6	55.2	55.9	56.6	58.1	59.5
x4.8	18.6	0.183	2 370	3.84	41.1	40.2	33.5	52.4	55.0	55.7	56.4	57.8	59.3
L102x102													
x19	55.4	0.544	7 060	6.48	93.1	30.3	32.4	44.4	47.4	48.1	48.9	50.5	52.1
x16	47.0	0.461	5 980	5.62	79.5	30.7	31.3	43.8	46.7	47.5	48.3	49.8	51.4
x13	38.1	0.374	4 860	4.69	65.3	31.1	30.2	43.3	46.2	46.9	47.7	49.2	50.8
x11	33.6	0.330	4 280	4.19	57.8	31.3	29.6	43.0	45.9	46.6	47.4	48.9	50.4
x9.5	29.1	0.285	3 710	3.68	50.4	31.5	29.0	42.8	45.6	46.4	47.1	48.6	50.1
x7.9	24.4	0.240	3 110	3.13	42.6	31.7	28.4	42.6	45.4	46.1	46.8	48.3	49.8
x6.4	19.7	0.193	2 510	2.56	34.5	31.9	27.9	42.4	45.1	45.8	46.5	48.0	49.5
x4.8	14.9	0.146	1 900	1.96	26.3	32.2	27.3	42.2	44.9	45.6	46.3	47.7	49.2
L89x89													
x16	40.4	0.396	5 150	3.61	59.3	26.5	28.0	38.5	41.5	42.3	43.1	44.7	46.3
x13	32.9	0.323	4 190	3.03	48.8	26.9	26.9	38.0	40.9	41.7	42.4	44.0	45.6
x11	29.1	0.285	3 700	2.71	43.3	27.1	26.3	37.7	40.6	41.4	42.1	43.7	45.3
x9.5	25.2	0.247	3 210	2.39	37.8	27.3	25.7	37.5	40.3	41.1	41.8	43.4	45.0
x7.9	21.2	0.208	2 700	2.04	32.0	27.5	25.2	37.3	40.1	40.8	41.6	43.1	44.6
x6.4	17.1	0.168	2 180	1.67	26.0	27.7	24.6	37.0	39.8	40.5	41.3	42.8	44.3
x4.8	12.9	0.127	1 650	1.29	19.8	28.0	24.0	36.8	39.6	40.3	41.0	42.5	44.0

*Designation consists of nominal leg sizes and thickness. For angle size, see Angles - Imperial Series.

Note: The properties of angles currently produced by cold forming are up to 7 percent less than the properties shown in the above tables. Check manufacturers' catalogue for the exact properties and dimensions. See also page 6-43.

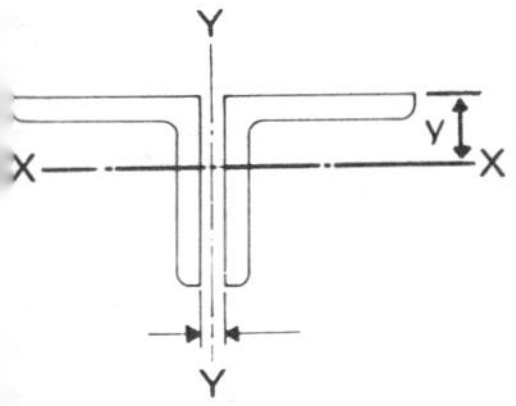

TWO ANGLES EQUAL LEGS - Imperial Series Back to Back

PROPERTIES OF SECTIONS

Designation*	Mass of 2 Angles	Dead Load	Area of 2 Angles	Axis X-X				Radii of Gyration about Axis Y-Y back to back of angles, millimetres					
				I	S	r	y	0	8	10	12	16	20
	kg/m	kN/m	mm	10^6 mm^4	10^3 mm^3	mm	mm						
L76x76													
x16	34.1	0.334	4 340	2.19	42.5	22.4	24.8	33.4	36.5	37.3	38.1	39.7	41.4
x13	27.9	0.273	3 550	1.85	35.1	22.8	23.7	32.9	35.9	36.6	37.4	39.0	40.7
x11	24.6	0.242	3 140	1.66	31.2	23.0	23.1	32.6	35.5	36.3	37.1	38.7	40.3
x9.5	21.4	0.210	2 720	1.47	27.3	23.2	22.5	32.3	35.3	36.0	36.8	38.4	40.0
x7.9	18.0	0.177	2 290	1.26	23.2	23.4	22.0	32.1	35.0	35.7	36.5	38.0	39.6
x6.4	14.6	0.143	1 850	1.04	18.9	23.6	21.4	31.9	34.7	35.4	36.2	37.7	39.3
x4.8	11.0	0.108	1 410	0.800	14.4	23.9	20.8	31.7	34.4	35.2	35.9	37.4	39.0
L64x64													
x13	22.8	0.224	2 900	1.02	23.7	18.8	20.5	27.8	30.8	31.6	32.4	34.1	35.8
x9.5	17.6	0.172	2 240	0.819	18.6	19.1	19.4	27.2	30.2	31.0	31.8	33.4	35.0
x7.9	14.8	0.146	1 890	0.707	15.8	19.3	18.8	27.0	29.9	30.7	31.4	33.0	34.7
x6.4	12.0	0.118	1 530	0.585	12.9	19.5	18.2	26.7	29.6	30.3	31.1	32.7	34.3
x4.8	9.14	0.0896	1 160	0.455	9.92	19.8	17.6	26.5	29.3	30.1	30.8	32.4	34.0
x3.2	6.18	0.0606	787	0.315	6.78	20.0	17.1	26.3	29.0	29.8	30.5	32.1	33.6
L51x51													
x9.5	13.8	0.135	1 750	0.399	11.5	15.1	16.2	22.1	25.2	26.0	26.8	28.5	30.2
x7.9	11.7	0.115	1 490	0.347	9.84	15.3	15.6	21.8	24.8	25.6	26.4	28.1	29.8
x6.4	9.50	0.0931	1 210	0.289	8.09	15.5	15.0	21.6	24.5	25.3	26.1	27.7	29.4
x4.8	7.24	0.0710	922	0.227	6.24	15.7	14.5	21.3	24.2	25.0	25.8	27.4	29.1
x3.2	4.91	0.0482	626	0.158	4.29	15.9	13.9	21.1	23.9	24.7	25.5	27.1	28.7
L44x44													
x9.5	11.9	0.117	1 510	0.259	8.65	13.1	14.6	19.6	22.7	23.5	24.4	26.1	27.8
x7.9	10.1	0.0991	1 290	0.226	7.42	13.3	14.0	19.3	22.4	23.2	24.0	25.7	27.4
x6.4	8.24	0.0808	1 050	0.190	6.11	13.4	13.4	19.0	22.0	22.8	23.6	25.3	27.0
x4.8	6.30	0.0618	802	0.150	4.73	13.7	12.9	18.8	21.7	22.5	23.3	24.9	26.6
x3.2	4.28	0.0420	546	0.105	3.26	13.9	12.3	18.5	21.4	22.2	23.0	24.6	26.3
L38x38													
x9.5	9.98	0.0978	1 270	0.155	6.18	11.1	12.9	17.0	20.2	21.1	21.9	23.7	25.5
x7.9	8.51	0.0835	1 080	0.136	5.31	11.2	12.4	16.7	19.9	20.7	21.5	23.3	25.0
x6.4	6.96	0.0683	887	0.115	4.39	11.4	11.8	16.4	19.5	20.3	21.2	22.9	24.6
x4.8	5.34	0.0524	680	0.0915	3.41	11.6	11.3	16.2	19.2	20.0	20.8	22.5	24.2
x4.0	4.50	0.0442	574	0.0786	2.90	11.7	11.0	16.1	19.0	19.8	20.6	22.3	24.0
x3.2	3.65	0.0358	464	0.0648	2.37	11.8	10.7	15.9	18.9	19.6	20.5	22.1	23.8
L32x32													
x9.5	8.09	0.0794	1 030	0.0852	4.16	9.09	11.3	14.5	17.8	18.7	19.6	21.3	23.2
x7.9	6.94	0.0681	884	0.0754	3.59	9.23	10.8	14.2	17.4	18.3	19.2	20.9	22.7
x6.4	5.71	0.0560	727	0.0642	2.98	9.40	10.2	13.9	17.1	17.9	18.8	20.5	22.3
x4.8	4.40	0.0431	560	0.0514	2.33	9.58	9.69	13.6	16.7	17.5	18.4	20.1	21.9
x3.2	3.02	0.0296	384	0.0368	1.62	9.79	9.12	13.4	16.4	17.2	18.0	19.7	21.5
L25x25													
x9.5	6.17	0.0606	787	0.0400	2.54	7.13	9.65	12.0	15.4	16.3	17.2	19.0	20.9
x7.9	5.34	0.0524	681	0.0357	2.19	7.24	9.14	11.7	15.0	15.9	16.8	18.6	20.5
x6.4	4.43	0.0435	565	0.0307	1.83	7.37	8.62	11.3	14.6	15.5	16.4	18.2	20.0
x4.8	3.44	0.0337	438	0.0249	1.44	7.54	8.07	11.0	14.2	15.1	16.0	17.8	19.6
x3.2	2.38	0.0233	303	0.0181	1.01	7.73	7.52	10.8	13.9	14.7	15.6	17.3	19.1

*Designation consists of nominal leg sizes and thickness. For angle size, see Angles - Imperial Series.

Note: The properties of angles currently produced by cold forming are up to 7 percent less than the properties shown in the above tables. Check manufacturers' catalogue for the exact properties and dimensions. See also page 6-43.

TWO ANGLES UNEQUAL LEGS - Imperial Series
Long Legs Back to Back

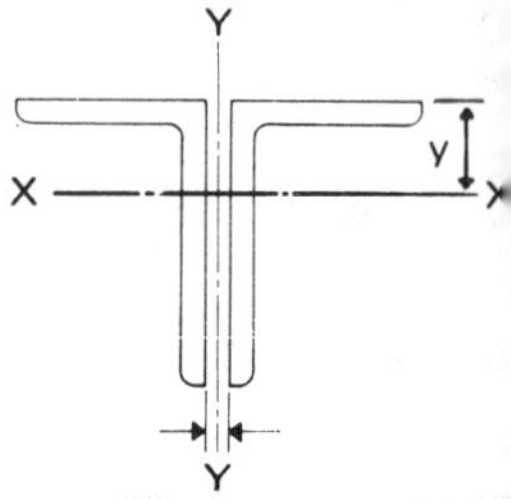

PROPERTIES OF SECTIONS

Designation*	Mass of 2 Angles	Dead Load	Area of 2 Angles	Axis X-X				Radii of Gyration about Axis Y-Y					
				I	S	r	y	back to back of angles, millimetres					
	kg/m	kN/m	mm	10^6 mm^4	10^3 mm^3	mm	mm	0	8	10	12	16	20
L203x152													
x25	131	1.29	16 700	67.0	494	63.3	67.4	60.6	63.4	64.1	64.9	66.3	67.9
x22	116	1.14	14 800	59.9	438	63.7	66.2	60.1	62.8	63.6	64.3	65.7	67.2
x19	101	0.988	12 800	52.7	382	64.1	65.1	59.6	62.3	63.0	63.7	65.2	66.7
x16	84.6	0.830	10 800	44.9	323	64.6	64.0	59.1	61.8	62.5	63.2	64.6	66.1
x14	76.5	0.750	9 740	40.9	293	64.8	63.4	58.9	61.6	62.3	63.0	64.4	65.8
x13	68.3	0.669	8 690	36.8	262	65.0	62.8	58.7	61.4	62.0	62.7	64.1	65.5
x11	59.9	0.588	7 630	32.5	231	65.3	62.2	58.5	61.1	61.8	62.5	63.9	65.3
L203x102													
x25	111	1.09	14 200	57.9	460	63.8	77.2	37.4	40.4	41.2	41.9	43.5	45.1
x19	85.7	0.841	10 900	45.7	357	64.7	74.8	36.3	39.1	39.8	40.5	42.1	43.6
x14	65.3	0.640	8 310	35.6	274	65.4	73.0	35.5	38.2	38.9	39.6	41.1	42.6
x13	58.3	0.572	7 420	32.0	245	65.7	72.4	35.3	37.9	38.6	39.3	40.7	42.2
L178x102													
x19	78.2	0.767	9 970	31.7	277	56.4	63.8	37.8	40.6	41.4	42.1	43.7	45.2
x16	65.9	0.647	8 400	27.1	235	56.8	62.6	37.3	40.0	40.8	41.5	43.0	44.5
x13	53.3	0.523	6 790	22.3	191	57.3	61.4	36.8	39.5	40.2	40.9	42.4	43.9
x9.5	40.5	0.397	5 160	17.2	146	57.8	60.2	36.4	39.0	39.7	40.4	41.8	43.3
L152x102													
x22	80.8	0.792	10 300	22.9	233	47.2	53.7	40.2	43.2	43.9	44.7	46.3	47.9
x19	70.4	0.691	8 970	20.3	204	47.6	52.5	39.7	42.6	43.3	44.1	45.6	47.2
x16	59.4	0.583	7 570	17.5	174	48.0	51.4	39.2	42.0	42.7	43.4	44.9	46.5
x14	53.8	0.528	6 860	16.0	158	48.2	50.8	38.9	41.7	42.4	43.1	44.6	46.2
x13	48.1	0.472	6 130	14.4	141	48.5	50.2	38.7	41.4	42.1	42.8	44.3	45.8
x11	42.3	0.415	5 390	12.8	125	48.7	49.6	38.5	41.1	41.9	42.6	44.0	45.5
x9.5	36.6	0.359	4 660	11.2	108	48.9	49.1	38.3	40.9	41.6	42.3	43.7	45.2
x7.9	30.7	0.301	3 910	9.44	91.2	49.2	48.5	38.1	40.7	41.3	42.0	43.5	44.9
x6.4	24.7	0.242	3 150	7.67	73.7	49.4	47.9	37.9	40.4	41.1	41.8	43.2	44.7
x4.8	18.6	0.183	2 370	5.85	55.8	49.6	47.3	37.7	40.2	40.9	41.6	42.9	44.4
L152x89													
x16	56.2	0.551	7 160	16.6	169	48.2	53.9	33.0	35.9	36.6	37.4	38.9	40.5
x13	45.5	0.446	5 800	13.7	138	48.6	52.7	32.5	35.3	36.0	36.7	38.2	39.8
x9.5	34.6	0.340	4 410	10.6	106	49.1	51.6	32.1	34.7	35.4	36.2	37.6	39.1
x7.9	29.0	0.285	3 700	9.01	89.1	49.3	51.0	31.9	34.5	35.2	35.9	37.3	38.8
x6.4	23.4	0.229	2 980	7.32	72.1	49.6	50.4	31.7	34.3	34.9	35.6	37.0	38.5
L127x89													
x19	59.0	0.579	7 520	11.6	140	39.3	44.4	35.5	38.4	39.2	40.0	41.5	43.2
x16	49.9	0.490	6 360	10.0	120	39.7	43.2	34.9	37.8	38.5	39.3	40.8	42.4
x13	40.5	0.397	5 160	8.31	97.9	40.1	42.1	34.4	37.2	37.9	38.7	40.2	41.8
x11	35.7	0.350	4 550	7.40	86.6	40.4	41.5	34.2	36.9	37.6	38.4	39.9	41.4
x9.5	30.9	0.303	3 930	6.48	75.2	40.6	40.9	33.9	36.6	37.4	38.1	39.6	41.1
x7.9	25.9	0.254	3 300	5.50	63.5	40.8	40.3	33.7	36.4	37.1	37.8	39.3	40.8
x6.4	20.9	0.205	2 660	4.48	51.4	41.0	39.7	33.5	36.2	36.8	37.5	39.0	40.5
x4.8	15.8	0.155	2 010	3.43	39.0	41.3	39.1	33.4	35.9	36.6	37.3	38.7	40.2

*Designation consists of nominal leg sizes and thickness. For angle size, see Angles - Imperial Series.

Note: The properties of angles currently produced by cold forming are up to 7 percent less than the properties shown in the above tables. Check manufacturers' catalogue for the exact properties and dimensions. See also page 6-43.

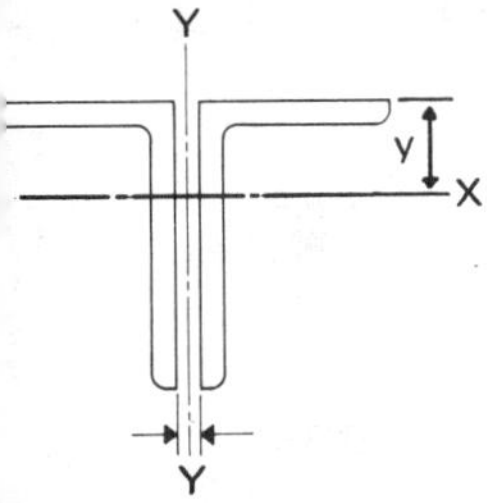

TWO ANGLES UNEQUAL LEGS - Imperial Series
Long Legs Back to Back

PROPERTIES OF SECTIONS

Designation*	Mass of 2 Angles	Dead Load	Area of 2 Angles	Axis X-X				Radii of Gyration about Axis Y-Y					
				I	S	r	y	back to back of angles, millimetres					
	kg/m	kN/m	mm	10^6 mm^4	10^3 mm^3	mm	mm	0	8	10	12	16	20
L127x76													
x16	46.8	0.459	5 960	9.48	116	39.9	45.6	28.9	31.9	32.6	33.4	35.0	36.6
x13	38.0	0.373	4 840	7.87	95.3	40.3	44.5	28.4	31.2	32.0	32.7	34.3	35.9
x11	33.5	0.328	4 260	7.01	84.4	40.6	43.9	28.2	30.9	31.7	32.4	33.9	35.5
x9.5	29.0	0.284	3 690	6.14	73.3	40.8	43.3	27.9	30.6	31.4	32.1	33.6	35.2
x7.9	24.3	0.239	3 100	5.21	61.9	41.0	42.7	27.7	30.4	31.1	31.8	33.3	34.8
x6.4	19.6	0.192	2 500	4.25	50.1	41.2	42.1	27.5	30.1	30.8	31.5	33.0	34.5
x4.8	14.8	0.145	1 890	3.25	38.0	41.5	41.5	27.3	29.9	30.6	31.3	32.7	34.2
L102x89													
x16	43.7	0.429	5 570	5.37	77.9	31.1	33.0	37.1	40.1	40.9	41.6	43.2	44.8
x13	35.5	0.349	4 530	4.48	63.9	31.5	31.9	36.6	39.5	40.2	41.0	42.6	44.1
x11	31.3	0.307	3 990	4.00	56.7	31.7	31.3	36.4	39.2	40.0	40.7	42.2	43.8
x9.5	27.1	0.266	3 460	3.52	49.4	31.9	30.8	36.1	38.9	39.7	40.4	41.9	43.5
x7.9	22.8	0.224	2 910	3.00	41.7	32.1	30.2	35.9	38.7	39.4	40.1	41.6	43.2
x6.4	18.4	0.180	2 340	2.45	33.9	32.3	29.6	35.7	38.4	39.1	39.9	41.3	42.9
x4.8	13.9	0.136	1 770	1.88	25.8	32.6	29.0	35.5	38.2	38.9	39.6	41.1	42.6
L102x76													
x16	40.5	0.397	5 160	5.09	75.9	31.4	35.0	30.9	33.9	34.6	35.4	37.0	38.7
x13	33.0	0.324	4 200	4.25	62.4	31.8	33.9	30.3	33.2	34.0	34.8	36.3	37.9
x11	29.1	0.286	3 710	3.80	55.3	32.0	33.3	30.1	32.9	33.7	34.4	36.0	37.6
x9.5	25.2	0.248	3 210	3.34	48.2	32.2	32.7	29.8	32.6	33.4	34.1	35.7	37.2
x7.9	21.2	0.208	2 700	2.85	40.7	32.4	32.1	29.6	32.4	33.1	33.8	35.4	36.9
x6.4	17.1	0.168	2 180	2.33	33.1	32.7	31.6	29.4	32.1	32.8	33.6	35.0	36.6
x4.8	13.0	0.127	1 650	1.79	25.2	32.9	31.0	29.2	31.9	32.6	33.3	34.8	36.3
L89x76													
x16	37.2	0.365	4 740	3.42	57.8	26.9	29.7	32.1	35.1	35.9	36.7	38.3	40.0
x13	30.4	0.298	3 870	2.87	47.7	27.3	28.6	31.5	34.5	35.2	36.0	37.6	39.2
x11	26.8	0.263	3 420	2.58	42.3	27.5	28.0	31.3	34.2	34.9	35.7	37.3	38.9
x9.5	23.3	0.228	2 970	2.27	36.9	27.7	27.4	31.0	33.9	34.6	35.4	36.9	38.5
x7.9	19.6	0.192	2 500	1.94	31.3	27.9	26.9	30.8	33.6	34.3	35.1	36.6	38.2
x6.4	15.8	0.155	2 020	1.59	25.4	28.1	26.3	30.6	33.3	34.1	34.8	36.3	37.9
x4.8	12.0	0.118	1 530	1.23	19.4	28.3	25.7	30.4	33.1	33.8	34.5	36.0	37.6
L89x64													
x16	34.1	0.334	4 340	3.21	56.1	27.2	31.7	25.9	29.0	29.8	30.6	32.2	33.9
x13	27.9	0.273	3 550	2.70	46.2	27.6	30.6	25.3	28.3	29.1	29.8	31.5	33.1
x11	24.6	0.242	3 140	2.42	41.1	27.8	30.0	25.0	28.0	28.7	29.5	31.1	32.8
x9.5	21.4	0.210	2 720	2.13	35.9	28.0	29.5	24.8	27.6	28.4	29.2	30.8	32.4
x7.9	18.0	0.177	2 290	1.82	30.4	28.2	28.9	24.5	27.4	28.1	28.9	30.4	32.0
x6.4	14.6	0.143	1 850	1.50	24.7	28.4	28.3	24.3	27.1	27.8	28.6	30.1	31.7
x4.8	11.0	0.108	1 410	1.15	18.9	28.7	27.7	24.1	26.8	27.5	28.3	29.8	31.3

*Designation consists of nominal leg sizes and thickness. For angle size, see Angles - Imperial Series.

Note: The properties of angles currently produced by cold forming are up to 7 percent less than the properties shown in the above tables. Check manufacturers' catalogue for the exact properties and dimensions. See also page 6-43.

TWO ANGLES UNEQUAL LEGS - Imperial Series
Long Legs Back to Back

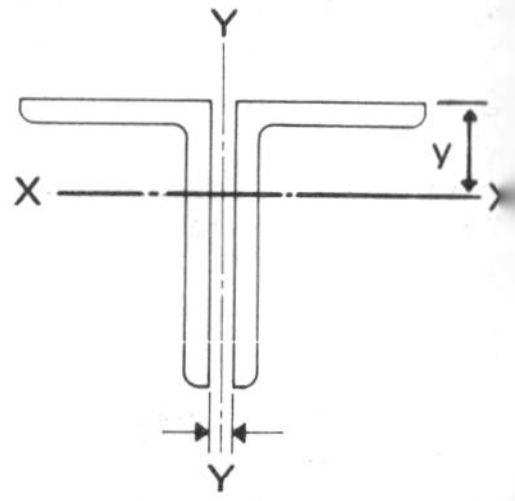

PROPERTIES OF SECTIONS

Designation*	Mass of 2 Angles	Dead Load	Area of 2 Angles	Axis X-X				Radii of Gyration about Axis Y-Y					
				I	S	r	y	back to back of angles, millimetres					
	kg/m	kN/m	mm	10^6 mm^4	10^3 mm^3	mm	mm	0	8	10	12	16	20
L76x64													
x16	30.9	0.303	3 940	2.05	41.3	22.8	26.5	27.0	30.1	30.9	31.8	33.4	35.1
x13	25.3	0.248	3 230	1.73	34.1	23.2	25.4	26.4	29.5	30.2	31.0	32.7	34.4
x11	22.4	0.220	2 850	1.56	30.4	23.4	24.8	26.2	29.1	29.9	30.7	32.3	34.0
x9.5	19.5	0.191	2 480	1.38	26.6	23.6	24.3	25.9	28.8	29.6	30.4	32.0	33.6
x7.9	16.4	0.161	2 090	1.18	22.6	23.8	23.7	25.7	28.5	29.3	30.0	31.6	33.3
x6.4	13.3	0.130	1 690	0.977	18.4	24.0	23.1	25.4	28.2	29.0	29.7	31.3	32.9
x4.8	10.1	0.0989	1 280	0.755	14.1	24.2	22.6	25.2	28.0	28.7	29.4	31.0	32.6
L76x51													
x16	27.7	0.272	3 530	1.89	39.7	23.1	28.6	21.0	24.1	25.0	25.8	27.5	29.3
x13	22.8	0.224	2 900	1.60	32.9	23.5	27.5	20.3	23.4	24.2	25.0	26.7	28.4
x11	20.2	0.198	2 570	1.44	29.3	23.7	27.0	20.0	23.0	23.8	24.6	26.3	28.0
x9.5	17.6	0.172	2 240	1.28	25.6	23.9	26.4	19.7	22.7	23.5	24.3	25.9	27.6
x7.9	14.8	0.146	1 890	1.10	21.8	24.1	25.8	19.5	22.4	23.1	23.9	25.6	27.2
x6.4	12.0	0.118	1 530	0.905	17.8	24.3	25.2	19.2	22.1	22.8	23.6	25.2	26.8
x4.8	9.14	0.0896	1 160	0.700	13.6	24.5	24.6	19.0	21.8	22.5	23.3	24.8	26.5
L64x51													
x9.5	15.7	0.154	2 000	0.760	17.9	19.5	21.1	20.8	23.8	24.6	25.4	27.1	28.8
x7.9	13.3	0.130	1 690	0.656	15.3	19.7	20.6	20.5	23.5	24.3	25.1	26.7	28.4
x6.4	10.8	0.106	1 370	0.544	12.5	19.9	20.0	20.3	23.2	23.9	24.7	26.3	28.0
x4.8	8.19	0.0803	1 040	0.423	9.60	20.1	19.4	20.1	22.9	23.6	24.4	26.0	27.6
x3.2	5.55	0.0544	707	0.293	6.57	20.4	18.8	19.9	22.6	23.3	24.1	25.7	27.3
L64x38													
x7.9	11.7	0.115	1 490	0.592	14.5	19.9	22.8	14.5	17.5	18.3	19.1	20.9	22.6
x6.4	9.50	0.0931	1 210	0.492	11.9	20.2	22.2	14.2	17.1	17.9	18.8	20.4	22.2
x4.8	7.24	0.0710	922	0.383	9.16	20.4	21.6	14.0	16.8	17.6	18.4	20.0	21.8
L51x38													
x9.5	11.9	0.116	1 510	0.361	11.0	15.4	18.0	15.7	18.9	19.7	20.6	22.3	24.1
x7.9	10.1	0.0990	1 290	0.314	9.41	15.6	17.4	15.5	18.5	19.4	20.2	21.9	23.7
x6.4	8.23	0.0807	1 050	0.263	7.74	15.8	16.9	15.2	18.2	19.0	19.8	21.5	23.3
x4.8	6.29	0.0617	801	0.206	5.97	16.0	16.3	14.9	17.9	18.6	19.5	21.1	22.8
x3.2	4.28	0.0420	545	0.144	4.11	16.3	15.7	14.7	17.5	18.3	19.1	20.7	22.4

*Designation consists of nominal leg sizes and thickness. For angle size, see Angles - Imperial Series.

Note: The properties of angles currently produced by cold forming are up to 7 percent less than the properties shown in the above tables. Check manufacturers' catalogue for the exact properties and dimensions. See also page 6-43.

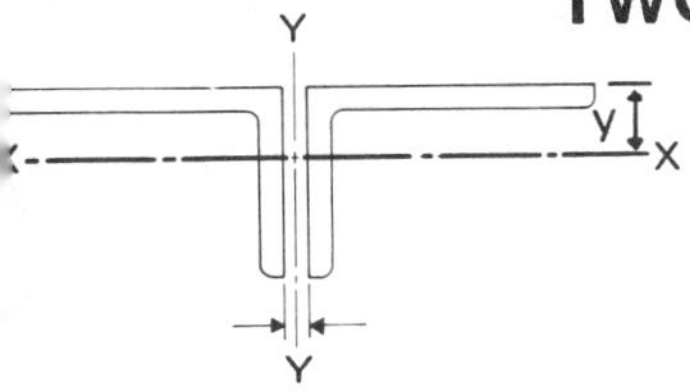

TWO ANGLES UNEQUAL LEGS - Imperial Series
Short Legs Back to Back

PROPERTIES OF SECTIONS

Designation*	Mass of 2 Angles	Dead Load	Area of 2 Angles	Axis X-X				Radii of Gyration about Axis Y-Y back to back of angles, millimetres					
				I	S	r	y	0	8	10	12	16	20
	kg/m	kN/m	mm	10^6 mm^4	10^3 mm^3	mm	mm						
L203x152													
x25	131	1.30	16 700	32.0	291	43.7	41.9	92.4	95.4	96.1	96.9	98.4	100
x22	116	1.10	14 800	28.8	259	44.1	40.7	91.9	94.8	95.6	96.3	97.8	99.3
x19	101	0.988	12 800	25.4	226	44.5	39.6	91.4	94.3	95.0	95.7	97.2	98.8
x16	84.6	0.830	10 800	21.8	192	44.9	38.5	90.9	93.7	94.5	95.2	96.7	98.2
x14	76.5	0.750	9 740	19.9	174	45.2	37.9	90.6	93.5	94.2	94.9	96.4	97.9
x13	68.3	0.669	8 690	17.9	156	45.4	37.3	90.4	93.2	93.9	94.6	96.1	97.6
x11	59.9	0.588	7 630	15.9	138	45.6	36.7	90.1	92.9	93.7	94.4	95.8	97.3
L203x102													
x25	111	1.10	14 200	9.81	130	26.3	26.7	100	103	104	105	106	108
x19	85.7	0.841	10 900	7.90	102	26.9	24.3	98.9	102	103	104	105	107
x14	65.3	0.640	8 310	6.26	78.8	27.4	22.5	98.0	101	102	103	104	106
x13	58.3	0.572	7 420	5.67	70.9	27.6	21.9	97.8	101	102	102	104	105
L178x102													
x19	78.2	0.767	9 970	7.64	100	27.7	25.8	85.1	88.1	88.9	89.7	91.3	92.8
x16	65.9	0.647	8 400	6.61	85.4	28.1	24.6	84.5	87.5	88.3	89.1	90.6	92.2
x13	53.3	0.523	6 790	5.50	69.9	28.5	23.4	84.0	86.9	87.7	88.5	90.0	91.5
x9.5	40.5	0.397	5 160	4.30	53.9	28.9	22.2	83.4	86.4	87.1	87.9	89.4	90.9
L152x102													
x22	80.8	0.792	10 300	8.20	112	28.2	28.7	71.5	74.5	75.3	76.1	77.6	79.2
x19	70.4	0.691	8 970	7.32	98.3	28.6	27.5	70.9	73.9	74.7	75.5	77.0	78.6
x16	59.4	0.583	7 570	6.34	83.8	28.9	26.4	70.3	73.3	74.1	74.8	76.4	77.9
x14	53.8	0.528	6 860	5.82	76.4	29.1	25.8	70.1	73.0	73.8	74.5	76.1	77.6
x13	48.1	0.472	6 130	5.28	68.7	29.3	25.2	69.8	72.7	73.5	74.2	75.8	77.3
x11	42.3	0.415	5 390	4.71	60.9	29.6	24.6	69.5	72.4	73.2	73.9	75.5	77.0
x9.5	36.6	0.359	4 660	4.13	53.0	29.8	24.1	69.3	72.2	72.9	73.7	75.2	76.7
x7.9	30.7	0.301	3 910	3.51	44.7	30.0	23.5	69.0	71.9	72.6	73.4	74.9	76.4
x6.4	24.7	0.242	3 150	2.87	36.3	30.2	22.9	68.8	71.6	72.4	73.1	74.6	76.1
x4.8	18.6	0.183	2 370	2.20	27.6	30.5	22.3	68.6	71.4	72.1	72.8	74.3	75.8
L152x89													
x16	56.2	0.551	7 160	4.23	63.6	24.3	22.4	72.3	75.3	76.1	76.9	78.5	80.1
x13	45.5	0.446	5 800	3.54	52.2	24.7	21.2	71.7	74.7	75.5	76.3	77.8	79.4
x9.5	34.6	0.340	4 410	2.78	40.4	25.1	20.0	71.2	74.2	74.9	75.7	77.2	78.7
x7.9	29.0	0.285	3 700	2.37	34.1	25.3	19.4	70.9	73.9	74.6	75.4	76.9	78.4
x6.4	23.4	0.229	2 980	1.94	27.7	25.5	18.8	70.7	73.6	74.3	75.1	76.6	78.1
L127x89													
x19	59.0	0.579	7 520	4.63	72.8	24.8	25.3	59.3	62.3	63.1	63.9	65.5	67.1
x16	49.9	0.490	6 360	4.03	62.2	25.2	24.2	58.7	61.7	62.5	63.2	64.8	66.4
x13	40.5	0.397	5 160	3.37	51.2	25.6	23.0	58.1	61.1	61.9	62.6	64.2	65.7
x11	35.7	0.350	4 550	3.02	45.4	25.8	22.4	57.9	60.8	61.6	62.3	63.9	65.4
x9.5	30.9	0.303	3 930	2.65	39.6	26.0	21.9	57.6	60.5	61.3	62.0	63.6	65.1
x7.9	25.9	0.254	3 300	2.26	33.5	26.2	21.3	57.4	60.3	61.0	61.7	63.3	64.8
x6.4	20.9	0.205	2 660	1.86	27.2	26.4	20.7	57.1	60.0	60.7	61.5	63.0	64.5
x4.8	15.8	0.155	2 010	1.43	20.7	26.6	20.1	56.9	59.7	60.4	61.2	62.7	64.2

*Designation consists of nominal leg sizes and thickness. For angle size, see Angles - Imperial Series.

Note: The properties of angles currently produced by cold forming are up to 7 percent less than the properties shown in the above tables. Check manufacturers' catalogue for the exact properties and dimensions. See also page 6-43.

TWO ANGLES UNEQUAL LEGS - Imperial Series
Short Legs Back to Back

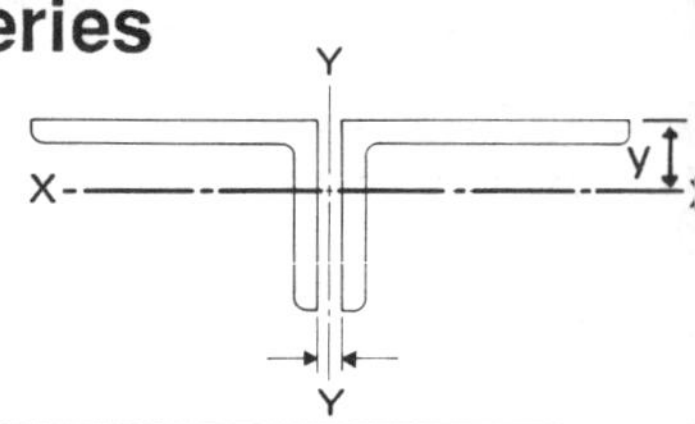

PROPERTIES OF SECTIONS

Designation*	Mass of 2 Angles	Dead Load	Area of 2 Angles	Axis X-X				Radii of Gyration about Axis Y-Y					
				I	S	r	y	back to back of angles, millimetres					
	kg/m	kN/m	mm	10^6 mm^4	10^3 mm^3	mm	mm	0	8	10	12	16	20
L127x76													
x16	46.8	0.459	5 960	2.55	45.6	20.7	20.2	60.6	63.7	64.4	65.2	66.8	68.4
x13	38.0	0.373	4 840	2.15	37.6	21.1	19.1	60.0	63.0	63.8	64.6	66.2	67.8
x11	33.5	0.328	4 260	1.93	33.4	21.3	18.5	59.7	62.7	63.5	64.3	65.8	67.4
x9.5	29.0	0.284	3 690	1.70	29.1	21.5	17.9	59.5	62.4	63.2	64.0	65.5	67.1
x7.9	24.3	0.239	3 100	1.45	24.7	21.7	17.3	59.2	62.1	62.9	63.7	65.2	66.8
x6.4	19.6	0.192	2 500	1.20	20.1	21.9	16.7	58.9	61.9	62.6	63.4	64.9	66.4
x4.8	14.8	0.145	1 890	0.923	15.4	22.1	16.1	58.7	61.6	62.3	63.1	64.6	66.1
L102x89													
x16	43.7	0.429	5 570	3.77	60.5	26.0	26.5	45.3	48.3	49.1	49.9	51.5	53.1
x13	35.5	0.349	4 530	3.16	49.7	26.4	25.4	44.8	47.7	48.5	49.3	50.8	52.4
x11	31.3	0.307	3 990	2.83	44.1	26.6	24.8	44.6	47.5	48.2	49.0	50.5	52.1
x9.5	27.1	0.266	3 460	2.49	38.5	26.8	24.2	44.3	47.2	47.9	48.7	50.2	51.8
x7.9	22.8	0.224	2 910	2.13	32.6	27.1	23.6	44.1	46.9	47.6	48.4	49.9	51.4
x6.4	18.4	0.180	2 340	1.74	26.5	27.3	23.1	43.8	46.6	47.4	48.1	49.6	51.1
x4.8	13.9	0.136	1 770	1.34	20.2	27.5	22.5	43.6	46.4	47.1	47.8	49.3	50.8
L102x76													
x16	40.5	0.397	5 160	2.40	44.3	21.6	22.1	47.0	50.1	50.9	51.6	53.2	54.9
x13	33.0	0.324	4 200	2.02	36.6	21.9	21.0	46.5	49.4	50.2	51.0	52.6	54.2
x11	29.1	0.286	3 710	1.81	32.5	22.1	20.4	46.2	49.1	49.9	50.7	52.2	53.8
x9.5	25.2	0.248	3 210	1.60	28.4	22.3	19.8	45.9	48.9	49.6	50.4	51.9	53.5
x7.9	21.2	0.208	2 700	1.37	24.1	22.5	19.2	45.7	48.6	49.3	50.1	51.6	53.2
x6.4	17.1	0.168	2 180	1.13	19.6	22.7	18.7	45.4	48.3	49.0	49.8	51.3	52.9
x4.8	13.0	0.127	1 650	0.872	15.0	23.0	18.1	45.2	48.0	48.8	49.5	51.0	52.6
L89x76													
x16	37.2	0.365	4 740	2.30	43.5	22.0	23.3	40.0	43.1	43.9	44.7	46.3	47.9
x13	30.4	0.298	3 870	1.94	35.9	22.4	22.2	39.5	42.5	43.2	44.0	45.6	47.2
x11	26.8	0.263	3 420	1.74	31.9	22.6	21.7	39.2	42.2	42.9	43.7	45.3	46.9
x9.5	23.3	0.228	2 970	1.54	27.9	22.8	21.1	39.0	41.9	42.6	43.4	45.0	46.6
x7.9	19.6	0.192	2 500	1.32	23.7	23.0	20.5	38.7	41.6	42.3	43.1	44.6	46.2
x6.4	15.8	0.155	2 020	1.09	19.3	23.2	19.9	38.5	41.3	42.1	42.8	44.3	45.9
x4.8	12.0	0.118	1 530	0.838	14.7	23.4	19.4	38.3	41.1	41.8	42.5	44.0	45.6
L89x64													
x16	34.1	0.334	4 340	1.34	30.1	17.6	19.0	41.8	44.9	45.7	46.5	48.1	49.8
x13	27.9	0.273	3 550	1.14	24.9	17.9	17.9	41.2	44.2	45.0	45.8	47.4	49.1
x11	24.6	0.242	3 140	1.02	22.2	18.1	17.3	40.9	43.9	44.7	45.5	47.1	48.7
x9.5	21.4	0.210	2 720	0.908	19.4	18.3	16.8	40.6	43.6	44.4	45.2	46.8	48.4
x7.9	18.0	0.177	2 290	0.782	16.5	18.5	16.2	40.4	43.3	44.1	44.9	46.4	48.0
x6.4	14.6	0.143	1 850	0.647	13.5	18.7	15.6	40.1	43.0	43.8	44.5	46.1	47.7
x4.8	11.0	0.108	1 410	0.502	10.3	18.9	15.0	39.9	42.7	43.5	44.2	45.8	47.4

*Designation consists of nominal leg sizes and thickness. For angle size, see Angles - Imperial Series.

Note: The properties of angles currently produced by cold forming are up to 7 percent less than the properties shown in the above tables. Check manufacturers' catalogue for the exact properties and dimensions. See also page 6-43.

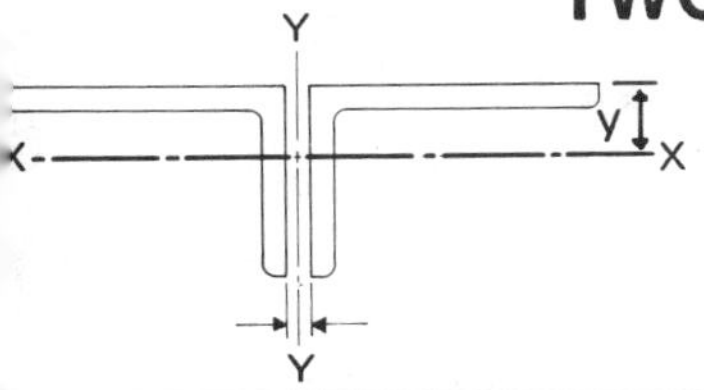

TWO ANGLES UNEQUAL LEGS - Imperial Series
Short Legs Back to Back

PROPERTIES OF SECTIONS

Designation*	Mass of 2 Angles	Dead Load	Area of 2 Angles	Axis X-X				Radii of Gyration about Axis Y-Y					
				I	S	r	y	back to back of angles, millimetres					
	kg/m	kN/m	mm	10^6 mm^4	10^3 mm^3	mm	mm	0	8	10	12	16	20
L76x64													
x16	30.9	0.303	3 940	1.28	29.4	18.0	20.2	35.0	38.1	38.9	39.7	41.4	43.1
x13	25.3	0.248	3 230	1.08	24.4	18.3	19.1	34.4	37.4	38.2	39.0	40.7	42.3
x11	22.4	0.220	2 850	0.978	21.7	18.5	18.5	34.1	37.1	37.9	38.7	40.3	42.0
x9.5	19.5	0.191	2 480	0.868	19.0	18.7	17.9	33.8	36.8	37.6	38.4	40.0	41.6
x7.9	16.4	0.161	2 090	0.748	16.2	18.9	17.4	33.6	36.5	37.3	38.1	39.6	41.3
x6.4	13.3	0.130	1 690	0.619	13.2	19.1	16.8	33.3	36.2	37.0	37.8	39.3	40.9
x4.8	10.1	0.0989	1 280	0.480	10.2	19.3	16.2	33.1	36.0	36.7	37.5	39.0	40.6
L76x51													
x16	27.7	0.272	3 530	0.655	18.8	13.6	15.9	36.8	40.0	40.8	41.6	43.3	45.0
x13	22.8	0.224	2 900	0.559	15.5	13.9	14.8	36.2	39.3	40.1	40.9	42.6	44.3
x11	20.2	0.198	2 570	0.507	13.9	14.0	14.3	35.9	39.0	39.8	40.6	42.2	43.9
x9.5	17.6	0.172	2 240	0.452	12.2	14.2	13.7	35.6	38.6	39.4	40.2	41.9	43.5
x7.9	14.8	0.146	1 890	0.392	10.4	14.4	13.1	35.3	38.3	39.1	39.9	41.5	43.2
x6.4	12.0	0.118	1 530	0.326	8.52	14.6	12.5	35.0	38.0	38.8	39.6	41.2	42.8
x4.8	9.14	0.0896	1 160	0.255	6.56	14.8	11.9	34.8	37.7	38.5	39.3	40.8	42.5
L64x51													
x9.5	15.7	0.154	2 000	0.428	11.9	14.6	14.8	28.7	31.8	32.6	33.4	35.0	36.7
x7.9	13.3	0.130	1 690	0.372	10.2	14.8	14.2	28.5	31.5	32.3	33.1	34.7	36.4
x6.4	10.8	0.106	1 370	0.310	8.34	15.0	13.6	28.2	31.2	32.0	32.7	34.3	36.0
x4.8	8.19	0.0803	1 040	0.242	6.42	15.2	13.1	28.0	30.9	31.6	32.4	34.0	35.6
x3.2	5.55	0.0544	707	0.169	4.42	15.5	12.5	27.7	30.6	31.3	32.1	33.7	35.3
L64x38													
x7.9	11.7	0.115	1 490	0.159	5.70	10.4	10.1	30.3	33.4	34.2	35.0	36.7	38.4
x6.4	9.50	0.0931	1 210	0.134	4.69	10.5	9.53	30.0	33.1	33.9	34.7	36.3	38.0
x4.8	7.24	0.0710	922	0.106	3.64	10.7	8.94	29.7	32.8	33.5	34.3	36.0	37.6
L51x38													
x9.5	11.9	0.116	1 510	0.171	6.45	10.6	11.6	23.7	26.9	27.7	28.5	30.2	32.0
x7.9	10.1	0.0990	1 290	0.150	5.53	10.8	11.1	23.4	26.5	27.3	28.2	29.8	31.6
x6.4	8.23	0.0807	1 050	0.126	4.57	11.0	10.5	23.1	26.2	27.0	27.8	29.5	31.2
x4.8	6.29	0.0617	801	0.0999	3.54	11.2	9.93	22.9	25.9	26.6	27.5	29.1	30.8
x3.2	4.28	0.0420	545	0.0706	2.46	11.4	9.35	22.6	25.5	26.3	27.1	28.7	30.4

*Designation consists of nominal leg sizes and thickness. For angle size, see Angles - Imperial Series.

Note: The properties of angles currently produced by cold forming are up to 7 percent less than the properties shown in the above tables. Check manufacturers' catalogue for the exact properties and dimensions. See also page 6-43.

TWO ANGLES EQUAL LEGS — Metric Series
Back to Back

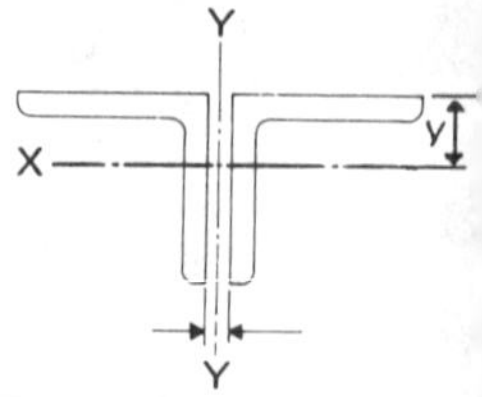

PROPERTIES OF SECTIONS

Size**	Thickness t	Mass of 2 Angles	Dead Load	Area of 2 Angles	Axis X-X				Radii of Gyration about Axis Y-Y back to back of angles, millimetres					
					I	S	r	y	0	8	10	12	16	20
mmxmm	mm	kg/m	kN/m	mm²	10⁶mm⁴	10³mm³	mm	mm						
200X200														
	30	174	1.71	22 200	80.6	580	60.3	60.9	85.7	88.6	89.3	90.1	91.6	93.1
	25	147	1.44	18 800	69.5	494	60.9	59.2	84.9	87.7	88.5	89.2	90.7	92.2
	20	119	1.17	15 200	57.6	404	61.6	57.4	84.1	86.9	87.6	88.3	89.8	91.3
	16	96.5	0.946	12 300	47.4	329	62.1	55.9	83.6	86.3	87.0	87.7	89.1	90.6
	13	79.0	0.775	10 100	39.4	271	62.6	54.8	83.2	85.9	86.6	87.2	88.7	90.1
	10	61.2	0.600	7 800	31.0	212	63.0	53.7	82.8	85.4	86.1	86.8	88.2	89.6
150X150														
	20	87.9	0.862	11 200	23.2	221	45.5	44.8	63.9	66.7	67.5	68.2	69.7	71.2
	16	71.3	0.700	9 090	19.3	181	46.0	43.4	63.3	66.1	66.8	67.5	69.0	70.5
	13	58.6	0.574	7 460	16.1	149	46.4	42.3	62.8	65.6	66.3	67.0	68.5	69.9
	10	45.5	0.446	5 800	12.7	117	46.9	41.2	62.4	65.1	65.8	66.5	68.0	69.4
125X125														
	16	58.8	0.576	7 490	10.8	123	38.0	37.1	53.1	56.0	56.7	57.5	59.0	60.5
	13	48.4	0.474	6 160	9.09	102	38.4	36.0	52.7	55.5	56.2	56.9	58.4	60.0
	10	37.7	0.370	4 800	7.23	80.3	38.8	34.9	52.2	55.0	55.7	56.4	57.9	59.4
	8	30.4	0.298	3 870	5.92	65.2	39.1	34.2	52.0	54.7	55.4	56.1	57.6	59.0
100X100														
	16	46.2	0.453	5 890	5.30	76.6	30.0	30.8	43.0	46.0	46.7	47.5	49.1	50.7
	13	38.2	0.374	4 860	4.49	63.9	30.4	29.8	42.5	45.4	46.2	46.9	48.5	50.0
	10	29.8	0.293	3 800	3.60	50.5	30.8	28.7	42.1	44.9	45.6	46.4	47.9	49.4
	8	24.1	0.236	3 070	2.96	41.1	31.1	28.0	41.8	44.6	45.3	46.0	47.5	49.0
	6	18.3	0.179	2 330	2.29	31.4	31.3	27.2	41.5	44.2	45.0	45.7	47.2	48.7
90X 90														
	13	34.1	0.334	4 340	3.21	51.1	27.2	27.2	38.5	41.4	42.2	42.9	44.5	46.1
	10	26.7	0.262	3 400	2.58	40.5	27.6	26.2	38.0	40.9	41.6	42.4	43.9	45.5
	8	21.6	0.212	2 750	2.13	33.0	27.8	25.5	37.7	40.5	41.3	42.0	43.5	45.1
	6	16.4	0.161	2 090	1.65	25.3	28.1	24.7	37.4	40.2	40.9	41.7	43.2	44.7
75X 75														
	13	28.0	0.274	3 560	1.78	34.6	22.4	23.5	32.4	35.4	36.2	37.0	38.6	40.3
	10	22.0	0.216	2 800	1.45	27.6	22.8	22.4	31.9	34.9	35.6	36.4	38.0	39.6
	8	17.8	0.175	2 270	1.20	22.6	23.0	21.7	31.6	34.5	35.2	36.0	37.6	39.2
	6	13.6	0.133	1 730	0.938	17.4	23.3	21.0	31.3	34.1	34.9	35.6	37.2	38.8
	5	11.4	0.112	1 450	0.797	14.6	23.4	20.6	31.2	34.0	34.7	35.5	37.0	38.5

**For angle designations, see page 6-19. See page 6-43 for more information on these tables.

Note: The properties of angles currently produced by cold forming are up to 7 percent less than the properties shown in the above tables. Check manufacturers' catalogue for the exact properties and dimensions.

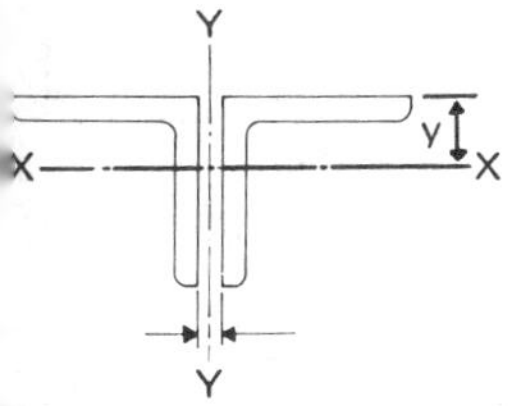

TWO ANGLES EQUAL LEGS — Metric Series
Back to Back

PROPERTIES OF SECTIONS

Size**	Thickness t	Mass of 2 Angles	Dead Load	Area of 2 Angles	Axis X-X				Radii of Gyration about Axis Y-Y back to back of angles, millimetres					
					I	S	r	y						
mmxmm	mm	kg/m	kN/m	mm²	$10^6 mm^4$	$10^3 mm^3$	mm	mm	0	8	10	12	16	20
65X 65														
	10	18.8	0.185	2 400	0.917	20.3	19.6	19.9	27.9	30.9	31.7	32.4	34.1	35.7
	8	15.3	0.150	1 950	0.766	16.7	19.8	19.2	27.6	30.5	31.3	32.0	33.6	35.3
	6	11.7	0.115	1 490	0.600	12.9	20.1	18.5	27.3	30.1	30.9	31.6	33.2	34.8
	5	9.81	0.096	1 250	0.511	10.9	20.2	18.1	27.1	30.0	30.7	31.5	33.0	34.6
55X 55														
	10	15.7	0.154	2 000	0.535	14.2	16.4	17.4	23.9	26.9	27.7	28.5	30.2	31.9
	8	12.8	0.126	1 630	0.450	11.7	16.6	16.7	23.5	26.5	27.3	28.1	29.7	31.4
	6	9.80	0.096	1 250	0.355	9.09	16.9	16.0	23.2	26.1	26.9	27.7	29.3	31.0
	5	8.24	0.081	1 050	0.303	7.70	17.0	15.6	23.1	25.9	26.7	27.5	29.1	30.7
	4	6.66	0.065	848	0.249	6.26	17.1	15.2	22.9	25.8	26.5	27.3	28.9	30.5
	3	5.04	0.049	642	0.192	4.78	17.3	14.9	22.8	25.6	26.3	27.1	28.7	30.3
45X 45														
	8	10.3	0.101	1 310	0.236	7.65	13.4	14.2	19.5	22.6	23.4	24.2	25.9	27.6
	6	7.91	0.078	1 010	0.188	5.95	13.7	13.4	19.2	22.2	22.9	23.8	25.4	27.1
	5	6.67	0.065	850	0.161	5.06	13.8	13.1	19.0	22.0	22.7	23.5	25.2	26.9
	4	5.40	0.053	688	0.133	4.13	13.9	12.7	18.9	21.8	22.5	23.3	25.0	26.7
	3	4.10	0.040	522	0.103	3.16	14.1	12.4	18.7	21.6	22.3	23.1	24.7	26.4
35X 35														
	6	6.03	0.059	768	0.084	3.49	10.5	10.9	15.1	18.2	19.1	19.9	21.6	23.4
	5	5.10	0.050	650	0.073	2.98	10.6	10.6	15.0	18.0	18.8	19.7	21.4	23.1
	4	4.14	0.041	528	0.060	2.44	10.7	10.2	14.8	17.8	18.6	19.4	21.1	22.9
	3	3.16	0.031	402	0.047	1.88	10.8	9.86	14.7	17.6	18.4	19.2	20.9	22.6
25X 25														
	5	3.53	0.035	450	0.024	1.45	7.39	8.06	10.9	14.1	15.0	15.9	17.7	19.5
	4	2.89	0.028	368	0.021	1.20	7.50	7.71	10.8	13.9	14.8	15.6	17.4	19.2
	3	2.21	0.022	282	0.016	0.930	7.63	7.35	10.6	13.7	14.5	15.4	17.1	19.0

**For angle designations, see page 6-19. See page 6-43 for more information on these tables.

Note: The properties of angles currently produced by cold forming are up to 7 percent less than the properties shown in the above tables. Check manufacturers' catalogue for the exact properties and dimensions.

TWO ANGLES UNEQUAL LEGS — Metric Series
Long Legs Back to Back

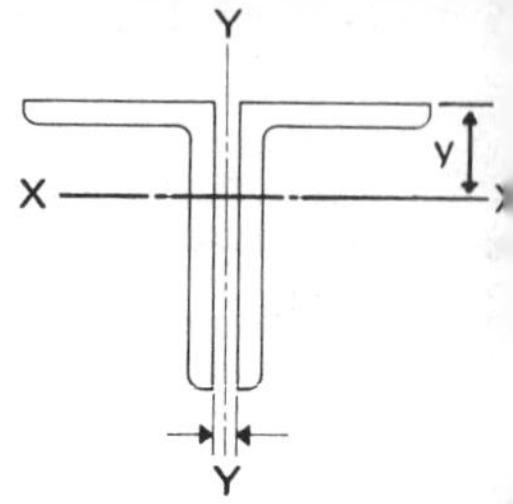

PROPERTIES OF SECTIONS

Size**	Thick-ness t	Mass of 2 Angles	Dead Load	Area of 2 Angles	Axis X-X				Radii of Gyration about Axis Y-Y					
					I	S	r	y	back to back of angles, millimetres					
mmxmm	mm	kg/m	kN/m	mm²	10⁶mm⁴	10³mm³	mm	mm	0	8	10	12	16	20
200X150														
	25	128	1.25	16 200	63.1	472	62.3	66.3	59.8	62.6	63.3	64.1	65.6	67.1
	20	104	1.02	13 200	52.4	387	63.0	64.5	59.0	61.8	62.5	63.2	64.6	66.1
	16	83.9	0.823	10 700	43.2	315	63.5	63.1	58.4	61.1	61.8	62.5	63.9	65.4
	13	68.8	0.675	8 760	35.9	260	64.0	62.0	58.0	60.7	61.4	62.0	63.4	64.9
200X100														
	20	87.9	0.862	11 200	45.3	360	63.6	74.3	35.7	38.6	39.3	40.0	41.6	43.1
	16	71.3	0.700	9 090	37.4	294	64.2	72.8	35.1	37.8	38.5	39.2	40.7	42.2
	13	58.6	0.574	7 460	31.1	243	64.6	71.7	34.6	37.2	37.9	38.7	40.1	41.6
	10	45.5	0.446	5 800	24.6	190	65.1	70.5	34.2	36.8	37.4	38.1	39.5	41.0
150X100														
	16	58.8	0.576	7 490	16.8	170	47.4	50.9	38.4	41.2	41.9	42.7	44.2	45.7
	13	48.4	0.474	6 160	14.1	140	47.8	49.9	37.9	40.7	41.4	42.1	43.6	45.1
	10	37.7	0.370	4 800	11.2	110	48.2	48.8	37.5	40.2	40.9	41.6	43.0	44.5
	8	30.4	0.298	3 870	9.11	89.3	48.5	48.0	37.3	39.9	40.6	41.3	42.7	44.2
125X 90														
	16	50.0	0.490	6 370	9.68	117	39.0	42.2	35.6	38.5	39.2	40.0	41.6	43.2
	13	41.2	0.404	5 250	8.15	97.1	39.4	41.2	35.1	37.9	38.7	39.4	40.9	42.5
	10	32.2	0.316	4 100	6.49	76.5	39.8	40.1	34.7	37.4	38.1	38.9	40.4	41.9
	8	26.0	0.255	3 310	5.32	62.1	40.1	39.3	34.4	37.1	37.8	38.5	40.0	41.5
125X 75														
	13	38.2	0.374	4 860	7.63	94.2	39.6	43.9	28.0	30.9	31.6	32.4	34.0	35.6
	10	29.8	0.293	3 800	6.09	74.2	40.0	42.8	27.6	30.3	31.0	31.8	33.3	34.9
	8	24.1	0.236	3 070	5.00	60.3	40.3	42.1	27.3	30.0	30.7	31.4	32.9	34.5
	6	18.3	0.179	2 330	3.84	46.0	40.6	41.3	27.1	29.6	30.3	31.1	32.5	34.0
100X 90														
	13	36.1	0.354	4 600	4.34	62.9	30.7	31.1	37.4	40.3	41.1	41.8	43.4	45.0
	10	28.3	0.277	3 600	3.48	49.7	31.1	30.0	37.0	39.8	40.5	41.3	42.8	44.3
	8	22.9	0.224	2 910	2.87	40.5	31.4	29.3	36.7	39.4	40.2	40.9	42.4	43.9
	6	17.3	0.170	2 210	2.21	31.0	31.7	28.5	36.4	39.1	39.8	40.6	42.0	43.6

**For angle designations, see page 6-19. See page 6-43 for more information on these tables.

Note: The properties of angles currently produced by cold forming are up to 7 percent less than the properties shown in the above tables. Check manufacturers' catalogue for the exact properties and dimensions.

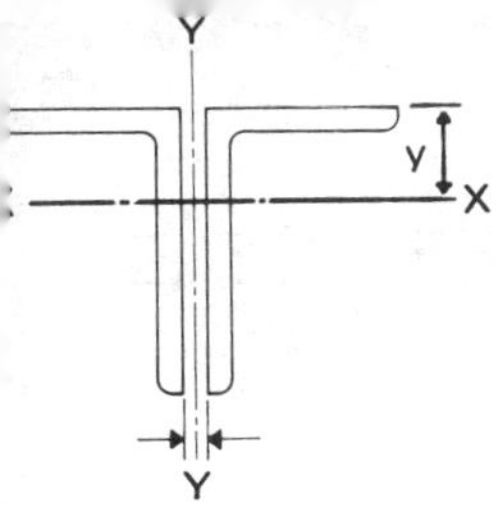

TWO ANGLES UNEQUAL LEGS — Metric Series
Long Legs Back to Back

PROPERTIES OF SECTIONS

Size**	Thickness t	Mass of 2 Angles	Dead Load	Area of 2 Angles	Axis X-X				Radii of Gyration about Axis Y-Y back to back of angles, millimetres					
					I	S	r	y						
mmxmm	mm	kg/m	kN/m	mm²	10⁶mm⁴	10³mm³	mm	mm	0	8	10	12	16	20
100X 75														
	13	33.1	0.324	4 210	4.07	61.1	31.1	33.4	30.0	32.9	33.6	34.4	36.0	37.6
	10	25.9	0.254	3 300	3.27	48.3	31.5	32.3	29.5	32.3	33.1	33.8	35.4	37.0
	8	21.0	0.206	2 670	2.70	39.4	31.8	31.5	29.2	32.0	32.7	33.4	35.0	36.5
	6	15.9	0.156	2 030	2.08	30.1	32.1	30.8	29.0	31.6	32.4	33.1	34.6	36.1
90X 75														
	13	31.0	0.304	3 950	3.02	49.7	27.6	29.3	30.9	33.8	34.6	35.4	37.0	38.6
	10	24.3	0.239	3 100	2.43	39.4	28.0	28.2	30.4	33.3	34.0	34.8	36.3	37.9
	8	19.7	0.193	2 510	2.01	32.2	28.3	27.5	30.1	32.9	33.6	34.4	35.9	37.5
	6	15.0	0.147	1 910	1.56	24.6	28.6	26.8	29.8	32.6	33.3	34.0	35.5	37.1
	5	12.6	0.123	1 600	1.32	20.8	28.7	26.4	29.7	32.4	33.1	33.9	35.4	36.9
90X 65														
	10	22.8	0.223	2 900	2.32	38.5	28.3	29.8	25.5	28.4	29.1	29.9	31.5	33.1
	8	18.5	0.181	2 350	1.92	31.5	28.5	29.1	25.2	28.0	28.7	29.5	31.1	32.7
	6	14.0	0.138	1 790	1.49	24.1	28.8	28.4	24.9	27.6	28.4	29.1	30.6	32.2
	5	11.8	0.115	1 500	1.26	20.3	29.0	28.0	24.8	27.5	28.2	28.9	30.4	32.0
80X 60														
	10	20.4	0.200	2 600	1.62	30.2	24.9	26.5	23.9	26.8	27.6	28.4	30.0	31.7
	8	16.6	0.163	2 110	1.34	24.7	25.2	25.8	23.6	26.5	27.2	28.0	29.6	31.2
	6	12.6	0.124	1 610	1.04	19.0	25.5	25.1	23.3	26.1	26.8	27.6	29.1	30.8
	5	10.6	0.104	1 350	0.886	16.0	25.6	24.7	23.2	25.9	26.6	27.4	28.9	30.5
75X 50														
	8	14.7	0.144	1 870	1.05	21.2	23.7	25.5	19.2	22.1	22.9	23.7	25.3	27.0
	6	11.2	0.110	1 430	0.820	16.3	24.0	24.7	18.9	21.7	22.5	23.2	24.8	26.5
	5	9.42	0.092	1 200	0.697	13.8	24.1	24.4	18.8	21.5	22.3	23.0	24.6	26.3
65X 50														
	8	13.4	0.132	1 710	0.701	16.1	20.2	21.3	20.0	23.0	23.7	24.5	26.2	27.9
	6	10.3	0.101	1 310	0.550	12.4	20.5	20.6	19.7	22.6	23.3	24.1	25.7	27.4
	5	8.63	0.085	1 100	0.469	10.5	20.7	20.2	19.6	22.4	23.1	23.9	25.5	27.2
	4	6.97	0.068	888	0.384	8.51	20.8	19.9	19.5	22.2	23.0	23.7	25.3	26.9
55X 35														
	6	7.91	0.078	1 010	0.304	8.46	17.4	19.0	13.3	16.3	17.1	17.9	19.6	21.4
	5	6.67	0.065	850	0.261	7.17	17.5	18.7	13.2	16.1	16.9	17.7	19.4	21.1
	4	5.40	0.053	688	0.214	5.84	17.7	18.3	13.0	15.9	16.7	17.5	19.1	20.9
	3	4.10	0.040	522	0.165	4.46	17.8	17.9	12.9	15.7	16.4	17.2	18.9	20.6
45X 30														
	6	6.50	0.064	828	0.163	5.58	14.0	15.7	11.7	14.8	15.6	16.5	18.2	20.0
	5	5.49	0.054	700	0.141	4.75	14.2	15.4	11.5	14.6	15.4	16.2	18.0	19.8
	4	4.46	0.044	568	0.116	3.88	14.3	15.0	11.4	14.3	15.2	16.0	17.7	19.5
	3	3.39	0.033	432	0.090	2.97	14.5	14.6	11.3	14.1	14.9	15.8	17.5	19.2

**For angle designations, see page 6-19. See page 6-43 for more information on these tables.

Note: The properties of angles currently produced by cold forming are up to 7 percent less than the properties shown in the above tables. Check manufacturers' catalogue for the exact properties and dimensions.

TWO ANGLES UNEQUAL LEGS — Metric Series
Short Legs Back to Back

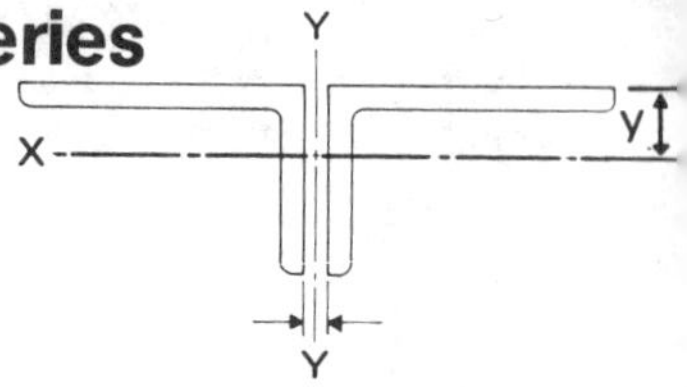

PROPERTIES OF SECTIONS

Size**	Thickness t	Mass of 2 Angles	Dead Load	Area of 2 Angles	Axis X-X I	Axis X-X S	Axis X-X r	Axis X-X y	Radii of Gyration about Axis Y-Y, back to back of angles, millimetres 0	8	10	12	16	20
mmxmm	mm	kg/m	kN/m	mm^2	10^6mm^4	10^3mm^3	mm	mm						
200X150*														
	25	128	1.25	16 200	30.3	279	43.2	41.3	91.0	94.0	94.7	95.5	97.0	98.6
	20	104	1.02	13 200	25.3	229	43.8	39.5	90.2	93.1	93.8	94.6	96.1	97.6
	16	83.9	0.823	10 700	21.0	188	44.3	38.1	89.5	92.4	93.1	93.9	95.4	96.9
	13	68.8	0.675	8 760	17.5	155	44.7	37.0	89.1	91.9	92.6	93.4	94.8	96.3
200X100*														
	20	87.9	0.862	11 200	7.69	102	26.2	24.3	97.8	101	102	102	104	106
	16	71.3	0.700	9 090	6.45	83.5	26.6	22.8	97.0	100	101	102	103	105
	13	58.6	0.574	7 460	5.44	69.4	27.0	21.7	96.5	99.5	100	101	103	104
	10	45.5	0.446	5 800	4.35	54.8	27.4	20.5	95.9	98.9	99.7	100	102	104
150X100														
	16	58.8	0.576	7 490	5.99	80.9	28.3	25.9	69.6	72.5	73.3	74.1	75.6	77.2
	13	48.4	0.474	6 160	5.06	67.3	28.7	24.9	69.0	72.0	72.7	73.5	75.0	76.6
	10	37.7	0.370	4 800	4.05	53.1	29.1	23.8	68.6	71.5	72.2	72.9	74.5	76.0
	8	30.4	0.298	3 870	3.33	43.3	29.3	23.0	68.2	71.1	71.8	72.6	74.1	75.6
125X 90														
	16	50.0	0.490	6 370	4.18	64.0	25.6	24.7	57.5	60.5	61.2	62.0	63.6	65.2
	13	41.2	0.404	5 250	3.54	53.4	26.0	23.7	57.0	59.9	60.7	61.4	63.0	64.6
	10	32.2	0.316	4 100	2.85	42.3	26.4	22.6	56.5	59.4	60.1	60.9	62.4	64.0
	8	26.0	0.255	3 310	2.35	34.5	26.6	21.8	56.2	59.0	59.8	60.5	62.0	63.6
125X 75														
	13	38.2	0.374	4 860	2.08	37.1	20.7	18.9	59.2	62.2	63.0	63.7	65.3	66.9
	10	29.8	0.293	3 800	1.68	29.4	21.0	17.8	58.6	61.6	62.4	63.1	64.7	66.3
	8	24.1	0.236	3 070	1.39	24.1	21.3	17.1	58.3	61.2	62.0	62.8	64.3	65.9
	6	18.3	0.179	2 330	1.08	18.5	21.6	16.3	58.0	60.9	61.6	62.4	63.9	65.5
100X 90														
	13	36.1	0.354	4 600	3.32	51.9	26.8	26.1	43.7	46.6	47.4	48.1	49.7	51.3
	10	28.3	0.277	3 600	2.67	41.1	27.2	25.0	43.2	46.1	46.8	47.6	49.1	50.7
	8	22.9	0.224	2 910	2.20	33.5	27.5	24.3	42.9	45.7	46.5	47.2	48.7	50.3
	6	17.3	0.170	2 210	1.71	25.7	27.8	23.5	42.6	45.4	46.1	46.9	48.4	49.9

**For angle designations, see page 6-19. See page 6-43 for more information on these tables.

Note: The properties of angles currently produced by cold forming are up to 7 percent less than the properties shown in the above tables. Check manufacturers' catalogue for the exact properties and dimensions.

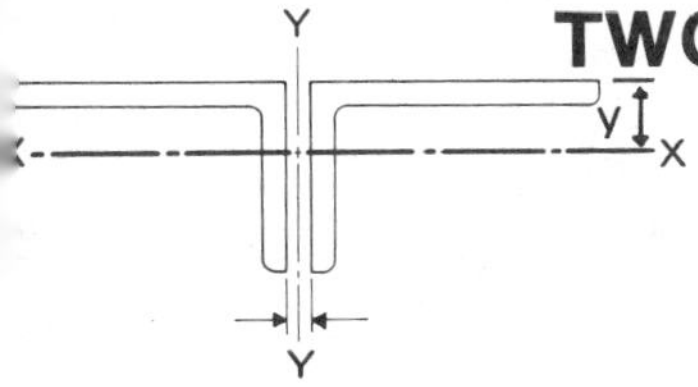

TWO ANGLES UNEQUAL LEGS — Metric Series
Short Legs Back to Back

PROPERTIES OF SECTIONS

Size**	Thick-ness t	Mass of 2 Angles	Dead Load	Area of 2 Angles	Axis X-X				Radii of Gyration about Axis Y-Y back to back of angles, millimetres					
					I	S	r	y	0	8	10	12	16	20
mmxmm	mm	kg/m	kN/m	mm²	10⁶mm⁴	10³mm³	mm	mm						
100X 75	13	33.1	0.324	4 210	1.95	36.1	21.5	20.9	45.6	48.6	49.4	50.2	51.7	53.3
	10	25.9	0.254	3 300	1.58	28.7	21.9	19.8	45.1	48.0	48.8	49.6	51.1	52.7
	8	21.0	0.206	2 670	1.31	23.5	22.2	19.0	44.8	47.7	48.4	49.2	50.7	52.3
	6	15.9	0.156	2 030	1.02	18.0	22.4	18.3	44.5	47.3	48.1	48.8	50.3	51.9
90X 75	13	31.0	0.304	3 950	1.89	35.6	21.9	21.8	40.3	43.3	44.0	44.8	46.4	48.0
	10	24.3	0.239	3 100	1.53	28.3	22.2	20.7	39.8	42.7	43.5	44.2	45.8	47.4
	8	19.7	0.193	2 510	1.27	23.1	22.5	20.0	39.5	42.3	43.1	43.9	45.4	47.0
	6	15.0	0.147	1 910	0.991	17.8	22.8	19.3	39.2	42.0	42.7	43.5	45.0	46.6
	5	12.6	0.123	1 600	0.841	15.0	22.9	18.9	39.0	41.8	42.6	43.3	44.8	46.4
90X 65	10	22.8	0.223	2 900	1.01	21.3	18.7	17.3	41.1	44.1	44.9	45.6	47.2	48.8
	8	18.5	0.181	2 350	0.844	17.4	18.9	16.6	40.8	43.7	44.5	45.2	46.8	48.4
	6	14.0	0.138	1 790	0.660	13.4	19.2	15.9	40.4	43.3	44.1	44.9	46.4	48.0
	5	11.8	0.115	1 500	0.562	11.4	19.4	15.5	40.3	43.2	43.9	44.7	46.2	47.8
80X 60	10	20.4	0.200	2 600	0.776	17.8	17.3	16.5	36.4	39.4	40.2	41.0	42.6	44.2
	8	16.6	0.163	2 110	0.648	14.7	17.5	15.8	36.1	39.0	39.8	40.6	42.2	43.8
	6	12.6	0.124	1 610	0.509	11.3	17.8	15.1	35.8	38.7	39.4	40.2	41.8	43.4
	5	10.6	0.104	1 350	0.434	9.58	17.9	14.7	35.6	38.5	39.2	40.0	41.6	43.2
75X 50	8	14.7	0.144	1 870	0.374	10.1	14.1	13.0	34.8	37.8	38.6	39.4	41.0	42.7
	6	11.2	0.110	1 430	0.296	7.84	14.4	12.2	34.4	37.4	38.2	39.0	40.6	42.2
	5	9.42	0.092	1 200	0.253	6.64	14.5	11.9	34.3	37.2	38.0	38.8	40.4	42.0
65X 50	8	13.4	0.132	1 710	0.359	9.93	14.5	13.8	29.4	32.4	33.2	34.0	35.6	37.3
	6	10.3	0.101	1 310	0.284	7.70	14.7	13.1	29.1	32.0	32.8	33.6	35.2	36.8
	5	8.63	0.085	1 100	0.243	6.53	14.9	12.7	28.9	31.8	32.6	33.4	35.0	36.6
	4	6.97	0.068	888	0.200	5.32	15.0	12.4	28.8	31.7	32.4	33.2	34.8	36.4
55X 35	6	7.91	0.078	1 010	0.096	3.70	9.77	9.04	25.8	28.9	29.7	30.5	32.1	33.8
	5	6.67	0.065	850	0.083	3.16	9.89	8.68	25.6	28.6	29.4	30.3	31.9	33.6
	4	5.40	0.053	688	0.069	2.59	10.0	8.31	25.4	28.4	29.2	30.0	31.7	33.4
	3	4.10	0.040	522	0.054	1.99	10.2	7.94	25.3	28.2	29.0	29.8	31.5	33.1
45X 30	6	6.50	0.064	828	0.058	2.65	8.35	8.22	21.1	24.2	25.0	25.9	27.6	29.3
	5	5.49	0.054	700	0.050	2.26	8.46	7.86	20.9	24.0	24.8	25.6	27.3	29.1
	4	4.46	0.044	568	0.042	1.86	8.58	7.49	20.7	23.8	24.6	25.4	27.1	28.8
	3	3.39	0.033	432	0.033	1.43	8.72	7.12	20.6	23.6	24.4	25.2	26.9	28.6

**For angle designations, see page 6-19. See page 6-43 for more information on these tables.

Note: The properties of angles currently produced by cold forming are up to 7 percent less than the properties shown in the above tables. Check manufacturers' catalogue for the exact properties and dimensions.

TWO CHANNELS
Toe to Toe

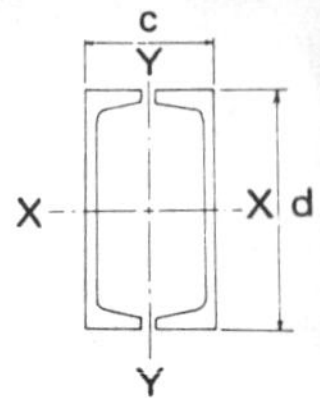

PROPERTIES OF SECTIONS

Channel Size	For Two Channels			Axis X–X			Axis Y–Y					
							Toe to Toe			c = d		
	Mass	Dead Load	Area	I_x	S_x	r_x	I_y	S_y	r_y	I_y	S_y	r_y
	kg/m	kN/m	mm^2	$10^6 mm^4$	$10^3 mm^3$	mm	$10^6 mm^4$	$10^3 mm^3$	mm	$10^6 mm^4$	$10^3 mm^3$	mm
MC460 x 86*	173	1.70	22 000	564	2 480	160	176	1 650	89.5	959	4 200	209
x 77.2*	155	1.52	19 700	522	2 280	163	149	1 430	87.0	861	3 770	209
x 68.2*	137	1.34	17 400	484	2 120	167	126	1 240	85.0	761	3 330	209
x 63.5*	127	1.25	16 200	462	2 020	169	112	1 120	83.2	707	3 090	209
C380 x 74*	149	1.46	19 000	336	1 760	133	112	1 190	76.9	558	2 930	172
x 60*	119	1.17	15 100	290	1 520	138	80.4	903	72.9	449	2 360	172
x 50*	101	0.990	12 900	262	1 370	143	62.8	730	69.9	381	2 000	172
C310 x 45*	89.4	0.876	11 400	135	884	109	49.4	618	65.9	213	1 400	137
x 37*	74.2	0.726	9 440	120	786	113	37.6	488	63.1	177	1 160	137
x 31*	61.6	0.604	7 840	107	702	117	28.2	381	60.0	146	958	136
C250 x 45*	89.0	0.874	11 300	85.6	674	86.9	43.6	574	62.0	142	1 120	112
x 37	74.6	0.730	9 500	75.8	598	89.4	34.0	466	59.8	120	949	113
x 30	59.2	0.582	7 560	65.4	514	93.0	24.1	350	56.5	96.6	761	113
x 23	45.2	0.442	5 760	55.6	438	98.2	15.7	242	52.3	72.9	574	113
C230 x 30	59.6	0.584	7 600	51.0	444	81.9	22.7	339	54.7	77.6	677	101
x 22	44.6	0.438	5 680	42.6	372	86.6	14.8	234	51.0	58.0	506	101
x 20	39.6	0.390	5 060	39.6	346	88.6	12.1	198	48.9	51.4	449	101
C200 x 28	55.8	0.548	7 120	36.4	360	71.6	19.2	299	51.9	55.7	548	88.4
x 21	40.8	0.400	5 200	29.8	294	75.8	11.8	200	47.6	41.1	405	88.9
x 17	34.0	0.167	4 340	27.0	266	78.8	8.93	157	45.4	33.9	334	88.4
C180 x 22*	43.8	0.428	5 560	22.6	254	63.7	12.1	209	46.7	32.8	369	76.8
x 18	36.4	0.356	4 620	20.0	226	65.9	9.02	164	44.2	27.5	309	77.1
x 15	29.0	0.284	3 700	17.7	199	69.3	6.50	123	41.9	21.7	244	76.6
C150 x 19	38.4	0.378	4 900	14.2	187	53.9	9.13	169	43.2	20.4	268	64.5
x 16	31.0	0.304	3 960	12.4	164	56.1	6.54	128	40.6	16.6	219	64.8
x 12	24.2	0.236	3 080	10.7	141	59.1	4.37	91.1	37.7	12.9	169	64.6
C130 x 17	34.2	0.335	4 340	8.66	136	44.7	7.28	140	41.0	11.8	185	52.0
x 13	26.6	0.262	3 400	7.32	115	46.5	4.69	99.8	37.2	9.56	150	53.0
x 10	19.8	0.194	2 520	6.18	97.2	49.5	2.94	66.8	34.1	7.02	111	52.8
C100 x 11	21.6	0.212	2 740	3.82	74.8	37.3	3.07	71.3	33.5	4.62	90.6	41.1
x 9	18.6	0.182	2 340	3.41	67.2	38.1	2.48	59.0	32.5	3.95	77.4	41.1
x 8	16.0	0.158	2 040	3.22	63.2	39.7	1.91	47.7	30.6	3.43	67.3	41.0
C 75 x 9	17.6	0.174	2 240	1.70	44.6	27.4	2.08	52.0	30.5	(1)	(1)	(1)
x 7	14.6	0.144	1 870	1.50	39.4	28.3	1.47	39.8	28.1	1.57	41.4	29.0
x 6	12.0	0.118	1 530	1.34	35.2	29.6	1.04	29.7	26.1	1.27	33.5	28.9

(1) The condition of c = d cannot be met for this section.
* Not available from Canadian Mills.

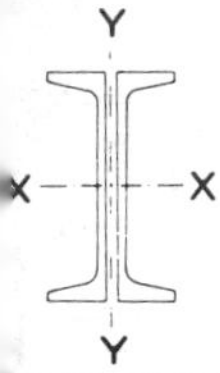

TWO CHANNELS
Back to Back

PROPERTIES OF SECTIONS

Channel Size	Mass of 2 Channels	Area of 2 Channels	Axis X–X			Radius of Gyration about Axis Y–Y					
			I_x	S_x	r_x	Back to Back Channels, millimetres					
	kg/m	mm^2	$10^6 mm^4$	$10^3 mm^3$	mm	0	8	10	12	16	20
MC460 x 86*	173	22 000	564	2 480	160	33.6	36.2	37.0	37.7	39.1	40.7
x 77.2*	155	19 700	522	2 280	163	33.8	36.5	37.2	37.9	39.3	40.8
x 68.2*	137	17 400	484	2 120	167	34.3	36.9	37.6	38.3	39.7	41.2
x 63.5*	127	16 200	462	2 020	169	34.7	37.3	37.9	38.7	40.1	41.6
C380 x 74*	149	19 000	336	1 760	133	30.0	32.8	33.5	34.3	35.9	37.5
x 60*	119	15 100	290	1 520	138	29.9	32.7	33.4	34.2	35.7	37.3
x 50*	101	12 900	262	1 370	143	30.5	33.2	33.9	34.7	36.2	37.8
C310 x 45*	89.4	11 400	135	884	109	25.7	28.5	29.3	30.0	31.6	33.2
x 37*	74.2	9 440	120	786	113	26.2	28.9	29.7	30.4	32.0	33.6
x 31*	61.6	7 840	107	702	117	26.7	29.5	30.2	30.9	32.5	34.1
C250 x 45*	89.0	11 300	85.6	674	86.9	23.4	26.3	27.1	27.9	29.5	31.2
x 37	74.6	9 500	75.8	598	89.4	23.3	26.1	26.9	27.7	29.3	30.9
x 30	59.2	7 560	65.4	514	93.0	23.3	26.1	26.8	27.6	29.2	30.8
x 23	45.2	5 760	55.6	438	98.2	23.9	26.8	27.5	28.3	29.9	31.5
C230 x 30	59.6	7 600	51.0	444	81.9	22.0	24.9	25.6	26.4	28.0	27.9
x 22	44.6	5 680	42.6	372	86.6	22.5	25.3	26.1	26.8	28.4	30.1
x 20	39.6	5 060	39.6	346	88.6	22.6	25.5	26.2	27.0	28.6	30.2
C200 x 28	55.8	7 120	36.4	360	71.6	21.0	23.9	24.7	25.5	27.1	28.8
x 21	40.8	5 200	29.8	294	75.8	20.9	23.8	24.5	25.3	26.9	28.6
x 17	34.0	4 340	27.0	266	78.8	21.5	24.4	25.1	25.9	27.5	29.2
C180 x 22*	43.8	5 560	22.6	254	63.7	19.7	22.6	23.4	24.2	25.8	27.5
x 18	36.4	4 620	20.0	226	65.9	19.5	22.4	23.2	24.0	25.6	27.3
x 15	29.0	3 700	17.7	199	69.3	20.2	23.1	23.9	24.7	26.3	28.0
C150 x 19	38.4	4 900	14.2	187	53.9	18.4	21.4	22.2	23.0	24.7	26.4
x 16	31.0	3 960	12.4	164	56.1	18.3	21.3	22.1	22.9	24.5	26.2
x 12	24.2	3 080	10.7	141	59.1	18.6	21.5	22.3	23.1	24.8	26.5
C130 x 17	34.2	4 340	8.66	136	44.7	18.1	21.1	21.9	22.8	24.5	26.2
x 13	26.6	3 400	7.32	115	46.5	17.0	20.0	20.8	21.6	23.3	25.1
x 10	19.8	2 520	6.18	97.2	49.5	17.4	20.4	21.2	22.0	23.7	25.4
C100 x 11	21.6	2 740	3.82	74.8	37.3	16.1	19.2	20.0	20.8	22.5	24.3
x 9	18.6	2 340	3.41	67.2	38.1	16.5	19.5	20.3	21.2	22.9	24.6
x 8	16.0	2 040	3.22	63.2	39.7	16.2	19.3	20.1	21.0	22.7	24.4
C 75 x 9	17.6	2 240	1.70	44.6	27.4	15.5	18.6	19.5	20.3	22.0	23.8
x 7	14.6	1 870	1.50	39.4	28.3	14.8	17.9	18.8	19.6	21.4	23.1
x 6	12.0	1 530	1.34	35.2	29.6	14.8	18.0	18.8	19.7	21.4	23.2

*Not available from Canadian Mills.

Note: For dead load of two channels refer to table on page 6–106.

W SHAPES AND CHANNELS

PROPERTIES OF SECTIONS

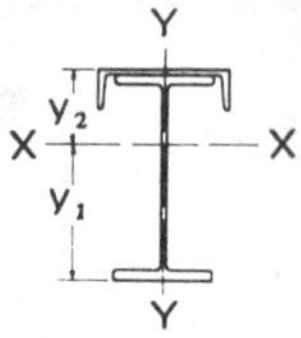

Beam	Channel	Mass	Dead Load	Total Area	Axis X–X					Axis Y–Y		
					I	$S_1 = \frac{I}{Y_1}$	$S_2 = \frac{I}{Y_2}$	r	Y_1	I	S	r
		kg/m	kN/m	mm^2	$10^6 mm^4$	$10^3 mm^3$	$10^3 mm^3$	mm	mm	$10^6 mm^4$	$10^3 mm^3$	mm
W920 x 289*	MC460 x 63.5*	352.2	3.45	44 900	6 410	11 800	16 300	378	545	387	1 690	92.8
	C380 x 50*	339.1	3.32	43 200	6 170	11 600	15 200	378	531	287	1 510	81.5
x 271*	MC460 x 63.5*	335.3	3.28	42 700	6 060	11 100	15 700	377	547	376	1 650	93.8
	C380 x 50*	322.2	3.20	41 000	5 830	11 000	14 500	377	532	276	1 450	82.0
x 253*	MC460 x 63.5*	317.3	3.10	40 400	5 680	10 300	14 900	375	550	365	1 600	95.1
	C380 x 50*	304.2	2.98	38 700	5 460	10 200	13 800	375	534	265	1 390	82.7
x 238*	MC460 x 63.5*	301.9	2.95	38 500	5 350	9 690	14 300	373	552	354	1 550	95.9
	C380 x 50*	288.8	2.82	36 800	5 130	9 570	13 200	373	536	254	1 330	83.0
x223*	MC460 x 63.5*	287.8	2.81	36 700	5 030	9 080	13 600	370	554	343	1 500	96.7
	C380 x 50*	274.7	2.68	35 000	4 820	8 960	12 500	371	537	243	1 280	83.3
W840 x 226*	MC460 x 63.5*	289.8	2.84	37 000	4 500	8 710	13 000	349	516	345	1 510	96.6
	C380 x 50*	277.1	2.72	35 300	4 310	8 600	12 000	349	501	245	1 290	83.3
x 210*	MC460 x 63.5*	274.4	2.68	34 900	4 180	8 050	12 300	346	519	334	1 460	97.8
	C380 x 50*	261.3	2.56	33 200	4 000	7 950	11 300	347	503	234	1 230	83.9
x 193*	MC460 x 63.5*	257.1	2.51	32 800	3 810	7 310	11 500	341	521	321	1 410	99.0
	C380 x 50*	244.0	2.38	31 100	3 640	7 220	10 500	342	505	221	1 160	84.3
W760 x 196*	MC460 x 63.5*	260.4	2.55	33 200	3 270	6 860	10 700	314	476	313	1 370	97.0
	C380 x 50*	247.3	2.42	31 500	3 120	6 770	9 800	315	462	213	1 120	82.1
x 185*	MC460 x 63.5*	248.4	2.43	31 600	3 070	6 420	10 300	312	479	306	1 340	98.4
	C380 x 50*	235.3	2.30	29 900	2 940	6 340	9 380	313	463	206	1 080	83.0
x 173*	MC460 x 63.5*	237.2	2.32	30 200	2 880	6 000	9 840	309	481	300	1 310	99.6
	C380 x 50*	224.1	2.20	28 500	2 750	5 920	8 940	310	465	200	1 050	83.7
x 161*	MC460 x 63.5*	224.0	2.19	28 500	2 660	5 490	9 300	305	484	292	1 280	101
	C380 x 50*	210.9	2.06	26 800	2 530	5 410	8 410	307	467	192	1 010	84.5
W690 x 170*	C380 x 50*	220.4	2.16	28 000	2 270	5 350	8 110	284	424	197	1 040	83.9
	C310 x 31*	200.7	1.96	25 500	2 080	5 220	6 870	285	398	120	785	68.5
x 152*	C380 x 50*	202.6	1.98	25 800	2 050	4 810	7 570	282	427	189	991	85.5
	C310 x 31*	182.9	1.79	23 300	1 870	4 690	6 350	284	400	111	730	69.1
x 140*	C380 x 50*	190.3	1.86	24 200	1 880	4 380	7 140	279	430	183	959	86.8
	C310 x 31*	170.6	1.67	21 700	1 720	4 270	5 930	281	402	105	690	69.6
W610 x 125	C380 x 50*	175.6	1.72	22 300	1 390	3 550	6 020	250	391	170	894	87.3
	C310 x 31*	155.9	1.52	19 800	1 260	3 460	4 950	252	364	92.8	609	68.4
x 113	C380 x 50*	163.9	1.60	20 800	1 260	3 200	5 650	246	395	165	868	89.1
	C310 x 31*	144.2	1.41	18 300	1 140	3 110	4 600	250	367	87.8	576	69.2
W530 x 101	C380 x 50*	151.9	1.49	19 300	908	2 560	4 710	217	355	158	829	90.4
	C310 x 31*	132.2	1.30	16 800	819	2 490	3 800	221	329	80.4	527	69.1
x 92	C380 x 50*	143.0	1.40	18 200	830	2 320	4 460	213	357	155	813	92.1
	C310 x 31*	123.3	1.21	15 700	747	2 260	3 560	218	330	77.3	507	70.1
W460 x 74	C380 x 50*	124.7	1.22	15 900	519	1 640	3 460	181	317	148	775	96.4
	C310 x 31*	105.0	1.03	13 400	467	1 600	2 720	187	292	70.1	460	72.4
W410 x 54	C380 x 50*	103.9	1.02	13 200	311	1 060	2 620	153	295	141	741	103
	C310 x 31*	84.2	0.826	10 700	279	1 030	2 010	161	271	63.6	417	77.0
W360 x 45	C310 x 31*	75.8	0.743	9 650	187	771	1 620	139	243	61.7	404	79.9
	C250 x 23	67.6	0.662	8 610	176	759	1 390	143	232	36.0	283	64.6
W310 x 39	C310 x 31*	69.5	0.682	8 860	132	605	1 350	122	219	60.8	398	82.8
	C250 x 23	61.3	0.601	7 820	124	597	1 160	126	208	35.1	276	67.0
W250 x 33	C250 x 23	55.3	0.542	7 050	74.0	417	857	102	178	32.5	256	67.9
	C200 x 17	49.7	0.488	6 340	70.0	412	749	105	170	18.2	180	53.6
W200 x 27	C250 x 23	49.2	0.482	6 270	40.4	276	607	80.3	147	31.1	245	70.4
	C200 x 17	43.6	0.428	5 560	38.2	272	529	82.9	140	16.8	166	55.0

*Not available from Canadian Mills.

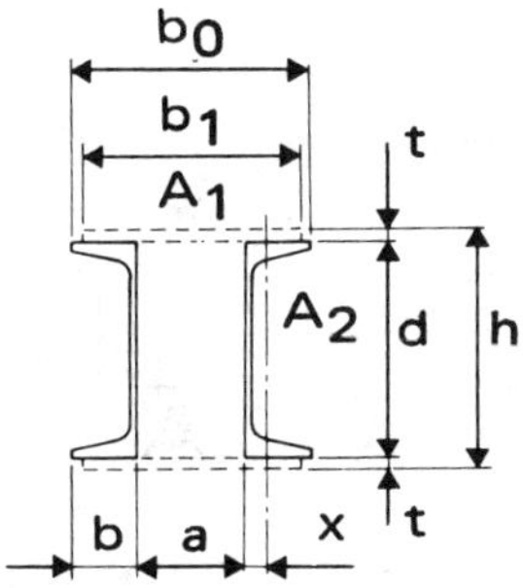

$h = d + 2t$ $\quad b_o = a + 2b$ $\quad S_{yy} = 2I_{yy}/b_o$ if $b_1 < b_o$

$A_1 = b_1 t$ $\quad S_{yy} = 2I_{yy}/b_1$ if $b_1 \geqslant b_o$

$A = 2(A_1 + A_2)$

$r_{yy} = \sqrt{I_{yy}/A}$

$$I_{xx} = 2I_{xc} + \frac{b_1}{12}\left[h^3 - d^3\right]$$

$S_{xx} = 2I_{xx}/h$

$r_{xx} = \sqrt{I_{xx}/A}$

$$I_{yy} = 2I_{yc} + \frac{A_1}{6}\; b_1^2 + 2A_2\;(x + a/2)^2$$

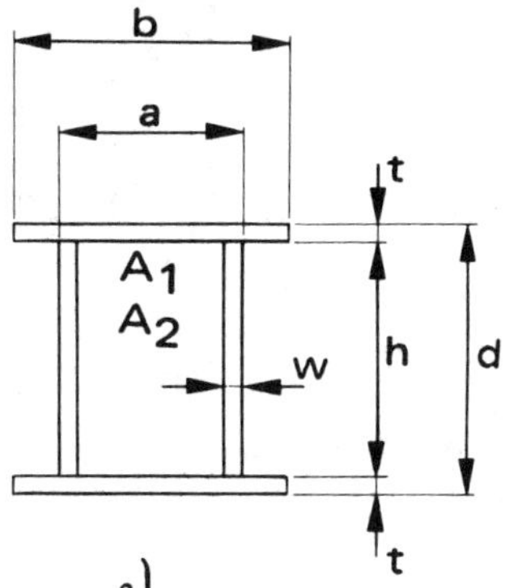

$h = d - 2t$

$A_1 = bt$

$A_2 = wh$

$A = 2(A_1 + A_2)$

$$I_{xx} = \frac{1}{12}\left\{b(d^3 - h^3) + 2A_2h^2\right\}$$

$S_{xx} = 2I_{xx}/d$

$$Z_{xx} = \frac{b}{4}\left\{d^2 - h^2\right\} + \frac{A_2h}{2} \qquad Z_{yy} = \frac{h}{4}\left\{a^2 - c^2\right\} + \frac{A_1b}{2}$$

$r_{xx} = \sqrt{I_{xx}/A}$

$c = a - 2w$ $\quad r_{yy} = \sqrt{I_{yy}/A}$

$$I_{yy} = \frac{1}{12}\left\{2A_1b^2 + h(a^3 - c^3)\right\}$$

$S_{yy} = 2I_{yy}/b$

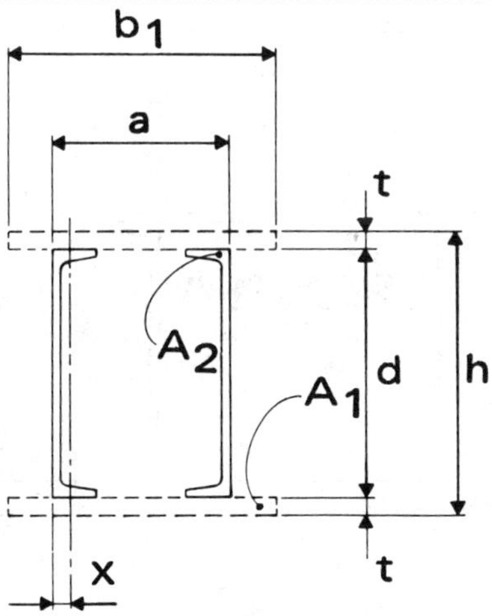

$h = d + 2t$

$A_1 = b_1 t$ $\quad I_{yy} = 2I_{yc} + \frac{A_1}{6}\, b_1^2 + 2A_2\,(a/2 - x)^2$

$A = 2(A_1 + A_2)$

$$I_{xx} = 2I_{xc} + \frac{b_1}{12}\,(h^3 - d^3) \qquad S_{yy} = 2I_{yy}/b_1 \text{ if } a < b_1$$

$S_{yy} = 2I_{yy}/a$ if $a \geqslant b_1$

$S_{xx} = 2I_{xx}/h$

$r_{yy} = \sqrt{I_{yy}/A}$

$r_{xx} = \sqrt{I_{xx}/A}$

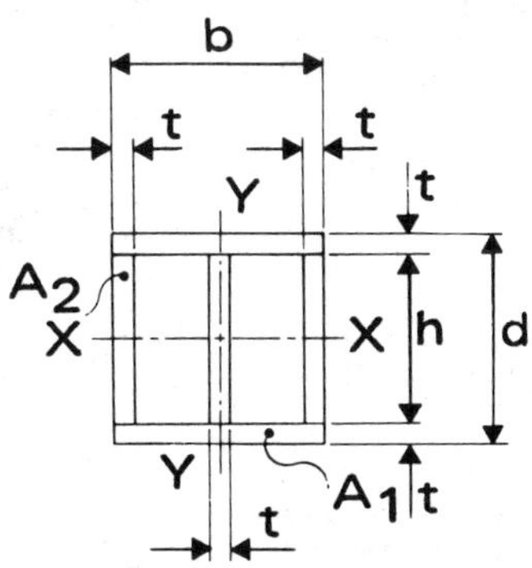

$h = d - 2t$

$A_1 = bt$

$A_2 = ht$

$A = 2A_1 + 3A_2$

$$I_{xx} = \frac{1}{12}\left[3A_2h^2 + b(d^3 - h^3)\right]$$

$$S_{xx} = 2I_{xx}/d \qquad Z_{yy} = \frac{A_1b}{2} + \frac{A_2t}{4} + A_2\;(b - t)$$

$r_{xx} = \sqrt{I_{xx}/A}$ $\quad S_{yy} = 2I_{yy}/b$

$$Z_{xx} = \frac{3A_2h}{4} + A_1(d - t) \qquad r_{yy} = \sqrt{I_{yy}/A}$$

$$I_{yy} = \frac{1}{12}\left[2A_1b^2 + A_2t^2 + 2t(b^3 - (b - 2t)^3)\right]$$

Elements of the shape which are shown in dotted outline are optional and if omitted the variable defining their size should be set equal to zero.

All elements of the shape are assumed to be continuous along the length of the shape.

BUILT-UP SECTIONS

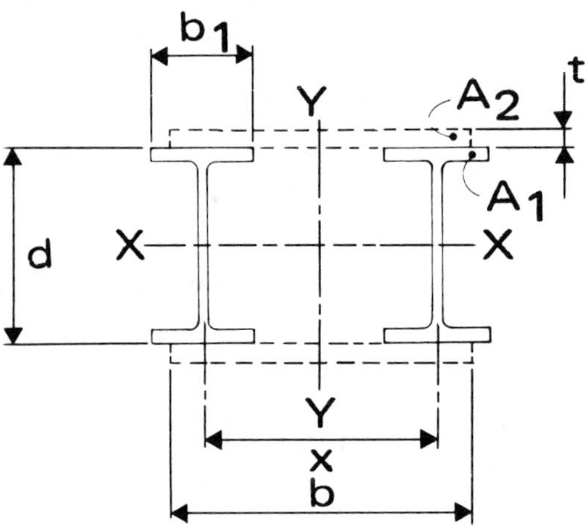

$A_2 = bt$

$A = 2(A_1 + A_2)$

$I_{xx} = 2I_{xw} + \frac{1}{12} b \left[(d + 2t)^3 - d^3\right]$

$S_{xx} = 2I_{xx}/(d + 2t)$

$r_{xx} = \sqrt{I_{xx}/A}$

$I_{yy} = 2I_{yw} + \frac{1}{6} A_2 t^2 + \frac{1}{2} A_1 x^2$

For $(x + b_1) > b$:

$S_{yy} = 2I_{yy}/(x + b_1)$

For $(x + b_1) \leqslant b$:

$S_{yy} = 2I_{yy}/b$

$r_{yy} = \sqrt{I_{yy}/A}$

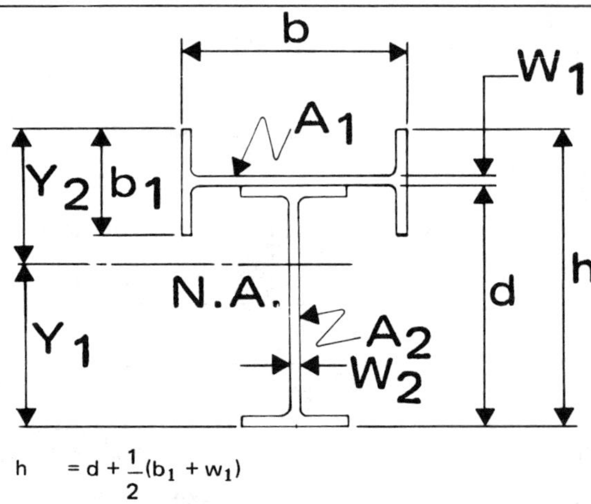

$h = d + \frac{1}{2}(b_1 + w_1)$

$Y_1 = \dfrac{A_1(d + W_1/2) + A_2 d/2}{A_1 + A_2}$

$Y_2 = h - Y_1$

$A = A_1 + A_2$

$I_{xx} = I_{y1} + I_{x2} + A_1(Y_2 - b_1/2)^2$

$S_{x1} = I_{xx}/Y_1$

$S_{x2} = I_{xx}/Y_2$

$r_{xx} = \sqrt{I_{xx}/A}$

$I_{yy} = I_{x1} + I_{y2}$

$^*I_{yT} = I_{x1} + I_{y2}/2 - (Y_1 - d/2)w^3$

$S_{yy} = 2I_{yy}/b$

$r_{yy} = \sqrt{I_{yy}/A}$

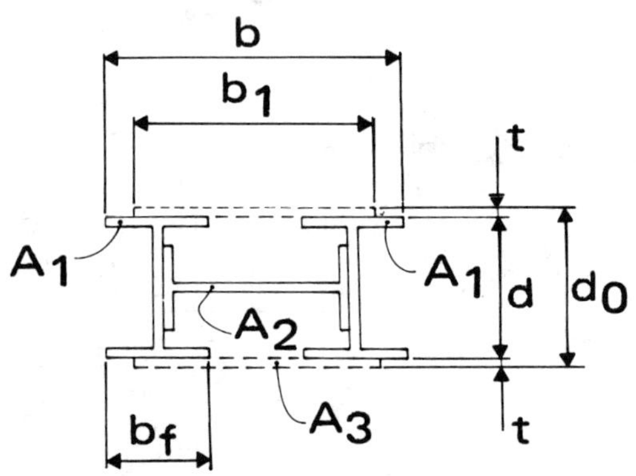

$d_o = d + 2t$

$A_3 = b_1 t$

$A = 2(A_1 + A_3) + A_2$

$I_{xx} = 2I_{x1} + I_{y2} + \frac{b_1}{12}(d_o^3 - d^3)$

$S_{xx} = 2I_{xx}/d_o$

$r_{xx} = \sqrt{I_{xx}/A}$

$I_{yy} = I_{x2} + 2I_{y1} + \frac{A_3}{6} b_1^2 + A_1 b_1^2/2$

$S_{yy} = 2I_{xx}/b \quad \text{if } b \geqslant b_1$

$S_{yy} = 2I_{xx}/b_1 \quad \text{if } b < b_1$

$r_{xx} = \sqrt{I_{xx}/A}$

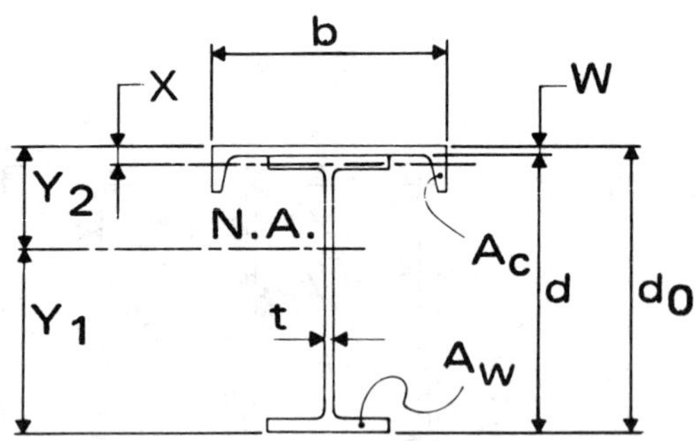

$Y_1 = \dfrac{A_w d/2 + A_c(d_o - x)}{A}$

$A = A_c + A_w$

$d_o = d + w$

$Y_2 = d_o - Y_1$

$I_{xx} = I_{xw} + I_{yc} + A_w(Y_1 - d/2)^2 + A_c(Y_2 - x)^2$

$S_{x1} = I_{xx}/Y_1$

$S_{x2} = I_{xx}/Y_2$

$r_{xx} = \sqrt{I_{xx}/A}$

$I_{yy} = I_{yw} + I_{xc}$

$^*I_{yT} = I_{xc} + \frac{I_{yw}}{2} - (Y_1 - d/2)w^3$

$S_{yy} = 2I_{yy}/b$

$r_{yy} = \sqrt{I_{yy}/A}$

$^*I_{yT}$ is the moment of inertia of the T section above the neutral axis.

Note: Centres of gravity of both channels are on the same vertical line.

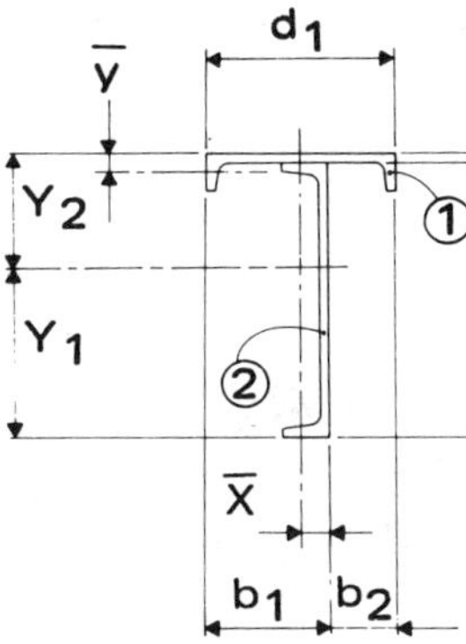

$d = d_2 + w_1$

$A = A_1 + A_2$

$b_1 = (d_1/2) + \bar{x}$

$b_2 = d_1 - b_1$

$$y_1 = \frac{A_1(d - \bar{y}) + \frac{A_2}{2}d_2}{A}$$

$$I_{xx} = I_{1y} + I_{2x} + A_1(y_2 - \bar{y})^2 + A_2(y_1 - \frac{d_2}{2})^2$$

$S_1 = \frac{I_{xx}}{y_1}$ $\qquad y_2 = d - y_1$

$I_{yy} = I_{1x} + I_{2y}$

$S_2 = \frac{I_{xx}}{y_2}$ $\qquad S_y = 2I_{yy}/d_1$

$r_{xx} = \sqrt{\frac{I_{xx}}{A}}$ $\qquad r_{yy} = \sqrt{\frac{I_{yy}}{A}}$

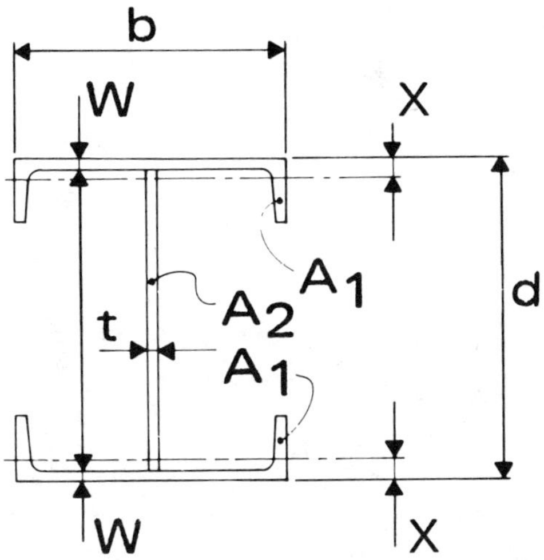

$h = d - 2w$

$A_2 = ht$

$A = 2A_1 + A_2$

$$I_{xx} = 2I_{yc} + \frac{1}{12}A_2h^2 + 2A_1(d/2 - x)^2$$

$$I_{yy} = 2I_{xc} + \frac{1}{12}A_2t^2$$

$S_{xx} = 2I_{xx}/d$

$S_{yy} = 2I_{yy}/b$

$r_{yy} = \sqrt{I_{yy}/A}$ $\qquad r_{xx} = \sqrt{I_{xx}/A}$

Note: a and b are the angle leg lengths and b_1 is the width of channel flange.

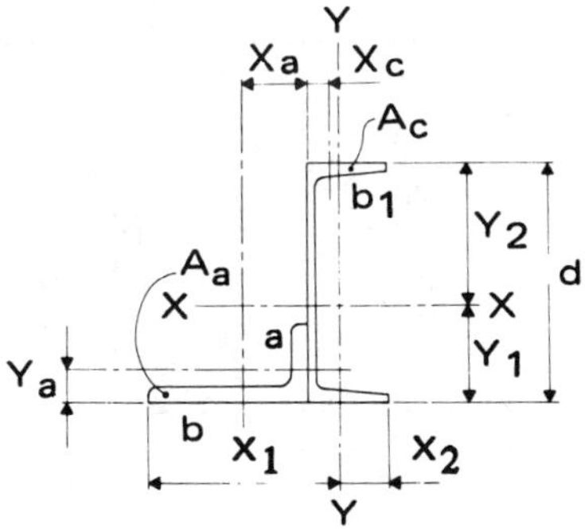

$A = A_a + A_c$

$x_1 = \frac{A_a(b - x_a) + A_c(b + x_c)}{A}$; $\qquad y_1 = \frac{A_ay_a + A_cd/2}{A}$

$x_2 = b_1 + b - x_1$ $\qquad y_2 = d - y_1$

$$I_{xx} = I_{ya} + I_{xc} + A_a(y_1 - y_a)^2 + A_c(\frac{d}{2} - y_1)^2$$

$$I_{yy} = I_{xa} + I_{yc} + A_a(x_1 - b + x_a)^2 + A_c(b_1 - x_2 - x_c)^2$$

$S_{y1} = I_{yy}/x_1$ $\qquad S_{x1} = I_{xx}/y_1$

$S_{y2} = I_{yy}/x_2$ $\qquad S_{x2} = I_{xx}/y_2$

$r_{yy} = \sqrt{I_{yy}/A}$ $\qquad r_{xx} = \sqrt{I_{xx}/A}$

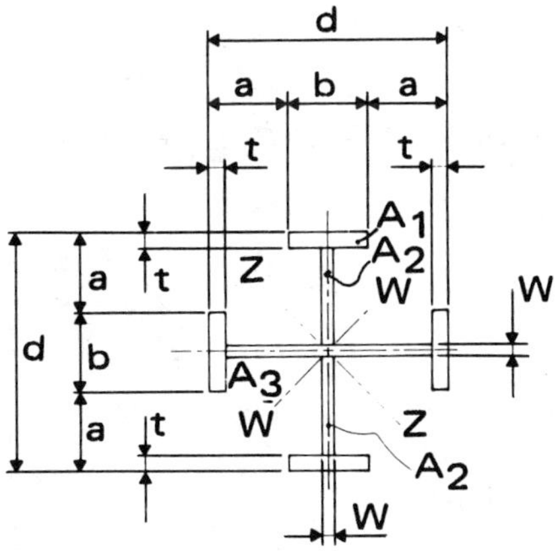

$A_1 = bt$ $\qquad A_3 = 2A_2 + w^2$

$A_2 = (d - w - 2t)w/2$ $\qquad A = 4A_1 + 2A_2 + A_3$

$$I_x = I_y = \frac{1}{12}\left\{b(d^3 - E^3) + wE^3 + 2tb^3 + Ew^3 - w^4\right\}$$

$S_x = S_y = 2I_x/d$

$E = (d - 2t)$

$r_x = r_y = \sqrt{I_x/A}$

BUILT-UP SECTIONS

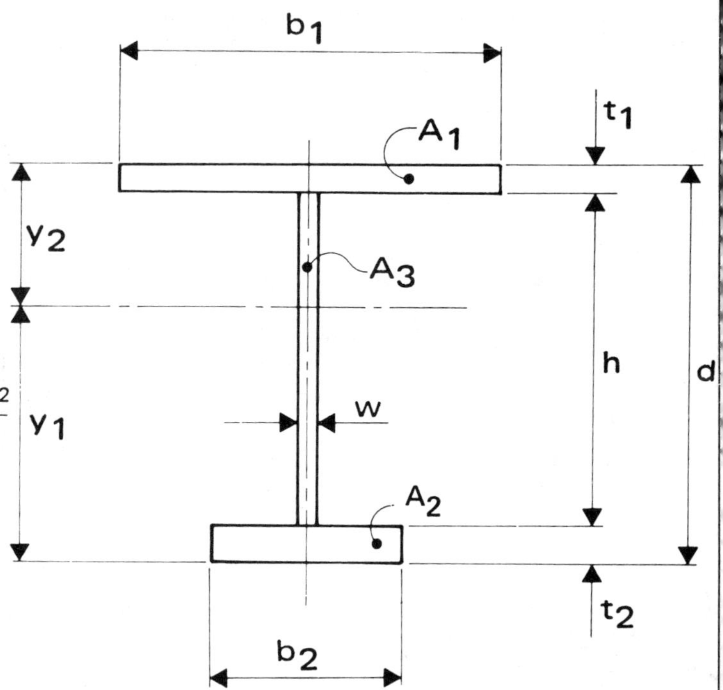

$A_1 = b_1 t_1 \qquad A_2 = b_2 t_2 \qquad A_3 = wh$

$d = h + t_1 + t_2$

$A = A_1 + A_2 + A_3$

$$y_1 = \frac{A_1(d - t_1/2) + A_3(t_2 + h/2) + A_2 t_2/2}{A}$$

$y_2 = d - y_1$

$$J = \frac{1}{3}\left\{A_1 t_1^2 + A_3 w^2 + A_2 t_2^2\right\}$$

$$C_w = \frac{\left(d - \dfrac{t_1 + t_2}{2}\right)^2 b_1^3 t_1}{12\left[1 + (b_1/b_2)^3 (t_1/t_2)\right]}$$

$$I_{yy} = \frac{1}{12}\left[A_1 b_1^2 + A_2 b_2^2 + A_3 w^2\right]$$

$$S_{x1} = \frac{I_{xx}}{y_1}; \qquad S_{x2} = \frac{I_{xx}}{y_2} \qquad S_{yy} = 2I_{yy}/b_1$$

$$^*I_{YT} = \frac{1}{12}\left[A_1 b_1^2 + (y_2 - t_1)w^3\right]$$

$$r_{xx} = \sqrt{\frac{I_{xx}}{A}} \qquad r_{yy} = \sqrt{\frac{I_{yy}}{A}}$$

$$I_{xx} = \frac{1}{12}\left[A_1 t_1^2 + A_2 t_2^2 + A_3 h^2\right] + A_1(y_2 - t_1/2)^2 + A_2(y_1 - t_2/2)^2 + A_3(y_1 - t_2 - h/2)^2$$

$A = A_1 + A_2 + A_s \quad ; \quad h = d + t_1 + t_2$

$$Y_1 = \frac{A_1(h - t_1/2) + A_s(t_2 + d/2) + A_2 t_2/2}{A}$$

$Y_2 = h - Y_1$

$$I_{xx} = I_{xs} + \frac{1}{12}(A_1 t_1^2 + A_2 t_2^2) + A_1(Y_2 - t_1/2)^2 + A_2(Y_1 - t_2/2)^2$$

$S_{x1} = I_{xx}/Y_1 \quad ; \quad S_{x2} = I_{xx}/Y_2$

$r_{xx} = \sqrt{I_{xx}/A}$

$$I_{yy} = I_{ys} + \frac{1}{12}(A_1 b_1^2 + A_2 b_2^2)$$

$S_{yy} = 2I_{yy}/b_1$ if $b_1 > b_2$

$S_{yy} = 2I_{yy}/b_2$ if $b_1 \leqslant b_2$

$r_{yy} = \sqrt{I_{yy}/A}$

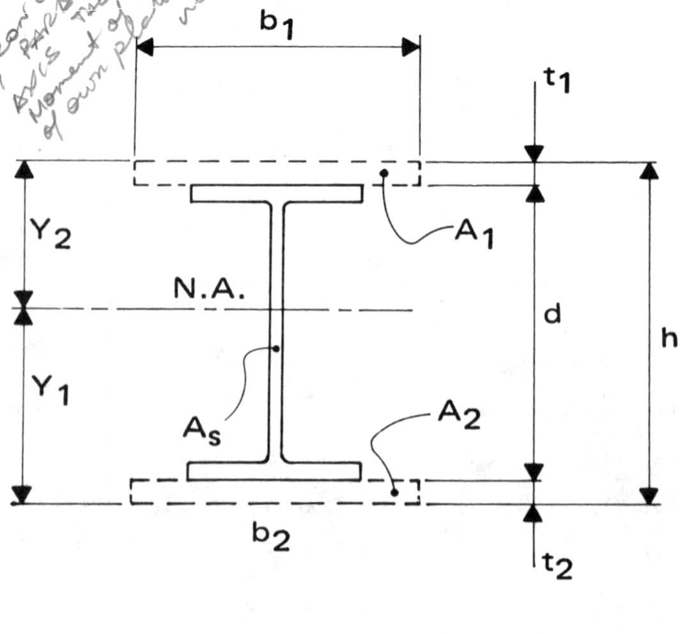

$^*I_{yt}$ is the moment of inertia of the T section above the neutral axis.

BARS AND PLATES

Bars

The term "Bars" means:

(a) Rounds, squares and hexagons of all sizes;

(b) Flat-rolled steel up to 200 mm inclusive in width and over 5.0 mm in thickness, except for widths from 150 mm to 200 mm up to 6.0 mm in thickness;

(c) Bar-size shapes under 75 mm in maximum dimension.

Hot rolled bar flat products are available in most widths and thickness according to CAN3-G312.1 and CAN3-G312.2-M Standards.

Plates

The term "Plates" means:

Flat hot-rolled steel, when ordered to thickness;

(a) Over 200 mm in width and over 6.0 mm in thickness, and

(b) Over 1200 mm in width and over 4.5 mm in thickness.

Slabs, sheet, bars, strip and skelp, although frequently falling within these size ranges, are not classified as plate. The following table, Standard Product Classification for Flat Hot-Rolled Steel Products and Bars, summarizes the ranges for plate, bar, strip and sheet products.

Plates may be further defined as "Universal Mill Plates" or "Sheared Plates". Sheared plates are rolled on a mill with horizontal rolls only, producing a product with uneven edges which must be sheared (or, at the option of the producer, flame cut) to ordered dimensions.

Universal mill plates are rolled to the ordered width on a mill having side rollers to control the width. Slab or ingot on a universal mill plate are not cross rolled, but are only elongated during the rolling process. The mill order must specify universal mill plate when it is required.

Extreme plate sizes produced by mills vary greatly with the size of various mills, and individual mills should be consulted for this information. Canadian mills can produce sheared and flame cut plate in widths ranging from 200 mm to 3900 mm and in thicknesses ranging from 4.5 mm to 150 mm. First and second preference thickness for plate according to CAN3-G312.1 is provided in the table on page 6-135. Where possible, designs should generally be based on first preference thickness for economy.

Standard mill practice is to invoice the purchaser for actual scale weight at point of shipment. Allowable overweight above theoretical that may be charged, is limited in accordance with either the CAN/CSA-G40.20-M or ASTM A6 Standards, depending upon whether the steel is furnished to Canadian or ASTM material specifications.

Various extras for thickness, width, length, cutting, quality, quantity, (or quantity discounts), and for other special requirements are added to the base price of plates. Particulars of these extras should be obtained from the producing mills.

Sketch Plates

Sketch plates of special or unusual shape usually require flame cutting, for which flame cutting extras apply. Some mills can supply sketch plates of certain shapes by shearing to size.

Floor Plates

Floor plates in different styles, patterns, and extreme dimensions are produced by different mills. The nominal, or ordered, thickness is that of the flat plate exclusive of the raised pattern. Individual producers should be consulted for more details.

Bearing Plates

Rolled steel bearing plates are used for column bases, and other bearing plates. Depending on the thickness required by design, bearing plates may require additional thickness for machining to ensure proper bearing. Plates up to and including 50 mm in thickness are rolled flat with surfaces sufficiently smooth to receive, without machining or flattening, the milled or machine cut ends of column shafts. Bearing plates 100 mm and under in thickness can sometimes be flattened by press to within required flatness tolerances. Rolled steel bearing plates greater than 100 mm thick usually require machining of the top surface in contact with the column shaft, but do not require flattening of the bottom surface when bearing on a grouted base. When bearing plates over 100 mm thick bear on a steel surface, both the top and bottom surfaces will require to be machined over the bearing areas.

Tables

The following Tables are included in this section:

Standard Product Classification of Flat Hot-Rolled Steel Products and Bars

Flat Metal Products - Sheet & Plate

Mass (kg/m) for Rectangular Steel Products

SI Wire Size - Wire Gauges Comparison

SI Thickness - Imperial Gauge Comparisons

Mass or Round Bars and Square Bars

STANDARD PRODUCT CLASSIFICATION of Flat Hot-Rolled Steel Products and Bars

Width W \ Thickness, t	$t > 6$	$6 \geq t > 5$	$5 \geq t > 4.5$	$4.5 \geq t > 1.2$	$1.2 \geq t > 0.9$	$0.9 \geq t > 0.65$
$W \leq 100$	BAR	BAR	STRIP	STRIP	STRIP	STRIP
$100 < W \leq 150$	BAR	BAR	STRIP	STRIP	STRIP	
$150 < W \leq 200$	BAR	STRIP	STRIP	STRIP		
$200 < W \leq 300$	PLATE	STRIP	STRIP	STRIP		
$300 < W \leq 1200$	PLATE	SHEET**	SHEET**	SHEET**		
$1200 < W$	PLATE	PLATE	PLATE	SHEET		

**For alloy steels, sheet begins at widths over 600 mm.

FLAT METAL PRODUCTS* – Sheet & Plate

Nominal Thickness,** mm		Mass† kg/m²	Dead Load kN/m²
First preference	Second preference		
0.050		0.393	0.003 85
0.060		0.471	0.004 62
0.080		0.628	0.006 16
0.10		0.785	0.007 70
0.12		0.942	0.009 24
	0.14	1.10	0.010 8
0.16		1.26	0.012 3
	0.18	1.41	0.013 9
0.20		1.57	0.015 4
	0.22	1.73	0.016 9
0.25		1.96	0.019 2
	0.28	2.20	0.021 6
0.30		2.36	0.023 1
	0.35	2.75	0.026 9
0.40		3.14	0.030 8
	0.45	3.53	0.034 6
0.50		3.93	0.038 5
	0.55	4.32	0.042 3
0.60		4.71	0.046 2
	0.65	5.10	0.050 0
	0.70	5.50	0.053 9
0.80		6.28	0.061 6
	0.90	7.07	0.069 3
1.0		7.85	0.077 0
	1.1	8.64	0.084 7
1.2		9.42	0.092 4
	1.4	11.0	0.108
1.6		12.6	0.123
	1.8	14.1	0.139
2.0		15.7	0.154
	2.2	17.3	0.169
2.5		19.6	0.192
	2.8	22.0	0.216
3.0		23.6	0.231
	3.2	25.1	0.246
3.5		27.5	0.269
	3.8	29.8	0.293
4.0		31.4	0.308
	4.2	33.0	0.323
4.5		35.3	0.346
	4.8	37.7	0.370
5.0		39.3	0.385
	5.5	43.2	0.423
6.0		47.1	0.462
7.0		55.0	0.539
8.0		62.8	0.616
	9.0	70.7	0.693
10		78.5	0.770
	11	86.4	0.847
12		94.2	0.924

Nominal Thickness,** mm		Mass† kg/m²	Dead Load kN/m²
First preference	Second preference		
	14	110	1.08
16		126	1.23
	18	141	1.39
20		157	1.54
	22	173	1.69
25		196	1.92
	28	220	2.16
30		236	2.31
	32	251	2.46
35		275	2.69
	38	298	2.93
40		314	3.08
	45	353	3.46
50		393	3.85
	55	432	4.23
60		471	4.62
	70	550	5.39
80		628	6.16
	90	707	6.93
100		785	7.70
	110	864	8.47
120		942	9.24
	130	1020	10.0
140		1100	10.8
	150	1180	11.5
160		1260	12.3
180		1410	13.9
200		1570	15.4
250		1960	19.2
300		2360	23.1

*Sizes are those listed in CAN3-G312.1-75

**For coated structural sheet, the nominal thickness applies to the base metal. For metric thickness dimensions for zinc coated structural quality sheet steel, see Part 7, Structural Sheet Steel Products.

† Computed using steel density of 7 850 kg/m³.

MASS (kg/m) for RECTANGULAR STEEL PRODUCTS*

Width millimetres		Thickness, millimetres														
Second							14		18		22		28		35	
	First	5	6	8	10	12		16		20		25		30		40
	25	0.981	1.18	1.57	1.96	2.36	2.75	3.14	3.53	3.93	4.32					
28		1.10	1.32	1.76	2.20	2.64	3.08	3.52	3.96	4.40	4.84	5.50				
	30	1.18	1.41	1.88	2.36	2.83	3.30	3.77	4.24	4.71	5.18	5.89	6.59			
35		1.37	1.65	2.20	2.75	3.30	3.85	4.40	4.95	5.50	6.04	6.87	7.69	8.24		
	40	1.57	1.88	2.51	3.14	3.77	4.40	5.02	5.65	6.28	6.91	7.85	8.79	9.42	11.0	
45		1.77	2.12	2.83	3.53	4.24	4.95	5.65	6.36	7.07	7.77	8.83	9.89	10.6	12.4	14.1
	50	1.96	2.36	3.14	3.93	4.71	5.50	6.28	7.07	7.85	8.64	9.81	11.0	11.8	13.7	15.7
55		2.16	2.59	3.45	4.32	5.18	6.04	6.91	7.77	8.64	9.50	10.8	12.1	13.0	15.1	17.3
	60	2.36	2.83	3.77	4.71	5.65	6.59	7.54	8.48	9.42	10.4	11.8	13.2	14.1	16.5	18.8
70		2.75	3.30	4.40	5.50	6.59	7.69	8.79	9.89	11.0	12.1	13.7	15.4	16.5	19.2	22.0
	80	3.14	3.77	5.02	6.28	7.54	8.79	10.0	11.3	12.6	13.8	15.7	17.6	18.8	22.0	25.1
90		3.53	4.24	5.65	7.07	8.48	9.89	11.3	12.7	14.1	15.5	17.7	19.8	21.2	24.7	28.3
	100	3.93	4.71	6.28	7.85	9.42	11.0	12.6	14.1	15.7	17.3	19.6	22.0	23.6	27.5	31.4
110		4.32	5.18	6.91	8.64	10.4	12.1	13.8	15.5	17.3	19.0	21.6	24.2	25.9	30.2	34.5
	120	4.71	5.65	7.54	9.42	11.3	13.2	15.1	17.0	18.8	20.7	23.6	26.4	28.3	33.0	37.7
140		5.50	6.59	8.79	11.0	13.2	15.4	17.6	19.8	22.0	24.2	27.5	30.8	33.0	38.5	44.0
	160		7.54	10.0	12.6	15.1	17.6	20.1	22.6	25.1	27.6	31.4	35.2	37.7	44.0	50.2
180			8.48	11.3	14.1	17.0	19.8	22.6	25.4	28.3	31.1	35.3	39.6	42.4	49.5	56.5
	200		9.42	12.6	15.7	18.8	22.0	25.1	28.3	31.4	34.5	39.3	44.0	47.1	55.0	62.8
220			10.4	13.8	17.3	20.7	24.2	27.6	31.1	34.5	38.0	43.2	48.4	51.8	60.4	69.1
	250		11.8	15.7	19.6	23.6	27.5	31.4	35.3	39.3	43.2	49.1	55.0	58.9	68.7	78.5
	300		14.1	18.8	23.6	28.3	33.0	37.7	42.4	47.1	51.8	58.9	65.9	70.7	82.4	94.2

Note: The mass has been computed using a steel density of 7 850 kg/m^3.

* Sizes are those listed in Table 2 of CAN3-G312.2-M76

SI WIRE SIZE – WIRE GAUGES COMPARISON

SI Wire Size Preferred Diam*. (mm)	The United States Steel Wire Gauge	American or Brown & Sharpe Wire Gauge	British Imperial or English Legal Standard Wire Gauge	Birmingham or Stubs Iron Wire Gauge
25.0				
24.0				
23.0				
22.0				
21.0				
20.0				
19.0				
18.0				
17.0				
16.0				
15.0				
		6/0's		
14.0				
		5/0's		
13.0				
			7/0's	5/0's
12.5				
	7/0's			
12.0				
11.8				
	6/0's	4/0's	6/0's	4/0's
11.2				
11.0				
	5/0's		5/0's	3/0's
10.6				
		3/0's	4/0's	
10.0	4/0's			
				2/0's
9.5				
	3/0's	2/0's	3/0's	
9.0				
			2/0's	1/0
8.5				
	2/0's	1/0	1/0	
8.0				
	1/0		1	1
7.5				
	1	1	2	2
7.0				
6.7	2			
		2		3
6.5				
			3	
6.3				
	3			4
6.0				
6.0				
	4	3	4	5
5.6				
			5	
5.3	5			
		4		6
5.0				
	6		6	
4.8				
		5		
4.6				
	7		7	7
4.4				
4.2				
	8	6	8	8
4.0				
3.8				
	9	7	9	9
3.6				
	10			
3.4				10
		8	10	
3.2				
	11			11
3.0				
		9	11	
2.8				
	12		12	12
2.6				
		10		13
2.4				
	13	11	13	
2.3				
2.2				
				14
2.1				
	14	12	14	
2.0				
1.90				
	15	13	15	15
1.80				
1.70				
		14	16	16
1.60				
	16			
1.50				

*From CAN3–G312.2-M76.

SI THICKNESS – IMPERIAL GAUGE COMPARISONS†

SI Preferred Thickness		United States Standard Gauge*				New Birmington Sheet Gauge		
		Weight	Ga. No.	Approximate thickness		Gauge Number	Thickness	
First mm	Second mm	Oz. per sq. ft.		Inches	mm		Inches	mm
	18							
						7/0's	0.6666	16.932
16								
						6/0's 5/0's	0.6250 0.5883	15.875 14.943
	14							
						4/0's 3/0's	0.5416 0.5000	13.757 12.700
12								
						2/0's	0.4452	11.562
	11							
						0	0.3964	10.069
10	9.0							
						1	0.3532	8.971
8.0						2	0.3147	7.993
						3	0.2804	7.122
7.0								
		160	3	0.2391	6.073	4	0.2500	6.350
6.0								
		150	4	0.2242	5.695	5	0.2225	5.652
	5.5							
		140	5	0.2092	5.314	6	0.1981	5.032
5.0								
		130	6	0.1943	4.935			
	4.8							
		120	7	0.1793	4.554			
4.5								
						7	0.1764	4.481
	4.2							
		110	8	0.1644	4.176			
4.0								
						8	0.1570	3.988
	3.8	100	9	0.1495	3.797			
						9	0.1398	3.551
3.5								
		90	10	0.1345	3.416			
	3.2							
		80	11	0.1196	3.038	10	0.1250	3.175
3.0								
						11	0.1113	2.827
	2.8							
		70	12	0.1046	2.657	12	0.0991	2.517
2.5								

†Preferred thicknesses are as per CAN3-G312.1-75.

*U.S. Standard Gauge is officially a weight gauge, in oz. per sq. ft. as tabulated. The Approx. thickness shown is the "Manufacturers' Standard" of the AISI based on a steel density of 501.81 lb. per ft.[3]

SI THICKNESS – IMPERIAL GAUGE COMPARISONS†

SI Preferred Thickness		United States Standard Gauge*				New Birmington Sheet Gauge		
		Weight	Ga. No.	Approximate thickness		Gauge Number	Thickness	
First mm	Second mm	Oz. per sq. ft.		Inches	mm		Inches	mm
		60	13	0.0897	2.278	13	0.0882	2.240
	2.2							
2.0						14	0.0785	1.994
		50	14	0.0747	1.897			
	1.8							
		45	15	0.0673	1.709	15	0.0699	1.775
1.6								
		40	16	0.0598	1.519	16 17	0.0625 0.0556	1.588 1.412
	1.4							
		36 32	17 18	0.0538 0.0478	1.367 1.214	18	0.0495	1.257
1.2								
						19	0.0440	1.118
	1.1							
		28	19	0.0418	1.062			
1.0						20	0.0392	0.996
		24	20	0.0359	0.912			
	0.90							
		22	21	0.0329	0.836	21	0.0349	0.886
0.80						22	0.0313	0.795
		20	22	0.0299	0.759			
	0.70					23	0.0278	0.706
		18	23	0.0269	0.683			
	0.65							
						24	0.0248	0.630
0.60		16	24	0.0239	0.607			
						25	0.0220	0.559
	0.55							
		14	25	0.0209	0.531			
0.50						26	0.0196	0.498
	0.45	12	26	0.0179	0.455	27	0.0175	0.445
		11	27	0.0164	0.417			
0.40						28	0.0156	0.396
		10	28	0.0149	0.378			
	0.35	9	29	0.0135	0.343	29	0.0139	0.353
						30	0.0123	0.312
0.30		8	30	0.0120	0.305			
	0.28					31	0.0110	0.279
		7	31	0.0105	0.267			
0.25								

†Preferred thicknesses are as per CAN3-G312.1-75.
*U.S. Standard Gauge is officially a weight gauge, in oz. per sq. ft. as tabulated. The Approx. thickness shown is the "Manufacturers' Standard" of the AISI based on a steel density of 501.81 lb. per ft.^{3}

MASS OF BARS
Round Bars

Preferred Nominal Size, mm		Mass*	Area	Preferred Nominal Size, mm		Mass*	Area	Preferred Nominal Size, mm		Mass*	Area
First	Second	kg/m	mm²	First	Second	kg/m	mm²	First	Second	kg/m	mm²
3.0		**0.055 5**	**7.07**		21	2.72	346		65	26.0	3 320
	3.5	0.075 5	9.62	**22**		**2.98**	**380**		70	30.2	3 850
4.0		**0.098 6**	**12.6**		23	3.26	415		72†	32.0	4 070
	4.5	0.125	15.9		24	3.55	452		75	34.7	4 420
5.0		**0.154**	**19.6**	**25**		**3.85**	**491**	**80**		**39.5**	**5 030**
	5.5	0.187	23.7		26	4.17	531		90	49.9	6 360
6.0		**0.223**	**28.3**		27†	4.49	573	**100**		**61.7**	**7 850**
	6.5	0.260	33.2		28	4.83	616		110	74.6	9 500
	7.0	0.302	38.5	**30**		**5.55**	**707**	**120**		**88.8**	**11 300**
8.0		**0.395**	**50.3**		32	6.31	804		130	104	13 300
	9.0	0.499	63.6	**35**		**7.55**	**962**	**140**		**121**	**15 400**
10		**0.617**	**78.5**		36†	7.99	1 020		150	139	17 700
	11	0.746	95.0		38	8.90	1 130	**160**		**158**	**20 100**
12		**0.888**	**113**	**40**		**9.86**	**1 260**		170	178	22 700
	13	1.04	133		42	10.9	1 390	**180**		**200**	**25 400**
14		**1.21**	**154**	**45**		**12.5**	**1 590**		190	223	28 400
	15	1.39	177		48	14.2	1 810	**200**		**247**	**31 400**
16		**1.58**	**201**	**50**		**15.4**	**1 960**		220	298	38 000
	17	1.78	227		55	18.7	2 380	**250**		**385**	**49 100**
18		**2.00**	**254**		56†	19.3	2 460		280	483	61 600
	19	2.23	284	**60**		**22.2**	**2 830**	**300**		**555**	**70 700**
20		**2.47**	**314**		64†	25.3	3 220		320	631	80 400

† Screw Stock, not listed in CAN3-G312.2-M76.

Square Bars

Preferred Nominal Size, mm		Mass*	Area	Preferred Nominal Size, mm		Mass*	Area	Preferred Nominal Size, mm		Mass*	Area
First	Second	kg/m	mm²	First	Second	kg/m	mm²	First	Second	kg/m	mm²
3.0		**0.070 6**	**9.00**	**25**		**4.91**	**625**	**100**		**78.5**	**10 000**
4.0		**0.126**	**16.0**		28	6.15	784		110	93.0	12 100
5.0		**0.196**	**25.0**	**30**		**7.06**	**900**	**120**		**113**	**14 400**
6.0		**0.283**	**36.0**		35	9.62	1 220		140	154	19 600
8.0		**0.502**	**64.0**	**40**		**12.6**	**1 600**	**160**		**201**	**25 600**
10		**0.785**	**100**		45	15.9	2 020		180	254	32 400
12		**1.13**	**144**	**50**		**19.6**	**2 500**	**200**		**314**	**40 000**
	14	1.54	196		55	23.7	3 020		220	380	48 400
16		**2.01**	**256**	**60**		**28.3**	**3 600**	**250**		**491**	**62 500**
	18	2.54	324		70	38.5	4 900	**300**		**706**	**90 000**
20		**3.14**	**400**	**80**		**50.2**	**6 400**				
	22	3.80	484		90	63.6	8 100				

*Computed using a steel density of 7 850 kg/m³

CRANE RAILS

General

Crane rails are designated by their mass in pounds per yard, with bolt sizes, hole diameters, and washer sizes dimensioned in inches. The SI metric dimensions and properties for crane rails and their accessories given on the following pages are soft converted from manufacturers' catalogues.

ASCE 40, 60 and 85 pound rails require special drilling and punching at time of purchase, while all other rail sizes are predrilled and punched by the manufacturer. The manufacturer should be consulted as to the availability of metric fasteners and metric size holes.

Rails listed in this handbook are the most popular sizes used for crane runways. For dimensions and properties not provided in the tables, consult the manufacturer.

ASCE 40 and 60 pound rails are usully supplied in 9140 mm lengths, ASCE 60 and 85 pound rails are available in 10 100 mm lengths and ASCE 85 pound and heavier rails are available in 11 900 mm lengths. If bolted rail bar splices are to be used, the number of rail lengths required, plus one short length in each run should be specified, to permit staggering of the joints. Orders must clearly specify that "THESE RAILS ARE INTENDED FOR CRANE SERVICE".

The ends of rails are often hardened to better resist rail end batter. Most manufacturers will chamfer the top and sides of the rail head at the ends, unless specified otherwise by the purchaser. Chamfering permits mild deformations to occur and minimizes chipping of the running surfaces.

When selecting a rail for crane service, the characteristics of operation must be considered. Some common variables which affect service life are:

- Frequency of operation
- Crane carriage speed and impact – rate of loading and unloading
- Corrosion – acidic mill conditions
- Abrasion
- Alignment of crane and supporting members
- Crane operating procedures

Crane rails are joined together end-to-end by either mechanical fasteners or welding. When bolting is used, special joint bars are employed as shown on page 6-144 and the rails are usually predrilled to 1/16-inch larger than the specified bolt size. If welded, manual arc welding is usually used and joint bars are not required. Welding has the advantage of eliminating mechanical joints, thus reducing the problem of aligning the top of rails.

PROPERTIES AND DIMENSIONS

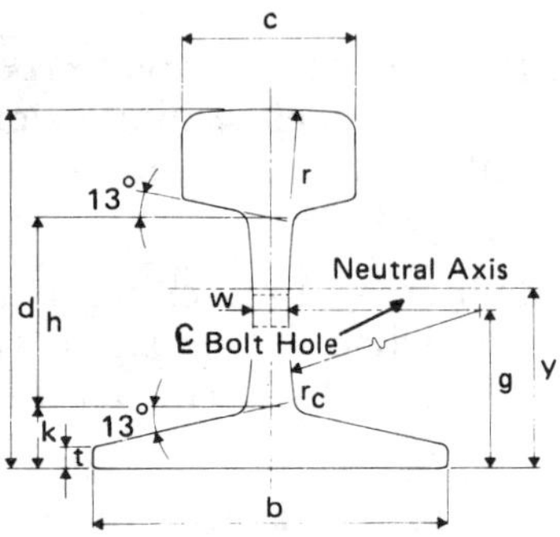

40 to 105 lbs.

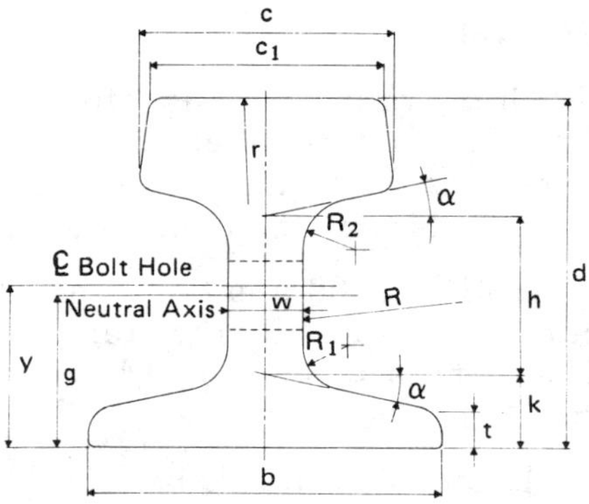

135 to 175 lbs.

Dimensions

Rail Type	Depth	Head	Base		Web						
						Gauge					
	d	c	b	t	w	g	k	h	r	r_c	R
	mm	mm	mm	mm	mm	mm	mm	mm	mm	mm	mm
ASCE 40	89	48	89	6	10	40	16	47	305	6.4	305
ASCE 60	108	60	108	7	12	52	19	58	305	6.4	305
ASCE 85	132	65	132	8	14	63	23	70	305	6.4	305
BETH 104	127	64	127	13	25	62	27	62	305	12.7	89
USS 105	132	65	132	10	24	56	25	61	305	6.4	305

Dimensions

Rail Type	Depth	Head		Base		Web								
							Gauge							
	d	c	c_1	b	t	w	g	k	h	r	R	R_1	R_2	α
	mm	mm	mm	mm	mm	mm	mm	mm	mm	mm	mm	mm	mm	deg
USS or BETH 135	146	87	76	132	12	32	63	27	71	356	305	19	19	13
BETH 171	152	109	102	152	16	32	67	32	70	FLAT	VERT	19	22	12
USS or BETH 175	152	108	102	152	13	38	68	29	79	457	VERT	29	51	12

Properties

Rail Type	Mass	Dead Load	Area	I_x	S_x Head	S_x Base	y
	kg/m	kN/m	mm^2	$10^6 mm^4$	$10^3 mm^3$	$10^3 mm^3$	mm
ASCE 40	19.8	0.195	2 540	2.72	58.8	63.7	42.7
60	29.8	0.292	3 830	6.06	108	116	52.1
85	42.2	0.413	5 370	12.5	182	199	62.7
BETH 104	51.6	0.506	6 640	12.4	175	221	56.1
USS 105	52.1	0.511	6 650	14.3	203	234	61.2
USS & BETH 135	67.0	0.657	8 590	21.1	282	295	71.1
BETH 171	84.9	0.832	10 800	30.6	402	400	76.5
USS & BETH 175	86.8	0.851	11 000	29.2	386	381	76.7

Rail fasteners

Hook bolts are primarily used when the flange of the crane beam is too narrow to permit the use of rail clamps. Hook bolts are used in groups of 2, located 75 to 100 mm apart, at 600 mm centres, and may be adjusted plus of minus 12 mm. Rails require special preparation either in the fabricator's shop or by the crane rail supplier.

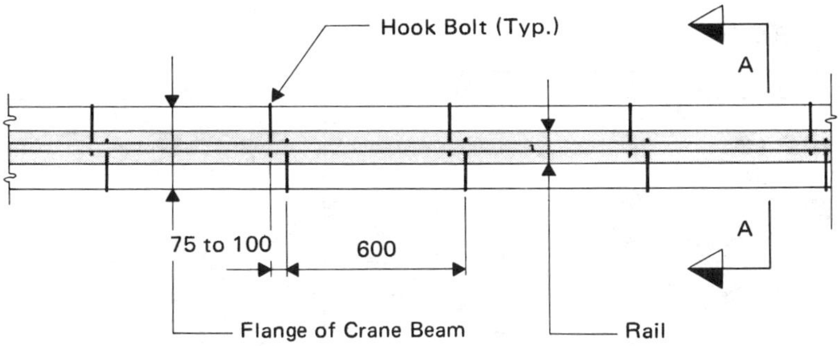

Suggested rail clamp dimensions are shown in Section B-B. For prefabricated rail clamps, reference should be made to Manufacturers' catalogues of track accessories. Two types of clamps are available; the tight clamp and the floating clamp. Floating clamps are used when longitudinal and controlled transverse movement is required for thermal expansion and alignment. Rail clamps are fabricated from pressed or forged steel and usually have single or double bolts.

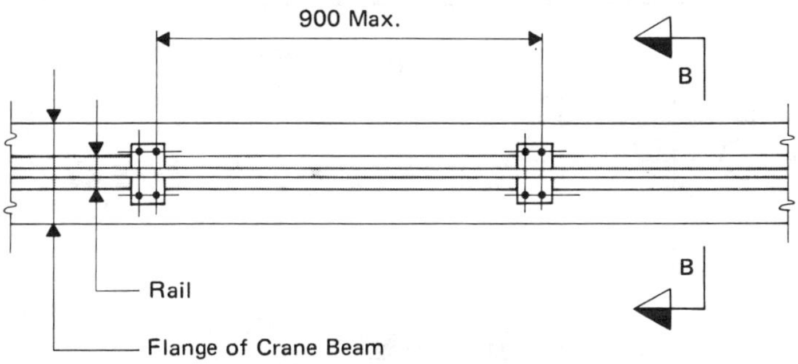

RAIL FASTENERS

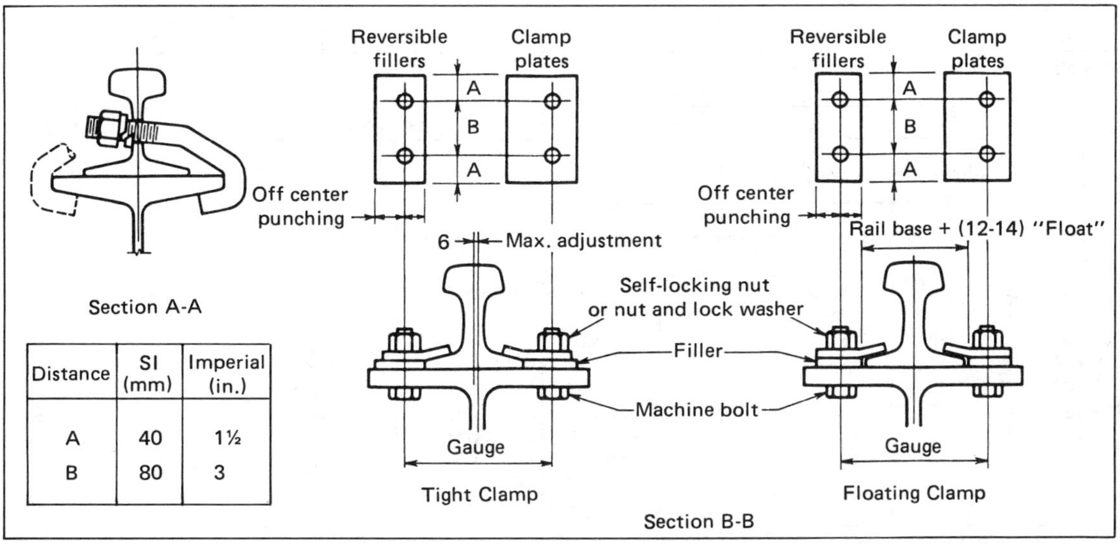

Distance	SI (mm)	Imperial (in.)
A	40	1½
B	80	3

RAIL SPLICES

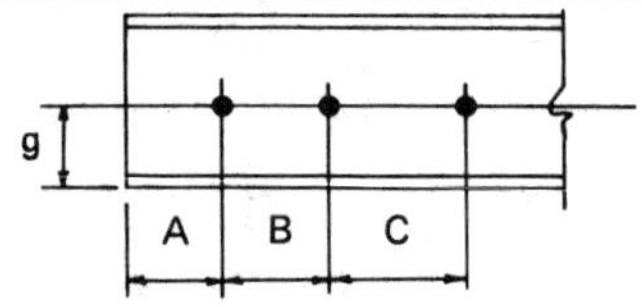

Rail End

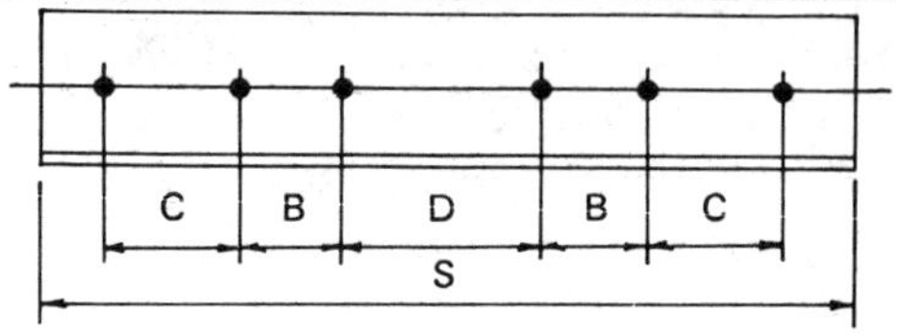

Joint Bar

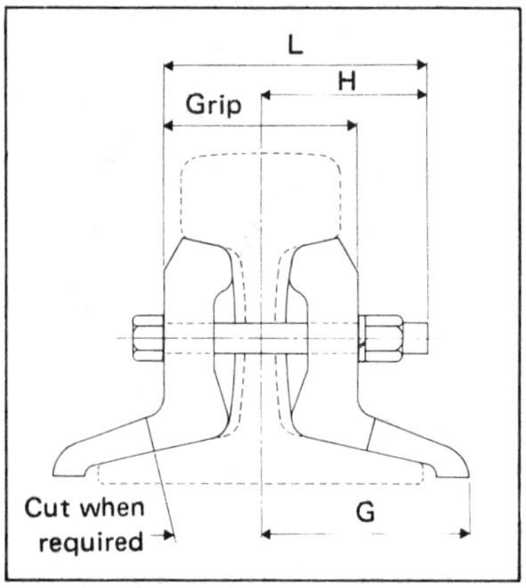

40 to 105 lbs.

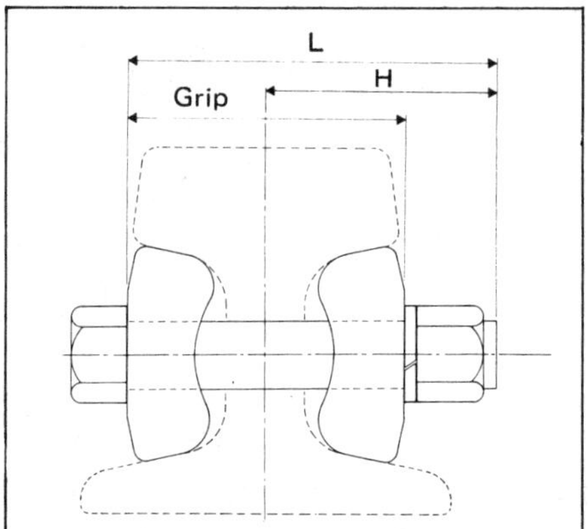

135 to 175 lbs.

Rail Type	Rail					Joint Bar					
	g	Hole dia.	A	B	C	Hole dia.	D	B	C	S	G
	mm	inch.	mm	mm	mm	inch.	mm	mm	mm	mm	mm
40	39.5	*13/16	63.5	127	–	*13/16	125	127	–	508	55.6
60	48.2	*13/16	63.5	127	–	*13/16	125	127	–	610	68.3
85	57.5	*15/16	63.5	127	–	*15/16	125	127	–	610	84.9
104	61.9	1-1/16	102	127	152	1-1/16	202	127	152	864	88.9
105	56.0	15/16	102	127	152	15/16	202	127	152	864	–
135	62.7	1-3/16	102	127	152	1-3/16	202	127	152	864	–
171	66.7	1-3/16	102	127	152	1-3/16	202	127	152	864	–
175	67.5	1-3/16	102	127	152	1-3/16	202	127	152	864	–

*special rail drilling and joint bar punching.

Rail Type	Bolt				Spring Washer		Mass of Ass'y	
	diam.	Grip	L	H	Hole dia.	Thk. & width	With Flg.	Without Flg.
	in.	mm	mm	mm	in.	in. in.	kg.	kg.
40	3/4	49.2	88.9	63.5	13/16	7/16 x 3/8	9.07	7.48
60	3/4	65.9	102	68.3	13/16	7/16 x 3/8	16.56	13.43
85	7/8	80.2	121	81.0	15/16	7/16 x 3/8	25.67	20.55
104	1	88.9	133	88.9	1-1/16	7/16 x 1/2	33.34	25.13
105	7/8	85.7	127	84.1	15/16	7/16 x 3/8	–	27.67
135	1-1/8	92.1	140	93.7	1-3/16	7/16 x 1/2	–	34.16
171	1-1/8	113	159	103	1-3/16	7/16 x 1/2	–	41.19
175	1-1/8	105	152	100	1-3/16	7/16 x 1/2	–	39.78

† For complete description of rail type, see Properties and Dimensions on page 6-142.

Splices

Rail drilling and joint bar punching as supplied for track work is not recommended for crane rails, since oversize holes may allow too much movement at the rail ends and result in failure. Tight joints which require special rail and joint bar drilling (see table on page 6-138) and squaring of the rail ends are recommended.

Light rails are not finished at the mill and are usually finished at the fabricator's shop or at the erection site. This may require reaming of holes for proper fit of bolts if dimensional tolerances are cumulative.

Joint bars are provided for crane service to match the rails ordered, and may be ordered blank. Under no circumstances should these joint bars be used as welding straps. Manufacturers' catalogues should be consulted for joint bar material specifications, dimensions and identification necessary to match the crane rail specified.

Joint bar bolts for crane service are readily identified from those used for track work, as they have straight shanks and are manufactured to ASTM A449 specification. Matching nuts are manufactured to ASTM A563 Grade B. The bolted assembly includes an alloy spring washer which is furnished to American Railway Engineers Association(AREA) specification. Bolts and nuts manufactured to ASTM A325 may also be acceptable.

To prolong the life of the runway, bolts should be retightened within 30 days after installation and every 3 months thereafter.

FASTENERS

General

The information on fasteners, provided herein, is based on standards, specifications and publications of the;

Canadian Standards Association

American National Standards Institute

Industrial Fasteners Institute

Research Council on Structural Connections

Additional fastener information can be obtained from the various manufacturers and from the Canadian Fasteners Institute.

Availability

The more commonly used fasteners for structural purposes in Canada have included the following:

5/8" ASTM A307 bolts for light steel framing such as girts, purlins, etc.

3/4" ASTM A325 bolts for building structures

7/8" ASTM A325 bolts for bridge structures

Larger sizes of ASTM A325 bolts, all sizes of ASTM A490 bolts, and all sizes of metric bolts (A325M and A490M) have not been in common use in Canada, and designers contemplating their use should first check for their availability.

Although the CISC Handbook has included data on metric bolts since publication of the Third Edition in 1980, demand for metric fasteners to date has not been sufficient to warrant their manufacture on a regular basis. However, with upcoming changes in Europe, and the introduction of the requirements for metric contained in the US Omnibus Trade Bill slated for 1992, it is expected that metric series bolts may start to become more readily available.

Clause 23.3.2(e) of CAN/CSA-S16.1-M89 permits the use of matching imperial size and metric size bolts in the appropriate metric size holes for the more commonly used high strength bolts. Designers and detailers should therefore base their calculations on the smaller size of matching fastener as appropriate, and allow substitutions as dictated by availability of metric or imperial size fasteners at the time of fabrication.

Definitions

Body Length means the distance from the underside of the head bearing surface to either the last scratch of thread or the top of the extrusion angle, whichever is the closest to the head.

Bolt Length means the length from the underside of the head bearing surface to the extreme point.

Finished Fastener means a fastener made to close tolerances and having surfaces other than the threads and bearing surface finished to provide a general high grade appearance.

Grip means the thickness of material or parts which the fastener is designed to secure when fully assembled.

Height of bolt Head means the overall distance, measured parallel to the fastener axis, from the extreme top (excluding raised identification marks) to the bearing surface and including the thickness of the washer face where provided.

Natural Finish means the as processed finish, unplated or uncoated of the bolt or nut.

Nominal size means the designation used for the purpose of general identification.

Proof Load means a specified test load which a fastener must withstand without any indication of significant deformation or failure.

Thickness of Nut means the overall distance from the top of the nut to the bearing surface, measured parallel to the axis of the nut.

Thread Length of a Bolt means the distance from the extreme point to the last complete thread.

Transition Thread Length means the distance from the last complete thread to either the last scratch of thread or the top of the extrusion angle, whichever is the closest to the head.

Washer Face means a circular boss on the bearing surface of a bolt or nut.

Tables

The following Tables are included in this section:

ASTM A325 and ASTM A490 High Strength Bolts and Nuts

ASTM A325M and ASTM A490M High Strength bolts and Nuts

Bolts Lengths for Various Grips — ASTM A325 and A490 bolts

Minimum and Maximum Grips for Metric Heavy Hex. Structural Bolts

Mass of ASTM A325 Bolts, Nuts and washers

Mass of ASTM A325M Bolts, Nuts and Washers

Metric Washer Dimensions

ASTM A307 Bolts and Nuts

Fasteners — Miscellaneous Detailing Data

Usual Gauges — W, M, S, C shapes, Angles

Erection Clearances — Bolt Impact Wrenches

ASTM A325 AND ASTM A490 HIGH STRENGTH BOLTS AND NUTS

REQUIRED MARKINGS

TYPE	A325		A490	
	BOLT	NUT	BOLT	NUT
1	(1) XYZ A325	Manufacturer's Identification Mark; XYZ; XYZ D; Arcs Indicate Grade C; Grade Mark (2) D, DH, 2 or 2H	XYZ A490	XYZ DH; DH or 2H (2)
2	XYZ A325; 3 Radial Lines at 60° Mandatory	Same as Type 1	XYZ A490; 6 Radial Lines at 30° Mandatory	Same as Type 1
3	(3) XYZ A325; Underline Mandatory	3; (3) XYZ DH3	(3) XYZ A490; Underline Mandatory	XYZ DH3

(1) Additional Optional 3 Radial Lines at 120° may be added.
(2) Type 3 also acceptable.
(3) Additional Optional Mark indicating Weathering Grade may be added.

DIMENSIONS

Dimensions						Coarse Thread Series—UNC**			
Nominal Bolt Size, Inches D	Bolt Dimensions*, Inches Heavy Hex Structural Bolts			Nut Dimensions*, Inches Heavy Hex nuts		Threads Per In.	Basic Pitch Dia	Section at Minor Dia	Tensile Stress Area
	Width across flats, F	Height, H	Thread length	Width across flats, W	Height, H		In.	In.^{2}	In.^{2}
1/2	7/8	5/16	1	7/8	31/64	13	0.4500	0.1257	0.1419
5/8	1 1/16	25/64	1 1/4	1 1/16	39/64	11	0.5660	0.202	0.226
3/4	1 1/4	15/32	1 3/8	1 1/4	47/64	10	0.6850	0.302	0.334
7/8	1 7/16	35/64	1 1/2	1 7/16	55/64	9	0.8028	0.419	0.462
1	1 5/8	39/64	1 3/4	1 5/8	63/64	8	0.9188	0.551	0.606
1 1/8	1 13/16	11/16	2	1 13/16	1 7/64	7	1.0322	0.693	0.763
1 1/4	2	25/32	2	2	1 7/32	7	1.1572	0.890	0.969
1 3/8	2 3/16	27/32	2 1/4	2 3/16	1 11/32	6	1.2667	1.054	1.155
1 1/2	2 3/8	15/16	2 1/4	2 3/8	1 15/32	6	1.3917	1.294	1.405

*Dimensions according to ANSI B18.2.1
**Thread dimensions according to ANSI B1.1

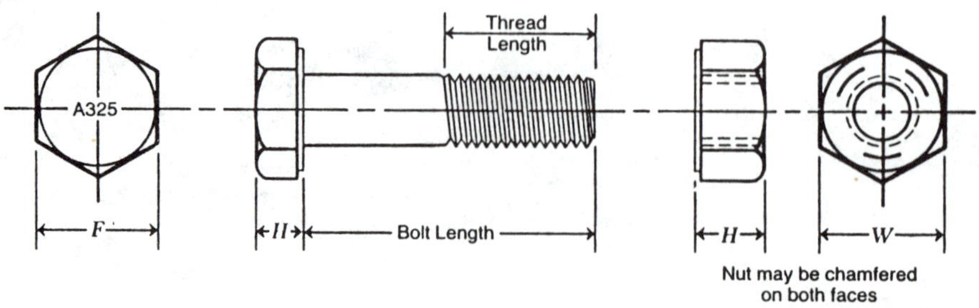

ASTM A325M AND ASTM A490M**
HIGH STRENGTH BOLTS AND NUTS†

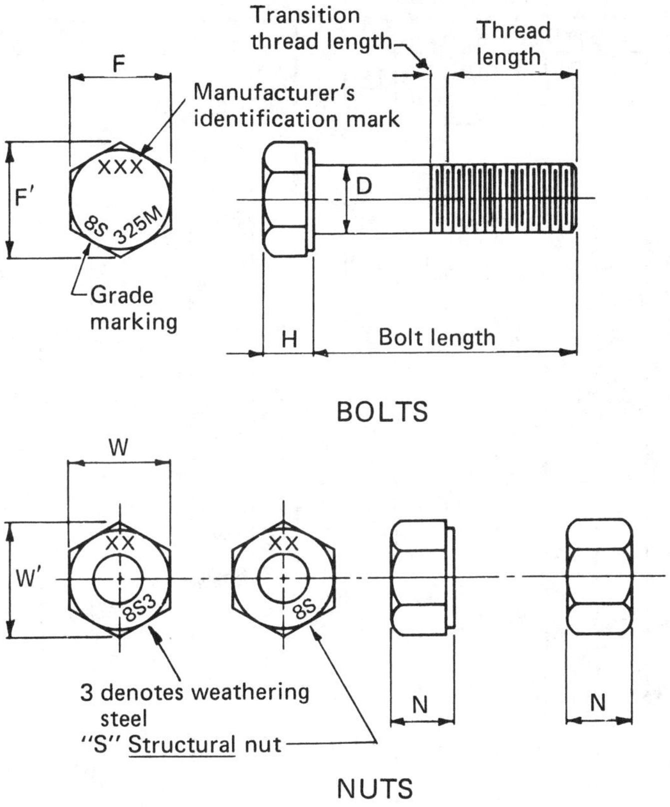

DIMENSIONS

Nominal Bolt Size	Heavy Hex Bolt or Nut Dimension				Heavy Hex Nut Max. Height N	Heavy Hex Structural Bolt			
	Across Flats F or W		Across Corners F' or W'			Max Head Height H	Thread Length*		Max. Transition Thread Length
	Max.	Min.	Max.	Min.			Bolt Lengths ⩽100	Bolt Lengths >100	
mm	mm	mm	mm	mm	mm	mm	mm	mm	mm
M16 x 2	27.00	26.16	31.18	29.56	17.1	10.75	31	38	6.0
M20 x 2.5	34.00	33.00	39.26	37.29	20.7	13.40	36	43	7.5
M22 x 2.5	36.00	35.00	41.57	39.55	23.6	14.90	38	45	7.5
M24 x 3	41.00	40.00	47.34	45.20	24.2	15.90	41	48	9.0
M27 x 3	46.00	45.00	53.12	50.85	27.6	17.90	44	51	9.0
M30 x 3.5	50.00	49.00	57.74	55.37	30.7	19.75	49	56	10.5
M36 x 4	60.00	58.80	69.28	66.44	36.6	23.55	56	63	12.0

*Does not include transition thread length.

Bolt dimensions conform to those listed in ANSI B18.2.3.7M-1979 "Metric Heavy Hex Structural Bolts", and the nut dimensions conform to those listed in CSA Standard B18.2.4.6-M1980 "Metric Heavy Hex Nuts".

** Strength requirements are based on ASTM Specifications A325M and A490M. See page 3–5.

BOLT LENGTHS* for VARIOUS GRIPS**
ASTM A325 and A490 Bolts

Grip mm	Grip in.	Bolt Diameter, in. ½	⅝	¾	⅞	1	1⅛	1¼	1⅜	1½
19	¾	1½	1¾		2			2½		2¾
21	13/16			2		2¼	2½		2¾	
22	⅞	1¾								
24	15/16		2		2¼			2¾		3
25	1									
27	1 1/16			2¼		2½	2¾		3	
29	1⅛	2								
30	1 3/16		2¼		2½			3		3¼
32	1¼									
33	1 5/16			2½		2¾	3		3¼	
35	1⅜	2¼								
37	1 7/16		2½		2¾			3¼		3½
38	1½									
40	1 9/16			2¾		3	3¼		3½	
41	1⅝	2½								
43	1 11/16		2¾		3			3½		3¾
44	1¾									
46	1 13/16			3		3¼	3½		3¾	
48	1⅞	2¾								
49	1 15/16		3		3¼			3¾		4
51	2									
52	2 1/16			3¼		3½	3¾		4	
54	2⅛	3								
56	2 3/16		3¼		3½			4		4¼
57	2¼									
59	2 5/16			3½		3¾	4		4¼	
60	2⅜	3¼								
62	2 7/16		3½		3¾			4¼		4½
64	2½									
65	2 9/16			3¾		4	4¼		4½	
67	2⅝	3½								
68	2 11/16		3¾		4			4½		4¾
70	2¾									
71	2 13/16			4		4¼	4½		4¾	
73	2⅞	3¾								
75	2 15/16		4		4¼			4¾		5
76	3									
78	3 1/16									

Grip mm	Grip in.	Bolt Diameter, in. ½	⅝	¾	⅞	1	1⅛	1¼	1⅜	1½
76	3		4		4¼			4¾		5
78	3 1/16			4¼		4½	4¾		5	
79	3⅛	4								
81	3 3/16		4¼		4½			5		5¼
83	3¼									
84	3 5/16			4½		4¾	5		5¼	
86	3⅜	4¼								
87	3 7/16		4½		4¾			5¼		5½
89	3½									
90	3 9/16			4¾		5	5¼		5½	
92	3⅝	4½								
94	3 11/16		4¾		5			5½		5¾
95	3¾									
97	3 13/16			5		5¼	5½		5¾	
98	3⅞	4¾								
100	3 15/16		5		5¼			5¾		6
102	4									
103	4 1/16			5¼		5½	5¾		6	
105	4⅛	5								
106	4 3/16		5¼		5½			6		6¼
108	4¼									
109	4 5/16			5½		5¾	6		6¼	
111	4⅜	5¼								
113	4 7/16		5½		5¾			6¼		6½
114	4½									
116	4 9/16			5¾		6	6¼		6½	
117	4⅝	5½								
119	4 11/16		5¾		6			6½		6¾
121	4¾									
122	4 13/16			6		6¼	6½		6¾	
124	4⅞	5¾								
125	4 15/16		6		6¼			6¾		7
127	5									
129	5 1/16			6¼		6½	6¾		7	
130	5⅛	6								
132	5 3/16		6¼		6½			7		7¼
133	5¼									
135	5 5/16									

*Bolt lengths must be specified in inches for ASTM A325 and A490 bolts.
**Grip is thickness of material to be connected exclusive of washers.
For each flat washer, add 5 mm (3/16-inch) to grip.
For each beveled washer, add 8 mm (5/16-inch) to grip.

MINIMUM AND MAXIMUM GRIPS FOR METRIC HEAVY HEX. STRUCTURAL BOLTS, IN MILLIMETRES

Nominal Bolt Size	M16		M20		M22		M24		M27		M30		M36	
L Nominal Length (mm)	Min. Grip	Max. Grip	Min. Grip	Max. Grip	Min. Grip	Max. Grip	Min. Grip	Max. Grip	Min. Grip	Max. Grip	Min. Grip	Max. Grip	Min. Grip	Max. Grip
45	14	26		23		20								
50	19	31	14	28		25		24						
55	24	36	19	32	17	29		29		25				
60	29	41	24	37	22	34	19	34		30		27		
65	34	46	29	42	27	39	24	39	21	35		32		
70	39	51	34	47	32	44	29	44	26	40	21	37		31
75	44	56	39	52	37	49	34	49	31	45	26	42		36
80	49	61	44	57	42	54	39	54	36	50	31	47	24	41
85	54	66	49	62	47	59	44	59	41	55	36	52	29	46
90	59	71	54	67	52	64	49	64	46	60	41	57	34	51
95	64	76	59	72	57	69	54	69	51	65	46	62	39	56
100	69	81	64	77	62	74	59	74	56	70	51	67	44	61
110	72	91	67	87	65	84	62	84	59	80	54	77	47	71
120	82	101	77	97	75	94	72	94	69	90	64	87	57	81
130	92	110	87	107	85	104	82	103	79	100	74	97	67	91
140	102	120	97	117	95	114	92	113	89	110	84	107	77	101
150	112	130	107	127	105	124	102	123	99	120	94	117	87	111
160	122	138	117	135	115	132	112	131	109	128	104	125	97	119
170	132	148	127	145	125	142	122	141	119	138	114	135	107	129
180	142	158	137	155	135	152	132	151	129	148	124	145	117	139
190	152	168	147	165	145	162	142	161	139	158	134	155	127	149
200	162	178	157	175	155	172	152	171	149	168	144	165	137	159
210	172	188	167	185	165	182	162	181	159	178	154	175	147	169
220	182	198	177	195	175	192	172	191	169	188	164	185	157	179
230	192	208	187	205	185	202	182	201	179	198	174	195	167	189
240	202	218	197	215	195	212	192	211	189	208	184	205	177	199
250	212	228	207	225	205	222	202	221	199	218	194	215	187	209
260	222	238	217	235	215	232	212	231	209	228	204	225	197	219
270	232	248	227	245	225	242	222	241	219	238	214	235	207	229
280	242	258	237	255	235	252	232	251	229	248	224	245	217	239
290	252	268	247	265	245	262	242	261	239	258	234	255	227	249
300	262	278	257	275	255	272	252	271	249	268	244	265	237	259

1. This table is based on ANSI B18.2.3.7M-1979.
2. Bolts with lengths above the heavy solid line are threaded full length.

MASS OF ASTM A325 BOLTS, NUTS AND WASHERS

APPROXIMATE WEIGHTS IN POUNDS PER 100 UNITS

HEAVY HEX STRUCTURAL BOLTS WITH HEAVY HEX NUTS (WITHOUT WASHERS)									
Length Under Head Inches	**Diameter of Bolt, Inches**								
	½	⅝	¾	⅞	1	1⅛	1¼	1⅜	1½
1½	19.2	33.1	52.2	78.0	109	148	197		
1¾	20.5	35.3	55.3	81.9	114	154	205	261	333
2	21.9	37.4	58.4	86.1	119	160	212	270	344
2¼	23.3	39.8	61.6	90.3	124	167	220	279	355
2½	24.7	41.7	64.7	94.6	130	174	229	290	366
2¾	26.1	43.9	67.8	98.8	135	181	237	300	379
3	27.4	46.1	70.9	103	141	188	246	310	391
3¼	28.8	48.2	74.0	107	146	195	255	321	403
3½	30.2	50.4	77.1	111	151	202	263	332	416
3¾	31.6	52.5	80.2	116	157	209	272	342	428
4	33.0	54.7	83.3	120	162	216	280	353	441
4¼	34.3	56.9	86.4	124	168	223	289	363	453
4½	35.7	59.0	89.5	128	173	230	298	374	465
4¾	37.1	61.2	92.7	133	179	237	306	384	478
5	38.5	63.3	95.8	137	184	244	315	395	490
5¼	39.9	65.5	98.9	141	190	251	324	405	503
5½	41.2	67.7	102	146	196	258	332	416	515
5¾	42.6	69.8	105	150	201	265	341	426	527
6	44.0	71.9	108	154	207	272	349	437	540
6¼	45.4	74.1	111	158	212	279	358	447	552
6½	46.8	76.3	114	163	218	286	367	458	565
6¾	48.1	78.5	118	167	223	293	375	468	577
7	49.5	80.6	121	171	229	300	384	479	589
7¼	50.9	82.8	124	175	234	307	392	489	602
7½	52.3	84.9	127	179	240	314	401	500	614
7¾	53.6	87.1	130	183	246	321	410	510	626
8	55.0	89.2	133	187	251	328	418	521	639
8¼	56.4	91.4	136	192	257	335	427	531	651
8½	57.8	93.5	139	196	262	342	435	542	664
8¾	59.1	95.7	142	200	268	349	444	552	676
9	60.5	97.8	145	204	273	356	453	563	689
9¼	62.0	100	149	209	279	363	461	574	701
9½	63.3	102	152	213	284	370	470	584	713
9¾	64.7	104	155	217	290	377	478	595	726
10	66.1	107	158	221	295	384	487	605	738
Per Inch Additional	5.5	8.6	12.4	16.9	22.1	28.0	34.4	42.5	49.7

Plain Round Washers	2.0	3.9	4.5	7.1	9.3	11.3	14.1	16.8	20.0
Bevelled Square Washers	24.9	23.8	22.6	21.0	19.2	34.0	31.6	—	—
Heavy Hex Nuts	6.54	11.9	19.3	29.7	42.5	59.2	78.6	102	131

MASS OF ASTM A325M BOLTS, NUTS AND WASHERS

APPROXIMATE MASS* OF 1000 UNITS (kg)

HEAVY HEX STRUCTURAL BOLTS WITH HEAVY HEX NUTS (WITHOUT WASHERS)

Bolt Length	Bolt Size (number following the letter M is the nominal bolt diameter in millimetres)						
mm	M16	M20	M22	M24	M27	M30	M36
40	170						
45	177	321	396				
50	184	331	409	539	744		
55	191	342	423	554	764		
60	198	353	436	570	784	1000	
65	204	364	449	586	804	1030	
70	211	375	462	601	824	1050	1710
75	218	386	476	617	845	1000	1750
80	225	397	489	632	865	1100	1780
85	232	407	502	648	885	1120	1820
90	239	418	515	664	905	1150	1850
95	246	429	529	679	925	1170	1890
100	253	440	542	695	945	1200	1920
110	267	462	569	726	985	1250	2000
120	281	483	595	757	1020	1300	2070
130	295	505	622	788	1070	1350	2140
140	309	527	648	820	1110	1400	2210
150	322	548	675	851	1150	1440	2280
160	336	570	701	882	1190	1490	2350
170	350	592	728	913	1230	1540	2420
180	364	613	754	945	1270	1590	2490
190	378	635	781	976	1310	1640	2570
Extra per 10 mm	13.9	21.7	26.5	31.2	40.1	49.2	71.3
Mass of Extra Heavy Hex Nuts	61.0	118	145	201	286	371	635
Mass of Plain Washers	18	27	33	45	57	65	100
Mass of Bevel Washers	101	91	85	78	154	142	115

*Computed theoretical mass using a steel density of 7 850 kg/m^3.

METRIC WASHER DIMENSIONS

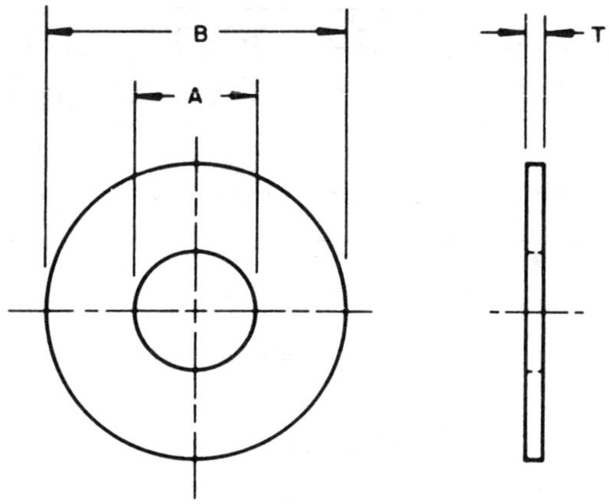

PLAIN CIRCULAR WASHERS

Metric Bolt Size	B Outside Diameter		A Hole Diameter		T Thickness	
	Max	Min	Max	Min	Max	Min
M16 x 2	34.0	32.4	18.4	18.0	4.6	3.1
M20 x 2.5	41.0	39.4	22.5	22.0	4.6	3.1
M22 x 2.5	44.0	42.4	24.5	24.0	4.6	3.4
M24 x 3	50.0	48.4	26.5	26.0	4.6	3.4
M27 x 3	56.0	54.1	30.5	30.0	4.6	3.4
M30 x 3.5	60.0	58.1	33.5	33.0	4.6	3.4
M36 x 4	72.0	70.1	39.5	39.0	4.6	3.4

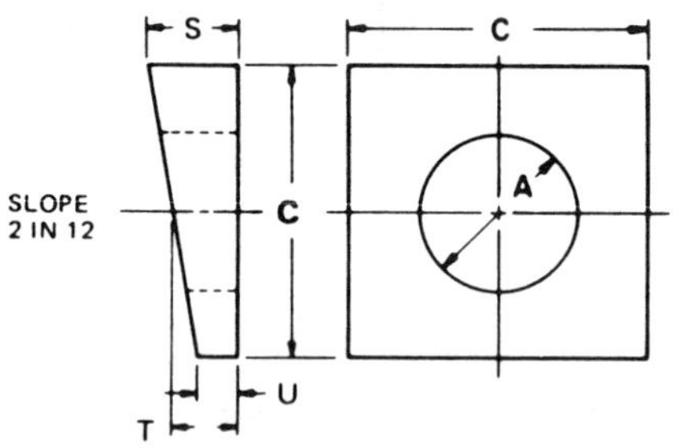

BEVELLED SQUARE WASHERS

Metric Bolt Size	C Width		A Hole Diameter		S Thickness: Thick Side ±0.5	T Thickness: Mean Nom.	U Thickness: Thin Side ±0.5
	Max	Min	Max	Min			
M16 x 2	45.0	43.0	18.4	18.0	11.7	8	4.3
M20 x 2.5	45.0	43.0	22.5	22.0	11.7	8	4.3
M22 x 2.5	45.0	43.0	24.5	24.0	11.7	8	4.3
M24 x 3	45.0	43.0	26.5	26.0	11.7	8	4.3
M27 x3	59.0	57.0	30.5	30.0	12.8	8	3.2
M30 x 3.5	59.0	57.0	33.5	33.0	12.8	8	3.2
M36 x 4	59.0	57.0	39.5	39.0	12.8	8	3.2

ASTM A307 BOLTS AND NUTS
DIMENSIONS IN IMPERIAL UNITS

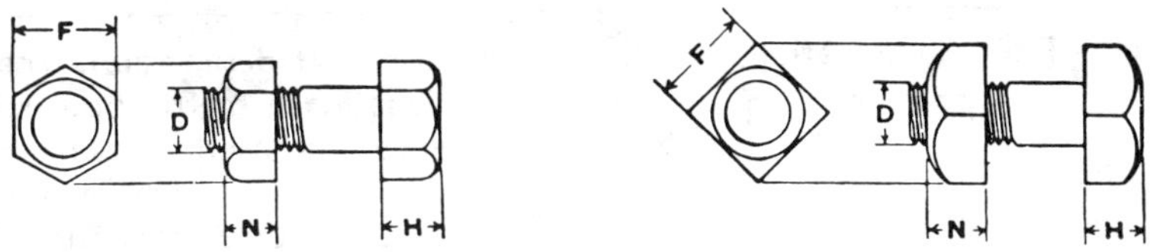

Nominal Diameter of Bolt	Regular Square Bolts		Finished Hex Bolts		Minimum Thread Lengths		Regular Square Nuts		Finished Hex Nuts	
	Nominal Width F	Nominal Height H	Basic Width F	Basic Height H	Length Under Head: 6 In. and Under	Length Under Head: Over 6 In.	Nominal Width F	Nominal Height N	Basic Width F	Basic Height N
Inches	Inches	Inches	Inches	Inches	Inches	Inches	Inches	Inches	Inches	Inches
1/4	3/8	11/64	7/16	5/32	¾	1	7/16	7/32	7/16	7/32
3/8	9/16	1/4	9/16	15/64	1	1¼	5/8	21/64	9/16	21/64
1/2	3/4	21/64	3/4	5/16	1¼	1½	13/16	7/16	3/4	7/16
5/8	15/16	27/64	15/16	25/64	1½	1¾	1	35/64	15/16	35/64
3/4	1 1/8	1/2	1 1/8	15/32	1¾	2	1 1/8	21/32	1 1/8	41/64
7/8	1 5/16	19/32	1 5/16	35/64	2	2¼	1 5/16	49/64	1 5/16	3/4
1	1 1/2	21/32	1 1/2	39/64	2¼	2½	1 1/2	7/8	1 1/2	55/64
1 1/8	1 11/16	3/4	1 11/16	11/16	2½	2¾	1 11/16	1	1 11/16	31/32
1 1/4	1 7/8	27/32	1 7/8	25/32	2¾	3	1 7/8	1 3/32	1 7/8	1 1/16

The dimensions for Regular Square Bolts conform to those for Square Head Bolts, Regular Series listed in CSA Standard B33.1-1961, "Square and Hexagon Bolts and Nuts, Studs and Wrench Openings", and to those for Square Bolts listed in ANSI Standard B18.2.1-1965, "Square and Hex Bolts and Screws".

The dimensions for Finished Hexagon Bolts (Hex Cap Screws) conform to those for Hexagon Head Bolts, Finished Grade, Regular Series listed in CSA Standard B33.1-1961, and to those for Hex Cap Screws (Finished Hex Bolts) listed in ANSI Standard B18.2.1-1965.

The minimum thread lengths are in agreement with the requirements of CSA Standard B33.1-1961 and with ANSI Standard B18.2.1-1965. In general, these requirements are as follows:

Bolts 6 inches or less in length — twice diameter plus ¼-inch.

Bolts longer than 6 inches — twice diameter plus ½-inch.

Bolts too short for the above thread lengths shall be threaded as close to the head as practicable.

The dimensions for Regular Square Nuts conform to those for Square Nuts, Regular Series listed in CSA Standard B33.1-1961, and to those for Square Nuts listed in ANSI Standard B18.2.2-1965, "Square and Hex Nuts"

The dimensions for Finished Hexagon Nuts conform to those for Hexagon Nuts, Finished Grade, Regular Series listed in CSA Standard B33.1-1961 and to those for Hex Nuts listed in ANSI Standard B18.2.2-1965.

Note: Square head bolts are used with either square or hexagon nuts. However, the use of Finished Hexagon Bolts and Finished Hexagon Nuts is gradually replacing the use of square head bolts and nuts in sizes up to one inch in diameter and six inches in length.
A307 bolts and nuts are manufactured in Imperial dimensions only.

FASTENERS — MISCELLANEOUS DETAILING DATA

Metric Fastener Designations

THREAD DATA

Diameter Pitch Combinations

Nominal dia. (mm)	Thread pitch (mm)	Nominal dia. (mm)	Thread pitch (mm)
1.6	0.35	20	2.5
2	0.4	22	2.5
2.5	0.45	24	3
3	0.5	27	3
3.5	0.6	30	3.5
4	0.7	36	4
5	0.8	42	4.5
6.0	1.0	48	5
8	1.25	56	5.5
10	1.5	64	6
12	1.75	72	6
14	2	80	6
16	2	90	6
		100	6

Basic Metric Thread designation: Metric screw threads are designated by the letter "M" followed by the nominal size (basic major diameter) in millimetres and the pitch in millimetres separated by the symbol "X".

M12	X	**1.75**	**−6g**
Size (mm)		Thread (pitch in mm)	Standard class of fit

Note: In the metric system, the pitch of the thread is given in mm instead of threads per inch — thus a M12 x 1.75 thread has a nominal diameter of 12 mm and the pitch of the thread is 1.75 mm.

PRODUCT DESIGNATION

Metric Bolt designation: The standard method of designating a metric bolt is by specifying (in sequence) the product name, nominal diameter and thread pitch, nominal length, type, steel property class, and protective coating (if required).

Heavy Hex Structural Bolt, M22x2.5x160, Type 2, ASTM A325M-79, Zinc Galvanized

Metric Nut designation: The standard method of designating a metric nut is by specifying (in sequence) the product name, nominal diameter and pitch, steel property class or material identification, and protective coating (if required).

Heavy Hex Nut, M30 x 3.5, ASTM A563M class 105, hot dipped galvanized

Note: It is common practice to omit the thread pitch from the product designation.

Slotted Hole Dimensions

See Clause 23.3.2, S16.1-M regarding use provisions.

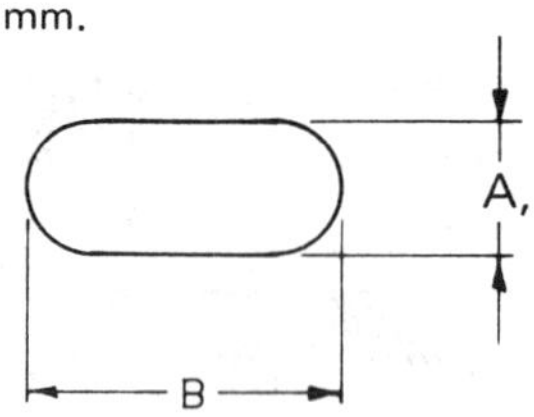

A, nominal diameter + 2mm

SHORT SLOT DIMENSIONS

Nominal Bolt Diameter	Slot Dimensions	
	Width, A	Length, B
mm	mm	mm
16	18	22
20	22	26
22	24	28
24	26	32
27	29	37
30	32	40
36	38	46

LONG SLOT DIMENSIONS

Nominal Bolt Diameter	Slot Dimensions	
	Width, A	Length, B
mm	mm	mm
16	18	40
20	22	50
22	24	55
24	26	60
27	29	67.5
30	32	75
36	38	90

FASTENERS — MISCELLANEOUS DETAILING DATA
Diagonal Distance for Staggered Fasteners

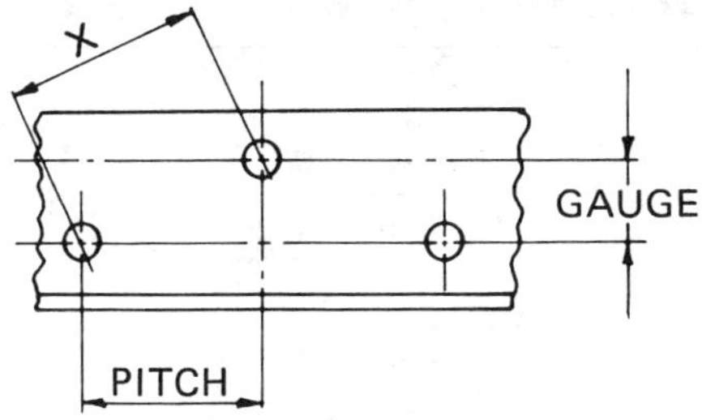

Pitch mm	GAUGE, mm																	
	25	30	35	40	45	50	55	60	65	70	75	80	85	90	95	100	105	110
5	25	30	35	40	45	50	55	60	65	70	75	80	85	90	95	100	105	110
10	27	32	36	41	46	51	56	61	66	71	76	81	86	91	96	100	105	110
15	29	34	38	43	47	52	57	62	67	72	76	81	86	91	96	101	106	111
20	32	36	40	45	49	54	59	63	68	73	78	82	87	92	97	102	107	112
25	35	39	43	47	51	56	60	65	70	74	79	84	89	93	98	103	108	113
30	39	42	46	50	54	58	63	67	72	76	81	85	90	95	100	104	109	114
35	43	46	49	53	57	61	65	69	74	78	83	87	92	97	101	106	111	115
40	47	50	53	57	60	64	68	72	76	81	85	89	94	98	103	108	112	117
45	51	54	57	60	64	67	71	75	79	83	87	92	96	101	105	110	114	119
50	56	58	61	64	67	71	74	78	82	86	90	94	99	103	107	112	116	121
55	60	63	65	68	71	74	78	81	85	89	93	97	101	105	110	114	119	123
60	65	67	69	72	75	78	81	85	88	92	96	100	104	108	112	117	121	125
65	70	72	74	76	79	82	85	88	92	96	99	103	107	111	115	119	123	128
70	74	76	78	81	83	86	89	92	96	99	103	106	110	114	118	122	126	130
75	79	81	83	85	87	90	93	96	99	103	106	110	113	117	121	125	129	133
80	84	85	87	89	92	94	97	100	103	106	110	113	117	120	124	128	132	136
85	89	90	92	94	99	99	101	104	107	110	113	117	120	124	127	131	135	139
90	93	95	97	98	101	103	105	108	111	114	117	120	124	127	131	135	138	142

BOLT LENGTH TOLERANCES

Nominal Length	Nominal Bolt Dia M16 thru M36
to 50 mm	± 1.2
over 50 to 80 mm	± 1.5
over 80 to 120 mm	± 1.8
over 120 to 150 mm	± 2.0
over 150 mm	± 4.0

MINIMUM EDGE DISTANCE FOR BOLT HOLES

Bolt Diameter mm	At sheared Edge mm	At Rolled or Gas Cut Edge † mm
16	28	22
20	34	26
22	38	28
24	42	30
27	48	34
30	52	38
36	64	46
over 36	1¾ x Diameter	1¼ x Diameter

† Gas cut edges shall be smooth and free from notches. Edge distance in this column may be decreased 3 mm when hole is at a point where computed stress under factored loads is not more than 0.3 of the yield stress.

USUAL GAUGES for W, M, S and C Shapes, Millimetres

S SHAPES

Flange Width	g†	
60 to 70	40	φ
75 to 90	45	φ
92 to 100	50	φ
102 to 120	60	
130 to 145	80	
150 to 185	100	
190 to 200	130	

STANDARD CHANNELS

Flange Width	g†	
35 to 40	22	φ
45	25	φ
47 to 50	30	φ
55	32	
57 to 60	35	
64 to 70	38	
73 to 80	45	
85 to 92	60	
95 to 110	65	

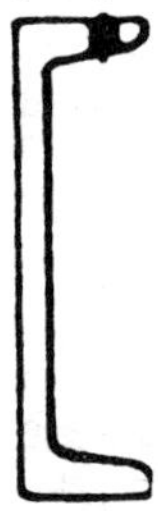

W AND M SHAPES

Flange Width	g†
100	60
130 to 145	80
150 to 180	100
190 and up	130

Extra gauges for W columns: 50 and 60; 70 and 80; 85 and 90

Depth of W or M column	150	200	250 to 350
g	60	100	130

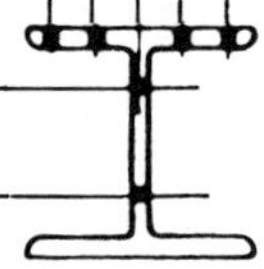

φ Holes usually drilled due to size of punch die block.
† Dependent on edge distance.

USUAL GAUGES for Angles, millimetres

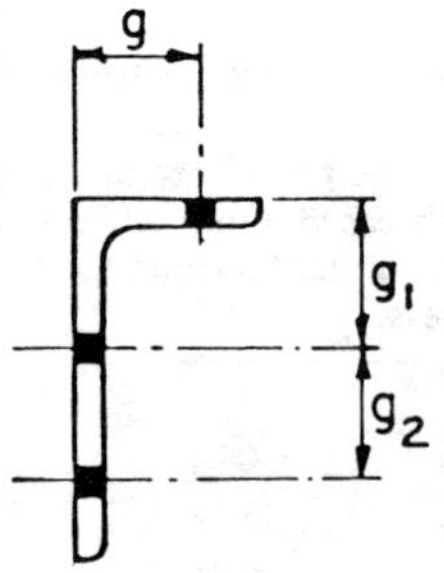

Notes:
Those values shown above the dashed line allow for full socket wrench clearance requirements.

The bolt sizes shown in italics to the left of g and g_1 are the maximum bolt sizes permissible for the dimensions shown.

$g_2 \geqslant$ 2-2/3 bolt diameters.

Leg \ Gauge	g		g_1		g_2	
200	*M36*	115	*M30*	80		80
150	*M36*	90	*M24*	55		65
125	*M30*	80	*M20*	45		54
100	*M27*	65				
90	*M24*	60				
80	*M24*	50				
75	*M24*	45				
65	*M24*	35				
60	*M24*	30				
55	*M22*	27				
50	*M16*	28				
45	*M16*	23				

ERECTION CLEARANCES
Bolt Impact Wrenches

METRIC

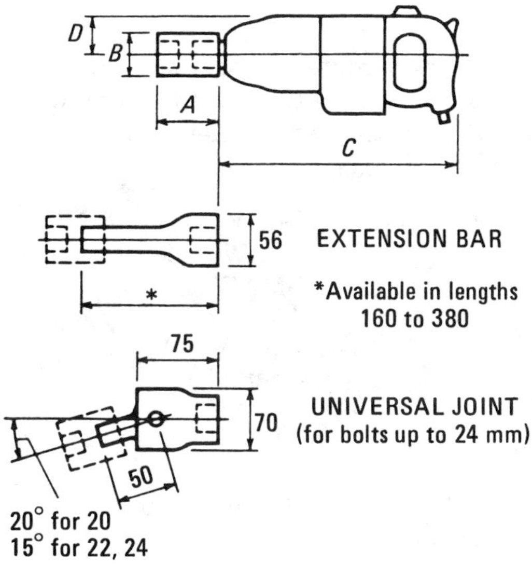

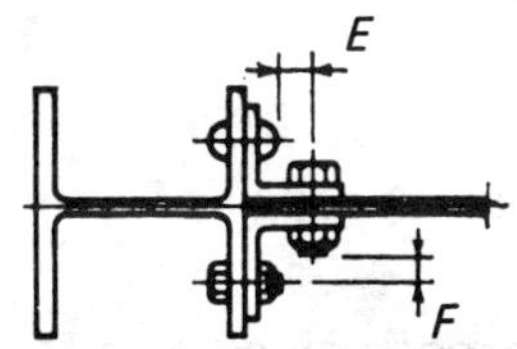

MINIMUM CLEARANCES

	Size	C	D
Light Wrenches	16 to 24	337 to 356	54
Heavy Wrenches	24 to 36	375 to 438	64

Sockets			Min. Clear.	
Bolt Size	A	B	E	F
16	80	45	25	28
20	85	54	30	34
22	90	57	32	36
24	95	60	34	38
27	100	70	38	42
30	110	75	41	45
36	130	90	48	52

IMPERIAL

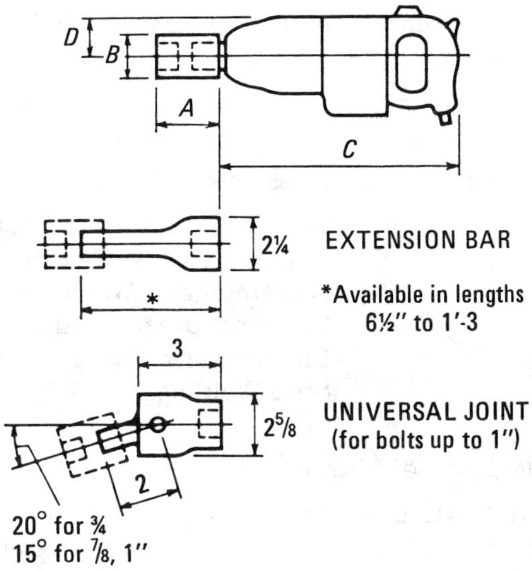

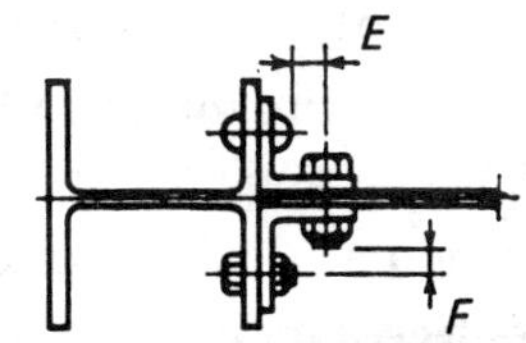

MINIMUM CLEARANCES

	Size	C	D
Light Wrenches	5/8 to 1	1-1¼ to 1-2	2 1/8
Heavy Wrenches	1 to 1½	1-2¾ to 1-5¼	2½

Sockets			Min. Clear.	
Bolt Size	A	B	E	F
5/8	2-5/8	1-3/4	1-1/8	1-1/4
3/4	3	2-1/4	1-1/4	1-3/8
7/8	3-1/4	2-1/2	1-5/16	1-7/16
1	3-1/2	2-5/8	1-7/16	1-9/16
1-1/8	3-3/4	2-7/8	1-9/16	1-11/16
1-1/4	4	3-1/8	1-5/8	1-3/4
1-3/8	4-1/4	3-1/4	2-1/8	2-1/4
1-1/2	4-3/8	4-1/4		

WELDING

The welding of steel shapes and plates for structural purposes is governed by CAN/CSA-S16.1-M, Limit States Design of Steel Structures, and CSA Standard W59, Welded Steel Construction (Metal-Arc Welding).

While both standards provide design information on the strength of welds, CSA Standard W59 extensively covers workmanship, inspection and acceptance criteria for welded joints in both statically and dynamically loaded structures.

Welding is a process used to join two or more pieces of material together. Arc welding is a process which produces coalescence of metals by heating them with an arc, with or without the application of pressure, and with or without the use of filler metal.

Welding processes used primarily for structural steel work are:

Shielded Metal Arc Welding	SMAW
Flux Cored Arc Welding	FCAW
Gas Metal Arc Welding	GMAW
Submerged Arc Welding	SAW
Electroslag Welding	ESW
Stud Welding	STW

Welding Definitions

Arc Cutting: means a group of cutting processes which melts the metal to be cut with the heat of an arc between an electrode and the base metal.

Arc Spot Weld: means a weld made by arc welding between or upon overlapping members in which coalescence may start and occur on the faying surfaces or may proceed from the surface of one member.

Base Metal: means the metal to be welded or cut.

Bevel Angle: means the angle formed between the prepared edge of a member and a plane perpendicular to the surface of the member.

Chain Intermittent Welds: means intermittent welds on both sides of a joint in which the weld increments on one side are approximately opposite those on the other side.

Coalescence: means the growing together or growth into one body of the material being welded.

Complete Joint Penetration: means when the weld metal completely fills the groove and is fused to the base metal throughout its total thickness.

Edge Joint: means a joint between the edge of two or more parallel or nearly parallel members.

Effective Weld Length: means the length of weld throughout which the correctly proportioned cross section exists. In a curved weld, it is measured along the axis of the weld.

Effective Throat: means the minimum distance from the root of a weld to its face less any reinforcement.

End Return (Boxing): means the continuation of a fillet weld around a corner of a member as an extension of the principal weld.

Face of Weld: means the exposed surface of a weld on the side from which the welding was done.

Fillet Weld: means a weld of approximately triangular cross section joining two surfaces approximately at right angles to each other in a lap joint, T-joint, or corner joint.

Groove Weld: means a weld made in a groove between two members to be joined.

Intermittent Weld: means a weld in which the continuity is broken by recurring unwelded spaces.

Joint Design: means the joint geometry together with the required dimensions of the welded joint.

Joint Penetration: means the minimum depth a groove weld extends from its face into a joint, exclusive of reinforcement, but including, if present, root penetration.

Leg of a Fillet Weld: is the distance from the root of the joint to the toe of the fillet weld.

Partial Joint Penetration: means a joint penetration which is less than complete.

Procedure Qualification: means a demonstration that welds made by a specific procedure can meet prescribed standards.

Root of Joint: means that portion of a joint to be welded where the members approach closest to each other. In cross section, the root of the joint may be either a point, a line or an area.

Root of Weld: means the points, as shown in cross section, at which the back of the weld intersects the base metal surfaces.

Root Penetration: means the depth that a weld extends into the root of a joint measured on the centreline of the root cross section.

Size of Weld:

> **Groove Weld:** means the joint penetration (depth of bevel plus the root penetration when specified). The size of a groove weld and its effective throat are one and the same.
>
> **Fillet Weld:**
>
> For equal leg fillet welds, the leg lengths of the largest isosceles right triangle which can be inscribed within the fillet weld cross section.
>
> For unequal leg fillet welds, the leg lengths of the largest right triangle which can be inscribed within the fillet weld cross section.
>
> Note: When one member makes an angle with the other member greater than 105 degrees, the leg length (size) is of less significance than the effective throat which is the controlling factor for the strength of a weld.

Tack Weld: means a weld made to hold parts of a weldment in proper alignment until the final welds are made.

Throat of a Fillet Weld.

> **Theoretical Throat:** means the distance from the beginning of the root of the joint perpendicular to the hypotenuse of the largest right triangle that can be inscribed within the fillet weld cross section. This dimension is based on the assumption that the root opening is equal to zero.
>
> **Actual Throat:** means the shortest distance from the root of weld to its face.
>
> **Effective Throat:** means the minimum distance minus any reinforcement from the root of weld to its face.

Fillet Welds

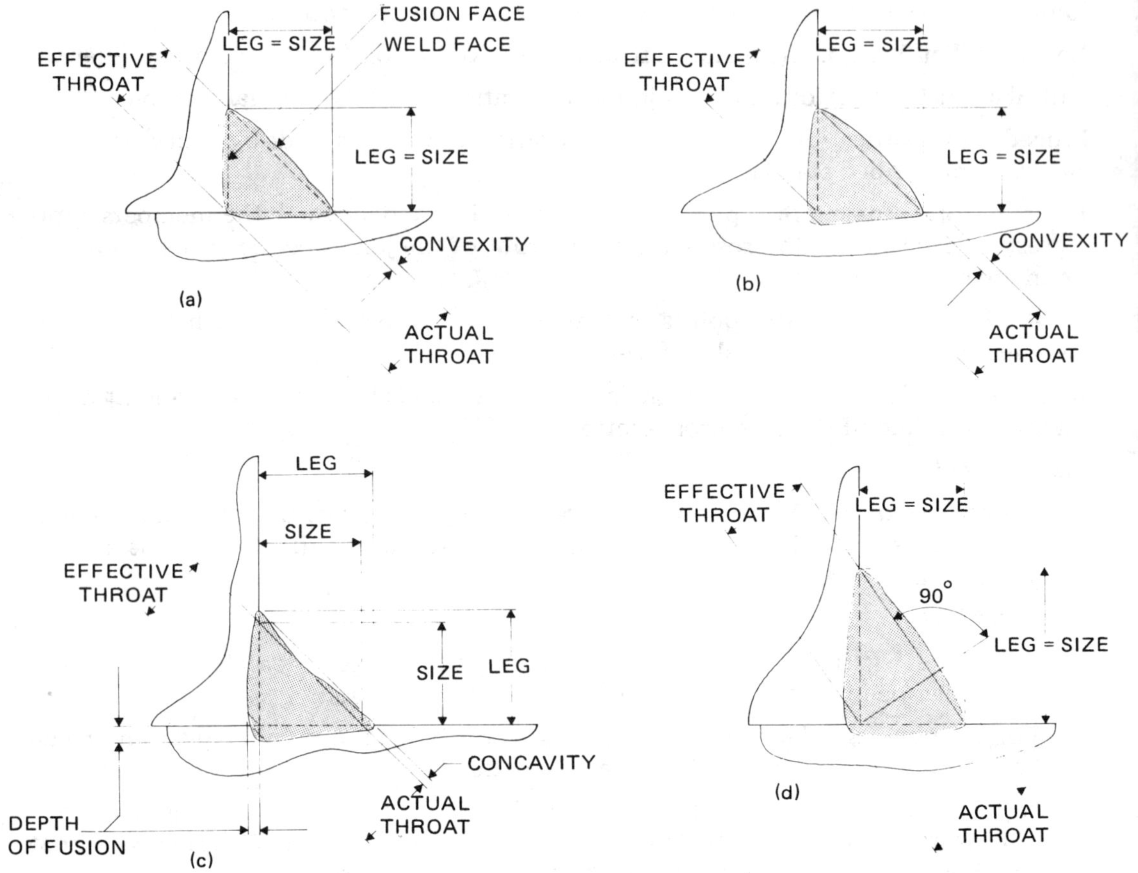

Minimum Size

- The minimum size of a fillet weld as measured should not be less than the values shown on the right, except that it need not exceed the thickness of the thinner part.

  When welding attachments to non-load carrying members the values on the right need not apply.

Material thickness of thicker part joined	Minimum Size of fillet weld, mm
to 12 mm incl.	5
over 12 mm to 20	6
over 20	8

- The minimum effective length of a fillet weld should be 40 mm or 4 times the size of the fillet, whichever is larger.

Maximum Size of Weld

- The maximum fillet weld size, D_{max}, recommended by good practice along a sheared edge is:

$$D_{max} \leqslant t \quad \text{when } t \leqslant 6 \text{ mm}$$

$$D_{max} = t - 2 \quad \text{when } t > 6 \text{ mm}$$

- Material with rolled edges:

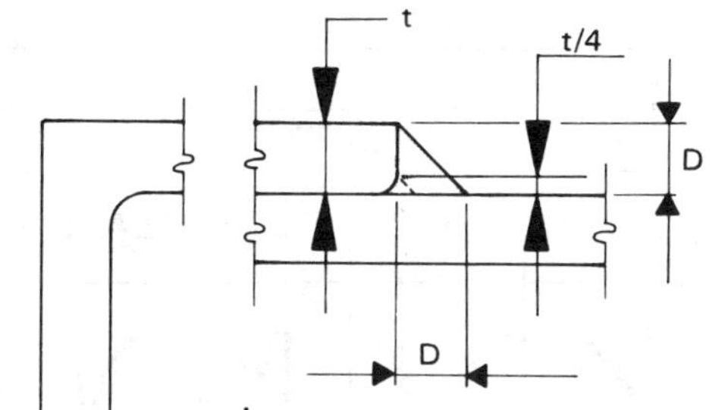

†Effective size = D − t/4
when D = t

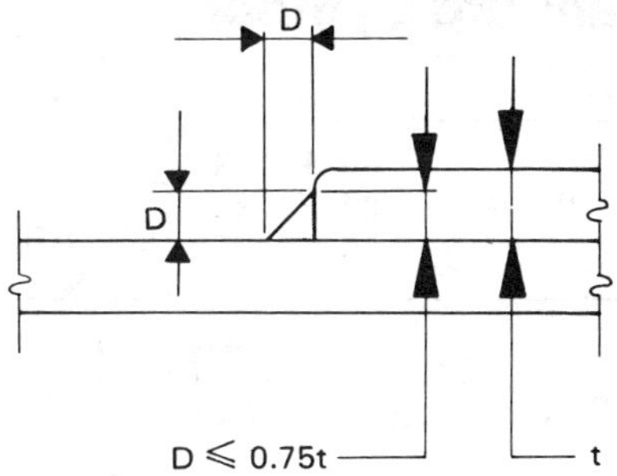

- When fillet welds are used in holes or slots, the diameter of the hole or the width of the slot should not be less than the thickness of the member containing it plus 8 mm.

Lap Joints

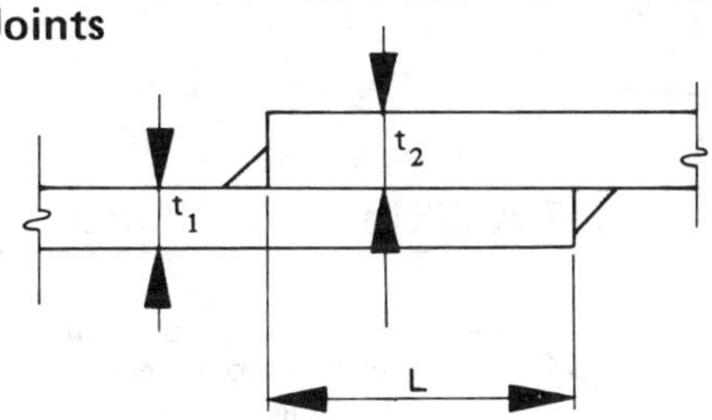

$L_{min} = 5t_1 \geqslant 25$ mm when $t_1 \leqslant t_2$

$L_{min} = 5t_2 \geqslant 25$ mm $\quad t_2 < t_1$

†This detail is not part of W59, but rather is a detail often used by the steel fabricating industry.

Partial Penetration Groove Welds

Minimum Groove Depth for Partial Joint Penetration Groove Welds†

Thickness of Thicker Part Joined (Millimetres)	Minimum Groove Depth*, mm	
	Groove Angle, α, at Root $45° \leqslant \alpha < 60°$ (V-, Bevel Grooves)	Groove Angle, α, at Root $\alpha \geqslant 60°$ (V-, Bevel, J-, U- Grooves)
Total 12 incl.	8	5
Over 12 – 20	10	6
Over 20 – 40	12	8
Over 40 – 60	14	10
Over 60	16	12

†Not combined with fillet welds.
*See Notes 2a and 2b of diagram on next page.

Flare Bevel and Flare V-Welds

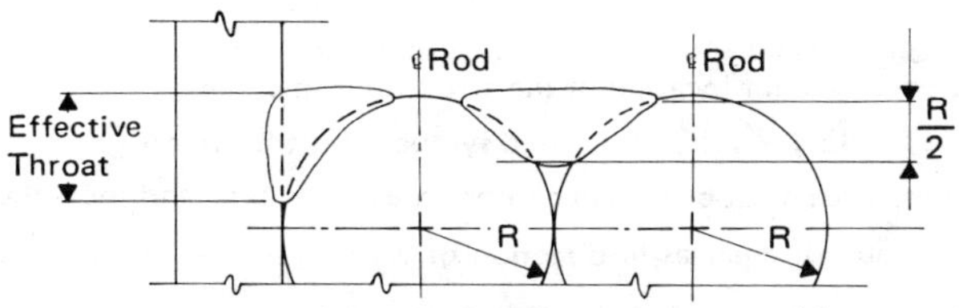

The effective throat thickness for flare groove welds on solid bars, when filled flush to the surface of the solid section of the bar is 5/16 R for Flare Bevel Groove welds and 1/2 R for Flare Vee Groove welds.

WELDED JOINTS
Standard Symbols

BASIC WELD SYMBOLS

BACK	FILLET	PLUG OR SLOT	GROOVE OR BUTT						
			SQUARE	V	BEVEL	U	J	FLARE V	FLARE BEVEL

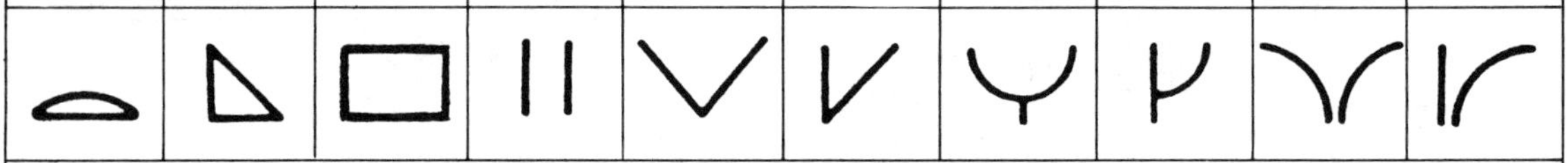

SUPPLEMENTARY WELD SYMBOLS

WELD ALL AROUND	*FIELD WELD	CONTOUR	
		FLUSH	CONVEX

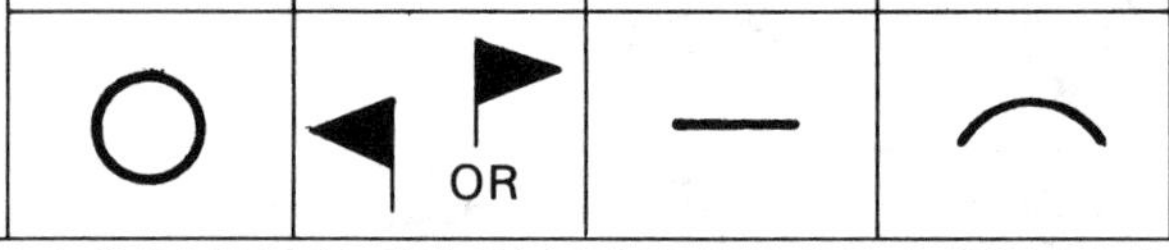

STANDARD LOCATION OF ELEMENTS OF A WELDING SYMBOL

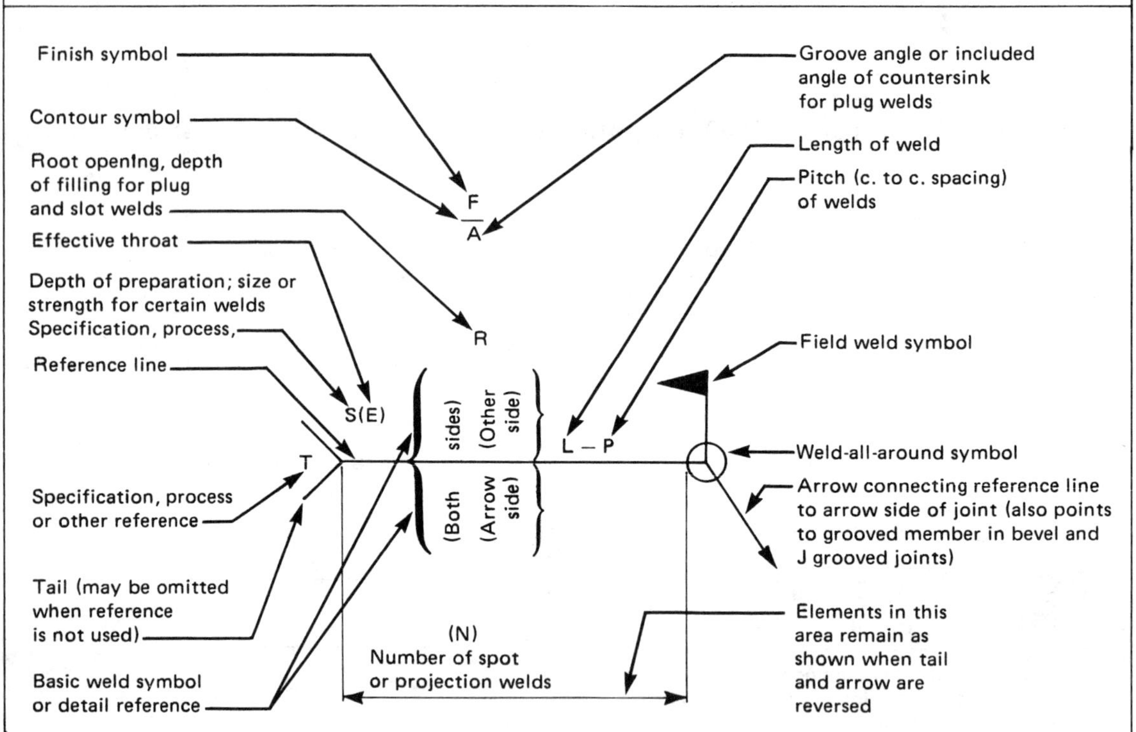

Note:

Size, weld symbol, length of weld and spacing must read in that order from left to right along the reference line. Neither orientation of reference line nor location of the arrow alter this rule.

The perpendicular leg of 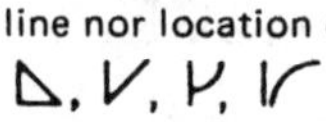 weld symbols must be at left.

Size and spacing of fillet welds must be shown on both the Arrow Side and the Other Side Symbol.

Symbols apply between abrupt changes in direction of welding unless governed by the "all around" symbol or otherwise dimensioned.

These symbols do not explicitly provide for the case that frequently occurs in structural work, where duplicate material (such as stiffeners) occurs on the far side of a web or gusset plate. The fabricating industry has adopted this convention; that when the billing of the detail material discloses the identity of far side with near side, the welding shown for the near side shall also be duplicated on the far side.

*Pennant points away from arrow.

WELDING SYMBOLS

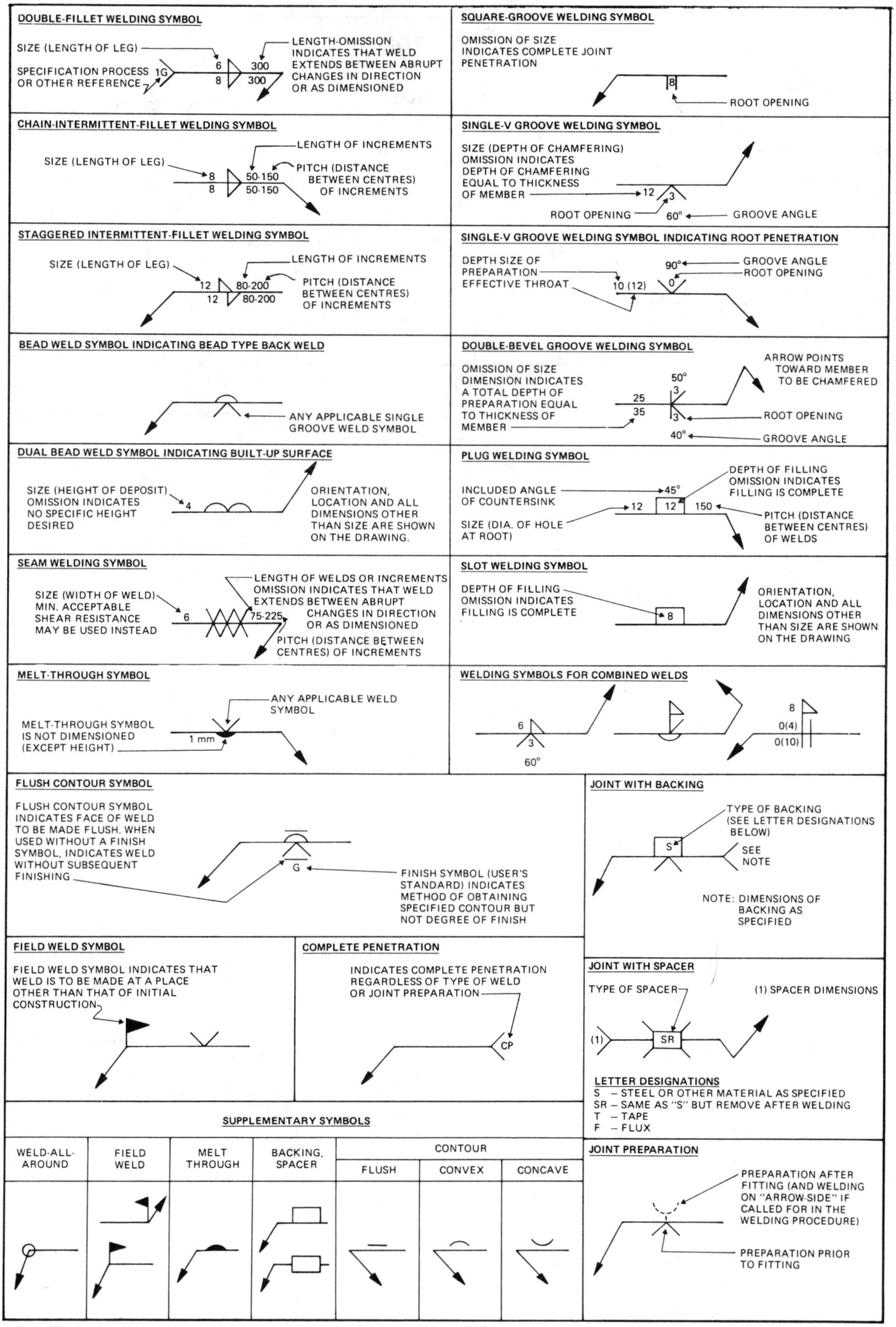

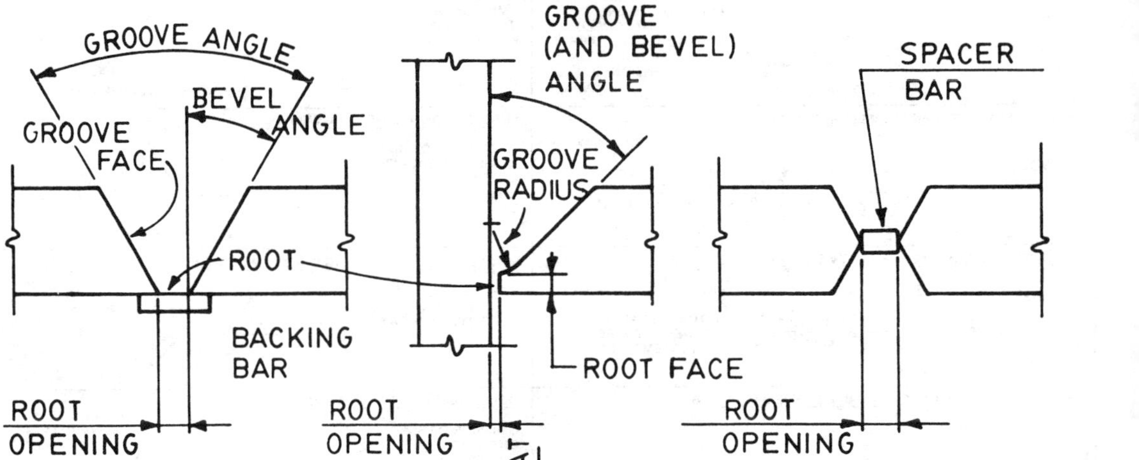

PREPARATION

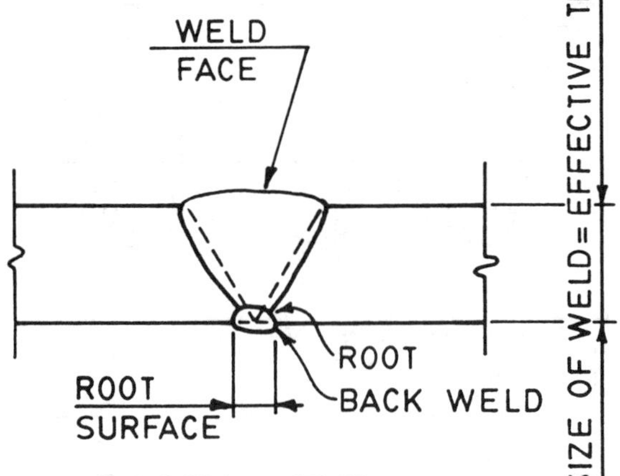

COMPLETE PENETRATION

notes:

1. FOR BEVEL GROOVES THE GROOVE ANGLE EQUALS THE ANGLE AT THE ROOT.(1)

2a. EFFECTIVE THROAT = DEPTH OF PREPARATION − 3mm WHEN, $45^\circ \leq$ ANGLE AT ROOT $< 60^\circ$ (2)

2b. EFFECTIVE THROAT = DEPTH OF PREPARATION WHEN ANGLE AT ROOT OF GROOVE $\geq 60^\circ$.(2)

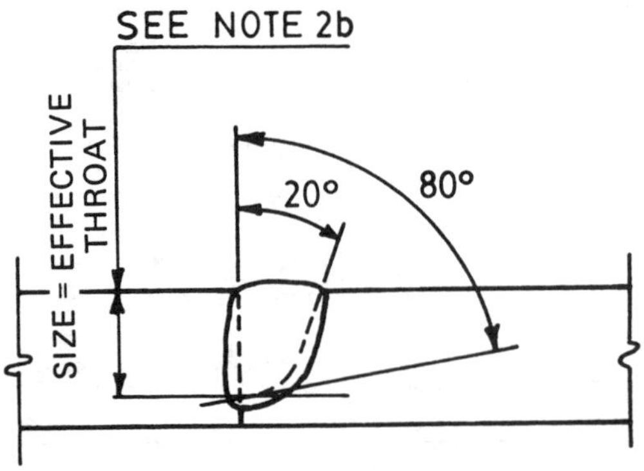

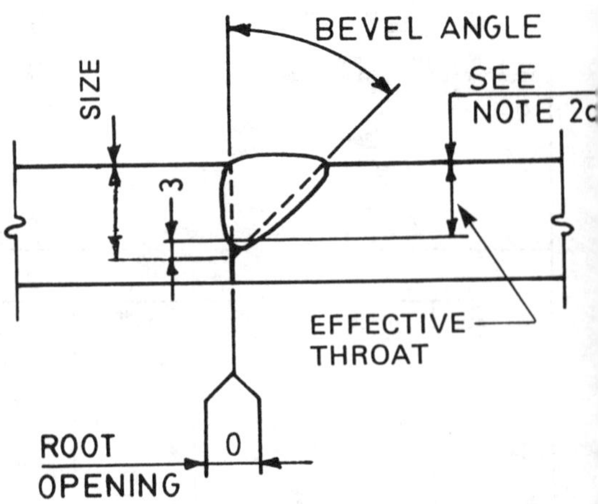

PARTIAL PENETRATION

(1) Not for J and U grooves.
(2) Applies only to PJPG welds.

STEEL PRODUCTS — RECORD OF CHANGES

Following is a chronological record of changes to the list of steel sections included in the CISC Handbook of Steel Construction since the first printing of the Third Edition.

1983 No longer produced by Algoma are —

M100 × 19
S150 × 26, and 19
S130 × 22, and 15
S100 × 11
S75 × 11, and 8
All angles except 8" × 8" leg sizes

1985 No longer produced by Algoma are —

WWF 550 × 217
WWF 350 × 385

New shapes and sections produced by Algoma —

WWF 1800 × 632, and 548
WWF 1600 × 579, and 495
WWF 1400 × 491, and 407
WWF 550 × 280

Welded Reduced Flange (WRF) shapes with top flanges narrower than the bottom flanges and intended primarily for composite bridge girders —

WRF 1800 × 543, 480, and 416
WRF 1600 × 491, 427, and 362
WRF 1400 × 413, 348, and 284
WRF 1200 × 373, 309, and 244
WRF 1000 × 340, 275, and 210

1986 New shapes and sections produced by Algoma —

W 610 × 91
W 610 × 84
W 530 × 72
W 310 × 31
W 250 × 24
W 200 × 21

1989 Sections produced by Algoma

Sections deleted

WWF1800 × 632 and 548
WWF1600 × 579 and 495
WWF1400 × 491 and 407
WWF1200 × 403 and 364
WWF1100 × 335, 291, 255, and 220
WWF1000 × 324, 280, and 244
WWF900 × 293, 249, and 213
WWF800 × 332—154
WWF700 × 222—141

Sections added

WWF2000 × 732—542
WWF1800 × 700—510
WWF1600 × 622—431
WWF1400 × 597—358
WWF1200 × 418, 380, and 333
WWF1100 × 351, 304, 273, and 234
WWF1000 × 340, 293, 262, and 223
WWF900 × 309, 262, and 231
WWF800 × 339—161
WWF700 × 245—152
WWF650 × 864—400
WWF600 × 793—369

Sections not available from Canadian mills added

W1000 — All
W920 × 1262—488
W840 × 922—392
W760 × 865—350
W760 × 134
W690 × 802—289
W610 × 732—262
W530 × 599—248
W460 × 464—193

1991 Sections no longer available from Canadian mills.

W310x283 and W310x253

C380x74–50 and C310x45–31

NOTES

PART SEVEN
MISCELLANEOUS

CISC
CODE OF STANDARD PRACTICE
for
Structural Steel

Published by the

CANADIAN INSTITUTE OF STEEL CONSTRUCTION

201 Consumers Road • Willowdale • Ontario

TABLE OF CONTENTS

CISC CODE OF STANDARD PRACTICE FOR STRUCTURAL STEEL

PREFACE

The CISC Code of Standard Practice for Structural Steel is a compilation of usual industry practices relating to the design, fabrication and erection of structural steel. These practices evolve over a period of time and are subject to change as improved methods replace those of an earlier period. The Code is revised whenever a sufficient number of changes have occurred to warrant a new edition.

The first edition of the Code was adopted and published in November 1958. A second edition incorporating minor revisions was published in October 1962. The third edition, published in September 1967 and revised in May 1970, incorporated minor changes throughout with principal changes in Section 2 — Definition of Structural Steel and Section 3 — Computation of Weights for Unit Price Bids.

The fourth edition adopted in June 1980, revised December 1980, broadened the scope to include bridges and other structures. It also incorporated the CISC "Guide to Tendering Procedures" into Section 3 and Appendices B & C. The Code was converted to SI (metric) units and provided conversion factors and Imperial units in Appendix E.

This fifth edition reflects the steel standard's recognition of the preparation of five types of fabrication and erection documents which may be produced in fulfilling a steel construction contract. These documents may be in the form of drawings, diagrams, sketches, computer output, hand calculations and other data which can be supplied by the fabricator/erector. This data is generally referred to in contract documents as "shop drawings". The computation of mass has been changed by deleting the mass of welds and the allowances for paint and other coatings.

Appendix B, Guideline for Unit Price Application for Changes, and Appendix C, A Suggested Format for Price Per Unit of Mass or Price Per Item Contracts have been substantially revised. To foster uniformity, two new appendices have been added; Miscellaneous Steel and A Suggested Format for a Monthly Progress Claim Form.

By documenting standard practices the Code aims to promote a clear understanding between the Canadian structural steel fabrication and erection industry and its clients.

Canadian Institute of Steel Construction

Adopted June 8, 1991

CISC CODE OF STANDARD PRACTICE FOR STRUCTURAL STEEL

1. General Provisions

1.1 Scope. This Code covers standard industry practice with respect to the furnishing of structural steel. In the absence of provisions to the contrary contained in contracts to which members of the Canadian Institute of Steel Construction are contracting parties, members will abide by the practices described herein.

1.2 Definitions.

Approved for Construction Drawings	Drawings and other documents approved by the client authorizing work to proceed.
Architect	As defined under the appropriate Architect's Act.
Client	A person, corporation, or authority with whom the fabricator has contracted.
Connection Design Details	Documents which provide details of typical and special types of connections and other data necessary for the preparation of shop details.
Construction Drawings	Drawings used to govern the construction of the works.
Construction Specifications	Specifications used to govern the construction of the works.
Contract	The agreement between the fabricator and/or erector, and the client.
Contract Documents	The documents which define the responsibilities of the parties involved in tendering, purchasing, supplying, fabricating and erecting structural steel, including tender drawings and tender specifications and applicable revisions in effect and agreed to at the time of contract award.
Cost Plus a Fee Contract	An Agreement whereby the fabricator and/or erector agrees to fulfil the contract for a consideration which is calculated on the basis of the fabricator's costs plus a specified fee as defined in the contract.
Designer	The designer of the structure. See Engineer of Record.
Engineer	As defined under the appropriate Professional Engineer's Act.
Engineer of Record	Professional Engineer who designs the structure, as defined under the appropriate Professional Engineer's Act.
Erection Bracing	Bracing materials or members which are used to plumb, align and stabilize structural members or the structure during construction and are removed when the structural members or the structure is secured by bolting or welding of structural members.
Erection Diagrams	Are general arrangement drawings showing the principal dimensions and elevations of the steel structure, sizes of the steel members, piece marks, size (diameter) and type of bolts, bolt installation requirements, elevations of column bases, all necessary dimensions and details for setting anchor bolts, and all other information necessary for the assembly of the structure.

Erection Procedures	Outline the construction methods, erection sequence, erection and temporary bracing requirements, and other engineering details necessary for shipping, handling, erecting, and maintaining the stability of the structural steel frame.
Erector	Means the party responsible for erection of the steelwork.
Fabricator	Means the party responsible for furnishing the structural or miscellaneous steel.
Field Work Details	Are details that provide complete information for modifying fabricated members in the field.
General Contractor, Constructor or Construction Manager	The person or corporation who constructs, coordinates, and supervises the construction of the work.
General Terminology e.g. Beams, Joists, Columns, etc.	These terms have the meanings stated or implied in CAN/CSA S16.1-M (latest edition), CAN/CSA-S6-M (latest edition) and Appendix A of this Code.
Lump Sum Price Contract	Also called Stipulated Price Contract. An agreement whereby the fabricator and/or erector contracts to fulfil the contract terms for a lump sum (stipulated price) consideration.
Miscellaneous Steel	Steel items described and listed in Appendix F of this code.
Others	Means a party or parties other than the fabricator and/or erector.
Owner	Means the owner of a structure and shall include his authorized agent and any person taking possession of a structure on the owner's behalf. Depending on the circumstances an authorized agent may be the architect, engineer, general contractor, construction manager, public authority or other designated representative of the owner.
Price Per Item Contract	An agreement whereby the fabricator and/or erector agrees to fulfil the contract terms for a consideration which is calculated on a "per item" basis, where the term "item" is defined in the contract.
Price Per Unit of Mass Contract	Also called Unit Price Contract. An agreement whereby the fabricator and/or erector contracts to fulfil the contract terms for a consideration which is based on the mass of steel calculated in accordance with the CISC Code of Standard Practice for Structural Steel.
Revision	A change in the contract documents.
Shop Details	Documents which provide complete information for the fabrication of various members and components of the structure, including the required material and product standards; the location, type, and size of all mechanical fasteners; bolt installation requirements; and welds.
Stipulated Price Contract	See Lump Sum Price Contract.
Structural Drawings	The tender drawings showing the structural steel required.
Structural Steel	Those items listed under Clause 2.1

Structural Steel Frame	An assemblage of structural steel components (beams, columns, purlins, girts, etc.) for the purpose of resisting loads and forces. See Clause 2.1.
Structural Steel Specifications	The portion of the tender specifications containing the requirements for the fabrication and erection of the structural steel.
Temporary Bracing	Members designed by the Engineer of Record to be removed at a later date at his or her instruction.
Tender Documents	Drawings, specifications, general conditions, addenda, etc., used as the basis for preparing a tender.
Tender Drawings	Drawings used as the basis for preparing a tender.
Tender Specifications	Specifications used as the basis for preparing a tender.
Unit Price Contract	See Price Per Unit of Mass Contract.

1.3 Governing Technical Standards. The provisions of the latest edition of CAN/CSA-S16.1-M "Limits States Design of Steel Structures", shall govern the design, fabrication and erection of steel structures except bridges. The provisions of the latest edition of CAN/CSA-S6-M "Design of Highway Bridges", "The Ontario Highway Bridge Design Code" (in Ontario) or the American Railway Engineering Association's "Specifications for Steel Railway Bridges" shall govern the design, fabrication and erection of structural steel for bridges. The provisions of the latest edition of CSA Standard W59 "Welded Steel Construction (Metal-Arc Welding)" shall govern arc welding design and practice. The provisions of other standards shall be applicable if called for in the tender drawings and tender specifications.

1.4 Responsibility for Design. When the client provides the structural drawings and specifications, the fabricator and the erector shall not be responsible for determining the adequacy of the design nor liable for the loss or damage resulting from an inadequate design. Should the client desire the fabricator to assume any responsibility for design beyond that of proposing adequate connections and details, and, when required, components, members, or assemblies standardized by the fabricator, the client shall state clearly his requirements in the invitation to tender or in the accompanying tender drawings and tender specifications. Even though proposed connections and design details may be prepared by the fabricator's technical staff, the over-all behaviour of the structure remains the responsibility of the designer of the structure. (See also Clause 5.6).

1.5 Responsibility for Erection Procedure. When the erection of structural steel is part of his contract, the fabricator shall be responsible for determining the erection procedure, for checking the adequacy of the connections for the uncompleted structure and for providing erection bracing or connection details. When the erection of the structural steel is not part of his contract, the fabricator shall not be responsible for determining the erection procedure, for checking the adequacy of the connections for the uncompleted structure, or for providing erection bracing or connection details not included in the contract documents, nor shall the fabricator be liable for loss or damage resulting from faulty erection. However, the steel fabricator shall be informed by the client of the erection sequence to be used which may influence the sequence and process of the manufacturing. (See also Clause 5.3 and 5.4).

1.6 Patented Devices. Except when the contract documents call for the design to be furnished by the fabricator or erector, the fabricator and erector assume that all necessary patent rights have been obtained by the client and that the fabricator and erector will be fully protected by the client in the use of patented designs, devices or parts required by the contract documents.

1.7 Scheduling. The client should provide a construction schedule in the tender documents. In the absence of such a schedule, one should be mutually agreed upon between the contracting parties, prior to the contract award.

2. Classification of Material

2.1 Structural Steel. Unless otherwise specified in the tender documents, a contract to supply, fabricate and deliver structural steel shall include only those items from the following list which are clearly indicated as being required by the structural drawings and tender specifications. (See Appendix A)

Anchors for structural steel.

Base plates and bearings for structural steel members.

Beams, purlins, girts forming part of the structural steel frame.

Bearing plates and angles for structural steel members and steel deck.

Bins and hoppers of 6 mm plate or heavier, attached to the structural steel frame.

Bracing for steel members, trusses or frames.

Brackets attached to the structural steel.

Bridge bearings, grating and railings, connected to the structural steel members.

Cables for permanent bracing or suspension systems.

Canopy framing if attached to the structural steel frame.

Cold formed channels when used as structural members as listed in the CISC Handbook of Steel Construction

Columns.

Conveyor galleries and supporting bents (exclusive of conveyor stringers, deck plate and supporting posts which are normally part of the conveyor assembly).

Crane rails and stops.

Curb angles and plates attached to the structural steel frame.

Diaphragms for bridges.

Door frames attached to the structural steel frame.

Embedded items connecting structural steel.

Expansion joints connected to the structural steel frame (excluding expansion joints for bridges).

Field bolts to connect structural steel components.

Floor plates, roof plates (raised pattern or plain) and steel grating connected to the structural steel frame.

Girders.

Grillage beams of structural steel.

Hangers supporting structural steel framing.

Jacking girders.

Lintels shown on the structural drawings.

Mechanical support framing shown on structural drawings.

Monorail beams of standard structural steel shapes.

Open-web steel joists, including anchors, bridging, headers and trimmers; also, when specified to be included in the structural steel contract documents, light-gauge forms and temperature reinforcement.

Sash angles attached to the structural steel frame.

Separators, angles, tees, clips and other detail fittings essential to the structural steel frame.

Shear connectors/studs, except when installed through sheet steel deck by deck installer.

Shelf angles attached to the structural steel frame.

Shop fasteners or welds, and fasteners required to assemble parts for shipment.

Steel tubes or cores for composite columns.

Steel stairs, walkways, ladders and handrails forming part of the structural steelwork. [See Appendix A]

Steel window sills attached to the structural steel frame.

Struts.

Suspended ceiling supports of structural steel shapes at least 75 mm in depth.

Temporary components to facilitate transportation to the site.

Tie, hanger and sag rods forming part of the structural steel frame.

Trusses.

2.2 Field Connection Material.

2.2.1 When the fabricator erects the structural steel, he shall supply all material required for temporary and for permanent connection of the component parts of the structural steel.

2.2.2 When the erection of the structural steel is not part of the fabricator's contract, unless otherwise specified, the fabricator shall furnish appropriate bolts and nuts (plus washers, if required) or special fasteners, of suitable size and in sufficient quantity for all field connections of steel to steel which are specified to be thus permanently connected, plus an over-allowance of two per cent of each size to cover waste.

Unless otherwise specified in the tender documents, welding electrodes, back-up bars, temporary shims, levelling plates, fitting-up bolts and drift pins required for the structural steel shall not be furnished by the fabricator when the erection of the structural steel is not part of the fabricator's contract.

2.3 Items Supplied by Others. Unless otherwise specified in the tender documents, the following steel or other items shall not be supplied by the structural steel fabricator.

Bolts for wood lagging.

Bins and hoppers not covered in Clause 2.1 of this Code.

Bridge bearings not connected to structural steel items.

Canopy framing not attached to structural steel.

Catch basin frames.

Concrete for filling HSS columns. Concrete is to be supplied and poured in columns by others in the shop or field with the co-operation of the fabricator/erector.

Connection material for other trades (e.g. precast concrete).

Conveyor stringers, deck plate and supporting posts.

Drain pipes.

Door and corner guards.

Door frames not covered in Clause 2.1 of this Code.

Drilling of holes into masonry or concrete, including core drilling of anchor bolts for bridges and drilling for deck support angles.

Edge forming less than 3.2 mm thick for steel deck and not covered in Clause 2.1 of this code.

Embedded steel parts not required for structural steel or deck.

Embedded steel parts in precast concrete.

Flagpoles and supports.

Floor plates, roof plates and grating not covered in Clause 2.1 of this Code.

Grout.

Hoppers and chutes.

Hose and tire storage brackets.

Installation of embedded parts.

Lag bolts, machine bolts and shields or inserts for attaching shelf angles, trimmer angles and channels to masonry or concrete.

Lintels over wall recesses.

Lintels which are an integral part of door frames not attached to the structural steel frame.

Machine bases, rollers and pulleys.

Members made from gauge material except cold-formed channels indicated in Clause 2.1.

Metal-clad doors and frames.

Miscellaneous Steel, See Appendix "F".

Shear connectors through sheet steel deck by deck installer.

Sheet steel cladding.

Sheet steel deck.

Sheet steel flashing.

Shelf angles not covered in Clause 2.1 of this Code.

Shoring under composite floors and stub girders.

Steel doors.

Steel sash.

Steel stacks.

Steel stairs, landings, walkways, ladders and handrails not covered in Clause 2.1 of this Code.

Steel tanks and pressure vessels.

Steel window sills not covered in Clause 2.1 of this Code.

Support for sheet steel deck at column cut outs and for openings not requiring framing connected to structural steel.

Temporary bracing for other trades.

Trench covers.

Trim angles, eave angles or fascia plates not directly attached to the structural steel frame.

3. Contracts

3.1 Standard Form of Contract. Unless otherwise agreed upon, a contract to fabricate, deliver and/or erect structural steel shall be based on the appropriate Standard Construction Documents approved by the Canadian Construction Documents Committee, or the Canadian Construction Association.

3.1.1 Progress Payment Claim Form. A suggested format for a progress payment claim form, is provided in Appendix G.

3.2 Types of Contracts.

3.2.1 For contracts stipulating a "lump sum price", the work required to be performed by the fabricator and/or erector must be completely defined by the contract documents.

3.2.2 For contracts stipulating a "price per unit of mass", the scope of the work, type of materials, character of fabrication, and conditions of erection are based upon the contract documents which must be representative of the work to be performed. See Appendix C of this Code.

3.2.3 For contracts stipulating "price per item", the work required to be performed by the fabricator and/or erector is based upon the quantity and character of the items described in the contract documents. See Appendix C of this Code.

3.2.4 For contracts stipulating "cost plus fee", the work required to be performed by the fabricator and/or erector is indefinite in nature at the time the contract documents are prepared. Consequently the contract documents should define the method of measurement of work performed, and the fee to be paid in addition to the fabricator's costs.

3.3 Revisions to Contract Documents.

3.3.1 Revisions to the contract shall be made by the issue of dated new or revised documents. All revisions shall be clearly indicated. Such revisions may be issued on a Detailed Change Notice.

3.3.2 The fabricator shall advise the client or client representative of any impact such revision or change will have on the existing agreement between the two parties.

3.3.3 Upon agreement between the fabricator and the client or client representative as to the revision's impact, the client or his representative shall issue a change order or extra work order authorizing the fabricator to proceed with the work.

3.3.4 Unless specifically stated to the contrary, the issue of a revision is authorization by the client to approve these documents for construction.

3.4 Computation of Mass

3.4.1 Unless another method is specified and fully described at the time tenders are requested, the computed mass of steel required for the structure shall be determined by the method of computation described herein. (Although the method of computation described does not result in the actual mass of fabricated structural steel and other items its relative simplicity results in low computational cost and it is based on quantities which can be readily computed and checked by all parties involved to establish the basis of payment. No additional mass allowance for welds, painting, galvanizing, and metallizing is to be included in the computation of mass.

(a) *Mass Density.* The mass density of steel is assumed to be 7850 kilograms per cubic metre.

(b) *Shapes, Bars and Hollow Structural Sections.* The mass of shapes, bars and hollow structural sections is computed using the finished dimensions shown on shop details. No deductions shall be made for holes created by cutting, punching or drilling, for material removed by coping or clipping, or for material removed by weld joint preparation. No cutting, milling or planing allowance shall be added to the finished dimensions. The mass per metre of length for shapes and hollow structural sections is the published mass. The mass per metre of length for bars is the published mass, or if no mass is published, the mass computed from the specified cross-sectional area.

(c) *Plates and Slabs.* The mass of plates and slabs is computed using the rectangular dimensions of plates or slabs from which the finished plate or slab pieces shown on the shop details can be cut. No burning, cutting, trimming or planing allowance shall be added. When it is practical and economical to do so, several irregularly-shaped pieces may be cut from the same plate or slab. In this case, the mass shall be computed using the rectangular dimensions of the plate or slab from which the pieces can be cut. No cutting or trimming allowance shall be added. In all cases, the specified plate or slab thickness is to be used to compute the mass. The mass of raised-pattern rolled plate is that published by the manufacturer.

(d) *Bolts.* The mass of shop and field bolts, nuts and washers is computed on the basis of the shop details and the nominal published mass of the applicable types and sizes of fastener.

(e) *Studs.* If not included in the contract on a "price per item basis", the mass of studs is computed on the basis of the shop details and/or erection diagrams and the published mass of the studs.

(f) *Grating.* If not included in the contract on the basis of a price per square metre, the mass of grating is computed on the basis of the shop details and/or erection diagrams and published mass of the grating. The area to be used is the minimum rectangular area from which the piece of grating can be cut.

(g) Where supplied, such items as shims, levelling plates, temporary connection material, back-up bars and certain field "consumables" shall be considered as part of the structural steel whether or not indicated specifically in the contract documents. Such items then will be added to, and become a part of, the computed mass of steel for the structure.

3.5 Contract Price Adjustments

3.5.1 When the responsibility of the fabricator and/or erector is changed from that previously established by the contract documents, an appropriate modification of the contract price shall be made. In computing the contract price adjustment, the fabricator and/or erector shall consider the quantity of work added or deleted, modifications in the character of the work, the timeliness of the change with respect to the status of material ordering, the detailing, fabrication and erection operations and related impact costs.

3.5.2 Requests for contract price adjustments shall be presented by the fabricator and/or erector and shall be accompanied by a description of the change in sufficient detail to permit evaluation and prompt approval by the client.

3.5.3 Price Per Unit of Mass and Price Per Item Contracts generally provide for minor revisions to the quantity of work prior to the time work is approved for construction. Minor revisions to the quantity of work should be limited to an increase or decrease in the mass of the steel for each category not exceeding ten percent. Should the mass of steel of any category vary by more than ten percent, then the contract unit price of that category may require adjustment. Changes to the character of the work or the mix of the work, at any time or changes to the quantity of the work after the work is approved for construction, may require a contract price adjustment.

3.5.4 A suggested format for accommodating contract price adjustments is contained in Appendix B.

3.6 Scheduling

3.6.1 The contract documents should specify the schedule for the performance of the work. This schedule should state when the approved for construction drawings will be issued and when the job site, foundations, piers and abutments will be ready, free from obstructions and accessible to the erector, so that erection can start at the designated time and continue without interference or delay caused by the client or other trades.

3.6.2 The fabricator and/or erector has the responsibility to advise the client of the effect any revision may have on the contract schedule.

3.6.3 If the fabrication and erection schedule is significantly delayed due to revisions, or for other reasons which are the client's responsibility, the fabricator and erector shall be compensated for additional costs incurred. Increases to the scope of the work shall provide additional time to the schedule.

4. Contract Documents

4.1 Tender Documents – Tender Drawings and Tender Specifications

4.1.1 At the time tenders are called, the steel fabricator shall receive a complete set of structural drawings and a complete set of tender specifications. In order to ensure adequate and complete tenders for Lump Sum Price Contracts*, these documents shall include complete structural drawings, conforming to the requirements for design drawings of the governing technical standard. Structural steel specifications should include any special requirements controlling the fabrication and erection of the structural steel, surface preparation and coating, and should indicate the extent of non-destructive examination, if any, to be carried out.

****For other types of contracts, it is desirable for the contract documents to be as complete as possible.***

4.1.2 Design drawings shall be drawn to a scale adequate to convey the required information. The drawings shall show a complete design of the structure with members suitably designated and located, including such dimensions and detailed description as necessary to permit the preparation of fabrication and erection documents. Floor levels, column centres, and offsets shall be dimensioned. The term "drawings" may include computer output and other data. Stiffeners and doubler plates required to maintain stability and which are an integral part of the main member shall be shown and dimensioned.

4.1.3 Design drawings shall designate the design standards used, shall show clearly the type or types of construction to be employed, shall show the category of the structural system used for seismic design, and shall designate the material or product standards applicable to the members and details depicted. Drawings shall give the governing combinations of shears, moments, and axial forces to be resisted by the connections.

4.1.4 Where connections are not shown, the connections shall be assumed to be in accordance with the requirements of the governing technical standard (see Clause 1.3).

4.2 Architectural, Electrical and Mechanical Drawings. Architectural, electrical and mechanical drawings may be used as a supplement to the structural drawing to define detail configurations and construction information, provided all requirements for the structural steel are noted on the structural drawings.

4.3 Construction Drawings and Construction Specifications

4.3.1 As soon as possible following award of the contract, the client shall furnish the fabricator with a plot plan of the construction site, and a set of complete drawings and specifications approved for construction consistent with the tender drawings and tender specifications. These construction drawings and specifications are required by the fabricator for ordering the material and for the preparation and completion of fabrication and erection documents. The approved for construction drawings shall show:

(a) all changes or revisions to the tender drawings, clearly indicated on the construction drawings,

(b) the complete design of the structure with members suitably designated and located, including such dimensions and detailed description as necessary to permit preparation of the fabrication and erection documents. Floor levels, column centres, and offsets shall be dimensioned;

(c) all materials to be furnished by the fabricator, together with sufficient information to prepare fabrication and erection documents, including the design standards used, the type or types of construction to be employed, the category of the system used for seismic design, the applicable material or product standards, and the governing combinations of shears, moments and axial forces to be resisted by connections.

The fabricator shall receive a complete set of the tender drawings and tender specifications.

5. Fabrication and Erection Documents

Note: *The term "shop drawings", frequently used in the construction industry, is replaced in this Code of Practice by the terms "fabrication and erection documents". These terms more correctly describe the following five separate and distinct documents that may be prepared by a fabricator/erector. See also Clause 1.2 for definitions. Not all of the five documents will be required for every project.*

5.1 Connection Design Details. Connection design details shall be prepared in advance of preparing shop details and submitted to the designer for confirmation that the intent of the design is met. Connection design details shall provide details of typical and special types of connections, and other data necessary for the preparation of shop details. Connection design details shall be referenced to the design drawings.

5.2 Shop Details. Unless provided by the client, shop details shall be prepared in advance of fabrication from the information on the approved construction drawings, the connection design details, and the erection diagrams. Shop details shall provide complete information for the fabrication of various members and components of the structure, including the required material and product standards; the location, type, and size of all attachments, mechanical fasteners, and welds. When shop details are required to be submitted for review and approval, only one reproducible copy of each shop detail will be submitted, unless a larger number of copies is required by the client as part of the tender specifications.

5.3 Erection Diagrams. Unless provided by the client, the fabricator will prepare erection diagrams from the approved construction drawings. In this regard, the fabricator may request reproducible copies of the structural drawings which may be altered for use as erection diagrams. When using reproducible copies of the structural drawings, the structural engineer's name and seal shall be removed. Erection diagrams shall be submitted to the designer for review and approval. Erection diagrams are general arrangement drawings showing the principal dimensions of the structure, piece marks, sizes of the members, size (diameter) and type of bolts, bolt installation requirements, elevations of column bases, all necessary dimensions and details for setting anchor bolts, and all other information necessary for the assembly of the structure. Only one reproducible copy of each diagram will be submitted for review and approval unless a larger number of copies is required by the client as part of the tender documents.

5.4 Erection Procedures. Erection procedures shall outline the construction methods, erection sequence, erection bracing, temporary bracing if required, and other engineering details necessary for shipping, erecting, and maintaining the stability of the steel frame. Erection procedures shall be supplemented by drawings and sketches to identify the location of stabilizing elements. Erection procedures shall be submitted for review when so specified.

5.5 Field Work Details. Field work details shall be submitted to the designer for review and approval. Field work details shall provide complete information for modifying fabricated members on the job site. All operations required to modify the member shall be shown on the field work details. If extra materials are necessary to make modifications, shop details may be required.

5.6 Review and Approval. Erection diagrams, connection design details, shop details and field work details, are normally submitted for review and approval. Approval, by the designer, of shop details prepared by the fabricator and/or erector, indicates that the fabricator has interpreted correctly the contract requirements. Approval by the designer of shop details prepared by the fabricator does not relieve the fabricator of the responsibility for accuracy of the detail dimensions on shop details, nor of the general fit up of parts to be assembled.

The preparation of fabrication and erection documents is governed by the following sequence of procedures.

(a) Sufficient information must be indicated in the construction drawings as stipulated in Clause 4.1 of this Code to permit the completion of fabrication and erection documents.

(b) The fabrication and erection documents are prepared by skilled technicians using industry and company standards, and represent the fabricator's interpretation of intent of the contract documents, particularly as described by the construction drawings. Connection design details are reviewed by the fabricator's technical staff prior to submission to the designer.

(c) The connection design details, shop details, erection diagrams and field work details are submitted to the designer for review and approval. Erection procedures are submitted when so specified.

(d) It is assumed by the fabricator that the fabrication and erection documents, when approved, have been reviewed by the client for accuracy in the interpretation of the contract requirements. Connection design details and shop details are reviewed and approved by the designer for structural adequacy and to ensure conformance with the loads, forces and special instructions contained in the contract documents.

(e) Shop details are prepared from the approved connection design details and erection diagrams.

If the client does not wish to review and approve the fabricator's fabrication and erection documents, the basis for interpreting the contract requirements, as well as the adequacy of connection details, is limited to the information contained in the structural steel specifications and shown on the structural drawings. This information shall be sufficient, as indicated in Clause 4.1, to permit proper execution of the work. However, the Engineer of Record is ultimately responsible for the structural integrity of the structure and the connections.

5.7 Additions, Deletions or Changes. Additions, deletions or changes, when approved, will be considered as contract revisions and constitute the client's authorization to release the additions, deletions or revisions for construction. See also clause 3.3 and 3.5.

5.8 Shop Details Furnished by the Client. When the shop details are furnished by the client he shall deliver them in time to permit fabrication to proceed in an orderly manner according to the time schedule agreed upon. The client shall prepare these shop details, in so far as practicable, in accordance with the detailing standards of the fabricator. The client shall be responsible for the completeness and accuracy of shop details so prepared.

5.9 Clipped Double Connections. Where two beams or girders, framing at right angles from opposite sides of a supporting member, share the same bolts, a clipped double connection shall be used unless a seated connection or other detail is used to facilitate safe erection of the beams or girders. A clipped double connection is not applicable to a two-bolt connection or when the beams are equal to or deeper than half the depth of the girder. For a description of a clipped double connection, see Appendix A.

6. Material, Fabrication, Inspection, Painting and Delivery

6.1 Stock Material. Materials taken from stock by the fabricator for structural use shall conform to structural steel material standards of the Canadian Standards Association, or the American Society for Testing and Materials, or to other published material specifications, in accordance with the requirements of the construction drawings and construction specifications.

6.2 Identification

6.2.1 The method of identification stipulated in CAN/CSA-S16.1-M shall form the basis for a fabricator's identification of stock material. Stock control and identification procedures may differ to some extent from fabricator to fabricator.

6.3 Preparation of Material

6.3.1 Flame cutting of structural steel may be by hand or mechanically guided means.

6.3.2 Surfaces noted as "finished" on the drawings are defined as having a roughness height rating not exceeding 500 (12.5 m) as defined in CSA Standard B95, Surface Texture (Roughness, Waviness and Lay), unless otherwise specified. Any fabricating technique such as friction sawing, cold sawing, milling, etc., that produces such a finish may be used.

6.4 Fitting and Fastening

6.4.1 Projecting elements of connection attachments need not be straightened in the connecting plane if it can be demonstrated that installation of the connectors or fitting aids will provide adequate contact between faying surfaces.

6.4.2 When runoff tabs are used, the fabricator or erector need not remove them unless specified in the contract documents or governing technical standard. When their removal is required, they may be hand flame-cut close to the edge of the finished member with no more finishing required, unless other finishing is specifically called for in the contract documents or governing technical standard.

6.5 Dimensional Tolerances. Tolerances on fabricated members shall be those prescribed in the applicable governing technical standard. Tolerances on steel material supplied from the mills or from fabricator's stock shall meet those prescribed in Canadian Standards Association Standard G40.20-M.

6.6 Inspection of Steelwork. Should the client wish to have an independent inspection and non-destructive examination of the steelwork, he shall reserve the right to do so in the tender documents. Arrangements should be made with the fabricator for inspection of steelwork at the fabrication shop by the client's inspectors. The cost of this inspection and testing is the responsibility of the client.

6.7 Surface Preparation. If paint is specified, the fabricator shall clean all steel surfaces to be painted of loose rust, loose mill scale, prominent spatter, slag or flux deposit, oil, dirt and other foreign matter by wire brushing or other suitable means. Unless specified, the fabricator shall not be obliged to blast-clean, pickle or perform any specific surface preparation operation aimed at total or near-total removal of tight mill scale, rust or non-deleterious matter.

6.8 Liability for Deterioration of Paint. When structural steel is specified to receive a shop coat of paint, the fabricator shall be responsible only to the extent of performing the surface preparation and painting in the specified manner. Unless otherwise agreed upon as part of the contract documents the fabricator shall not be responsible for the deterioration of the paint that may result from exposure to the weather for more than ninety days after completion of the painting.

6.9 Marking and Shipping

6.9.1 Except for weathering steel surfaces exposed to view and for architecturally exposed steel, erection marks shall be painted or otherwise legibly marked on the members. Preferably, members which are heavy enough to require special erection equipment shall be marked to indicate the computed or scale mass and the centre of gravity for lifting.

6.9.2 Bolts of the same length and diameter, and loose nuts and washers of each size shall be packaged separately. Pins, bolts, nuts, washers and other small parts shall be shipped in boxes, crates, kegs or barrels, none of which exceed 135 kg gross mass. A list and description of material contained therein shall be marked plainly on the outside of each container.

6.9.3 When requested by the erector, long girders shall be loaded and marked so that they will arrive at the job site in position for handling without turning. Instructions for such delivery shall be given to the carrying agency when required.

6.9.4 For each shipment, the fabricator shall furnish a shipping bill listing the items in the shipment. Such bill shall show the erection mark, the approximate length, the description (whether beam, column, angle, etc.) of each item. Such bill shall be signed by the receiver and returned to the fabricator within 48 hours of receipt of the shipment with a note regarding shortages or damages, if any, and the bill shall act as a receipt for the shipment. When the shipments are made by truck transport, the bills should accompany the shipment. When shipments are made by rail or water, the bills shall be sent to the receiver to arrive on or before receipt of the shipment.

6.9.5 Unless otherwise specified at time of tender, steel during shipment will not be covered by tarpaulins or otherwise protected. When such protection is specified the shipper is to notify the carrier of the protection requirements.

6.10 Delivery of Materials

6.10.1 Fabricated structural steel shall be delivered in a sequence which will permit the most efficient and economical performance of shop fabrication and erection. If the client contracts separately for delivery and erection he must coordinate planning between the fabricator, erector and general contractor.

6.10.2 Anchor bolts, washers and other anchorages, grillages, or materials to be built into masonry or concrete should be shipped so that they will be on hand when needed. The client must give the fabricator sufficient notice to permit fabrication and shipping of materials before they are needed.

6.10.3 The quantities of material shown by the shipping bill are customarily accepted by the client, fabricator and erector as correct. If any shortage or damage is claimed, the client or erector should, within 48 hours, notify the carrier and the fabricator in order that the claim may be investigated.

6.10.4 The size and mass of structural steel assemblies may be limited by the shop capabilities, the permissible mass and clearance dimensions of available transportation or government regulations and the job site conditions. The fabricator limits the number of field splices to those consistent with economy.

6.10.5 On supply only contracts the unloading of steel is the responsibility of others. Unless stated otherwise the unloading of steel is part of the steel erection.

7. Erection

7.1 Method of Erection. Unless otherwise specified or agreed upon, erection shall proceed according to the most efficient and economical method available to the erector on the basis of continuous operation consistent with the drawings and specifications.

7.1.1 Temporary Bracing. Temporary bracing of the steel frame shall only be removed on instruction from the Engineer of Record.

7.2 Erection Safety. Erection shall be done in a safe manner and in accordance with applicable provincial legislation.

7.3 Site Conditions. The client shall provide and maintain adequate access roads cleared of snow and ice into and through the site for the safe delivery of derricks, cranes, other necessary equipment, and the material to be erected. The client shall provide for the erector a firm, properly graded, drained, convenient and adequate space of sufficient load carrying capacity at the site for the operation of erection equipment and shall remove at the client's cost all overhead obstructions such as power lines, telephone lines, etc., in order to provide a safe and adequate working area for erection of the steelwork. The erector shall provide and install the safety protection required for his own operations or for his work forces to meet the safety requirements of applicable Acts or Codes. The general contractor shall install protective covers to all protruding rebar, machinery anchor bolts, etc., which are a hazard to workers. Any protection for pedestrians, property, other trades, etc., not essential to the steel erection activity is the responsibility of the client. When the structure does not occupy the full available site, the client shall provide adequate storage space to enable the fabricator and erector to operate at maximum practicable speed and efficiency.

7.4 Foundations. The tender specifications preferably shall specify the time that foundations will be ready, free from obstruction and accessible to the erector. Unless otherwise agreed upon, the work of erection shall be tendered on the basis that it will start at the time designated in the tender specifications without interference or delay caused by others. Neither the fabricator nor the erector shall be responsible for the accurate location, strength and suitability of foundations. The finished tops of all footing shall be at a specified level to predetermine the amount of shimming that may be required.

7.5 Building Lines and Bench Marks. The erector shall be provided with a plot plan accurately locating building lines and bench marks at the site of the structure.

7.6 Installation of Anchor Bolts and Embedded Items

7.6.1 Anchor bolts and foundation bolts shall be set by the client in accordance with the erection diagrams. They must not vary from the dimensions shown on the erection diagrams by more than the following (see also Appendix D):

(a) 3 mm centre to centre of any two bolts within an anchor bolt group, where an anchor bolt group is defined as the set of anchor bolts which receives a single fabricated steel shipping piece;

(b) 6 mm centre-to-centre of adjacent anchor bolt groups;

(c) Maximum accumulation of 6 mm per 30 000 mm along the established column line of multiple anchor bolt groups, but not to exceed a total of 25 mm. The established column line is the actual field line most representative of the centres of the as-built anchor bolt groups along a line of columns;

(d) 6 mm from the centre of any anchor bolt group to the established column line through that group.

The tolerances of paragraphs b,c, and d also apply to offset dimensions, shown on the construction drawings, measured parallel and perpendicular to the nearest established column line for individual columns shown on the drawings to be offset from established column lines.

7.6.2 Unless shown otherwise, anchor bolts shall be set perpendicular to the theoretical bearing surface, threads shall be protected and free of concrete and nuts should run freely on the threads. Shear pockets shall be cleaned of debris, formwork, ice and snow by the client prior to steel erection.

7.6.3 Other embedded items or connection materials between the structural steel and the work of others shall be located and set by the client in accordance with approved erection diagrams. Accuracy of these items must satisfy the erection tolerance requirements of Clause 7.11.

7.6.4 All work performed by the client shall be completed so as not to delay or interfere with the erection of the structural steel.

7.7 Bearing Devices. The client shall set to lines and grades all levelling plates and loose bearing plates which can be handled without a derrick or crane. All other bearing devices supporting structural steel shall be set and wedged, shimmed or adjusted with levelling screws by the erector to lines and grades established by the client. The fabricator and/or erector shall provide the wedges, shims or levelling screws that are required, and shall scribe clearly the bearing devices with working lines to facilitate proper alignment. Promptly after the setting of any bearing devices, the client shall check lines and grades, and grout as required. The final location and proper grouting of bearing devices are the responsibility of the client.

When steel columns, girders or beams which will be supported on concrete or masonry have base plates or bearing plates fabricated as an integral part of the member, the bearing area of the support shall be suitably prepared by others so as to be at exact grade and level to receive the steelwork.

7.8 Examination by Erector. Prior to field erection, the erector shall examine the work of all others on which his work is in any way dependent and shall report to the client any errors or discrepancies that may affect erection of structural steel.

7.9 Adjustable Shelf Angles and Sash Angles. The erector shall position at time of erection all adjustable shelf angles and sash angles attached to the steel frame true and level within the tolerances permitted by the governing technical standard. Any subsequent adjustment that may be necessary to accommodate the work of others shall be performed by others.

7.10 Loose Lintels and Wall Bearing Members. Unless otherwise specified, loose lintels, shelf angles, wall bearing members and other pieces not attached to the structural steel frame shall be received and set by others.

7.11 Tolerances. Unless otherwise specified, tolerances on erected structural steel shall be those prescribed in the applicable governing technical standard.

7.12 Checking Erected Steelwork. Prior to placing or applying any other materials, the owner is responsible for determining that the location of the structural steel is acceptable for plumbness, level and alignment within tolerances with bolts correctly installed and welds inspected. The erector is given timely notice of acceptance by the owner or a listing of specific items to be corrected in order to obtain acceptance. Such notice is rendered immediately upon completion of any part of the work and prior to the start of work by other trades that may be supported, attached or applied to the structural steelwork.

7.13 Removal of Bracing.

7.13.1 Removal of Erection Bracing. Guys, braces and falsework or cribbing supplied by the erector shall remain the property of the erector. The erector shall remove them when the steel structure is otherwise adequately braced unless other arrangements are made. Guys and braces temporarily left in place under such other arrangements shall be removed by others provided prior permission by the erector for their removal has been given, and returned to the erector in good condition, see Clause 7.13.2.

7.13.2 Removal of Temporary Bracing. Temporary bracing required by the designer shall only be removed on instruction from the Engineer of Record.

7.14 Correction of Errors when Material is Not Erected by the Fabricator. Correction of minor misfits and a moderate amount of cutting, welding, and reaming shall be considered a part of the erection; in the same manner as if the Fabricator would be erecting the work. Any major rework required due to incorrect shop work shall be immediately reported to the Fabricator, before rework commences. The Fabricator shall then either correct the error or approve the method of correction including applicable costs.

7.15 Field Assembly. Unless otherwise specified, the fabricator shall provide for suitable field connections that will, in his opinion, afford the greatest overall economy.

7.16 Accommodation of Other Trades. Neither the fabricator nor the erector shall cut, drill or otherwise alter the work of others or his own work to accommodate other trades unless such work is clearly defined in the tender drawings and tender specifications and unless detailed information is provided before the erection diagrams are approved. Any subsequent cutting, drilling or other alteration of the structural steel performed by the fabricator or the erector for the accommodation of other trades, shall be specifically agreed upon and authorized by the client before such work is commenced.

7.17 Temporary Floors and Access Stairs. Unless otherwise required by law, all temporary access stairs shall be provided by others, except for the floor upon which erecting equipment is located. On this floor the erector shall provide such temporary flooring as he requires, moving his planking, etc., as the work progresses.

7.18 Touch-Up of Shop Paint. Unless so specified, the erector will not spot-paint field fasteners and field welds nor touch-up abrasions to the shop paint.

7.19 Final Painting. Unless so specified, the erector will not be responsible for cleaning the steel after erection in preparation for field painting, nor for any general field painting that may be required.

7.20 Final Clean-Up. Except as provided in Clause 7.13, upon completion of erection and before final acceptance, the erector shall remove all falsework, rubbish and temporary building furnished by him.

Notes

(Appendix A follows on facing pages 7-24 and 7-25.)

APPENDIX A - Structural Steel in Buildings

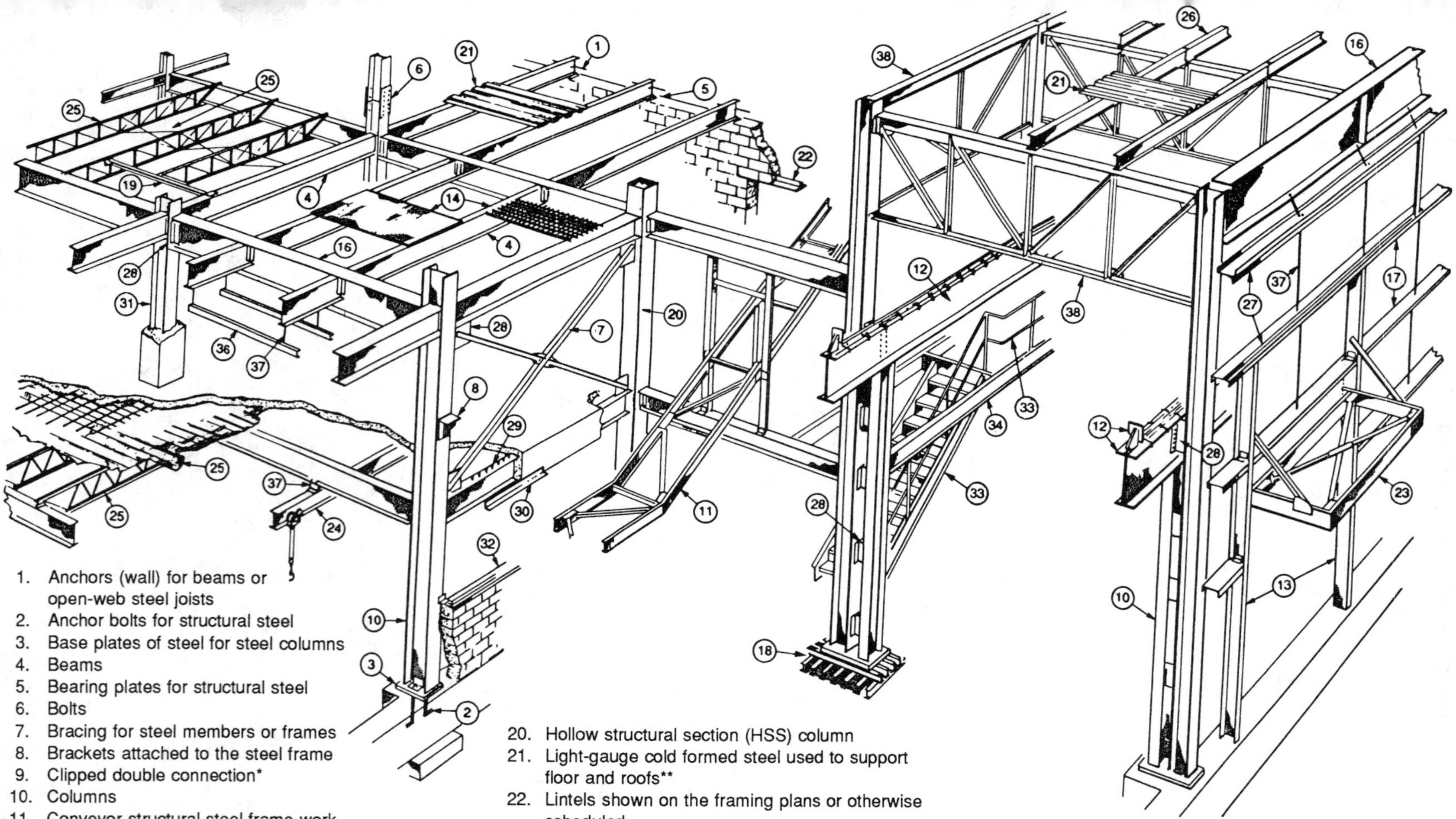

1. Anchors (wall) for beams or open-web steel joists
2. Anchor bolts for structural steel
3. Base plates of steel for steel columns
4. Beams
5. Bearing plates for structural steel
6. Bolts
7. Bracing for steel members or frames
8. Brackets attached to the steel frame
9. Clipped double connection*
10. Columns
11. Conveyor structural steel frame work
12. Crane, girders, rails and stops
13. Door frames constituting part of and connected to the steel frame
14. Floor and roof plates (raised pattern or plain), grating, connected to steel frame
15. Gerber girder*
16. Girders
17. Girts
18. Grillage beams of steel
19. Headers or trimmers for support of open-web steel joists where such headers or trimmers frame into structural steel members
20. Hollow structural section (HSS) column
21. Light-gauge cold formed steel used to support floor and roofs**
22. Lintels shown on the framing plans or otherwise scheduled
23. Marquee or canopy (structural frame only) when forming an integral part of the steel frame
24. Monorail beams of standard structural shapes, attached to the steel frame
25. Open-web steel joists, bridging, and accessories when supplied with steel joists
26. Purlins
27. Sash angles connected to the steel frame
28. Separators, angles, tees, clips, and other detail fitting essential to the structural steel frame
29. Shear connectors
30. Shelf angles attached to the steel frame
31. Steel cores for composite columns
32. Steel window sills attached to the steel frame
33. Steel stairs and handrails
34. Struts
35. Stub girder*
36. Suspended ceiling supports of structural steel shapes 75 mm or greater in depth
37. Ties, hangers and sag rods forming part of the structural frame
38. Trusses and brace frames

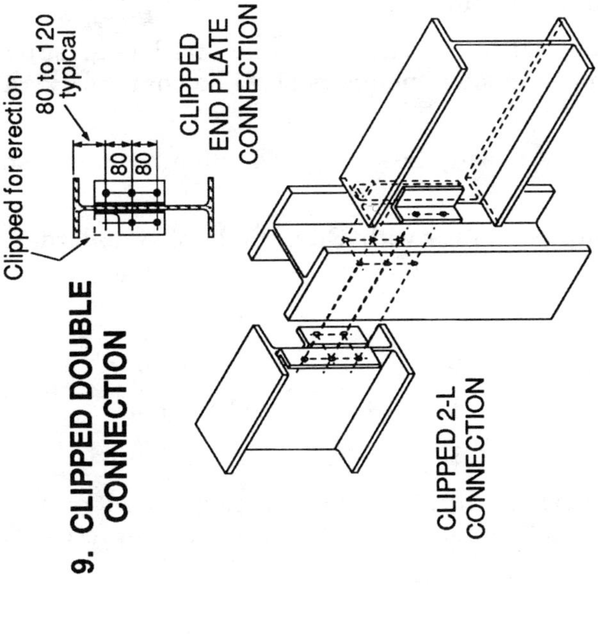

9. CLIPPED DOUBLE CONNECTION

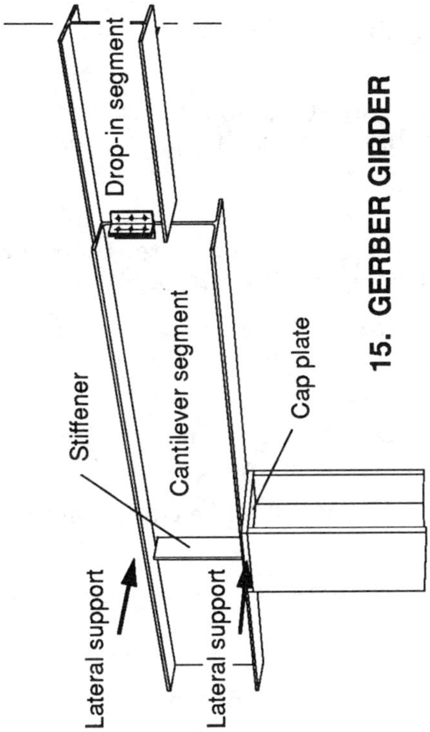

15. GERBER GIRDER

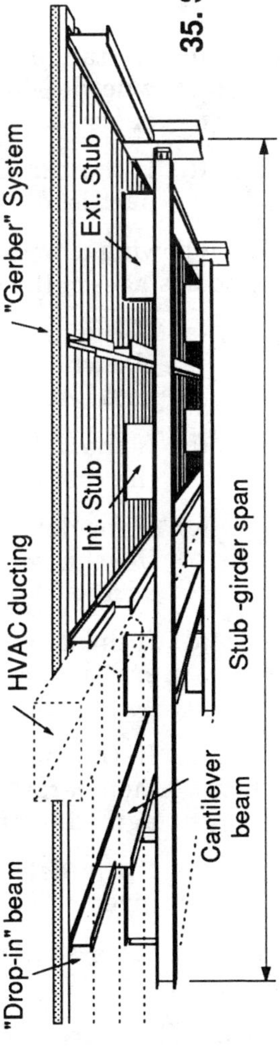

35. STUB-GIRDER

APPENDIX B
Guideline for Unit Price Application for Changes

1. Unit rates shall apply on their own, only up until commencement of material order or shop detail drawings, whichever is the earlier.
2. The following amounts, additional to the unit rate, shall be charged on additions at the various stages of the contract.
   (a) If the addition affects shop drawings (e.g. of support members) already in progress or complete, then the changes to such drawings or redetailing shall be charged extra at an agreed hourly rate.
   (b) If the addition requires additional work to material manufacture or erection (e.g. supporting members) in progress or complete, then such additional work shall be charged extra at an agreed hourly rate.
   (c) "Detail" or "Connection" materials added to existing or supporting members, whether due to an additional member or not, shall be charged on a cost plus basis.
   (d) If the timing of the addition causes the added material to be shipped as a part load, then transportation shall be charged extra at cost plus 10%.
3. The following amounts, additional to the unit rate, shall be charged for deletions at the various stages of the contract.
   (a) If the deleted material has been ordered or delivered and cannot be used elsewhere then a restocking charge shall be levied (charge varies between 15% - 40% depending on section size).
   (b) If the deleted member has been detailed or drawings are in progress, then the cost of such drawings shall be charged extra at an agreed hourly rate.
   (c) If the deletion affects shop drawings already completed or in progress, then the changes to such drawings or the redetailing shall be charged extra at an agreed hourly rate.
   (d) If the deleted member has been manufactured or erected or manufacture or erection is in progress, then the cost of such manufacture or erection shall be charged extra at an agreed hourly rate.
   (e) If the deletion affects members already manufactured (e.g. supporting members) then the changes to such members shall be charged extra at an agreed hourly rate.
   (f) If the deleted member has already been shipped, then no credit shall be given.
4. All unit rates shall be applied in accordance with CISC Code of Standard Practice Clause 3.4.1.
5. Hourly Rates are as follows:
   (a) Drawing Office Labour - $ /labour hour
   (b) Shop Labour - $ /labour hour
   (c) Field Labour - $ /labour hour
   (d) Equipment used for revisions will be charged at negotiated rental rates, according to Canadian Construction Association standard practice.

6. Revisions involving the use of grades of steel, sources of supply, or types of sections, other than specified, will be subject to price adjustments.
7. Mass will be computed in accordance with Clause 3.4 of the CISC Code of Standard Practice for Structural Steel.

APPENDIX C
A Suggested Format for Price Per Unit of Mass or Price Per Item Contracts

The following is a list of suggested categories for which unit prices could be asked, *such categories being selected or added to,* depending upon the nature of the project.

For payment purposes, the connection material required to connect an individual member to its supporting member is assumed to be part of the member to which it is attached for shipping purposes.

A. STRUCTURAL STEEL

1. COLUMNS

(a) Rolled Shapes
1. Up to 90 kg/m
2. 90-150 kg/m
3. Over 150 kg/m

(b) WWF or Plate Fabricated
1. Up to 90 kg/m
2. 90-150 kg/m
3. Over 150 kg/m

(c) USA Rolled Shapes
(F_y=350 MPa)
(Alternate to Item (b) above)
1. Up to 30 kg/m
2. 30-90 kg/m
3. 90-150 kg/m
4. Over 150 kg/m

(d) Circular Hollow Sections
1. Up to 30 kg/m
2. Over 30 kg/m

(e) Square or Rectangular Hollow Sections
1. Up to 30 kg/m
2. Over 30 kg/m

2. BEAMS

(a) Rolled Shapes
1. Up to 30 kg/m
2. 30-90 kg/m
3. 90-150 kg/m
4. Over 150 kg/m

(b) WWF or Plate Fabricated
1. Up to 90 kg/m
2. 90-150 kg/m
3. Over 150 kg/m

(c) USA Rolled Shapes
(F_y=350 MPa)
(Alternate to Item (b) above)

1. Up to 30 kg/m
2. 30-90 kg/m
3. 90-150 kg/m
4. Over 150 kg/m

(d) Stud Shear Connectors

___diam. ___mm long

3. CRANE RAILS

(a) Rail complete with Rail Clips

1. 30 kg/m
2. 42 kg/m

(b) Stops

4.1 TRUSSES - where principal members are:

(a) Tees, Angles or W Shapes

(b) Hollow Structural Sections

4.2 OPEN WEB STEEL JOISTS

5. BRACING

(a) Rolled Shapes

1. Under 30 kg/m
2. Over 30 kg/m

(b) Circular Hollow Sections

1. Under 30 kg/m
2. Over 30 kg/m

6. PURLINS

(a) Rolled Shapes

7. GIRTS

(a) Rolled Shapes

(b) Hollow Structural Sections

1. Under 30 kg/m
2. Over 30 kg/m

(c) Sag Rods

8.FRAMING

Wall and Roof Openings

9. STAIR STRINGERS

10. LADDERS (Galvanized or Painted)

(a) Without safety cage

(b) With safety cage

11. UTILITY BRIDGES (Plus Interior Framing)

(a) Tees, Angles or W Shapes

(b) Hollow Structural Sections

12. CONVEYOR GALLERIES

(a) Tees, Angles or W Shapes

(b) Hollow Structural Sections

(c) Open Conveyor Trusses

13. **MISCELLANEOUS PLATFORMS (Tees, Angles or W Shapes)**
    1. Up to 30 kg/m
    2. Over 30 kg/m

14. **PREMIUM FOR ERECTION OF LEAVE-OUT STEEL AFTER EQUIPMENT INSTALLATION**

15. **MEMBERS LESS THAN 1200mm IN LENGTH**

B. MISCELLANEOUS STEEL & GRATING

1. GRATING

(a) Floors
1. Galvanized
2. Fiberglass
3. Painted

(b) Stair Landings
1. Galvanized
2. Fiberglass
3. Painted

2. CHECKERED PLATE (6mm thick)

(a) Steel – Galvanized

(b) Steel – Painted

3.STAIR TREADS

(Maximum 900 mm long)

(a) Grating – Galvanized

(b) Grating – Fiberglass

(c) Grating – Painted

(d) Channel

4. HANDRAIL (without kickplate)

(a) Horizontal
1. Aluminum Pipe
2. Steel Pipe – Galvanized
3. Steel Pipe – Painted

(b) Sloping
1. Aluminum Pipe
2. Steel Pipe – Galvanized
3. Steel Pipe – Painted

5 KICKPLATE

(a) Plate 100 mm x 6 mm
1. Aluminum
2. Steel – Galvanized
3. Steel – Painted

(b) Banding - 140 mm x 6 mm
1. Steel – Galvanized
2. Steel – Painted

C. MOBILIZATION AND DEMOBILIZATION

1. Mobilization for erection and demobilization

D. UNIT RATES FOR EXTRA WORK

1. For extra detailing work ____________/hr. all inclusive (composite rate)
2. For extra shop labour ____________/hr. all inclusive (composite rate)
3. For extra field erection ____________/hr. all inclusive (composite rate)
4. Field labour work week ____________/hrs. per week

NOTE: "All inclusive" shall mean, all labour cost including overheads and profits, and in the field all small tools (up to $1000 value).

APPENDIX D
Tolerances on Anchor Bolt Placement

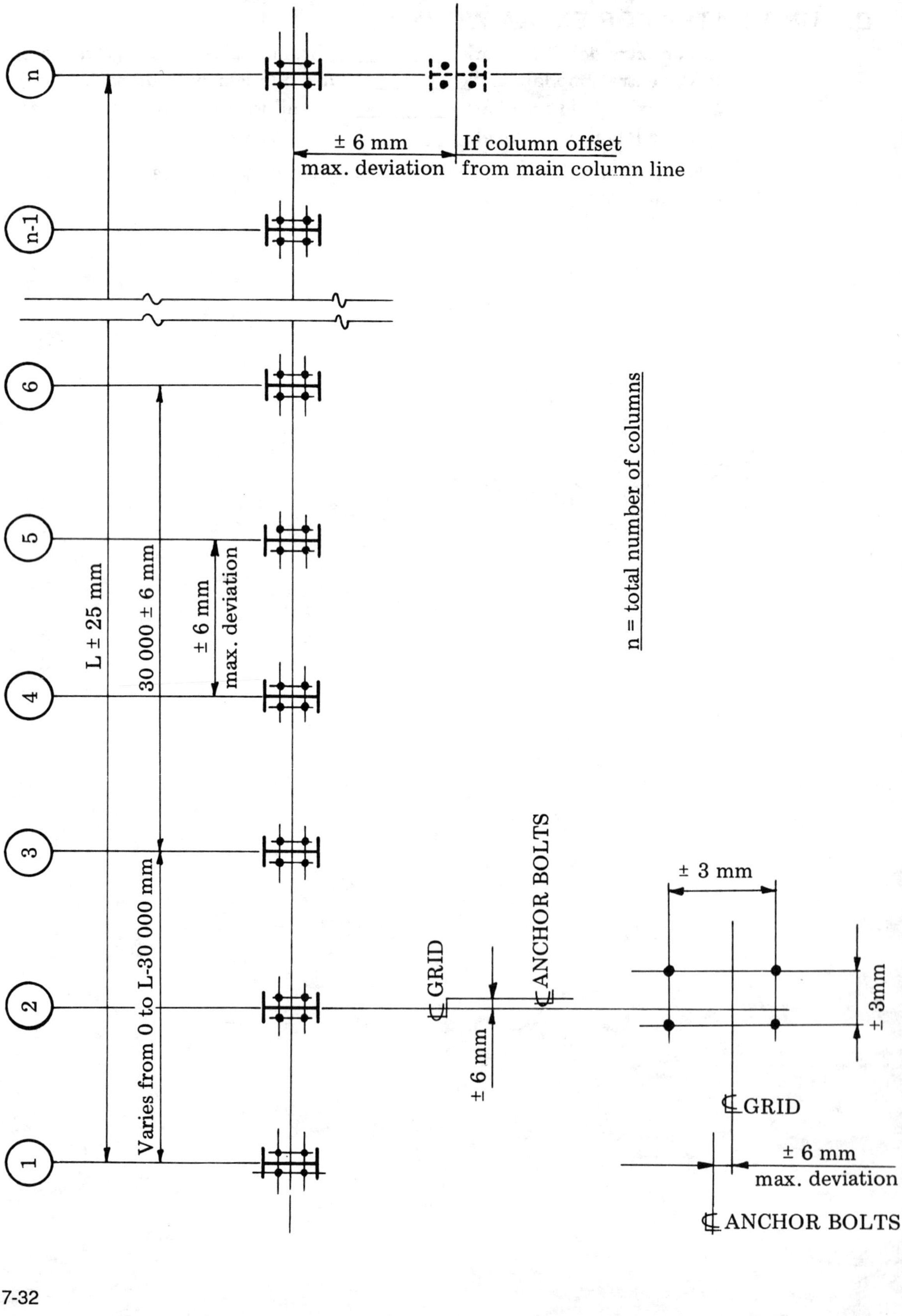

APPENDIX E
Conversion of SI Units to Imperial Units

When Imperial units are used in contract documents, unless otherwise stipulated, the SI units used in the CISC Code of Standard Practice for Structural Steel shall be replaced by the Imperial units shown, for the clause as noted.

Clause 3.2(a). Unit Weight. The unit weight of steel is assumed to be 0.2833 pounds per cubic inch.

For other clauses, the standard conversion factors (for length, mass, etc.) stipulated in CISC handbooks should be used.

APPENDIX F
Miscellaneous Steel

Miscellaneous Steel. Unless otherwise specified in the tender documents, the following items are considered miscellaneous steel of ferrous metal only, fabricated from 2.0 mm (14 ga.) and more metal, including galvanizing, cadmium and chrome plating, but not stainless steel and cast iron items.

This list of items is to be read in conjunction with Clause 2.1 Structural Steel and Clause 2.2 Items Supplied by Others, and shall include all steel items not included in Clauses 2.1 and 2.3 unless specified otherwise.

Access doors and frames — except trade-name items and those required for servicing mechanical and electrical equipment.

Angles and channel frames for doors and wall openings — drilling and tapping to be specified as being done by others.

Benches and brackets.

Bolts — only includes those bolts and anchors required for anchoring miscellaneous steel supplied under this list.

Bumper posts and rails.

Burglar/security bars.

Clothes line poles, custom fabricated types only.

Coat rods, custom fabricated types only.

Corner protection angles.

Expansion joint angles, plates, etc., including types made from steel, or a combination of steel and non-ferrous metal.

Fabricated convector frames and enclosures.

Fabricated items where clearly detailed or specified and made from 2.0 mm (14 ga.) and heavier steel, except where included in another division.

Fabricated steel framing for curtain walls and storefronts where not detailed on structural drawings and not enclosed by architectural metal.

Fabricated wire mesh and expanded metal partitions and screens.

Fire escapes.

Flag poles — steel.

Custom fabricated footscrapers, mud and foot grilles, including pans, but less drains.

Frames, grating and plate covers for manholes, catch basins, sumps, trenches, hatches, pits, etc., except cast iron, frames and covers and trade-name floor and roof drains.

Gates, grilles, grillwork and louvres, excluding baked enamel or when forming part of mechanical system.

Grating type floors and catwalks — excluding those forming part of mechanical system.

Handrails, balusters and any metal brackets attached to steel rail including plastic cover, excluding steel handrails forming part of structural steel framing.

Joist hangers, custom fabricated types only.

Joist strap anchors.

Lintels, unless shown on structural drawings.

Mat recess frames, custom fabricated types only.

Mobile chalk and tackboard frames, custom fabricated types only.

Monorail beams of standard shapes, excluding trade name items, unless shown on structural drawings.

Shop drawings and/or erection diagrams.

Shop preparation and/or priming.

Sleeves if specified, except for mechanical and electrical division.
Stair nosings, custom fabricated types only.
Steel ladders and ladder rungs not forming part of structural steel or mechanical work.
Steel stairs and landings not forming part of structural steel.
Table and counter legs, frames and brackets, custom fabricated types only.
Thresholds and sills, custom fabricated types only.
Vanity and valance brackets, custom fabricated types only.
Weatherbars — steel.

Miscellaneous Steel Items Excluded

Bases and supports for mechanical and electrical equipment where detailed on mechanical or electrical drawings.
Bolts other than for anchoring items of miscellaneous steel.
Cast iron frames and covers for manhole and catch basins.
Chain link and woven wire mesh.
Glulam connections and anchorages except when attached to miscellaneous steel, e.g. columns.
Joist hangers, trade-name types
Metal cladding and covering, less than 2.0 mm (14 ga.)
Precast concrete connections and anchorages in building structure.
Reinforcing steel or mesh.
Roof and floor hatches when trade-name items.
Sheet metal items, steel decking and siding and their attachments, closures, etc., less than 2.0 mm (14 ga.)
Shoring under composite floors and stub-girders.
Steel stacks.
Steel reinforcement for architectural metal storefronts, curtainwalls and windows.
Stud shear connectors when used with steel deck.
Stone anchors.
Temporary bracing for other trades.
Thimbles and breeching, also mechanical fire dampers.
Window and area wells.

When miscellaneous steel fabricator erects miscellaneous steel, all material required for temporary and/or permanent connections of the component parts of the miscellaneous steel shall be supplied.

APPENDIX G
A Suggested Format for a Monthly Progress Payment Claim Form

MONTHLY PROGRESS PAYMENT CLAIM FORM

CISC MPPC FORM

____
(FIRM NAME)

PROJECT: ____________________
CONTRACT NO: ____________________
PROGRESS CLAIM NO: ____________________
DATE: ____________________

ITEM	ORIGINAL BASE CONTRACT	APPROVED CHANGES TO-DATE	REVISED BASE CONTRACT	PROGRESS TO-DATE	PREVIOUS AMOUNT CLAIMED	THIS PROGRESS CLAIM	% COMPLETE
1. ENGINEERING & DRAWING PREPARATION 2. RAW MATERIALS IN YARD 3. FABRICATION 4. FREIGHT TO SITE 5. ERECTION 6. PLUMB / BOLT / CLEAN UP							
7. TOTAL GROSS AMOUNT							
8. HOLDBACK ___ %							
9. NET AMOUNT							
10. G S T ___ % OF LINE 9							
11. TOTAL AMOUNT DUE							

APPROVED CHANGE ORDER(S) TO-DATE: ____________________

STRUCTURAL SHEET STEEL PRODUCTS

General

Structural sheet steel products such as roof deck, floor deck and cladding complement the structural steel frame of a building. These large-surface elements often

TYPICAL STEEL DECK AND CLADDING PROFILES

Roof Deck

Can be supplied perforated for acoustical applications.

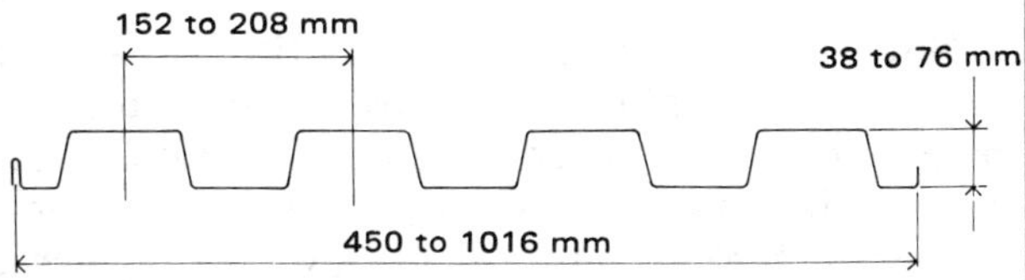

Floor Deck

Available in composite and non-composite profiles and as cellular or non-cellular units.

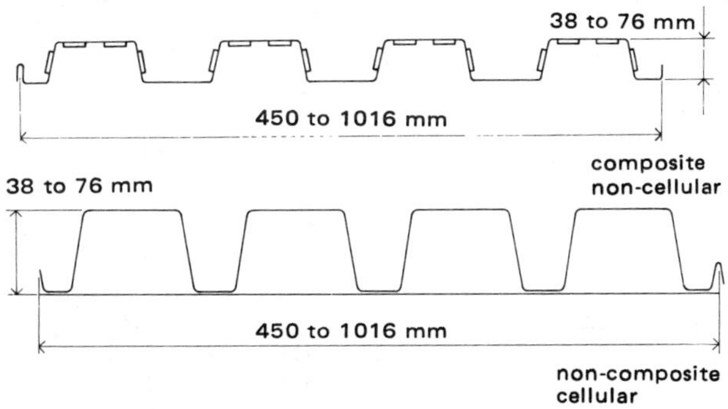

Architectural Cladding

Available in various profiles widths, coatings and colours.

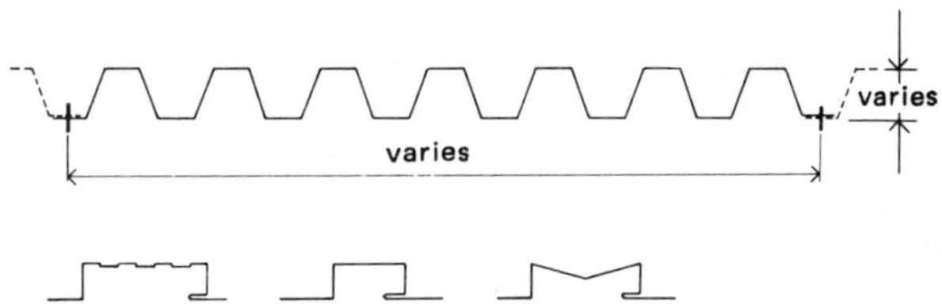

Steel Building Systems Cladding

Available in various profiles, widths, coatings and colours.

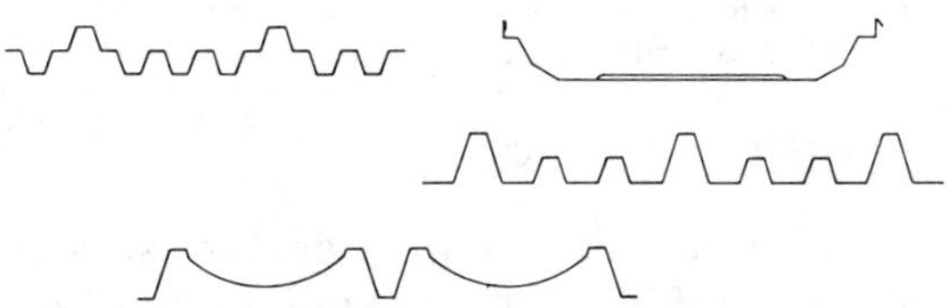

Farm Building Cladding

Available in various profiles, widths, coatings and colours.

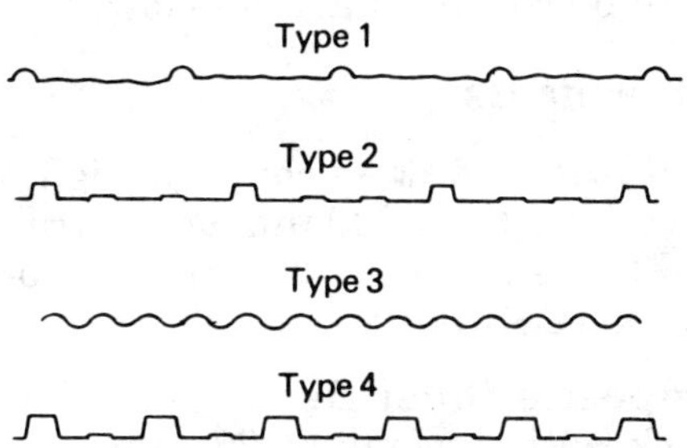

Figure 1: Typical Cladding Profiles

Figure 1 is for general information only, and manufacturers may produce additional profiles which are not represented by any of the types shown.

perform both structural and non-structural functions, thereby enhancing the overall economy of the design.

Many of the sheet steel products used in Canada are supplied by members of the Canadian Sheet Steel Building Institute, a national association of steel building producers, zinc producers, coil coaters, insulation manufacturers and fabricators of steel building products, steel building systems and lightweight steel framing components. The Institute promotes the use of sheet steel in building construction by encouraging good design, pleasing form and greater economy.

Sheet steel materials for building construction are metallic coated (zinc or aluminum-zinc alloy) and can be prefinished for extra corrosion protection and aesthetics. Consult fabricators' catalogues for details of available products, profiles, widths, lengths, thicknesses, load capacities and other characteristics. The table below relates base steel thickness increments with the corresponding overall zinc coated thickness for various zinc coating designations applicable to structural quality sheets. For structural design calculations the base steel design thickness is used.

METALLIC COATED STRUCTURAL QUALITY STEEL SHEET THICKNESSES

Base Steel Nominal Thickness (1) (mm)	Overall Metallic Coated Nominal Thickness (mm)					
	Metallic Coating Designation (2)					
	(3) ZF75 102-C	Z275 AZ150	(4) Z350 AZ180	(4) Z450	(4) Z600	(4) Z700
2.67	2.67	2.71	2.72	2.74	2.76	2.77
1.91	1.91	1.95	1.96	1.98	2.00	2.01
1.52	1.52	1.56	1.57	1.59	1.61	1.62
1.22	1.22	1.26	1.27	1.29	1.31	1.32
0.91	0.91	0.95	0.96	0.98	1.00	1.01
0.76	0.76	0.80	0.81	0.83	0.85	0.86
0.61	0.61	0.65	0.66	0.68	0.70	0.71
0.46	0.46	0.50	0.51	0.53	0.55	0.56

(5) line: below 0.98 in Z450 column; below 1.32 in Z600 and Z700 columns.

Notes:

(1) Base steel nominal thickness is used to establish section properties and for structural design calculations.

(2) The listed metallic coating designations apply to metric material from Canadian producers.

(3) The small thickness increment for ZF75 (wiped coat) and 102-C (electrogalvanized, chromate treated) is usually disregarded.

(4) Enquire as to delivery date, if time is critical.

(5) Enquire as to availability of thicknesses below the line.

CSSBI PUBLICATIONS

CSSBI publications include industry product standards, informational bulletins and special publications as well as non-technical promotional material. A selection of current publications is listed below.

CSSBI Standards

Steel Roof Deck — covers design, fabrication and erection of steel roof deck with flutes not more than 200 mm on centre and a nominal 77 mm maximum profile depth, intended for use with built-up roofing or other suitable weather-resistant cover on top of the deck. (CSSBI 10M)

Composite Steel Deck — covers design, fabrication and erection of composite steel deck with a nominal 77 mm maximum profile depth, intended for use with a concrete cover slab on top of the deck to create a composite slab. (CSSBI 12M)

Sheet Steel Cladding for Architectural/Industrial Applications — covers design, fabrication and erection of weather-tight wall and roof cladding made from metallic coated, prefinished sheet steel for use on buildings with low internal humidity. (CSSBI 20M)

Steel Building Systems — covers the design, fabrication and erection of steel building systems (SBS). Includes definitions, classification of SBS by type, checklist of items normally furnished, criteria for load combinations, design standards, and certification by a registered engineer. (CSSBI 30M)

Steel Farm Cladding — covers the manufacture, load carrying capacity, handling and installation of sheet steel cladding intended for application to walls and/or roofs of farm buildings. (CSSBI 21M)

Bulletins and Special Publications

Criteria for the Testing of Composite Slabs — provides the criteria for conducting a series of shear-bond tests necessary to determine the structural capacity of a composite slab. (CSSBI S2)

Criteria for the Design of Composite Slabs — contains design criteria, based on limit states design, for composite slabs made of a structural concrete placed permanently over a composite steel deck. (CSSBI S3)

Design of Steel Deck Diaphragms — offers a simple and practical approach to the design of steel deck diaphragms supported by horizontal steel framing. (CSSBI B14)

Criteria for the Design and Installation of Double Skin Insulated Steel Roofs — provides guidelines on good design and installation practices for roof systems consisting of a steel liner and profiled exterior sheet separated by sub-purlins and containing thermal insulation. (CSSBI B11)

Snow Load Design Criteria for Steel Building Systems — illustrates NBCC roof snow load magnitudes, distribution and special considerations for which Steel Building Systems would be designed. (CSSBI B7)

Wind Load Design Criteria for Steel Building Systems — illustrates NBCC wind load provisions used for the design of all structural components of a Steel Building System. (CSSBI B10)

Lightweight Steel Framing Manual — is an introduction to Lightweight Steel Framing (LSF) and describes the applications for LSF in floors, roofs, curtain walls and axial load bearing walls. A suggested Guide Specification is also included. (CSSBI 50M)

Contact CSSBI at the address below for a complete listing of publications, copies of publications, or other information concerning sheet steel in construction.

Canadian Sheet Steel Building Institute
305 — 201 Consumers Road
Willowdale, Ontario M2J 4G8
Tel (416) 493 - 8780
Fax (416) 493 - 8838

MASS AND FORCES FOR MATERIALS

MATERIAL	Mass (kg/m³)	Force (kN/m³)
METALS, ALLOYS, ORES		
Aluminum	2 640	25.9
Brass	8 550	83.8
Bronze, 7.9-14% tin	8 150	79.9
Bronze, aluminum	7 700	75.5
Copper	8 910	87.4
Copper ore, pyrites	4 200	41.2
Gold	19 300	189
Iron, cast, pig	7 210	70.7
Iron, wrought	7 770	76.2
Iron, spiegel-eisen	7 500	73.5
Iron, ferro-silicon	7 000	68.6
Iron ore, hematite	5 210	51.1
Iron ore, hematite in bank	2 560-2 880	25.1-28.2
Iron ore, hematite, loose	2 080-2 560	20.4-25.1
Iron ore, limonite	3 800	37.3
Iron ore, magnetite	5 050	49.5
Iron slag	2 760	27.1
Lead	11 400	112
Lead ore, galena	7 450	73.1
Magnesium	1 790	17.6
Manganese	7 610	74.6
Manganese ore	4 150	40.7
Mercury	13 600	133
Monel	8 910	87.4
Nickel	9 050	88.8
Platinum	21 300	209
Silver	10 500	103
Steel, rolled	7 850	77.0
Tin	7 350	72.1
Tin ore, cassiterite	6 700	65.7
Zinc	7 050	69.1
Zinc ore, blende	4 050	39.7
MASONRY		
Ashlar	2 240-2 560	22.0-25.1
Brick, soft	1 760	17.3
Brick, common	2 000	19.6
Brick, pressed	2 240	22.0
Clay tile, average	961	9.42
Rubble	2 080-2 480	20.4-24.3
Concrete, cinder, haydite	1 600-1 760	15.7-17.3
Concrete, slag	2 080	20.4
Concrete, stone	2 310	22.7
Concrete, stone, reinforced	2 400	23.5
SOLID FUELS		
Coal, anthracite, piled	753-929	7.38-9.11
Coal, bituminous, piled	641-865	6.29-8.48
Coke, piled	368-513	3.61-5.03
Charcoal, piled	160-224	1.57-2.20
Peat, piled	320-416	3.14-4.08
ICE AND SNOW		
Ice	897	8.80
Snow, dry, fresh fallen	128	1.26
Snow, dry, packed	192-400	1.88-3.92
Snow, wet	432-641	4.24-6.29

MATERIAL	Mass (kg/m³)	Force (kN/m³)
TIMBER, AIR-DRY		
Birch	689	6.76
Cedar	352	3.45
Fir, Douglas, seasoned	545	5.34
Fir, Douglas, unseasoned	641	6.29
Fir, Douglas, wet	801	7.86
Fir, Douglas, glue laminated	545	5.34
Hemlock	481	4.72
Larch, tamarack	561	5.50
Larch, western	609	5.97
Maple	737	7.23
Oak, red	689	6.76
Oak, white	753	7.38
Pine, jack	481	4.72
Pine, ponderosa	513	5.03
Pine, red	449	4.40
Pine, white	416	4.08
Poplar	481	4.72
Spruce	449	4.40
For pressure treated timber add retention to mass of air-dry material.		
LIQUIDS		
Alcohol, pure	785	7.70
Gasoline	673	6.60
Oils	929	9.11
Water, fresh at 4°C (max. density)	1 000	9.81
Water, fresh at 100°C	961	9.42
Water, salt	1 030	10.1
EARTH, ETC. EXCAVATED		
Earth, wet	1 600	15.7
Earth, dry	1 200	11.8
Sand and gravel, wet	1 920	18.8
Sand and gravel, dry	1 680	16.5
VARIOUS BUILDING MATERIALS		
Cement, portland, loose	1 510	14.8
Cement, portland, set	2 930	28.7
Lime, gypsum, loose	849-1 030	8.33-10.1
Mortar, cement-lime, set	1 650	16.2
Quarry stone, piled	1 440-1 760	14.1-17.3
MISCELLANEOUS		
Asphaltum	1 300	12.7
Tar, bituminous	1 200	11.8
Glass, common	2 500	24.5
Glass, plate or crown	2 580	25.3
Glass, crystal	2 950	28.9
Paper	929	9.11

DESIGN DEAD LOAD (kPa) OF MATERIALS

Material	Load (kPa)
STEEL DECKS	
Steel deck* 38 mm deep	
(up to 0.91 mm thick)	0.10
(1.22 to 1.52 mm thick)	0.15
Steel deck* 76 mm deep (Narrow-Rib)	
(up to 0.91 mm thick)	0.15
(1.22 to 1.91 mm thick)	0.30
Steel deck* 76 mm deep (Wide-Rib)	
(up to 0.91 mm thick)	0.10
(1.22 to 1.52 mm thick)	0.15
* for cellular deck, add	0.08
CONCRETE, per 100 mm	
- 2350 kg/m^3 (N.D.)	2.31
- 2000 kg/m^3 (slag aggregate)	1.96
- 1850 kg/m^3 (S.L.D.)	1.82
HOLLOW CORE PRECAST (no topping)	
- 200 mm deep (N.D.)	2.60
- 300 mm deep (N.D.)	3.50
WOOD JOISTS (at 400 mm centres)	
- 38 mm x 184 mm joists	0.09
- 38 mm x 235 mm joists	0.12
- 38 mm x 286 mm joists	0.14
PLYWOOD	
- 11 mm thick	0.06
- 14 mm thick	0.08
- 19 mm thick	0.11
CHIPBOARD	
- 12.7 mm thick	0.07
- 15.9 mm thick	0.09
- 19.0 mm thick	0.11
WALLS AND CLADDING	
- Solid brick wall (concrete)	
- 100 mm thick (S.L.D.)	1.40
- 100 mm thick (N.D.)	1.90
- Hollow block (S.L.D.)	
- 100 mm thick	1.10
- 200 mm thick	1.60
- 300 mm thick	2.30
- Hollow block (N.D.)	
- 100 mm thick	1.40
- 200 mm thick	2.10
- 300 mm thick	2.90
- P.C. wall plus glazing	2.40 - 3.80
- Metal curtain wall	0.74 - 1.50
- Insulated sheet steel wall (exclude girts)	0.25 - 0.40
- 38 x 89 wood studs @ 400 mm	0.05
- Gypsum wallboard per 10 mm	0.08
- Stone veneer per 25 mm	0.40

Material	Load (kPa)
FLOOR FINISHING	
- Vinyl, linoleum or asphalt tile	0.07
- Softwood subfloor per 10 mm	0.06
- Hardwood per 10 mm	0.08
- Carpeting	0.10
- Asphaltic concrete per 10 mm	0.23
- 20 mm Ceramic or quarry tiles on 12 mm mortar bed	0.80
- Terrazzo per 10 mm	0.24
- Mastic floor (20 mm)	0.45
ROOFING	
- 3 ply asphalt, no gravel	0.15
- 4 ply asphalt, no gravel	0.20
- 3 ply asphalt and gravel	0.27
- 4 ply asphalt and gravel	0.32
- Asphalt strip shingles	0.15
- Gypsum wallboard per 10 mm	0.08
INSULATION (per 100 mm thick)	
- Glass fibre, batts	0.05
- Glass fibre, blown	0.04
- Glass fibre, rigid	0.07
- Urethane, rigid foam	0.03
- Insulating concrete	0.06
CEILINGS	
- Gypsum wallboard per 10 mm	0.08
- Tiled ceiling & suspension system, with fixtures, average	0.20
- 20 mm plaster on lath/furring	0.40
- Sprayed fire protect'n, aver'g	0.07
- Ducts/pipes/wiring allowance (average condition)	0.25
DECK-SLABS (average condition)	
- 38 mm deck with	
- 65 mm N.D. cover#	1.95
- 90 mm N.D. cover#	2.55
- 65 mm S.L.D. cover##	1.55
- 85 mm S.L.D. cover##	1.90
- 75 mm (or 76 mm) "wide-rib" deck with	
- 65 mm N.D. cover#	2.55
- 90 mm N.D. cover#	3.15
- 65 mm S.L.D. cover##	2.15
- 85 mm S.L.D. cover##	2.50
- 76 mm "narrow-rib" deck with	
- 65 mm N.D. cover#	2.20
- 90 mm N.D. cover#	2.80
- 65 mm S.L.D. cover##	1.90
- 85 mm S.L.D. cover##	2.25

assume 2350 kg/m^3 concrete
assume 1850 kg/m^3 concrete

M/D RATIOS

Designation	SI (kg/m)/m		Imperial (lb./ft.)/in.	
	Beam[1]	Column[2]	Beam[1]	Column[2]
WWF1200				
X487	121		2.07	
X403	100		1.71	
X364	94.1		1.61	
X302	84.6		1.44	
X263	80.4		1.37	
WWF1100				
X458	120		2.05	
X388	106		1.80	
X335	91.3		1.56	
X291	86.4		1.47	
X255	75.7		1.29	
X220	71.7		1.22	
WWF1000				
X447	123		2.11	
X377	109		1.85	
X324	93.4		1.59	
X280	88.3		1.51	
X244	77.0		1.31	
X200	69.7		1.19	
WWF900				
X417	122		2.08	
X347	106		1.81	
X293	89.3		1.52	
X249	83.6		1.43	
X213	71.5		1.22	
X192	71.6		1.22	
X169	63.1		1.08	
WWF800				
X332	108		1.84	
X279	90.6		1.55	
X235	84.5		1.44	
X198	71.2		1.22	
X164	66.1		1.13	
X154	62.1		1.06	
WWF700				
X222	86.0		1.47	
X203	78.7		1.34	
X185	71.7		1.22	
X164	71.9		1.23	
X151	66.2		1.13	
X141	61.8		1.06	
WWF550				
X721		227		3.87
X620		191		3.27
X503		154		2.63
X420		129		2.20
X280		85.6		1.46
WWF500				
X651		226		3.86
X561		191		3.26
X456		154		2.63
X381		129		2.20
X343		116		1.98
X306		103		1.76
X276		92.9		1.59
X254		85.5		1.46
X223		75.1		1.28
X197		66.1		1.13

Designation	SI (kg/m)/m		Imperial (lb./ft.)/in.	
	Beam[1]	Column[2]	Beam[1]	Column[2]
WWF450				
X503		191		3.25
X409		154		2.62
X342		129		2.19
X308		116		1.98
X274		103		1.76
X248		92.9		1.59
X228		85.4		1.46
X201		75.3		1.28
X177		66.0		1.13
WWF400				
X444		190		3.24
X362		153		2.62
X303		128		2.19
X273		116		1.97
X243		103		1.76
X220		92.8		1.58
X202		85.2		1.45
X178		75.1		1.28
X157		66.0		1.13
WWF350				
X315		153		2.61
X263		128		2.18
X238		116		1.97
X212		103		1.76
X192		92.8		1.58
X176		85.0		1.45
X155		74.9		1.28
X137		65.9		1.12
W920				
X446	144		2.47	
X417	135		2.31	
X387	126		2.16	
X365	120		2.04	
X342	112		1.92	
W920				
X313	114		1.94	
X289	105		1.80	
X271	99.2		1.69	
X253	92.9		1.59	
X238	87.7		1.50	
X223	82.4		1.41	
X201	74.8		1.28	
W840				
X359	123		2.11	
X329	114		1.94	
X299	104		1.78	
W840				
X226	88.4		1.51	
X210	82.8		1.41	
X193	76.3		1.30	
X176	69.9		1.19	
W760				
X314	117		2.00	
X284	106		1.82	
X257	96.7		1.65	

[1] M/D = mass/(surface area—top of top flange)

[2] M/D = mass/surface area

Designation	SI (kg/m)/m		Imperial (lb./ft.)/in.	
	Beam[1]	Column[2]	Beam[1]	Column[2]
W760				
X196	84.8		1.45	
X185	80.3		1.37	
X173	75.4		1.29	
X161	70.5		1.20	
X147	64.6		1.10	
W690				
X265	108		1.84	
X240	98.6		1.68	
X217	89.5		1.53	
W690				
X170	80.0		1.37	
X152	71.8		1.23	
X140	66.5		1.13	
X125	59.9		1.02	
W610				
X241	109		1.85	
X217	98.5		1.68	
X195	88.9		1.52	
X174	80.0		1.37	
X155	71.6		1.22	
W610				
X140	73.7		1.26	
X125	66.1		1.13	
X113	60.0		1.02	
X101	54.0		0.921	
X91	48.9		0.833	
X84	45.4		0.769	
W610				
X92	53.5		0.912	
X82	47.9		0.818	
W530				
X219	107		1.83	
X196	96.8		1.65	
X182	90.3		1.54	
X165	82.2		1.40	
X150	75.1		1.28	
W530				
X138	80.9		1.38	
X123	72.4		1.24	
X109	64.5		1.10	
X101	60.1		1.03	
X92	55.1		0.940	
X82	49.4		0.843	
X72	43.6		0.739	
W530				
X85	55.1		0.940	
X74	48.2		0.823	
X66	43.3		0.739	
W460				
X177	99.2		1.69	
X158	89.0		1.52	
X144	81.5		1.39	
X128	72.8		1.24	
X113	64.9		1.11	
W460				
X106	70.9		1.21	
X97	65.2		1.11	
X89	60.2		1.03	
X82	55.8		0.953	
X74	50.3		0.859	
X67	45.9		0.783	
X61	42.0		0.718	
W460				
X68	49.8		0.850	
X60	44.2		0.755	
X52	38.9		0.663	
W410				
X149	91.7		1.57	
X132	81.6		1.39	
X114	71.3		1.22	
X100	62.9		1.07	
W410				
X85	62.5		1.07	
X74	54.8		0.936	
X67	50.0		0.853	
X60	45.0		0.769	
X54	40.8		0.697	
W410				
X46	38.0		0.649	
X39	32.2		0.550	
W360				
X1086		388		6.62
X990		360		6.14
X900		333		5.69
X818		309		5.27
X744		285		4.87
X677		262		4.48
W360				
X634		249		4.24
X592		235		4.01
X551		220		3.76
X509		205		3.50
X463		189		3.23
X421		174		2.97
X382		159		2.72
X347		146		2.49
X314		134		2.28
X287		123		2.09
X262		113		1.93
X237		103		1.76
X216		94.3		1.61
W360				
X196		88.7		1.51
X179		81.4		1.39
X162	89.1	74.0	1.52	1.26
X147	81.2	67.4	1.39	1.15
X134	74.4	61.8	1.27	1.05
W360				
X122	82.8	70.5	1.41	1.20
X110	75.1	64.0	1.28	1.09
X101	69.4	59.1	1.18	1.01
X91	62.9	53.5	1.07	0.914
W360				
X79	60.5	52.3	1.03	0.893
X72	55.6	48.0	0.948	0.819
X64	49.7	43.0	0.849	0.733
W360				
X57	46.8		0.799	
X51	42.2		0.720	
X45	37.5		0.641	
W360				
X39	36.4		0.621	
X33	31.0		0.530	

[1] M/D = mass/(surface area—top of top flange)

[2] M/D = mass/surface area

M/D RATIOS

Designation	SI (kg/m)/m		Imperial (lb./ft.)/in.	
	Beam[1]	Column[2]	Beam[1]	Column[2]
W310				
X500		236		4.03
X454		217		3.71
X415		201		3.44
X375		185		3.15
X342		170		2.90
X313		157		2.68
W310				
X283		144		2.46
X253		130		2.23
X226		118		2.01
X202		106		1.81
X179		95.2		1.63
X158		84.9		1.45
X143		77.3		1.32
X129	84.2	70.1	1.44	1.20
X118	77.5	64.5	1.32	1.10
X107	70.7	58.8	1.21	1.00
X97	64.0	53.3	1.09	0.910
W310				
X86	63.0	53.1	1.07	0.906
X79	58.3	49.1	0.994	0.838
W310				
X74	60.9	52.1	1.04	0.889
X67	55.6	47.5	0.948	0.811
X60	50.1	42.9	0.856	0.731
W310				
X52	46.3		0.790	
X45	40.4		0.689	
X39	35.3		0.602	
X31	28.4		0.490	
W310				
X33	35.9		0.614	
X28	30.8		0.526	
X24	26.7		0.456	
X21	23.4		0.399	
W250				
X167		104		1.78
X149		94.3		1.61
X131		84.0		1.43
X115		74.2		1.27
X101	79.3	66.0	1.35	1.13
X89	70.4	58.6	1.20	0.999
X80	63.7	53.0	1.09	0.904
X73	58.6	48.7	1.00	0.831
W250				
X67	60.6	51.1	1.03	0.873
X58	52.9	44.6	0.902	0.761
X49	45.0	38.0	0.769	0.648
W250				
X45	46.8		0.798	
X39	40.9		0.698	
X33	35.0		0.597	
X24	25.8		0.437	
W250				
X28	34.2		0.584	
X25	30.9		0.528	
X22	27.6		0.471	
X18	22.5		0.385	
W200				
X100	94.3	78.7	1.61	1.34
X86	82.6	68.8	1.41	1.17
X71	68.7	57.3	1.17	0.977
X59	58.1	48.4	0.992	0.825
X52	51.7	43.0	0.882	0.734
X46	46.1	38.3	0.787	0.654

Designation	SI (kg/m)/m		Imperial (lb./ft.)/in.	
	Beam[1]	Column[2]	Beam[1]	Column[2]
W200				
X42	47.0	39.6	0.802	0.676
X36	40.7	34.3	0.694	0.585
W200				
X31	38.5	33.0	0.656	0.563
X27	33.9	29.0	0.578	0.496
X21	26.4		0.449	
W200				
X22	31.1		0.530	
X19	27.2		0.465	
X15	21.7		0.371	
W150				
X37	48.3	40.2	0.824	0.686
X30	39.6	33.0	0.676	0.563
X22	29.4	24.4	0.502	0.417
W150				
X24	39.5	33.8	0.674	0.577
X18	30.1	25.7	0.514	0.439
X14	23.7	20.3	0.405	0.346
W130				
X28		36.8		0.629
X24		32.0		0.546
W100				
X19		31.1		0.532
S610				
X180	99.1		1.69	
X158	87.3		1.49	
S610				
X149	85.8		1.46	
X134	77.5		1.32	
X119	68.7		1.17	
S510				
X143	93.0		1.59	
X128	83.6		1.43	
S510				
X112	76.3		1.30	
X98	66.6		1.14	
S460				
X104	77.0		1.31	
X81.4	60.4		1.03	
S380				
X74	63.4		1.08	
X64	55.2		0.942	
S310				
X74	74.7		1.27	
S310				
X60.7	61.5		1.05	
S310				
X52	53.6		0.914	
X47	48.3		0.824	
S250				
X52	60.9		1.04	
X38	45.1		0.770	
S200				
X34	48.3		0.824	
X27	38.7		0.660	
S180				
X30	47.5		0.810	
X22.8	36.4		0.621	
S150				
X26	47.4		0.808	
X19	34.9		0.595	
M100				
X19		32.1		0.550

[1] M/D = mass/(surface area—top of top flange)

[2] M/D = mass/surface area

COEFFICIENTS OF THERMAL EXPANSION

(Linear, per degree x 10^{-6})

METALS	c per °C	c per °F
Aluminum	**23**	13
Brass	**19**	10.4
Bronze	**18**	10.1
Copper	**16.7**	9.3
Iron, Gray Cast	**11**	5.9
Iron, Wrought	**12.0**	6.7
Lead	**28.7**	15.9
Magnesium	**28.8**	16
Nickel	**12.6**	7
Steel, Cast	**11.3**	6.3
Steel, Stainless	**17.8**	9.9
Steel, Structural	**11.7**	6.5
Zinc, Rolled	**31**	17.3

NON-METALS	c per °C	c per °F
Cement, Portland	**13**	7
Concrete, Stone	**10**	5.7
Glass	**7**	4
Granite	**8.3**	4.6
Limestone	**7.9**	4.4
Marble	**9**	5
Masonry, Ashlar	**6.3**	3.5
Masonry, Brick	**6.1**	3.4
Masonry, Rubble	**6.3**	3.5
Plaster	**16**	9
Sandstone	**11**	6
Slate	**10**	5.8
Fir (parallel to fibre)	**3.8**	2.1
Fir (perpendicular to fibre)	**58**	32

NOTE: Coefficients of thermal expansion indicated are average values from various sources. Minor variations may be expected in metals. Large variations may be expected in concrete and masonry due to the many combinations of constituents possible.

Coefficients apply in general to a temperature range from 0 to 100 degrees Celsius.

The coefficient of linear thermal expansion (c) is the change in length per unit of length for a change of one degree of temperature. The coefficient for surface expansion is approximately two times, and the coefficient of volume expansion is approximately three times, the linear coefficient.

Change in length = cL x change in temperature, if member is free to elongate or contract.

Change in unit stress = cE x change in temperature, if member is not permitted to elongate or contract.

ELECTRONIC AIDS

General

Electronic aids now range from the simple pocket calculator, through the desk top and mini computers, to the powerful main frame large core computer in datacentres. An increasing number of reasonably priced but powerful computers are available to designers which can be programmed for steel design. In various parts of this book, for example, Built Up Sections in Part 6, expressions or flow charts are provided for rapid use on a calculator or to assist in the preparation of programs.

The CISC together with the Canadian Steel Construction Council (CSCC) have developed computer programs on calculators (HP97, 41C), personal computers (i.e. IBM), desk top computers (HP9000 Series 200) and large scale computers (FORTRAN language) for a broad range of steel design situations. Further information on current programs for steel design may be obtained from the CISC.

Electronic Calculators and Metric

Converting to, and using SI has been made easier by the availability of a broad range of electronic calculators. Although even an inexpensive model will be of some help, a more sophisticated model will be more effective in dealing with engineering problems.

Handling All Those Zeros

Both Hewlett-Packard (HP) and Texas Instruments (TI) market calculators that offer display format control in what is called "engineering notation". The display value in this format consists of a mantissa (the significant digits) and an exponent (power of ten) that have been adjusted so that exponent is a multiple of three (10^9, 10^{-6}, etc.) and the mantissa has 1, 2, or 3 digits to the left of the decimal point. Some models also allow the user to control the number of digits in the mantissa, that is, the number of significant digits displayed may be limited.

Since the properties of steel shapes in this book are organized so that the exponents are also in multiples of three, stress calculations are simplified by being able to use the number (mantissa and exponent) as tabulated.

This exploits a basic feature of the SI system. For example, to determine the extreme fibre stress of a W250 X 33 beam subject to 60 kN•m moment the following operations are performed:

Operation	Display	
1. Enter 60 kN·m	60.0	00
2. Multiply by 1 000 once to bring m to mm and again to bring kN to N	60.0	06
3. Obtain S_x for W250 × 33 beam (See p. 6—50), 379 x 10^3 mm^3	379.0	03
4. Divide to obtain stress in MPa ($N/mm^2 \equiv MN/m^2 \equiv MPa$)	158.3	00

Data Storage

A calculator with at least 6 data storage registers is very convenient in that it can be used to hold a number of conversion factors which pertain to structural design, such as the number of kN per kip, kN·m per ft-kip, MPa per ksi etc. Thus, easy comparisons between systems of units can be made, if required. Calculators with preprogrammed conversions are generally of marginal value as they are designed for the US household, not the engineering market, and they convert US Customary units instead of Imperial units.

Programmable vs. Non-programmable

Although price may determine the choice, the programmable models do offer some unique advantages when dealing with metric conversion. One such application is the manipulation of the display to show values of linear measure in the Imperial system in feet, inches and sixteenth of an inch, as shown

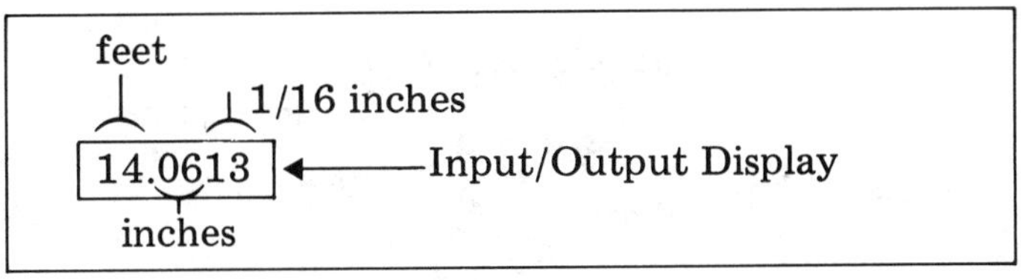

and have them converted to metric or vice versa. Subsequent measurements can then be added to or subtracted from a previous value.

Of course, any repetitive calculation is greatly simplified with the programmable models.

Aids

The following expressions are provided in addition to those found in other Parts of this book.

1. Mass of Any Cross-Section in kilograms per metre

$$7\,850 \times 10^{-6} \times A\ (mm^2)$$

2. Force Due to Gravity on Any Cross-Section in kilonewtons per metre

$$77 \times 10^{-6} \times A\ (mm^2)$$

3. Factored Moment Resistance of Composite Beam

The following expressions are structured for easy programming on a programmable calculator. The majority of the symbols used are defined in Part One, CAN3-S16.1-M78 or in Composite Beams, Part Five of this book.

a) Plastic Neutral Axis in the Slab

When $\phi\, A_s\, F_y \leqslant 0.85\, \phi_c\, b_1 t_c\, F_c'$

$$M_{rc} = \phi\, A_s\, F_y \left(\frac{d}{2} + t_d + t_c - \frac{\phi\, A_s\, F_y}{1.7 \phi_c b_1 F_c'} \right)$$

b) Plastic Neutral Axis in the Steel Flange

When $\phi\, A_s\, F_y > 0.85\, \phi_c\, b_1\, t_c\, F_c'$ and

$$(\phi\, A_s\, F_y - 0.85\, \phi_c\, b_1\, t_c\, F_c')/2 \leqslant \phi\, b\, t\, F_y$$

$$M_{rc} = \phi\, b\, t_1\, F_y \left[d + \frac{A_s(t_1 - d) + 2\, t_1\, b\, (d - t_1)}{2\, (A_s - b\, t_1)} \right] +$$

$$0.85\, \phi_c b_1 t_c F_c' \left[d + t_d + \frac{t_c}{2} - \frac{A_s d - b t_1 (2d - t_1)}{2(A_s - b t_1)} \right]$$

Where $t_1 = \dfrac{\phi A_s F_y - 0.85 \phi_c b_1 t_c F'_c}{2 \phi F_y b}$

c) Plastic Neutral Axis in the Steel Web

When $\phi A_s F_y > 0.85 \phi_c b_1 t_c F'_c$ and

$$(\phi A_s F_y - 0.85 \phi_c b_1 t_c F'_c)/2 > \phi b t F_y$$

$$M_{rc} = \phi F_y (bt + d_1 w)\left[d - \frac{bt^2 + 2d_1 tw + d_1{}^2 w}{2(bt + d_1 w)} - d_2\right] +$$

$$0.85 \phi_c b_1 t_c F'_c \left[d + t_d + \frac{t_c}{2} - d_2\right]$$

Where $d_1 = \dfrac{\left[\dfrac{\phi A_s F_y - 0.85 \phi_c b_1 t_c F'_c}{2 \phi F_y}\right] - b t}{w}$;

and, $d_2 = \dfrac{\dfrac{A_s d}{2} - (b t + d_1 w)\left[d - \dfrac{bt^2 + 2d_1 tw + d_1{}^2 w}{2(b t + d_1 w)}\right]}{A_s - (b t + d_1 w)}$

For partial shear connection the above expressions are applicable when the following substitutions are made

t_d = Deck thickness + Slab thickness $- t_c$

$t_c = nq_r/\phi_{sc}/(0.85 b_1 F'_c)$

NOTES

PROPERTIES OF GEOMETRIC SECTIONS
Definitions

Neutral Axis

The line, in any given section of a member subject to bending, on which there is neither tension nor compression.

For pure elastic bending of a straight beam, the neutral axis at any cross-section is coincident with the centroidal axis of the cross-section.

In the case of fully plastic bending, the neutral axis divides the sectional area equally. Therefore, the neutral axis for elastic and plastic bending coincide only in the case of sections symmetrical about the neutral axis.

Moment of Inertia I

The sum of the products obtained by multiplying each of the elementary areas, of which the section is composed, by the square of its perpendicular distance from the axis about which the moment of inertia is being calculated.

Elastic Section Modulus S

The moment of inertia divided by the perpendicular distance from the axis about which the moment of inertia has been calculated to the most remote part of the section.

The elastic section modulus is used to determine the bending stress in the extreme fibre of a section by dividing the bending moment by the section modulus, referred to the neutral axis perpendicular to the plane of bending, both values being expressed in like units of measure.

Radius of Gyration r

The perpendicular distance from a neutral axis to the centre of gyration (i.e., the point where the entire area is considered to be concentrated so as to have the same moment of inertia as the actual area). The square of the radius of gyration of a section is equal to the moment of inertia (referred to the appropriate axis) divided by the area.

The radius of gyration of a section is used to ascertain the load this section will sustain when used in compression as a strut or column. The ratio of the effective unsupported length of the section divided by the least radius of gyration applicable to this length is called the slenderness ratio.

Plastic Modulus Z

The modulus of resistance to bending of a completely yielded cross-section, calculated by taking the combined statical moment, about the neutral axis, of the cross-sectional areas above and below that axis.

In general, the plastic modulus is calculated by simple statics and has been included for only a few of the shapes listed.

PROPERTIES OF GEOMETRIC SECTIONS

SQUARE

Axis of moments through center

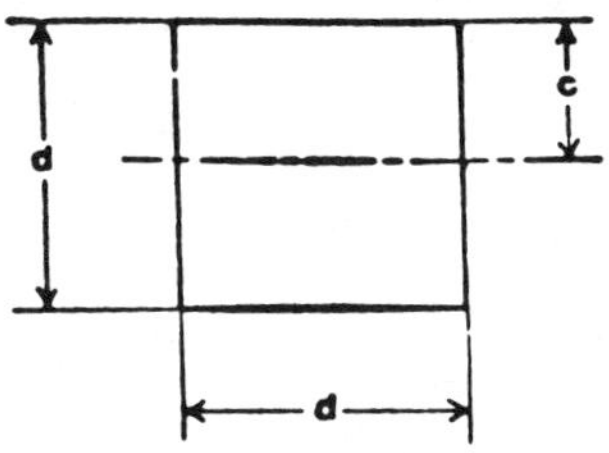

$$A = d^2$$

$$c = \frac{d}{2}$$

$$I = \frac{d^4}{12}$$

$$S = \frac{d^3}{6}$$

$$r = \frac{d}{\sqrt{12}}$$

$$Z = \frac{d^3}{4}$$

SQUARE

Axis of moments on base

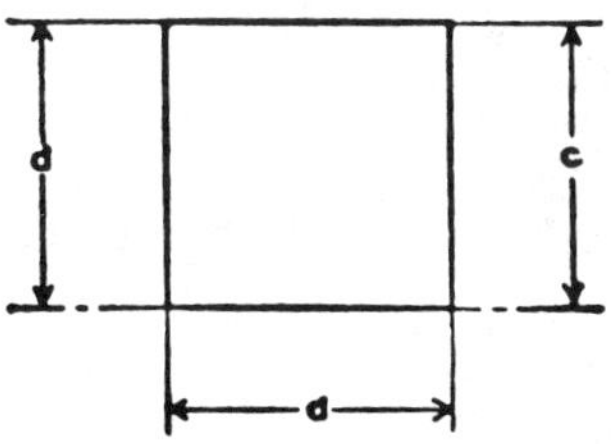

$$A = d^2$$

$$c = d$$

$$I = \frac{d^4}{3}$$

$$S = \frac{d^3}{3}$$

$$r = \frac{d}{\sqrt{3}}$$

SQUARE

Axis of moments on diagonal

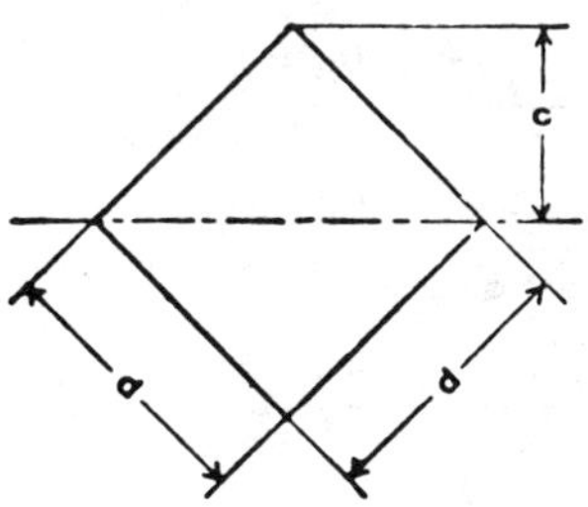

$$A = d^2$$

$$c = \frac{d}{\sqrt{2}}$$

$$I = \frac{d^4}{12}$$

$$S = \frac{d^3}{6\sqrt{2}}$$

$$r = \frac{d}{\sqrt{12}}$$

$$Z = \frac{2c^3}{3} = \frac{d^3}{3\sqrt{2}}$$

RECTANGLE

Axis of moments through center

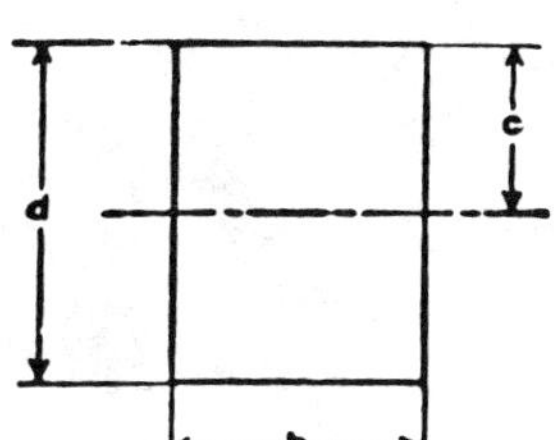

$$A = bd$$

$$c = \frac{d}{2}$$

$$I = \frac{bd^3}{12}$$

$$S = \frac{bd^2}{6}$$

$$r = \frac{d}{\sqrt{12}}$$

$$Z = \frac{bd^2}{4}$$

PROPERTIES OF GEOMETRIC SECTIONS

RECTANGLE

Axis of moments on base

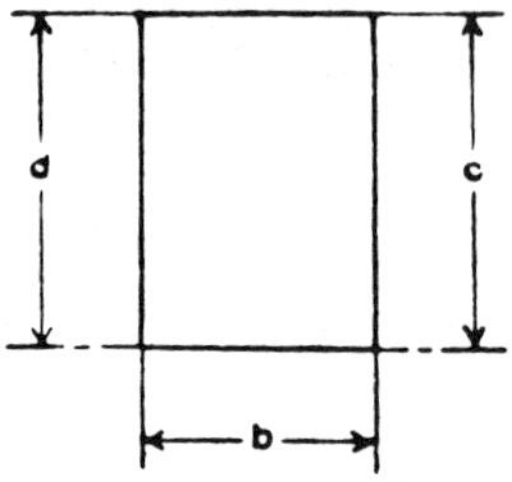

$$A = bd$$

$$c = d$$

$$I = \frac{bd^3}{3}$$

$$S = \frac{bd^2}{3}$$

$$r = \frac{d}{\sqrt{3}}$$

RECTANGLE

Axis of moments on diagonal

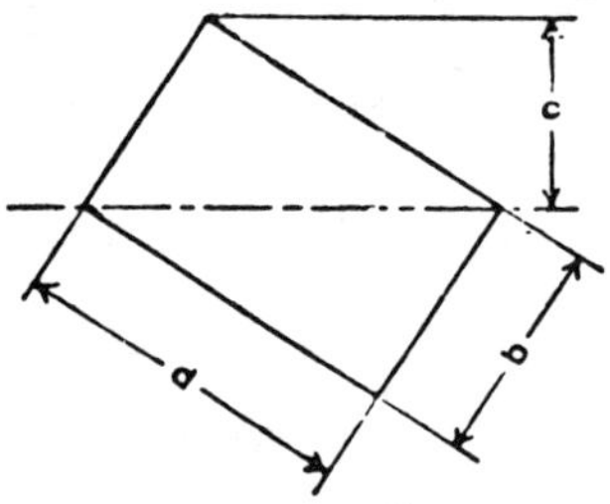

$$A = bd$$

$$c = \frac{bd}{\sqrt{b^2 + d^2}}$$

$$I = \frac{b^3 d^3}{6\,(b^2 + d^2)}$$

$$S = \frac{b^2 d^2}{6\sqrt{b^2 + d^2}}$$

$$r = \frac{bd}{\sqrt{6\,(b^2 + d^2)}}$$

RECTANGLE

Axis of moments any line through center of gravity

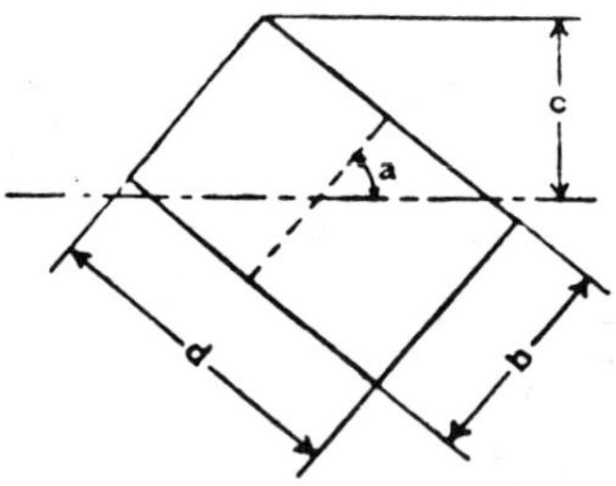

$$A = bd$$

$$c = \frac{b \sin a + d \cos a}{2}$$

$$I = \frac{bd\,(b^2 \sin^2 a + d^2 \cos^2 a)}{12}$$

$$S = \frac{bd\,(b^2 \sin^2 a + d^2 \cos^2 a)}{6\,(b \sin a + d \cos a)}$$

$$r = \sqrt{\frac{b^2 \sin^2 a + d^2 \cos^2 a}{12}}$$

HOLLOW RECTANGLE

Axis of moments through center

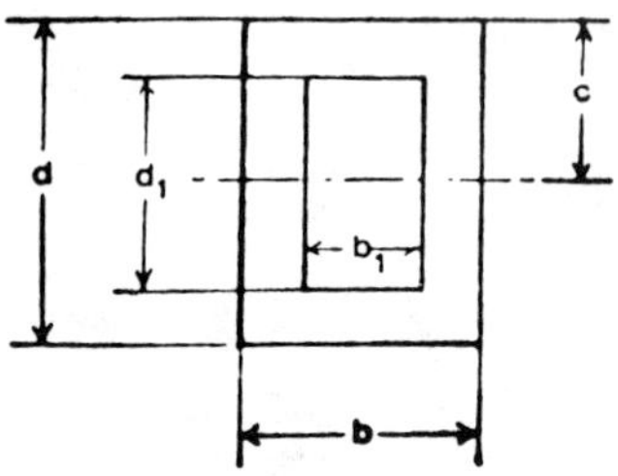

$$A = bd - b_1 d_1$$

$$c = \frac{d}{2}$$

$$I = \frac{bd^3 - b_1 d_1{}^3}{12}$$

$$S = \frac{bd^3 - b_1 d_1{}^3}{6d}$$

$$r = \sqrt{\frac{bd^3 - b_1 d_1{}^3}{12A}}$$

$$Z = \frac{1}{4}(bd^2 - b_1 d_1^2)$$

EQUAL RECTANGLES

Axis of moments through center of gravity

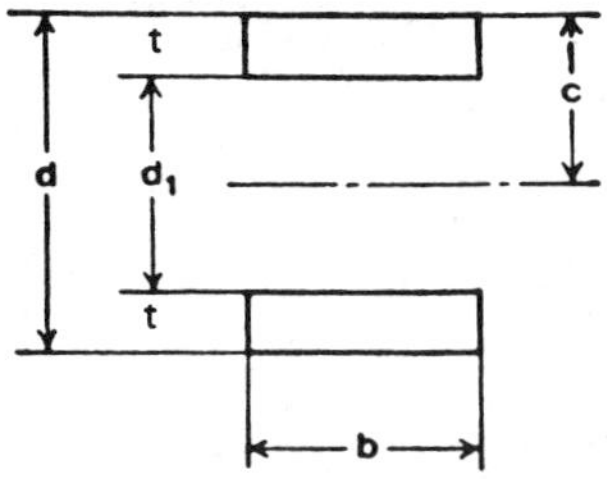

$$A = b\,(d - d_1)$$

$$c = \frac{d}{2}$$

$$I = \frac{b\,(d^3 - d_1{}^3)}{12}$$

$$S = \frac{b\,(d^3 - d_1{}^3)}{6d}$$

$$r = \sqrt{\frac{d^3 - d_1{}^3}{12(d - d_1)}}$$

$$Z = \frac{b}{4}(d^2 - d_1^2) = bt(d - t)$$

UNEQUAL RECTANGLES

Axis of moments through center of gravity

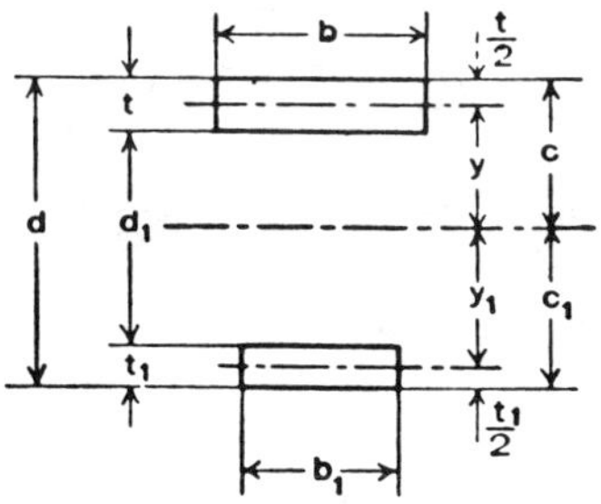

$$A = bt + b_1 t_1$$

$$c = \frac{\tfrac{1}{2}\,bt^2 + b_1 t_1\,(d - \tfrac{1}{2}\,t_1)}{A}$$

$$I = \frac{bt^3}{12} + bty^2 + \frac{b_1 t_1{}^3}{12} + b_1 t_1 y_1{}^2$$

$$S = \frac{I}{c} \qquad S_1 = \frac{I}{c_1}$$

$$r = \sqrt{\frac{I}{A}}$$

TRIANGLE

Axis of moments through center of gravity

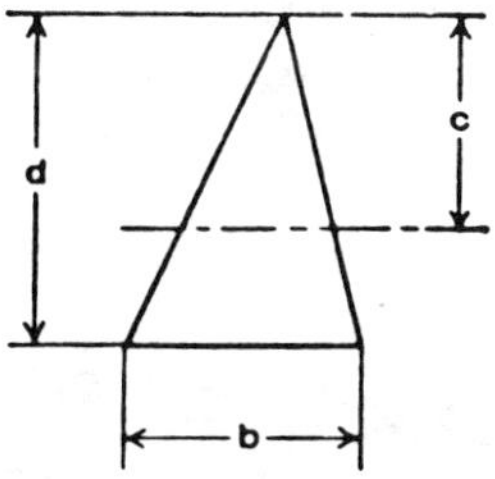

$$A = \frac{bd}{2}$$

$$c = \frac{2d}{3}$$

$$I = \frac{bd^3}{36}$$

$$S = \frac{bd^2}{24}$$

$$r = \frac{d}{\sqrt{18}}$$

TRIANGLE

Axis of moments on base

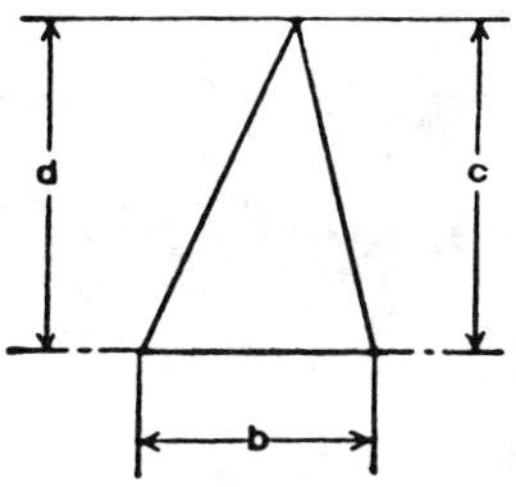

$$A = \frac{bd}{2}$$

$$c = d$$

$$I = \frac{bd^3}{12}$$

$$S = \frac{bd^2}{12}$$

$$r = \frac{d}{\sqrt{6}}$$

PROPERTIES OF GEOMETRIC SECTIONS

TRAPEZOID

Axis of moments through center of gravity

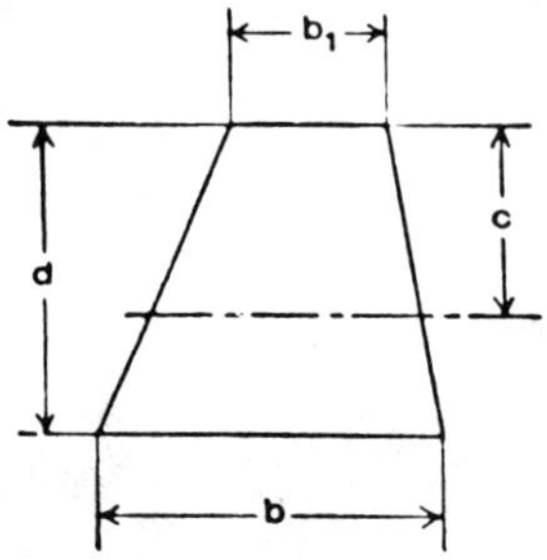

$$A = \frac{d(b + b_1)}{2}$$

$$c = \frac{d(2b + b_1)}{3(b + b_1)}$$

$$I = \frac{d^3\,(b^2 + 4\,bb_1 + b_1^2)}{36\,(b + b_1)}$$

$$S = \frac{d^2\,(b^2 + 4\,bb_1 + b_1^2)}{12\,(2b + b_1)}$$

$$r = \frac{d}{6(b + b_1)}\sqrt{2\,(b^2 + 4\,bb_1 + b_1^2)}$$

CIRCLE

Axis of moments through center

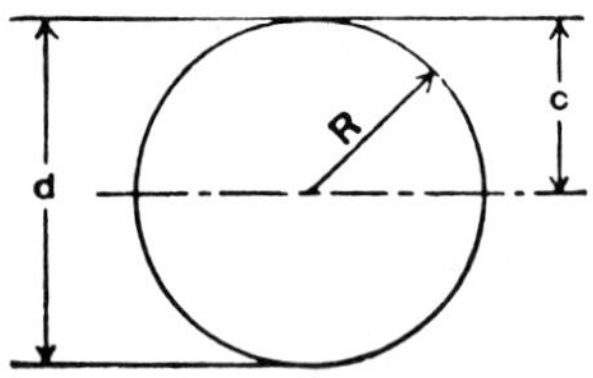

$$A = \frac{\pi d^2}{4} = \pi R^2$$

$$c = \frac{d}{2} = R$$

$$I = \frac{\pi d^4}{64} = \frac{\pi R^4}{4}$$

$$S = \frac{\pi d^3}{32} = \frac{\pi R^3}{4}$$

$$r = \frac{d}{4} = \frac{R}{2}$$

$$Z = \frac{d^3}{6}$$

HOLLOW CIRCLE

Axis of moments through center

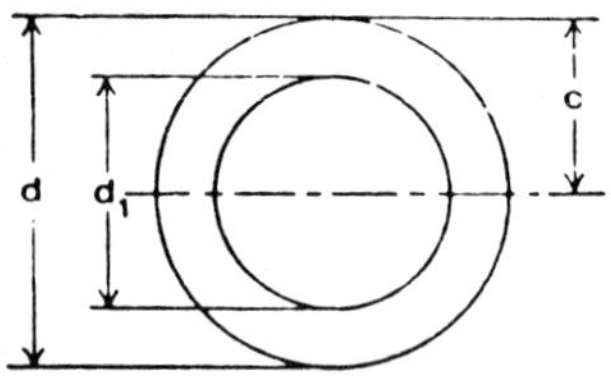

$$A = \frac{\pi(d^2 - d_1^2)}{4}$$

$$c = \frac{d}{2}$$

$$I = \frac{\pi(d^4 - d_1^4)}{64}$$

$$S = \frac{\pi(d^4 - d_1^4)}{32d}$$

$$r = \frac{\sqrt{d^2 + d_1^2}}{4}$$

$$Z = \frac{1}{6}(d^3 - d_1^3)$$

HALF CIRCLE

Axis of moments through center of gravity

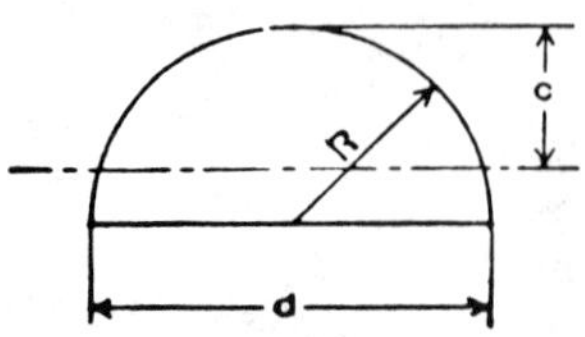

$$A = \frac{\pi R^2}{2}$$

$$c = R\left(1 - \frac{4}{3\pi}\right)$$

$$I = R^4\left(\frac{\pi}{8} - \frac{8}{9\pi}\right)$$

$$S = \frac{R^3}{24}\,\frac{(9\pi^2 - 64)}{(3\pi - 4)}$$

$$r = R\,\frac{\sqrt{9\pi^2 - 64}}{6\pi}$$

PARABOLA

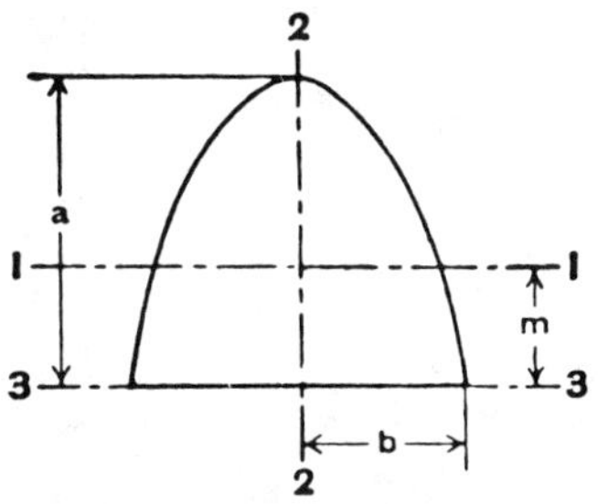

$$A = \frac{4}{3}ab$$

$$m = \frac{2}{5}a$$

$$I_1 = \frac{16}{175}a^3L$$

$$I_2 = \frac{4}{15}ab^3$$

$$I_3 = \frac{32}{105}a^3b$$

HALF PARABOLA

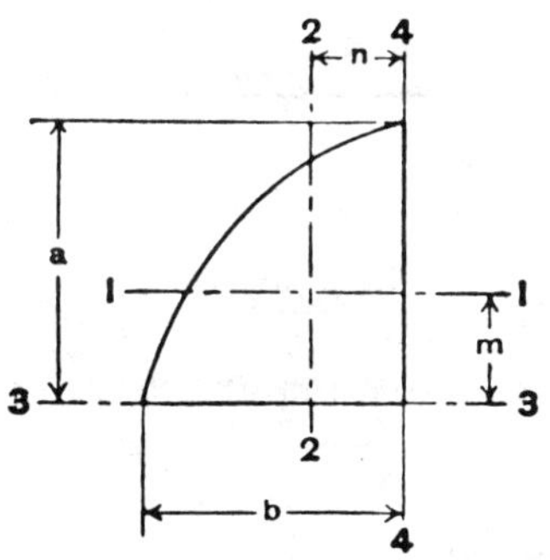

$$A = \frac{2}{3}ab$$

$$m = \frac{2}{5}a$$

$$n = \frac{3}{8}b$$

$$I_1 = \frac{8}{175}a^3b$$

$$I_2 = \frac{19}{480}ab^3$$

$$I_3 = \frac{16}{105}a^3b$$

$$I_4 = \frac{2}{15}ab^3$$

COMPLEMENT OF HALF PARABOLA

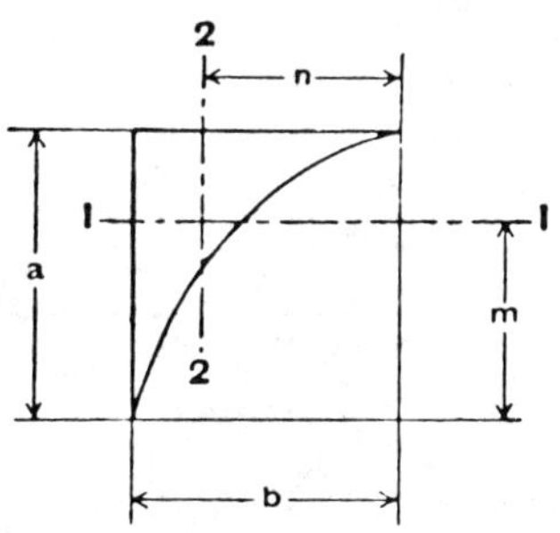

$$A = \frac{1}{3}ab$$

$$m = \frac{7}{10}a$$

$$n = \frac{3}{4}b$$

$$I_1 = \frac{37}{2100}a^3b$$

$$I_2 = \frac{1}{80}ab^3$$

PARABOLIC FILLET IN RIGHT ANGLE

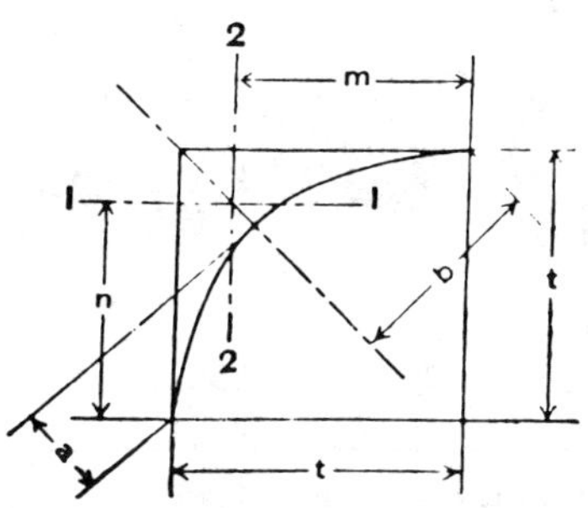

$$a = \frac{t}{2\sqrt{2}}$$

$$b = \frac{t}{\sqrt{2}}$$

$$A = \frac{1}{6}t^2$$

$$m = n = \frac{4}{5}t$$

$$I_1 = I_2 = \frac{11}{2100}t^4$$

PROPERTIES OF GEOMETRIC SECTIONS

* HALF ELLIPSE

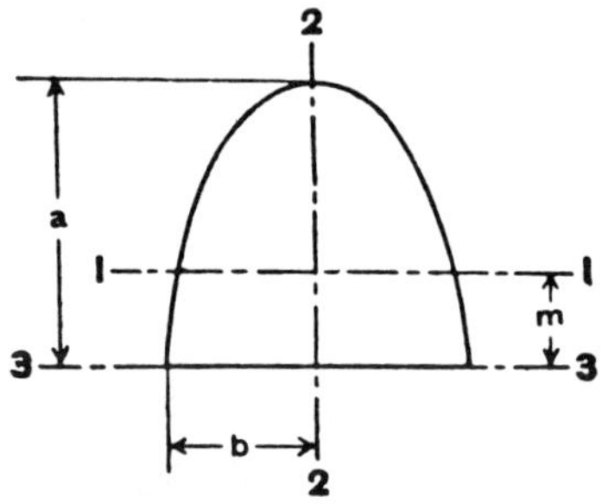

$$A = \frac{1}{2}\pi ab$$

$$m = \frac{4a}{3\pi}$$

$$I_1 = a^3 b\left(\frac{\pi}{8} - \frac{8}{9\pi}\right)$$

$$I_2 = \frac{1}{8}\pi ab^3$$

$$I_3 = \frac{1}{8}\pi a^3 b$$

* QUARTER ELLIPSE

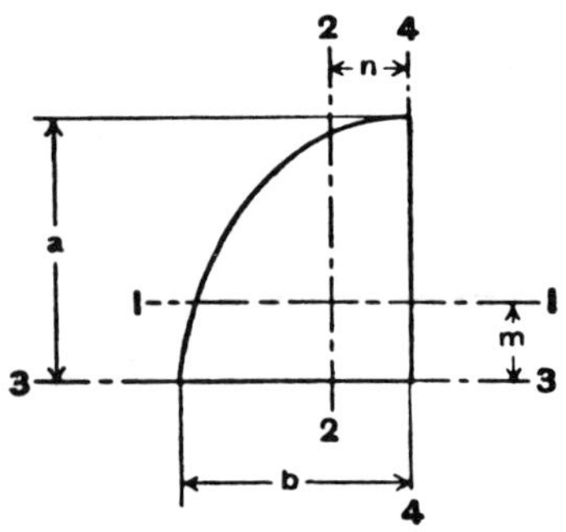

$$A = \frac{1}{4}\pi ab$$

$$m = \frac{4a}{3\pi}$$

$$n = \frac{4b}{3\pi}$$

$$I_1 = a^3 b\left(\frac{\pi}{16} - \frac{4}{9\pi}\right)$$

$$I_2 = ab^3\left(\frac{\pi}{16} - \frac{4}{9\pi}\right)$$

$$I_3 = \frac{1}{16}\pi a^3 b$$

$$I_4 = \frac{1}{16}\pi ab^3$$

* ELLIPTIC COMPLEMENT

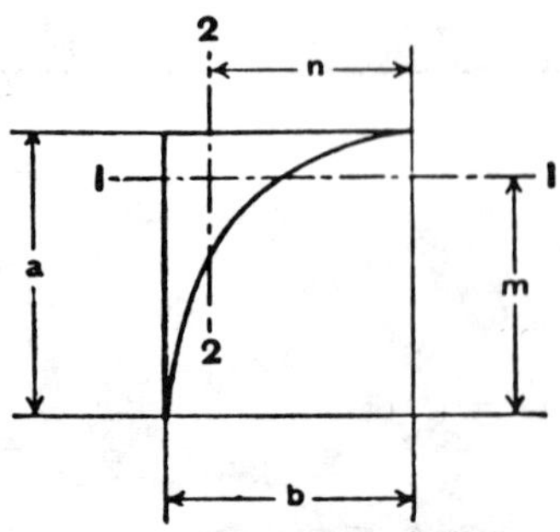

$$A = ab\left(1 - \frac{\pi}{4}\right)$$

$$m = \frac{a}{6\left(1 - \frac{\pi}{4}\right)}$$

$$n = \frac{b}{6\left(1 - \frac{\pi}{4}\right)}$$

$$I_1 = a^3 b\left(\frac{1}{3} - \frac{\pi}{16} - \frac{1}{36\left(1 - \frac{\pi}{4}\right)}\right)$$

$$I_2 = ab^3\left(\frac{1}{3} - \frac{\pi}{16} - \frac{1}{36\left(1 - \frac{\pi}{4}\right)}\right)$$

* To obtain properties of half circle, quarter circle and circular complement substitute a = b = R.

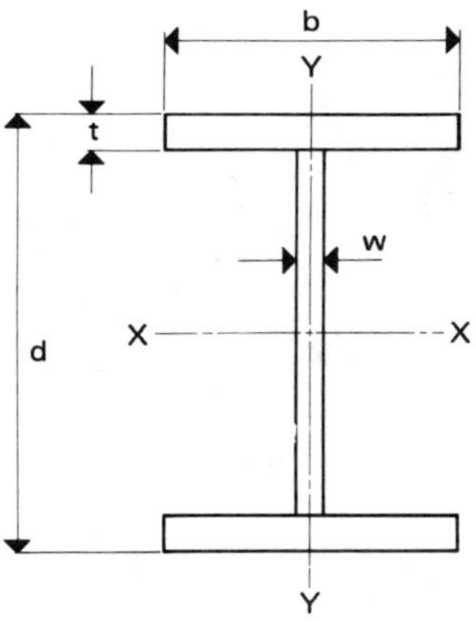

$$A = 2bt + (d - 2t)\,w$$

$$I = \frac{1}{12}\left[bd^3 - (b - w)(d - 2t)^3\right]$$

$$S = \frac{1}{6d}\left[bd^3 - (b - w)(d - 2t)^3\right]$$

$$r = \sqrt{\frac{I}{A}}$$

$$Z = \frac{1}{4}\left[bd^2 - (b - w)(d - 2t)^2\right]$$

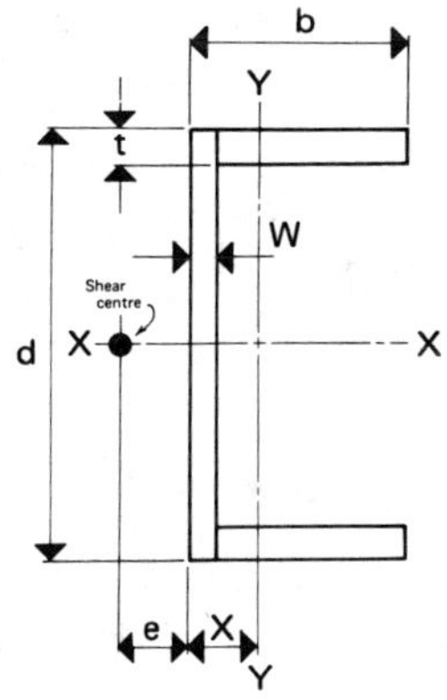

$$A = dw + 2\,(b - w)\,t$$

$$I = \frac{1}{12}\left[bd^3 - (b - w)(d - 2t)^3\right]$$

$$S = \frac{1}{6d}\left[bd^3 - (b - w)(d - 2t)^3\right]$$

$$r = \sqrt{\frac{I}{A}}$$

$$e = \frac{b^2d^2t}{4I} - \frac{w}{2}$$

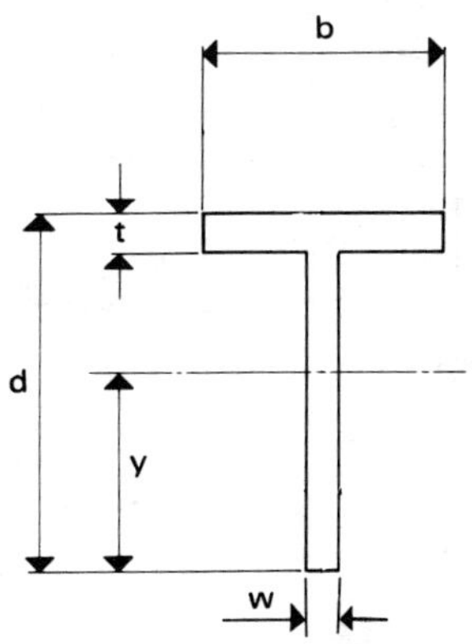

$$A = bt + w\,(d - t)$$

$$y = \frac{1}{2}\left(\frac{bdt}{A} + d - t\right)$$

$$I = \frac{1}{12}\left[bt^3 + w\,(d - t)^3 + \frac{3\,bwtd^2\,(d - t)}{A}\right]$$

$$S_1 = \frac{I}{y}$$

$$S_2 = \frac{I}{d - y}$$

$$r = \sqrt{\frac{I}{A}}$$

PROPERTIES OF GEOMETRIC SECTIONS AND STRUCTURAL SHAPES

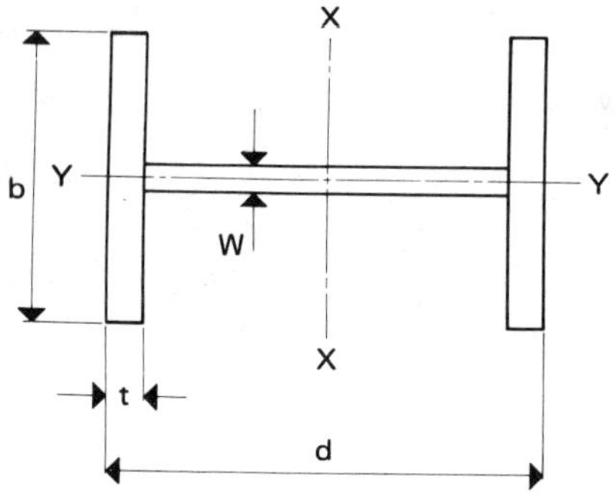

$$A = 2bt + w(d - 2t)$$

$$I = \frac{1}{12}\left[2tb^3 + (d - 2t)w^3\right]$$

$$S = \frac{1}{6b}\left[2tb^3 + (d - 2t)w^3\right]$$

$$r = \sqrt{\frac{I}{A}}$$

$$Z = \frac{1}{4}\left[2t\,(b^2 - w^2) + d\,w^2\right]$$

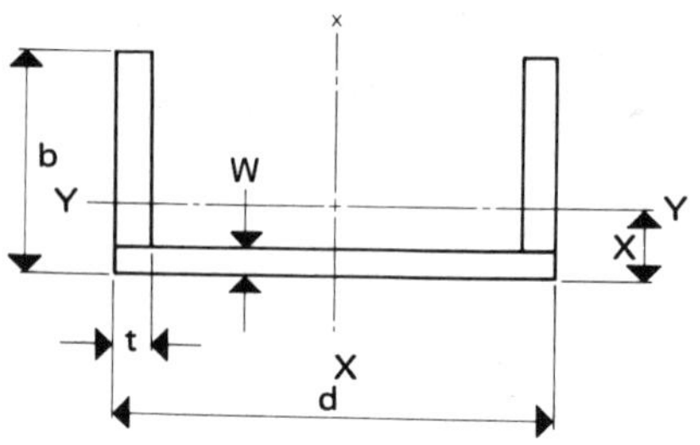

$$A = dw + 2(b - w)t$$

$$x = \frac{1}{2A}\left[(d - 2t)\,w^2 + 2tb^2\right]$$

$$I = \frac{1}{3}\left[dx^3 + 2t\,(b-x)^3 - (d-2t)\,(x-w)^3\right]$$

$$S_1 = \frac{I}{b - x}\,;\; S_2 = \frac{I}{x}$$

$$r = \sqrt{\frac{I}{A}}$$

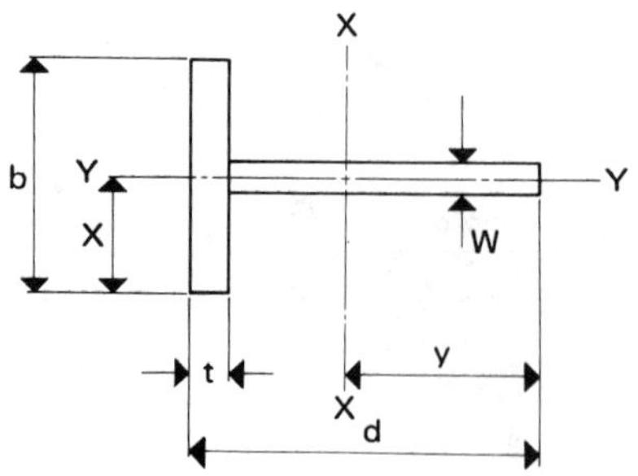

$$A = bt + (d - t)w$$

$$x = b/2$$

$$I = \frac{1}{12}\left[tb^3 + (d - t)\,w^3\right]$$

$$S = \frac{2I}{b}$$

$$r = \sqrt{\frac{I}{A}}$$

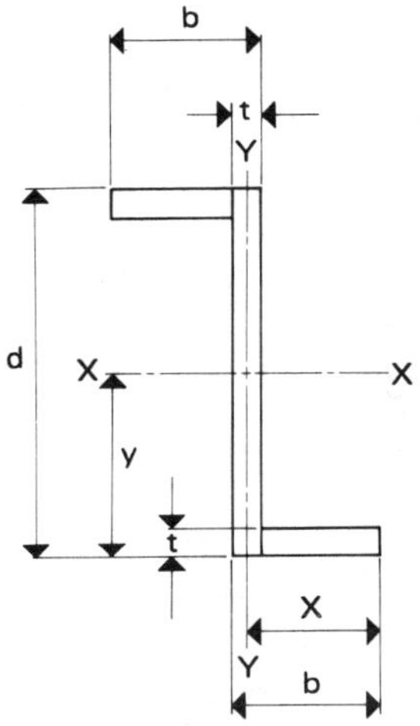

$$A = t\left[d + 2(b - t)\right]$$

$$y = d/2$$

$$I = \frac{bd^3 - (b - t)(d - 2t)^3}{12}$$

$$S = \frac{I}{y}$$

$$r = \sqrt{\frac{bd^3 - (b - t)(d - 2t)^3}{12t\left[d + 2(b - t)\right]}}$$

ANGLE

Axis of moments through center of gravity

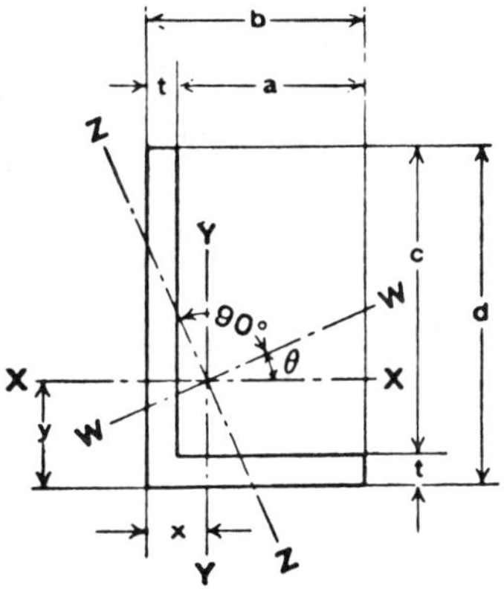

Z-Z is axis of minimum I

$$\tan 2\theta = \frac{2K}{I_y - I_x}$$

$$A = t(b + c) \qquad x = \frac{b^2 + ct}{2(b + c)} \qquad y = \frac{d^2 + at}{2(b + c)}$$

K = Product of Inertia about X-X & Y-Y

$$= \mp \frac{abcdt}{4(b + c)}$$

$$I_x = \frac{1}{3}\left(t(d - y)^3 + by^3 - a(y - t)^3\right)$$

$$I_y = \frac{1}{3}\left(t(b - x)^3 + dx^3 - c(x - t)^3\right)$$

$$I_z = I_x \sin^2\theta + I_y \cos^2\theta + K \sin 2\theta$$

$$I_w = I_x \cos^2\theta + I_y \sin^2\theta - K \sin 2\theta$$

K is negative when heel of angle, with respect to c. g., is in 1st or 3rd quadrant, positive when in 2nd or 4th quadrant.

BEAMS AND CHANNELS

Transverse force oblique through center of gravity

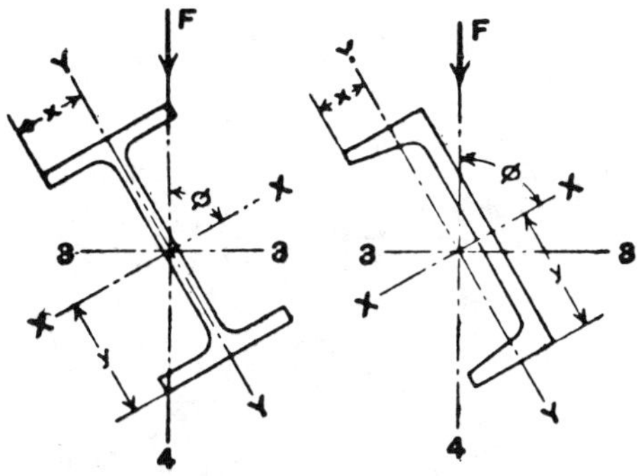

$$I_3 = I_x \sin^2\phi + I_y \cos^2\phi$$

$$I_4 = I_x \cos^2\phi + I_y \sin^2\phi$$

$$f = M\left(\frac{y}{I_x}\sin\phi + \frac{x}{I_y}\cos\phi\right)$$

where M is bending moment due to force F.

PROPERTIES OF GEOMETRIC SECTIONS AND STRUCTURAL SHAPES

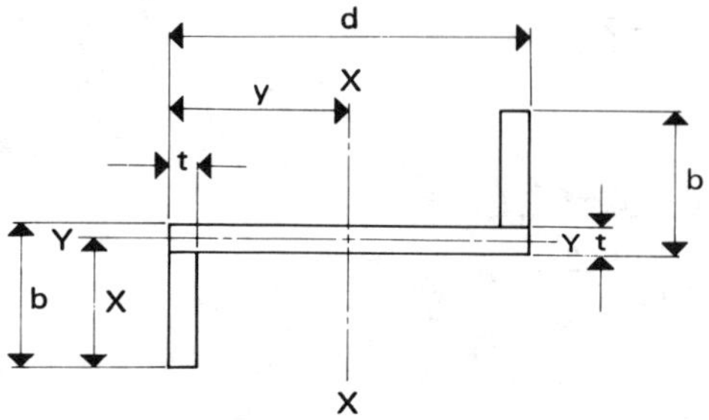

$$A = t\,[d + 2(b - t)]$$

$$x = b - t/2$$

$$I = \frac{1}{12}\left[(2b - t)^3\, t + (d - 2t)\, t^3\right]$$

$$S = \frac{I}{x}$$

$$r = \sqrt{\frac{I}{A}}$$

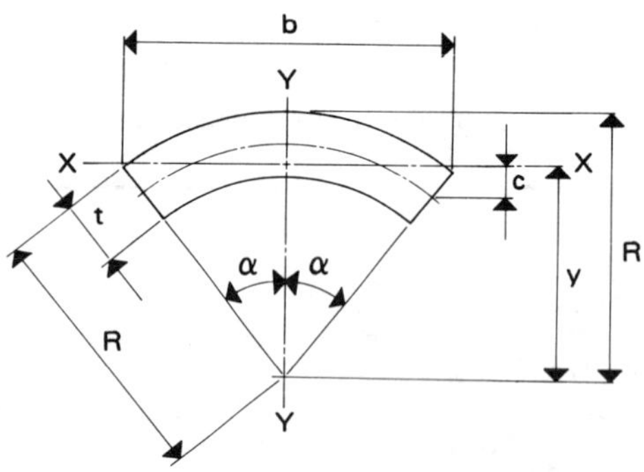

$$A = \frac{\pi \alpha t}{180}(2R - t)$$

$$c = \left(\frac{180 \sin \alpha}{\pi \alpha} - \cos \alpha\right)\left(R - \frac{t}{2}\right)$$

$$n = \left(R - \frac{t}{2}\right) \sin \alpha$$

$$I_x = \left(\frac{\pi \alpha}{180} + \sin \alpha \cos \alpha - \frac{360 \sin^2 \alpha}{\pi \alpha}\right)\left(R - \frac{t}{2}\right)^3 t$$

$$I_y = \left(\frac{\pi \alpha}{180} - \sin \alpha \cos \alpha\right) R^3\, t$$

$$y = \frac{180 \sin \alpha}{\pi \alpha}\left(R - \frac{t}{2}\right)$$

PROPERTIES OF THE CIRCLE

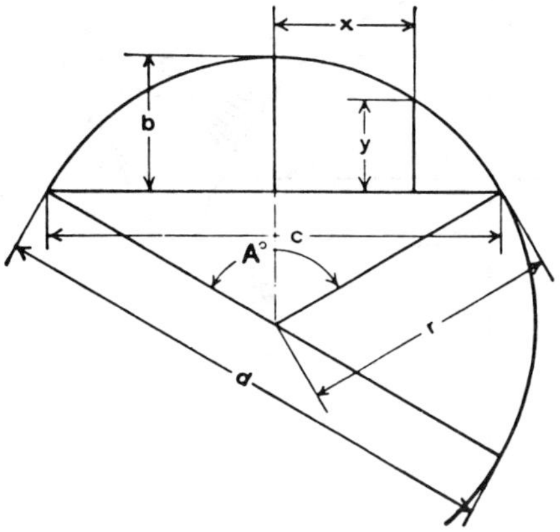

Circumference $= 6.28318\ r = 3.14159\ d$

Diameter $= 0.31831$ circumference

Area $= 3.14159\ r^2$

Arc a $= \dfrac{\pi r A^\circ}{180^\circ} = 0.017453\ r A^\circ$

Angle A° $= \dfrac{180^\circ a}{\pi r} = 57.29578 \dfrac{a}{r}$

Radius r $= \dfrac{4b^2 + c^2}{8b}$

Chord c $= 2\sqrt{2br - b^2} = 2r \sin \dfrac{A}{2}$

Rise b $= r - ½\sqrt{4r^2 - c^2} = \dfrac{c}{2} \tan \dfrac{A}{4}$

$= 2r \sin^2 \dfrac{A}{4} = r + y - \sqrt{r^2 - x^2}$

y $= b - r + \sqrt{r^2 - x^2}$

x $= \sqrt{r^2 - (r + y - b)^2}$

Diameter of circle of equal periphery as square	= 1.27324 side of square
Side of square of equal periphery as circle	= 0.78540 diameter of circle
Diameter of circle circumscribed about square	= 1.41421 side of square
Side of square inscribed in circle	= 0.70711 diameter of circle

CIRCULAR SECTOR

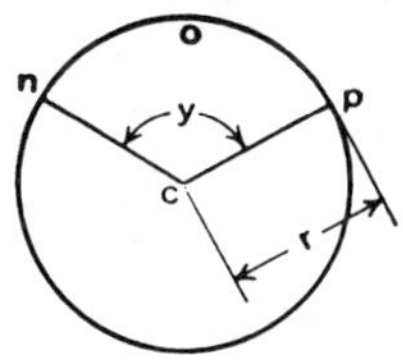

r = radius of circle   y = angle ncp in degrees

Area of Sector ncpo = ½ (length of arc nop X r)

$= \text{Area of Circle} \times \dfrac{y}{360}$

$= 0.0087266 \times r^2 \times y$

CIRCULAR SEGMENT

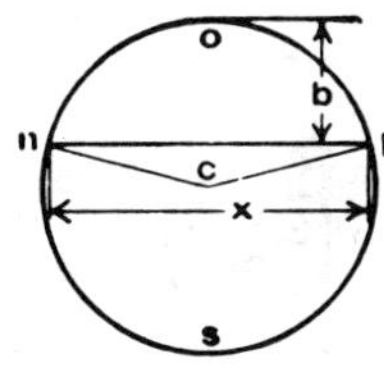

r = radius of circle   x = chord   b = rise

Area of Segment nop = Area of Sector ncpo − Area of triangle ncp

$= \dfrac{(\text{Length of arc nop} \times r) - x(r - b)}{2}$

Area of Segment nsp = Area of Circle − Area of Segment nop

VALUES FOR FUNCTIONS OF π

$\pi = 3.14159265359$, log = 0.4971499

$\pi^2 = 9.8696044$, log = 0.9942997 $\dfrac{1}{\pi} = 0.3183099$, log = $\bar{1}.5028501$ $\sqrt{\dfrac{1}{\pi}} = 0.5641896$, log = $\bar{1}.7514251$

$\pi^3 = 31.0062767$, log = 1.4914496 $\dfrac{1}{\pi^2} = 0.1013212$, log = $\bar{1}.0057003$ $\dfrac{\pi}{180} = 0.0174533$, log = $\bar{2}.2418774$

$\sqrt{\pi} = 1.7724539$, log = 0.2485749 $\dfrac{1}{\pi^3} = 0.0322515$, log = $\bar{2}.5085500$ $\dfrac{180}{\pi} = 57.2957795$, log = 1.7581226

Note: Logs of fractions such as $\bar{1}.5028501$ and $\bar{2}.5085500$ may also be written 9.5028501 − 10 and 8.5085500 − 10 respectively.

PROPERTIES OF PARABOLA AND ELLIPSE

PARABOLA

When $H \div B = 0.1$ or less, approximate $\frac{1}{2}$ perimeter $= \sqrt{B^2 + 4/3H^2}$ or use formulas for circular arcs

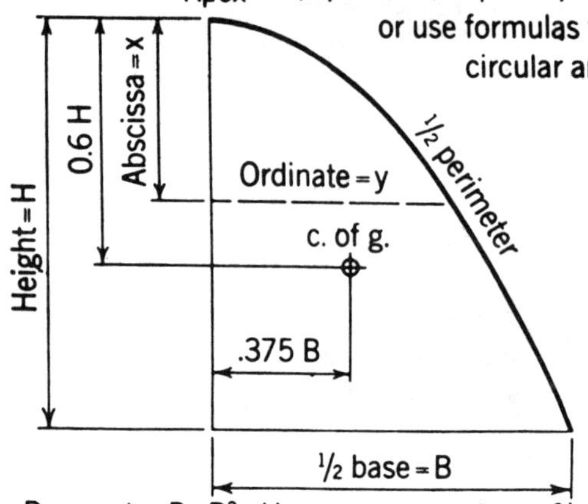

Parameter $P = B^2 \div H$

Area $= \frac{2}{3}HB$

$x = y^2 \div P$

$y = \sqrt{xP}$

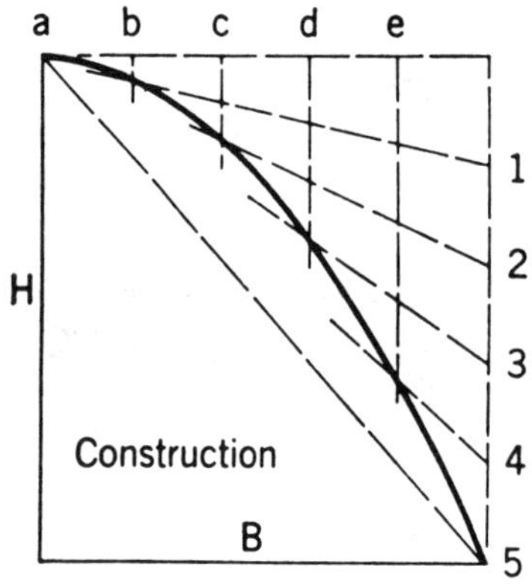

ELLIPSE

$(x^2 \div H^2) + (y^2 \div B^2) = 1$

$x = (H \div B)\sqrt{B^2 - y^2}$

$y = (B \div H)\sqrt{H^2 - x^2}$

Approximate $\frac{1}{4}$ perimeter $= \frac{\pi}{4}\sqrt{2(H^2+B^2)}$

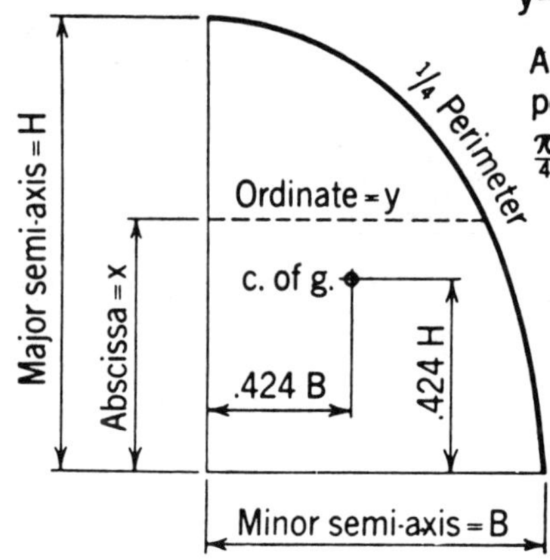

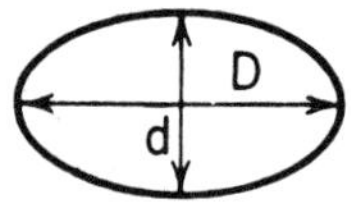

Area = .7854 Dd

Construction

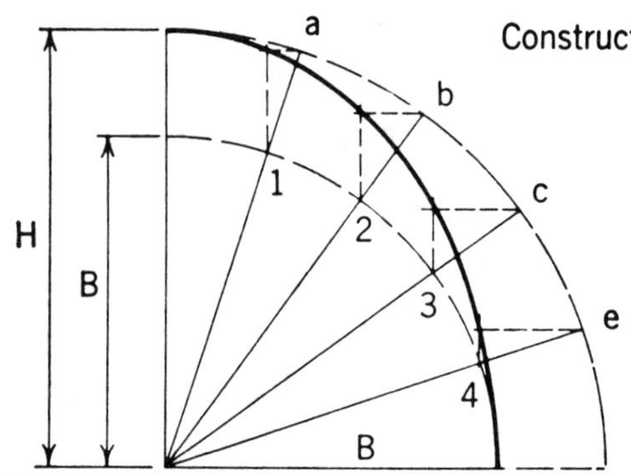

AREA BETWEEN PARABOLIC CURVE AND SECANT

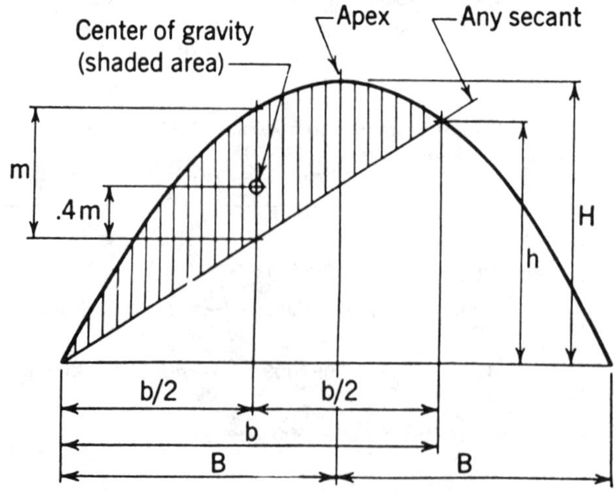

Length b may vary from 0 to 2B

$$h = Hb\left(\frac{2B-b}{B^2}\right)$$

$$m = \frac{Hb^2}{4B^2}$$

Shaded area $= \frac{2}{3}bm$

$$= \frac{Hb^3}{6B^2}$$

RECTANGULAR PARALLELEPIPED

Volume = abc

Surface area = $2(ab + ac + bc)$

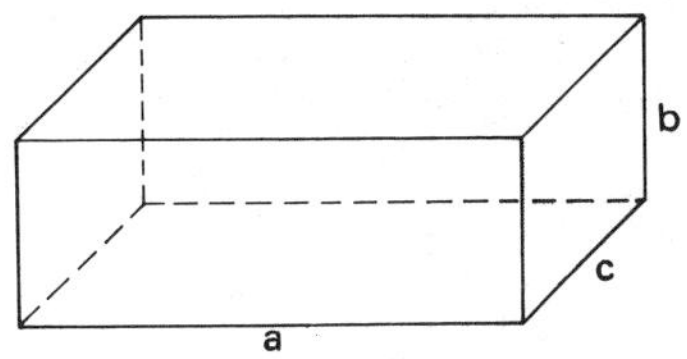

PARALLELEPIPED

Volume = $Ah = abc \sin \theta$

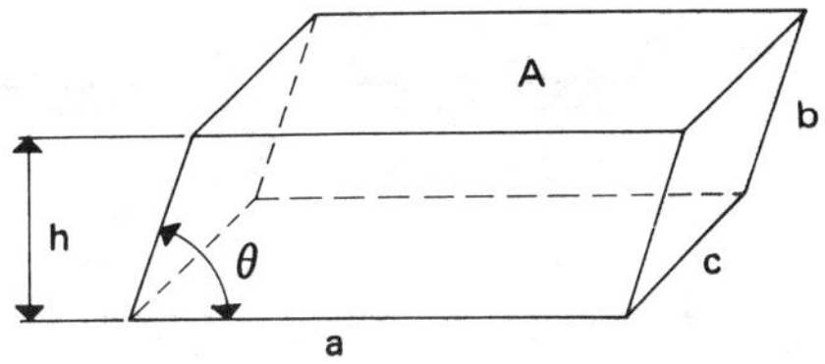

PYRAMID

$$\text{Volume} = \frac{1}{3} Ah$$

The centroid of a pyramid is located y-distance from the base on the line joining the centre of gravity of area A and the apex.

$$y = \frac{h}{4}$$

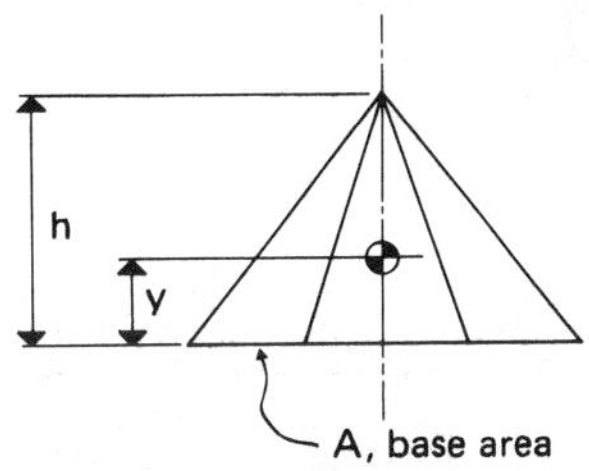

FRUSTUM OF PYRAMID

$$V = \frac{h}{3}(A_1 + A_2 + \sqrt{A_1 A_2})$$

The centroid is located y-distance up from area A_2 on the line joining the centres of gravity of areas A_1 and A_2

$$y = \frac{h\,(A_1 + 2\sqrt{A_1 A_2} + 3A_2)}{4\,(A_1 + \sqrt{A_1 A_2} + A_2)}$$

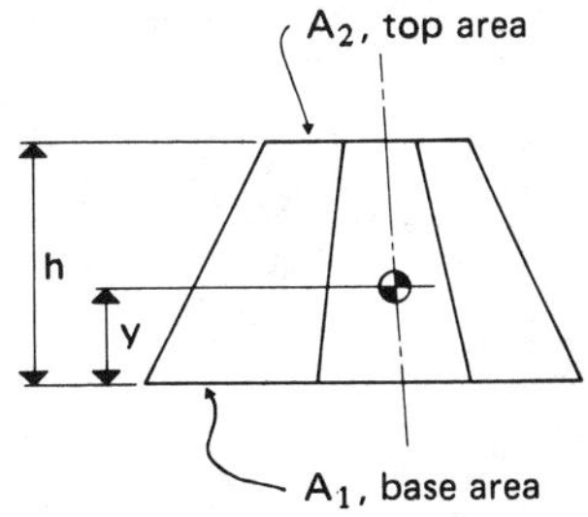

WEDGE

$$V = \frac{(2a + c)\,bh}{6}$$

The centroid is located y-distance from the base on the line joining the centre of gravity of the base area and the mid point of edge, c.

$$y = \frac{h\,(a + c)}{2\,(2a + c)}$$

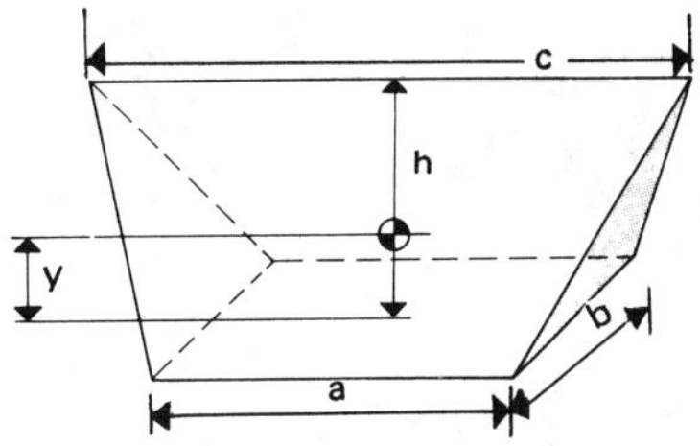

PROPERTIES OF SOLIDS

RIGHT CIRCULAR CYLINDER

Volume = $\pi r^2 h$

Lateral surface area = $2\pi rh$

$$y = \frac{h}{2}$$

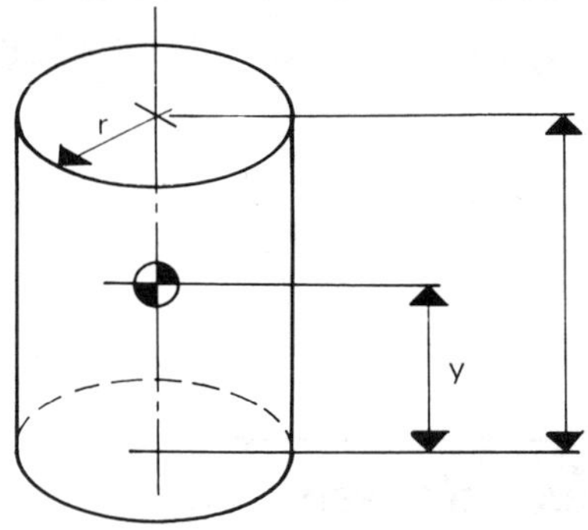

RIGHT CIRCULAR CONE

Volume = $\frac{1}{3}\pi r^2 h$

Lateral surface area = $\pi r\sqrt{r^2 + h^2} = \pi rl$

$$y = \frac{h}{4}$$

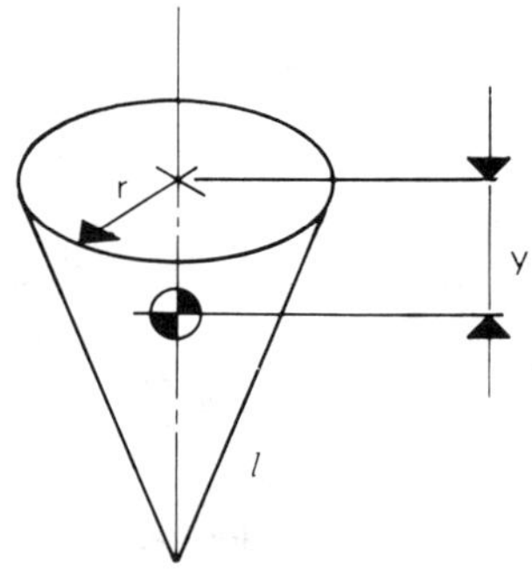

FRUSTRUM OF RIGHT CIRCULAR CONE

Volume = $\frac{1}{3}\pi h(a^2 + ab + b^2)$

Lateral surface area = $\pi(a + b)\sqrt{h^2 + (b - a)^2}$

$= \pi(a + b)l$

$$y = \frac{h\ (b^2 + 2ab + 3a^2)}{4\ \ (b^2 + ab + a^2)}$$

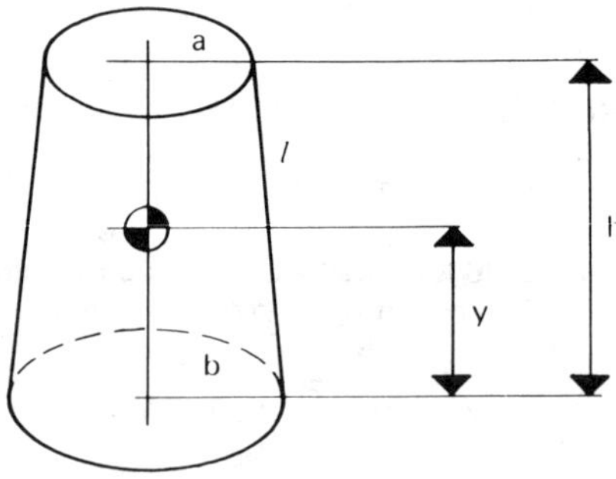

SPHERE

Volume = $\frac{4}{3}\pi r^3$

Surface area = $4\pi r^2$

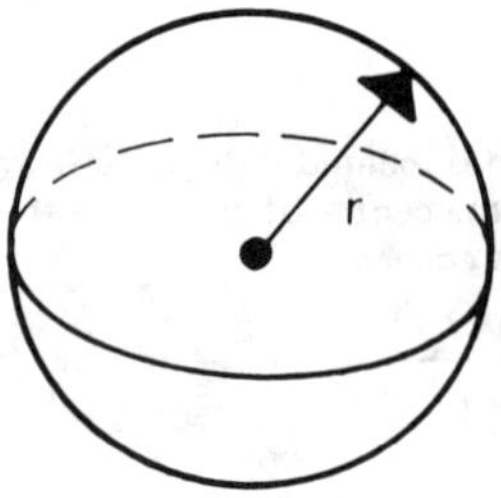

TRIGONOMETRIC FUNCTIONS

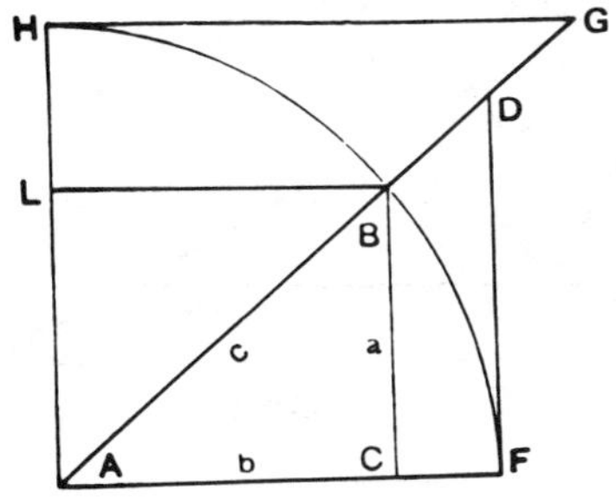

Radius AF $= 1$

$= \sin^2 A + \cos^2 A = \sin A \operatorname{cosec} A$

$= \cos A \sec A = \tan A \cot A$

Sine A $= \dfrac{\cos A}{\cot A} = \dfrac{1}{\operatorname{cosec} A} = \cos A \tan A = \sqrt{1-\cos^2 A} = BC$

Cosine A $= \dfrac{\sin A}{\tan A} = \dfrac{1}{\sec A} = \sin A \cot A = \sqrt{1-\sin^2 A} = AC$

Tangent A $= \dfrac{\sin A}{\cos A} = \dfrac{1}{\cot A} = \sin A \sec A = FD$

Cotangent A $= \dfrac{\cos A}{\sin A} = \dfrac{1}{\tan A} = \cos A \operatorname{cosec} A = HG$

Secant A $= \dfrac{\tan A}{\sin A} = \dfrac{1}{\cos A} = AD$

Cosecant A $= \dfrac{\cot A}{\cos A} = \dfrac{1}{\sin A} = AG$

RIGHT ANGLED TRIANGLES

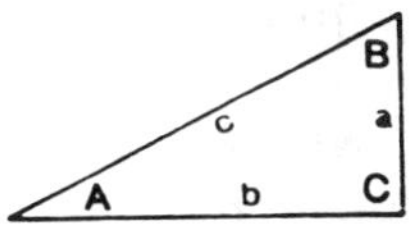

$$a^2 = c^2 - b^2$$
$$b^2 = c^2 - a^2$$
$$c^2 = a^2 + b^2$$

Known	Required					
	A	B	a	b	c	Area
a, b	$\tan A = \frac{a}{b}$	$\tan B = \frac{b}{a}$			$\sqrt{a^2 + b^2}$	$\frac{ab}{2}$
a, c	$\sin A = \frac{a}{c}$	$\cos B = \frac{a}{c}$		$\sqrt{c^2 - a^2}$		$\frac{a\sqrt{c^2 - a^2}}{2}$
A, a		$90° - A$		$a \cot A$	$\frac{a}{\sin A}$	$\frac{a^2 \cot A}{2}$
A, b		$90° - A$	$b \tan A$		$\frac{b}{\cos A}$	$\frac{b^2 \tan A}{2}$
A, c		$90° - A$	$c \sin A$	$c \cos A$		$\frac{c^2 \sin 2A}{4}$

OBLIQUE ANGLED TRIANGLES

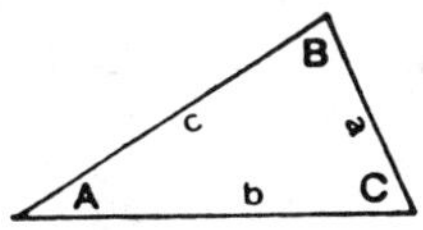

$$s = \frac{a + b + c}{2}$$

$$K = \sqrt{\frac{(s-a)(s-b)(s-c)}{s}}$$

$$a^2 = b^2 + c^2 - 2bc \cos A$$
$$b^2 = a^2 + c^2 - 2ac \cos B$$
$$c^2 = a^2 + b^2 - 2ab \cos C$$

Known	Required					
	A	B	C	b	c	Area
a, b, c	$\tan \frac{1}{2} A = \frac{K}{s-a}$	$\tan \frac{1}{2} B = \frac{K}{s-b}$	$\tan \frac{1}{2} C = \frac{K}{s-c}$			$\sqrt{s(s-a)(s-b)(s-c)}$
a, A, B			$180° - (A+B)$	$\frac{a \sin B}{\sin A}$	$\frac{a \sin C}{\sin A}$	
a, b, A		$\sin B = \frac{b \sin A}{a}$			$\frac{b \sin C}{\sin B}$	
a, b, C	$\tan A = \frac{a \sin C}{b - a \cos C}$				$\sqrt{a^2 + b^2 - 2ab \cos C}$	$\frac{ab \sin C}{2}$

BRACING FORMULAE

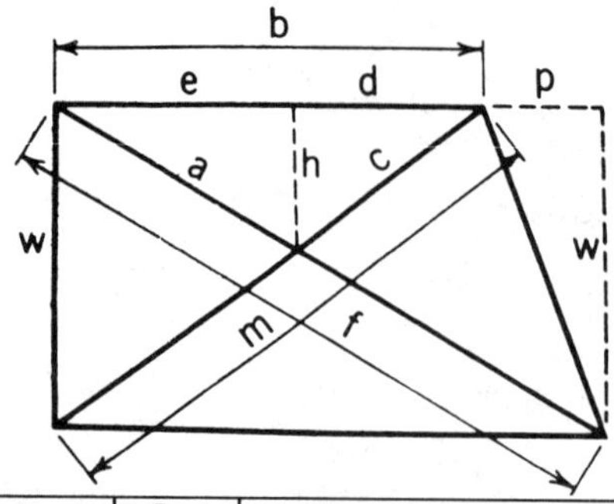

Given	To Find	Formula
bpw	f	$\sqrt{(b+p)^2+w^2}$
bw	m	$\sqrt{b^2+w^2}$
bp	d	$b^2 \div (2b+p)$
bp	e	$b(b+p) \div (2b+p)$
bfp	a	$bf \div (2b+p)$
bmp	c	$bm \div (2b+p)$
bpw	h	$bw \div (2b+p)$
afw	h	$aw \div f$
cmw	h	$cw \div m$

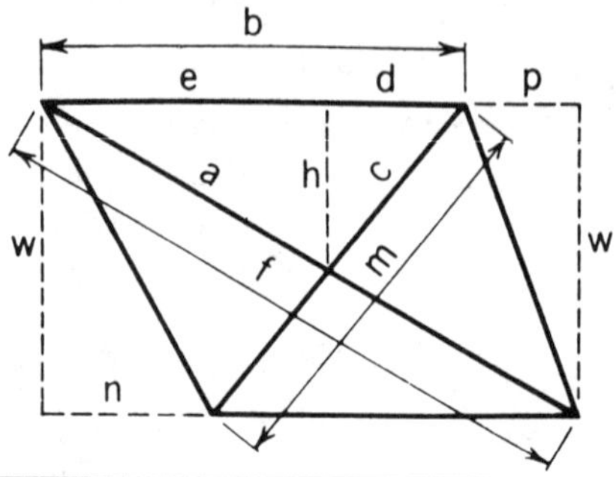

Given	To Find	Formula
bpw	f	$\sqrt{(b+p)^2+w^2}$
bnw	m	$\sqrt{(b-n)^2+w^2}$
bnp	d	$b(b-n) \div (2b+p-n)$
bnp	e	$b(b+p) \div (2b+p-n)$
bfnp	a	$bf \div (2b+p-n)$
bmnp	c	$bm \div (2b+p-n)$
bnpw	h	$bw \div (2b+p-n)$
afw	h	$aw \div f$
cmw	h	$cw \div m$

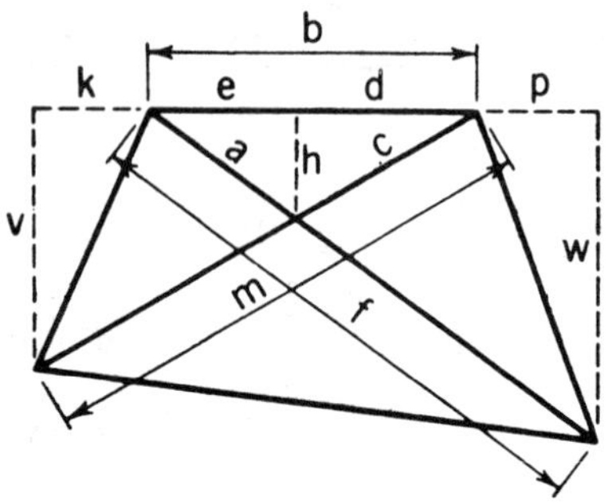

Given	To Find	Formula
bpw	f	$\sqrt{(b+p)^2+w^2}$
bkv	m	$\sqrt{(b+k)^2+v^2}$
bkpvw	d	$bw(b+k) \div [v(b+p)+w(b+k)]$
bkpvw	e	$bv(b+p) \div [v(b+p)+w(b+k)]$
bfkpvw	a	$fbv \div [v(b+p)+w(b+k)]$
bkmpvw	c	$bmw \div [v(b+p)+w(b+k)]$
bkpvw	h	$bvw \div [v(b+p)+w(b+k)]$
afw	h	$aw \div f$
cmv	h	$cv \div m$

PARALLEL BRACING

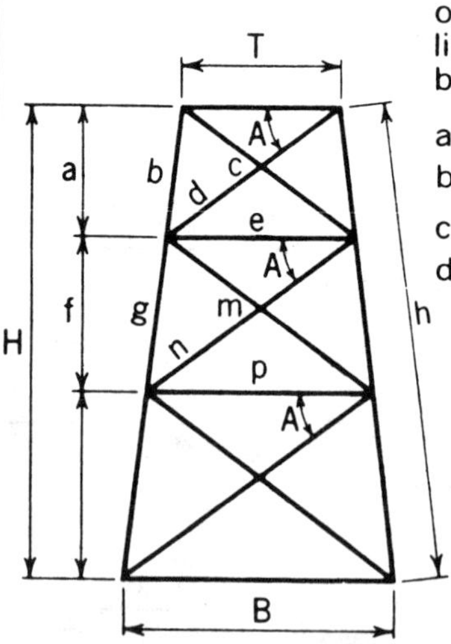

$k = (\log B - \log T) \div$ no. of panels. Constant k plus the logarithm of any line equals the log of the corresponding line in the next panel below.

$a = TH \div (T+e+p)$

$b = Th \div (T+e+p)$

$c = \sqrt{(\frac{1}{2}T + \frac{1}{2}e)^2 + a^2}$

$d = ce \div (T+e)$

$\log e = k + \log T$

$\log f = k + \log a$

$\log g = k + \log b$

$\log m = k + \log c$

$\log n = k + \log d$

$\log p = k + \log e$

The above method can be used for any number of panels.
In the formulas for "a" and "b" the sum in parenthesis, which in the case shown is (T + e + p), is always composed of all the horizontal distances except the base.

LENGTH OF CIRCULAR ARCS FOR UNIT RADIUS

By the use of this table, the length of any arc may be found if the length of the radius and the angle of the segment are known.

Example: Required the length of arc of segment 32° 15′ 27″ with radius of 8 000 mm.

From table: Length of arc (Radius 1) for 32° = .5585054
15′ = .0043633
27″ = .0001309

.5629996

.5629996 X 8 000 (length of radius) = 4504 mm

For the same arc but with the radius expressed as 24 feet 3 inches, the length of arc would be 0.5629996 X 24.25 = 13.65 feet

DEGREES						MINUTES		SECONDS	
1	.017 4533	61	1.064 6508	121	2.111 8484	1	.000 2909	1	.000 0048
2	.034 9066	62	1.082 1041	122	2.129 3017	2	.000 5818	2	.000 0097
3	.052 3599	63	1.099 5574	123	2.146 7550	3	.000 8727	3	.000 0145
4	.069 8132	64	1.117 0107	124	2.164 2083	4	.001 1636	4	.000 0194
5	.087 2665	65	1.134 4640	125	2.181 6616	5	.001 4544	5	.000 0242
6	.104 7198	66	1.151 9173	126	2.199 1149	6	.001 7453	6	.000 0291
7	.122 1730	67	1.169 3706	127	2.216 5682	7	.002 0362	7	.000 0339
8	.139 6263	68	1.186 8239	128	2.234 0214	8	.002 3271	8	.000 0388
9	.157 0796	69	1.204 2772	129	2.251 4747	9	.002 6180	9	.000 0436
10	.174 5329	70	1.221 7305	130	2.268 9280	10	.002 9089	10	.000 0485
11	.191 9862	71	1.239 1838	131	2.286 3813	11	.003 1998	11	.000 0533
12	.209 4395	72	1.256 6371	132	2.303 8346	12	.003 4907	12	.000 0582
13	.226 8928	73	1.274 0904	133	2.321 2879	13	.003 7815	13	.000 0630
14	.244 3461	74	1.291 5436	134	2.338 7412	14	.004 0724	14	.000 0679
15	.261 7994	75	1.308 9969	135	2.356 1945	15	.004 3633	15	.000 0727
16	.279 2527	76	1.326 4502	136	2.373 6478	16	.004 6542	16	.000 0776
17	.296 7060	77	1.343 9035	137	2.391 1011	17	.004 9451	17	.000 0824
18	.314 1593	78	1.361 3568	138	2.408 5544	18	.005 2360	18	.000 0873
19	.331 6126	79	1.378 8101	139	2.426 0077	19	.005 5269	19	.000 0921
20	.349 0659	80	1.396 2634	140	2.443 4610	20	.005 8178	20	.000 0970
21	.366 5191	81	1.413 7167	141	2.460 9142	21	.006 1087	21	.000 1018
22	.383 9724	82	1.431 1700	142	2.478 3675	22	.006 3995	22	.000 1067
23	.401 4257	83	1.448 6233	143	2.495 8208	23	.006 6904	23	.000 1115
24	.418 8790	84	1.466 0766	144	2.513 2741	24	.006 9813	24	.000 1164
25	.436 3323	85	1.483 5299	145	2.530 7274	25	.007 2722	25	.000 1212
26	.453 7856	86	1.500 9832	146	2.548 1807	26	.007 5631	26	.000 1261
27	.471 2389	87	1.518 4364	147	2.565 6340	27	.007 8540	27	.000 1309
28	.488 6922	88	1.535 8897	148	2.583 0873	28	.008 1449	28	.000 1357
29	.506 1455	89	1.553 3430	149	2.600 5406	29	.008 4358	29	.000 1406
30	.523 5988	90	1.570 7963	150	2.617 9939	30	.008 7266	30	.000 1454
31	.541 0521	91	1.588 2496	151	2.635 4472	31	.009 0175	31	.000 1503
32	.558 5054	92	1.605 7029	152	2.652 9005	32	.009 3084	32	.000 1551
33	.575 9587	93	1.623 1562	153	2.670 3538	33	.009 5993	33	.000 1600
34	.593 4119	94	1.640 6095	154	2.687 8070	34	.009 8902	34	.000 1648
35	.610 8652	95	1.658 0628	155	2.705 2603	35	.010 1811	35	.000 1697
36	.628 3185	96	1.675 5161	156	2.722 7136	36	.010 4720	36	.000 1745
37	.645 7718	97	1.692 9694	157	2.740 1669	37	.010 7629	37	.000 1794
38	.663 2251	98	1.710 4227	158	2.757 6202	38	.011 0538	38	.000 1842
39	.680 6784	99	1.727 8760	159	2.775 0735	39	.011 3446	39	.000 1891
40	.698 1317	100	1.745 3293	160	2.792 5268	40	.011 6355	40	.000 1939
41	.715 5850	101	1.762 7825	161	2.809 9801	41	.011 9264	41	.000 1988
42	.733 0383	102	1.780 2358	162	2.827 4334	42	.012 2173	42	.000 2036
43	.750 4916	103	1.797 6891	163	2.844 8867	43	.012 5082	43	.000 2085
44	.767 9449	104	1.815 1424	164	2.862 3400	44	.012 7991	44	.000 2133
45	.785 3982	105	1.832 5957	165	2.879 7933	45	.013 0900	45	.000 2182
46	.802 8515	106	1.850 0490	166	2.897 2466	46	.013 3809	46	.000 2230
47	.820 3047	107	1.867 5023	167	2.914 6999	47	.013 6717	47	.000 2279
48	.837 7580	108	1.884 9556	168	2.932 1531	48	.013 9626	48	.000 2327
49	.855 2113	109	1.902 4089	169	2.949 6064	49	.014 2535	49	.000 2376
50	.872 6646	110	1.919 8622	170	2.967 0597	50	.014 5444	50	.000 2424
51	.890 1179	111	1.937 3155	171	2.984 5130	51	.014 8353	51	.000 2473
52	.907 5712	112	1.954 7688	172	3.001 9663	52	.015 1262	52	.000 2521
53	.925 0245	113	1.972 2221	173	3.019 4196	53	.015 4171	53	.000 2570
54	.942 4778	114	1.989 6753	174	3.036 8729	54	.015 7080	54	.000 2618
55	.959 9311	115	2.007 1286	175	3.054 3262	55	.015 9989	55	.000 2666
56	.977 3844	116	2.024 5819	176	3,071 7795	56	.016 2897	56	.000 2715
57	.994 8377	117	2.042 0352	177	3.089 2328	57	.016 5806	57	.000 2763
58	1.012 2910	118	2.059 4885	178	3.106 6861	58	.016 8715	58	.000 2812
59	1.029 7443	119	2.076 9418	179	3.124 1394	59	.017 1624	59	.000 2860
60	1.047 1976	120	2.094 3951	180	3.141 5927	60	.017 4533	60	.000 2909

SI SUMMARY

General

The following information on SI units is provided to assist those involved in the planning, design, fabrication and erection of steel structures prepared in SI units. Information related to the metric system in general is to be found in CAN3-Z234.1-79, "Canadian Metric Practice Guide" and for terms related to the steel industry in the "Industry Practice Guide for SI Metric Units in the Canadian Iron and Steel Industry". The latter is available from the Task Force for Metric Conversion in the Canadian Iron and Steel Industry, P.O. Box 4248, Station "D", Hamilton, Ontario, L8V 4L6.

The eleventh General Conference of Weights and Measures, in 1960, adopted the name International System of Units for a coherent system which includes the metre as the base unit of length and the kilogram as the base unit of mass. The international abbreviation of the name of this system, in all languages, is SI.

Canada is a signatory to the General Conference on Weights and Measures, and in 1970, the Canadian government stated that the eventual conversion to the metric system is an objective of Canadian policy. Since that time, metric conversion activity in Canada has developed to the point where material and design standards, building codes and technical literature are available in SI units.

The SI system is based on the seven base units listed in Table 7—1. Decimal multiples and sub-multiples of the SI base units are formed by the addition of the prefixes given in Table 7-2.

SI BASE UNITS
Table 7-1

Quantity	Name	Symbol
length	metre	m
mass	kilogram	kg
time	second	s
electric current	ampere	A
thermodynamic temperature	kelvin	K
amount of substance	mole	mol
luminous intensity	candela	cd

SI PREFIXES
Table 7-2

Multiplying Factor	Prefix	Symbol
1 000 000 000 000 = 10^{12}	tera	T
1 000 000 000 = 10^{9}	giga	G
1 000 000 = 10^{6}	mega	M
1 000 = 10^{3}	kilo	k
100 = 10^{2}	hecto	h
10 = 10^{1}	deca	da
0.1 = 10^{-1}	deci	d
0.01 = 10^{-2}	centi	c
0.001 = 10^{-3}	milli	m
0.000 001 = 10^{-6}	micro	μ
0.000 000 001 = 10^{-9}	nano	n
0.000 000 000 001 = 10^{-12}	pico	p
0.000 000 000 000 001 = 10^{-15}	femto	f
0.000 000 000 000 000 001 = 10^{-18}	atto	a

In choosing the appropriate decimal multiple or sub-multiple, the Canadian Metric Practice Guide recommends the use of prefixes representing 10 raised to a power that is a multiple of 3, a ternary power. Thus, common structural steel design units would be:

Force — newton (N), kilonewton (kN)
Stress — pascal (Pa), kilopascal (kPa), megapascal (MPa)
Length — millimetre (mm), metre (m)
Mass — kilogram (kg), megagram (Mg)

The tonne is a special unit, equal to 1 000 kg (or 1 Mg) that will be used in the basic steel industry, but should not be used in structural design calculations.

Designers using SI units must transform loads given in mass (kilograms) to forces, using the relationship force = mass times acceleration. In the design of structures on earth, acceleration is the acceleration due to gravity, designated by "g" and established as 9.806 65 metres per second per second at the third General Conference on Weights and Measures in 1901.

The unit of force to be used in design is the newton (N) (or multiples thereof) where a newton is defined as the force that, when applied to a body having a mass of one kilogram (kg), gives the body an acceleration of one metre (m) per second squared (s^2). The unit of stress is the pascal (Pa), which is one newton per square metre (m^2). Since this is a very small unit, designers of steel structures will generally use megapascals (MPa), where one megapascal is one million pascals and equals one newton per square millimetre (N/mm^2). See also "Structural Loads, Mass and Force".

Properties and dimensions of steel sections are given, in this book, in millimetre units, tabulated to an appropriate ternary power of 10, and millimetres should be used for dimensioning steel structures. Some relationships and values of interest to steel designers are shown below:

SI PREFIXES
Table 7-3

Density of Steel		7 850	kg/m^3
Modulus of Elasticity	E	200 000	MPa
Shear Modulus of Steel	G	77 000	MPa
Coefficient of Thermal Expansion		11.7×10^{-6}	/°C
Acceleration due to Earth's Gravity	g	9.806 65	m/s^2

For a more complete description of SI, the Canadian Metric Practice Guide should be consulted; however, Table 7—4 provides a convenient summary listing selected SI units, the quantity represented, the unit name and typical application.

Structural Loads, Mass and Force

Since most civil engineers have been accustomed to designing structures on earth to withstand loads more variable than the acceleration due to gravity, the pound-force and the kilogram-force have been used as standard units of force. These units were assumed to be numerically equal to their mass counter-parts, the pound-mass and the kilogram-mass respectively.

In SI, the units of mass and force, the kilogram and the newton respectively, are distinctly different both in name and in value. The two are related through the famous Newtonian equation, force = mass times acceleration, or

$$F = ma.$$

Thus a newton (N) is defined as the force required to give one kilogram (kg) mass an acceleration of one metre (m) per second (s) squared, or

$$1\ N = 1\ kg \cdot m/s^2 .$$

The standard international value of acceleration due to gravity is 9.806 65 m/s^2.

However, for hand calculations in Canada a value of

$$g = 9.81\ m/s^2$$

may be more acceptable as it retains three significant figures (adequate for most structural design) and produces a numerical value of force distinctly different from the value of mass. Thus, whether or not the mass has been converted to a force will be readily apparent, and errors will tend to be reduced.

SELECTED SI UNITS
Table 7-4

Quantity	Preferred Units	Unit Name	Typical Applications	Remarks
Area	mm^2	square millimetre	Area of cross section for structural sections	Avoid cm^2
	m^2	square metre	Areas in general	
Bending Moment	kN•m	kilonewton metre	Bending moment in structural sections	
Coating mass	g/m^2	gram per square metre	Mass of zinc coating on steel deck	
Coefficient of Thermal Expansion	1/°C*	reciprocal (of) degree Celsius	Expansion of materials subject to temperature change (generally expressed as a ratio per degree Celsius)	11.7×10^{-6}/°C for steel
Density, mass	kg/m^3	kilogram per cubic metre	Density of materials in general; mass per unit volume	7 850 kg/m^3 for steel
Force	N	newton	Unit of force used in structural calculations	$1N = 1\ kg \cdot m/s^2$
	kN	kilonewton	Force in structural elements such as columns; concentrated forces; axial forces; reactions; shear force; gravitational force	
Force per Unit Length	N/m	newton per metre	Unit for use in calculations	$1\ kg/m \times 9.81\ m/s^2 = (9.81\ kg \cdot m/s^2) \times \frac{1}{m} = 9.81\ N/m$
	kN/m	kilonewton per metre	Transverse force per unit length on a beam, column etc.; dead load of a beam for stress calculations	$(1\ kg/m \times 9.81\ m/s^2) \times \frac{1\ 000}{1\ 000} = (9.81\ kg \cdot m/s^2) \times \frac{1}{m} \times \frac{1\ 000}{1\ 000} = (9.81\ N/m) \times \frac{1\ 000}{1\ 000} = 9.81\ kN/m \times 1/1\ 000 = 0.009\ 81\ kN/m$
Force per Unit Area (See Pressure)				
Frequency	Hz	hertz	Frequency of vibration	$1\ Hz = 1/s = s^{-1}$ replaces cycle per second (cps)
Impact energy	J	joule	Charpy V-notch test	1 N•m = 1 J
Length	mm	millimetre	Dimensions on all drawings; dimensions of sections, spans, deflection, elongations, eccentricity	
	m	metre	Overall dimensions; in calculations; contours; surveys	
	km	kilometre	Distances for transportation purposes	
	μm	micrometre	Thickness of coatings (paint)	
Mass	kg	kilogram	Mass of materials, structural elements and machinery	A metric tonne, t $1t = 10^3\ kg = 1Mg = 1\ 000\ kg$
Mass per Unit Length	kg/m	kilogram per metre	Mass per unit length of section, bar, or similar items of uniform cross section.	Also known as "linear density"
Mass per Unit Area	kg/m^2	kilogram per square metre	Mass per unit area of plates, slabs, or similar items of uniform thickness; rating for load-carrying capacities on floors (display on notices only)	DO NOT USE IN STRESS CALCULATION
Mass Density	kg/m^3	kilogram per cubic metre	Density of materials in general; mass per unit volume	7 850 kg/m^3 for steel
Modulus of Elasticity (Young's)	MPa	megapascal	Modulus of elasticity; Young's modulus	200 000 MPa for carbon, high-strength low alloy and low-alloy wrought steels
Modulus, Shear	MPa	megapascal	Shear Modulus	77 000 MPa assumed for steel
Modulus, Section	mm^3	millimetre to third power	First moment of area of cross section of structural section, such as plastic section modulus, elastic section modulus	

* The preferred unit is 1/K, however 1/°C is an acceptable unit for the construction industry.

SELECTED SI UNITS
Table 7-4

Quantity	Preferred Units	Unit Name	Typical Applications	Remarks
Moment of Inertia	mm^4	millimetre to fourth power	Second moment of area; moment of inertia of a section; torsional constant of cross section	
Moment of Force	kN•m	kilonewton metre	Bending moment (in structural sections); overturning moment	
	N•m	newton metre		
Pressure (see also Stress)	Pa	pascal	Unit used in calculation	1 Pa = 1 N/m^2
	kPa	kilopascal	Uniformly distributed loads on floors; soil pressure; wind loads; snow loads; dead loads; live loads.	1 kPa = 1 kN/m^2
Section Modulus (see Modulus)				
Stress	MPa	megapascal	Stress (yield, ultimate, permitted, calculated) in structural steel	1 MPa = 1 MN/m^2 = 1 N/mm^2
Structural Load (see Force)				
Temperature	°C	degree Celsius	Ambient temperature	0°C ≃ 273.15K However, for temperature intervals 1°C = 1K
Thickness	mm	millimetre	Thickness of web, flange, plate, etc.	
	μm	micrometre	Thickness of paint	
Torque	kN•m	kilonewton metre	Torsional moment on a cross section; inspection torque for high strength bolts.	
Volume	m^3	cubic metre	Volume; volume of earthworks, excavation, concrete, sand, all bulk materials.	1 m^3 = 1 000 L The cubic metre is the preferred unit of volume for engineering purposes
	L	litre	Volume of fluids and containers for fluids	
Work, Energy	J	joule	Energy absorbed in impact testing of materials; energy in general	1 kWh = 3.6 MJ where kWh is a kilowatt hour.

There are two common areas where the designer of a structure must be alert to the distinction between mass and force:

1. dead loads due to the mass of the structural elements, permanent equipment etc.,
2. superimposed, or live loads due to storage of materials.

In these, and other cases where mass is well known since it is the unit of commerce, the designer must convert mass to force by multiplying by g.

COMMON CONVERSION FACTORS
Table 7-5

Item	Imperial — SI		SI — Imperial	
Acceleration	1 ft./s^2	= 0.304 8 m/s^2	1 m/s^2	= 3.2808 ft./s^2
Area	1 acre	= 0.404 685 6 ha	1 ha	= 2.471 acres
	1 ft.^{2}	= 0.092 903 04 m^2	1 m^2	= 10.764 ft.^{2}
	1 in.^{2}	= 645.16 mm^2	1 mm^2	= $1.55\text{x}10^{-3}$ in.^{2}
	1 mi.^{2}	= 2.589 988 km^2	1 km^2	= 0.3861 mi.^{2}
	1 yd.^{2}	= 0.836 127 4 m^2	1 m^2	= 1.20 yd.^{2}
Capacity (Canadian Legal Units)	1 oz.	= 28.413 062 mL	1 mL	= $35.2\text{x}10^{-3}$ oz.
	1 gal.	= 4.546 090 L	1 L	= 0.220 gal.
	1 pt.	= 0.568 261 L	1 L	= 1.76 pt.
	1 qt.	= 1.136 522 L	1 L	= 0.880 qt.
Density, Mass	1 lb./ft.	= 1.488 16 kg/m	1 kg/m	= 0.672 lb./ft.
	1 lb./yd.	= 0.496 055 kg/m	1 kg/m	= 2.016 lb./yd.
	1 oz./ft.^{2}	= 305.152 g/m^2	1 g/m^2	= $3.277\text{x}10^{-3}$ oz./ft.^{2}
	1 lb./ft.^{2}	= 4.882 43 kg/m^2	1 kg/m^2	= 0.205 lb./ft.^{2}
	1 lb./in.^{2}	= 703.069 6 kg/m^2	1 kg/m^2	= $1.42\text{x}10^{-3}$ lb./in.^{2}
	1 lb./ft.^{3}	= 16.018 46 kg/m^3	1 kg/m^3	= $62.4\text{x}10^{-3}$ lb./ft.^{3}
	1 lb./in.^{3}	= 27.679 90 Mg/m^3	1 Mg/m^3	= 0.0361 lb./in.^{3}
Force	1 kip	= 4.448 222 kN	1 kN	= 0.225 kip
Length	1 ft.	= 0.304 8 m = 304.8 mm	1 m	= 3.28 ft.
	1 in.	= 25.4 mm	1 mm	= 0.0394 in.
	1 mile	= 1.609 344 km	1 km	= 0.622 mi.
	1 yd.	= 0.914 4 m	1 m	= 1.09 yd.
Mass	1 lb.	= 0.453 592 37 kg	1 kg	= 2.20 lb.
	1 ton (2000 lb.)	= 0.907 184 74 Mg	1 Mg	= 1.10 ton = 2200 lb.
Mass per Unit Area	1 lb./ft.^{2}	= 4.882 43 kg/m^2	1 kg/m^2	= 0.205 lb./ft.^{2}
Mass per Unit Length	1 lb./ft.	= 1.488 16 kg/m	1 kg/m	= 0.672 lb./ft.
Moment of Inertia				
a) Second Moment of Area	1 in.^{4}	= 416 231.4 mm^4	1 mm^4	= $2.4\text{x}10^{-6}$ in.^{4}
b) Section Modulus	1 in.^{3}	= 16 387.064 mm^3	1 mm^3	= $0.061\text{x}10^{-3}$ in.^{3}
Pressure or Stress	1 ksi	= 6.894 757 MPa	1 MPa	= 0.145 ksi
	1 psf	= 47.880 26 Pa	1 Pa	= 0.0209 psf
	1 psi	= 6.894 757 kPa	1 kPa	= 0.145 psi
Torque or Moment of Force	1 ft.•kipf	= 1.355 818 kN•m	1 kN•m	= 0.738 ft.•kipf
Volume	1 in.^{3}	= 16 387.064 mm^3	1 mm^3	= $0.061\text{x}10^{-3}$ in.^{3}
	1 ft.^{3}	= 28.316 85 dm^3	1 dm^3	= 0.0353 ft.^{3}
	1 yd.^{3}	= 0.764 555 m^3	1 m^3	= 1.308 yd.^{3}
Costs	1 \$/ft.	= 3.28 \$/m	1 \$/m	= 0.305 \$/ft.
	1 \$/ft.^{2}	= 10.764 \$/m^2	1 \$/m^2	= 0.0929 \$/ft.^{2}
	1 \$/yd.^{2}	= 1.20 \$/m^2	1 \$/m^2	= 0.836 \$/yd.^{2}
	1 \$/ft.^{3}	= 35.34 \$/m^3	1 \$/m^3	= 0.0283 \$/ft.^{3}
	1 \$/yd.^{3}	= 1.307 \$/m^3	1 \$/m^3	= 0.765 \$/yd.^{3}

MILLIMETRE EQUIVALENTS
DECIMALS AND EACH 64TH OF AN INCH

FRACTIONS		INCHES	mm
1/64		.015625	.397
	1/32	.03125	.794
		.03937	1
3/64		.046875	1.191
	1/16	.0625	1.588
5/64		.078125	1.984
		.07874	2
	3/32	.09375	2.381
7/64		.109375	2.778
		.11811	3
	1/8	.125	3.175
9/64		.140625	3.572
	5/32	.15625	3.969
		.15748	4
11/64		.171875	4.366
	3/16	.1875	4.763
		.19685	5
13/64		.203125	5.159
	7/32	.21875	5.556
15/64		.234375	5.953
		.23622	**6**
	¼	.25	6.350
17/64		.265625	6.747
		.27559	**7**
	9/32	.28125	7.144
19/64		.296875	7.541
	5/16	.3125	7.938
		.31496	**8**
21/64		.328125	8.334
	11/32	.34375	8.731
		.35433	**9**
23/64		.359375	9.128
	3/8	.375	9.525
25/64		.390625	9.922
		.3937	10
	13/32	.40625	10.319
27/64		.421875	10.716
		.43307	**11**
	7/16	.4375	11.113
29/64		.453125	11.509
	15/32	.46875	11.906
		.47244	**12**
31/64		.484375	12.303
	½	.5	12.700

FRACTIONS		INCHES	mm
		.51181	**13**
33/64		.515625	13.097
	17/32	.53125	13.494
35/64		.546875	13.891
		.55118	**14**
	9/16	.5625	14.288
37/64		.578125	14.684
		.59055	15
	19/32	.59375	15.081
39/64		.609375	15.478
	5/8	.625	15.875
		.62992	**16**
41/64		.640625	16.272
	21/32	.65625	16.669
		.66929	**17**
43/64		.671875	17.066
	11/16	.6875	17.463
45/64		.703125	17.859
		.70866	**18**
	23/32	.71875	18.256
47/64		.734375	18.653
		.74893	**19**
	¾	.75	19.050
49/64		.765625	19.447
	25/32	.781250	19.844
		.7874	20
51/64		.796875	20.241
	13/16	.8125	20.638
		.82677	**21**
53/64		.828125	21.034
	27/32	.84375	21.431
55/64		.859375	21.828
		.86614	**22**
	7/8	.875	22.225
57/64		.890625	22.622
		.90551	**23**
	29/32	.90625	23.019
59/64		.921875	23.416
	15/16	.9375	23.813
		.94488	**24**
61/64		.953125	24.209
	31/32	.96875	24.606
		.98425	25
63/64		.984375	25.003
	1	1.000	25.4

MISCELLANEOUS CONVERSION FACTORS

Area

1 acre	= 0.404 685 6 ha
1 hectare	= **1** hm^2
1 legal subdivision (40 acres)	= 0.161 874 2 km^2
1 section (1 mile square, 640 acres)	= 2.589 988 km^2
1 square foot	= **929.0304** cm^2
1 square inch	= **645.16** mm^2
1 square mile	= 2.589 988 km^2
1 square yard	= 0.836 127 4 m^2
1 township (36 sections)	= 93.239 57 km^2

Linear Density (Mass per Unit Length)

1 pound per inch	= 17.858 kg/m
1 pound per foot	= 1.488 16 kg/m
1 pound per yard	= 0.496 055 kg/m

Area Density (Mass per Unit Area)

1 ounce per square foot	= 305.152 g/m^2
1 pound per square foot	= 4.882 43 kg/m^2
1 pound per square inch	= 703.0696 kg/m^2

Mass Density (Mass per Unit Volume)

1 pound per cubic foot	= 16.018 46 kg/m^3
1 pound per cubic inch	= 27.679 90 Mg/m^3
1 ton (long) per cubic yard	= 1.328 939 Mg/m^3
1 ton (short) per cubic yard	= 1.186 553 Mg/m^3

Energy

1 British thermal unit (Btu) (International Table)	= **1.055 056 kJ**
1 foot pound-force	= 1.355 818 J
1 horsepower hour	= 2.684 52 MJ
1 kilowatt hour	= **3.6** MJ

Force

1 kilogram-force	= **9.806 65** N
1 kip (thousand pounds force)	= 4.448 222 kN
1 pound-force	= 4.448 222 N

Heat

1 Btu *foot per (square foot hour °F)	= 1.730 74 W/(m•K)	k - value
1 Btu per (square foot hour °F)	= 5.678 29 W/(m^2•K)	U - value
1 square foot hour °F per Btu	= 0.176 109 m^2•K/W	R - value

* *Based on the Btu IT.*

Length

1 chain (66 feet)	= **20.1168** m
1 foot	= **0.3048** m
1 inch	= **25.4** mm
1 microinch	= **25.4** nm
1 micron	= **1** μm
1 mil (0.001 inch)	= **25.4** μm
1 mile	= **1.609 344** km
1 mile (International nautical)	= **1.852** km
1 mile (UK nautical)	= **1.853 184** km
1 mile (US nautical)	= **1.852** km
1 yard	= **0.9144** m

MISCELLANEOUS CONVERSION FACTORS

Mass

1 hundredweight (100 lb)	= **45.359 237** kg
1 hundredweight (long) (112 lb, UK)	= 50.802 345 kg
1 pennyweight	= 1.555 174 g
1 pound (avoirdupois)	= **0.453 592 37** kg
1 ton (long, 2240 lb, UK)	= **1.016 046 908 8** Mg
1 ton (short, 2000 lb)	= **0.907 184 74** Mg

Mass Concentration

1 pound per cubic foot	= 16.018 46 kg/m^3

Second Moment of Area (Moment of Inertia)

1 $inch^4$	= 0.416 231 4 x $10^6 mm^4$

Section Modulus

1 $inch^3$	= **16.387 064 x $10^3 mm^3$**

Momentum

1 pound foot per second	= 0.138 255 kg•m/s

Power. See also Energy.

1 Btu (IT)* per hour	= 0.293 072 W
1 foot pound-force per hour	= 0.376 616 1 mW
1 foot pound-force per minute	= 22.596 97 mW
1 foot pound-force per second	= 1.355 818 W
1 horsepower (550 ft•lbf/s)	= 745.6999 W

* *International Tables.*

Pressure or Stress (Force per Area)

1 atmosphere, standard	= **101.325** kPa
1 inch of mercury (conventional, 32°F)	= 3.386 39 kPa
1 inch of water (conventional)	= 249.089 Pa
1 ksi (1000 lbf/in^2)	= 6.894 757 MPa
1 mm mercury (conventional, 0°C)	= 133.322 Pa
1 pound-force per square foot	= 47.880 26 Pa
1 pound-force per square inch (psi)	= 6.894 757 kPa
1 ton-force per square inch	= 13.789 514 MPa
1 ton-force (UK) per square inch	= 15.4443 MPa

Temperature

Scales

Celsius * temperature	= temperature in kelvins – **273.15**
Fahrenheit temperature	= **1.8** (Celsius temperature) + **32**
Fahrenheit temperature	= **1.8** (temperature in kelvins) – **459.67**
Rankine temperature	= **1.8** (temperature in kelvins)

Intervals

1 degree Celsius*	= **1** K
1 degree Fahrenheit	= **5/9** K
1 degree Rankine	= **5/9** K

**"Celsius" replaced "Centigrade" in 1948 to eliminate confusion with the word centigrade, associated with centesimal angular measure.*

MISCELLANEOUS CONVERSION FACTORS

Time	
1 day (mean solar	= **86.4** ks
1 hour (mean solar)	= **3.6** ks
1 minute (mean solar)	= **60** s
1 month (mean calendar, 365/12 days)	= **2.628** Ms
1 year (calendar, 365 days)	= **31.536** Ms
Torque (Moment of Force)	
1 pound-force foot	= 1.355 818 N•m
1 pound-force inch	= 0.112 985 N•m
Volume	
1 acre foot	= 1233.482 m^3
1 barrel (oil, 42 US gallons)	= **0.158 987 3** m^3
1 board foot*	= 2.359 737 dm^3
1 cubic foot	= 28.316 85 dm^3
1 cubic inch	= **16.387 064** cm^3
1 cubic yard	= 0.764 555 m^3
1 gallon‡	= **4.546 09** dm^3
1 gallon (UK)§	= 4.546 092 dm^3
1 gallon (US)	= 3.785 412 dm^3

**The board foot is nominally 1 x 12 x 12 = 144 in^3. However, the actual volume of wood is about 2/3 of the nominal quantity.*

§ Also referred to as the "Imperial gallon."

Volume Rate of Flow	
1 cubic foot per minute	= 0.471 947 4 dm^3/s
1 cubic foot per second	= 28.316 85 dm^3/s
1 cubic yard per minute	= 12.742 58 dm^3/s
1 gallon per minute	= 75.768 17 cm^3/s
1 gallon (UK) per minute	= 75.7682 cm^3/s
1 gallon (US) per minute	= 63.0902 cm^3/s
1 million gallons per day	= 52.6168 dm^3/s

Notes:

1. The conversion factors give the relationship between SI units and other Canadian legal units as well as commonly encountered units of measure of United Kingdom and USA origin. The yard and the pound are the same throughout the world; by definition they are specified fractions of the metre and the kilogram. The gallons of Canada and Australia, which are identical, differ by a relatively insignificant amount from the gallon of the United Kingdom, whereas that of the USA is a much smaller measure.

2. The conversion factors given in tables apply to Canadian units unless stated otherwise.

3. Conversion factors that are exact are shown in boldface type. Other factors are given to more than sufficient accuracy for most general and scientific work.

4. Conversions are those listed in CAN3-Z234.1-79

PART EIGHT
GENERAL INDEX